CONCENTRATED ACIDS AND BASES

NAME	APPROXIMATE WEIGHT PERCENT	APPROXIMATE MOLARITY	mL OF REAGENT NEEDED TO PREPARE 1 L OF ~1.0 M SOLUTION
Acid			
Acetic	99.8	17.4	57.5
Hydrochloric	37.2	12.1	82.6
Hydrofluoric	49.0	28.9	34.6
Nitric	70.4	15.9	62.9
Perchloric	70.5	11.7	85.5
Phosphoric	85.5	14.8	67.6
Sulfuric	96.0	18.0	55.6
Base			
Ammonia[†]	28.0	14.5	69.0
Sodium hydroxide	50.5	19.4	51.5
Potassium hydroxide	45.0	11.7	85.5

[†] 28.0% ammonia is the same as 56.6% ammonium hydroxide.

QUANTITATIVE CHEMICAL ANALYSIS

Third Edition

QUANTITATIVE CHEMICAL ANALYSIS

Third Edition

Daniel C. Harris
Michelson Laboratory
China Lake, California

W. H. Freeman and Company
New York

Cover Image: Permanent magnet levitates above superconducting disk of yttrium barium copper oxide in pool of liquid nitrogen. Box 16-1 describes the chemical analysis of the superconductor and Experiment 25-11 gives a procedure for the analysis. [Photo courtesy D. Cornelius, Michelson Laboratory, with materials from T. Vanderah.]

Library of Congress Cataloging-in-Publication Data

Harris, Daniel C. 1948–
   Quantitative chemical analysis / Daniel C. Harris.—3d ed.
     p.   cm.
   Includes index.
   ISBN 0-7167-2170-8 (hardcover)   ISBN 0-7167-2171-6
   1. Chemistry, Analytic—Quantitative.   I. Title.
QD101.2.H37   1991
545—dc20   90-38694
                                                CIP

Printed in the United States of America

2 3 4 5 6 7 8 9 0 KP 9 9 8 7 6 5 4 3 2 1

Contents

Paper mill on the Potomac River near Westernport, Maryland. [Photo courtesy C. Dalpra, Potomac River Basin Commission.]

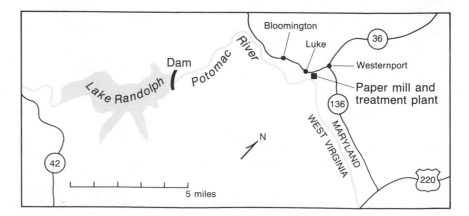

The Potomac River above the paper mill is acidic and lifeless. After passing the mill and water treatment plant, the neutralized river teems with life.

Preface

The North Branch of the Potomac River runs crystal clear through a beautiful section of the Appalachian Mountains. But for much of its length, the river is dead—a victim of acid drainage from abandoned coal mines. As the river flows past a paper mill and waste water treatment plant at Westernport, Maryland, the pH rises from a lifeless 4.5 to a fertile 7.2, in which fish and plants thrive. Something at Westernport is neutralizing the acid in the river. In the early 1980s a great dam was built at Bloomington, on the acidic part of the river, to serve as a reservoir and to provide flood control. The question arose of how to regulate the release water from the reservoir so as not to overcome the neutralization occurring at Westernport.

As in many environmental quality issues, analytical chemistry played a key role in solving this problem. Combining many of the techniques you will study in this book—including acid–base titrations, ion-exchange chromatography, atomic absorption spectroscopy, potentiometric titrations, and even gravimetric analysis—the chemistry at Westernport was unraveled and quantified.[†] The key source of neutralization is suspended solid calcium carbonate—from chemicals used to treat wood pulp—exiting the paper mill. Part of this solid is trapped at the treatment plant and does not enter the river. However, respiration of bacteria in the treatment plant produces a massive quantity of carbon dioxide that reacts with solid calcium carbonate to make soluble calcium bicarbonate:

$$CaCO_3(s) + CO_2(aq) + H_2O(l) \rightarrow \underbrace{Ca^{2+}(aq) + 2HCO_3^{-}(aq)}$$

Calcium   Dissolved calcium bicarbonate
carbonate

The bicarbonate from the treatment plant neutralizes the acidic river, permitting life to thrive immediately downstream of the plant:

$$HCO_3^{-}(aq) + H^{+}(aq) \rightarrow CO_2(g) + H_2O(l)$$

[†] D. P. Sheer and D. C. Harris, *J. Water Pollution Control Federation,* **54,** 1441 (1982).

By identifying and quantifying the chemistry at Westernport, operating rules for release of water from the reservoir at Bloomington were devised so that acidic water behind the dam would not overcome the neutralizing capacity of the paper mill.

In one way or another, almost every branch of science relies on analytical chemistry to answer the questions "What is it?" (qualitative analysis) and "How much is there?" (quantitative analysis). This text provides a foundation for understanding how the latter question is approached in the laboratory, although the two questions are usually intertwined.

In writing this text, my goals have been to provide a sound physical understanding of the principles of analytical chemistry and to show how these principles are applied in chemistry and related disciplines, especially the life sciences. I have attempted to present the subject matter of a traditional analytical chemistry course in a rigorous, readable, and interesting manner that will appeal to students whether or not their primary interest is chemistry. I intend the material presented here to be lucid enough for nonmajors, yet to contain the depth required by advanced undergraduates. My goals are reflected in the book's origins: it grew out of an introductory analytical chemistry course taught mainly for nonchemistry majors at the University of California, Davis, and out of a course taken by third-year chemistry majors at Franklin and Marshall College in Lancaster, Pennsylvania.

Although there is no way for one course to cover every topic in this book, the text is designed to satisfy the diverse needs of many analytical chemistry courses. Topics include statistics, chemical equilibrium, acid–base chemistry, electroanalytical methods, spectroscopic techniques, methods of separation, and laboratory procedures. Chapter 19 discusses spectrophotometric methods and can be easily covered out of its order in this book. Some problems in earlier chapters use spectrophotometry and refer the student to the beginning of Chapter 19.

Several major changes have occurred in this edition. The introduction to acids and bases has been moved forward to Chapter 5, to help pave the way for some acid–base chemistry used in Chapter 7. The introduction to electrochemistry in Chapter 14 was rewritten with the hope of reducing confusion about signs in the calculation of cell voltages. Instructors who have taught with the previous edition should pay particular attention to how signs are treated in this edition. Changes in the discussion of EDTA complex equilibrium are also intended to improve clarity. Chapters on chromatographic methods and spectrophotometric instrumentation were rewritten to incorporate a decade of advances. A new chapter on sample preparation appears at the end of the book. It collects aspects of sample preparation that were previously scattered through the book, and includes much material new to this edition. Numerous sections of the book were updated in small ways, new Boxes were added, a new experiment was added, and tables of formation constants were added to the Appendix. New Color Plates are included at the center of the book. The Computer Projects of the previous edition were deleted to help make room for the additions.

To help lighten the load of a very dense subject, the text is laced with interesting Boxes and Demonstrations and includes a color insert to illustrate the Demonstrations. There is an extensive Glossary at the back of the book. The collections of Exercises and Problems at the end of each chapter are vital, since only by attacking them can the student master this subject. The Exercises are the smallest set of problems; they cover all the most important, sometimes complex, topics. Complete solutions to Exercises and short answers to Problems appear at the back of the book. Complete solutions to Problems appear in the *Solutions Manual to Accompany Quantitative Chemical Analysis*, available to instructors upon request. The manual may be made available to students through bookstores.

I am grateful to James P. Rybarczyk of Ball State University, Jerome W. O'Laughlin of the University of Missouri, and Timothy C. Donnelly of the University of California, Davis, who reviewed the Second Edition and then reviewed the entire manuscript for the Third Edition in excruciating detail. Detailed reviews of the previous edition used to plan this revision were also received from Joe P. Foley (Lousiana State University), Harry B. Mark (University of Cincinnati), Joseph Wang (New Mexico State University), John P. Walters (St. Olaf College), William E. Kurtin (Trinity University), Ned A. Daugherty (Colorado State University), Anna Brajter-Toth (University of Florida), and Dale Hawley (Kansas State University). A published review by Royce C. Engstrom (University of South Dakota) and a survey by David C. Lock (Queens College) and W. E. L. Grossman (Hunter College) were also helpful. Comments received from John G. Dorsey (University of Cincinnati), Lowell M. Schwartz (University of Massachusetts, Boston), Rollie J. Myers (University of California, Berkeley), George F. Atkinson (University of Waterloo), Robert de Levie (Georgetown University), Richard Ulsh (Elmira College), Gerald Seebach (Transylvania University), Jack Penciner (Tel Aviv University), Kenneth Sauer (University of California, Berkeley), Truman D. Turnquist (Mount Union College), and Donald C. Jackman (Pfeiffer College) were all appreciated. I especially savored letters from students.

At Michelson Laboratory I consulted a great deal with Eric Erickson, Mike Seltzer, Wayne Weimer, and Alice Harper, all of whom also helped check the solutions to problems for accuracy. The fine people at W. H. Freeman and Company, Gary Carlson, Stephen Wagley, Alice Fernandes-Brown, Julia De Rosa, and José Fonfrias, helped this edition to take form. New Color Plates were obtained with the help of Klaus Grohmann (Hunter College) and Ruth Anderson (Ohio State University) and were photographed by Ken Karp. No edition of this book would have gotten off the ground without the participation of my wife Sally, who typed the Solutions Manual, prepared the manuscript for production, compiled the index, proofread, and helped in numerous indescribable ways.

This book is dedicated to the students who use it, who occasionally smile when they read it, who gain new insight, and who feel satisfaction after struggling to solve a problem. I have been successful if this book helps you develop critical, independent reasoning that you can apply to new problems. I truly appreciate any comments, criticisms, suggestions and corrections from students and teachers. Please address correspondence to me at the Chemistry Division, Research Department, Michelson Laboratory, China Lake CA 93555.

Dan Harris
August 1990

QUANTITATIVE CHEMICAL ANALYSIS

Third Edition

["The Experiment" by Sempé. Copyright C. Charillon, Paris.]

1 Units and Concentrations

The immediate goal of quantitative analysis is to answer the query "How much?" How much vanadium is contained in the ore? How much phosphate is bound to the enzyme? How much pesticide is contained in the groundwater? Some of the principles and methods that enable us to measure "how much" are the subject of this text. The ultimate goal of analytical chemistry is not just to measure "how much," but to use that knowledge for some greater purpose. This could be a scientific investigation, a policy decision, a cost analysis, philosophical satisfaction, or myriad other reasons. We begin with a brief discussion of units of measurement.

1-1 SI UNITS

Scientists around the world use a uniform system of measurement known as the *Système International d'Unités,* whose units are called **SI units.** The fundamental units, from which all others are derived, are listed in Table 1-1. The standards of length, mass, and time are the familiar metric units of *meter* (m), *kilogram* (kg), and *second* (s). The other fundamental units of most concern to us measure electric current (*ampere,* A), temperature (*kelvin,* K), and amount of substance (*mole,* mol).

All the remaining physical quantities, such as energy, force, and charge, can be expressed in terms of the fundamental units. Some of these derived quantities are shown in Table 1-2, along with their names and symbols. Conversion factors relating some common non-SI units to SI units are listed in Table 1-3.

You can find the definitions of many terms used in this text in the Glossary at the back of the book.

1

Table 1-1
Fundamental SI units

Quantity	Unit	Symbol	Definition
Length	meter	m	The meter is the distance light travels in a vacuum during $\frac{1}{299\,792\,458}$ of a second. This definition fixes the speed of light at exactly 299 792 458 m/s.
Mass	kilogram	kg	One kilogram is the mass of the prototype kilogram kept at Sèvres, France. This is the only SI unit whose primary standard is not defined in terms of physical constants.
Time	second	s	The second is the duration of 9 192 631 770 periods of the radiation corresponding to the two hyperfine levels of the ground state of ^{113}Cs.
Electric current	ampere	A	One ampere is the amount of constant current that will produce a force of 2×10^{-7} N/m (newtons per meter of length) when maintained in two straight, parallel conductors of infinite length and negligible cross section, separated by one meter.
Temperature	kelvin	K	The thermodynamic temperature is defined such that the triple point of water (at which solid, liquid, and gaseous water are in equilibrium) is 273.16 K and the temperature of absolute zero is 0 K.
Luminous intensity	candela	cd	One candela is the luminous intensity, in a given direction, of a source that emits monochromatic radiation of frequency 540 THz and that has a radiant intensity of $\frac{1}{683}$ W/sr in that direction.
Amount of substance	mole	mol	One mole of substance contains as many molecules (or atoms, if the substance is a monatomic element) as there are atoms of carbon in exactly 0.012 kg of ^{12}C. The number of particles in a mole is approximately $6.022\,136\,7 \times 10^{23}$.
Plane angle	radian	rad	The radian is such that there are 2π radians in a complete circle.
Solid angle	steradian	sr	The steradian is defined such that there are 4π steradians in a complete sphere.

Table 1-2
Some SI-derived units with special names

Quantity	Units	Symbol	Expression in terms of other units	Expression in terms of SI base units
Frequency	hertz	Hz		$1/s$
Force	newton	N		$m \cdot kg/s^2$
Pressure	pascal	Pa	N/m^2	$kg/(m \cdot s^2)$
Energy, work, quantity of heat	joule	J	$N \cdot m$	$m^2 \cdot kg/s^2$
Power, radiant flux	watt	W	J/s	$m^2 \cdot kg/s^3$
Quantity of electricity, electric charge	coulomb	C		$s \cdot A$
Electric potential, potential difference electromotive force	volt	V	W/A	$m^2 \cdot kg/(s^3 \cdot A)$
Capacitance	farad	F	C/V	$s^4 \cdot A^2/(m^2 \cdot kg)$
Electric resistance	ohm	Ω	V/A	$m^2 \cdot kg/(s^3 \cdot A^2)$
Conductance	siemens	S	A/V	$s^3 \cdot A^2/(m^2 \cdot kg)$
Magnetic flux	weber	Wb	$V \cdot s$	$m^2 \cdot kg/(s^2 \cdot A)$
Magnetic flux density	tesla	T	Wb/m^2	$kg/(s^2 \cdot A)$
Inductance	henry	H	Wb/A	$m^2 \cdot kg/(s^2 \cdot A^2)$
Luminous flux	lumen	lm		$cd \cdot sr$
Illuminance	lux	lx		$cd \cdot sr/m^2$

$J/A \cdot s$

$\dfrac{J}{C} \cdot s$

Table 1-3
Some conversion factors

Quantity	Unit	Symbol	SI equivalent
Volume	liter	L	$*10^{-3} \ m^3$
	milliliter	mL	$*10^{-6} \ m^3$
Length	angstrom	Å	$*10^{-10} \ m$
	inch	in.	$*0.025 \ 4 \ m$
Mass	pound	lb	$*0.453 \ 592 \ 37 \ kg$
Force	dyne	dyn	$*10^{-5} \ N$
Pressure	atmosphere	atm	$*101 \ 325 \ N/m^2$
	torr	1 mm Hg	$133.322 \ N/m^2$
	pound/in.^{2}	psi	$6 \ 894.76 \ N/m^2$
Energy	erg	erg	$*10^{-7} \ J$
	electron volt	eV	$1.602 \ 177 \ 33 \times 10^{-19} \ J$
	calorie, thermochemical	cal	$*4.184 \ J$
	British thermal unit	Btu	$1 \ 055.06 \ J$
Power	horsepower		$745.700 \ W$

Note: An asterisk (*) indicates that the conversion is exact (by definition).

The recommended way to write numbers is to leave a space between digits after every third digit on either side of the decimal point. An example is

$$1\ 032.971\ 35$$

Commas are *not* to be used for spacing. In Europe the decimal point is usually written as a comma, and the number above would appear as

$$1\ 032,971\ 35$$

EXAMPLE: Conversion Between Units

The most common unit of pressure is the *atmosphere* (atm). The SI unit of pressure is the *pascal* (Pa), which equals a force of one newton per square meter (N/m^2). What pressure in pascals corresonds to a pressure of 0.268 atm?

Table 1-3 tells us that 1 atm is exactly 101 325 N/m^2 = 101 325 Pa. We can write

$$(0.268\ \cancel{atm})\left(101\ 325\ \frac{Pa}{\cancel{atm}}\right) = 27\ 200\ Pa$$

The units should be written beside each quantity, and the answer should be displayed with its units.

Table 1-4
Prefixes

Prefix	Symbol	Factor
exa	E	10^{18}
peta	P	10^{15}
tera	T	10^{12}
giga	G	10^{9}
mega	M	10^{6}
kilo	k	10^{3}
hecto	h	10^{2}
deca	da	10^{1}
deci	d	10^{-1}
centi	c	10^{-2}
milli	m	10^{-3}
micro	μ	10^{-6}
nano	n	10^{-9}
pico	p	10^{-12}
femto	f	10^{-15}
atto	a	10^{-18}

Various prefixes employed to indicate fractions or multiples of units are given in Table 1-4. It is inconvenient continually to write a number such as 3.2×10^{-11} s, so we write 32 ps instead. One *picosecond* (ps) is 10^{-12} s. To express 3.2×10^{-11} s in picoseconds, we perform the conversion as follows:

$$\frac{3.2 \times 10^{-11}\ \cancel{s}}{10^{-12}\ \dfrac{\cancel{s}}{ps}} = 32\ ps$$

$1\ \mu L = 10^{-3}\ mL = 10^{-6}\ L$

The SI unit of volume (which has the dimensions length^3) is the *cubic meter* (m^3). The common unit of volume is the *liter* (L), which is defined as the volume of a cube 0.1 m on each edge. The *milliliter* (mL; 1 mL = 10^{-3} L) is exactly 1 cm^3. Small-scale work, especially in biochemistry, often employs *microliter* (μL; 1 μL = 10^{-6} L) volumes.

1-2 EXPRESSIONS OF CONCENTRATION

Concentration signifies how much of a substance is contained in a specified volume or mass. This section describes most of the common ways to express concentration. Normality and titer, which are not used in this text, are defined in the Glossary, and normality is further discussed in Appendix E.

Molarity

The most common unit of concentration is **molarity** (moles per liter), abbreviated M. Molarity can also be expressed as millimoles per milliliter, where one millimole (mmol) is 10^{-3} mol. A **mole** is defined as the number of atoms of ^{12}C in exactly 12 g of ^{12}C. This number of atoms is called *Avogadro's number* and its best value at this time is $6.022\ 136\ 7 \times 10^{23}$. The term "gram-atom," is sometimes used for Avogadro's number of atoms, while "mole" is sometimes reserved for Avogadro's number of molecules. We will not make such a distinction. A mole is simply $6.022\ 136\ 7 \times 10^{23}$ of anything.

The **molecular weight** (M.W.) of a substance is the number of grams that contain Avogadro's number of molecules. The molecular weight is simply the sum of the atomic weights of the constituent atoms. The terms "mass" and "weight" are sometimes used interchangeably. Weight actually refers to the force exerted by a mass in a gravitational field.

$$\text{Molarity (M)} = \frac{\text{moles of solute}}{\text{liters of solution}}$$

$$= \frac{\text{millimoles of solute}}{\text{milliliters of solution}}$$

A mole is $6.022\ 136\ 7 \times 10^{23}$ of anything.

EXAMPLE: Solution Concentration

A solution is made by dissolving 12.00 g of benzene, C_6H_6, in enough hexane to give 250.0 mL of solution. Find the molarity of the benzene.

The molecular weight of benzene is 6 (atomic weight of carbon) + 6 (atomic weight of hydrogen) = 6 (12.011) + 6 (1.008) = 78.114 g/mol. The units of molecular weight, grams per mole, are often not written but simply understood. The number of moles in 12.00 g is

$$\frac{12.00\ g}{78.114\ g/mol} = 0.153\ 6\ mol$$

Atomic weights appear on the inside back cover of this text.

The molarity (moles per liter) is found by dividing the number of moles by the number of liters:

$$\frac{0.153\ 6\ mol}{0.250\ 0\ L} = 0.614\ 4\ \text{M}$$

The small capital M is read "moles per liter" or "molar."

Note that milliliters must first be converted to liters by dividing the number of milliliters by 1 000 mL/L:

$$\frac{250.0\ mL}{1\ 000\ mL/L} = 0.250\ 0\ L$$

Students have a propensity for confusing **moles** with **moles per liter**, especially on tests. When you want to write moles, the proper abbreviation is "mol." When you want to write moles per liter, use a capital M. Do not use the symbol "m," which means neither mole nor M. Writing the units beside all numbers is the best way to keep track of your calculations and to reduce the occurrence of silly errors.

Formality

HBr is a **strong electrolyte,** which means that it is virtually completely dissociated into H^+ and Br^- ions in aqueous solution. By contrast, acetic acid is a **weak electrolyte,** being only partially dissociated into $CH_3CO_2^-$ and H^+ in water.

If a solution is made by diluting 1.000 mol of HBr to 1.000 L with water, the **formal concentration,** F, of HBr is 1.000 mole per liter. But the actual concentration of HBr molecules is nearly zero, because the HBr molecules have dissociated. The formal concentration refers to the amount of substance dissolved, without regard to its actual composition in solution. Rather than calling the HBr solution 1.000 M, it would be more correct to call it 1.000 F. The small capital F is read "formal." Many texts use the terms "formality" and "molarity" interchangeably, and we will be guilty of the same simplification.

AlCl$_3$ has the structure below in many solvents.

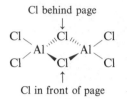

Cl behind page

Cl in front of page

Unless you are fully aware of the chemistry of a particular compound, you rarely know its true molarity, but you can know its formal concentration from the amount weighed into a solution or measured by an analytical procedure. For this reason, the formal concentration also is called the **analytical concentration.**

The **formula weight** (F.W.) of a substance is the mass of one formula unit. For example, the formula weight of AlCl$_3$ is $[26.982 + 3(35.453)] = 133.341$. The true molecular weight in many nonpolar organic solvents is twice the formula weight, since the molecule exists as a dimer (pronounced DIE-mer) with the composition Al$_2$Cl$_6$. The formula weight refers to the species that has been written. We could refer to the formula weight of Al$_2$Cl$_6$, which is twice as much as the formula weight of AlCl$_3$.

Percent Composition

The percentage of a substance in a solution is most often expressed as a **weight percent,** which is defined as

$$\text{Weight percent} = \frac{\text{mass of substance}}{\text{mass of total solution}} \times 100$$

Weight percent is commonly abbreviated as wt/wt percent. A solution labeled 40% (wt/wt) aqueous ethanol contains 40 g of ethanol per 100 g (not 100 mL) of solution. It is made by mixing 40 g of ethanol with 60 g of H$_2$O.

Other common percent units are volume percent (vol/vol percent) and weight-volume percent (wt/vol percent):

$$\text{Volume percent} = \frac{\text{volume of substance}}{\text{volume of total solution}} \times 100$$

$$\text{Weight/volume percent} = \frac{\text{mass of substance (in grams)}}{\text{volume of total solution (in milliliters)}} \times 100$$

Although the units of wt or vol should always be specified, wt/wt is usually implied when units are absent.

Density (also called specific gravity) is mass/volume = g/mL.

EXAMPLE: Concentration from Weight Percent
Commercial concentrated HCl is labeled 37.0%, which you may assume means weight percent. Its **density,** sometimes called **specific gravity,** is 1.18 g/mL. Find (a) the molarity of the HCl; (b) the mass of solution containing 0.100 mol of HCl; and (c) the volume of solution containing 0.100 mol of HCl.

(a) A 37.0% solution contains 37.0 g of HCl per 100 g of solution. The mass of one liter of solution is

$$(1\,000\,\text{mL})\left(1.18\,\frac{\text{g}}{\text{mL}}\right) = 1\,180\,\text{g}$$

The mass of HCl in 1 180 g of solution is

$$\left(0.370\,\frac{\text{g HCl}}{\text{g solution}}\right)(1\,180\,\text{g solution}) = 437\,\text{g HCl}$$

Since the molecular weight of HCl is 36.461, the molarity of HCl is

$$\frac{437 \text{ g/L}}{36.461 \text{ g/mol}} = 12.0 \frac{\text{mol}}{\text{L}} = 12.0 \text{ M}$$

(b) Since 0.100 mol of HCl equals 3.65 g, the mass of solution containing 0.100 mol is

$$\frac{3.65 \text{ g HCl}}{0.370 \text{ g HCl/g solution}} = 9.85 \text{ g solution}$$

(c) The volume of solution containing 0.100 mol of HCl is

$$\frac{9.85 \text{ g solution}}{1.18 \text{ g solution/mL}} = 8.35 \text{ mL}$$

Parts per Million and Its Relatives

Composition is often expressed as **parts per million** (ppm), **parts per billion** (ppb), or **parts per thousand** (ppt). The term one part per million, for example, indicates that one gram of the substance of interest is present per million grams of total solution or mixture. For an aqueous solution whose density is close to 1.00 g/mL, 1 ppm corresponds to 1 μg/mL or 1 mg/L.

$$\text{ppt} = \frac{\text{g of substance}}{\text{g of sample}} \times 10^3$$

$$\text{ppm} = \frac{\text{g of substance}}{\text{g of sample}} \times 10^6$$

$$\text{ppb} = \frac{\text{g of substance}}{\text{g of sample}} \times 10^9$$

EXAMPLE: Concentration from Parts per Million

A sample of salt water with a density of 1.02 g/mL contains 17.8 ppm nitrate, NO_3^-. Calculate the molarity of nitrate in the water.

Molarity is moles per liter, and 17.8 ppm means that the water contains 17.8 μg of NO_3^- per gram of solution. One liter of solution weighs

Mass of solution = volume (mL) × density (g/mL) = 1 000 × 1.02 = 1 020 g

One liter therefore contains

$$\text{Grams of } NO_3^- = \frac{17.8 \times 10^{-6} \text{ g } NO_3^-}{\text{g solution}} \times 1\,020 \text{ g solution} = 0.018\,2 \text{ g } NO_3^-$$

The molarity of nitrate is therefore

$$\frac{\text{mol } NO_3^-}{\text{L solution}} = \frac{0.018\,2 \text{ g } NO_3^-/(62.065 \text{ g } NO_3^-/\text{mol})}{\text{L solution}}$$

$$= 2.93 \times 10^{-4} \text{ M}$$

Parts per million is a popular unit, widely used in the press and television. Saying that drinking water contains 0.1 ppm trichloroethylene means that the water has 0.1 μg of trichloroethylene per gram of water. Sometimes ppm is used ambiguously. The statement that air contains 6 ppm ozone probably refers to volume (6 μL ozone per liter of air), not mass (6 μg ozone per gram of air). If there may be confusion, you should qualify your use of ppt, ppm, and ppb.

$$\text{Molality } (m) = \frac{\text{moles of solute}}{\text{kilograms of solvent}}$$

$$\text{Osmolarity} = \frac{\text{moles of particles}}{\text{liters of solution}}$$

Other Concentration Units

Molality

The **molality,** m, defined as the number of moles of solute per kilogram of solvent, is useful for precise physical measurements. The reason is that molality is not temperature-dependent, whereas molarity is temperature-dependent. A dilute aqueous solution expands approximately 0.02% per degree Celsius when heated near 20°C. Therefore, the moles of solute per liter (molarity) decreases by this same percent.

Osmolarity

Osmolarity, which is encountered in biochemical and clinical literature, is defined as the total number of moles of particles dissolved in one liter of solution. For nonelectrolytes such as glucose, the osmolarity is equal to the molarity. For the strong electrolyte $CaCl_2$, the osmolarity is three times the molarity, since each formula weight of $CaCl_2$ provides three moles of ions in solution ($Ca^{2+} + 2Cl^-$). Blood plasma is 0.308 osmolar.

1-3 PREPARING SOLUTIONS

If a pure solid or liquid reagent is to be used to prepare a solution of a given molarity, we simply weigh out the correct mass of reagent, dissolve it in the solvent, and dilute the solution to the desired final volume. The dilution is usually done in a volumetric flask, as described in the next chapter. To prepare 1.00 M NaCl, we would not weigh out 1.00 mol of NaCl and mix it with 1.00 L of water, because the total volume of the mixture would not be 1.00 L.

EXAMPLE: Solution Preparation

What quantity of $H_2C_2O_4 \cdot 2H_2O$ (oxalic acid dihydrate) should be used to prepare 250 mL of 0.150 M aqueous oxalic acid?

The formula weight of oxalic acid dihydrate ($C_2H_6O_6$) is 126.07. If we want to prepare 250 mL of 0.150 M oxalic acid, we will need

$$\left(\frac{250 \text{ mL}}{1\,000 \text{ mL}/\text{L}} \right)\left(0.150 \, \frac{\text{mol}}{\text{L}} \right) = 0.037\,5 \text{ mol}$$

This is equivalent to

$$(0.037\,5 \text{ mol oxalic acid})\left(126.07 \, \frac{\text{g of } H_2C_2O_4 \cdot 2H_2O}{\text{mol oxalic acid}} \right) = 4.73 \text{ g}$$

So 4.73 g of oxalic acid dihydrate should be dissolved in water and diluted to 250 mL.

It is frequently necessary to prepare a dilute solution of a reagent from a more concentrated solution. A useful equation for calculating the required volume of concentrated reagent is

$$M_{conc}V_{conc} = M_{dil}V_{dil} \tag{1-1}$$

Since $M \cdot V = (\text{moles/L})(L) = \text{moles}$, Equation 1-1 simply states that the moles of solute in both solutions are equal. Dilution occurs because the volume has changed.

where conc refers to the concentrated solution and dil refers to the dilute solution.

EXAMPLE: Dilution Calculation

A solution of ammonia in water is also called "ammonium hydroxide" because of the equilibrium

$$NH_3 + H_2O \rightleftharpoons NH_4^+ + OH^-$$

Ammonia   Ammonium   Hydroxide

The density of concentrated ammonium hydroxide, which contains 28.0% (wt/wt) NH_3, is 0.899 g/mL. What volume of this reagent should be diluted to 500 mL to make 0.100 M NH_3?

We begin by calculating the molarity of the concentrated reagent. Since the solution contains 0.899 g of solution per milliliter and there are 0.280 g of NH_3 per gram of solution (28.0% wt/wt), we can write

$$\text{Molarity of } NH_3 = \frac{\left(899 \, \frac{\text{g solution}}{L}\right)\left(0.280 \, \frac{\text{g } NH_3}{\text{g solution}}\right)}{17.03 \, \frac{\text{g } NH_3}{\text{mol } NH_3}} = 14.8 \text{ M}$$

To find the volume of 14.8 M NH_3 required to prepare 500 mL of 0.100 M NH_3, Equation 1-1 may be used:

$$M_{\text{conc}} V_{\text{conc}} = M_{\text{dil}} V_{\text{dil}}$$

$$\left(14.8 \, \frac{\text{mol}}{L}\right) V_{\text{conc}}(L) = \left(0.100 \, \frac{\text{mol}}{L}\right)(0.500 \text{ L}) \qquad (1-2)$$

$$V_{\text{conc}} = 3.38 \times 10^{-3} \text{ L} = 3.38 \text{ mL}$$

Note that both volumes in Equation 1-2 could be expressed in milliliters instead of liters.

Summary

The SI base units are meter (m), kilogram (kg), second (s), ampere (A), kelvin (K), candela (cd), and mole (mol). Quantities such as force, pressure, and energy are measured in units derived from the base units. In calculations, the dimensions should be carried along with the numbers. Prefixes such as kilo- and milli- are used in the SI system to denote multiples of units. Common expressions of con- centration are molarity (moles of solute per liter of so- lution), formality (formula units per liter), molality (moles of solute per kilogram of solvent), percent composition, and parts per million. You should be able to calculate the quantities of reagents needed to prepare a given so- lution, and the equality $M_{\text{conc}} V_{\text{conc}} = M_{\text{dil}} V_{\text{dil}}$ is useful for this purpose.

Terms to Understand

analytical concentration
concentration
density
formal concentration
formula weight
molality
molarity

mole
molecular weight
osmolarity
parts per billion
parts per million
parts per thousand
SI units

specific gravity
strong electrolyte
volume percent
weak electrolyte
weight percent

Exercises†

1-A. A solution with a final volume of 500.0 mL was prepared by dissolving 25.00 mL of methanol (CH_3OH, density = 0.791 4 g/mL) in chloroform.
  (a) Calculate the *molarity* of methanol in the solution.
  (b) If the solution has a density of 1.454 g/mL, find the *molality* of methanol.

1-B. A 48.0% (wt/wt) solution of HBr in water has a density of 1.50 g/mL.
  (a) What is the formal concentration (mol/L) of the solution?
  (b) What mass of solution contains 36.0 g of HBr?
  (c) What volume (mL) of solution contains 233 mmol of HBr?
  (d) How much solution is needed to prepare 0.250 L of 0.160 M HBr?

1-C. A solution contains 12.6 ppt of dissolved $MgCl_2$ (which is actually dissociated into $Mg^{2+} + 2Cl^-$). What is the concentration of Cl^- in parts per thousand?

Problems‡

A1-1. State the fundamental quantities and their units in the SI system. Give one example of a derived quantity.

A1-2. Write the name and number represented by each symbol. For example, for kW you should write kW = kilowatt = 10^3 watts.
  (a) mW (b) pm (c) KΩ (d) μF
  (e) TJ (f) ns (g) fg (h) dPa

A1-3. Write each quantity using an appropriate prefix. For example, the quantity 1.01×10^5 Pa is written 101 kPa.
  (a) 10^{-13} J   (b) $4.317\,28 \times 10^{-8}$ H
  (c) $2.997\,9 \times 10^{14}$ Hz (d) 10^{-10} m
  (e) 2.1×10^{13} W   (f) 48.3×10^{-20} mol

A1-4. How many joules per second (J/s) are used by a device that requires 5.00×10^3 British thermal units per hour (Btu/h)? How many watts (W) does the device use?

A1-5. What is the formal concentration (mol/L) of NaCl when 32.0 g is dissolved in water and diluted to 0.500 L?

A1-6. How many grams of perchloric acid, $HClO_4$, are contained in 37.6 g of 70.5% (wt/wt) aqueous perchloric acid? How many grams of water are in the same solution?

A1-7. Any dilute aqueous solution has a density near 1.00 g/mL. If the solution contains 1 ppm of solute, express the solute concentration in g/L, μg/L, μg/mL, and mg/L.

A1-8. The density of 70.5% (wt/wt) aqueous perchloric acid is 1.67 g/mL. Recall that grams refers to grams of *solution* (= g $HClO_4$ + g H_2O).
  (a) How many grams of solutions are in 1.00 L?
  (b) How many grams of $HClO_4$ are in 1.00 L?
  (c) How many moles of $HClO_4$ are in 1.00 L?

A1-9. What is the formal concentration of acetic acid when 2.67 g is dissolved in butanol to give 0.100 L of solution? The formula for acetic acid can be found in Appendix G.

A1-10. Find the molarity of pyridine, C_5H_5N, if 5.00 g is dissolved in water to give a total volume of 457 mL.

A1-11. It is recommended that drinking water contain 1.6 ppm of fluoride, F^-, for prevention of tooth decay. How many grams of fluoride should be in 1.00×10^6 kg of water? How many grams of NaF contain this much fluoride?

A1-12. How many grams of methanol, CH_3OH, are contained in 0.100 L of 1.71 M aqueous methanol (i.e., 1.71 mol CH_3OH/L solution)?

A1-13. What is the maximum volume of 0.25 M sodium hypochlorite solution, NaOCl (laundry bleach), that can be prepared by dilution of 1.00 L of 0.80 M NaOCl?

---

† Detailed Solutions to Exercises are provided at the end of the book.
‡ Brief numerical answers to the Problems are given after the Solutions to Exercises. Detailed solutions to Problems appear in the *Solutions Manual to Accompany Quantitative Chemical Analysis*. Problems preceded by A are intended to be easier to solve than the other Problems and Exercises.

1-14. Newton's law tells us that force = mass × acceleration. We also know that energy = force × distance and that pressure = force/area. From these relations, derive the dimensions of newtons, joules, and pascals in terms of the fundamental SI units in Table 1-1.

1-15. If 0.250 L of aqueous solution with a density of 1.00 g/mL contains 13.7 μg of pesticide, express the concentration of pesticide in (a) parts per million and (b) parts per billion.

1-16. A bottle of concentrated aqueous sulfuric acid, labeled 98.0% (wt/wt) H_2SO_4, has a concentration of 18.0 M.

(a) How many milliliters of reagent should be diluted to 1.00 L to give 1.00 M H_2SO_4?

(b) Calculate the density of 98.0% H_2SO_4.

1-17. Find the osmolarity of 1.00 L of solution containing 3.15 g $CaCl_2$, 0.153 g KCl, 1.57 g K_2SO_4, and 0.994 g of sucrose (table sugar, formula = $C_{12}H_{22}O_{11}$, a nonionic compound).

1-18. An aqueous solution containing 20.0% (wt/wt) KI has a density of 1.168 g/mL. Find the molality (not molarity) of the KI.

1-19. What is the density of 53.4% (wt/wt) aqueous NaOH if 16.7 mL of the solution produces 0.169 M NaOH when diluted to 2.00 L?

2 Tools of the Trade

Much of this text deals with fundamental "wet" chemical procedures, although elaborate instrumental techniques are discussed in later chapters. The principles developed in these early chapters are essential to the understanding of sophisticated techniques. In this chapter we describe some of the basic laboratory apparatus and manipulations associated with chemical measurements.

2-1 LAB NOTEBOOK

The critical functions of a lab notebook are to state *what what was done* and *what you observed*. The greatest flaw, found even with experienced scientists, is that notebooks are difficult to understand. Hard as it is to believe, even the author of a notebook cannot understand his or her own notes after a few years. The problem is not usually one of legibility, but rather of poorly labeled entries and incomplete descriptions. The habit of writing in *complete sentences* is an excellent way to avoid incomplete descriptions.

Beginning students often find it useful to write a very complete description of an experiment, with formal sections dealing with purpose, methods, results, and conclusions. Arranging a notebook to accept numerical data prior to coming to the lab is an excellent way to prepare for an experiment.

The measure of scientific "truth" is the ability of different people to reproduce an experiment. Sometimes two scientists in different labs cannot reproduce each other's work, and neither has complete enough notebooks to understand why. Details that seem unimportant on the day of an experiment may become important months or years later. A good lab notebook will state everything that was done and will allow you or anyone else to repeat the experiment in exactly the same manner at any future date.

A complete notebook should also contain all your observations. Long after you have forgotten what happened, you should be able to rely on your

The lab notebook must
1. State what was done.
2. State what was observed.
3. Be understandable to someone else.

Without a doubt, somebody reading this book today is going to make an important discovery in the future and will seek a patent. The lab notebook is your legal record of your discovery. For this purpose each page in your notebook should be signed and dated. Anything of potential importance should also be signed and dated by a second person.

notes to tell what happened and perhaps provide a key observation needed to interpret an experiment. It could be that you do not understand what you observe during an experiment, but at some later time new knowledge will enable you to interpret old observations.

It is good practice to write a balanced chemical equation for every reaction you use. This helps you understand what you are doing and may point out what you do not understand about what you are doing.

2-2 ANALYTICAL BALANCE

The most common balance is the single-pan, semi-micro balance, with a total capacity of 100 to 200 g and a sensitivity of 0.01 or 0.1 mg. A typical balance is shown in Figure 2-1.

The usual weighing operation consists of first weighing a sheet of weighing paper or a receiving vessel on the balance pan. Then the substance to be weighed is added and a second reading is taken. The difference between the two masses corresponds to the mass of added substance. The mass of the empty weighing vessel is called the **tare.** Many balances can tare the receiving vessel. That is, with the receiver on the pan, the scale can be set to read zero. Then the substance to be weighed is added and its mass is read directly. No chemical should ever be placed directly on the balance pan. This protects the balance from corrosion and also permits you to recover all the chemical being weighed.

Alternatively, weighing "by difference" is sometimes convenient. First a small vessel containing a reagent is weighed. Then some of the reagent is delivered to a receiver and the vessel is weighted again. The difference equals the mass of reagent delivered. Weighing by difference is especially useful for **hygroscopic reagents** (ones that rapidly absorb moisture from the air), since the weighing bottle can be kept closed during the weighing operations.

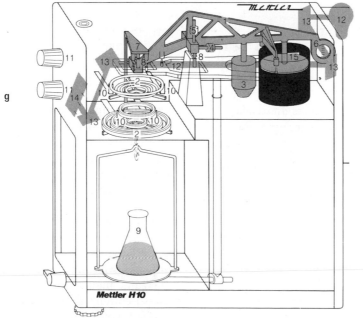

Figure 2-1
Cross section of a Mettler analytical balance, showing the projection of the optical scale from the back of the balance beam to the front of the instrument. (1) Balance beam. (2) Set of weights. (3) Fixed counterweight. (4) Zero-point adjustment weight. (5) Scale-deflection adjustment weight. (6) Graduation plate. (7) Parallelogram suspension. (8) Knife edges. (9) Sample to be weighed. (10) Weight-lifting mechanism. (11) Weight-control knobs. (12) Light bulb (light path shaded). (13) Mirrors. (14) Readout panel. (15) Air damping. [Courtesy Mettler Instrument Corp., Hightstown, N.J.]

Principle of Operation

The principle of most balances is illustrated in Figure 2-2. The sample is suspended on one side of a fulcrum, and a weight is placed on the opposite side. The two will be in perfect balance when

$$m_1 l_1 = m_2 l_2 \qquad (2\text{-}1)$$

where m_1 and m_2 are the masses on each side and l_1 and l_2 are the distances of the balance point from the fulcrum on each side. Normally, $l_1 = l_2$, and m_2 is a known mass.

When one mass in Figure 2-2 is heavier than the other, the balance beam tilts toward the heavier mass. The *sensitivity* of the balance is the deflection of the pointer divided by the mass difference between m_1 and m_2. The greater the sensitivity, the more will be the deflection for a given mass difference. The balance points and fulcrum are typically agate prisms in contact with agate plates. As the agate knife edges wear out with use (and less-than-gentle handling), the sensitivity of the balance decreases. Placing the two balance points and the fulcrum in the same horizontal plane causes the sensitivity of the balance to remain constant with varying loads.

The schematic diagram of a single-pan balance in Figure 2-3 shows the empty balance pan and a set of *removable* weights suspended from the balance point to the left of the fulcrum. The counterweight at the right exactly balances the empty pan and the removable weights. When a sample is placed on the pan, the left side becomes heavier. The dials on the front of the balance are then used to *remove* some of the weights attached to the balance pan until the two sides are almost back in balance. The remaining deflection from horizontal of the balance beam is measured by an optical scale, located at the back of the balance, that is projected to the front of the instrument, as shown in Figure 2-1. The sum of the weights removed plus the reading on the optical scale equals the mass of the sample.

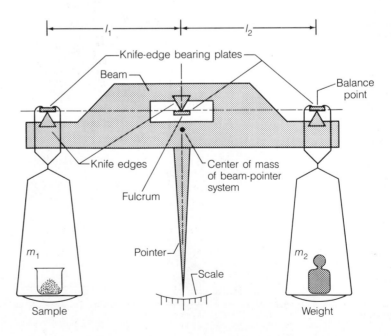

Figure 2-2
Principle of operation of a double-pan balance.

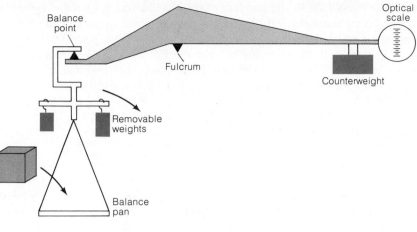

Figure 2-3
Principle of operation of a single-pan balance. [Courtesy Mettler Instrument Corp., Hightstown, N.J.]

Effect of Buoyancy

For accurate work, the buoyant effect of air must be considered. When a sample is placed on the balance pan, it displaces a certain amount of air. Removing this mass of air from the pan makes the object seem lighter than it actually is, because the pan was set to zero *with* the air on the pan. A similar effect applies to the removable weights on the balance. There will be a net effect of **buoyancy** whenever the density of the object being weighed is not equal to the density of the standard weights. The true mass of the object is given by[†]

$$m - \frac{m'\left(1 - \dfrac{d_a}{d_w}\right)}{1 - \dfrac{d_a}{d}} \qquad (2\text{-}2)$$

where m = true mass of object weighed in vacuum

m' = mass read from balance

d_a = density of air (0.001 2 g/mL near 1 atm and 25°C)

d_w = density of weights (typically 8.0 g/mL)

d = density of object being weighted

For accurate work the correct air density at the particular values of temperature, pressure, and humidity should be used.[‡]

[†] R. Batting and A. G. Williamson, *J. Chem. Ed.,* **61,** 51 (1984); J. E. Lewis and L. A. Woolf, *J. Chem. Ed.,* **48,** 639 (1971); F. F. Cantwell, B. Kratochvil, and W. E. Harris, *Anal. Chem.,* **50,** 1010 (1978). The notation *J. Chem. Ed.,* **61,** 51 (1984) means *Journal of Chemical Education,* volume 61, page 51, in the year 1984.

[‡] The density of air (g/L) is given by $d = 0.464\,68\left(\dfrac{B - 0.378\,3V}{T}\right)$, where B is the barometric pressure (in torr), V is the vapor pressure (in torr) of water in the air, and T is the air temperature (in kelvins).

The magnitude of the buoyancy factor is illustrated by a few examples. Suppose that the air density is 0.001 2 g/mL and the density of the balance weights is 8.0 g/mL. When we weigh water, with a density of 1.00 g/mL, the true mass is 1.001 1 g when the balance reads 1.000 0 g. The error is 0.11% in this case. For NaCl, with a density of 2.16 g/mL, the error would be 0.04%. For $AgNO_3$, whose density is 4.35 g/mL, the error is only 0.01%.

Errors in Weighing

Some care should be exercised to minimize weighing errors. A vessel being weighed should not be touched with your hands, since your fingerprints will change its mass. A sample should always be at ambient temperature before weighing to avoid errors due to convective air currents. Cooling a sample that has been dried in an oven usually requires about half an hour in a desiccator at room temperature. The balance pan should be in its arrested position when you are adding a load and in its half-arrested position when you are dialing weights. This protects against abrupt forces that wear down the knife edges serving as fulcrum and balance point (Figure 2-3). The glass doors of the analytical balance must be closed during readings to protect against oscillations due to air currents. Top-loading balances should be provided with a low solid "fence" around the pan to minimize the effect of air currents. Often a sensitive balance is placed on a heavy base (such as a large marble slab) to minimize the effect of room vibrations on the reading. Adjustable feet and a bubble meter at the top allow the balance to be maintained in a level position.

Electronic Balance[†]

The instrument in Figure 2-1 is a **mechanical balance,** even though it uses an electric light bulb to illuminate the scale deflection. A weighing is made by substituting the unknown for the built-in weights. The balance beam is restored to a position near its original position, and residual deflection is measured on the illuminated scale.

The **electronic balance,** which is rapidly replacing the mechanical balance, has no built-in weights. It uses electromagnetic force to restore the balance beam to its original position. The electric current needed to generate the force is proportional to the mass of the object being weighed.

The principle is best understood by first focusing on the electromagnetic restoring system in Figure 2-4. When a mass is placed on the pan the null detector senses a displacement and sends an error signal to the circuit that generates a correction current. This current flows through the coil attached to the base of the balance pan, creating a magnetic field. The magnetic field of the coil is attracted or repelled by the permanent magnet mounted beneath the pan. As the deflection decreases, the output of the null detector decreases. The correction current needed to restore the system to its initial position is proportional to the mass of the unknown on the balance. The instrument is calibrated to read in units of mass.

[†] R. M. Schoonover, *Anal. Chem.,* **54,** 973A (1982); B. B. Johnson and J. D. Wells, *J. Chem. Ed.,* **63,** 86 (1986).

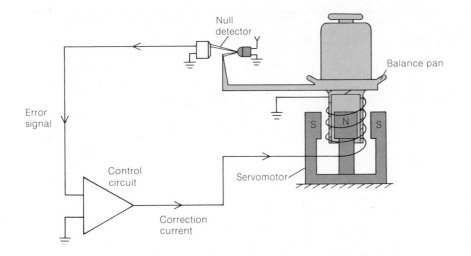

Figure 2-4
Principle of electromagnetic servo system. [R. M. Schoonover, *Anal. Chem.,* **54,** 973A (1982).]

A typical top-loading electronic balance is shown in cross section in Figure 2-5. The most notable feature is the absence of the balance beam and knife edges of the mechanical balance (Figure 2-3). The weighing pan is firmly attached to a solid parallelogram load-constraint system. This whole system flexes (bends) when a mass is placed on the pan. The load coupler transfers the motion to the beam at the center of the diagram. A null detector and servomotor similar to that in Figure 2-4 then restore the system to its initial position by applying an electric current. The servomotor is driven by the output of the null detector. When the detector senses that the original position has been restored, the motor stops correcting the position.

The electronic balance has potential errors not encountered with mechanical balances. There may be errors in weighing magnetic materials. You can check for this by moving the magnetic object near the empty pan and looking for changes in the zero reading. Electromagnetic radiation from nearby equipment might affect the balance reading. Dust must not be allowed to enter the gap between the coil and the permanent magnet of the servomotor.

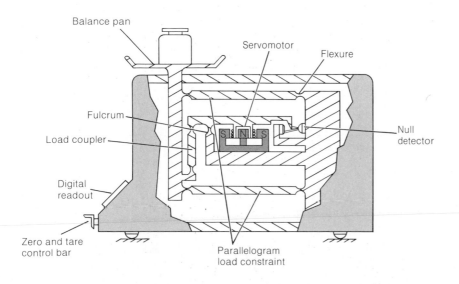

Figure 2-5
Cross section of electronic balance. [R. M. Schoonover, *Anal. Chem.,* **54,** 973A (1982).]

Box 2-1 PIEZOELECTRIC CRYSTAL DETECTORS

One of the most sensitive detectors of a change in mass is a vibrating quartz crystal.[†] When pressure is applied to the crystal, an electric voltage develops between certain surfaces of the crystal. This is called the *piezoelectric effect*. Conversely, a voltage applied to the crystal causes a distortion of the crystal. Your electric wristwatch and microcomputer use a sinusoidal voltage to set a quartz crystal into a precisely defined oscillation whose period serves as a clock in these devices.

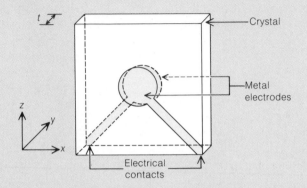

A typical piezoelectric quartz crystal. A sinusoidal voltage applied between the metal electrodes (along the *y* direction) causes the crystal to vibrate with a shearing motion along the *x* direction. The vibrational frequency, *F*, is given by $F = N/t$, where *t* is the thickness of the quartz (in meters) and $N = 1\,679$ Hz · m. A thickness of 0.17 mm gives an oscillation frequency of 10 MHz.

If a thin layer of mass *m* is spread over area *A* on the quartz surface, the oscillation frequency, *F*, decreases by an amount

$$\Delta F = 2.3 \times 10^{-10} \, F^2 \frac{m}{A}$$

where ΔF and *F* are given in Hz, *m* is measured in grams, and *A* is given in m^2. A 10-MHz crystal coated over 1 cm^2 of its surface with 1 μg of material will change its vibrational frequency by $\Delta F = 230$ Hz, a very measurable change.

[†] J. Hlavay and G. G. Guilbault, *Anal. Chem.,* **49,** 1890 (1977). For a fascinating account of how these crystals are grown artificially from superheated aqueous solution (called *hydrothermal* synthesis) see R. A. Laudise, *Chem. Eng. News,* 28 Sept. 1987, p. 30. Another sensitive chemical sensor based on piezoelectric crystals uses *surface acoustic waves.* To learn about this, see D. S. Ballantine, Jr., and H. Wohltjen, *Anal. Chem.,* **61,** 704A (1989).

A piezoelectric crystal coated with a water-absorbing material was sent to Mars with the Viking spacecraft to measure the concentration of water vapor in the Martian atmosphere. Another application of piezoelectric detectors has been to measure ozone (O_3) levels in the air around arc welders. Too much ozone in a poorly ventilated workspace is a health hazard. To detect ozone, a quartz crystal was coated with polybutadiene, which reacts irreversibly with O_3 to increase the mass of the coating. Air containing ozone was passed over the crystal and the vibrational frequency decreased at a constant rate as the reaction occurred. The figure below shows that the more ozone in the atmosphere, the greater was the rate of change of the oscillation frequency. Such a device was used as a monitor at the Danish Welding Institute.

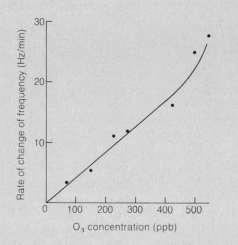

Rate of change of frequency of a polybutadiene-coated quartz oscillator when exposed to a 1 L/min flow of air containing ozone. [Data from H. M. Fog and B. Rietz, *Anal. Chem., 57,* 2634 (1985).]

Antibodies are proteins manufactured by living organisms to bind to foreign molecules and foreign cells and mark them for elimination. Antibodies are one of the primary defensive means by which we protect our bodies from infection. The molecule bound by an antibody is called an *antigen.* By chemically attaching antibodies to a piezoelectric device, the quartz crystal becomes a sensitive detector for such biological antigens as enzymes (proteins that catalyze biochemical reactions) and even whole microbes.[†]

[†] R. C. Ebersole and M. D. Ward, *J. Am. Chem. Soc.,* **110,** 8623 (1988); S. Borman, *Anal. Chem., 59,* 1161A (1987); H. Muramatsu, K. Kajiwara, E. Tamiya, and I. Karube, *Anal. Chim. Acta,* **188,** 257 (1986).

Table 2-1
Tolerances for laboratory balance weights

Denomination	Weight classification		
	S	S-1	P
Grams			
200	0.50	2.0	4.0
100	0.25	1.0	2.0
50	0.12	0.60	1.2
30	0.074	0.45	0.90
20	0.074	0.35	0.70
10	0.074	0.25	0.50
5	0.054	0.18	0.36
3	0.054	0.15	0.30
2	0.054	0.13	0.26
1	0.054	0.10	0.20
Milligrams			
500	0.025	0.080	0.16
300	0.025	0.070	0.14
200	0.025	0.060	0.12
100	0.025	0.050	0.10
50	0.014	0.042	0.085
30	0.014	0.038	0.075
20	0.014	0.035	0.070
10	0.014	0.030	0.060
5	0.014	0.028	0.055
3	0.014	0.026	0.052
2	0.014	0.025	0.050
1	0.014	0.025	0.050

Note: All tolerances are in milligrams.

Perhaps the most important limitation of the electronic balance is that it is calibrated with a standard mass at a factory where the force of gravity is not the same as the force of gravity in your lab. Gravitational acceleration varies over a range of $\sim 0.1\%$ among different locations in the United States. It is therefore essential that you calibrate the balance with a standard mass in your own lab. The buoyancy correction for subsequent weighings uses Equation 2-2 with d_w set to the density of the standard mass. The manufacturer uses a standard mass of $d = 8.0\,g/mL$ for the factory calibration. The tolerances of standard masses that can be used for balance calibration are listed in Table 2-1.

2-3 BURETS

A **buret** is a precisely bored glass tube with graduations enabling you to measure the volume of liquid delivered. This is done by reading the level before and after draining liquid from the buret. The typical glass buret in Figure 2-6a has a Teflon stopcock. A loosely fitting cap is used to keep dust out and vapors in. The graduations of Class A burets are certified to meet the tolerances in Table 2-2.

When you read the height of liquid in a buret, it is important that your eye be at the same level as the top of the liquid. This minimizes the **parallax** error associated with reading the position of the liquid. If your eye is above this level, the liquid seems to be higher than it actually is. If your eye is too low, there appears to be less liquid than is actually present.

The surface of most liquids forms a concave **meniscus,** as shown in Figure 2-7. It is useful to use a piece of black tape on a white card as a background for locating the precise position of the meniscus. To use this card, align the top of the black tape with the bottom of the meniscus and read

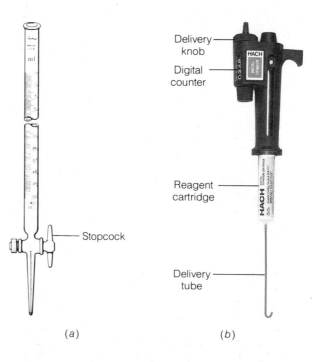

Figure 2-6
(a) Glass buret. [Courtesy A. H. Thomas Co., Philadelphia, Pa.] (b) Digital titrator with plastic cartridge containing reagent solution. [Courtesy Hach Co., Loveland, Co.]

(a)   (b)

Table 2-2
Tolerances of Class A burets

Buret volume (mL)	Smallest graduation (mL)	Tolerance (mL)
5	0.01	±0.01
10	0.05 *or* 0.02	±0.02
25	0.1	±0.03
50	0.1	±0.05
100	0.2	±0.10

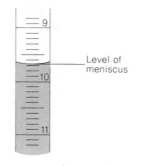

Level of meniscus

Figure 2-7
A portion of a buret with the level of the meniscus at 9.68 mL. You should always estimate the reading of any scale to the nearest tenth of a division. This corresponds to 0.01 mL for the buret in this figure.

the position on the buret. Some solutions, especially highly colored ones, appear to have two meniscuses. In such cases, either one may be used. The important point is to perform the readings reproducibly.

The markings of a buret normally have a thickness that is *not* negligible in comparison with the distance between markings. The thickness of the markings corresponds to about 0.02 mL, for a 50-mL buret. To most accurately use the buret, you should select one portion of the marking to be called zero. For example, say that the liquid level is at the mark when the bottom of the meniscus just touches the *top* of the mark on the glass. When the bottom of the meniscus is at the *bottom* of the mark, the reading is 0.02 mL greater.

Near the end point of a titration, it is desirable to deliver less than one drop at a time from the buret. This permits a finer location of the end point, since the volume of a drop is about 0.05 mL (from a 50-mL buret). To deliver a fraction of a drop, carefully open the stopcock until part of a drop is hanging from the buret tip. (Alternatively, you can allow just a fraction of a drop to emerge from the buret by very rapidly turning the stopcock through the open position.) Then touch the inside glass wall of the receiving flask to the buret tip to transfer the liquid to the wall of the flask. Carefully tip the flask so that the main body of liquid washes over the newly added liquid, and mix the contents. Near the end of a titration, the flask should be tipped and rotated to ensure that droplets on the walls containing unreacted analyte come in contact with the bulk solution. Liquid should drain evenly down the walls of a buret. If it does not, the buret should be cleaned with detergent and a buret brush. If this is insufficient, the buret should be soaked in peroxydisulfate–sulfuric acid cleaning solution.[†] Volumetric glassware should never be soaked in an alkaline cleaning solution, since glass is slowly attacked by base. The tendency of liquid to stick to the walls of a buret can be diminsihed by draining the buret slowly. A slowly drained buret will also provide greater reproducibility of results. *A drainage rate of no more than 20 mL/min should be used.*

One of the most common errors in using a buret is caused by failure to expel the bubble of air that often forms directly beneath the stopcock (Figure

Buret reading tips:

1. Read the bottom of the concave meniscus.
2. Avoid parallax.
3. Account for the thickness of the marking lines in your readings.

[†] Cleaning solution is prepared by dissolving 36 g of ammonium peroxydisulfate, $(NH_4)_2S_2O_8$, in 2.2 L (a "one-gallon" bottle) of 98% sulfuric acid. More ammonium peroxydisulfate is added every few weeks, as necessary, to maintain the oxidizing capability of the solution. This mixture is an extremely powerful oxidant that eats clothing and people, as well as dirt and grease. It should be stored and used in a fume hood. Peroxydisulfate cleaning solution replaces chromic acid, which is a carcinogen. [H. M. Stahr, W. Hyde, and L. Sigler, *Anal. Chem.,* **54,** 1456A (1982).]

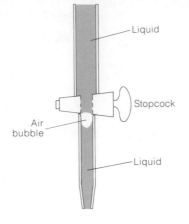

Figure 2-8
The air bubble often trapped beneath the stopcock of a buret should be expelled before you use the apparatus.

2-8). If an air bubble is present at the start of a titration, it may fill in during the titration, causing an error in the volume of liquid delivered from the buret. Usually the bubble can be dislodged by draining the buret for a second or two with the stopcock wide open. Sometimes a tenacious bubble must be expelled by carefully shaking the buret while draining liquid into a sink.

When you fill a buret with fresh solution, it is a good idea to rinse the buret several times with the new solution, discarding each wash. It is not necessary to fill the whole buret with the wash solution. Simply tilt the buret to allow the whole surface to come in contact with a small volume of liquid. This same washing technique can be applied to any vessel (such as a spectrophotometer cuvet or a pipet) that must be reused without opportunity for drying.

The digital titrator in Figure 2-6b is more convenient and portable, but less accurate, than the customary glass buret in Figure 2-6a. The digital titrator is especially useful for conducting measurements in the field where samples are collected. The counter tells how much reagent from the disposable cartridge has been dispensed by rotation of the delivery knob. Prepackaged reagents are available, or you can fill an empty cartridge with your own solution. The accuracy of 1% is about ten times poorer than the accuracy of a glass buret, but many field or quality control operations do not require greater accuracy.

> Washing any glassware with a new solution is a good idea.

> Section 9-3 describes a method that uses a syringe to deliver a known *mass* of reagent instead of a buret to deliver a kown *volume*.

2-4 VOLUMETRIC FLASKS

A **volumetric flask** is calibrated to contain a particular volume of water at 20°C when the bottom of the meniscus is adjusted to the center of the line marked on the neck of the flask (Figure 2-9, Table 2-3). Most flasks bear the label "TC 20°C," which means that the flask is calibrated *to contain* the indicated volume at 20°C. (Other types of glassware may be calibrated *to deliver,* "TD," their indicated volume.) The temperature is important because both the liquid and the glass expand as they are heated.

To adjust the liquid level to the *center* of the mark of a volumetric flask (or of a pipet, which is calibrated in the same way), look at the mark from *above* or *below* the level of the mark. The front and back of the mark will not be aligned with each other and will describe an ellipse. Drain the liquid

> Thermal expansion of water and glass is discussed in Section 2-9.

> Although glass expands when it is heated, modern laboratory glassware made of Pyrex or other low expansion glass can be safely dried in an oven to at least 320°C without harm. [D. R. Burfield and G. Hefter, *J. Chem. Ed.,* **64,** 1054 (1987).] Glass is normally dried at 110–150°C and there is rarely reason to heat it to a higher temperature.

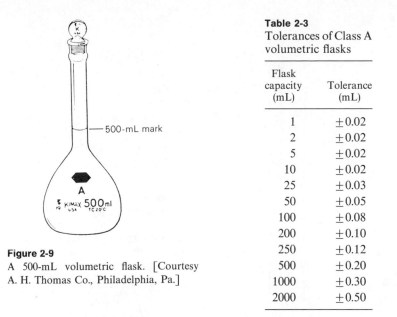

Figure 2-9
A 500-mL volumetric flask. [Courtesy
A. H. Thomas Co., Philadelphia, Pa.]

Table 2-3
Tolerances of Class A
volumetric flasks

Flask capacity (mL)	Tolerance (mL)
1	±0.02
2	±0.02
5	±0.02
10	±0.02
25	±0.03
50	±0.05
100	±0.08
200	±0.10
250	±0.12
500	±0.20
1000	±0.30
2000	±0.50

until the bottom of the meniscus is at the *center* of the ellipse (Figure 2-10). This places the bottom of the meniscus exactly at the center of the calibration mark, if viewed edge on.

A volumetric flask is used to prepare a solution of a known volume. Most commonly, a reagent is weighed into a volumetric flask, dissolved, and diluted to the mark. This way the mass of reagent and the final volume of the solution are both known. The solid should first be dissolved in *less* liquid than the flask is calibrated to contain. Then more liquid is added, and the solution is again mixed. The final adjustment to the mark should be done with as much well-mixed solution as possible in the flask. This minimizes the change in volume upon mixing pure liquid with solution already in the flask. The final drops of liquid should be added with a pipet, *not a squirt bottle,* for best control. After adjusting the solution to its final volume the cap should be held firmly in place and the flask *inverted 20 or more times* to assure complete mixing.

Figure 2-10
Proper position of the meniscus: at the center of the ellipse formed by the front and back of the calibration mark when viewed from above or below. Volumetric flasks and pipets are calibrated to this position.

2-5 PIPETS AND SYRINGES

Pipets are used to deliver known volumes of liquid. Four common types are shown in Figure 2-11. The transfer pipet, which is the most accurate, is calibrated to deliver one fixed volume. The last drop of liquid does not drain out of the pipet; it is meant to be left in the pipet. *It should not be blown out.* The measuring pipet is calibrated to deliver a variable volume, as indicated by the difference between the volumes indicated before and after delivery. The measuring pipet in Figure 2-11 could, for example, be used to deliver 5.6 mL by starting delivery at the 1-mL mark and ending at the 6.6-mL mark. The Ostwald–Folin pipet is similar to the transfer pipet, except that the last drop *should* be blown out. When in doubt about whether the pipet is made for blowout, consult the manufacturer's catalog. Serological pipets are measuring pipets calibrated all the way to the tip. The serological pipet in Figure 2-11 will deliver 10 mL when the last drop is blown out.

Do not blow out the last drop from a transfer pipet.

Calibration mark

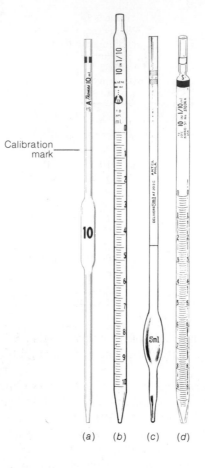

(a)   (b)   (c)   (d)

Figure 2-11
Some common pipets. (a) Transfer. (b) Measuring (Mohr). (c) Ostwald–Folin (blow out last drop). (d) Serological (blow out last drop). [Courtesy A. H. Thomas Co., Philadelphia, Pa.]

Table 2-4
Tolerances of Class A transfer pipets

Volume (mL)	Tolerance (mL)
0.5	±0.006
1	±0.006
2	±0.006
3	±0.01
4	±0.01
5	±0.01
10	±0.02
15	±0.03
20	±0.03
25	±0.03
50	±0.05
100	±0.08

In general, transfer pipets are more accurate than measuring pipets. The tolerances for Class A transfer pipets are given in Table 2-4. To deliver the calibrated volume, bring the bottom of the meniscus to the center of the calibration mark, as shown in Figure 2-10.

Using a Transfer Pipet

Using a rubber bulb, *not your mouth,* suck liquid up past the mark. Quickly put your index finger over the end of the pipet in place of the bulb. The liquid should still be above the mark at this time. Pressing the pipet against the bottom of the vessel while removing the rubber bulb helps prevent liquid from draining while you get your finger in place.[†] Wipe the excess liquid off the outside of the pipet with a clean tissue. *Touch the tip of the pipet to the side of a beaker,* and drain the liquid until the bottom of the meniscus just reaches the center of the mark. The pipet must touch the beaker during the draining so that extra liquid is not hanging from the tip of the pipet when the meniscus reaches the mark. Any liquid outside of the pipet will be drawn

---

[†] G. Deckey (*J. Chem. Ed.,* **57,** 526 (1980)) describes a way to use a rubber bulb and an Eppendorf pipet tip to apply suction to a pipet whose outer diameter is smaller than the hole in the rubber bulb.

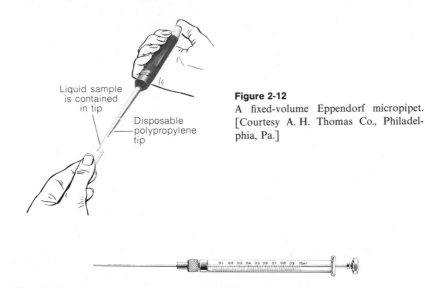

Figure 2-12
A fixed-volume Eppendorf micropipet. [Courtesy A. H. Thomas Co., Philadelphia, Pa.]

Figure 2-13
A Hamilton syringe with a total volume of 1 μL and divisions separated by 0.01 μL. The barrel is made of glass, and the needle is stainless steel. In this particular syringe, the entire sample is contained in the needle. In larger syringes, the liquid is held in the barrel. [Courtesy A. H. Thomas Co., Philadelphia. Pa.]

onto the wall of the beaker. Now transfer the pipet to the desired receiving vessel and drain it *while holding the tip to the wall of the vessel.* After the pipet stops draining, hold it against the wall for a few seconds more to be sure that everything that can drain has drained. *Do not blow out the last drop.* The pipet should be nearly vertical to ensure that the proper amount of liquid has drained. When you finish with a pipet, it should be rinsed with distilled water or placed in a pipet container to soak until it is cleaned. Solutions should never be allowed to dry inside a pipet because removing deposits is very difficult.

Delivering Small Volumes

Plastic micropipets, such as that shown in Figure 2-12, are used to deliver microliter volumes (1 μL = 10^{-6} L). The liquid is contained entirely in a disposable plastic tip. Pipets are available covering the range 1–1000 μL, in fixed or variable volumes. The accuracy is 1–2%, and the precision may reach 0.5%. Accuracy refers to how close the delivered volume is to the desired volume. Precision refers to the reproducibility of replicate deliveries.

For delivering very small variable volumes of liquid, a microliter **syringe** (Figure 2-13) is excellent. Syringes are available in a wide range of volumes, with accuracy and precision close to 1%.

For maximum accuracy an individual operator can calibrate an individual pipet tip as described by B. Kratochvil and N. Motkosky, *Anal. Chem.,* **59,** 1064 (1987).

The composition of the container (glass, plastic, steel, etc.) is important when you are handling small volumes. It takes very little impurity from the container wall to significantly affect the composition of 1 μL of solution.

2-6 FILTRATION

In **gravimetric analysis,** the mass of product from a reaction is measured to determine how much unknown was present. Precipitates for gravimetric analysis are collected by filtration, washed, and then dried. If the precipitate does

Figure 2-14
A fritted-glass Gooch filter crucible [Courtesy A. H. Thomas Co., Philadelphia, Pa.]

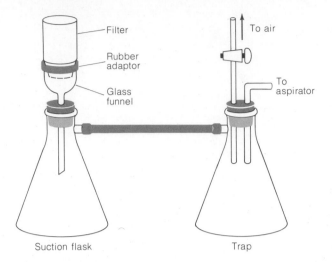

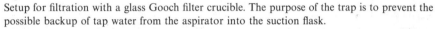

Figure 2-15
Setup for filtration with a glass Gooch filter crucible. The purpose of the trap is to prevent the possible backup of tap water from the aspirator into the suction flask.

not need to be ignited, it is most conveniently collected in a fritted-glass funnel, such as that shown in Figure 2-14. This funnel has a porous glass disk that allows liquid to pass through, but retains solid particles. In gravimetric analysis, the filter crucible is dried and weighed before the precipitate is collected. After the product is collected and dried in the crucible, the crucible and its contents are weighed together to determine the mass of precipitate. The filter crucible is usually used with suction provided by a water aspirator, as shown in Figure 2-15. The liquid from which a substance precipitates or crystallizes is called the **mother liquor.** The liquid that passes through a filter is called the **filtrate.**

Some gravimetric procedures require isolating a precipitate, then **igniting** (heating) it at high temperature to convert it to a well-defined product of known composition. For example, Fe^{3+} is precipitated as a poorly defined hydrated form of $Fe(OH)_3$ and ignited to Fe_2O_3 before weighing. When a gravimetric precipitate is to be ignited, it is collected in *ashless* filter paper that leaves little residue when oxidized at high temperature. The filter paper is folded in quarters, one corner torn off, and the paper placed in a conical glass funnel (Figure 2-16). The paper should fit snugly and be seated with a little distilled water. When liquid is poured into the funnel, an unbroken stream of liquid should fill the stem of the funnel. The weight of the liquid in the stem helps to speed the filtration.

The correct procedure for filtration is shown in Figure 2-17. The liquid containing suspended precipitate is poured down a glass rod into the filter. The rod helps prevent splattering or dripping down the side of the beaker.

Figure 2-16
Folding filter paper for a conical funnel. (a) Fold the paper in half. (b) Then fold it in half again. (c) Tear off a corner. (d) Open the side that has not been torn when fitting the paper in the funnel.

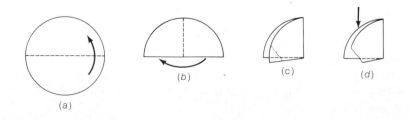

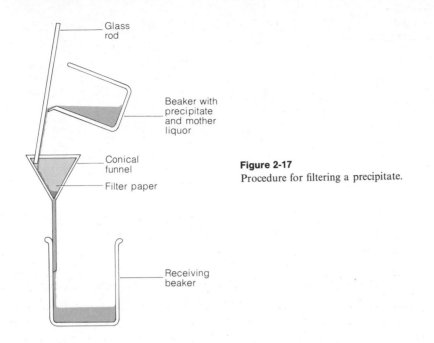

Glass
rod

Beaker with
precipitate
and mother
liquor

Conical
funnel

Filter paper

Receiving
beaker

Figure 2-17
Procedure for filtering a precipitate.

Particles adhering to the beaker or rod can be dislodged with a **rubber policeman** (a glass rod with a flattened piece of rubber attached to the end) and transferred with a jet of liquid from a squirt bottle containing distilled water or an appropriate wash liquid. Particles that remain in the beaker may be wiped onto a small piece of moist filter paper and the paper placed in the filter, to be ignited with the bulk of the precipitate.

2-7 DRYING

Reagents, precipitates, and glassware are conveniently dried in an electric oven, usually maintained at about 110°C. (Some reagents or precipitates require other drying temperatures.) When a filter crucible is to be brought to constant mass prior to a gravimetric analysis, the crucible should be dried for one hour or longer and then cooled in a desiccator. The crucible is weighed and then heated again for about 30 minutes. When successive weighings agree to within 0.3 mg, the filter is considered to have been dried "to constant mass." Drying solid reagents or filter crucibles in an oven should be done using a beaker and watchglass (Figure 2-18) to prevent dust from falling into the reagent.

Dust is an important source of contamination in analytical chemistry. It is good practice to cover all vessels on the benchtop whenever possible.

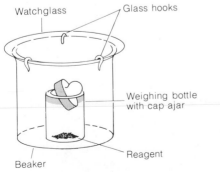

Watchglass   Glass hooks

Weighing bottle
with cap ajar

Reagent

Beaker

Figure 2-18
Use of a beaker and watchglass to keep dust out of a reagent while it is drying in the oven.

Table 2-5

Efficiencies of drying agents

Agent	Formula	Water left in atmosphere (μg H_2O/L)
Magnesium perchlorate, anhydrous	$Mg(ClO_4)_2$	0.2
"Anhydrone"	$Mg(ClO_4)_2 \cdot 1-1.5H_2O$	1.5
Barium oxide	BaO	2.8
Alumina	Al_2O_3	2.9
Phosphorus pentoxide	P_4O_{10}	3.6
Lithium perchlorate, anhydrous	$LiClO_4$	13
Calcium chloride (dried at 127°C)	$CaCl_2$	67
Calcium sulfate ("Drierite")	$CaSO_4$	67
Silica gel	SiO_2	70
Ascarite	NaOH on asbestos	93
Sodium hydroxide	$NaOH$	513
Barium perchlorate	$Ba(ClO_4)_2$	599
Calcium oxide	CaO	656
Magnesium oxide	MgO	753
Potassium hydroxide	KOH	939

Note: Moist nitrogen was passed over each desiccant, and the water remaining in the gas was condensed and weighed.

SOURCE: A. I. Vogel, *A Textbook of Quantitative Inorganic Analysis,* 3rd ed. (New York: Wiley, 1961), p. 178.

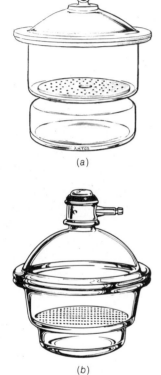

Figure 2-19

Desiccators. (a) Ordinary desiccator. (b) Vacuum desiccator. The vacuum desiccator can be evacuated through the sidearm at the top and then sealed by rotating the joint with the sidearm. Drying is more efficient at low pressure. [Courtesy A. H. Thomas Co., Philadelphia, Pa.]

After a regent or crucible has been dried at high temperature, it is cooled to room temperature in a **desiccator,** a closed chamber that contains a drying agent, or **desiccant.** The interface between the lid and the body of the desiccator is greased to make an airtight seal. Two common types of desiccators are shown in Figure 2-19. The desiccant is placed beneath the perforated disk near the bottom of the chamber. Data on the efficacy of various common drying agents are given in Table 2-5. In addition to those listed, 98% sulfuric acid is also a common and efficient desiccant. After placing a hot object in a desiccator, leave the lid cracked open for a minute or two until the object has cooled slightly. This prevents the lid from popping open when the air inside is warmed by the hot object. An object being cooled prior to weighing should be given about 30 minutes to reach room temperature. The correct way to open a desiccator is to slide the lid sideways until it can be removed. The greased seal prevents you from opening the desiccator with a direct upward pull on the lid.

2-8 IGNITION

The object of the ignition is to convert the moist $Fe(OH)_3$ to dry, pure Fe_2O_3.

If a precipitate such as $Fe(OH)_3$ is to be ignited, it is first collected in ashless filter paper as described in the Section 2-6. The filter should be allowed to drain thoroughly—preferably overnight—with the contents protected from dust by a watchglass. Carefully lift the paper out of the funnel, fold it as shown in Figure 2-20, and transfer it to a porcelain crucible that has been brought to constant mass by several cycles of heating, cooling in a desiccator,

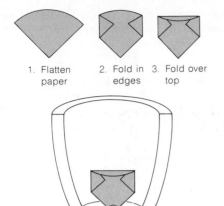

1. Flatten paper   2. Fold in edges   3. Fold over top

4. Place inside crucible with point pushed against bottom

Figure 2-20
Folding filter paper and placing it inside a crucible for ignition. Continued folding of the paper so that the entire packet fits on the bottom of the crucible may be necessary. Be careful not to puncture the paper.

and weighing. The lid of the crucible should also have been heated cooled, and weighed together with the crucible.

With the filter paper and precipitate inside the crucible, dry the contents cautiously with a small flame, as shown in Figure 2-21. The flame should be directed at the top of the crucible, and the lid should be off. Avoid spattering. After it is dry, *char* the filter paper by increasing the flame temperature. For precipitates such as $Fe(OH)_3$, the crucible should have free access to air to avoid reduction of the product by carbon from the filter paper. The lid should be kept handy to smother the crucible if the paper inflames. Use tongs, not your hands, to manipulate the crucible and lid. Any carbon left on the crucible or lid should be burned by directing the burner flame at it. Finally, complete the ignition by heating the bottom of the crucible at the maximum flame temperature (the bluest flame) for 15 minutes.

After ignition the crucible and lid are cooled briefly in air and then in a desiccator for 30 minutes. Bring the crucible and its contents and lid to constant mass by repeated heatings.

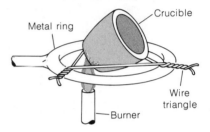

Metal ring   Crucible   Wire triangle   Burner

Figure 2-21
Positioning of a crucible above a burner.

2-9 CALIBRATION OF VOLUMETRIC GLASSWARE

When greatest accuracy is desired, it is necessary to calibrate the individual volumetric glassware that you are using. This is usually done by measuring the mass of water delivered or contained in the vessel, and using the density of water to convert mass to volume. By this means you could determine, for example, that your particular 10-mL pipet delivers 10.016 mL, not 10.000 mL.

Table 2-6 shows that pure water expands ~0.02% per degree near 20°C. This change has a practical implication for the calibration of glassware and for the temperature dependence of reagent concentrations.

For small or odd-shaped vessels, mercury can be used instead of water. Mercury is easier to pour out of glass and weighs 13.5 times as much as an equal volume of water.

EXAMPLE: Effect of Temperature on Solution Concentration

A dilute aqueous solution with a molarity of 0.031 46 M was prepared in the winter when the lab temperature was 17°C. What will be the molarity of that solution on a warm spring day when the temperature is 25°C?

We assume that the thermal expansion of the solution is similar to the thermal expansion of pure water. Since the concentration of a solution is proportional to its density, we can write

$$\frac{c'}{d'} = \frac{c}{d} \tag{2-3}$$

where c' and d' are the concentration and density at temperature T', and c and d apply at temperature T. Using the densities in column 2 of Table 2-6, we can write

Note that the concentration of the solution decreases when the temperature increases.

$$\frac{c \text{ at } 25°}{0.997\,05} = \frac{0.031\,46}{0.998\,78} \quad \Rightarrow \quad c = 0.031\,41 \text{ M}$$

The concentration has decreased by 0.16%.

Table 2-6
Density of water

Temperature (°C)	Density of water (g/cm³)	Volume of 1 g of water (cm³)	
		At temperature shown†	Corrected to 20°C‡
0	0.999 842 5	—	—
4	975 0	—	—
5	966 8	—	—
10	702 6	1.001 4	1.001 5
11	608 4	1.001 5	1.001 6
12	500 4	1.001 6	1.001 7
13	380 1	1.001 7	1.001 8
14	247 4	1.001 8	1.001 9
15	102 6	1.002 0	1.002 0
16	0.998 946 0	1.002 1	1.002 1
17	777 9	1.002 3	1.002 3
18	598 6	1.002 5	1.002 5
19	408 2	1.002 7	1.002 7
20	207 1	1.002 9	1.002 9
21	0.997 995 5	1.003 1	1.003 1
22	773 5	1.003 3	1.003 3
23	541 5	1.003 5	1.003 5
24	299 5	1.003 8	1.003 8
25	047 9	1.004 0	1.004 0
26	0.996 786 7	1.004 3	1.004 2
27	516 2	1.004 6	1.004 5
28	236 5	1.004 8	1.004 7
29	0.995 947 8	1.005 1	1.005 0
30	650 2	1.005 4	1.005 3
35	0.994 034 9	—	—
37	0.993 331 6	—	—
40	0.992 218 7	—	—
100	0.958 366 5	—	—

† Corrected for buoyancy with Equation 2-2, using 0.001 2 g/mL air density and density of weights = 8.0 g/mL.
‡ Corrected for expansion of borosilicate glass (0.001 0% per degree Celsius).

Box 2-2 DISPOSAL OF CHEMICAL WASTE

Many of the chemicals we commonly use are harmful to plants, animals, and people if carelessly discarded. While many solutions can be safely poured down the drain, others will put toxic waste into rivers and groundwater. For each experiment you do, your instructor needs to determine in advance how reagents will be disposed. Options include (1) pouring solutions down the drain and diluting with tap water, (2) saving the waste for proper disposal in an approved landfill, (3) chemically treating the waste to make it less hazardous and then pouring it down the drain or saving it for disposal at a landfill, and (4) recycling the chemical so that little waste is produced. It is critical that chemically incompatible wastes not be combined with each other and that *each waste container be properly labeled to identify the type and approximate quantity of waste.* Practical methods for dealing with different wastes can be found in the book *Prudent Practices for Disposal of Chemicals from Laboratories* (Washington: National Academy, 1983).

A few examples illustrate different approaches to managing lab waste.[†] Waste dichromate ($Cr_2O_7^{2-}$) is first reduced to Cr(III) with sodium bisulfite ($NaHSO_3$), precipitated with hydroxide to make insoluble $Cr(OH)_3$, and evaporated to a small volume of solid waste for a landfill. Waste solutions of acids can be mixed with waste solutions of base until they are nearly neutral (as determined with pH paper) and then poured down the drain. Waste iodate (IO_3^-) solutions can be treated with $NaHSO_3$ to reduce the IO_3^- to I^-. This generates acid that must be neutralized with base. Then the whole solution can be safely poured down the drain. Waste Pb^{2+} solution is treated with sodium metasilicate (Na_2SiO_3) solution to precipitate insoluble $PbSiO_3$ that can be packaged for a landfill. Waste silver or gold should be chemically treated to recover these valuable metals. A toxic gas used in a fume hood should be bubbled through a chemical trap to prevent its escape from the hood. Some toxic gases can be passed through a burner that converts them to harmless products.

[†] M.-A. Armour, *J. Chem. Ed.,* **65**, A64 (1988); W. A. Walton, *J. Chem. Ed.,* **64**, A69 (1987).

Glass itself expands when it is heated. Pyrex and other borosilicate glasses, which are the most common types, expand by about $1.0 \times 10^{-3}\%$ per degree Celsius near room temperature. This means that if a glass container's temperature is increased 10°C, its volume will increase by about $(10)(0.001\ 0\%) = 0.01\%$. For all but the most accurate work, this expansion is insignificant. Soft glass expands approximately two to three times as much as borosilicate glass.

Suppose that you want to calibrate a 25-mL pipet. You should first weigh an empty weighing bottle such as the one in Figure 2-18. Then carefully fill the pipet to the 25-mL mark with distilled water. Carefully drain the pipet into the weighing bottle and put the cap on to prevent evaporation from the jar. Weigh the jar again to find the mass of water delivered from the pipet. Use Table 2-6 to convert the mass of water into volume of water.

EXAMPLE: Calibration of a Pipet

The empty weighing bottle weighed 10.283 g. After filling it from a 25-mL pipet, the mass was 35.225 g. If the lab temperature was 23°C, find the volume of water delivered from the pipet. What volume would be delivered if the temperature were 20°C?

The mass of water in the pipet is 35.225 - 10.283 = 24.942 g. From column 3 of Table 2-6, the volume of water is (24.942 g)(1.003 5 mL/g) = 25.029 mL at 23°C. This is the volume delivered by the pipet at 23°C. If the pipet and water were at 20°C

Column 3 of Table 2-6 includes the effect of buoyancy.

instead of 23°C, the pipet would contract slightly and contain less water. However, columns 3 and 4 of Table 2-6 show that the correction from 23°C to 20°C is not significant to four decimal places. The volume delivered at 20°C would still be 25.029 mL.

In the most careful work it is necessary to account for thermal expansion and contraction of solutions and glassware. It is therefore necessary to know the laboratory temperature at the time solutions are made and when they are used.

Summary

"Wet" chemical analyses require the use of equipment and techniques with which you should become familiar. The analytical balance should be treated as a delicate piece of equipment, and buoyancy corrections should be employed in exact work. Burets should be read in a reproducible manner and drained slowly for best results. Always interpolate between the markings to obtain accuracy one decimal place beyond the graduations. Volumetric flasks are used to prepare solutions with a known volume. Transfer pipets are used to deliver a fixed volume of liquid, while measuring pipets, which are less accurate, deliver variable volumes. Filtering solutions or collecting precipitates requires proper technique, as does the drying of reagents, precipitates, and glassware in ovens and desiccators. Ignition is used to convert a gravimetric precipitate to a known, stable composition useful for measurement by weighing. In the most careful work the concentrations of solutions and volumes delivered or contained in glassware should be corrected for changes in temperature. The importance of maintaining a complete, accurate, and intelligible notebook cannot be overstated. Chemical waste must be disposed in a deliberate, safe, and legal manner.

Terms to Understand

ashless filter paper	desiccator	ignition	pipet
buoyancy	electronic balance	mechanical balance	rubber policeman
buret	filtrate	meniscus	syringe
constant mass	gravimetric analysis	mother liquor	tare
desiccant	hygroscopic	parallax	volumetric flask

Exercises[†]

2-A. What is the true mass of a sample of water whose mass measured in the atmosphere is 5.397 4 g? When you look up the density of water, assume that the lab temperature is (a) 15°C and (b) 25°C. Assume that the density of air is 0.001 2 g/mL at both temperatures and the density of balance weights is 8.0 g/mL.

[†] In case you missed this note in Chapter 1, detailed solutions to Exercises appear at the end of this book. Brief numerical answers to the Problems appear after the solutions to Exercises. Detailed solutions to the Problems are given in the *Solutions Manual to Accompany Quantitative Chemical Analysis*. The A Problems are intended to be simpler than the other Problems and Exercises.

2-B. The density of ferric oxide, Fe_2O_3, is 5.24 g/mL. A sample obtained from ignition of a gravimetric precipitate weighed 0.296 1 g in the atmosphere. What is the true mass in vacuum?

2-C. A solution of potassium permanganate ($KMnO_4$) was found by titration to have a molarity of 0.051 38 M at 24°C on a warm day. What was the molarity of the same solution on a cool night when the lab temperature dropped to 16°C?

2-D. Water was drained from a buret between the 0.12 mL and 15.78 mL marks. The apparent volume delivered was $15.78 - 0.12 = 15.66$ mL. Measured in the air at 22°C, the mass of water delivered was 15.569 g. What was the true volume delivered from the buret?

Problems

A2-1. State three essential attributes of a lab notebook.

A2-2. What is the true mass of benzene ($d = 0.88$ g/mL) weighed into a flask when a balance says that 9.947 g is present? Assume that the air density is $d = 0.001\ 2$ g/mL and the reference masses in the balance have a density of $d = 8.0$ g/mL.

A2-3. What do the symbols "TD" and "TC" mean on volumetric glassware?

A2-4. Describe the proper procedure for preparing 250.0 mL of 0.150 0 M K_2SO_4 with a volumetric flask.

A2-5. Describe the correct procedure for transferring 5.00 mL of a liquid reagent using a transfer pipet.

A2-6. Which pipet is more accurate: a transfer pipet or a measuring pipet?

A2-7. What would you do differently when delivering 1.00 mL of liquid from a 1-mL serological pipet, as compared to using a 1-mL measuring pipet?

A2-8. Which drying agent is more efficient, "Drierite" or phosphorus pentoxide?

A2-9. What is the purpose of the trap in Figure 2-15? What is the purpose of the watchglass in Figure 2-18?

A2-10. What is the purpose of igniting a gravimetric precipitate? List the steps in ignition.

A2-11. By what percentage does a dilute aqueous solution expand when heated from 15°C to 25°C? If a 0.500 0 M solution is prepared at 15°C, what would its molarity be at 25°C?

A2-12. An empty 10-mL volumetric flask weighs 10.263 4 g. When filled to the mark with distilled water and weighed again in the air at 20°C, the mass is 20.214 4 g. What is the true volume of the flask at 20°C?

2-13. The true volume of a particular 50 mL volumetric flask is 50.037 mL at 20°C. What mass of water measured in vacuum at 20°C would be contained in the flask? What mass of water measured in air at 20°C would be contained in the flask?

2-14. You would like to prepare 500.0 mL of solution containing exactly 1.000 M KNO_3 at 20°C, but the lab (and water) temperature is 24°C at the time of preparation. How many grams of KNO_3 should be dissolved in a volume of 500.0 mL at 24°C so that the concentration will be 1.000 M at 20°C? The density of KNO_3 is 2.109 g/mL. What apparent mass of KNO_3 weighed in air is required?

2-15. Prepare a graph showing the buoyancy correction (expressed as a percent of the sample weight) versus the sample density. Calculate correction factors for the following densities (g/mL), assuming $d_a = 0.001\ 2$ g/mL and $d_w = 8.0$ g/mL.
(a) 0.5 (b) 1 (c) 2 (d) 3 (e) 4
(f) 6 (g) 8 (h) 10 (i) 12 (j) 14
For comparison, look up the densities of the following substances: pentane, acetic acid, CCl_4, sulfur, sodium acetate, $AgNO_3$, Hg, Pb, PbO_2. For which substance will the buoyancy correction be least?

2-16. (a) What is the vapor pressure of water in the air when the temperature is 20°C and the humidity is 42%? The vapor pressure of water at 20°C at equilibrium is 17.5 torr. (*Hint:* Humidity refers to the fraction of the maximum water vapor pressure present in the air.)

(b) Find the air density (g/mL, not g/L) under the conditions of part a if the barometric pressure is 705 torr.

(c) What is the true mass of water under the conditions of part b if the balance indicates that 1.000 0 g is present? Assume $d_w = 8.0$ g/mL.

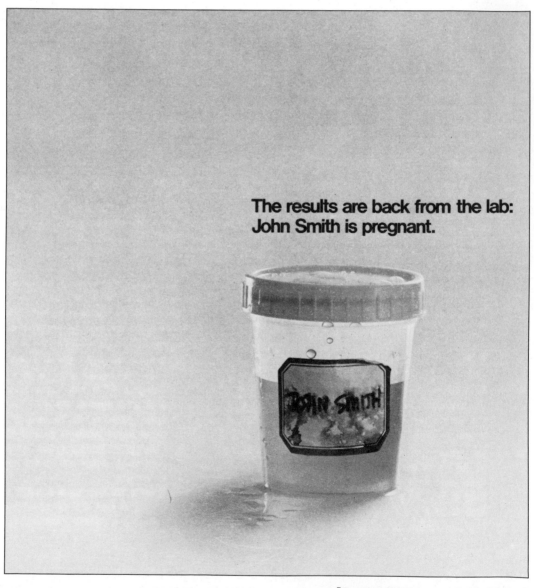

The results are back from the lab:
John Smith is pregnant.

3 Experimental Error

Not all laboratory errors are so monumental as that depicted on the opposite page, but there is error associated with every kind of measurement. There is no way to measure the "true value" of anything. The best we can do in a chemical analysis is to apply carefully a technique that experience tells us is reliable. Also, we can measure a quantity in several ways using different methods to see if the measurements agree. You should always be cognizant of the uncertainty associated with a result and, therefore, of how reliable that result may be. In this chapter we deal with the relationship between uncertainties in the individual measurements made during an experiment and the reliability of the final result.

3-1 SIGNIFICANT FIGURES

The number of **significant figures** is the minimum number of digits needed to write a given value in scientific notation without loss of accuracy. The number 142.7 has four significant figures, since it can be written as 1.427×10^2, and all four figures are needed to express fully the value. If you wrote 1.4270×10^2, you would be implying that you know the value of the digit after 7, which is not the case for the number 142.7. The number 1.4270×10^2 therefore has five significant figures.

The number 6.302×10^{-6} has four significant figures, since all four digits are necessary. You could write the same number as 0.000 006 302, which also has just *four* significant figures. The zeros to the left of the 6 are all merely holding decimal places. Since 0.000 006 302 may also be written as 6.302×10^{-6}, only four figures are necessary, and we say only four are significant. The number 92 500 is ambiguous with regard to significant figures.

Definition of significant figures.

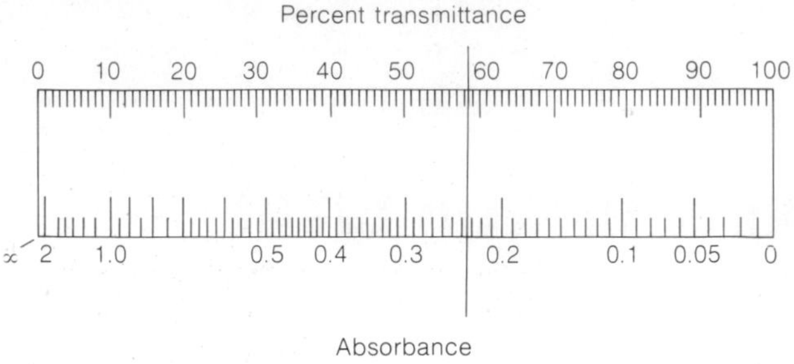

Percent transmittance

Absorbance

Figure 3-1
The scale of a Bausch and Lomb Spectronic 20 spectrophotometer. The percent transmittance is a linear scale, and the absorbance is a logarithmic scale.

It could mean any of the following:

9.25×10^4	3 significant figures
9.250×10^4	4 significant figures
9.2500×10^4	5 significant figures

It is preferable to write one of the three numbers above, instead of 92 500, to indicate how many figures are actually known.

Zeros are significant when they occur (1) in the middle of a number or (2) at the end of a number on the right-hand side of a decimal point.

Significant zeros below are **bold:**

106   0.010**6**   0.106   0.106**0**

The last significant figure in a measured quantity always has some associated uncertainty. The minimum uncertainty would be ± 1 in the last digit. The scale of a Spectronic 20 spectrophotometer is drawn in Figure 3-1. The needle shown in the figure appears to be at an absorbance value of 0.234. We say that there are three significant figures because the numbers 2 and 3 are completely certain and the number 4 is an estimate. The value could be read 0.233 or 0.235 by different people. The percent transmittance is near 58.3. Because the transmittance scale is smaller than the absorbance scale at this point, there is probably more uncertainty in the last digit of transmittance. A reasonable estimate of the uncertainty might be 58.3 ± 0.2. There are three significant figures in the number 58.3.

Always estimate any reading to the nearest tenth of the distance between scale divisions.

In general, when reading the scale of any apparatus, you should interpolate between the markings. It is usually possible to estimate to the nearest tenth of the distance between two marks. Thus on a 50-mL buret, which is graduated to 0.1 mL, you should read the level to the nearest 0.01 mL. When using a ruler calibrated in millimeters, estimate distances to the nearest tenth of a millimeter.

Arithmetic Operations

This section deals with determining the number of significant figures to keep in the answer after you have performed various arithmetic operations with your data.

Addition and subtraction

Often, the numbers to be added or subtracted have equal numbers of digits. In this case, the answer should go to the *same decimal place* as in the individual numbers. For example,

$$\begin{array}{r} 1.362 \times 10^{-4} \\ +\,3.111 \times 10^{-4} \\ \hline 4.473 \times 10^{-4} \end{array}$$

The number of significant figures in the answer can sometimes either exceed or be less than that in the original data:

$$\begin{array}{r} 5.345 \\ +\,6.728 \\ \hline 12.073 \end{array} \qquad \begin{array}{r} 7.26 \times 10^{14} \\ -\,6.69 \times 10^{14} \\ \hline 0.57 \times 10^{14} \end{array}$$

If the numbers being added do not have the same number of significant figures, we are usually limited by the least certain one. For example, in calculating the formula weight of KrF_2, the answer is known only to the second decimal place, as we are limited by our knowledge of the atomic weight of Kr.

$$\begin{array}{rl} 18.998\,403\,2 & \text{(F)} \\ +\,18.998\,403\,2 & \text{(F)} \\ +\,83.80 & \text{(Kr)} \\ \hline 121.796\,806\,4 \end{array}$$

The number 121.796 806 4 should be rounded to 121.80 as the final answer.

    When rounding off, look at *all* the digits *beyond* the last place desired. In the example above, the digits 6 806 4 lie beyond the last significant decimal place. Since this number is more than half way to the next higher digit, we round the 9 up to 10 (i.e., we round up to 121.80 instead of down to 121.79). If the insignificant figures were less than half way, we would round down. For example, 121.794 8 is correctly rounded to 121.79.

    In the special case where the number is exactly halfway, we round to the nearest *even* digit. Thus, 43.550 00 is rounded to 43.6, if we can only have three significant figures. If we are retaining only three figures, 1.425×10^{-9} becomes 1.42×10^{-9}. The number $1.425\,01 \times 10^{-9}$ would become 1.43×10^{-9}, since 501 is more than half way to the next digit. The rationale for rounding to the nearest even digit is that it avoids systematically increasing or decreasing results through successive round-off errors. On the average, half our round-offs will be up and half down.

    In adding or subtracting numbers expressed in scientific notation, all numbers should first be expressed with the same exponent. For example, to do the following addition we could write

$$\begin{array}{r} 1.632 \times 10^{5} \\ +\,4.107 \times 10^{3} \\ +\,0.984 \times 10^{6} \\ \hline \end{array} \Rightarrow \begin{array}{r} 1.632 \times 10^{5} \\ +\,0.041\,07 \times 10^{5} \\ +\,9.84 \times 10^{5} \\ \hline 11.51 \times 10^{5} \end{array}$$

The sum $11.51\,307 \times 10^{5}$ is rounded to 11.51×10^{5} because the number 9.84×10^{5} limits us to two decimal places when all numbers are expressed as multiples of 10^{5}.

Rules for rounding off numbers.

For addition and subtraction, express all numbers using the same exponent, and align all numbers with respect to the decimal point. Round off the answer according to the number of decimal places in the number with the fewest decimal places.

Challenge: Show that the answer would still have four significant figures if all numbers were expressed as multiples of 10^4 instead of 10^5.

A more complete discussion of multiplication and division is reserved for the end of the chapter, after we have looked at relative uncertainties.

In the operations of multiplication and division we are normally limited to the number of digits contained in the number with the fewest significant figures. For example:

$$3.26 \times 10^{-5}$$
$$\times 1.78$$
$$\overline{5.80 \times 10^{-5}}$$

$$4.317\,9 \times 10^{12}$$
$$\times 3.6 \quad \times 10^{-19}$$
$$\overline{1.6 \quad \times 10^{-6}}$$

$$34.60$$
$$\div 2.462\,87$$
$$\overline{14.05}$$

The power of ten has no influence on the number of figures that should be retained.

Logarithms and antilogarithms

A refresher on the algebra of logarithms and exponents can be found in Appendix A.

The **logarithm** of a is the number b, whose value is such that

$$a = 10^b \qquad (3\text{-}1)$$

$$\log a = b \qquad (3\text{-}2)$$

The number a is said to be the **antilogarithm** of b. A logarithm is composed of a **mantissa** and a **character**:

$$\log 339 = 2.530$$

Character   Mantissa

Number of figures in *mantissa* of $\log x$ = number of significant figures in x:

$$\log (5.403 \times 10^{-8}) = -7.267\,4$$
4 digits   4 digits

The number 339 can be written 3.39×10^2. *The number of figures in the mantissa of log 339 should equal the number of significant figures in 339.* The logarithm of 339 is properly expressed as 2.530. The character, 2, corresponds to the exponent in 3.39×10^2.

To see that the third decimal place is the last significant place, consider the following results:

$$10^{2.531} = 340 \;(339.6)$$
$$10^{2.530} = 339 \;(338.8)$$
$$10^{2.529} = 338 \;(338.1)$$

The numbers in parentheses are the results prior to rounding to three figures. Changing the exponent by one digit in the third decimal place changes the answer by one digit in the last place of 339.

In converting a logarithm to its antilogarithm, *the number of significant figures in the antilogarithm should equal the number of figures in the **mantissa**.* Thus

Number of figures in antilog x ($= 10^x$) = number of figures in *mantissa* of x:

$$10^{6.142} = 1.39 \times 10^6$$
3 digits   3 digits

$$\text{antilog}(-3.42) = 10^{-3.42} = 3.8 \times 10^{-4}$$
2 digits   2 digits 2 digits

The following examples show the proper use of significant figures for logs and antilogs:

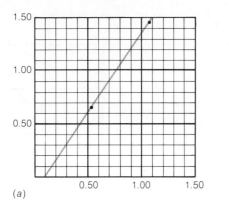
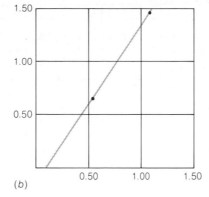

Figure 3-2
Graphs demonstrating choice of rulings in relation to significant figures in the data. The graph in part b does not have fine enough divisions to plot data accurate to the hundredths place.

$$\log 0.001\,237 = -2.907\,6 \qquad \text{antilog } 4.37 = 2.3 \times 10^4$$

$$\log 1\,237 = 3.092\,4 \qquad 10^{4.37} = 2.3 \times 10^4$$

$$\log 3.2 = 0.51 \qquad 10^{-2.600} = 2.51 \times 10^{-3}$$

3-2 SIGNIFICANT FIGURES AND GRAPHS

The rulings on a sheet of graph paper should be compatible with the number of significant figures of the coordinates. The graph in Figure 3-2a has reasonable rulings on which to graph the points (0.53, 0.65) and (1.08, 1.47). There are rulings corresponding to every 0.1 unit, and it is easy to estimate the position of the 0.01 unit. The graph in Figure 3-2b is the same size but does not have fine enough rulings for estimating the position of the 0.01 unit.

In general, a graph must be at least as accurate as the data being plotted. Therefore, it is a good practice to use the most finely ruled graph paper available. Paper ruled with 10 lines per centimeter or 20 lines per inch is usually best. Plan the coordinates so that the data are spread over as much of the sheet of paper as possible.

> The accuracy of a graph should be consistent with the accuracy of the data being plotted.

3-3 TYPES OF ERROR

Experimental error can be classified as either **systematic** or **random.**

Systematic Error

A systematic error, also called a **determinate** error, can in principle be discovered and corrected. One example would be using a pH meter that has been standardized incorrectly. Suppose you think that the pH of the buffer used to standardize the meter is 7.00, but it is really 7.08. If the meter is otherwise working properly, all of your pH readings will be 0.08 pH unit too low. When you read a pH of 5.60, the actual pH of the sample is 5.68. This is a simple example of systematic error. It is always in the same direction and could be discovered by using another buffer of known pH to test the meter.

> Systematic error is a consistent error that may be detected and corrected.

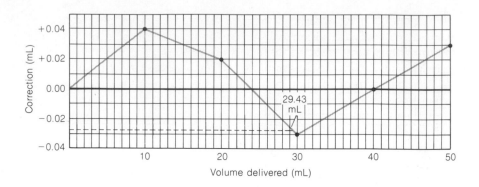

Figure 3-3
Calibration curve for a 50-mL buret.

Ways to detect systematic error:
1. Analyze samples of known composition. Your method should agree with the known composition. (See Box 15-2 for an example.)
2. Analyze "blank" samples containing none of the substance being sought. If you observe a nonzero result, your method responds to more than what you intend.
3. Use different analytical methods to measure the same quantity. If the methods do not agree, there is error associated with one (or more) of the methods.
4. Samples of the same material can be analyzed by different people in different laboratories (using the same or different methods). Disagreements beyond the expected random error indicate systematic errors.

Random error is due to the limitations of physical measurement and cannot be eliminated. A better experiment may reduce the magnitude of the random error, but cannot eliminate it entirely.

A slightly more complicated example of systematic error occurs with an uncalibrated buret. The manufacturer's tolerance for a Class A 50-mL buret is ±0.05 mL. That is, when you believe that you have delivered 29.43 mL, the actual volume could be 29.40 mL and still be within the manufacturer's tolerance. One way to correct for this type of error is by constructing an experimental calibration curve, such as that shown in Figure 3-3. In this procedure distilled water is delivered from the buret into a flask and weighed. By knowing the density of water, we can determine the volume of water from its mass. The graph of correction factors assumes that the level of liquid is always close to the zero mark at the start of the titration. Using Figure 3-3, you would apply a correction factor of −0.03 mL to the measured value of 29.43 mL to reach the correct value of 29.40 mL.

The systematic error associated with using the buret whose calibration is shown in Figure 3-3 is a positive error in some regions and a negative error in others. The key feature of systematic error is that, with care and cleverness, you can detect it and correct it.

Random Error

Random error is also called **indeterminate** error. It arises from natural limitations on our ability to make physical measurements. As the name implies, random error is sometimes positive and sometimes negative. It is always present, cannot be corrected, and is the ultimate limitation on the determination of a quantity. One type of random error is that associated with reading a scale. Different people reading the absorbance or transmittance in Figure 3-1 would report a range of values reflecting their subjective interpolations between the markings. One person reading the same instrument several times probably would also report several different readings. Another type of indeterminate error may result from electrical noise in an instrument. In measuring a voltage, for example, the reading generally has a small fluctuation resulting from electrical instability of the meter itself. This sort of instability is normally random. Positive and negative fluctuations occur with approximately equal frequency and cannot be completely eliminated.

Precision and Accuracy

Precision: reproducibility.
Accuracy: nearness to the "truth."

Precision is a measure of the reproducibility of a result. **Accuracy** refers to how close a measured value is to the "true" value.

The result of an experiment may be very reproducible, but wrong. For example, if you made an error in preparing a solution for a titration, the solution would not have the desired concentration. You might then do a series of highly reproducible titrations, but report an incorrect result because the concentration of solution was not as you intended. In such a case we would say that the precision of the result is high, but the accuracy is poor. Conversely, it is possible to make a series of poorly reproducible measurements clustered around the correct value. In this case, the precision is low but the accuracy is high. An ideal procedure will provide both precision and accuracy.

Accuracy is defined as nearness to the "true" value. The word "true" is in quotes because somebody had to *measure* the "true" value, and there is error associated with *every* measurement. The "true" value is best obtained by an experienced worker using a well-tested procedure. It is desirable to test the result by using different procedures, because, while each method might be precise, systematic error could lead to poor agreement between methods. Good agreement among several methods affords us some confidence, but never proof, that the results are correct.

Absolute and Relative Uncertainty

Absolute uncertainty is an expression of the margin of uncertainty associated with a measurement. If the estimated uncertainty in reading a perfectly calibrated buret is ± 0.02 mL, we call the quantity ± 0.02 mL the absolute uncertainty associated with the reading.

Relative uncertainty is an expression comparing the size of the absolute uncertainty to the size of its associated measurement. The relative uncertainty of a buret reading of 12.35 ± 0.02 mL is

$$\text{Relative uncertainty} = \frac{\text{absolute uncertainty}}{\text{magnitude of measurement}} \qquad (3\text{-}3)$$

$$= \frac{0.02 \text{ mL}}{12.35 \text{ mL}} = 0.002$$

The percent relative uncertainty is simply

$$\text{Percent relative uncertainty} = 100 \times \text{relative uncertainty} \qquad (3\text{-}4)$$

In the above example the percent relative uncertainty is 0.2%.

A constant absolute uncertainty leads to a smaller relative uncertainty as the magnitude of the measurement increases. If the uncertainty in reading a buret is constant at ± 0.02 mL, the relative uncertainty is 0.2% for a volume of 10 mL and 0.1% for a volume of 20 mL.

3-4 PROPAGATION OF UNCERTAINTY

It is usually possible to estimate or measure the random error associated with a particular measurement, such as the length of an object or the temperature of a solution. The uncertainty could be based on your estimate of

Standard deviation is defined and discussed in Chapter 4.

how well you can read an instrument or on your experience with a particular method. When possible, uncertainty is usually expressed as the *standard deviation* of a series of replicate measurements. The discussion that follows applies only to random error; it is assumed that any systematic error has been detected and corrected.

In most experiments, it is necessary to perform arithmetic operations on several numbers, each of which has its associated random error. The most likely uncertainty in the result is not simply the sum of the individual errors, because some of these are likely to be positive and some negative. We expect a certain amount of cancellation of errors.

Addition and Subtraction

Suppose you wish to perform the following arithmetic, in which the experimental uncertainties are given in parentheses:

$$\begin{array}{r} 1.76\,(\pm 0.03) \leftarrow e_1 \\ +1.89\,(\pm 0.02) \leftarrow e_2 \\ -0.59\,(\pm 0.02) \leftarrow e_3 \\ \hline 3.06\,(\pm e_4) \end{array} \qquad (3\text{-}5)$$

The arithmetic answer is 3.06, but what is the uncertainty associated with this result?

We start by calling the three uncertainties e_1, e_2, and e_3. For addition and subtraction, the uncertainty in the answer is obtained by manipulating the *absolute uncertainties* of the individual terms:

For addition and subtraction, use absolute uncertainty.

$$e_4 = \sqrt{e_1^2 + e_2^2 + e_3^2} \qquad (3\text{-}6)$$

For the sum in Equation 3-5 we can write

$$e_4 = \sqrt{(0.03)^2 + (0.02)^2 + (0.02)^2} = 0.04_1 \qquad (3\text{-}7)$$

The absolute uncertainty associated with the sum is ± 0.04, and we can write the answer as 3.06 ± 0.04. Although there is only one significant figure in the uncertainty, we wrote it initially as 0.04_1, with the first insignificant figure subscripted. The reason for retaining one or more insignificant figures is to avoid introducing round-off errors into later calculations through the number 0.04_1. The insignificant figure was subscripted to remind us where the last significant figure should be at the conclusion of the calculations.

If we wish to express the percent relative uncertainty in the sum of Equation 3-5, we may write

$$\text{Percent relative uncertainty} = \frac{0.04_1}{3.06} \times 100 = 1._3\% \qquad (3\text{-}8)$$

The uncertainty, 0.04_1, is $1._3\%$ of the result, 3.06. The subscript 3 in $1._3\%$ is not significant. It would now be sensible to drop the insignificant figures and express the final result as

$$3.06 \, (\pm 0.04) \quad \text{(absolute uncertainty)}$$

or

$$3.06 \, (\pm 1\%) \quad \text{(relative uncertainty)}$$

For addition and subtraction, use absolute uncertainties. The relative uncertainty can be found at the end of the calculation.

Multiplication and Division

For multiplication and division we first convert all uncertainties to percent relative uncertainties (or relative uncertainties). Then we calculate the error of the product or quotient as follows:

$$\%e_4 = \sqrt{(\%e_1)^2 + (\%e_2)^2 + (\%e_3)^2} \tag{3-9}$$

For multiplication and division, use relative uncertainty.

For example, consider the following operations:

$$\frac{1.76 \, (\pm 0.03) \times 1.89 \, (\pm 0.02)}{0.59 \, (\pm 0.02)} = 5.6_4 \pm ? \tag{3-10}$$

Some advice: Retain one or more extra insignificant figures until you have finished your entire calculation. Then round the final answer to the correct number of figures. If you are using a calculator, just keep all of the digits in the calculator until you need to express the final answer.

First convert all the absolute uncertainties to percent relative uncertainties:

$$\frac{1.76 \, (\pm 1._7\%) \times 1.89 \, (\pm 1._1\%)}{0.59 \, (\pm 3._4\%)} = 5.6_4 \pm ? \tag{3-11}$$

Then find the relative uncertainty of the answer using Equation 3-9:

$$\%e_4 = \sqrt{(1._7)^2 + (1._1)^2 + (3._4)^2} = 4._0\% \tag{3-12}$$

The answer is $5.6_4. \, (\pm 4._0\%)$.

To convert the relative uncertainty to absolute uncertainty, find $4._0\%$ of the answer:

$$4._0\% \times 5.6_4 = 0.04_0 \times 5.6_4 = 0.2_3 \tag{3-13}$$

The answer is $5.6_4 \, (\pm 0.2_3)$. Finally, we drop all the figures that are not significant. The result may be expressed as

$$5.6 \, (\pm 0.2) \quad \text{(absolute uncertainty)}$$

$$5.6 \, (\pm 4\%) \quad \text{(relative uncertainty)}$$

For multiplication and division, use relative uncertainties. The absolute uncertainty can be found at the end of the calculation.

There are only two significant figures because we are limited by the denominator, 0.59, in the original problem.

Mixed Operations

As a final example, consider the following mixed operations:

$$\frac{[1.76 \, (\pm 0.03) - 0.59 \, (\pm 0.02)]}{1.89 \, (\pm 0.02)} = 0.619_0 \pm ? \tag{3-14}$$

A general treatment of propagation of uncertainty in complex calculations (including logs and exponents) can be found in Appendix C.

First work out the difference in brackets, using absolute uncertainties:

$$1.76\,(\pm 0.03) - 0.59\,(\pm 0.02) = 1.17 \pm 0.03_6 \tag{3-15}$$

since $\sqrt{(0.03)^2 + (0.02)^2} = 0.03_6$.

Then convert to relative uncertainties:

$$\frac{1.17\,(\pm 0.03_6)}{1.89\,(\pm 0.02)} = \frac{1.17\,(\pm 3._1\%)}{1.89\,(\pm 1._1\%)} = 0.619_0\,(\pm 3._3\%) \tag{3-16}$$

since $\sqrt{(3._1)^2 + (1._1)^2} = 3._3$.

The relative uncertainty in the result is $3._3\%$. The absolute uncertainty is $0.03_3 \times 0.619_0 = 0.02_0$. The final answer could be written as

$$0.619\,(\pm 0.02_0) \qquad \text{(absolute uncertainty)}$$

or

$$0.619\,(3._3\%) \qquad \text{(relative uncertainty)}$$

The result of a calculation ought to be written in a manner consistent with the uncertainty in the result.

Since the uncertainty spans the last *two* places of the result, it would also be reasonable to write the result as

$$0.62\,(\pm 0.02)$$

or

$$0.62\,(\pm 3\%)$$

Comment on Significant Figures

The number of figures used to express a calculated result should be consistent with the uncertainty in that result. For example, the quotient

$$\frac{0.002\,364\,(\pm 0.000\,003)}{0.025\,00\,(\pm 0.000\,05)} = 0.094\,6\,(\pm 0.000\,2)$$

is properly expressed with *three* significant figures, even though the original data have four figures. *The first uncertain figure of the answer is the last significant figure.* The quotient

$$\frac{0.002\,664\,(\pm 0.000\,003)}{0.025\,00\,(\pm 0.000\,05)} = 0.106\,6\,(\pm 0.000\,2)$$

is expressed with *four* figures because the uncertainty occurs in the fourth place. The quotient

$$\frac{0.821\,(\pm 0.002)}{0.803\,(\pm 0.002)} = 1.022\,(\pm 0.004)$$

The first uncertain figure should be the last significant figure.

is expressed with *four* figures even though the dividend and divisor each have *three* figures.

Summary

The number of significant figures in a value is the minimum number of digits needed to write the value in scientific notation. The first uncertain digit in a calculated result should be the last significant digit. In addition and subtraction, the last significant figure is determined by the decimal place of the least certain number. In multiplication and division, the number of figures is usually limited by the factor with the least number of digits. The number of figures in the mantissa of the logarithm of a quantity should equal the number of significant figures in the quantity. Random error mainly affects the precision (reproducibility) of a result, while systematic error mainly affects the accuracy (nearness to the "true" value). To analyze propagation of uncertainty in addition and subtraction, we use absolute uncertainties: $e_3 = \sqrt{e_1^2 + e_2^2}$. In multiplication and division, propagation of uncertainty is analyzed by using relative uncertainties: $\%e_3 = \sqrt{\%e_1^2 + \%e_2^2}$. Always retain more digits than necessary during a calculation, and round off to the appropriate number of digits at the end.

Terms to Understand

absolute uncertainty
accuracy
antilogarithm
character
determinate error
indeterminate error
logarithm

mantissa
precision
random error
relative uncertainty
significant figure
systematic error

Exercises

3-A. Write each answer with a reasonable number of figures. Find the absolute uncertainty and percent relative uncertainty for each answer.
   (a) $[12.41\,(\pm 0.09) \div 4.16\,(\pm 0.01)]$
       $\times 7.068\,2\,(\pm 0.000\,4) = ?$
   (b) $3.26\,(\pm 0.10) \times 8.47\,(\pm 0.05)$
       $- 0.18\,(\pm 0.06) = ?$
   (c) $6.843\,(\pm 0.008) \times 10^4$
       $\div [2.09\,(\pm 0.04) - 1.63\,(\pm 0.01)] = ?$

3-B. Suppose that you have a bottle of aqueous solution labeled "53.4 $(\pm 0.4)\%$ (wt/wt) NaOH—density = 1.52 (± 0.01) g/mL."
   (a) How many milliliters of 53.4% NaOH are needed to prepare 2.000 L of 0.169 M NaOH?
   (b) If the uncertainty in delivering the NaOH is ± 0.10 mL, calculate the absolute uncertainty in

the molarity (0.169 M). You may assume negligible uncertainty in the molecular weight of NaOH and in the final volume, 2.000 L.

3-C. Consider a solution containing 37.0 $(\pm 0.5)\%$ (wt/wt) HCl in water. The density of the solution is 1.18 (± 0.01) g/mL. To deliver 0.050 0 $(\pm 2\%)$ mol of HCl requires 4.18 $(\pm x)$ mL of solution. Find x.
   Caution: In this problem you have been given the uncertainty in the *answer* to a calculation. You need to find the uncertainty in one factor along the way in the calculation. Be sure to propagate uncertainties in the right direction. For example, if $a = b \cdot c$, then $\%e_a^2 = \%e_b^2 + \%e_c^2$. If you are given $\%e_a$, the value of $\%e_c$ must be *smaller* than $\%e_a$, and $\%e_c^2 = \%e_a^2 - \%e_b^2$.

Problems

A3-1. Indicate how many significant figures there are in
   (a) 1.903 0   (b) 0.039 10   (c) 1.40×10^4

A3-2. Round each number to the number of significant figures indicated.
   (a) 1.236 7 to 4 figures
   (b) 1.238 4 to 4 figures
   (c) 0.135 2 to 3 figures
   (d) 2.051 to 2 figures
   (e) 2.005 0 to 3 figures

A3-3. Round each number to three significant figures.
   (a) 0.216 74   (b) 0.216 5   (c) 0.216 500 3

A3-4. Write each answer with the correct number of significant figures.
(a) $1.021 + 2.69 = 3.711$
(b) $12.3 - 1.63 = 10.67$
(c) $4.34 \times 9.2 = 39.928$
(d) $0.060\ 2 \div (2.113 \times 10^4) = 2.849\ 03 \times 10^{-6}$

A3-5. Write the answers with the correct number of figures.
(a) $\log(4.218 \times 10^{12}) = ?$
(b) $\text{antilog}(-3.22) = ?$
(c) $10^{2.384} = ?$

A3-6. Using the correct number of significant figures, calculate the formula weight of (a) $BaCl_2$ and (b) $C_{31}H_{32}O_8N_2$.

A3-7. Rewrite the number 3.123 56 ($\pm 0.167\ 89\%$) in the forms (a) number ($\pm$ absolute uncertainty) and (b) number ($\pm$ percent relative uncertainty). Use a reasonable number of figures in each expression.

A3-8. Each target in the figure shows where arrows have struck. Match the letter of the target with the description below.
(a) accurate and precise   – c
(b) accurate but not precise – b
(c) precise but not accurate   – d
(d) neither precise nor accurate – a

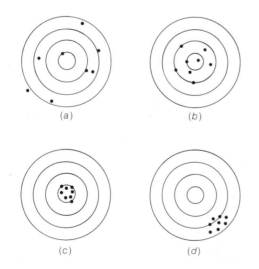

(a)   (b)

(c)   (d)

A3-9. Find the absolute uncertainty and percent relative uncertainty for each calculation. Express the answers with a reasonable number of significant figures.
(a) $6.2\ (\pm 0.2) - 4.1\ (\pm 0.1) = ?$
(b) $9.43\ (\pm 0.05) \times 0.016\ (\pm 0.001) = ?$

3-10. Write each answer with the correct number of significant figures.
(a) $1.0 + 2.1 + 3.4 + 5.8 = 12.300\ 0$
(b) $106.9 - 31.4 = 75.500\ 0$
(c) $107.868 - (2.113 \times 10^2) + (5.623 \times 10^3) = 5\ 519.568$
(d) $(26.14/37.62) \times 4.38 = 3.043\ 413$
(e) $(26.14/37.62 \times 10^8) \times (4.38 \times 10^{-2}) = 3.043\ 413 \times 10^{-10}$
(f) $(26.14/3.38) + 4.2 = 11.933\ 7$
(g) $\log(3.98 \times 10^4) = 4.599\ 9$
(h) $10^{-6.31} = 4.897\ 79 \times 10^{-7}$

3-11. Find the absolute uncertainty and percent relative uncertainty for each calculation. Express the answers with a reasonable number of significant figures.
(a) $9.23\ (\pm 0.03) + 4.21\ (\pm 0.02) - 3.26\ (\pm 0.06) = ?$
(b) $91.3\ (\pm 1.0) \times 40.3\ (\pm 0.2)/21.1\ (\pm 0.2) = ?$
(c) $[4.97\ (\pm 0.05) - 1.86\ (\pm 0.01)]/21.1\ (\pm 0.2) = ?$
(d) $2.016\ 4\ (\pm 0.000\ 8) + 1.233\ (\pm 0.002) + 4.61\ (\pm 0.01) = ?$
(e) $2.016\ 4\ (\pm 0.000\ 8) \times 10^3 + 1.233\ (\pm 0.002) \times 10^2 + 4.61\ (\pm 0.01) \times 10^1 = ?$

3-12. (a) Show that the formula weight of NaCl is $58.442\ 5 \pm 0.000\ 9$ g/mol.
(b) To prepare a solution of NaCl, you weigh out 2.634 (± 0.002) g and dissolve it in a volumetric flask whose volume is 100.00 ± 0.08 mL. Express the molarity of the resulting solution, and its uncertainty, with the correct number of significant figures.

3-13. Consider a buoyancy correction using Equation 2-2. What is the true mass in vacuum of water weighed at 24°C in the air if the apparent mass is $1.034\ 6 \pm 0.000\ 2$ g? Assume that the density of air is $0.001\ 2 \pm 0.000\ 1$ g/mL and the density of the balance weights is 8.0 ± 0.5 g/mL. The uncertainty in the density of water in Table 2-6 is negligible in comparison to the uncertainty in the density of air.

3-14. The constant $\hbar$ (read "h bar") is defined as $h/2\pi$, where h is Planck's constant [$6.626\ 075\ 5\ (\pm 0.000\ 004\ 0) \times 10^{-34}$ J·s]. Calculate the value and absolute uncertainty of $\hbar$. The number 2 is an integer (infinitely accurate) and π is also an exact number. The first ten digits of π are 3.141 592 653.

3-15. The value of Boltzmann's constant (k) listed on the inside front cover of the book is calculated from the quotient R/N, where R is the gas constant and N is Avogadro's number. If the uncertainty in R is 0.000 070 J/(mol·K) and the uncertainty in N is $0.000\ 003\ 6 \times 10^{23}$/mol, find the uncertainty in k.

4 Statistics

Since all real measurements contain experimental error, it is never possible to be completely certain of a result. Nevertheless, we still seek to answer such questions as "Is my red blood cell count today higher than its usual value?" If today's count is twice as high as its usual value, answering this question is trivial. But what if the "high" count does not seem excessively high in comparison with counts on "normal" days? Consider the following values:

"Normal" days	Today
5.1 ⎫	5.6 × 10⁶ cells/μL
5.3 ⎪	
4.8 ⎬ × 10⁶ cells/μL	
5.4 ⎪	
5.2 ⎭	

The number 5.6 is higher than the five normal values, but the random variation in normal values might lead us to expect that 5.6 will be observed on some "normal" days.

The study of statistics addresses the question of how to deal with the variations in experimental results. We will never be able to answer this question with complete certainty, but we are able to say that the value 5.6×10^6 cells/μL is expected to be observed on one out of 20 normal days. There is a 5% probability that 5.6×10^6 represents a normal count and a 95% probability that it represents an elevated count.

The best we can do is assess the probability that today's count is elevated. We cannot say that it *is* or is *not* elevated.

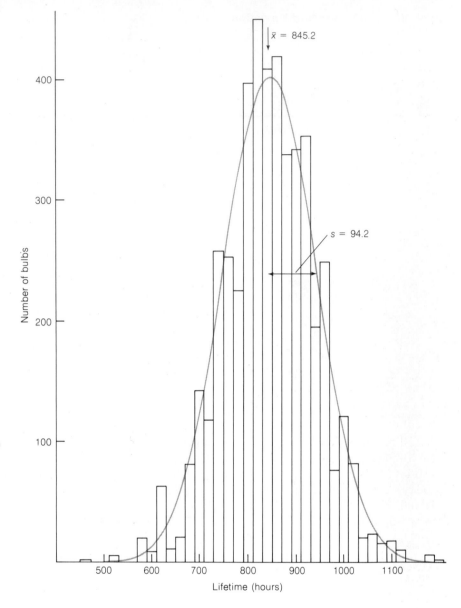

Figure 4-1
Bar graph and normal error curve describing the lifetime of a hypothetical set of electric bulbs. The smooth curve has the same mean, standard deviation, and area as those for the bar graph.

4-1 GAUSSIAN ERROR CURVE

When the variation in a set of experimental data is strictly random, a line joining the graphed values approximates a bell-shaped curve, as illustrated in Figure 4-1. In this hypothetical case, a manufacturer has tested the lifetimes of 4 768 electric light bulbs. The bar graph in Figure 4-1 shows the number of bulbs having a lifetime in each 20-hour interval. The smooth curve is the **Gaussian curve,** or **normal error curve,** that best fits the data. In the graph of any finite set of data, there will be variation from the Gaussian curve. As the number of data points increases, the line connecting them should more closely approach the smooth curve.

Mean Value and Standard Deviation

The data for light bulb lifetimes, and the corresponding Gaussian curve, are characterized by two parameters. The arithmetic **mean,** $\bar{x}$, is defined as

$$\text{Mean} = \bar{x} = \frac{\sum_i x_i}{n} \qquad (4\text{-}1)$$

The mean gives the center of the distribution. The standard deviation measures the width of the distribution.

where each x_i is the lifetime of an individual bulb. The symbol $\sum$ means summation. Therefore, $\sum_i x_i = x_1 + x_2 + x_3 + \cdots + x_n$. The mean, also called the **average,** is the sum of the measured values divided by n, the total number of values. In Figure 4-1, the mean value is indicated by the arrow at a value of 845.2 hours.

The **standard deviation,** s, measures how closely the data are clustered about the mean.

$$\text{Standard deviation} = s = \sqrt{\frac{\sum_i (x_i - \bar{x})^2}{n-1}} \qquad (4\text{-}2)$$

An experimental technique that produces a small standard deviation is more reliable (precise) than one that produces a large standard deviation, provided that they are equally accurate.

For the data in Figure 4-1, $s = 94.2$ h. The significance of s is that the smaller the standard deviation, the more closely the data are clustered about the mean (Figure 4-2). A set of light bulbs having a small standard deviation in lifetime must be more uniformly manufactured than a set with a large standard deviation.

For an *infinite* set of data, the mean is called μ (the population mean) and the standard deviation is called σ (the population standard deviation). We can never measure μ and σ, but the values of $\bar{x}$ and s approach μ and σ as the number of measurements increases.

The **degrees of freedom** of the system are given by the quantity $n - 1$ in Equation 4-2. The square of the standard deviation is called the **variance.**

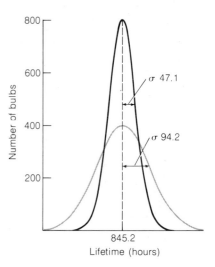

Figure 4-2
Gaussian curves for two sets of light bulbs, one having a standard deviation one-half that of the other. The number of bulbs described by each curve is the same.

EXAMPLE: Mean and Standard Deviation

To illustrate the use of Equations 4-1 and 4-2, suppose that four measurements are made: 821, 783, 834, and 855 hours. The average is

$$\bar{x} = \frac{821 + 783 + 834 + 855}{4} = 823._2 \text{ h}$$

To avoid round-off errors, we generally retain one more significant figure for the average and the standard deviation than was present in the original data. The standard deviation is

$$s = \sqrt{\frac{(821 - 823.2)^2 + (783 - 823.2)^2 + (834 - 823.2)^2 + (855 - 823.2)^2}{(4-1)}}$$

$$= 30._3 \text{ h}$$

The average and the standard deviation should both end at the *same decimal place.* For $\bar{x} = 823.2$, the value $s = 30.3$ is reasonable, but $s = 30.34$ is not.

Suppose you obtain experimental values having many significant digits, such as 1.638 451 2, 1.638 460 2, 1.638 457 1, and 1.638 449 7. If you use a calculator or computer to find standard deviation, the computer may not retain enough digits during the computation to produce a correct answer. In such cases it is wise to subtract the constant part from each of the numbers and just use the variable parts. To find the standard deviation of these four values, subtract 1.638 4 from each, leaving 0.000 051 2, 0.000 060 2, 0.000 057 1, and 0.000 049 7. The standard deviation of these three-significant-figure numbers is 0.000 004 9, which is also the standard deviation of the original eight-significant-figure numbers. A computer or calculator will correctly process the three-digit numbers but may not be able to handle the eight-digit numbers.

Other terms

We now define some quantities that do not apply directly to the normal error curve, but whose definitions you ought to know. The **median** is the value above and below which there is an equal number of data points. For an odd number of points, the median is the middle one. Therefore 3 is the median of 1, 2, 3, 7, 8. For an even number of points, the median is halfway between the two center values. Thus the median of 1, 2, 3, and 6 is 2.5. The **range,** or **spread,** is the difference between the highest and lowest values. The range of 126.2, 127.5, 127.1, 125.9, and 126.4 is $(127.5 - 125.9) = 1.6$. The **geometric mean** of n numbers is

$$\text{Geometric mean} = \sqrt[n]{\prod_i x_i} \qquad (4\text{-}3)$$

The symbol $\prod$ means the product of all the values. The geometric mean of 2, 4, 9, 13, and 29 is $(2 \cdot 4 \cdot 9 \cdot 13 \cdot 29)^{1/5} = 7.7$.

Standard Deviation and Probability

The formula for a Gaussian curve is

$$y = \frac{1}{\sigma\sqrt{2\pi}} e^{-(x-\mu)^2/2\sigma^2} \qquad (4\text{-}4)$$

where $e\ (= 2.718\ 28\ldots)$ is the base of the natural logarithm. To describe the curve for a finite set of data, we approximate μ by $\bar{x}$ and σ by s. A graph of Equation 4-4 is shown in Figure 4-3, in which we have set $\sigma = 1$ and $\mu = 0$ for simplicity. In general, the most probable value of x is $x = \mu$ and the curve is symmetric about $x = \mu$. The relative probability of making a particular measurement is proportional to the ordinate (y value) for the value of x. Thus in Figure 4-3 the maximum probability for any measurement occurs at $x = \mu = 0$. The probability of measuring the value $x = 1$ is $0.242/0.399 = 0.607$ times the probability of measuring the value $x = 0$.

In dealing with a Gaussian curve, it is especially useful to express deviations from the mean value in multiples of the standard deviation. That is, we transform x into z, given by

$$z = \frac{x - \mu}{\sigma} \approx \frac{x - \bar{x}}{s} \qquad (4\text{-}5)$$

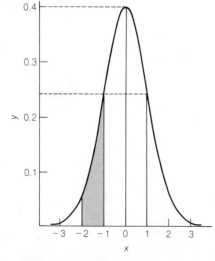

Figure 4-3
A Gaussian error curve in which $\mu = 0$ and $\sigma = 1$.

Table 4-1
Ordinate and area for the Gaussian error curve

$$y = \frac{1}{\sqrt{2\pi}} e^{-z^2/2}$$

$|z|$	y	Area	$|z|$	y	Area	$|z|$	y	Area
0.0	0.398 9	0.000 0	1.4	0.149 7	0.419 2	2.8	0.007 9	0.497 4
0.1	0.397 0	0.039 8	1.5	0.129 5	0.433 2	2.9	0.006 0	0.498 1
0.2	0.391 0	0.079 3	1.6	0.110 9	0.445 2	3.0	0.004 4	0.498 7
0.3	0.381 4	0.117 9	1.7	0.094 1	0.455 4	3.1	0.003 3	0.499 0
0.4	0.368 3.	0.155 4	1.8	0.079 0	0.464 1	3.2	0.002 4	0.499 3
0.5	0.352 1	0.191 5	1.9	0.065 6	0.471 3	3.3	0.001 7	0.499 5
0.6	0.333 2	0.225 8	2.0	0.054 0	0.477 3	3.4	0.001 2	0.499 7
0.7	0.312 3	0.258 0	2.1	0.044 0	0.482 1	3.5	0.000 9	0.499 8
0.8	0.289 7	0.288 1	2.2	0.035 5	0.486 1	3.6	0.000 6	0.499 8
0.9	0.266 1	0.315 9	2.3	0.028 3	0.489 3	3.7	0.000 4	0.499 9
1.0	0.242 0	0.341 3	2.4	0.022 4	0.491 8	3.8	0.000 3	0.499 9
1.1	0.217 9	0.364 3	2.5	0.017 5	0.493 8	3.9	0.000 2	0.500 0
1.2	0.194 2	0.384 9	2.6	0.013 6	0.495 3	4.0	0.000 1	0.500 0
1.3	0.171 4	0.403 2	2.7	0.010 4	0.496 5			

Note: The area refers to the area between $z = 0$ and $z =$ the value in the table. Thus the area from $z = 0$ to $z = 1.4$ is 0.419 2. The area from $z = -0.7$ to $z = 0$ is the same as from $z = 0$ to $z = 0.7$. The area from $z = -0.5$ to $z = +0.3$ is (0.191 5 + 0.117 9) = 0.309 4. The total area between $z = -\infty$ and $z = +\infty$ is unity. A more complete table can be found in any edition of the *Handbook of Chemistry and Physics* (Boca Raton, Fla.: CRC Press).

Table 4-1 gives ordinate and area values for the Gaussian curve in Figure 4-3.

The probability of measuring z in a certain *range* is proportional to the *area* of that range. For example, the probability of observing z between -2 and -1 is 0.136. This corresponds to the shaded area in Figure 4-3. The area under each portion of the Gaussian curve is given in Table 4-1. Since the sum of the probabilities of all the measurements must be unity, the area under the whole curve from $z = -\infty$ to $z = +\infty$ must be unity. The number $1/\sigma\sqrt{2\pi}$ in Equation 4-4 is called the *normalization factor*. It guarantees that the area under the entire curve is unity.

EXAMPLE: Gaussian Curve

Suppose the manufacturer offers to replace free of charge any bulb that burns out in less than 600 hours. What fraction of the bulbs should be kept available as replacements?

To answer this question we express the desired interval in multiples of the standard deviation. Then we find the area of the interval by using Table 4-1. Since $\bar{x} = 845.2$ and $s = 94.2$, $z = (600 - 845.2)/94.2 = -2.60$. The area under the curve between the mean value and $z = -2.60$ is given as 0.495 3 in Table 4-1. Since the entire area from $-\infty$ to the mean value is 0.500 0, the area from $-\infty$ to -2.60 must be 0.004 7. That is, the area to the left of 600 hours in Figure 4-1 is only 0.47% of the entire area under the normal error curve. Only 0.47% of the bulbs are expected to fail in less than 600 hours.

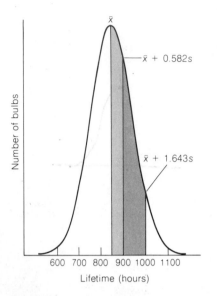

Figure 4-4
Use of the Gaussian curve to find the fraction of bulbs with a lifetime between 900 and 1 000 hours.

Range	Percent of measurements
$\mu \pm 1\sigma$	68.3
$\mu \pm 2\sigma$	95.5
$\mu \pm 3\sigma$	99.7

EXAMPLE: More of the Gaussian Curve

What fraction of the bulbs is expected to have a lifetime between 900 and 1 000 hours?

To answer this question *we need to find the fraction of the area of the Gaussian curve between x = 900 and x = 1 000 hours.* When $x = 900$, $z = (900 - 845.2)/94.2 = 0.582$. When $x = 1\,000$, $z = (1\,000 - 845.2)/94.2 = 1.643$. Figure 4-4 shows that we can find the area between 900 and 1 000 hours by first finding the area from $\bar{x}$ to 1 000 hours and then subtracting the area from $\bar{x}$ to 900 hours.

Table 4-1 can be used to find the area from $\bar{x}$ to 900 hours ($z = 0.582$) as follows. The area up to $z = 0.5$ is 0.191 5. The area up to $z = 0.6$ is 0.225 8. We can estimate the area up to 0.582 by *linear interpolation:*

$$\begin{array}{c} \text{Area between } z = 0.500 \\ \text{and } z = 0.582 \end{array} = \underbrace{\left(\frac{0.582 - 0.500}{0.600 - 0.500} \right)}_{\substack{\text{Fraction of} \\ \text{interval between} \\ z = 0.5 \text{ and } z = 0.6}} \underbrace{(0.225\,8 - 0.191\,5)}_{\substack{\text{Area between} \\ z = 0.5 \text{ and } z = 0.6}} = 0.028\,1$$

Area between 0 and 0.582 = area between 0 and 0.5 + area between 0.5 and 0.582

$$= 0.191\,5 + 0.028\,1 = 0.219\,6$$

To find the area between $\bar{x}$ and 1 000 hours ($z = 1.643$) we interpolate between $z = 1.6$ and $z = 1.7$ to find

$$\text{Area between 0 and } 1.643 = 0.449\,6$$

The area between 900 and 1 000 hours is $0.449\,6 - 0.219\,6 = 0.230\,0$. That is, 23% of the bulbs are expected to have a lifetime between 900 and 1 000 hours.

The significance of the standard deviation is that it measures the width of the Gaussian error curve. The larger the value of σ, the broader the curve. In a Gaussian curve (such as in Figure 4-3), 68.3% of the area falls under $\mu \pm 1\sigma$. That is, more than two-thirds of the measurements are expected to lie within one standard deviation of the mean. Also, 95.5% of the area lies within $\mu \pm 2\sigma$, and 99.7% of the area lies within $\mu \pm 3\sigma$ (Table 4-1).

Suppose that you use two different experimental methods to measure the percent of sulfur in coal: method A has a standard deviation of 0.4%, and method B has a standard deviation of 1.1%. You can expect that two-thirds of measurements from method A will lie within 0.4% of the mean. For method B, two-thirds of measurements will lie within 1.1% of the mean.

4-2 STUDENT'S *t*

Student's *t* is a valuable statistical tool used to measure probability. We use it most frequently to express confidence intervals and for comparing results from different experiments.

Confidence Intervals

From a limited number of measurements it is impossible to find the true population mean, μ, or the true standard deviation, σ. What we can determine

Table 4-2
Values of Student's *t*

Degrees of freedom	Confidence level (%)				
	50	80	90	95	99
1	1.000	3.078	6.314	12.706	63.657
2	0.816	1.886	2.920	4.303	9.925
3	0.765	1.638	2.353	3.182	5.841
4	0.741	1.533	2.132	2.776	4.604
5	0.727	1.476	2.015	2.571	4.032
6	0.718	1.440	1.943	2.447	3.707
7	0.711	1.415	1.895	2.365	3.500
8	0.706	1.397	1.860	2.306	3.355
9	0.703	1.383	1.833	2.262	3.250
10	0.700	1.372	1.812	2.228	3.169
15	0.691	1.341	1.753	2.131	2.947
20	0.687	1.325	1.725	2.086	2.845
∞	0.674	1.282	1.645	1.960	2.576

Note: In calculating confidence intervals, σ may be substituted for s in Equation 4-6 if you have a great deal of experience with a particular method and have therefore determined its "true" population standard deviation. If σ is used instead of s, the value of t to use in Equation 4-6 comes from the bottom row of Table 4-2.

are $\overline{x}$ and s, the sample mean and the sample standard deviation. The **confidence interval** is an expression stating that the true mean, μ, is likely to lie within a certain distance of the measured mean, $\overline{x}$. The confidence interval of μ is given by

$$\mu_r = \overline{x} \pm \frac{ts}{\sqrt{n}} \qquad (4\text{-}6)$$

where s is the measured standard deviation, n is the number of observations, and t is a number called Student's t. Values of t for various confidence levels are given in Table 4-2.

EXAMPLE: Confidence Interval

Suppose that the percent carbohydrate content of a glycoprotein (a protein with sugars attached to it) is determined to be 12.6, 11.9, 13.0, 12.7, and 12.5 in replicate analyses. Find the 50% and 95% confidence intervals for the carbohydrate content.

First we calculate $\overline{x} = 12.5_4\%$ and $s = 0.4_0\%$ for the five measurements. To calculate the 50% confidence interval, look up t in Table 4-2 under 50% and across from *four* degrees of freedom. (Recall that degrees of freedom $= n - 1$.) The value of t is 0.741, so the 50% confidence interval is

$$\mu = \overline{x} \pm \frac{ts}{\sqrt{n}} = 12.5_4 \pm \frac{(0.741)(0.4_0)}{\sqrt{5}} = 12.5_4 \pm 0.1_3$$

The 95% confidence interval is

$$\mu = \overline{x} \pm \frac{ts}{\sqrt{n}} = 12.5_4 \pm \frac{(2.776)(0.4_0)}{\sqrt{5}} = 12.5_4 \pm 0.5_0$$

"Student" was the pseudonym of W. S. Gossett, who published the classic paper on this subject (*Biometrika*, **6**, 1 [1908]). Gossett's employer, the Guinness Breweries of Ireland, restricted publications for proprietary reasons. Because of the importance of Gossett's work, he was allowed to publish it, but under an assumed name.

These calculations mean that there is a 50% chance that the true mean, μ, lies within the range 12.4_1–12.6_7. There is a 95% chance that μ lies within the range 12.0_4–13.0_4.

EXAMPLE: Are Results "Too Far" from the Expected Value?

Suppose that you wish to test the validity of a new analytical technique. You prepare a sample containing 0.031 9% Ni and analyze it four times with your new method. You observe values of 0.032 9, 0.032 2, 0.033 0, and 0.032 3. Does your method yield a value "significantly" higher than the known value?

To answer this question, we can calculate the value of μ at several confidence intervals. First we calculate $\bar{x} = 0.032\,6_0$ and $s = 0.000\,4_1$. Plugging into Equation 4-6 gives

$$90\% \text{ confidence:} \quad \mu = 0.032\,6_0 \pm \frac{(2.353)(0.000\,4_1)}{\sqrt{4}} = 0.032\,6_0 \pm 0.000\,4_8$$

$$95\% \text{ confidence:} \quad \mu = 0.032\,6_0 \pm \frac{(3.182)(0.000\,4_1)}{\sqrt{4}} = 0.032\,6_0 \pm 0.000\,6_5$$

$$99\% \text{ confidence:} \quad \mu = 0.032\,6_0 \pm \frac{(5.841)(0.000\,4_1)}{\sqrt{4}} = 0.032\,6_0 \pm 0.001\,1_9$$

The known value, 0.031 9%, lies approximately at the limit of the 95% confidence interval (which covers the range $0.032\,6_0 \pm 0.000\,6_5 = 0.031\,9_5$ to $0.033\,2_5$). Therefore you can be 95% confident that the new method produces a high value. That is, if $\mu = 0.031\,9$, the value $\bar{x} = 0.032\,6$ will be observed in only about 5% of experiments. The chances are 19 out of 20 that μ is really greater than 0.031 9 for your new method. Because 0.031 9 is within the 99% confidence interval ($(0.031)\,4_1$ to $0.033\,7_9$), you cannot be 99% confidence that your new method produces a high result.

Statistical tests only give us probabilities. They do not relieve us of the responsibility of interpreting our results. You may be interested in Box 4-1 now.

Statistical tests do not relieve you of the ultimate subjective decision to accept or reject a conclusion. The tests only provide guidance in the form of probabilities. In the foregoing example, it would probably be reasonable to conclude that values obtained with the new method are systematically high. However, with only four values, it would be wise to run the analysis several more times to confirm your conclusion.

Comparison of Means

We sometimes need to compare the results of two tests to see whether they are "the same" or "different" from each other. For this purpose, we perform the **t test** using Student's t. The discussion that follows assumes that the population standard deviation (σ) for each method is essentially the same.

Comparing replicate measurements

For the two sets of data we calculate a value of t using the formula

$$t = \frac{\bar{x}_1 - \bar{x}_2}{s} \sqrt{\frac{n_1 n_2}{n_1 + n_2}} \tag{4-7}$$

Box 4-1   STUDENT'S *t* AND THE LAW

By this point you should begin to appreciate that knowledge of the uncertainty associated with a result is as important as the result itself. Clearly, the meanings of 1.083 ± 0.007 and $1.0_{83} \pm 0.4_{17}$ are very different.

As a person who will either derive or use analytical results, you should be aware of this warning published in a report entitled "Principles of Environmental Analysis":[†]

> Analytical chemists must always emphasize to the public that **the single most important characteristic of any result obtained from one or more analytical measurements is an adequate statement of its uncertainty interval.** Lawyers usually attempt to dispense with uncertainty and try to obtain unequivocal statements; therefore, an uncertainty interval must be clearly defined in cases involving litigation and/or enforcement proceedings. Otherwise, a value of 1.001 without a specified uncertainty, for example, may be viewed as legally exceeding a permissible level of 1.

[†] L. H. Keith, W. Crummett, J. Deegan, Jr., R. A. Libby, J. K. Taylor, and G. Wentler, *Anal. Chem.*, **55**, 2210 (1983).

where

$$s = \sqrt{\frac{\sum_{\text{set 1}} (x_i - \overline{x}_1)^2 + \sum_{\text{set 2}} (x_j - \overline{x}_2)^2}{n_1 + n_2 - 2}} \qquad (4\text{-}8)$$

The value of s is a *pooled* standard deviation making use of both sets of data. The value of t from Equation 4-7 is to be compared to the value of t in Table 4-2 for $n_1 + n_2 - 2$ degrees of freedom. *If the calculated t is greater than the tabulated t, the two results are significantly different at the confidence level in question.*

If $t_{\text{calculated}} > t_{\text{table}}$, difference is significant.

EXAMPLE: Are Two Means Different at the 95% Confidence Level?

The radioactive compound $^{14}CO_2$ may be used as a tracer to study metabolism in plants. Suppose that a compound isolated from a plant exhibited 28, 32, 27, 39, and 40 counts of radioactive decays per minute. A blank sample used to measure the background counts of the radiation counter (due to electrical noise and background radiation) gave 28, 21, 28, and 20 counts per minute. It appears that the isolated compound gives more counts than those from background. Can we be 95% confident that the compound is indeed radioactive?

The averages for the two sets of data are, respectively, $33._2$ and $24._2$ counts per minute. The pooled s is calculated from Equation 4-8 as follows:

$$s = \sqrt{\frac{\begin{array}{c}(28 - 33._2)^2 + (32 - 33._2)^2 + (27 - 33._2)^2 + (39 - 33._2)^2 + (40 - 33._2)^2 \\ + (28 - 24._2)^2 + (21 - 24._2)^2 + (28 - 24._2)^2 + (20 - 24._2)^2\end{array}}{5 + 4 - 2}}$$

$$= 5._4$$

Equation 4-7 then gives

$$t = \frac{33._2 - 24._2}{5._4} \sqrt{\frac{5 \times 4}{5 + 4}} = 2._{48}$$

The calculated value of t is bigger than the t value 2.365 listed in Table 4-2 for the seven degrees of freedom and 95% confidence. Therefore, we can be 95% confident that the observed counts are higher than background and that radioactive ^{14}C has been incorporated into the compound. Note that the calculated t (2.48) is less than the t for 99% confidence (3.500). Therefore, the probability that the two sets of counts are different lies between 95% and 99%.

Comparing individual differences

Suppose that the cholesterol content of six sets of human blood plasma is measured by two different techniques. The results are listed in Table 4-3. Each of the six plasma samples is a different sample with a different cholesterol content. Method B gives a lower result than method A in five out of the six samples. Is method B systematically different from method A?

To answer this question, we perform a t test on the individual *differences* between results for each sample:

$$t = \frac{\bar{d}}{s_d} \sqrt{n} \tag{4-9}$$

where

$$s_d = \sqrt{\frac{\sum (d_i - \bar{d})^2}{n - 1}} \tag{4-10}$$

The quantity $\bar{d}$ is the average difference between methods A and B, and n is the number of pairs of data (six in this case). For the results in Table 4-3, the standard deviation, s_d, of the differences is calculated to be

$$s_d = \sqrt{\frac{\begin{array}{c}(0.04 - 0.06_0)^2 + (-0.16 - 0.06_0)^2 + (0.17 - 0.06_0)^2 \\ + (0.17 - 0.06_0)^2 + (0.04 - 0.06_0)^2 + (0.10 - 0.06_0)^2\end{array}}{6 - 1}}$$

$$= 0.12_2$$

Table 4-3
Comparison of two methods for measuring cholesterol

Plasma sample	Cholesterol content (g/L)		Difference (d_i)
	Method A	Method B	
1	1.46	1.42	0.04
2	2.22	2.38	−0.16
3	2.84	2.67	0.17
4	1.97	1.80	0.17
5	1.13	1.09	0.04
6	2.35	2.25	0.10
			$\bar{d} = +0.06_0$

Putting this value into Equation 4-9 gives

$$t = \frac{0.06_0}{0.12_2} \sqrt{6} = 1.20$$

The calculated value of t ($= 1.20$) lies between the tabulated values of t at 50% and 80% confidence levels for five degrees of freedom in Table 4-2. That is, there is more than a 50% chance, but less than an 80% chance, that the two methods are systematically different. It would be reasonable to conclude that the two techniques are *not* significantly different.

When applying t tests, most people regard differences as significant when they occur in the 90–95% confidence range. A confidence level of 99% is considered highly significant.

From Table 4-2, for $6 - 1 = 5$ degrees of freedom, we find:

$$t(50\% \text{ confidence}) = 0.727$$
$$t(80\% \text{ confidence}) = 1.476$$

4-3 DEALING WITH BAD DATA

Sometimes one datum appears to be inconsistent with the remaining data. When this happens, you are faced with the decision of whether to retain the questionable point or to throw it away as unreliable. The Q test is used to help make this decision.

Consider the five results 12.53, 12.56, 12.47, 12.67, and 12.48. Is 12.67 a "bad point"? To apply the Q test, we arrange the data in order of increasing value and calculate Q defined as

$$Q = \frac{\text{gap}}{\text{range}} \qquad (4\text{-}11)$$

Gap = 0.11

| 12.47 | 12.48 | 12.53 | 12.56 | | (12.67) | —Questionable value (too high?) |

Range = 0.20

The **gap** is the difference between the questionable point and the nearest value.

If Q (observed) $> Q$ (tabulated), the questionable point should be discarded. For the numbers above, $Q = 0.11/0.20 = 0.55$. Referring to Table 4-4, we see that the critical value of Q at the 90% confidence limit is 0.64. *Since the observed Q is smaller than the tabulated Q, the questionable point should be retained.* That is, there is more than a 10% chance that the value 12.67 is a member of the same population as the other four numbers.

If Q (observed) $> Q$ (tabulated), discard the questionable point.

Table 4-4
Values of Q for rejection of data

Q (90% confidence)	0.94	0.76	0.64	0.56	0.51	0.47	0.44	0.41
Number of observations	3	4	5	6	7	8	9	10

Note: $Q = $ gap/range. If Q (observed) $> Q$ (tabulated), the value in question may be rejected with 90% confidence.
Source: R. B. Dean and W. J. Dixon, *Anal. Chem.*, **23**, 636 (1951). This reference also provides useful recipes for the rapid estimation of statistical parameters for small sets of data ($n \leq 10$).

Table 4-5

Spectrophotometer readings for protein analysis by the Lowry method

Sample (μg)	Absorbance of three independent samples			Range	Average with all data
0	0.099	0.099	0.100	0.001	0.099_3
5	0.185	0.187	0.188	0.003	0.188_7
10	0.282	0.272	0.272	0.010	0.275_3
15	0.392	0.345	0.347	0.047	0.361_2
20	0.425	0.425	0.430	0.005	0.426_7
25	0.483	0.488	0.496	0.013	0.489_0

The Q test is fairly stringent and not particularly helpful for small sets of data ($n < 5$). If you strive to retain anything that has more than a 10% chance of being real, you may accumulate some "bad" data. *Good common sense and the fortitude to repeat a questionable experiment are usually more valuable than any statistical test.*

Some real data from a spectrophotometric analysis are given in Table 4-5. In this analysis a color is developed that is in proportion to the amount of protein present in the sample. Scanning across the three absorbance values for each size of sample, we note that the number 0.392 seems clearly out of line and should be rejected. The absorbance values in Table 4-5 show that the range for the 15-μg sample is much bigger than the range for the other samples. The number 0.392 is inconsistent with the other values observed for the same sample. The nearly linear relation between the average values of absorbance up to the 20-μg sample indicates that the value 0.392 is in error (Figure 4-5).

It is reasonable to ask whether all three absorbances for the 25-μg sample are low for some unknown reason, since this point falls below the straight line in Figure 4-5. The answer to this question can be discovered only by repetition of the experiment. Many repetitions of this particular analysis show that the 25-μg point is consistently below the straight line and there is nothing "wrong" with the data in Table 4-5.

What is the difference between the Q test and a confidence interval? The confidence interval applies to the *mean,* whereas the Q test applies to individual data. For the five points 12.53, 12.56, 12.47, 12.67, and 12.48, we find $\bar{x} = 12.54_2$, $s = 0.08_0$, and

$$90\% \text{ confidence:} \quad \mu = 12.54_2 \pm 0.07_6$$

Student's t tells us that there is *less* than a 10% chance that the mean value lies above $12.54_2 + 0.07_6 = 12.61_8$. The Q test tells us that there is *more* than a 10% chance that the individual value 12.67 is a normal part of the data set, differing from the other points through purely random error.

Use a common-sense approach to rejection of data.

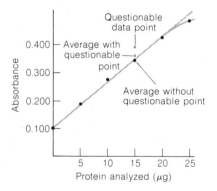

Figure 4-5

Graph of the average values of absorbance in Table 4-5.

4-4 FINDING THE "BEST" STRAIGHT LINE

Often we seek to draw the "best" straight line through a graphed set of experimental data points, as in Figure 4-5. This section provides a very important recipe for finding that "best" line.

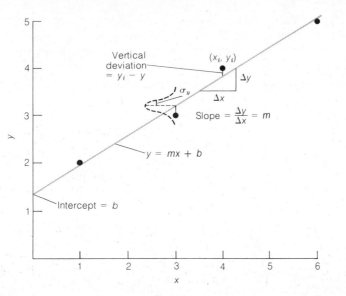

Figure 4-6
An example of least-squares curve fitting. The points (1, 2) and (6, 5) do not fall exactly on the solid line, but they are too close to the line to show their deviations. The Gaussian curve drawn over the point (3, 3) is a schematic indication of the fact that each value of y_i is normally distributed about the straight line. That is, the most probable value of y will fall on the line, but there is a finite probability of measuring y some distance from the line.

Method of Least Squares

The method of least squares assumes that the errors in the y values are substantially greater than the errors in the x values.[†] This condition is often true and applies to Figure 4-5, since the variation in absorbance is normally much greater than the uncertainty in concentration. A second assumption is that the uncertainties (standard deviations) in all of the y values are similar.

Suppose we seek to draw the best straight line through the points in Figure 4-6 by minimizing the vertical deviations between the points and the line. The reason for minimizing only the vertical deviations is the assumption that uncertainties in the y values are much greater than uncertainties in the x values.

Let the equation of the line be

$$y = mx + b \qquad (4\text{-}12)$$

in which m is the **slope** and b is the **y-intercept.** The vertical deviation for the point (x_i, y_i) will be given by $y_i - y$, where y is the ordinate of the straight line when $x = x_i$.

The equation of a straight line is discussed in Appendix B.

[†] An outstanding, readable article showing how to apply a simple least-squares procedure to many complicated nonlinear problems is R. de Levie, *J. Chem. Ed., 63,* 11 (1986). The case in which both the x and y coordinates have substantial uncertainty is treated in very readable articles by D. York, *Can. J. Phys., 44,* 1079 (1966) and J. A. Irvin and T. I. Quickenden, *J. Chem. Ed., 60,* 711 (1983). A general approach to least-squares curve fitting using a computer can be found in an article by W. E. Wentworth, *J. Chem. Ed., 42,* 96, 162 (1965). Articles dealing with the statistical significance of least-squares parameters can be found in M. D. Pattengill and D. E. Sands, *J. Chem. Ed., 56,* 244 (1979) and E. Heilbronner, *J. Chem. Ed., 56,* 240 (1979).

$$\text{Vertical deviation} = d_i = y_i - y = y_i - (mx_i + b) \tag{4-13}$$

Some of the deviations are positive and some are negative. Since we wish to minimize the magnitude of the deviations irrespective of their signs, we can square all the deviations so that we are dealing only with positive numbers:

$$d_i^2 = (y_i - y)^2 = (y_i - mx_i - b)^2 \tag{4-14}$$

Because we seek to minimize the squares of the deviations, this is called the method of least squares. It can be shown that minimizing the squares of the deviations (rather than simply their magnitudes) corresponds to assuming that the set of y values is the most probable set.

Finding the values of m and b that minimize the sum of the squares of the vertical deviations involves some calculus, which we will omit. We will express the final solution for slope and intercept in terms of **determinants,** which are a graphic way of expressing certain products. The determinant $\begin{vmatrix} e & f \\ g & h \end{vmatrix}$ has the value $eh - fg$. So, for example,

$$\begin{vmatrix} 6 & 5 \\ 4 & 3 \end{vmatrix} = 6 \times 3 - 5 \times 4 = -2 \tag{4-15}$$

The slope and the intercept of the "best" straight line are found to be

Least-squares slope and intercept
$$m = \begin{vmatrix} \sum x_i y_i & \sum x_i \\ \sum y_i & n \end{vmatrix} \div D \tag{4-16}$$

$$b = \begin{vmatrix} \sum (x_i^2) & \sum x_i y_i \\ \sum x_i & \sum y_i \end{vmatrix} \div D \tag{4-17}$$

where the number D is given by

$$D = \begin{vmatrix} \sum (x_i^2) & \sum x_i \\ \sum x_i & n \end{vmatrix} \tag{4-18}$$

and n is the number of points.

Now we apply Equations 4-16 and 4-17 to the points in Figure 4-6 to find the slope and the intercept of the best straight line through the four points. The work is set out in Table 4-6. Noting that $n = 4$ and putting the various sums into the determinants in Equations 4-16, 4-17, and 4-18 gives

$$m = \begin{vmatrix} 57 & 14 \\ 14 & 4 \end{vmatrix} \div \begin{vmatrix} 62 & 14 \\ 14 & 4 \end{vmatrix} = \frac{32}{52} = 0.615\,38 \tag{4-19}$$

$$b = \begin{vmatrix} 62 & 57 \\ 14 & 14 \end{vmatrix} \div \begin{vmatrix} 62 & 14 \\ 14 & 4 \end{vmatrix} = \frac{70}{52} = 1.346\,15 \tag{4-20}$$

The equation of the best straight line through the points in Figure 4-6 is therefore

$$y = 0.615\,38x + 1.346\,15 \tag{4-21}$$

Table 4-6
Calculations for least-squares analysis

x_i	y_i	$x_i y_i$	x_i^2	$d_i (= y_i - mx_i - b)$	d_i^2
1	2	2	1	0.038 47	0.001 479 9
3	3	9	9	−0.192 29	0.036 975
4	4	16	16	0.192 33	0.036 991
6	5	30	36	−0.038 43	0.001 476 9
$\sum x_i = 14$	$\sum y_i = 14$	$\sum x_i y_i = 57$	$\sum (x_i^2) = 62$		$\sum (d_i^2) = 0.076\ 923$

We tackle the question of how many significant figures should be associated with m and b in the next section.

How Reliable Are Least-Squares Parameters?

To estimate the uncertainties (expressed as standard derivations) in the slope and the intercept, an uncertainty analysis must be performed on Equations 4-16 and 4-17. Since the uncertainties in m and b are related to the uncertainty in measuring each value of y, we need first to estimate the standard deviation that describes the population of y values. This standard deviation, σ_y, characterizes the little Gaussian curve inscribed in Figure 4-6.

We will estimate σ_y, the population standard deviation of all y values, by calculating s_y, the standard deviation, for the four measured values of y. The deviation of each value of y_i from the center of its Gaussian curve is just $d_i = y_i - y = y_i - (mx_i + b)$ (Equation 4-13). The standard deviation of these vertical deviations will be given by an equation analogous to Equation 4-2:

$$\sigma_y \approx s_y = \sqrt{\frac{\sum(d_i - \bar{d})^2}{\text{(degrees of freedom)}}} \qquad (4\text{-}22)$$

But the average deviation, $\bar{d}$, is zero for the best straight line. So the numerator of Equation 4-22 reduces to $\sum(d_i^2)$.

The degrees of freedom is the number of independent pieces of information available. In Equation 4-2 we set the degrees of freedom equal to $n - 1$. The rationale for so doing is that we began with n degrees of freedom, but one degree of freedom is "lost" in determining the average value ($\bar{x}$). That is, only $n - 1$ pieces of information are available in addition to the average value. If you know $n - 1$ values and you also know the average value, then you can calculate the nth value.

How does this apply to Equation 4-22? We began with n points. But two degrees of freedom were "used up" in determining the slope and the intercept of the best line. Therefore, the degrees of freedom is $n - 2$. Equation 4-22 should be written

If you know $\bar{x}$ and $n - 1$ of the individual values, you can calculate the nth value. Therefore the problem has just $n - 1$ degrees of freedom once $\bar{x}$ is known.

$$\sigma_y \approx s_y = \sqrt{\frac{\sum(d_i^2)}{n - 2}} \qquad (4\text{-}23)$$

where d_i is given by Equation 4-13.

The uncertainty analysis for Equations 4-16 and 4-17 uses the methods of Appendix C and leads to the following results:

$$\text{Standard deviation} \atop \text{of slope and} \atop \text{intercept} \quad \begin{cases} \sigma_m^2 = \dfrac{\sigma_y^2 n}{D} & (4\text{-}24) \\[4mm] \sigma_b^2 = \dfrac{\sigma_y^2 \sum(x_i^2)}{D} & (4\text{-}25) \end{cases}$$

where σ_m is our estimate of the standard deviation of the slope, σ_b is our estimate of the standard deviation of the intercept, σ_y is given by Equation 4-23, and D is given by Equation 4-18.

Now we can finally address the question of significant figures for the slope and the intercept of the line in Figure 4-6. In Table 4-6 it can be seen that $\sum(d_i^2) = 0.076\,923$. Putting this number into Equation 4-23 gives

$$\sigma_y^2 \approx s_y^2 = \frac{0.076\,923}{4-2} = 0.038\,462 \qquad (4\text{-}26)$$

Now, we can plug numbers into Equations 4-24 and 4-25 to find

$$\sigma_m^2 = \frac{\sigma_y^2 n}{D} = \frac{(0.038\,462)(4)}{52} = 0.002\,958\,6 \qquad (4\text{-}27)$$

$$\sigma_b^2 = \frac{\sigma_y^2 \sum(x_i^2)}{D} = \frac{(0.038\,462)(62)}{52} = 0.045\,859 \qquad (4\text{-}28)$$

or

$$\sigma_m = 0.054\,39 \qquad \text{and} \qquad \sigma_b = 0.214\,15$$

Combining the results for m, σ_m, b, and σ_b, we can write

$$\text{Slope:} \qquad \frac{0.615\,38}{\pm 0.054\,39} = 0.62 \pm 0.05$$

$$\text{Intercept:} \qquad \frac{1.346\,15}{\pm 0.214\,15} = 1.3 \pm 0.2$$

where the uncertainties represent one standard deviation. The choice of the last significant figure of slope and intercept is dictated by the decimal place of the first digit of the standard deviation. *The first uncertain figure is the last significant figure.*

> The first digit of the uncertainty is the last significant figure.

A Practical Example of the Method of Least Squares

What good is all of this? One real application of a least-squares analysis involves the determination of protein concentration by using the absorbance values in Table 4-5. To obtain the numbers in this table, analysis of a series of known protein standards was performed. Each standard leads to a certain absorbance measured with a spectrophotometer. Figure 4-5 shows that the standards containing from 0 to 20 μg of protein appear to fall on a straight line. These standards provide 14 absorbance values in Table 4-5, with the value 0.392 omitted. Using all 14 values, we calculate the least-squares parameters for Figure 4-5 to be

$$m = 0.016\,3_0 \qquad \sigma_m = 0.000\,2_2$$

$$b = 0.104_0 \qquad \sigma_b = 0.002_6$$

$$\sigma_y = 0.005_9$$

Now suppose that the absorbance of an unknown sample is found to be 0.246. How many micrograms of protein does it contain, and what uncertainty is associated with the answer? Note that in Figure 4-5 the y axis is absorbance, and the x axis is micrograms of protein. Solving for concentration gives

$$x = \frac{y - b}{m}$$

Follow the rules regarding propagation of uncertainty for subtraction and division.

$$= \frac{0.246\,(\pm 0.005_9) - 0.104_0\,(\pm 0.002_6)}{0.016\,3_0\,(\pm 0.000\,2_2)}$$

$$= \frac{0.142_0\,(\pm 0.006_4)}{0.016\,3_0\,(\pm 0.000\,2_2)}$$

$$= \frac{0.142_0\,(\pm 4._5\%)}{0.016\,3_0\,(\pm 1._3\%)}$$

$$= 8.7_1\,(\pm 4._7\%) = 8.7\,(\pm 0.4)\ \mu g \text{ of protein} \qquad (4\text{-}29)$$

Operating on the standard deviations with the usual rules for propagation of uncertainty leads us to calculate that the unknown has 8.7 (± 0.4) μg of protein. The estimate of uncertainty is the standard deviation associated with the number 8.7.[†]

A final caution is in order. Many people have calculators or computer programs that perform least-squares curve fitting automatically. If you do not first make a graph of your data, you will not have the opportunity to reject bad data. Mindless use of a computer program may not do justice to your hard-earned data. *The operation in which a human being evaluates his or her data should never be sacrificed.*

You eyes are smarter than your calculator!

[†] The calculation of uncertainty in Equation 4-29 is approximately correct. It neglects the fact that the calculated slope and intercept are not independent of each other. A full treatment (which is beyond the scope of this text) gives

$$\text{(Uncertainty in } x)^2 = \frac{\sigma_y^2}{m^2}\left[1 + \left(\frac{y - b}{m}\right)^2 \left(\frac{n}{D}\right) + \frac{\sum(x_i^2)}{D} - 2\left(\frac{y - b}{m}\right)\left(\frac{\sum(x_i)}{D}\right)\right]$$

where D is given by Equation 4-18. If you measure several values of y and take their average, the uncertainty in x will be reduced. In this case change the first term in brackets from 1 to $1/k$, where k is the number of y values that have been averaged. For the example in Equation 4-29, one value of y (0.246) has been measured, and application of the equation in this footnote gives uncertainty in $x = \pm 0.3_7\ \mu g$ instead of the $\pm 0.4_1\ \mu g$ calculated using Equation 4-29.

Summary

The results of many measurements of an experimental quantity follow a Gaussian distribution, provided the errors are purely random. The measured mean, $\bar{x}$, approaches the true mean, μ, as the number of measurements becomes very large. The broader the distribution, the greater is σ, the standard deviation. For a limited number of measurements, the standard deviation is given by the formula $s = \sqrt{[\sum(x_i - \bar{x})^2]/(n-1)}$. About two-thirds of all measurements lie within $\pm 1\sigma$ and 95% lie within $\pm 2\sigma$. The probability of observing a value within a certain interval is proportional to the area of that interval, given in Table 4-1.

Student's t is used to find confidence intervals ($\mu = \bar{x} \pm ts/\sqrt{n}$) and to compare means. To compare two sets of replicate measurements, we use the formula $t = (1/s)(\bar{x}_1 - \bar{x}_2)\sqrt{n_1 n_2/(n_1 + n_2)}$. To compare individual differences in a series of measurements made by two methods, we use the formula $t = \bar{d}\sqrt{n}/s_d$. The Q test is used to reject bad data, but common sense and repetition of a questionable experiment are even better ideas.

The method of least squares is used to find the slope and the intercept of the best straight line through a series of points. The standard deviations of slope and intercept are used in analyzing the error associated with an experiment measurement.

Terms to Understand

average
confidence interval
determinant
Gaussian curve
geometric mean
intercept
mean
median

normal error curve
Q test
range
slope
spread
standard deviation
t test
variance

Exercises

4-A. For the numbers 116.0, 97.9, 114.2, 106.8, and 108.3, find the mean, standard deviation, median, geometric mean, range, and 90% confidence interval for the mean. Using the Q test, decide whether the number 97.9 should be discarded.

4-B. Suppose that data were collected for 10 000 sets of automobile brakes. The mileage at which each set had been 80% worn through was recorded. The average was 62 700, and the standard deviation was 10 400 miles.
   (a) What fraction of brakes is expected to be 80% worn in less than 45 800 miles?
   (b) What fraction is expected to be 80% worn at a mileage between 60 000 and 70 000 miles?

4-C. It is found from a reliable assay that the ATP (adenosine triphosphate) content of a certain type of cell is 112 μmol/100 mL. You have developed a new assay, which gave the following values for replicate analyses: 117, 119, 111, 115, 120 μmol/100 mL. The average value is 116.4. Can you be 90% confident that your method produces a high value? Can you be 99% confident?

4-D. The Ca content of a powdered mineral sample was analyzed five times by each of two methods, with similar standard deviations:

	Ca (percent composition)
Method 1:	0.027 1, 0.028 2, 0.027 9, 0.027 1, 0.027 5
Method 2:	0.027 1, 0.026 8, 0.026 3, 0.027 4, 0.026 9

Are the mean values significantly different at the 90% confidence level?

4-E. A common procedure for protein determination is the dye-binding assay of Bradford.[†] In this method, a dye binds to the protein, and the color of the dye changes from brown to blue. The amount of blue color is proportional to the amount of protein present. Some real data are given above right.

[†] M. Bradford, *Anal. Biochem.*, **72**, 248 (1976).

Protein (μg): 0.00   9.36   18.72   28.08   37.44
Absorbance
at 595 nm: 0.466 0.676 0.883 1.086 1.280

(a) Using the method of least squares, determine the equation of the best straight line through these points. Use the standard deviation of the slope and intercept to express the equation in the form

$$y = [m(\pm\sigma_m)]x + [b(\pm\sigma_b)]$$ with a reasonable number of significant figures.

[handwritten: $y = [0.0218\,(\pm\,0.0002)]x + [0.471\,(\pm\,0.004)]$]

(b) Make a graph showing the experimental data and the straight line calculated in part a.

(c) An unknown protein sample gave an absorbance of 0.973. Calculate the number of micrograms of protein in the unknown, and estimate its uncertainty (expressed as a standard deviation).

Problems

A4-1. A population of results with purely random variation follows a Gaussian curve. Use Table 4-1 to state the fraction of such a population that lies within the following intervals:
(a) $\mu \pm \sigma$ (b) $\mu \pm 2\sigma$ (c) μ to $+\sigma$
(d) μ to $+0.5\sigma$ (e) $-\sigma$ to -0.5σ

A4-2. The ratio of the number of atoms of the isotopes ^{69}Ga and ^{71}Ga in samples from different sources was measured in an effort to understand differences in reported values of the atomic weight of gallium.[†] Results for eight samples were as follows:

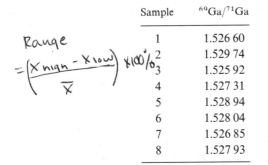

[handwritten: Range $= \left(\dfrac{x_{high} - x_{low}}{\bar{x}}\right) \times 100\%$]

Sample	^{69}Ga/^{71}Ga
1	1.526 60
2	1.529 74
3	1.525 92
4	1.527 31
5	1.528 94
6	1.528 04
7	1.526 85
8	1.527 93

(a) Find the mean value of ^{69}Ga/^{71}Ga.
(b) Find the standard deviation.
(c) Find the variance.
(d) Find the median value.
(e) What is the range of the data?

A4-3. Sample 8 of Problem A4-2 was analyzed seven times, with $\bar{x} = 1.527\,93$ and $s = 0.000\,07$. The value of s is a measure of the reproducibility of the analytical procedure.
(a) If sample 8 is analyzed many more times, what fraction of the results is expected to lie within $\bar{x} \pm 3s = 1.527\,93 \pm 0.000\,21$?
(b) How many of the first seven samples in Problem A4-2 have values lying within the range $1.527\,93 \pm 0.000\,21$? Is the variation among samples due to irreproducibility in the analysis or real differences among samples?

A4-4. The percent of an additive in gasoline was measured six times with the following results: 0.13, 0.12, 0.16, 0.17, 0.20, 0.11%. Find the 90% and 99% confidence intervals for the percent of the additive.

A4-5. The Ti content of five different ore samples (each with a different Ti content) was measured by each of two methods:

	Sample				
	A	B	C	D	E
Method 1:	0.013 4	0.014 4	0.012 6	0.012 5	0.013 7
Method 2:	0.013 5	0.015 6	0.013 7	0.013 7	0.013 6

Do the two analytical techniques give results that are significantly different at the 90% confidence level?

A4-6. The calcium content of a person's urine was determined on two different days:

Day	Average Ca (mg/L)	Number of measurements
1	238	4
2	255	5

The analytical method applied to many samples yields a standard deviation known to be 14 mg/L. Are the two average values significantly different at the 90% confidence level?

A4-7. Using the Q test, decide whether the value 216 should be rejected from the set of results 192, 216, 202, 195, 204.

[handwritten: (value − closest) / (high − low)]

† J. W. Gramlich and L. A. Machlan, *Anal. Chem.*, **57**, 1788 (1985).

A4-8. A straight line is drawn through the points

$$(3.0, -3.87 \times 10^4)$$
$$(10.0, -12.99 \times 10^4)$$
$$(20.0, -25.93 \times 10^4)$$
$$(30.0, -3.889 \times 10^4)$$
$$(40.0, -51.96 \times 10^4)$$

using the method of least squares. The results are $m = -1.298\,72 \times 10^4$, $b = 256.695$, $\sigma_m = 13.190$, and $\sigma_b = 323.57$. Express the slope and intercept and their uncertainties with the correct significant figures.

4-9. (a) Calculate the fraction of bulbs in Figure 4-1 expected to have a lifetime greater than 1 000 hours.
(b) Calculate the fraction expected to have a lifetime between 800 and 900 hours.

4-10. Write the equation of the smooth Gaussian curve in Figure 4-1. Use the equation to calculate the value of y when $x = 1\,000$ h. See if your calculated value agrees with the value on the graph. Remember that the curve represents the results of 4 768 measurements, and each bar on the graph corresponds to a 20-hour interval.

4-11. The time needed for a certain process to occur was measured five times and found to be 14.48, 14.57, 14.59, 14.32, and 14.52 s.
(a) Can the number 14.32 be rejected as bad data at the 90% confidence level?
(b) Including the value 14.32 in the data set, calculate the fraction of measurements expected between 14.55 and 14.60 s if a large number of measurements is made.

4-12. If you measure a quantity four times and the standard deviation is 1.0% of the average, can you be at least 90% confident that the true value is within 1.2% of the measured average?

4-13. Two methods were used to measure the specific activity (units of enzyme activity per milligram of protein) of an enzyme. One unit of enzyme activity is defined as the amount of enzyme that catalyzes the formation of one micromole of product per minute under specified conditions

Enzyme activity (five replications)

| Method 1: | 139 | 147 | 160 | 158 | 135 |
| Method 2: | 148 | 159 | 156 | 164 | 159 |

Is the mean value of method 1 significantly different from the mean value of method 2 at the 90% confidence level? at the 80% confidence level?

4-14. Using the Q test, determine the largest number (n) that should be retained in the set 63, 65, 68, 72, n.

4-15. Use the method of least squares to calculate the equation of the best straight line going through the points (1, 3), (3, 2), and (5, 0). Express your answer in the form $y = [m(\pm\sigma_m)]x + [b(\pm\sigma_b)]$, with a reasonable number of significant figures.

4-16. Consider a suspension of cells containing 4.13 $(\pm 0.09) \times 10^{-13}$ mol of cells per liter. (This is $\sim 2.5 \times 10^8$ cells per milliliter.) Call this total concentration of cells C_t. Each cell has n equivalent binding sites to which the hormone H can bind. The dissociation constant, K, is the equilibrium constant for the reaction below:

$$CH \underset{}{\overset{K}{\rightleftharpoons}} C + H$$

where C is a cell, H is the hormone, and CH is hormone attached to the cell. To determine the number of binding sites, n, per cell, one usually varies the concentration of H while maintaining a fixed concentration of cells C_t. The concentrations of free hormone, H_f, and bound hormone, H_b, are measured in each experiment. A graph of H_b/H_f versus H_b is then drawn. This is called a *Scatchard plot*. It can be shown that

$$\frac{H_b}{H_f} = -\left(\frac{1}{K}\right)H_b + \frac{nC_t}{K}$$

That is, a graph of H_b/H_f versus H_b should give a straight line.

The results of a hypothetical experiment are shown below. The slope and the intercept were calculated by the method of least squares and have *not* been reduced to a reasonable number of significant figures. The uncertainties shown in parentheses are one standard deviation.

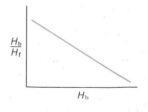

Intercept: $b = 3.138\,6\,(\pm 0.289\,9) \times 10^{-3}$
Slope: $m = -1.364\,5\,(\pm 0.132\,8) \times 10^6\ \text{M}^{-1}$

Find the value of n and the absolute uncertainty (standard deviation) of n. Express your answer with a reasonable number of significant figures. Do not try to assign significant figures until the *end* of the calculation.

4-17. Consider the least-squares problem illustrated in Figure 4-6. The slope and intercept and their standard deviations are computed in Equations 4-19, 4-20, 4-27, and 4-28 and the work is set out in Table 4-6. Suppose that a single new measurement produces a y value of 2.58.

(a) Use the method of Equation 4-29 to calculate the x value (and its uncertainty) associated with $y = 2.58$.

(b) Use the equation in the footnote on page 63 to calculate the uncertainty of the x value associated with $y = 2.58$.

(c) Suppose you measure y four times and the average is 2.58. Use the equation in the footnote on page 63 to calculate the uncertainty of the x value given that 2.58 is the average of four measurements instead of one measurement.

5 Chemical Equilibrium

In this chapter we will study some of the common types of equilibria in aqueous solution, including solubility of ionic compounds, complex formation, and acid–base reactions. Detailed studies of acid–base equilibria and complex formation are reserved for Chapters 10–13. Oxidation–reduction equilibria will be introduced in Chapter 14.

5-1 THE EQUILIBRIUM CONSTANT

For the reaction

$$a\text{A} + b\text{B} \rightleftharpoons c\text{C} + d\text{D} \tag{5-1}$$

The equilibrium constant is more correctly expressed as a ratio of *activities,* rather than concentrations. We reserve a discussion of activity for the next chapter.

it is customary to write the **equilibrium constant,** K, in the form

$$K = \frac{[\text{C}]^c[\text{D}]^d}{[\text{A}]^a[\text{B}]^b} \tag{5-2}$$

where the small superscript letters denote stoichiometric coefficients and each capital letter stands for a chemical species. The symbol [A] stands for the concentration of A.

In the thermodynamic derivation of the equilibrium constant, each quantity in Equation 5-2 is expressed as the *ratio* of the concentration of each species to its concentration in its **standard state.** For solutes the standard state is 1 M. For gases the standard state is 1 atm, and for solids and liquids the standard states are the pure solid or liquid. It is understood (but rarely written) that the term [A] in Equation 5-2 really means [A]/(1 M) if A is a solute. If D is a gas, [D] really means (pressure of D in atmospheres)/(1 atm). To emphasize that [D] means pressure of D, we will usually write P_D in place of [D]. The terms of Equation 5-2 are actually dimensionless; therefore all equilibrium constants are dimensionless.

Equilibrium constants are dimensionless.

For the ratios [A]/(1 M) and [D]/(1 atm) to be dimensionless, [A] *must* be expressed in moles per liter (M), and [D] *must* be expressed in atmospheres. If C were a pure liquid or solid, the ratio [C]/(concentration of C in its standard state) would be unity (1) because the standard state is the pure liquid or solid. If [C] is a solvent, the concentration is so close to that of pure liquid C that the value of [C] is still essentially 1.

The take-home lesson is this. In evaluating an equilibrium constant,

1. The concentrations of solutes should be expressed as moles per liter.
2. The concentrations of gases should be expressed in atmospheres.
3. The concentrations of pure solids, pure liquids, and solvents are omitted because they are unity.

These conventions are arbitrary, but you must use them if you wish to use tabulated values of equilibrium constants, standard reduction potentials, and free energies.

Equilibrium constants are dimensionless, but in specifying concentrations you must use units of M for solutes and atm for gases.

Manipulation of Equilibrium Constants

Consider the reaction

$$HA \rightleftharpoons H^+ + A^- \qquad K_1 = \frac{[H^+][A^-]}{[HA]} \qquad (5\text{-}3)$$

Throughout this text it is understood that all species in chemical equations are in aqueous solution, unless otherwise specified.

If the direction of a reaction is reversed, the new value of K is simply the reciprocal of the original value of K.

$$H^+ + A^- \rightleftharpoons HA \qquad K_1' = \frac{[HA]}{[H^+][A^-]} = 1/K_1 \qquad (5\text{-}4)$$

If two reactions are added, the new value of K is the product of the two individual values:

$$\begin{array}{lll} HA & \rightleftharpoons \cancel{H^+} + A^- & K_1 \\ \cancel{H^+} + C \rightleftharpoons CH^+ & K_2 \\ \hline HA + C \rightleftharpoons A^- + CH^+ & K_3 \end{array}$$

$$K_3 = K_1 K_2 = \frac{[\cancel{H^+}][A^-]}{[HA]} \cdot \frac{[CH^+]}{[\cancel{H^+}][C]} = \frac{[A^-][CH^+]}{[HA][C]} \qquad (5\text{-}5)$$

If a reaction is reversed, $K' = 1/K$. If two reactions are added, $K = K_1 K_2$.

If *n* reactions are added, the overall equilibrium constant is the product of all *n* individual equilibrium constants.

EXAMPLE: Combining Equilibrium Constants

The equilibrium constant for the reaction

$$H_2O \rightleftharpoons H^+ + OH^-$$

is called $K_w (=[H^+][OH^-])$ and has the value 1.0×10^{-14} at 25°C.

Given that

$$NH_3(aq) + H_2O \rightleftharpoons NH_4^+ + OH^- \qquad K_{NH_3} = 1.8 \times 10^{-5}$$

find the equilibrium constant for the reaction

$$NH_4^+ \rightleftharpoons NH_3(aq) + H^+$$

The third reaction can be obtained by reversing the second reaction and adding it to the first reaction:

$$\begin{array}{llll}
\cancel{H_2O} \rightleftharpoons H^+ & + \cancel{OH^-} & K_1 = K_w \\
NH_4 + \cancel{OH^-} \rightleftharpoons NH_3(aq) + \cancel{H_2O} & & K_2 = 1/K_{NH_3} \\
\hline
NH_4^+ \qquad\qquad \rightleftharpoons H^+ & + NH_3(aq) & K_3 = K_w \cdot \dfrac{1}{K_{NH_3}} = 5.6 \times 10^{-10}
\end{array}$$

Equilibrium and Thermodynamics

Enthalpy

$$\Delta H = (+)$$

Heat is absorbed.
Endothermic.

$$\Delta H = (-)$$

Heat is liberated.
Exothermic.

The change in **enthalpy**, ΔH, for a reaction is the heat absorbed when the reaction takes place under constant applied pressure. The *standard enthalpy* change, $\Delta H°$, refers to the heat absorbed when all reactants and products are in their standard states.[†] For the following reaction, $\Delta H° = -75.15$ kJ/mol at 25°C.

$$HCl(g) \rightleftharpoons H^+(aq) + Cl^-(aq) \tag{5-6}$$

The negative sign of $\Delta H°$ indicates that heat is given off by the products when Reaction 5-6 occurs. This means that the solution becomes warmer when the reaction occurs. For other reactions, ΔH is positive, which means that heat is absorbed by reactants as they react. That is, the solution gets colder during the reaction. A reaction for which ΔH is positive is said to be **endothermic.** If ΔH is negative, the reaction is **exothermic.**

Entropy

$$\Delta S = (+)$$

Products more disordered
than reactants.

$$\Delta S = (-)$$

Products more ordered
than reactants.

The **entropy,** S, of a substance is a measure of its "disorder," which we will not attempt to define in a quantitative way. The greater the disorder, the greater the entropy. In general, a gas is more disordered (has higher entropy) than a liquid, which is more disordered than a solid. Ions in aqueous solution are normally more disordered than in their solid salt. For example, for Reaction 5-7, we find $\Delta S° = +76$ J/(K·mol) at 25°C.

$$KCl(s) \rightleftharpoons K^+(aq) + Cl^-(aq) \tag{5-7}$$

[†] The precise definition of the standard state contains subtleties beyond the scope of this text. For Reaction 5-6 the standard state of H^+ or Cl^- is the hypothetical state in which each ion is present at a concentration of 1 M but behaves as if it were in an infinitely dilute solution. That is, the standard concentration is 1 M, but the standard physical behavior is what would be observed in a very dilute solution in which each ion is unaffected by surrounding ions.

This means that a mole of $K^+(aq)$ plus a mole of $Cl^-(aq)$ is more disordered than a mole of $KCl(s)$ plus the solvent water. (The dimensions of entropy are energy/temperature per mole of reactant.) For Reaction 5-6, $\Delta S° = -131.5$ J/(K·mol) at 25°C. The aqueous ions are less disordered than gaseous HCl plus the solvent water.

Free energy

There is a tendency in nature for a system to lower its enthalpy and raise its entropy. A chemical reaction is driven toward the formation of products by a *negative* value of ΔH (heat given off) and/or a *positive* value of ΔS (entropy increases). If ΔH is negative and ΔS is positive, the reaction is clearly favored. If ΔH is positive and ΔS is negative, the reaction is clearly disfavored.

If ΔH and ΔS are both positive or both negative, what decides whether a reaction will be favored? The change in **Gibbs free energy**, ΔG, is the arbiter between opposing tendencies of ΔH and ΔS. At constant temperature

$$\Delta G = \Delta H - T\,\Delta S \tag{5-8}$$

The effects of entropy and enthalpy are combined in Equation 5-8. *A reaction is favored if ΔG is negative.*

For Reaction 5-6, $\Delta H°$ favors the reaction and $\Delta S°$ disfavors it. To find the net result, we must evaluate $\Delta G°$:

$$\begin{aligned}
\Delta G° &= \Delta H° - T\,\Delta S° \\
&= (-75.15 \times 10^3 \text{ J}) - (298.15 \text{ K})(-131.5 \text{ J/K}) \\
&= -35.94 \text{ kJ/mol} \tag{5-9}
\end{aligned}$$

Note that 25.00°C = 298.15 K.

$\Delta G°$ is negative and the reaction is favored under standard conditions. The favorable enthalpy change is greater than the unfavorable entropy change in this case.

The relation between $\Delta G°$ and the equilibrium constant for a reaction is

$$K = e^{-\Delta G°/RT} \tag{5-10}$$

where R is the gas constant [$= 8.314\,510$ J/(K·mol)] and T is in kelvins. For Reaction 5-6, we find

$$K = e^{-(-35.94 \times 10^3 \text{J/mol})/[8.314\,510\text{J/(K·mol)}](298.15\text{K})}$$

$$= 1.98 \times 10^6 \tag{5-11}$$

Since the equilibrium constant is large, HCl(g) is very soluble in water and is nearly completely ionized to H^+ and Cl^- when it dissolves.

To summarize, ΔG takes into account the tendency for heat to be liberated (ΔH negative) and disorder to increase (ΔS positive). A reaction is favored ($K > 1$) if $\Delta G°$ is negative. A reaction is disfavored ($K < 1$) if $\Delta G°$ is positive.

$\Delta G = (+)$
_____
Reaction is disfavored.

$\Delta G = (-)$
_____
Reaction is favored.

Le Châtelier's Principle

Suppose that a system at equilibrium is subjected to a change that disturbs the system. **Le Châtelier's principle** states that the direction in which the system proceeds back to equilibrium is such that the change is partially offset.[†]

To see what this means, consider the following reaction:

$$BrO_3^- + 2Cr^{3+} + 4H_2O \rightleftharpoons Br^- + Cr_2O_7^{2-} + 8H^+ \qquad (5\text{-}12)$$

for which the equilibrium constant is given by

$$K = \frac{[Br^-][Cr_2O_7^{2-}][H^+]^8}{[BrO_3^-][Cr^{3+}]^2} = 1 \times 10^{11} \text{ at } 25°C \qquad (5\text{-}13)$$

In one particular equilibrium state of this system, the following concentrations exist:

$$[H^+] = 5.0 \text{ M} \qquad [Cr_2O_7^{2-}] = 0.10 \text{ M} \qquad [Cr^{3+}] = 0.003\,0 \text{ M}$$

$$[Br^-] = 1.0 \text{ M} \qquad [BrO_3^-] = 0.043 \text{ M}$$

The reaction quotient has the same form as the equilibrium constant, but the concentrations are generally not the equilibrium concentrations.

Suppose that the equilibrium is disturbed by adding dichromate to the solution to increase the concentration of $[Cr_2O_7^{2-}]$ from 0.10 to 0.20 M. In what direction will the reaction proceed to reach equilibrium? According to the principle of Le Châtelier, the reaction should go back to the left to partially offset the increase in dichromate, which appears on the right side of Reaction 5-12. We can verify this algebraically by setting up the **reaction quotient,** Q, which has the same form as the equilibrium constant. The only difference is that Q is evaluated with whatever concentrations happen to exist in the solution. When the system reaches equilibrium, $Q = K$. For Reaction 5-12,

$$Q = \frac{(1.0)(0.20)(5.0)^8}{(0.043)(0.003\,0)^2} = 2 \times 10^{11} > K \qquad (5\text{-}14)$$

If $Q < K$, the reaction must proceed to the right to reach equilibrium. If $Q > K$, the reaction must proceed to the left to reach equilibrium.

Since $Q > K$, the reaction must go to the left to decrease the numerator and increase the denominator, until $Q = K$.

In general,

1. If a reaction is at equilibrium and products are added (or reactants are removed), the reaction goes to the left.

2. If a reaction is at equilibrium and reactants are added (or products are removed), the reaction goes to the right.

When the temperature of a system is changed, so is the equilibrium constant. Equations 5-8 and 5-10 can be combined to predict the effect of temperature on K:

$$K = e^{-\Delta G°/RT} = e^{-(\Delta H° - T\,\Delta S°)/RT} = e^{(-\Delta H°/RT + \Delta S°/R)}$$

$$= e^{-\Delta H°/RT} \cdot e^{\Delta S°/R} \qquad (5\text{-}15)$$

[†] All that you never wanted to know about Le Châtelier's principle can be found in several detailed articles, including those by F. G. Helfferich, *J. Chem. Ed.,* **62,** 305 (1985); R. Fernández-Prini, *J. Chem. Ed.,* **59,** 550 (1982); and L. K. Brice, *J. Chem. Ed.,* **60,** 387 (1983).

The term $e^{\Delta S°/R}$ is independent of T. The term $e^{-\Delta H°/RT}$ increases with increasing temperature if $\Delta H°$ is positive. The term $e^{-\Delta H°/RT}$ decreases with increasing temperature if $\Delta H°$ is negative. Equation 5-15 therefore tells us

1. The equilibrium constant of an endothermic reaction increases if the temperature is raised.
2. The equilibrium constant for an exothermic reaction decreases if the temperature is raised.

These statements can be understood in terms of Le Châtelier's principle as follows. Consider an endothermic reaction:

$$\text{Heat} + \text{reactants} \rightleftharpoons \text{products} \qquad (5\text{-}16)$$

If the temperature is raised, heat is added to the system. Reaction 5-16 proceeds to the right to partially offset this change.[†]

In dealing with equilibrium problems, we are making thermodynamic predictions, not kinetic predictions. We are calculating what must happen for a system to reach equilibrium, but not how long it will take. Some reactions are over in an instant; others will not reach equilibrium in a million years.

Heat can be treated as if it were a reactant in an endothermic reaction and a product in an exothermic reaction.

5-2 SOLUBILITY PRODUCT

The **solubility product** is the equilibrium constant for the reaction in which a solid salt dissolves to give its constituent ions in solution.[‡] The concentration of the solid is omitted from the equilibrium constant because the solid is in its standard state. A table of solubility products can be found in Appendix F.

As a typical example, consider the dissolution of mercurous chloride (Hg_2Cl_2) in water. The reaction is

$$Hg_2Cl_2(s) \rightleftharpoons Hg_2^{2+} + 2Cl^- \qquad (5\text{-}17)$$

for which the solubility product K_{sp} is

$$K_{sp} = [Hg_2^{2+}][Cl^-]^2 = 1.2 \times 10^{-18} \qquad (5\text{-}18)$$

A solution containing excess, undissolved solid is said to be **saturated** with that solid. The solution contains all the solid capable of being dissolved under the prevailing conditions. What will be the concentration of Hg_2^{2+} in a solution saturated with Hg_2Cl_2?

The equilibrium constant omits the pure solid because its concentration is unchanged.

Box 5-1 contains some information about Hg_2^{2+}.

[†] The solubility of the vast majority of ionic compounds increases with temperature, despite the fact that the standard heat of solution ($\Delta H°$) is negative for about half of them. Discussions of this seeming contradiction can be found in G. M. Bodner, *J. Chem. Ed.,* **57,** 117 (1980) and R. S. Treptow, *J. Chem. Ed.,* **61,** 499 (1984).

[‡] The solubility product does not tell the entire story on the solubility of slightly soluble salts. For many, the dissolution of *undissociated* species is as important as dissolution accompanied by dissociation. That is, the salt MX(s) can give MX(aq) as well as M⁺(aq) and X⁻(aq). In the case of $CaSO_4$, the total concentration of dissolved material is 15–19 mM [A. K. Sawyer, *J. Chem. Ed.,* **60,** 416 (1983)]. Using activity coefficients as described for LiF in the next chapter, we estimate that the concentration of dissociated Ca^{2+} should be 9 mM. This suggests that about half of the dissolved $CaSO_4$ is undissociated.

Box 5-1 THE MERCUROUS ION

Mercury has three common oxidation states:

1. $Hg(0)$ is metallic mercury. Hg is one of only two elements (the other being Br) that is a liquid at 298 K. Ga metal melts just above room temperature, at 303 K.
2. $Hg(I)$ is the mercurous ion, which is a dimer with a bond length close to 250 pm (2.50 Å).

$$[Hg—Hg]^{2+}$$

When mercurous salts dissolve, they do not give monatomic Hg^+ in solution. Hg_2^{2+} remains present as a diatomic ion.
3. $Hg(II)$ is the mercuric ion, which exists as Hg^{2+}.

$Hg(I)$ and $Hg(II)$ exist in equilibrium in the presence of $Hg(l)$:

$$Hg_2^{2+} \rightleftharpoons Hg^{2+} + Hg(l)$$

$$K = [Hg^{2+}]/[Hg_2^{2+}] = 0.011$$

In the presence of excess $Hg(l)$, Hg^{2+} is reduced to Hg_2^{2+}. In the absence of $Hg(l)$, Hg_2^{2+} **disproportionates** to give Hg^{2+} and $Hg(l)$. A disproportionation is a reaction in which an element in a particular oxidation state reacts with itself to produce products in higher and lower oxidation states. In the reaction above $Hg(I)$ gives $Hg(II)$ and $Hg(0)$.

The mercurous ion is stable in the presence of only a limited number of anions. Anions such as OH^-, S^{2-}, and CN^- stabilize $Hg(II)$, causing $Hg(I)$ to disproportionate:

$$Hg_2^{2+} + 2CN^- \rightarrow Hg(CN)_2(aq) + Hg(l)$$

In Reaction 5-17 we see that two Cl^- ions are produced for each Hg_2^{2+} ion. If the concentration of dissolved Hg_2^{2+} is x M, the concentration of dissolved Cl^- must be $2x$ M.

$$Hg_2Cl_2(s) \rightleftharpoons Hg_2^{2+} + 2Cl^-$$

Initial concentration:	solid	0	0
Final concentration:	solid	x	$2x$

Putting these values of concentration into the solubility product gives

To find the cube root of a number with a calculator, raise the number to the 0.333 333 33 ... power.

$$[Hg_2^{2+}][Cl^-]^2 = (x)(2x)^2 = 1.2 \times 10^{-18}$$

$$4x^3 = 1.2 \times 10^{-18} \tag{5-19}$$

$$x = 6.7 \times 10^{-7} \text{ M}$$

The concentration of Hg_2^{2+} ion is calculated to be 6.7×10^{-7} M, and the concentration of Cl^- is $(2)(6.7 \times 10^{-7}) = 13.4 \times 10^{-7}$ M.

The physical meaning of the solubility product is this: If an aqueous solution is left in contact with excess solid Hg_2Cl_2, the solid will dissolve until the condition $[Hg_2^{2+}][Cl^-]^2 = K_{sp}$ is satisfied. Thereafter, the amount of undissolved solid remains constant. Unless excess solid remains, there is no guarantee that $[Hg_2^{2+}][Cl^-]^2 = K_{sp}$. If Hg_2^{2+} and Cl^- are mixed together (with appropriate counterions) such that the product $[Hg_2^{2+}][Cl^-]^2$ exceeds K_{sp}, then Hg_2Cl_2 will precipitate.

5-3 COMMON ION EFFECT

What will be the concentration of Hg_2^{2+} in a solution containing 0.030 M NaCl saturated with Hg_2Cl_2? The concentration table now looks like this:

$$Hg_2Cl_2(s) \rightleftharpoons Hg_2^{2+} + 2Cl^-$$

Initial concentration:	solid	0	0.030
Final concentration:	solid	x	$2x + 0.030$

The initial concentration of Cl^- is due to the dissolved NaCl (which is completely dissociated into Na^+ and Cl^-). The final concentration of Cl^- is due to contributions from NaCl and Hg_2Cl_2.

The proper solubility equation is

$$[Hg_2^{2+}][Cl^-]^2 = (x)(2x + 0.030)^2 = K_{sp} \tag{5-20}$$

But think about the size of x. In the previous example, $x = 6.7 \times 10^{-7}$ M, which is pretty small compared with 0.030 M. In the present example, we anticipate that x will be even smaller than 6.7×10^{-7}, by Le Châtelier's principle. Addition of a product (Cl^- in this case) to Reaction 5-17 displaces the reaction toward the left. There will be less dissolved Hg_2^{2+} than in the absence of a second source of Cl^-. This application of Le Châtelier's principle is called the **common ion effect**. *A salt will be less soluble if one of its constituent ions is already present in the solution.*

Getting back to Equation 5-20, we expect that $2x \ll 0.030$. As a good approximation, we will ignore $2x$ in comparison to 0.030. The equation simplifies to

$$(x)(0.030)^2 = K_{sp} = 1.2 \times 10^{-18}$$
$$x = 1.3 \times 10^{-15} \tag{5-21}$$

The answer shows that it was clearly justified to omit $2x$ ($= 2.6 \times 10^{-15}$) to solve the problem. The answer also illustrates the common ion effect. In the absence of Cl^-, the solubility of Hg_2^{2+} was 6.7×10^{-7} M. In the presence of 0.030 M Cl^-, the solubility of Hg_2^{2+} is reduced to 1.3×10^{-15} M.

What is the Cl^- concentration in a solution in which the concentration Hg_2^{2+} is somehow *fixed* at 1.0×10^{-9} M? Our concentration table looks like this now:

$$Hg_2Cl_2(s) \rightleftharpoons Hg_2^{2+} + 2Cl^-$$

Initial concentration:	0	1.0×10^{-9}	0
Final concentration:	solid	1.0×10^{-9}	x

$[Hg_2^{2+}]$ is not x in this example, so there is no reason to set $[Cl^-] = 2x$.

Common ion effect: A salt is less soluble if one of its ions is already present in the solution

It is important to confirm at the end of the calculation that the approximation $2x \ll 0.030$ is valid. Look at Box 5-2 for more on approximations.

Box 5-2   THE LOGIC OF APPROXIMATIONS

It is hopelessly complicated to solve most real problems in chemistry (and other sciences) without the aid of judicious numerical approximations. For example, in solving the equation

$$(x)(2x + 0.030)^2 = 1.2 \times 10^{-18}$$

we used the approximation that $2x \ll 0.030$, and therefore solved the much simpler equation

$$(x)(0.030)^2 = 1.2 \times 10^{-18}$$

Whenever you use an approximation you are *assuming* that the approximation is true. *If the assumption is true, it will not create a contradiction. If the assumption is false, it will lead to a contradiction.*

You may object to this reasoning, feeling "How can the truth of an assumption be tested by using the assumption?" Suppose you wish to test the statement "Gail can swim 100 meters." To see if the statement is true you can *assume* that it is true. If Gail can swim 100 meters, then you could dump her in the middle of a lake with a radius of 100 meters and expect her to swim to shore. If she comes ashore alive, then your assumption was correct and no contradiction is created. If she does not make it to shore, there is a contradiction. Your assumption must have been wrong. There are only two possibilities: Either the assumption is correct and using it is correct, or the assumption is wrong and using it is wrong. You can test an assumption by using it and seeing whether you are right or wrong when you are done.

Gail's lake

Here are some numerical examples illustrating the logic of assumptions used to solve numerical equations:

Example 1. $(x)(3x + 0.01)^3 = 10^{-12}$
$(x)(0.01)^3 = 10^{-12}$ (assuming $3x \ll 0.01$)
$x = 10^{-12}/(0.01)^3 = 10^{-6}$
No contradiction: $3x = 3 \times 10^{-6} \ll 0.01$.
The assumption is true.

The problem is solved by plugging each concentration into the solubility product:

$$[Hg_2^{2+}][Cl^-]^2 = K_{sp}$$

$$(1.0 \times 10^{-9})(x)^2 = 1.2 \times 10^{-18} \tag{5-22}$$

$$x = [Cl^-] = 3.5 \times 10^{-5} \text{ M}$$

5-4   SEPARATION BY PRECIPITATION

It is sometimes useful to separate one substance from another by precipitating one from solution.[†] As an example, consider a solution containing plumbous

---

[†] A nice classroom demonstration of selective precipitation by addition of Pb^{2+} to a solution containing carbonate and iodide is described by T. P. Chirpich, *J. Chem. Ed.,* **65,** 359 (1988).

Example 2. $(x)(3x + 0.01)^3 = 10^{-8}$
$\qquad\quad (x)(0.01)^3 = 10^{-8}$ (assuming $3x \ll 0.01$)
$\qquad\quad x = 10^{-8}/(0.01)^3 = 0.01.$
$\qquad\quad$ A contradiction: $3x = 0.03 > 0.01.$
$\qquad\quad$ The assumption is false.

In Example 2 the assumption leads to a contradiction, so the assumption cannot be correct. When this happens you must solve the quartic equation $x(3x + 0.01)^3 = 10^{-8}$.

You can try to solve the quartic equation exactly, but approximate methods are usually easier. For the equation $x(3x + 0.01)^3 = 10^{-8}$, one quick procedure is trial-and-error guessing. The first guess, $x = 0.01$, comes from the (incorrect) assumption that $3x \ll 0.01$ made in Example 2.

Guess	$x(3x + 0.01)^3$	
$x = 0.01$	6.4×10^{-7}	
$x = 0.005$	7.8×10^{-8}	
$x = 0.002$	8.2×10^{-9}	
$x = 0.002\ 4$	1.22×10^{-8}	
$x = 0.002\ 2$	1.006×10^{-8}	
$x = 0.002\ 18$	9.86×10^{-9}	
$x = 0.002\ 19$	9.96×10^{-9}	best value of x with 3 figures

Example 3. $(x)(3x + 0.01)^3 = 10^{-9}$
$\qquad\quad (x)(0.01)^3 = 10^{-9}$ (assuming $3x \ll 0.01$)
$\qquad\quad x = 10^{-9}/(0.01)^3 = 10^{-3}$

In this case $3x = 0.003$. This is less than 0.01, but not a great deal less. *Whether or not the assumption is adequate depends on your purposes.* If you need an answer accurate to within a factor of 2, the approximation is acceptable. If you need an answer accurate to 1%, the approximation is not adequate. The correct answer is 6.06×10^{-4}.

(Pb^{2+}) and mercurous (Hg_2^{2+}) ions, each at a concentration of 0.010 M. Each forms an insoluble iodide, but the mercurous iodide is considerably less soluble, as indicated by the smaller value of K_{sp}:

$$PbI_2(s) \rightleftharpoons Pb^{2+} + 2I^- \qquad K_{sp} = 7.9 \times 10^{-9} \qquad (5\text{-}23)$$

$$Hg_2I_2(s) \rightleftharpoons Hg_2^{2+} + 2I^- \qquad K_{sp} = 1.1 \times 10^{-28} \qquad (5\text{-}24)$$

Is it possible to *completely* separate the Pb^{2+} and Hg_2^{2+} by selectively precipitating the latter with iodide?

To answer this, we must first define complete separation. "Complete" can mean anything we choose. Let us ask whether 99.99% of the Hg_2^{2+} can be precipitated without causing Pb^{2+} to precipitate. That is, we seek to reduce the Hg_2^{2+} concentration to 0.01% of its original value without precipitating Pb^{2+}. The concentration of Hg_2^{2+} will be 0.01% of 0.010 M $= 1.0 \times 10^{-6}$ M.

Complete separation means whatever you want it to. *You* must define what you mean by complete.

The minimum concentration of I^- needed to effect this precipitation is calculated as follows:

	$Hg_2I_2(s) \rightleftharpoons$	Hg_2^{2+}	$+ \; 2I^-$
Initial concentration:	0	0.010	0
Final concentration:	solid	$\underbrace{1.0 \times 10^{-6}}$	x
		0.01% of 0.010 M	

The quantity x is the concentration of I^- needed to reduce $[Hg_2^{2+}]$ to 10^{-6} M.

$$[Hg_2^{2+}][I^-]^2 = K_{sp}$$

$$(1.0 \times 10^{-6})(x)^2 = 1.1 \times 10^{-28} \tag{5-25}$$

$$x = [I^-] = 1.0 \times 10^{-11} \; M$$

Will this concentration of I^- cause Pb^{2+} to precipitate? We answer this by seeing whether the solubility product of PbI_2 is exceeded.

$$Q = [Pb^{2+}][I^-]^2 = (0.010)(1.0 \times 10^{-11})^2$$

$$= 1.0 \times 10^{-24} < K_{sp} \quad \text{(for } PbI_2) \tag{5-26}$$

The reaction quotient, Q, is 1.0×10^{-24}, which is less than K_{sp} ($= 7.9 \times 10^{-9}$). Therefore, the Pb^{2+} will not precipitate and the "complete" separation of Pb^{2+} and Hg_2^{2+} is feasible. We predict that adding I^- to a solution of Pb^{2+} and Hg_2^{2+} will precipitate virtually all of the mercurous ion before any plumbous ion precipitates.

Life should be so easy. What we have just done is to make a (valid) thermodynamic prediction. If the system comes to equilibrium, we can achieve the desired separation. However, occasionally one substance **coprecipitates** with the other. For example, some Pb^{2+} might become attached to the surface of the Hg_2I_2 crystal, or might even occupy sites within the crystal. The foregoing calculation says that the separation is worth trying. *But only an experiment can show whether the separation will actually work.*

Question: If you want to know whether a small amount of Pb^{2+} coprecipitates with Hg_2I_2, should you measure the Pb^{2+} concentration in the mother liquor (the solution) or the Pb^{2+} concentration in the precipitate? Which measurement is more sensitive? By "sensitive," we mean responsive to a small amount of coprecipitation.

5-5 COMPLEX FORMATION

If the anion, X, precipitates the metal, M, it is often observed that at a high concentration of X, MX redissolves. One example is the formation of Pb^{2+} in the presence of excess I^-:

$$PbI_2(s) \xrightarrow{K_{sp}} Pb^{2+} + 2I^- \qquad K_{sp} = [Pb^{2+}][I^-]^2 = 7.9 \times 10^{-9} \tag{5-27}$$

Besides forming $PbI_2(s)$, Pb^{2+} and I^- can react to form the **complex ions** PbI_n^{2-n}:

The notation for these equilibrium constants is discussed in Box 5-3.

$$Pb^{2+} + I^- \xrightarrow{K_1} PbI^+ \qquad K_1 = [PbI^+]/[Pb^{2+}][I^-] = 1.0 \times 10^2 \tag{5-28}$$

$$Pb^{2+} + 2I^- \xrightarrow{\beta_2} PbI_2(aq) \qquad \beta_2 = [PbI_2(aq)]/[Pb^{2+}][I^-]^2 = 1.4 \times 10^3 \tag{5-29}$$

$$Pb^{2+} + 3I^- \xrightarrow{\beta_3} PbI_3^- \qquad \beta_3 = [PbI_3^-]/[Pb^{2+}][I^-]^3 = 8.3 \times 10^3 \tag{5-30}$$

Box 5-3 NOTATION FOR FORMATION CONSTANTS

Formation constants are the equilibrium constants for complex ion formation. The **stepwise formation constants,** designated K_i, are defined as follows:

$$M + X \xrightleftharpoons{K_1} MX \qquad K_1 = [MX]/[M][X]$$

$$MX + X \xrightleftharpoons{K_2} MX_2 \qquad K_2 = [MX_2]/[MX][X]$$

$$MX_{n-1} + X \xrightleftharpoons{K_n} MX_n \qquad K_n = [MX_n]/[MX_{n-1}][X]$$

The **overall,** or **cumulative, formation constants** are denoted β_i:

$$M + 2X \xrightleftharpoons{\beta_2} MX_2 \qquad \beta_2 = [MX_2]/[M][X]^2$$

$$M + nX \xrightleftharpoons{\beta_n} MX_n \qquad \beta_n = [MX_n]/[M][X]^n$$

A useful relationship is that $\beta_n = K_1 K_2 \cdots K_n$.

$$Pb^{2+} + 4I^- \xrightleftharpoons{\beta_4} PbI_4^{2-} \qquad \beta_4 = [PbI_4^{2-}]/[Pb^{2+}][I^-]^4 = 3.0 \times 10^4 \qquad (5\text{-}31)$$

The species $PbI_2(aq)$ in Reaction 5-29 is *dissolved* PbI_2, containing two iodine atoms bound to a lead atom. Reaction 5-29 is *not* the reverse of Reaction 5-27, since the latter deals with solid PbI_2.

In the four species PbI^+, PbI_2, PbI_3^-, and PbI_4^{2-}, iodide is said to be the **ligand** of Pb. A ligand is any atom or group of atoms attached to the species of interest.

At low I^- concentrations, the solubility of lead is governed by the solubility of $PbI_2(s)$. However, at high I^- concentrations, Reactions 5-28 through 5-31 are driven to the right (Le Châtelier's principle), and the total concentration of dissolved lead is considerably greater than that of Pb^{2+} alone (Figure 5-1).

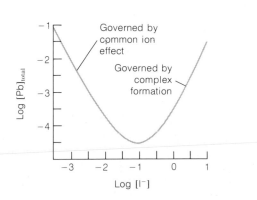

Figure 5-1

Total concentration of dissolved lead as a function of the concentration of free iodide. To the left of the minimum, $[Pb]_{total}$ is governed by the solubility product for $PbI_2(s)$. As $[I^-]$ is increased, $[Pb]_{total}$ decreases because of the common ion effect. At high values of $[I^-]$, $PbI_2(s)$ begins to redissolve because of its reaction with I^- to form soluble complex ions.

A most useful characteristic of chemical equilibrium is that *all equilibrium conditions are satisfied simultaneously.* If we somehow know the concentration of I^-, we can calculate the concentration of Pb^{2+} by substituting this value into the equilibrium constant expression given with Reaction 5-27, regardless of whether there are zero, four, or 1 006 other reactions involving Pb^{2+}. *The concentration of Pb^{2+} that satisfies any one of the equilibria must satisfy all of the equilibria. There can be only one concentration of Pb^{2+} in the solution.*

It is instructive to consider the composition of the solution when the final concentration of I^- is at 0.001 0 M and again at 1.0 M. From the K_{sp} expression given with Reaction 5-27, we calculate

$$[Pb^{2+}] = K_{sp}/[I^-]^2 = 7.9 \times 10^{-3} \text{ M} \tag{5-32}$$

if $[I^-] = 10^{-3}$ M. From Reactions 5-28 through 5-31 we then calculate the concentrations of the other lead-containing species:

$$[PbI^+] = K_1[Pb^{2+}][I^-] = (1.0 \times 10^2)(7.9 \times 10^{-3})(1.0 \times 10^{-3})$$

$$= 7.9 \times 10^{-4} \text{ M} \tag{5-33}$$

$$[PbI_2(aq)] = \beta_2[Pb^{2+}][I^-]^2 = 1.1 \times 10^{-5} \text{ M} \tag{5-34}$$

$$[PbI_3^-] = \beta_3[Pb^{2+}][I^-]^3 = 6.6 \times 10^{-8} \text{ M} \tag{5-35}$$

$$[PbI_4^{2-}] = \beta_4[Pb^{2+}][I^-]^4 = 2.4 \times 10^{-10} \text{ M} \tag{5-36}$$

If, instead, we take $[I^-] = 1.0$ M, then the concentrations are calculated to be

$$[Pb^{2+}] = 7.9 \times 10^{-9} \text{ M} \qquad [PbI_3^-] = 6.6 \times 10^{-5} \text{ M}$$

$$[PbI^+] = 7.9 \times 10^{-7} \text{ M} \qquad [PbI_4^{2-}] = 2.4 \times 10^{-4} \text{ M}$$

$$[PbI_2(aq)] = 1.1 \times 10^{-5} \text{ M}$$

The total concentration of dissolved lead is

$$[Pb]_{total} = [Pb^{2+}] + [PbI^+] + [PbI_2(aq)] + [PbI_3^-] + [PbI_4^{2-}] \tag{5-37}$$

When $[I^-] = 10^{-3}$ M, $[Pb]_{total} = 8.7 \times 10^{-3}$ M, of which 91% is Pb^{2+}. As $[I^-]$ increases, $[Pb]_{total}$ decreases by the common ion effect operating in Reaction 5-27. However, at sufficiently high $[I^-]$, complex formation takes over and $[Pb]_{total}$ increases (Figure 5-1). When $[I^-] = 1.0$ M, $[Pb]_{total} = 3.2 \times 10^{-4}$ M, of which 76% is PbI_4^{2-}.

Comment on Equilibrium Constants

If you look up the equilibrium constant of a chemical reaction in two different books, there is an excellent chance that the values will be different (sometimes by a factor of 10 or more). This happens because the constant may have been determined under different conditions by different investigators, perhaps using different techniques.

One of the most common sources of variation in the reported value of K is the ionic composition of the solution. It is important to note whether K is reported for a particular, stated ionic composition or whether the value has been extrapolated to zero ionic strength. If you need to use an equilibrium constant for your own work, you should choose a value of K measured under conditions as close as possible to those you will employ. If the value of K is critical to your experiment, it is best to measure K yourself under the precise conditions of your experiment.[†]

5-6 WHAT ARE ACIDS AND BASES?

In aqueous chemistry, an **acid** is most usefully defined as a substance that increases the concentration of H_3O^+ when added to water. Conversely, a **base** decreases the concentration of H_3O^+ in aqueous solution. The formula H_3O^+, which represents the **hydronium ion,** is a more accurate description of what we have been calling H^+ so far in this chapter. The hydronium ion is a combination of H^+ (a proton; i.e., a hydrogen atom that has lost an electron) with H_2O. As we shall see shortly, a decrease in H_3O^+ concentration necessarily requires an increase in OH^- concentration. Therefore a base is a substance that increases the concentration of OH^- in aqueous solution.

Two Classifications of Acid–Base Behavior

In 1923 Brønsted and Lowry classified acids as proton donors and bases as proton acceptors. These definitions include the one given above. For example, HCl is an acid (a proton donor) and it increases the concentration of H_3O^+ in water:

$$HCl + H_2O \rightleftharpoons H_3O^+ + Cl^- \qquad (5\text{-}38)$$

The Brønsted–Lowry definition does not require that H_3O^+ be formed. This definition can therefore be extended to nonaqueous solvents and even to the gas phase:

$$\underset{\substack{\text{Hydrochloric} \\ \text{acid} \\ \text{(acid)}}}{HCl(g)} + \underset{\substack{\text{Ammonia} \\ \text{(base)}}}{NH_3(g)} \rightleftharpoons \underset{\substack{\text{Ammonium} \\ \text{chloride} \\ \text{(salt)}}}{NH_4^+Cl^-(s)} \qquad (5\text{-}39)$$

Brønsted–Lowry definitions:

    acid: proton donor

    base: proton acceptor

Any ionic solid, such as ammonium chloride, is called a **salt.** In a formal sense, a salt can be thought of as the product of an acid–base reaction. When

[†] Some of the most useful compilations of equilibrium constants are found in L. G. Sillén and A. E. Martell, *Stability Constants of Metal-Ion Complexes* (Chemical Society, London, Special Publications No. 17 and 25, 1964 and 1971), and A. E. Martell and R. M. Smith, *Critical Stability Constants* (New York: Plenum Press, 1974), a multivolume collection. Books dealing with the experimental measurement of equilibrium constants include A. Martell and R. Motekaltis, *Stability Constants: Determination and Use* (New York: VCH Publishers, 1989); K. A. Connors, *Binding Constants: The Measurement of Molecular Complex Stability* (New York: John Wiley and Sons, 1987); M. Meloun, *Computation of Solution Equilibria* (New York: John Wiley and Sons, 1988); D. J. Leggett, ed., *Computational Methods for the Determination of Formation Constants* (New York: Plenum Press, 1985).

an acid and a base react, they are said to **neutralize** each other. Most salts are **strong electrolytes.** This means that they dissociate completely into their component ions when dissolved in water. Thus, ammonium chloride gives NH_4^+ and Cl^- in aqueous solution:

$$NH_4^+Cl^-(s) \rightarrow NH_4^+(aq) + Cl^-(aq) \qquad (5\text{-}40)$$

Another definition of acid–base behavior was put forth by G. N. Lewis in the 1920s. He defined an acid as an electron-pair acceptor and a base as an electron-pair donor. The Lewis definitions are more general and include those given previously here, extending the concept of acid and base to species that need not have a reactive H^+. The simplest application of this classification is to the reaction between H_3O^+ and OH^- in aqueous solution:

Lewis definitions:

acid: electron pair acceptor

base: electron pair donor

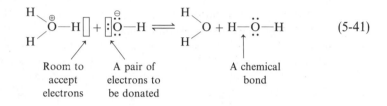

$$(5\text{-}41)$$

The pair of electrons donated by OH^- is accepted by H^+ (which dissociates from H_3O^+), thus forming a bond.

The Lewis definitions can apply to reactions that have little resemblance to aqueous acid–base reactions. For example, the reaction of antimony pentachloride with chloride in certain nonaqueous solvents can be thought of as an acid–base reaction:

These drawings represent geometry, not bonding, in the complexes. $SbCl_5$ is a trigonal bipyramid in which all five Cl atoms are bound to Sb. There are no Cl—Cl bonds. $SbCl_6^-$ is octahedral, with six Sb—Cl bonds and no Cl—Cl bonds. Reaction 5-42 is an example of *complex ion formation*, discussed in Section 5-5.

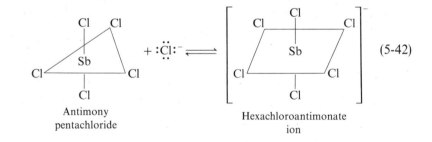

$$(5\text{-}42)$$

Antimony pentachloride

Hexachloroantimonate ion

The antimony pentachloride is acting as a **Lewis acid,** accepting Cl^-. The chloride ion donates a pair of electrons and is therefore a **Lewis base.**

For the remainder of this text, when we speak of acids and bases, we are speaking of **Brønsted–Lowry acids and bases.** *The acids we speak of are proton donors, and the bases are proton acceptors.* Reactions such as 5-42 are important, but we usually think of them as complex ion equilibria rather than as acid–base reactions.

Conjugate Acids and Bases

The products of any reaction between an acid and base may also be classified as acids and bases:

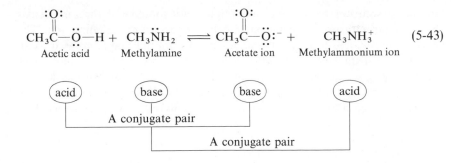

$$CH_3C\overset{:O:}{\underset{}{\|}}—\overset{..}{\underset{..}{O}}—H + CH_3\overset{..}{N}H_2 \rightleftharpoons CH_3C\overset{:O:}{\underset{}{\|}}—\overset{..}{\underset{..}{O}}:^- + CH_3NH_3^+ \qquad (5\text{-}43)$$

Acetic acid Methylamine Acetate ion Methylammonium ion

acid — base — base — acid

A conjugate pair

A conjugate pair

Conjugate acids and bases are related by the gain or loss of one proton.

Acetate is a base because it can accept a proton to make acetic acid. The methylammonium ion is an acid because it can donate a proton and become methylamine. Acetic acid and the acetate ion are said to be a **conjugate acid–base pair.** Methylamine and the methylammonium ion are likewise conjugate. *Conjugate acids and bases are related to each other by the gain or loss of one* H^+.

Nature of H^+ and OH^-

For simplicity, we will generally write H^+, not H_3O^+, as the formula for an ionized hydrogen atom (a proton) in aqueous solution. It is certain that the proton does not exist by itself in water. The simplest formula, which is found in some crystalline salts, is H_3O^+. For example, crystals of the compound perchloric acid monohydrate consist of tetrahedral perchlorate ions and pyramidal hydronium (also called hydroxonium) ions.

We will write H^+ when we really mean H_3O^+.

$$HClO_4 \cdot H_2O \quad \text{is really} \quad \left[\begin{array}{c} H\overset{..}{\underset{}{O}} \\ H \quad H \end{array} \right]^+ \left[\begin{array}{c} O \\ O—Cl \\ O \quad O \end{array} \right]^-$$

Hydronium Perchlorate

Recall that a solid wedge is a bond coming out of the plane of the page and a dashed wedge is a bond to an atom behind the page.

The formula $HClO_4 \cdot H_2O$ is a way of specifying the composition of the substance when we are ignorant of its structure. A more accurate formula would be $H_3O^+ClO_4^-$.

The average dimensions of the H_3O^+ cation that occurs in many crystals are shown in Figure 5-2. The bond enthalpy of the OH bond of H_3O^+ is 544 kJ/mol, some 84 kJ/mol greater than the OH bond enthalpy in H_2O.

Bond enthalpy refers to the heat needed to break a chemical bond in the gas phase.

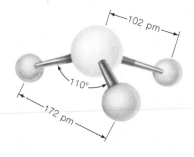

Figure 5-2
Typical structure of the H_3O^+ ion, as found in several crystals.

102 pm
110°
172 pm

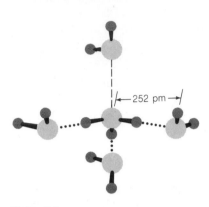

Figure 5-3
Typical coordination environment of H_3O^+ in dilute aqueous solution. Three molecules of water are bound by very strong hydrogen bonds, and one molecule (at the top) is held by much weaker ion-dipole interactions.

In aqueous solution the H_3O^+ cation is tightly associated with three molecules of water through exceptionally strong hydrogen bonds (Figure 5-3). The $O—H \cdots O$ hydrogen-bonded distance of 252 pm may be compared to an $O—H \cdots O$ distance of 283 pm between hydrogen-bonded water molecules. In liquid SO_2, where the H_3O^+ cation is not surrounded by H_2O molecules, H_3O^+ appears to be planar, or nearly so.[†] The $H_5O_2^+$ cation is another simple species in which a hydrogen ion is shared equally by two water molecules.[‡]

$$\begin{array}{c} H \qquad\qquad\qquad H \\ \diagdown O \cdots H \cdots O \diagup \\ H \diagup \quad |\!\!\leftarrow 243\ pm \rightarrow\!\!| \quad \diagdown H \end{array}$$

Less is known of the structure of OH^- in solution, but the ion $H_3O_2^-$ ("$OH^- \cdot H_2O$") has been observed by x-ray crystallography.[*] The central $O—H—O$ linkage contains the shortest hydrogen bond involving H_2O that has ever been observed.

$$\begin{array}{c} \qquad\qquad\qquad H \\ O—H—O \diagup \\ H \diagup \ |\!\!\leftarrow 229\ pm \rightarrow\!\!| \end{array}$$

$H_3O_2^-$ has also been observed as a ligand, forming metal $—O \cdots H \cdots O—$ metal bridges in several compounds. The $O \cdots H \cdots O$ distances range from 244 to 252 pm.[§]

We will ordinarily write H^+ in most chemical equations, though we really mean H_3O^+. To emphasize certain chemistry of water, we may write H_3O^+. For example, water can be either an acid or a base. Water is an acid with respect to methoxide:

$$H—\overset{..}{\underset{..}{O}}—H + CH_3—\overset{..}{\underset{..}{O}}:^- \rightleftharpoons H—\overset{..}{\underset{..}{O}}:^- + CH_3—\overset{..}{\underset{..}{O}}—H \qquad (5\text{-}44)$$

$\quad$ Water $\qquad$ Methoxide $\qquad$ Hydroxide $\qquad$ Methanol

But with respect to hydrogen bromide, water is a base:

$$\underset{\text{Water}}{H_2O} + \underset{\substack{\text{Hydrogen}\\\text{bromide}}}{HBr} \rightleftharpoons \underset{\substack{\text{Hydronium}\\\text{ion}}}{H_3O^+} + \underset{\text{Bromide}}{Br^-} \qquad (5\text{-}45)$$

Water also undergoes self-ionization, also called **autoprotolysis,** in which it acts as both an acid and a base:

$$H_2O + H_2O \rightleftharpoons H_3O^+ + OH^- \qquad (5\text{-}46)$$

or

$$H_2O \rightleftharpoons H^+ + OH^- \qquad (5\text{-}47)$$

Reaction 5-47 should be taken to mean the same as Reaction 5-46.

[†] G. D. Mateescu and G. M. Benedikt, *J. Amer. Chem. Soc.,* **101,** 3959 (1979).
[‡] F. A. Cotton, C. K. Fair, G. E. Lewis, G. N. Mott, F. K. Ross, A. J. Schultz, and J. M. Williams, *J. Amer. Chem. Soc.,* **106,** 5319 (1984).
[*] K. Abu-Dari, K. N. Raymond, and D. P. Freyberg, *J. Amer. Chem. Soc.,* **101,** 3688 (1979).
[§] A. Bino and D. Gibson, *Inorg. Chem.,* **23,** 109 (1984).

Other **protic solvents** (solvents with a reactive H^+) also undergo autoprotolysis. An example is acetic acid:

$$2CH_3COH \rightleftharpoons CH_3C\!\!\begin{array}{c}+ \\ \diagup OH \\ \diagdown OH \end{array} + CH_3C\!\!-\!\!O^- \text{ (in acetic acid)} \qquad (5\text{-}48)$$

The extent of these reactions of water or of acetic acid is very small. The *autoprotolysis constants* (equilibrium constants) for Reactions 5-47 and 5-48 are 1.0×10^{-14} and 3.5×10^{-15}, respectively, at 25°C.

5-7 pH

The autoprotolysis constant for H_2O is given the special symbol K_w, where the subscript w denotes water.

$$H_2O \overset{K_w}{\rightleftharpoons} H^+ + OH^- \qquad (5\text{-}49)$$

The equilibrium constant K_w is

$$K_w = [H^+][OH^-] \qquad (5\text{-}50)$$

As is true of all equilibrium constants, K_w varies with temperature (Table 5-1). The value of K_w at 25.00°C (298.15 K) is 1.01×10^{-14}.

Examples of protic solvents:

$$H_2O \quad \text{and} \quad CH_3CH_2OH$$

Examples of **aprotic** solvents (solvents without acidic protons):

$$CH_3CH_2OCH_2CH_3 \quad \text{and} \quad CH_3CN$$

Recall that H_2O (the solvent) is omitted from the equilibrium constant. The value $K_w = 1.0 \times 10^{-14}$ at 25°C will be accurate enough for our purposes in this text.

Table 5-1
Temperature dependence of K_w

Temperature (°C)	K_w	$pK_w = -\log K_w$
0	1.14×10^{-15}	14.944
5	1.85×10^{-15}	14.734
10	2.92×10^{-15}	14.535
15	4.51×10^{-15}	14.346
20	6.81×10^{-15}	14.167
24	1.00×10^{-14}	14.000
25	1.01×10^{-14}	13.996
30	1.47×10^{-14}	13.833
35	2.09×10^{-14}	13.680
40	2.92×10^{-14}	13.535
45	4.02×10^{-14}	13.396
50	5.47×10^{-14}	13.262
100	5.45×10^{-13}	12.264
150	2.31×10^{-12}	11.637
200	5.15×10^{-12}	11.288
250	6.43×10^{-12}	11.192
300	3.93×10^{-12}	11.406
350	5.07×10^{-13}	12.295

EXAMPLE: Concentration of H⁺ and OH⁻ in Pure Water at 25°C

Calculate the concentration of H^+ and OH^- in pure water at 25°C.

The stoichiometry of Reaction 5-49 tells us that H^+ and OH^- are produced in a 1:1 mole ratio. Their concentrations must be equal. Calling each concentration x, we can write

$$K_w = 1.0 \times 10^{-14} = [H^+][OH^-] = [x][x]$$

$$\Rightarrow \quad x = 1.0 \times 10^{-7} \text{ M}$$

The concentrations of H^+ and OH^- are both 1.0×10^{-7} M.

EXAMPLE: Concentration of OH⁻ When [H⁺] Is Known

What is the concentration of OH^- if $[H^+] = 1.0 \times 10^{-3}$ M?

Putting $[H^+] = 1.0 \times 10^{-3}$ M into Equation 5-50 gives

$$K_w = 1.0 \times 10^{-14} = (1.0 \times 10^{-3})[OH^-]$$

$$\Rightarrow \quad [OH^-] = 1.0 \times 10^{-11} \text{ M}$$

A concentration $[H^+] = 1.0 \times 10^{-3}$ M gives $[OH^-] = 1.0 \times 10^{-11}$ M. *As the concentration of H^+ increases, the concentration of OH^- necessarily decreases, and vice versa.* A concentration $[OH^-] = 1.0 \times 10^{-3}$ M gives $[H^+] = 1.0 \times 10^{-11}$ M.

An approximate definition of **pH** is the negative logarithm of the H^+ concentration:

$pH = -\log[H^+]$

$$pH \approx -\log[H^+] \qquad (5\text{-}51)$$

In the next chapter we will define pH more precisely in terms of *activities,* but for most purposes Equation 5-51 is a good working definition of pH. In pure water at 25°C with $[H^+] = 1.0 \times 10^{-7}$ M, the pH is $-\log(1.0 \times 10^{-7}) = 7.00$. If the concentration of OH^- is 1.0×10^{-3} M, then $[H^+] = 1.0 \times 10^{-11}$ M and the pH is 11.00.

A useful relation between the concentrations of H^+ and OH^- is

Take the log of both sides of Equation 5-50 to derive Equation 5-52:

$K_w = [H^+][OH^-]$

$\log K_w = \log[H^+] + \log[OH^-]$

$-\log K_w = pH + pOH$

$\qquad = 14.00 \quad$ at 25°C

$$pH + pOH = -\log(K_w) = 14.00 \quad \text{at 25°C} \qquad (5\text{-}52)$$

where $pOH = -\log[OH^-]$, just as $pH = -\log[H^+]$. This is a fancy way of saying that if pH = 3.58, then pOH = 14.00 − 3.58 = 10.42, or $[OH^-] = 10^{-10.42} = 3.8 \times 10^{-11}$ M.

A solution is **acidic** if $[H^+] > [OH^-]$. A solution is **basic** if $[H^+] < [OH^-]$. At 25°C, an acidic solution has a pH below 7, and a basic solution has a pH above 7.

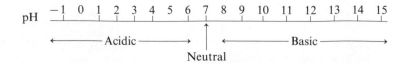

Although the pH of most solutions is in the range 0–14, these are not the limits of pH. A pH of -1.00, for example, means $-\log[H^+] = -1.00$, or $[H^+] = 10$. This concentration is easily attained in a concentrated solution of a strong acid such as HCl.

Is there such a thing as pure water? In most labs, the answer is "No." Distilled water from the tap in most labs is acidic because it contains CO_2 from the atmosphere. The CO_2 is an acid by virtue of the reaction

$$CO_2 + H_2O \rightleftharpoons \underset{\text{Bicarbonate}}{HCO_3^-} + H^+ \qquad (5\text{-}53)$$

Carbon dioxide can be largely removed by first boiling the water and then protecting it from the atmosphere. Alternatively, the water may be distilled under an inert atmosphere, such as pure N_2.

About a hundred years ago, careful measurements of the conductivity of water were made by Friedrich Kohlrausch and his students. In removing ionic impurities from their water, they found it necessary to distill the water *42 consecutive times* under vacuum to reduce conductivity to a limiting value.

5-8 STRENGTHS OF ACIDS AND BASES

Acids and bases are commonly classified as strong or weak, depending on whether they react "completely" or only "partly" to produce H^+ or OH^-. Since there is a continuous range of possibilities for "partial" reaction, there is no sharp distinction between weak and strong. However, some compounds react so completely that they are easily considered strong acids or bases, and, by convention, anything else is termed weak.

Strong Acids and Bases

The most common strong acids and bases are listed in Table 5-2. By definition, a strong acid or base is completely dissociated in aqueous solution. That is, the equilibrium constants for the following reactions are very large.

$$HCl(aq) \rightleftharpoons H^+ + Cl^- \qquad (5\text{-}54)$$

$$KOH(aq) \rightleftharpoons K^+ + OH^- \qquad (5\text{-}55)$$

Virtually no undissociated HCl or KOH exists in aqueous solution. Demonstration 5-1 shows one consequence of the strong acid behavior of HCl.

Notice that, although the hydrogen halides HCl, HBr, and HI are strong acids, HF is *not* a strong acid. Box 5-4 gives an explanation of this unexpected observation. For most practical purposes, the hydroxides of the alkaline earth metals (Mg^{2+}, Ca^{2+}, Sr^{2+}, and Ba^{2+}) may be considered strong bases, although they are far less soluble than alkali metal hydroxides and also have some tendency to form MOH^+ complexes (Table 5-3).

pH is usually measured with a glass electrode, whose operation is described in Chapter 15. Some methods of extending pH measurements to very low or very high values (where the glass electrode is useless) are discussed in Box 11-2.

Table 5-2
Common strong acids and bases

Formula	Name
	Acids
HCl	Hydrochloric acid (hydrogen chloride)
HBr	Hydrogen bromide
HI	Hydrogen iodide
H_2SO_4[†]	Sulfuric acid
HNO_3	Nitric acid
$HClO_4$	Perchloric acid
	Bases
LiOH	Lithium hydroxide
NaOH	Sodium hydroxide
KOH	Potassium hydroxide
RbOH	Rubidium Hydroxide
CsOH	Cesium hydroxide
R_4NOH[‡]	Quaternary ammonium hydroxide

[†] For H_2SO_4, only the first proton ionization is complete. Dissociation of the second proton has an equilibrium constant of 1.0×10^{-2}.
[‡] This is a general formula for any hydroxide salt of an ammonium cation containing four organic groups. An example is tetrabutylammonium hydroxide: $(CH_3CH_2CH_2CH_2)_4N^+OH^-$.

Table 5-3
Equilibria of alkaline earth metal hydroxides

$$M(OH)_2(s) \rightleftharpoons M^{2+} + 2OH^-$$
$$K_{sp} = [M^{2+}][OH^-]^2$$
$$M^{2+} + OH^- \rightleftharpoons MOH^+$$
$$K_1 = [MOH^+]/[M^{2+}][OH^-]$$

Metal	$\log K_{sp}$	$\log K_1$
Mg^{2+}	-11.15	2.58
Ca^{2+}	-5.19	1.30
Sr^{2+}	—	0.82
Ba^{2+}	—	0.64

Note: 25°C and ionic strength = 0.

Demonstration 5-1   THE HCl FOUNTAIN

The complete dissociation of HCl into H^+ and Cl^- makes HCl(g) extremely soluble in water.

$$HCl(g) \rightleftharpoons HCl(aq) \tag{a}$$

$$HCl(aq) \rightleftharpoons H^+(aq) + Cl^-(aq) \tag{b}$$

Net reaction: $$HCl(g) \rightleftharpoons H^+(aq) + Cl^-(aq) \tag{c}$$

Since the equilibrium of Reaction b lies far to the right, it pulls Reaction a to the right as well.

Challenge: The standard free energy change ($\Delta G°$) for Reaction c is -36.0 kJ/mol. Show that the equilibrium constant is 2.0×10^6.

   The extreme solubility of HCl(g) in water is the basis for the HCl fountain, assembled as shown below. In Figure a an inverted 250-mL round-bottom flask containing air is set up with its inlet tube leading to a source of HCl(g) and its outlet tube directed into an inverted bottle of water. As HCl is admitted to the flask, air is displaced. When the bottle is filled with air the flask is filled mostly with HCl(g).
   The hoses are disconnected and replaced with a beaker of indicator and a rubber bulb (Figure b). For an indicator we use slightly alkaline methyl purple, which is green above pH 5.4 and purple below pH 4.8. When ~1 mL of water is squirted from the rubber bulb into the flask, a vacuum is created and indicator solution is drawn up into the flask, making a fascinating fountain (Color Plate 1).

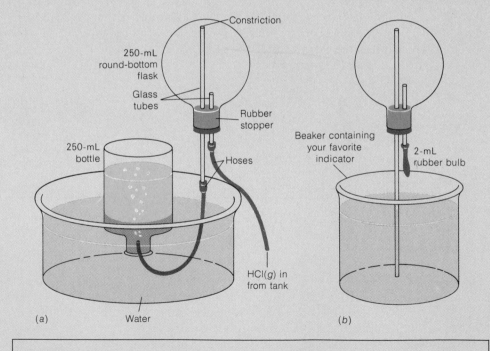

(a)   Water   (b)

Question: Why is a vacuum created when water is squirted into the flask, and why does the indicator change color when it enters the flask?

Box 5-4   THE STRANGE BEHAVIOR OF HYDROFLUORIC ACID[†]

The hydrogen halides HCl, HBr, and HI are all strong acids, which means that the reactions

$$HX(aq) + H_2O \rightarrow H_3O^+ + X^-$$

(X = Cl, Br, I) all go to completion. Why, then, does HF behave as a weak acid?

The answer is curious. First, HF *does* completely give up its proton to H_2O:

$$\underset{\substack{\text{Hydronium} \\ \text{ion}}}{HF(aq) \rightarrow \quad H_3O^+} + \underset{\substack{\text{Fluoride} \\ \text{ion}}}{F^-}$$

But fluorine forms the strongest hydrogen bonds of any element. The hydronium ion remains tightly associated with the fluoride ion through a hydrogen bond. We call such an association an **ion pair.**

$$\underset{\text{An ion pair}}{H_3O^+ + F^- \rightleftharpoons F^- \cdots H_3O^+}$$

Ion pairs are common in nonaqueous solvents, which cannot promote ion dissociation as well as water. But the hydronium–fluoride ion pair is unusual for aqueous solutions.

Thus HF does not behave as a strong acid because the F^- and H_3O^+ ions remain associated with each other. Dissolving one mole of the strong acid HCl in water creates one mole of free H_3O^+. Dissolving one mole of the "weak" acid HF in water creates very little free H_3O^+.

Hydrofluoric acid is not unique in its propensity to form ion pairs. Many moderately strong acids, such as those below, are thought to exist predominantly as ion pairs in aqueous solution $(HA + H_2O \rightleftharpoons A^- \cdots H_3O^+).[‡]$

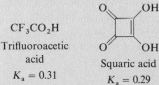

CF₃CO₂H
Trifluoroacetic
acid
$K_a = 0.31$

Squaric acid
$K_a = 0.29$

[†] P. A. Giguère, *J. Chem. Ed.,* **56,** 571 (1979); P. A. Giguère and S. Turrell, *J. Chem. Soc.,* **102,** 5473 (1980).

[‡] R. I. Gelb, L. M. Schwartz, and D. A. Laufer, *J. Chem. Soc.,* **103,** 5664 (1981).

Weak Acids and Bases

All weak acids, HA, react with water according to Equation 5-56:

$$HA + H_2O \xrightleftharpoons{K_a} H_3O + A^- \tag{5-56}$$

which means exactly the same as

$$HA \xrightleftharpoons{K_a} H^+ + A^- \tag{5-57}$$

Acid dissociation constant:

The equilibrium constant for both of these equations is called K_a, the **acid dissociation constant.**

$$K_a = \frac{[H^+][A^-]}{[HA]}$$

$$K_a = \frac{[H^+][A^-]}{[HA]} \tag{5-58}$$

By definition, a weak acid is one that is only partially dissociated in water. This means that K_a is "small" for a weak acid.

Weak bases, B, react with water according to Equation 5-59

$$B + H_2O \overset{K_b}{\rightleftharpoons} BH^+ + OH^- \tag{5-59}$$

The equilibrium constant K_b is usually called the **base hydrolysis constant** or **base "dissociation" constant.**

$$K_b = \frac{[BH^+][OH^-]}{[B]} \tag{5-60}$$

By definition, a weak base is one for which K_b is "small."

Common classes of weak acids and bases

Acetic acid and methylamine are typical of common weak acids and bases.

$$\underset{\substack{\text{Acetic} \\ \text{acid (HA)}}}{CH_3\overset{\displaystyle O}{\overset{\|}{C}}OH} \rightleftharpoons \underset{\text{Acetate (A}^-)}{CH_3\overset{\displaystyle O}{\overset{\|}{C}}O^-} + H^+ \qquad K_a = 1.75 \times 10^{-5} \tag{5-61}$$

$$\underset{\text{Methylamine (B)}}{CH_3\overset{\displaystyle ..}{N}H_2} + H_2O \rightleftharpoons \underset{\substack{\text{Methylammonium} \\ \text{ion (BH}^+)}}{CH_3\overset{\displaystyle +}{N}H_3} + OH^- \qquad K_b = 4.4 \times 10^{-4} \tag{5-62}$$

Acetic acid is representative of the carboxylic acids, which have the general structure

$$\underset{\text{A carboxylic acid}}{R-\overset{\displaystyle O}{\overset{\|}{C}}-OH}$$

where R is an organic substituent. *Most* **carboxylic acids** *are weak acids, and most* **carboxylate anions** *are weak bases.*

$$\underset{\text{A carboxylate anion}}{R-\overset{\displaystyle O}{\overset{\|}{C}}-O^-}$$

Methylamine is a representative amine, a nitrogen-containing compound.

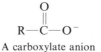

$R\overset{..}{N}H_2$	a primary amine	RNH_3^+
$R_2\overset{..}{N}H$	a secondary amine	$R_2NH_2^+$
$R_3\overset{..}{N}$	a tertiary amine	R_3NH^+

ammonium ions

Base hydrolysis constant:

$$K_b = \frac{[BH^+][OH^-]}{[B]}$$

Carboxylic acids (RCO_2H) and ammonium ions (R_3NH^+) are weak acids. Carboxylate anions (RCO_2^-) and amines (R_3N) are weak bases.

Amines *are weak bases, and* **ammonium ions** *are weak acids.* The "parent" of all amines is ammonia, NH_3. When a base such as methylamine reacts with water, the product formed is the conjugate acid. That is, the methylammonium ion produced in Reaction 5-62 is a weak acid:

$$\underset{BH^+}{CH_3\overset{+}{N}H_3} \overset{K_a}{\rightleftharpoons} \underset{B}{CH_3\overset{\cdot\cdot}{N}H_2} + H^+ \qquad K_a = 2.3 \times 10^{-11} \qquad (5\text{-}63)$$

The methylammonium ion is the conjugate acid of methylamine.

You should learn to recognize whether a compound will have acidic or basic properties. Note carefully that the salt methylammonium chloride, for example, will dissociate completely in aqueous solution to give the methylammonium cation and the chloride anion.

$$\underset{\substack{\text{Methylammonium} \\ \text{chloride}}}{CH_3\overset{+}{N}H_3Cl^-(s)} \rightarrow CH_3\overset{+}{N}H_3(aq) + Cl^-(aq) \qquad (5\text{-}64)$$

The methylammonium ion, being the conjugate acid of methylamine, is a weak acid (Reaction 5-63). The chloride ion is neither an acid nor a base. It is the conjugate base of HCl, a strong acid. This means that Cl^- *has virtually no tendency to associate with* H^+, or else HCl would not be a strong acid. We can predict that a solution of methylammonium chloride will be acidic, since the methylammonium ion is an acid and Cl^- is not a base.

Polyprotic acids and bases

Polyprotic acids and bases are compounds that can donate or accept more than one proton. For example, oxalic acid is diprotic and phosphate is tribasic:

Although we will usually write a **base** as **B** and an **acid** as **HA,** it is important to realize that **BH$^+$** is also an **acid** and **A$^-$** is also a **base.**

Methylammonium chloride is a weak acid because
1. It dissociates into $CH_3NH_3^+$ and Cl^-.
2. $CH_3NH_3^+$ is a weak acid, being conjugate to CH_3NH_2, a weak base.
3. Cl^- has no basic properties. It is conjugate to HCl, a strong acid. That is, HCl dissociates completely.

Challenge: Phenol is a weak acid. Show that a solution of the ionic compound potassium phenolate ($C_6H_5O^-K^+$) will be basic.

$$\underset{\substack{\text{Oxalic} \\ \text{acid}}}{HO\overset{\overset{O}{\|}}{C}\overset{\overset{O}{\|}}{C}OH} \rightleftharpoons H^+ + \underset{\substack{\text{Monohydrogen} \\ \text{oxalate}}}{^-O\overset{\overset{O}{\|}}{C}\overset{\overset{O}{\|}}{C}OH} \qquad K_{a1} = 5.60 \times 10^{-2} \qquad (5\text{-}65)$$

$$^-O\overset{\overset{O}{\|}}{C}\overset{\overset{O}{\|}}{C}OH \rightleftharpoons H^+ + \underset{\substack{\text{Oxalate}}}{^-O\overset{\overset{O}{\|}}{C}\overset{\overset{O}{\|}}{C}O^-} \qquad K_{a2} = 5.42 \times 10^{-5} \qquad (5\text{-}66)$$

Notation for acid and base equilibrium constants: K_{a1} *refers to the acidic species with the most protons and* K_{b1} *refers to the basic species with the least protons. The subscript* a *in acid dissociation constants will usually be omitted.*

$$\underset{\text{Phosphate}}{PO_4^{3-}} + H_2O \rightleftharpoons \underset{\substack{\text{Monohydrogen} \\ \text{phosphate}}}{HPO_4^{2-}} + OH^- \qquad K_{b1} = 1.4 \times 10^{-2} \qquad (5\text{-}67)$$

$$\underset{}{HPO_4^{2-}} + H_2O \rightleftharpoons \underset{\substack{\text{Dihydrogen} \\ \text{phosphate}}}{H_2PO_4^-} + OH^- \qquad K_{b2} = 1.59 \times 10^{-7} \qquad (5\text{-}68)$$

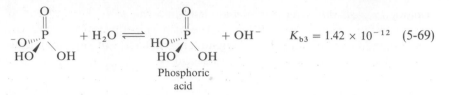

$$K_{b3} = 1.42 \times 10^{-12} \quad (5\text{-}69)$$

Phosphoric
acid

The standard notation for successive acid dissociation constants of a polyprotic acid is K_1, K_2, K_3, and so on. That is, the subscript a is usually omitted. We will retain or omit the subscript as dictated by clarity. For successive base hydrolysis constants, we will retain the subscript b. The examples above illustrate that K_{a1} *(or K_1) refers to the acidic species with the most protons, and K_{b1} refers to the basic species with the least number of protons.* Carbonic acid, the diprotic carboxylic acid derived from CO_2, is described in Box 5-5.

Box 5-5   CARBONIC ACID[†]

Carbonic acid is the acidic form of dissolved carbon dioxide.

$$CO_2(g) \rightleftharpoons CO_2(aq) \qquad\qquad K = \frac{[CO_2(aq)]}{P_{CO_2}} = 0.0344$$

$$CO_2(aq) + H_2O \rightleftharpoons \underset{\text{Carbonic acid}}{\text{HO}-\overset{\overset{\displaystyle O}{\|}}{C}-\text{OH}} \qquad K = \frac{[H_2CO_3]}{[CO_2(aq)]} \approx 1.3 \times 10^{-3}$$

$$H_2CO_3 \rightleftharpoons \underset{\text{Bicarbonate}}{HCO_3^-} \quad H^+ \qquad K_{a1}$$

$$HCO_3^- \rightleftharpoons \underset{\text{Carbonate}}{CO_3^{2-}} + H^+ \qquad K_{a2}$$

Its behavior as a diprotic acid appears anomalous at first, because the value of K_{a1} is about 10^2 to 10^4 times smaller than K_a for other carboxylic acids.

$$H_2CO_3 \qquad K_{a1} = 4.45 \times 10^{-7}$$

CH_3CO_2H	$K_a = 1.75 \times 10^{-5}$	HCO_2H	$K_a = 1.80 \times 10^{-4}$
Acetic acid		Formic acid	
$N\equiv CCH_2CO_2H$	$K_a = 3.37 \times 10^{-3}$	$HOCH_2CO_2H$	$K_a = 1.48 \times 10^{-4}$
Cyanoacetic acid		Glycolic acid	

The reason for this seeming anomaly is not that H_2CO_3 is unusual but rather that the value commonly given for K_{a1} applies to the equation

---

[†] M. Kern, *J. Chem. Ed.,* **37,** 14 (1960); H. S. Harned and R. Davis, Jr., *J. Amer. Chem. Soc.,* **65,** 2030 (1943).

A most important relationship exists between K_a and K_b of a conjugate acid–base pair in aqueous solution. We will derive this result using the acid HA and its conjugate base A^-.

$$\text{HA} \rightleftharpoons \text{H}^+ + \text{A}^- \qquad K_a = \frac{[\text{H}^+][\text{A}^-]}{[\text{HA}]} \tag{5-70}$$

$$\text{A}^- + \text{H}_2\text{O} \rightleftharpoons \text{HA} + \text{OH}^- \qquad K_b = \frac{[\text{HA}][\text{OH}^-]}{[\text{A}^-]} \tag{5-71}$$

$$\overline{\text{H}_2\text{O} \rightleftharpoons \text{H}^+ + \text{OH}^-} \qquad K_w = K_a \cdot K_b$$

$$= \frac{[\text{H}^+][\text{A}^-]}{[\text{HA}]} \frac{[\text{HA}][\text{OH}^-]}{[\text{A}^-]} \tag{5-72}$$

All dissolved $\text{CO}_2 \rightleftharpoons \text{HCO}_3^- + \text{H}^+$
$(= \text{CO}_2(aq) + \text{H}_2\text{CO}_3)$

$$K_{a1} = \frac{[\text{HCO}_3^-][\text{H}^+]}{[\text{CO}_2(aq) + \text{H}_2\text{CO}_3]} = 4.45 \times 10^{-7}$$

Only about 0.2% of dissolved CO_2 is in the form H_2CO_3. When the true value of $[\text{H}_2\text{CO}_3]$ is used, instead of the value $[\text{H}_2\text{CO}_3 + \text{CO}_2(aq)]$, the value of the equilibrium constant becomes

$$K_{a1} = \frac{[\text{HCO}_3^-][\text{H}^+]}{[\text{H}_2\text{CO}_3]} = 2 \times 10^{-4}$$

The hydration of CO_2 and dehydration of H_2CO_3 are surprisingly slow reactions, which can be demonstrated readily in a classroom.[†] Living cells utilize the enzyme *carbonic anhydrase* to speed the rate at which H_2CO_3 and CO_2 equilibrate, in order to process this key metabolite.

Very careful titrations reveal another species in the carbonic acid system.[†] It is thought to be the product of the reaction of H_2CO_3 and HCO_3^-:

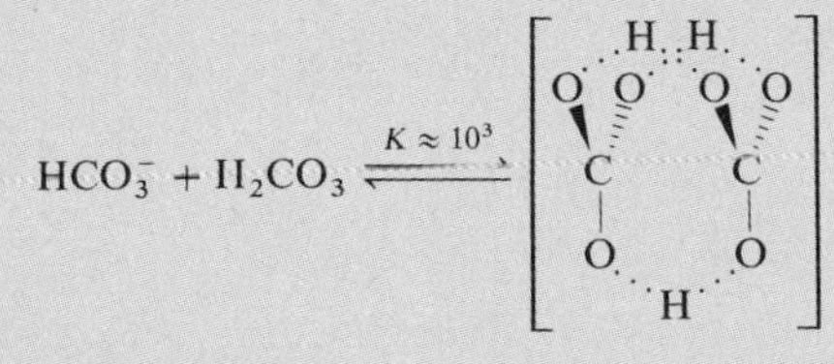

The nature of the species is speculative at this time, but it may be the triply hydrogen-bonded species drawn above.

[†] A. K. Covington, *Chem. Soc. Rev.*, **14**, 265 (1985).

$K_a \cdot K_b = K_w$ for a conjugate acid–base pair in aqueous solution.

When Reactions 5-70 and 5-71 are added, their equilibrium constants must be multiplied, giving a most useful result:

$$K_a \cdot K_b = K_w \qquad (5\text{-}73)$$

Equation 5-73 applies to any acid and its conjugate base in aqueous solution.

EXAMPLE: Finding K_b for the Conjugate Base

The value of K_a for acetic acid is 1.75×10^{-5} (Reaction 5-61). Find K_b for the acetate ion.

The solution is trivial:

$$K_b = \frac{K_w}{K_a} = \frac{1.0 \times 10^{-14}}{1.75 \times 10^{-5}} = 5.7 \times 10^{-10}$$

EXAMPLE: Finding K_a for the Conjugate Acid

The value of K_b for methylamine is 4.4×10^{-4} (Reaction 5-62). Find K_a for the methylammonium ion.

Once again:

$$K_a = \frac{K_w}{K_b} = 2.3 \times 10^{-11}$$

For a diprotic acid, we can derive results relating each of two acids and their conjugate bases:

$$\cancel{H_2A} \rightleftharpoons H^+ + \cancel{HA^-} \quad K_{a1} \qquad\qquad \cancel{HA^-} \rightleftharpoons H^+ + \cancel{A^{2-}} \quad K_{a2}$$

$$\underline{\cancel{HA^-} + H_2O \rightleftharpoons \cancel{H_2A} + OH^- \quad K_{b2}} \qquad \underline{\cancel{A^{2-}} + H_2O \rightleftharpoons \cancel{HA^-} + OH^- \quad K_{b1}}$$

$$H_2O \rightleftharpoons H^+ + OH^- \quad K_w \qquad\qquad H_2O \rightleftharpoons H^+ + OH^- \quad K_w$$

The final results are

$$K_{a1} \cdot K_{b2} = K_w \qquad (5\text{-}74)$$

$$K_{a2} \cdot K_{b1} = K_w \qquad (5\text{-}75)$$

Challenge: Derive the following results for a triprotic acid:

$$K_{a1} \cdot K_{b3} = K_w \qquad (5\text{-}76)$$

$$K_{a2} \cdot K_{b2} = K_w \qquad (5\text{-}77)$$

$$K_{a3} \cdot K_{b1} = K_w \qquad (5\text{-}78)$$

Summary

At equilibrium, the forward and reverse rates of a chemical reaction are equal. For the reaction $aA + bB \rightleftharpoons cC + dD$, the equilibrium constant is $K = [C]^c[D]^d/[A]^a[B]^b$. Solute concentrations should be expressed in moles per liter; gas concentrations should be in atmospheres; and the concentrations of pure solids, liquids, and solvents are omitted. If the direction of a reaction is changed, $K' = 1/K$. If two reactions are added, $K_3 = K_1K_2$. The value of the equilibrium constant can be calculated from the free energy change for a chemical reaction: $K = e^{-\Delta G°/RT}$. Le Châtelier's principle predicts the effect on a chemical reaction when reactants or products are added, or temperature is changed. The reaction quotient, Q, is used to tell whether a system is at equilibrium, or how it must change to reach equilibrium.

The solubility product, which is the equilibrium constant for the dissolution of a solid salt, is used to calculate the solubility of a salt in aqueous solution. If one of the ions of that salt is already present in the solution, the solubility of the salt is decreased (the common ion effect). You should be able to evaluate the feasibility of selectively precipitating one ion from a solution containing other ions. At high concentration of ligand, a precipitated metal ion may redissolve because of the formation of soluble complex ions. Setting up and solving equilibrium problems are important skills, as are the use and testing of numerical approximations.

Acids are proton donors, and bases are proton acceptors. An acid increases the concentration of H_3O^+ in aqueous solution, and a base increases the concentration of OH^-. An acid–base pair related through the gain or loss of a single proton is described as conjugate. When a proton is transferred from one molecule to another molecule of a protic solvent, the reaction is called autoprotolysis.

The definition of pH is: $pH = -\log[H^+]$ (which will be modified to include activity in the next chapter). K_a refers to the equilibrium constant for the dissociation of an acid: $HA + H_2O \rightleftharpoons H_3O^+ + A^-$ K_b is the base hydrolysis constant for the reaction $B + H_2O \rightleftharpoons BH^+ + OH^-$. When K_a or K_b is large, the acid or base is said to be strong; otherwise the acid or base is weak. The common strong acids and bases are listed in Table 5-2. The most common weak acids are carboxylic acids (RCO_2H), and the most common weak bases are amines ($R_3N:$). Carboxylate anions (RCO_2^-) are weak bases, and ammonium ions (R_3NH^+) are weak acids. For a conjugate acid–base pair, $K_a \cdot K_b = K_w$. For polyprotic acids we denote the successive acid dissociation constants as $K_{a1}, K_{a2}, K_{a3}, \ldots$, or just $K_1, K_2, K_3, \ldots$. For polybasic species we denote successive hydrolysis constants $K_{b1}, K_{b2}, K_{b3}, \ldots$. For a polyprotic system with n protons, the relation between the ith K_a and the $(n-i)$th K_b is $K_{ai} \cdot K_{b(n-i)} = K_w$.

Terms to Understand

acid
acidic solution
acid dissociation constant (K_a)
amine
ammonium ion
aprotic solvent
autoprotolysis
base
base "dissociation" constant (K_b)
base hydrolysis constant (K_b)
basic solution
Brønsted–Lowry acid
Brønsted–Lowry base
carboxylate anion
carboxylic acid
common ion effect
complex ion
conjugate acid–base pair
coprecipitation
disproportionation
endothermic
enthalpy

entropy
equilibrium constant
exothermic
Gibbs free energy
hydronium ion
ion pair
Le Châtelier's principle
Lewis acid
Lewis base
ligand
neutralization
overall formation constant
pH
polyprotic acids and bases
protic solvent
reaction quotient
salt
saturated solution
solubility product
standard state
stepwise formation constant
strong electrolyte

Exercises

5-A. Consider the following equilibria, in which all ions are aqueous:

(1) $Ag^+ + Cl^- \rightleftharpoons AgCl(aq)$ $K = 2.0 \times 10^3$
(2) $AgCl(aq) + Cl^- \rightleftharpoons AgCl_2^-$ $K = 9.3 \times 10^1$
(3) $AgCl(s) \rightleftharpoons Ag^+ + Cl^-$ $K = 1.8 \times 10^{-10}$

(a) Calculate the numerical value of the equilibrium constant for the reaction $AgCl(s) \rightleftharpoons AgCl(aq)$.
(b) Calculate the concentration of $AgCl(aq)$ in equilibrium with excess undissolved solid AgCl.
(c) Find the numerical value of K for the reaction $AgCl_2^- \rightleftharpoons AgCl(s) + Cl^-$.

5-B. Reaction 5-12 is allowed to come to equilibrium in a solution initially containing 0.010 0 M BrO_3^-, 0.010 0 M Cr^{3+}, and 1.00 M H^+.

(a) Set up an equation analogous to Equation 5-20 to find the concentrations at equilibrium.
(b) Noting that $K = 1 \times 10^{11}$ for Reaction 5-12, it is reasonable to assume that the reaction will go nearly all the way "to completion." That is, we expect both the concentration of Br^- and of $Cr_2O_7^{2-}$ to be close to 0.005 00 M at equilibrium. (Why?) Using these concentrations for Br^- and $Cr_2O_7^{2-}$, find the concentrations of BrO_3^- and Cr^{3+} at equilibrium. Note that Cr^{3+} is the *limiting reagent* in this problem. The reaction uses up the Cr^{3+} before consuming all of the BrO_3^-.

5-C. How many grams of lanthanum iodate, $La(IO_3)_3$ (which dissociates into La^{3+} and $3IO_3^-$), will dissolve in 250.0 mL of (a) water and (b) 0.050 M $LiIO_3$?

5-D. Which will be more soluble (moles of metal dissolved per liter of solution) in water?

(a) $Ba(IO_3)_2$ ($K_{sp} = 1.5 \times 10^{-9}$) or
$Ca(IO_3)_2$ ($K_{sp} = 7.1 \times 10^{-7}$)
(b) $TlIO_3$ ($K_{sp} = 3.1 \times 10^{-6}$) or
$Sr(IO_3)_2$ ($K_{sp} = 3.3 \times 10^{-7}$)

5-E. Fe(III) can be precipitated from acidic solution by addition of OH^- to form $Fe(OH)_3(s)$. At what concentration of OH^- will the concentration of Fe(III) be reduced to 1.0×10^{-10} M? If Fe(II) is used instead, what concentration of OH^- is necessary to reduce the Fe(II) concentration to 1.0×10^{-10} M?

5-F. It is desired to perform a 99% complete separation of 0.010 M Ca^{2+} and 0.010 M Ce^{3+} by precipitation with oxalate ($C_2O_4^{2-}$). Given the solubility products below, decide whether this is feasible.

$$CaC_2O_4 \qquad K_{sp} = 1.3 \times 10^{-8}$$
$$Ce_2(C_2O_4)_3 \qquad K_{sp} = 3 \times 10^{-29}$$

5-G. For a solution of Ni^{2+} and ethylenediamine, the following equilibrium constants apply at 20°C:

$$Ni^{2+} + H_2NCH_2CH_2NH_2 \rightleftharpoons Ni(en)^{2+}$$

Ethylenediamine $\log K_1 = 7.52$
(abbreviated en)

$Ni(en)^{2+} + en \rightleftharpoons Ni(en)_2^{2+}$ $\log K_2 = 6.32$

$Ni(en)_2^{2+} + en \rightleftharpoons Ni(en)_3^{2+}$ $\log K_3 = 4.49$

Calculate the concentration of free Ni^{2+} in a solution prepared by mixing 0.100 mol of en plus 1.00 mL of 0.010 0 M Ni^{2+} and diluting to 1.00 L. *Hint:* Start by assuming that nearly all of the Ni is in the form $Ni(en)_3^{2+}$ (i.e., $[Ni(en)_3^{2+}] = 1.00 \times 10^{-5}$ M).

5-H. If each of the following is dissolved in water, will the solution be acidic, basic, or neutral?

(a) Na^+Br^- (b) $Na^+CH_3CO_2^-$
(c) $NH_4^+Cl^-$ (d) K_3PO_4
(e) $(CH_3)_4N^+Cl^-$ (f) $(CH_3)_4N^+ \bigcirc\!\!\!-\!CO_2^-$

5-I. Succinic acid dissociates as follows:

$$\underset{O}{\overset{O}{HOCCH_2CH_2COH}} \overset{K_1}{\rightleftharpoons}$$

$$\underset{O}{\overset{O}{HOCCH_2CH_2CO^-}} + H^+ \qquad K_1 = 6.2 \times 10^{-5}$$

$$\underset{O}{\overset{O}{HOCCH_2CH_2CO^-}} \overset{K_2}{\rightleftharpoons}$$

$$\underset{O}{\overset{O}{{}^-OCCH_2CH_2CO^-}} + H^+ \qquad K_2 = 2.3 \times 10^{-6}$$

Calculate K_{b1} and K_{b2} for the following reactions:

$$\underset{O}{\overset{O}{{}^-OCCH_2CH_2CO^-}} + H_2O \overset{K_{b1}}{\rightleftharpoons}$$

$$\underset{O}{\overset{O}{HOCCH_2CH_2CO^-}} + OH^-$$

$$\underset{O}{\overset{O}{HOCCH_2CH_2CO^-}} + H_2O \overset{K_{b2}}{\rightleftharpoons}$$

$$\underset{O}{\overset{O}{HOCCH_2CH_2COH}} + OH^-$$

5-J. Histidine is a triprotic amino acid:

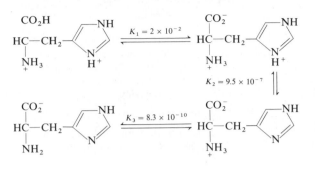

$$K_1 = 2 \times 10^{-2}$$
$$K_2 = 9.5 \times 10^{-7}$$
$$K_3 = 8.3 \times 10^{-10}$$

What is the value of the equilibrium constant for the reaction

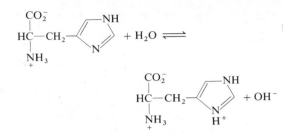

5-K. Using the values of K_w in Table 5-1, calculate the pH of distilled water at 0°C, 20°C, and 40°C.

Problems

A5-1. Write the expression for the equilibrium constant for each of the following reactions. Write the pressure of a gaseous molecule, X, as P_X.
   (a) $3Ag^+(aq) + PO_4^{3-}(aq) \rightleftharpoons Ag_3PO_4(s)$
   (b) $C_6H_6(l) + \frac{15}{2}O_2(g) \rightleftharpoons 3H_2O(l) + 6CO_2(g)$
   (c) $Cl_2(g) + 2OH^-(aq) \rightleftharpoons Cl^-(aq) + OCl^-(aq) + H_2O(l)$
   (d) $Hg(l) + I_2(g) \rightleftharpoons HgI_2(s)$

A5-2. For the reaction $2A(g) + B(aq) + 3C(l) \rightleftharpoons D(s) + 3E(g)$, the concentration at equilibrium are found to be

   A: 2.8×10^3 Pa   D: 16.5 M

   B: 1.2×10^{-2} M   E: 3.6×10^4 torr

   C: 12.8 M

   Find the numerical value of the equilibrium constant that would appear in a conventional table of equilibrium constants.

A5-3. From the equations

   $HOCl \rightleftharpoons H^+ + OCl^-$ $K = 3.0 \times 10^{-8}$

   $HOCl + OBr^- \rightleftharpoons HOBr + OCl^-$ $K = 15$

   find the value of K for the reaction $HOBr \rightleftharpoons H^+ + OBr^-$. All species are aqueous.

A5-4. From the equilibrium constants below, calculate the equilibrium constant for the reaction $HO_2CCO_2H \rightleftharpoons 2H^+ + C_2O_4^{2-}$. All species in this problem are aqueous.

$$HO\overset{\overset{O}{\|}}{C}\overset{\overset{O}{\|}}{C}OH \rightleftharpoons H^+ + HO\overset{\overset{O}{\|}}{C}\overset{\overset{O}{\|}}{C}O^-$$
Oxalic acid $K_1 = 5.6 \times 10^{-2}$

$$HO\overset{\overset{O}{\|}}{C}\overset{\overset{O}{\|}}{C}O^- \rightleftharpoons H^+ + {}^-O\overset{\overset{O}{\|}}{C}\overset{\overset{O}{\|}}{C}O^-$$
Oxalate $K_2 = 5.4 \times 10^{-5}$

A5-5. (a) A favorable entropy change occurs when ΔS is positive. Does the order of the system increase or decrease when ΔS is positive?
   (b) A favorable enthalpy change occurs when ΔH is negative. Does the system absorb heat or give off heat when ΔH is negative?
   (c) Write the relation between ΔG, ΔH, and ΔS. Use the results of parts a and b to state whether ΔG must be positive or negative for a spontaneous change.

A5-6. For the reaction $HCO_3^- \rightleftharpoons H^+ + CO_3^{2-}$, $\Delta G^\circ = +59.0$ kJ/mol at 298.15 K. Find the value of K for the reaction.

A5-7. The formation of tetrafluoroethylene from its elements is highly exothermic.

$$2F_2(g) + 2C(s) \rightleftharpoons F_2C{=}CF_2(g)$$
Fluorine   Graphite   Tetrafluoroethylene

   (a) If a mixture of F_2, graphite, and C_2F_4 is at equilibrium in a closed container, will the reaction go to the right or to the left if F_2 is added?
   (b) Will it go to the right or left if rare bacteria from the planet Telfon are added? These bacteria eat C_2F_4 and make Telfon for their cell walls. Telfon has the structure.

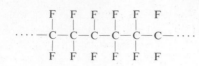

(c) Will it go to the right or left if solid graphite is added? (Neglect any effect of increased pressure due to the decreased volume in the vessel when solid is added.)

(d) Will it go to the right or left if the container is crushed to one-eighth its original volume?

(e) Will it go to the right or left if the container is heated?

A5-8. Suppose that the following reaction has come to equilibrium:

$$Br_2(l) + I_2(s) + 4Cl^-(aq) \rightleftharpoons$$
$$2Br^-(aq) + 2ICl_2^-(aq)$$

If more $I_2(s)$ is added, will the concentration of ICl_2^- in the aqueous phase increase, decrease, or remain unchanged?

A5-9. For the reaction $H_2O(l) \rightleftharpoons H^+(aq) + OH^-(aq)$, $K = 1.0 \times 10^{-14}$ at 25°C. The concentrations in a system out of equilibrium are $[H^+] = 3.0 \times 10^{-5}$ M and $[OH^-] = 2.0 \times 10^{-7}$ M. Will the reaction proceed to the left or to the right to reach equilibrium?

A5-10. Will $CaSO_4$ be more soluble in distilled water or in 0.1 M $CaCl_2$?

A5-11. Use the solubility product to calculate the solubility of CuBr in water expressed as (a) moles per liter and (b) g/100 mL.

A5-12. Use the solubility product to calculate the solubility of Ag_2CrO_4 ($\rightarrow 2Ag^+ + CrO_4^{2-}$) in water expressed as (a) moles per liter and (b) g/100 mL.

A5-13. How many grams of PbI_2 will dissolve in 0.500 L of (a) water and (b) 0.063 4 M NaI?

A5-14. What concentration of carbonate must be added to 0.10 M Zn^{2+} to precipitate 99.9% of the Zn^{2+}?

A5-15. Explain why the solubility of lead first decreases and then increases as the concentration of I^- is increased in Figure 5-1.

A5-16. Write the definitions of acids and bases advanced (a) by Brønsted and Lowry; (b) by Lewis.

A5-17. Make a list of the common strong acids and strong bases. Memorize this list.

A5-18. Write the formulas and names for two classes of weak acids and two classes of weak bases.

A5-19. Why does "pure" distilled water (at 25°C) from the tap in most labs have a pH below 7?

A5-20. Identify the Lewis acids in the following reactions:

(a) $BF_3 + NH_3 \rightleftharpoons F_3\overset{-}{B}-\overset{+}{N}H_3$

(b) $F^- + AsF_5 \rightleftharpoons AsF_6^-$

A5-21. Identify the Brønsted–Lowry acids among the reactants in the following reactions:

(a) $NaHSO_3 + NaOH \rightleftharpoons Na_2SO_3 + H_2O$

(b) $KCN + HI \rightleftharpoons HCN + KI$

(c) $PO_4^{3-} + H_2O \rightleftharpoons HPO_4^{2-} + OH^-$

A5-22. Write the autoprotolysis reaction of H_2SO_4.

A5-23. Write the K_a reaction for trichloroacetic acid, Cl_3CCO_2H, and for the anilinium ion,

A5-24. Write the K_b reactions for pyridine and for sodium 2-mercaptoethanol.

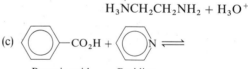

Pyridine   Sodium 2-mercaptoethanol

A5-25. Write the K_a *and* K_b reactions of $NaHCO_3$.

A5-26. Identify the conjugate acid–base pairs in the following reactions:

(a) $H_2NCH_2CH_2NH_2 + H_2O \rightleftharpoons$
   Ethylenediamine

$$H_3\overset{+}{N}CH_2CH_2NH_2 + OH^-$$

(b) $H_3\overset{+}{N}CH_2CH_2\overset{+}{N}H_3 + H_2O \rightleftharpoons$

$$H_3\overset{+}{N}CH_2CH_2NH_2 + H_3O^+$$

(c)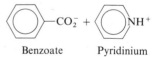

Benzoic acid   Pyridine

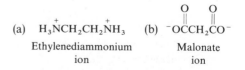

Benzoate   Pyridinium

A5-27. Write the equations for the stepwise acid–base reactions for the following ions in water. Write the correct symbol (e.g., K_{b1}) for the equilibrium constant for each reaction.

(a) $H_3\overset{+}{N}CH_2CH_2\overset{+}{N}H_3$
   Ethylenediammonium ion

(b) $^-OCCH_2CO^-$ (with two $\overset{O}{\underset{||}{}}$ groups)
   Malonate ion

A5-28. Which is a stronger acid, a or b?

(a) $\underset{\text{Dichloroacetic acid}}{Cl_2HCCOH}$ with O double bonded above C, $K_a = 5.0 \times 10^{-2}$

Dichloroacetic acid

(b) $\underset{\text{Chloroacetic acid}}{ClH_2CCOH}$ with O double bonded above C, $K_a = 1.36 \times 10^{-3}$

Chloroacetic acid

Which is a stronger base, c or d?

(c) H_2NNH_2 $K_b = 3.0 \times 10^{-6}$

Hydrazine

(d) H_2NCNH_2 with O double bonded above C, $K_b = 1.5 \times 10^{-14}$

Urea

A5-29. Write the K_b reaction of CN^-. Given that the K_a value for HCN is 6.2×10^{-10}, calculate K_b for CN^-.

A5-30. Write the K_{a2} reaction of phosphoric acid (H_3PO_4) and the K_{b2} reaction of disodium oxalate ($Na_2C_2O_4$).

A5-31. From the K_b values for phosphate in Equations 5-67 through 5-69, calculate the three K_a values of phosphoric acid.

5-32. The cumulative formation constant for $SnCl_2(aq)$ in 1.0 M $NaNO_3$ is $\beta_2 = 12$. Find the concentration of $SnCl_2(aq)$ for a solution in which the concentrations of Sn^{2+} and Cl^- are both somehow fixed at 0.20 M.

5-33. The solubility product for CuCl is 1.9×10^{-7}. The equilibrium constant for the reaction

$$Cu(s) + Cu^{2+} \rightleftharpoons 2Cu^+$$

is 9.6×10^{-7}. Calculate the equilibrium constant for the reaction

$$Cu(s) + Cu^{2+} + 2Cl^- \rightleftharpoons 2CuCl(s)$$

5-34. For the sum of two reactions, we know that $K_3 = K_1K_2$.

$$\begin{array}{ll} A + B \rightleftharpoons C + D & K_1 \\ \underline{D + E \rightleftharpoons B + F} & K_2 \\ A + E \rightleftharpoons C + F & K_3 \end{array}$$

Show that this implies that $\Delta G_3^\circ = \Delta G_1^\circ + \Delta G_2^\circ$.

5-35. The equilibrium constant for the reaction $H_2O \rightleftharpoons H^+ + OH^-$ is 1.0×10^{-14} at 25°C. What is the value of K for the reaction $4H_2O \rightleftharpoons 4H^+ + 4OH^-$?

5-36. For the reaction $Mg^{2+} + Cu(s) \rightleftharpoons Mg(s) + Cu^{2+}$, $K = 10^{-92}$ and $\Delta S^\circ = +18$ J/(K·mol).
   (a) Under standard conditions, is ΔG° positive or negative? The term "standard conditions" means that reactants and products are in their standard states.
   (b) Under standard conditions, is the reaction endothermic or exothermic?

5-37. When $BaCl_2 \cdot H_2O(s)$ is dried in an oven, it loses gaseous water:

$$BaCl_2 \cdot H_2O(s) \rightleftharpoons BaCl_2(s) + H_2O(g)$$

$$\Delta H^\circ = 63.11 \text{ kJ/mol at } 25°C$$

$$\Delta S^\circ = +148 \text{ J/(K·mol) at } 25°C$$

   (a) Calculate the vapor pressure of gaseous H_2O above $BaCl_2 \cdot H_2O$ at 298 K.
   (b) Assuming that ΔH° and ΔS° are not temperature-dependent (a poor assumption), estimate the temperature at which the vapor pressure of $H_2O(g)$ above $BaCl_2 \cdot H_2O(s)$ will be 1 atm.

5-38. The equilibrium constant for the reaction of ammonia with water has the following values in the range 5–10°C:

$$NH_3 + H_2O \rightleftharpoons NH_4^+ + OH^-$$

$$K = 1.479 \times 10^{-5} \text{ at } 5°C$$

$$K = 1.570 \times 10^{-5} \text{ at } 10°C$$

   (a) Assuming that ΔH° and ΔS° are constant in the interval 5–10°C, use Equation 5-15 to find the value of ΔH° for the reaction in this temperature range.
   (b) Describe how Equation 5-15 could be used to make a linear graph to determine ΔH°, if ΔH° and ΔS° were constant over some range of temperature.

5-39. For the reaction $H_2(g) + Br_2(g) \rightleftharpoons 2HBr(g)$, $K = 7.2 \times 10^{-4}$ at 1 362 K and ΔH° is positive. A vessel is charged with 48.0 Pa HBr, 1 370 Pa H_2, and 3 310 Pa Br_2 at 1 362 K.
   (a) Will the reaction proceed to the left or right to reach equilibrium?
   (b) Calculate the pressure (in pascals) of each species in the vessel at equilibrium.
   (c) If the mixture at equilibrium is compressed to half its original volume, will the reaction proceed to the left or right to reestablish equilibrium?
   (d) If the mixture at equilibrium is heated from 1 362 to 1 407 K, will HBr be formed or consumed in order to reestablish equilibrium?

5-40. Express the solubility of $AgIO_3$ in 10.0 mM KIO_3 as a fraction of its solubility in pure water.

5-41. Ag^+ at a concentration of 10–100 ppb (ng/mL) is an effective disinfectant for swimming pool water. However, the concentration should not exceed a few hundred parts per billion for human health. One way to maintain an appropriate concentration of Ag^+ is to add a slightly soluble silver salt to the pool. For each salt below, calculate the concentration of Ag^+ (in parts per billion) that would exist at equilibrium.

$$AgCl: \quad K_{sp} = 1.8 \times 10^{-10}$$

$$AgBr: \quad K_{sp} = 5.0 \times 10^{-13}$$

$$AgI: \quad K_{sp} = 8.3 \times 10^{-17}$$

5-42. If a solution containing 0.10 M Cl^-, Br^-, I^-, and CrO_4^{2-} is treated with Ag^+, in what order will the anions precipitate?

5-43. A given solution contains 0.050 0 M Ca^{2+} and 0.030 0 M Ag^+. Can 99% of either ion be precipitated by addition of sulfate, without precipitating the other metal ion? What will be the concentration of Ca^{2+} when Ag_2SO_4 begins to precipitate?

5-44. (a) Calculate the solubility (milligrams per liter) of $Zn_2Fe(CN)_6$ in distilled water. This salt dissociates as follows:

$$Zn_2Fe(CN)_6(s) \rightleftharpoons 2Zn^{2+} + Fe(CN)_6^{4-}$$
$$\text{Ferrocyanide}$$

$$K_{sp} = 2.1 \times 10^{-16}$$

(b) Calculate the concentration of ferrocyanide in a solution of 0.040 M $ZnSO_4$ saturated with $Zn_2Fe(CN)_6$. $ZnSO_4$ will be completely dissociated into Zn^{2+} and SO_4^{2-}.

(c) What concentration of $K_4Fe(CN)_6$ should be added to a suspension of solid $Zn_2Fe(CN)_6$ in water to give $[Zn^{2+}] = 5.0 \times 10^{-7}$ M?

5-45. Given the equilibria below, calculate the concentrations of each zinc-containing species in a solution saturated with $Zn(OH)_2(s)$ and containing $[OH^-]$ at a fixed concentration of 3.2×10^{-7} M.

$$Zn(OH)_2(s): \quad K_{sp} = 3.0 \times 10^{-16}$$

$$Zn(OH)^+: \quad \beta_1 = 2.5 \times 10^4$$

$$Zn(OH)_3^-: \quad \beta_3 = 7.2 \times 10^{15}$$

$$Zn(OH)_4^{2-}: \quad \beta_4 = 2.8 \times 10^{15}$$

5-46. For the reaction below, $K = 6.9 \times 10^{-18}$.

$$Cu(OH)_{1.5}(SO_4)_{0.25}(s) \rightleftharpoons$$
$$Cu^{2+} + \tfrac{3}{2}OH^- + \tfrac{1}{4}SO_4^{2-}$$

What will be the concentration of Cu^{2+} in a solution saturated with $Cu(OH)_{1.5}(SO_4)_{0.25}$?

5-47. Consider the following equilibria:

$$AgCl(s) \rightleftharpoons Ag^+ + Cl^- \quad K_{sp} = 1.8 \times 10^{-10}$$
$$AgCl(s) + Cl^- \rightleftharpoons AgCl_2^- \quad K_2 = 1.5 \times 10^{-2}$$
$$AgCl_2^- + Cl^- \rightleftharpoons AgCl_3^- \quad K_3 = 0.49$$

Find the total concentration of silver-containing species in a silver-saturated, aqueous solution containing the following concentrations of Cl^-: (a) 0.010 M (b) 0.20 M (c) 2.0 M

5-48. Use electron dot structures to show why tetramethylammonium hydroxide is an ionic compound. That is, show why the hydroxide is not covalently bound to the rest of the molecule.

5-49. (a) Using only K_{sp} in Table 5-3, calculate how many moles of $Ca(OH)_2$ will dissolve in 1.00 L of water.

(b) How will the solubility calculated in part a be affected by the K_1 reaction in Table 5-3?

5-50. Although KOH, RbOH, and CsOH show no evidence of association between metal and hydroxide in aqueous solution, Li^+ and Na^+ do form complexes with OH^-:

$$Li^+ + OH^- \rightleftharpoons LiOH(aq) \quad K_1 = \frac{[LiOH(aq)]}{[Li^+][OH^-]}$$
$$= 0.83$$

$$Na^+ + OH^- \rightleftharpoons NaOH(aq) \quad K_1 = 0.20$$

Calculate the fraction of sodium in the form $NaOH(aq)$ in 1 F NaOH.

5-51. The diagram at the top of the next page gives pH values of common substances.

(a) Explain why the theoretical pH of "pure" rain is not 7.

(b) Explain how SO_2 in the atmosphere makes acidic rain. SO_2 is emitted by combustion of sulfur-containing fuels, especially coal.

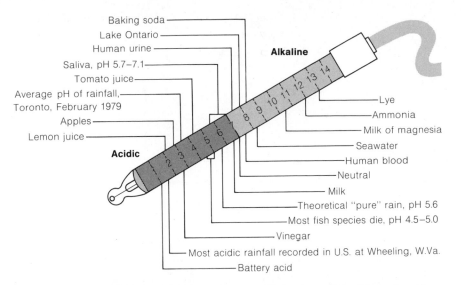

The most acidic rainfall in the United States is more acid than lemon juice. [Reproduced from *Chemical and Engineering News*, September 14, 1981.]

5-52. The planet Aragonose (which is made mostly of the mineral aragonite, whose composition is $CaCO_3$) has an atmosphere containing methane and carbon dioxide, each at a pressure of 0.10 atm. The oceans are saturated with aragonite and have a concentration of H^+ equal to 1.8×10^{-7} M. Given the following equilibria, calculate how many grams of calcium are contained in 2.00 L of Aragonose seawater.

$$CaCO_3(s, \text{aragonite}) \rightleftharpoons Ca^{2+}(aq) + CO_3^{2-}(aq)$$
$$K_{sp} = 6.0 \times 10^{-9}$$

$$CO_2(g) \rightleftharpoons CO_2(aq)$$
$$K_{CO2} = 3.4 \times 10^{-2}$$

$$CO_2(aq) + H_2O(l) \rightleftharpoons HCO_3^-(aq) + H^+(aq)$$
$$K_1 = 4.4 \times 10^{-7}$$

$$HCO_3^-(aq) \rightleftharpoons H^+(aq) + CO_3^{2-}(aq)$$
$$K_2 = 4.7 \times 10^{-11}$$

Hint: Don't be too upset. Reverse the first reaction, add all the reactions together, and see what cancels.

5-53. The solubility product for $BaF_2(s)$ is given by $[Ba^{2+}][F^-]^2 = K_{sp}$. This equation is plotted on the graph at the top of the next column.

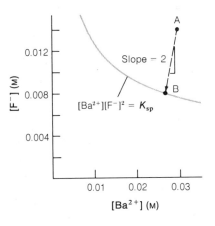

At the point A, $[Ba^{2+}][F^-]^2 > K_{sp}$ and the solution is said to be *supersaturated*. It contains more solute than should be present at equilibrium. Suppose that a supersaturated solution with composition A on the graph is prepared by some means. Is the following statement true or false? "The composition of the solution at equilibrium will be given by the point B, which is connected to A by a line of slope = 2."

5-54. Consider the values of K_w in Table 5-1. Use Le Châtelier's principle to decide if the autoprotolysis of water is endothermic or exothermic at (a) 25°C, (b) 100°C, and (c) 300°C.

6 Activity

In Chapter 5 we wrote the equilibrium constant for the reaction $aA + bB \rightleftharpoons cC + dD$ in the form

$$K = \frac{[C]^c[D]^d}{[A]^a[B]^b} \tag{6-1}$$

This statement is not strictly correct, because the quotient $[C]^c[D]^d/[A]^a[B]^b$ is not constant under all conditions. In this chapter we will see how the concentrations should be replaced by *activities*, and we will see how activities are used.

6-1 ROLE OF IONIC STRENGTH IN IONIC EQUILIBRIA

Equilibria involving ionic species are affected by the presence of all ions in a solution. The most useful measure of the total concentration of ions in a solution is the **ionic strength,** μ, defined as

$$\mu = \frac{1}{2}\sum_i c_i z_i^2 \tag{6-2}$$

where c_i is the concentration of the ith species and z_i is its charge. The sum extends over *all* ions in solution.

EXAMPLE: Calculation of Ionic Strength
Find the ionic strength of (a) 0.10 M $NaNO_3$, (b) 0.10 M Na_2SO_4, and (c) 0.020 M KBr plus 0.030 M $ZnSO_4$.

(a) $\mu = \frac{1}{2}\{[Na^+] \cdot (+1)^2 + [NO_3^-] \cdot (-1)^2\}$
$\quad = \frac{1}{2}\{(0.10) \cdot 1 + (0.10) \cdot 1\} = 0.10 \text{ M}$

(b) $\mu = \frac{1}{2}\{[Na^+] \cdot (+1)^2 + [SO_4^{2-}] \cdot (-2)^2\}$
$\quad = \frac{1}{2}\{(0.20) \cdot 1 + (0.10) \cdot 4\} = 0.30 \text{ M}$

Note that $[Na^+] = 0.20$ M because there are two moles of Na^+ per mole of Na_2SO_4.

(c) $\mu = \frac{1}{2}\{[K^+] \cdot (+1)^2 + [Br^-] \cdot (-1)^2 + [Zn^{2+}] \cdot (+2)^2 + [SO_4^{2-}] \cdot (-2)^2\}$
$\quad = \frac{1}{2}\{(0.020) \cdot 1 + (0.020) \cdot 1 + (0.030) \cdot 4 + (0.030) \cdot 4\} = 0.14 \text{ M}$

Electrolyte	Molarity	Ionic strength
1:1	M	M
2:1	M	3M
3:1	M	6M
2:2	M	4M

$NaNO_3$ is called a 1:1 electrolyte because the cation and anion both have a charge of 1. For 1:1 electrolytes the ionic strength equals the molarity. For any other stoichiometry (such as the 2:1 electrolyte, Na_2SO_4), the ionic strength is greater than the molarity.

Effect of Ionic Strength on Solubility of Salts

Consider a saturated solution of $Hg_2(IO_3)_2$ in distilled water. Based on the solubility product, we expect the concentration of mercurous ion to be 6.9×10^{-7} M:

$$Hg_2(IO_3)_2(s) \rightleftharpoons Hg_2^{2+} + 2IO_3^- \qquad K_{sp} = 1.3 \times 10^{-18} \qquad (6\text{-}3)$$

$$K_{sp} = [Hg_2^{2+}][IO_3^-]^2 = x(2x)^2$$

$$x = [Hg_2^{2+}] = 6.9 \times 10^{-7} \text{ M}$$

This concentration is indeed observed when $Hg_2(IO_3)_2$ is dissolved in distilled water.

However, a seemingly strange effect is observed when a salt such as KNO_3 is added to the solution. Neither K^+ nor NO_3^- reacts with either Hg_2^{2+} or IO_3^-. Yet, if 0.050 M KNO_3 is added to the saturated solution of $Hg_2(IO_3)_2$, more solid dissolves until the concentration of Hg_2^{2+} has increased by about 50% (to 1.0×10^{-6} M).

It turns out to be a general observation that adding any "inert" salt (such as KNO_3) to any sparingly soluble salt (such as $Hg_2(IO_3)_2$) increases the solubility of the latter. By "inert" we mean a salt whose ions do not react with the compound of interest.

Addition of an "inert" salt increases the solubility of an ionic compound.

Explanation of Increased Solubility

How can we explain the increased solubility induced by the addition of ions to the solution? Consider one particular Hg_2^{2+} ion and one particular IO_3^- ion in the solution. The IO_3^- ion is surrounded by cations (K^+, Hg_2^{2+}) and anions (NO_3^-, IO_3^-) in its region of solution. However, on the average, there will be more cations than anions around any one chosen anion. This is because cations are attracted to the anion, but anions are repelled. This gives rise to a region of net positive charge around any particular anion. We call this region the **ionic atmosphere.** Ions are continually diffusing into and out of

An anion is surrounded by more cations than anions. A cation is surrounded by more anions than cations.

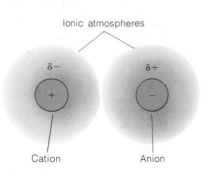

Ionic atmospheres

$\delta-$ $\delta+$

+   −

Cation   Anion

Figure 6-1
Schematic representation of the ionic atmospheres surrounding any ions in solution. The charges of the atmospheres ($\delta+$ and $\delta-$) are less than the charges of the central ions.

Ion dissociation is increased by increasing the ionic strength. Demonstration 6-1 shows this effect in the case of Reaction 6-5

the ionic atmosphere. The net charge in the atmosphere, averaged over time, is less than the charge of the anion at the center. A similar phenomenon accounts for a region of negative charge surrounding any cation in solution. This is shown schematically in Figure 6-1.

The ionic atmosphere attenuates (decreases) the attraction between ions in solution. The cation plus its negative atmosphere has less positive charge than the cation alone. The anion plus its atmosphere has less negative charge than the anion alone. The net attraction between the cation-plus-atmosphere and the anion-plus-atmosphere is smaller than it would be between pure cation and anion in the absence of ionic atmospheres. *The greater the ionic strength of a solution, the higher the charge in the ionic atmosphere. This means that each ion-plus-atmosphere contains less charge, and there is less attraction between any particular cation and anion.*

Increasing the ionic strength therefore reduces the attraction between any particular Hg_2^{2+} ion and any IO_3^- ion, as compared to their attraction for each other in distilled water. The effect is to reduce their tendency to come together, thereby increasing the solubility of $Hg_2(IO_3)_2$.

Increasing ionic strength promotes the dissociation of compounds into ions. Thus, each of the following reactions is driven to the right if the ionic strength is raised from, say, 0.01 to 0.1 M:

$$Hg_2(IO_3)_2(s) \rightleftharpoons Hg_2^{2+} + 2IO_3^- \qquad (6\text{-}3)$$

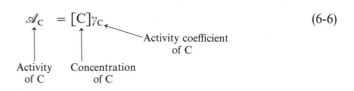

$$Fe(SCN)^{2+} \rightleftharpoons Fe^{3+} + \underset{\text{Thiocyanate}}{SCN^-} \qquad (6\text{-}5)$$

6-2 ACTIVITY COEFFICIENTS

The equilibrium "constant" in Equation 6-1 does not predict that there should be any effect of ionic strength on ionic reactions. To account for the effect of ionic strength, the concentrations should be replaced by **activities:**

$$\mathscr{A}_C = [C]\gamma_C \qquad (6\text{-}6)$$

Activity   Concentration   Activity coefficient
of C   of C   of C

Do not confuse the terms "activity" and "activity coefficient."

where C is any species. The activity of each species is its concentration multiplied by its **activity coefficient.**

The correct form of the equilibrium constant is

This is the "real" equilibrium constant.

$$K = \frac{\mathscr{A}_C^c \mathscr{A}_D^d}{\mathscr{A}_A^a \mathscr{A}_B^b} = \frac{[C]^c\gamma_C^c[D]^d\gamma_D^d}{[A]^a\gamma_A^a[B]^b\gamma_B^b} \qquad (6\text{-}7)$$

Equation 6-7 allows for the effect of ionic strength on a chemical equilibrium because the activity coefficients depend on ionic strength.

For Reaction 6-3 the equilibrium constant is

$$K = \mathscr{A}_{Hg_2^{2+}} \mathscr{A}_{IO_3^-}^2 = [Hg_2^{2+}]\gamma_{Hg_2^{2+}}[IO_3^-]^2\gamma_{IO_3^-}^2 \qquad (6-8)$$

If the concentrations of Hg_2^{2+} and IO_3^- are to increase with increasing ionic strength, the values of the activity coefficients must decrease with increasing ionic strength.

At low ionic strength, activity coefficients approach unity, and the thermodynamic equilibrium constant, Equation 6-7, approaches the "concentration" equilibrium constant, Equation 6-1. One way to measure thermodynamic equilibrium constants is to measure the concentration ratio, Equation 6-1, at successively lower ionic strenths and then extrapolate to zero ionic strength. Very commonly, tabulated equilibrium constants are not true thermodynamic constants but just the concentration ratio, Equation 6-1, measured under a particular set of conditions.

Activity Coefficients of Ions

Detailed treatment of the ionic atmosphere model leads to the **extended Debye-Hückel equation,** relating activity coefficients to ionic strength:

$$\log \gamma = \frac{-0.51z^2\sqrt{\mu}}{1 + (\alpha\sqrt{\mu}/305)} \quad \text{(at 25°C)} \qquad (6-9)$$

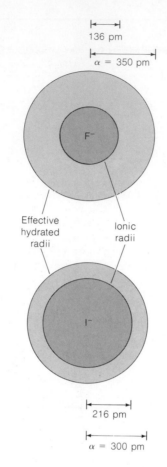

Effective hydrated radii

Ionic radii

Figure 6-2
Ionic and hydrated radii of fluoride and iodide.

1 pm (picometer) = 10^{-12} m

In Equation 6-9, γ is the activity coefficient of an ion of charge $\pm z$ and size α (pm) in an aqueous medium of ionic strength μ. The equation works fairly well for $\mu \leq 0.1$ M.

The value of α is the effective **hydrated radius** of the ion plus its tightly bound sheath of water molecules. Small, highly charged ions bind solvent molecules more tightly and have *larger* hydrated radii than do larger or less highly charged ions. F^-, for example, has a hydrated radius greater than that of I^- (Figure 6-2). Each anion attracts solvent molecules mainly by electrostatic interaction between the negative ion and the positive pole of the H_2O dipole:

Note that H^+ has a size of 900 pm and Li^+ has a size of 600 pm. No species with sizes of 800 or 700 are listed in this table.

Table 6-1 lists the sizes and calculated activity coefficients of many common ions. The table is arranged according to the size (α) and charge of the ion. All ions of the same size and charge appear on the same line and have the same activity coefficients. At the end of the table is a list giving the sizes of some organic ions. To find the activity coefficient of these ions, refer to the appropriate size and charge in the main body of the table.

Table 6-1
Activity coefficients for aqueous solution at 25°C

Ion	Ion size (α, pm)	Ionic strength (μ, M)				
		0.001	0.005	0.01	0.05	0.1
Charge = ± 1						
H^+	900	0.967	0.933	0.914	0.86	0.83
	800	0.966	0.931	0.912	0.85	0.82
	700	0.965	0.930	0.909	0.845	0.81
Li^+	600	0.965	0.929	0.907	0.835	0.80
	500	0.964	0.928	0.904	0.83	0.79
Na^+, $CdCl^+$, ClO_2^-, IO_3^-, HCO_3^-, $H_2PO_4^-$, HSO_3^-, $H_2AsO_4^-$, $Co(NH_3)_4(NO_2)_2^+$	450	0.964	0.928	0.902	0.82	0.775
	400	0.964	0.927	0.901	0.815	0.77
OH^-, F^-, SCN^-, OCN^-, HS^-, ClO_3^-, ClO_4^-, BrO_3^-, IO_4^-, MnO_4^-	350	0.964	0.926	0.900	0.81	0.76
K^+, Cl^-, Br^-, I^-, CN^-, NO_2^-, NO_3^-	300	0.964	0.925	0.899	0.805	0.755
Rb^+, Cs^+, NH_4^+, Tl^+, Ag^+	250	0.964	0.924	0.898	0.80	0.75
Charge = ± 2						
Mg^{2+}, Be^{2+}	800	0.872	0.755	0.69	0.52	0.45
	700	0.872	0.755	0.685	0.50	0.425
Ca^{2+}, Cu^{2+}, Zn^{2+}, Sn^{2+}, Mn^{2+}, Fe^{2+}, Ni^{2+}, Co^{2+}	600	0.870	0.749	0.675	0.485	0.405
Sr^{2+}, Ba^{2+}, Cd^{2+}, Hg^{2+}, S^{2-}, $S_2O_4^{2-}$, WO_4^{2-}	500	0.868	0.744	0.67	0.465	0.38
Pb^{2+}, CO_3^{2-}, SO_3^{2-}, MoO_4^{2-}, $Co(NH_3)_5Cl^{2+}$, $Fe(CN)_5NO^{2-}$	450	0.867	0.742	0.665	0.455	0.37
Hg_2^{2+}, SO_4^{2-}, $S_2O_3^{2-}$, $S_2O_6^{2-}$, $S_2O_8^{2-}$, SeO_4^{2-}, CrO_4^{2-}, HPO_4^{2-}	400	0.867	0.740	0.660	0.445	0.355
Charge = ± 3						
Al^{3+}, Fe^{3+}, Cr^{3+}, Sc^{3+}, Y^{3+}, In^{3+}, lanthanides†	900	0.738	0.54	0.445	0.245	0.18
	500	0.728	0.51	0.405	0.18	0.115
PO_4^{3-}, $Fe(CN)_6^{3-}$, $Cr(NH_3)_6^{3+}$, $Co(NH_3)_6^{3+}$, $Co(NH_3)_5H_2O^{3+}$	400	0.725	0.505	0.395	0.16	0.095
Charge = ± 4						
Th^{4+}, Zr^{4+}, Ce^{4+}, Sn^{4+}	1100	0.588	0.35	0.255	0.10	0.065
$Fe(CN)_6^{4-}$	500	0.57	0.31	0.20	0.048	0.021

Sizes of some organic ions (α, pm): Charge = ± 1

$HCOO^-$, $H_2citrate^-$, $CH_3NH_3^+$, $(CH_3)_2NH_2^+$	350
$NH_3^+CH_2COOH$, $(CH_3)_3NH^+$, $C_2H_5NH_3^+$	400
CH_3COO^-, CH_2ClCOO^-, $(CH_3)_4N^+$, $(C_2H_5)_2NH_2^+$, $NH_2CH_2COO^-$	450
$CHCl_2COO^-$, CCl_3COO^-, $(C_2H_5)_3NH^+$, $(C_3H_7)NH_3^+$	500
$C_6H_5COO^-$, $C_6H_4OHCOO^-$, $C_6H_4ClCOO^-$, $C_6H_5CH_2COO^-$, $CH_2{=}CHCH_2COO^-$, $(CH_3)_2CHCH_2COO^-$, $(C_2H_5)_4N^+$,	

(Continued)

Table 6-1 (*continued*)

Ion	Ion size (α, pm)	Ionic strength (μ, M)				
		0.001	0.005	0.01	0.05	0.1
$(C_3H_7)_2NH_2^+$	600					
$[OC_6H_2(NO_3)_3]^-$, $(C_3H_7)_3NH^+$, $CH_3OC_6H_4COO^-$	700					
$(C_6H_5)_2CHCOO^-$, $(C_3H_7)_4N^+$	800					
Charge = ± 2						
$(COO)_2^{2-}$, H citrate^{2-}	450					
$H_2C(COO)_2^{2-}$, $(CH_2COO)_2^{2-}$, $(CHOHCOO)_2^{2-}$	500					
$C_6H_4(COO)_2^{2-}$, $H_2C(CH_2COO)_2^{2-}$, $(CH_2CH_2COO)_2^{2-}$	600					
$[OOC(CH_2)_5COO]^{2-}$, $[OOC(CH_2)_6COO]^{2-}$, Congo red anion^{2-}	700					
Charge = ± 3						
Citrate^{3-}	500					

† Lanthanides are elements 57–71 in the periodic table.
SOURCE: Data from J. Kielland, *J. Amer. Chem. Soc.*, **59**, 1675 (1937).

Over the range of ionic strength from zero to 0.1 M, the effect of each variable on activity coefficients is as follows:

1. As the ionic strength increases, the activity coefficient decreases. For all ions, γ approaches unity as μ approaches zero.
2. As the charge of the ion increases, the departure of the activity coefficient from unity increases. Activity corrections are much more important for an ion with a charge of ± 3 than for one with a charge of ± 1. Note that the activity coefficients in Table 6-1 depend on the magnitude of the charge, but not on its sign.
3. The smaller the hydrated radius of the ion, the more important activity effects become.

Effects 1 and 2 for ions of hydrated radius 500 pm are shown in Figure 6-3.

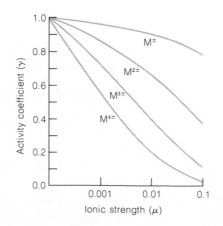

Figure 6-3
Dependence of activity coefficient on ionic strength for ions with a hydrated radius of 500 pm. Note that the abscissa is logarithmic.

EXAMPLE: Using Table 6-1

Find the activity coefficient of Hg_2^{2+} in a solution of 0.033 M $Hg_2(NO_3)_2$.

The ionic strength is

$$\mu = \tfrac{1}{2}([Hg_2^{2+}] \cdot 2^2 + [NO_3^-] \cdot (-1)^2)$$
$$= \tfrac{1}{2}([0.033] \cdot 4 + [0.066] \cdot 1) = 0.10 \text{ M}$$

In Table 6-1, Hg_2^{2+} is listed under the charge ± 2 and has a size of 400 pm. Thus $\gamma = 0.355$ when $\gamma = 0.1$ M.

EXAMPLE: How to Interpolate

Calculate the activity coefficient of H^+ when $\mu = 0.025$ M.

We can do this by two methods. The one we will use for the remainder of this text is to interpolate between values listed in Table 6-1. H^+ is the first entry in Table 6-1.

$$H^+: \quad \mu = 0.01 \qquad 0.025 \qquad 0.05$$
$$\gamma = 0.914 \qquad ? \qquad 0.86$$

By linear interpolation we find the value of γ when $\mu = 0.025$ M as follows:

$$\gamma = 0.914 - \left(\frac{0.025 - 0.01}{0.05 - 0.01}\right)(0.914 - 0.86) = 0.89_4$$

| Value of μ when $\gamma = 0.01$ | Fraction of interval between 0.01 and 0.05 | Difference in γ between 0.01 and 0.05 |

A more correct and more tedious calculation uses Equation 6-9. Table 6-1 gives a value of 900 pm for the size of H^+:

$$\log \gamma_{H^+} = \frac{(-0.51)(1^2)\sqrt{0.025}}{1 + (900\sqrt{0.025}/305)} = -0.054\,98$$
$$\gamma_{H^+} = 0.88_1$$

The difference between this calculated value and the interpolated value is less than 2%.

Activity Coefficients of Nonionic Compounds

Neutral molecules, such as benzene or acetic acid, are not surrounded by an ionic atmosphere because they have no charge. To a good approximation, their activity coefficients are unity when the ionic strength is less than 0.1 M. For all problems in this text, we will set $\gamma = 1$ for neutral molecules. That is, *the activity of a neutral molecule will be assumed to be equal to its concentration.*

For neutral species, $\mathscr{A}_C \approx [C]$.

For gaseous reactants such as H_2, the activity is written

$$\mathscr{A}_{H_2} = P_{H_2}\gamma_{H_2} \qquad (6\text{-}10)$$

For gases, $\mathscr{A} \approx P$ (atm).

where P_{H_2} is pressure in atmospheres. For most gases at or below 1 atm, $\gamma \approx 1$. For all gases, *we will assume that* $\mathscr{A} = P$ *(atm)*. The activity of a gas is called its **fugacity,** and the activity coefficient is called the *fugacity coefficient*. Deviation of gas behavior from the ideal gas law results in deviation of the fugacity coefficient from unity.

High Ionic Strengths

For the dissociation of acetic acid, we can write

$$CH_3\overset{\overset{\text{O}}{\|}}{C}OH(aq) \rightleftharpoons CH_3\overset{\overset{\text{O}}{\|}}{C}O^- + H^+ \tag{6-11}$$

$$K = \frac{[CH_3CO_2^-]\gamma_{CH_3CO_2^-}[H^+]\gamma_{H^+}}{[CH_3CO_2H]} \tag{6-12}$$

where $\gamma_{CH_3CO_2H}$ has been set equal to unity. Equation 6-12 can be rearranged to show the quotient of concentrations:

$$R = \frac{[CH_3CO_2^-][H^+]}{[CH_3CO_2H]} = \frac{K}{\gamma_{CH_3CO_2^-}\gamma_{H^+}} \tag{6-13}$$

The extended Debye–Hückel equation 6-9 predicts that each γ will decrease as μ increases. The quotient R should therefore increase as the ionic strength increases.

Experimental values of R are shown in Figure 6-4. As the ionic strength is increased by addition of KCl, R increases up to an ionic strength near 0.5 M. Above this ionic strength the quotient R decreases. The only way this can happen is if the activity coefficients in Equation 6-13 *increase* above an ionic strength of 0.5 M.

At high ionic strength, γ increases with increasing μ.

This effect is general. Above an ionic strength of approximately 1 M, most activity coefficients increase. The ionic atmosphere model cannot account for this. A model that is successful for ionic strengths as high as 3 M is based on the change in dielectric constant of the solution in the immediate vicinity of each ion in the solution.[†] We will not attempt any activity coefficient calculations for ionic strengths above 0.1 M.

[†] L. W. Bahe, *J. Phys. Chem.,* **76**, 1062 (1972); L. W. Bahe and D. Parker, *J. Amer. Chem. Soc.,* **97**, 5664 (1975).

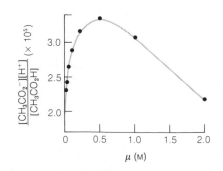

Figure 6-4

Value of R in Equation 6-13 as a function of increasing KCl concentration in an aqueous solution of acetic acid. [Data from H. S. Harned and F. C. Hickey, *J. Amer. Chem. Soc.,* **59**, 2303 (1937).]

Mean Activity Coefficient

The theories dealing with activity coefficients derive coefficients for individual ions. In most experiments, ions are available only in pairs, and the **mean activity coefficient,** $\gamma_\pm$, is derived from various measurements. For a salt with the stoichiometry (cation)$_m$(anion)$_n$, the mean activity coefficient is related to the individual coefficients by the equation.

$$\gamma_\pm^{m+n} = \gamma_+^m \gamma_-^n \qquad (6\text{-}14)$$

where γ_+ is the activity coefficient of the cation and γ_- is the activity coefficient of the anion.

Prediction of $\gamma_\pm$ for La(NO$_3$)$_3$ from the individual coefficients for La^{3+} and NO$_3^-$ in Table 6-1 at $\mu = 0.1$ M:

$$\gamma_\pm = [(0.18)^1(0.755)^3]^{1/4} = 0.53$$

Observed value $= 0.59$.

6-3 USING ACTIVITY COEFFICIENTS

This section presents three examples illustrating the proper use of activity coefficients in equilibrium calculations. The general prescription is trivial: *Write each equilibrium constant with activities in place of concentrations. Use the ionic strength of the solution to find the correct values of activity coefficients.*

   In the first example, we will calculate the concentration of Ca^{2+} in a 0.012 5 M solution of MgSO$_4$ saturated with CaF$_2$. The relevant equilibrium is

$$CaF_2(s) \rightleftharpoons Ca^{2+} + 2F^- \qquad K_{sp} = 3.9 \times 10^{-11} \qquad (6\text{-}15)$$

The equilibrium expression is set up with the aid of a little table:

	CaF$_2$(s) $\rightleftharpoons$ Ca^{2+} +	2F$^-$
Initial concentration:	solid	0	0
Final concentration:	solid	x	$2x$

$$K = \mathscr{A}_{Ca^{2+}}\mathscr{A}_{F^-}^2 = [Ca^{2+}]\gamma_{Ca^{2+}}[F^-]^2\gamma_{F^-}^2$$
$$= [x]\gamma_{Ca^{2+}}[2x]^2\gamma_{F^-}^2 \qquad (6\text{-}16)$$

To find the values of γ to use in Equation 6-16, we need to calculate the ionic strength. The ionic strength is due to the dissolved MgSO$_4$ *and* the dissolved CaF$_2$. However, K_{sp} for CaF$_2$ is quite small, and we will start by guessing that the contribution of CaF$_2$ to the ionic strength will be negligible. For 0.012 5 M MgSO$_4$, we calculate

$$\mu = \tfrac{1}{2}([0.012\,5] \cdot 2^2 + [0.012\,5] \cdot (-2)^2) = 0.050\,0 \text{ M} \qquad (6\text{-}17)$$

Using $\mu = 0.050\,0$ M, we look at Table 6-1 to find $\gamma_{Ca^{2+}} = 0.485$ and $\gamma_{F^-} = 0.81$. Substituting into Equation 6-16 gives

$$3.9 \times 10^{-11} = [x](0.485)[2x]^2(0.81)^2 \qquad (6\text{-}18)$$
$$x = [Ca^{2+}] = 3.1 \times 10^{-4} \text{ M}$$

Note that [F$^-$] and γ_{F^-} are both squared.

Question: What is the value of μ, including the contribution of CaF_2?

Our assumption was correct; the contribution of CaF_2 to the ionic strength is negligible.

As the next example, let's find the concentration of Ca^{2+} in a 0.050 M NaF solution saturated with CaF_2. The ionic strength is again 0.050 M, this time owing to the presence of NaF.

	$CaF_2(s) \rightleftharpoons Ca^{2+} +$		$2F^-$
Initial concentration:	solid	0	0.050
Final concentration:	solid	x	$2x + 0.050$

Assuming that $2x \ll 0.050$, we solve the problem as follows:

Question: Why is the solubility of CaF_2 lower in this solution than in $MgSO_4$ solution?

$$K_{sp} = [Ca^{2+}]\gamma_{Ca^{2+}}[F^-]^2\gamma_{F^-}^2$$

$$3.9 \times 10^{-11} = (x)(0.485)(0.050)^2(0.81)^2 \qquad (6\text{-}19)$$

$$x = [Ca^{2+}] = 4.9 \times 10^{-8}$$

As a final example, we calculate the solubility of LiF in distilled water:

$$LiF(s) \rightleftharpoons Li^+ + F^- \qquad K_{sp} = 1.7 \times 10^{-3} \qquad (6\text{-}20)$$

The ionic strength is determined by the concentration of LiF *dissolved in the solution.* We do not know this concentration yet. As a first approximation, we will calculate the concentrations of Li^+ and F^- by neglecting activity coefficients. Calling x_1 our first approximation for $[Li^+]$ (and $[F^-]$), we can write

$$K_{sp} \approx [Li^+][F^-] = x_1^2 \quad \Rightarrow \quad x_1 = [Li^+] = [F^-] = 0.041 \qquad (6\text{-}21)$$

This is a method of *successive approximations*. Each cycle of calculations is called one *iteration*.

As a second approximation, we will assume that $\mu = 0.041$ M, which is the result of our first approximation. Interpolating in Table 6-1 gives $\gamma_{Li^+} = 0.851$ and $\gamma_{F^-} = 0.830$ for $\mu = 0.041$ M. Putting these values into the expression for K_{sp} gives

$$K_{sp} = [Li^+]\gamma_{Li^+}[F^-]\gamma_{F^-}$$

$$= [x_2](0.851)[x_2](0.830) \quad \Rightarrow \quad x_2 = 0.049 \text{ M} \qquad (6\text{-}22)$$

As a third approximation, we assume that $\mu = 0.049$ M. Using this to find new values of activity coefficients in Table 6-1 gives

$$K_{sp} = [x_3](0.837)[x_3](0.812) \quad \Rightarrow \quad x_3 = 0.050 \text{ M} \qquad (6\text{-}23)$$

Using $\mu = 0.050$ M gives a fourth approximation:

$$K_{sp} = [x_4](0.835)[x_4](0.81) \quad \Rightarrow \quad x_4 = 0.050 \text{ M} \qquad (6\text{-}24)$$

As LiF dissolves, it increases the ionic strength and increases its own solubility.

The fourth answer is the same as the third answer. We have reached a self-consistent result, which must therefore be correct.

The definition of pH given in Chapter 5, pH $\approx -\log[H^+]$, is not quite correct. The real definition is

$$pH = -\log \mathscr{A}_{H^+} = -\log[H^+]\gamma_{H^+} \qquad (6\text{-}25)$$

The real definition of pH!

When we measure pH with a pH meter, we are measuring the negative logarithm of the hydrogen ion *activity,* not the concentration.

EXAMPLE: pH of Pure Water at 25°C

Let's calculate the pH of pure water by using activity coefficients correctly.

The relevent equilibrium is

$$H_2O \overset{K_w}{\rightleftharpoons} H^+ + OH^- \qquad (6\text{-}26)$$

$$K_w = \mathscr{A}_{H^+}\mathscr{A}_{OH^-} = [H^+]\gamma_{H^+}[OH^-]\gamma_{OH^-} \qquad (6\text{-}27)$$

The stoichiometry of the reaction tells us that H^+ and OH^- are produced in a 1:1 mole ratio. Their concentrations must therefore be equal. Calling each concentration x, we can write

$$K_w = 1.0 \times 10^{-14} = (x)\gamma_{H^+}(x)\gamma_{OH^-}$$

But the ionic strength of pure water is so small that it is reasonable to expect that $\gamma_{H^+} = \gamma_{OH^-} = 1$. Using these values in the preceding equation gives

$$1.0 \times 10^{-14} = x^2 \quad \Rightarrow \quad x = 1.0 \times 10^{-7} \text{ M}$$

The concentrations of H^+ and OH^- are both 1.0×10^{-7} M. Their activities are also both 1.0×10^{-7} because each activity coefficient is very close to 1.00. The pH is

$$pH = -\log[H^+]\gamma_{H^+} = -\log(1.0 \times 10^{-7})(1.00) = 7.00$$

EXAMPLE: pH of Water Containing a Salt

Now let's calculate the pH of water containing 0.10 M KCl at 25°C.

Reaction 6-26 tells us that $[H^+] = [OH^-]$. However, the values of γ in Equation 6-27 are not equal. The ionic strength of 0.10 M KCl is 0.10 M. According to Table 6-1, the activity coefficients of H^+ and OH^- are 0.83 and 0.76, respectively, when the ionic strength is 0.10 M. Putting these values into Equation 6-27 gives

$$K_w = [H^+]\gamma_{H^+}[OH^-]\gamma_{OH^-}$$

$$1.0 \times 10^{-14} = (x)(0.83)(x)(0.76)$$

$$x = 1.26 \times 10^{-7} \text{ M}$$

The concentrations of H^+ and OH^- are equal and are both greater than 1.0×10^{-7} M. The activities of H^+ and OH^- are not equal in this solution:

$$\mathscr{A}_{H^+} = [H^+]\gamma_{H^+} = (1.26 \times 10^{-7})(0.83) = 1.05 \times 10^{-7}$$

$$\mathscr{A}_{OH^-} = [OH^-]\gamma_{OH^-} = (1.26 \times 10^{-7})(0.76) = 0.96 \times 10^{-7}$$

Finally, we calculate the pH:

$$pH = -\log \mathscr{A}_{H^+} = -\log(1.05 \times 10^{-7}) = 6.98$$

These two examples show that the pH of water changes from 7.00 to 6.98 when we add 0.10 M KCl. KCl is not an acid or a base. The small change in pH arises because KCl affects the activities of H^+ and OH^-. The pH change of 0.02 units lies at the current limit of accuracy of pH measurements and is hardly important. However, the *concentration* of H^+ in 0.10 M KCl (1.26×10^{-7} M) is 26% greater than the concentration of H^+ in pure water (1.00×10^{-7} M).

The change in pH of water when a salt is added is an example of a *matrix effect*. The **matrix** is the medium containing whatever we are interested in. In this case, we are interested in H^+ in a matrix of 0.10 M KCl. The matrix changes the activity of H^+, even though there is no direct chemical reaction between H^+ and KCl. We will find instances throughout this book where the matrix can have significant effects on chemical analysis.

Summary

The true thermodynamic equilibrium constant for the reaction $aA + bB \rightleftharpoons cC + dD$ is $K = \mathscr{A}_C^c \mathscr{A}_D^d / \mathscr{A}_A^a \mathscr{A}_B^b$, where $\mathscr{A}_i$ is the activity of the ith species. The activity is the product of the concentration and the activity coefficient: $\mathscr{A}_i = c_i \gamma_i$. For nonionic compounds and gases, we will assume that $\gamma_i = 1$. For ionic species the activity coefficient depends on the ionic strength, defined as $\mu = \frac{1}{2}\sum c_i z_i^2$. The activity coefficient decreases as the ionic strength increases, at least for low ionic strengths. The extent of dissociation of ionic compounds increases with ionic strength because the ionic atmosphere of each ion diminishes the attraction of ions for each other. You should be able to perform equilibrium calculations correctly using activities instead of concentrations. You should be able to estimate activity coefficients by interpolation in Table 6-1.

Terms to Understand

activity	hydrated radius	matrix
activity coefficient	ionic atmosphere	mean activity coefficient
extended Debye–Hückel equation	ionic strength	pH
fugacity		

Exercises

6-A. Calculate the ionic strength of
  (a) 0.02 M KBr
  (b) 0.02 M Cs_2CrO_4
  (c) 0.02 M $MgCl_2$ plus 0.03 M $AlCl_3$

6-B. Find the activity (not the activity coefficient) of the $(CH_3CH_2CH_2)_4N^+$ ion in a solution containing 0.005 0 M $(CH_3CH_2CH_2)_4N^+Br^-$ plus 0.005 0 M $(CH_3)_4N^+Cl^-$.

6-C. Using activities, find the solubility of AgSCN (moles of Ag^+/L) in (a) 0.060 M KNO_3 (b) 0.060 M KSCN

6-D. Using activities, find the concentration of OH^- in a solution of 0.025 M $CaCl_2$ saturated with $Mn(OH)_2$.

6-E. Calculate the mean activity coefficient for $MgCl_2$ at a concentration of 0.020 M from the data in Table 6-1.

6-F. Using activities correctly, calculate the pH and concentration of H^+ in pure water containing 0.050 M LiBr at 25°C.

Problems

A6-1. Which statements are true: In the ionic strength range 0–0.1 M, activity coefficients *decrease* with
(a) increasing ionic strength
(b) increasing ionic charge
(c) decreasing hydrated radius

A6-2. Calculate the ionic strength of
(a) $0.000\,2$ M $La(IO_3)_3$
(b) 0.02 M $CuSO_4$
(c) 0.01 M $CuCl_2 + 0.05$ M $(NH_4)_2SO_4$ $+ 0.02$ M $Gd(NO_3)_3$

A6-3. Find the activity coefficient of each ion at the indicated ionic strength.
(a) SO_4^{2-} $(\mu - 0.01$ M$)$
(b) Sc^{3+} $(\mu = 0.005$ M$)$
(c) Eu^{3+} $(\mu = 0.1$ M$)$
(d) $(CH_3CH_2)_3NH^+$ $(\mu = 0.05$ M$)$

A6-4. Use interpolation in Table 6-1 to find the activity coefficient of H^+ when $\mu = 0.030$ M.

A6-5. Calculate the solubility of Hg_2Br_2 (expressed as moles of Hg_2^{2+} per liter) in
(a) $0.000\,33$ M $Mg(NO_3)_2$
(b) $0.003\,3$ M $Mg(NO_3)_2$
(c) 0.033 M $Mg(NO_3)_2$

A6-6. Find the activity coefficient of H^+ in a solution containing 0.010 M HCl plus 0.040 M $KClO_4$. What is the pH of the solution?

6-7. Suggest a reason why the hydrated radii decrease in the order $Sn^{4+} > In^{3+} > Cd^{2+} > Rb^+$.

6-8. Calculate the activity coefficient of Zn^{2+} when $\mu = 0.083$ M by
(a) using Equation 6-9
(b) using linear interpolation with Table 6-1

6-9. The observed mean activity coefficient $(\gamma_\pm)$ for HCl at a concentration of 0.005 M is 0.93. What is the value of $\gamma_\pm$ calculated from Table 6-1?

6-10. Find the concentration of Ba^{2+} in a $0.033\,3$ M $Mg(IO_3)_2$ solution saturated with $Ba(IO_3)_2$. 1.5×10^{-9}

6-11. Calculate the concentration of Pb^{2+} in a saturated solution of PbF_2 in water.

6-12. Using activities correctly, calculate the pH of a solution containing 0.010 M NaOH plus $0.004\,0$ M $Ca(NO_3)_2$. What would be the pH if you neglected activities?

6-13. The equilibrium constant for dissolution in water of a nonionic compound, such as diethyl ether $(CH_3CH_2OCH_2CH_3)$, can be written

$$\text{ether } (l) \rightleftharpoons \text{ether } (aq) \qquad K = [\text{ether } (aq)]\gamma_{\text{ether}}$$

At low ionic strength, $\gamma \approx 1$ for all nonionic compounds. At high ionic strength, ether and most other neutral molecules can be *salted out* of aqueous solution. That is, when a high concentration (typically >1 M) of an ionic compound (such as NaCl) is added to an aqueous solution, neutral molecules usually become *less* soluble. Does the activity coefficient, γ_{ether}, increase or decrease at high ionic strength?

6-14. The temperature-dependent form of the extended Debye–Hückel equation 6-9 is

$$\log \gamma = \frac{(-1.825 \times 10^6)(\varepsilon T)^{-3/2}z^2\sqrt{\mu}}{1 + \alpha\sqrt{\mu}/(2.00\sqrt{\varepsilon T})}$$

where ε is the (dimensionless) dielectric constant[†] of water, T is kelvins, z is the charge of the ion of interest, μ is the ionic strength of the solution (mol/L), and α is the size of the ion in picometers. The dependence of ε on temperature is given by

$$\varepsilon = 79.755e^{(-4.6 \times 10^{-3})(T - 293.15)}$$

Calculate the activity coefficient of SO_4^{2-} at $50.00°C$ when $\mu = 0.100$ M. Compare your value with the one in Table 6-1.

6-15. An empirical equation describing activity coefficients at ionic strengths higher than 0.1 M is

$$\log \gamma = \frac{-A\sqrt{\mu}}{1 + B\sqrt{\mu}} + C\mu$$

where γ is the activity coefficient, μ is the ionic strength, and A, B, and C are constants characteristic of a particular solution. An approximate measure of single ion activity coefficients can be made with ion-selective electrodes. One set of measurements for Cu^{2+} [I. Uemasu and Y. Umezawa, *Anal. Chem.*, **55**, 386 (1983)] gives $A = 3.23$ (± 0.32), $B = 2.57$ (± 0.32), and $C = 0.198$ (± 0.012). Calculate the activity coefficient of Cu^{2+} at the following ionic strengths: 0.001, 0.01, 0.1, 0.5, 1.0, 1.5, 2.0, and 3.0 M. Make a graph of γ versus μ and compare your values of γ with those in Table 6-1.

[†] The dielectric constant of a solvent is a measure of how well that solvent can separate oppositely charged ions. The force of attraction (in newtons) between two ions of charge q_1 and q_2 (in coulombs) separated by a distance r (in meters) is

$$\text{Force} = -(8.988 \times 10^9)\frac{q_1q_2}{\varepsilon r^2}$$

where ε is the (dimensionless) dielectric constant. The larger the value of ε, the smaller the attraction between ions. Water, with $\varepsilon \approx 80$, separates ions very well. The value of ε for benzene is 2, and the value of ε for vacuum is 1.

7 Systematic Treatment of Equilibrium

Many chemical equilibrium problems are exceedingly complex because of the large number of chemical reactions and species involved. This text focuses on some of the most common (and simplest) types of equilibria. Even so, there are times when we must resort to a powerful procedure that allows us to deal with a range of equilibrium problems from the simplest to the very difficult. In favorable cases, this *systematic treatment of equilibrium* can be applied with little arithmetic difficulty. In complicated cases, computers must be employed to solve the equations.

The basic procedure in the systematic treatment is to write as many algebraic equations as there are unknowns (species) in the problem. By working with n equations and n unknowns, the problem can, in principle, be solved. Sometimes the solution is simple, but more often it is not. The n equations are generated by writing all the chemical equilibrium conditions plus two more: the balances of charge and of mass. The following two sections describe these latter relationships.

7-1 CHARGE BALANCE

Solutions must have zero total charge.

The **charge balance** is an algebraic statement of electroneutrality of the solution. That is, *the sum of the positive charges in solution equals the sum of the negative charges in solution.*

Consider a sulfate ion with concentration 0.016 7 M. Since the charge on SO_4^{2-} is -2, the charge contributed by sulfate is $(-2)(0.016\ 7) = -0.033\ 4$ M. In general, an ion with a charge of $\pm n$ and a concentration $[A]$ will contribute $\pm n[A]$ to the charge of the solution. The charge balance equates the magnitude of the total positive charge to the magnitude of the total negative charge.

Suppose that a solution contains the following ionic species: H^+, OH^-, K^+, $H_2PO_4^-$, HPO_4^{2-}, and PO_4^{3-}. The charge balance is written

$$[H^+] + [K^+] = [OH^-] + [H_2PO_4^-] + 2[HPO_4^{2-}] + 3[PO_4^{3-}] \quad (7\text{-}1)$$

This statement says that the total charge contributed by H^+ and K^+ equals the magnitude of the charge contributed by all of the anions on the right side of the equation. *The coefficient in front of each species always equals the magnitude of the charge on the ion.* This is because a mole of, say, PO_4^{3-} contributes three moles of negative charge. If $[PO_4^{3-}] = 0.01$ M, the negative charge is $3[PO_4^{3-}] = 3(0.01) = 0.03$ M.

The charge balance, Equation 7-1, appears unbalanced to many people. "The right side of the equation has much more charge than the left side!" you may think. But that is wrong.

For example, consider a solution prepared by weighing out 0.025 0 mol of KH_2PO_4 plus 0.030 0 mol of KOH and diluting to 1.00 L. The concentrations of the species at equilibrium are calculated to be

$$[H^+] = 3.9 \times 10^{-12} \text{ M} \qquad [H_2PO_4^-] = 1.4 \times 10^{-6} \text{ M}$$

$$[K^+] = 0.055\,0 \text{ M} \qquad [HPO_4^{2-}] = 0.022\,6 \text{ M}$$

$$[OH^-] = 0.002\,6 \text{ M} \qquad [PO_4^{3-}] = 0.002\,4 \text{ M}$$

This calculation, which you should be able to do when you have finished studying acids and bases, takes into account the reaction of OH^- with $H_2PO_4^-$ to produce HPO_4^{2-} and PO_4^{3-}.

Are the charges balanced? Yes indeed. Plugging into Equation 7-1, we find

$$[H^+] + [K^+] = [OH^-] + [H_2PO_4^-] + 2[HPO_4^{2-}] + 3[PO_4^{3-}]$$

$$3.9 \times 10^{-12} + 0.055\,0 = 0.002\,6 + 1.4 \times 10^{-6} + 2(0.022\,6) + 3(0.002\,4)$$

$$0.055\,0 = 0.055\,0 \qquad (7\text{-}2)$$

The total positive charge (to three figures) is 0.055 0 M, and the total negative charge is also 0.055 0 M. The charges must be balanced in every solution. Otherwise your beaker with excess positive charge would glide across the lab bench and smash into another beaker with excess negative charge.

The general form of the charge balance for any solution is

$$n_1[C_1] + n_2[C_2] + \cdots = m_1[A_1] + m_2[A_2] + \cdots \qquad (7\text{-}3)$$

where $[C_i]$ = concentration of the *i*th cation

$\quad n_i$ = charge of the *i*th cation

$\quad [A_i]$ = concentration of the *i*th anion

$\quad m_i$ = magnitude of the charge of the *i*th anion

The coefficient of each term in the charge balance equals the magnitude of the charge on each ion.

For the force between beakers of "charged solutions," see Problem 7-23.

$\sum$ [positive charges] $= \sum$ [negative charges]. *Activity coefficients do not appear in the charge balance. The charge contributed by 0.1 M H^+ is exactly 0.1 M. Think about this.*

Write the charge balance for a solution containing H_2O, H^+, OH^-, ClO_4^-, $Fe(CN)_6^{3-}$, CN^-, Fe^{3+}, Mg^{2+}, CH_3OH, HCN, NH_3, and NH_4^+.

The correct equation is

$$[H^+] + 3[Fe^{3+}] + 2[Mg^{2+}] + [NH_4^+] = [OH^-] + [ClO_4^-] + 3[Fe(CN)_6^{3-}] + [CN^-]$$

Neutral species (H_2O, CH_3OH, HCN, and NH_3) do not appear in the charge balance.

7-2 MASS BALANCE

The mass balance is a statement of the conservation of matter. It really refers to conservation of atoms, not to mass.

The **mass balance,** also called the material balance, is a statement of the conservation of matter. The mass balance states that *the sum of the amounts of all species containing a particular atom (or group of atoms) must equal the amount of that atom (or group) delivered to the solution.* It is easier to see this through particular examples than by a general statement.

Suppose that a solution is prepared by dissolving 0.050 mol of acetic acid in water to give a total volume of 1.00 L. The acetic acid will partially dissociate into acetate:

$$CH_3CO_2H \rightleftharpoons CH_3CO_2^- + H^+ \qquad (7\text{-}4)$$
$$\text{Acetic acid} \qquad \text{Acetate}$$

The mass balance is simply a statement that the sum of the amount of dissociated and undissociated acid must equal the amount of acid put into the solution.

Mass balance: $\quad 0.050 \text{ M} = [CH_3CO_2H] + [CH_3CO_2^-] \qquad (7\text{-}5)$

Activity coefficients do not appear in the mass balance. The concentration of each species gives an exact count of the number of atoms of that species.

Phosphoric acid (H_3PO_4) can dissociate into $H_2PO_4^-$, HPO_4^{2-}, and PO_4^{3-}. The mass balance for a solution prepared by dissolving 0.025 0 mol of H_3PO_4 in 1.00 L is

$$0.025\ 0 \text{ M} = [H_3PO_4] + [H_2PO_4^-] + [HPO_4^{2-}] + [PO_4^{3-}] \qquad (7\text{-}6)$$

Write the mass balances for K^+ and for phosphate in a solution prepared by mixing 0.025 0 mol KH_2PO_4 plus 0.030 0 mol KOH and diluting to 1.00 L.

The total concentration of K^+ is 0.025 0 M + 0.030 0 M, so one trivial mass balance is

$$[K^+] = 0.055\ 0 \text{ M}$$

The total concentration of *all forms* of phosphate is 0.025 0M. The mass balance for phosphate is

$$[H_3PO_4] + [H_2PO_4^-] + [HPO_4^{2-}] + [PO_4^{3-}] = 0.025\ 0 \text{ M}$$

Now consider a solution prepared by dissolving $La(IO_3)_3$ in water.

$$La(IO_3)_3(s) \xrightleftharpoons{K_{sp}} La^{3+} \rightarrow 3IO_3^- \qquad (7\text{-}7)$$

$$\text{Iodate}$$

We do not know how much La^{3+} or IO_3^- is dissolved, but we do know that there must be three iodate ions for each lanthanum ion dissolved. The mass balance is

$$[IO_3^-] = 3[La^{3+}] \qquad (7\text{-}8)$$

We have already used this sort of relation in solubility problems. Perhaps the following table will jog your memory:

	$La(IO_3)_3(s) \rightleftharpoons La^{3+} + 3IO_3^-$		
Initial concentration:	solid	0	0
Final concentration:	solid	x	$3x$

When we set $[IO_3^-] = 3x$, we are saying that $[IO_3^-] = 3[La^{3+}]$.

EXAMPLE: Mass Balance When the Total Concentration Is Unknown

Write the mass balance for a saturated solution of the slightly soluble salt Ag_3PO_4, which produces PO_4^{3-} plus $3Ag^+$ when it dissolves.

If the phosphate in solution remained as PO_4^{3-}, we could write

$$[Ag^+] = 3[PO_4^{3-}]$$

because three silver ions are produced for each phosphate ion. However, since phosphate reacts with water to give HPO_4^{2-}, $H_2PO_4^-$, and H_3PO_4, the correct mass balance is

$$[Ag^+] = 3\{[PO_4^{3-}] + [HPO_4^{2-}] + [H_2PO_4^-] + [H_3PO_4]\} \qquad \text{Atoms of Ag} = 3 \text{ (atoms of P)}$$

That is, the number of atoms of Ag^+ must equal three times the total number of atoms of phosphorus, regardless of how many species contain phosphorus atoms.

7-3 SYSTEMATIC TREATMENT OF EQUILIBRIUM

General Prescription

STEP 1. Write all the pertinent chemical reactions.

STEP 2. Write the charge balance.

STEP 3. Write the mass balance.

STEP 4. Write the equilibrium constant for each chemical reaction. This is the only step in which activity coefficients figure.

Step 6 does not necessarily imply doom and gloom. In this text we will emphasize simplifications that allow you to deal with equilibrium problems by hand. In real life you can use commercially available computer software such as *Eureka, MathCAD,* or *TK Solver Plus*[†] to solve complicated systems of simultaneous equations. For an example of how to use equation-solving software, see F. T. Chau and A. S. W. Chik, *J. Chem. Ed.,* **66,** A61 (1989).

STEP 5. Count the equations and unknowns. At this point you should have as many equations as unknowns (chemical species). If not, you must either write more equilibria or fix some concentrations at known values.

STEP 6. By hook or by crook, solve for all the unknowns.

Steps 1 and 6 are usually the heart of the problem. Knowing (or guessing) what chemical equilibria exist in a given solution requires a fair degree of chemical intuition. In this text you will usually be given some help with Step 1. Unless we know all the relevant equilibria, it is not possible to correctly calculate the composition of a solution. Because of not knowing all the chemical reactions, we undoubtedly oversimplify many equilibrium problems.

Step 6 is a mathematical problem, not a chemical problem. In some cases it is easy to solve for all the unknowns, but for most problems a computer must be employed and/or approximations must be made. How the systematic treatment of equilibrium works is best understood by studying some examples.

Ionization of Water

The dissociation of water into H^+ and OH^- occurs in every aqueous solution:

$$H_2O \xrightleftharpoons{K_w} H^+ + OH^- \qquad K_w = 1.0 \times 10^{-14} \text{ at } 25°C \qquad (7\text{-}9)$$

Let us apply the systematic treatment of equilibrium to find the concentrations of H^+ and OH^- in pure water.

STEP 1. Pertinent reactions. The only one is Reaction 7-9.

STEP 2. Charge balance. The only ions are H^+ and OH^-, so the charge balance is

$$[H^+] = [OH^-] \qquad (7\text{-}10)$$

STEP 3. Mass balance. In Reaction 7-9 we see that one H^+ ion is generated each time one OH^- ion is made. The mass balance is simply

$$[H^+] = [OH^-]$$

which is the same as the charge balance for this particularly trivial system.

STEP 4. Write the equilibrium constants. The only one is

$$K_w = [H^+]\gamma_{H^+}[OH^-]\gamma_{OH^-} = 1.0 \times 10^{-14} \qquad (7\text{-}11)$$

This is the only step in which activity coefficients are introduced into the problem.

[†] For information about these programs contact, for *Eureka,* Borland International, 4585 Scotts Valley Drive, Scotts Valley, CA 95066; for *MathCAD,* Mathsoft, Inc., One Kendall Square, Cambridge MA 02139; and for *TK Solver Plus,* Technical Systems, 1220 Rock Street, Rockford IL 61101.

STEP 5. Count the equations and unknowns. We have two equations, 7-10 and 7-11, and two unknowns, $[H^+]$ and $[OH^-]$.

It requires n equations to solve for n unknowns.

STEP 6. Solve.

Now we must stop and decide what to do about the activity coefficients. Later, it will be our custom to ignore them unless the calculation requires an accurate result. In the present problem, we anticipate that the ionic strength of pure water will be very low (since the only ions are small amounts of H^+ and OH^-). Therefore, it is reasonable to suppose that γ_{H^+} and γ_{OH^-} are both unity, since $\mu \approx 0$.

Putting the equality $[H^+] = [OH^-]$ into Equation 7-11 gives

Recall that γ approaches 1 as μ approaches 0.

$$[H^+]\gamma_{H^+}[OH^-]\gamma_{OH^-} = 1.0 \times 10^{-14} \qquad (7\text{-}12)$$

$$[H^+] \cdot 1 \cdot [H^+] \cdot 1 = 1.0 \times 10^{-14} \qquad (7\text{-}13)$$

$$[H^+] = 1.0 \times 10^{-7} \text{ M}$$

Since $[H^+] = [OH^-]$, $[OH^-] = 1.0 \times 10^{-7}$ M also. We have solved the problem. Just as a reminder, the pH is given by

The ionic strength is 10^{-7} M. The assumption that $\gamma_{H^+} = \gamma_{OH^-} = 1$ is good.

$$pH = -\log \mathscr{A}_{H^+} = -\log[H^+]\gamma_{H^+} \qquad (7\text{-}14)$$

$$= -\log(1.0 \times 10^{-7})(1) \qquad (7\text{-}15)$$

$$= 7.00$$

This is the correct definition of pH. When we neglect activity coefficients, we will write $pH = -\log[H^+]$.

A note about activity coefficients

Although it is proper to write all equilibrium constants in terms of activities, the algebraic complexity of manipulating the activity coefficients often obscures the chemistry of a problem. For the remainder of this text we will omit activity coefficients unless there is a particular point to be made with them. There will be occasional problems in which activity is used. This provides you with an occasional reminder of activities. Alternatively, there is no loss in continuity if the activity problems are skipped.

Solubility of Hg_2Cl_2

Let's study one application of the systematic treatment of equilibrium by calculating the concentration of Hg_2^{2+} in a saturated solution of Hg_2Cl_2.

STEP 1. Pertinent reactions. The two reactions that come to mind are

$$Hg_2Cl_2(s) \underset{}{\overset{K_{sp}}{\rightleftharpoons}} Hg_2^{2+} + 2Cl^- \qquad (7\text{-}16)$$

$$H_2O \underset{}{\overset{K_w}{\rightleftharpoons}} H^+ + OH^- \qquad (7\text{-}17)$$

The equilibrium given by Reaction 7-17 exists in every aqueous solution.

Multiply $[Hg_2^{2+}]$ by two because one mole of this ion has two moles of charge.

STEP 2. Charge balance. Equating positive and negative charges gives

$$[H^+] + 2[Hg_2^{2+}] = [Cl^-] + [OH^-] \qquad (7\text{-}18)$$

STEP 3. Mass balance. There are two mass balances in this system. One is the trivial statement that $[H^+] = [OH^-]$, since both arise only from the ionization of water. If there were any other reactions involving H^+ or OH^-, we could not automatically say $[H^+] = [OH^-]$. A slightly more interesting mass balance is

Two Cl^- ions are produced for each Hg_2^{2+} ion.

$$[Cl^-] = 2[Hg_2^{2+}] \qquad (7\text{-}19)$$

STEP 4. The equilibrium constants are

$$[Hg_2^{2+}][Cl^-]^2 = 1.2 \times 10^{-18} \qquad (7\text{-}20)$$

$$[H^+][OH^-] = 1.0 \times 10^{-14} \qquad (7\text{-}21)$$

We have neglected the activity coefficients in these equations.

STEP 5. There are four equations (7-18 to 7-21) and four unknowns: $[H^+]$, $[OH^-]$, $[Hg_2^{2+}]$, and $[Cl^-]$.

STEP 6. Since we have not written any chemical reactions between the ions produced by H_2O and the ions produced by Hg_2Cl_2, there are really two separate problems. One is the trivial problem of water ionization, which we already solved:

$$[H^+] = [OH^-] = 1.0 \times 10^{-7} \text{ M} \qquad (7\text{-}22)$$

A slightly more interesting problem is the Hg_2Cl_2 equilibrium. Noting that $[Cl^-] = 2[Hg_2^{2+}]$, we can write

$$[Hg_2^{2+}][Cl^-]^2 = [Hg_2^{2+}](2[Hg_2^{2+}])^2 = K_{sp} \qquad (7\text{-}23)$$

$$[Hg_2^{2+}] = (K_{sp}/4)^{1/3} = 6.7 \times 10^{-7} \text{ M}$$

Satisfy yourself that the calculated concentrations fulfill the charge balance condition.

This result is exactly what you would have found by writing a table of concentrations and solving by the methods of Chapter 5.

7-4 DEPENDENCE OF SOLUBILITY ON pH

We are now in a position to study some cases that are not trivial. In this section we present two examples in which there are *coupled equilibria*. That is, the product of one reaction is a reactant in the next reaction.

Solubility of CaF$_2$

Let's set up the equations needed to find the solubility of CaF_2 in water. There are three pertinent reactions in this sytem. First $CaF_2(s)$ dissolves:

$$CaF_2(s) \underset{}{\overset{K_{sp}}{\rightleftharpoons}} Ca^{2+} + 2F^- \qquad (7\text{-}24)$$

The fluoride ion can then react with water to give HF(*aq*):

$$F^- + H_2O \xrightleftharpoons{K_b} HF(aq) + OH^- \qquad (7\text{-}25)$$

Finally, for every aqueous solution we can write

$$H_2O \xrightleftharpoons{K_w} H^+ + OH^- \qquad (7\text{-}26)$$

The equilibrium constant is called K_b because F^- is functioning as a base when it removes H^+ from H_2O.

If Reaction 7-25 occurs, the solubility of CaF_2 is greater than what is predicted by the solubility product. The reason is that a product of Reaction 7-24 is consumed in Reaction 7-25. According to Le Châtelier's principle, Reaction 7-24 will be driven to the right. The systematic treatment of equilibrium allows us to find the net effect of all three reactions.

STEP 1. The pertinent reactions are 7-24 through 7-26.

STEP 2. Charge balance:

$$[H^+] + 2[Ca^{2+}] = [OH^-] + [F^-] \qquad (7\text{-}27)$$

STEP 3. Mass balance: If all fluoride remained in the form F^-, we could write $[F^-] = 2[Ca^{2+}]$ from the stoichiometry of Reaction 7-24. But some F^- reacts to give HF. The total moles of fluorine atoms is equal to the sum of F^- plus HF. The mass balance is

$$\underbrace{[F^-] + [HF]}_{\substack{\text{Total concentration} \\ \text{of fluorine}}} = 2[Ca^{2+}] \qquad (7\text{-}28)$$

STEP 4. Equilibria:

$$K_{sp} = [Ca^{2+}][F^-]^2 = 3.9 \times 10^{-11} \qquad (7\text{-}29)$$

$$K_b = \frac{[HF][OH^-]}{[F^-]} = 1.5 \times 10^{-11} \qquad (7\text{-}30)$$

$$K_w = [H^+][OH^-] = 1.0 \times 10^{-14} \qquad (7\text{-}31)$$

STEP 5. We have five equations (7-27 through 7-31) and five unknowns: $[H^+]$, $[OH^-]$, $[Ca^{2+}]$, $[F^-]$, and $[HF]$.

The final step is to solve the problem, which is no simple matter for these five equations. Instead, let us ask a simpler question: "What will be the concentrations of $[Ca^{2+}]$, $[F^-]$, and $[HF]$ if the pH is somehow *fixed* at the value 3.00?" To fix the pH at 3.00 means that $[H^+] = 1.0 \times 10^{-3}$ M.

Once we know the value of $[H^+]$, a simple procedure for solving all of the equations is the following: Setting $[H^+] = 1.0 \times 10^{-3}$ M in Equation 7-31 gives

If you insist on solving the problem, see Box 7-1.

The pH could be fixed at a desired value by adding a buffer, which is discussed in Chapter 10.

$$[OH^-] = \frac{K_w}{[H^+]} = 1.0 \times 10^{-11} \qquad (7\text{-}32)$$

Box 7-1 ALL RIGHT, DAN, HOW WOULD YOU REALLY SOLVE THE CaF₂ PROBLEM?

Suppose that we did not simplify the CaF_2 problem by specifying a fixed pH. How can we find the composition of the system if it simply contains CaF_2 dissolved in water? A systematic guessing procedure is a pretty good way to do it.

First let's use a little intuition. F^- is a pretty weak base ($K_b = 1.5 \times 10^{-11}$, Equation 7-25) and there will not be very much F^- dissolved (since $K_{sp} = 3.9 \times 10^{-11}$, Equation 7-24). Therefore, a first guess is that the pH is in the neighborhood of 7.

Guessing that the pH is 7.00, we can calculate the concentration of each species and see if the charge balance, Equation 7-27, is satisfied. It is valid to use the charge balance now because we have not fixed the pH by addition of some other reagent. Assuming pH = 7.00, the concentrations of Ca^{2+} and F^- are calculated to be 2.14×10^{-4} M and 4.27×10^{-4} M, respectively.

$$[H^+] + 2[Ca^{2+}] = [OH^-] + [F^-] \tag{7-27}$$

$$10^{-7} + 2(2.14 \times 10^{-4}) > 10^{-7} + 4.27 \times 10^{-4}$$

A pH of 7.00 almost satisfies the charge balance. Raising the pH will increase the right side of the equation above and bring it into balance. A little trial-and-error guessing shows that a pH of 7.11 comes closest to satisfying the charge balance. When the charge balance is satisfied we *must* have found the correct pH and composition of the solution. Although this procedure is cumbersome to do by hand, it is well-suited to a computer.

Putting this value of $[OH^-]$ into Equation 7-30 gives

$$\frac{[HF]}{[F^-]} = \frac{K_b}{[OH^-]} = 1.5$$

$$[HF] = 1.5\,[F^-] \tag{7-33}$$

Substituting this expression for $[HF]$ into the mass balance (Equation 7-28) gives

$$[F^-] + [HF] = 2[Ca^{2+}]$$

$$[F^-] + 1.5[F^-] = 2[Ca^{2+}]$$

$$[F^-] = 0.80[Ca^{2+}] \tag{7-34}$$

Finally, we use this value of $[F^-]$ in the solubility product (Equation 7-29):

$$[Ca^{2+}][F^-]^2 = K_{sp}$$

$$[Ca^{2+}](0.80[Ca^{2+}])^2 = K_{sp}$$

$$[Ca^{2+}] = 3.9 \times 10^{-4} \text{ M} \tag{7-35}$$

Challenge: Use the concentration of Ca^{2+} that we just calculated to show that the concentrations of $[F^-]$ and $[HF]$ are 3.1×10^{-4} M and 4.7×10^{-4} M, respectively.

A graph of the pH dependence of the concentrations of Ca^{2+}, F^-, and HF is shown in Figure 7-1. At high pH there is very little HF, so

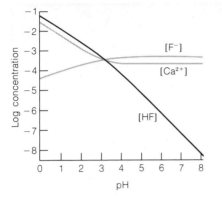

Figure 7-1
pH dependence shown by the concentrations of Ca^{2+}, F^-, and HF in a saturated solution of CaF_2. Note the logarithmic scale.

$[F^-] \approx 2[Ca^{2+}]$. At low pH there is very little F^-, so $[HF] \approx 2[Ca^{2+}]$. The concentration of Ca^{2+} increases at low pH, because Reaction 7-24 is drawn to the right by the reaction of F^- with H^+ to make HF. It is a general phenomenon that salts of basic ions will become more soluble at low pH, because the basic ions react with H^+. Some examples of basic ions are F^-, OH^-, S^{2-}, CO_3^{2-}, $C_2O_4^{2-}$, and PO_4^{3-}. For an example using the interaction of pH, solubility, and tooth decay, see Box 7-2.

High pH means there is a low concentration of H^+. Low pH means that there is a high concentration of H^+.

Finally, you should realize that the charge balance equation (7-27) is no longer valid if the pH is fixed by external means. To adjust the pH, an ionic compound must necessarily have been added to the solution. Equation 7-27 is therefore incomplete, since it omits those ions. We did not use Equation 7-27 to solve the problem.

Fixing the pH invalidates the original charge balance. There exists a new charge balance, but we do not know enough to write an equation expressing this fact.

Solubility of HgS

The mineral cinnabar consists of red HgS. When it dissolves in water, some reactions that may occur are

$$HgS(s) \rightleftharpoons Hg^{2+} + S^{2-} \qquad K_{sp} = 5 \times 10^{-54} \qquad (7\text{-}36)$$

$$S^{2-} + H_2O \rightleftharpoons HS^- + OH^- \qquad K_{b1} = 0.80 \qquad (7\text{-}37)$$

$$HS^- + H_2O \rightleftharpoons H_2S(aq) + OH^- \qquad K_{b2} = 1.1 \times 10^{-7} \qquad (7\text{-}38)$$

$$H_2O \rightleftharpoons H^+ + OH^- \qquad K_w = 1.0 \times 10^{-14} \qquad (7\text{-}39)$$

We are ignoring the equilibrium $H_2S(aq) \rightleftharpoons H_2S(g)$.

Because S^{2-} is a strong base, it reacts with H_2O to give HS^-, thereby drawing Reaction 7-36 to the right and increasing the solubility of HgS. Now let's set up the equations to find the composition of a saturated solution of HgS.

STEP 1. The pertinent reactions are 7-36 through 7-39. It is worth restating that if there are any other significant reactions, the calculated composition will be wrong. We are necessarily limited by our knowledge of the chemistry of the system.

STEP 2. Charge balance. If the pH is not adjusted by external means, the charge balance is

$$2[Hg^{2+}] + [H^+] = 2[S^{2-}] + [HS^-] + [OH^-] \qquad (7\text{-}40)$$

Box 7-2 pH AND TOOTH DECAY

The enamel covering of teeth contains the mineral *hydroxyapatite*, a calcium hydroxyphosphate. This slightly soluble mineral will dissolve in acid, because both the PO_4^{3-} and OH^- react with H^+:

$$Ca_{10}(PO_4)_6(OH)_2 + 14H^+ \rightleftharpoons 10Ca^{2+} + 6H_2PO_4^- + 2H_2O$$

Hydroxyapatite

The decay-causing bacteria that adhere to teeth produce lactic acid from the metabolism of sugar.

$$\underset{\text{Lactic acid}}{\overset{\overset{\displaystyle OH}{|}}{CH_3CHCO_2H}}$$

The lactic acid lowers the pH at the surface of the tooth to less than 5. At any pH below about 5.5, hydroxyapatite dissolves and tooth decay occurs, as shown in the electron micrographs below:

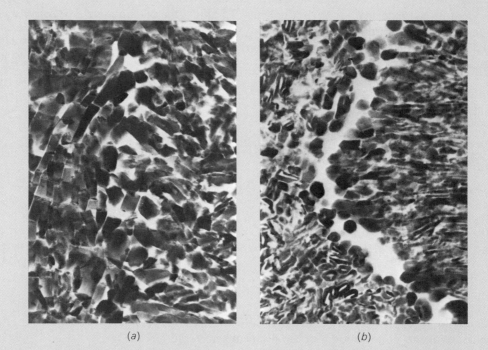

(a)   *(b)*

(a) Transmission electron micrograph of normal human tooth enamel, showing crystals of hydroxyapatite. (b) Transmission electron micrograph of decayed enamel, showing regions where the mineral has been dissolved by acid. [Courtesy D. B. Scott, J. W. Simmelink, and V. K. Nygaard, Case Western Reserve University, School of Dentistry. Originally published in *J. Dent. Res., 53,* 165 (1974).]

Fluoride inhibits tooth decay because it forms *fluorapatite*, $Ca_{10}(PO_4)_6F_2$, which is less soluble and more acid-resistant than hydroxyapatite.

STEP 3. Mass balance. For each atom of Hg in solution, there is one atom of S. If all of the S remained as S^{2-}, the mass balance would be $[Hg^{2+}] = [S^{2-}]$. But S^{2-} reacts with water to give HS^- and H_2S. The mass balance is

$$[Hg^{2+}] = [S^{2-}] + [HS^-] + [H_2S] \qquad (7\text{-}41)$$

STEP 4. Equilibria:

$$K_{sp} = [Hg^{2+}][S^{2-}] = 5 \times 10^{-54} \qquad (7\text{-}42)$$

$$K_{b1} = \frac{[HS^-][OH^-]}{[S^{2-}]} = 0.80 \qquad (7\text{-}43)$$

$$K_{b2} = \frac{[H_2S][OH^-]}{[HS^-]} = 1.1 \times 10^{-7} \qquad (7\text{-}44)$$

$$K_w = [H^+][OH^-] = 1.0 \times 10^{-14} \qquad (7\text{-}45)$$

STEP 5. There are six equations (7-40 through 7-45) and six unknowns: $[Hg^{2+}]$, $[S^{2-}]$, $[HS^-]$, $[H_2S]$, $[H^+]$, and $[OH^-]$.

STEP 6. Solve.

Once again the problem is very difficult to solve. We will simplify matters by assuming that the pH is *fixed* at 8.00 by some external means. This means that the charge balance (7-40) is invalid, but we can use the remaining equations to find the composition of the system.

The charge balance is invalid because we are adding new ions to adjust the pH to 8.00.

If the pH is 8.00, Equation 7-45 tells us that $[OH^-] = 1.0 \times 10^{-6}$ M. Putting this value into Equation 7-44 gives

$$[H_2S] = \frac{K_{b2}[HS^-]}{[OH^-]} = 0.11[HS^-] \qquad (7\text{-}46)$$

Substituting the value of $[OH^-]$ into Equation 7-43 gives

$$[HS^-] = \frac{K_{b1}[S^{2-}]}{[OH^-]} = 8.0 \times 10^5[S^{2-}] \qquad (7\text{-}47)$$

Now we can use these values of $[H_2S]$ and $[HS^-]$ in the mass balance (Equation 7-41) to write

$$[Hg^{2+}] = [S^{2-}] + [HS^-] + [H_2S]$$

$$[Hg^{2+}] = [S^{2-}] + 8.0 \times 10^5[S^{2-}] + 0.11[HS^-]$$

$$[Hg^{2+}] = [S^{2-}] + 8.0 \times 10^5[S^{2-}] + (0.11)(8.0 \times 10^5)[S^{2-}]$$

$$[Hg^{2+}] = [S^{2-}](8.88 \times 10^5) \qquad (7\text{-}48)$$

Putting this relation between $[Hg^{2+}]$ and $[S^{2-}]$ into the solubility product solves the problem:

$$K_{sp} = [Hg^{2+}][S^{2-}]$$

$$K_{sp} = [Hg^{2+}]\left(\frac{[Hg^{2+}]}{8.88 \times 10^5}\right) \quad \Rightarrow \quad [Hg^{2+}] = 2.1 \times 10^{-24} \text{ M} \qquad (7\text{-}49)$$

Challenge: Show that the other concentrations are
$[S^{2-}] = 2.4 \times 10^{-30}$ M,
$[HS^-] = 1.9 \times 10^{-24}$ M, and
$[H_2S] = 2.1 \times 10^{-25}$ M.

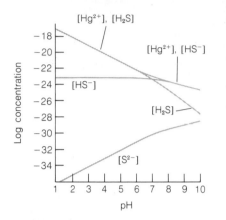

Figure 7-2
pH dependence shown by the concentrations of species in a saturated solution of HgS. Note the logarithmic scale.

A graph showing the effect of pH on the composition of a saturated solution of HgS is shown in Figure 7-2. Below pH ≈ 6, H_2S is the predominant form of sulfur in solution, so $[Hg^{2+}] \approx [H_2S]$. Above pH ≈ 8, HS^- is the predominant form, and $[Hg^{2+}] \approx [HS^-]$. The solubility of HgS increases as the pH is lowered since S^{2-} reacts with H^+, drawing Reaction 7-36 to the right.

We can use the procedure above to find the composition of the solution at any given pH. If we happen to guess the true pH of the system, then the charge balance (Equation 7-40) will be satisfied. Since HgS has such a small solubility product, so little will be dissolved that it will have very little effect on the pH of the water. A reasonable first guess for the pH is 7.00. Successive guesses with the aid of a computer show that the pH of a saturated solution of HgS is, in fact, 7.00.

Recapitulation

In the previous section we dealt with the solubility of salts of the type MA, where M is a metal and A is a basic anion. A can react with water to give HA, $H_2A, \ldots$, and so on. A general procedure for finding the composition of the system at a given pH is the following:

STEP 1. Set up all the equations (except the charge balance) using the systematic treatment of equilibrium.

STEP 2. Use the concentration of OH^- to write expressions for each protonated form of the anion (H_nA) in terms of A.

STEP 3. Substitute the values of H_nA from Step 2 into the mass balance to derive a relation between M and A.

STEP 4. Put the relation between M and A into the solubility product to solve for the concentrations of M and A and other species.

Comments

Although the systematic treatment is always a valid approach to equilibrium problems, it is not always the handiest approach. In this text we will use whatever method proves to be most convenient. It will be extremely helpful for you to learn the shortcuts as they are introduced. When all else fails, the systematic treatment cannot lead you astray, even though it may not be the easiest method.

In all approaches to all equilibrium problems, we are ultimately limited by how much of the system's chemistry is understood. Unless we know all of the relevant equilibria, it is not possible to correctly calculate the composition of a solution. From ignorance of some of the chemical reactions, we undoubtedly oversimplify many equilibrium problems.

Summary

In the systematic treatment of equilibrium, we write down all the pertinent equilibrium expressions, as well as the charge and mass balances. The charge balance states that the sum of all positive charge in solution equals the sum of all negative charge. The mass balance states that the sum of the moles of all forms of an element in solution must equal the moles of that element delivered to the solution. We make certain that we have as many equations as unknowns and then set about to solve for the concentration of each species, using algebra, insight, approximations, computers, magic, or anything else. Application of this procedure to salts of basic anions shows that their solubilities increase at low pH because the anion is protonated in acidic solution

Terms to Understand

charge balance
mass balance

Exercises

7-A. Write the charge balance for a solution prepared by dissolving CaF_2 in H_2O. Consider that CaF_2 may give Ca^{2+}, F^-, and CaF^+.

7-B. (a) Write the mass balance for a solution of $CaCl_2$ in water, where the aqueous species are Ca^{2+} and Cl^-.

(b) Write a mass balance where the aqueous species are Ca^{2+}, Cl^-, and $CaCl^+$.

7-C. (a) Write a mass balance for a saturated solution of CaF_2 in water in which the reactions are

$$CaF_2(s) \rightleftharpoons Ca^{2+} + 2F^-$$

$$F^- + H^+ \rightleftharpoons HF(aq)$$

(b) Write a mass balance for CaF_2 in water where, in addition to the previous reactions, the following reaction occurs:

$$HF(aq) + F^- \rightleftharpoons HF_2^-$$

7-D. Write a mass balance for an aqueous solution of $Ca_3(PO_4)_2$, where the aqueous species are Ca^{2+}, PO_4^{3-}, HPO_4^{2-}, $H_2PO_4^-$, and H_3PO_4.

7-E. (a) Find the concentrations of Ag^+, CN^-, and HCN in a saturated solution of AgCN whose pH is somehow fixed at 9.00. Consider the following equilibria:

$$AgCN(s) \rightleftharpoons Ag^+ + CN^-$$
$$K_{sp} = 2.2 \times 10^{-16}$$

$$CN^- + H_2O \rightleftharpoons HCN(aq) + OH^-$$
$$K_h = 1.6 \times 10^{-5}$$

(b) *Activity problem:* Do as directed in Exercise 7-E(a), but use activity coefficients to solve it correctly. Assume that the ionic strength is fixed at 0.10 M by addition of an inert salt. When using activities, the statement that the pH is 9.00 means that $-\log[H^+]\gamma_{H^+} = 9.00$.

7-F. Calculate the solubility of ZnC_2O_4 (g/L) in a solution held at pH 3.00. Consider the equilibria

$$ZnC_2O_4(s) \rightleftharpoons Zn^{2+} + C_2O_4^{2-}$$
$$\text{Oxalate}$$
$$K_{sp} = 7.5 \times 10^{-9}$$

$$C_2O_4^{2-} + H_2O \rightleftharpoons HC_2O_4^- + OH^-$$
$$K_{b1} = 1.8 \times 10^{-10}$$

$$HC_2O_4^- + H_2O \rightleftharpoons H_2C_2O_4 + OH^-$$
$$K_{b2} = 1.8 \times 10^{-13}$$

Problems

A7-1. State in words the meaning of the charge balance equation.

A7-2. State in words the meaning of a mass balance equation.

A7-3. Why are activity coefficients excluded from the charge and mass balances?

A7-4. Write a charge balance for a solution containing H^+, OH^-, Ca^{2+}, HCO_3^-, CO_3^{2-}, $Ca(HCO_3)^+$, $Ca(OH)^+$, K^+, and ClO_4^-.

A7-5. Write a charge balance for a solution of H_2SO_4 in water if the H_2SO_4 ionizes to HSO_4^- and SO_4^{2-}.

A7-6. Write the charge balance for an aqueous solution of arsenic acid, H_3AsO_4, in which the acid can dissociate to $H_2AsO_4^-$, $HAsO_4^{2-}$, and AsO_4^{3-}. Just for practice look up the structure of arsenic acid in Appendix G and write the structure of $HAsO_4^{2-}$.

A7-7. Suppose that $MgBr_2$ dissolves to give Mg^{2+} and Br^-.

(a) Write a mass balance equation for Mg^{2+} for a 0.20 M solution of $MgBr_2$ in water.

(b) Write a mass balance equation for Br^- for a 0.20 M solution of $MgBr_2$ in water.

Now suppose that in addition to Mg^{2+} and Br^-, $MgBr^+$ can be formed.

(c) Write a mass balance equation for Mg^{2+} for a 0.20 M solution of $MgBr_2$ in water.

(d) Write a mass balance equation for Br^- for a 0.20 M solution of $MgBr_2$ in water.

A7-8. (a) Suppose that $MgBr_2$ dissolves to give Mg^{2+} and Br^-. Write a charge balance equation for this aqueous solution.

(b) Now suppose that in addition to Mg^{2+} and Br^-, $MgBr^+$ can be formed. Write a charge balance equation for this aqueous solution.

A7-9. For a 0.1 F aqueous solution of sodium acetate, $Na^+CH_3CO_2^-$, one mass balance is simply $[Na^+] = 0.1$ M. Write a mass balance involving acetate.

A7-10. Use the systematic treatment of equilibrium to calculate the concentration of Zn^{2+} in a saturated aqueous solution of $Zn_2[Fe(CN)_6]$. This salt dissociates into Zn^{2+} and $Fe(CN)_6^{4-}$ (ferrocyanide). Assume that neither zinc ion nor ferrocyanide ion reacts with H_2O.

A7-11. Use the procedure in Section 7-4 to calculate the concentrations of Ca^{2+}, F^-, and HF in a saturated aqueous solution of CaF_2 held at pH 2.00.

7-12. Consider the dissolution of the compound X_2Y_3, which might give the following species: $X_2Y_2^{2+}$, X_2Y^{4+}, $X_2Y_3(aq)$, and Y^{2-}. Use the mass balance for this solution to find an expression for $[Y^{2-}]$ in terms of the other concentrations. Simplify your answer as much as possible.

7-13. Calculate the concentration of each ion in a solution of 4.0×10^{-8} M $Mg(OH)_2$, which is completely dissociated to Mg^{2+} and OH^-.

7-14. A certain metal salt of acrylic acid has the formula $M(H_2C=CHCO_2)_2$. Find the concentration of M^{2+} in a saturated aqueous solution of this salt in which $[OH^-]$ is maintained at the value 1.8×10^{-10} M. The equilibria are

$$M(CH_2=CHCO_2)_2(s) \rightleftharpoons$$
$$M^{2+} + 2H_2C=CHCO_2^-$$
$$K_{sp} = 6.3 \times 10^{-14}$$

$$H_2C=CHCO_2^- + H_2O \rightleftharpoons$$
$$H_2C=CHCO_2H + OH^-$$

Acrylic acid

$$K_b = 1.8 \times 10^{-10}$$

7-15. Consider a saturated aqueous solution of the slightly soluble salt $R_3NH^+Br^-$, where R is an organic group.

$$R_3NH^+Br^-(s) \rightleftharpoons R_3NH^+ + Br^-$$
$$K_{sp} = 4.0 \times 10^{-8}$$

$$R_3NH^+ \rightleftharpoons R_3N + H^+$$
$$K_b = 2.3 \times 10^{-9}$$

Calculate the solubility (moles per liter) of $R_3NH^+Br^-$ in a solution maintained at pH 9.50.

7-16. (a) Use the systematic treatment of equilibrium to find how many moles of PbO will dissolve in a 1.00-L solution in which the pH is fixed at 10.50. Consider the equilibrium involving Pb^{2+} to be

$$PbO(s) + H_2O \rightleftharpoons Pb^{2+} + 2OH^-$$
$$K = 5.0 \times 10^{-16}$$

(b) Answer the same question as in part a, but also consider the reaction

$$Pb^{2+} + H_2O \rightleftharpoons PbOH^+ + H^+$$
$$K_a = 1.3 \times 10^{-18}$$

(c) *Activity problem:* Answer the same question as in part a, now using activity coefficients. Assume μ is fixed at 0.050 M.

7-17. Calculate the molarity of Ag^+ in a saturated aqueous solution of Ag_3PO_4 at pH 6.00 if the equilibria are

$$Ag_3PO_4(s) \rightleftharpoons 3Ag^+ + PO_4^{3-}$$
$$K_{sp} = 2.8 \times 10^{-18}$$

$$PO_4^{3-} + H_2O \rightleftharpoons HPO_4^{2-} + OH^-$$
$$K_{b1} = 2.3 \times 10^{-2}$$

$$HPO_4^{2-} + H_2O \rightleftharpoons H_2PO_4^- + OH^-$$
$$K_{b2} = 1.6 \times 10^{-7}$$

$$H_2PO_4^- + H_2O \rightleftharpoons H_3PO_4 + OH^-$$
$$K_{b3} = 1.4 \times 10^{-12}$$

7-18. (a) Calculate the ratio $[Pb^{2+}]/[Sr^{2+}]$ in a solution of distilled water saturated with PbF_2 *and* SrF_2.
   (b) Calculate the concentrations of Pb^{2+}, Sr^{2+}, and F^- in the solution above.

7-19. When ammonium sulfate dissolves, both the anion and cation have acid–base reactions in water:

$$(NH_4)_2SO_4(s) \rightleftharpoons 2NH_4^+ + SO_4^{2-}$$
$$K_{sp} = 276$$

$$NH_4^+ \rightleftharpoons NH_3(aq) + H^+$$
$$K_a = 5.70 \times 10^{-10}$$

$$SO_4^- + H_2O \rightleftharpoons HSO_4^- + OH^-$$
$$K_b = 9.80 \times 10^{-13}$$

(a) Write a charge balance for this system.
(b) Write a mass balance for this system.
(c) Find the concentration of $NH_3(aq)$ if the pH is somehow fixed at 9.25.

7-20. Consider the following simultaneous equilibria:

$$FeG^+ + G^- \rightleftharpoons FeG_2(aq)$$
$$K_2 = 3.2 \times 10^3$$

$$G^- + H_2O \rightleftharpoons HG + OH^-$$
$$K_b = 6.0 \times 10^{-5}$$

where G is the amino acid glycine, $H_3N^+CH_2CO_2^-$. Suppose that 0.050 0 mol of FeG_2 is dissolved in 1.00 L.
(a) Write the charge balance for the solution.
(b) Write two independent mass balances for the solution.

(c) *Using activity coefficients,* find the concentration of FeG^+ if the pH is fixed at 8.50 and the ionic strength is 0.10 M. For FeG^+ use $\gamma = 0.79$ and for G^- use $\gamma = 0.78$.

7-21. Consider an aqueous system in which the following equilibria can occur:

$$M^{2+} + X^- \rightleftharpoons MX^- \qquad K_1$$
$$MX^- + X^- \rightleftharpoons MX_2(aq) \qquad K_2$$

Derive an equation giving the concentration of M^{2+} when 0.10 mol of MX_2 is dissolved in 1.00 L. Your equation should contain $[M^{2+}]$, K_1, and K_2 as the only variables.

7-22. *Ion Pairing.* When an ionic compound dissolves in water, one of the possible equilibria involves tight association of the cations and anions to give an *ion pair*. In Box 5-4 this occurred by hydrogen bonding between H_3O^+ and F^-, but hydrogen bonding need not be involved in all ion pairs. In a solution of zinc sulfate, for example, some fraction of the cations and anions are associated as $Zn^{2+}SO_4^{2-}$, which behaves as a single species in the solution. Consider a solution in which the formal concentration of $ZnSO_4$ is 0.010 M. In the absence of ion pairing this means that $[Zn^{2+}] = [SO_4^{2-}] = 0.010$ M. In fact, ion pairing is significant:

$$Zn^{2+} + SO_4^{2-} \rightleftharpoons Zn^{2+}SO_4^{2-} \quad \text{(ion pair)}$$
$$K = \frac{[Zn^{2+}SO_4^{2-}]\gamma_{Zn^{2+}SO_4^{2-}}}{[Zn^{2+}]\gamma_{Zn^{2+}}[SO_4^{2-}]\gamma_{SO_4^{2-}}} = 200$$

Since the species $Zn^{2+}SO_4^{2-}$ is neutral, its activity coefficient can be taken as unity in this equilibrium constant. The mass balance for the solution is

$$0.010 \text{ M} = [Zn^{2+}] + [Zn^{2+}SO_4^{2-}]$$

and we also know that $[Zn^{2+}] = [SO_4^{2-}]$, even though we do not know the value of either concentration.

(a) Neglecting activity coefficients, use the ion pair equilibrium to calculate the concentration of Zn^{2+} in the solution.
(b) Use the answer from part a to compute the ionic strength of the solution and the activity coefficients of Zn^{2+} and SO_4^{2-}. Repeat the calculation of part a, using activity coefficients this time.
(c) Repeat the procedure in part b two more times to find a good estimate of $[Zn^{2+}]$. What percent of the salt is ion paired?

(d) With the ionic strength from your final iteration in part c, calculate the ionic strength of the solution. Calculate the mean activity coefficient, $\gamma_\pm$, of zinc sulfate in this solution using Equation 6-14. In the illustration below the upper curve was calculated neglecting ion pairing. The lower curve was calculated with ion pairing and the dots are experimental values of $\gamma_\pm$. [From S. O. Russo and G. I. H. Hanania, *J. Chem. Ed.*, **66**, 148 (1989), which includes other good examples of ion pairing.]

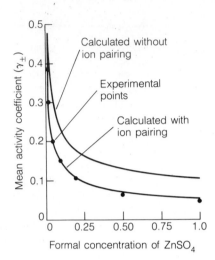

Calculated without ion pairing

Experimental points

Calculated with ion pairing

Formal concentration of $ZnSO_4$

7-23. This problem demonstrates what would happen if charge balance did not exist in a solution. The force (in newtons) between two charges q_1 and q_2 (in coulombs) is given by

$$\text{Force} = -(8.988 \times 10^9) \frac{q_1 q_2}{r^2}$$

where r is the distance (meters) between the two charges. What is the force (in newtons and pounds) between two beakers separated by 1.5 m if one beaker contains 250 mL of solution with 1.0×10^{-6} M excess negative charge and the other has 250 mL of 1.0×10^{-6} M excess positive charge? Note that there are 9.648×10^4 coulombs per mole of charge and 0.224 8 pound per newton.

8 Gravimetric Analysis

Gravimetric analysis encompasses a variety of techniques in which the mass of a product is used to determine the quantity of the original **analyte** (the species being analyzed). Because mass can be measured very accurately, gravimetric methods are among the most accurate in analytical chemistry.

8-1 EXAMPLES OF GRAVIMETRIC ANALYSIS

A familiar example of gravimetric analysis is the determination of Cl^- by precipitation with Ag^+:

$$Ag^+ + Cl^- \rightarrow AgCl(s) \tag{8-1}$$

The weight of AgCl produced tells us how much Cl^- was originally present.

EXAMPLE: A Simple Gravimetric Calculation

A 10.00-mL solution containing Cl^- was treated with excess $AgNO_3$ to precipitate 0.436 8 g of AgCl. What was the molarity of Cl^- in the unknown?

The formula weight of AgCl is 143.321. A precipitate weighing 0.463 8 g contains

$$\frac{0.436\ 8\ \text{g AgCl}}{143.321\ \text{g AgCl/mol AgCl}} = 3.048 \times 10^{-3}\ \text{mol AgCl}$$

Since one mole of AgCl contains one mole of Cl^-, there must have been 3.048×10^{-3} mol of Cl^- in the unknown.

$$[Cl^-] = \frac{3.048 \times 10^{-3}\ \text{mol}}{0.010\ 00\ \text{L}} = 0.304\ 8\ \text{M}$$

Exceedingly careful gravimetric analysis employing AgCl was used by T. W. Richards and his colleagues to determine the atomic weights of Ag, Cl, and N to six-figure accuracy.[†] This Nobel Prize-winning research formed the basis for accurate determination of the atomic weights of many other elements.

Combustion Analysis

Combustion methods are useful for the analysis of carbon, hydrogen, nitrogen, sulfur, and halogens.

Historically, one of the most widely used forms of gravimetric analysis has been **combustion analysis,** used to determine the carbon and hydrogen content of organic compounds.[‡] Because the technique is accurate and applicable to a very wide range of substances, most chemists consider this form of elemental analysis to be a necessary step in the characterization and identification of a new compound. The procedure requires specialized equipment and is commonly carried out in commercial laboratories. A simplified diagram of the apparatus is shown in Figure 8-1.

The sample is heated in an oxygen atmosphere, and the partially combusted product is passed through catalysts at elevated temperature to complete the oxidation of the material to CO_2 and H_2O. Useful catalysts include Pt gauze, CuO, PbO_2, and MnO_2. After the catalyst, sometimes additional reagents are added to remove halogen or sulfur compounds. The combustion products are flushed through a chamber containing P_4O_{10} ("phosphorus pentoxide"), which absorbs the water, and then through a chamber of Ascarite (NaOH on asbestos), which absorbs the CO_2. The increase in mass of each chamber tells how much hydrogen and carbon, respectively, was produced. A guard tube downstream of the two chambers prevents atmospheric H_2O or CO_2 from entering the chambers to be weighed.

[†] T. W. Richards, *Chem. Rev.,* **1,** 1 (1925).

[‡] A comprehensive discussion of this technique can be found in N. H. Furman, ed., *Standard Methods of Chemical Analysis,* 6th ed., Vol. 1 (Princeton, N.J.: van Nostrand, 1962), pp. 279–287. Modern combustion methods use gas chromatography for measurement of the product gases. For details, see E. Pella, *Amer. Lab.,* February 1990, p. 116.

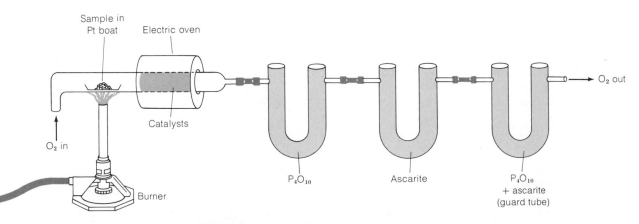

Figure 8-1
Combustion analysis of carbon and hydrogen.

EXAMPLE: Determination of Masses of Carbon and Hydrogen by Combustion Analysis

An organic compound weighing 5.714 mg produced 14.414 mg of CO_2 and 2.529 mg of H_2O upon combustion. Find the weight percent of C and H in the sample.

One mole of CO_2 contains one mole of carbon. Therefore,

Moles of C in sample = moles of CO_2 produced

$$= \frac{14.414 \times 10^{-3} \text{ g } CO_2}{44.010 \text{ g/mol } CO_2} = 3.275 \times 10^{-4} \text{ mol}$$

Mass of C in sample = $(3.275 \times 10^{-4} \text{ mol C})(12.011 \text{ g/mol C}) = 3.934 \text{ mg}$

Weight percent of C = $\dfrac{3.934 \text{ mg C}}{5.714 \text{ mg sample}} \times 100 = 68.84\%$

One mole of H_2O contains *two* moles of H. Therefore,

Moles of H in sample = 2(moles of H_2O produced)

$$= 2\left(\frac{2.529 \times 10^{-3} \text{ g } H_2O}{18.015\,2 \text{ g/mol } H_2O}\right) = 2.808 \times 10^{-4} \text{ mol}$$

Mass of H in sample = $(2.808 \times 10^{-4} \text{ mol H})(1.007\,9 \text{ g/mol H}) = 2.830 \times 10^{-4} \text{ g}$

Weight percent of H = $\dfrac{0.283\,0 \text{ mg H}}{5.714 \text{ mg sample}} \times 100 = 4.952\%$

The last calculation uses the conversion $(2.830 \times 10^{-4} \text{ g})(1\,000 \text{ mg/g}) = 0.283\,0 \text{ mg}$.

The nitrogen content of the compound may also be determined by a modification known as the Dumas method. In this procedure the sample is mixed with powdered CuO and ignited in a stream of CO_2, decomposing the compound to H_2O, CO_2, N_2, and some nitrogen oxides. The latter are reduced to N_2 by a hot Cu catalyst downstream. The gases then enter a gas buret filled with concentrated KOH solution to absorb all CO_2. The volume of gas finally collected is due to the evolved N_2.

Problem 9-22 describes halogen analysis by a combustion method.

8-2 PRECIPITATION PROCESS

The ideal product of a gravimetric analysis should be very insoluble, easily filterable, and very pure and possess a known and constant composition. Although few substances meet all these requirements, appropriate techniques can help optimize the properties of gravimetric precipitates.

Gravimetric analyses will give rise to systematic error if any of these criteria are not met.

Solubility

The solubility of a precipitate may be decreased by cooling the solution, since the solubility of most compounds decreases as the temperature is lowered. Alternatively, the solvent may be changed. The solubility of many organic salts can be decreased by decreasing the polarity of the solvent to less than that of pure water.

See Demonstration 8-1 for an experiment with colloids and dialysis.

Filterability

The particles of the product need to be large enough so that the precipitate does not clog the filter or, worse, pass through the filter. The most desirable particles are, of course, crystals. At the other extreme is a **colloidal suspension** of particles so small that they remain indefinitely suspended and pass right through most filters. Colloids are particles with diameters in the range 1–100 nm. The conditions of the precipitation have much to do with the resulting particle size.

Crystal growth

Crystallization is generally considered as occurring in two phases: nucleation and particle growth. **Nucleation** is the process whereby molecules in solution randomly come together and form small aggregates. Particle growth involves the addition of more molecules to the nucleus to form a crystal. When a solution contains more solute than should be present at equilibrium, the solution is said to be **supersaturated. Relative supersaturation** is expressed as

---

Demonstration 8-1   COLLOIDS AND DIALYSIS

Colloids are particles with diameters in the range 1–100 nm. They are larger than what we usually think of as molecules, but too small to precipitate. Colloids remain in solution indefinitely, suspended by the Brownian motion (random movement) of the solvent molecules.

Heat one beaker containing 200 mL of distilled water to 70–90°C, and leave an identical beaker of water at room temperature. Add 1 mL of 1 M $FeCl_3$ to each beaker and stir. The warm solution turns brown-red in a few seconds, while the cold solution remains yellow (Color Plate 3). The yellow color is characteristic of low molecular weight Fe(III) compounds. The red color results from colloidal aggregates of Fe(III) ions held together by hydroxide, oxide, and some chloride ions. These particles have a molecular weight of $\sim 10^5$, a diameter of ~ 10 nm and contain $\sim 10^3$ atoms of Fe.[†]

To demonstrate the size of colloidal particles, we can perform a **dialysis** experiment. In dialysis, two solutions are separated by a *semipermeable membrane*. A semipermeable membrane is one with holes through which some molecules, but not others, can diffuse. The common dialysis tubing available from most scientific supply houses is made of cellulose and has pore diameters of 1–5 nm.[‡] Small molecules can diffuse through these pores, but large molecules (such as proteins or colloids) cannot.

Pour some of the brown-red colloidal Fe solution into a dialysis tube knotted at one end, then tie off the other end. Drop this into a flask of distilled water to show that the color remains entirely within the bag even after several days (Color Plate 3). For comparison, an identical bag containing a dark blue solution of 1 M $CuSO_4 \cdot 5H_2O$ can be left in another flask of water. (The notation $CuSO_4 \cdot 5H_2O$ means that the crystal contains five water molecules for every $CuSO_4$.) The blue color of the Cu^{2+} will diffuse out of the bag, and the entire solution in the flask will be a uniform light blue color in 24 hours.

[†] R. N. Silva, *Rev. Pure and Appl. Chem.,* **22,** 115 (1972); K. M. Towe and W. F. Bradley, *J. Colloid Interface Sci.,* **24,** 384 (1967); T. G. Spiro, S. E. Allerton, J. Renner, A. Terzis, R. Bils, and P. Saltman, *J. Amer. Chem. Soc.,* **88,** 2721 (1966).

[‡] Tubing such as catalog number 3787, sold by A. H. Thomas Co., P. O. Box 99, Swedesboro, NJ 08085-0099, is adequate for this demonstration.

$(Q - S)/S$, where Q is the concentration of solute actually present and S is the concentration at equilibrium. The more substance that is dissolved, the greater the supersaturation.

The rate of nucleation has been found to depend more on relative supersaturation than does the rate of particle growth. That is, in a highly supersaturated solution, nucleation proceeds faster than particle growth. The result is a suspension of very tiny particles or, worse, a colloid. In a less supersaturated solution, nucleation is not so rapid, and the nuclei have a chance to grow into larger, more tractable particles.

To decrease supersaturation, and thereby promote particle growth, three techniques may be employed:

1. The temperature is raised to increase S and thereby decrease the relative supersaturation. (Most substances are more soluble in warm solution than in cold solution.)

2. The precipitant is added slowly with vigorous mixing, to avoid a local, highly supersaturated condition where the stream of precipitant first enters the analyte.

3. The volume of solution is kept large so that the concentrations of analyte and precipitant are low.

Low relative supersaturation promotes increased particle size.

The **precipitant** is the reagent that is added to cause the precipitation to occur.

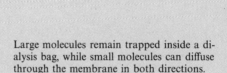

Dialysis tubing knotted at each end

2.5-nm-diameter pores

Large molecules

Small molecules

Large molecules remain trapped inside a dialysis bag, while small molecules can diffuse through the membrane in both directions.

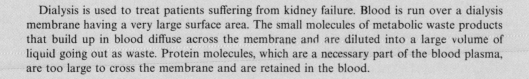

Dialysis is used to treat patients suffering from kidney failure. Blood is run over a dialysis membrane having a very large surface area. The small molecules of metabolic waste products that build up in blood diffuse across the membrane and are diluted into a large volume of liquid going out as waste. Protein molecules, which are a necessary part of the blood plasma, are too large to cross the membrane and are retained in the blood.

Controlled precipitation

It is sometimes possible to control the solubility of a precipitate through chemical means, such as pH or complexing ion control. For example, CaC_2O_4 is commonly precipitated from hot acidic solution, in which the salt is more soluble by virtue of the reaction of $C_2O_4^{2-}$ with H^+:

$$\underset{\text{Calcium oxalate}}{CaC_2O_4(s)} \underset{}{\overset{K_{sp}}{\rightleftharpoons}} Ca^{2+} + C_2O_4^{2-} \qquad (8\text{-}2)$$

$$C_2O_4^{2-} + H^+ \rightleftharpoons HC_2O_4^- \qquad (8\text{-}3)$$

Gradually raising the pH drives Reaction 8-3 to the left. This, in turn, drives Reaction 8-2 to the left.

Homogeneous precipitation

Homogeneous precipitation is one of the nicest tricks for controlling supersaturation. In this technique, the precipitating agent is generated slowly by means of a chemical reaction. The most commonly employed reagent is urea, which slowly decomposes in boiling water to produce OH^-:

$$\underset{\text{Urea}}{\underset{H_2N \qquad NH_2}{\overset{\overset{\textstyle O}{\overset{\|}{C}}}{}}} + 3H_2O \xrightarrow{\text{heat}} CO_2 + 2NH_4^+ + 2OH^- \qquad (8\text{-}4)$$

By this means the OH^- concentration of a solution can be raised very gradually. For example, one precipitation where slow hydroxide formation greatly enhances the particle size is that of ferric formate:

$$(H_2N)_2CO + 3H_2O \xrightarrow{\text{heat}} CO_2 + 2NH_4^+ + 2OH^- \qquad (8\text{-}5)$$

$$OH^- + \underset{\text{Formic Acid}}{\overset{\overset{\textstyle O}{\overset{\|}{HCOH}}}{}} \rightarrow \underset{\text{Formate}}{HCO_2^-} + H_2O \qquad (8\text{-}6)$$

$$3HCO_2^- + Fe^{3+} \rightarrow \underset{\text{Ferric formate}}{Fe(HCO_2)_3 \cdot nH_2O} \qquad (8\text{-}7)$$

Box 8-1 shows one application of homogeneous precipitation.

Urea hydrolysis can also be used to drive Reactions 8-3 and 8-2 very slowly to the left, producing easily filterable crystals of CaC_2O_4. Table 8-1 lists some reagents employed in homogeneous precipitation.

Precipitation in the presence of an electrolyte

Recall that an *electrolyte* is a compound that dissociates into ions when it dissolves.

It is desirable to precipitate most ionic compounds in the presence of added electrolyte. To understand the reason for this, we must discuss how tiny colloidal crystallites *coagulate* (come together) into larger particles (crystals). To illustrate this, we will discuss the formation of AgCl, which is commonly formed in the presence of ~ 0.1 M HNO_3.

Figure 8-2 is a schematic drawing of a colloidal particle of AgCl growing in a solution containing excess Ag^+, H^+, and NO_3^-. The surface of the particle

Table 8-1
Some common reagents used for homogeneous precipitation

Precipitant	Reagent	Reaction	Some elements precipitated
OH^-	Urea	$(H_2N)_2CO + 3H_2O \rightarrow CO_2 + 2NH_4^+ + 2OH^-$	Al, Ga, Th, Bi, Fe, Sn
OH^-	Potassium cyanate	$HOCN + 2H_2O \rightarrow NH_4^+ + CO_2 + OH^-$ Hydrogen cyanate	Cr, Fe
S^{2-}	Thioacetamide†	$\overset{\overset{\displaystyle S}{\|\|}}{CH_3CNH_2} + H_2O \rightarrow \overset{\overset{\displaystyle O}{\|\|}}{CH_3CNH_2} + H_2S$	Sb, Mo, Cu, Cd
SO_4^{2-}	Sulfamic acid	$H_3\overset{+}{N}SO_3^- + H_2O \rightarrow NH_4^+ + SO_4^{2-} + H^+$	Ba, Ca, Sr, Pb
$C_2O_4^{2-}$	Dimethyl oxalate	$\overset{\overset{\displaystyle O\,O}{\|\|\ \|\|}}{CH_3OCCOCH_3} + 2H_2O \rightarrow 2CH_3OH + C_2O_4^{2-} + 2H^+$	Ca, Mg, Zn
PO_4^{3-}	Trimethyl phosphate	$(CH_3O)_3P{=}O + 3H_2O \rightarrow 3CH_3OH + PO_4^{3-} + 3H^+$	Zr, Hf
CrO_4^{2-}	Chromic ion plus bromate	$2Cr^{3+} + BrO_3^- + 5H_2O \rightarrow 2CrO_4^{2-} + Br^- + 10H^+$	Pb
8-Hydroxyquinoline	8-Acetoxyquinoline	[structure] CH_3CO ... $+ H_2O \rightarrow$ ... OH ... $+ CH_3CO_2H$	Al, U, Mg, Zn

† Hydrogen sulfide is volatile and toxic; it should be handled only in a well-vented hood. Thioacetamide is a carcinogen that should be handled with gloves. If thioacetamide contacts your skin, wash yourself thoroughly immediately. Leftover reagent is destroyed by heating at 50°C with 5 moles of NaOCl per mole of thioacetamide, and then washing down the drain. [H. Elo, *J. Chem. Ed.*, **64**, A144 (1987).]

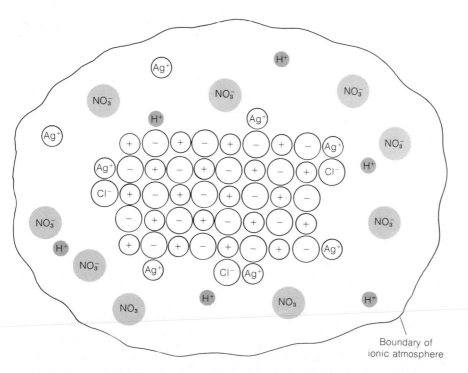

Figure 8-2
Schematic diagram showing a colloidal particle of AgCl growing in a solution containing excess Ag^+, H^+, and NO_3^-. The particle has a net positive charge because of adsorbed Ag^+ ions. The region of solution surrounding the particle is called the ionic atmosphere. It has a net negative charge, since the particle attracts anions and repels cations.

Boundary of ionic atmosphere

Box 8-1 MICROSCOPIC STUDY OF HOMOGENEOUS PRECIPITATION

Zinc sulfide is used as a high-temperature window material for transmission of infrared light. One way to fabricate ceramic windows is to *hot press* ZnS powder (applying both heat and pressure) until the particles weld themselves together into a fully dense solid body containing no voids. It is generally thought that microscopic, spherical particles of one size (not a distribution of sizes) are optimal for hot pressing. Irregular shapes and sizes are not expected to pack together as well as uniform spheres.

In an effort to produce spherical particles of a single size (said to be "monosized"), the homogeneous precipitation of Zn^{2+} by thioacetamide was studied:[†]

$$\underset{\text{Thioacetamide}}{CH_3\overset{\overset{\displaystyle S}{\|}}{C}NH_2} + H_2O \xrightarrow{\text{heat}} \underset{\text{Acetamide}}{CH_3\overset{\overset{\displaystyle O}{\|}}{C}NH_2} + H_2S$$

$$H_2S + Zn^{2+} \rightarrow ZnS(precipitate) + 2H^+$$

(a)

Scanning electron micrographs of ZnS particles produced by homogeneous precipitation. The magnifications are different: the white bar in each photograph represents a length of 1 μm. The monosized particles in Figure a were produced at a slower rate than the bimodal distribution in Figure b. [Courtesy Mufit Akinc, Iowa State University.]

Precipitation commenced when a *relative supersaturation* of approximately 10 was reached. At this point small crystallites of ZnS, 14 nm in size, were produced throughout the solution. These crystallites diffused together to form larger spherical aggregates that grew larger with increased time. In Figure a the particles have reached a uniform 200-nm (0.2-μm) diameter. This collection of monosized spherical particles is the desired form of ZnS for hot pressing of ceramic windows.

By varying the temperature, pH, initial concentration of thioacetamide, and the anion of the zinc salt, the rate of production of 14-nm crystallites could be increased relative to the rate of growth of large spherical aggregates. In Figure b one population of spherical aggregates grew to a fairly large size. Then when the concentration of 14-nm crystallites reached a critical value, a second population of smaller aggregates began to form. The resulting size distribution is said to be *bimodal* (pronounced by-MODE-all), with two particle sizes. Careful control of the homogeneous precipitation process therefore allows the size and size distribution of ZnS particles to be selected.

---

† A. Celikkaya and M. Akinc, *J. Amer. Ceramic Soc.*, **73**, 245 (1990); *Ibid.*, **73**, 2360 (1990).

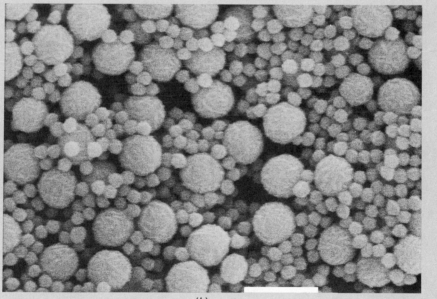

(b)

While it is common to find the excess common ion adsorbed on the crystal surface, it is also possible to find other ions selectively adsorbed. In the presence of citrate and sulfate there is more citrate than sulfate adsorbed on a particle of $BaSO_4$.

has an excess positive charge due to the **adsorption** of extra silver ions on exposed chloride ions. (To be adsorbed means to be attached to the surface. In contrast, **absorption** involves penetration beyond the surface, to the inside.) The positively charged surface attracts anions and repels cations from the *ionic atmosphere* surrounding the particle. (Compare this to Figure 6-1.) The positively charged particle and the negatively charged ionic atmosphere together are called the **electric double layer.**

In order for two colloidal particles of AgCl to coalesce, they must collide with each other. However, the negatively charged ionic atmospheres of the particles repel each other. The particles, therefore, must have enough kinetic energy to overcome this electrostatic repulsion before they can coalesce.

Two steps can be taken to promote coalescence of the particles. One is to heat the solution, thereby increasing the particles' kinetic energy. The other is to increase the concentration of electrolyte (HNO_3 in this case). The greater the concentration of electrolyte, the less volume the ionic atmospheres occupy, and the closer together the two particles can come before their electrostatic repulsion becomes significant. This is why most gravimetric precipitations are done in the presence of an electrolyte.

Digestion

Following precipitation, most gravimetric procedures call for a period of standing in the presence of the mother liquor, usually with heating. This treatment, called **digestion,** promotes slow recrystallization of the precipitate. During this process the particle size usually increases, and impurities tend to be expelled from the crystal.

Purity

Impurities bound to the surface of a crystal are said to be *adsorbed*. Impurities held within the crystal (*absorbed* impurities) are classified as either **inclusions** or **occlusions.** Inclusions are impurity ions that randomly occupy sites in the crystal lattice normally occupied by ions that belong in the crystal. Inclusions are more likely when the impurity ion has a similar size and charge to one of the ions that belongs to the product. Occlusions are pockets of impurity that are literally trapped inside the growing crystal.

Reprecipitation improves the purity of some precipitates.

Adsorbed, occluded, and included impurities are said to be **coprecipitated.** That is, the impurity is precipitated along with the desired product, even though the solubility of the impurity has not been exceeded. Coprecipitation tends to be worst in colloidal precipitates (which have a large surface area), such as $BaSO_4$, $Al(OH)_3$, and $Fe(OH)_3$. Many procedures call for washing away the mother liquor, redissolving the precipitate, and **reprecipitating** the product. During the second precipitation the concentration of impurities in the solution is lower than during the first precipitation, and the degree of coprecipitation therefore tends to be lower. Occasionally a trace component is intentionally isolated by being coprecipitated with a major component of the solution. The precipitate used to collect the trace component is said to be a **gathering agent,** and the process is called **gathering.**

As an example, Se(IV) can be *gathered* by coprecipitation with $Fe(OH)_3$. This allows Se(IV) at a concentration of 25 ng/L to be analyzed with a precision of 6%. [K. W. M. Siu and S. S. Berman, *Anal. Chem.,* **56,** 1806 (1984).]

Sometimes impurity species can be prevented from precipitating by addition of a **masking agent.** In the gravimetric analysis of Be^{2+}, Mg^{2+}, Ca^{2+}, or Ba^{2+} with the reagent N-*p*-chlorophenylcinnamohydroxamic acid, impurities such as Ag^+, Mn^{2+}, Zn^{2+}, Cd^{2+}, Hg^{2+}, Cu^{2+}, Fe^{2+}, and Ga^{3+} can be maintained in solution by excess KCN.

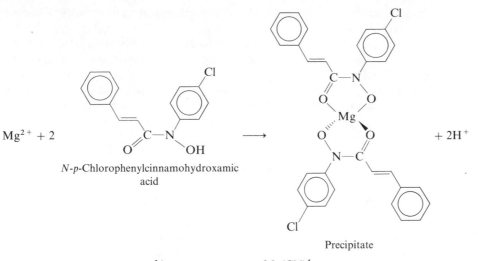

$Mg^{2+} + 2$ [N-p-Chlorophenylcinnamohydroxamic acid] $\longrightarrow$ [Precipitate] $+ 2H^+$

$$Mn^{2+} + 6CN^- \rightarrow Mn(CN)_6^{4-}$$
(from KCN)   (stays in solution)

In this reaction, cyanide *masks* Mn^{2+} from the precipitating agent.

The ions Pb^{2+}, Pd^{2+}, Sb^{3+}, Sn^{2+}, Bi^{3+}, Zr^{4+}, Ti^{4+}, V^{5+}, and Mo^{6+} are masked with a mixture of citrate and oxalate.[†]

Sometimes a precipitate forms in a pure state, but, while it is standing in the mother liquor, impurities form on the product. This is called **postprecipitation** and usually involves a supersaturated impurity that does not readily crystallize. After a period of time the impurity crystallizes and contaminates the desired product. An example is the postprecipitation of MgC_2O_4 in the presence of CaC_2O_4.

Washing is an important step in a gravimetric analysis. When the precipitate is collected on a filter, there are always droplets of liquid containing excess solute adhering to the solid. If the solution containing $AgNO_3$ were not washed away from the solid AgCl, the $AgNO_3$ would still be present after drying and the precipitate would weigh too much.

Some precipitates can simply be washed with water to remove the solute. However, many precipitates require electrolyte to maintain their coherence. For these, the ionic atmospheres are still needed to neutralize the surface charges of the tiny particles. If the electrolyte is washed away with water, the charged solid particles repel each other and the product breaks up. This breaking up is called **peptization** and can actually result in loss of the product through the filter. Silver chloride is an example of a precipitate that will peptize if washed with water. Therefore, AgCl is washed with dilute HNO_3 to remove excess $AgNO_3$ and to prevent peptization. The electrolyte used for washing must be volatile, so that it will be lost during drying. Some common volatile electrolytes are HNO_3, HCl, NH_4NO_3, NH_4Cl, and $(NH_4)_2CO_3$.

Ammonium chloride, for example, decomposes as follows when it is heated:

$$NH_4Cl(s) \rightarrow NH_3(g) + HCl(g)$$

Composition of Product

The final product must have a known, stable composition. If it is **hygroscopic** (picks up water from the air), it will be difficult to weigh accurately. Many precipitates contain a variable quantity of water and must be dried to achieve a constant composition. Some precipitates are **ignited** (heated strongly) to

[†] Y. K. Agrawal and D. R. Roshania, *J. Indian Chem. Soc.,* **61**, 248 (1984)

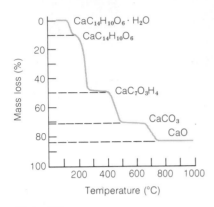

Figure 8-3
Thermogravimetric curve for calcium salicylate. [G. Liptay, ed., *Atlas of Thermoanalytical Curves* (London: Heyden and Son, 1976).]

change the chemical form. Conditions of heating or ignition vary with each gravimetric analysis.

An example of how the composition depends on heating temperature is shown in Figure 8-3. The curve in this figure was obtained by a *thermobalance,* and the technique is called **thermogravimetric analysis.** In this procedure a sample is heated, and its change in mass is measured as a function of temperature. We see that calcium salicylate decomposes in the following stages:

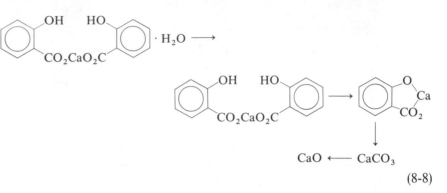

$$(8\text{-}8)$$

Clearly, the composition of the product to be weighed depends on the temperature and, usually, the duration of heating.

Many gravimetric precipitates are made to produce a stable product of definite composition by being ignited at high temperature in an oven or over a flame. Two examples are $Fe(HCO_2)_3 \cdot nH_2O$, which is ignited at 850°C for one hour to give Fe_2O_3; and $Mg(NH_4)PO_4 \cdot 6H_2O$, which is ignited at 1 100°C to give $Mg_2P_2O_7$.

8-3 SCOPE OF GRAVIMETRIC ANALYSIS

The only major equipment needed for gravimetric analysis is an accurate balance. It is one of the oldest analytical techniques, and gravimetric analyses were developed for many elements and compounds long before alternative methods were available. Gravimetric procedures are usually very accurate, but more tedious than other methods.

To indicate the scope of gravimetric analysis, some analytically useful precipitations of common cations and anions are listed in Table 8-2. A few common organic precipitating agents are listed in Table 8-3. Because most precipitants (agents that cause precipitation) are not specific, conditions usually need to be carefully controlled in order to selectively precipitate one species in the presence of interfering substances. Often the potentially interfering substance must be separated from the analyte prior to analysis. It is important to be cognizant of potential interference in all analytical techniques, not just gravimetric analysis.

Separation of interfering species from the analyte is a necessary step in many analytical procedures.

8-4 CALCULATIONS OF GRAVIMETRIC ANALYSIS

We will now examine some examples that illustrate how to relate the mass of a gravimetric precipitate to the quantity of the original analyte. The general approach is to relate the moles of product to the moles of reactant.

Table 8-2
Representative gravimetric analyses

Species analyzed	Precipitated form	Form weighed	Some interfering species
K^+	$KB(C_6H_5)_4$	$KB(C_6H_5)_4$	NH_4^+, Ag^+, Hg_2^{2+}, Tl^+, Rb^+, Cs^+
Mg^{2+}	$Mg(NH_4)PO_4 \cdot 6H_2O$	$Mg_2P_2O_7$	Many metals except Na^+ and K^+
Ca^{2+}	$CaC_2O_4 \cdot H_2O$	$CaCO_3$ or CaO	Many metals except Mg^{2+}, Na^+, K^+
Ba^{2+}	$BaSO_4$	$BaSO_4$	Na^+, K^+, Li^+, Ca^{2+}, Al^{3+}, Cr^{3+}, Fe^{3+}, Sr^{2+}, Pb^{2+}, NO_3^-
Ti^{4+}	$TiO(5,7\text{-dibromo-8-hydroxyquinoline})_2$	Same	Fe^{3+}, Zr^{4+}, Cu^{2+}, $C_2O_4^{2-}$, citrate, HF
VO_4^{3-}	Hg_3VO_4	V_2O_5	Cl^-, Br^-, I^-, SO_4^{2-}, CrO_4^{2-}, AsO_4^{3-}, PO_4^{3-}
Cr^{3+}	$PbCrO_4$	$PbCrO_4$	Ag^+, NH_4^+
Mn^{2+}	$Mn(NH_4)PO_4 \cdot H_2O$	$Mn_2P_2O_7$	Many metals
Fe^{3+}	$Fe(HCO_2)_3$	Fe_2O_3	Many metals
Co^{2+}	$Co(1\text{-nitroso-2-naphtholate})_3$	$CoSO_4$ (by reaction with H_2SO_4)	Fe^{3+}, Pd^{2+}, Zr^{4+}
Ni^{2+}	$Ni(dimethylglyoximate)_2$	Same	Pd^{2+}, Pt^{2+}, Bi^{3+}, Au^{3+}
Cu^{2+}	$CuSCN$	$CuSCN$	NH_4^+, Pb^{2+}, Hg^{2+}, Ag^+
Zn^{2+}	$Zn(NH_4)PO_4 \cdot H_2O$	$Zn_2P_2O_7$	Many metals
Ce^{4+}	$Ce(IO_3)_4$	CeO_2	Th^{4+}, Ti^{4+}, Zr^{4+}
Al^{3+}	$Al(8\text{-hydroxyquinolate})_3$	Same	Many metals
Sn^{4+}	$Sn(cupferron)_4$	SnO_2	Cu^{2+}, Pb^{2+}, $As(III)$
Pb^{2+}	$PbSO_4$	$PbSO_4$	Ca^{2+}, Sr^{2+}, Ba^{2+}, Hg^{2+}, Ag^+, HCl, HNO_3
NH_4^+	$NH_4B(C_6H_5)_4$	$NH_4B(C_6H_5)_4$	K^+, Rb^+, Cs^+
Cl^-	$AgCl$	$AgCl$	Br^-, I^-, SCN^-, S^{2-}, $S_2O_3^{2-}$, CN^-
Br^-	$AgBr$	$AgBr$	Cl^-, I^-, SCN^-, S^{2-}, $S_2O_3^{2-}$, CN^-
I^-	AgI	AgI	Cl^-, Br^-, SCN^-, S^{2-}, $S_2O_3^{2-}$, CN^-
SCN^-	$CuSCN$	$CuSCN$	NH_4^+, Pb^{2+}, Hg^{2+}, Ag^+
CN^-	$AgCN$	$AgCN$	Cl^-, Br^-, I^-, SCN^-, S^{2-}, $S_2O_3^{2-}$
F^-	$(C_6H_5)_3SnF$	$(C_6H_5)_3SnF$	Many metals (except alkali metals), SiO_4^{4-}, CO_3^{2-}
ClO_4^-	$KClO_4$	$KClO_4$	Na^+, K^+, Li^+, Ca^{2+}, Al^{3+}, Cr^{3+}, Fe^{3+}, Sr^{2+}, Pb^{2+}, NO_3^-
SO_4^{2-}	$BaSO_4$	$BaSO_4$	Many metals except Na^+, K^+
PO_4^{3-}	$Mg(NH_4)PO_4 \cdot 6H_2O$	$Mg_2P_2O_7$	Many metals except Na^+, K^+
NO_3^-	Nitron nitrate	Nitron nitrate	ClO_4^-, I^-, SCN^-, CrO_4^{2-}, ClO_3^-, NO_2^-, Br^-, $C_2O_4^{2-}$
CO_3^{2-}	CO_2 (by acidification)	CO_2	(The liberated CO_2 is trapped with Ascarite and weighed.)

Table 8-3
Common organic precipitating agents

Name	Structure	Some ions precipitated
Dimethylglyoxime		Ni^{2+}, Pd^{2+}, Pt^{2+}
Cupferron		Fe^{3+}, VO_2^+, Ti^{4+}, Zr^{4+}, Ce^{4+}, Ga^{3+}, Sn^{4+}
8-Hydroxyquinoline (oxine)		Mg^{2+}, Zn^{2+}, Cu^{2+}, Cd^{2+}, Pb^{2+}, Al^{3+}, Fe^{3+}, Bi^{3+}, Ga^{3+}, Th^{4+}, Zr^{4+}, UO_2^{2+}, TiO^{2+}
Salicylaldoxime		Cu^{2+}, Pb^{2+}, Bi^{3+}, Zn^{2+}, Ni^{2+}, Pd^{2+}
1-Nitroso-2-naphthol		Co^{2+}, Fe^{3+}, Pd^{2+}, Zr^{4+}
Nitron		NO_3^-, ClO_4^-, BF_4^-, WO_4^{2-}
Sodium tetraphenylborate	$Na^+B(C_6H_5)_4^-$	K^+, Rb^+, Cs^+, NH_4^+, Ag^+, organic ammonium ions
Tetraphenylarsonium chloride	$(C_6H_5)_4As^+Cl^-$	$Cr_2O_7^{2-}$, MnO_4^-, ReO_4^-, MoO_4^{2-}, WO_4^{2-}, ClO_4^-, I_3^-

---

EXAMPLE: Relating Mass of Product to Mass of Reactant

The piperazine content of an impure commercial grade of piperazine can be determined by precipitating and weighing the diacetate:[†]

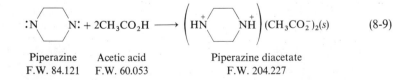

$$\text{(8-9)}$$

Piperazine	Acetic acid	Piperazine diacetate
F.W. 84.121	F.W. 60.053	F.W. 204.227

In one experiment 0.312 6 g of the sample was dissolved in 25 mL of acetone, and 1 mL of acetic acid was added. After five minutes, the precipitate was filtered, washed with acetone, dried at 110°C, and found to weigh 0.712 1 g. What is the weight percent of piperazine in the commercial material?

[†] G. W. Latimer, Jr., *J. Chem. Ed.*, **43**, 148 (1966); G. R. Bond, *Anal. Chem.*, **32**, 1332 (1962).

For each mole of piperazine in the reaction, one mole of product will be formed. The moles of product will therefore be

$$\text{Moles of product} = \frac{0.712\ 1\ \cancel{g}}{204.227\ \cancel{g}/\text{mol}} = 3.487 \times 10^{-3}\ \text{mol}$$

This many moles of piperazine corresponds to

$$\text{Grams of piperazine} = (3.487 \times 10^{-3}\ \cancel{\text{mol}})\left(84.121\ \frac{g}{\cancel{\text{mol}}}\right) = 0.293\ 3\ g$$

which gives

$$\text{Percent of piperazine in analyte} = \frac{0.293\ 3\ \cancel{g}}{0.312\ 6\ \cancel{g}} \times 100 = 93.83\%$$

An alternative (but equivalent) way to work this problem is to realize that 204.227 g (1 mol) of product will be formed for every 84.121 g (1 mol) of piperazine analyzed. Since 0.712 1 g of product was formed, the amount of reactant is given by

$$\frac{x\ \text{g piperazine}}{0.712\ 1\ \text{g product}} = \frac{84.121\ \text{g piperazine}}{204.227\ \text{g product}}$$

or

$$x\ \text{g piperazine} = \left(\frac{84.121}{204.227}\right)0.712\ 1\ g = 0.293\ 3\ g$$

The quantity 84.121/204.227 is often referred to as a *gravimetric factor* relating the amount of starting material to the amount of product.

As a practical note, if you were performing this analysis, it would be important to know or to determine that the impurities in the piperazine are not precipitated under the conditions of this experiment.

EXAMPLE: Calculating How Much Precipitant to Use

The nickel content in steel can be determined gravimetrically. First dissolve the alloy in 12 M HCl and neutralize in the presence of citrate ion, which maintains the iron in solution near neutral pH. Warming the slightly basic solution in the presence of dimethylglyoxime (DMG) causes the red DMG–nickel complex to precipitate quantitatively. The product is filtered, washed with cold water, and dried at 110°C.

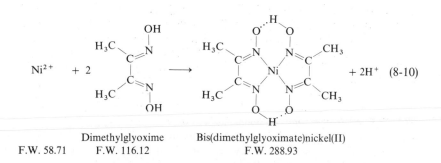

Dimethylglyoxime   Bis(dimethylglyoximate)nickel(II)
F.W. 58.71   F.W. 116.12   F.W. 288.93

Suppose that the nickel content of the steel is known to be near 3% and you wish to analyze 1.0 g of the steel. What volume of 1.0% alcoholic DMG solution should be used to give a 50% excess of DMG for the analysis? Assume that the density of the alcohol solution is 0.79 g/mL.

Since the Ni content is around 3%, 1.0 g of steel will contain about 0.03 g of Ni, which corresponds to

$$\frac{0.03 \text{ g}}{58.71 \text{ g/mol}} = 5.11 \times 10^{-4} \text{ mol Ni}$$

This amount of metal requires

$$2(5.11 \times 10^{-4} \text{ mol})(116.12 \text{ g DMG/mol}) = 0.119 \text{ g DMG}$$

since one mole of Ni^{2+} requires two moles of DMG. A 50% excess of DMG would be $(1.5)(0.119 \text{ g}) = 0.178 \text{ g}$. This much DMG is contained in

$$\left(\frac{0.178 \text{ g DMG}}{0.010 \text{ g DMG/g solution}} \right) = 17.8 \text{ g solution}$$

which occupies a volume of

$$\frac{17.8 \text{ g}}{0.79 \text{ g/mL}} = 23 \text{ mL}$$

In the previous calculation we made use of the fact that 1.0% DMG means that 1.0 g (not 1.0 mL) of solution contains 0.010 g of DMG.

If the analysis of 1.163 4 g of steel resulted in formation of 0.179 5 g of precipitate, what is the percentage of Ni in the steel?

For each mole of Ni in the steel, one mole of precipitate will be formed. Therefore, 0.179 5 g of precipitate corresponds to

$$\frac{0.179 \text{ 5 g}}{288.93 \text{ g/mol}} = 6.213 \times 10^{-4} \text{ mol Ni(DMG)}_2$$

The Ni in the alloy must therefore be

$$(6.213 \times 10^{-4} \text{ mol Ni})(58.71 \text{ g/mol Ni}) = 0.036 \text{ 47 g}$$

The weight percent of Ni in steel is

$$\frac{0.036 \text{ 47 g Ni}}{1.163 \text{ 4 g steel}} \times 100 = 3.135\%$$

A slightly simpler way to approach this problem comes from realizing that 58.71 g of Ni (1 mol) would give 288.93 g (1 mol) of product. Calling the mass of Ni in the sample x, we can write

$$\frac{\text{Grams of Ni analyzed}}{\text{Grams of product formed}} = \frac{x}{0.179 \text{ 5}} = \frac{58.71}{288.93} \quad \Rightarrow \quad \text{Ni} = 0.036 \text{ 47 g}$$

EXAMPLE: A Problem with Two Components

A mixture made up of the 8-hydroxyquinoline complexes of aluminum and magnesium weighed 1.084 3 g. We do not know how much of each complex is in the mixture.

When ignited in a Bunsen burner flame, the mixture decomposed, leaving a residue of Al_2O_3 and MgO weighing 0.134 4 g.

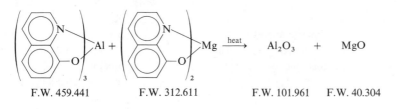

$$\text{F.W. 459.441} \qquad \text{F.W. 312.611} \qquad \text{F.W. 101.961} \qquad \text{F.W. 40.304}$$

Find the weight percent of $Al(C_9H_6NO)_3$ in the original mixture.

We will abbreviate the 8-hydroxyquinoline anion as Q. Letting the mass of AlQ_3 be x and the mass of MgQ_2 be y, we can write

$$\underset{\substack{\text{Mass of}\\AlQ_3}}{x} + \underset{\substack{\text{Mass of}\\MgQ_2}}{y} = 1.084\ 3\ \text{g}$$

The moles of Al will be $x/459.441$, and the moles of Mg will be $y/312.611$. The moles of Al_2O_3 will be one-half of the total moles of Al, since it takes two moles of Al to make one mole of Al_2O_3.

$$\text{Moles of } Al_2O_3 = \left(\frac{1}{2}\right)\frac{x}{459.441}$$

The moles of MgO will equal the moles of Mg $= y/312.611$. Now we can write

$$\overbrace{\underbrace{\left(\frac{1}{2}\right)\frac{x}{459.441}}_{\text{mol}}\underbrace{(101.961)}_{\text{g/mol}}}^{\text{Mass of } Al_2O_3} + \overbrace{\underbrace{\frac{y}{312.611}}_{\text{mol}}\underbrace{(40.304)}_{\text{g/mol}}}^{\text{Mass of MgO}} = 0.134\ 4\ \text{g}$$

Substituting $y = 1.084\ 3 - x$ into the equation above gives

$$\left(\frac{1}{2}\right)\left(\frac{x}{459.441}\right)(101.961) + \left(\frac{1.084\ 3 - x}{312.611}\right)(40.304) = 0.134\ 4\ \text{g}$$

from which we find $x = 0.300\ 3$ g, which is 27.70% of the original mixture.

Summary

Gravimetric analysis is based on the formation of a weighable product whose mass can be related to the mass of analyte. Most commonly, one ion is precipitated by the addition of a suitable counterion. The ideal product should be insoluble, easily filterable, and pure and possess a known composition. Measures taken to reduce supersaturation and promote the formation of large, easily filtered particles include (1) raising the temperature during precipitation, (2) slow addition and vigorous mixing of reagents, (3) maintaining a large sample volume, and (4) use of homogeneous precipitation. Colloid formation is the extreme of undesirability in gravimetric analysis. After a precipitate forms, it is usually digested in the mother liquor at an elevated temperature to promote particle growth and recrystallization. All precipitates are then filtered and washed; some must be washed with a volatile electrolyte to prevent peptization. In the final step, the precipitate is heated to dryness or ignited to achieve a reproducible, stable composition. All gravimetric calculations seek to relate moles of product to moles of analyte.

Terms to Understand

absorption
adsorption
colloid
combustion analysis
coprecipitation
dialysis
digestion
electric double layer

gathering
gravimetric analysis
homogeneous precipitation
hygroscopic
ignition
inclusion
masking agent
nucleation

occlusion
peptization
postprecipitation
precipitant
relative supersaturation
reprecipitation
thermogravimetric analysis

Exercises

8-A. An organic compound with a molecular weight of 417 was analyzed for ethoxyl (CH_3CH_2O—) groups using the reactions

$$ROCH_2CH_3 + HI \rightarrow ROH + CH_3CH_2I$$

$$CH_3CH_2I + Ag^+ + H_2O \rightarrow$$
$$AgI(s) + CH_3CH_2OH$$

A 25.42-mg sample of compound produced 29.03 mg of AgI. How many ethoxyl groups are there in each molecule?

8-B. A 0.649-g sample containing only K_2SO_4 (F.W. 174.27) and $(NH_4)_2SO_4$(F.W. 132.14) was dissolved in water and treated with $Ba(NO_3)_2$ to precipitate all of the SO_4^{2-} as $BaSO_4$ (F.W. 233.40). Find the weight percent of K_2SO_4 in the sample if 0.977 g of precipitate was formed.

8-C. Consider a mixture of the two solids $BaCl_2 \cdot 2H_2O$ and KCl, in an unknown ratio. (The notation $BaCl_2 \cdot 2H_2O$ means that a crystal is formed with two water molecules for each $BaCl_2$.) When the unknown is heated to 160°C for one hour, the water of crystallization is driven off:

$$BaCl_2 \cdot 2H_2O(s) \xrightarrow{160°C} BaCl_2(s) + 2H_2O(g)$$

A sample originally weighing 1.783 9 g weighed 1.562 3 g after heating. Calculate the weight percent of Ba, K, and Cl in the original sample.

8-D. A mixture containing only aluminum tetrafluoroborate, $Al(BF_4)_3$(F.W. 287.39), and magnesium nitrate, $Mg(NO_3)_2$ (F.W. 148.31), weighed 0.282 8 g. It was dissolved in 1% aqueous HF solution and treated with nitron solution to precipitate a mixture of nitron tetrafluoroborate and nitron nitrate weighing 1.322 g. Calculate the weight percent of Mg in the original solid mixture.

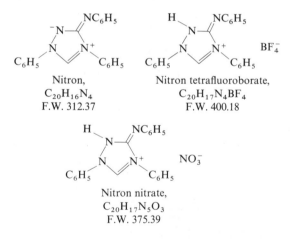

Nitron,
$C_{20}H_{16}N_4$
F.W. 312.37

Nitron tetrafluoroborate,
$C_{20}H_{17}N_4BF_4$
F.W. 400.18

Nitron nitrate,
$C_{20}H_{17}N_5O_3$
F.W. 375.39

Problems

A8-1. (a) What is the difference between absorption and adsorption?
(b) How is an inclusion different from an occlusion?

A8-2. State four desirable properties of a gravimetric precipitate.

A8-3. Why is a high relative supersaturation undesirable in a gravimetric precipitation?

A8-4. What measures can be taken to decrease the relative supersaturation during a precipitation?

A8-5. Why are many ionic precipitates washed with electrolyte solution instead of pure water?

A8-6. Why is it less desirable to wash AgCl precipitate with aqueous $NaNO_3$ than with HNO_3 solution?

A8-7. A 50.00-mL solution containing NaBr was treated with excess $AgNO_3$ to precipitate 0.214 6 g of

AgBr. What was the molarity of NaBr in the solution?

A8-8. To find the Ce(IV) content of a solid, 4.37 g was dissolved and treated with excess iodate to precipitate $Ce(IO_3)_4$. The precipitate was collected, washed well, dried, and ignited to produce 0.104 g of CeO_2. What was the weight percent of Ce in the original solid?

A8-9. Write a balanced equation for the combustion of benzoic acid, $C_6H_5CO_2H$, to give CO_2 and H_2O. How many milligrams of CO_2 and of H_2O will be produced by the combustion of 4.635 mg of benzoic acid?

A8-10. A 0.050 02-g sample of impure piperazine contained 71.29% piperazine. How many grams of product will be formed when this sample is analyzed by Reaction 8-9?

A8-11. Describe what is done in thermogravimetric analysis.

A8-12. Why would a reprecipitation be employed in a gravimetric analysis?

A8-13. A 1.000-g sample of unknown gave 2.500 g of bis(dimethylglyoximate)nickel(II) when analyzed by Reaction 8-10. Find the mass of Ni in the unknown.

A8-14. Referring to Figure 8-3, name the product when calcium salicylate monohydrate is heated to 550°C or 1 000°C. Using the formula weights of these products, calculate what mass is expected to remain when 0.635 6 g of calcium salicylate monohydrate is heated to 550°C or 1 000°C.

A8-15. A method for measurement of soluble organic carbon in seawater involves oxidation of the organic materials to CO_2 with $K_2S_2O_8$, followed by gravimetric determination of the CO_2 trapped by a column of NaOH-coated asbestos.[†] A water sample weighing 6.234 g produced 2.378 mg of CO_2. Calculate the ppm carbon in the seawater.

8-16. How many milliliters of 2.15% alcoholic dimethylglyoxime should be used to provide a 50.0% excess for Reaction 8-10 with 0.998 4 g of steel containing 2.07% Ni? Assume that the density of the dimethylglyoxime solution is 0.790 g/mL.

8-17. Twenty dietary iron tablets with a total mass of 22.131 g were ground and mixed thoroughly. Then 2.998 g of the powder was dissolved in HNO_3 and heated to convert all the iron to Fe(III). Addition of NH_3 caused quantitative precipitation of $Fe_2O_3 \cdot xH_2O$, which was ignited to give 0.264 g of Fe_2O_3 (F.W. 159.69). What is the average mass of $FeSO_4 \cdot 7H_2O$ (F.W. 278.01) in each tablet?

[†] P. Hannaker and A. S. Buchanan, *Anal. Chem.*, **55**, 1922 (1985).

8-18. A mixture weighing 7.290 mg contained only cyclohexane, C_6H_{12} (F.W. 84.161), and oxirane, C_2H_4O (F.W. 44.053). When the mixture was analyzed by combustion analysis, 21.999 mg of CO_2 (F.W. 44.010) was produced. Find the weight percent of oxirane in the mixture.

8-19. A 1.475-g sample containing NH_4Cl, K_2CO_3, and inert ingredients was dissolved to give 0.100 L of solution. A 25.0-mL aliquot was acidified and treated with excess sodium tetraphenylborate, $Na^+B(C_6H_5)_4^-$, to precipitate K^+ and NH_4^+ ions completely:

$$(C_6H_5)_4B^- + K^+ \rightarrow (C_6H_5)_4BK(s)$$
$$(C_6H_5)_4B^- + NH_4^+ \rightarrow (C_6H_5)_4BNH_4(s)$$

The resulting precipitate amounted to 0.617 g. A fresh 50.0-mL aliquot of the original solution was made alkaline and heated to drive off all the NH_3:

$$NH_4^+ + OH^- \rightarrow NH_3(g) + H_2O$$

It was then acidified and treated with sodium tetraphenylborate to give 0.554 g of precipitate. Find the weight percent of NH_4Cl and of K_2CO_3 in the original solid.

8-20. A mixture containing only Al_2O_3 and Fe_2O_3 weighs 2.019 g. When heated under a stream of H_2, the Al_2O_3 is unchanged, but the Fe_2O_3 is converted to metallic Fe plus $H_2O(g)$. If the residue weighs 1.774 g, what is the weight percent of Fe_2O_3 in the original mixture?

8-21. A solid mixture weighing 0.548 5 g contained only ferrous ammonium sulfate and ferrous chloride. The sample was dissolved in 1 M H_2SO_4, oxidized to Fe(III) with H_2O_2, and precipitated with cupferron. The ferric cupferron complex was ignited to produce 0.167 8 g of ferric oxide, Fe_2O_3 (F.W. 159.69). Calculate the weight percent of Cl in the original sample.

$FeSO_4 \cdot (NH_4)_2SO_4 \cdot 6H_2O$	$FeCl_2 \cdot 6H_2O$
Ferrous ammonium sulfate	Ferrous chloride
F.W. 392.13	F.W. 234.84

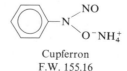

Cupferron
F.W. 155.16

8-22. A 2.000-g sample of a solid mixture containing only $PbCl_2$ (F.W. 278.1), $CuCl_2$ (F.W. 134.45), and KCl (F.W. 74.55) was dissolved in water to give 100.0 mL of solution. First 50.00 mL of the unknown solution was treated with sodium piperidine dithiocarbamate to precipitate 0.726 8 g of lead piperidine dithiocarbamate:

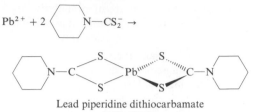

Lead piperidine dithiocarbamate
F.W. 527.74

Then 25.00 mL of the unknown solution was treated with iodic acid to precipitate 0.838 8 g of $Pb(IO_3)_2$ and $Cu(IO_3)_2$.

$$Cu^{2+} + 2IO_3^- \rightarrow Cu(IO_3)_2$$
F.W. 413.35

$$Pb^{2+} + 2IO_3^- \rightarrow Pb(IO_3)_2$$
F.W. 557.0

Calculate the weight percent of Cu in the unknown mixture.

8-23. *Propagation of error.* A mixture containing only silver nitrate and mercurous nitrate was dissolved in water and treated with excess sodium cobalticyanide, $Na_3[Co(CN_6)]$ to precipitate both cobalticyanide salts:

$$AgNO_3 \quad \text{F.W. 169.873}$$
$$Ag_3Co(CN)_6 \quad \text{F.W. 538.643}$$

$$Hg_2(NO_3)_2 \quad \text{F.W. 525.19}$$
$$(Hg_2)_3[Co(CN)_6]_2 \quad \text{F.W. 1 633.62}$$

(a) The unknown weighed 0.432 1 g and the product weighed 0.451 5 g. Find the weight percent of silver nitrate in the unknown. *Caution:* In this type of arithmetic you should keep all the digits in your calculator or else serious rounding errors may occur. Do not round off until the end of the calculation.

(b) Even a good analyst is not likely to have less than a 0.3% error in isolating the precipitate. Suppose that there is negligible (zero) error in all quantities in this experiment, except the mass of product. Suppose that the mass of product has an uncertainty of 0.30%. Calculate the *relative uncertainty* in the mass of silver nitrate in the unknown.

8-24. The *superconductor* $YBa_2Cu_3O_{7-x}$ has a variable oxygen content, where x is in the range 0 to $\frac{1}{2}$. One way to measure the value of x is by ther-

mogravimetric analysis in a stream of flowing H_2. When heated to 1 000°C the following reaction goes to completion:

$$YBa_2Cu_3O_{7-x} + (3.5 - x)H_2(g) \rightarrow$$
$$\tfrac{1}{2}Y_2O_3 + 2BaO + 3Cu + (3.5 - x)H_2O(g)$$

The starting material is a solid and the first three products are also solids.

(a) *Thermogravimetric analysis.* When 34.397 mg of superconductor was subjected to this analysis, 31.661 mg of solid remained after heating to 1 000°C. Find the value of x in the formula $YBa_2Cu_3O_{7-x}$ of the original material.

(b) *Propagation of Error.* Suppose that the uncertainty in each mass in part a is ± 0.002 mg. Find the uncertainty in the value of x.

8-25. *The Man in the Vat Problem.*[†] Long ago a workman at a dye factory fell into a vat containing a hot concentrated mixture of sulfuric and nitric acids. He dissolved completely! Since nobody witnessed the accident, it was necessary to prove that he fell in so that the man's wife could collect his insurance money. The man weighed 70 kg, and a human body contains about 6.3 ppt (parts per thousand) phosphorus. The acid in the vat was analyzed for phosphorus to see if it contained a dissolved human.

(a) The vat contained 8.00×10^3 L of liquid and 100.0 mL was analyzed. If the man did fall into the vat, what is the expected quantity of phosphorus in 100.0 mL?

(b) The 100.0 mL was treated with a molybdate reagent that caused ammonium phosphomolybdate, $(NH_4)_3[P(Mo_{12}O_{40})] \cdot 12H_2O$, to precipitate. This substance was dried at 110°C to remove waters of hydration and heated to 400°C until it reached a constant composition corresponding to the formula $P_2O_5 \cdot 24MoO_3$, which weighed 0.371 8 g. When a fresh mixture of the same acids (not from the vat) was treated in the same manner, 0.033 1 g of $P_2O_5 \cdot 24MoO_3$ was produced. This *blank determination* gives the amount of phosphorus in the starting reagents. The $P_2O_5 \cdot 24MoO_3$ that could have come from the dissolved man is therefore $0.371\ 8 - 0.033\ 1 = 0.338\ 7$ g. How much phosphorus was present in the 100.0-mL sample? Is this quantity consistent with a dissolved man?

[†] R. W. Ramette, *J. Chem. Ed.,* **65,** 800 (1988).

9 Precipitation Titrations

The use of precipitation reactions for gravimetric analysis is not very popular because of the time, labor, and skill required. However, many precipitation reactions can be adapted for accurate, easily performed titrations in which we measure the volume of precipitant needed for complete reaction. This tells us how much analyte was present. This chapter begins with a general discussion of titrations and then proceeds to the details of precipitation titrations. We will illustrate titration calculations for many kinds of reactions, not just precipitations, because the principles of the calculations apply to every reaction.

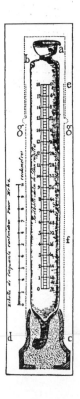

Figure 9-1
Earliest known buret, invented by F. Descroizilles in the early 1800s, was used in the same manner as a graduated cylinder is used today. This device and its progeny have terrorized generations of analytical chemistry students. From E. R. Madsen, *The Development of Titrimetric Analysis 'till 1806* (Copenhagen: C. E. C. Gad Publishers, 1958). For more history, see C. Duval, *J. Chem. Ed.,* **28,** 508 (1951) and A. Johansson, *Anal. Chim. Acta,* **206,** 97 (1988).

9-1 PRINCIPLES OF VOLUMETRIC ANALYSIS

In **volumetric analysis** the volume of reagent needed to react with the **analyte** (that being analyzed) is measured. In a **titration,** increments of the reagent solution—the **titrant**—are added to the analyte until their reaction is complete. The most common procedure is to deliver titrant from a buret, as shown in Figures 9-1 and 9-2.

Titrations can be based on any kind of chemical reaction. The principal requirements for the reaction are that it be *complete* (have a large equilibrium

Titration reactions should ideally be complete and rapid.

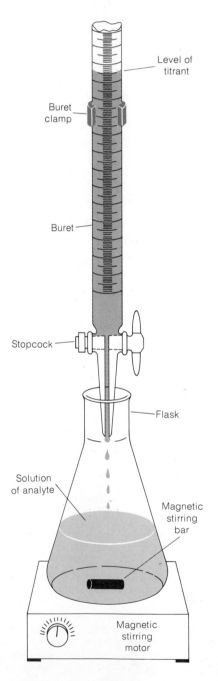

Figure 9-2
Typical setup for a titration. The analyte is contained in the flask, and the titrant in the buret. The magnetic stirring bar is a magnet coated with Teflon, which is inert to almost all solutions. The bar is spun by a rotating magnet inside the stirring motor.

Level of titrant

Buret clamp

Buret

Stopcock

Flask

Solution of analyte

Magnetic stirring bar

Magnetic stirring motor

constant) and *rapid*. The most common titrations are based on acid–base, oxidation–reduction, complex formation, or precipitation reactions.

Methods of determining when the reaction is complete are discussed in several places in this text. The usual methods involve observing an indicator color change, monitoring a spectrophotometric absorbance change, or detecting changes in current or voltage between a pair of electrodes immersed in the analyte solution. An **indicator** is a compound that has a physical property (usually color) that changes abruptly near the equivalence point. The change is caused by the sudden disappearance of analyte or appearance of titrant at the equivalence point.

The **equivalence point** is that point in the titration at which the quantity of titrant added is the exact amount necessary for stoichiometric reaction with the analyte. For example, oxalic acid reacts with permanganate in hot acidic solution as follows:

Indicators are discussed in the following sections:

precipitation indicators: 9-7

acid–base indicators: 11-5

metal ion indicators: 13-4

redox indicators: 16-3

$$5\text{HOCCOH} + 2\text{MnO}_4^- + 6\text{H}^+ \longrightarrow 10\text{CO}_2 + 2\text{Mn}^{2+} + 8\text{H}_2\text{O} \quad (9\text{-}1)$$

Oxalic acid (colorless) Permanganate (purple) (Colorless) (Colorless)

Suppose the analyte contains 5.000 mmol of oxalate. The equivalence point is reached when 2.000 mmol of MnO_4^- has been added, since *two moles of permanganate are required for five moles of oxalic acid.*

The equivalence point is the ideal result we seek in a titration. What we actually measure is the **end point.** The end point is marked by a sudden change in a physical or chemical property of the solution. In this particular case, the most convenient end point is the appearance of the purple color of permanganate in the flask. Up to the equivalence point, all the added permanganate is consumed by the oxalate, and the analyte solution remains colorless. After the equivalence point, unreacted MnO_4^- ion builds up until there is enough to color the solution purple. The appearance of a *trace* of purple color marks the end point. The better your eyes are, the closer will be your measured end point to the true equivalence point. Here the end point cannot exactly equal the equivalence point because extra MnO_4^-, beyond that needed to react with oxalic acid, is required to show the purple color.

The difference between the end point and the equivalence point is an inescapable **titration error.** By choosing an appropriate physical property, in which a change is easily observed (such as indicator color, optical absorbance of a reactant or product, pH, or conductivity), it is possible to observe and record the end point very close to the true equivalence point. It is also usually possible to estimate the titration error by using a **blank titration.** In the foregoing example, a solution containing no oxalic acid could be titrated with MnO_4^- under conditions otherwise identical to those of the original experiment. The amount of MnO_4^- needed to show a purple color in the blank titration is subtracted from the needed amount observed in the oxalic acid titration. Subtracting the blank reading from the observed end point reading accurately locates the equivalence point.

The validity of an analytical procedure depends on knowing the amount of one of the reactants used. In our discussion so far, we have assumed that we know the concentration of the titrant in the buret. A known concentration is possible if the titrant was prepared by dissolving a weighed amount of pure reagent in a known volume of solution. In such a case, we call the reagent a **primary standard,** since it is pure enough to be weighed and used directly. To be useful, a **primary standard** should be 99.9% pure, or better. It should not decompose under ordinary storage, and it should be stable when dried (by heating or vacuum).[†]

In most cases, however, the titrant is not available as a primary standard. Instead, a solution having the approximate desired concentration is used to titrate a weighed, primary standard. By this procedure, called **standardization,** we determine the precise concentration of the solution to be used in the analysis. We then say that the solution is a **standard solution.** In all cases, the validity of an analytical result ultimately depends on knowing the composition of some primary standard. The National Institute of Standards and Technology sells many analyzed reference materials that can be used to test the accuracy of analytical procedures (Box 9-1).

In a **direct titration,** titrant is added to the analyte until the reaction is complete. Occasionally it is more convenient to perform a **back titration.** This is done by *adding* a known *excess* of reagent to the analyte. Then a second reagent is used to titrate the *excess* of the first reagent. Back titrations are most useful when the end point of the back titration is clearer than the end point of the direct titration or when an excess of the first reagent is necessary for complete reaction with the analyte.

> A reagent is *standardized* by titration against a primary standard.

9-2 CALCULATIONS OF VOLUMETRIC ANALYSIS

In volumetric analysis our calculations serve to *relate the moles of titrant to the moles of analyte.* If the reaction can be written

$$tT + aA \rightarrow \text{products} \tag{9-2}$$

then t moles of titrant (T) are required to react with a moles of analyte (A).

The moles of titrant delivered are

$$\text{Moles of T} = V_T M_T \tag{9-3}$$

where V_T is the volume of T and M_T is the molarity of T in the buret. The moles of A that must have been present to react with this many moles of T are

$$\text{Moles of A} = \frac{a}{t} (\text{moles of T}) \tag{9-4}$$

since a moles of A react with t moles of T. For example, if the reaction were $2T + 3A \rightarrow$ products, then at the equivalence point

$$\text{Moles of A} = \tfrac{3}{2} (\text{moles of T}) \tag{9-5}$$

The following examples illustrate calculations of volumetric analysis for precipitation, oxidation–reduction, and acid–base reactions. The principles of the calculations are the same for all types of reactions.

> *Calculations:*
>
> $tT + aA \rightarrow$ products
>
> $\text{Moles of T} = V_T M_T$
>
> $\text{Moles of A} = \dfrac{a}{t} (\text{moles of T})$

[†] Recommended primary standards for many elements are given in Table 24-8.

Box 9-1   STANDARD REFERENCE MATERIALS

Inaccurate laboratory measurements can mean wrong medical diagnosis and treatment, lost production time, wasted energy and materials, manufacturing rejects, and product liability problems. To minimize errors in laboratory measurements, the U.S. National Institute of Standards and Technology distributes more than 1 000 *standard reference materials*, such as metals, chemicals, rubber, plastics, engineering materials, radioactive substances, and environmental and clinical standards. These are carefully assayed samples that can be used to test the accuracy of analytical procedures used in different laboratories.[†]

For example, in the management of patients with epilepsy, physicians depend on laboratory tests to ascertain that the blood serum concentrations of anticonvulsant drugs are in the correct range. Low levels lead to seizures and high levels to drug toxicity. Tests of identical serum specimens revealed unacceptably large differences in results among different laboratories.[‡] The National Institute of Standards and Technology was asked to develop a standard reference material containing known levels of antiepilepsy drugs in serum. The reference material would allow different laboratories to detect and correct errors in their assay procedures.

Before introduction of this reference material, called SRM 900, five different laboratories analyzing identical samples reported a range of results with relative errors of 40–110% of the expected value. The graph below shows that after SRM 900 distribution the error in the analyses was reduced to 20–40%.

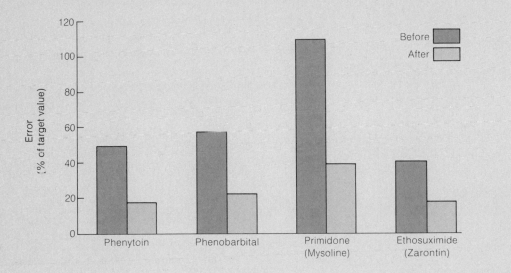

Drug	Before use of SRM 900		After use of SRM 900	
	Target	Range Reported	Target	Range reported
Phenytoin	23.9	17.9–30.0	16.7	14.1–17.0
Phenobarbital	19.2	14.2–25.0	21.6	20.0–24.4
Primidone	3.0	1.0– 4.3	8.1	6.6– 9.5
Ethosuximide	38.9	33.1–48.0	75.9	70.2–83.7

[†] A brochure describing standard reference materials can be obtained from the Office of Standard Reference Materials, Room B-311, Chemistry Building, National Institute of Standards and Technology, Gaithersburg MD 20899.

[‡] C. E. Peppinger, J. K. Penny, B. G. White, D. D. Daly, and R. Buddington, *Arch. Neurol.*, **33**, 351 (1976).

EXAMPLE: Volumetric Chloride Determination

The chloride content of blood serum, cerebrospinal fluid, or urine can be measured by titration of the chloride with mercuric ion:

$$Hg^{2+} + 2Cl^- \rightarrow HgCl_2(aq) \qquad (9\text{-}6)$$

When the reaction is complete, excess Hg^{2+} appears in the solution. This is detected with the indicator diphenylcarbazone, which forms a violet-blue complex with Hg^{2+}.

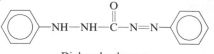

Diphenylcarbazone

Mercuric nitrate was standardized by titrating a solution containing 147.6 mg of NaCl in about 25 mL of water. It required 28.06 mL of mercuric nitrate solution to reach the violet-blue indicator end point. When this same Hg^{2+} solution was used to titrate 2.000 mL of urine, 22.83 mL was required. (a) Calculate the molarity of Hg^{2+} in the titrant and (b) find the concentration of Cl^- (mg/mL) in the urine.

(a) The moles of Cl^- in 147.6 mg of NaCl are

$$\frac{147.6 \times 10^{-3}\text{g NaCl}}{58.443 \text{ g NaCl/mol}} = 2.526 \times 10^{-3} \text{ mol}$$

According to Reaction 9-6, two moles of Cl^- are required for each mole of Hg^{2+}. The moles of Hg^{2+} used in the reaction must have been

$$\text{Moles of } Hg^{2+} = \tfrac{1}{2}(\text{moles of } Cl^-) = 1.263 \times 10^{-3} \text{ mol}$$

This many moles are contained in 28.06 mL, so the molarity of Hg^{2+} is

$$\text{Molarity of } Hg^{2+} = \frac{1.263 \times 10^{-3} \text{ mol}}{28.06 \times 10^{-3} \text{ L}} = 0.045\,01 \text{ M}$$

As a point of useful information, you should be able to use the units mg, mL, and mmol together in the same way you use g, L, and mol together:

$$\frac{147.6 \text{ mg}}{58.443 \text{ mg/mmol}} = 2.526 \text{ mmol}$$

$$\frac{(\tfrac{1}{2})(2.526 \text{ mmol})}{28.06 \text{ mL}} = 0.045\,01 \, \frac{\text{mmol}}{\text{mL}} = 0.045\,01 \, \frac{\text{mol}}{\text{L}}$$

(b) In the titration of 2.000 mL of urine, 22.83 mL of Hg^{2+} was required. The moles of Hg^{2+} used were

$$\text{Moles of } Hg^{2+} = (22.83 \times 10^{-3} \text{ L})(0.045\,01 \text{ M}) = 1.028 \times 10^{-3} \text{ mol}$$

Since one mole of Hg^{2+} reacts with two moles of Cl^-, the moles of Cl^- in the 2.000-mL sample must have been

$$\text{Moles of } Cl^- = 2(\text{moles of } Hg^{2+}) = 2.056 \times 10^{-3} \text{ mol}$$

The quantity of Cl^- in one milliliter is half of this value, or 1.028×10^{-3} mol.

This much Cl⁻ weighs

$$(1.028 \times 10^{-3} \text{ mol})(35.453 \text{ g Cl/mol}) = 36.45 \text{ mg}$$

so the concentration of Cl^- is 36.45 mg/mL of urine.

EXAMPLE: Volumetric Calcium Determination

The calcium content of urine can be determined by the following procedure:

1. Ca^{2+} is precipitated as calcium oxalate in basic solution:

$$Ca^{2+} + C_2O_4^{2-} \rightarrow Ca(C_2O_4) \cdot H_2O(s)$$

$$\text{Oxalate} \qquad \text{Calcium oxalate}$$

2. After the precipitate is washed to remove any free oxalate, the solid is dissolved in acid.

3. The dissolved oxalic acid is heated to 60°C and titrated with standardized potassium permanganate until the purple end point is observed:

$$5H_2C_2O_4 + 2MnO_4^- + 6H^+ \rightarrow 10CO_2 + 2Mn^{2+} + 8H_2O$$

$$\text{(Colorless)} \quad \text{(Purple)} \qquad \text{(Colorless)} \quad \text{(Colorless)}$$

Suppose that 0.356 2 g of $Na_2C_2O_4$ is dissolved in a total volume of 250.0 mL. If 10.00 mL of this solution requires 48.36 mL of $KMnO_4$ solution for titration, find the molarity of the permanganate solution.

The concentration of the oxalate solution is

$$\frac{0.356\ 2 \text{ g Na}_2\text{C}_2\text{O}_4/(134.00 \text{ g Na}_2\text{C}_2\text{O}_4/\text{mol})}{0.250\ 0 \text{ L}} = 0.010\ 63 \text{ M}$$

The moles of $C_2O_4^{2-}$ in 10.00 mL are

$$\left(0.010\ 63 \frac{\text{mol}}{\text{L}}\right)(0.010\ 0 \text{ L}) = 1.063 \times 10^{-4} \text{ mol} = 0.106\ 3 \text{ mmol}$$

Since two moles of MnO_4^- are required for five moles of oxalate, the moles of MnO_4^- delivered must have been

$$\text{Moles of MnO}_4^- = \tfrac{2}{5}(\text{moles of C}_2\text{O}_4^{2-}) = 0.042\ 53 \text{ mmol}$$

The concentration of MnO_4^- is

$$\text{Molarity of MnO}_4^- = \frac{0.042\ 53 \text{ mmol}}{48.36 \text{ mL}} = 8.795 \times 10^{-4} \text{ M}$$

The key step in this calculation is to note that five moles of oxalate require two moles of permanganate. You should look at the answer to be sure that the moles of MnO_4^- are *less* than the moles of oxalate. If not, you probably used the factor $\tfrac{5}{2}$ instead of $\tfrac{2}{5}$ in the calculation.

Suppose that the calcium in a 5.00-mL urine sample is precipitated with excess oxalate, and the excess oxalate is washed away from the solid. After the precipitate is dissolved in acid, it requires 16.17 mL of the standard MnO_4^- solution to titrate the oxalate. Find the molar concentration of Ca^{2+} in the urine.

In 16.17 mL of MnO_4^- there are $(0.016\ 17\ L)(8.795 \times 10^{-4}\ mol/L) = 1.422 \times 10^{-5}$ mol of MnO_4^-. This will react with

$$\text{Moles of oxalate} = \tfrac{5}{2}(\text{moles of } MnO_4^-) = 3.555 \times 10^{-5}\ mol$$

Since there is one oxalate ion for each calcium ion in $CaC_2O_4 \cdot H_2O$, there must have been 3.555×10^{-5} mol of Ca^{2+} in 5.00 mL of urine. The concentration of Ca^{2+} is

$$\frac{3.555 \times 10^{-5}\ mol}{5.00 \times 10^{-3}\ L} = 0.007\ 11\ \text{M}$$

EXAMPLE: Kjeldahl Nitrogen Determination

The most accurate and reliable analysis of protein uses the Kjeldahl (pronounced KEL-dall) nitrogen determination, which is applicable to a wide variety of organic compounds. The sample is first chemically digested with boiling H_2SO_4 containing a catalyst that converts all the organic nitrogen to NH_4^+. The solution is made basic, converting the ammonium ion to NH_3, which is distilled out into a *known excess* of aqueous HCl. The HCl in excess of the amount needed to react with NH_3 is then *back-titrated* with NaOH to determine how much NH_3 was collected.

Digestion:	$N \text{ (in protein)} \rightarrow NH_4^+$	(9-7)
Distillation of NH_3:	$NH_4^+ + OH^- \rightarrow NH_3(g) + H_2O$	(9-8)
Collection of NH_3 in HCl:	$NH_3 + H^+ \rightarrow NH_4^+$	(9-9)
Titration of unreacted HCl:	$H^+ + OH^- \rightarrow H_2O$	(9-10)

Suppose that a 0.500-mL aliquot of protein solution is analyzed by the Kjeldahl procedure. (**Aliquot** is a fancy word for *portion*.) The protein is known to contain 16.2% (wt/wt) nitrogen (a typical figure for proteins). The liberated ammonia was collected in 10.00 mL of 0.021 40 M HCl, and the unreacted acid required 3.26 mL, of 0.0198 M NaOH for complete titration. Find the concentration of protein (milligrams of protein per milliliter) in the original solution.

The total moles of HCl in the flask receiving the NH_3 is

$$(10.00\ mL)(0.021\ 40\ mmol/mL) = 0.214\ 0\ mmol\ HCl$$

The NaOH required for titration of the unreacted HCl is

$$(3.26\ mL)(0.019\ 8\ mmol/mL) = 0.064\ 6\ mmol\ NaOH$$

The difference between the moles of HCl and the moles of NaOH must equal the moles of NH_3 participating in Reaction 9-9.

$$0.214\ 0\ mmol\ HCl - 0.064\ 6\ mmol\ NaOH = 0.149\ 4\ mmol\ NH_3$$

Since one mole of nitrogen in the protein gives rise to one mole of NH_3, there must have been 0.149 4 mmol of nitrogen in the protein sample. This much nitrogen weighs

$$(0.149\ 4\ mmol)\left(14.006\ 7\ \frac{mg\ N}{mmol}\right) = 2.093\ mg\ nitrogen$$

If the protein contains 16.2% (wt/wt) nitrogen, there must be

$$\frac{2.093\ mg\ N}{0.162\ mg\ N/mg\ protein} = 12.9\ mg\ protein$$

This corresponds to

$$\frac{12.9 \text{ mg protein}}{0.500 \text{ mL}} = 25.8 \text{ mg protein/mL}$$

9-3 MASS TITRATIONS

An alternative way to conduct titrations is to deliver known *masses* of solutions instead of known *volumes*. Such **mass titrations** afford comparable or greater precision than ordinary volumetric titrations and use less reagent. At a time when waste disposal is a critical problem, mass titrations allow us to perform typical analyses with one-tenth as much reagent as that needed for a 50-mL buret.

A convenient device for mass titrations in student laboratories is the buret–syringe–whose construction is shown in Figure 9-3. The body is made from a "disposable" 5-mL plastic syringe. The shaft and handle of the plunger are cut away from the rubber end and discarded. The shaft is replaced by a long, finely threaded bolt mounted on the aluminium base, whose central hole is threaded to accommodate the bolt. Two nuts and bolts join the aluminum to the plastic base of the syringe.

Operation of the buret–syringe in Figure 9-4 begins by removing the metal bolt and base assembly. A larger syringe is used to suck reagent solution

Advantages of mass titrations:

1. Smaller quantities of reagents required
2. High precision

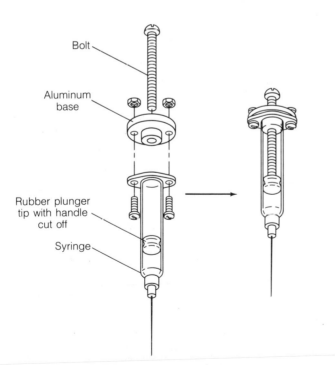

Figure 9-3
Construction of buret–syringe for mass titrations. Adapted from D. D. Siemer, S. D. Reeder, and M. A. Wade, *J. Chem. Ed.*, **65**, 467 (1988).

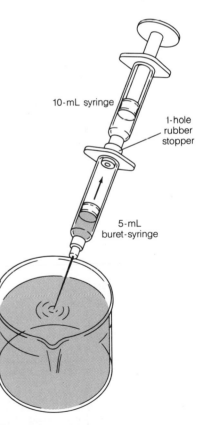

Figure 9-4
Use of second syringe to load buret–syringe. From D. D. Siemer, S. D. Reeder, and M. A. Wade, *J. Chem. Ed.*, **65**, 467 (1988).

into the buret–syringe, with the aid of a rubber stopper to make an air-tight seal between the two syringes. The large syringe is then replaced by the aluminum base and bolt. The buret–syringe is inverted and air bubbles are gently expelled by advancing the bolt to push the plunger. The mass of the loaded buret–syringe is measured to the nearest milligram on a balance. To do this, place the buret–syringe upside down in a beaker or flask on the balance pan. The pan should have a wind screen to prevent air currents in the lab from affecting the reading. The buret–syringe is then used to deliver titrant to a solution and is reweighed to find the mass of reagent delivered. Careful advancement of the bolt allows you to deliver increments smaller than 5 mg.

How does this method compare to a conventional titration with a 50-mL buret? The 50-mL buret can be read to 0.01 mL, which is a reading precision of 0.01 mL/50 mL = 0.02%. A full drop is usually 0.05 mL, which is 0.05/50 = 0.1% of the buret volume. A careful person can deliver a fraction of a drop. The precision of reading the buret–syringe is 1 mg. If the syringe volume is 5 mL, the contents weigh about 5 g. Therefore the buret–syringe reading precision is 0.001 g/5 g = 0.02%, the same as that of a 50-mL buret. With care, you can deliver increments approaching 1 mg. The 5-mL buret–syringe therefore allows you to obtain the same precision as a 50 mL buret, but using one-tenth as much reagent.

Thermal expansion of solutions is discussed in Section 2-9.

The concentrations of solutions used for mass titrations are conveniently expressed as moles of solute per kilogram of *solution*. (This is not the same as molality, which is moles of solute per kilogram of *solvent*.) *An added bonus is that mol/kg does not change when the temperature of the solution changes.* In contrast, since solutions expand or contract when temperature changes, the molarity of a reagent on a warm day is different from the molarity of the same solution on a cool day.

The concentrations of solutions used for small-scale experiments should be similar to those used for large-scale experiments. Only the total volumes or masses are decreased. This practice maintains the sharpness of end points in small-scale experiments. If reagents were diluted, end points would be harder to observe. The syringe is compatible with most aqueous reagents. The steel needle can be used for HNO_3, H_2SO_4, and $HClO_4$, but HCl or HBr attack the metal. Dilute bases, most oxidants (Chapter 16), and most complexing agents (Chapter 13) are compatible with a steel needle.

EXAMPLE: Preparing Titrant for a Mass Titration

A solution of potassium dichromate, $K_2Cr_2O_7$, was needed for titration of an unknown containing ferrous ion, Fe^{2+}. The desired $K_2Cr_2O_7$ concentration was about 0.03 M. Therefore approximately 0.003 mol was dissolved in approximately 0.1 L. The exact amount of $K_2Cr_2O_7$ was 0.893 8 g and the total mass of solution (water plus $K_2Cr_2O_7$) was 102.346 g. What is the concentration of this solution expressed as mol solute/kg solution?

Since the formula weight of $K_2Cr_2O_7$ is 294.185 g/mol, the concentration is

$$\text{concentration} = \frac{\text{mol of reagent}}{\text{kg of solution}} = \frac{(0.893\ 8\ \text{g})/(294.185\ \text{g/mol})}{0.102\ 346\ \text{kg}} = 0.029\ 69\ \text{mol/kg}$$

EXAMPLE: A Mass Titration

An unknown solid containing Fe^{2+} was dissolved in sulfuric plus phosphoric acid. (In this solution Fe^{2+} is not rapidly oxidized to Fe^{3+} by atmospheric oxygen.) Then 0.308 5 g of unknown was titrated with the solution from the previous example:

$$Cr_2O_7^{2-} + 6Fe^{2+} + 14H^+ \rightarrow 2Cr^{3+} + 6Fe^{3+} + 14H_2O$$

If 3.622 g of dichromate solution was required to reach the end point (detected by an indicator color change), what was the weight percent of Fe^{2+} in the unknown?

The quantity of dichromate delivered was

$$(3.622 \times 10^{-3} \text{ kg})(0.029 \text{ } 69 \text{ mol/kg}) = 0.107 \text{ } 5 \text{ mmol } Cr_2O_7^{2-}$$

Since each mole of $Cr_2O_7^{2-}$ reacts with 6 moles of Fe^{2+}, there must have been $6(0.107 \text{ } 5 \text{ mmol}) = 6.452 \text{ mmol of } Fe^{2+}$ present, or 0.036 03 g of Fe^{2+} (using the atomic weight of Fe). This corresponds to a weight percent of $100 \times (0.036 \text{ } 03 \text{ g of } Fe^{2+}/0.308 \text{ } 5 \text{ g of unknown}) = 11.68\%$.

9-4 EXAMPLE OF A PRECIPITATION TITRATION

A sensitive (but not very precise) determination of sulfate is by precipitation titration with Ba^{2+}:

$$SO_4^{2-} + Ba^{2+} \rightarrow BaSO_4(s) \qquad (9\text{-}11)$$

The best means of detecting the end point is to measure the light scattered by particles of precipitate. A liquid containing very fine particles that scatter light is said to be *turbid* (Box 9-2).

In the $BaSO_4$ precipitation, the **turbidity** (light scattering) increases until the equivalence point is reached. The abrupt leveling-off of the scattering marks the end point. Since scattering is very dependent on particle size, the reaction must be carried out in a very reproducible manner to obtain precise results. A glycerol–alcohol mixture is used to help stabilize the particles and prevent the rapid settling of a mass of solid.

Alternatively, the end point can be detected by using an indicator that forms a colored precipitate with Ba^{2+}. In this procedure, excess standard $BaCl_2$ is added to precipitate all the sulfate. The indicator disodium rhodizonate is added to give a red precipitate with excess Ba^{2+}:

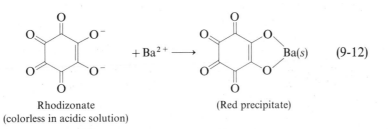

Rhodizonate
(colorless in acidic solution)

(Red precipitate)

$(9\text{-}12)$

Back titration with standard SO_4^{2-} converts the red precipitate to colorless $BaSO_4$. The disappearance of the last trace of color marks the end point.

The back titration is necessary because the free indicator is not stable under the conditions of the titration.

Box 9-2   TURBIDIMETRY AND NEPHELOMETRY

A suspension of particles scatters light and is said to be *turbid*. In **turbidimetry** the suspension is placed in a spectrophotometer cuvet and the "absorbance" is measured with an ordinary spectrophotometer.[†] Light is not really absorbed by the solution, but is scattered in all directions and does not reach the detector.

The apparent absorption usually obeys an equation analogous to Beer's law over some limited range of concentration:

$$\text{"}A\text{"} = \log_{10} \frac{P_0}{P} = kbc$$

where "A" is the apparent absorbance, P_0 is the radiant power of the incident light, P is the radiant power of emergent light, k is a constant, b is the path length, and c is the "concentration" of precipitate. The value of k is determined empirically with a series of standards. The equation above can be rearranged to the form

$$\text{"}T\text{"} = \frac{P}{P_0} = e^{-\tau b}$$

where "T" is the "transmittance" of the turbid solution and τ is called the **turbidity coefficient.**

In **nephelometry** the light scattered at 90° to the incident beam by the turbid solution is measured. The power of the scattered beam is usually proportional to the concentration of particles over some limited range of concentrations. An empirical curve is used to relate the intensity of the scattered beam to the concentration of particles.

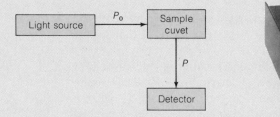

(a) Schematic diagram of nephelometer.

(b) Monitek digital nephelometer. [Courtesy A. H. Thomas Co., Philadelphia, Pa.]

In turbidimetry we measure the *fraction* of light scattered (the "transmittance"). This is independent of the intensity of the light source. In nephelometry we measure the *absolute intensity* of the scattered light. Even though the fraction of scattered light is constant, the observed intensity of scattered light increases if the source intensity is increased. Therefore, the sensitivity of nephelometry can be raised simply by increasing the power of the light source or by using a more sensitive detector. Being more sensitive than turbidimetry, nephelometry is more useful at very low concentrations of analyte.

Turbidimetric and nephelometric titrations are less precise than others because end-point detection depends on particle size, which is not very reproducible. However, the sensitivity is excellent; sulfate can be analyzed at the parts-per-millon level.

[†] The basic concepts of absorption spectroscopy, including Beer's law, will be used from time to time in this text before its formal introduction in Chapter 19. We assume that most readers are familiar with this materials. If you are not, Sections 19-1 and 19-2 will give you enough background to understand all that we do with spectroscopy prior to Chapter 19.

Deriving a theoretical titration curve will help you to appreciate what is happening in precipitation titrations. The titration curve is a graph showing how the concentration of one of the reactants varies as titrant is added. Rather than plotting concentration directly, it is more useful to plot the p function, defined as

$$pX = -\log_{10} \mathscr{A}_X = -\log[X]\gamma_X \qquad (9\text{-}13)$$

where $\mathscr{A}_X$ is the activity of X, $[X]$ is the concentration of X, and γ_X is the activity coefficient of X. If we neglect activity coefficients, as will usually be our custom, we can consider the p function to be given by

$$pX \approx -\log[X] \qquad (9\text{-}14)$$

Consider the titration of 25.00 mL of 0.100 0 M I^- with 0.050 00 M Ag^+

$$I^- + Ag^+ \rightarrow AgI(s) \qquad (9\text{-}15)$$

and suppose that we are monitoring the Ag^+ concentration with an electrode. Reaction 9-15 is the reverse of the dissolution of $AgI(s)$, whose solubility product is rather small:

$$AgI(s) \rightleftharpoons Ag^+ + I^- \qquad K_{sp} = [Ag^+][I^-] = 8.3 \times 10^{-17} \qquad (9\text{-}16)$$

Since the equilibrium constant for the titration reaction 9-15 is large ($K = 1/K_{sp} = 1.2 \times 10^{16}$), the equilibrium lies far to the right. It is reasonable to say that each aliquot of Ag^+ reacts completely with I^-, leaving a very tiny amount of Ag^+ in solution. At the equivalence point, there will be a sudden increase in the Ag^+ concentration because all the I^- has been consumed and we are now adding Ag^+ directly to the solution.

What volume of Ag^+ solution is needed to reach the equivalence point? To calculate this volume, which we designate V_e (volume of titrant at the equivalence point), first note that one mole of Ag^+ is required for each mole of I^-.

$$mol\ I^- = mol\ Ag^+$$

$$(0.025\,00\ L)(0.100\,0\ mol\ I^-/L) = (V_e)(0.050\,00\ mol\ Ag^+/L) \qquad (9\text{-}17)$$

$$V_e = 0.050\,00\ L = 50.00\ mL$$

An easy way to find V_e in this example is to note that the concentration of Ag^+ is half the concentration of I^-. Therefore, the volume of Ag^+ required will be twice as great as the volume of I^-, and 25.00 mL of I^- will require 50.00 mL of Ag^+.

The titration curve has three distinct regions. The calculations are different depending on whether we are before, at, or after the equivalence point. We will consider each region separately.

This is the correct definition of the p function.

This is our usual working definition of the p function.

V_e = volume of titrant at equivalence point

Before the Equivalence Point

Consider the point at which the volume of Ag^+ added is 10.00 mL. Since there are more moles of I^- than Ag^+ at this point, virtually all the Ag^+ is "used up" to make $AgI(s)$. We want to find the very small concentration of Ag^+ remaining in solution after reaction with I^-. One way to do this is to imagine that Reaction 9-15 has gone to completion and that some AgI redissolves (Reaction 9-16). The solubility of Ag^+ will be determined by the concentration of free I^- remaining in the solution:

When $V < V_e$, the concentration of unreacted I^- regulates the solubility of AgI.

$$[Ag^+] = \frac{K_{sp}}{[I^-]} \tag{9-18}$$

The free I^- is overwhelmingly due to the I^- that has not been precipitated by 10.00 mL of Ag^+. By comparison, the I^- due to dissolution of $AgI(s)$ is negligible.

So let's find the concentration of unprecipitated I^-. The moles of I^- remaining in solution will be

Moles of I^- = original moles of I^- − moles of Ag^+ added

$$= (0.025\ 00\ L)(0.100\ mol/L) - (0.010\ 00\ L)(0.050\ 00\ mol/L)$$

$$= 0.002\ 000\ mol\ I^- \tag{9-19}$$

Since the volume is 0.035 00 L (25.00 mL + 10.00 mL), the concentration is

$$[I^-] = \frac{0.002\ 000\ mol\ I^-}{0.035\ 00\ L} = 0.057\ 14\ \text{M} \tag{9-20}$$

The concentration of Ag^+ in equilibrium with this much I^- is given by

$$[Ag^+] = \frac{K_{sp}}{[I^-]} = \frac{8.3 \times 10^{-17}}{0.057\ 14} = 1.4_5 \times 10^{-15}\ \text{M} \tag{9-21}$$

The p function we seek is given by

$$pAg^+ = -\log[Ag^+] = 14.84 \tag{9-22}$$

Two details are worth noting. First, we have neglected activity coefficients in this calculation. A more rigorous calculation would include activity coefficients in Equations 9-21 and 9-22. Second, there are two significant figures in the concentration of Ag^+ because there are two significant figures in K_{sp}. The two figures in $[Ag^+]$ translate into two figures in the *mantissa* of the p function. The p function is therefore correctly written as 14.84.

The step-by-step calculation just outlined is a safe, but tedious, way to find the concentration of I^- at this point in the titration. We will now examine a streamlined procedure that is well worth learning. The streamlined calculation requires us to bear in mind that $V_e = 50.00$ mL. When 10.00 mL of Ag^+ has been added, the reaction is one-fifth complete because 10.00 mL out of the 50.00 mL of Ag^+ needed for complete reaction has been added. There-

$$\log(1.4_5 \times 10^{-15}) = \underbrace{14}_{\text{Two significant figures}}.\underbrace{84}_{\text{Two digits in mantissa}}$$

If you need a review of significant figures in logarithms, see Section 3-1.

fore, four-fifths of the I^- remains unreacted. If there were no dilution, the concentration of I^- would be four-fifths of its original value. However, the original volume of 25.00 mL has been increased to 35.00 mL. If no I^- had been consumed, the concentration would be the original value of $[I^-]$ times (25.00/35.00). Accounting for both the reaction and the dilution, we can write

$$[I^-] = \underbrace{\left(\frac{4.000}{5.000}\right)}_{\substack{\text{Fraction}\\\text{remaining}}} \underbrace{(0.100\ 0\ \text{M})}_{\substack{\text{Original}\\\text{concentration}}} \underbrace{\left(\frac{25.00}{35.00}\right)}_{\substack{\text{Dilution}\\\text{factor}}} = 0.057\ 14\ \text{M} \tag{9.23}$$

Original volume of I^-

Total volume of solution

This streamlined calculation is *well worth using.*

This is the same result found from Equation 9-20.

Just to reinforce the streamlined procedure, let's calculate pAg^+ when $V_{Ag^+} = 49.00$ mL. (V_{Ag^+} is the volume of Ag^+ added from the buret.) Since $V_e = 50.00$ mL, the fraction of I^- reacted is 49.00/50.00, and the fraction remaining is 1.00/50.00. The total volume is $25.00 + 49.00 = 74.00$ mL.

$$[I^-] = \left(\frac{1.00}{50.00}\right)(0.100\ 0\ \text{M})\left(\frac{25.00}{74.00}\right) = 6.76 \times 10^{-4}\ \text{M} \tag{9-24}$$

$$[Ag^+] = K_{sp}/[I^-] = 1.2_3 \times 10^{-13}\ \text{M} \tag{9-25}$$

$$pAg^+ = -\log[Ag^+] = 12.91 \tag{9-26}$$

The value of $pAg^+ = 12.91$ shows that the concentration of Ag^+ is negligible compared with the concentration of unreacted I^-. This is true even though the titration is 98% complete.

At the Equivalence Point

Now we have added exactly enough Ag^+ to react with all the I^-. You may imagine that all the AgI precipitates and some redissolves to give equal concentrations of Ag^+ and I^-. The value of pAg^+ is found very simply:

$$[Ag^+][I^-] = K_{sp}$$

$$(x)(x) = 8.3 \times 10^{-17} \quad \Rightarrow \quad x = 9.1 \times 10^{-9} \tag{9-27}$$

$$pAg^+ = -\log x = 8.04 \tag{9-28}$$

When $V = V_e$, the concentration of Ag^+ is determined by the solubility of pure AgI.

This value of pAg^+ is independent of the original concentration or volumes.

After the Equivalence Point

Now the concentration of Ag^+ is determined almost completely by the amount of Ag^+ added after the equivalence point. Virtually all the Ag^+ added before the equivalence point has precipitated as AgI. Suppose that $V_{Ag^+} = 52.00$ mL.

When $V > V_e$, the concentration of Ag^+ is determined by the excess Ag^+ added from the buret.

The amount added past the equivalence point is 2.00 mL. The calculation proceeds as follows:

$$\text{mol } Ag^+ = (0.002\,00 \text{ L})(0.050\,00 \text{ mol } Ag^+/\text{L}) = 0.000\,100 \text{ mol} \quad (9\text{-}29)$$

$$[Ag^+] = (0.000\,100 \text{ mol})/(0.077\,00 \text{ L}) = 1.30 \times 10^{-3} \text{M} \quad (9\text{-}30)$$

Total volume = 77.00 mL

$$pAg^+ = 2.89 \quad (9\text{-}31)$$

It would be justified to use three significant figures for the mantissa of pAg^+, since there are now three significant figures in the value of $[Ag^+]$. To be consistent with our earlier results, we will retain only two figures. *For consistency, we will generally express all p functions in this text with two decimal places.*

A somewhat streamlined calculation can save time. The concentration of Ag^+ in the buret is 0.050 00 M, and 2.00 mL of this solution is being diluted to $(25.00 + 52.00) = 77.00$ mL. Hence $[Ag^+]$ is given by

This is the streamlined calculation.

$$[Ag^+] = (0.050\,00 \text{ M})\left(\frac{2.00}{77.00}\right) = 1.30 \times 10^{-3} \text{M} \quad (9\text{-}32)$$

Volume of excess Ag^+

Total volume of solution

Original concentration of Ag^+ — Dilution factor

The Shape of the Titration Curve

Figure 9-5 shows the complete titration curve and also illustrates the effect of reactant concentrations on the titration. The equivalence point is the steepest point of the curve. It is the point of maximum slope (a negative slope in this case) and is therefore an inflection point:

Steepest slope: $\dfrac{dy}{dx}$ reaches its greatest value

Inflection point: $\dfrac{d^2y}{dx^2} = 0$

A *complexometric* titration involves complex formation between titrant and analyte.

In titrations involving 1:1 stoichiometry of reactants, the equivalence point is the steepest point of the titration curve. This is true of acid–base, complexometric, and redox titrations as well. For stoichiometries other than 1:1, such as that for the reaction $2Ag^+ + CrO_4^{2-} \rightarrow Ag_2CrO_4(s)$, the curve is not symmetric near the equivalence point. The equivalence point is not at the center of the steepest section of the curve, and it is not an inflection point. In practice, titration curves are steep enough that the steepest point is almost always taken as a good estimate of the equivalence point, regardless of the stoichiometry.

Figure 9-6 illustrates the role of the magnitude of K_{sp} in the titration of halide ions. It is clear that the least-soluble product, AgI, gives the sharpest

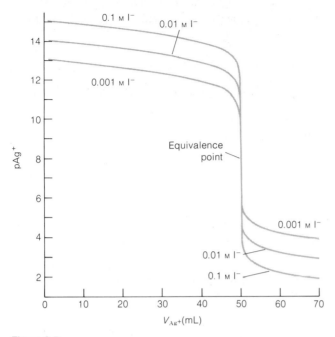

Figure 9-5
Titration curves showing the effect of concentration.

OUTER CURVE: 25.00 mL of 0.100 0 M I^- titrated with 0.050 00 M Ag^+.

MIDDLE CURVE: 25.00 mL of 0.010 00 M I^- titrated with 0.005 000 M Ag^+.

INNER CURVE: 25.00 mL of 0.001 000 M I^- titrated with 0.000 500 0 M Ag^+.

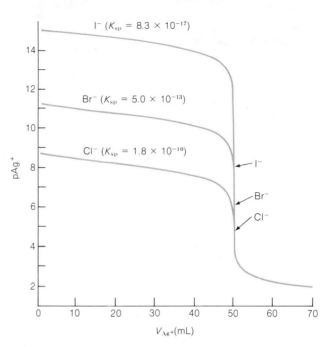

Figure 9-6
Titration curves showing the effect of K_{sp}. Each curve is calculated for 25.00 mL of 0.100 0 M halide titrated with 0.050 00 M Ag^+. Equivalence points are marked by arrows.

change at the equivalence point. However, even for AgCl, the curve is steep enough to locate the equivalence point with very little error. The larger the equilibrium constant for any titration reaction, the more pronounced will be the change in concentration near the equivalence point.

At the equivalence point, the titration curve is steepest for the least-soluble precipitates.

EXAMPLE: Calculating Concentrations during a Precipitation Titration
A solution containing 25.00 mL of 0.041 32 M $Hg_2(NO_3)_2$ was titrated with 0.057 89 M KIO_3.

$$Hg_2^{2+} + 2IO_3^- \rightarrow Hg_2(IO_3)_2(s)$$
$$\text{Iodate}$$

The solubility product for $Hg_2(IO_3)_2$ is 1.3×10^{-18}. Calculate the concentration of Hg_2^{2+} ion in the solution (a) after addition of 34.00 mL of KIO_3, (b) after addition of 36.00 mL of KIO_3, and (c) at the equivalence point.

The reaction requires two moles of IO_3^- per mole of Hg_2^{2+}. The volume of iodate needed to reach the equivalence point is found as follows:

$$\text{Moles of } IO_3^- = 2(\text{moles of } Hg_2^{2+}) \quad \left\{ \begin{array}{l} \text{Notice which side of} \\ \text{the equation has the 2} \end{array} \right.$$

$$(V_e)(0.057\ 89\ \text{M}) = 2(25.00\ \text{mL})(0.041\ 32\ \text{M}) \quad \Rightarrow \quad V_e = 35.69\ \text{mL}$$

(a) When $V = 34.00$ mL, the precipitation of Hg_2^{2+} is not yet complete.

$$[Hg_2^{2+}] = \underbrace{\left(\frac{35.69 - 34.00}{35.69}\right)}_{\substack{\text{Fraction} \\ \text{remaining}}} \underbrace{(0.041\ 32)}_{\substack{\text{Original} \\ \text{concentration} \\ \text{of } Hg_2^{2+}}} \underbrace{\left(\frac{25.00}{25.00 + 34.00}\right)}_{\substack{\text{Dilution} \\ \text{factor}}} = 8.29 \times 10^{-4}\ \text{M}$$

Original volume of Hg_2^{2+}

Total volume of solution

(b) When $V = 36.00$ mL, the precipitation is complete. We have gone $(36.00 - 35.69) = 0.31$ mL *past* the equivalence point. The concentration of excess IO_3^- in the solution is

$$[IO_3^-] = \underbrace{(0.057\ 89)}_{\substack{\text{Original} \\ \text{concentration} \\ \text{of } IO_3^-}} \underbrace{\left(\frac{0.31}{25.00 + 36.00}\right)}_{\substack{\text{Dilution} \\ \text{factor}}} = 2.9 \times 10^{-4}\ \text{M}$$

Volume of excess IO_3^-

Total volume of solution

The concentration of Hg_2^{2+} in equilibrium with solid $Hg_2(IO_3)_2$ plus this much IO_3^- is

$$[Hg_2^{2+}] = \frac{K_{sp}}{[IO_3^-]^2} = \frac{1.3 \times 10^{-18}}{(2.9 \times 10^{-4})^2} = 1.5 \times 10^{-11}\ \text{M}$$

(c) At the equivalence point, we have added only enough IO_3^- to react with all the Hg_2^{2+}. We can write

$$Hg_2(IO_3)_2(s) \rightleftharpoons Hg_2^{2+} + 2IO_3^-$$

$$ x \qquad 2x$$

$$(x)(2x)^2 = K_{sp} \quad \Rightarrow \quad x = [Hg_2^{2+}] = 6.9 \times 10^{-7}\ \text{M}$$

9-6 TITRATION OF A MIXTURE

When a mixture is titrated, the product with the *smaller* K_{sp} precipitates first, if the stoichiometry of the different possible precipitates is the same.

The precipitation of I^- and Cl^- with Ag^+ produces two distinct breaks in the titration curve. The first corresponds to the reaction of I^- and the second to the reaction of Cl^-.

If a mixture of two precipitable ions is titrated, the less-soluble precipitate will be formed first. If the two solubility products are sufficiently different, the first precipitation will be nearly complete before the second commences.

Consider the titration with $AgNO_3$ of a solution containing KI and KCl. Since $K_{sp}(AgI) \ll K_{sp}(AgCl)$, the first Ag^+ added will precipitate AgI. Further addition of Ag^+ continues to precipitate I^- with no effect on Cl^-. When precipitation of I^- is almost complete, the concentration of Ag^+ abruptly increases. Then, when the concentration of Ag^+ is high enough, AgCl begins to precipitate and $[Ag^+]$ levels off again. Finally, when the Cl^- is consumed, another abrupt change in $[Ag^+]$ occurs. Qualitatively, we expect to see two breaks in the titration curve. The first corresponds to the AgI equivalence point, and the second to the AgCl equivalence point.

Figure 9-7 shows an experimental curve for this titration. The apparatus used to measure the curve is shown in Figure 9-8, and the theory of how this system measures Ag^+ concentration is discussed in Chapter 15.

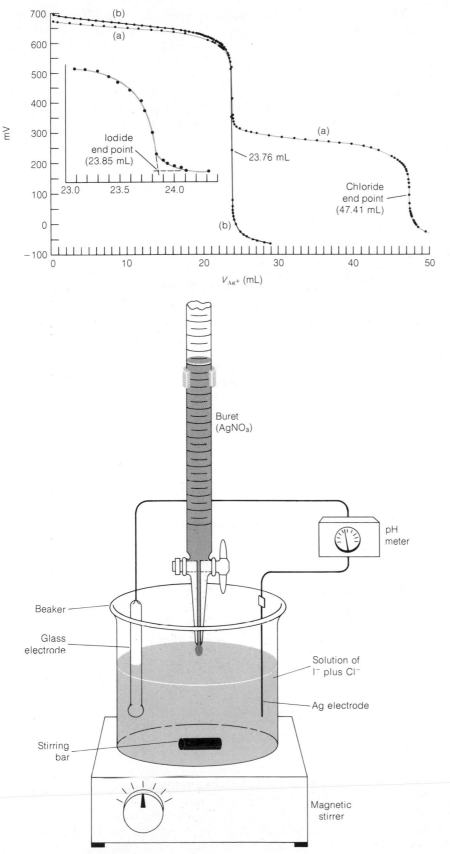

Figure 9-7
Experimental titration curves. (a) Titration curve for 40.0 mL of 0.050 2 M KI plus 0.050 0 M KCl titrated with 0.084 5 M AgNO₃. The inset is an expanded view of the region near the first equivalence point. (b) Titration curve for 20.00 mL of 0.100 4 M I⁻ titrated with 0.084 5 M Ag⁺.

Figure 9-8
Apparatus for measuring the titration curves in Figure 9-7. The silver electrode responds to changes in the Ag⁺ concentration, and the glass electrode provides a constant reference potential in this experiment. The measured voltage changes by approximately 59 mV for each factor-of-10 change in [Ag⁺]. All solutions, including AgNO₃, were maintained at pH 2.0 using 0.010 M sulfate buffer prepared from H₂SO₄ and KOH.

Before Cl^- precipitates, the calculations for AgI precipitation are just as they were in Section 9-5.

The I^- end point is taken as the intersection of the steep and nearly horizontal curves shown in the inset of Figure 9-7. The reason for using the intersection is that the precipitation of I^- is not quite complete as Cl^- begins to precipitate. Therefore, the end of the steep portion (the intersection) is a better approximation of the equivalence point than is the middle of the steep section. The Cl^- end point is taken as the midpoint of the second steep section, at 47.41 mL. The moles of Cl^- in the sample correspond to the moles of Ag^+ delivered between the first and second end points. That is, it requires 23.85 mL of Ag^+ to precipitate I^-, and $(47.41 - 23.85)$ mL of Ag^+ to precipitate Cl^-.

Comparing the I^-/Cl^- and pure I^- titration curves in Figure 9-7 shows that the iodide end point is 0.38% too high in the I^-/Cl^- titration. We expect the first end point at 23.76 mL, but it is observed at 23.85 mL. There are two factors contributing to this high value. One is experimental error, which is always present. This is as likely to be positive as negative. However, the end point in some titrations, especially Br^-/Cl^- titrations, is found to be systematically 0–3% high, depending on conditions. This has been attributed to a small amount of **coprecipitation** of AgCl with AgBr. Even though the solubility product of AgCl has not been exceeded, a little Cl^- precipitates with the Br^- and carries down a corresponding amount of Ag^+. A high concentration of an anion such as nitrate has been found to minimize the coprecipitation.

The second end point in Figure 9-7 corresponds to the total precipitation of both halides. It is observed at the expected value of V_{Ag^+}. The concentration of Cl^-, found from the *difference* between the two end points, will be slightly low in Figure 9-7, because the first end point is slightly high.

9-7 END-POINT DETECTION

Three techniques are commonly employed to detect the end point in precipitation titration:

1. potentiometric methods—using electrodes, as in Figure 9-8 (these techniques are discussed in Chapter 15)
2. indicator methods—discussed in this section
3. light-scattering methods—turbidimetry and nephelometry (see Box 9-2)

In the remainder of this section we will discuss three types of indicator methods applied to the titration of Cl^- with Ag^+. Titrations with Ag^+ are called **argentometric titrations.** The three indicator methods are

1. **Mohr titration**—formation of a colored precipitate at the end point
2. **Volhard titration**—formation of a soluble, colored complex at the end point
3. **Fajans titration**—adsorption of a colored indicator on the precipitate at the end point

Mohr Titration

A nice classroom demonstration of the Mohr titration is described by R. J. Stolzberg, *J. Chem, Ed.*, **65**, 621 (1988).

In the Mohr titration, Cl^- is titrated with Ag^+ in the presence of CrO_4^{2-} (chromate, dissolved as Na_2CrO_4).

Titration reaction: $\quad Ag^+ + Cl^- \underset{K_{sp}}{\overset{K_f}{\rightleftharpoons}} AgCl(s)$ $\quad$ (9-33)

(White)

End-point reaction: $\quad 2Ag^+ + CrO_4^{2-} \rightarrow Ag_2CrO_4(s)$ $\quad$ (9-34)

(Red)

The AgCl precipitates before Ag_2CrO_4. The color of AgCl is white; dissolved CrO_4^{2-} is yellow; and Ag_2CrO_4 is red. The AgCl end point is indicated by the first appearance of red Ag_2CrO_4. Reasonable control of CrO_4^{2-} concentration and pH is required for Ag_2CrO_4 precipitation to occur at the desired point in the titration.

Since a certain amount of excess Ag_2CrO_4 is necessary for visual detection, the color is not seen until after the true equivalence point. We can correct for this titration error in two ways. One is by means of a blank titration with no chloride present. The volume of Ag^+ needed to form detectable red color is then subtracted from V_{Ag^+} in the Cl^- titration. Alternatively, we can standardize the $AgNO_3$ by the Mohr method, using a standard NaCl solution and conditions similar to those for the titration of unknown. The Mohr method is useful for Cl^-, Br^-, and CN^-, but not for I^- or SCN^- (thiocyanate).

$$K_{sp} = [Ag^+][Cl^-]$$

$$K_f = \frac{1}{[Ag^+][Cl^-]}$$

$$= \frac{1}{K_{sp}}$$

Titration error arises because some excess Ag^+ is needed to form a detectable amount of red precipitate.

5.6×10^9

9.1×10^{11}

Volhard Titration

The Volhard titration is actually a procedure for the titration of Ag^+. To determine Cl^-, a back titration is necessary. First, the Cl^- is precipitated by a known, excess quantity of standard $AgNO_3$.

$$Ag^+ + Cl^- \rightarrow AgCl(s) \qquad (9\text{-}35)$$

The AgCl is isolated, and the excess Ag^+ is titrated with standard KSCN in the presence of Fe^{3+}.

$$Ag^+ + SCN^- \rightarrow AgSCN(s) \qquad (9\text{-}36)$$

When all the Ag^+ has been consumed, the SCN^- reacts with Fe^{3+} to form a red complex.

$$Fe^{3+} + SCN^- \rightarrow FeSCN^{2+} \qquad (9\text{-}37)$$

(Red)

The appearance of the red color signals the end point. Knowing how much SCN^- was required for the back titration tells us how much Ag^+ was left over from the reaction with Cl^-. Since the total amount of Ag^+ is known, the amount consumed by Cl^- can then be calculated.

In the analysis of Cl^- by the Volhard method, the end point slowly fades because AgCl is more soluble than AgSCN. The AgCl slowly dissolves and is replaced by AgSCN. To prevent this from happening, two techniques are commonly used. One is to filter off the AgCl and titrate only the Ag^+ in the filtrate. An easier procedure is to shake a few millimeters of nitrobenzene,

Since the Volhard method is a titration of Ag^+, it can be adapted for the determination of any anion that forms an insoluble silver salt.

$C_6H_5NO_2$, with the precipitated AgCl prior to the back titration. Nitrobenzene coats the AgCl and effectively isolates it from attack by SCN^-. Br^- and I^-, whose silver salts are *less* soluble than AgSCN, may be titrated by the Volhard method without isolating the silver halide precipitate.

Fajans Titration

The Fajans titration uses an **adsorption indicator.** To see how this works, we must recall the electrical phenomena associated with precipitate formation. When Ag^+ is added to Cl^-, there will be excess Cl^- ions in solution prior to the equivalence point. Some Cl^- is selectively adsorbed on the AgCl surface, imparting a negative charge to the crystal surface. After the equivalence point, there is excess Ag^+ in solution. Adsorption of the Ag^+ cations on the crystal surface creates a positive charge on the particles of precipitate. The abrupt change from negative charge to positive charge occurs at the equivalence point.

See Figure 8-2 and its associated discussion for a description of the *electric double layer* surrounding a particle of precipitate.

The common adsorption indicators are anionic dyes, which are attracted to the positively charged particles of precipitate produced immediately after the equivalence point. The adsorption of the negatively charged dye on the positively charged surface changes the color of the dye by interactions that are not well understood. The color change signals the end point in the titration. Because the indicator reacts with the precipitate surface, it is desirable to have as much surface area as possible. This means performing the titration under conditions that tend to keep the particles as small as possible. (Small particles have more surface area than an equal volume of large particles.) This condition is the opposite of that required for gravimetric analysis, in which large, easily filterable particles are desired. Low electrolyte concentration helps to prevent coagulation of the precipitate and maintain small particle size.

The indicator most commonly used for AgCl is dichlorofluorescein.

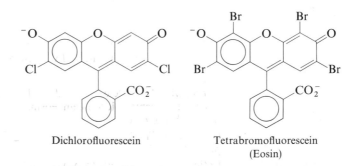

Dichlorofluorescein   Tetrabromofluorescein
(Eosin)

This dye has a greenish-yellow color in solution but turns pink when it is adsorbed on AgCl (see Demonstration 9-1). Because the indicator is a weak acid, and must be present in its anionic form, the pH of the reaction must be controlled. The dye eosin is useful in the titration of Br^-, I^-, and SCN^-. It gives a sharper end point than dichlorofluorescein and is more sensitive (i.e., less halide is required for titration). It cannot be used for AgCl because the eosin anion is more strongly bound to AgCl than is Cl^- ion. Eosin will bind to the AgCl crystallites even before the particles become positively charged.

In all argentometric titrations, but especially with adsorption indicators, strong light (such as daylight through a window) should be avoided. Light causes decomposition of the silver salts, and adsorbed indicators are especially light-sensitive.

Scope of Indicator Methods

To indicate the scope of indicator methods in precipitation titrations, a few applications are listed in Table 9-1. Whereas the Mohr and Volhard methods are specifically applicable to argentometric titrations, the Fajans method is used for a wider variety of titrations. Because the Volhard titration is carried out in acidic solution (typically 0.2 M HNO_3), it avoids certain interferences that would affect other titrations. The silver salts of such anions as CO_3^{2-}, $C_2O_4^{2-}$, and AsO_4^{3} are soluble in acidic solution, so these anions do not interfere with the analysis.

Table 9-1
Some applications of precipitation titrations

Species analyzed	Notes
Cl^-, Br^-	MOHR METHOD Ag_2CrO_4 end point is used.
Br^-, I^-, SCN^-, CNO^-, AsO_4^{3-}	VOLHARD METHOD Precipitate removal is unnecessary.
Cl^-, PO_4^{3-}, CN^-, $C_2O_4^{2-}$, CO_3^{2-}, S^{2-}, CrO_4^{2-}	Precipitate removal required.
BH_4^-	Back titration of Ag^+ left after reaction with BH_4^-: $$BH_4^- + 8Ag^+ + 8OH^- \rightarrow 8Ag(s) + H_2BO_3^- + 5H_2O$$
K^+	K^+ is first precipitated with a known excess of $(C_6H_5)_4B^-$. Remaining $(C_6H_5)_4B^-$ is precipitated with a known excess of Ag^+. Unreacted Ag^+ is then titrated with SCN^-.
Cl^-, Br^-, I^-, SCN^-, $Fe(CN)_6^{4-}$	FAJANS METHOD Titration with Ag^+. Detection with such dyes as fluorescein, dichlorofluorescein, eosin, bromophenol blue.
F^-	Titration with $Th(NO_3)_4$ to produce ThF_4. End point detected with alizarin red S.
Zn^{2+}	Titration with $K_4Fe(CN)_6$ to produce $K_2Zn_3[Fe(CN)_6]_2$. End-point detection with diphenylamine.
SO_4^{2-}	Titration with $Ba(OH)_2$ in 50% (vol/vol) aqueous methanol using alizarin red S as indicator.
Hg_2^{2+}	Titration with NaCl to produce Hg_2Cl_2. End point detected with bromophenol blue.
PO_4^{3-}, $C_2O_4^{2-}$	Titration with $Pb(CH_3CO_2)_2$ to give $Pb_3(PO_4)_2$ or PbC_2O_4. End point detected with dibromofluorescein (PO_4^{3-}) or fluorescein ($C_2O_4^{2-}$).

Demonstration 9-1   FAJANS TITRATION

The Fajans titration of Cl^- with Ag^+ convincingly demonstrates the utility of indicator end points in precipitation titrations. Dissolve 0.5 g of NaCl plus 0.15 g of dextrin in 400 mL of water. The purpose of the dextrin is to retard coagulation of the AgCl precipitate. Add 1 mL of dichlorofluorescein indicator. The indicator solution contains 1 mg/mL of dichlorofluorescein in 95% aqueous ethanol or 1 mg/mL of the sodium salt in water. Titrate the NaCl solution with a solution containing 2 g of $AgNO_3$ in 30 mL. About 20 mL are required to reach the end point.

Color Plate 4a shows the yellow color of the indicator in the NaCl solution prior to the titration. Color Plate 4b shows the milky-white appearance of the AgCl suspension during titration, before the end point is reached. The pink suspension in Color Plate 4c appears at the end point, when the anionic indicator becomes adsorbed to the cationic particles of precipitate.

Summary

In a titration the volume of titrant required for complete reaction is used to calculate the quantity of analyte present. If the stoichiometry of the reaction is $tT + aA \rightarrow$ products, the moles of titrant (T) delivered are $V_T M_T$, and the moles of analyte (A) are (a/t)(moles of T). We approximate the equivalence point by measuring the end point, at which a sudden change in a physical property (such as indicator color, pH, electrode potential, conductivity, or absorbance) occurs. The difference between the measured end point and the true equivalence point, called the titration error, can be estimated by performing a blank titration. The utility of any titration depends on knowledge of the titrant concentration. This is determined by dissolving a known mass of primary standard in a known volume of solution. Alternatively, titrant can be standardized by titrating a known mass of primary standard. In a mass titration, the mass of titrant is used to find the quantity of unknown. Mass titrations require less reagent than is usually used in volumetric titrations. In a direct titration, titrant is added to analyte and the volume of titrant is measured. In a back titration, a known excess of reagent is added to analyte, and the excess is titrated with a second standard reagent.

In a precipitation titration, analyte is precipitated by the titrant. The end point can be detected by potentiometric, light-scattering, or indicator methods. The most general indicator method, the Fajans titration, is based on the adsorption of a charged indicator by the charged surface of the precipitate after the equivalence point. The Volhard titration, used to measure Ag^+, is based on the reaction of Fe^{3+} with SCN^- after the precipitation of AgSCN is complete. The Mohr titration utilizes the precipitation of colored Ag_2CrO_4 after the argentometric titration of Cl^- or Br^-.

The concentrations of reactants and products during a precipitation titration are calculated in three ways. Before the end point, there is excess analyte. The concentration of titrant can be found from the solubility product of the precipitate and the known concentration of excess analyte. At the equivalence point, the concentrations of both reactants are governed by the dissociation equilibrium of product. After the equivalence point, the concentration of analyte can be determined from the solubility product of precipitate and the known concentration of excess titrant.

Terms to Understand

adsorption indicator
aliquot
analyte
argentometric titration
back titration
blank titration
coprecipitation

direct titration
end point
equivalence point
Fajans titration
indicator
mass titration
Mohr titration

nephelometry
primary standard
standard solution
standardization
titrant
titration error
turbidimetry

turbidity
turbidity coefficient
Volhard titration
volumetric analysis

Exercises

9-A. Ascorbic acid (vitamin C) reacts with I_3^- according to

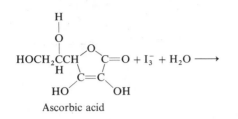

Ascorbic acid

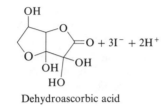

Dehydroascorbic acid

Starch is used as an indicator in the reaction. The end point is marked by the appearance of a deep blue starch–iodine complex when the first drop of unreacted I_3^- remains in the solution.
(a) If 29.41 mL of I_3^- solution is required to react with 0.197 0 g of pure ascorbic acid, what is the molarity of the I_3^- solution?
(b) A vitamin C tablet containing ascorbic acid plus an inert binder was ground to a powder, and 0.424 2 g was titrated by 31.63 mL of I_3^-. Find the weight percent of ascorbic acid in the tablet.

9-B. A solid sample weighing 0.237 6 g contained only malonic acid (F.W. 104.06) and anilinium chloride (F.W. 129.59). If it required 34.02 mL of 0.087 71 M NaOH to neutralize the sample, find the weight percent of each component in the solid mixture. The reactions are

$$CH_2(CO_2H)_2 + 2OH^- \rightarrow CH_2(CO_2^-)_2 + H_2O$$
Malonic acid   Malonate

Anilinium chloride

—$NH_3^+Cl^- + OH^- \longrightarrow$

—$NH_2 + H_2O + Cl^-$

Aniline

9-C. A solution of NaOH was standardized by titration of a known quantity of the primary standard, potassium acid phthalate:

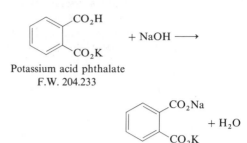

Potassium acid phthalate
F.W. 204.233

The NaOH was then used to find the concentration of an unknown solution of H_2SO_4:

$$H_2SO_4 + 2NaOH \rightarrow Na_2SO_4 + H_2O$$

(a) Titration of 0.824 g of potassium acid phthalate required 38.314 g of NaOH solution to reach the end point detected by phenolphthalein indicator. Find the concentration of NaOH (mol NaOH/kg solution).
(b) A 10.00 mL aliquot of H_2SO_4 solution required 57.911 g of NaOH solution to reach the phenolphthalein end point. Find the molarity of H_2SO_4.

9-D. Suppose that 50.0 mL of 0.080 0 M KSCN is titrated with 0.040 M Cu^+. The solubility product of CuSCN is 4.8×10^{-15}. At each of the following volumes of titrant, calculate pCu^+, and construct a graph of pCu^+ versus milliliters of Cu^+ added: 0.10, 10.0, 25.0, 50.0, 75.0, 95.0, 99.0, 99.9, 100.0, 100.1, 101.0, 110.0 mL.

9-E. A 40.0-ml solution of 0.040 0 M $Hg_2(NO_3)_2$ was titrated with 60.0 mL of 0.100 M KI to precipitate $Hg_2I_2(s)$ ($K_{sp} = 4.5 \times 10^{-29}$).
(a) What volume of KI is needed to reach the equivalence point?
(b) Calculate the ionic strength of the solution when 60.0 mL of KI has been added.
(c) Without ignoring activity, calculate pHg_2^{2+} ($= -\log \mathcal{A}_{Hg_2^{2+}}$) when 60.0 mL of KI has been added.

9-F. Construct a graph of pAg^+ versus milliliters of Ag^+ for the titration of 40.00 mL of solution containing 0.050 00 M Br^- and 0.050 00 M Cl^-. The titrant is 0.084 54 M $AgNO_3$. Calculate pAg^+ at the following volumes: 2.00, 10.00, 22.00, 23.00, 24.00, 30.00, 40.00, second equivalence point, 50.00 mL.

9-G. Consider the titration of 50.00 (± 0.05) mL of a mixture of I^- and SCN^- with 0.068 3 (± 0.000 1) M Ag^+. The first equivalence point is observed at 12.6 (± 0.4) mL, and the second occurs at 27.7 (± 0.3) mL.

(a) Find the molarity and the uncertainty in molarity of thiocyanate in the original mixture.

(b) Suppose that the uncertainties listed above are all the same, except that the uncertainty of the first equivalence point ($12.6 \pm$? mL) is variable. What is the maximum uncertainty (milliliters) of the first equivalence point if the uncertainty in SCN^- molarity is to be $\leq 4.0\%$?

9-H. Sketch graphs showing how (a) the turbidimetric "transmittance" and (b) the nephelometric scattered light intensity would vary during the titration of SO_4^{2-} by Ba^{2+}. Indicate the end point in each case.

Problems

A9-1. Explain the following statement: "The validity of an analytical result ultimately depends on knowing the composition of some primary standard."

A9-2. Distinguish between the terms *end point* and *equivalence point*.

A9-3. Why does the surface charge of a precipitate particle change sign at the equivalence point?

A9-4. Examine the procedure in Table 9-1 for the Fajans titration of Zn^{2+}. Do you expect the charge on the precipitate to be positive or negative after the equivalence point?

A9-5. What is the difference between a direct titration and a back titration?

A9-6. Describe how to analyze a solution of NaI using the Volhard titration.

A9-7. Why is nitrobenzene used in the Volhard titration of chloride?

A9-8. Why is a blank titration useful in the Mohr titration?

A9-9. How many milliliters of 0.100 M KI are needed to react with 40.0 mL of 0.040 0 M $Hg_2(NO_3)_2$ if the reaction is $Hg_2^{2+} + 2I^- \rightarrow Hg_2I_2(s)$?

A9-10. For Reaction 9-1, how many milliliters of 0.165 0 M $KMnO_4$ are needed to react with 108.0 mL of 0.165 0 M oxalic acid? How many milliliters of 0.165 0 M oxalic acid are required to react with 108.0 mL of 0.165 0 M $KMnO_4$?

A9-11. How many milligrams of oxalic acid dihydrate, $H_2C_2O_4 \cdot 2H_2O$, will react with 1.00 mL of 0.027 3 M ceric sulfate ($Ce(SO_4)_2$) if the reaction is $H_2C_2O_4 + 2Ce^{4+} \rightarrow 2CO_2 + 2Ce^{3+} + 2H^+$?

A9-12. A solution of $HClO_4$ with a mass of 4.618 g required 3.123 g of KOH solution (containing 0.064 88 mol KOH/kg solution) for complete titration. Find the concentration of the $HClO_4$ solution (mol $HClO_4$/kg solution).

A9-13. Consider the titration of 25.00 mL of 0.082 30 M KI with 0.051 10 M $AgNO_3$. Calculate pAg^+ at the following volumes of added $AgNO_3$:
(a) 39.00 mL (b) V_e (c) 44.30 mL

9-14. Sulfur can be determined by combustion analysis, which produces a mixture of SO_2 and SO_3. The gas stream is passed through H_2O_2 to convert these products to H_2SO_4, which is titrated with standard base. When 6.123 mg of a substance was burned, the resulting H_2SO_4 required 3.01 mL of 0.015 76 M NaOH for titration. What is the weight percent of sulfur in the sample?

9-15. The Kjeldahl procedure (Reactions 9-7 through 9-10) was used to analyze 256 μL of a solution containing 37.9 mg protein/mL. The liberated NH_3 was collected in 5.00 mL of 0.033 6 M HCl, and the remaining acid required 6.34 mL of 0.010 M NaOH for complete titration. What is the weight percent of nitrogen in the protein?

9-16. Ammonia reacts with hypobromite, OBr^-, according to

$$2NH_3 + 3OBr^- \rightarrow N_2 + 3Br^- + 3H_2O$$

What is the molarity of a hypobromite solution if 1.00 mL of the OBr^- solution reacts with 1.69 mg of NH_3?

9-17. Arsenious oxide (As_2O_3) is available in pure form and is a useful (and poisonous) primary standard for standardizing many oxidizing agents, such as MnO_4^-. The As_2O_3 is first dissolved in base and then titrated with MnO_4^- in acidic solution. A small amount of iodide (I^-) or iodate (IO_3^-) is used to catalyze the reaction between H_3AsO_3 and MnO_4^-. The reactions are

$$As_2O_3 + 4OH^- \rightleftharpoons 2HAsO_3^{2-} + H_2O$$
$$HAsO_3^{2-} + 2H^+ \rightleftharpoons H_3AsO_3$$
$$5H_3AsO_3 + 2MnO_4^- + 6H^+ \rightleftharpoons$$
$$5H_3AsO_4 + 2Mn^{2+} + 3H_2O$$

(a) A 3.214-g aliquot of $KMnO_4$ (F.W. 158.034) was dissolved in 1.000 L of water, heated to cause any reactions with impurities to occur, cooled, and filtered. What is the theoretical molarity of this solution if no MnO_4^- was consumed by impurities?

(b) What mass of As_2O_3 (F.W. 197.84) would be just sufficient to react with 25.00 mL of the $KMnO_4$ solution in part a?

(c) It was found that 0.146 8 g of As_2O_3 required 29.98 mL of $KMnO_4$ solution for the faint color of unreacted MnO_4^- to appear. In a blank titration, 0.03 mL of MnO_4^- was required to produce enough color to be seen. Calculate the molarity of the permanganate solution.

9-18. A cyanide solution with a volume of 12.73 mL was treated with 25.00 mL of Ni^{2+} solution (containing excess Ni^{2+}) to convert the cyanide to tetracyanonickelate:

$$4CN^- + Ni^{2+} \rightarrow Ni(CN)_4^{2-}$$

The excess Ni^{2+} was then titrated with 10.15 mL of 0.013 07 M ethylenediaminetetraacetic acid (EDTA). One mole of this reagent reacts with one mole of Ni^{2+}:

$$Ni^{2+} + EDTA^{4-} \rightarrow Ni(EDTA)^{2-}$$

$Ni(CN)_4^{2-}$ does not react with EDTA. If 39.35 mL of EDTA was required to react with 30.10 mL of the original Ni^{2+} solution, calculate the molarity of CN^- in the 12.73-mL cyanide sample.

9-19. A 0.238 6-g sample contained only NaCl and KBr. It was dissolved in water and required 48.40 mL of 0.048 37 M $AgNO_3$ for complete titration of both halides. Calculate the weight percent of Br in the solid sample.

9-20. A 30.00-mL solution containing an unknown amount of I^- was treated with 50.00 mL of 0.365 0 M $AgNO_3$. The precipitated AgI was filtered off, and the filtrate (plus Fe^{3+}) was titrated with 0.287 0 M KSCN. When 37.60 mL had been added, the solution turned red. How many milligrams of I^- were present in the original solution?

9-21. A convenient procedure[†] for determining the halogen content of organic compounds makes use of the argentometric titration in Figure 9-8. To 50 mL of anhydrous ether is added a carefully weighed sample (10–100 mg) of unknown, plus 2 mL of sodium dispersion and 1 mL of methanol. (Sodium dispersion is finely divided solid sodium suspended in oil. With methanol it makes sodium methoxide, which attacks the organic compound, liberating halides.) Excess sodium is destroyed by slow addition of 2-propanol, after which 100 mL of water is added. (Sodium should not be treated directly with water, since the H_2 produced can explode in the presence of O_2: $2Na + 2H_2O \rightarrow 2NaOH + H_2$) This gives a two-phase mixture, with an ether layer floating on top of the aqueous layer that contains the halide salts. The aqueous layer is adjusted to pH 4 and titrated with Ag^+ using the electrodes in Figure 9-8. How many grams of $AgNO_3$ solution (containing 0.043 66 g $AgNO_3$/kg solution) will be required to reach each equivalence point when 82.67 mg of 1-bromo-4-chlorobutane ($BrCH_2CH_2CH_2CH_2Cl$, F.W. 171.464) is analyzed?

9-22. *Pregl halogen analysis.* Another way to measure halogens in organic compounds is the Pregl method, which is a combustion procedure analogous to that used for C, H, and N (Section 8-1). The sample is burned in a stream of oxygen over a red-hot Pt catalyst. Carbon and hydrogen are converted to CO_2 and H_2O, respectively. Halogens (X) are converted to a mixture of HX and X_2. The gas stream is bubbled through a carbonate solution containing sodium sulfite to reduce X_2:

$$\underset{\text{Sulfite}}{SO_3^{2-}} + X_2 + H_2O \rightarrow \underset{\text{Sulfate}}{SO_4^{2-}} + 2X^- + 2H^+$$

After neutralizion of the solution with HNO_3, the halide can be titrated. An 8.463-mg sample of unknown containing C, H, Br, and Cl subjected to the Pregl procedure required 0.405 g and 1.192 g of $AgNO_3$ titrant (containing 0.093 84 mol $AgNO_3$/kg solution) to reach the two end points in a mass titration analogous to that in Figure 9-7. (The second end point is $1.192 - 0.405 = 0.787$ g past the first end point.) Find the weight percent of Br and Cl in the compound and the Br/Cl atomic ratio.

9-23. Co^{2+} can be analyzed by treatment with a known excess of thiocyanate in the presence of pyridine:

$$Co^{2+} + 4\, \underset{\text{Pyridine}}{\bigcirc N:} + \underset{\text{Thiocyanate}}{2SCN^-} \longrightarrow$$

$$Co(C_5H_5N)_4(SCN)_2(s)$$

The precipitate is filtered off, and the SCN^- content of the filtrate is determined by Volhard titration. A 25.00-mL unknown solution was treated with 3 mL of pyridine and 25.00 mL of 0.102 8 M KSCN in a 250-mL volumetric flask. The solution was diluted to the mark, mixed, and filtered. After the first few milliliters of filtrate were discarded, 50.0 mL of filtrate was acidified with HNO_3 and treated with 5.00 mL of 0.105 5 M $AgNO_3$. After addition of Fe^{3+} indicator, the excess Ag^+ required 3.76 mL of 0.102 8 M KSCN to reach the Volhard end point. Calculate the Co^{2+} concentration in the unknown.

[†] M. L. Ware, M. D. Argentine, and G. W. Rice, *Anal. Chem.,* **60**, 383 (1988).

9-24. Derive an algebraic relation between the turbidity coefficient and the constant k in Box 9-2. What values of τ give apparent transmittances of 90.0% and 10.0% if $b = 1.00$ cm? What value of τ gives an absorbance of 1.00? (You may wish to use the relation $\ln x = (\log x)(\ln 10)$ derived in Appendix A.)

9-25. A solution of volume 25.00 mL containing 0.031 1 M $Na_2C_2O_4$ was titrated with 0.025 7 M $La(ClO_4)_3$ to precipitate lanthanum oxalate:

$$2La^{3+} + 3C_2O_4^{2-} \rightarrow La_2(C_2O_4)_3(s)$$

Lanthanum   Oxalate

(a) What volume of $La(ClO_4)_3$ is required to reach the equivalence point?
(b) Calculate the value of pLa^{3+} when 10.00 mL of $La(ClO_4)_3$ has been added to the sodium oxalate.

9-26. A mixture having a volume of 10.00 mL and containing 0.100 0 M Ag^+ and 0.100 0 M Hg_2^{2+} was titrated with 0.100 0 M KCN to precipitate $Hg_2(CN)_2$ and AgCN.

(a) Calculate pCN^- at each of the following volumes of added KCN: 5.00, 10.00, 15.00, 19.90, 20.10, 25.00, 30.00, 35.00 mL.
(b) Should any AgCN be precipitated at 19.90 mL?

9-27. A solution containing 10.00 mL of 0.100 M LiF was titrated with 0.010 0 M $Th(NO_3)_4$ to precipitate ThF_4.

(a) What volume of $Th(NO_3)_4$ is needed to reach the equivalence point?
(b) What is the ionic strength of the solution when 1.00 mL of $Th(NO_3)_4$ has been added?
(c) Neglecting activities and using the value of K_{sp} for ThF_4 in Appendix F, calculate pTh^{4+} when 1.00 mL of $Th(NO_3)_4$ has been added to the LiF.

9-28. Suppose that 20.00 mL of 0.100 0 M Br^- is titrated with 0.080 0 M $AgNO_3$ and the end point is detected by the Mohr method. The equivalence point occurs at 25.00 mL. If the titration error is to be $\leq 0.1\%$, then the Ag_2CrO_4 should precipitate between 24.975 and 25.025 mL. Calculate the upper and lower limits for the CrO_4^{2-} concentration in the titration solution at the equivalence point if the precipitation of Ag_2CrO_4 is to commence between 24.975 and 25.025 mL.

9-29. *Activity problem.* A 0.010 0 M $Na_3[Co(CN)_6]$ solution with a volume of 50.0 mL was treated with 0.010 0 M $Hg_2(NO_3)_2$ solution to precipitate $(Hg_2)_3[Co(CN)_6]_2$. Using activity coefficients, find $pCo(CN)_6$ when 90.0 mL of $Hg_2(NO_3)_2$ was added.

10 Acid–Base Equilibria

The next three chapters present a fairly detailed description of acid–base chemistry. Understanding the behavior of acids and bases is essential in virtually every field of science involving chemistry. It would be difficult to have a meaningful discussion of any process, from protein biosynthesis to the weathering of rocks, without understanding the chemistry of acids and bases. If your major field of interest is *not* chemistry, Chapters 10–12 are probably the most important ones in this book for you. We assume at this point that you are already familiar with the basic concepts of acids and bases presented in Sections 5-6 through 5-8.

10-1 STRONG ACIDS AND BASES

What could be easier than calculating the pH of 0.10 M HBr? Since HBr is a **strong acid,** it is completely dissociated. So $[H^+] = 0.10$ M and

$$pH = -\log[H^+] = \log(0.10) = 1.00 \qquad (10\text{-}1)$$

Table 5-2 gave a list of strong acids and bases.

EXAMPLE: Activity Coefficient in a Strong Acid Calculation

Equation 10-1 uses our customary tactic of ignoring activity coefficients. Now calculate the pH of 0.10 M HBr properly, using activity coefficients.

The ionic strength of 0.10 M HBr is 0.10 M, at which the activity coefficient of H^+ is 0.83 (Table 6-1). The pH is given by

$$pH = -\log[H^+]\gamma_{H^+} = -\log(0.10)(0.83) - 1.08$$

The activity correction is not very large, and we will generally continue to ignore activity coefficients in our calculations.

If you know $[OH^-]$, you can always find $[H^+]$, since $[H^+] = K_w/[OH^-]$.

How do we calculate the pH of 0.10 M KOH? Since KOH is a strong base, it is completely dissociated, and $[OH^-] = 0.10$ M. Using $K_w = [H^+][OH^-]$, we write

$$[H^+] = \frac{K_w}{[OH^-]} = \frac{1.0 \times 10^{-14}}{0.10} = 1.0 \times 10^{-13} \qquad (10\text{-}2)$$

$$pH = -\log[H^+] = 13.00$$

To find the pH of other concentrations of KOH is pretty trivial:

$[OH^-]$ (M)	$[H^+]$ (M)	pH
$10^{-3.00}$	$10^{-11.00}$	11.00
$10^{-4.00}$	$10^{-10.00}$	10.00
$10^{-5.00}$	$10^{-9.00}$	9.00

A generally useful relation is that

pH + pOH = 14.00 at 25°C

$$pH + pOH = -\log K_w = 14.00 \qquad \text{at } 25°C \qquad (10\text{-}3)$$

The Dilemma

What is the pH of 1.0×10^{-8} M KOH? Applying our usual reasoning, we calculate

$$[H^+] = K_w/(1.0 \times 10^{-8}) = 1.0 \times 10^{-6} \quad \Rightarrow \quad pH = 6.00 \qquad (10\text{-}4)$$

But how can the base KOH produce an acidic solution (pH < 7) when dissolved in pure water? It's impossible.

The Cure

Clearly, there is something wrong with our calculation. In particular, we have not considered the contribution of OH^- from the ionization of water. In pure water, $[OH^-] = 1.0 \times 10^{-7}$ M, which is greater than the amount of KOH added to the solution.

To handle this problem, we resort to the systematic treatment of equilibrium discussed in Chapter 7. The procedure is to write the charge and mass balances, as well as all relevant equilibria. The species in the solution are K^+, OH^-, and H^+. All of the H^+ and some of the OH^- come from dissociation of water. The remainder of the OH^- comes from KOH. The charge balance is

$$[K^+] + [H^+] = [OH^-] \qquad (10\text{-}5)$$

The mass balance is rather trivial in this case:

$$[K^+] = 1.0 \times 10^{-8} \text{ M} \qquad (10\text{-}6)$$

The only equilibrium equation to consider is

$$[H^+][OH^-] = K_w \qquad (10\text{-}7)$$

There are three equations and three unknowns ($[H^+]$, $[OH^-]$, $[K^+]$), so we have enough information to solve the problem.

Since we are seeking the pH, let's set $[H^+] = x$. Using the values $[K^+] = 1.0 \times 10^{-8}$ M and $[H^+] = x$, and substituting into Equation 10-5, we find

$$[OH^-] = [K^+] + [H^+] = 1.0 \times 10^{-8} + x \qquad (10\text{-}8)$$

Using this value of $[OH^-]$ in the K_w equilibrium allows us to solve the problem:

$$[H^+][OH^-] = K_w$$

$$(x)(1.0 \times 10^{-8} + x) = 1.0 \times 10^{-14} \qquad (10\text{-}9)$$

$$x^2 + (1.0 \times 10^{-8})x - (1.0 \times 10^{-14}) = 0$$

$$x = \frac{-1.0 \times 10^{-8} \pm \sqrt{(10^{-8})^2 - 4(1)(-1.0 \times 10^{-14})}}{2(1)} \qquad (10\text{-}10)$$

$$= 9.6 \times 10^{-8} \quad \text{or} \quad -1.1 \times 10^{-7} \text{ M}$$

Solution of a quadratic equation:

$$ax^2 + bx + c = 0$$

$$x = \frac{-b \pm \sqrt{b^2 - 4ac}}{2a}$$

Rejecting the negative solution (because a concentration cannot be negative), we conclude that

$$[H^+] = 9.6 \times 10^{-8} \text{ M} \qquad (10\text{-}11)$$

$$pH = -\log[H^+] = 7.02 \qquad (10\text{-}12)$$

This pH is eminently reasonable, since a solution of 1.0×10^{-8} M KOH should be very slightly basic.

Figure 10-1 shows the pH calculated for different concentrations of a strong base or a strong acid dissolved in water. We can think of these curves in terms of three regions:

1. If the concentration is "high" ($\gtrsim 10^{-6}$ M), the pH has the value we would calculate by just considering the concentration of added H^+ or OH^-. This is, the pH of $10^{-5.00}$ M KOH *is* 9.00.

2. If the concentration is "low" ($\lesssim 10^{-8}$ M) the pH is 7.00. We have not added enough acid or base to significantly affect the pH of the water itself.

3. At intermediate concentrations ($\sim 10^{-6}$–10^{-8} M), the effects of water ionization and the added acid or base are comparable. Only in this region is it necessary to do a systematic equilibrium calculation.

Case 1 is the only practical case. Unless you protected, say, a 10^{-7} M KOH solution from the air, the pH would be overwhelmingly governed by the dissolved CO_2, not the 10^{-7} M KOH. To make a pH near 7, you should use a buffer, not a strong acid or base. You do not need to use the systematic treatment of equilibrium to find the pH of any practical concentration of a strong acid or base.

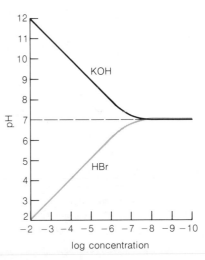

Figure 10-1
Graph showing the calculated pH as a function of the concentration of a strong acid or strong base dissolved in water.

Any acid or base suppresses water ionization. This is an application of Le Châtelier's principle.

The common misconception that dissociation of water always produces 10^{-7} M H^+ and 10^{-7} M OH^- is true *only* in pure water with no added acids or bases. In a $10^{-4.00}$ M solution of HBr, for example, the pH is 4.00. The concentration of OH^- is

$$[OH^-] = K_w/[H^+] = 10^{-10.00} \text{ M} \qquad (10\text{-}13)$$

Question: What concentrations of H^+ and OH^- are produced by H_2O dissociation in 10^{-2} M NaOH?

But the only source of $[OH^-]$ is the dissociation of water. If water produces only $10^{-10.00}$ M OH^-, it must also be producing only $10^{-10.00}$ M H^+, since it makes one H^+ for every OH^-. In a $10^{-4.00}$ M HBr solution, water dissociation produces only $10^{-10.00}$ M OH^- and $10^{-10.00}$ M H^+.

10-2 WEAK ACIDS AND BASES

We begin by reviewing what is meant by the **acid dissociation constant K_a**, for the acid HA:

$$HA \rightleftharpoons H^+ + A^- \qquad (10\text{-}14)$$

Of course you know that K_a is really $\mathscr{A}_{H^+}\mathscr{A}_{A^-}/\mathscr{A}_{HA}$.

$$K_a = \frac{[H^+][A^-]}{[HA]} \qquad (10\text{-}15)$$

A **weak acid** is one that is not completely dissociated. That is, Reaction 10-14 does not go to completion. For a base, B, the **base "dissociation" constant**, or **base hydrolysis constant**, K_b, is defined by the reaction

$$B + H_2O \rightleftharpoons BH^+ + OH^- \qquad (10\text{-}16)$$

$$K_b = \frac{[BH^+][OH^-]}{[B]} \qquad (10\text{-}17)$$

A **weak base** is one for which Reaction 10-16 does not go to completion.

The term **pK** refers to the negative logarithm of the equilibrium constant. So we can write

$$pK_w = -\log K_w = -\log [H^+][OH^-] \qquad (10\text{-}18)$$

$$pK_a = -\log K_a = -\log \frac{[A^-][H^+]}{[HA]} \qquad (10\text{-}19)$$

$$pK_b = -\log K_b = -\log \frac{[BH^+][OH^-]}{[B]} \qquad (10\text{-}20)$$

As K_a increases, pK_a decreases. The smaller pK_a is, the stronger the acid is.

As a K value gets bigger, its p function decreases, and vice versa. Comparing formic and benzoic acids, we see that the former is a stronger acid, with a bigger dissociation constant and a smaller pK_a than benzoic acid.

$$\underset{\text{Formic acid}}{\overset{\overset{\displaystyle O}{\overset{\displaystyle \|}{}}}{HCOH}} \rightleftharpoons H^+ + \underset{\text{Formate}}{HCO_2^-} \qquad \begin{array}{l} K_a = 1.80 \times 10^{-4} \\ \mathbf{pK_a = 3.745} \end{array} \qquad (10\text{-}21)$$

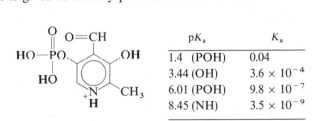

$$K_a = 6.28 \times 10^{-5} \quad (10\text{-}22)$$
$$pK_a = 4.202$$

The acid HA and its corresponding base, A⁻, are said to be a **conjugate** acid–base pair, because they are related by the gain or loss of a single proton. Similarly, B and BH⁺ are a conjugate pair. An important relationship between K_a and K_b for a conjugate acid–base pair was derived in Section 5-8:

$$K_a \cdot K_b = K_w \quad (10\text{-}23)$$

HA and A⁻ are a *conjugate acid–base pair.* B and BH⁺ are also conjugate.

See Reactions 5-70 through 5-72 if you need to familiarize yourself with the derivation of Equation 10-23.

Weak Is Conjugate to Weak

The conjugate base of a weak acid is a weak base. The conjugate acid of a weak base is a weak acid. Let's examine these statements in some detail. Consider a weak acid, HA, with $K_a = 10^{-4}$. The conjugate base, A⁻, has $K_b = K_w/K_a = 10^{-10}$. That is, if HA is a weak acid, A⁻ is a weak base. If the K_a value were 10^{-5}, then the K_b value would be 10^{-9}. We see that as HA becomes a weaker acid, A⁻ becomes a stronger base. Conversely, the greater the acid strength of HA, the less the base strength of A⁻. However, if either A⁻ or HA is weak, so is its conjugate. If HA is strong (such as HCl), its conjugate base (Cl⁻) is *so* weak that it is not a base at all.

The conjugate base of a weak acid is a weak base. The conjugate acid of a weak base is a weak acid. *Weak is conjugate to weak.*

Using Appendix G

A table of acid dissociation constants appears in Appendix G. Each compound is shown in its *fully protonated form.* Methylamine, for example, is shown as $CH_3NH_3^+$, which is really the methylammonium ion. The value of K_a (2.3×10^{-11}) given for methylamine is actually K_a for the methylammonium ion. To find K_b for methylamine, we write $K_b = K_w/K_a = 1.0 \times 10^{-14}/2.3 \times 10^{-11} = 4.3 \times 10^{-4}$.

For polyprotic acids and bases, several K_a values are given. Pyridoxal phosphate is given in its fully protonated form as follows:[†]

pK_a	K_a
1.4 (POH)	0.04
3.44 (OH)	3.6×10^{-4}
6.01 (POH)	9.8×10^{-7}
8.45 (NH)	3.5×10^{-9}

This means that pK_1 (1.4) is for dissociation of one of the phosphate protons, and pK_2 (3.44) is for the hydroxyl proton. The third most acidic proton is the other phosphate proton, for which p$K_3 = 6.01$. Finally, the NH group is the least acidic (p$K_4 = 8.45$).

[†] Measuring acid dissociation constants does not tell us which protons dissociate in each step. The assignments for pyridoxal phosphate were done with nuclear magnetic resonance spectroscopy by B. Szpoganicz and A. E. Martell, *J. Amer. Chem. Soc.,* **106**, 5513 (1984).

The acetyl ($CH_3\overset{\displaystyle O}{\overset{\displaystyle \|}{C}}$—) derivative of
o-hydroxybenzoic acid is the active
ingredient in aspirin.

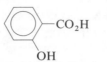 Acetylsalicylic acid

Let's compare the ionization of *o*-hydroxybenzoic acid and *p*-hydroxybenzoic acid.

o-Hydroxybenzoic acid   *p*-Hydroxybenzoic acid
(salicylic acid)

$pK_a = 2.97$ $pK_a = 4.58$

It is thought that the reason that the *o*-hydroxy acid is more than an order of magnitude stronger than the *p*-hydroxy acid is that the conjugate base of the *o*-hydroxy acid is stabilized by strong intramolecular hydrogen bonding.

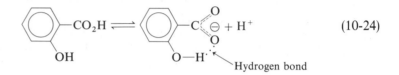

(10-24)

Such intramolecular hydrogen bonding between the hydroxyl and carboxyl groups is not possible in the para isomer because the two functional groups are too far apart. Since the ortho isomer is a stronger acid than the para isomer, we expect that a solution of the former should have a lower pH than an equimolar solution of the latter.

A Typical Problem

Formal concentration is the total number of moles of a compound dissolved in a liter. The formal concentration of a weak acid refers to the total amount of HA placed in the solution, regardless of the fact that some has changed into A^-.

Our problem is to find the pH of a solution of the weak acid HA, given the formal concentration of HA and the value of K_a. Let us call the formal concentration F. One way to attack this problem is by the systematic treatment of equilibrium discussed in Chapter 7.

Charge balance: $[H^+] = [A^-] + [OH^-]$ (10-25)

Mass balance: $F = [A^-] + [HA]$ (10-26)

Equilibria: $HA \rightleftharpoons H^+ + A^-$ $K_a = \dfrac{[H^+][A^-]}{[HA]}$ (10-27)

$H_2O \rightleftharpoons H^+ + OH^-$ $K_w = [H^+][OH^-]$ (10-28)

There are four independent equations and four unknowns ($[A^-]$, $[HA]$, $[H^+]$, $[OH^-]$), so the problem is solved if we can just do the necessary algebra.

The problem is that it's not so easy to solve these four simultaneous equations. If you combine them, you will discover that a cubic equation results. At this point the chemist steps in and cries, "Wait! There is no reason to solve a cubic equation. We can make an excellent, simplifying approximation. (Besides, I have trouble solving cubic equations.)"

In a solution of any respectable weak acid, the concentration of H^+ due to acid dissociation will be much greater than the concentration due to water dissociation. When HA dissociates, it produces A^-. When H_2O dissociates, it produces OH^-. If the acid dissociation is much greater than the water dissociation, we can say $[A^-] \gg [OH^-]$, and Equation 10-25 reduces to

$$[H^+] \approx [A^-] \qquad (10\text{-}29)$$

Using Equations 10-26, 10-27, and 10-29, we can set up a solution. First set $[H^+] = x$. Equation 10-29 says that $[A^-] = x$. Equation 10-26 says that $[HA] = F - [A^-] = F - x$. Putting these values into Equation 10-27 gives

$$K_a = \frac{[H^+][A^-]}{[HA]} = \frac{(x)(x)}{F - x} \qquad (10\text{-}30)$$

It will now be useful to put real numbers into the problem. Let $F = 0.0500$ M and $K_a = 1.07 \times 10^{-3}$ for o-hydroxybenzoic acid. Equation 10-30 can be readily solved, since it is just a quadratic equation.

$$\frac{x^2}{F - x} = 1.07 \times 10^{-3} \qquad (10\text{-}31)$$

$$x^2 + (1.07 \times 10^{-3})x - 5.35 \times 10^{-5} = 0$$

$$x = 6.80 \times 10^{-3} \text{ (negative root rejected)}$$

$$[H^+] = [A^-] = x = 6.80 \times 10^{-3} \text{ M} \qquad (10\text{-}32)$$

$$[HA] = F - x = 0.0432 \text{ M} \qquad (10\text{-}33)$$

$$pH = -\log x = 2.17 \qquad (10\text{-}34)$$

Was the approximation $[H^+] \approx [A^-]$ justified? The calculated pH is 2.17, which means that $[OH^-] = K_w/[H^+] = 1.47 \times 10^{-12}$ M.

$[A^-]$ (from HA dissociation) $= 6.80 \times 10^{-3}$ M
    (implies that $[H^+]$ from HA dissociation $= 6.80 \times 10^{-3}$ M)

$[OH^-]$ (from H_2O dissociation) $= 1.47 \times 10^{-12}$ M
    (implies that $[H^+]$ from H_2O dissociation $= 1.47 \times 10^{-12}$ M)

> In a solution of a weak acid, H^+ is derived almost entirely from the weak acid, not from H_2O dissociation.

The assumption that H^+ is derived mainly from HA is excellent.

For uniformity, we will calculate pH values to the 0.01 decimal place, regardless of what is justified by significant figures. It is important to retain all the digits in your calculator during the solution of a quadratic equation. In the quadratic formula, the term b^2 is often nearly equal to $4ac$, and, if you do not keep all the digits, the subtraction $b^2 - 4ac$ may generate garbage instead of a real answer. Box 10-1 provides some good ideas about quadratic equations.

> We will express pH values to the 0.01 decimal place in this text.

Solving Equation 10-31 by using the quadratic formula requires rearrangement to the form

$$ax^2 + bx + c = 0$$

whose two solutions are

$$x_+ = \frac{-b + \sqrt{b^2 - 4ac}}{2a} \quad \text{and} \quad x_- = \frac{-b - \sqrt{b^2 - 4ac}}{2a} \tag{a}$$

It is not very difficult to use these formulas, but for many problems the method of *successive approximations* is even simpler. The first step is to rearrange Equation 10-31 to the form

$$x = \sqrt{(1.07 \times 10^{-3})(0.050\,0 - x)} \tag{b}$$

As a first approximation, we will neglect x on the right-hand side. That is, we are supposing that $x \ll 0.050\,0$, which might be a good or bad approximation. Neglecting x on the right gives

$$x_1 = \sqrt{(1.07 \times 10^{-3})(0.050\,0)} = 7.31 \times 10^{-3}$$

Now we have a value for x_1, our first approximation. We then plug this value into the right side of Equation b to get a second approximation, x_2.

$$x_2 = \sqrt{(1.07 \times 10^{-3})(0.050\,0 - 7.31 \times 10^{-3})} = 6.76 \times 10^{-3}$$

Continuing, we find that

$$x_3 = \sqrt{(1.07 \times 10^{-3})(0.050\,0 - 6.76 \times 10^{-3})} = 6.80 \times 10^{-3}$$

$$x_4 = \sqrt{(1.07 \times 10^{-3})(0.050\,0 - 6.80 \times 10^{-3})} = 6.80 \times 10^{-3}$$

In four iterations we have come to an answer that is constant to three figures. The first guess (ignoring x on the right-hand side) gave an error of 7.6% in the value of x.

Fraction of Dissociation

α is the fraction of HA that has dissociated:

$$\alpha = \frac{[A^-]}{[A^-] + [HA]}$$

The **fraction of dissociation,** α, of a weak acid is defined as the fraction that is in the form A^-. It is given by

$$\alpha = \frac{[A^-]}{[A^-] + [HA]} = \frac{x}{x + (F - x)} = \frac{x}{F} \tag{10-35}$$

For 0.050 0 M o-hydroxybenzoic acid, we find

$$\alpha = \frac{6.80 \times 10^{-3} \text{ M}}{0.050\,0 \text{ M}} = 0.136 \tag{10-36}$$

That is, the acid is 13.6% dissociated at a formal concentration of 0.050 0 M.
    The variation of α with formal concentration is shown in Figure 10-2. All **weak electrolytes** (compounds that are only partially dissociated) dissociate more as they are diluted. (Demonstration 10-1 illustrates some properties of weak electrolytes.) It can be seen in Figure 10-2 that o-hydroxybenzoic acid

You may find that you make fewer mistakes solving a quadratic equation by successive approximations than with the quadratic formula. If the approximations do not converge rapidly, you can always fall back on the quadratic formula.

When using the quadratic formula (Equation a), you will occasionally encounter the condition $b^2 \gg 4ac$. In this case the root x_+ cannot be computed accurately unless your calculator has many significant digits.

For example, consider the equation

$$x^2 + 4x + 10^{-10} = 0$$

The solution x_- is easy to find:

$$x_- = \frac{-b - \sqrt{b^2 - 4ac}}{2a} = \frac{-4 - \sqrt{4^2 - 4(1)(10^{-10})}}{2(1)} \approx \frac{-4 - 4}{2} = -4$$

However, the solution x_+ is difficult because the difference $4^2 - 4 \times 10^{-10}$ inside the radical is equal to 4 on most calculators. This gives the false result

$$x_+ = \frac{-b + \sqrt{b^2 - 4ac}}{2a} = \frac{-4 + \sqrt{4^2 - 4(1)(10^{-10})}}{2(1)} = \frac{-4 + 4}{2} = 0 \quad \text{(not true)}$$

To find the correct value of x_+ we make use of a property of all quadratic equations, which is

$$x_+ \cdot x_- = \frac{c}{a}$$

Inserting the easily calculated value $x_- = -4$ gives

$$x_+ \cdot (-4) = \frac{10^{-10}}{1} \quad \Rightarrow \quad x_+ = -2.5 \times 10^{-11} \quad \text{(the right value)}$$

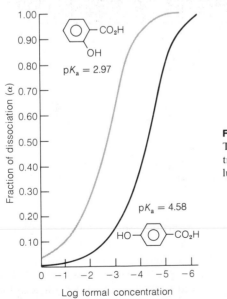

Figure 10-2
The fraction of dissociation of a weak electrolyte increases as the electrolyte is diluted.

Demonstration 10-1 CONDUCTIVITY OF WEAK ELECTROLYTES

The relative conductivity of strong and weak acids is directly related to their different degrees of dissociation in aqueous solution. To demonstrate conductivity we use the equipment shown below, but any kind of buzzer or light bulb could easily be substituted for the electric horn.[†] The voltage required will depend on the buzzer or light chosen.

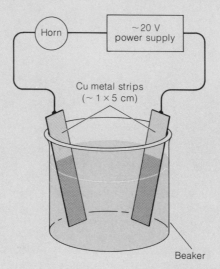

Apparatus for demonstrating conductivity of electrolyte solutions.

When a conducting solution is placed in the beaker, the horn can be heard. We can first show that distilled water or sucrose solutions are nonconductive. Solutions of the strong electrolytes NaCl or HCl are conductive. We compare strong and weak electrolytes by demonstrating that 1 mM HCl gives a loud sound, while 1 mM acetic acid gives little or no sound. With 10 mM acetic acid the strength of the sound varies noticeably as the electrodes are moved away from each other in the beaker.[‡]

[†] The horn used in this demonstration is a Radio Shack piezo alerting buzzer.
[‡] Inexpensive apparatus for a quantitative conductivity experiment is described by T.-R. Rettich, *J. Chem. Ed.*, **66**, 168 (1989).

is more dissociated than *p*-hydroxybenzoic acid at the same formal concentration. This is reasonable, since the ortho isomer is a stronger acid than the para isomer.

Essence of Weak-Acid Problems

When faced with finding the pH of a weak acid, you should immediately realize that $[H^+] = [A^-] = x$ and proceed to set up and solve the equation

$$\frac{[H^+][A^-]}{[HA]} = \frac{x^2}{F - x} = K_a \qquad (10\text{-}37)$$

The way to do it.

where F is the formal concentration of HA. The approximation $[H^+] = [A^-]$ would be poor only if the acid were outrageously dilute ($\leq 10^{-6}$ M) or ridiculously weak. Neither of these conditions constitutes a practical problem.

EXAMPLE: A Weak-Acid Problem

Find the pH of 0.100 M trimethylammonium chloride.

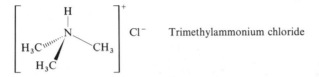

Trimethylammonium chloride

We must first realize that salts of this type are *completely dissociated* to give $(CH_3)_3NH^+$ and Cl^-. We then recognize that the trimethylammonium ion is a weak acid, being the conjugate acid of trimethylamine, $(CH_3)_3N$, a typical weak organic base. Cl^- has no basic or acidic properties and should be ignored. Looking in Appendix G, we find the trimethylammonium ion listed under the name trimethylamine, but drawn as the trimethylammonium ion. The value of pK_a is 9.800, so

$$K_a = 10^{-pK_a} = 1.58 \times 10^{-10}$$

From here everything is downhill.

$$(CH_3)_3NH^+ \rightleftharpoons (CH_3)_3N + H^+$$
$$F - x \qquad\qquad x \qquad\quad x$$

$$\frac{x^2}{0.100 - x} = 1.58 \times 10^{-10}$$

$$x = 3.97 \times 10^{-6} \text{ M} \quad \Rightarrow \quad pH = 5.40$$

10-4 WEAK-BASE EQUILIBRIA

The treatment of weak bases is almost the same as that of weak acids. The usual notation for the base reaction is

$$B + H_2O \rightleftharpoons BH^+ + OH^- \qquad (10\text{-}16)$$

$$K_b = \frac{[BH^+][OH^-]}{[B]} \qquad (10\text{-}17)$$

If we suppose that the dissociation of H_2O is negligible compared with that in Reaction 10-16, then we can say that just about all the OH^- in the solution comes from Reaction 10-16. Setting $[OH^-] = x$, we must also set $[BH^+] = x$, since one BH^+ is produced for each OH^-. If the formal concentration of base ($= [B] + [BH^+]$) is called F, we can set

$$[B] = F - [BH^+] = F - x \qquad (10\text{-}38)$$

A weak-base problem has the same algebra as a weak-acid problem, except $K = K_b$ and $x = [OH^-]$.

Plugging these values into Equation 10-17, we get

$$K_b = \frac{(x)(x)}{F - x} \tag{10-39}$$

which looks a lot like a weak-acid problem, except that now $x = [OH^-]$.

Standard Weak-Base Problem

Let's work one problem using the weak base cocaine as an example.

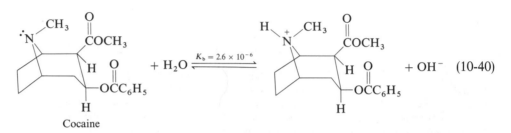

Cocaine

If the formal concentration of cocaine is 0.037 2 M, the problem can be formulated and solved as follows:

$$\begin{array}{ccccc} \text{B} & + \text{H}_2\text{O} \rightleftharpoons & \text{BH}^+ & + & \text{OH}^- \\ 0.0372 - x & & x & & x \end{array} \tag{10-41}$$

$$\frac{x^2}{0.037\,2 - x} = 2.6 \times 10^{-6} \quad \Rightarrow \quad x = 3.1_0 \times 10^{-4} \tag{10-42}$$

Since $x = [OH^-]$, we can write

$$[H^+] = K_w/[OH^-] = 1.0 \times 10^{-14}/3.1_0 \times 10^{-4} = 3.2_2 \times 10^{-11} \tag{10-43}$$

$$pH = -\log[H^+] = 10.49 \tag{10-44}$$

This is a reasonable pH for a weak base.

What fraction of cocaine has reacted with water in this solution? We can formulate α for a base, called the **fraction of association,** as the fraction that has reacted with water:

> *Question:* Based on the final answer, was it justified to neglect water dissociation as a source of OH^-? What concentration of OH^- is produced by H_2O dissociation in this solution?

For a base, α is the fraction that has reacted with water:

$$\alpha = \frac{[BH^+]}{[B] + [BH^+]}$$

$$\alpha = \frac{[BH^+]}{[BH^+] + [B]} = \frac{x}{F} = 0.008\,3 \tag{10-45}$$

Only 0.83% of the base has reacted.

Conjugate Acids and Bases, Revisited

HA and A^- are a conjugate acid–base pair. So are BH^+ and B.

Earlier we noted that **the conjugate base of a weak acid is a weak base,** and that **the conjugate acid of a weak base is a weak acid.** We also derived an exceedingly important relation between the equilibrium constants for a **conjugate acid–base pair**:

$$K_a \cdot K_b = K_w \qquad (10\text{-}23)$$

In Section 10-3 we considered *o*- and *p*-hydroxybenzoic acids, designated HA. Now consider their conjugate bases. For example, the salt sodium *o*-hydroxybenzoate will dissolve to give the Na^+ cation (which has no acid–base chemistry) and the *o*-hydroxybenzoate anion, which is a weak base.

The acid–base chemistry is the reaction of *o*-hydroxybenzoate with water:

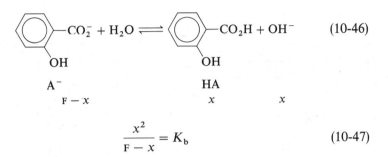

$$\frac{x^2}{F-x} = K_b \qquad (10\text{-}47)$$

Using the values of K_a for the ortho and para isomers, we can calculate the K_b values of the conjugate bases.

Isomer of hydroxybenzoic acid	K_a	K_b
ortho	1.07×10^{-3}	9.35×10^{-12}
para	2.63×10^{-5}	3.80×10^{-10}

Putting each value of K_b into Equation 10-47, and letting F $= 0.050\,0$ M, we find

pH of $0.050\,0$ M *o*-hydroxybenzoate $= 7.83$

pH of $0.050\,0$ M *p*-hydroxybenzoate $= 8.64$

These are reasonable pH values for solutions of weak bases. Furthermore, as expected, the conjugate base of the stronger acid is the weaker base.

EXAMPLE: A Weak-Base Problem
Find the pH of 0.10 M ammonia.
When ammonia is dissolved in water, its reaction is

$$NH_3 + H_2O \overset{K_b}{\rightleftharpoons} NH_4^+ + OH^-$$

Ammonia   Ammonium
                                                ion

F $-x$ x x

In Appendix G we find the ammonium ion, NH_4^+, listed next to ammonia. The pK_a value for the ammonium ion is given as 9.244. Therefore, the K_b value for NH_3 is

$$K_b = \frac{K_w}{K_a} = \frac{10^{-14.00}}{10^{-9.244}} = 1.75 \times 10^{-5}$$

To find the pH of 0.10 M NH_3 we set up and solve the equation

$$\frac{[NH_4^+][OH^-]}{[NH_3]} = \frac{x^2}{0.10 - x} = K_b = 1.75 \times 10^{-5}$$

$$x = [OH^-] = 1.3_1 \times 10^{-3} \text{ M}$$

$$[H^+] = \frac{K_w}{[OH^-]} = 7.6_1 \times 10^{-12} \text{ M}$$

$$pH = -\log[H^+] = 11.12$$

10-5 BUFFERS

A buffered solution is one that resists changes in pH when acids or bases are added or when dilution occurs. The **buffer** consists of a mixture of an acid and its conjugate base. The importance of buffers in all areas of science is overwhelming. Biochemists and other life scientists are particularly concerned with buffers because the proper functioning of any biological system is critically dependent upon pH. For example, Figure 10-3 shows how the rate of the enzyme-catalyzed Reaction 10-48 varies with pH.

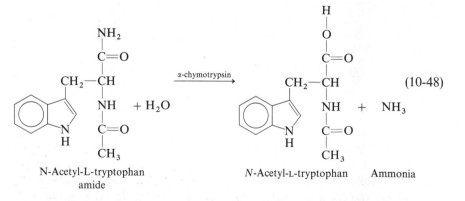

$$(10\text{-}48)$$

N-Acetyl-L-tryptophan amide   *N*-Acetyl-L-tryptophan   Ammonia

Figure 10-3
Graph showing the pH dependence of the rate of cleavage of the C—N bond in Reaction 10-48. [M. L. Bender, G. E. Clement, F. J. Kézdy, and H. A. Heck, *J. Amer. Chem. Soc.*, **86**, 3680 (1964).]

In the absence of the enzyme α-chymotrypsin, the rate of this reaction is negligible over the same pH range, 5–10. The efficiency of the enzyme is highly dependent upon pH. For any organism to survive, it must control the pH of each subcellular compartment so that each of its enzyme-catalyzed reactions may proceed at the proper rate.

Mixing a Weak Acid and Its Conjugate Base

When you mix a weak acid with a weak base, you get what you mix!

The purpose of this section is to show that *if you mix A moles of a weak acid with B moles of its conjugate base, the moles of acid remain close to A and the moles of base remain close to B.* Very little reaction occurs to change either concentration.

To understand why this should be so, look at the K_a and K_b reactions in terms of Le Châtelier's principle. Consider an acid with $pK_a = 4.00$ and its conjugate base with $pK_b = 10.00$. We will calculate the fraction of acid that dissociates in a 0.100 M solution of HA.

$$HA \rightleftharpoons H^+ + A^- \qquad pK_a = 4.00 \qquad \text{(10-49)}$$

$$0.100 - x \qquad x \qquad x$$

$$\frac{x^2}{F - x} = K_a \quad \Rightarrow \quad x = 3.11 \times 10^{-3} \qquad \text{(10-50)}$$

$$\text{Fraction of dissociation} = \alpha = \frac{x}{F} = 0.031\ 1 \qquad \text{(10-51)}$$

The acid is only 3.11% dissociated under these conditions. In a solution containing 0.100 mol of A^- dissolved in 1.00 L, the extent of reaction of A^- with water is even smaller.

$$A^- + H_2O \rightleftharpoons HA + OH^- \qquad pK_b = 10.00 \qquad \text{(10-52)}$$

$$0.100 - x \qquad\qquad x \qquad x$$

$$\frac{x^2}{F - x} = K_b \quad \Rightarrow \quad x = 3.16 \times 10^{-6} \qquad \text{(10-53)}$$

$$\text{Fraction of association} = \alpha = \frac{x}{F} = 3.16 \times 10^{-5} \qquad \text{(10-54)}$$

Pure HA dissociates very little, and adding extra A^- to the solution will make the HA dissociate even less. Similarly, A^- does not react very much with water, and adding extra HA makes A^- react even less. If 0.0050 mol of A^- plus 0.036 mol of HA is added to water, there will be close to 0.050 mol of A^- and close to 0.036 mol of HA in the solution at equilibrium.

This approximation breaks down for dilute solutions or at extremes of pH. We will test the validity of the approximation later in this section.

Henderson–Hasselbalch Equation

The central equation dealing with buffers is the **Henderson–Hasselbalch equation,** which is merely a rearranged form of the K_a equilibrium expression.

$$HA \rightleftharpoons H^+ + A^-$$

$$K_a = \frac{[H^+][A^-]}{[HA]}$$

$$\log K_a = \log \frac{[H^+][A^-]}{[HA]} = \log[H^+] + \log \frac{[A^-]}{[HA]}$$

Remember that $\log xy = \log x + \log y$.

$$\underbrace{-\log[H^+]}_{pH} = \underbrace{-\log K_a}_{pK_a} + \log \frac{[A^-]}{[HA]}$$

The names of Henderson and Hasselbalch appear to be associated with Equation 10-55 not because they invented the equation (it is just the rearranged equilibrium equation), but because they recognized that the concentrations $[A^-]$ and $[HA]$ can be set equal to their formal concentrations. They were among the first people to apply Equation 10-55 to practical problems.

$$pH = pK_a + \log \frac{[A^-]}{[HA]} \qquad (10\text{-}55)$$

The Henderson–Hasselbalch equation, 10-55, tells us the pH of a solution, provided we know the ratio of the concentrations of conjugate acid and base, as well as pK_a for the acid. If a solution is prepared from the weak base B and its conjugate acid, the analogous equation is

$$pH = pK_a + \log \frac{[B]}{[BH^+]} \qquad \begin{array}{c} pK_a \text{ applies to} \\ \textit{this} \text{ acid} \end{array} \qquad (10\text{-}56)$$

where pK_a is the acid dissociation constant of the weak acid BH^+. The important features of Equations 10-55 and 10-56 are that the base (A^- or B) appears in the numerator of both equations, and the equilibrium constant is K_a of the acid in the denominator.

Challenge: Show that if activities are not neglected, the correct form of the Henderson–Hasselbalch equation is

$$pH = pK_a + \log \frac{[A^-]\gamma_{A^-}}{[HA]\gamma_{HA}} \qquad (10\text{-}57)$$

Properties of the Henderson–Hasselbalch equation

In Equation 10-55 you can see that if $[A^-] = [HA]$, then $pH = pK_a$.

$$pH = pK_a + \log \frac{[A^-]}{[HA]} = pK_a + \log 1 = pK_a \qquad (10\text{-}58)$$

When $[A^-] = [HA]$, $pH = pK_a$

Regardless of how complex a solution may be, whenever $pH = pK_a$, $[A^-]$ must equal $[HA]$. This is true because *all equilibria must be satisfied simultaneously in any solution at equilibrium.* If there are 10 different acids and bases in the solution, the 10 forms of Equation 10-55 must all give the same pH, because **there can be only one concentration of H^+ in a solution.**

Another feature of the Henderson–Hasselbalch equation is that for every power-of-10 change in the ratio $[A^-]/[HA]$, the pH changes by one unit. As the base (A^-) increases, the pH goes up. As the acid (HA) increases, the pH goes down. This is shown in Table 10-1. For any conjugate acid–base pair, you can say, for example, that if $pH = pK_a - 1$, ten-elevenths is in the form HA and one-eleventh is in the form A^-.

Table 10-1
Change of pH with change of $[A^-]/[HA]$

$[A^-]/[HA]$	pH
100:1	$pK_a + 2$
10:1	$pK_a + 1$
1:1	pK_a
1:10	$pK_a - 1$
1:100	$pK_a - 2$

EXAMPLE: Using the Henderson–Hasselbalch Equation
Sodium hypochlorite (NaOCl, the active ingredient of almost all bleaches) was dissolved in a solution buffered to pH 6.20. Find the ratio $[OCl^-]/[HOCl]$ in this solution.

In Appendix G we find that $pK_a = 7.53$ for hypochlorous acid, HOCl. Since the pH is known, the ratio $[OCl^-]/[HOCl]$ can be calculated from the Henderson–Hasselbalch equation.

$$HOCl \rightleftharpoons H^+ + OCl^-$$

$$pH = pK_a + \log \frac{[OCl^-]}{[HOCl]}$$

$$6.20 = 7.53 + \log \frac{[OCl^-]}{[HOCl]}$$

$$-1.33 = \log \frac{[OCl^-]}{[HOCl]}$$

$$10^{-1.33} = 10^{\log([OCl^-]/[HOCl])} = \frac{[OCl^-]}{[HOCl]}$$

$$0.047 = \frac{[OCl^-]}{[HOCl]}$$

Note that finding the ratio $[OCl^-]/[HOCl]$ only requires knowing the pH. We do not need to know what else is in the solution, how much NaOCl was added, or the volume of the solution.

A Buffer in Action

For illustration, we will work with a very widely used buffer called "tris," which is short for tris(hydroxymethyl)aminomethane.

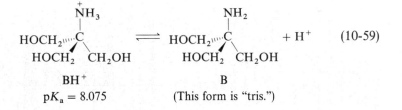

$$\qquad\qquad\qquad\qquad (10\text{-}59)$$

BH$^+$ $\qquad\qquad\qquad\qquad$ B

$pK_a = 8.075$ $\qquad\qquad$ (This form is "tris.")

In Appendix G we find pK_a for the conjugate acid of tris to be 8.075. An example of a salt containing the BH$^+$ cation is tris hydrochloride, which is really BH$^+$Cl$^-$. The formula weight of tris is 121.136, and the formula weight of tris hydrochloride is 157.597. When BH$^+$Cl$^-$ is dissolved in water, it dissociates completely to BH$^+$ plus Cl$^-$.

EXAMPLE: A Buffer Solution

Find the pH of a solution prepared by dissolving 12.43 g of tris plus 4.67 g of tris hydrochloride in 1.00 L of water.

The concentrations of B and BH$^+$ added to the solution are

$$[B] = \frac{12.43 \text{ g/L}}{121.136 \text{ g/mol}} = 0.102\ 6 \text{ M}$$

$$[BH^+] = \frac{4.67 \text{ g/L}}{157.597 \text{ g/mol}} = 0.029\ 6 \text{ M}$$

Assuming that what we mixed stays in the same form, we can simply plug these concentrations into the Henderson–Hasselbalch equation to find the pH.

$$pH = pK_a + \log \frac{[B]}{[BH^+]} = 8.075 + \log \frac{0.102\ 6}{0.029\ 6} = 8.61$$

The pH of a buffer is nearly independent of volume.

Notice that *the volume of solution is irrelevant to finding the pH,* since the volume cancels in the numerator and denominator of the log term:

$$pH = pK_a + \log \frac{\text{moles of B/L of solution}}{\text{moles of BH}^+/\text{L of solution}}$$

$$= pK_a + \log \frac{\text{moles of B}}{\text{moles of BH}^+} \tag{10-60}$$

The pH in the preceding example would be 8.61 whether the volume were 1.00, 0.63, or 2.41 L.

EXAMPLE: Effect of Adding Acid to a Buffer

If we add 12.0 mL of 1.00 M HCl to the solution used in the previous example, what will be the new pH?

The key to this problem is to realize that *when a strong acid is added to a weak base, both react completely to give* BH^+. (You can read more about this important statement in Box 10-2.) In the present example we are adding 12.0 mL of 1.00 M HCl, which contains $(0.012\ 0\ \text{L})(1.00\ \text{mol/L}) = 0.012\ 0$ mol of H^+. This much H^+ will consume 0.012 0 mol of B to create 0.012 0 mol of BH^+. This is shown conveniently in a little table:

	B (tris)	+	H^+ (from HCl)	→	BH^+
Initial moles:	0.102 6		0.012 0		0.029 6
Final moles:	0.090 6		—		0.041 6
	(0.102 6 − 0.012 0)				(0.029 6 + 0.012 0)

The table contains enough information for us to calculate the pH.

$$pH = pK_a + \log \frac{\text{moles of B}}{\text{moles of BH}^+}$$

$$= 8.075 + \log \frac{0.090\ 6}{0.041\ 6} = 8.41$$

The volume of the solution is irrelevant, as usual.

Question: Does the pH change in the right direction when HCl is added?

Box 10-2 STRONG PLUS WEAK REACTS COMPLETELY

A strong acid reacts with a weak base essentially "completely" because the equilibrium constant is large.

$$\underset{\substack{\text{Weak} \\ \text{base}}}{B} + \underset{\substack{\text{Strong} \\ \text{acid}}}{H^+} \rightleftharpoons BH^+ \qquad K = \frac{1}{K_a} \text{(for } BH^+\text{)}$$

If B is tris(hydroxymethyl)aminomethane, the equilibrium constant for reaction with HCl is

$$K = \frac{1}{K_a} = \frac{1}{10^{-8.075}} = 1.2 \times 10^8$$

A strong base reacts "completely" with a weak acid because the equilibrium constant is, again, very large.

$$\underset{\substack{\text{Strong} \\ \text{base}}}{OH^-} + \underset{\substack{\text{Weak} \\ \text{acid}}}{HA} \rightleftharpoons A^- + H_2O \qquad K = \frac{1}{K_b} \text{(for } A^-\text{)}$$

If HA is acetic acid, the equilibrium constant for reaction with NaOH is

$$K = \frac{1}{K_b} = \frac{K_a \text{ for HA}}{K_w} = 1.7 \times 10^9$$

The reaction of a strong acid with a strong base is even more complete than a strong + weak reaction:

$$\underset{\substack{\text{Strong} \\ \text{acid}}}{H^+} + \underset{\substack{\text{Strong} \\ \text{base}}}{OH^-} \rightleftharpoons H_2O \qquad K = \frac{1}{K_w} = 10^{14}$$

If you mix a strong acid, a strong base, a weak acid, and a weak base, the strong acid and base will react with each other until one is used up. The remainder of the strong acid or base will then react with the weak base or weak acid.

The foregoing example illustrates that *the pH of a buffer does not change very much when a strong acid or base is added.* Addition of 12.0 mL of 1.00 M HCl changed the pH from 8.61 to 8.41. Addition of 12.0 mL of 1.00 M HCl to 1.00 L of unbuffered solution would have lowered the pH to 1.93.

But *why* does a buffer resist changes in pH? It does so because the strong acid or base is consumed by B or BH^+. If you add HCl to tris, B is converted to BH^+. If you add NaOH, BH^+ is converted to B. As long as we don't use up the B or BH^+ (by adding too much HCl or NaOH) the log term of the Henderson–Hasselbalch equation does not change very much and the pH

A buffer resists changes in pH . . .

. . . . because the buffer consumes the added acid or base.

Demonstration 10-2   HOW BUFFERS WORK

A buffer resists changes in pH because the added acid or base is consumed by the buffer. As the buffer is used up it becomes less resistant to changes in pH.

In this demonstration,[†] a mixture containing approximately a 10:1 mole ratio of HSO_3^-: SO_3^{2-} is prepared. Since pK_a for HSO_3^- is 7.2, the pH should be approximately

$$pH = pK_a + \log \frac{[SO_3^{2-}]}{[HSO_3^-]} = 7.2 + \log \frac{1}{10} = 6.2$$

When formaldehyde is added, the net reaction is the consumption of HSO_3^-, but not of SO_3^{2-}.

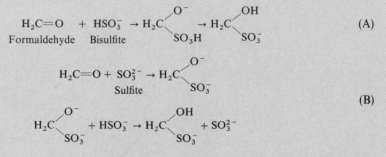

(A)

(B)

(In sequence A bisulfite is consumed directly. In sequence B the net reaction is destruction of HSO_3^-, with no change in the SO_3^{2-} concentration.)

We can prepare a table showing how the pH should change as the HSO_3^- reacts.

Percent of reaction completed	$[SO_3^{2-}]:[HSO_3^-]$	Calculated pH
0	1:10	6.2
90	1:1	7.2
99	1:0.1	8.2
99.9	1:0.01	9.2
99.99	1:0.001	10.2

[†] F. B. Dutton and G. Gordon in H. N. Alyea and F. B. Dutton, eds., *Tested Demonstrations in Chemistry* 6th ed. (Easton, Pa.: Journal of Chemical Education, 1965), p. 147; R. L. Barrett, *J. Chem. Ed.*, **32**, 78 (1955). A kinetics experiment and more detailed discussions of the mechanism of this reaction are given by M. G. Burnett, *J. Chem. Ed.*, **59**, 160 (1982) and P. Warneck, *J. Chem. Ed.*, **66**, 334 (1989).

does not change very much. Demonstration 10-2 provides a nice illustration of how buffers work. The buffer has its maximum capacity to resist changes of pH when $pH = pK_a$. We will return to this point later.

EXAMPLE: Calculating How to Prepare a Buffer Solution
How many milliliters of 0.500 M NaOH should be added to 10.0 g of tris hydrochloride to give a pH of 7.60 in a final volume of 250 mL?

You can see that through 90% completion the pH should rise by just 1 pH unit. In the next 9% of the reaction the pH will rise by another unit. At the end of the reaction the change in pH should be very abrupt.

In the formaldehyde clock reaction, formaldehyde is added to a solution containing HSO_3^-, SO_3^{2-}, and phenolphthalein indicator. Phenolphthalein is colorless below a pH of ~ 8.5 and red above this pH. What is observed is that the solution remains colorless for more than a minute. Suddenly the pH shoots up and the liquid turns pink. Monitoring the pH with a glass electrode gave the results below.

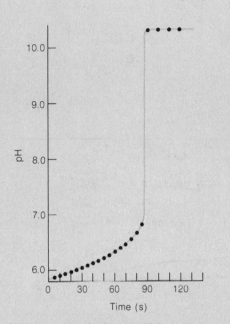

Graph of pH versus time in the formaldehyde clock reaction.

Procedure: All solutions should be fresh. Prepare a solution of formaldehyde by diluting 9 mL of 37%(wt/wt) formaldehyde to 100 mL. Dissolve 1.5 g of $NaHSO_3$ and 0.18 g of Na_2SO_3 in 400 mL of water, and add ~ 1 mL of phenolphthalein indicator solution (Table 11-3). Add 23 mL of the formaldehyde solution to the well-stirred buffer solution to initiate the clock reaction. The time of reaction can be adjusted by changing the temperature, concentrations, or volume.

The number of moles of tris hydrochloride in 10.0 g is $(10.0\ \text{g})/(157.597\ \text{g/mol}) =$ 0.063 5. We can make a table to help solve the problem.

Reaction with OH^-:	BH^+	$+\ OH^- \rightarrow$	B
Initial moles:	0.063 5	x	—
Final moles:	0.063 5 $-\ x$	—	x

The Henderson–Hasselbalch equation allows us to find x, since we know pH and pK_a.

$$pH = pK_a + \log \frac{mol\ B}{mol\ BH^+}$$

$$7.60 = 8.075 + \log \frac{x}{0.063\ 5 - x}$$

$$-0.475 = \log \frac{x}{0.063\ 5 - x}$$

$$10^{-0.475} = \frac{x}{0.063\ 5 - x} \quad \Rightarrow \quad x = 0.015\ 9\ mol$$

This many moles of NaOH is contained in

$$\frac{0.015\ 9\ mol}{0.500\ mol/L} = 0.031\ 8\ L = 31.8\ mL$$

Notice that the volume of buffer solution (250 mL) was not used anywhere in answering the question.

Preparing a Buffer in Real Life!

Reasons why a calculation would be wrong:

1. You might have ignored activity coefficients.

2. The temperature might not be just right.

3. The approximations that $[HA] = F_{HA}$ and $[A^-] = F_{A^-}$ could be in error.

4. The pK_a reported for tris in your favorite table is probably not what you would measure in your lab.

5. You will probably make an arithmetic error anyway.

If you really wanted to prepare a tris buffer of pH 7.60, you would *not* do it by calculating what to mix. Suppose that you wish to prepare 1.00 L of buffer containing 0.100 M tris at a pH of 7.60. You have available solid tris hydrochloride and approximately 1 M NaOH. Here's how to do it:

1. Weigh out 0.100 mol of tris hydrochloride and dissolve it in a beaker containing about 800 mL of water.

2. Place a pH electrode in the solution and monitor the pH.

3. Add NaOH until the pH is exactly 7.60.

4. Transfer the solution to a volumetric flask and wash the beaker a few times. Add the washings to the volumetric flask.

5. Dilute to the mark and mix.

You do not mix calculated quantities, though a quick calculation is helpful so that you have some idea of how much will be needed.

Buffer Capacity

The **buffer capacity,** β, also called **buffer intensity,** is defined as

$$\beta = \frac{dC_b}{dpH} = -\frac{dC_a}{dpH} \tag{10-61}$$

where C_a and C_b are the number of moles of strong acid or base per liter needed to produce a unit change in pH. Buffer capacity is a positive number. The larger the value of β, the more resistant the solution is to pH change.

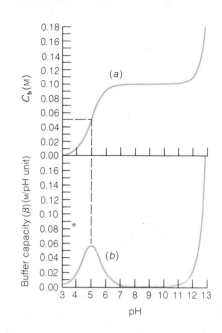

Figure 10-4
(a) C_b versus pH for a solution containing 0.100 F HA with $pK_a = 5.00$. (b) Buffer capacity versus pH for the same system. This curve is the derivative of the curve in part a.

At the top of Figure 10-4 is a graph showing C_b versus pH for a solution containing 0.100 F HA with $pK_a = 5.00$. The ordinate (C_b) is the formal concentration of strong base needed to be mixed with 0.100 F HA to give the indicated pH. For example, a solution containing 0.050 F OH^- plus 0.100 F HA would have a pH of 5.00 (neglecting activities).

The lower graph in Figure 10-4 shows the buffer capacity as a function of pH for the same system of HA plus strong base. The lower curve in Figure 10-4 is the derivative of the upper curve. The most notable feature of buffer capacity is that it reaches a maximum when $pH = pK_a$. That is, a buffer is most effective in resisting changes in pH when $pH = pK_a$ (i.e., when $[HA] = [A^-]$).

In choosing a buffer for an experiment, you should seek one whose, pK_a *is as close as possible to the desired pH. The useful pH range of a buffer is usually considered to be* $pK_a \pm 1$ *pH unit.* Outside this range there is not enough of either the weak acid or the weak base to react with added base or acid. Clearly, the buffer capacity can be increased by increasing the concentration of the buffer.

Choose a buffer whose pK_a is close to the desired pH.

The buffer capacity curve in Figure 10-4 continues upward at high pH (and at low pH, which is not shown in the figure) simply because there is a high concentration of OH^-. Addition of a small amount of acid or base to a large amount of OH^- (or H^+) will not have a very great effect on pH. A solution of high pH is well buffered by the H_2O/OH^- conjugate acid–conjugate base pair. A solution of low pH is well buffered by the H_3O^+/H_2O conjugate acid–conjugate base pair.

Table 10-2 lists the pK_a values for several common buffers. These are widely used in biochemistry. The measurement of pH with glass electrodes, and the buffers used by the U.S. National Institute of Standards and Technology to define the pH scale, are described in Chapter 15.

Table 10-2
Structures and pK_a values for some commonly used buffers

Name	Structure	pK_a ($\sim 25°C$)	Formula weight
Phosphoric acid	H_3PO_4	2.15 (pK_1)	97.995
Citric acid	$HO_2CCH_2\overset{\displaystyle OH}{\underset{\displaystyle CO_2H}{C}}CH_2CO_2H$	3.13 (pK_1)	160.013
Formic acid	HCO_2H	3.74	46.026
Succinic acid	$HO_2CCH_2CH_2CO_2H$	4.21 (pK_1)	118.089
Citric acid	$H_2(citrate)^-$	4.76 (pK_2)	160.013
Acetic acid	CH_3CO_2H	4.76	68.029
Succinic acid	$H(succinate)^-$	5.64 (pK_2)	118.089
2-(N-Morpholino)ethanesulfonic acid (MES)	$O\underset{}{\bigcirc}\overset{+}{N}HCH_2CH_2SO_3^-$	6.15	195.240
Cacodylic acid	$(CH_3)_2AsO_2H$	6.19	153.033
Citric acid	$H(citrate)^{2-}$	6.40 (pK_3)	160.013
N-2-Acetamidoiminodiacetic acid (ADA)	$H_2NCCH_2\overset{+}{N}H \begin{smallmatrix} CH_2CO_2^- \\ \\ CH_2CO_2H \end{smallmatrix}$	6.60	190.156
1,3-Bis[tris(hydroxymethyl)methylamino]propane (BIS-TRIS propane)	$(HOCH_2)_3C\overset{+}{N}H_2(CH_2)_3NHC(CH_2OH)_3$	6.80	283.345
Piperazine-N,N'-bis(2-ethanesulfonic acid) (PIPES)	$^-O_3SCH_2CH_2\overset{+}{N}H\bigcirc\overset{+}{H}NCH_2CH_2SO_3^-$	6.80	302.373
N-2-Acetamido-2-aminoethanesulfonic acid (ACES)	$H_2NCCH_2\overset{+}{N}H_2CH_2CH_2SO_3^-$	6.90	182.200
3-(N-Morpholine)-2-hydroxypropanesulfonic acid (MOPSO)	$O\bigcirc\overset{+}{N}H CH_2\overset{OH}{C}HCH_2SO_3^-$	6.95	225.265
Imidazole hydrochloride	$H\overset{+}{N}\underset{\underset{H}{N}}{\bigcirc} Cl^-$	6.99	104.539
3-(N-Morpholine)propanesulfonic acid (MOPS)	$O\bigcirc\overset{+}{N}HCH_2CH_2CH_2SO_3^-$	7.20	209.266
Phosphoric acid	$H_2PO_4^-$	7.20 (pK_2)	97.995
N-Tris(hydroxymethyl)methyl-2-aminoethanesulfonic acid (TES)	$(HOCH_2)_3C\overset{+}{N}HCH_2CH_2SO_3^-$	7.50	229.254
N-2-Hydroxyethylpiperazine-N'-2-ethanesulfonic acid (HEPES)	$HOCH_2CH_2N\bigcirc\overset{+}{N}HCH_2CH_2SO_3^-$	7.56†	238.308
N-2-Hydroxyethylpiperazine-N'-3-propanesulfonic acid (HEPPS)	$HOCH_2CH_2N\bigcirc\overset{+}{N}HCH_2CH_2CH_2SO_3^-$	8.00	252.335
N-Tris(hydroxymethyl)methylglycine (TRICINE)	$(HOCH_2)_3C\overset{+}{N}H_2CH_2CO_2^-$	8.15	179.173

(*continued*)

Table 10-2 (*continued*)

Name	Structure	pK_a ($\sim25°C$)	Formula weight
Glycine amide, hydrochloride	$\overset{\displaystyle O}{\overset{\displaystyle \parallel}{H_3\overset{+}{N}CH_2CNH_2}}$ $\quad Cl^-$	8.20	110.543
Tris(hydroxymethyl)aminomethane hydrochloride (TRIS hydrochloride)	$(HOCH_2)_3C\overset{+}{N}H_3$ $\quad Cl^-$	8.08	157.597
N,N-Bis(2-hydroxyethyl)glycine (BICINE)	$(HOCH_2CH_2)_2\overset{+}{N}HCH_2CO_2^-$	8.35	163.174
Glycylglycine	$\overset{\displaystyle O}{\overset{\displaystyle \parallel}{H_3\overset{+}{N}CH_2CNHCH_2CO_2^-}}$	8.40	132.119
Boric acid	$B(OH)_3$	9.24 (pK_1)	61.833
Cyclohexylaminoethanesulfonic acid (CHES)	$\overset{+}{N}H_2CH_2CH_2SO_3^-$	9.50	207.294
3-(Cyclohexylamino)propanesulfonic acid (CAPS)	$\overset{+}{N}H_2CH_2CH_2CH_2SO_2^-$	10.40	221.321
Phosphoric acid	HPO_4^{2-}	12.35 (pK_3)	97.995
Boric acid	$OB(OH)_2^-$	12.74 (pK_2)	61.833

Note: The protonated form of each molecule is shown. Acidic hydrogen atoms are shown in **bold** type. Several buffers in this table are widely used in biomedical research because of their relatively weak binding of metal ions and physiologic inertness [C. L. Bering, *J. Chem. Ed., 64,* 803 (1987)]. However, the buffers ADA, BICINE, ACES, and TES have greater metal-binding ability than formerly thought [R. Nakon and C. R. Krishnamoorthy, *Science, 221,* 749 (1983)]. A set of lutidine buffers for the pH range 3–8 with very limited metal ion coordination power has been described by U. Bips, H. Elias, M. Hauröder, G. Kleinhans, S. Pfeifer, and K. J. Wannowius, *Inorg. Chem., 22,* 3862 (1983).
[†] Temperature and ionic strength dependence of HEPES acid dissociation are given by D. Feng, W. F. Koch, and Y. C. Wu, *Anal. Chem., 61,* 1400 (1989).

Limitations of Buffers

Activity coefficients

The *correct* Henderson–Hasselbalch equation, 10-57, includes activity coefficients. Failure to include activity coefficients is the principal reason why our calculated pH values will not be in perfect agreement with measured pH values. The activity coefficients also predict that the pH of a buffer will vary with ionic strength. Adding an inert salt (such as NaCl) will change the pH of a buffer. This effect can be significant for highly charged species, such as citrate or phosphate. When a 0.5 M stock solution of phosphate buffer at pH 6.6 is diluted to 0.05 M, the pH rises to 6.9. This is a rather significant effect of changing ionic strength.

Changing ionic strength changes pH.

Temperature

Most buffers exhibit a noticeable dependence of pK_a on temperature. Tris has an exceptionally large dependence, approximately -0.031 pK_a unit per degree, near room temperature. A solution of tris made up to pH 8.08 at 25°C will have pH ≈ 8.7 at 4° and pH ≈ 7.7 at 37°C.

Changing temperature changes pH.

What you mix is not what you get in dilute solutions or at extremes of pH.

In a dilute solution, or at extremes of pH, the molar concentrations of HA and A^- in solution are not equal to their formal concentrations. This can be seen as follows. Suppose we mix F_{HA} moles of HA and F_{A^-} moles of A^-. The equilibria are

$$HA \rightleftharpoons H^+ + A^- \qquad K_a \qquad (10\text{-}62)$$

$$A^- + H_2O \rightleftharpoons HA + OH^- \qquad K_b \qquad (10\text{-}63)$$

Reaction 10-62 decreases the concentration of HA, and Reaction 10-63 increases the concentration of HA. For each mole of H^+ made by Reaction 10-62, HA decreases by one mole. For each mole of OH^- made in Reaction 10-63, HA increases by one mole. The total concentration of HA in the solution is therefore

$$[HA] = F_{HA} - [H^+] + [OH^-] \qquad (10\text{-}64)$$

(Equation 10-64 neglects any contribution of H_2O dissociation to the concentrations of H^+ and OH^-.) By similar reasoning we can write

$$[A^-] = F_{A^-} + [H^+] - [OH^-] \qquad (10\text{-}65)$$

In our work so far, we have assumed that $[HA] \approx F_{HA}$ and $[A^-] \approx F_{A^-}$, and we used these values in the Henderson–Hasselbalch equation. A more rigorous procedure is to use the values given by Equations 10-64 and 10-65. We see that if F_{HA} or F_{A^-} is small, or if $[H^+]$ or $[OH^-]$ is large, the approximations $[HA] \approx F_{HA}$ and $[A^-] \approx F_{A^-}$ are not good. In acidic solutions $[H^+] \gg [OH^-]$, so $[OH^-]$ can be ignored in Equations 10-64 and 10-65. In basic solutions $[H^+]$ can be neglected.

EXAMPLE: A Dilute Buffer Prepared from a Moderately Strong Acid

What will be the pH if 0.010 0 mol of HA (with $pK_a = 2.00$) and 0.010 0 mol of A^- are dissolved in water to make 1.00 L of solution?

Since the solution will be acidic (pH $\approx pK_a = 2.00$), we can neglect the $[OH^-]$ terms in Equations 10-64 and 10-65. Setting $[H^+] = x$, we can use the K_a equation to calculate the value of $[H^+]$.

$$\begin{array}{ccccc} HA & \rightleftharpoons & H^+ & + & A^- \\ 0.010\,0 - x & & x & & 0.010\,0 + x \end{array}$$

$$\frac{[H^+][A^-]}{[HA]} = \frac{(x)(0.010\,0 + x)}{(0.010\,0 - x)} = 10^{-2.00} \quad \Rightarrow \quad x = 0.004\,14 \text{ M}$$

$$pH = -\log[H^+] = 2.38$$

The concentrations of HA and A^- are not what we mixed:

$$[HA] = F_{HA} - [H^+] = 0.005\,86 \text{ M}$$

$$[A^-] = F_{A^-} + [H^+] = 0.014\,1 \text{ M}$$

In this example, HA is too strong and the concentrations are too low for HA and A^- to be equal to their formal concentrations.

The Henderson–Hasselbalch equation is a true statement. It is just a rearrangement of the K_a equilibrium expression, which is always true. What is an approximation is the pair of statements $[HA] \approx F_{HA}$ and $[A^-] \approx F_{A^-}$.

The Henderson–Hasselbalch equation (with activity coefficients) is *always* true.

Summary of Buffers

A buffer consists of a mixture of a weak acid and its conjugate base. The buffer is most useful when $pH \approx pK_a$. Over a reasonable range of concentration, the pH of a buffer is nearly independent of concentration. A buffer resists changes in pH because it reacts with added acids or bases. If too much acid or base is added, the buffer will be consumed and no longer resist changes in pH.

10-6 DIPROTIC ACIDS AND BASES

Polyprotic acids and bases are those that can donate or accept more than one proton. The 20 common amino acids, which are the building blocks of proteins, are all polyprotic. Most are **diprotic,** which means that their acid–base chemistry involves two protons.

The general structure of the natural **amino acids** is

Ammonium group $\longrightarrow$ $H_3\overset{+}{N}$
$\diagdown$
$CH-R$
Carboxyl group $\longrightarrow$ $^-O-C\diagup$
$\parallel$
O

where R is a different group for each compound. The carboxyl group, drawn here in its ionized (basic) form, is a stronger acid than the ammonium group. Therefore, the nonionized form rearranges spontaneously to the **zwitterion:**

A *zwitterion* is a molecule with positive and negative groups.

H_2N
$\diagdown$
$CH-R$ $\longrightarrow$
HO_2C

$H_3\overset{+}{N}$
$\diagdown$
$CH-R$
^-O_2C
Zwitterion

At low pH, both the ammonium group and the carboxyl group are protonated. At high pH, neither is protonated. The acid dissociation constants of the amino acids are listed in Table 10-3, where each compound is drawn in its fully protonated form.

In our discussion, we will take as a specific example the amino acid leucine, designated HL.

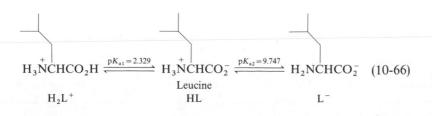

$$H_3\overset{+}{N}CHCO_2H \underset{}{\overset{pK_{a1}=2.329}{\rightleftharpoons}} H_3\overset{+}{N}CHCO_2^- \underset{}{\overset{pK_{a2}=9.747}{\rightleftharpoons}} H_2NCHCO_2^- \quad (10\text{-}66)$$

H_2L^+ Leucine L^-
           HL

The side chain in leucine is $(CH_3)_2CHCH_2-$.

Table 10-3
Acid dissociation constants of amino acids

Amino acid	Structure[†]	Carboxylic acid[‡]	Ammonium group[‡]	Substituent[‡]
Alanine	$\overset{NH_3^+}{\underset{CO_2H}{CH}}$—CH₃	$pK_a = 2.348$	$pK_a = 9.867$	
Arginine	$\overset{NH_3^+}{\underset{CO_2H}{CH}}$—CH₂CH₂CH₂NHC$\overset{\overset{+}{NH_2}}{\underset{NH_2}{}}$	$pK_a = 1.823$	$pK_a = 8.991$	$(pK_a = 12.48)$
Asparagine	$\overset{NH_3^+}{\underset{CO_2H}{CH}}$—CH₂$\overset{O}{\overset{\|}{C}}$NH₂	$pK_a = 2.14^\S$	$pK_a = 8.72^\S$	
Aspartic acid	$\overset{NH_3^+}{\underset{CO_2H}{CH}}$—CH₂CO₂H	$pK_a = 1.990$	$pK_a = 10.002$	$pK_a = 3.900$
Cysteine	$\overset{NH_3^+}{\underset{CO_2H}{CH}}$—CH₂SH	$(pK_a = 1.71)$	$pK_a = 10.77$	$pK_a = 8.36$
Glutamic acid	$\overset{NH_3^+}{\underset{CO_2H}{CH}}$—CH₂CH₂CO₂H	$pK_a = 2.23$	$pK_a = 9.95$	$pK_a = 4.42$
Glutamine	$\overset{NH_3^+}{\underset{CO_2H}{CH}}$—CH₂CH₂$\overset{O}{\overset{\|}{C}}$NH₂	$pK_a = 2.17^\S$	$pK_a = 9.01^\S$	
Glycine	$\overset{NH_3^+}{\underset{CO_2H}{CH}}$—H	$pK_a = 2.350$	$pK_a = 9.778$	
Histidine	$\overset{NH_3^+}{\underset{CO_2H}{CH}}$—CH₂ (imidazole)	$pK_a = 1.7^\S$	$pK_a = 9.08^\S$	$pK_a = 6.02^\S$
Isoleucine	$\overset{NH_3^+}{\underset{CO_2H}{CH}}$—CH$\overset{CH_3}{\underset{CH_2CH_3}{}}$	$pK_a = 2.319$	$pK_a = 9.754$	
Leucine	$\overset{NH_3^+}{\underset{CO_2H}{CH}}$—CH₂CH(CH₃)₂)	$pK_a = 2.329$	$pK_a = 9.747$	
Lysine	$\overset{NH_3^+}{\underset{CO_2H}{CH}}$—CH₂CH₂CH₂CH₂NH₃⁺	$pK_a = 2.04^\S$	$pK_a = 9.08^\S$	$pK_a = 10.69^\S$

(*continued*)

Table 10-3 (*continued*)   **209**

Amino acid	Structure[†]	Carboxylic acid[‡]	Ammonium group[‡]	Substituent[‡]
Methionine	$\overset{NH_3^+}{\underset{CO_2H}{CH}}-CH_2CH_2SCH_3$	$pK_a = 2.20^§$	$pK_a = 9.05^§$	
Phenylalanine	$\overset{NH_3^+}{\underset{CO_2H}{CH}}-CH_2-\bigcirc$	$pK_a = 2.20$	$pK_a = 9.31$	
Proline	$H_2\overset{+}{N}-$ pyrrolidine ring HO_2C-	$pK_a = 1.952$	$pK_a = 10.640$	
Serine	$\overset{NH_3^+}{\underset{CO_2H}{CH}}-CH_2OH$	$pK_a = 2.187$	$pK_a = 9.209$	
Threonine	$\overset{NH_3^+}{\underset{CO_2H}{CH}}-\underset{OH}{\overset{CH_3}{CH}}$	$pK_a = 2.088$	$pK_a = 9.100$	
Tryptophan	$\overset{NH_3^+}{\underset{CO_2H}{CH}}-CH_2-$ indole	$pK_a = 2.35^§$	$pK_a = 9.33^§$	
Tyrosine	$\overset{NH_3^+}{\underset{CO_2H}{CH}}-CH_2-\bigcirc-OH$	$pK_a = 2.17^§$	$pK_a = 9.19$	$pK_a = 10.47$
Valine	$\overset{NH_3^+}{\underset{CO_2H}{CH}}-CH(CH_3)_2$	$pK_a = 2.286$	$pK_a = 9.718$	

[†] The acidic protons are shown in **bold** type. Each amino acid is written in its fully protonated form.
[‡] pK_a values refer to 25°C and zero ionic strength unless marked by [§]. Values considered to be uncertain are enclosed in parentheses.
[§] For these entries the ionic strength is 0.1 M, and the constant refers to a product of concentrations instead of activities.
SOURCE: A. E. Martell and R. M. Smith, *Critical Stability Constants*, Vol. 1 (New York: Plenum Press 1974).

The equilibrium constants refer to the following reactions:

$$H_2L^+ \rightleftharpoons HL + H^+ \qquad K_{a1} = K_1 \qquad (10\text{-}67)$$

$$HL \rightleftharpoons L^- + H^+ \qquad K_{a2} = K_2 \qquad (10\text{-}68)$$

$$L^- + H_2O \rightleftharpoons HL + OH^- \qquad K_{b1} \qquad (10\text{-}69)$$

$$HL + H_2O \rightleftharpoons H_2L^+ + OH^- \qquad K_{b2} \qquad (10\text{-}70)$$

We customarily omit the subscript a in K_{a1} and K_{a2}. We will always write the subscript b in K_{b1} and K_{b2}.

You recall that the relations between the acid and base equilibrium constants are

These are Equations 5-74 and 5-75

$$K_{a1} \cdot K_{b2} = K_w \tag{10-71}$$

$$K_{a2} \cdot K_{b1} = K_w \tag{10-72}$$

We now set out to calculate the pH and composition of individual solutions of 0.050 0 M H_2L^+, 0.050 0 M HL, and 0.050 0 M L^-. Our methods are general. They do not depend on the charge type of the acids and bases. That is, we would use the same procedure to find the pH of the diprotic acids H_2A, where A is anything, or H_2L^+, where L is leucine.

Easy stuff.

Acidic Form, H_2L^+

A salt such as leucine hydrochloride contains the protonated species, H_2L^+, which can dissociate twice, as indicated in Reaction 10-66. Since $K_1 = 4.69 \times 10^{-3}$, H_2L^+ is a weak acid. HL is an even weaker acid, since $K_2 = 1.79 \times 10^{-10}$. It appears that the H_2L^+ will dissociate only partly, and the resulting HA^- will hardly dissociate at all. For this reason, we make the (superb) approximation that a solution of H_2L^+ behaves as a monoprotic acid, with $K_a = K_1$.

With this approximation, the calculation of the pH of 0.050 0 M H_2L^+ is a trivial matter.

H_2L^+ can be treated as monoprotic, with $K_a = K_{a1}$.

$$H_3NCHCO_2H \rightleftharpoons H_3NCHCO_2^- + H^+$$

$$\underset{0.050\,0-x}{H_2L^+} \rightleftharpoons \underset{x}{HL} + \underset{x}{H^+}$$

$$K_a = K_1 = 4.69 \times 10^{-3}$$

$$\frac{x^2}{F - x} = K_a \quad \Rightarrow \quad x = 1.31 \times 10^{-2} \text{ M} \tag{10-73}$$

$$[HL] = x = 1.31 \times 10^{-2} \text{ M} \tag{10-74}$$

$$[H^+] = x = 1.31 \times 10^{-2} \text{ M} \quad \Rightarrow \quad pH = 1.88 \tag{10-75}$$

$$[H_2L^+] = F - x = 3.69 \times 10^{-2} \text{ M} \tag{10-76}$$

What is the concentration of L^- in the solution? We have already assumed that it is very small, but it cannot be zero. We can calculate $[L^-]$ using the K_{a2} equation (10-68), with the concentrations of HL and H^+ from Equations 10-74 and 10-75.

$$K_{a2} = \frac{[H^+][L^-]}{[HL]}$$

$$[L^-] = \frac{K_{a2}[HL]}{[H^+]} \qquad (10\text{-}77)$$

$$[L^-] = \frac{(1.79 \times 10^{-10})(1.31 \times 10^{-2})}{(1.31 \times 10^{-2})} = 1.79 \times 10^{-10} \text{ M } (= K_{a2})$$

Since we have made the approximation that $[H^+] = [HL]$, it follows from Equation 10-77 that $[L^-] = K_{a2} = 1.79 \times 10^{-10}$ M.

Our approximation is confirmed by this last result. The concentration of L^- is about eight orders of magnitude smaller than that of HL. As a source of protons, the dissociation of HL is indeed negligible compared to the dissociation of H_2L^+. For most diprotic acids, K_1 is sufficiently larger than K_2 for this approximation to be valid. In the example just treated, even if K_2 were ten times less than K_1, the value of $[H^+]$ calculated by ignoring the second ionization would be in error by only 4%. The error in pH would be only 0.01 pH unit. *In summary, a solution of a diprotic acid behaves as a solution of a monoprotic acid, with $K_a = K_{a1}$.*

Basic Form, L^-

More easy stuff.

The fully basic species, L^-, would be found in a salt such as sodium leucinate, which could be prepared by treating leucine with an equimolar quantity of NaOH. Dissolving sodium leucinate in water gives a solution of L^-, the fully basic species. The two K_b values for this dibasic anion are

$$K_{b1} = K_w/K_{a2} = 5.59 \times 10^{-5} \qquad (10\text{-}78)$$

$$K_{b2} = K_w/K_{a1} = 2.13 \times 10^{-12} \qquad (10\text{-}79)$$

These values tell us that L^- will not hydrolyze very much to give HL. Furthermore, the resulting HL is such a weak base that hardly any further reaction to make H_2L^+ will occur.

As before, then, it is reasonable to treat L^- as a monobasic species, with $K_b = K_{b1}$. The results of this (fantastic) approximation can be outlined as follows.

Hydrolysis refers to the reaction of anything with water. Specifically, the reaction $L^- + H_2O \rightleftharpoons HL + OH^-$ is called hydrolysis.

$$H_2NCHCO_2^- + H_2O \rightleftharpoons \overset{+}{H_3N}CHCO_2^- + OH^-$$

$$\begin{array}{cccc} L^- & + H_2O \rightleftharpoons & HL & + OH^- \\ 0.050\,0 - x & & x & x \end{array}$$

L^- can be treated as monobasic, with $K_b = K_{b1}$.

$$K_b = K_{b1} = \frac{K_w}{K_{a2}} = 5.59 \times 10^{-5} \qquad (10\text{-}80)$$

$$\frac{x^2}{F - x} = 5.59 \times 10^{-5} \quad \Rightarrow \quad x = 1.64 \times 10^{-3} \text{ M}$$

$$[HL] = x = 1.64 \times 10^{-3} \text{ M} \qquad (10\text{-}81)$$

$$[H^+] = K_w/x = 6.08 \times 10^{-12} \text{ M} \quad \Rightarrow \quad \text{pH} = 11.22 \qquad (10\text{-}82)$$

$$[L^-] = F - x = 4.84 \times 10^{-2} \text{ M} \qquad (10\text{-}83)$$

The concentration of H_2L^+ can be found from the K_{b2} (or K_{a1}) equilibrium.

$$K_{b2} = \frac{[H_2L^+][OH^-]}{[HL]} = \frac{[H_2L^+]x}{x} = [H_2L^+] \qquad (10\text{-}84)$$

We find that $[H_2L^+] = K_{b2} = 2.13 \times 10^{-12}$ M, and the approximation that $[H_2L^+]$ is insignificant compared with $[HL]$ is well justified. In summary, if there is any reasonable separation between K_{a1} and K_{a2} (and therefore between K_{b1} and K_{b2}), *a solution of the fully basic form of a diprotic acid can be treated as monobasic, with $K_b = K_{b1}$.*

A tougher problem.

Intermediate Form, HL

A solution prepared from leucine, HL, is more complicated than either H_2L^+ or L^-, because HL is both an acid and a base.

HL is both an acid and a base.

$$HL \rightleftharpoons H^+ + L^- \qquad K_a = K_{a2} = 1.79 \times 10^{-10} \quad (10\text{-}85)$$

$$HL + H_2O \rightleftharpoons H_2L^+ + OH^- \qquad K_b = K_{b2} = 2.13 \times 10^{-12} \quad (10\text{-}86)$$

A molecule that can both donate and accept a proton is said to be **amphiprotic.** The acid dissociation reaction (10-85) has a larger equilibrium consant than the base association reaction (10-86), so we expect that a solution of leucine will be acidic.

However, we cannot simply ignore Reaction 10-86. It turns out that both reactions proceed to a nearly equal extent for the following reason: Reaction 10-85 produces one mole of H^+ for each mole of L^-. The mole of H^+ reacts with a mole of OH^- from Reaction 10-86, thereby driving Reaction 10-86 to the right. The number of molecules of HL reacting by each path is nearly equal, with Reaction 10-85 dominating just slightly.

To treat this case correctly, we must resort to the systematic method of Chapter 7. The procedure is applied to leucine, whose intermediate form (HL) has no net charge. However, the results we obtain apply to the intermediate form of *any* diprotic acid, regardless of its charge.

Our problem deals with 0.050 0 M leucine, in which both Reactions 10-85 and 10-86 can happen. The charge balance is

$$[H^+] + [H_2L^+] = [L^-] + [OH^-] \qquad (10\text{-}87)$$

which can be rearranged to

$$[H_2L^+] - [L^-] + [H^+] - [OH^-] = 0 \qquad (10\text{-}88)$$

We see from Equation 10-67 that we can replace $[H_2L^+]$ with $[HL][H^+]/K_1$, and from Equation 10-68 that we can replace $[L^-]$ with $[HL]K_2/[H^+]$. The K_w equation tells us that we can always write $[OH^-] = K_w/[H^+]$.

Putting all these values into Equation 10-88 gives

$$\frac{[HL][H^+]}{K_1} - \frac{[HL]K_2}{[H^+]} + [H^+] - \frac{K_w}{[H^+]} = 0 \qquad (10\text{-}89)$$

Equation 10-89 can now be solved for $[H^+]$. First we multiply all terms by $[H^+]$:

$$\frac{[HL][H^+]^2}{K_1} - [HL]K_2 + [H^+]^2 - K_w = 0$$

Then we factor out $[H^+]^2$ and rearrange:

$$[H^+]^2\left(\frac{[HL]}{K_1} + 1\right) = K_2[HL] + K_w$$

$$[H^+]^2 = \frac{K_2[HL] + K_w}{\dfrac{[HL]}{K_1} + 1} \qquad (10\text{-}90)$$

Multiplying the numerator and denominator of Equation 10-90 by K_1 and then taking the square root of both sides gives

$$[H^+] = \sqrt{\frac{K_1K_2[HL] + K_1K_w}{K_1 + [HL]}} \qquad (10\text{-}91)$$

Up to this point we have made no approximations, except to neglect activity coefficients in the equilibria. We have solved for $[H^+]$ in terms of known constants plus the single unknown, $[HL]$. Where do we proceed from here?

A chemist promptly gallops down from the mountains on her white stallion to provide the missing insight: "The major species will be HL, because it is both a weak acid and a weak base. Neither Reaction 10-85 nor Reaction 10-86 will go very far. For the concentration of HL in Equation 10-91, you can simply substitute the value 0.050 0 M."

The missing insight!

Taking the chemist's advice, we write Equation 10-91 in its most useful form:

$$\boxed{[H^+] \approx \sqrt{\frac{K_1K_2F + K_1K_w}{K_1 + F}} \qquad (10\text{-}92)}$$

K_1 and K_2 in this equation are both *acid* dissociation constants (K_{a1} and K_{a2}).

where F is the formal concentration of HL ($= 0.050\,0$ M in the present case).

At long last we can calculate the pH of 0.050 0 M leucine with Equation 10-92

$$[H^+] = \sqrt{\frac{(4.69 \times 10^{-3})(1.79 \times 10^{-10})(0.050\,0) + (4.69 \times 10^{-3})(1.0 \times 10^{-14})}{4.69 \times 10^{-3} + 0.050\,0}}$$

$$= 876 \times 10^{-7}\ \text{M} \quad \Rightarrow \quad \text{pH} = 6.06 \qquad (10\text{-}93)$$

The concentrations of H_2L^+ and L^- can be found from Equations 10-67 and 10-68, using $[H^+] = 8.76 \times 10^{-7}$ M and $[HL] = 0.050\,0$ M. From Equation 10-67 we get

$$[H_2L^+] = \frac{[H^+][HL]}{K_1} = \frac{(8.76 \times 10^{-7})(0.050\,0)}{4.69 \times 10^{-3}} = 9.34 \times 10^{-6} \text{ M} \quad (10\text{-}94)$$

and from Equation 10-68 we get

$$[L^-] = \frac{K_2[HL]}{[H^+]} = \frac{(1.79 \times 10^{-10})(0.050\,0)}{8.76 \times 10^{-7}} = 1.02 \times 10^{-5} \text{ M} \quad (10\text{-}95)$$

If $[H_2L^+] + [L^-]$ is not much less than $[HL]$, and if you wish to refine your values of $[H_2L^+]$ and $[L^-]$, the method in Box 10-3 can be used.

Was the approximation $[HL] \approx 0.050\,0$ M a good one? It certainly was, because $[H_2L^+]$ (9.34×10^{-6} M) and $[L^-]$ (1.02×10^{-5} M) are quite small in comparison to 0.050 0 M. Nearly all the leucine remained in the form HL. Note also that $[H_2L^+]$ is nearly equal to $[L^-]$; this confirms that Reactions

Box 10-3 SUCCESSIVE APPROXIMATIONS

The method of *successive approximations* is a good way to deal with difficult equations that do not have simple solutions. Consider a case in which the concentration of the intermediate (amphiprotic) species is not very close to F, the formal concentration of the solution. This happens when K_1 and K_2 are not very far apart, and F is small. Consider a solution of 1.00×10^{-3} M HM^-, where HM^- is the intermediate form of malic acid.

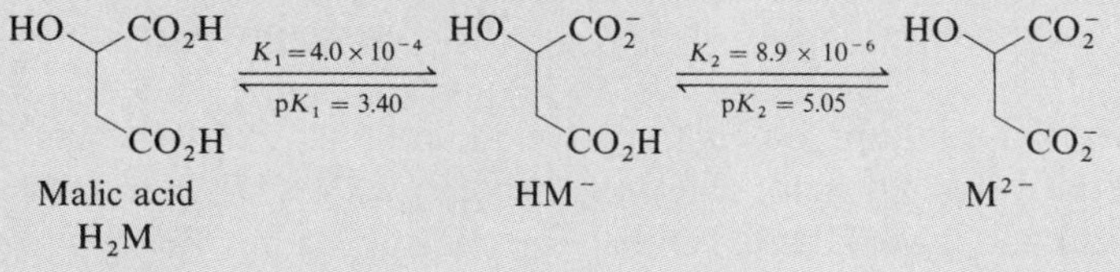
HO
CO2H
CO2H
K1 = 4.0 × 10−4
pK1 = 3.40
HO
CO2−
CO2H
K2 = 8.9 × 10−6
pK2 = 5.05
HO
CO2−
CO2−
Malic acid
H2M
HM−
M2−

As a first approximation, we assume that $[HM^-] \approx 1.00 \times 10^{-3}$ M. Plugging this value into Equation 10-92, we calculate first approximations for $[H^+]$, $[H_2M]$, and $[M^{2-}]$.

$$[H^+]_1 = \sqrt{\frac{K_1K_2(0.001\,00) + K_1K_w}{K_1 + (0.001\,00)}}$$

$$= 5.04 \times 10^{-5} \text{ M} \Rightarrow [H_2M]_1 = 1.26 \times 10^{-4} \text{ M} \quad \text{and}$$

$$[M^{2-}]_1 = 1.77 \times 10^{-4} \text{ M}$$

Clearly, $[H_2M]$ and $[M^{2-}]$ are not negligible in comparison to F $= 1.00 \times 10^{-3}$ M, so we need to revise our estimate of $[HM^-]$. As a second approximation, use

$$[HM^-]_2 = \text{F} - [H_2M]_1 - [M^{2-}]_1$$

$$= 0.001\,00 - 0.000\,126 - 0.000\,177$$

$$= 0.000\,697 \text{ M}$$

10-85 and 10-86 proceed almost equally, even though K_a is 84 times bigger than K_b for leucine.

We will usually find that Equation 10-92 is a fair-to-excellent approximation. It applies to the intermediate form of any diprotic acid, regardless of its charge type.

An even simpler form of Equation 10-92 results from two conditions that usually exist. First, if $K_2F \gg K_w$, the second term in the numerator of Equation 10-92 can be dropped.

$$[H^+] \approx \sqrt{\frac{K_1 K_2 F + \cancel{K_1 K_w}}{K_1 + F}}$$

Then, if $K_1 \ll F$, the first term in the denominator can also be neglected.

$$[H^+] \approx \sqrt{\frac{K_1 K_2 F}{K_1 + F}}$$

Using the value $[HM^-]_2 = 0.000\,697$ in Equation 10-91 gives

$$[H^+]_2 = \sqrt{\frac{K_1 K_2(0.000\,697) + K_1 K_w}{K_1 + (0.000\,697)}}$$

$$= 4.76 \times 10^{-5}\,\text{M} \quad \Rightarrow \quad [H_2M]_2 = 8.29 \times 10^{-5}\,\text{M} \qquad \text{and}$$

$$[M^{2-}]_2 = 1.30 \times 10^{-4}\,\text{M}$$

The values of $[H_2M]_2$ and $[M^{2-}]_2$ can be used to calculate a third approximation for $[HM^-]$:

$$[HM^-]_3 = F - [H_2M]_2 - [M^{2-}]_2 = 0.000\,787\,\text{M}$$

Plugging $[HM^-]_3$ into Equation 10-91 gives

$$[H^+]_4 = 4.85 \times 10^{-5}$$

and the procedure can be repeated to get

$$[H^+]_5 = 4.83 \times 10^{-5}$$

We are homing in on an estimate of $[H^+]$ in which the uncertainty is already less than 1%. This is more accuracy than is justified by the accuracy of the equilibrium constants, K_1 and K_2. The fifth approximation for $[H^+]$ gives pH = 4.32, compared to pH = 4.30 for the first approximation and pH = 4.23 using the formula pH $\approx (pK_1 + pK_2)/2$. Considering the uncertainty in pH measurements, all this calculation was hardly worth the effort. However, the concentration of $[HM^-]$ is 0.000 768 M, which is 23% less than the original estimate ($[HM^-] \approx F = 0.001\,00$ M).

Canceling F in the numerator and denominator gives

$$[H^+] \approx \sqrt{K_1 K_2} \qquad (10\text{-}96)$$

or

Recall that $\log(x^{1/2}) = \frac{1}{2}\log x$ and $\log xy = \log x + \log y$.

$$\log[H^+] \approx \tfrac{1}{2}(\log K_1 + \log K_2)$$

$$-\log[H^+] \approx -\tfrac{1}{2}(\log K_1 + \log K_2)$$

The pH of the intermediate form of a diprotic acid is roughly midway between the two pK_a values and is almost independent of concentration.

$$\boxed{pH \approx \frac{pK_1 + pK_2}{2}} \qquad (10\text{-}97)$$

Equation 10-97 is a good one to keep in your head. It is not as exact as Equation 10-92, but it is usually pretty close. Equation 10-97 gives a pH of 6.04 for leucine, compared to pH = 6.06 from Equation 10-92. Equation 10-97 says that *the pH of the intermediate form of a diprotic acid is roughly midway between pK_1 and pK_2, regardless of the formal concentration.*

EXAMPLE: pH of the Intermediate Form of a Diprotic Acid

Potassium acid phthalate, KHP, is a salt of the intermediate form of phthalic acid. Calculate the pH of 0.10 M and of 0.010 M KHP.

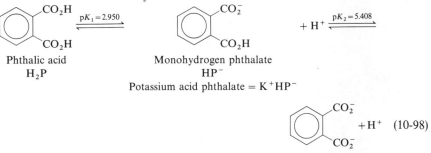

Phthalic acid
H_2P

Monohydrogen phthalate
HP^-
Potassium acid phthalate = K^+HP^-

Phthalate
P^{2-}

$(10\text{-}98)$

The pH of potassium acid phthalate is calculated to be $(pK_1 + pK_2)/2 = 4.18$, regardless of concentration, by using Equation 10-97. Using Equation 10-92, we calculate pH = 4.18 for 0.10 M K^+HP^- and pH = 4.20 for 0.010 M K^+HP^-.

A good way to do it.

Good advice

When faced with the intermediate form of a diprotic acid, use Equation 10-92 to calculate the pH. As a check on your calculation, the answer should be very close to $(pK_1 + pK_2)/2$.

Table 10-4
Summary of calculations for leucine

Solution	pH	$[H^+]$ (M)	$[H_2L^+]$ (M)	$[HL]$ (M)	$[L^-]$ (M)
0.050 0 M H_2A	1.88	1.31×10^{-2}	3.69×10^{-2}	1.31×10^{-2}	1.79×10^{-10}
0.050 0 M HA^-	6.06	8.76×10^{-7}	9.34×10^{-6}	5.00×10^{-2}	1.02×10^{-5}
0.050 0 M A^{2-}	11.22	6.08×10^{-12}	2.13×10^{-12}	1.64×10^{-3}	4.84×10^{-2}

Summary of Diprotic Acids

The way to calculate the pH and composition of solutions prepared from different forms of a diprotic acid (H_2A, HA^-, or A^{2-}) is outlined below. A summary of calculations for leucine is given in Table 10-4.

Solution of H₂A

(a) Treat H_2A as a monoprotic acid with $K_a = K_1$. This gives $[H^+]$, $[HA^-]$ and $[H_2A]$.

$$H_2A \underset{F-x}{\overset{K_1}{\rightleftharpoons}} \underset{x}{H^+} + \underset{x}{HA^-}$$

$$\frac{x^2}{F-x} = K_1$$

(b) Use the K_2 equilibrium to solve for $[A^{2-}]$ using values of $[H^+]$ and $[HA^-]$ from part a.

$$[A^{2-}] = \frac{K_2[\cancel{HA^-}]}{[\cancel{H^+}]} = K_2$$

Solution of HA⁻

(a) Use the approximation $[HA^-] \approx F$ and find the pH with Equation 10-92.

$$[H^+] = \sqrt{\frac{K_1 K_2 F + K_1 K_w}{K_1 + F}}$$

The pH should be close to $(pK_1 + pK_2)/2$.

(b) Using $[H^+]$ from part a and $[HA^-] \approx F$, solve for H_2A and A^{2-} using the K_1 and K_2 equilibria.

$$[H_2A] = \frac{[HA^-][H^+]}{K_1}$$

$$[A^{2-}] = \frac{K_2[HA^-]}{[H^+]}$$

Solution of A²⁻

(a) Treat A^{2-} as monobasic, with $K_b = K_{b1} = K_w/K_{a2}$. This gives $[A^{2-}]$, $[HA^-]$, and $[H^+]$.

$$A^{2-} + H_2O \xrightleftharpoons{K_{b1}} HA^- + OH^-$$
$$\phantom{A^{2-} + H_2O} F - x \qquad\qquad x \qquad x$$

$$\frac{x^2}{F - x} = K_{b1} = \frac{K_w}{K_{a2}}$$

$$[H^+] = \frac{K_w}{[OH^-]} = \frac{K_w}{x}$$

(b) Use the K_1 equilibrium to solve for $[H_2A]$, using the values of $[HA^-]$ and $[H^+]$ from part a.

$$[H_2A] = \frac{[HA^-][H^+]}{K_{a1}} = \frac{[HA^-]K_w}{K_{a1}[OH^-]} = K_{b2}$$

Diprotic Buffers

A buffer made from a diprotic (or polyprotic) acid is treated just as a buffer made from a monoprotic acid. For the acid H_2A we can write the two Henderson–Hasselbalch equations below, both of which are *always* necessarily true. We can use either one whenever we please.

$$pH = pK_1 + \log \frac{[HA^-]}{[H_2A]} \tag{10-99}$$

All Henderson–Hasselbalch equations (with activity coefficients) are always true for a solution at equilibrium.

$$pH = pK_2 + \log \frac{[A^{2-}]}{[HA^-]} \tag{10-100}$$

EXAMPLE: A Diprotic Buffer System
Find the pH of a solution prepared by dissolving 1.00 g of potassium acid phthalate and 1.20 g of disodium phthalate in 50.0 mL of water.

The structures of monohydrogen phthalate and phthalate were shown in Reaction 10-98. The formula weights are $KHP = C_8H_5O_4K = 204.223$ and $Na_2P = C_8H_4O_4Na_2 = 210.097$. The pH is given by

$$pH = pK_2 + \log \frac{[P^{2-}]}{[HP^-]} = 5.408 + \log \frac{1.20/210.097}{1.00/204.223} = 5.47$$

We used pK_2 because K_2 is the acid dissociation constant of HP^-, which appears in the denominator of the Henderson–Hasselbalch equation. Notice that the volume of solution was not used in answering the question.

EXAMPLE: Preparing a Buffer in a Diprotic System
How many milliliters of 0.800 M KOH should be added to 3.38 g of oxalic acid to give a pH of 4.40 when diluted to 500 mL?

OO
‖ ‖
HOCCOH

Oxalic acid
(H_2Ox)

Formula weight = 90.036

$pK_1 = 1.252$

$pK_2 = 4.266$

We know that a 1:1 mole ratio of $HOx^-:Ox^{2-}$ would have pH = pK_2 = 4.266. If the pH is to be 4.40 there must be more Ox^{2-} than HOx^- present. We must add enough base to convert all of the H_2Ox to HOx^-, plus enough additional base to convert the right amount of HOx^- into Ox^{2-}.

$$H_2Ox + OH^- \rightarrow HOx^- + H_2O$$
$$pH \approx \frac{pK_1 + pK_2}{2} = 2.76$$

$$HOx^- + OH^- \rightarrow Ox^{2-} + H_2O$$

A 1:1 mixture would have
pH = pK_2 = 4.27

In 3.38 g of H_2Ox there are 0.037 5$_4$ mol. The volume of 0.800 M KOH needed to react with this much H_2Ox to make HOx^- is

$$(\text{Volume of KOH})(0.800 \text{ M}) = 3.75_4 \times 10^{-2} \text{ mol}$$

$$\text{Volume} = 46.9_2 \text{ mL}$$

To produce a pH of 4.40 requires

	HOx^-	+ $OH^- \rightarrow$	Ox^{2-}
Initial moles:	0.037 5$_4$	x	—
Final moles:	0.037 5$_4$ − x	—	x

$$pH = pK_2 + \log \frac{[Ox^{2-}]}{[HOx^-]}$$

$$4.40 = 4.266 + \log \frac{x}{0.037 5_4 - x} \quad \Rightarrow \quad x = 0.021 6_4 \text{ mol}$$

The volume of KOH needed to deliver 0.021 6$_4$ mole is 0.021 6$_4$ mol/0.800 M = 27.0$_5$ mL. The total volume of KOH needed to bring the pH up to 4.40 is 46.9$_2$ + 27.0$_5$ = 73.9$_7$ mL.

10-7 POLYPROTIC ACIDS AND BASES

The treatment of diprotic acids and bases can be extended to polyprotic systems. By way of review let's write the pertinent equilibria for a triprotic system.

$$H_3A \rightleftharpoons H_2A^- + H^+ \qquad K_{a1} = K_1 \qquad (10\text{-}101)$$

$$H_2A^- \rightleftharpoons HA^{2-} + H^+ \qquad K_{a2} = K_2 \qquad (10\text{-}102)$$

$$HA^{2-} \rightleftharpoons A^{3-} + H^+ \qquad K_{a3} = K_3 \qquad (10\text{-}103)$$

$$A^{3-} + H_2O \rightleftharpoons HA^{2-} + OH^- \qquad K_{b1} = \frac{K_w}{K_{a3}} \qquad (10\text{-}104)$$

The acid–base equilibria for a triprotic system:

$$K_{bi} = \frac{K_w}{K_{a(3-i)}}$$

$$HA^{2-} + H_2O \rightleftharpoons H_2A^- + OH^- \qquad K_{b2} = \frac{K_w}{K_{a2}} \qquad (10\text{-}105)$$

$$H_2A^- + H_2O \rightleftharpoons H_3A + OH^- \qquad K_{b3} = \frac{K_w}{K_{a1}} \qquad (10\text{-}106)$$

Triprotic systems are treated as follows:

1. H_3A is treated as a monoprotic weak acid, with $K_a = K_1$.
2. H_2A^- is treated as the intermediate form of a diprotic acid.

The K values in Equations 10-107 and 10-108 are K_a values for the triprotic acid.

$$[H^+] \approx \sqrt{\frac{K_1 K_2 F + K_1 K_w}{K_1 + F}} \qquad (10\text{-}107)$$

3. HA^{2-} is also treated as the intermediate form of a diprotic acid. However, HA^{2-} is "surrounded" by H_2A^- and A^{3-}, so the equilibrium constants to use in Equation 10-92 are K_2 and K_3, instead of K_1 and K_2.

$$[H^+] \approx \sqrt{\frac{K_2 K_3 F + K_2 K_w}{K_2 + F}} \qquad (10\text{-}108)$$

4. A^{3-} is treated as monobasic, with $K_b = K_{b1} = K_w/K_{a3}$.

EXAMPLE: A Triprotic System

Find the pH of 0.10 M H_3His^{2+}, 0.10 M H_2His^+, 0.10 M HHis, and 0.10 M His^-, where His stands for the amino acid histidine.

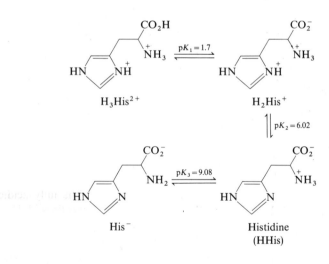

His⁻   Histidine (HHis)

0.10 M H_3His^{2+}: Treating this as a monoprotic acid we write

$$H_3His^{2+} \rightleftharpoons H_2His^+ + H^+$$

$$\quad F - x \qquad\qquad x \qquad\quad x$$

$$\frac{x^2}{F - x} = K_1 = 2 \times 10^{-2} \quad \Rightarrow \quad x = 3._{58} \times 10^{-2} M \quad \Rightarrow \quad pH = 1.45$$

0.10 M H_2His^+: Using Equation 10-107, we write

$$[H^+] = \sqrt{\frac{(2 \times 10^{-2})(9.5 \times 10^{-7})(0.10) + (2 \times 10^{-2})(1.0 \times 10^{-14})}{2 \times 10^{-2} + 0.10}}$$

$$= 1.2_6 \times 10^{-4} \text{ M} \quad \Rightarrow \quad pH = 3.90$$

Note that $(pK_1 + pK_2)/2 = 3.86$.

0.10 M HHis: Using Equation 10-108, we write

$$[H^+] = \sqrt{\frac{(9.5 \times 10^{-7})(8.3 \times 10^{-10})(0.10) + (9.5 \times 10^{-7})(1.0 \times 10^{-14})}{9.5 \times 10^{-7} + 0.10}}$$

$$= 2.8_1 \times 10^{-8} \quad \Rightarrow \quad pH = 7.55$$

Note that $(pK_2 + pK_3)/2 = 7.55$.

0.10 M His^-: Treating this as monobasic, we can write

$$His^- + H_2O \rightleftharpoons HHis + OH^-$$

$$\text{F} - x \qquad\qquad x \qquad x$$

$$\frac{x^2}{\text{F} - x} = K_{b1} = \frac{K_w}{K_{a3}} = 1.2 \times 10^{-5} \quad \Rightarrow \quad x = 1.0_9 \times 10^{-3} \text{ M}$$

$$pH = -\log\left(\frac{K_w}{x}\right) = 11.04$$

In this chapter we have reduced most acid–base problems to a few common types. When you encounter an acid or base you should first write down (or think about) the acid–base chemistry of that species. Decide whether you are dealing with an acidic, basic, or intermediate form. Then do the appropriate arithmetic to answer the question at hand.

There are only three forms of acids and bases: acidic, basic, and intermediate (amphiprotic) species.

Summary

Acid–base systems fall into several categories:

Strong acids or bases. For practical concentrations ($\gtrsim 10^{-6}$ M), pH or pOH can be found by inspection. When the concentration is near 10^{-7} M, we use the systematic treatment of equilibrium to calculate pH. At still lower concentrations, the pH is 7.00, set by autoprotolysis of the solvent.

Weak acids. For the reaction $HA \rightleftharpoons H^+ + A^-$, we set up and solve the equation $K_a = x^2/(\text{F} - x)$, where $[H^+] = [A^-] = x$, and $[HA] = \text{F} - x$. The fraction of dissociation is given by $\alpha = [A^-]/([HA] + [A^-]) = x/\text{F}$. The term pK_a is defined as $pK_a = -\log K_a$.

Weak bases. For the reaction $B + H_2O \rightleftharpoons BH^+ + OH^-$, we set up and solve the equation $K_b = x^2/(\text{F} - x)$, where $[OH^-] = [BH^+] = x$, and $[B] = \text{F} - x$. The conjugate acid of a weak base is a weak acid, and the conjugate base of a weak acid is a weak base. For a conjugate acid–base pair, $K_a \cdot K_b = K_w$.

Diprotic systems. These problems are divided into three categories:

1. The fully acidic form, H_2A, behaves as a monoprotic acid, $H_2A \rightleftharpoons H^+ + HA^-$, for which we solve the equation $K_{a1} = x^2/(\text{F} - x)$, where $[H^+] = [HA^-] = x$, and $[H_2A] = \text{F} - x$. After finding $[HA^-]$ and $[H^+]$, the value of $[A^{2-}]$ can be found from the K_{a2} equilibrium condition.

2. The fully basic form, A^{2-}, behaves as a base, $A^{2-} + H_2O \rightleftharpoons HA^- + OH^-$, for which we solve the equation $K_{b1} = x^2/(\text{F} - x)$, where $[OH^-] = [HA^-] = x$, and $[A^{2-}] = \text{F} - x$. After finding these concentrations, $[H_2A]$ can be found from the K_{a1} or K_{b2} equilibrium conditions.

3. The intermediate (amphiprotic) form, HA^-, is both an acid and a base. Its pH is given by

$$[H^+] = \sqrt{\frac{K_1 K_2 F + K_1 K_w}{K_1 + F}}$$

where K_1 and K_2 are acid dissociation constants for H_2A, and F is the formal concentration of the intermediate. In most cases this equation reduces to the form $pH \approx (pK_1 + pK_2)/2$, in which pH is independent of concentration.

In triprotic systems there are two intermediate forms. The pH of each is found with an equation analogous to that for the intermediate form of a diprotic system. Triprotic systems also have one fully acidic and one fully basic form; these can be treated as monoprotic for the purpose of calculating pH.

Buffers. A buffer is a mixture of a weak acid and its conjugate base. It resists changes in pH because it reacts with added acid or base. The pH is given by the Henderson–Hasselbalch equation.

$$pH = pK_a + \log \frac{[A^-]}{[HA]}$$

where pK_a applies to the species in the denominator. The concentrations of HA and A^- are essentially unchanged from those used to prepare the solution. The pH of a buffer is nearly independent of dilution, but the buffer capacity increases as the concentration of buffer increases. The maximum buffer capacity is found at $pH = pK_a$, and the useful range of a buffer is approximately $pH = pK_a \pm 1$. The Henderson–Hasselbalch equation can be used for polyprotic systems, as long as a conjugate acid–base pair is used in the log term.

Terms to Understand

acid dissociation constant, K_a
amino acid
amphiprotic
base "dissociation" constant, K_b
base hydrolysis constant, K_b
buffer
buffer capacity (buffer intensity)
conjugate acid–base pair
diprotic
fraction of association, α (of a base)

fraction of dissociation, α (of an acid)
Henderson–Hasselbalch equation
hydrolysis
leveling effect
pK
polyprotic
strong acid
weak acid
weak electrolyte
zwitterion

Exercises

10-A. Using activity coefficients correctly, find the pH of 1.0×10^{-2} M NaOH.

10-B. Calculate the pH of
(a) 1.0×10^{-8} M HBr
(b) 1.0×10^{-8} M H_2SO_4 (The H_2SO_4 dissociates completely to $2H^+$ plus SO_4^{2-} at this low concentration.)

10-C. What is the pH of a solution prepared by dissolving 1.23 g of 2-nitrophenol in 0.250 L?

10-D. The pH of 0.010 M o-cresol is 6.05. Find pK_a for this weak acid.

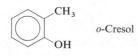
o-Cresol

10-E. Calculate the limiting value of the fraction of dissociation (α) of a weak acid ($pK_a = 5.00$) as the concentration of HA approaches zero. Repeat the same calculation for $pK_a = 9.00$.

10-F. Find the pH of 0.050 M sodium butanoate (the sodium salt of butanoic acid, also called butyric acid.)

10-G. The pH of 0.10 M ethylamine is 11.80.
(a) Without using Appendix G, find K_b for ethylamine.
(b) Using the results of part a, calculate the pH of 0.10 M ethylammonium chloride.

10-H. Which of the following bases would be most suitable for preparing a buffer of pH 9.00?
(a) NH_3 (ammonia, $K_b = 1.75 \times 10^{-5}$)

(b) $C_6H_5NH_2$ (aniline, $K_b = 3.99 \times 10^{-10}$)
(c) H_2NNH_2 (hydrazine, $K_b = 3.0 \times 10^{-6}$)
(d) C_5H_5N (pyridine, $K_b = 1.69 \times 10^{-9}$)

10-I. A solution contains 63 different conjugate acid–base pairs. Among them is acrylic acid and acrylate ion, with the ratio [acrylate]/[acrylic acid] = 0.75. What is the pH of the solution?

$$H_2C=CHCO_2H \qquad pK_a = 4.25$$
Acrylic acid

10-J. (a) How many grams of $NaHCO_3$ (F.W. 84.007) must be added to 4.00 g of K_2CO_3 (F.W. 138.206) to give a pH of 10.80 in 500 mL of water?

(b) What will be the pH if 100 mL of 0.100 M HCl are added to the solution in part a?

(c) How many milliliters of 0.320 M HNO_3 should be added to 4.00 g of K_2CO_3 to give a pH of 10.00 in 250 mL?

10-K. How many milliliters of 0.800 M KOH should be added to 3.38 g of oxalic acid to give a pH of 2.40 when diluted to 500 mL?

10-L. Find the pH and the concentrations of H_2SO_3, HSO_3^-, and SO_3^{2-} in each of the following solutions:

(a) 0.050 M H_2SO_3
(b) 0.050 M $NaHSO_3$
(c) 0.050 M Na_2SO_3

10-M. Calculate the pH of a 0.010 M solution of each amino acid in the form drawn above right.

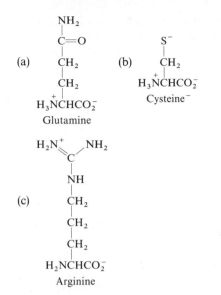

Glutamine (a), Cysteine$^-$ (b), Arginine (c)

10-N. A solution with an ionic strength of 0.10 M containing 0.010 0 M phenylhydrazine has a pH of 8.13. Using activity coefficients correctly, find pK_a for the phenylhydrazinium ion found in phenylhydrazine hydrochloride. Assume that $\gamma_{BH^+} = 0.80$.

Phenylhydrazine
B

Phenylhydrazine hydrochloride
BH^+Cl^-

Problems

A10-1. Calculate the pH of the following solutions.
(a) 1.0×10^{-3} M HBr
(b) 1.0×10^{-2} M KOH

A10-2. Write the chemical reaction whose equilibrium constant is
(a) K_a for benzoic acid
(b) K_b for benzoate ion
(c) K_b for aniline
(d) K_a for anilinium ion

Benzoic acid

Potassium benzoate

Aniline

Anilinium bromide

A10-3. Which of the following acids would be most suitable for preparing a buffer of pH 3.10?
(a) hydrogen peroxide
(b) propanoic acid
(c) cyanoacetic acid
(d) 4-aminobenzenesulfonic acid

A10-4. Phosphate, present to an extent of 0.01 M, is one of the main buffers in blood plasma, whose pH is 7.45. Would phosphate be as useful if the plasma pH were 8.5?

A10-5. Explain the following statement: The Henderson–Hasselbalch equation is *always* true; what may not be correct are the values of $[A^-]$ and $[HA]$ that we choose to use in the equation.

A10-6. Write the general structure of an amino acid. Why do some amino acids in Table 10-3 have two pK values while others have three values?

A10-7. Starting with the fully protonated species, write the stepwise acid dissociation reactions of the amino acids glutamic acid and tyrosine. Be sure to remove the protons in the correct order.

A10-8. Write the chemical reactions whose equilibrium constants are K_{b1} and K_{b2} for the amino acid proline. Find the values of K_{b1} and K_{b2}.

A10-9. Find the pH and fraction of dissociation (α) of a 0.100 M solution of the weak acid HA with $K_a = 1.00 \times 10^{-5}$.

A10-10. Find the pH and fraction of association (α) of a 0.100 M solution of the weak base B with $K_b = 1.00 \times 10^{-5}$.

A10-11. A buffer was prepared by dissolving 0.100 mol of the weak acid HA ($K_a = 1.00 \times 10^{-5}$) plus 0.050 mol of its conjugate base Na^+A^- in 1.00 L. Find the pH.

A10-12. Consider the diprotic acid H_2A with $K_1 = 1.00 \times 10^{-4}$ and $K_2 = 1.00 \times 10^{-8}$. Find the pH and concentrations of H_2A, HA^-, and A^{2-} in each case of the following solutions:
(a) 0.100 M H_2A
(b) 0.100 M NaHA
(c) 0.100 M Na_2A

A10-13. $BH^+ClO_4^-$ is a salt formed from the base B ($K_b = 1.00 \times 10^{-4}$) and perchloric acid. It dissociates into BH^+, a weak acid, and ClO_4^-, which is neither an acid nor a base. Find the pH of 0.100 M $BH^+ClO_4^-$.

A10-14. Derive the Henderson–Hasselbalch equation, including activity coefficients, from the K_a equation: $K_a = [H^+]\gamma_{H^+}[A^-]\gamma_{A^-}/[HA]\gamma_{HA}$.

10-15. Find the pH and concentrations of $(CH_3)_3N$ and $(CH_3)_3NH^+$ in a 0.060 M solution of trimethylamine.

10-16. Find the pH and concentrations of $(CH_3)_3N$ and $(CH_3)_3NH^+$ in a 0.060 M solution of trimethylammonium chloride.

10-17. Find the pH of 0.050 M NaCN.

10-18. Calculate the pH of 0.085 0 M pyridinium bromide, $C_5H_5NH^+Br^-$.

10-19. Calculate the fraction of association (α) for 1.00×10^{-1}, 1.00×10^{-2}, and 1.00×10^{-12} M sodium acetate. Does α increase or decrease with dilution?

10-20. *When is a weak acid weak and when is a weak acid strong?* Show that the weak acid HA will be 92% dissociated when dissolved in water if the formal concentration is one tenth of the K_a ($F = K_a/10$). Show that the fraction of dissociation is 27% when $F = 10K_a$. At what formal concentration will the acid be 99% dissociated? Compare your answer to the upper curve in Figure 10-2.

10-21. Write the Henderson–Hasselbalch equation for a solution of formic acid. Calculate the quotient $[HCO_2^-]/[HCO_2H]$ at the following values of pH: (a) 3.000 (b) 3.745 (c) 4.000

10-22. Given that pK_b for nitrite ion (NO_2^-) is 10.85, find the quotient $[HNO_2]/[NO_2^-]$ in a solution of sodium nitrite at
(a) pH 2.00 (b) pH 10.00

10-23. We will abbreviate malonic acid, $CH_2(CO_2H)_2$, as H_2M. Find the pH and concentrations of H_2M, HM^-, and M^{2-} in each of the following solutions:
(a) 0.100 M H_2M
(b) 0.100 M NaHM
(c) 0.100 M Na_2M

10-24. A 0.045 0 M solution of benzoic acid has a pH of 2.78. Calculate pK_a for this acid.

10-25. If a 0.10 M solution of a base has pH = 9.28, find K_b for the base.

10-26. Compound A reacts with H_2O as follows:

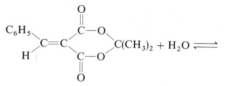

Compound A

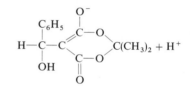

The equilibrium constant (in aqueous methanol solution) is $10^{-5.4}$. Suppose that this same equilibrium constant applies in pure water. Find the pH of a 0.020 M solution of compound A.

10-27. (a) Describe the correct procedure for preparing 0.250 L of 0.050 0 M HEPES (Table 10-2), pH 7.45.
(b) Would you need NaOH or HCl to bring the pH to 7.45?

10-28. (a) Calculate the pH of a solution prepared by dissolving 10.0 g of tris (Equation 10-59) plus 10.0 g of tris hydrochloride in 0.250 L of water.
(b) What will be the pH if 10.5 mL of 0.500 M NaOH is added?

10-29. How many milliliters of 0.246 M HNO_3 should be added to 213 mL of 0.006 66 M ethylamine to give a pH of 10.52?

10-30. (a) Write the chemical reactions whose equilibrium constants are K_b and K_a for imidazole and imidazole hydrochloride, respectively.

(handwritten notes in left margin):

$\dfrac{x^2}{F-x} = K_b$

$\dfrac{x^2}{(5.71 \times 10^{-9})} = \dfrac{0.100}{0.100 - x}$

$x^2 = 9x^2 + x - 0.100$

$5.71 \times 10^{-9}x^2 + x - 0.100$

$x = \dfrac{-b \pm \sqrt{b^2 - 4ac}}{2a}$

$x = 9.98 \times 10^{-2}$

$x = 9.98 \times 10^{-2}$

$= 1.75 \times 10^{?}$

$x = 0.02785$

(handwritten near 10-18): $Ka = 1.75 \times 10^{-5}$ $K = 9.74 \times 10^{?}$

(b) Calculate the pH of a solution prepared by mixing 1.00 g of imidazole with 1.00 g of imidazole hydrochloride and diluting to 100.0 mL.

(c) Calculate the pH of the solution if 2.30 mL of 1.07 M $HClO_4$ is added to the solution.

(d) How many milliliters of 1.07 M $HClO_4$ should be added to 1.00 g of imidazole to give a pH of 6.993?

10-31. Calculate the pH of a solution prepared by mixing 0.080 0 mol of chloroacetic acid plus 0.040 0 mol of sodium chloroacetate in 1.00 L of water.

(a) First do the calculation assuming that the concentrations of HA and A^- equal their formal concentrations.

(b) Then do the calculation using the real values of [HA] and $[A^-]$ in the solution.

(c) Using first your head, and then the Henderson–Hasselbalch equation, find the pH of a solution prepared by dissolving all of the following in one beaker containing a total volume of 1.00 L: 0.180 mol of $ClCH_2CO_2H$, 0.020 mol of $ClCH_2CO_2Na$, 0.080 mol HNO_3, and 0.080 mol $Ca(OH)_2$. Assume that $Ca(OH)_2$ dissociates completely.

10-32. Calculate the pH of a solution prepared by mixing 0.010 0 mol of the base B ($K_b = 10^{-2.00}$) with 0.020 0 mol of BH^+Br^- and diluting to 1.00 L. First calculate the pH assuming [B] = 0.010 0 and $[BH^+] = 0.020 0$ M. Compare this answer with the pH calculated without making such an assumption. Which calculation is more correct?

10-33. Calculate the pH of 0.300 M piperazine. Calculate the concentration of each form of piperazine in this solution.

10-34. (a) Calculate the quotient $[H_3PO_4]/[H_2PO_4^-]$ in 0.050 0 M KH_2PO_4.

(b) Find the same ratio for 0.050 0 M K_2HPO_4.

10-35. Barbituric acid dissociates as follows:

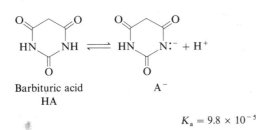

Barbituric acid A^-
HA

$$K_a = 9.8 \times 10^{-5}$$

(a) Calculate the pH and fraction of dissociation of $10^{-2.00}$ M barbituric acid.

(b) Calculate the pH and fraction of dissociation of $10^{-10.00}$ M barbituric acid.

(c) Calculate the pH and fraction of association of $10^{-2.00}$ M potassium barbiturate.

10-36. Calculate how many milliliters of 0.626 M KOH should be added to 5.00 g of HEPES (Table 10-2) to give a pH of 7.40.

30-37. Use Equations 10-64 and 10-65 to find the concentrations of HA and A^- in a solution prepared by mixing 0.002 000 mol of acetic acid plus 0.004 000 mol of sodium acetate in 1.000 L of water.

10-38. Find the pH and the concentration of each species of lysine in a solution of 0.010 0 M lysine · HCl, lysine monohydrochloride.

10-39. (a) Calculate how many milliliters of 0.100 M HCl should be added to how many grams of sodium acetate dihydrate (NaOAc · $2H_2O$, F.W. 118.06) at 5°C to prepare 250.0 mL of 0.100 M buffer, pH 5.00. At 5°C, $pK_w = 14.734$ and pK_a for acetic acid is 4.770.

(b) If you mixed what you calculated in part a, the pH would not be 5.00. Describe how you would actually prepare this buffer in the lab.

10-40. The temperature dependence of pK_a for acetic acid is given in the following table. Is the dissociation of this acid endothermic or exothermic (a) at 5°C? (b) at 45°C?

Temperature (°C)	pK_a	Temperature (°C)	pK_a
0	4.781	30	4.757
5	4.770	35	4.762
10	4.762	40	4.769
15	4.758	45	4.777
20	4.756	50	4.787
25	4.756		

10-41. *Thermodynamics and propagation of uncertainty.* The bisulfite ion exists in the following equilibrium forms:

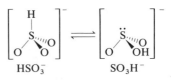

HSO_3^- SO_3H^-

The temperature dependence of the equilibrium constant at an ionic strength of 1.0 M is $\ln K = -3.23 \ (\pm 0.53) + 1.44 \ (\pm 0.15) \times 10^3 \times (1/T)$, where T is kelvins. Since $\ln K$ must be dimensionless, the number -3.23 is dimensionless and the number 1.44×10^3 has the unit of kelvins.

(a) Using Equation 5-15, calculate the enthalpy change, $\Delta H°$, and the entropy change, $\Delta S°$, for the isomerization reaction. Include uncertainties in your answers.

(b) Calculate the quotient $[SO_3H^-]/[HSO_3^-]$ at 298 K, including estimated uncertainty.

10-42. Using activity coefficients correctly, calculate the pH of
(a) 0.050 M HBr   (b) 0.050 M NaOH

10-43. Using activity coefficients correctly, calculate the fraction of dissociation, α, of 50.0 mM hydroxybenzene (phenol) in 0.050 M LiBr. Assume that the size of $C_6H_5O^-$ is 600 pm.

10-44. (a) Using activity coefficients correctly, calculate the pH of a solution containing a 2.00:1.00 mole ratio of $HC^{2-}:C^{3-}$, where H_3C is citric acid. Assume that the ionic strength is 0.010 M.

(b). What will be the pH if the ionic strength is raised to 0.10 M and the mole ratio $HC^{2-}:C^{3-}$ is kept constant?

10-45. Use the method of Box 10-3 to calculate the concentrations of H^+, H_2Ox, HOx^-, and Ox^{2-} in 0.001 00 M monosodium oxalate, NaHOx.

10-46. In this problem we will calculate the pH of the intermediate form of a diprotic acid correctly, taking activities into account.

(a) Derive Equation 10-91 for a solution of potassium acid phthalate (KHP in Reaction 10-98). Do not neglect activity coefficients in this derivation.

(b) Calculate the pH of 0.050 M KHP using the results of part a. Assume that the sizes of both HP^- and P^{2-} are 600 pm.

A10-47. Write down, but do not attempt to solve, the exact equations needed to calculate the composition of one liter of solution containing F_1 mol of HCl, F_2 mol of disodium ascorbate (Na_2A, the salt of a weak acid whose two K_a values may be called K_1 and K_2), and F_3 mol of trimethylamine (a weak base, B, whose equilibrium constant should be called K_b). Include activity coefficients wherever appropriate.

11 Acid–Base Titrations

Acid–base titrations are used routinely in virtually every field of chemistry. Figure 11-1 shows experimental data for the titration of the enzyme ribonuclease, with either strong acid or strong base. Ribonuclease is a protein with 124 amino acids. Its function is to cleave molecules of ribonucleic acid (RNA). Near pH 9.6, the enzyme has no net charge. Of the 124 amino acids of the neutral enzyme, 16 can be protonated by titration with acid, and 20 can lose protons through titration with base.

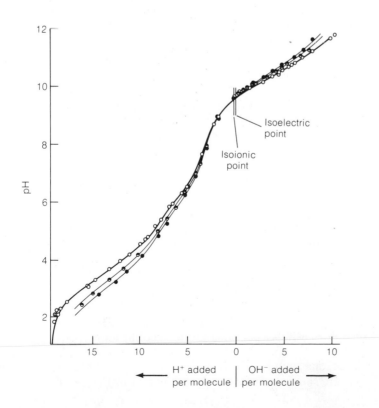

Figure 11-1
Acid–base titration of the enzyme ribonuclease. Circles are experimental points for the following ionic strengths: ● = 0.01 M, ◖ = 0.03 M, ○ = 0.15 M. The abscissa represents moles of acid or base added per mole of enzyme. The isoelectric and isoionic points are discussed in Chapter 12. [C. T. Tanford and J. D. Hauenstein, *J. Amer. Chem. Soc.,* **78,** 5287 (1956).]

From the shape the titration curve in Figure 11-1 it is possible to deduce the approximate value for pK_a for each titratable group. This, in turn, provides insight into the immediate environment of that amino acid in its place in the protein. In the case of ribonuclease it was found that three tyrosine residues exhibit "normal" values of pK_a (~ 9.95), and three others have $pK_a > 12$. The interpretation is that three tyrosine groups are accessible to solvent and OH^-, while three are buried inside the protein where they cannot be easily titrated. The solid lines in Figure 11-1 are calculated from the values of pK_a deduced for all the titratable groups.

In this chapter we will discuss the shapes of titration curves. We will not deal with molecules as complicated as ribonuclease, but the principles we discuss can be applied to any complex molecule.

11-1 TITRATION OF STRONG ACID WITH STRONG BASE

For each type of titration studied in this chapter *our goal is to construct a graph showing how the pH changes as titrant is added.* If you can do this, then you understand what is happening during the titration, and you will be able to interpret an experimental titration curve.

First write the reaction between titrant and analyte.

The first step in each case is to write the chemical reaction between titrant and analyte. We will then use that reaction to calculate the composition and pH after each addition of titrant. As a simple example, let's focus our attention on the titration of 50.00 mL of 0.020 00 M KOH with 0.100 0 M HBr. The chemical reaction between titrant and analyte is merely

The titration reaction.

$$H^+ + OH^- \rightarrow H_2O \qquad (11\text{-}1)$$

Since the equilibrium constant for this reaction is $1/K_w = 10^{14}$, it is fair to say that it "goes to completion". *Any amount of H^+ added will consume a stoichiometric amount of OH^-.*

One useful beginning is to calculate the volume of HBr needed to reach the equivalence point (V_e).

$$\underbrace{(V_e(\text{mL}))(0.100\ 0\ \text{M})}_{\substack{\text{mmol of HBr} \\ \text{at equivalence point}}} = \underbrace{(50.00\ \text{mL})(0.020\ 00\ \text{M})}_{\substack{\text{mmol of } OH^- \\ \text{being titrated}}} \quad \Rightarrow \quad V_e = 10.00\ \text{mL} \quad (11\text{-}2)$$

It is helpful to bear in mind that when 10.00 mL of HBr has been added, the titration is complete. Prior to this point there will be excess, unreacted OH^- present. After V_e has been obtained there will be excess H^+ in the solution.

In the titration of any strong base with any strong acid, there are three regions of the titration curve that represent different kinds of calculations:

1. Before reaching the equivalence point the pH is determined by excess OH^- in the solution.

2. At the equivalence point the H^+ is just sufficient to react with all of the OH^- to make H_2O. The pH is determined by the dissociation of water.

3. After the equivalence point, pH is determined by the excess H^+ in the solution.

We will do one sample calculation for each region. The results of the calculations are shown in Table 11-1 and plotted in Figure 11-2.

Table 11-1
Calculation of the titration curve for 50.00 mL of 0.020 00 M KOH treated with 0.100 0 M HBr

	mL HBr added (V_a)	Concentration of unreacted OH^- (M)	Concentration of excess H^+ (M)	pH
Region 1	0.00	0.020 0		12.30
	1.00	0.017 6		12.24
	2.00	0.015 4		12.18
	3.00	0.013 2		12.12
	4.00	0.011 1		12.04
	5.00	0.009 09		11.95
	6.00	0.007 14		11.85
	7.00	0.005 26		11.72
	8.00	0.003 45		11.53
	9.00	0.001 69		11.22
	9.50	0.000 840		10.92
	9.90	0.000 167		10.22
	9.99	0.000 016 6		9.22
Region 2	10.00	—	—	7.00
Region 3	10.01		0.000 016 7	4.78
	10.10		0.000 166	3.78
	10.50		0.000 826	3.08
	11.00		0.001 64	2.79
	12.00		0.003 23	2.49
	13.00		0.004 76	2.32
	14.00		0.006 25	2.20
	15.00		0.007 69	2.11
	16.00		0.009 09	2.04

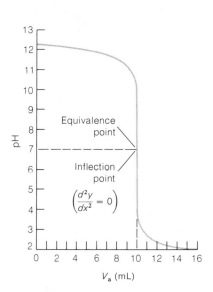

Figure 11-2
Calculated titration curve for the reaction of 50.00 mL of 0.020 00 M KOH with 0.100 0 M HBr. The equivalence point is also an inflection point.

Region 1: Before the Equivalence Point

When 3.00 mL of HBr has been added the reaction is three-tenths complete, since 10.00 mL of HBr is required to reach the equivalence point. The fraction of OH^- left unreacted is seven-tenths. The concentration of OH^- remaining in the flask is

Before the equivalence point, there is excess OH^-.

$$[OH^-] = \left(\frac{10.00 - 3.00}{10.00}\right)(0.020\,00)\left(\frac{50.00}{50.00 + 3.00}\right) = 0.013\,2 \text{ M} \quad (11\text{-}3)$$

Fraction of OH^- remaining — Initial concentration of OH^- — Dilution factor — Initial volume of OH^- — Total volume of solution

$$[H^+] = \frac{K_w}{[OH^-]} = \frac{1.0 \times 10^{-14}}{0.013\,2}$$

$$= 7.5_8 \times 10^{-13} \text{ M} \quad \Rightarrow \quad pH = 12.12 \quad (11\text{-}4)$$

Equation 11-3 is an example of a type of calculation that was introduced in Section 9-5 in connection with precipitation titrations. This equation tells us that the concentration of OH^- is equal to a certain fraction of the initial concentration, with a correction for dilution. The dilution factor equals the initial volume of OH^- divided by the total volume of solution.

Challenge: Using a setup similar to Equation 11-3, calculate $[OH^-]$ when 6.00 mL of HBr have been added. Check your pH against the value in Table 11-1.

In Table 11-1 and Figure 11-2 the volume of acid added is designated V_a. The values of pH in Table 11-1 are all expressed to the 0.01 decimal place, regardless of what is justified by significant figures. We will do this throughout the text for the sake of consistency and also because that is the usual limit of accuracy in pH measurements.

Region 2: At the Equivalence Point

Region 2 is the equivalence point, where enough H^+ has been added to react with all the OH^-. We could prepare the same solution by dissolving KBr in water. The pH is determined by the dissociation of water:

$$H_2O \rightleftharpoons H^+ + OH^-$$

(11-5)

$$ x \qquad x$$

At the equivalence point, pH = 7.00, but *only* in a strong acid–strong base reaction.

$$K_w = x^2 \quad \Rightarrow \quad x = 1.00 \times 10^{-7} \, \text{M} \quad \Rightarrow \quad \text{pH} = 7.00$$

(11-6)

The pH at the equivalence point in the titration of any strong base (or acid) with acid (or base) will be 7.00 at 25°C.

As we will discover in Section 11-2, *the pH is **not** 7.00 at the equivalence point in the titration of weak acids or bases*. The pH is 7.00 only if the titration involves a strong acid and a strong base.

Region 3: After the Equivalence Point

Beyond the equivalence point we are adding excess HBr to the solution. The concentration of excess H^+ at, say, 10.50 mL is given by

After the equivalence point, there is excess H^+.

$$[H^+] = \underbrace{(0.100\,0)}_{\substack{\text{Initial} \\ \text{concentration} \\ \text{of } H^+}} \underbrace{\left(\frac{\overset{\text{Volume of excess } H^+}{0.50}}{\underset{\text{Total volume of solution}}{50.00 + 10.50}}\right)}_{\substack{\text{Dilution} \\ \text{factor}}} = 8.26 \times 10^{-4} \, \text{M}$$

(11-7)

$$\text{pH} = -\log[H^+] = 3.08$$

(11-8)

At $V_a = 10.50$ mL, there is an excess of just $V_a - V_e = 10.50 - 10.00 = 0.50$ mL of HBr. That is the reason why 0.50 appears in the dilution factor.

Titration Curve

The complete titration curve is shown in Figure 11-2. Characteristic of all analytically useful titrations is a sudden change in pH near the equivalence point. The equivalence point is where the slope $(d\text{pH}/dV_a)$ is greatest. It is therefore an inflection point. To repeat an important statement, the pH at the equivalence point is 7.00 *only* in a strong acid–strong base titration. If one or both of the reactants is weak, the equivalence point pH is *not* 7.00, and might be quite far from 7.00.

11-2 TITRATION OF WEAK ACID WITH STRONG BASE

The titration of a weak acid with a strong base allows us to put all our knowledge of acid–base chemistry to work. The example we will examine is the titration of 50.00 mL of 0.020 00 M MES with 0.100 0 M NaOH. MES is an abbreviation for 2-(*N*-morpholino)ethanesulfonic acid, which is a weak acid with $pK_a = 6.15$.

The *titration reaction* is

$$O\hspace{-0.3em}\bigcirc\hspace{-0.3em}\overset{+}{N}HCH_2CH_2SO_3^- + OH^- \longrightarrow O\hspace{-0.3em}\bigcirc\hspace{-0.3em}NCH_2CH_2SO_3^- + H_2O \quad (11\text{-}9)$$

HA A^-

MES, $pK_a = 6.15$

Always start by writing the titration reaction.

A look at Reaction 11-9 will show you that it is the reverse of the K_b reaction for the base A^-. Therefore the equilibrium constant for Reaction 11-9 is $K = 1/K_b = 1/(K_w/K_a \text{ (for HA)}) = 7.1 \times 10^7$. The equilibrium constant is so large that we may say that the reaction goes to completion after each addition of OH^-. As we saw in Box 10-2, *strong plus weak react completely.*

Strong + weak → complete reaction

It is very helpful first to calculate the volume of base, V_b, needed to reach the equivalence point.

$$\underbrace{(V_b(\text{mL}))(0.100\ 0\ \text{M})}_{\text{mmol of base}} = \underbrace{(50.00\ \text{mL})(0.020\ 00\ \text{M})}_{\text{mmol of HA}} \quad \Rightarrow \quad V_b = 10.00\ \text{mL} \quad (11\text{-}10)$$

The titration calculations for this problem are of four types:

1. Before any base is added, the solution contains just HA in water. This is a weak-acid problem in which the pH is determined by the equilibrium

$$HA \overset{K_a}{\rightleftharpoons} H^+ + A^- \quad (11\text{-}11)$$

2. From the first addition of NaOH until immediately before the equivalence point, there is a mixture of unreacted HA plus the A^- produced by Reaction 11-9. *A buffer!* We can use the Henderson–Hasselbalch equation to find the pH.

3. At the equivalence point "all" of the HA has been converted to A^-. The problem is the same as if the solution had been made by merely dissolving A^- in water. We have a weak-base problem in which the pH is determined by Reaction 11-12.

$$A^- + H_2O \overset{K_b}{\rightleftharpoons} HA + OH^- \quad (11\text{-}12)$$

4. Beyond the equivalence point, excess NaOH is being added to a solution of A^-. To a good approximation, the pH is determined by the strong base. We will calculate the pH as if we had simply added the excess NaOH to water. We will neglect the very small effect from having A^- present as well.

Region 1: Before Base Is Added

The initial solution contains just the weak acid HA.

Before adding any base, we have a solution of 0.020 00 M HA with $pK_a = 6.15$. This is simply a weak-acid problem.

$$HA \rightleftharpoons H^+ + A^- \qquad K_a = 10^{-6.15}$$

$$F - x \qquad x \qquad x$$

$$\frac{x^2}{0.020\,00 - x} = K_a \quad \Rightarrow \quad x = 1.19 \times 10^{-4} \quad \Rightarrow \quad pH = 3.93 \quad (11\text{-}13)$$

Region 2: Before the Equivalence Point

Before the equivalence point, there is a mixture of HA and A^-, which is a buffer.

Once we begin to add OH^-, a mixture of HA and A^- is created by the titration reaction (11-9). This mixture is a buffer whose pH can be calculated with the Henderson–Hasselbalch equation (10-55) once we know the quotient $[A^-]/[HA]$.

Suppose we wish to calculate the quotient $[A^-]/[HA]$ when 3.00 mL of OH^- have been added. Since $V_e = 10.00$ mL, we have added enough base to react with three-tenths of the HA. We can make a table showing the relative concentrations before and after the reaction:

Titration reaction:	HA	+ OH$^-$	→ A$^-$	+ H$_2$O
Relative initial quantities (HA ≡ 1):	1	$\frac{3}{10}$	—	
Relative final quantities:	$\frac{7}{10}$	—	$\frac{3}{10}$	

Once we know the *quotient* $[A^-]/[HA]$ in any solution, we know its pH:

$$pH = pK_a + \log \frac{[A^-]}{[HA]} = 6.15 + \log \frac{\frac{3}{10}}{\frac{7}{10}} = 5.78 \qquad (11\text{-}14)$$

The point at which the volume of titrant is $\frac{1}{2}V_e$ is a special one in any titration.

	HA	+ OH$^-$	→ A$^-$	+ H$_2$O
Relative initial quantities:	1	$\frac{1}{2}$	—	
Relative final quantities:	$\frac{1}{2}$	—	$\frac{1}{2}$	

$$pH = pK_a + \log \frac{\frac{1}{2}}{\frac{1}{2}} = pK_a$$

When $V_b = \frac{1}{2}V_e$, pH = pK_a. This is a landmark point in any titration.

When the volume of titrant is $\frac{1}{2}V_e$, $pH = pK_a$ for the acid HA (neglecting activity coefficients). If you have an experimental titration curve you can find the approximate value of pK_a by reading the pH when $V_b = \frac{1}{2}V_e$, where V_b is the volume of added base. (To find the true value of pK_a requires a knowledge of the ionic strength and activity coefficients.)

In considering the amount of OH^- needed to react with HA, you need not worry about the small amount of HA that dissociates to A^- in the absence of OH^-. The reason for this is explained in Box 11-1.

Advice. As soon as you recognize a mixture of HA and A^- in any solution, *you have a buffer!* You can calculate the pH if you can find the quotient $[A^-]/[HA]$.

$$pH = pK_a + \log \frac{[A^-]}{[HA]}$$

Learn to recognize buffers! They lurk in every corner of acid–base chemistry.

Box 11-1 THE ANSWER TO A NAGGING QUESTION

Consider the titration of 100 mL of a 1.00 M solution of the weak acid HA with 1.00 M NaOH. The equivalence point occurs at $V_b = 100$ mL. At any point between $V_b = 0$ and $V_b = 100$ mL part of the HA will be converted to A^- and part will be left as HA. If 10 mL of NaOH had been added, we would calculate the pH as follows:

Titration reaction: $HA + OH^- \rightarrow A^- + H_2O$

Initial mmol:	100	10	—	—
Final mmol:	90	—	10	—

$$pH = pK_a + \log \frac{[A^-]}{[HA]} = pK_a + \log \frac{10}{90}$$

Upon some reflection you might think "Wait a minute! The initial solution contains some A^- in equilibrium with the HA. There must be less than 100 mmol of HA and more than 0 mmol of A^- at the start. Doesn't this make our answer wrong?"

It does not, and the reason is easy to see with some numbers. Suppose that the HA dissociates to give 1 mmol of A^- before any NaOH is added.

$$HA \xrightleftharpoons{K_a} H^+ + A^- \qquad (A)$$

Initial solution: 99 mmol   1 mmol   1 mmol

The solution contains 1 mmol of H^+, 1 mmol of A^-, and 99 mmol of HA.

When 10 mmol of OH^- is added, 1 mmol reacts with the strong acid, H^+, and 9 mmol is left to react with the weak acid, HA.

$$H^+ + OH^- \rightarrow H_2O \qquad (B)$$

1 mmol   1 mmol   1 mmol

$$HA + OH^- \rightarrow A^- \qquad (C)$$

9 mmol   9 mmol   9 mmol

The resulting solution contains 1 mmol of A^- from Reaction A and 9 mmol of A^- from Reaction C. The total A^- is 10 mmol. The total HA is 90 mmol. These are exactly the same numbers that we calculated by ignoring the initial dissociation of HA.

Moral: You can treat a mixture of the weak acid HA plus OH^- as if no dissociation of HA occurred before adding the NaOH.

Region 3: At the Equivalence Point

At the equivalence point, HA has been converted to A^-, a *weak base.*

At the equivalence point the quantity of NaOH is exactly enough to consume the HA.

Titration reaction:			HA + OH$^-$ → A$^-$ + H$_2$O
Initial relative quantities:	1	1	—
Final relative quantities:	—	—	1

The resulting solution contains "just" A^-. We could have prepared the same solution by dissolving the salt Na^+A^- in distilled water. *A solution of Na^+A^- is merely a solution of a weak base.*

To compute the pH of a weak base we write the reaction of the weak base with water:

$$A^- + H_2O \rightleftharpoons HA + OH^- \qquad K_b = \frac{K_w}{K_a} \qquad (11\text{-}15)$$
$$\text{F} - x \qquad\qquad x \qquad x$$

The only tricky point is that the formal concentration of A^- is no longer $0.020\,00$ M, which was the initial concentration of HA. The A^- has been diluted by NaOH from the buret:

$$\text{F}' = (0.020\,00)\left(\frac{50.00}{50.00 + 10.00}\right) = 0.016\,7 \text{ M} \qquad (11\text{-}16)$$

Initial concentration of HA   Dilution factor   Initial volume of HA   Total volume of solution

With this value of F' we can solve the problem:

$$\frac{x^2}{\text{F}' - x} = K_b = \frac{K_w}{K_a} = 1.43 \times 10^{-8} \quad \Rightarrow \quad x = 1.54 \times 10^{-5} \text{ M} \quad (11\text{-}17)$$

$$\text{pH} = -\log[\text{H}^+] = -\log\frac{K_w}{x} = 9.18 \qquad (11\text{-}18)$$

The pH is higher than 7 at the equivalence point in the titration of a weak acid with a strong base.

The pH at the equivalence point in this titration is 9.18. **It is not 7.00.** The equivalence point pH will *always* be above 7 for the titration of a weak acid, since the acid is converted to its conjugate base at the equivalence point.

Region 4: After the Equivalence Point

Here we assume the pH is governed by the excess OH$^-$.

Now we are adding NaOH to a solution of A^-. The NaOH is so much stronger a base than A^- that it is a fair approximation to say that the pH is determined by the concentration of excess OH$^-$ in the solution.

Let's calculate the pH when $V_b = 10.10$ mL. This is just 0.10 mL past V_e. The concentration of excess OH^- is

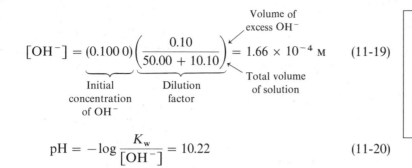

$$[OH^-] = \underbrace{(0.100\ 0)}_{\substack{\text{Initial}\\\text{concentration}\\\text{of }OH^-}} \underbrace{\left(\frac{0.10}{50.00 + 10.10}\right)}_{\substack{\text{Dilution}\\\text{factor}}} = 1.66 \times 10^{-4}\ \text{M} \qquad (11\text{-}19)$$

Volume of excess OH^-

Total volume of solution

$$pH = -\log \frac{K_w}{[OH^-]} = 10.22 \qquad (11\text{-}20)$$

> *Challenge:* Compare the concentration of OH^- due to excess titrant at $V_b = 10.10$ mL to the concentration of OH^- due to hydrolysis of A^-. Satisfy yourself that it is fair to neglect the contribution of A^- to the pH after the equivalence point.

Titration Curve

A summary of the calculations for the titration of MES with NaOH is shown in Table 11-2. The calculated titration curve appears in Figure 11-3. The curve has two easily identified points. One is the equivalence point, which is the steepest part of the curve. The other landmark is the point where $V_b = \frac{1}{2}V_e$ and $pH = pK_a$. This latter point is also an inflection point, having the minimum slope.

Landmark points in a titration:

At $V_b = V_e$, curve is steepest.
At $V_b = \frac{1}{2}V_e$, $pH = pK_a$ and the slope is minimal.

Table 11-2
Calculation of the titration curve for 50.00 mL of 0.020 00 M MES treated with 0.100 0 M NaOH

	mL base added (V_b)	pH
Region 1	0.00	3.93
Region 2	0.50	4.87
	1.00	5.20
	2.00	5.55
	3.00	5.78
	4.00	5.97
	5.00	6.15
	6.00	6.33
	7.00	6.52
	8.00	6.75
	9.00	7.10
	9.50	7.43
	9.90	8.15
Region 3	10.00	9.18
Region 4	10.10	10.22
	10.50	10.91
	11.00	11.21
	12.00	11.50
	13.00	11.67
	14.00	11.79
	15.00	11.88
	16.00	11.95

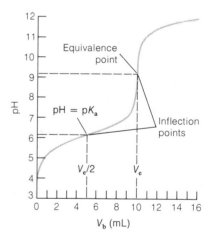

Figure 11-3
Calculated titration curve for the reaction of 50.00 mL of 0.020 00 M MES with 0.100 0 M NaOH.

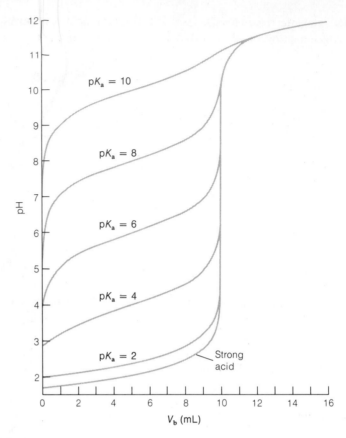

Figure 11-4
Calculated curves showing the titration of 50.0 mL of 0.020 0 M HA with 0.100 M NaOH. As the acid becomes weaker, the change in slope at the equivalence point becomes less abrupt.

The *buffer capacity* measures the ability to resist changes in pH.

If you look back at Figure 10-4, you will note that the maximum *buffer capacity* occurs when pH = pK_a. That is, the solution is most resistant to pH changes when pH = pK_a (and $V_b = \frac{1}{2}V_e$); the slope ($d\text{pH}/dV_b$) is therefore at its minimum.

Figure 11-4 shows how the titration curve depends upon the acid dissociation constant of HA. As K_a decreases, the abruptness of the inflection near the equivalence point decreases, until the equivalence point becomes too shallow to detect. A similar phenomenon occurs as the concentration of analyte and/or titrant decreases. *It is not practical to titrate an acid or base when it is too weak or too dilute.*

11-3 TITRATION OF WEAK BASE WITH STRONG ACID

The titration of a weak base with a strong acid is just the reverse of the titration of a weak acid with a strong base. The *titration reaction* is

$$B + H^+ \rightarrow BH^+ \tag{11-21}$$

Since the reactants are a weak base and a strong acid, the reaction goes essentially to completion after each addition of acid. There are four distinct regions of the titration curve:

1. Before acid is added, the solution contains just the weak base, B, in water. The pH is determined by the K_b reaction:

$$B + H_2O \overset{K_b}{\rightleftharpoons} BH^+ + OH^- \qquad (11\text{-}22)$$
$$\phantom{B + H_2O \overset{K_b}{\rightleftharpoons}}F - x x x$$

When $V_a = 0$, we have a *weak-base* problem.

2. Between the initial point and the equivalence point, there is a mixture of B and BH^+—*a buffer!* The pH is computed by using

$$pH = pK_a \text{ (for } BH^+) + \log \frac{[B]}{[BH^+]} \qquad (11\text{-}23)$$

When $0 < V_a < V_e$, we have a *buffer*.

In adding acid (increasing V_a), we reach the special point where $V_a = \frac{1}{2}V_e$ and $pH = pK_a$ (for BH^+). As before, pK_a (and therefore pK_b) can be determined easily from the titration curve.

3. At the equivalence point, B has been converted into BH^+, a weak acid. The pH is calculated by considering the acid dissociation reaction of BH^+.

$$BH^+ \rightleftharpoons B + H^+ \qquad K_a = \frac{K_w}{K_b} \qquad (11\text{-}24)$$
$$F' - x \quad x \quad x$$

When $V_a = V_e$, the solution contains the *weak acid* BH^+.

The formal concentration of BH^+, F′, is not the same as the original formal concentration of B, since some dilution has occurred. Since the solution contains BH^+ at the equivalence point, it is acidic. *The pH at the equivalence point must be below* 7.

4. After the equivalence point, there is excess H^+ in the solution. We treat this problem by considering only the concentration of excess H^+ and neglecting the contribution of a weak acid, BH^+.

For $V_a > V_e$, there is excess *strong acid.*

EXAMPLE: Titration of Pyridine with HCl

Consider the titration of 25.00 mL of 0.083 64 M pyridine with 0.106 7 M HCl.

$$K_b = 1.69 \times 10^{-9}$$

Pyridine

The titration reaction is

and the equivalence point occurs at 19.60 mL:

$$\underbrace{(V_e(\text{mL}))(0.106\ 7\ \text{M})}_{\text{mmol of HCl}} = \underbrace{(25.00\ \text{mL})(0.083\ 64\ \text{M})}_{\text{mmol of pyridine}} \Rightarrow V_e = 19.60\ \text{mL}$$

Find the pH when $V_a = 4.63$ mL.

Part of the pyridine has been neutralized, so there is a mixture of pyridine and pyridinium ion–*a buffer*. The fraction of pyridine that has been titrated is $4.63/19.60 = 0.236$, since it takes 19.60 mL to titrate the whole sample. The fraction of pyridine remaining is $(19.60 - 4.63)/19.60 = 0.764$. The pH is

$$pH = pK_a\left(= -\log\frac{K_w}{K_b}\right) + \log\frac{[B]}{[BH^+]}$$

$$= 5.23 + \log\frac{0.764}{0.236} = 5.74$$

11-4 TITRATIONS IN DIPROTIC SYSTEMS

The principles developed for titrations of monoprotic acids and bases are readily extended to titrations of polyprotic acids and bases. These principles are demonstrated in this section by a few sample calculations.

A Typical Case

The upper curve in Figure 11-5 is calculated for the titration of 10.0 mL of 0.100 M base (B) with 0.100 M HCl. The base is dibasic, with $pK_{b1} = 4.00$ and $pK_{b2} = 9.00$. The titration curve has reasonably sharp breaks at both equivalence points, corresponding to the reactions

$$B + H^+ \rightarrow BH^+ \tag{11-25}$$

$$BH^+ + H^+ \rightarrow BH_2^{2+} \tag{11-26}$$

The volume at the first equivalence point is 10.00 mL because

$$\underbrace{V_e(mL)(0.100\ \text{M})}_{\text{mmol of HCl}} = \underbrace{(10.0\ mL)(0.100\ 0\ \text{M})}_{\text{mmol of B}} \tag{11-27}$$

$$V_e = 10.0\ mL$$

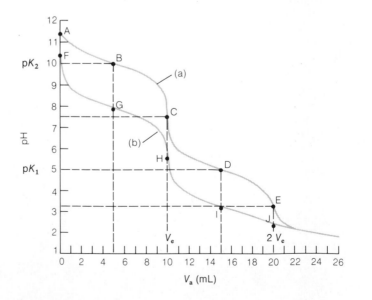

Figure 11-5

(a) Titration curve for the reaction of 10.0 mL of 0.100 M base ($pK_{b1} = 4.00$, $pK_{b2} = 9.00$) with 0.100 M HCl. The two equivalence points are labeled C and E. Points B and D are the half-neutralization points, whose pH values equal pK_{a2} and pK_{a1}, respectively. (b) Titration curve for the reaction of 10.0 mL of 0.100 M nicotine ($pK_{b1} = 6.15$, $pK_{b2} = 10.85$) with 0.100 M HCl.

The volume at the second equivalence point must be $2V_e$, since Reaction 11-26 requires exactly the same number of moles of HCl as those for Reaction 11-25.

The calculation of pH at each point along the curve is very similar to the corresponding point in the titration of a monobasic compound. We will examine each point, A through E in Figure 11-5.

$V_{e2} = 2V_{e1}$, always.

Point A

Before any acid is added, the solution contains just B, a weak base, whose pH is governed by the following reaction:

$$B \quad + H_2O \xrightleftharpoons{K_{b1}} BH^+ + OH^- \qquad (11\text{-}28)$$

$$0.100 - x \qquad\qquad x \qquad x$$

Recall that the fully basic form of a dibasic compound can be treated as if it were monobasic. (The K_{b2} reaction can be neglected.)

$$\frac{x^2}{0.100 - x} = 1.00 \times 10^{-4} \quad\Rightarrow\quad x = 3.11 \times 10^{-3}$$

$$[H^+] = \frac{K_w}{x} \quad\Rightarrow\quad pH = 11.49$$

Point B

At any point between A (the initial point) and C (the first equivalence point), we have a buffer containing B and BH$^+$. Point B is halfway to the equivalence point, so $[B] = [BH^+]$. The pH is calculated by using the Henderson–Hasselbalch equation *for the weak acid*, BH^+, whose acid dissociation constant is K_{a2} for BH_2^{2+}. The value of K_{a2} is

Of course, you remember that

$$K_{a2} = \frac{K_w}{K_{b1}} = 10^{-10.00} \qquad (11\text{-}29)$$

$$K_{b1} = \frac{K_w}{K_{a2}}$$

$$K_{b2} = \frac{K_w}{K_{a1}}$$

To calculate the pH at point B, we write

$$pH = pK_{a2} + \log \frac{[B]}{[BH^+]} = 10.00 + \log 1 = 10.00 \qquad (11\text{-}30)$$

So the pH at point B is just pK_{a2}.

To calculate the quotient $[B]/[BH^+]$ at any point in the buffer region, just find what fraction of the way from point A to point C the titration has progressed. For example, if $V_a = 1.5$ mL, then

$$\frac{[B]}{[BH^+]} = \frac{8.5}{1.5} \qquad (11\text{-}31)$$

because 10.0 mL are required to reach the equivalence point and we have added just 1.5 mL. The pH at $V_a = 1.5$ mL is given by

$$pH = 10.00 + \log \frac{8.5}{1.5} = 10.75 \qquad (11\text{-}32)$$

BH$^+$ is the *intermediate form* of a diprotic acid.

pH $\approx \frac{1}{2}(pK_1 + pK_2)$

Point C

At the first equivalence point, B *has been converted to* BH$^+$, *the intermediate form of the diprotic acid,* BH$_2^{2+}$. BH$^+$ *is both an acid and a base.* As described in Section 10-6, the pH is given by

$$[H^+] \approx \sqrt{\frac{K_1 K_2 F + K_1 K_w}{K_1 + F}} \tag{11-33}$$

where K_1 and K_2 are the acid dissociation constants of BH$_2^{2+}$.

The formal concentration of BH$^+$ is calculated by considering the dilution of the original solution of B.

$$F = \underbrace{(0.100\ \text{M})}_{\substack{\text{Original} \\ \text{concentration} \\ \text{of B}}} \underbrace{\left(\frac{10.0}{20.0}\right)}_{\substack{\text{Dilution} \\ \text{factor}}} = 0.050\ 0\ \text{M} \tag{11-34}$$

Initial volume of B

Total volume of solution

Plugging all the numbers into Equation 11-33 gives

$$[H^+] = \sqrt{\frac{(10^{-5})(10^{-10})(0.050\ 0) + (10^{-5})(10^{-14})}{10^{-5} + 0.050\ 0}} = 3.16 \times 10^{-8} \tag{11-35}$$

$$pH = 7.50$$

Note that in this example pH $= (pK_{a1} + pK_{a2})/2$.

Point D

At any point between C and E, we can consider the solution to be a buffer containing BH$^+$ (the base) and BH$_2^{2+}$ (the acid). When $V_a = 15.0$ mL, $[BH^+] = [BH_2^{2+}]$ and

$$pH = pK_{a1} + \log \frac{[BH^+]}{[BH_2^{2+}]} = 5.00 + \log 1 = 5.00 \tag{11-36}$$

Challenge: Show that if V_a were 17.2 mL, the ratio in the log term would be

$$\frac{[BH^+]}{[BH_2^{2+}]} = \frac{20.0 - 17.2}{17.2 - 10.0} = \frac{2.8}{7.2}$$

Point E

Point E is the second equivalence point, at which the solution is formally the same as one prepared by dissolving BH$_2$Cl$_2$ in water. The formal concentration of BH$_2^{2+}$ is

$$F = (0.100 \text{ M})\left(\frac{10.0}{30.0}\right) = 0.033\ 3 \text{ M} \qquad (11\text{-}37)$$

Original volume of B

Total volume of solution

The pH is determined by the acid dissociation reaction of BH_2^{2+}.

$$BH_2^{2+} \rightleftharpoons BH^+ + H^+ \qquad K_{a1} = \frac{K_w}{K_{b2}} \qquad (11\text{-}38)$$

$$F - x \qquad x \qquad x$$

At the second equivalence point, we have made BH_2^{2+}, which can be treated as a monoprotic weak acid.

$$\frac{x^2}{0.033\ 3 - x} = 1.0 \times 10^{-5} \quad \Rightarrow \quad x - 5.72 \times 10^{-4} \quad \Rightarrow \quad pH = 3.24$$

Beyond the second equivalence point ($V_a > 20.0$ mL), the pH of the solution can be calculated from the volume of strong acid added to the solution. For example, at $V_a = 25.00$ mL, there is an excess of 5.00 mL of 0.100 M HCl in a total volume of $10.00 + 25.00 = 35.00$ mL. The pH is found by writing

$$[H^+] = (0.100 \text{ M})\left(\frac{5.00}{35.00}\right) = 1.43 \times 10^{-2} \text{ M} \quad \Rightarrow \quad pH = 1.85 \quad (11\text{-}39)$$

Blurred End Points

The titrations of many diprotic acids or bases do show two clear end points, as in the upper curve in Figure 11-5. However, other titrations may not show both end points, as in the lower curve of Figure 11-5. This latter curve is calculated for the titration of 10.0 mL of 0.100 M nicotine ($pK_{b1} = 6.15$, $pK_{b2} = 10.85$) with 0.100 M HCl. The two reactions are

Under some conditions, the end point is obscured.

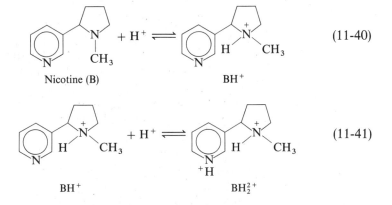

Nicotine (B) BH^+ (11-40)

BH^+ BH_2^{2+} (11-41)

Both reactions occur, but there is almost no perceptible break at the second equivalence point, simply because BH_2^{2+} is too strong an acid (or, equivalently, BH^+ is too weak a base).

The footnote at the beginning of Section 12-4 gives references to articles that describe how to use computer programs to solve acid–base equilibrium problems without introducing the simplifying (but sometimes incorrect) approximations that we have been using.

The pH values at points F, G, H, I, and J in Figure 11-5 are calculated in just the same manner used for points A, B, C, D, and E. However, as the acidic end of the titration is approached (say, pH $\lesssim$ 3), the approximation that all the HCl reacts with BH^+ to give BH_2^{2+} is not true. The acidity of BH_2^{2+} is great enough for it to exist partially dissociated in equilibrium with substantial free H^+. To calculate the pH correctly, we use the systematic treatment of equilibrium. It was in this manner, using a computer, that the lower portion of the titration curve was calculated.

The pH at point J is correctly calculated by using the same procedure as for point E. That is, at point J the solution contains the salt $[BH_2^{2+}][Cl^-]_2$, regardless of whether it was created by a titration or by dissolving BH_2Cl_2 in water. The pH calculated for point J is 2.35.

As an example of what should alert you to a poor approximation, suppose we try calculating the pH at $V_a = 21.0$ mL. We will use the approximation that the pH is determined by the excess HCl.

$$[H^+] \approx \underbrace{(0.100 \text{ M})}_{\substack{\text{Initial} \\ \text{concentration} \\ \text{of HCl}}} \underbrace{\left(\frac{1.0}{31.0}\right)}_{\substack{\text{Dilution} \\ \text{factor}}} = 3.2 \times 10^{-3} \text{ M} \quad \Rightarrow \quad \text{pH} = 2.49 \quad (11\text{-}42)$$

where the 1.0 is the Volume of excess HCl and 31.0 is the Total volume of solution.

We calculate that pH = 2.49 for this value of V_a, beyond the second equivalence point. But the pH at the equivalence point was correctly calculated to be 2.35, *which is lower than 2.49.* The pH cannot turn around and start climbing. If the pH at point J is correct, the pH at $V_a = 21.0$ mL must be wrong. The problem is that we have neglected the contribution to $[H^+]$ made by dissociation of BH_2^{2+}, which is a fairly strong acid ($pK_{a1} = 3.15$). The systematic treatment of equilibrium gives a value of pH = 2.19 at $V_a = 21.0$ mL.

One moral of this section is that whenever an acid is too strong (low pK_a) or a base too weak (high pK_b), the titration curve may show no evident break. This is true for monoprotic or polyprotic systems. A second lesson is that the approximations used in our calculations tend to be misleading at very high or very low pH. The approximations also become poorer as the formal concentration of acid or base being titrated becomes smaller.

In the titration curve for ribonuclease shown in Figure 11-1, there is a continual change in pH, with no clear breaks. The reason is that there are 29 groups being titrated in the pH interval shown. The 29 end points are so close together that a nearly uniform rise results. The curve can be analyzed to find the many pK_a values, but this requires a computer, and the individual pK values will not be determined very precisely.

11-5 FINDING THE END POINT

Titrations are most commonly performed either to find out how much analyte is present or to measure equilibrium constants of the analyte. To find out how much analyte is present requires a knowledge of V_e, the volume of titrant at the equivalence point. The most popular ways of determining the equivalence point involve indicators or pH measurement with the glass electrode.

What is an indicator?

An acid–base **indicator** is itself an acid or base whose different protonated species have different colors. An example is thymol blue, which has two useful color changes.

An indicator is an acid or a base whose different protonated forms have different colors.

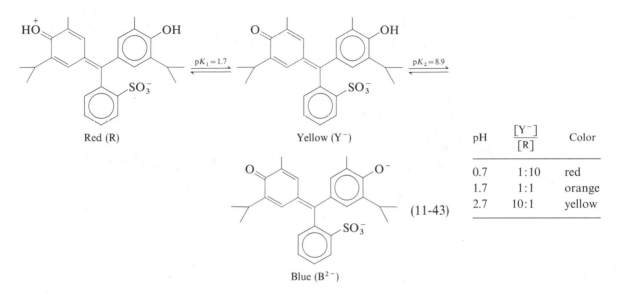

pH	$\dfrac{[Y^-]}{[R]}$	Color
0.7	1:10	red
1.7	1:1	orange
2.7	10:1	yellow

(11-43)

Blue (B^{2-})

Below pH 1.7, the predominant species is red; between pH 1.7 and pH 8.9, the predominant species is yellow; and above pH 8.9, the predominant species is blue. For simplicity, we designate the three species R, Y^-, and B^{2-}, respectively. The sequence of color changes for thymol blue is shown in Color Plate 5.

One of the most common indicators is phenolphthalein, usually used for its colorless ↔ pink transition at pH 8.0–9.6.

The equilibrium between R and Y^- can be written

$$R \rightleftharpoons Y^- + H^+$$

and

$$K_1 = \frac{[Y^-][H^+]}{[R]}$$

$$pH = pK_1 + \log \frac{[Y^-]}{[R]} \qquad (11\text{-}44)$$

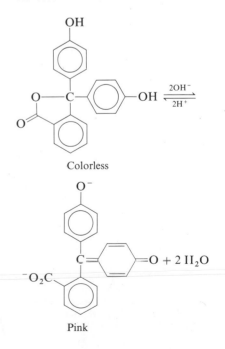

Colorless

At pH = 1.7 ($= pK_1$), there will be a 1:1 mixture of the yellow and red species, which will appear orange. As a very crude rule of thumb we may say that the solution will appear red when $[Y^-]/[R] \lesssim 1/10$ and yellow when $[Y^-]/[R] \gtrsim 10/1$. From the Henderson–Hasselbalch equation (11-44) we can see that the solution will be red when pH ≈ $pK_1 - 1$ and yellow when pH ≈ $pK_1 + 1$. In tables of indicator colors, thymol blue is listed as red below pH 1.2 and yellow above pH 2.8. By comparison, the pH values predicted by our rule of thumb are 0.7 and 2.7. Between pH 1.2 and pH 2.8, the indicator exhibits varying shades of orange. While most indicators have a single color

Pink

Demonstration 11-1   INDICATORS AND THE ACIDITY OF CO₂

This demonstration is just plain fun to watch.[†] Fill two 1-L graduated cylinders with 900 mL of water and place a magnetic stirring bar in each. Add 10 mL of 1 M NH_3 to each. Then put 2 mL of phenolphthalein indicator solution in one and 2 mL of bromothymol blue indicator solution in the other. Both indicators will have the color of their basic species.

Drop a few chunks of Dry Ice (solid CO_2) in each cylinder. As the CO_2 bubbles through each cylinder the solutions become more acidic. First the pink phenolphthalein color disappears. After some time the pH drops just low enough for bromothymol blue to change from blue to its green intermediate color. The pH does not go low enough to turn the indicator to its yellow color.

Add about 20 mL of 6 M HCl to *the bottom* of each cylinder, using a length of Tygon tubing attached to a funnel. Then stir each solution for a few seconds on a magnetic stirrer. Explain what happens. The sequence of events in this demonstration is shown in Color Plate 6.

[†] A set of fascinating indicator demonstrations using universal indicator (a mixed indicator with many color changes) is described in J. T. Riley, *J. Chem. Ed.*, **54**, 29 (1977).

In strong acid the colorless form of phenolphthalein turns orange-red. In strong base the red species loses its color [G. Wittke, *J. Chem. Ed.*, **60**, 239 (1983)].

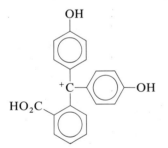

Orange-red
(formed in 65–98% H_2SO_4)

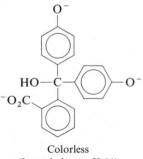

Colorless
(formed above pH 11)

Choose an indicator whose color change comes as close as possible to the theoretical pH of the equivalence point.

change, thymol blue undergoes another transition, from yellow to blue, between pH 8.0 and pH 9.6. In this range, various shades of green would be seen. The interpretation of color varies among individuals, and indicator colors are no exception. Acid–base indicator color changes form the basis of Demonstration 11-1.

Choosing an indicator

A titration curve for which pH = 5.54 at the equivalence point is shown in Figure 11-6. An indicator with a color change near this pH would be useful

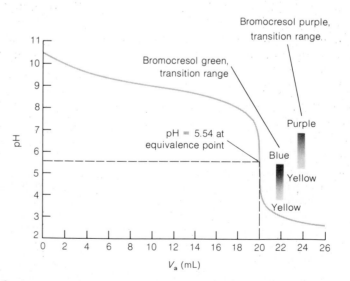

Figure 11-6
Calculated titration curve for the reaction of 100 mL of 0.010 0 M base (pK_b = 5.00) with 0.050 0 M HCl.

in determining the end point of the titration. You can see on the graph in Figure 11-6 that the pH drops steeply (from 7 to 4) over a small volume interval. Therefore, any indicator with a color change in this pH interval would provide a fair approximation to the equivalence point. The closer the point of color changes is to pH 5.54, the more accurate will be the end point. The difference between the observed end point (color change) and the true equivalence point is called the **indicator error**.

A list of some common indicators is given in Table 11-3. A rather large number of indicators in the table would provide a useful end point for the

[handwritten: $pK_b = 3.925$]

[handwritten: HCl THAM $pK_a = 8.075$]

[handwritten: $K_a = 8.41 \times 10^{-9}$]

Table 11-3
Some common indicators

Indicator	Transition range (pH)	Acid color	Base color	Preparation
Methyl violet	0.0–1.6	Yellow	Blue	0.05% in H_2O
Cresol red	0.2–1.8	Red	Yellow	0.1 g in 26.2 mL 0.01 M NaOH. Then add $\sim$225 mL H_2O.
Thymol blue	1.2–1.8	Red	Yellow	0.1 g in 21.5 mL 0.01 M NaOH. Then add $\sim$225 mL H_2O.
Cresol purple	1.2–2.8	Red	Yellow	0.1 g in 26.2 mL 0.01 M NaOH. Then add $\sim$225 mL H_2O.
Erythrosine, disodium	2.2–3.6	Orange	Red	0.1% in H_2O
Methyl orange	3.1–4.4	Red	Orange	0.01% in H_2O
Congo red	3.0–5.0	Violet	Red	0.1% in H_2O
Ethyl orange	3.4–4.8	Red	Yellow	0.1% in H_2O
Bromocresol green	3.8–5.4	Yellow	Blue	0.1 g in 14.3 mL 0.01 M NaOH. Then add $\sim$225 mL H_2O.
Methyl red	4.8–6.0	Red	Yellow	0.02 g in 60 mL ethanol. Then add 40 mL H_2O.
Chlorophenol red	4.8–6.4	Yellow	Red	0.1 g in 23.6 mL 0.01 M NaOH. Then add $\sim$225 mL H_2O.
Bromocresol purple	5.2–6.8	Yellow	Purple	0.1 g in 18.5 mL 0.01 M NaOH. Then add $\sim$225 mL H_2O.
p-Nitrophenol	5.6–7.6	Colorless	Yellow	0.1% in H_2O
Litmus	5.0–8.0	Red	Blue	0.1% in H_2O
Bromothymol blue	6.0–7.6	Yellow	Blue	0.1 g in 16.0 mL 0.01 M NaOH. Then add $\sim$225 mL H_2O.
Phenol red	6.4–8.0	Yellow	Red	0.1 g in 28.2 mL 0.01 M NaOH. Then add $\sim$225 mL H_2O.
Neutral red	6.8–8.0	Red	Orange	0.01 g in 50 mL ethanol. Then add 50 mL H_2O.
Cresol red	7.2–8.8	Yellow	Red	See above.
α-Naphtholphthalein	7.3–8.7	Yellow	Blue	0.1 g in 50 mL ethanol. Then add 50 mL H_2O.
Cresol purple	7.6–9.2	Yellow	Purple	See above.
Thymol blue	8.0–9.6	Yellow	Blue	See above.
Phenolphthalein	8.0–9.6	Colorless	Red	0.05 g in 50 mL ethanol. Then add 50 mL H_2O.
Thymolphthalein	8.3–10.5	Colorless	Blue	0.04 g in 50 mL ethanol. Then add 50 mL H_2O.
Alizarin yellow	10.1–12.0	Yellow	Orange-red	0.01% in H_2O
Nitramine	10.8–13.0	Colorless	Orange-brown	0.1 g in 70 mL ethanol. Then add 30 mL H_2O.
Tropaeolin O	11.1–12.7	Yellow	Orange	0.1% in H_2O

Box 11-2   WHAT DOES A NEGATIVE pH MEAN?

In the 1930s, Louis Hammett and his students devised a means to measure the basicity of very weak bases and the acidity of very strong acids. They began with a weak reference base, B, whose base strength could be measured in aqueous solution. An example is *p*-nitroaniline, whose protonated form has a pK_a of 0.99.

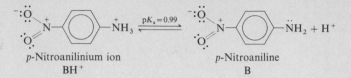

p-Nitroanilinium ion   *p*-Nitroaniline
BH$^+$ B

Suppose that some *p*-nitroaniline and a second base, C, are dissolved in a strong acid, such as 2 M HCl. The pK_a of CH$^+$ can be measured relative to that of BH$^+$ by first writing a Henderson–Hasselbalch equation for each acid:

$$pH = pK_a(\text{for BH}^+) + \log \frac{[\text{B}]\gamma_\text{B}}{[\text{BH}^+]\gamma_{\text{BH}^+}}$$

$$pH = pK_a(\text{for CH}^+) + \log \frac{[\text{C}]\gamma_\text{c}}{[\text{CH}^+]\gamma_{\text{CH}^+}}$$

Setting the two equations equal (since there is only one pH) gives

$$\underbrace{pK_a(\text{for CH}^+) - pK_a(\text{for BH}^+)}_{\Delta pK_a} = \log \frac{[\text{B}][\text{CH}^+]}{[\text{C}][\text{BH}^+]} + \log \frac{\gamma_\text{B}\gamma_{\text{CH}^+}}{\gamma_\text{c}\gamma_{\text{BH}^+}}$$

In solvents of high dielectric constant, the second term on the right, above, is close to zero because the ratio of activity coefficients is close to unity. Neglecting this last term gives the operationally useful result:

$$\Delta pK_a = \log \frac{[\text{B}][\text{CH}^+]}{[\text{C}][\text{BH}^+]}$$

That is, if you have a way to find the concentrations of B, BH$^+$, C, and CH$^+$, and you know the pK_a for BH$^+$, then you can find the pK_a for CH$^+$.

In practice, the concentrations can be measured spectrophotometrically, so pK_a for CH$^+$ can be determined. Then, using CH$^+$ as the reference, the pK_a for another compound, DH$^+$, can be measured. This procedure can be extended to measure the strengths of successively weaker bases, far too weak to be protonated in water.

The acidity of the solvent used to protonate the weak base, B, can be defined as H_0, which is analogous to the pH of an aqueous solution:

$$H_0 = pK_a(\text{for BH}^+) + \log \frac{[\text{B}]}{[\text{BH}^+]}$$

For dilute aqueous solutions, H_0 approaches pH. For concentrated solutions of strong acids, H_0 is considered to be a measure of the acid strength. H_0 is called the **Hammett acidity function.**

A graph of H_0 versus acid concentration for several strong acids is shown below.

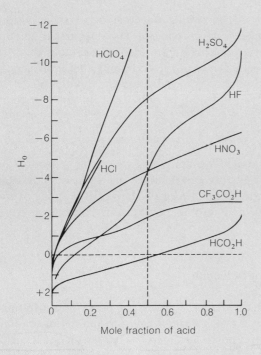

Hammett acidity function, H_0, for aqueous solutions of acids. [Data from R. A. Cox and K. Yates, *Can. J. Chem.,* **61**, 2225 (1983), which provides an informative review of acidity functions.]

When we refer to negative pH values, we are usually referring to H_0 values. For example, as measured by its ability to protonate very weak bases, 8 M $HClO_4$ has a "pH" close to -4. The figure shows why $HClO_4$ is considered to be a stronger acid than the other common strong acids. The values of H_0 for several powerfully acidic solvents are given below.

Acid	Name	H_0
H_2SO_4 (100%)	Sulfuric acid	-11.93
$H_2SO_4 \cdot SO_3$	Fuming sulfuric acid (oleum)	-14.14
HSO_3F	Fluorosulfuric acid	-15.07
$HSO_3F + 10\%$ SbF_5	"Super acid"	-18.94
$HSO_3F + 7\%$ $SbF_5 \cdot 3SO_3$	—	-19.35

The idea of using indicators to measure acidity can be extended to environments in which conventional chemical measurements are difficult. For example, the pH inside microscopic vesicles (compartments) within living cells can be estimated by infusion of an appropriate indicator and measuring the spectrum of the indicator inside the vesicle. The "pH" of very acidic surfaces of solid catalysts can be measured by observing the color of Hammet indicators adsorbed on the catalyst surface.

titration in Figure 11-6. For example, if bromocresol purple were used, we would use the purple-to-yellow color change as the end point. The last trace of purple should disappear near pH 5.2, which is quite close to the true equivalence point in Figure 11-6. If bromocresol green were used as the indicator, a color change from blue to green (= yellow + blue) would mark the end point.

In general, *we seek an indicator whose transition range overlaps the steepest part of the titration curve as closely as possible.* The steepness of the titration curve near the equivalence point in Figure 11-6 ensures that the indicator error caused by the noncoincidence of the end point and equivalence point will not be large. For example, if the indicator end point were at pH 6.4 (instead of 5.54), the error in V_e would be only 0.25% in this particular case.

Other uses of indicators

Indicators can be used to measure pH.

Equation 11-44 implies that if we know the pK of an indicator and the concentrations of both forms, we can measure pH with the indicator. This approach allows us to extend the idea of pH beyond the range over which glass electrodes function, and also to nonaqueous solvents, as Box 11-2 explains.

Potentiometric End-point Detection

An alternative way to measure the equivalence point of a titration is to monitor the pH continuously with a glass electrode and pH meter. Since the meter measures the electric potential difference across the glass membrane of the electrode, we refer to this as a *potentiometric measurement*. The theory of such measurements is described in Chapter 15. Box 11-3 describes an important potentiometric measurement in environmental analysis.

Figure 11-7 shows experimental results for the titration of a hexaprotic weak acid, H_6A, with NaOH. Because the compound is very difficult to purify, only a tiny amount was available for titration. Just 1.430 mg was dissolved in 1.00 mL of water and titrated with microliter quantities of 0.065 92 M NaOH, delivered with a Hamilton syringe.

The curve in Figure 11-7 shows two clear breaks, near 90 and 120 μL, which correspond to titration of the *third* and *fourth* protons of H_6A.

$$H_4A^{2-} + OH^- \rightarrow H_3A^{3-} + H_2O \qquad (\sim 90 \ \mu L \text{ equivalence point})$$

$$H_3A^{3-} + OH^- \rightarrow H_2A^{4-} + H_2O \qquad (\sim 120 \ \mu L \text{ equivalence point})$$

The first two and last two equivalence points give unrecognizable end points, because they occur at pH values that are too low or too high.

The end point is the point of maximum slope.

The end point of an acid–base titration is taken as the point where the slope (dpH$/dV$) of the titration curve is greatest. How can we locate the steepest sections in Figure 11-7? A surprisingly accurate (and simple) way is to take a pencil and run it along the curve, feeling how the slope changes as you go. Most people can feel the inflection point at which the slope stops increasing and begins to decrease. A better, but much more laborious method, is to prepare a Gran plot.

Box 11-3 ALKALINITY AND ACIDITY

Alkalinity is defined as the capacity of a natural water specimen to react with H^+ to reach pH 4.5. To a good approximation, alkalinity is determined by OH^-, CO_3^{2-} and HCO_3^-:

$$\text{alkalinity} \approx [OH^-] + 2[CO_3^{2-}] + [HCO_3^-]$$

The titration curve for CO_3^{2-} with H^+ in Problem 11-34 shows that the second equivalence point is near pH 4.5. When water whose pH is greater than 4.5 is titrated with acid to pH 4.5 (measured with a pH meter), all OH^-, CO_3^{2-} and HCO_3^- will have reacted. Other basic species also react, but OH^-, CO_3^{2-} and HCO_3^- account for most of the alkalinity in most water samples. Alkalinity is normally expressed as mmol H^+ needed to bring 1 L of water to pH 4.5.

Alkalinity and *hardness* (dissolved Ca^{2+} and Mg^{2+}, Box 13-2) are important characteristics of irrigation water. Alkalinity in excess of the $Ca^{2+} + Mg^{2+}$ content is called "residual sodium carbonate." Water with a residual sodium carbonate content chemically equivalent to ≥ 2.5 mmol H^+/L is not suitable for irrigation. A residual sodium carbonate content between 1.25 and 2.5 mmol H^+/L is marginal, while a content ≤ 1.25 mmol H^+/L is suitable for irrigation.

Acidity of natural waters refers to the total acid content that can be titrated to pH 8.3 with NaOH. Problem 11-34 shows that pH 8.3 is near the second equivalence point for titration of carbonic acid (H_2CO_3) with OH^-. Almost all weak acids that might be present in the water will also be titrated in this procedure. The acidity is expressed as mmol OH^- needed to bring 1 L of water to pH 8.3.

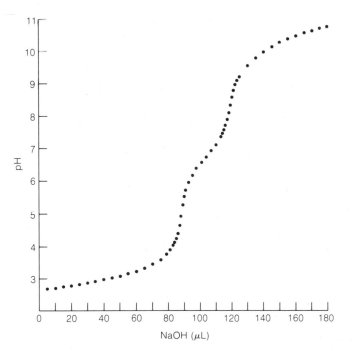

Figure 11-7
Experimental points in the titration of 1.430 mg of xylenol orange, a hexaprotic acid, dissolved in 1.00 mL of aqueous 0.10 M $NaNO_3$. The titrant was 0.065 92 M NaOH.

Gran plot[†]

Consider the titration of a weak acid, HA, whose acid dissociation constant can be written as follows:

$$HA \rightleftharpoons H^+ + A^- \qquad K_a = \frac{[H^+]\gamma_{H^+}[A^-]\gamma_{A^-}}{[HA]\gamma_{HA}} \qquad (11\text{-}45)$$

It will be necessary to include activity coefficients in this discussion because a pH electrode responds to hydrogen ion *activity,* not concentration.

Strong plus weak react completely.

At any point between the initial point and the end point of the titration it is usually a good approximation to say that each mole of NaOH converts one mole of HA into one mole of A^-. If we have titrated V_a mL of HA (whose formal concentration is F_a) with V_b mL of NaOH (whose formal concentration is F_b), we can write

$$[A^-] = \frac{\text{moles of } OH^- \text{ delivered}}{\text{total volume}} = \frac{V_b F_b}{V_b + V_a} \qquad (11\text{-}46)$$

$$[HA] = \frac{\text{original moles of HA} - \text{moles of } OH^-}{\text{total volume}} = \frac{V_a F_a - V_b F_b}{V_a + V_b} \qquad (11\text{-}47)$$

Substituting the values of $[A^-]$ and $[HA]$ from Equations 11-46 and 11-47 into Equation 11-45 gives

$$K_a = \frac{[H^+]\gamma_{H^+}V_b F_b \gamma_{A^-}}{(V_a F_a - V_b F_b)\gamma_{HA}} \qquad (11\text{-}48)$$

which can be rearranged to

$$\mathscr{A}_{H^+} = [H^+]\gamma_{H^+} = 10^{-pH}$$

$$\underbrace{V_b[H^+]\gamma_{H^+}}_{10^{-pH}} = \frac{\gamma_{HA}}{\gamma_{A^-}} K_a \left(\frac{V_a F_a - V_b F_b}{F_b} \right) \qquad (11\text{-}49)$$

The term on the left can be written $V_b \cdot 10^{-pH}$, since $[H^+]\gamma_{H^+} = 10^{-pH}$. The term in parentheses on the right is

$$V_a F_a = V_e F_b \quad \Rightarrow \quad V_e = \frac{V_a F_a}{F_b}$$

$$\frac{V_a F_a}{F_b} - V_b = V_e - V_b \qquad (11\text{-}50)$$

Equation 11-49 can therefore be written in the form

$$V_b \cdot 10^{-pH} = \frac{\gamma_{HA}}{\gamma_{A^-}} K_a (V_e - V_b) \qquad (11\text{-}51)$$

A graph of $V_b \cdot 10^{-pH}$ versus V_b is called a **Gran plot.** If γ_{HA}/γ_{A^-} is constant, the graph should be a straight line with a slope of $-K_a \gamma_{HA}/\gamma_{A^-}$.

[†] More extensive discussions of the use and applications of the Gran plot can be found in G. Gran, *Anal. Chim. Acta,* **206,** 111 (1988), and F. J. C. Rossotti and H. Rossotti, *J. Chem. Ed.,* **42,** 375 (1965). Useful extensions of the Gran equations deal with cases in which Equations 11-51 and 11-52 produce curved lines instead of straight lines. See L. M. Schwartz, *J. Chem. Ed.,* **64,** 947 (1987).

The intercept on the V_b axis will be V_e. A Gran plot for the titration in Figure 11-7 is shown in Figure 11-8. Any units can be used for V_b, but the same units should be used on both axes. In Figure 11-8, V_b was expressed in microliters on both axes.

The beauty of a Gran plot is that it enables us to use data taken before the end point to find the end point. Other graphic methods (which we will not discuss) require carefully collected data close to V_e. This is the hardest region of the titration curve in which to get accurate data, so these other graphic methods are often no better than estimating the steepest slope by eye (or "feel"). The slope of the Gran plot enables us to find K_a for the acid, HA.

The Gran function, $V_b \cdot 10^{-pH}$, does not actually go to zero, because 10^{-pH} is never zero. The curve must be extrapolated to find V_e. The reason the function does not reach zero is that we have used the approximation that every mole of OH^- generates one mole of A^-. This approximation breaks down as V_b approaches V_e. The approximation that HA is undissociated can also be poor for acids of moderate strength in the early part of the titration. As a practical matter, only the linear portion of the Gran plot is used.

Another source of curvature in the Gran plot is changing ionic strength, which causes γ_{HA}/γ_{A^-} to vary. In Figure 11-7 this variation was avoided by having enough $NaNO_3$ present to maintain an essentially constant ionic strength throughout the titration. Even without added salt, the last 10–20% of data before V_e gives a fairly straight line because the value of γ_{HA}/γ_{A^-} does not change very much.

Although we derived the Gran function for a monoprotic acid, the same plot ($V_b \cdot 10^{-pH}$ versus V_b) applies to polyprotic acids (such as H_6A in Figure 11-7). An analogous function can also be derived for the titration of a weak base with a strong acid.

Gran plot:

Plot $V_b \cdot 10^{-pH}$ versus V_b

x intercept $= V_e$

Slope $= -K_a \gamma_{HA}/\gamma_{A^-}$

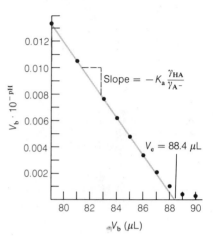

Figure 11-8
Gran plot for the data near the first equivalence point of Figure 11-7. The last 10–20% of volume prior to V_e is normally the useful region for a Gran plot.

Challenge: When a weak base, B, is titrated with a strong acid, the appropriate Gran function is

$$V_H \cdot 10^{+pH} = \left(\frac{1}{K_a} \cdot \frac{\gamma_{A^-}}{\gamma_{HA}} \right)(V_e - V_H) \qquad (11\text{-}52)$$

where V_H is the volume of strong acid added, and K_a is the acid dissociation constant of BH^+. A graph of $V_H \cdot 10^{+pH}$ versus V_H should be a straight line with a slope of $-\gamma_{A^-}/\gamma_{HA}K_a$ and an intercept of V_e on the V_H axis. Use a procedure similar to the one above to derive Equation 11-52.

11-6 PRACTICAL NOTES

Several acids and bases can be obtained in pure enough form to be used as **primary standards**.[†] Some of them are listed in Table 11-4. Note that NaOH

[†] Instructions for purifying and using primary standards may be found in the following books: L. Meites, *Handbook of Analytical Chemistry* (New York: McGraw-Hill, 1963), pp. 3-32–3-35; J. Bassett, R. C. Denney, G. H. Jeffery, and J. Mendham, *Vogel's Textbook of Quantitative Inorganic Analysis* (Essex: Longman, 1978), 4th ed., pp. 296–306; I. M. Kolthoff and V. A. Stenger, *Volumetric Analysis*, Vol. 2 (New York: Wiley-Interscience, 1947).

Table 11-4
Some primary standards

Compound	Formula weight	Notes

ACIDS

Potassium acid phthalate

204.233 — The pure commercial material is dried at 105°C and used to standardize base. A phenolphthalein end point is satisfactory.

HCl
Hydrochloric acid

36.461 — HCl and water distill as an *azeotrope* (a mixture) whose composition (~ 6 M) depends upon pressure. The composition is tabulated as a function of the pressure during distillation. See Problem 11-46 for more information.

$KH(IO_3)_2$
Potassium hydrogen iodate

389.912 — This is a strong acid, so any indicator with an end point between ~ 5 and ~ 9 is adequate.

BASES

$H_2NC(CH_2OH)_3$
tris(hydroxymethyl)aminomethane
(also called "Tris" or "Tham")

121.136 — The pure commercial material is dried at 100–103°C and titrated with strong acid. The end point is in the range pH 4.5–5.

$$H_2NC(CH_2OH)_3 + H^+ \rightarrow H_3\overset{+}{N}C(CH_2OH)_3$$

HgO
Mercuric oxide

216.59 — Pure HgO is dissolved in a large excess of I^- or Br^-, whereupon $2OH^-$ are liberated:

$$HgO + 4I^- + H_2O \rightarrow HgI_4^{2-} + 2OH^-$$

The base is titrated using an indicator end point.

Na_2CO_3
Sodium carbonate

105.989 — Primary standard grade Na_2CO_3 is commercially available. Alternatively, recrystallized $NaHCO_3$ can be heated for 1 hour at 260–270°C to produce pure Na_2CO_3. Sodium carbonate is titrated with acid to an end point of pH 4–5. Just before the end point the solution is boiled to expel CO_2.

$Na_2B_4O_7 \cdot 10H_2O$
Borax

381.367 — The recrystallized material is dried in a chamber containing an aqueous solution saturated with NaCl and sucrose. This gives the decahydrate in pure form. The standard is titrated with acid to a methyl red end point.

$$"B_4O_7 \cdot 10H_2O^{2-}" + 2H^+ \rightarrow 4B(OH)_3 + 5H_2O$$

NaOH and KOH must be standardized with primary standards.

and KOH are not primary standards, for the reagent-grade materials contain carbonate (from reaction with atmospheric CO_2) and adsorbed water. Solutions of NaOH and KOH must be standardized against a primary standard. Potassium acid phthalate is among the most convenient compounds for this purpose. Dilute solutions of NaOH for titrations are prepared by diluting a stock solution of 50% (wt/wt) aqueous NaOH. Sodium carbonate is relatively insoluble in this stock solution and settles to the bottom.

Alkaline solutions (e.g., 0.1 M NaOH) must be protected from the atmosphere, or else they absorb CO_2:

$$OH^- + CO_2 \rightarrow HCO_3^- \tag{11-53}$$

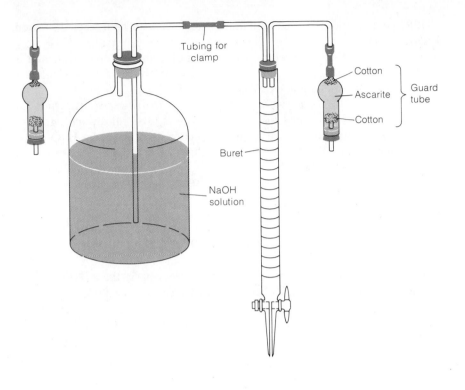

Cotton

Ascarite

Cotton

Guard
tube

Buret

NaOH
solution

Figure 11-9
Setup for protecting alkaline solutions
from atmospheric CO_2. The buret is filled
by applying suction to the guard tube at
the right. In this manner solution is transferred without exposure to air.

CO_2 absorption changes the concentration of strong base over a period of
time and decreases the extent of reaction near the end point in the titration
of weak acids. If the solutions are kept in tightly capped polyethylene bottles,
they may be used for about a week with little change. For longer periods,
or for frequent use, the setup shown in Figure 11-9 is convenient. The Ascarite
(NaOH coated on asbestos) is a powerful CO_2 absorbent that protects the
solution from air. A Gran plot can be used to estimate the carbonate content
of an alkaline titrant.

Strongly basic solutions attack glass and are best stored in plastic containers. Such solutions should not be kept in a buret longer than necessary.
Boiling 0.01 M aqueous NaOH in a flask for 1 h causes a 10% decrease in
molarity, owing to the reaction of the base with glass.[†]

Strong base slowly dissolves glass.

11-7 TITRATIONS IN NONAQUEOUS SOLVENTS

After reading this far, you may think that water is the only solvent in the
world. In fact, there are a great many chemical reactions that cannot be
studied with aqueous solutions. In acid–base chemistry there are three common reasons why we might choose a nonaqueous solvent:

1. The reactants or products might be insoluble in water.
2. The reactants or products might react with water.
3. The analyte is too weak an acid or base to be titrated in water.

[†] A. A. Smith, *J. Chem. Ed.*, **63**, 85 (1986).

Reasons 1 and 2 are self-explanatory and will not be further discussed. Figure 11-4 gives an example illustrating reason 3. An acid with $pK_a \gtrsim 8$ does not give a distinct potentiometric end point and cannot be titrated in water. However, a very sharp end point might be observed if the same acid were titrated in a nonaqueous solvent.

To see why this is so, we must realize that *the strongest acid that can exist in water is* H_3O^+, *and the strongest base is* OH^-. If an acid stronger than H_3O^+ is dissolved in water, it protonates H_2O to make H_3O^+. If a base stronger than OH^- is dissolved in water, it deprotonates H_2O to make OH^-. Because of this **leveling effect**, $HClO_4$ and HCl behave as if they had the same acid strength; both are *leveled* to H_3O^+:

The leveling effect.

$$HClO_4 + H_2O \rightarrow H_3O^+ + ClO_4^- \tag{11-54}$$

$$HCl \quad + H_2O \rightarrow H_3O^+ + Cl^- \tag{11-55}$$

In a solvent (such as acetic acid) that is less basic than H_2O, $HClO_4$ and HCl are not be leveled to the same strength:

In acetic acid solution, $HClO_4$ is a stronger acid than HCl, but in aqueous solution they are leveled to the strength of H_3O^+.

$$HClO_4 + CH_3CO_2H \rightleftharpoons CH_3CO_2H_2^+ + ClO_4^- \qquad K = 1.3 \times 10^{-5}$$
$$\underset{\substack{\text{Acetic acid} \\ \text{(solvent)}}}{} \tag{11-56}$$

$$HCl \quad + CH_3CO_2H \rightleftharpoons CH_3CO_2H_2^+ + Cl^- \qquad K = 2.8 \times 10^{-9} \tag{11-57}$$

The equilibrium constants show that $HClO_4$ is a stronger acid than HCl in acetic acid solvent.

Figure 11-10 shows a titration curve for a mixture of five acids titrated with 0.2 M tetrabutylammonium hydroxide in methyl isobutyl ketone solvent.

Question: Where do you think the end point for the acid $H_3O^+ClO_4^-$ would come in Figure 11-10? (*Answer:* Between HCl and 2-hydroxybenzoic acid)

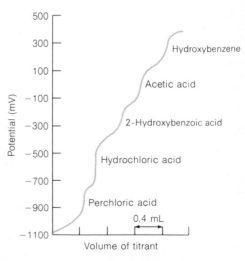

Figure 11-10

Titration of a mixture of acids with tetrabutylammonium hydroxide in methyl isobutyl ketone solvent. The measurements were made with a glass electrode and a platinum reference electrode. [D. B. Bruss and G. E. A. Wyld, *Anal. Chem.*, **29**, 232 (1957).]

This solvent is not protonated to a great extent by any of the acids. It can be seen that perchloric acid is a stronger acid than HCl in this solvent as well.

Consider a base too weak to give a distinct end point when titrated with a strong acid in water.

$$B + H_3O^+ \rightleftharpoons BH^+ + H_2O \tag{11-58}$$

The reason no end point is recognized is that the equilibrium constant for the titration reaction (11-58) is not large enough. If a stronger acid than H_3O^+ were available, the titration reaction might have a large enough equilibrium constant to give a distinct end point. If the same base were dissolved in acetic acid and titrated with $HClO_4$ in acetic acid, a clear end point might be observed. The reaction

$$B + HClO_4 \xrightarrow{\text{in acetic acid}} \underbrace{BH^+ClO_4^-}_{\text{An ion pair}} \tag{11-59}$$

might have a large equilibrium constant, because $HClO_4$ is a much stronger acid than H_3O^+. (The product in Reaction 11-59 is written as an ion pair because acetic acid has too low a dielectric constant to allow ions to separate extensively.)

A wide variety of very weak acids and bases, or compounds insoluble or unstable in water, can be titrated in nonaqueous solution. An extensive literature of nonaqueous titrations exists.[†]

A base too weak to be titrated by H_3O^+ in water might be titrated by $HClO_4$ in acetic acid solvent.

[†] J. S. Fritz, *Acid-Base Titrations in Nonaqueous Solvents* (Boston: Allyn and Bacon, 1973); J. Kucharsky and L. Safarik, *Titrations in Non-Aqueous Solvents* (New York: Elsevier, 1963); W. Huber, *Titrations in Nonaqueous Solvents* (New York: Academic Press, 1967); I. Gyenes, *Titration in Non-Aqueous Media* (Princeton, N.J.: van Nostrand, 1967).

Summary

In the titration of a strong acid with a strong base (or vice versa), the titration reaction is $H^+ + OH^- \rightarrow H_2O$, and the pH is determined by the concentration of excess unreacted analyte or titrant. The pH at the equivalence point is 7.00.

The titration curve of a weak acid with a strong base $(HA + OH^- \rightarrow A^- + H_2O)$ can be divided into four regions:

1. Before any base is added, the pH is determined by the acid dissociation reaction of the weak acid $(HA \rightleftharpoons H^+ + A^-)$.

2. Between the initial point and the equivalence point, there is a buffer made from A^- (which is equivalent to the moles of added base) and the excess unreacted HA. $pH = pK_a + \log([A^-]/[HA])$. At the special point when $V_b = \frac{1}{2}V_e$, $pH = pK_a$ (neglecting activity coefficients).

3. At the equivalence point, the weak acid has been converted to its conjugate base, A^-, whose pH is governed by hydrolysis $(A^- + H_2O \rightleftharpoons HA + OH^-)$. The pH is necessarily above 7.00.

4. After the equivalence point, the pH is determined by the concentration of excess strong base.

The titration curve for the reaction of a weak base with a strong acid has four regions analogous to those above. Before beginning, the pH is governed by hydrolysis of base, B. Between the initial and equivalence points, there is a buffer consisting of B plus BH^+. When $V_a = \frac{1}{2}V_e$, $pH = pK_a$ (for BH^+). At the equivalence point, B has been converted to its conjugate acid, BH^+, whose pH must be below 7.00. Beyond the equivalence point, the pH is governed by the concentration of excess strong acid titrant.

A titration curve for the reaction of a diprotic acid, H_2A, with a strong base has six distinct regions:

1. The initial pH is determined by the dissociation of H_2A, which behaves as a monoprotic weak acid ($H_2A \rightleftharpoons H^+ + HA^-$).

2. Between the initial point and the first equivalence point, the solution is buffered by H_2A plus HA^-, whose pH is given by $pH = pK_{a1} + \log([HA^-]/[H_2A])$. When $V_b = \frac{1}{2}V_e$, $pH = pK_{a1}$.

3. At the first equivalence point, H_2A has been converted to HA^-, whose pH is given by

$$[H^+] = \sqrt{\frac{K_1 K_2 F' + K_1 K_w}{K_1 + F'}}$$

$$pH \approx \frac{1}{2}(pK_{a1} + pK_{a2})$$

where F' contains a correction for dilution of starting material.

4. Between the first and second equivalence points, there is a buffer consisting of HA^- and A^{2-}, whose pH is given by $pH = pK_{a2} + \log([A^{2-}]/[HA^-])$. When $V_b = \frac{3}{2}V_e$, $pH = pK_{a2}$.

5. At the second equivalence point, HA^- has been converted to its conjugate base, A^{2-}, whose pH is governed by hydrolysis ($A^{2-} + H_2O \rightleftharpoons HA^- + OH^-$).

6. Beyond the second equivalence point, the pH is governed by the concentration of excess strong base titrant.

In any titration, if the reactants are too dilute, or if the equilibrium constant for the titration reaction is not large enough, no sharp end point will be observed. When choosing an indicator, select one whose color transition range matches the pH of the equivalence point of the titration as closely as possible. A Gran plot, such as a graph of $V_b \cdot 10^{-pH}$ versus V_b, is a useful way to find the end point and the acid or base equilibrium constant. Acids or bases too weak to be titrated in H_2O may be titrated in a nonaqueous solvent in which the titrant is not leveled to the strength of H_3O^+ or OH^-.

Terms to Understand

Gran plot
Hammett acidity function
indicator

indicator error
leveling effect
primary standard

Exercises

11-A. Calculate the pH at each of the following points in the titration of 50.00 mL of 0.010 0 M NaOH with 0.100 M HCl. Volume of acid added: 0.00, 1.00, 2.00, 3.00, 4.00, 4.50, 4.90, 4.99, 5.00, 5.01, 5.10, 5.50, 6.00, 8.00, and 10.00 mL. Make a graph of pH versus volume of HCl added.

11-B. Calculate the pH at each point listed for the titration of 50.0 mL of 0.050 0 M formic acid with 0.050 0 M KOH. The points to calculate are $V_b = 0.0$, 10.0, 20.0, 25.0, 30.0, 40.0, 45.0, 48.0, 49.0, 49.5, 50.0, 50.5, 51.0, 52.0, 55.0, and 60.0 mL. Draw a graph of pH versus V_b.

11-C. Calculate the pH at each point listed for the titration of 100.0 mL of 0.100 M cocaine (Equation 10-40) with 0.200 M HNO_3. The points to calculate are $V_a = 0.0$, 10.0, 20.0, 25.0, 30.0, 40.0, 49.0, 49.9, 50.0, 50.1, 51.0, and 60.0 mL. Draw a graph of pH versus V_a.

11-D. Consider the titration of 50.0 mL of 0.050 0 M malonic acid with 0.100 M NaOH. Calculate the pH at each point listed and sketch the titration curve:

$V_b = 0.0$, 8.0, 12.5, 19.3, 25.0, 37.5, 50.0, and 56.3 mL.

11-E. Write the chemical reactions (including structures of reactants and products) that occur when histidine is titrated with perchloric acid. (Histidine is a molecule with no net charge.) A solution containing 25.0 mL of 0.050 0 M histidine was titrated with 0.050 0 M $HClO_4$. Calculate the pH at the following values of V_a: 0, 4.0, 12.5, 25.0, 26.0, 50.0 mL.

11-F. Select indicators from Table 11-3 that would be useful for the titrations in Figures 11-2, 11-3, and the second highest curve in Figure 11-4. Select a different indicator for each titration and state what color change you would use as the end point.

11-G. Acid–base indicators are themselves acids or bases. Consider an indicator, HIn, which dissociates according to the equation

$$HIn \xrightleftharpoons{K_a} H^+ + In^-$$

Suppose that the molar absorptivity, ε, is $2\,080\ \mathrm{M^{-1}\,cm^{-1}}$ for HIn and is $14\,200\ \mathrm{M^{-1}\,cm^{-1}}$ for In$^-$, at a wavelength of 440 nm.[†]

(a) Write the expression giving the absorbance at 440 nm of a solution containing HIn at a concentration [HIn] and In$^-$ at a concentration [In$^-$]. Assume the cell path-length is 1.00 cm. Note that absorbance is additive. The total absorbance is the sum of absorbances of all components.

(b) A solution containing the indicator at a formal concentration of 1.84×10^{-4} M is adjusted to pH 6.23 and found to exhibit an absorbance of 0.868 at 440 nm. Calculate pK_a for this indicator.

11-H. When 100.0 mL of a weak acid was titrated with 0.093 81 M NaOH, 27.63 mL was required to reach the equivalence point. The pH at the equivalence point was 10.99. What was the pH when only 19.47 mL of NaOH had been added?

11-I. A 0.100 M solution of the weak acid HA was titrated with 0.100 M NaOH. The pH measured when $V_b = \frac{1}{2}V_e$ was 4.62. Using activity coefficients correctly, calculate pK_a. The size of the A$^-$ anion is 450 pm.

[†] This problem is an application of Beer's law, which you can read about in Sections 19-1 and 19-2.

Problems

A11-1. Distinguish the terms *end point* and *equivalence point*.

A11-2. Give the name and formula of a primary standard used to standardize (a) HCl and (b) NaOH.

A11-3. For what is a Gran plot used?

A11-4. Write the formula of a compound with a negative pK_a.

A11-5. What is meant by the leveling effect?

A11-6. Consider the titration of 100.0 mL of 0.100 M NaOH with 1.00 M HBr. Find the pH at the following volumes of acid added and make a graph of pH versus V_a: $V_a = 0, 1, 5, 9, 9.9, 10, 10.1$, and 12 mL.

A11-7. A weak acid HA (p$K_a = 5.00$) was titrated with 1.00 M KOH. The acid solution had a volume of 100.0 mL and a molarity of 0.100 M. Find the pH at the following volumes of base added and make a graph of pH versus V_b: $V_b = 0, 1, 5, 9, 9.9, 10, 10.1$, and 12 mL.

A11-8. A 100.0-mL aliquot of 0.100 M weak base B (p$K_b = 5.00$) was titrated with 1.00 M HClO$_4$. Find the pH at the following volumes of acid added and make a graph of pH versus V_a: $V_a = 0, 1, 5, 9, 9.9, 10, 10.1$, and 12 mL.

A11-9. The dibasic compound B (p$K_{b1} = 4.00$, p$K_{b2} = 8.00$) was titrated with 1.00 M HCl. The initial solution of B was 0.100 M and had a volume of 100.0 mL. Find the pH at the following volumes of acid added and make a graph of pH versus V_a: $V_a = 0, 1, 5, 9, 10, 11, 15, 19, 20$, and 22 mL.

A11-10. A 100.0-mL aliquot of 0.100 M diprotic acid H$_2$A (p$K_1 = 4.00$, p$K_2 = 8.00$) was titrated with 1.00 M NaOH. Find the pH at the following volumes of base added and make a graph of pH versus V_b: $V_b = 0, 1, 5, 9, 10, 11, 15, 19, 20$, and 22 mL.

A11-11. A solution was prepared from 1.023 g of the primary standard Tris (Table 11-4) plus 99.367 g of water; 4.963 g of the solution was titrated with 5.262 g of aqueous HNO$_3$ to reach the methyl red end point. Calculate the concentration of the HNO$_3$ (expressed as mol HNO$_3$/kg solution).

A11-12. Consider the titration in Figure 11-3, for which the pH at the equivalence point is calculated to be 9.18. If thymol blue is used as an indicator, what color will be observed through most of the titration prior to the equivalence point? at the equivalence point? after the equivalence point?

A11-13. What color do you expect to observe for cresol purple indicator (Table 11-3) at the following pH values?
(a) 1.0 (b) 2.0 (c) 3.0

A11-14. Cresol red has *two* transition ranges listed in Table 11-3. What color would you expect it to be at the following pH values?
(a) 0 (b) 1 (c) 6 (d) 9

A11-15. Explain how to use potassium acid phthalate to standardize a solution of NaOH.

A11-16. At what point in the titration of a weak base with a strong acid is the maximum buffer capacity reached? This is the point at which a given small addition of acid causes the least pH change.

A11-17. What is the equilibrium constant for the reaction between benzylamine and HCl?

A11-18. Explain why sodium methoxide (NaOCH$_3$) and sodium ethoxide (NaOCH$_2$CH$_3$) are leveled to the same base strength in aqueous solution. Write the chemical reactions that occur when these bases are added to water.

11-19. A solution containing 50.0 mL of 0.031 9 M benzylamine was titrated with 0.050 0 M HCl. Calculate the pH at the following volumes of added acid: $V_a = 0$, 12.0, $\frac{1}{2}V_e$, 30.0, V_e, and 35.0 mL.

11-20. Consider the titration of the weak acid HA with NaOH. At what fraction of V_e does pH = $pK_a - 1$? At what fraction of V_e does pH = $pK_a + 1$? Use these two points, plus $V_b = 0$, $\frac{1}{2}V_e$, V_e, and $1.2V_e$ to sketch the titration curve for the reaction of 100 mL of 0.100 M anilinium bromide ("aminobenzene · HBr") with 0.100 M NaOH.

11-21. A solution of 100.00 mL of 0.040 0 M sodium propanoate (the sodium salt of propanoic acid) was titrated with 0.083 7 M HCl. Calculate the pH at the points $V_a = 0$, $\frac{1}{4}V_e$, $\frac{1}{2}V_e$, $\frac{3}{4}V_e$, V_e, and $1.1\ V_e$.

11-22. What is the pH at the equivalence point when 0.100 M hydroxyacetic acid is titrated with 0.050 0 M KOH?

11-23. Calculate the pH when 25.0 mL of 0.020 0 M 2-aminophenol has been titrated with 10.9 mL of 0.015 0 M $HClO_4$.

11-24. Calculate the pH at 10.0-mL intervals (from 0 to 100 mL) in the titration of 40.0 mL of 0.100 M piperazine with 0.100 M HCl. Make a graph of pH versus V_a.

11-25. Consider the titration of 50.0 mL of 0.100 M sodium glycinate with 0.100 M HCl.

   (a) Calculate the pH at the second equivalence point.

   (b) Show that our approximate method of calculations gives incorrect (physically unreasonable) values of pH at $V_a = 90.0$ and $V_a = 101.0$ mL.

11-26. A solution containing 0.100 M glutamic acid (the molecule with no net charge) was titrated with 0.025 0 M RbOH.

   (a) Draw the structures of reactants and products.

   (b) Calculate the pH at the first equivalence point.

11-27. A solution containing 0.010 0 M tyrosine was titrated to the first equivalence point with 0.004 00 M KOH.

   (a) Draw the structures of reactants and products.

   (b) Calculate the pH at the first equivalence point.

11-28. Find the pH of the solution when 0.010 0 M tyrosine is titrated to the equivalence point with 0.004 00 M $HClO_4$.

11-29. This problem deals with the amino acid cysteine, which we will abbreviate H_2C.

   (a) A 0.030 0 M solution was prepared by dissolving dipotassium cysteine, K_2C, in water. Then 40.0 mL of this solution was titrated with 0.060 0 M $HClO_4$. Calculate the pH at the first equivalence point.

   (b) Calculate the quotient $[C^{2-}]/[HC^-]$ in a solution of 0.050 0 M cysteinium bromide (the salt $H_3C^+Br^-$).

11-30. How many milliliters of 0.043 1 M NaOH should be added to 59.6 mL of 0.122 M leucine to obtain a pH of 8.00?

11-31. Would the indicator bromocresol green, with a transition range of pH 3.8–5.4, ever be useful in the titration of a weak acid with a strong base?

11-32. Why would an indicator end point not be very useful in the titration curve for $pK_a = 10.00$ in Figure 11-4?

11-33. (a) What is the pH at the equivalence point when 0.030 0 M NaF is titrated with 0.060 0 M $HClO_4$?

   (b) Why would an indicator end point probably not be useful in this titration?

11-34. A titration curve for Na_2CO_3 titrated with HCl follows this problem. Suppose that *both* phenolphthalein and bromocresol green are present in the titration solution. State what colors you expect to observe at the following volumes of added HCl:

(a) 2 mL (b) 10 mL (c) 19 mL

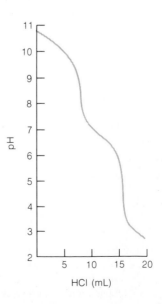

11-35. Consider the titration of 0.10 M pyridinium bromide (the salt of pyridine plus HBr) by 0.10 M NaOH. Sketch the titration curve using calculated pH values for the volumes $0.99\,V_e$, V_e, and $1.01\,V_e$. Select an indicator from Table 11-3 that would be suitable for this titration and state what color change will be used.

11-36. In the Kjeldahl nitrogen determination (Reactions 9-7 through 9-10), the final product is a solution of NH_4^+ ion in HCl solution. It is necessary to titrate the HCl without titrating the NH_4^+ ion.

    (a) Calculate the pH of pure 0.010 M NH_4Cl.

    (b) Select an indicator from Table 11-3 that would allow you to titrate HCl but not NH_4^+.

11-37. A 10.231-g sample of window cleaner containing ammonia was diluted with 39.466 g of water. Then 4.373 g of solution was titrated with 14.22 mL of 0.106 3 M HCl to reach a bromocresol green end point. Find the weight percent of NH_3 in the cleaner.

11-38. A solution was prepared by dissolving 0.194 7 g of HgO (Table 11-4) in 20 mL of water containing 4 g of KBr. Titration with HCl required 17.98 mL to reach a phenolphthalein end point. Calculate the molarity of the HCl.

11-39. How many grams of potassium acid phthalate should be weighed into a flask to standardize ~0.05 M NaOH if you wish to use ~30 mL of base for the titration?

11-40. Find the equilibrium constant for the reaction of MES (Table 10-2) with NaOH.

11-41. Calculate the pH of a solution made by mixing 50.00 mL of 0.100 M NaCN with

    (a) 4.20 mL of 0.438 M $HClO_4$

    (b) 11.82 mL of 0.438 M $HClO_4$

    (c) What is the pH at the equivalence point with 0.438 M $HClO_4$?

11-42. When 22.63 mL of aqueous NaOH was added to 1.214 g of cyclohexylaminoethanesulfonic acid (F.W. 207.29, structure in Table 10-2) dissolved in 41.37 mL of water, the pH was 9.24. Calculate the molarity of the NaOH.

11-43. How many grams of dipotassium oxalate (F.W. 166.22) should be added to 20.0 mL of 0.800 M $HClO_4$ to give a pH of 4.40 when the solution is diluted to 500 mL?

11-44. Data for the titration of 100.00 mL of a weak acid by NaOH are given above right. Find the equivalence point by preparing a Gran plot using data for the last 10% of the volume prior to V_e.

mL NaOH	pH	mL NaOH	pH
0.00	4.14	21.61	6.27
1.31	4.30	21.77	6.32
2.34	4.44	21.93	6.37
3.91	4.61	22.10	6.42
5.93	4.79	22.27	6.48
7.90	4.95	22.37	6.53
11.35	5.19	22.48	6.58
13.46	5.35	22.57	6.63
15.50	5.50	22.70	6.70
16.92	5.63	22.76	6.74
18.00	5.71	22.80	6.78
18.35	5.77	22.85	6.82
18.95	5.82	22.91	6.86
19.43	5.89	22.97	6.92
19.93	5.95	23.01	6.98
20.48	6.04	23.11	7.11
20.75	6.09	23.17	7.20
21.01	6.14	23.21	7.30
21.10	6.15	23.30	7.49
21.13	6.16	23.32	7.74
21.20	6.17	23.40	8.30
21.30	6.19	23.46	9.21
21.41	6.22	23.55	9.86
21.51	6.25		

11-45. Borax (Table 11-4) was used to standardize a solution of HNO_3. Titration of 0.261 9 g of borax required 21.61 mL. What is the molarity of the HNO_3?

11-46. Constant boiling aqueous HCl can be used as a primary standard for acid–base titrations. When $\sim20\%$ (wt/wt) HCl is distilled, the composition of the distillate varies in a regular manner with the barometric pressure:

P (torr)	HCl[†] (g/100 g solution)
770	20.196
760	20.220
750	20.244
740	20.268
730	20.292

[†] The composition of distillate is from the paper of C. W. Foulk and M. Hollingsworth, *J. Am. Chem. Soc.,* **45,** 1223 (1923), with numbers corrected for the current values of atomic weights.

Suppose that constant boiling HCl was collected at a pressure of 746 torr.

(a) Make a graph of the data in the table to find the weight percent of HCl collected at 746 torr.

(b) What mass of distillate (weighed in air, using weights whose density is 8.0 g/mL) should be dissolved in 1.000 0 L to give 0.100 00 M HCl? The density of distillate over the whole range in the table is close to 1.096 g/mL. You will need this density to change the mass measured in vacuum to mass measured in air. See Section 2-2 for buoyancy corrections.

11-47. The base B is too weak to titrate in aqueous solution.

(a) Which solvent, pyridine or acetic acid, would be more suitable for titration of B with $HClO_4$?

(b) Which solvent would be more suitable for the titration of a very weak acid with tetra-butylammonium hydroxide?

11-48. Explain why sodium amide ($NaNH_2$) and phenyl lithium (C_6H_5Li) are leveled to the same base strength in aqueous solution. Write the chemical reactions that occur when these are added to water.

11-49. Some properties of a particular indicator are given below:[†]

$$HIn \underset{\xleftarrow{\hspace{1cm}}}{\overset{pK_a = 7.95}{\rightleftharpoons}}$$

$\lambda_{max} = 395$ nm

$\varepsilon_{395} = 1.80 \times 10^4$ M^{-1} cm^{-1}

$\varepsilon_{604} = 0$

$$In^- + H^+$$

$\lambda_{max} = 604$ nm

$\varepsilon_{604} = 4.97 \times 10^4$ M^{-1} cm^{-1}

A solution with a volume of 20.0 mL containing 1.40×10^{-5} M indicator plus 0.050 0 M benzene-1,2,3-tricarboxylic acid was treated with 20.0 mL of aqueous KOH. The resulting solution had an absorbance at 604 nm of 0.118 in a 1.00-cm cell. Calculate the molarity of the KOH solution.

11-50. A very weakly basic aromatic amine ($pK_b = 14.79$) was used to measure the pH of a concentrated acid. A solution was prepared by dissolving 6.390 mg of the amine (M.W. 278.16) in 100.0 mL of the acid. The absorbance measured at 385 nm in a 1.000-cm cell was 0.350.[†] Find the pH of the solution.

$$B \qquad\qquad + H^+ \rightleftharpoons$$

$\varepsilon_{385} = 2 860$ M^{-1} cm^{-1}

$$BH^+$$

$\varepsilon_{385} = 937$ M^{-1} cm^{-1}

11-51. An aqueous solution containing ~1 g of oxo-butanedioic acid per 100 mL was titrated with 0.094 32 M NaOH to measure the acid molarity.

(a) What will be the pH at each equivalence point?

(b) Which equivalence point would be best to use in this titration?

(c) You have the indicators erythrosine, ethyl orange, bromocresol green, bromothymol blue, thymolphthalein, and alizarin yellow. Which indicator will you use and what color change will you look for?

11-52. Consider the neutral form of the amino acid histidine, which we will abbreviate HA.

(a) Write the sequence of reactions that occurs when HA is titrated with $HClO_4$. Draw structures of reactants and products.

(b) How many mL of 0.050 0 M $HClO_4$ should be added to 25.0 mL of 0.040 0 M HA to give a pH of 3.00?

11-53. *Use activity coefficients correctly* to calculate the pH after 10.0 mL of 0.100 M trimethylam-monium bromide was titrated with 4.0 mL of 0.100 M NaOH.

11-54. When 5.00 mL of 0.103 2 M NaOH was added to 0.112 3 g of alanine (M.W. 89.094) in 100.0 mL of 0.10 M KNO_3, the measured pH was 9.57. *Use activity coefficients correctly* to find pK_2 for alanine. Consider the ionic strength of the solution to be 0.10 M and consider each ionic form of alanine to have an activity coefficient of 0.77.

[†] This problem makes use of Beer's law, which you can read about in Sections 19-1 and 19-2.

12 Advanced Topics in Acid–Base Chemistry

This chapter contains a collection of topics that build upon the two previous chapters. We begin with a discussion of how to think about the composition of an acid–base mixture without doing any calculations at all.

12-1 WHICH IS THE PRINCIPAL SPECIES?

We are often faced with the problem of identifying which species of acid, base, or intermediate is predominant under given conditions. A simple example is this: What is the principal form of benzoic acid in an aqueous solution at pH 8?

Benzoic acid

$pK_a = 4.20$

The pK_a for benzoic acid is 4.20. This means that at pH 4.20 there would be a 1:1 mixture of benzoic acid (HA) and benzoate ion (A$^-$). At pH = pK_a + 1 ($=5.20$) the quotient [A$^-$]/[HA] is 10:1. At pH = pK_a + 2 ($=6.20$), the quotient [A$^-$]/[HA] is 100:1. As the pH increases the quotient [A$^-$]/[HA] increases still further.

It is easy to see that for a monoprotic system the basic species, A$^-$, is the predominant form when pH > pK_a. The acidic species, HA, is the predominant form when pH < pK_a. The predominant form of benzoic acid at pH 8 is the benzoate anion, $C_6H_5CO_2^-$.

At pH = pK_a, [A$^-$] = [HA]. This follows from the Henderson–Hasselbalch equation:

$$pH = pK_a + \log \frac{[A^-]}{[HA]}$$

pH	Major species
$<pK_a$	HA
$>pK_a$	A$^-$

EXAMPLE: Principal Species—Which One and How Much?

What is the predominant form of ammonia in a solution at pH 7.0? Approximately what fraction is in this form?

In Appendix G we find that $pK_a = 9.24$ for the ammonium ion (NH_4^+, the conjugate acid of ammonia, NH_3). At pH = 9.24, $[NH_4^+] = [NH_3]$. Below pH 9.24, NH_4^+ will be the predominant form. Since pH = 7.0 is about 2 pH units below pK_a, the quotient $[NH_4^+]/[NH_3]$ will be around 100:1. More than 99% is in the form NH_4^+.

For polyprotic systems the reasoning is the same, but there are several values of pK_a. Consider oxalic acid, H_2Ox, with $pK_1 = 1.25$ and $pK_2 = 4.27$. At pH = pK_1, $[H_2Ox] = [HOx^-]$. At pH = pK_2, $[HOx^-] = [Ox^{2-}]$. We can make a little chart showing the major species in each pH region:

pH	Major species
$pH < pK_1$	H_2A
$pK_1 < pH < pK_2$	HA^-
$pH > pK_2$	A^{2-}

pH range	Predominant species
$pH < 1.25$	H_2Ox
$1.25 < pH < 4.27$	HOx^-
$pH > 4.27$	Ox^{2-}

EXAMPLE: Principal Species in a Polyprotic System

The amino acid arginine has the following forms:

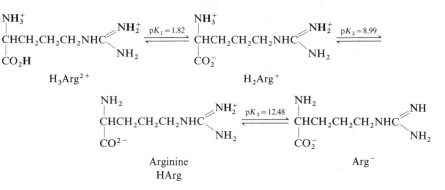

Arginine
HArg

Note that the α-ammonium group (the one next to the carboxyl group) is more acidic than the substituent. This information comes from Appendix G. What is the principal form of arginine at pH 10.0? Approximately what fraction is in this form? What is the second most abundant form at this pH?

We know that at pH = $pK_2 = 8.99$, $[H_2Arg^+] = [HArg]$. At pH = $pK_3 = 12.48$, $[HArg] = [Arg^-]$. At pH = 10.0, the major species is HArg. Since pH 10.0 is about one pH unit higher than pK_2, we can say that $[HArg]/[H_2Arg^+] \approx 10:1$. About 90% of arginine is in the form HArg. The second most important species is H_2Arg^+, which makes up about 10% of the arginine.

EXAMPLE: More on Polyprotic Systems

In the pH range 1.82 to 8.99, H_2Arg^+ is the principal form of arginine. Which is the second most prominent species at pH 6.0? at pH 5.0?

We know that the pH of the pure intermediate (amphiprotic) species, H_2Arg^+, is given by Equation 10-97:

$$\text{pH of } H_2Arg^+ \approx \tfrac{1}{2}(pK_1 + pK_2) = 5.40$$

Above pH 5.40 (and below $pH = pK_2$), we expect that HArg, the conjugate base of H_2Arg^+, will be the second most important species. Below pH 5.40 (and above $pH = pK_1$), we anticipate that H_3Arg^{2+} will be the second most important species.

The features of a triprotic system are sketched qualitatively in Figure 12-1. We determine the principal species by comparing the pH of the solution to the various pK_a values.

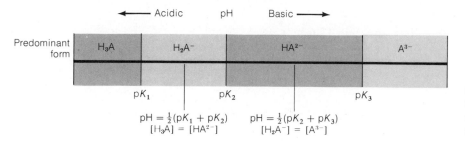

Figure 12-1
Schematic diagram showing the predominant form of a triprotic system (H_3A) in each pH interval.

12-2 FRACTIONAL COMPOSITION EQUATIONS

We will now derive equations that give the fraction of acid or base present in each possible form at a given pH. These equations are useful for the detailed solution of equilibrium problems involving acids and bases. They will also be used later in sections on EDTA titrations and electrochemistry.

Monoprotic Systems

First consider a solution of a monoprotic acid with a formal concentration F.

$$HA \underset{}{\overset{K_a}{\rightleftharpoons}} H^+ + A^- \tag{12-1}$$

$$K_a = \frac{[H^+][A^-]}{[HA]} \tag{12-2}$$

$$\text{Mass balance:} \quad F = [HA] + [A^-] \tag{12-3}$$

Equation 12-3 can be rearranged to give

$$[A^-] = F - [HA] \tag{12-4}$$

which can be plugged into Equation 12-2 to give

$$K_a = \frac{[H^+](F - [HA])}{[HA]} \tag{12-5}$$

or

$$[HA] = \frac{[H^+]F}{[H^+] + K_a} \tag{12-6}$$

In a similar manner, we can derive the relationship

$$[A^-] = \frac{K_aF}{[H^+] + K_a} \tag{12-7}$$

α_0 = fraction of species in the form HA

α_1 = fraction of species in the form A^-

$$\boxed{\alpha_0 + \alpha_1 = 1}$$

The *fraction* of molecules in the form HA is called $\boldsymbol{\alpha_0}$.

$$\alpha_0 = \frac{[HA]}{[HA] + [A^-]} = \frac{[HA]}{F} \tag{12-8}$$

Dividing Equation 12-6 by F gives

$$\alpha_0 = \frac{[HA]}{F} = \frac{[H^+]}{[H^+] + K_a} \tag{12-9}$$

The fraction denoted here as α_1 is the same thing we called the *fraction of dissociation* (α) on page 188.

In a similar manner, the fraction in the form A^-, designated $\boldsymbol{\alpha_1}$, can be obtained:

$$\alpha_1 = \frac{K_a}{[H^+] + K_a} \tag{12-10}$$

Figure 12-2 is a graph of Equations 12-9 and 12-10 for a system with $pK_a = 5.00$. At low pH almost all of the acid is in the form HA. At high pH almost everything is in the form A^-.

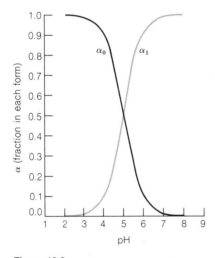

Figure 12-2
Fractional composition diagram of a monoprotic system with $pK_a = 5.00$.

Diprotic Systems

Now we will derive the fractional composition equations for a diprotic system. The derivation follows the same pattern used for the monoprotic system.

$$H_2A \overset{K_1}{\rightleftharpoons} H^+ + HA^- \tag{12-11}$$

$$HA^- \overset{K_2}{\rightleftharpoons} H^+ + A^{2-} \tag{12-12}$$

$$K_1 = \frac{[H^+][HA^-]}{[H_2A]} \quad \Rightarrow \quad [HA^-] = [H_2A]\frac{K_1}{[H^+]} \tag{12-13}$$

$$K_2 = \frac{[H^+][A^{2-}]}{[HA^-]} \quad \Rightarrow \quad [A^{2-}] = [HA^-]\frac{K_2}{[H^+]} \tag{12-14}$$

Equality 12-15 comes from using Equation 12-13 to express $[HA^-]$ in terms of $[H_2A]$.

$$[A^{2-}] = [H_2A]\frac{K_1K_2}{[H^+]^2} \tag{12-15}$$

Mass balance: $F = [H_2A] + [HA^-] + [A^{2-}]$ (12-16)

$$F = [H_2A] + \frac{K_1}{[H^+]}[H_2A] + \frac{K_1K_2}{[H^+]^2}[H_2A]$$ (12-17)

$$F = [H_2A]\left(1 + \frac{K_1}{[H^+]} + \frac{K_1K_2}{[H^+]^2}\right)$$ (12-18)

For a diprotic system we designate the fraction in the form H_2A as α_0, the fraction in the form HA^- as α_1, and the fraction in the form A^{2-} as α_2. From the definition of α_0 we can write

$$\alpha_0 = \frac{[H_2A]}{F} = \frac{[H^+]^2}{[H^+]^2 + [H^+]K_1 + K_1K_2}$$ (12-19)

In a similar manner, we can derive the following equations:

$$\alpha_1 = \frac{[HA^-]}{F} = \frac{K_1[H^+]}{[H^+]^2 + [H^+]K_1 + K_1K_2}$$ (12-20)

$$\alpha_2 = \frac{[A^{2-}]}{F} = \frac{K_1K_2}{[H^+]^2 + [H^+]K_1 + K_1K_2}$$ (12-21)

Figure 12-3 shows the curves for the fractions α_0, α_1, and α_2 for fumaric acid, whose two pK_a values are only 1.5 units apart. The value of α_1 rises only to 0.72 because the two pK values are so close together. There is a substantial amount of both H_2A and A^{2-} in the region $pK_1 < pH < pK_2$.

Equations 12-19 through 12-21 apply equally well to B, BH^+, and BH_2^{2+} obtained by dissolving the base B in water. The fraction α_0 applies to the acidic form, BH_2^{2+}. Similarly, α_1 applies to BH^+ and α_2 applies to B. The constants K_1 and K_2 are the acid dissociation constants of BH_2^{2+} ($K_1 = K_w/K_{b2}$ and $K_2 = K_w/K_{b1}$).

α_0 = fraction of species in the form H_2A

α_1 = fraction of species in the form HA^-

α_2 = fraction of species in the form A^{2-}

$$\boxed{\alpha_0 + \alpha_1 + \alpha_2 = 1}$$

The general form of α for the polyprotic acid H_nA is

$$\alpha_0 = \frac{[H^+]^n}{D}$$

$$\alpha_1 = \frac{K_1[H^+]^{n-1}}{D}$$

$$\alpha_j = \frac{K_1K_2\cdots K_j[H^+]^{n-j}}{D}$$

where $D = [H^+]^n + K_1[H^+]^{n-1} + K_1K_2[H^+]^{n-2} + \cdots + K_1K_2K_3\cdots K_n$

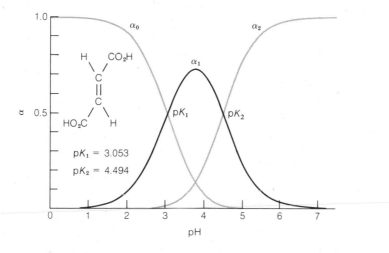

Figure 12-3

Fractional composition diagram for fumaric acid (*trans*-butenedioic acid).

12-3 ISOELECTRIC AND ISOIONIC pH

Biochemists often refer to the isoelectric or isoionic pH of a polyprotic species, such as a protein. These terms can be readily understood in terms of a simple diprotic system, such as the amino acid alanine.

$$\underset{\substack{\text{Alanine cation}\\ H_2A^+}}{\overset{\overset{\displaystyle CH_3}{\overset{+\ |}{H_3NCHCO_2H}}}{}} \rightleftharpoons \underset{\substack{\text{Neutral zwitterion}\\ HA}}{\overset{\overset{\displaystyle CH_3}{\overset{+\ |}{H_3NCHCO_2^-}}}{}} + H^+ \qquad pK_1 = 2.35 \qquad (12\text{-}22)$$

$$\overset{\overset{\displaystyle CH_3}{\overset{+\ |}{H_3NCHCO_2^-}}}{} \rightleftharpoons \underset{\substack{\text{Alanine anion}\\ A^-}}{\overset{\overset{\displaystyle CH_3}{\overset{|}{H_2NCHCO_2^-}}}{}} + H^+ \qquad pK_2 = 9.87 \qquad (12\text{-}23)$$

Isoionic pH is the pH of the pure, neutral, polyprotic acid.

The **isoionic point** (or isoionic pH) is the pH obtained when the pure, neutral polyprotic acid HA (the neutral zwitterion) is dissolved in water. The only ions in such a solution are H_2A^+, A^-, H^+, and OH^-. Most of the alanine is in the form HA.

Isoelectric pH is the pH at which the average charge of the polyprotic acid is zero.

The **isoelectric point** (or isoelectric pH) is the pH at which the *average* charge of the polyprotic acid is zero. At this pH most of the molecules are in the uncharged form HA. The concentrations of H_2A^+ and A^- are equal; therefore, the *average* charge of alanine will be zero. It is important to realize that the molecule is never present in only its neutral form. There will always be some H_2A^+ and some A^- in equilibrium with HA.

The isoionic pH for alanine occurs when pure alanine is dissolved in water. Since alanine (HA) is the intermediate form of a diprotic acid (H_2A^+), the pH is given by

Alanine is the intermediate form of a polyprotic acid, so we use Equation 12-24 (which is the same as Equation 10-92) to find the pH.

$$[H^+] = \sqrt{\frac{K_1K_2F + K_1K_w}{K_1 + F}} \qquad (12\text{-}24)$$

where F is the formal concentration of alanine. For 0.10 M alanine, the isoionic pH is

$$[H^+] = \sqrt{\frac{K_1K_2(0.10) + K_1K_w}{K_1 + (0.10)}} = 7.6 \times 10^{-7}\ \text{M} \quad \Rightarrow \quad pH = 6.12 \quad (12\text{-}25)$$

The isoionic pH shows a slight concentration dependence. For 0.010 M alanine, the isoionic pH is 6.19.

The isoelectric point is the pH at which the concentrations of H_2A^+ and A^- are equal, and the average charge of alanine is therefore zero. We can calculate the isoelectric pH by first writing expressions for the concentration of the cation and anion.

$$[H_2A^+] = \frac{[HA][H^+]}{K_1} \qquad (12\text{-}26)$$

$$[A^-] = \frac{K_2[HA]}{[H^+]} \qquad (12\text{-}27)$$

Setting $[H_2A^+] = [A^-]$, we find

$$\frac{[HA][H^+]}{K_1} = \frac{K_2[HA]}{[H^+]} \qquad (12\text{-}28)$$

or

$$[H^+] = \sqrt{K_1 K_2} \qquad \text{(at the isoelectric pH)} \qquad (12\text{-}29)$$

which gives

$$\text{Isoelectric pH} = \frac{pK_1 + pK_2}{2} \qquad (12\text{-}30)$$

For a diprotic amino acid, the isoelectric pH is halfway between the two pK_a values. The isoelectric pH of alanine is $\frac{1}{2}(2.35 + 9.87) = 6.11$. The isoelectric pH is not concentration-dependent.

You can see that the isoelectric and isoionic points for a polyprotic acid are very nearly the same. At the isoelectric pH, the average charge of the molecule is zero; thus $[H_2A^+] = [A^-]$ and $pH = \frac{1}{2}(pK_1 + pK_2)$. At the isoionic point, the pH is given by Equation 12-24, and $[H_2A^+]$ is not exactly equal to $[A^-]$. However, the pH is quite close to the isoelectric pH.

The relevance of isoelectric and isoionic points with respect to proteins is this: The *isoionic* pH of a protein is the pH of a solution containing the pure protein with no counterions except H^+ or OH^-. Proteins are usually isolated in a charged form together with various counterions (such as Na^+, NH_4^+, or Cl^-). If such a protein is subjected to intensive **dialysis** against pure water, the pH in the protein compartment will approach the isoionic point provided the counterions are free to pass through the dialysis membrane. The *isoelectric* point is the pH at which the protein has no net charge. This property forms the basis for a very sensitive and important technique used in the preparative and analytical separation of proteins, as described in Box 12-1.

> The isoelectric point is midway between the two pK_a values "surrounding" the neutral, intermediate species.

> Dialysis was discussed in Demonstration 8-1.

12-4 REACTIONS OF WEAK ACIDS WITH WEAK BASES

In Chapters 10 and 11 we systematically dealt with the reactions between acids and bases in the combinations of strong-plus-strong and strong-plus-weak. These reactions were easy to treat because their equilibrium constants are large. Such reactions "go to completion." The final coup would be the ability to deal with reactions between weak acids and weak bases.[†]

Calling the acid HA and the base B, we can write the reaction

$$HA + B \rightleftharpoons A^- + BH^+ \qquad (12\text{-}31)$$

[†] Although details are beyond the scope of this text, it should be mentioned that the most complicated acid–base systems can be analyzed with a computer. Readable articles telling how to approach such problems include the following: C. J. Willis, *J. Chem. Ed.*, **58**, 659 (1981); R. A. Stavis, *J. Chem. Ed.*, **55**, 99 (1978); M. J. D. Brand, *J. Chem. Ed.*, **53**, 771 (1976); J. A. Devore, *J. Chem. Ed.*, **65**, 868 (1988); D. P. Herman, K. K. Booth, O. J. Parker, and G. L. Breneman, *J. Chem. Ed.*, **67**, 501 (1990); and E. R. Malinowski, *J. Chem. Ed.*, **67**, 502 (1990). An article describing how to use commercial equation-solving software to set up and solve acid–base problems is F. T. Chau and A. S. W. Chik, *J. Chem. Ed.*, **66**, A61 (1989).

Box 12-1 ISOELECTRIC FOCUSING

At its *isoelectric* point, a protein has no net charge. It will therefore not migrate in an electric field at its isoelectric pH. This is the basis of a very sensitive technique of protein separation called **isoelectric focusing.** A mixture of proteins is subjected to a strong electric field in a medium having a pH gradient. Positively charged molecules move toward the negative pole and negatively charged molecules move toward the positive pole. The proteins migrate in one direction or the other until they reach the region of their isoelectric pH. In this region they have no net charge and no longer move. Thus each protein in the mixture is focused in one small region at its isoelectric pH.

An example of isoelectric focusing is shown on the following page. In this experiment a mixture of seven proteins (and, apparently, a host of impurities) was applied to a polyacrylamide gel containing a mixture of polyprotic compounds called *ampholytes.* Each end of the gel is placed in contact with a conducting solution and several hundred volts are applied across the length of the gel. The ampholytes migrate until they form a stable pH gradient (ranging from about pH 3 at one end of the gel to pH 10 at the other).[†] The proteins migrate until they reach their isolectric pH, at which point they have no net charge and cease migrating. If a molecule diffuses out of its isoelectric region, it becomes charged and immediately migrates back to its isoelectric zone in the gel. When the proteins finish migrating, the field is removed. The proteins are precipitated in place on the gel and stained with a dye to make their positions visible.

The stained gel is shown at the bottom of the figure. A spectrophotometer scan of the dye peaks is shown on the graph, and a profile of measured pH is also plotted. Each dark band of stained protein gives an absorbance peak. The instrument that measures absorbance as a function of position along the gel is called a *densitometer.*

[†] For a review of advances in pH gradient formation, including the use of an immobilized pH gradient and the structures of gradient-forming buffers, see P. G. Righetti, E. Gianazza, C. Gelfi, M. Chiari, and P. K. Sinha, *Anal. Chem.,* **61,** 1602 (1989).

Isoelectric focusing patterns. (1) Soybean trypsin inhibitor. (2) β-Lactoglobulin A. (3) β-Lactoglobulin B. (4) Ovotransferrin. (5) Horse myoglobin. (6) Whale myoglobin. (7) Cytochrome *c.* [Courtesy BioRad Laboratories, Richmond. Calif.]

whose equilibrium constant can be calculated as follows:

$$HA \rightleftharpoons H^+ + A^- \qquad K = K_a$$

$$B + H^+ \rightleftharpoons BH^+ \qquad K = \frac{1}{K_a \text{ (for } BH^+)} = \frac{K_b}{K_w}$$

$$HA + B \rightleftharpoons A^- + BH^+ \qquad K = \frac{K_a K_b}{K_w} \qquad (12\text{-}32)$$

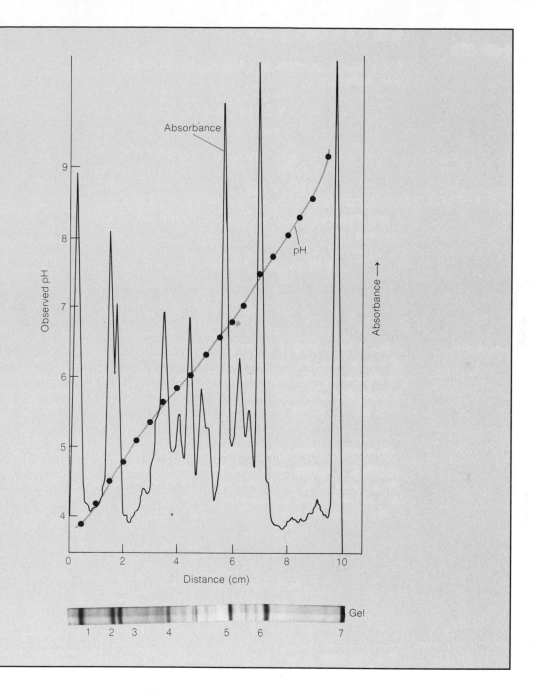

For weak acids and bases, K for Reaction 12-32 might be large or small.

It proves useful to divide the problems of weak bases and weak bases into three cases:

Case 1: K is large ($K \gg 1$).

Case 2: K is not large.

Case 3: Equimolar mixture of HA and B, regardless of the size of K.

Case 1: K Is Large ($K \gg 1$)

For a specific example, let's look at the pH of a mixture of chloroacetic acid and methylamine.

$$\text{ClCH}_2\text{CO}_2\text{H (HA)} \qquad \text{CH}_3\text{NH}_2 \text{ (B)}$$

Chloroacetic acid   Methylamine

$$K_\text{a} = 1.36 \times 10^{-3} \qquad K_\text{b} = 4.4 \times 10^{-4}$$

$$\text{p}K_\text{a} = 2.86 \qquad \text{p}K_\text{b} = 3.36$$

Using Equation 12-32, we find the equilibrium constant for the acid–base reaction:

$$\text{ClCH}_2\text{CO}_2\text{H} + \text{CH}_3\text{NH}_2 \rightleftharpoons \text{ClCH}_2\text{CO}_2^- + \text{CH}_3\text{NH}_3^+ \qquad (12\text{-}33)$$

$$K = \frac{K_\text{a}K_\text{b}}{K_\text{w}} = 6.0 \times 10^7$$

Case 1 is just like the weak-plus-strong combination.

Since K is so large, it is fair to say that the reaction will go to completion. *When the reactants are mixed, they will proceed to make products until one of the reactants has been consumed.* That is, if K is large, the reaction between a weak acid and weak base can be treated just like the case in which one reactant is a strong acid or base.

EXAMPLE: A Reaction with a Large Equilibrium Constant

What would be the pH if 100.0 mL of 0.050 0 M chloroacetic acid were mixed with 60.0 mL of 0.060 0 M methylamine?

We proceed as follows:

	HA	+ B	$\rightleftharpoons$ A$^-$	+ BH$^+$
Initial mmol:	5.00	3.60	—	—
Final mmol:	1.40	—	3.60	3.60

This little table shows that we have created a known mixture of HA and A$^-$–**Aha! A buffer!** So the pH is easily calculated from the appropriate Henderson–Hasselbalch equation:

$$\text{pH} = \text{p}K_\text{a} + \log\frac{[\text{A}^-]}{[\text{HA}]} = 2.86 + \log\frac{3.60}{1.40} = 3.27$$

Our analysis of case 1 indicates that the reaction of a sufficiently strong "weak" base with a sufficiently strong "weak" acid can go "to completion." The lower curve in Figure 12-4 is calculated for the titration of chloroacetic

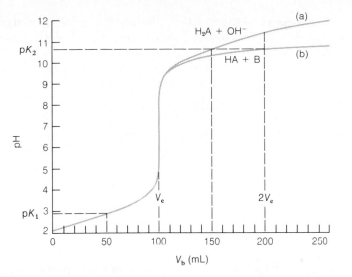

Challenge: Figure 12-4 compares the titration of chloroacetic acid with methylamine to the titration of a diprotic acid (H_2A) with NaOH. Why does pK_2 intersect the upper curve at $\frac{3}{2}V_e$ and the lower curve at $2V_e$? On the lower curve, "pK_2" is pK_a for the acid, BH^+. If you can answer this question, you've come a long way.

Figure 12-4
(a) Titration of 100 mL of 0.50 M H_2A ($pK_1 = 2.86$, $pK_2 = 10.64$) with 0.050 M NaOH. (b) Titration of 100 mL of the weak acid HA (0.50 M, $pK_a = 2.86$) with the weak base B (0.050 M, $pK_b = 3.36$).

acid by methylamine. Up to the equivalence point, the curve is identical to that of the titration of a weak acid with a strong base, shown in the upper curve.

Case 2: *K* Is Not Large

Suppose that 10.0 mL of 0.020 0 M ammonium chloride and 10.0 mL of 0.032 0 M trimethylamine were mixed together. What would be the pH?

$$NH_4^+ \qquad\qquad (CH_3)_3N$$

Ammonium ion   Trimethylamine

$$pK_a = 9.244 \qquad\qquad pK_b = 4.200$$

$$K_a = 5.70 \times 10^{-10} \qquad K_b = 6.31 \times 10^{-5}$$

The acid–base reaction is

$$NH_4^+ + (CH_3)_3N \rightleftharpoons NH_3 + (CH_3)_3NH^+ \qquad (12\text{-}34)$$

$$K = \frac{K_a K_b}{K_w} = 3.60$$

and the equilibrium constant is only 3.60. This is not a large number. We cannot say that the reaction between this acid and base goes "to completion." A substantial quantity of unreacted starting material will exist in equilibrium with products.

To find the composition at equilibrium we can set up a little table of concentrations. When 10.0 mL of acid and 10.0 mL of base are mixed, they

In case 2 the limiting reactant is not used up.

dilute each other by a factor of two. Therefore, the initial molarities are 0.010 0 M and 0.016 0 M . Since the products are formed in a 1:1 mole ratio, we can write

	NH_4^+	+	$(CH_3)_3N$	$\rightleftharpoons$	NH_3	+	$(CH_3)_3NH^+$
Initial concentration (M):	0.010 0		0.016 0		—		—
Final concentration (M):	0.010 0 − x		0.016 0 − x		x		x

Putting these values into the equilibrium expression for Reaction 12-34 gives

$$\frac{x^2}{(0.010\,0 - x)(0.016\,0 - x)} = 3.60 \quad \Rightarrow \quad x = 0.007\,88 \qquad (12\text{-}35)$$

This value of x tells us that the concentration of each species is

$$[NH_4^+] = 0.002\,12 \text{ M} \qquad [(CH_3)_3N] = 0.008\,12$$

$$[NH_3] = 0.007\,88 \text{ M} \qquad [(CH_3)_3NH^+] = 0.007\,88$$

Aha! Two buffers!

The pH can be calculated by using either of the buffer systems present:

There are two buffer systems:

1. NH_4^+/NH_3
2. $(CH_3)_3NH^+/(CH_3)_3N$

$$pH = pK_a \text{ (for } NH_4^+\text{)} + \log \frac{[NH_3]}{[NH_4^+]}$$

$$= 9.244 + \log \frac{0.007\,88}{0.002\,12} = 9.81 \qquad (12\text{-}36)$$

$$pH = pK_a \text{ (for } (CH_3)_3NH^+\text{)} + \log \frac{[(CH_3)_3N]}{[(CH_3)_3NH^+]}$$

$$= 9.800 + \log \frac{0.008\,12}{0.007\,88} = 9.81 \qquad (12\text{-}37)$$

If the value of x was calculated correctly in Equation 12-35, then Equations 12-36 and 12-37 must produce the same pH.

Case 3: Equimolar Mixture of HA and B

Before we treat this case, let's take another look at a diprotic acid, which has three forms: H_2A contains two protons, HA^- has one proton, and A^{2-} has no protons. The intermediate form, HA^-, can donate *or* accept one proton. This behavior led to the equation

$$[H^+] \approx \sqrt{\frac{K_1 K_2 F + K_1 K_w}{K_1 + F}} \qquad (10\text{-}92)$$

to describe a solution of pure HA^-.

But an equimolar mixture of HA and B is very much like the intermediate form of a diprotic system. Let's draw the diprotic acid as $HA_x \sim A_y H$, where A_x and A_y are attached to each other (i.e., part of the same molecule). The three states of each system can be drawn as follows:

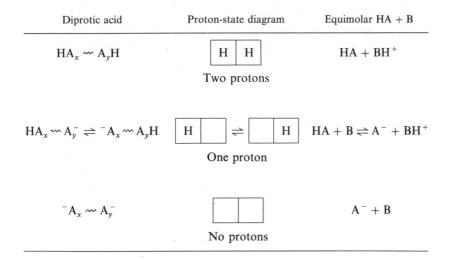

| Diprotic acid | Proton-state diagram | Equimolar HA + B |

The analogy between the intermediate species of a diprotic acid and an equimolar mixture of HA + B suggests that the pH of a mixture of HA + B can be calculated from the equation

$$[H^\cdot] \approx \sqrt{\frac{K_1 K_2 F + K_1 K_w}{K_1 + F}} \qquad (12\text{-}38)$$

where $K_1 = K_a$ for the stronger of the two acids HA and BH^+

$K_2 = K_a$ for the weaker of the two acids HA and BH^+

F = formal concentration of either HA or B (they are equimolar)

An equimolar mixture of HA and B behaves as the intermediate form of the imaginary diprotic acid $HA \sim BH^+$.

With this in mind, consider a solution prepared by mixing 100 mL of 0.050 0 M chloroacetic acid with 100 mL of 0.050 0 M methylamine (Equation 12-33). This is an equimolar mixture of HA and B and should behave *just as the intermediate form of a diprotic acid*. The values of K_a for the two acids are

$$K_a \text{ for } ClCH_2CO_2H = 1.36 \times 10^{-3}$$

$$K_a \text{ for } CH_3NH_3^+ = 2.3 \times 10^{-11}$$

Calling the larger value K_1 and the smaller value K_2, we write

$$[H^+] = \sqrt{\frac{(1.36 \times 10^{-3})(2.3 \times 10^{-11})(0.025\,0) + (1.36 \times 10^{-3})(1.0 \times 10^{-14})}{1.36 \times 10^{-3} + 0.025\,0}}$$

$$= 1.74 \times 10^{-7} \text{ M} \implies pH = 6.76 \qquad (12\text{-}39)$$

We used F = 0.025 0 M because each solution was diluted by the other when they were mixed.

To summarize, an equimolar mixture of a weak acid and a weak base behaves as the intermediate form of a polyprotic acid. *The pH of the equimolar mixture should be very nearly midway between pK_a for HA and pK_a for BH^+.*

EXAMPLE: Analogy to a Triprotic System

Find the pH of 0.10 M ammonium bicarbonate.

Ammonium bicarbonate can be thought of as the first intermediate species of the *triprotic* system consisting of NH_4^+ and H_2CO_3:

$$NH_4^+: \qquad pK_a = 9.244$$
$$H_2CO_3: \qquad pK_1 = 6.352$$
$$pK_2 = 10.329$$

We can think of this triprotic system as follows:

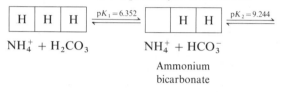

$$NH_4^+ + H_2CO_3 \qquad\qquad NH_4^+ + HCO_3^-$$

Ammonium
bicarbonate

$$NH_3 + HCO_3^- \qquad\qquad NH_3 + CO_3^{2-}$$

Since ammonium bicarbonate is "surrounded" by pK_1 and pK_2,

$$[H^+] = \sqrt{\frac{(10^{-6.352})(10^{-9.244})(0.10) + (10^{-6.352})(K_w)}{10^{-6.352} + 0.10}}$$

$$= 1.59 \times 10^{-8} \text{ M} \quad \Rightarrow \quad pH = 7.80$$

The pH is midway between pK_1 (6.352) and pK_2 (9.244) for the imaginary triprotic system.

Summary

The principal species of a monoprotic or polyprotic system is found by comparing the pH with the various pK_a values. For $pH < pK_1$, the fully protonated species is the predominant form. For $pK_1 < pH < pK_2$ the form $H_{n-1}A^-$ is favored and at each successive pK value the next deprotonated species becomes the principal species. Finally, at pH values higher than the highest pK, the fully basic form (A^{n-}) is dominant. The fractional composition of a solution is expressed by α, given in Equations 12-9 and 12-10 for a monoprotic system and Equations 12-19 through 12-21 for a diprotic system.

The isoelectric pH of a polyprotic species is that pH at which its net charge is zero. For a diprotic amino acid whose amphiprotic form is neutral, the isoelectric pH is given by $pH = \frac{1}{2}(pK_1 + pK_2)$. The isoionic pH of a polyprotic species is the pH that would exist in a solution containing only the ions derived from the neutral polyprotic species and from H_2O.

The reaction of a weak acid with a weak base can be written $HA + B \rightleftharpoons A^- + BH^+$, and the equilibrium constant expressed as K_aK_b/K_w. Such reactions fall into three classes:

1. If the equilibrium constant is large, the reaction proceeds to completion and can be treated as a strong-plus-weak titration.

2. If the equilibrium constant is not large, we must set up and solve an equilibrium problem to find the composition of the solution. Knowing the composition, we can use either of two Henderson–Hasselbalch equations to find the pH. Both equations are satisfied simultaneously.

3. An equimolar mixture of a weak acid and a weak base can be treated as the intermediate form of a polyprotic system, regardless of the magnitude of the equilibrium constant.

Terms to Understand

dialysis
isoelectric focusing

isoelectric point
isoionic point

Exercises

12-A. (a) Draw the structure of the predominant form (principal species) of 1,3-dihydroxybenzene at pH 9.00 and at pH 11.00.
(b) What is the second most prominent species at each pH?
(c) Calculate the percent in the major form at each pH.

12-B. Draw the structures of the predominant forms of glutamic acid and tyrosine at pH 9.0 and pH 10.0. What is the second most abundant species at each pH?

12-C. Calculate the isoionic pH of 0.010 M lysine.

12-D. Neutral lysine can be written HL. The other forms of lysine are H_3L^{2+}, H_2L^+, and L^-. The isoelectric point is the pH at which the *average* charge of lysine is zero. Therefore, at the isoelectric point $2[H_3L^{2+}] + [H_2L^+] = [L^-]$. Use this condition to calculate the isoelectric pH of lysine.

12-E. Solution A contains 0.050 0 M HF. Solution B contains 0.050 0 M ammonia.
(a) Find the pH of a solution containing 20.0 mL of A plus 14.0 mL of B.
(b) Find the pH of a solution containing 20.0 mL of B plus 14.0 mL of A.

12-F. Calculate the pH of a 0.100 M solution of
(a) tetraethylammonium formate
(b) triethylammonium formate

12-G. What is the pH when 25.0 mL of 0.010 M tyrosine is treated with 25.0 mL of 0.009 0 M ammonia?

12-H. Which solution will have the highest pH?
(a) 0.010 M potassium acid phthalate
(b) 0.025 M monosodium malonate
(c) 0.030 M glycine
(d) 0.016 M $(NH_4)(HCO_3)$
(e) 0.013 M K_2HPO_4

12-I. A certain acid–base indicator exists in three colored forms:

$$H_2In \xrightleftharpoons{pK_1 = 1.00} HIn^- \xrightleftharpoons{pK_2 = 7.95} In^{2-}$$

$\lambda_{max} = 520$ nm $\qquad \lambda_{max} = 435$ nm $\qquad \lambda_{max} = 572$ nm

$\varepsilon_{520} = 5.00 \times 10^4$ $\quad \varepsilon_{435} = 1.80 \times 10^4$ $\quad \varepsilon_{572} = 4.97 \times 10^4$

Red	Yellow	Red
$\varepsilon_{435} = 1.67 \times 10^4$	$\varepsilon_{520} = 2.13 \times 10^3$	$\varepsilon_{520} = 2.50 \times 10^4$
$\varepsilon_{572} = 2.03 \times 10^4$	$\varepsilon_{572} = 2.00 \times 10^2$	$\varepsilon_{435} = 1.15 \times 10^4$

A solution containing 10.0 mL of 5.00×10^{-4} M indicator was mixed with 90.0 mL of 0.1 M phosphate buffer (pH 7.50). Calculate the absorbance of this solution at 435 nm in a 1.00-cm cell.[†]

_____

[†] This problem makes use of Beer's law, which you can read about in Sections 19-1 and 19-2.

Problems

A12-1. The acid HA has $pK_a = 7.00$.
(a) Which is the principal species, HA or A^-, at pH 6.00?

(b) Which is the principal species at pH 8.00?
(c) What is the quotient $[A^-]/[HA]$ at pH 7.00? at pH 6.00?

A12-2. The diprotic acid H_2A has $pK_1 = 4.00$ and $pK_2 = 8.00$.
  (a) At what pH is $[H_2A] = [HA^-]$?
  (b) At what pH is $[HA^-] = [A^{2-}]$?
  (c) Which is the principal species, H_2A, HA^-, or A^{2-}, at pH 2.00?
  (d) Which is the principal species at pH 6.00?
  (e) Which is the principal species at pH 10.00?

A12-3. The base B has $pK_b = 5.00$.
  (a) What is the value of pK_a for the acid BH^+?
  (b) At what pH is $[BH^+] = [B]$?
  (c) Which is the principal species, B or BH^+, at pH 7.00?
  (d) What is the quotient $[B]/[BH^+]$ at pH 12.00?

A12-4. The acid HA has $pK_a = 4.00$. Use Equations 12-9 and 12-10 to find the fraction in the form HA and the fraction in the form A^- at pH = 5.00. Does your answer agree with what you expect for the quotient $[A^-]/[HA]$ at pH 5.00?

A12-5. A dibasic compound, B, has $pK_{b1} = 4.00$ and $pK_{b2} = 6.00$. Find the fraction in the form BH_2^{2+} at pH 7.00 using Equation 12-19. Note that K_1 and K_2 in Equation 12-19 are acid dissociation constants for BH_2^{2+} ($K_1 = K_w/K_{b2}$ and $K_2 = K_w/K_{b1}$).

A12-6. What is wrong with the following statement: At its isoelectric point, the charge on all molecules of a particular protein is zero.

A12-7. The acid HA has $pK_a = 5.00$ and the base B has $pK_b = 5.00$.
  (a) Calculate the equilibrium constant for the reaction $HA + B \rightleftharpoons A^- + BH^+$.
  (b) What is the pH if 0.100 mol of HA plus 0.050 mol of B are mixed in 1.00 L of solution?

A12-8. Answer the same questions as those in Problem A12-7 if $pK_b = 9.00$.

A12-9. A mixture of 0.100 mol of HA ($pK_a = 5.00$) and 0.100 mol of B ($pK_b = 10.00$) in 1.00 L of water is prepared.
  (a) Which is the stronger acid, HA or BH^+?
  (b) Use Equation 12-38 to find the pH.

12-10. Calculate the isoelectric and isoionic pH of 0.010 M threonine.

12-11. Draw the structure of the predominant form of pyridoxal-5-phosphate at pH 7.00.

12-12. What fraction of ethane-1,2-dithiol is in each form (H_2A, HA^-, A^{2-}) at pH 8.00? at pH 10.00?

12-13. Calculate α_0, α_1, and α_2 for *cis*-butenedioic acid at pH 1.00, 1.91, 6.00, 6.33, and 10.00.

12-14. (a) Derive equations for α_0, α_1, α_2, and α_3 for a triprotic system.
  (b) Calculate the values of these fractions for phosphoric acid at pH 7.00.

12-15. (a) Write the acid–base reaction that occurs when pyridinium bromide is mixed with potassium 4-methylphenolate.
  (b) Calculate the equilibrium constant for this reaction.

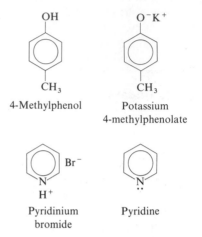

4-Methylphenol   Potassium 4-methylphenolate

Pyridinium bromide   Pyridine

12-16. Calculate the pH that results from mixing 20.0 mL of 0.010 0 M pyridinium bromide with 23.0 mL of 0.010 0 M potassium 4-methylphenolate. (See Problem 12-15 for structures.)

12-17. Find the pH of a solution prepared by mixing 212 mL of 0.200 M acetic acid with 325 mL of 0.050 0 M sodium benzoate.

12-18. Calculate the pH of a solution prepared by mixing 50.0 mL of 0.100 M acetic acid with 50.0 mL of 0.100 M sodium benzoate.

12-19. Butanoic acid (0.100 M) was titrated with 0.100 M ethylamine. Calculate the pH when $V_b = \frac{1}{2}V_e$.

12-20. Butanoic acid (0.100 M) was titrated with 0.100 M aminobenzene. Calculate the pH when $V_b = \frac{1}{2}V_e$.

12-21. What is the pH of 0.100 M ethylammonium butanoate, $(CH_3CH_2NH_3^+)$ $(CH_3CH_2CH_2CO_2^-)$?

12-22. What is the pH of 0.050 M pyridinium bisulfite, $C_5H_5NH^+HSO_3^-$?

12-23. A solution was prepared using 10.0 mL of 0.100 M cacodylic acid plus 10.0 mL of 0.080 0 M NaOH. To this was added 1.00 mL of 1.27×10^{-6} M morphine. Calling morphine B, calculate the fraction of morphine present in the form BH^+.

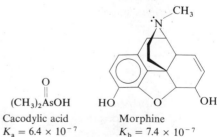

$(CH_3)_2AsOH$ HO   Morphine   OH
Cacodylic acid
$K_a = 6.4 \times 10^{-7}$ Morphine
$K_b = 7.4 \times 10^{-7}$

12-24. Calculate the pH of a 0.100 M solution of
(a) sodium bicarbonate   (b) ammonium acetate

12-25. What is the pH of a solution prepared by mixing 0.050 0 mol aspartic acid, 0.030 0 mol LiOH, and 0.030 0 mol H_2SO_4 in a total volume of 1.00 L?

12-26. What is the charge of the predominant form of citric acid at pH 5.00?

12-27. What is the pH of 0.060 M ammonium azide? (Azide is the anion N_3^- derived from hydrazoic acid, HN_3.)

12-28. The base Na^+A^-, whose anion is dibasic, was titrated with HCl to give the lower curve in Figure 11-5. Is point H, the first equivalence point, the isoelectric point, or the isoionic point?

12-29. Figure 11-1 gives the titration curve for an enzyme. Is the molecule positively charged, negatively charged, or neutral at its isoionic point? Explain.

12-30. *Separation by capillary zone electrophoresis.* In *electrophoresis,* charged molecules are separated by their ability to migrate in an electric field. Benzoic acid containing ordinary ^{16}O can be separated from benzoic acid containing heavy ^{18}O by electrophoresis at a suitable pH because they have slightly different acid dissociation constants.

$$^{16}O \qquad\qquad ^{18}O$$
$$\| \qquad\qquad\quad \|$$
$$C_6H_5C\text{-}^{16}OH \qquad C_6H_5C\text{-}^{18}OH$$

Benzoic acid-^{16}O Benzoic acid-^{18}O

$$H^{16}A \qquad\qquad H^{18}A$$

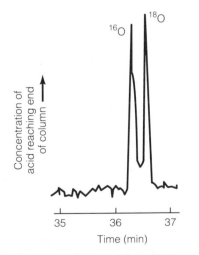

Emergence of isotopic benzoic acids from capillary electrophoresis column. From S. Terebe, T. Yashima, N. Tanaka, and M. Araki, *Anal. Chem.,* **60,** 1673 (1988).

The mixture of isotopic acids is applied to one end of a 50-μm-diameter × 75-cm-long fused silica tube filled with a buffer solution. Application of a 40-kV electric field causes negatively charged solute to migrate from the application end of the tube to the positive pole of the electric field at the other end of the tube. For isotopic acids, the difference in mobility is caused by the different fraction of each acid in the anionic form, A^-. Calling this fraction α, we can write

$$H^{16}A \overset{^{16}K}{\rightleftharpoons} H^+ + {}^{16}A^- \qquad H^{18}A \overset{^{18}K}{\rightleftharpoons} H^+ + {}^{18}A^-$$

$$^{16}\alpha = \frac{^{16}K}{^{16}K + [H^+]} \qquad\qquad ^{18}\alpha = \frac{^{18}K}{^{18}K + [H^+]}$$

where K is the equilibrium constant. The greater the fraction of acid in the form A^-, the faster it will migrate in the electric field. It can be shown that for electrophoresis the maximum separation will occur when $\Delta\alpha/\sqrt{\bar\alpha}$ is a maximum. In this expression, $\Delta\alpha = {}^{16}\alpha - {}^{18}\alpha$ and $\bar\alpha$ is the average fraction of dissociation $(=\frac{1}{2}[{}^{16}\alpha + {}^{18}\alpha])$.

(a) Let us denote the ratio of acid dissociation constants as $R = {}^{16}K/{}^{18}K$. In general, R will be close to unity. For benzoic acid $R = 1.020$. Abbreviate ^{16}K as K and write $^{18}K = K/R$. Derive an expression for $\Delta\alpha/\sqrt{\bar\alpha}$ in terms of K, $[H^+]$, and R. Since both equilibrium constants are nearly equal (R is close to unity), set $\bar\alpha$ equal to $^{16}\alpha$ in your expression.

(b) Find the maximum value of $\Delta\alpha/\sqrt{\bar\alpha}$ by taking the derivative with respect to $[H^+]$ and setting it equal to zero. Show that the maximum difference in mobility of isotopic benzoic acids occurs when

$$[H^+] = \frac{K + K\sqrt{1 + 8R}}{2R}$$

(c) Show that for $R \approx 1$, this simplifies to $[H^+] = 2K$, or pH = pK $- 0.30$. That is, the maximum electrophoretic separation should occur when the column buffer has pH = pK $- 0.30$, regardless of the exact value of R.

12-31. Consider the titration of 50.0 mL of 0.050 M aspartic acid with 0.100 M ammonia.
(a) Write the sequence of titration reactions, including structures of reactants and products.
(b) Calculate the pH when 37.5 mL of base has been added.
(c) The pH at the first equivalence point is 6.61 and the titration curve is steep enough to use an indicator to mark the end point. You have available bromocresol green, chlorophenol red, litmus, cresol red, and thymol blue. Which will you use and what color change will you look for?

12-32. A solution containing acetic acid, oxalic acid, ammonia, and pyridine has a pH of 9.00. What fraction of ammonia is not protonated?

12-33. A solution was prepared by mixing 25.00 mL of 0.080 0 M aniline, 25.00 mL of 0.060 0 M sulfanilic acid, and 1.00 mL of 1.23×10^{-4} M HIn and then diluting to 100.0 mL. (HIn stands for protonated indicator.)

Anilinium ion $pK_a = 4.601$

Sulfanilic acid $pK_a = 3.232$

$$HIn \rightleftharpoons H^+ + In^-$$

$\varepsilon_{325} = 2.45 \times 10^4$ $\varepsilon_{325} = 4.39 \times 10^3 \ M^{-1} \ cm^{-1}.$

$\varepsilon_{550} = 2.26 \times 10^4$ $\varepsilon_{550} = 1.53 \times 10^4$

The absorbance measured at 550 nm in a *5.00-cm* cell was 0.110. Find pK_a for HIn.[†]

_____

[†] This problem is an application of Beer's law, which you can read about in Sections 19-1 and 19-2.

13 EDTA Titrations

Any chemical reaction that proceeds rapidly and has a well-defined stoichiometry and a large equilibrium constant is of potential use for a titration. We have seen how precipitation and acid–base reactions meet these requirements. In this chapter we will examine how the formation of metal ion complexes can be used for analytical purposes.

13-1 METAL CHELATE COMPLEXES

Metal ions are **Lewis acids,** in that they can share electron pairs donated by ligands, which are therefore **Lewis bases.** Cyanide is termed a **monodentate** ligand because it binds to a metal ion through only one atom (the carbon atom). Most transition metal ions have room to bind six ligand atoms. A ligand that can bind to a metal ion through more than one ligand atom is said to be **multidentate.** It is also called a **chelating ligand** (pronounced KEE late ing).

One simple chelating ligand is ethylenediamine ($H_2NCH_2CH_2NH_2$, also called 1,2-diaminoethane), whose binding to a metal ion is shown in the margin. We say that ethylenediamine is *bidentate* because it binds to the metal through two ligand atoms.

An important *tetradentate* ligand is adenosine triphosphate (ATP), which binds to divalent metal ions (such as Mg^{2+}, Mn^{2+}, Co^{2+}, and Ni^{2+}) through four of their six coordination positions (Figure 13-1). The fifth and sixth positions are occupied by water molecules. The biologically active form of ATP is generally the Mg^{2+} complex.

$$Ag^{\oplus} + 2:\overset{\ominus}{C}\equiv N: \rightleftharpoons$$

Lewis acid (electron-pair acceptor)   Lewis base (electron-pair donor)

$$[N\equiv C-Ag-C\equiv N]^{\ominus}$$

Complex ion

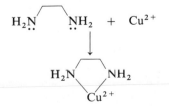

Bidentate bonding

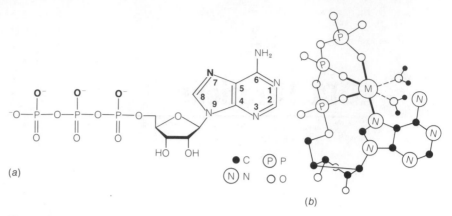

Figure 13-1

(a) Structure of adenosine triphosphate (ATP), with ligand atoms shown in **bold** type. (b) Possible structure of a metal–ATP complex, with four bonds to ATP and two bonds to H_2O ligands. There is controversy as to whether N_7 is bound directly to the metal or whether a molecule of water forms a hydrogen-bonded bridge between N_7 and the metal ion.

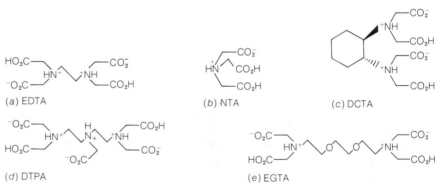

Figure 13-2

Structures of some analytically useful chelating agents. (a) Ethylenediaminetetraacetic acid (also called ethylenedinitrilotetraacetic acid). (b) Nitrilotriacetic acid. (c) *trans*-1,2 Diaminocyclohexanetetraacetic acid. (d) Diethylenetriaminepentaacetic acid. (e) Bis-(aminoethyl)glycolether-*N*,*N*,*N′*,*N′*-tetraacetic acid.

Although ATP forms some strong metal complexes, it is not analytically useful because it is very expensive and unstable. The synthetic aminocarboxylic acids shown in Figure 13-2 are some of the more common chelating agents for titrations based on complex formation. The nitrogen atoms and carboxylate oxygen atoms are the potential ligand atoms in these molecules (Figure 13-3). When these atoms bind to a metal ion, the ligand atoms lose their protons. Ethylenediaminetetraacetic acid (EDTA) is by far the most widely used chelating agent for titrations.

A titration based on formation of a complex ion is called a **complexometric titration.** The ligands other than NTA in Figure 13-2 are especially useful because they form strong 1:1 complexes with many metal ions. The equilibrium constant for the reaction of a metal with a ligand is called the

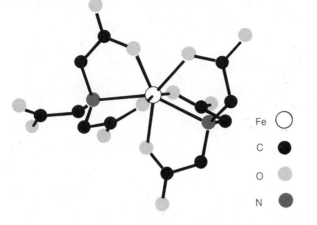

Figure 13-3
Structure of $Fe(NTA)_2^{3-}$ in the salt $Na_3[Fe(NTA)_2] \cdot 5H_2O$. Note that one ligand binds to Fe through three oxygen atoms and one nitrogen atom. The other ligand uses two oxygen atoms and one nitrogen atom. Its third carboxyl group is uncoordinated. The Fe atom is seven-coordinate. [W. Clegg, A. K. Powell, and M. J. Ware, *Acta Cryst.*, **C40**, 1822 (1984).]

formation constant, K_f, also termed the **stability constant.** The reaction between *trans*-1,2-diaminocyclohexanetetraacetic acid (DCTA) and Ni^{2+}, for example, has a formation constant of $10^{19.4}$.

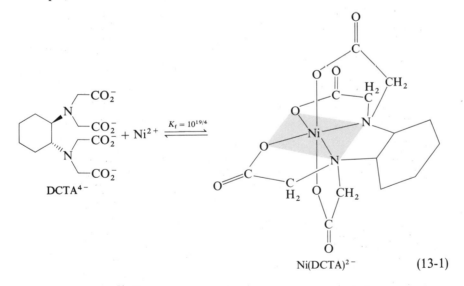

(13-1)

The stoichiometry of the reaction between DCTA and a metal ion is 1:1, *regardless of the charge of the metal.* The only common ions that do not form strong complexes with the ligands in Figure 13-2 are univalent metal ions (Li^+, Na^+, K^+, etc.).

Chelate Effect

Multidentate ligands usually form stronger metal complexes than those formed by similar monodentate ligands. For example, the reaction of Cd^{2+}

with two molecules of ethylenediamine is more favorable than its reaction with four molecules of methylamine:

$$Cd^{2+} + 2H_2\ddot{N}CH_2CH_2\ddot{N}H_2 \rightleftharpoons \left[\begin{array}{cc} NH_2 & H_2N \\ & Cd \\ NH_2 & H_2N \end{array} \right]^{2+} \quad K = 2 \times 10^{10}$$

Ethylenediamine

(13-2)

$$Cd^{2+} + 4CH_3\ddot{N}H_2 \rightleftharpoons \left[\begin{array}{cc} CH_3NH_2 & H_2NCH_3 \\ & Cd \\ CH_3NH_2 & H_2NCH_3 \end{array} \right]^{2+} \quad K = 3 \times 10^6$$

Methylamine

(13-3)

At pH 12 in the presence of 2 M ethylenediamine and 4 M methylamine, the quotient $[Cd(\text{ethylenediamine})_2^{2+}]/[Cd(\text{methylamine})_4^{2+}]$ is 200.

This can be understood on thermodynamic grounds. The two tendencies that drive a chemical reaction are decreasing enthalpy (negative ΔH, liberation of heat) and increasing entropy (positive ΔS, more disorder). In Reactions 13-2 and 13-3, four Cd—N bonds are formed, and ΔH is about the same for both reactions.

However, Reaction 13-2 represents the coming together of *three* molecules (Cd^{2+} + 2 ethylenediamine), whereas in Reaction 13-3 *five* molecules (Cd^{2+} + 4 methylamine) are involved. There is more disorder associated with five molecules than with three molecules. If the enthalpy change (ΔH) of each reaction is about the same, the entropy change (ΔS) will favor Reaction 13-2 over 13-3.

The **chelate effect,** then, is the observation that multidentate ligands form more stable metal complexes than those formed by similar monodentate ligands. The stability of the multidentate complex is mainly an entropy effect.[†] The chelate effect is most pronounced for ligands such as EDTA or DCTA, which can occupy all six coordination sites about a metal ion. An important medical use of the chelate effect is discussed in Box 13-1.

A reaction is favorable if $\Delta G < 0$.

$$\Delta G = \Delta H - T\Delta S$$

The reaction is favored by negative ΔH and positive ΔS.

Chelate effect: A multidentate ligand forms stronger complexes than a similar monodentate ligand.

13-2 EDTA

One mole of EDTA reacts with *one* mole of metal ion.

Ethylenediaminetetraacetic acid (EDTA) is by far the most widely used chelator in analytical chemistry. It forms strong 1:1 complexes with most metal ions. By direct titration or through an indirect sequence of reactions, virtually every element of the periodic table can be analyzed with EDTA.

Acid–Base Properties

EDTA is a hexaprotic system, which we will designate H_6Y^{2+}. Its acid–base properties are summarized by the following pK_a values:[‡]

[†] More detailed discussion of the chelate effect and the role of factors other than entropy can be found in articles by J. J. R. Fraústo da Silva, *J. Chem. Ed.,* **60,** 390 (1983), and C.-S. Chung, *J. Chem. Ed.,* **61,** 1062 (1984).

[‡] pK_1 applies at 25°C, $\mu = 1.0$ M. The other values shown apply at 20°C, $\mu = 0.1$ M. [A. E. Martell and R. M. Smith, *Critical Stability Constants,* Vol. 1 (New York: Plenum Press, 1974), p. 204.]

$$pK_1 = 0.0$$
$$pK_2 = 1.5$$
$$pK_3 = 2.0$$
$$pK_4 = 2.66$$
$$pK_5 = 6.16$$
$$pK_6 = 10.24$$

$$K_1 = \frac{[H_5Y^-][H^+]}{[H_6Y^{2-}]}$$

HO₂CCH₂ and CH₂CO₂H structure with $\overset{+}{H}NCH_2CH_2\overset{+}{N}H$

H_6Y^{2+}

* Think in terms of their K_b
then it does make sense.

Box 13-1 CHELATION THERAPY AND THALASSEMIA

Oxygen is carried in the human circulatory system by the iron-containing protein hemoglobin, which consists of two pairs of subunits, designated α and β. β-Thalassemia major is a genetic disease in which the β subunits of hemoglobin are not synthesized in adequate quantities. Children afflicted with this disease can survive only with frequent transfusions of normal red blood cells.

The problem with this treatment is that the patient accumulates 4–8 g of iron per year from the hemoglobin in the transfused cells. The body has no mechanism for excreting such large quantities of iron, so iron builds up in all tissues. Most thalassemia victims die by age 20 from the toxic effects of iron overload.

To help the body excrete excess iron, intensive chelation therapy is used. The most successful drug so far is the chelator *desferrioxamine B*, isolated from bacteria. The structure of the iron complex (ferrioxamine B) is shown below.

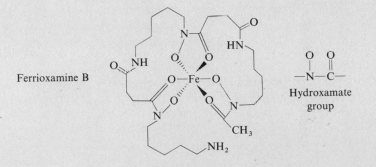

Ferrioxamine B

Hydroxamate group

The ligand contains three hydroxamate groups, which occupy all six positions about the ferric ion. The formation constant for ferrioxamine B is $10^{30.6}$.

Used in conjunction with ascorbic acid—vitamin C, a reducing agent that reduces Fe(III) to the more soluble Fe(II)—desferrioxamine is able to remove several grams of iron per year from an overloaded patient. The ferrioxamine complex is excreted in the urine.

A decade of clinical trials has demonstrated that desferrioxamine reduces the incidence of heart and liver disease in thalassemia patients and maintains approximately correct iron balance.[†] However, desferrioxamine is very expensive and must be taken by continuous injection for maximum effect. The molecule is not absorbed through the intestine. A multitude of potent iron chelators has been tested to find an effective one that can be taken orally, but to date none has replaced desferrioxamine.[‡]

[†] L. Wolfe, D. Sallan, and D. G. Nathan, *Anal. N. Y. Acad. Sci.*, **445**, 248 (1985).
[‡] A. E. Martel, R. J. Motekaitis, I. Murase, L. F. Sala, R. Stoldt, C. Y. Ng, H. Rosenkrantz, and J. J. Metterville, *Inorg. Chim. Acta*, **138**, 215 (1987).

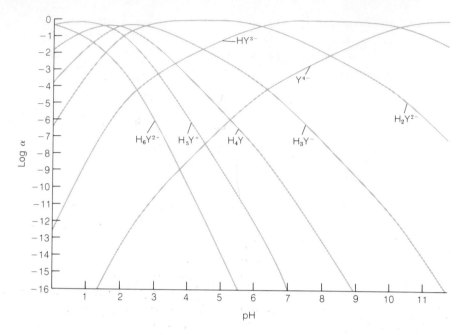

Figure 13-4

Fractional composition diagram for EDTA. Note that the ordinate is logarithmic.

Table 13-1
Values of $\alpha_{Y^{4-}}$ for EDTA at 20°C and $\mu = 0.10$ M

pH	$\alpha_{Y^{4-}}$
0	1.3×10^{-23}
1	1.9×10^{-18}
2	3.3×10^{-14}
3	2.6×10^{-11}
4	3.8×10^{-9}
5	3.7×10^{-7}
6	2.3×10^{-5}
7	5.0×10^{-4}
8	5.6×10^{-3}
9	5.4×10^{-2}
10	0.36
11	0.85
12	0.98
13	1.00
14	1.00

The first four pK values apply to carboxyl protons, and the last two are for the ammonium protons. The neutral acid is tetraprotic, with the formula H_4Y. A commonly used reagent is the disodium salt, $Na_2H_2Y \cdot 2H_2O$.[†]

The logarithm of the fraction of EDTA in each of its protonated forms is plotted in Figure 13-4. As in Section 12-2, we may define α for each species as the fraction of EDTA in that form. For example, $\alpha_{Y^{4-}}$ is defined as

$$\alpha_{Y^{4-}} = \frac{[Y^{4-}]}{[H_6Y^{2+}] + [H_5Y^+] + [H_4Y] + [H_3Y^-] \\ + [H_2Y^{2-}] + [HY^{3-}] + [Y^{4-}]}$$

$$\alpha_{Y^{4-}} = \frac{[Y^{4-}]}{[EDTA]} \tag{13-4}$$

where $[EDTA]$ is the total concentration of all *free* EDTA species in the solution. By "free" we mean EDTA not complexed to metal ions. Following the derivation in Section 12-2, it can be shown that $\alpha_{Y^{4-}}$ is given by

$$\alpha_{Y^{4-}} = \frac{K_1K_2K_3K_4K_5K_6}{[H^+]^6 + [H^+]^5K_1 + [H^+]^4K_1K_2 + [H^+]^3K_1K_2K_3 \\ + [H^+]^2K_1K_2K_3K_4 + [H^+]K_1K_2K_3K_4K_5 + K_1K_2K_3K_4K_5K_6} \tag{13-5}$$

Table 13-1 gives values for $\alpha_{Y^{4-}}$ as a function of pH.

[†] H_4Y can be dried at 140°C for two hours and used as a primary standard. It can be dissolved by adding NaOH solution from a plastic container. NaOH solution from a glass bottle should not be used because it contains alkaline earth metals leached from the glass. Reagent-grade $Na_2H_2Y \cdot 2H_2O$ contains ~0.3% excess water. It may be used in this form with suitable correction for the mass of excess water, or dried to the composition $Na_2H_2Y \cdot 2H_2O$ at 80°C.

The formation constant, K_f, of a metal–EDTA complex is the equilibrium constant for the reaction

$$M^{n+} + Y^{4-} \rightleftharpoons MY^{n-4} \qquad K_f = \frac{[MY^{n-4}]}{[M^{n+}][Y^{4-}]} \qquad (13\text{-}6)$$

Note that K_f is defined for reaction of the species Y^{4-} with the metal ion. This is only one of the seven different forms of free EDTA present in the solution. Table 13-2 shows that the formation constants for most EDTA complexes are quite large and tend to be larger for more positively charged metal ions.

In many complexes the EDTA ligand completely engulfs the metal ion, forming the six-coordinate species shown in Figure 13-5. In this structure, two nitrogen atoms occupy adjacent positions of the octahedrally coordinated metal ion. The other four positions are occupied by carboxyl oxygen atoms.

If you try to build a space-filling model of a six-coordinate metal–EDTA complex, you will find that there is considerable strain in the chelate rings.

Equation 13-6 does not imply that Y^{4-} is the only species that reacts with M^{n+}. It simply says that the equilibrium constant is expressed in terms of the concentration of Y^{4-}.

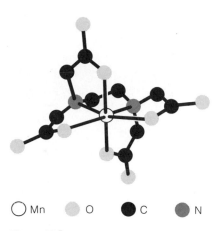

○ Mn   ◒ O   ● C   ◓ N

Figure 13-5
Six-coordinate geometry of a metal–EDTA complex found in the compound $KMnEDTA \cdot 2H_2O$. [J. Stein, J. P. Fackler, Jr., G. J. McClune, J. A. Fee, and L. T. Chan, *Inorg. Chem.*, **18**, 3511 (1979).]

Table 13-2
Formation constants for metal–EDTA complexes

Ion	$\log K_f$	Ion	$\log K_f$	Ion	$\log K_f$
Li^+	2.79	Mn^{3+}	25.3 (25°C)	Ce^{3+}	15.98
Na^+	1.66	Fe^{3+}	25.1	Pr^{3+}	16.40
K^+	0.8	Co^{3+}	41.4 (25°C)	Nd^{3+}	16.61
Be^{2+}	9.2	Zr^{4+}	29.5	Pm^{3+}	17.0
Mg^{2+}	8.79	Hf^{4+}	29.5 ($\mu = 0.2$)	Sm^{3+}	17.14
Ca^{2+}	10.69	VO^{2+}	18.8	Eu^{3+}	17.35
Sr^{2+}	8.73	VO_2^+	15.55	Gd^{3+}	17.37
Ba^{2+}	7.86	Ag^+	7.32	Tb^{3+}	17.93
Ra^{2+}	7.1	Tl^+	6.54	Dy^{3+}	18.30
Sc^{3+}	23.1	Pd^{2+}	18.5 (25°C, $\mu = 0.2$)	Ho^{3+}	18.62
Y^{3+}	18.09			Er^{3+}	18.85
La^{3+}	15.50			Tm^{3+}	19.32
V^{2+}	12.7	Zn^{2+}	16.50	Yb^{3+}	19.51
Cr^{2+}	13.6	Cd^{2+}	16.46	Lu^{3+}	19.83
Mn^{2+}	13.87	Hg^{2+}	21.7	Am^{3+}	17.8 (25°C)
Fe^{2+}	14.32	Sn^{2+}	18.3 ($\mu = 0$)	Cm^{3+}	18.1 (25°C)
Co^{2+}	16.31	Pb^{2+}	18.04	Bk^{3+}	18.5 (25°C)
Ni^{2+}	18.62	Al^{3+}	16.3	Cf^{3+}	18.7 (25°C)
Cu^{2+}	18.80	Ga^{3+}	20.3	Th^{4+}	23.2
Ti^{3+}	21.3 (25°C)	In^{3+}	25.0	U^{4+}	25.8
V^{3+}	26.0	Tl^{3+}	37.8 ($\mu = 1.0$)	Np^{4+}	24.6 (25°C, $\mu = 1.0$)
Cr^{3+}	23.4	Bi^{3+}	27.8		

Note: The stability constant is the equilibrium constant for the reaction $M^{n+} + Y^{4-} \rightleftharpoons MY^{n-4}$. Values in table apply at 20°C, and ionic strength 0.1 M, unless otherwise noted.
SOURCE: A. E. Martell and R. M. Smith, *Critical Stability Constants*, Vol. 1 (New York: Plenum Press, 1974), pp. 204–211.

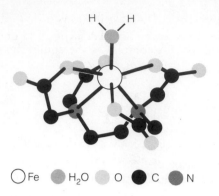

Figure 13-6
Seven-coordinate geometry of Fe(EDTA)
$(H_2O)^-$. Other metal ions that form seven-
coordinate EDTA complexes include
Ru^{3+}, Cr^{3+}, Co^{3+}, Mg^{2+}, Mn^{2+}, and
Ti^{4+}. [M. D. Lind, J. L. Hoard, M. J. Hamor, and T. A. Hamor, *Inorg. Chem.*, **3**,
34 (1964); X. Solans, M. Font Altaba, and
J. Garcia-Oricain, *Acta. Cryst.*, **C40**, 573
(1984); J. P. Fackler, Jr; F. J. Kristine, A.
M. Mazany, T. J. Moyer, and R. E. Shep-
herd, *Inorg. Chem.*, **24**, 1857 (1985).]

$\bigcirc$ Fe   ● H_2O ● O   ● C   ● N

This strain is relieved if the oxygen ligands are drawn back toward the nitrogen atoms. Such distortion opens up a seventh coordination position, which becomes occupied by a water molecule, as shown in Figure 13-6. In some complexes, such as $Ca(EDTA)^{2-}$, the metal ion is so large that two coordination positions are available and the metal ion becomes eight-coordinate.[†]

The formation constant can still be formulated as in Equation 13-6, even if there are water molecules attached to the product. This is true because the solvent (H_2O) is omitted from the reaction quotient.

Conditional Formation Constant

The formation constant in Equation 13-6 describes the reaction between Y^{4-} and a metal ion. As you can see in Figure 13-4, most of the EDTA is not Y^{4-} below pH 10.24. The species HY^{3-}, H_2Y^{2-}, etc. predominate at lower pH. It is convenient to express the fraction of free EDTA by using the equation

$$[Y^{4-}] = \alpha_{Y^{4-}}[\text{EDTA}] \tag{13-7}$$

where [EDTA] refers to the total concentration of all EDTA species not bound to metal ion. Equation 13-7 is just a rearranged form of Equation 13-4.

The equilibrium constant for Reaction 13-6 can now be rewritten as

> Only some of the free EDTA is in the form Y^{4-}.

$$K_f = \frac{[MY^{n-4}]}{[M^{n+}][Y^{4-}]} = \frac{[MY^{n-4}]}{[M^{n+}]\alpha_{Y^{4-}}[\text{EDTA}]} \tag{13-8}$$

If the pH is fixed by a buffer, then $\alpha_{Y^{4-}}$ is a constant that can be combined with K_f:

$$K_f' = \alpha_{Y^{4-}}K_f = \frac{[MY^{n-4}]}{[M^{n+}][\text{EDTA}]} \tag{13-9}$$

The number $K_f' = \alpha_{Y^{4-}}K_f$ is called the **conditional formation constant** or *effective formation constant*. It describes the formation of MY^{n-4} at any particular pH.

[†] R. L. Barnett and V. A. Uchtman, *Inorg. Chem.*, **18**, 2674 (1979).

The conditional formation constant is useful because it allows us to look at EDTA complex formation as if the uncomplexed EDTA were all in one form:

$$M^{n+} + EDTA \rightleftharpoons MY^{n-4} \qquad K'_f = \alpha_{Y^{4-}} K_f \qquad (13\text{-}10)$$

At any given pH, we can find $\alpha_{Y^{4-}}$ and evaluate K'_f.

With the conditional formation constant, we can treat EDTA complex formation as if all the free EDTA were in one form.

EXAMPLE: Using the Conditional Formation Constant

The formation constant in Table 13-2 for FeY^- is $10^{25.1} = 1.3 \times 10^{25}$. Calculate the concentration of free Fe^{3+} in solutions of 0.10 M FeY^- at pH 8.00 and at pH 2.00.

The complex formation reaction is

$$Fe^{3+} + EDTA \rightleftharpoons FeY^- \qquad K'_f = \alpha_{Y^{4-}} K_f$$

where EDTA on the left side of the equation refers to all forms of unbound EDTA ($= Y^{4-}$, HY^{3-}, H_2Y^{2-}, H_3Y^-, etc.). Using values of $\alpha_{Y^{4-}}$ from Table 13-1 we find

At pH 8.00: $\quad K'_f = (5.6 \times 10^{-3})(1.3 \times 10^{25}) = 7.3 \times 10^{22}$

At pH 2.00: $\quad K'_f = (3.3 \times 10^{-14})(1.3 \times 10^{25}) = 4.3 \times 10^{11}$

Since dissociation of FeY^- must produce equal quantities of Fe^{3+} and EDTA, we can write

	Fe^{3+}	$+$ EDTA $\rightleftharpoons$	FeY^-
Initial concentration (M):	0	0	0.10
Final concentration (M):	x	x	$0.10 - x$

$$\frac{0.10 - x}{x^2} = K'_f = 7.3 \times 10^{22} \quad \text{at pH 8.00}$$

$$= 4.3 \times 10^{11} \quad \text{at pH 2.00}$$

Solving for $x (= [Fe^{3+}])$, we find $[Fe^{3+}] = 1.2 \times 10^{-12}$ at pH 8.00 and 4.8×10^{-7} M at pH 2.00. *Using the conditional formation constant, we treat the dissociated EDTA as if it were a single species.*

You can see from the example that a metal–EDTA complex becomes less stable at lower pH. For a titration reaction to be effective, it must "go to completion." That is, its equilibrium constant must be large. Figure 13-7 shows how pH affects the titration of Ca^{2+} with EDTA. Below pH $\approx$ 8, the break at the end point is not sharp enough to allow accurate determination. This is because the conditional formation constant for CaY^{2-} is too small below pH 8.

Figure 13-8 gives the minimum pH needed for titration of many metal ions. The figure provides a strategy for the selective titration of one ion in the presence of another. For example, a solution containing both Fe^{3+} and Ca^{2+} could be titrated with EDTA at pH 4. At this pH, Fe^{3+} is titrated without interference from the Ca^{2+} ion.

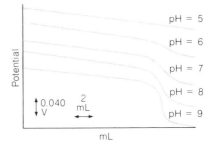

Figure 13-7

Titration of Ca^{2+} with EDTA as a function of pH. As the pH is lowered, the end point becomes less distinct. The potential was measured with mercury and calomel electrodes as described in Problem 15-B. [C. N. Reilley and R. W. Schmid, *Anal. Chem.*, **30**, 947 (1958).]

Control of pH can be used to select which metals will be titrated by EDTA and which will not.

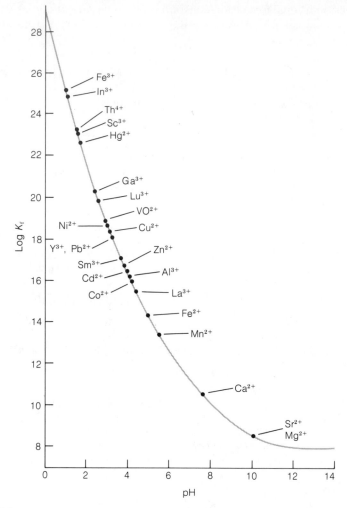

Figure 13-8
Minimum pH for effective titration of various metal ions by EDTA. The minimum pH was specified arbitrarily as the pH at which the conditional formation constant for each metal–EDTA complex is 10^6. [C. N. Reilley and R. W. Schmid, *Anal. Chem.*, **30**, 947 (1958).]

EXAMPLE: What Is the Minimum K'_f for "Complete" Titration?

Suppose we define "complete" to mean 99.9% complexation. What value of K'_f is required for complete reaction at the equivalence point in a titration?

Let the formal concentration of MY^{n-4} be F at the equivalence point. For 99.9% complexation, $[M^{n+}] = [EDTA] = 10^{-3}$ F.

$$M^{n+} + EDTA \rightleftharpoons MY^{n-4}$$

$$10^{-3}\,F \quad 10^{-3}\,F \quad F - 10^{-3}\,F$$

$$K'_f = \frac{[MY^{n-4}]}{[M^{n+}][EDTA]} = \frac{F - 10^{-3}\,F}{(10^{-3}\,F)(10^{-3}\,F)} \approx \frac{F}{10^{-6}\,F^2} = \frac{10^6}{F}$$

For F $= 10^{-2}$ M, K'_f would have to be $10^6/10^{-2} = 10^8$ for 99.9% complete reaction. The points in Figure 13-8 give $K'_f = 10^6$. This corresponds to $\sim 99\%$ complete reaction of 10^{-2} M metal ion at the equivalence point.

$$[H^+] \qquad MY \longrightarrow H_3Y^-$$

In this section we will calculate the concentration of free metal ion during the course of the titration of metal with EDTA. This is analogous to the titration of a strong acid by a weak base. The metal ion plays the role of H^+, and EDTA is the base. The titration reaction is

$$M^{n+} + EDTA \rightleftharpoons MY^{n-4} \qquad K_f' = \alpha_{Y4-} K_f \qquad (13\text{-}11)$$

K_f' is the effective formation constant at the fixed pH of the solution.

If K_f' is large, we can consider the reaction to be complete at each point in the titration.

The titration curve has three natural regions (Figure 13-9).

Region 1: Before the equivalence point

In this region there is excess M^{n+} left in solution after the EDTA has been consumed. The concentration of free metal ion is equal to the concentration of excess, unreacted M^{n+}. The dissociation of MY^{n-4} is negligible.

Region 2: At the equivalence point

There is exactly as much EDTA as metal in the solution. We can treat the solution as if it had been made by dissolving pure MY^{n-4}. Some free M^{n+} is generated by the slight dissociation of MY^{n-4}:

$$MY^{n-4} \rightleftharpoons M^{n+} + EDTA \qquad (13\text{-}12)$$

In Reaction 13-12, EDTA refers to the total concentration of free EDTA in all of its forms. At the equivalence point, $[M^{n+}] = [EDTA]$.

Region 3: After the equivalence point

Now there is excess EDTA, and virtually all the metal ion is in the form MY^{n-4}. Reaction 13-11 still governs the concentration of M^{n+}. However, the concentration of free EDTA can be equated to the concentration of excess EDTA added after the equivalence point.

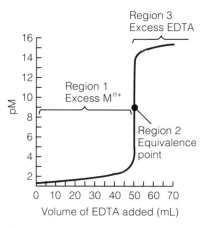

Figure 13-9
Three regions in an EDTA titration illustrated for reaction of 50.0 mL of 0.050 0 M M^{n+} with 0.050 0 M EDTA, assuming $K_f' = 1.15 \times 10^{16}$. The concentration of free M^{n+} decreases as the titration proceeds.

Titration Calculations

Let's calculate the shape of the titration curve for the reaction of 50.0 mL of 0.050 0 M Mg^{2+} with 0.050 0 M EDTA. Suppose that the Mg^{2+} solution is buffered to pH 10.00. The titration reaction is

$$Mg^{2+} + EDTA \rightarrow MgY^{2-} \qquad (13\text{-}13)$$

and

$$K_f' = \alpha_{Y4-} K_f = (0.36)(6.2 \times 10^8) = 2.2 \times 10^8 \qquad (13\text{-}14)$$

The value of $\alpha_{Y^{4-}}$ comes from Table 13-1.

The equivalence point will be 50.0 mL. Since K_f' is large, it is reasonable to say that the reaction goes to completion with each addition of titrant. What we seek is to make a graph on which pMg^{2+} $(= -\log[Mg^{2+}])$ is plotted versus mL of EDTA added.

Region 1: Before the equivalence point

Before the equivalence point, there is excess unreacted M^{n+}.

Consider the case in which 5.0 mL of EDTA has been added. Since the equivalence point is 50.0 mL, one-tenth of the Mg^{2+} will be consumed and nine-tenths should remain.

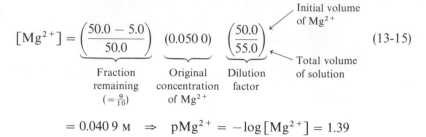

$$[Mg^{2+}] = \underbrace{\left(\frac{50.0 - 5.0}{50.0}\right)}_{\substack{\text{Fraction} \\ \text{remaining} \\ (=\frac{9}{10})}} \underbrace{(0.050\,0)}_{\substack{\text{Original} \\ \text{concentration} \\ \text{of } Mg^{2+}}} \underbrace{\left(\frac{50.0}{55.0}\right)}_{\substack{\text{Dilution} \\ \text{factor}}} \quad (13\text{-}15)$$

Initial volume of Mg^{2+}

Total volume of solution

$$= 0.040\,9 \text{ M} \quad \Rightarrow \quad pMg^{2+} = -\log[Mg^{2+}] = 1.39$$

In a similar manner, we could calculate pMg^{2+} for any volume of EDTA less than 50.0 mL.

Region 2: At the equivalence point

At the equivalence point, the major species is MY^{n-4}. There are small and equal amounts of M^{n+} and EDTA in equilibrium with the complex.

Virtually all the metal is in the form MgY^{2-}. Assuming negligible dissociation, the concentration of MgY^{2-} is equal to the original concentration of Mg^{2+}, with a correction for dilution.

Initial volume of Mg^{2+}

$$[MgY^{2-}] = \underbrace{(0.050\,0 \text{ M})}_{\substack{\text{Original} \\ \text{concentration} \\ \text{of } Mg^{2+}}} \underbrace{\left(\frac{50.0}{100.0}\right)}_{\substack{\text{Dilution} \\ \text{factor}}} = 0.025\,0 \text{ M} \quad (13\text{-}16)$$

Total volume of solution

The concentration of free Mg^{2+} is small and unknown. We can write

	Mg^{2+} +	EDTA $\rightleftharpoons$	MgY^{2-}
Initial concentration (M):	—	—	0.025 0
Final concentration (M):	x	x	0.025 0 − x

In Equation 13-17, [EDTA] refers to the total concentration of all forms of EDTA not bound to metal.

$$\frac{[MgY^{2-}]}{[Mg^{2+}][EDTA]} = K_f' = 2.2 \times 10^8 \quad (13\text{-}17)$$

$$\frac{0.025\,0 - x}{x^2} = 2.2 \times 10^8 \quad \Rightarrow \quad x = 1.06 \times 10^{-5} \text{ M} \quad (13\text{-}18)$$

$$pMg^{2+} = -\log x = 4.97$$

Region 3: After the equivalence point

After the equivalence point, virtually all the metal is present as MY^{n-4}. There is a known excess of EDTA present. A small amount of free M^{n+} exists in equilibrium with the MY^{n-4} and EDTA.

In this region virtually all the metal is in the form MgY^{2-}, and there is excess, unreacted EDTA. The concentrations of MgY^{2-} and excess EDTA are easily calculated. For example, at 51.0 mL there is 1.0 mL of excess EDTA.

$$[EDTA] = \underbrace{(0.050\,0)}_{\substack{\text{Original} \\ \text{concentration} \\ \text{of EDTA}}} \underbrace{\left(\frac{1.0}{101.0}\right)}_{\substack{\text{Dilution} \\ \text{factor}}} = 4.95 \times 10^{-4} \text{ M} \qquad (13\text{-}19)$$

Volume of excess EDTA

Total volume of solution

$$[MgY^{2-}] = \underbrace{(0.050\,0)}_{\substack{\text{Original} \\ \text{concentration} \\ \text{of Mg}^{2+}}} \underbrace{\left(\frac{50.0}{101.0}\right)}_{\substack{\text{Dilution} \\ \text{factor}}} = 2.48 \times 10^{-2} \text{ M} \qquad (13\text{-}20)$$

Original volume of Mg^{2+}

Total volume of solution

The concentration of Mg^{2+} is still governed by Equation 13-17:

$$\frac{[MgY^{2-}]}{[Mg^{2+}][EDTA]} = K_f' = 2.2 \times 10^8 \qquad (13\text{-}17)$$

$$\frac{[2.48 \times 10^{-2}]}{[Mg^{2+}][4.95 \times 10^{-4}]} = 2.2 \times 10^8 \qquad (13\text{-}21)$$

$$[Mg^{2+}] = 2.2 \times 10^{-7} \text{ M} \quad \Rightarrow \quad pMg^{2+} = 6.65$$

The same sort of calculation can be used for any volume past the equivalence point.

Titration curve

The calculated titration curve is shown in Figure 13-10. As in previous titrations, a distinct break occurs at the equivalence point, where the slope is greatest. For comparison, the titration curve for 0.050 0 M Zn^{2+} is also shown in Figure 13-10. The change in pZn^{2+} at the equivalence point is even greater than the change in pMg^{2+}, because the formation constant for ZnY^{2-} is greater than that of MgY^{2-}.

The completeness of reaction (and hence the sharpness of the equivalence point) is determined by the conditional formation constant, $\alpha_{Y^{4-}} \cdot K_f$, which is pH-dependent. Since $\alpha_{Y^{4-}}$ decreases drastically as the pH is lowered, the pH is an important variable determining whether a titration is feasible. The end point is more distinct at high pH. However, the pH of a titration must not be so high that the metal hydroxide precipitates. The effect of pH on the titration of Ca^{2+} was shown in Figure 13-7.

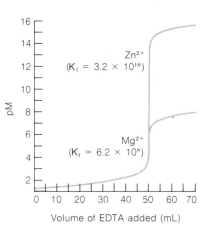

Figure 13-10
Theoretical titration curves for the reaction of 50.0 mL of 0.050 0 M metal ion with 0.050 0 M EDTA at pH 10.00.

The lower the pH, the less distinct is the end point.

13-4 AUXILIARY COMPLEXING AGENTS

The titration curve for Zn^{2+} in Figure 13-10 is not realistic because the pH is high enough to precipitate $Zn(OH)_2$ ($K_{sp} = 3.0 \times 10^{-16}$) before EDTA is added. The titration of Zn^{2+} is usually carried out in ammonia buffer, which not only fixes the pH, but serves to complex the metal ion and keep it in solution. The ammonia is called an **auxiliary complexing agent,** because it complexes the metal ion until EDTA is introduced. Let's see how this works.

Metal–Ligand Equilibria

Consider a metal ion, M, that can form two complexes with the ligand L:

$$M + L \rightleftharpoons ML \qquad \beta_1 = \frac{[ML]}{[M][L]} \qquad (13\text{-}22)$$

$$M + 2L \rightleftharpoons ML_2 \qquad \beta_2 = \frac{[ML_2]}{[M][L]^2} \qquad (13\text{-}23)$$

Overall (β) and stepwise (K) formation constants were distinguished in Box 5-3. The relation between β_n and the stepwise formation constants is $\beta_n = K_1 K_2 K_3 \cdots K_n$.

The equilibrium constants, β_i, are called **overall** or **cumulative formation constants**. The fraction of metal ion in the uncomplexed state, M, can be expressed as

$$\alpha_M = \frac{[M]}{C_M} \qquad (13\text{-}24)$$

where C_M refers to the total concentration of all forms of M ($= M$, ML, and ML_2 in this case).

Now let's find a useful expression for α_M. The mass balance for metal is simply

$$C_M = [M] + [ML] + [ML_2] \qquad (13\text{-}25)$$

Equations 13-22 and 13-23 allow us to say $[ML] = \beta_1[M][L]$ and $[ML_2] = \beta_2[M][L]^2$. Therefore

$$C_M = [M] + \beta_1[M][L] + \beta_2[M][L]^2$$

$$= [M]\{1 + \beta_1[L] + \beta_2[L]^2\} \qquad (13\text{-}26)$$

If the metal forms more than two complexes, Equation 13-27 takes the form

$$\alpha_M = \frac{1}{1 + \beta_1[L] + \beta_2[L]^2 + \beta_3[L]^3 + \cdots + \beta_n[L]^n}$$

Substituting Equation 13-26 into Equation 13-24 gives the desired result:

$$\alpha_M = \frac{[M]}{[M]\{1 + \beta_1[L] + \beta_2[L]^2\}} = \frac{1}{1 + \beta_1[L] + \beta_2[L]^2} \qquad (13\text{-}27)$$

EXAMPLE: Ammonia Complexes of Zinc

In a solution containing Zn^{2+} and NH_3 the complexes $Zn(NH_3)^{2+}$, $Zn(NH_3)_2^{2+}$, $Zn(NH_3)_3^{2+}$, and $Zn(NH_3)_4^{2+}$ are all present. If the concentration of free NH_3 is 0.10 M, find the fraction of zinc in the form Zn^{2+}. This question means that the concentration of *unprotonated* NH_3 is 0.10 M. (At any pH there will also be some concentration of NH_4^+ in equilibrium with NH_3.) The question also implies that there are no other complexing agents present.

Appendix I gives four stepwise formation constants (K_1, K_2, K_3, and K_4) for the complexes $Zn(NH_3)^{2+}$, $Zn(NH_3)_2^{2+}$, $Zn(NH_3)_3^{2+}$, and $Zn(NH_3)_4^{2+}$. The overall formation constants are related to the stepwise formation constants as follows:

$$\beta_1 = K_1 = 10^{2.18}$$

$$\beta_2 = K_1 K_2 = 10^{2.18} 10^{2.25} = 10^{4.43}$$

$$\beta_3 = K_1 K_2 K_3 = 10^{2.18} 10^{2.25} 10^{2.31} = 10^{6.74}$$

$$\beta_4 = K_1 K_2 K_3 K_4 = 10^{2.18} 10^{2.25} 10^{2.31} 10^{1.96} = 10^{8.70}$$

The constants β_1 through β_4 refer to four equilibria analogous to Reactions 13-22 and 13-23. The appropriate form of Equation 13-27 is

$$\alpha_{Zn^{2+}} = \frac{1}{1 + \beta_1[L] + \beta_2[L]^2 + \beta_3[L]^3 + \beta_4[L]^4} \qquad (13\text{-}28)$$

Putting in $[L] = 0.10$ M and the four values of β_i gives $\alpha_{Zn^{2+}} = 1.8 \times 10^{-5}$, which means that very little zinc is in the form Zn^{2+} in the presence of 0.10 M NH_3.

EDTA Titration in the Presence of an Auxiliary Complexing Agent

Now consider a titration of Zn^{2+} by EDTA in the presence of NH_3. The extension of Equation 13-11 requires a new, effective formation constant that accounts for the fact that only some of the EDTA is in the form Y^{4-} and only some of the zinc is in the form Zn^{2+}:

$$K_f'' = \alpha_{Zn^{2+}}\alpha_{Y^{4-}}K_f \qquad (13\text{-}29)$$

K_f'' is the effective formation constant at a given fixed pH and given fixed concentration of auxiliary complexing agent.

In this expression $\alpha_{Zn^{2+}}$ is given by Equation 13-28 and $\alpha_{Y^{4-}}$ is given by Equation 13-5. For particular values of pH and NH_3 concentration we can calculate a value of K_f'' and proceed with titration calculations as in Section 13-3, substituting K_f'' for K_f'. An assumption in this process is that EDTA is a much stronger complexing agent than ammonia, and essentially all of the EDTA added at any point in the titration is bound to Zn^{2+}, until the Zn^{2+} is consumed.

EXAMPLE: EDTA Titration of Zinc in the Presence of Ammonia

Consider the titration of 50.0 mL of 1.00×10^{-3} M Zn^{2+} with 1.00×10^{-3} M EDTA at pH 10.00 in the presence of 0.10 M NH_3. (This is the concentration of NH_3. There is also NH_4^+ in the solution.) The equivalence point is at 50.0 mL. Find pZn^{2+} after addition of 20.0, 50.0, and 60.0 mL of EDTA.

In Equation 13-28 we found that $\alpha_{Zn^{2+}} = 1.8 \times 10^{-5}$. Table 13-1 tells us that $\alpha_{Y^{4-}} = 0.36$. Therefore the effective formation constant is

$$K_f'' = \alpha_{Zn^{2+}}\alpha_{Y^{4-}}K_f = (1.8 \times 10^{-5})(0.36)(10^{16.50}) = 2.0_5 \times 10^{11}$$

(a) *Before the equivalence point—20.0 mL:* The concentration of zinc not bound to EDTA ($C_{Zn^{2+}}$) is found by a calculation analogous to Equation 13-15. Since the equivalence point is 50.0 mL, the fraction of Zn^{2+} remaining is 30.0/50.0. The dilution factor is 50.0/70.0. Therefore

$$C_{Zn^{2+}} = (30.0/50.0)(1.00 \times 10^{-3})(50.0/70.0) = 4.3 \times 10^{-4} \text{ M}$$

However, nearly all of the zinc not bound to EDTA is bound to NH_3. The concentration of free Zn^{2+} is

$$[Zn^{2+}] = \alpha_{Zn^{2+}}C_{Zn^{2+}} = (1.8 \times 10^{-5})(4.3 \times 10^{-4}) = 7.7 \times 10^{-9} \text{ M}$$

$$\Rightarrow pZn^{2+} = -\log[Zn^{2+}] = 8.11$$

$[Zn^{2+}] = \alpha_{Zn^{2+}}C_{Zn^{2+}}$. This follows from Equation 13-24.

(b) *At the equivalence point—50.0 mL:* At the equivalence point the dilution factor is 50.0/100.0, so $[ZnY^{2-}] = (50.0/100.0)(1.00 \times 10^{-3}) = 5.00 \times 10^{-4}$ M. By analogy with Equation 13-17 we can write

$$C_{Zn^{2+}} + EDTA \rightleftharpoons ZnY^{2-}$$

Initial concentration (M):	0	0	5.00×10^{-4}
Final concentration (M):	x	x	$5.00 \times 10^{-4} - x$

$$K_f'' = 2.0_5 \times 10^{11} = \frac{[ZnY^{2-}]}{[C_{Zn^{2+}}][EDTA]} = \frac{5.00 \times 10^{-4} - x}{x^2}$$

$$\Rightarrow \quad x = C_{Zn^{2+}} = 4.9 \times 10^{-8} \text{ M}$$

$$[Zn^{2+}] = \alpha_{Zn^{2+}} C_{Zn^{2+}} = (1.8 \times 10^{-5})(4.9 \times 10^{-8}) = 8.9 \times 10^{-13} \text{ M}$$

$$\Rightarrow \quad pZn^{2+} = -\log[Zn^{2+}] = 12.05$$

(c) *After the equivalence point—60.0 mL:* Now we are past the equivalence point, so almost all of the zinc is in the form ZnY^{2-}. By analogy to Equation 13-20 with a dilution factor of 50.0/110.0, we can write

$$[ZnY^{2-}] = (50.0/110.0)(1.00 \times 10^{-3}) = 4.5 \times 10^{-4} \text{ M}$$

We also know the concentration of excess EDTA. By analogy to Equation 13-19, with a dilution factor of 10.0/110.0, we can write

$$[EDTA] = (10.0/110.0)(1.00 \times 10^{-3}) = 9.1 \times 10^{-5} \text{ M}$$

Once we know $[ZnY^{2-}]$ and $[EDTA]$, we can use an equation analogous to Equation 13-17 to find $[Zn^{2+}]$:

$$\frac{[ZnY^{2-}]}{[Zn^{2+}][EDTA]} = \alpha_{Y^{4-}} K_f = K_f' = (0.36)(10^{16.50}) = 1.1 \times 10^{16}$$

$$\frac{[4.5 \times 10^{-4}]}{[Zn^{2+}][9.1 \times 10^{-5}]} = 1.1 \times 10^{16} \quad \Rightarrow \quad [Zn^{2+}] = 4.3 \times 10^{-16} \quad \Rightarrow \quad pZn^{2+} = 15.36$$

Note that past the equivalence point the problem did not depend on the presence of NH_3, because we knew the concentrations of both $[ZnY^{2-}]$ and $[EDTA]$.

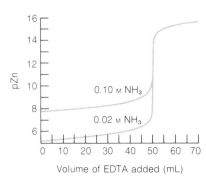

Figure 13-11
Titration curves for the reaction of 50.0 mL of 1.00×10^{-3} M Zn^{2+} with 1.00×10^{-3} M EDTA at pH 10.00 in the presence of two different concentrations of NH_3.

Figure 13-11 compares the calculated titration curves for Zn^{2+} in the presence of different concentrations of auxiliary complexing agent. The greater the concentration of NH_3, the smaller the change of pZn^{2+} near the equivalence point. When an auxiliary ligand is used, its amount must be kept below the level that would obliterate the end point of the titration.

13-5 METAL ION INDICATORS

Several methods can be used to detect the end point in EDTA titrations. The most common technique is to use a metal ion indicator. Alternatively, a mercury electrode (described in Problem 15-B) can be used to produce titration curves such as those in Figure 13-7. A glass pH electrode can also be used to follow the course of the titration in an unbuffered solution, since H_2Y^{2-} releases $2H^+$ when it forms HY^{2-}. An ion-selective electrode (Section 15-5) for the metal being titrated is also a good end-point detector.

A **metal ion indicator** is a compound whose color changes when it binds to a metal ion. Several common indicators are shown in Table 13-3. *For an indicator to be useful, it must bind metal less strongly than EDTA does.*

End-point detection methods:

1. metal ion indicators
2. mercury electrode
3. glass (pH) electrode
4. ion-selective electrode

Table 13-3
Some common metal ion indicators

295

Name	Structure	pK_a	Color of free indicator	Color of metal ion complex
Eriochrome black T	(H_2In^-)	$pK_2 = 6.3$ $pK_3 = 11.6$	H_2In^- —red HIn^{2-} —blue In^{3-} —orange	Wine-red
Calmagite	(H_2In^-)	$pK_2 = 8.1$ $pK_3 = 12.4$	H_2In^- —red HIn^{2-} —blue In^{3-} —orange	Wine-red
Murexide	(H_4In^-)	$pK_2 = 9.2$ $pK_3 = 10.9$	H_4In^- —red-violet H_3In^{2-} —violet H_2In^{3-} —blue	Yellow (with Co^{2+}, Ni^{2+}, Cu^{2+}) Red with Ca^{2+}
Xylenol orange	(H_3In^{3-})	$pK_2 = 2.32$ $pK_3 = 2.85$ $pK_4 = 6.70$ $pK_5 = 10.47$ $pK_6 = 12.23$	H_5In^- —yellow H_4In^{2-} —yellow H_3In^{3-} —yellow H_2In^{4-} —violet HIn^{5-} —violet In^{6-} —violet	Red
Pyridylazo-naphthol (PAN)	(HIn)	$pK_a = 12.3$	HIn —orange-red In^- —pink	Red
Pyrocatechol violet	(H_3In^-)	$pK_1 = 0.2$ $pK_2 = 7.8$ $pK_3 = 9.8$ $pK_4 = 11.7$	H_4In —red H_3In^- —yellow H_2In^{2-} —violet HIn^{3-} —red-purple	Blue

A typical analysis is illustrated by the titration of Mg^{2+} with EDTA using eriochrome black T as the indicator. We can write the reaction as follows:

The indicator must release its metal to EDTA.

$$MgIn \; + \; EDTA \; \rightarrow \; MgEDTA \; + \; In \qquad (13\text{-}30)$$

$$\text{(Red)} \quad \text{(Colorless)} \quad \text{(Colorless)} \quad \text{(Blue)}$$

A small amount of indicator (In) is added to the Mg^{2+} to form a red complex. As EDTA is added, it reacts first with free, colorless Mg^{2+} and then with the small amount of red MgIn complex. (The EDTA must therefore bind to Mg^{2+} better than the indicator binds to Mg^{2+}.) The change from the red of MgIn to the blue of unbound In signals the end point of the titration. Demonstration 13-1 illustrates this titration.

Most metal ion indicators are also acid–base indicators. Some pertinent pK_a values are listed in Table 13-3. Because the color of free indicator is pH-dependent, most indicators can be used only in certain pH ranges. For example, xylenol orange (pronounced ZY-leen-ol) changes from yellow to red when it binds to a metal ion at pH 5.5. This is an easy color change to observe. At pH 7.5 the change is from violet to red and rather difficult to see. A spectrophotometer can be used to measure an indicator color change, but it is more convenient if we can see it.

Demonstration 13-1 METAL ION INDICATOR COLOR CHANGES

We will demonstrate the color change associated with Reaction 13-30 and also show how a second dye can be added to a solution to produce a more easily detected color change.
    Prepare the following solutions:

  eriochrome black T: Dissolve 0.1 g of the solid in 7.5 mL of triethanolamine plus 2.5 mL of absolute ethanol.

  methyl red: Dissolve 0.02 g in 60 mL of ethanol. Then add 40 mL of water.

  buffer: Add 142 mL of concentrated (14.5 M) aqueous ammonia to 17.5 g of ammonium chloride and dilute to 250 mL with water.

  $MgCl_2$: 0.05 M

  EDTA: 0.05 M $Na_2EDTA \cdot 2H_2O$

    Prepare a solution containing 25 mL of $MgCl_2$, 5 mL of buffer and 300 mL of water. Add 6 drops of eriochrome black T indicator and titrate with EDTA. Note the color change from wine-red to pale blue at the end point (Color Plate 8A).
    For some observers the change of indicator color is not as sharp as desired. The colors can be affected by adding an "inert" dye whose color alters the appearance of the solution before and after the titration. Adding 3 mL of methyl red (or many other yellow dyes) produces an orange color prior to the end point and a green color after it. This sequence of colors is shown in Color Plate 8B.

Some metal ion indicators are unstable. Solutions of azo indicators (compounds with —N≡N— bonds) deteriorate rapidly and should probably be prepared each week. Murexide solution should be prepared fresh each day.

For an indicator to be useful in the titration of a metal with EDTA, the indicator must give up its metal ion to the EDTA. If a metal does not freely dissociate from an indicator, the metal is said to **block** the indicator. Eriochrome black T is blocked by Cu^{2+}, Ni^{2+}, Co^{2+}, Cr^{3+}, Fe^{3+}, and Al^{3+}. It cannot be used as an indicator for the direct titration of any of these metals. Eriochrome black T can be used for a back titration, however. For example, excess standard EDTA can be added to Cu^{2+}. Then indicator is added and the excess EDTA is back-titrated with Mg^{2+}.

> *Question:* What will be the color change when the back titration is performed?

13-6 EDTA TITRATION TECHNIQUES

Because so many elements can be analyzed by titration with EDTA, there is an extensive literature dealing with many variations of the basic procedure.[†] In this section we discuss several important techniques.

Direct Titration

In a **direct titration,** analyte is titrated with standard EDTA. The analyte is buffered to an appropriate pH at which the conditional formation constant for the metal–EDTA complex is large enough to produce a sharp end point. Since most metal ion indicators are also acid–base indicators, they have different colors at different values of pH. An appropriate pH must be one at which the free indicator has a distinctly different color from the metal–indicator complex.

The larger the effective formation constant, the more abrupt is the change in metal ion concentration at the end point.

In many titrations an *auxiliary complexing agent,* such as ammonia, tartrate, citrate, or triethanolamine, is employed to prevent the metal ion from precipitating in the absence of EDTA. For example, the direct titration of Pb^{2+} is carried out in ammonia buffer at pH 10 in the presence of tartrate, which complexes the metal ion and does not allow $Pb(OH)_2$ to precipitate. The lead–tartrate complex must be less stable than the lead–EDTA complex, or the titration would not be feasible.

Back Titration

In a **back titration** a known excess of EDTA is added to the analyte. The excess EDTA is then titrated with a standard solution of a second metal ion. A back titration is necessary if the analyte precipitates in the absence of EDTA, if it reacts too slowly with EDTA under titration conditions, or if it blocks the indicator. The metal ion used in the back titration must not displace the analyte metal ion from its EDTA complex.

[†] Some good sources for reading about EDTA titration techniques are G. Schwarzenbach and H. Flaschka, *Complexometric Titrations* (H. M. N. H. Irving, trans.) (London: Methuen, 1969); H. A. Flaschka, *EDTA Titrations* (New York: Pergamon Press, 1959); and C. N. Reilley, A. J. Barnard, Jr., and R. Puschel in L. Meites, ed., *Handbook of Analytical Chemistry* (New York: McGraw-Hill, 1963), pp. 3-76 to 3-234.

EXAMPLE: A Back Titration

Ni^{2+} can be analyzed by a back titration using standard Zn^{2+} at pH 5.5 with xylenol orange indicator. A solution containing 25.00 mL of Ni^{2+} in dilute HCl was treated with 25.00 mL of 0.052 83M Na_2EDTA. The solution was neutralized with NaOH, and the pH was adjusted to 5.5 with acetate buffer. The solution turned yellow when a few drops of indicator were added. It was then titrated with 17.61 mL of 0.022 99 M Zn^{2+} to reach the red end point. What was the molarity of Ni^{2+} in the unknown?

The unknown was treated with 25.00 mL of 0.052 83 M EDTA, which contains

$$(25.00 \text{ mL})(0.052\ 83 \text{ M}) = 1.320\ 8 \text{ mmol of EDTA}$$

Back titration required 17.61 mL of 0.022 99 M Zn^{2+}:

$$(17.61 \text{ mL})(0.022\ 99 \text{ M}) = 0.404\ 9 \text{ mmol of } Zn^{2+}$$

Since one mole of EDTA reacts with one mole of any metal ion, there must have been $(1.320\ 8 \text{ mmol EDTA} - 0.404\ 9 \text{ mmol } Zn^{2+}) = 0.915\ 9 \text{ mmol } Ni^{2+}$. The concentration of Ni^{2+} was 0.915 9 mmol/25.00 mL = 0.036 64 M.

Preventing precipitation

Al^{3+} precipitates as $Al(OH)_3$ at pH 7 in the absence of EDTA. An acidic solution of Al^{3+} can be treated with excess EDTA, adjusted to pH 7–8 with sodium acetate, and boiled to ensure complete complexation of the ion. The Al^{3+}–EDTA complex is stable in solution at this pH. The solution is then cooled; eriochrome black T indicator is added; and back titration with standard Zn^{2+} is performed.

Displacement Titrations

For metal ions that do not have a satisfactory indicator, a **displacement titration** may be feasible. In this procedure the analyte usually is treated with excess $Mg(EDTA)^{2-}$ to displace Mg^{2+}, which is later titrated with standard EDTA.

$$M^{n+} + MgY^{2-} \rightarrow MY^{n-4} + Mg^{2+} \qquad (13\text{-}31)$$

Challenge: Calculate the equilibrium constant for Reaction 13-31 if M = Hg. Why is $Mg(EDTA)^{2-}$ used for a displacement titration?

Hg^{2+} is determined in this manner. The formation constant of $Hg(EDTA)^{2-}$ must be greater than the formation constant for $Mg(EDTA)^{2-}$, or else Reaction 13-31 will not work.

There is no suitable indicator for Ag^+. However, Ag^+ will displace Ni^{2+} from the tetracyanonickelate ion:

$$2Ag^+ + Ni(CN)_4^{2-} \rightarrow 2Ag(CN)_2^- + Ni^{2+} \qquad (13\text{-}32)$$

The liberated Ni^{2+} can then be titrated with EDTA to find out how much Ag^+ was added.

Indirect Titrations

Anions that form precipitates with certain metal ions may be analyzed with EDTA by **indirect titration.** For example, sulfate can be analyzed by precipitation with excess Ba^{2+} at pH 1. The $BaSO_4$ precipitate is filtered and washed. Boiling the precipitate with excess EDTA at pH 10 brings the Ba^{2+} back into solution as $Ba(EDTA)^{2-}$. The excess EDTA is back-titrated with Mg^{2+}.

Alternatively, an anion may be precipitated with excess metal ion. The precipitate is filtered and washed, and the excess metal ion in the filtrate is titrated with EDTA. Anions such as CO_3^{2-}, CrO_4^{2-}, S^{2-}, and SO_4^{2-} can be determined by indirect titration with EDTA.

Masking

A **masking agent** is a reagent that protects some component of the analyte from reaction with EDTA. For example, Al^{3+} reacts with F^- to form the very stable complex AlF_6^{3-}. The Mg^{2+} in a mixture of Mg^{2+} and Al^{3+} can be titrated by first masking the Al^{3+} with F^-, leaving only the Mg^{2+} to react with EDTA.

Masking is used to prevent one element from interfering in the analysis of another element. Masking is not restricted to EDTA titrations. See Box 13-2 for an important application of masking.

Box 13-2   WATER HARDNESS

Hardness refers to the total concentration of alkaline earth ions in water. Since the concentrations of Ca^{2+} and Mg^{2+} are usually much greater than the concentrations of other alkaline earth ions, hardness can be equated to $[Ca^{2+}] + [Mg^{2+}]$. Hardness is commonly expressed as the equivalent number of milligrams of $CaCO_3$ per liter. Thus if $[Ca^{2+}] + [Mg^{2+}] = 1$ mM, we would say that the hardness is 100 mg $CaCO_3$ per liter because 100 mg $CaCO_3 = 1$ mmol $CaCO_3$. Water whose hardness is less than 60 mg $CaCO_3$ per liter is considered to be "soft." *Individual hardness* refers to the individual concentration of each alkaline earth ion.

Hard water reacts with soap to form insoluble curds.

$$Ca^{2+} + 2RSO_3^- \rightarrow Ca(RSO_3)_2(s) \qquad \text{(a)}$$
$$\text{soap} \qquad \text{precipitate}$$

Enough soap to consume the Ca^{2+} and Mg^{2+} must be used before the soap will be useful for cleaning. It is not believed that hard water is unhealthy for man. Hardness is beneficial in irrigation water because the alkaline earth ions tend to flocculate (cause to aggregate) colloidal particles in soil and thereby increase the permeability of the soil to water.

To measure total hardness, the sample is treated with ascorbic acid (or hydroxylamine) to reduce Fe^{3+} to Fe^{2+} and with cyanide to mask Fe^{2+}, Cu^+, and several other minor metal ions. Titration with EDTA at pH 10 in ammonia buffer then gives the total concentrations of Ca^{2+} and Mg^{2+}. The concentration of Ca^{2+} may be determined separately if the titration is carried out at pH 13 without ammonia. At this pH, $Mg(OH)_2$ precipitates and is inaccessible to the EDTA.

Insoluble carbonates are converted to soluble bicarbonates by excess carbon dioxide:

$$CaCO_3(s) + CO_2 + H_2O \rightarrow Ca(HCO_3)_2(aq) \qquad \text{(b)}$$

Heating converts bicarbonate to carbonate (driving off CO_2) and causes $CaCO_3$ to precipitate. The reverse of Reaction b forms a solid scale that clogs boiler pipes. The fraction of hardness due to $Ca(HCO_3)_2(aq)$ is called *temporary hardness* because this calcium is lost (by precipitation of $CaCO_3$) upon heating. Hardness arising from other salts (mainly dissolved $CaSO_4$) is called *permanent hardness*, because it is not removed by heating.

$N(CH_2CH_2OH)_3$

Triethanolamine

$$HOCH_2\overset{\overset{\displaystyle SH}{|}}{C}HCH_2SH$$

2,3-Dimercaptopropanol

Cyanide is a common masking agent that forms complexes with Cd^{2+}, Zn^{2+}, Hg^{2+}, Co^{2+}, Cu^+, Ag^+, Ni^{2+}, Pd^{2+}, Pt^{2+}, Fe^{2+}, and Fe^{3+}, but not with Mg^{2+}, Ca^{2+}, Mn^{2+}, or Pb^{2+}. If cyanide is first added to a solution containing Cd^{2+} and Pb^{2+}, only the Pb^{2+} is then able to react with EDTA. Fluoride can mask Al^{3+}, Fe^{3+}, Ti^{4+}, and Be^{2+}. Triethanolamine masks Al^{3+}, Fe^{3+}, and Mn^{2+}, and 2,3-dimercaptopropanol masks Bi^{3+}, Cd^{2+}, Cu^{2+}, Hg^{2+}, and Pb^{2+}.

Demasking refers to the release of a metal ion from a masking agent. Cyanide complexes can be demasked by treatment with formaldehyde:

$$M(CN)_m^{n-m} + mH_2CO + mH^+ \longrightarrow mH_2C\overset{\overset{\displaystyle OH}{\diagup}}{\underset{\underset{\displaystyle CN}{\diagdown}}{}} + M^{n+} \qquad (13\text{-}33)$$

Thiourea reduces Cu^{2+} to Cu^+ and masks Cu^+. Cu^{2+} can be liberated from the Cu(I)–thiourea complex by demasking with H_2O_2. The selectivity afforded by masking, demasking, and pH control allows individual components of complex mixtures of metal ions to be analyzed by EDTA titration.

Summary

In a complexometric titration the reaction between analyte and titrant is one of complex ion formation, the equilibrium constant for which is called the formation constant, K_f. Chelating ligands (termed multidentate because they bind to a metal through more than one ligand atom) form more stable complexes than those formed by monodentate ligands. The reason for this chelate effect is that the entropy of complex formation favors the binding of one large ligand, rather than many small ligands. Synthetic multidentate aminocarboxylic acids, such as EDTA, have large metal-binding constants and are widely used in analytical chemistry.

Although EDTA is a hexaprotic system, the formation constant for complex formation is defined in terms of the form Y^{4-}. Since the fraction ($\alpha_{Y^{4-}}$) of free EDTA in the form Y^{4-} depends on pH, we define an effective (or conditional) formation constant as $K_f' = \alpha_{Y^{4-}}K_f = [MY^{n-4}]/[M^{n+}][EDTA]$. This constant describes the hypothetical reaction $M^{n+} + EDTA \rightleftharpoons MY^{n-4}$, where EDTA refers to all forms of EDTA not bound to metal ion. Titration calculations fall into three categories. When excess unreacted M^{n+} is present, pM is calculated directly from $pM = -\log[M^{n+}]$. When excess EDTA is present, we know both $[MY^{4-n}]$ and [EDTA], so $[M^{n+}]$ can be calculated from the conditional formation constant. At the equivalence point the condition $[M^{n+}] = [EDTA]$ allows us to solve for $[M^{n+}]$. EDTA titration curves become sharper as the formation constant increases and as the pH is raised. Auxiliary complexing agents, which compete with EDTA for the metal ion and thereby limit the sharpness of the titration curve, are sometimes necessary to keep the metal in solution. Calculations for a solution containing EDTA and an auxiliary complexing agent utilize a conditional formation constant defined as $K_f'' = \alpha_M\alpha_{Y^{4-}}K_f$, where α_M is the fraction of free metal ion not complexed by the auxiliary ligand.

For end-point detection we commonly use metal ion indicators, a glass electrode, an ion-selective electrode, or a mercury electrode. When a direct titration is not suitable, because the analyte is unstable, reacts slowly with EDTA, or has no suitable indicator, a back titration of excess EDTA or a displacement titration of $MgEDTA^{2-}$ may be feasible. Masking is commonly used to prevent interference by certain species in complex solutions. Indirect EDTA titration procedures are available for the analysis of many anions or other species that do not react directly with the reagent.

Terms to Understand

auxiliary complexing agent

back titration

blocking

chelate effect

chelating ligand

complexometric titration

conditional formation constant
demasking
direct titration
displacement titration
indirect titration
Lewis acid

Lewis base
masking agent
metal ion indicator
monodentate
multidentate
stability constant

Exercises

13-A. The potassium ion in a 250.0 ($\pm$ 0.1)-mL water sample was precipitated with sodium tetraphenylborate:

$$K^+ + (C_6H_5)_4B^- \rightarrow KB(C_6H_5)_4(s)$$

The precipitate was filtered, washed, and dissolved in an organic solvent. Treatment of the organic solution with an excess of Hg(II)–EDTA then gave the following reaction:

$$4HgY^{2-} + (C_6H_5)_4B^- + 4H_2O \rightarrow$$

$$H_3BO_3 + 4C_6H_5Hg^+ + 4HY^{3-} + OH^-$$

The liberated EDTA was titrated with 28.73 ($\pm$ 0.03) mL of 0.043 7 ($\pm$0.000 1) M Zn^{2+}. Find the concentration (and absolute uncertainty) of K^+ in the original sample.

13-B. A 25.00-mL sample of unknown containing Fe^{3+} and Cu^{2+} required 16.06 mL of 0.050 83 M EDTA for complete titration. A 50.00-mL sample of the unknown was treated with NH_4F to protect the Fe^{3+}. Then the Cu^{2+} was reduced and masked by addition of thiourea. Upon addition of 25.00 mL of 0.050 83 M EDTA, the Fe^{3+} was liberated from its fluoride complex and formed an EDTA complex. The excess EDTA required 19.77 mL of 0.018 83 M Pb^{2+} to reach an end point using xylenol orange. Find the concentration of Cu^{2+} in the unknown.

13-C. Calculate pGa^{3+} (to the 0.01 decimal place) at each of the following points in the titration of 50.0 mL of 0.040 0 M EDTA with 0.080 0 M $Ga(NO_3)_3$ at pH 4.00:
(a) 0.1 mL   (b) 5.0 mL   (c) 10.0 mL
(d) 15.0 mL   (e) 20.0 mL   (f) 24.0 mL

(g) 25.0 mL   (h) 26.0 mL   (i) 30.0 mL
Make a graph of pGa^{3+} versus volume of titrant.

13-D. Calculate the concentration of H_2Y^{2-} at the equivalence point in Problem 13-C.

13-E. Suppose that 0.010 0 M Fe^{3+} is titrated with 0.005 00 M EDTA at pH 2.00.
(a) What is the concentration of free Fe^{3+} at the equivalence point?
(b) What is the quotient $[H_3Y^-]/[H_2Y^{2-}]$ in the solution when the titration is just 63.7% of the way to the equivalence point?

13-F. A solution containing 20.0 mL of 1.00×10^{-3} M Co^{2+} in the presence of 0.10 M $C_2O_4^{2-}$ at pH 9.00 was titrated with 1.00×10^{-2} M EDTA. Using equilibrium constants in Table 13-2 and Appendix I, calculate pCo^{2+} for the following volumes of added EDTA: 0, 1.00, 2.00, and 3.00 mL. Consider the concentration of $C_2O_4^{2-}$ to be fixed at 0.10 M. Sketch a graph of pCo^{2+} versus mL of added EDTA.

13-G. Iminodiacetic acid, abbreviated H_2X in this problem, forms 2:1 complexes with many metal ions:

$$H_2\overset{+}{N}\underset{CH_2CO_2H}{\overset{CH_2CO_2H}{\diagdown}} = H_3X^+$$

$$\alpha_{X^{2-}} = \frac{[X^{2-}]}{[H_3X^+] + [H_2X] + [HX^-] + [X^{2-}]}$$

$$Cu^{2+} + 2X^{2-} \rightleftharpoons CuX_2^{2-} \qquad K = 3.5 \times 10^{16}$$

A solution of volume 25.0 mL containing 0.120 M iminodiacetic acid buffered to pH 7.00 was titrated with 25.0 mL of 0.050 0 M Cu^{2+}. Given that $\alpha_{X^{2-}} = 4.6 \times 10^{-3}$ at pH 7.00, calculate the concentration of Cu^{2+} in the resulting solution.

Problems

A13-1. Explain why the change from red to blue in Reaction 13-30 occurs suddenly at the equivalence point instead of gradually throughout the entire titration.

A13-2. List four methods for detecting the end point of an EDTA titration.

A13-3. Give three circumstances in which an EDTA back titration might be necessary.

A13-4. State (in words) what $\alpha_{Y^{4-}}$ means. Calculate $\alpha_{Y^{4-}}$ for EDTA at
(a) pH 3.50 (b) pH 10.50

A13-5. Calcium ion was titrated with EDTA at pH 11 using calmagite as indicator (Table 13-3). Which is the principal species of calmagite at pH 11? What color was observed before the equivalence point? after the equivalence point?

A13-6. Describe what is done in a displacement titration and give an example.

A13-7. State the purpose of an auxiliary complexing agent and give an example of its use.

A13-8. Give an example of the use of a masking agent.

A13-9. How many milliliters of 0.050 0 M EDTA are required to react with 50.0 mL of 0.010 0 M Ca^{2+}? with 50.0 mL of 0.010 0 M Al^{3+}?

A13-10. A 50.0-mL sample containing Ni^{2+} was treated with 25.0 mL of 0.050 0 M EDTA to complex all the Ni^{2+} and leave excess EDTA in solution. The excess EDTA was then back-titrated, requiring 5.00 mL of 0.050 0 M Zn^{2+}. What was the concentration of Ni^{2+} in the original solution?

A13-11. The ion M^{n+} was titrated with EDTA as in Equation 13-11. The initial solution contained 100.0 mL of 0.050 0 M metal ion buffered to pH 9.00. The titrant was 0.050 0 M EDTA.
(a) What is the equivalence volume, V_e, in milliliters?
(b) Calculate the concentration of free metal ion at $V = \frac{1}{2}V_e$.
(c) What fraction ($\alpha_{Y^{4-}}$) of free EDTA is in the form Y^{4-} at pH 9.00?
(d) The formation constant (K_f in Equation 13-6) is $10^{12.00}$. Calculate the value of the conditional formation constant K_f' ($= \alpha_{Y^{4-}} \cdot K_f$).
(e) Calculate the concentration of free metal ion at $V = V_e$.
(f) What is the concentration of free metal ion at $V = 1.100V_e$?

A13-12. According to Appendix I, Cu^{2+} forms two complexes with acetate:

$VOSO_4 + EDTA$
$\Rightarrow$.

$$Cu^{2+} + CH_3CO_2^- \rightleftharpoons Cu(CH_3CO_2)^+$$
$$K_1 (= \beta_1)$$

$$Cu(CH_3CO_2)^+ + CH_3CO_2^- \rightleftharpoons$$
$$Cu(CH_3CO_2)_2(aq) \qquad K_2$$

(a) Find the value of β_2 for the reaction

$$Cu^{2+} + 2CH_3CO_2^- \rightleftharpoons$$
$$Cu(CH_3CO_2)_2(aq) \quad \beta_2 = K_1K_2$$

(b) Consider 1.00 L of solution prepared by mixing 1.00×10^{-4} mol $Cu(ClO_4)_2$ and 0.100 mol CH_3CO_2Na. Use Equation 13-27 to find the fraction of copper in the form Cu^{2+}.

13-13. Calculate pCo^{2+} at each of the following points in the titration of 25.00 mL of 0.020 26 M Co^{2+} by 0.038 55 M EDTA at pH 6.00:
(a) 12.00 mL (b) V_e (c) 14.00 mL

13-14. A 50.0-mL aliquot of solution containing 0.450 g of $MgSO_4$ in 0.500 L required 37.6 mL of EDTA solution for titration. How many milligrams of $CaCO_3$ will react with 1.00 mL of this EDTA solution?

13-15. A 1.000-mL sample of unknown containing Co^{2+} and Ni^{2+} was treated with 25.00 mL of 0.038 72 M EDTA. Back titration with 0.021 27 M Zn^{2+} at pH 5 required 23.54 mL to reach the xylenol orange end point. A 2.000-mL sample of unknown was passed through an ion-exchange column that retards Co^{2+} more than Ni^{2+}. The Ni^{2+} that passed through the column was treated with 25.00 mL of 0.038 72 M EDTA and required 25.63 mL of 0.021 27 M Zn^{2+} for back titration. The Co^{2+} emerged from the column later. It, too, was treated with 25.00 mL of 0.038 72 M EDTA. How many milliliters of 0.021 27 M Zn^{2+} will be required for back titration?

13-16. Consider the titration of 25.0 mL of 0.020 0 M $MnSO_4$ with 0.010 0 M EDTA in a solution buffered to pH 8.00. Calculate pMn^{2+} at the following volumes of added EDTA:
(a) 0 mL (b) 20.0 mL (c) 40.0 mL
(d) 49.0 mL (e) 49.9 mL (f) 50.0 mL
(g) 50.1 mL (h) 55.0 mL (i) 60.0 mL

13-17. Using the same volumes as in Problem 13-16, calculate pCa^{2+} for the titration of 25.00 mL of 0.020 00 M EDTA with 0.010 00 M $CaSO_4$ at pH 10.00.

13-18. Calculate the molarity of HY^{3-} in a solution prepared by mixing 10.00 mL of 0.010 0 M $VOSO_4$, 9.90 mL of 0.010 0 M EDTA, and 10.0 mL of buffer with a pH of 4.00.

13-19. Consider the derivation of the fraction α_M in Equations 13-22 through 13-27.
(a) Derive expressions analogous to Equation 13-27 for the fractions α_{ML} and α_{ML_2}.
(b) Calculate the values of α_{ML} and α_{ML_2} for the conditions in Problem A13-12.

13-20. Calculate pCu^{2+} at each of the following points in the titration of 50.00 mL of 0.001 00 M Cu^{2+} with 0.001 00 M EDTA at pH 11.00 in a solution whose NH_3 concentration is somehow *fixed* at 0.100 M:

(a) 0 mL (b) 1.00 mL (c) 45.00 mL
(d) 50.00 mL (e) 55.00 mL

13-21. A 50.0-mL solution containing Ni^{2+} and Zn^{2+} was treated with 25.0 mL of 0.045 2 M EDTA to bind all the metal. The excess unreacted EDTA required 12.4 mL of 0.012 3 M Mg^{2+} for complete reaction. An excess of the reagent 2,3-dimercapto-1-propanol was then added to displace the EDTA from zinc. Another 29.2 mL of Mg^{2+} was required for reaction with the liberated EDTA. Calculate the molarity of Ni^{2+} and Zn^{2+} in the original solution.

13-22. Cyanide can be determined indirectly by EDTA titration. A known excess of Ni^{2+} is added to the cyanide to form tetracyanonickelate:

$$4CN^- + Ni^{2+} \rightarrow Ni(CN)_4^{2-}$$

When the excess Ni^{2+} is titrated with standard EDTA, $Ni(CN)_4^{2-}$ does not react. In a cyanide analysis 12.7 mL of cyanide solution was treated with 25.0 mL of standard solution containing excess Ni^{2+} to form tetracyanonickelate. The excess Ni^{2+} required 10.1 mL of 0.013 0 M EDTA for complete reaction. In a separate experiment, 39.3 mL of 0.013 0 M EDTA was required to react with 30.0 mL of the standard Ni^{2+} solution. Calculate the molarity of CN^- in the 12.7-mL sample of unknown.

13-23. Sulfide ion was determined by indirect titration with EDTA. To a solution containing 25.00 mL of 0.043 32 M $Cu(ClO_4)_2$ plus 15 mL of 1 M acetate buffer (pH 4.5) was added 25.00 mL of unknown sulfide solution with vigorous stirring. The CuS precipitate was filtered and washed with hot water. Then ammonia was added to the filtrate (which contains excess Cu^{2+}) until the blue color of $Cu(NH_3)_4^{2+}$ was observed. Titration with 0.039 27 M EDTA required 12.11 mL to reach the murexide end point. Calculate the molarity of sulfide in the unknown.

13-24 A mixture of Mn^{2+}, Mg^{2+}, and Zn^{2+} was analyzed as follows: The 25.00-mL sample was treated with 0.25 g of $NH_3OH^+Cl^-$ (hydroxylammonium chloride, a reducing agent that maintains manganese in the +2 state), 10 mL of ammonia buffer (pH 10), and a few drops of eriochrome black T indicator and then diluted to 100 mL. It was warmed to 40°C and titrated with 39.98 mL of 0.045 00 M EDTA to the blue end point. Then 2.5 g of NaF was added to displace Mg^{2+} from its EDTA complex. The liberated EDTA required 10.26 mL of standard 0.020 65 M Mn^{2+} for complete titration. After this second end point was reached, 5 mL of 15% (wt/wt) aqueous KCN was added to displace

Zn^{2+} from its EDTA complex. This time the liberated EDTA required 15.47 mL of standard 0.020 65 M Mn^{2+}. Calculate the number of milligrams of each metal (Mn^{2+}, Zn^{2+}, and Mg^{2+}) in the 25.00-mL sample of unknown.

13-25. Pyrocatechol violet (Table 13-3) is to be used as a metal ion indicator in an EDTA titration. The procedure is as follows:

1. Add a known excess of EDTA to the unknown metal ion.

2. Adjust the pH with a suitable buffer.

3. Back-titrate the excess chelate with standard Al^{3+}.

From the following available buffers, select the best buffer, and then state what color change will be observed at the end point. Explain your answer.
(a) pH 6–7 (b) pH 7–8
(c) pH 8–9 (d) pH 9–10

13-26. *Indirect EDTA determination of cesium.*[†] Cesium ion does not form a strong EDTA complex, but it can be analyzed by adding a known excess of $NaBiI_4$ in cold concentrated acetic acid containing excess NaI. Solid $Cs_3Bi_2I_9$ is precipitated, filtered, and removed. The excess yellow BiI_4^- is then titrated with EDTA. The end point occurs when the yellow color disappears. (Sodium thiosulfate is used in the reaction to prevent the liberated I^- from being oxidized to yellow aqueous I_2 by O_2 from the air.) The precipitation is fairly selective for Cs^+. The ions Li^+, Na^+, K^+, and low concentrations of Rb^+ do not interfere, although Tl^+ does. Suppose that 25.00 mL of unknown containing Cs^+ was treated with 25.00 mL of 0.086 40 M $NaBiI_4$ and the unreacted BiI_4^- required 14.24 mL of 0.043 7 M EDTA for complete titration. Find the concentration of Cs^+ in the unknown.

13-27. *Metal ion buffers.* By analogy to a hydrogen ion buffer, a metal ion buffer tends to maintain a particular metal ion concentration in solution. A mixture of the acid HA and its conjugate base A^- is a hydrogen ion buffer that maintains a pH defined by the equation $K_a = [A^-][H^+]/[HA]$. A mixture of CaY^{2-} and Y^{4-} serves as a Ca^{2+} buffer governed by the equation $1/K_f' = [EDTA][Ca^{2+}]/[CaY^{2-}]$. How many grams of $Na_2EDTA \cdot 2H_2O$ (F.W. 372.23) should be

[†] Numerous indirect EDTA titrations of monovalent ions are described in a review article by I. M. Yurist, M. M. Talmud, and P. M. Zaitsev, *J. Anal. Chem. U.S.S.R.*, **42**, 911 (1987).

mixed with 1.95 g of $Ca(NO_3)_2 \cdot 2H_2O$ (F.W. 200.12) in a 500-mL volumetric flask to give a buffer with $pCa^{2+} = 9.00$ at pH 9.00?

13-28. *Allosteric interactions.* The molecule drawn below contains two large rings with oxygen atoms capable of binding metal atoms, one on each ring.

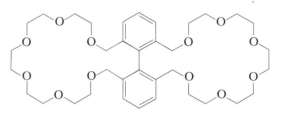

Calling the molecule L, we can represent the metal-binding reactions as

$$L + M \rightleftharpoons LM \qquad K_1 = \frac{[LM]}{[L][M]} \qquad (A)$$

$$LM + M \rightleftharpoons LM_2 \qquad K_2 = \frac{[LM_2]}{[LM][M]} \qquad (B)$$

If binding at one site influences binding at the other site, there is said to be an *allosteric interaction* between the sites. If binding at one site makes binding at the other site more favorable than it was in the absence of the first binding, there is said to be *positive cooperativity* between the sites. If binding at one site makes binding at the second site less favorable, there is *negative cooperativity* between the sites. If there is no interaction between sites, binding is said to be *noncooperative*. This means that a metal at one site has no effect on metal binding at the other site.

The binding of $Hg(CF_3)_2$ to the molecule above in benzene solution was found to have $K_1 = 4.0 \, (\pm 0.1) \, K_2$.[†] Show that $K_1 = 4K_2$ corresponds to noncooperative binding. *Hint:* If the two binding sites are represented as ⬜⬜, we can represent the equilibria as follows:

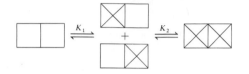

where ⬛ represents metal bound at one site. In noncooperative binding, the four populations ⬜⬜, ⬛⬛, ⬜⬛, and ⬛⬛ must be equal when the ligand is 50% saturated with metal.

13-29. *A brainbuster!* What is the quotient $[MgY^{2-}]/[NaY^{3-}]$ in a solution prepared by mixing 0.100 M Na_2EDTA with an equal volume of 0.100 M $Mg(NO_3)_2$? Assume that the pH is high enough that there is a negligible amount of unbound EDTA. You can approach this problem by realizing that nearly all Mg^{2+} will be bound to EDTA and nearly all Na^+ will be free.

[†] J. Rebek, Jr., T. Costello, L. Marshall, R. Wattley, R. C. Gadwood, and K. Onan, *J. Amer. Chem. Soc., 107,* 7481 (1985).

14 Fundamentals of Electrochemistry

A major branch of analytical chemistry makes use of electrical measurements for analytical purposes. In this chapter we will review fundamental concepts of electricity and electrochemical cells. The principles we develop will lay a foundation for the discussion of potentiometric measurements, electrogravimetric and coulometric analysis, polarography, and amperometric methods in the next four chapters.

14-1 BASIC CONCEPTS

A **redox reaction** involves transfer of electrons from one species to another. A species is said to be **oxidized** when it *loses electrons*. It is **reduced** when it *gains electrons*. An **oxidizing agent,** also called an **oxidant,** takes electrons from another substance and becomes reduced. A **reducing agent,** also called a **reductant,** gives electrons to another substance and is oxidized in the process. In the reaction.

Oxidation: loss of electrons
Reduction: gain of electrons
Oxidizing agent: takes electrons
Reducing agent: gives electrons

$$\underset{\substack{\text{Oxidizing} \\ \text{agent}}}{Fe^{3+}} + \underset{\substack{\text{Reducing} \\ \text{agent}}}{V^{2+}} \rightarrow Fe^{2+} + V^{3+} \qquad (14\text{-}1)$$

Fe^{3+} is the oxidizing agent, since it takes an electron from V^{2+}. V^{2+} is the reducing agent, since it gives an electron to Fe^{3+}. Fe^{3+} is reduced, and V^{2+} is oxidized as the reaction proceeds from left to right. A review of oxidation numbers and redox equation balancing can be found in Appendix D.

305

The *quantity* of electrons that flow
from a reaction is proportional to
the quantity of analyte that reacts.

The electric force (voltage) is related
to the identity and concentrations of
reactants and products.

Faraday constant (F):
96 485.309 C/mol e^-

Relation Between Chemistry and Electricity

If the electrons involved in a redox reaction can be made to flow through
an electric circuit, we can learn something about the reaction by studying
the circuit. In Reaction 14-1, one electron must be transferred to oxidize one
atom of V^{2+} and to reduce one atom of Fe^{3+}. If we know how many moles
of electrons are transferred from V^{2+} to Fe^{3+}, then we know how many
moles of product have been formed.

The voltage produced by a cell in which Reaction 14-1 is taking place
is related to the free energy change occurring when electrons flow from V^{2+}
to Fe^{3+}. The free energy change would be different if electrons were flowing,
for example, from Sn^{2+} to Fe^{3+}. In techniques such as polarography, the
voltage can be used to identify the reactants. Voltage is also related to the
concentrations of reactants and products, as we shall see when we study the
Nernst equation (Section 14-4).

Electrical Measurements

Charge

Electric charge, denoted by q, is measured in **coulombs,** abbreviated C. The
charge of a single electron is $1.602\,177\,33 \times 10^{-19}$ C. One mole of electrons
has a charge of $9.648\,530\,9 \times 10^4$ C; this number is called the **Faraday
constant** and given the symbol F. The relation between coulombs and moles
is therefore

$$q = n \cdot F$$
$$\text{Coulombs} = \text{moles} \cdot \frac{\text{coulombs}}{\text{mole}}$$

(14-2)

EXAMPLE: Relating Coulombs to Quantity of Reaction

If 5.585 g of Fe^{3+} was reduced in Reaction 14-1, how many coulombs of charge must
have been transferred from V^{2+} to Fe^{3+}?

First, we find that 5.585 g of Fe^{3+} equals 0.100 0 mol of Fe^{3+}. Since each Fe^{3+}
ion requires one electron in Reaction 14-1, 0.100 0 mol of electrons must have been
transferred. Using the Faraday constant, we find that 0.100 0 mol of electrons corresponds to

$$(0.100\,0 \text{ mol e}^-)\left(9.649 \times 10^4 \frac{C}{\text{mol e}^-}\right) = 9.649 \times 10^3 \text{ C}$$

Current

The quantity of charge flowing each second through a circuit is called the
current. The unit of current is the **ampere,** abbreviated A. A current of one

$1 \text{ A} = 1 \text{ C/s}$

$$\left(1 \text{C} \right)\left(\frac{1 \text{mole } e^-}{96 \times 10^3 \text{ C}} \right) = 1.04 \times 10^{-5} \text{ mole } e^-$$

ampere represents a charge flowing past a point in a circuit at a rate of one coulomb per second.

$$1 \text{ A} \not\Rightarrow 1 \times 10^{-5} \text{ mole } e^-$$
per second.

EXAMPLE: Relating Current to Rate of Reaction

Suppose that electrons are forced into a platinum wire immersed in a solution containing Sn^{4+} (Figure 14-1), which is reduced to Sn^{2+} at a constant rate of 4.24 mmol/h. How much current flows into the solution?

Two electrons are required to reduce *one* Sn^{4+} ion:

$$Sn^{4+} + 2e^- \rightarrow Sn^{2+}$$

If Sn^{4+} is reacting at a rate of 4.24 mmol/h, electrons flow at a rate of 2(4.24) = 8.48 mmol/h, which corresponds to

$$\frac{8.48 \text{ mmol/h}}{3\,600 \text{ s/h}} = 2.356 \times 10^{-3} \text{ mmol/s} = 2.356 \times 10^{-6} \text{ mol/s}$$

To find the current, we convert moles of electrons per second to coulombs per second:

$$\text{Current} = \frac{\text{coulombs}}{\text{second}} = \frac{q}{s} = \frac{nF}{s} = \frac{n}{s} F = \frac{\text{moles}}{\text{second}} \cdot F$$

$$= \left(2.356 \times 10^{-6} \frac{\text{mol}}{\text{s}} \right)\left(9.649 \times 10^4 \frac{\text{C}}{\text{mol}} \right) = 0.227 \frac{\text{C}}{\text{s}} = 0.227 \text{ A}$$

A current of 0.227 A can also be expressed as 227 mA.

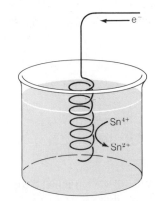

Figure 14-1
Electrons flowing into a coil of Pt wire at which Sn^{4+} ions in solution are reduced to Sn^{2+} ions. This process could not happen by itself because there is not a complete circuit. If Sn^{4+} is to be reduced at this Pt electrode, some other species must be oxidized at some other place.

In the foregoing example, we encountered a Pt electrode. An **electrode** is any device that conducts electrons into or out of the chemical species involved in a redox reaction. Platinum is commonly used as an *inert* conductor; it does not participate in the redox chemistry except as a conductor of electrons.

Voltage, work, and free energy

The difference in **electric potential** between two points is a measure of the work that is needed (or can be done) when an electric charge moves from one point to the other. The symbol E customarily denotes a potential difference, measured in **volts** (V). Work has the dimensions of energy, whose units are joules (J).

When a charge, q, moves through a potential difference, E, the work done is

It costs energy to move like charges toward each other. Energy is released when opposite charges move toward each other.

$$\boxed{\begin{array}{c} \text{Work} = E \cdot q \\[2mm] \text{Joules} = \text{volts} \cdot \text{coulombs} \end{array}}$$

(14-3)

One **joule** of energy is gained or lost when one *coulomb* of charge is moved through a potential difference of one volt. Equation 14-3 tells us that the dimensions of volts are joules/coulomb.

1 volt = 1 J/C

EXAMPLE: Electrical Work

How much work is required to move 2.36 mmol of electrons through a potential difference of 1.05 V?

To use Equation 14-3, we must convert moles of electrons to coulombs of charge. The relation is simply

$$q = nF = (2.36 \times 10^{-3} \text{ mol})(9.649 \times 10^4 \text{ C/mol}) = 2.277 \times 10^2 \text{ C}$$

The work required is

$$\text{Work} = E \cdot q = (1.05 \text{ V})(2.277 \times 10^2 \text{ C}) = 239 \text{ J}$$

The greater the potential difference between two points, the stronger will be the "push" on a charged particle traveling between those points. A 12-V battery will "push" electrons through a circuit eight times harder than a 1.5-V dry cell.

The free energy change, ΔG, for a chemical reaction conducted reversibly at constant temperature and pressure equals the maximum possible electrical work that can be done by the reaction on its surroundings.

You might review Chapter 5 for a brief discussion of ΔG.

$$\text{Work} = -\Delta G \tag{14-4}$$

The negative sign in Equation 14-4 indicates that the free energy of a system is considered as decreasing when the work done on the surroundings is positive.

Combining Equations 14-2, 14-3, and 14-4 produces a relationship of utmost importance to chemistry:

$$\Delta G = -\text{work} = -E \cdot q$$

Relation between free energy and electric potential:

$$\Delta G = -nFE$$

$$\boxed{\Delta G = -nFE} \tag{14-5}$$

Equation 14-5 relates the free energy change of a chemical reaction to the electrical potential difference (i.e., the voltage) that can be generated by the reaction.

Ohm's law

The current, I, flowing through a circuit is directly proportional to the voltage and inversely proportional to the **resistance, R,** of the circuit.

Ohm's law: $I = E/R$. The greater the voltage, the more current will flow. The greater the resistance, the less current will flow.

$$\boxed{I = \frac{E}{R}} \tag{14-6}$$

The units of resistance are ohms, assigned the Greek symbol Ω. A current of one ampere will flow through a circuit with a potential difference of one volt if the resistance of the circuit is one ohm.

Power, P, is the work done per unit time. The SI unit of power is J/s, better known as the **watt** (W).

$$P = \frac{\text{work}}{s} = \frac{E \cdot q}{s} = E \cdot \frac{q}{s} \qquad (14\text{-}7)$$

Since q/s is the current, I, we can write

$$P = E \cdot I \qquad (14\text{-}8)$$

A cell capable of delivering one ampere at a potential of one volt has a power output of one watt.

Power (watts) = work per second

$$P = E \cdot I = (IR) \cdot I = I^2 R$$

$$P = E \cdot I = E \cdot \frac{E}{R} = \frac{E^2}{R}$$

EXAMPLE: Using Ohm's Law

A schematic diagram of a very simple circuit is shown in Figure 14-2. The battery generates a potential difference of 3.0 V, and the resistor has a resistance of 100 Ω. We assume that the resistance of the wire connecting the battery and the resistor is negligible. How much current and how much power are delivered by the battery in this circuit?

The current flowing through this circuit is

$$I = \frac{E}{R} = \frac{3.0\ \text{V}}{100\ \Omega} = 0.030\ \text{A} = 30\ \text{mA}$$

The power produced by the battery must be

$$P = E \cdot I = (3.0\ \text{V})(0.030\ \text{A}) = 90\ \text{mW}$$

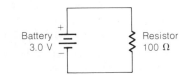

Figure 14-2
A simple electric circuit with a battery and a resistor.

What happens to the energy needed to push electrons through the circuit in Figure 14-2? Ideally, the only place at which energy is lost is in the resistor. *The energy appears as heat in the resistor.* The power (90 mW) equals the rate at which heat is produced in the resistor.

14-2 GALVANIC CELLS

A **galvanic cell** (also called a *voltaic cell*) is one that uses a *spontaneous* chemical reaction to generate electricity. In order to accomplish this, one reagent must be oxidized and another must be reduced. The two cannot be in contact, or electrons would simply flow directly from the reducing agent to the oxidizing agent. Instead, the oxidizing and reducing agents are physically separated, and electrons are forced to flow through an external circuit in order to go from one reactant to the other.

A galvanic cell uses a spontaneous chemical reaction to generate electricity.

A Cell in Action

Figure 14-3 shows a galvanic cell containing two electrodes suspended in an aqueous solution of $CdCl_2$. One electrode is a strip of cadmium metal; the other is a piece of metallic silver coated with solid AgCl. The chemical reactions occurring in this cell are

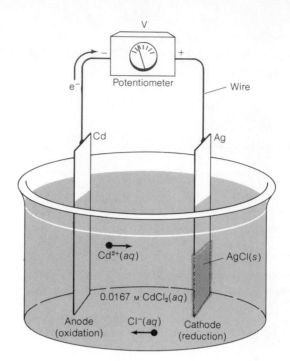

Figure 14-3
A simple galvanic cell. The potentiometer is a device for measuring voltage.

Oxidation: $$Cd(s) \rightleftharpoons Cd^{2+}(aq) + 2e^-$$ (14-9)

Reduction: $$2AgCl(s) + 2e^- \rightleftharpoons 2Ag(s) + 2Cl^-(aq)$$ (14-10)

Net reaction: $$Cd(s) + 2AgCl(s) \rightleftharpoons Cd^{2+}(aq) + 2Ag(s) + 2Cl^-(aq)$$ (14-11)

The net reaction is composed of an oxidation reaction and a reduction reaction, each of which is called a **half-reaction.** The two half-reactions are written with equal numbers of electrons so that their sum includes no free electrons.

Oxidation of Cd metal—to produce $Cd^{2+}(aq)$—provides electrons that flow through the circuit to the Ag electrode in Figure 14-3. At the surface of the Ag electrode, Ag^+ (from AgCl) is reduced to $Ag(s)$. The chloride from AgCl is left in solution. The free energy change for Reaction 14-11 is -150 kJ per mole of Cd. It is the energy liberated by this spontaneous reaction that provides the driving force pushing electrons through the circuit.

> Recall that ΔG is *negative* for a spontaneous reaction.

EXAMPLE: Voltage Produced by a Chemical Reaction

Calculate the voltage that would be measured by the potentiometer in Figure 14-3.

Since $\Delta G = -150$ kJ/mol of Cd, we can use Equation 14-5 (where n is the number of moles of electrons transferred in the balanced net reaction) to write

$$E = -\frac{\Delta G}{nF} = -\frac{-150 \times 10^3 \text{ J}}{(2 \text{ mol})\left(9.649 \times 10^4 \dfrac{C}{mol}\right)}$$

$$= +0.777 \text{ J/C} = +0.777 \text{ V}$$

A spontaneous chemical reaction (negative ΔG) produces a *positive voltage.*

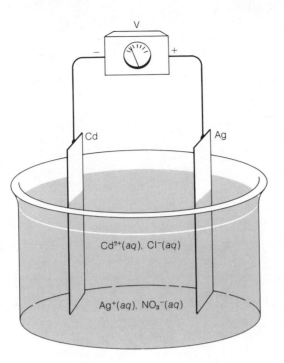

Figure 14-4
A cell that will not work. The solution contains $CdCl_2$ and $AgNO_3$.

Anode and Cathode

Chemists define the electrode at which *oxidation* occurs as the **anode**. The **cathode** is the electrode at which *reduction* occurs. In Figure 14-3, Cd is the anode because it is oxidized ($Cd \rightarrow Cd^{2+} + 2e^-$) and Ag is the cathode because reduction takes place at its surface ($2AgCl + 2e^- \rightarrow 2Ag + 2Cl^-$).

Anode ↔ oxidation

Cathode ↔ reduction

Salt Bridge

Consider the cell in Figure 14-4, in which the reactions are intended to be

Anode:
$$Cd(s) \rightleftharpoons Cd^{2+}(aq) + 2e^- \qquad (14\text{-}12)$$

Cathode:
$$2Ag^+(aq) + 2e^- \rightleftharpoons 2Ag(s) \qquad (14\text{-}13)$$

Net reaction:
$$Cd(s) + 2Ag^+(aq) \rightleftharpoons Cd^{2+}(aq) + 2Ag(s) \quad (14\text{-}14)$$

Reaction 14-14 is spontaneous, but little current will flow through the circuit because Ag^+ ions are not forced to be reduced at the Ag electrode. The Ag^+ ions in solution can react directly at the Cd(s) surface, giving Reaction 14-14 with no flow of electrons through the circuit.

The cell in Figure 14-4 is *short-circuited*.

Okay. Let's try separating the two reactants, as in Figure 14-5. This time electrons will flow through the circuit for an instant as Reaction 14-14 begins. But after an instant, there will be a negative charge in the right half-cell (since electrons flowed into it) and a positive charge in the left half-cell (since electrons flowed out of it). The excess negative charge on the right-hand side will repel electrons trying to gain entry through the circuit. In an instant, the charge repulsion exactly counterbalances the driving force of the chemical reaction and no current can flow.

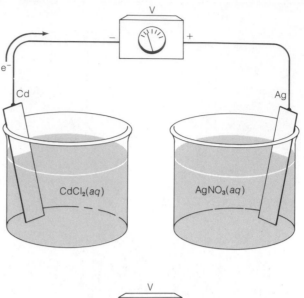

Figure 14-5
Another cell that will not work.

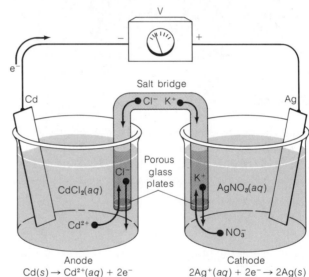

Anode   Cathode
$Cd(s) \rightarrow Cd^{2+}(aq) + 2e^-$ $2Ag^+(aq) + 2e^- \rightarrow 2Ag(s)$

Figure 14-6
At last! A cell that works!

The purpose of a salt bridge is to maintain electroneutrality (no charge buildup) throughout the cell. See Demonstration 14-1.

To fix the cell in Figure 14-5, we can insert a **salt bridge,** as shown in Figure 14-6. The salt bridge consists of a U-shaped tube filled with a gel containing KCl (or any other electrolyte not involved in the cell reaction).[†] The ends of the bridge are covered with porous glass disks that allow ions to diffuse, but that minimize mixing of the solutions inside and outside the bridge. When the galvanic cell is operating, K^+ from the bridge migrates into the cathode compartment and NO_3^- migrates from the cathode into the bridge. The ion migration exactly offsets the charge buildup that would otherwise occur as electrons flow into the silver electrode. In the other half-cell, Cd^{2+} migrates into the bridge while Cl^- migrates into the anode compartment to compensate for the buildup of positive charge that would otherwise occur.

[†] A typical salt bridge is prepared by heating 3 g of agar with 30 g of KCl in 100 ml of water until a clear solution is obtained. The solution is poured into the U-tube and allowed to solidify to a gel. The bridge is stored in a solution of saturated aqueous KCl.

Demonstration 14-1   THE HUMAN SALT BRIDGE

A salt bridge consists of any ionic medium with a semipermeable barrier on each end. You can demonstrate a "proper" salt bridge by filling a U-tube with agar and KCl as described in the footnote on the facing page. A suitable demonstration cell is shown below.

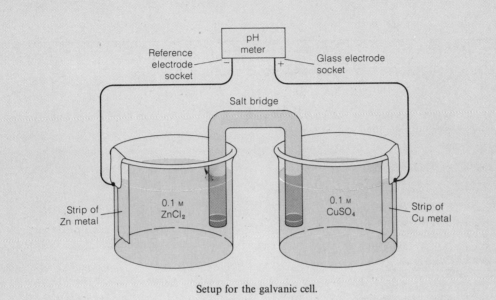

Setup for the galvanic cell.

The pH meter is a potentiometer whose negative terminal is the reference electrode socket.

You should be able to write the two half-reactions for this cell and use the Nernst equation in Section 14-4 to calculate the theoretical voltage. Measure the voltage with a conventional salt bridge. Then replace the salt bridge with one made of filter paper freshly soaked in NaCl solution and measure the voltage again. Finally, replace the filter-paper salt bridge with two fingers of the same hand and measure the voltage again. The human body is really just a bag of salt housed in a semipermeable membrane. The small differences in voltage observed when the salt bridge is replaced can be attributed to the junction potential discussed in Section 15-3.

Line Notation

A shorthand notation is commonly used to describe electrochemical cells.
Only two symbols are needed:

| phase boundary   ‖   salt bridge

The cell in Figure 14-3 is represented by the *line diagram*

$$Cd(s)\,|\,CdCl_2(aq)\,|\,AgCl(s)\,|\,Ag(s)$$

Boundary between separate phases: |

Each phase boundary is indicated by a vertical line. The electrodes are shown at the extreme left- and right-hand sides of the diagram. The cell in Figure 14-6 can be written as follows:

Salt bridge: ‖

$$Cd(s) | CdCl_2(aq) \| AgNO_3(aq) | Ag(s)$$

The contents of the salt bridge need not be specified.

Box 14-1   SOLID IONIC CONDUCTORS: SOLID SALT BRIDGES[†]

An active area of solid-state research is the development and study of solid ionic conductors such as sodium β-alumina, shown below.

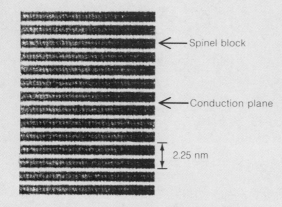

← Spinel block

← Conduction plane

2.25 nm

Electron micrograph of sodium β-alumina, showing rigid spinel blocks (*dark bands*) and conduction planes (*light bands*) in which Na⁺ ions can migrate in two dimensions. [S. Horiuchi, *Am. Ceramic Soc. Bull.*, **64**, 1590 (1985).]

This material has the approximate composition $Na_2O \cdot 11Al_2O_3$ and contains layers of aluminum oxide sandwiching thin layers of sodium oxide. The aluminum oxide layers have a rigid construction similar to the mineral spinel. The thin *conduction planes* contain rigid oxide anions bridging aluminum ions of the spinel blocks. In the conduction planes are also sodium ions and many vacant sites. The nearest Na—O distance is 0.287 nm, which is much greater than the sum of the ionic radii of Na^+ and O^{2-} (0.235 nm). The combination of loose packing and vacant sites allows the sodium ions to migrate from one site to another in this plane with a small energy barrier (16 kJ/mol). Sodium β-alumina is therefore a good *two-dimensional conductor* of Na^+ ions. Conduction in the third dimension (into the spinel blocks) is prohibited.

This solid sodium ion conductor is used as a "salt bridge" in the sodium–sulfur battery shown at the right. At the operating temperature of 300–350°C, both the sodium and sulfur electrodes are molten. Sodium is oxidized at the sodium/sodium β-alumina interface, and electrons go out the top of the battery.

$$Na \rightarrow Na^+ + e^-$$

The sodium ions are conducted into the sulfur compartment through the sodium β-alumina "salt bridge." Electrons from the sodium oxidation reenter the battery at the molten sulfur

[†] D. F. Shriver and G. C. Farrington, *Chem. Eng. News,* May 20, 1985, pp. 42–57; M. D. Ingram and C. A. Vincent, *Chem. Brit.,* March 1984, pp. 235–239; E. C. Subbarao, ed., *Solid Electrolytes and Their Applications* (New York: Plenum Press, 1980).

The voltage measured in the experiment in Figure 14-6 is the difference in electric potential between the Ag electrode on the right and the Cd electrode on the left. That is, the voltage tells us how much work can be done by electrons flowing from one side to the other (Equation 14-3). The potentiometer (voltmeter) used to measure the voltage has "positive" and "negative"

electrode (which contains some carbon to increase its conductivity). The sulfur is reduced to polysulfide ions.

$$xS + 2e^- \rightarrow S_x^{2-}$$

The net reaction converts elemental sodium and sulfur to sodium polysulfide. The entire process is reversed when the battery is recharged.

This battery stores several times as much energy per unit mass as the conventional lead–acid battery. This is because of the low atomic masses of Na and S and because there is so little of any other material in the cell.

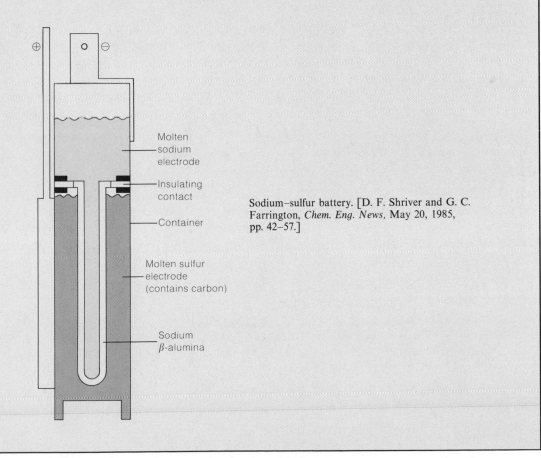

Sodium–sulfur battery. [D. F. Shriver and G. C. Farrington, *Chem. Eng. News*, May 20, 1985, pp. 42–57.]

terminals (connectors) labeled "+" and "−" in our diagrams.[†] The potentiometer indicates a positive voltage when electrons flow into the negative terminal and out of the positive terminal, as in Figure 14-6. If electrons flow the other way, the voltage will be negative. Thus the potentiometer tells us which way electrons flow in a galvanic cell.

To predict the voltage that will be observed when different half-cells are connected to each other, the **standard reduction potential** ($E°$) for each half-cell is measured by an experiment shown in an idealized form in Figure 14-7. The half-reaction of interest in this diagram is

$$\text{Ag}^+ + \text{e}^- \rightleftharpoons \text{Ag}(s) \qquad (14\text{-}15)$$

which occurs in the right half-cell. We use the term "standard" to describe this cell because the activities of all species are unity. For Reaction 14-15 this means that $\mathscr{A}_{\text{Ag}^+} = 1$ and, by definition, the activity of Ag(s) is also unity.

> *Question:* What is the pH of the standard hydrogen electrode?
> *Answer:* Since $\mathscr{A}_{\text{H}^+} = 1$, pH = $-\log \mathscr{A}_{\text{H}^+} = 0$.

The left half-cell is called the **standard hydrogen electrode** (S.H.E.). It consists of a catalytic Pt surface in contact with an acidic solution in which $\mathscr{A}_{\text{H}^+} = 1$. A stream of $\text{H}_2(g)$ is bubbled through the electrode compartment so that the solution is saturated with $\text{H}_2(aq)$. The activity of $\text{H}_2(g)$ is unity if the pressure of $\text{H}_2(g)$ is 1 atm. The reaction that comes to equilibrium at the surface of the Pt electrode is

$$\tfrac{1}{2}\text{H}_2(g, \mathscr{A} = 1) \rightleftharpoons \text{H}^+(aq, \mathscr{A} = 1) + \text{e}^- \qquad (14\text{-}16)$$

By convention, $E° = 0$ for S.H.E.

We *arbitrarily* assign a potential of zero to the standard hydrogen electrode. The voltage meaured by the meter in Figure 14-7 can therefore be *assigned* to Reaction 14-15 that occurs in the right half-cell. The measured value

---

[†] Sometimes the negative terminal of a voltmeter is labeled "common" or "ground." The negative terminal may also be color coded black, while the positive terminal is color coded red. When a pH meter is used as a potentiometer, the positive terminal is the wide receptacle to which the glass pH electrode is connected. The negative terminal is the narrow receptacle to which the reference electrode is connected.

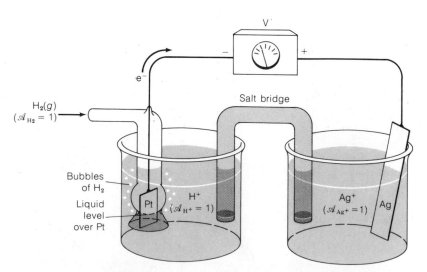

Figure 14-7

Cell used to measure the standard potential of the reaction $\text{Ag}^+ + \text{e}^- \rightleftharpoons \text{Ag}(s)$. This cell is hypothetical because it is usually not possible to adjust the activity of a species to 1.

$\text{Pt}(s)|\text{H}_2(g, \mathscr{A} = 1)|\text{H}^+(aq, \mathscr{A} = 1)\|\text{Ag}^+(aq, \mathscr{A} = 1)|\text{Ag}(s)$

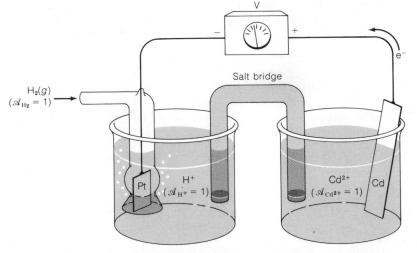

Pt(s)|H₂(g, $\mathscr{A}$ = 1)|H⁺(aq, $\mathscr{A}$ = 1)‖Cd²⁺(aq, $\mathscr{A}$ = 1)|Cd(s)

Figure 14-8
Cell used to measure the standard reduction potential for the reaction $Cd^{2+} + 2e^- \rightleftharpoons Cd(s)$.

$E° = +0.799$ V is the standard reduction potential for Reaction 14-15. The positive sign tells us that electrons flow from left to right through the meter.

Figure 14-8 shows the hypothetical cell that would be used to measure the standard reduction potential for the reaction

$$Cd^{2+} + 2e^- \rightleftharpoons Cd(s) \tag{14-17}$$

In this case the measured voltage is $E° = -0.402$ V. The negative sign means that electrons flow from right to left (not left to right) through the meter in Figure 14-8. Therefore Cd behaves as an anode in this experiment, and the reaction that actually occurs is the reverse of Reaction 14-17.

Table of Standard Potentials

The two galvanic cells we have just constructed allow us to establish a rudimentary table of standard reduction potentials.

	$E°$ (V)
$Ag^+ + e^- \rightleftharpoons Ag(s)$	+0.799
$2H^+ + 2e^- \rightleftharpoons H_2(g)$	0 (by convention)
$Cd^{2+} + 2e^- \rightleftharpoons Cd(s)$	−0.402

The potentials in this table tell us that electrons will flow from the standard hydrogen electrode to the $Ag^+|Ag(s)$ cell under standard conditions. The table also tells us that electrons flow from the standard $Cd^{2+}|Cd(s)$ cell to the S.H.E. It follows that electrons would also flow from the $Cd^{2+}|Cd(s)$ cell to the $Ag^+|Ag(s)$ cell if the two were connected (as in Figure 14-6).

Appendix H contains standard reduction potentials for many different reactions, arranged alphabetically by element. If the table were arranged according to descending value of $E°$ (as in Table 14-1), we would find the strongest oxidizing agents at the upper left and the strongest reducing agents at the lower right.

Remember that "standard conditions" means that all activities are unity.

Table 14-1
Ordered redox potentials

	Oxidizing agent	Reducing agent		$E°$ (V)
	$F_2(g) + 2e^- \rightleftharpoons 2F^-$			2.890
	$O_3(g) + 2H^+ + 2e^- \rightleftharpoons O_2(g) + H_2O$			2.075
	$MnO_4^- + 8H^+ + 5e^- \rightleftharpoons Mn^{2+} + 4H_2O$			1.507
	$Ag^+ + e^- \rightleftharpoons Ag(s)$			0.799
	$Cu^{2+} + 2e^- \rightleftharpoons Cu(s)$			0.339
	$2H^+ + 2e^- \rightleftharpoons H_2(g)$			0.000
	$Cd^{2+} + 2e^- \rightleftharpoons Cd(s)$			−0.402
	$K^+ + e^- \rightleftharpoons K(s)$			−2.936
	$Li^+ + e^- \rightleftharpoons Li(s)$			−3.040

Oxidizing power increases ↑ Reducing power increases ⇓

Question: The potential for the reaction $K^+ + e^- \rightleftharpoons K(s)$ is −2.936 V. This means that K^+ is a very poor oxidizing agent. (It does not readily accept electrons.) Does this imply that K^+ is therefore a good reducing agent?
Answer. No! To be a good reducing agent, K^+ would have to give up electrons easily (forming K^{2+}), which it cannot.

How We Use $E°$ Values

The half-reactions in Appendix H can be combined to produce balanced redox reactions. We simply reverse one half-reaction to turn it into an oxidation, and add the oxidation half-reaction to the reduction half-reaction. *When you reverse a half-reaction, you should change the sign of $E°$.* When you add the resulting reactions, you also add the values of $E°$.

When you reverse the direction of a reaction, reverse the sign of $E°$

EXAMPLE: Combining Half-Reactions

Find the standard potential ($E°$) for the reaction

$$Cu(s) + PbF_2(s) \rightleftharpoons Cu^{2+} + Pb(s) + 2F^- \tag{14-18}$$

We can break the reaction into two half-reactions that appear in Appendix H:

Anode (oxidation): $\quad Cu(s) \rightleftharpoons Cu^{2+} + 2e^- \tag{14-19}$

Cathode (reduction): $\quad PbF_2(s) + 2e^- \rightleftharpoons Pb(s) + 2F^- \tag{14-20}$

The standard potential for the reduction 14-20 is given as −0.350 V in Appendix H. The standard potential for the reduction $Cu^{2+} + 2e^- \rightleftharpoons Cu(s)$ is +0.339 V. When we reverse the copper reaction to write the oxidation 14-9, we also reverse the sign of $E°$ to −0.339 V. The net reaction 14-18 is just the sum of 14-19 and 14-20:

$$Cu(s) \rightleftharpoons Cu^{2+} + 2e^- \qquad E° = -0.339 \text{ V}$$
$$\underline{PbF_2(s) + 2e^- \rightleftharpoons Pb(s) + 2F^- \qquad E° = -0.350 \text{ V}}$$
$$Cu(s) + PbF_2(s) \rightleftharpoons Cu^{2+} + Pb(s) + 2F^- \qquad E° = -0.339 + (-0.350) = -0.689 \text{ V}$$

The negative value of $E°$ means that the reaction would be spontaneous in the reverse direction under standard conditions.

Frequently we must multiply one half-reaction by an integer before adding it to the other half-reaction, so that no electrons appear in the balanced equation. *When you multiply a half-reaction, you do not multiply the value of* $E°$. The reason can be seen in Equation 14-3. The potential difference between two points is the work done *per coulomb of charge* carried through that potential difference ($E = \text{work}/q$). The work per coulomb is the same whether 0.1, 2.3, or 10^4 coulombs have been transferred. The total work is different in each case, but the work per coulomb is constant. Therefore we do not double $E°$ if we multiply a half-reaction by 2.

When you multiply a reaction, *do not multiply $E°$.*

EXAMPLE: Don't Multiply $E°$ by n

Find the standard potential for the reaction

$$Cd(s) + 2Ag^+ \rightleftharpoons Cd^{2+} + 2Ag(s)$$

In Appendix H we find the following data:

$$Ag^+ + e^- \rightleftharpoons Ag(s) \qquad E° = +0.799 \text{ V} \qquad (14\text{-}21)$$

$$Cd^{2+} + 2e^- \rightleftharpoons Cd(s) \qquad E° = -0.402 \text{ V} \qquad (14\text{-}22)$$

After reversing Reaction 14-22 and multiplying Reaction 14-21 by 2, we can add them together:

$$2Ag^+ + 2e^- \rightleftharpoons 2Ag(s) \qquad E° = +0.799 \text{ V}$$

$$\underline{Cd(s) \rightleftharpoons Cd^{2+} + 2e^- \qquad E° = +0.402 \text{ V}}$$

$$Cd(s) + 2Ag^+ \rightleftharpoons Cd^{2+} + 2Ag(s) \qquad E° = 0.799 + 0.402 = +1.201 \text{ V}$$

Now that you know how to find $E°$ for *balanced reactions*, Box 14-2 tells you how to find $E°$ for a *half-reaction* that is the sum of two or more half-reactions in Appendix H.

14-4 NERNST EQUATION

For reactions run under standard conditions (unit activities), the more positive the value of $E°$, the greater is the driving force. We know that changing the concentrations of reactants and products influences the free energy change of the reaction. Le Châtelier's principle tells us that increasing reactant concentrations drives a reaction to the right and increasing the product concentrations drives a reaction to the left. The net driving force for a reaction is expressed by the **Nernst equation**, whose two terms include the driving force under standard conditions ($E°$) and a term that shows the dependence on concentrations.

For the balanced reaction

$$aA + bB \rightleftharpoons cC + dD \qquad (14\text{-}23)$$

the Nernst equation giving the cell potential, E, is

$$E = E° - \frac{RT}{nF} \ln \frac{\mathscr{A}_C^c \mathscr{A}_D^d}{\mathscr{A}_A^a \mathscr{A}_B^b} \qquad (14\text{-}24)$$

A reaction is spontaneous if ΔG is negative and E is positive. $\Delta G°$ and $E°$ refer to the free energy change and potential when the activities of all reactants and products are unity. $\Delta G° = -nFE°$.

Challenge: Show that Le Châtelier's principle requires a negative sign in front of the log term in the Nernst equation.

where E° = standard reduction potential ($\mathscr{A}_A = \mathscr{A}_B = \mathscr{A}_C = \mathscr{A}_D = 1$)

R = gas constant [8.314 510 (V·C)/(K·mol)]

T = temperature (K)

n = number of electrons in each half-reaction

F = Faraday constant ($9.648\ 530\ 9 \times 10^4$ C/mol)

$\mathscr{A}_i$ = activity of species i

The logarithmic term in the Nernst equation is the **reaction quotient**, Q.

$$Q = \frac{\mathscr{A}_C^c \mathscr{A}_D^d}{\mathscr{A}_A^a \mathscr{A}_B^b} \qquad (14\text{-}25)$$

Q has the same form as the equilibrium constant, but the activities need not have their equilibrium values. That is, $Q \neq K$ unless the system happens to be at equilibrium. Recall that pure solids, pure liquids, and solvents are omitted from Q because their activities are unity (or close to unity), while concentrations of solutes are expressed as mol/L and concentrations of gases are expressed as pressure in atm. When all activities are unity, $Q = 1$ and ln $Q = 0$, giving $E = E^\circ$.

Appendix A tells how to convert ln to log.

Converting the natural logarithm in Equation 14-24 to the base 10 logarithm, and inserting $T = 298.15$ K ($25.00°C$) gives the most useful form of the Nernst equation:

$$E = E^\circ - \frac{0.059\ 16\ \text{V}}{n} \log \frac{\mathscr{A}_C^c \mathscr{A}_D^d}{\mathscr{A}_A^a \mathscr{A}_B^b} \qquad \text{(at 25°C)} \qquad (14\text{-}26)$$

We see that the potential changes by $59.16/n$ mV for each factor-of-10 change in the value of Q.

Using the Nernst Equation

Let's apply the Nernst equation to the cell in Figure 14-6, in which the reactions are

anode:	$Cd(s) \rightleftharpoons Cd^{2+} + 2e^-$	$E^\circ = 0.402$ V
cathode:	$2Ag^+ + 2e^- \rightleftharpoons 2Ag(s)$	$E^\circ = 0.799$ V

$$Cd(s) + 2Ag^+ \rightleftharpoons Cd^{2+} + 2Ag(s) \qquad E^\circ = 1.201\ \text{V} \qquad (14\text{-}27)$$

In general, we will write concentrations instead of activities in the Nernst equation, unless there is a specific point to be made with activities. With this approximation, the Nernst equation for Reaction 14-27 is

Remember that pure solids, pure liquids, and solvents are omitted from Q.

$$E = E^\circ - \frac{0.059\ 16}{2} \log \frac{[Cd^{2+}]}{[Ag^+]^2} \qquad (14\text{-}28)$$

Box 14-2   LATIMER DIAGRAMS

A **Latimer diagram** summarizes the standard reduction potential ($E°$) connecting various oxidation states of an element. For example, in acidic solution the following standard reduction potentials are observed:

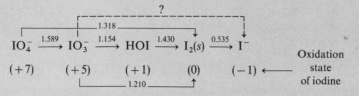

As an example of what each arrow means, let's write the balanced equation represented by the arrow connecting IO_3^- and HOI:

$$IO_3^- \xrightarrow{1.154} HOI$$

means

$$IO_3^- + 5H^+ + 4e^- \rightleftharpoons HOI + 2H_2O \qquad E° = +1.154 \text{ V}$$

It is possible to derive reduction potentials for arrows that are not shown in the diagram. Suppose we wish to determine $E°$ for the reaction shown by the dashed line in the Latimer diagram, which is

$$IO_3^- + 6H^+ + 6e^- \rightleftharpoons I^- + 3H_2O$$

We can use a thermodynamic cycle to find $E°$ for this reaction by expressing the desired reaction as a sum of reactions whose potentials are known.

The standard free energy change, $\Delta G°$, for a reaction is given by

$$\Delta G° = -nFE°$$

When two reactions are added to give a third reaction, the sum of the individual $\Delta G°$ values must equal the overall value of $\Delta G°$.

To apply free energy to the problem above, we write two reactions whose sum is the desired reaction:

$$IO_3^- + 6H^+ + 5e^- \xrightarrow{E_1° = 1.210} \tfrac{1}{2}I_2(s) + 3H_2O \qquad \Delta G_1° = -5F(1.210)$$

$$\tfrac{1}{2}I_2(s) + e^- \xrightarrow{E_2° = 0.535} I^- \qquad \Delta G_2° = -1F(0.535)$$

$$\overline{IO_3^- + 6H^+ + 6e^- \xrightarrow{E_3° = ?} I^- + 3H_2O} \qquad \Delta G_3° = -6FE_3°$$

But since $\Delta G_1° + \Delta G_2° = \Delta G_3°$, we may solve for $E_3°$:

$$\Delta G_3° = \Delta G_1° + \Delta G_2°$$

$$-6FE_3° = -5F(1.210) - 1F(0.535)$$

$$E_3° = \frac{5(1.210) - 1(0.535)}{6} = 1.098 \text{ V}$$

If the concentrations happen to be $[Cd^{2+}] = 0.010$ M and $[Ag^+] = 0.50$ M, the cell voltage is

$$E = 1.201 - \frac{0.059\ 16}{2} \log \frac{[0.010]}{[0.50]^2} = 1.242\ \text{V} \qquad (14\text{-}29)$$

The voltage is increased from its standard value (1.242 versus 1.201) because the product, Cd^{2+}, is at a relatively low concentration compared to the reactant, Ag^+. The more positive the voltage, the more favorable the reaction is.

We used the value $n = 2$ in the denominator of Equation 14-28 because two electrons appear in each half-reaction whose sum is Reaction 14-27. Box 14-3 shows that if we had written the half-reactions with only one electron, the cell voltage would still be 1.242 V. The cell voltage is a measurable physical quantity whose value cannot possibly depend on how we write the chemical reaction.

The cell voltage cannot depend on how we write the reaction!

Two Different Descriptions of the Same Reaction

Now we turn our attention to the cell in Figure 14-3, whose half-reactions can be expressed as follows:

$$Cd(s) \rightleftharpoons Cd^{2+} + 2e^- \qquad E° = 0.402\ \text{V}$$

$$\underline{2AgCl(s) + 2e^- \rightleftharpoons 2Ag(s) + 2Cl^- \qquad E° = 0.222\ \text{V}}$$

$$Cd(s) + 2AgCl(s) \rightleftharpoons Cd^{2+} + 2Ag(s) + 2Cl^- \qquad E° = 0.624\ \text{V} \quad (14\text{-}30)$$

The Nernst equation for Reaction 14-30 looks like this:

$$E = 0.624 - \frac{0.059\ 16}{2} \log [Cl^-]^2[Cd^{2+}] \qquad (14\text{-}31)$$

Since the cell in Figure 14-3 contains 0.016 7 M $CdCl_2$, the Nernst equation becomes

$$E = 0.624 - \frac{0.059\ 16}{2} \log (0.033\ 4)^2(0.016\ 7) = 0.764\ \text{V} \qquad (14\text{-}32)$$

We expect a voltage of 0.764 V.

Suppose that a different author had written this book and had chosen to describe the cell in Figure 14-3 in a different way:

$$Cd(s) \rightleftharpoons Cd^{2+} + 2e^- \qquad E° = 0.402\ \text{V}$$

$$\underline{2Ag^+ + 2e^- \rightleftharpoons 2Ag(s) \qquad E° = 0.799\ \text{V}}$$

$$Cd(s) + 2Ag^+ \rightleftharpoons Cd^{2+} + 2Ag(s) \qquad E° = 1.201\ \text{V} \qquad (14\text{-}33)$$

Equation 14-33 is just as valid as Equation 14-30. One person chose to describe the silver half-reaction as $AgCl(s) + e^- \rightleftharpoons Ag(s) + Cl^-$, and the other chose

Box 14-3   THE CELL VOLTAGE DOES NOT DEPEND ON HOW YOU WRITE THE CELL REACTION

The cell reaction for Figure 14-6 was written with two electrons in Equations 14-12 and 14-13, giving

$$Cd(s) + 2Ag^+ \rightleftharpoons Cd^{2+} + 2Ag(s)$$

$$E = E^\circ - \frac{0.059\,16}{2} \log \frac{[Cd^{2+}]}{[Ag^+]^2} \tag{A}$$

For 0.010 M $CdCl_2$ and 0.50 M $AgNO_3$, we calculate E = 1.242 V in Equation 14-29.

Suppose that the cell reaction were written with just one electron being transferred, instead of two:

$$\tfrac{1}{2}Cd(s) \rightleftharpoons \tfrac{1}{2}Cd^{2+} + e^- \qquad E^\circ = 0.402 \text{ V}$$

$$Ag^+ + e^- \rightleftharpoons Ag(s) \qquad E^\circ = 0.799 \text{ V}$$

$$\overline{\tfrac{1}{2}Cd(s) + Ag^+ \rightleftharpoons \tfrac{1}{2}Cd^{2+} + Ag(s)} \qquad E^\circ = 1.201 \text{ V}$$

For 0.010 M $CdCl_2$ and 0.50 M $AgNO_3$ we calculate

$$E = E^\circ - \frac{0.059\,16}{1} \log \frac{[Cd^{2+}]^{1/2}}{[Ag^+]} \tag{B}$$

$$E = 1.201 - \frac{0.059\,16}{1} \log \frac{\sqrt{(0.010)}}{(0.50)} = 1.242 \text{ V}$$

This is the same voltage calculated with Equation A, which was based on a balanced reaction in which two electrons are transferred.

Why do Equations A and B give the same voltage? The reason is that the factor $1/n$ in front of the log term is related to the exponents of the concentrations within the log term. Making use of the identity

$$\log a^b = b \log a$$

we can write

$$\underbrace{\frac{1}{2} \log \frac{[Cd^{2+}]}{[Ag^+]^2}}_{\substack{\text{The form in} \\ \text{Equation A}}} = \frac{1}{2} \log \frac{[Cd^{2+}]^{(1/2) \cdot 2}}{[Ag^+]^{1 \cdot 2}} = \underbrace{\frac{2}{2} \log \frac{[Cd^{2+}]^{1/2}}{[Ag^+]}}_{\substack{\text{The form in} \\ \text{Equation B}}}$$

The cell voltage does not depend on how you write the cell reaction.

to look at it as $Ag^+ + e^- \rightleftharpoons Ag(s)$. In both equations, Ag(I) is being reduced to Ag(0).

If the two descriptions are equally valid, they should predict the same voltage. The Nernst equation for Reaction 14-33 is

$$E = 1.201 - \frac{0.059\,16}{2} \log \frac{[Cd^{2+}]}{[Ag^+]^2} \tag{14-34}$$

$$K_{sp} = [Ag^+][Cl^-]$$

To find the concentration of Ag^+, we must use the solubility product for AgCl. Since the cell contains 0.033 4 M Cl^- and solid AgCl, we can say

$$[Ag^+] = \frac{K_{sp} \text{ (for AgCl)}}{[Cl^-]} = \frac{1.8 \times 10^{-10}}{0.033\ 4} = 5.4 \times 10^{-9}\ \text{M} \quad (14\text{-}35)$$

Putting this value into Equation 14-34 gives

$$E = 1.201 - \frac{0.059\ 16}{2} \log \frac{0.016\ 7}{(5.4 \times 10^{-9})^2} = 0.764\ \text{V} \quad (14\text{-}36)$$

Lo and behold! Equations 14-36 and 14-32 give the same voltage. They certainly should, because they describe the same cell.

The cell voltage cannot depend on how we write the reaction!

Sign of the Cell Voltage

We write an oxidation for the left half-cell and a reduction for the right half-cell. If the cell voltage is positive, our guess was correct. If the voltage is negative, the reaction proceeds spontaneously in the reverse direction.

For each cell that we have analyzed, we have written an oxidation for the left half-cell and a reduction for the right half-cell. This is an arbitrary convention that we will continue to follow throughout this book. If the calculated cell voltage is positive, then we have guessed correctly that the left-hand electrode is actually the anode. If the voltage is negative, however, it means that the left-hand electrode is in fact the cathode. The sign of the voltage simply tells us which way electrons flow through the circuit.

With this in mind, let's apply the Nernst equation to the cell in Figure 14-8. We will write an oxidation for the left half-cell and a reduction for the right half-cell:

Anode: $\qquad\qquad H_2(g) \rightleftharpoons 2H^+ + 2e^- \qquad E° = 0\ V$

Cathode: $\qquad \dfrac{Cd^{2+} + 2e^- \rightleftharpoons Cd(s) \qquad\quad E° = -0.402\ V}{}$

$\qquad\qquad H_2(g) + Cd^{2+} \rightleftharpoons 2H^+ + Cd(s) \qquad E° = -0.402\ V \quad (14\text{-}37)$

$$E = E° - \frac{0.059\ 16}{2} \log \frac{[H^+]^2}{[Cd^{2+}]P_{H_2}} \quad (14\text{-}38)$$

If $[H^+] = [Cd^{2+}] = 1$ M and $P_{H_2} = 1$ atm, then $E = E° = -0.402$ V. That is, the cell reaction proceeds spontaneously in the *reverse* direction from that in Reaction 14-37.

A positive cell voltage means that the reaction is spontaneous in the forward direction. A negative voltage means that the reaction is spontaneous in the reverse direction. A voltage of zero means that the cell is at equilibrium.

$E = (+) \Rightarrow$ reaction goes right

$E = (-) \Rightarrow$ reaction goes left

$E = 0 \Rightarrow$ reaction at equilibrium

By Le Châtelier's principle, we should be able to force Reaction 14-37 to go forward by decreasing the concentration of the product, H^+. At what pH will the cell reaction be in equilibrium if we maintain $[Cd^{2+}] = 1$ M and $P_{H_2} = 1$ atm? Setting $E = 0$ in Equation 14-38 and solving for $[H^+]$, we find

$$0 = -0.402 - \frac{0.059\ 16}{2} \log \frac{[H^+]^2}{(1)(1)} \quad (14\text{-}39)$$

$$[H^+] = 1.6 \times 10^{-7}\ \text{M} \quad \Rightarrow \quad pH = 6.80 \quad (14\text{-}40)$$

At a pH of 6.80 (*in the left half-cell*), the reaction will be in equilibrium. If the pH is higher than 6.80, the concentration of $[H^+]$ will be even lower than the equilibrium value, and Reaction 14-37 will proceed spontaneously to the right.

Question: Why does the pH apply to the left half-cell?

Some Advice

When you are faced with a cell drawing or a line diagram, the first step toward understanding the cell is to write an anode reaction for the left half-cell and a cathode reaction for the right half-cell. To write these reactions, *look for an element in the cell in two oxidation states.* For the cell

How to figure out the cell reaction.

$$Pb(s)\,|\,PbF_2(s)\,|\,F^-(aq)\,\|\,Cu^{2+}(aq)\,|\,Cu(s)$$

we see Pb in two oxidation states, as $Pb(s)$ and $PbF_2(s)$, and Cu in two oxidation states, as Cu^{2+} and $Cu(s)$. Thus, the half-reactions are

$$\text{Anode:}\qquad Pb(s) + 2F^- \rightleftharpoons PbF_2(s) + 2e^- \qquad (14\text{-}41)$$

$$\text{Cathode:}\qquad Cu^{2+} + 2e^- \rightleftharpoons Cu(s) \qquad (14\text{-}42)$$

If, perhaps, the phase of the moon were different, you might choose to write the Pb half-reaction as

$$\text{Anode:}\qquad Pb(s) \rightleftharpoons Pb^{2+} + 2e^- \qquad (14\text{-}43)$$

because you know that if $PbF_2(s)$ is present, there *must* be some Pb^{2+} in the solution. As we saw previously in the $AgCl\,|\,Ag$ example, Reactions 14-41 and 14-43 are equally valid descriptions of the cell, and each should predict the same cell voltage. The decision to use Reaction 14-41 or 14-43 depends on whether the F^- or Pb^{2+} concentration is more easily available to you.

We described the left half-cell in terms of a redox reaction involving Pb because Pb is the element that appears in two oxidation states. We would *not* write a reaction such as

Don't invent species not shown in the cell. Use what is shown in the line diagram to select the half-reactions.

$$2F^- \rightleftharpoons F_2(g) + 2e^- \qquad (14\text{-}44)$$

because $F_2(g)$ is not shown in the line diagram of the cell.

The Nernst Equation Is Used in Measuring Standard Reduction Potentials

The standard reduction potential is defined as the potential that would be observed if the half-cell of interest (with unit activities) were connected to a standard hydrogen electrode, as in Figures 14-7 and 14-8. However, it is not easy to construct these cells because we have no way to adjust concentrations and ionic strength to give unit activities. Remember that a 1 M solution does not have unit activity. In reality, activities less than unity are used in each half-cell and the Nernst equation is used to extract the value of $E°$ from the

Problem 14-23 gives an example of the use of the Nernst equation to find $E°$.

cell voltage measured under nonstandard conditions.[†] In the hydrogen electrode, standard buffers with known pH (Table 15-3) are used to obtain known values of $\mathscr{A}_{H^+}$.

14-5 RELATION OF $E°$ AND THE EQUILIBRIUM CONSTANT

A galvanic cell produces electricity because the cell reaction is not at equilibrium. When we use a potentiometer to measure the cell voltage, we allow negligible current flow so that the concentrations in each half-cell remain unchanged. If we replaced the potentiometer with a wire, much more current would flow, the concentrations of reactants would decrease, and the concentrations of products would increase. This process would continue until the cell reached equilibrium. At that point, there would be no more force driving the reaction, and E would be zero.

At equilibrium, E (not $E°$) = 0.

When $E = 0$, a cell is at equilibrium. Therefore, the reaction quotient, Q, is equal to the equilibrium constant, K, when $E = 0$. This allows us to derive a most important relation between K and $E°$ for a chemical reaction:

$$E = E° - \frac{0.059\ 16}{n} \log Q \qquad \text{(at any time)} \qquad (14\text{-}45)$$

$$0 = E° - \frac{0.059\ 16}{n} \log K \qquad \text{(at equilibrium)} \qquad (14\text{-}46)$$

Rearranging Equation 14-46 gives

To go from Equation 14-47 to 14-48:

$$\frac{0.059\ 16}{n} \log K = E°$$

$$\log K = \frac{n}{0.059\ 16} E°$$

$$10^{\log K} = 10^{nE°/0.059\ 16}$$

$$K = 10^{nE°/0.059\ 16}$$

$$\boxed{\frac{0.059\ 16}{n} \log K = E°} \qquad \text{at } 25°\text{C} \qquad (14\text{-}47)$$

or

$$\boxed{K = 10^{nE°/0.059\ 16}} \qquad \text{at } 25°\text{C} \qquad (14\text{-}48)$$

Equation 14-48 allows us to deduce the equilibrium constant for any reaction for which $E°$ is known. Alternatively, knowing the equilibrium constant for a reaction lets us find $E°$, using Equation 14-47.

The correct form of Equation 14-48 at any temperature is

The right-hand column of Appendix H allows us to calculate $E°$ at temperatures other than 25°C. See the footnote to Appendix H on page AP34 for instructions on this calculation.

$$K = 10^{nFE°/RT \ln 10} \qquad (14\text{-}49)$$

If you want to determine K at some particular temperature, you must know $E°$ at that temperature. The value of $E°$ is temperature-dependent.

[†] A student experiment using the Nernst equation to measure $E°$ has been described by A. Arévalo and G. Pastor, *J. Chem. Ed.,* **62,** 882 (1985).

For the reaction

$$Cu(s) + 2Fe^{3+} \rightleftharpoons 2Fe^{2+} + Cu^{2+} \qquad E° = 0.432 \text{ V} \qquad (14\text{-}50)$$

we can use Equation 14-48 to evaluate K as follows:

$$K = 10^{(2)(0.432)/(0.059\ 16)} = 4 \times 10^{14} \qquad (14\text{-}51)$$

Significant figures for logs and exponents were discussed on page 38.

Note that a very large value of $E°$ is not required to produce a very large equilibrium constant. The value of K in Equation 14-51 is correctly expressed with one significant figure. The value of $E°$ has three figures. Two are used for the exponent (14), and only one is left for the multiplier (4).

By judicious choice of half-reactions, we can evaluate equilibrium constants for reactions that need not be redox reactions. For example, the sum of the equations for the two half-reactions below gives the solubility equation for ferrous carbonate:

$$FeCO_3(s) + 2e^- \rightleftharpoons Fe(s) + CO_3^{2-} \qquad E° = -0.756 \text{ V}$$

$$\underline{Fe(s) \rightleftharpoons Fe^{2+} + 2e^- \qquad\qquad\qquad E° = +0.44 \text{ V}}$$

$$FeCO_3(s) \rightleftharpoons Fe^{2+} + CO_3^{2-} \quad (K = K_{sp}) \qquad E° = -0.31_6 \text{ V}$$

Ferrous
carbonate

$$K_{sp} = 10^{(2)(-0.31_6)/(0.059\ 16)} = 2 \times 10^{-11} \qquad (14\text{-}52)$$

The net reaction *need not be a redox reaction.* We can still use $E°$ to find K.

Potentiometric measurements provide one of the most useful means of measuring equilibrium constants that are too small or too large to measure by determining concentrations of reactants and products directly.

The general form of a problem involving the relation between $E°$ values for half-reactions and K for a net reaction is

Cathode:	$E_1°$
Anode:	$E_2°$
Net reaction:	$E_3°, K$

If you know $E_1°$ and $E_2°$, you can find $E_3° (= E_1° + E_2°)$ and $K (= 10^{nE_3°/0.059\ 16})$. Alternatively, if you know $E_3°$ and either $E_1°$ or $E_2°$, you can find the missing $E°$. If you know K, you can calculate $E_3°$ and use it to find either $E_1°$ or $E_2°$, provided you know one of them.

Any two pieces of information allow us to calculate the third piece.

EXAMPLE: Relating $E°$ and K

Suppose we have the following information:

$$Ni^{2+} + 2 \text{ glycine} \rightleftharpoons Ni(glycine)_2^{2+} \qquad K = \beta_2 = 1.2 \times 10^{11}$$

$$Ni^{2+} + 2e^- \rightleftharpoons Ni(s) \qquad\qquad\qquad E° = -0.236 \text{ V}$$

From the value of the overall formation constant of $Ni(glycine)_2^{2+}$ plus the value of $E°$ for the $Ni^{2+}|Ni(s)$ couple, deduce the value of $E°$ for the reaction

$$Ni(glycine)_2^{2+} + 2e^- \rightleftharpoons Ni(s) + 2 \text{ glycine} \tag{14-53}$$

To accomplish this task, we need to see the relation among the three reactions:

$$Ni^{2+} + 2e^- \rightleftharpoons Ni(s) \qquad\qquad E_1° = -0.236 \text{ V}$$

$$\underline{Ni(s) + 2 \text{ glycine} \rightleftharpoons Ni(glycine)_2^{2+} + 2e^- \qquad E_2° = ?}$$

$$Ni^{2+} + 2 \text{ glycine} \rightleftharpoons Ni(glycine)_2^{2+} \qquad E_3° = ? \qquad K = 1.2 \times 10^{11}$$

We know that $E_1° + E_2°$ must equal $E_3°$, so we can deduce the value of $E_2°$ if we can find $E_3°$. But $E_3°$ can be determined from the equilibrium constant for the net reaction.

$$K = 10^{nE_3°/0.059\,16} \quad \Rightarrow \quad E_3° = \frac{0.059\,16}{n} \log K$$

$$E_3° = \frac{0.059\,16}{2} \log(1.2 \times 10^{11}) = 0.328 \text{ V}$$

Hence,

$$E_2° = E_3° - E_1° = 0.564 \text{ V}$$

Since $E_2°$ applies to an oxidation, the standard reduction potential for Reaction 14-53 is -0.564 V.

14-6 USING CELLS AS CHEMICAL PROBES

It is essential to distinguish two classes of equilibria associated with galvanic cells:

1. equilibrium *between* the two half-cells
2. equilibrium *within* each half-cell

If a galvanic cell is producing a nonzero voltage (either positive or negative), then the net cell reaction cannot be at equilibrium. We say that equilibrium *between* the two half-cells has not been established.

> A chemical reaction that can occur *within one half-cell* will reach equilibrium and is assumed to remain at equilibrium. Such a reaction is not the net cell reaction.

The electrodes of cells used as chemical probes are connected by a potentiometer that allows negligible flow of current (Box 14-4). When we use cells as chemical probes, *we allow the half-cells to stand long enough to come to chemical equilibrium **within** each half-cell*. For example, in the right-hand half-cell in Figure 14-9, the equilibrium

$$AgCl(s) \rightleftharpoons Ag^+(aq) + Cl^-(aq) \tag{14-54}$$

exists with or without the presence of another half-cell. Reaction 14-54 is not part of the net cell reaction. It is simply a chemical reaction whose equilibrium will be established when $AgCl(s)$ is in contact with an aqueous solution. In the left half-cell, the reaction

$$CH_3CO_2H \rightleftharpoons CH_3CO_2^- + H^+ \tag{14-55}$$

has also come to equilibrium. Neither of these is a redox reaction involved in the net cell reaction.

Box 14-4 CONCENTRATIONS IN THE OPERATING CELL

Doesn't operation of a cell change the concentrations in the cell? Yes, but cell voltage is measured under conditions of *negligible current flow*. For example, the resistance of a high-quality pH meter is 10^{13} Ω. If you use this meter to measure a potential of 1 V, the current is

$$I = \frac{E}{R} = \frac{1}{10^{13}} = 10^{-13} \text{ A}$$

If the cell in Figure 14-6 produces 50 mV, the current through the circuit is 0.050 V/10^{13} $\Omega = 5 \times 10^{-15}$ A. This corresponds to a flow of

$$\frac{5 \times 10^{-15} \text{ C/s}}{9.648 \times 10^{4} \text{ C/mol}} = 5 \times 10^{-20} \text{ mol e}^{-}/\text{s}$$

The rate at which Cd^{2+} is produced is only 2.5×10^{-20} mol/s. Clearly, this will not have any effect on the cadmium concentration in the cell. *The purpose of the potentiometer is to measure the voltage of the cell without affecting the concentrations in the cell.*

If the salt bridge were left in a real cell for very long, the concentrations and ionic strength would change because of diffusion between each compartment and the salt bridge. We assume that the cells are set up for a short enough time that this does not happen.

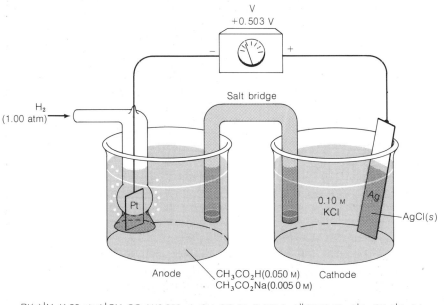

Figure 14-9

This galvanic cell can be used to measure the pH of the left half-cell.

The half-reaction for the right-hand cell of Figure 14-9 is

$$AgCl(s) + e^- \rightleftharpoons Ag(s) + Cl^-(aq) \qquad E° = 0.222 \text{ V} \qquad (14\text{-}56)$$

But what is the half-reaction in the left-hand cell? The only element we find in two oxidation states is hydrogen. We see that $H_2(g)$ is being bubbled into the cell, and we also realize that every aqueous solution contains H^+. Therefore, hydrogen is present in two oxidation states, and the half-reaction can be written as

$$H_2(g) \rightleftharpoons 2H^+(aq) + 2e^- \qquad E° = 0 \qquad (14\text{-}57)$$

The net cell reaction must be

$$2AgCl(s) + H_2(g, 1.00 \text{ atm}) \rightleftharpoons 2H^+(x \text{ M}) + 2Cl^-(0.10 \text{ M}) + 2Ag(s) \qquad (14\text{-}58)$$

The reaction cannot be at equilibrium, because the measured voltage is 0.503 V.

When we write the Nernst equation for the net cell reaction

$$E = 0.222 - \frac{0.059\,16}{2} \log \frac{[H^+]^2[Cl^-]^2}{P_{H_2}} \qquad (14\text{-}59)$$

we discover that all quantities except $[H^+]$ are known ($[Cl^-] = 0.10$ M and $P_{H_2} = 1.00$ atm). *The measured voltage therefore allows us to find the concentration of H^+ in the left half-cell.*

$$0.503 = 0.222 - \frac{0.059\,16}{2} \log \frac{[H^+]^2[0.10]^2}{(1.00)}$$

$$\Rightarrow \quad [H^+] = 1.76 \times 10^{-4} \text{ M} \qquad (14\text{-}60)$$

This, in turn, allows us to evaluate the equilibrium constant for the acid–base reaction that has come to equilibrium in the left half-cell:

<table>
<tr><td>

Question: Why can we assume that the concentrations of acetic acid and acetate ion are equal to their initial (formal) concentrations?

</td></tr>
</table>

$$K_a = \frac{[CH_3CO_2^-][H^+]}{[CH_3CO_2H]} = \frac{(0.005\,0)(1.8 \times 10^{-4})}{0.050} = 1.8 \times 10^{-5} \qquad (14\text{-}61)$$

The voltage of a galvanic cell may serve as a probe, providing information on the concentration of an unknown within the cell. This particular cell behaves as a pH meter, a probe for H^+.

The cell in Figure 14-9 may be thought of as a *probe* to measure the unknown H^+ concentration in the left half-cell. Using this type of cell, we could determine the equilibrium constant for dissociation of any acid or the hydrolysis of any base placed in the left half-cell. The use of electrochemical cells as probes will be explored further in Chapter 15.

The problems at the end of this chapter include several brainbusters designed to bring together your knowledge of electrochemistry, chemical equilibrium, solubility, complex formation, and acid–base chemistry. They require you to find the equilibrium constant for a reaction that occurs in only one half-cell. The reaction of interest is *not* the net cell reaction and is not even a redox reaction. A good approach to such problems is outlined below:

The half-reactions *must* involve species that appear in two oxidation states in the cell.

STEP 1. Write the anode and cathode half-reactions and their standard potentials. If you choose a half-reaction for which you cannot find $E°$, find another way to write the reaction.

STEP 2. Write a Nernst equation for the net reaction, and put in all the known quantities. If all is well, there will be only one unknown in the equation.

STEP 3. Solve for the unknown concentration, and use that concentration to solve the chemical equilibrium problem that was originally posed.

EXAMPLE: Analyzing a Very Complicated Cell

The cell in Figure 14-10 can be used to measure the formation constant (K_f) of $Hg(EDTA)^{2-}$. The solution in the cathode compartment contains 0.500 mmol of Hg^{2+} and 2.00 mmol of EDTA in a volume of 0.100 L buffered to pH 6.00. If the measured voltage is $+0.331$ V, find the value of K_f for $Hg(EDTA)^{2-}$.

STEP 1. The left half-cell is a standard hydrogen electrode for which we can say

$$\text{Anode:} \qquad H_2(g,\ 1.00\ \text{atm}) \rightleftharpoons 2H^+(1.00\ \text{M}) + 2e^- \qquad E° = 0$$

Mercury is the element in two oxidation states in the right half-cell, so let us write the cathode reaction as

$$\text{Cathode:} \qquad Hg^{2+} + 2e^- \rightleftharpoons Hg(l) \qquad E° = 0.852\ \text{V}$$

In the right half-cell, the reaction between Hg^{2+} and EDTA is

$$Hg^{2+} + Y^{4-} \xrightleftharpoons{K_f} HgY^{2-}$$

Since we expect K_f to be large, we will assume that virtually all the Hg^{2+} has reacted to make HgY^{2-}. Therefore, the concentration of HgY^{2-} is 0.500 mmol/100 mL = 0.005 00 M. The remaining EDTA has a total concentration of (2.00 − 0.50) mmol/100 mL = 0.015 0 M. The cathode compartment therefore contains 0.005 00 M HgY^{2-}, 0.015 0 M EDTA, and a very small, unknown concentration of Hg^{2+}.

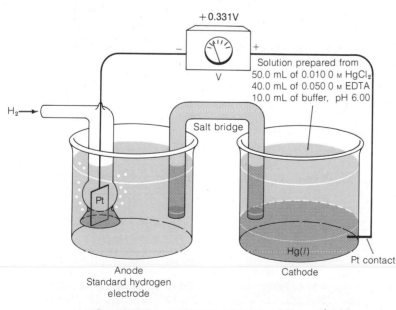

Anode
Standard hydrogen
electrode

Cathode

S.H.E.‖$Hg(EDTA)^{2-}(aq,\ 0.005\ 00\ \text{M})$, $EDTA(aq,\ 0.015\ 0\ \text{M})$|$Hg(l)$

Figure 14-10
A galvanic cell that can be used to measure the formation constant for $Hg(EDTA)^{2-}$.

Recall that $[Y^{4-}] = \alpha_{Y^{4-}}[EDTA]$.

The formation constant for HgY^{2-} can be written

$$K_f = \frac{[HgY^{2-}]}{[Hg^{2+}][Y^{4-}]} = \frac{[HgY^{2-}]}{[Hg^{2+}]\alpha_{Y^{4-}}[EDTA]}$$

where $[EDTA]$ is the formal concentration of EDTA not bound to metal. In this cell, $[EDTA] = 0.015\ 0$ M. The fraction of EDTA in the form Y^{4-} is $\alpha_{Y^{4-}}$ discussed in Section 13-2. Since we know that $[HgY^{2-}] = 0.005\ 00$ M, all we need to find is $[Hg^{2+}]$ in order to evaluate K_f.

STEP 2. The net cell reaction is

$$Hg^{2+} + H_2(g) \rightleftharpoons Hg(l) + 2H^+ \qquad E^\circ = 0.852$$

and the Nernst equation is

The values of $[H^+]$ and P_{H_2} apply to the left half-cell, where H^+ and H_2 participate in the net cell redox reaction.

$$E = E^\circ - \frac{0.059\ 16}{2} \log \frac{[H^+]^2}{[Hg^{2+}]P_{H_2}}$$

STEP 3. Putting in all the known values allows us to solve for $[Hg^{2+}]$:

$$0.331 = 0.852 - \frac{0.059\ 16}{2} \log \frac{(1.00)^2}{[Hg^{2+}](1.00)}$$

$$[Hg^{2+}] = 2.4 \times 10^{-18}\ \text{M}$$

This value of $[Hg^{2+}]$ can be used to evaluate the formation constant for HgY^{2-}:

The value of $\alpha_{Y^{4-}}$ comes from Table 13-1.

$$K_f = \frac{[HgY^{2-}]}{[Hg^{2+}]\alpha_{Y^{4-}}[EDTA]} = \frac{(0.005\ 00)}{(2.4 \times 10^{-18})(2.3 \times 10^{-5})(0.015\ 0)}$$

$$= 6 \times 10^{21}$$

The mixture of EDTA plus $Hg(EDTA)^{2-}$ in the cathode serves as a mercuric ion "buffer" that fixes the concentration of Hg^{2+}. This, in turn, determines the cell voltage.

14-7 NERNST EQUATION FOR HALF-REACTIONS

In the next chapter it will be very useful to write Nernst equations for half-reactions, just as we have been writing such equations for complete reactions in this chapter. The form of the Nernst equation is the same in both cases. For the reaction

$$aA + ne^- \rightleftharpoons bB \qquad E^\circ \tag{14-62}$$

the Nernst equation is

$$E = E^\circ - \frac{0.059\ 16}{n} \log \frac{\mathscr{A}_B^b}{\mathscr{A}_A^a} \tag{14-63}$$

EXAMPLE: Nernst Equation for a Half-Reaction

Write the Nernst equation for the reaction

$$FeO_4^{2-} + 3H_2O + 3e^- \rightleftharpoons FeOOH(s) + 5OH^- \qquad E^\circ = 0.80 \text{ V}$$

Using concentrations instead of activities, we write

$$E = 0.80 \text{ V} - \frac{0.059\ 16}{3} \log \frac{[OH^-]^5}{[FeO_4^{2-}]}$$

14-8 BIOCHEMISTS USE $E^{\circ\prime}$

Many of the reactions in living organisms are redox reactions. Perhaps the most important redox reactions are involved in the respiratory process, in which molecules of food are oxidized by O_2 to yield energy or metabolic intermediates. The standard reduction potentials that we have been using so far apply to systems in which all activities of reactants and products are unity. If H^+ is involved in the reaction, E° applies when pH = 0 ($\mathscr{A}_{H^+} = 1$). *Whenever H^+ appears in a redox reaction, or whenever reactants or products are acids or bases, reduction potentials are pH-dependent.*

Since the pH inside a cell is in the neighborhood of 7, the reduction potentials that apply at pH 0 are not particularly appropriate and may be misleading. For example, at pH 0, ascorbic acid (vitamin C) is a more powerful reducing agent than is succinic acid. However, at pH 7, this order is reversed. It is the reducing strength at pH 7, not at pH 0, that is relevant to the chemistry of a living cell.

The *standard potential* for a redox reaction is defined for a galvanic cell in which all activities are unity. The **formal potential** is the reduction potential that applies under a *specified* set of conditions (including pH, ionic strength, concentration of complexing agents, etc.). Biochemists call the formal potential at pH 7 $E^{\circ\prime}$ (read "E zero prime"). Table 14-2 lists $E^{\circ\prime}$ values for a few biologically important redox couples.

The formal potential at pH = 7 is called $E^{\circ\prime}$.

Relation Between E° and $E^{\circ\prime}$

Consider the half-reaction

$$aA + ne^- \rightleftharpoons bB + mH^+ \qquad E^\circ \qquad \text{(14-64)}$$

in which A is an oxidized species and B is a reduced species. Both A and B might be acids or bases, as well. The Nernst equation for the reduction is written

$$E = E^\circ - \frac{0.059\ 16}{n} \log \frac{[B]^b[H^+]^m}{[A]^a} \qquad \text{(14-65)}$$

Table 14-2
Reduction potentials of biological interest

Reaction	$E°$ (V)	$E°'$ (V)
$O_2 + 4H^+ + 4e^- \rightleftharpoons 2H_2O$	+1.229	+0.816
$Fe^{3+} + e^- \rightleftharpoons Fe^{2+}$	+0.771	+0.771
$I_2 + 2e^- \rightleftharpoons 2I^-$	+0.535	+0.535
Cytochrome a (Fe^{3+}) $\rightleftharpoons$ cytochrome a (Fe^{2+})	+0.290	+0.290
$O_2(g) + 2H^+ + 2e^- \rightleftharpoons H_2O_2$	+0.695	+0.281
Cytochrome c (Fe^{3+}) $\rightleftharpoons$ cytochrome c (Fe^{2+})	—	+0.254
2,6-Dichlorophenolindophenol + $2H^+ + 2e^- \rightleftharpoons$ reduced 2,6-dichlorophenolindophenol	—	+0.22
Dehydroascorbate + $2H^+ + 2e^- \rightleftharpoons$ ascorbate + H_2O	+0.390	+0.058
Fumarate + $2H^+ + 2e^- \rightleftharpoons$ succinate	+0.433	+0.031
Methylene blue + $2H^+ + 2e^- \rightleftharpoons$ reduced product	+0.532	+0.011
Glyoxylate + $2H^+ + 2e^- \rightleftharpoons$ glycolate	—	−0.090
Oxalacetate + $2H^+ + 2e^- \rightleftharpoons$ malate	+0.330	−0.102
Pyruvate + $2H^+ + 2e^- \rightleftharpoons$ lactate	+0.224	−0.190
Riboflavin + $2H^+ + 2e^- \rightleftharpoons$ reduced riboflavin	—	−0.208
FAD + $2H^+ + 2e^- \rightleftharpoons FADH_2$	—	−0.219
(Glutathione-S)$_2$ + $2H^+ + 2e^- \rightleftharpoons$ 2 glutathione-SH	—	−0.23
Safranine T + $2e^- \rightleftharpoons$ leucosafranine T	−0.235	−0.289
$(C_6H_5S)_2 + 2H^+ + 2e^- \rightleftharpoons 2C_6H_5SH$	—	−0.30
$NAD^+ + H^+ + 2e^- \rightleftharpoons NADH$	−0.105	−0.320
$NADP^+ + H^+ + 2e^- \rightleftharpoons NADPH$	—	−0.324
Cystine + $2H^+ + 2e^- \rightleftharpoons$ 2 cysteine	—	−0.340
Acetoacetate + $2H^+ + 2e^- \rightleftharpoons$ L-β-hydroxybutyrate	—	−0.346
Xanthine + $2H^+ + 2e^- \rightleftharpoons$ hypoxanthine + H_2O	—	−0.371
$2H^+ + 2e^- \rightleftharpoons H_2$	0.000	−0.414
Gluconate + $2H^+ + 2e^- \rightleftharpoons$ glucose + H_2O	—	−0.44
$SO_4^{2-} + 2e^- + 2H^+ \rightleftharpoons SO_3^{2-} + H_2O$	—	−0.454
$2SO_3^{2-} + 2e^- + 4H^+ \rightleftharpoons S_2O_4^{2-} + 2H_2O$	—	−0.527

Note: A more complete table can be found in the *Handbook of Biochemistry and Molecular Biology*, 3d ed. (Cleveland: CRC Press, 1976), *Physical and Chemical Data*, Vol. 1, pp. 122–129.

To find $E°'$, we must rearrange the Nernst equation to a form in which the log term contains only the *formal concentrations* of A and B raised to the powers a and b, respectively.

The recipe for finding $E°'$.

$$E = \underbrace{E° + \text{other terms}}_{\substack{\text{All of this is} \\ \text{called } E°' \\ \text{when pH} = 7}} - \frac{0.059\,16}{n} \log \frac{F_B^b}{F_A^a} \qquad (14\text{-}66)$$

The entire collection of terms over the brace, evaluated at pH = 7, is called $E°'$.

To convert [A] or [B] to F_A or F_B, we make use of equations such as 12-6, 12-7, 12-19, 12-20, or 12-21, which relate the formal (i.e., total) concentration of *all* forms of an acid or a base to its concentration in a *particular* form. These useful equations are repeated below:

For a monoprotic acid:

$$F = [HA] + [A^-]$$

For a diprotic acid:

$$F = [H_2A] + [HA^-] + [A^{2-}]$$

Monoproptic system:

$$[HA] = \alpha_0 F = \frac{[H^+]F}{[H^+] + K_a} \quad (14\text{-}67)$$

$$[A^-] = \alpha_1 F = \frac{K_a F}{[H^+] + K_a} \quad (14\text{-}68)$$

Diprotic system:

$$[H_2A] = \alpha_0 F = \frac{[H^+]^2 F}{[H^+]^2 + [H^+]K_1 + K_1 K_2} \quad (14\text{-}69)$$

$$[HA^-] = \alpha_1 F = \frac{K_1 [H^+]F}{[H^+]^2 + [H^+]K_1 + K_1 K_2} \quad (14\text{-}70)$$

$$[A^{2-}] = \alpha_2 F = \frac{K_1 K_2 F}{[H^+]^2 + [H^+]K_1 + K_1 K_2} \quad (14\text{-}71)$$

where K_a is the acid dissociation constant for HA and K_1 and K_2 are the acid dissociation constants for H_2A.

EXAMPLE: Finding the Formal Potential

Find $E^{\circ\prime}$ for the reaction[†]

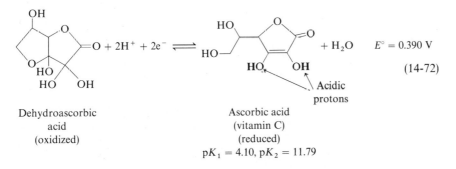

Dehydroascorbic acid (oxidized)

Ascorbic acid (vitamin C) (reduced)

$pK_1 = 4.10, pK_2 = 11.79$

$E^\circ = 0.390$ V

$(14\text{-}72)$

Acidic protons

Abbreviating dehydroascorbic acid as D, and ascorbic acid as H_2A, we can rewrite the reduction as

$$D + 2H^+ + 2e^- \rightleftharpoons H_2A + H_2O$$

for which the Nernst equation is

$$E = E^\circ - \frac{0.059\,16}{2} \log \frac{[H_2A]}{[D][H^+]^2} \quad (14\text{-}73)$$

D is not an acid or a base, so $F_D = [D]$. For the diprotic acid H_2A, we can use Equation 14-69 to express $[H_2A]$ in terms of F_{H_2A}:

$$[H_2A] = \frac{[H^+]^2 F_{H_2A}}{[H^+]^2 + [H^+]K_1 + K_1 K_2}$$

[†] The chemistry of ascorbic acid is discussed by D. T. Sawyer, G. Chiericato, Jr., and T. Tsuchiya, *J. Amer. Chem. Soc.*, **104**, 6273 (1982).

Putting these values into Equation 14-73 gives

$$E = E° - \frac{0.059\ 16}{2} \log \left(\frac{\dfrac{[H^+]^2 F_{H_2A}}{[H^+]^2 + [H^+]K_1 + K_1K_2}}{F_D[H^+]^2} \right)$$

which can be rearranged to the form

$$E = \underbrace{E° - \frac{0.059\ 16}{2} \log \frac{1}{[H^+]^2 + [H^+]K_1 + K_1K_2}}_{\text{Formal potential } (= E°' \text{ if pH} = 7)} - \frac{0.059\ 16}{2} \log \frac{F_{H_2A}}{F_D} \tag{14-74}$$

Putting the values of $E°$, K_1, and K_2 into Equation 14-74 and setting $[H^+] = 10^{-7.00}$, we find $E°' = +0.062$ V.

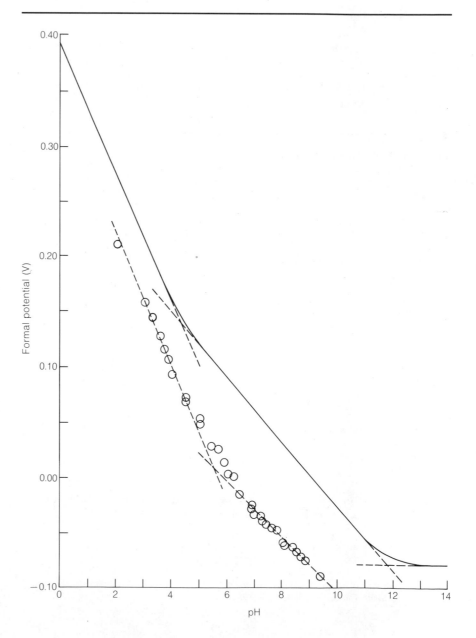

Figure 14-11
Formal reduction potential of ascorbic acid, showing its dependence on pH. *Solid line:* Graph of the function labeled formal potential in Equation 14-74. *Circles:* Experimental polarographic half-wave reduction potential of ascorbic acid in a medium of ionic strength = 0.2 M. The half-wave potential, discussed in Section 18-2, is nearly the same as the formal potential. At high pH (>12) the half-wave potential does not level off to a slope of zero, as Equation 14-74 predicts. Instead, a hydrolysis reaction of ascorbic acid occurs and the chemistry is more complex than Reaction 14-72. [J. J. Ruiz, A. Aldaz, and M. Dominguez, *Can. J. Chem.,* **55,** 2799 (1977); *ibid.,* **56,** 1533 (1978).]

The solid line in Figure 14-11 shows how the calculated formal potential for Reaction 14-72 depends on pH. The potential decreases as the pH increases, until $pH \approx pK_2$. Above pK_2, A^{2-} is the dominant form of ascorbic acid, and no protons are involved in the net redox reaction. Therefore, the potential becomes independent of pH.

Summary

The work done when a charge of q coulombs passes through a potential difference of E volts is $W = E \cdot q$. The maximum work that can be done by a spontaneous chemical change is related to the free energy change: $W = -\Delta G$. If the chemical change produces a potential difference E, the relation between free energy and the potential difference is $\Delta G = -nFE$. Ohm's law ($I = E/R$) describes the relation among current, voltage, and resistance in an electric circuit. It can be combined with the definitions of work and power ($P =$ work per second) to give $P = E \cdot I = I^2 R$.

A galvanic cell uses a spontaneous redox reaction to produce electricity. The electrode at which oxidation occurs is the anode and the electrode at which reduction occurs is the cathode. The two half-cells are usually separated by a salt bridge that allows ions to migrate from one side to the other to maintain charge neutrality, but prevents reactants in the two half-cells from mixing. The standard reduction potential of a half-reaction is measured by pairing that half-reaction with a standard hydrogen electrode. The term "standard" means that activities of reactants and products are unity. When you reverse the direction of a reaction, you change the sign of its potential. A complete reaction is the sum of an oxidation half-reaction and a reduction half-reaction. The standard potential for the complete reaction is the sum of the potentials for each half-reaction. If several half-reactions are added to give another half-reaction, the standard potential of the net half-reaction can be found by equating the free energy of the net half-reaction to the sum of free energies of the component half-reactions.

When the activities of reactants and products are not unity, the potential of a complete reaction or half-reaction is given by the Nernst equation: $E = E^\circ - (0.059\ 16/n) \log Q$ (at 25°C), where Q is the reaction quotient. The reaction quotient has the same form as the equilibrium constant, but it is evaluated with concentrations existing at the time of interest.

Complex equilibria can be studied by making the system part of an electrochemical cell. If we measure the voltage and know the concentrations (activities) of all but one of the reactants and products, the Nernst equation allows us to compute the concentration of that one unknown species. In this way the electrochemical cell serves as a probe for that species.

Biochemists prefer to use the formal potential of a half-reaction at pH 7 ($E^{\circ\prime}$) instead of the standard potential (E°), which applies at pH 0. The value of $E^{\circ\prime}$ is found by writing the Nernst equation for the desired half-reaction and grouping together all terms except the logarithm containing the formal concentrations of reactant and product. The combination of terms, evaluated at pH 7, is $E^{\circ\prime}$.

Terms to Understand

ampere	half-reaction	
anode	joule	redox reaction
cathode	Latimer diagram	reducing agent
coulomb	Nernst equation	reductant
current	Ohm's law	reduction
$E^{\circ\prime}$	oxidant	resistance
electric potential	oxidation	salt bridge
electrode	oxidizing agent	standard hydrogen electrode
Faraday constant	potentiometer	standard reduction potential
formal potential	power	volt
galvanic cell	reaction quotient	watt

Exercises

14-A. A mercury cell used to power heart pacemakers runs on the following reaction:

$$Zn(s) + HgO(s) \rightarrow ZnO(s) + Hg(l) \quad E° = 1.35 \text{ V}$$

If the power required to operate the pacemaker is 0.010 0 W, how many kilograms of HgO will be consumed in 365 days? How many pounds of HgO is this? (1 pound = 453.6 g)

14-B. Calculate $E°$ and K for each of the following reactions.
(a) $I_2(s) + 5Br_2(aq) + 6H_2O \rightleftharpoons$
$2IO_3^- + 10Br^- + 12H^+$
(b) $Cr^{2+} + Fe(s) \rightleftharpoons Fe^{2+} + Cr(s)$
(c) $Mg(s) + Cl_2(g) \rightleftharpoons Mg^{2+} + 2Cl^-$
(d) $5MnO_2(s) + 4H^+ \rightleftharpoons$
$3Mn^{2+} + 2MnO_4^- + 2H_2O$
(e) $Ag^+ + 2S_2O_3^{2-} \rightleftharpoons Ag(S_2O_3)_2^{3-}$
(f) $CuI(s) \rightleftharpoons Cu^+ + I^-$

14-C. Calculate the voltage of each of the following cells.
(a) $Fe(s)|FeBr_2(0.010 \text{ M})\|NaBr(0.050 \text{ M})$
$|Br_2(l)|Pt(s)$
(b) $Cu(s)|Cu(NO_3)_2(0.020 \text{ M})$
$\|Fe(NO_3)_2(0.050 \text{ M})|Fe(s)$
(c) $Hg(l)|Hg_2Cl_2(s)|KCl(0.060 \text{ M})\|KCl(0.040 \text{ M})$
$|Cl_2(g, 0.50 \text{ atm})|Pt(s)$

14-D. Consider the cell pictured below.
The anode reaction can be written in *either* of two ways:

$$Ag(s) + I^- \rightarrow AgI(s) + e^- \qquad (1)$$

or

$$Ag(s) \rightarrow Ag^+ + e^- \qquad (2)$$

The cathode reaction is

$$H^+ + e^- \rightarrow \tfrac{1}{2}H_2(g) \qquad (3)$$

(a) Using Reactions 2 and 3, calculate $E°$ and write the Nernst equation for the cell.
(b) Use the value of K_{sp} for AgI to compute $[Ag^+]$ and find the cell voltage.
(c) Suppose, instead, that you wish to describe the cell with Reactions 1 and 3. We know that the cell voltage (E, not $E°$) must be the same, no matter which description we use. Write the Nernst equation for Reactions 1 and 3 and use *it* to solve the $E°$ in Reaction 1. Compare your answer with the value in Appendix H.

14-E. Calculate the voltage of the cell

$$Cu(s)|Cu^{2+}(0.030 \text{ M})\|K^+Ag(CN)_2^-(0.010 \text{ M}),$$
$$HCN(0.10 \text{ F}), \text{ buffer to pH } 8.21|Ag(s)$$

You may wish to refer to the following reactions:

$$Ag(CN)_2^- + e^- \rightleftharpoons Ag(s) + 2CN^- \quad E° = -0.310 \text{ V}$$
$$HCN \rightleftharpoons H^+ + CN^- \quad pK_a = 9.21$$

14-F. (a) Write a balanced equation for the reaction $PuO_2^+ \rightarrow Pu^{4+}$ and calculate $E°$ for the reaction.

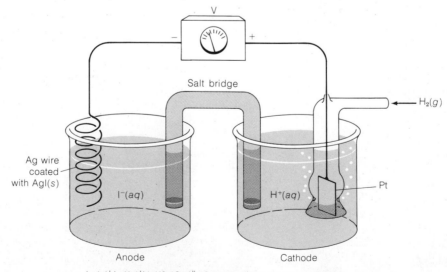

$Ag(s)|AgI(s)|NaI(0.10 \text{ M})\|HCl(0.10 \text{ M})|H_2(g, 0.20 \text{ atm})|Pt(s)$

$$PuO_2^{2+} \xrightarrow{+0.966} PuO_2^+ \xrightarrow{\text{?}} Pu^{4+} \xrightarrow{+1.006} Pu^{3+}$$

$$\underset{1.021}{\underline{\hspace{11cm}}}$$

(b) Predict whether or not an equimolar mixture of PuO_2^{2+} and PuO_2^+ will oxidize H_2O to O_2 at a pH of 2.00. You may assume that $P_{O_2} = 0.20$ atm. Will O_2 be liberated at pH 7.00?

14-G. Calculate the voltage of the cell below, in which KHP is potassium acid phthalate, the monopotassium salt of phthalic acid.

$$Hg(l)\,|\,Hg_2Cl_2(s)\,|\,KCl(0.10\ \text{M})\,\|\,KHP(0.050\ \text{M})$$
$$|\,H_2(g,\ 1.00\ \text{atm})\,|\,Pt(s)$$

14-H. The cell below has a voltage of -0.321 V:

$$Hg(l)\,|\,Hg(NO_3)_2(0.001\ 0\ \text{M}),\ KI(0.010\ \text{M})\,\|\,\text{S.H.E.}$$

Calculate the equilibrium constant for the reaction

$$Hg^{2+} + 4I^- \rightleftharpoons HgI_4^{2-}$$

You may assume that the only forms of mercury in solution are Hg^{2+} and HgI_4^{2-}.

14-I. The formation constant for $Cu(EDTA)^{2-}$ is 6.3×10^{18}, and the value of $E°$ for the reaction $Cu^{2+} + 2e^- \rightleftharpoons Cu(s)$ is $+0.339$ V. From this information, find $E°$ for the reaction

$$CuY^{2-} + 2e^- \rightleftharpoons Cu(s) + Y^{4-}$$

14-J. Using the reaction below, state which compound, $H_2(g)$ or glucose, is the more powerful reducing agent at pH $= 0.00$.

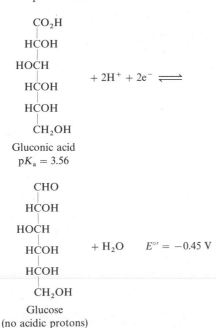

14-K. Living cells convert energy derived from sunlight or combustion of food into energy-rich ATP (adenosine triphosphate) molecules. For ATP synthesis, $\Delta G = +34.5$ kJ/mol. This energy is then made available to the cell when ATP is hydrolyzed to ADP (adenosine diphosphate). In animals, ATP is synthesized when protons pass through a complex enzyme in the mitochondrial membrane.[†] Two factors account for the movement of protons through this enzyme into the mitochondrion (see the figure below). One factor is the gradient of concentration of H^+, higher outside the mitochondrion than inside. This gradient arises because protons are *pumped* out of the mitochondrion by enzymes involved in the oxidation of food molecules. A second factor is that the inside of the mitochondrion is negatively charged with respect to the outside. The synthesis of one ATP molecule requires two protons to pass through the phosphorylation enzyme.

(a) The difference in free energy when a molecule travels from a region of high activity to a region of low activity is given by

$$\Delta G = -RT \ln \frac{\mathscr{A}_{\text{high}}}{\mathscr{A}_{\text{low}}}$$

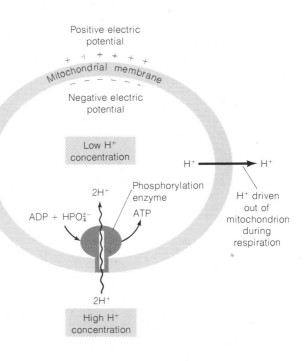

[†] For an interesting discussion of this topic, see "How Cells Make ATP," by P. C. Hinkle and R. E. McCarty, *Scientific American* (March 1978), p. 104. An article describing how gradients of pH and electric potential across biological membranes are measured is D. F. Wilson and N. G. Forman, *Biochem.*, **21**, 1438 (1982).

How big must the pH difference be if the passage of two protons is to provide enough energy to synthesize one ATP molecule?

(b) pH differences this large have not been observed in mitochondria. How great an electric potential difference between inside and outside is necessary for the movement of two protons to provide energy to synthesize ATP? In answering this question, neglect any contribution from the pH difference.

(c) It is thought that the energy for ATP synthesis is provided by *both* the pH difference and the electric potential. If the pH difference is 1.00 pH unit, how many millivolts must be the potential difference?

Problems

A14-1. (a) How many electrons are in one coulomb?

(b) How many coulombs are there in one mole of charge?

A14-2. The basal rate of consumption of O_2 by a 70-kg human is about 16 mol of O_2 per day. This O_2 oxidizes food and is reduced to H_2O, providing energy for the organism:

$$O_2 + 4H^+ + 4e^- \rightleftharpoons 2H_2O$$

(a) To what current (in amperes = C/s) does this respiration rate correspond? (Current is defined by the flow of electrons from food to O_2.)

(b) Compare your answer in part a with the current drawn by a refrigerator using 500 W at 115 V. Remember that power (in watts) = work/s = $E \cdot I$.

(c) If the electrons flow from nicotinamide adenine dinucleotide (NADH) to O_2, they experience a potential drop of 1.1 V. What is the power output (in watts) of our human friend?

A14-3. A 6.00-V battery is connected across a 2.00-kΩ resistor.

(a) How many electrons per second flow through the circuit?

(b) How many joules of heat are produced for each electron?

(c) If the circuit operates for 30.0 min, how many moles of electrons will have flowed through the resistor?

(d) What voltage would the battery need to deliver for the power to be 100 W?

A14-4. Consider the redox reaction

$$I_2 + 2S_2O_3^{2-} \rightleftharpoons 2I^- + \quad S_4O_6^{2-}$$
$$\text{Thiosulfate} \qquad \text{Tetrathionate}$$

(a) Identify the oxidizing agent on the left side of the reaction and write a balanced half-reaction for the oxidant.

(b) Identify the reducing agent on the left side of the rection and write a balanced half-reaction for the reductant.

(c) How many coulombs of charge are passed from reductant to oxidant when 1.00 g of thiosulfate reacts?

(d) If the rate of reaction is 1.00 g of thiosulfate consumed per minute, what current (in amperes) flows from reductant to oxidant?

A14-5. The free energy change for the reaction $CO + \frac{1}{2}O_2 \rightleftharpoons CO_2$ is $\Delta G° = -257$ kJ per mole of CO at 298 K.

(a) Find $E°$ for the reaction.

(b) Find the equilibrium constant for the reaction.

A14-6. Which will be the strongest oxidizing agent under standard conditions (all activities = 1): HNO_2, Se, UO_2^{2+}, Cl_2, H_2SO_3, or MnO_2?

A14-7. Which will be the strongest reducing agent under standard conditions (all activities = 1): Se, Sn^{4+}, Cr^{2+}, Mg^{2+}, or $Fe(CN)_6^{4-}$?

A14-8. For each picture at the top of the next page, write the line notation to describe the cell. Write an anode reaction for the left electrode and a cathode reaction for the right electrode.

A14-9. (a) Draw a picture of the following cell, showing the location of each species:

$$Pt(s) | Fe^{3+}(aq), Fe^{2+}(aq) \| Cr_2O_7^{2-}(aq),$$
$$Cr^{3+}(aq), HA(aq) | Pt(s)$$

(b) Write an oxidation half-reaction for the left electrode and a reduction half-reaction for the right electrode.

(c) Write a balanced equation for the cell reaction.

A14-10. A light-weight rechargeable battery developed for electric automobiles uses the following cell:

$$Zn(s) | ZnCl_2(aq) \| Cl^-(aq) | Cl_2(l) | C(s)$$

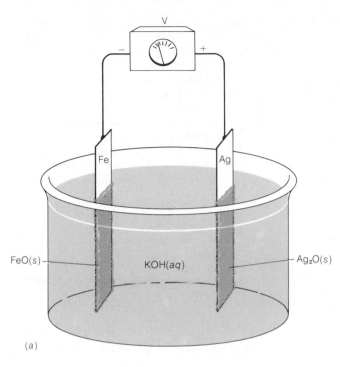

(a)

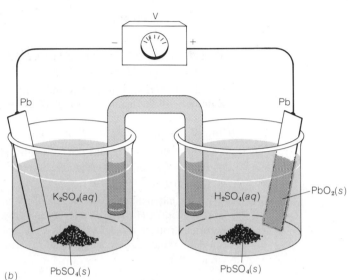

(b)

(a) Write an oxidation half-reaction for the left electrode and a reduction half-reaction for the right electrode.

(b) If the battery delivers a constant current of 1.00×10^3 A for 1.00 h, how many kilograms of Cl_2 will be consumed?

A14-11. Draw a picture of each of the following cells, showing the location of each chemical species. For each cell, write an anode reaction for the left electrode and a cathode reaction for the right electrode.

(a) $Au(s) | Fe(CN)_6^{4-}(aq). Fe(CN)_6^{3-}(aq)$

$\| Ag(CN)_2^-(aq), KCN(aq) | Ag(s)$

(b) $Pt(s) | Hg(l) | Hg_2Cl_2(s) | KCl(aq)$

$\| ZnCl_2(aq) | Zn(s)$

A14-12. Consider the half-reaction

$$As(s) + 3H^+ + 3e^- \rightleftharpoons AsH_3(g) \quad E° = -0.238 \text{ V}$$
$$\text{Arsine}$$

(a) Write the Nernst equation for the half-reaction.

(b) Find E (not $E°$) when pH = 3.00 and $P_{AsH_3} = 1.00$ torr. (Remember that 1 atm = 760 torr.)

A14-13. Consider the cell

$$Pt(s) | H_2(g, 0.100 \text{ atm}) | H^+(aq, \text{pH} = 2.54)$$

$$\| Cl^-(aq, 0.200 \text{ M}) | Hg_2Cl_2(s) | Hg(l) | Pt(s)$$

(a) Write a half-reaction for each electrode, with oxidation at the left.

(b) Write the net cell reaction and find E. For the Hg_2Cl_2 half-reaction, $E° = 0.268$ V.

A14-14. From the half-reactions below, calculate the solubility product of $Mg(OH)_2$.

$$Mg^{2+} + 2e^- \rightleftharpoons Mg(s)$$
$$E° = -2.360 \text{ V}$$
$$Mg(OH)_2(s) + 2e^- \rightleftharpoons Mg(s) + 2OH^-$$
$$E° = -2.690 \text{ V}$$

A14-15. In this problem we will use a cell similar to that in Figure 14-9 as a probe to find the concentration of Cl^- in the right compartment. The cell is

$$Pt(s) | H_2(g, 1.00 \text{ atm}) | H^+(aq, \text{pH} = 3.60)$$

$$\| Cl^-(aq, x \text{ M}) | AgCl(s) | Ag(s)$$

(a) Write an anode reaction for the left half-cell.

(b) Write a cathode reaction for the right half-cell.

(c) Write a balanced net cell reaction and write the Nernst equation for this reaction.

(d) Given a measured cell voltage of 0.485 V, find $[Cl^-]$ in the right compartment.

A14-16. In this problem we will find $E°'$ for the reaction $C_2H_2(g) + 2H^+ + 2e^- \rightleftharpoons C_2H_4(g)$.

(a) Write the Nernst equation for the half-reaction, using $E°$ from Appendix H.

(b) Rearrange the Nernst equation to the form

$$E = E° + \text{other terms} - \frac{0.059\,16}{2} \log \frac{P_{C_2H_4}}{P_{C_2H_2}}$$

(c) The quantity $(E° + \text{other terms})$ is $E°'$. Evaluate $E°'$ for pH $= 7.00$.

A14-17. Use the numerical values of R and E to derive Equation 14-26 from Equation 14-24. What would be the value of the numerical constant in front of the log term at 0°C? at 37°C?

14-18. Use the right-hand column of Appendix H to find the value of $E°$ for the reaction $Al^{3+} + 3e^- \rightleftharpoons Al(s)$ at 50°C.

14-19. Use Le Châtelier's principle and half-reactions from Appendix H to find which of the following become stronger oxidizing agents as the pH is lowered. Which are unchanged, and which become weaker?

$$Cl_2 \qquad Cr_2O_7^{2-} \qquad Fe^{3+}$$
$$\text{Chlorine} \qquad \text{Dichromate} \qquad \text{Ferric}$$

$$MnO_4^- \qquad IO_3^-$$
$$\text{Permanganate} \qquad \text{Iodate}$$

14-20. Calculate $E°$, $\Delta G°$, and K for each of the following reactions.

(a) $4Co^{3+} + 2H_2O \rightleftharpoons 4Co^{2+} + O_2(g) + 4H^+$

(b) $Cu(s) + Cu^{2+} \rightleftharpoons 2Cu^+$

(c) $Ag(S_2O_3)_2^{3-} + Fe(CN)_6^{4-} \rightleftharpoons$
$Ag(s) + 2S_2O_3^{2-} + Fe(CN)_6^{3-}$

(d) $2Cu^{2+} + 2I^- + HO$—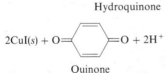—$OH \rightleftharpoons$

Hydroquinone

$2CuI(s) + O$=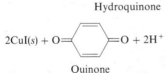=$O + 2H^+$

Quinone

A14-21. Consider a circuit in which the left half-cell was prepared by dipping a Pt wire in a beaker containing an equimolar mixture of Cr^{2+} and Cr^{3+}. The right half-cell contained a Tl rod immersed in 1.00 M $TlClO_4$.

(a) Use line notation to describe this cell.

(b) Calculate the cell voltage.

(c) Write the spontaneous net cell reaction.

(d) When the two electrodes are connected by a salt bridge and a wire, which terminal (Pt or Tl) will be the anode?

14-22. A solution contains 0.100 M Ce^{3+}, 1.00×10^{-4} M Ce^{4+}, 1.00×10^{-4} M Mn^{2+}, 0.100 M MnO_4^-, and 1.00 M $HClO_4$.

(a) Write a balanced net reaction that can occur between the species in this solution.

(b) Calculate $\Delta G°$ and K for the reaction.

(c) Calculate E for the conditions given above.

(d) Calculate ΔG for the conditions given above.

(e) At what pH would the concentrations of Ce^{4+}, Ce^{3+}, Mn^{2+}, and MnO_4^- listed above be in equilibrium at 298 K? That is, at what pH would there be no net reaction?

14-23. *Measuring a standard reduction potential.* The following cell was set up to measure the standard reduction potential of the $Ag^+|Ag$ couple:

$Pt(s)|HCl(0.010\,00$ M$), H_2(g)\|$

$$AgNO_3(0.010\,00 \text{ M})|Ag(s)$$

The temperature was 25°C (the standard condition) and atmospheric pressure was 751.0 torr. Since the vapor pressure of water is 23.8 torr at 25°C, P_{H_2} in the cell was $751.0 - 23.8 = 727.2$ torr. The Nernst equation for the cell, including activity coefficients, is

$$\tfrac{1}{2}H_2(g) + Ag^+ \rightleftharpoons H^+ + Ag(s)$$

$$E = E°_{Ag^+|Ag} - 0.059\,16 \log \frac{[H^+]\gamma_{H^+}}{P_{H_2}^{1/2}[Ag^+]\gamma_{Ag^+}}$$

Given a measured cell voltage of $+0.798\,3$ V, and using activity coefficients from Table 6-1, find $E°_{Ag^+|Ag}$.

14-24. (a) In the presence of cyanide ion, the reduction potential of Fe(III) is decreased from 0.771 to 0.356 V.

$$Fe^{3+} + e^- \rightleftharpoons Fe^{2+}$$
$$\text{Ferric} \qquad \text{Ferrous}$$

$$E° = 0.771 \text{ V}$$

$$Fe(CN)_6^{3-} + e^- \rightleftharpoons Fe(CN)_6^{4-}$$
$$\text{Ferricyanide} \qquad \text{Ferrocyanide}$$

$$E° = 0.356 \text{ V}$$

Which ion, Fe^{3+} or Fe^{2+}, is stabilized more by complexing with CN^-?

(b) Using Appendix H, answer the same question when the ligand is phenanthroline instead of cyanide.

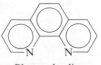

Phenanthroline

14-25. Consider the cell

$$\text{S.H.E.} \,\|\, \text{Ag(S}_2\text{O}_3)_2^{3-}(aq, 0.010 \text{ M}),$$

$$\text{S}_2\text{O}_3^{2-}(aq, 0.050 \text{ M}) \,|\, \text{Ag}(s)$$

(a) Using the half-reaction $\text{Ag(S}_2\text{O}_3)_2^{3-} + e^- \rightleftharpoons \text{Ag}(s) + 2\text{S}_2\text{O}_3^{2-}$, calculate the cell voltage (E, not $E°$).

(b) Alternatively, the cell could have been described with the half-reaction $\text{Ag}^+ + e^- \rightleftharpoons \text{Ag}(s)$. Using the cell voltage from part a, calculate $[\text{Ag}^+]$ in the right half-cell.

(c) Use the answer to part b to find the formation constant for the reaction

$$\text{Ag}^+ + 2\text{S}_2\text{O}_3^{2-} \overset{K_f}{\rightleftharpoons} \text{Ag(S}_2\text{O}_3)_2^{3-}$$
$$\text{Thiosulfate}$$

14-26. (a) Write the line notation for the cell below.

(b) Calculate the cell voltage (E, not $E°$), and state the direction in which electrons will flow through the potentiometer.

(c) The left half-cell was loaded with 14.3 mL of $\text{Br}_2(l)$ (density = 3.12 g/mL). The aluminum electrode contains 12.0 g of Al. Which element, Br_2 or Al, is the limiting reagent in this cell? (That is, which reagent will be used up first?)

(d) If the cell is somehow operated under conditions in which it produces a constant voltage of 1.50 V, how much electrical work will have been done when 0.231 mL of $\text{Br}_2(l)$ has been consumed?

(e) If the potentiometer is replaced by a 1.20-kΩ resistor, and if the heat dissipated by the resistor is 1.00×10^{-4} J/s, at what rate (grams per second) is Al(s) dissolving? (In this question the voltage is not 1.50 V.)

14-27. Suppose that the concentrations of NaF and KCl were each 0.10 M in the cell $\text{Pb}(s) \,|\, \text{PbF}_2(s) \,|\, \text{F}^-(aq) \,\|\, \text{Cl}^-(aq) \,|\, \text{AgCl}(s) \,|\, \text{Ag}(s)$.

(a) Using the half-reactions $\text{Pb}(s) + 2\text{F}^- \rightleftharpoons \text{PbF}_2(s) + 2e^-$ and $\text{AgCl}(s) + e^- \rightleftharpoons \text{Ag}(s) + \text{Cl}^-$, calculate the cell voltage.

(b) Now calculate the cell voltage using the reactions $\text{Pb}(s) \rightleftharpoons \text{Pb}^{2+} + 2e^-$ and $\text{Ag}^+ + e^- \rightleftharpoons \text{Ag}(s)$. For this part, you will need the solubility products for PbF_2 and AgCl.

14-28. The quinhydrone electrode was introduced in 1921 as a means of measuring pH. A cell employing this electrode is shown on the following page. The solution whose pH is to be measured is placed in the left half-cell, which also contains a 1:1 mole ratio of quinone and hydroquinone. The cell reaction is

$$\text{HO}-\bigcirc-\text{OH} + \text{Hg}_2\text{Cl}_2(s) \rightleftharpoons$$
$$\text{Hydroquinone}$$

$$\text{O}=\bigcirc=\text{O} + 2\text{H}^+ + 2\text{Hg}(l) + 2\text{Cl}^-$$
$$\text{Quinone}$$

(a) Ignoring activities and using the relation $\text{pH} = -\log[\text{H}^+]$, the Nernst equation can be changed into the form

$$E(\text{cell}) = A + B \cdot \text{pH}$$

where A and B are constants. Find the numerical values of A and B at 25°C,

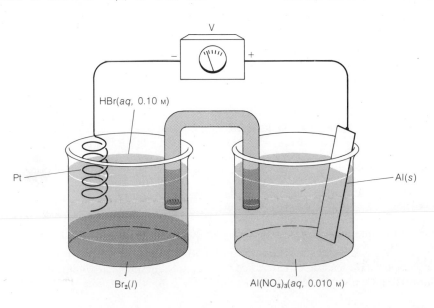

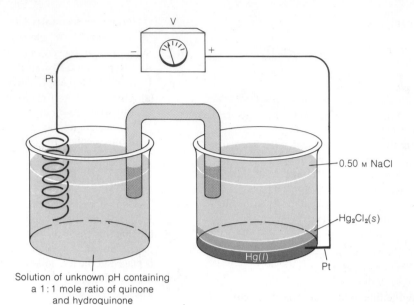

Solution of unknown pH containing
a 1:1 mole ratio of quinone
and hydroquinone

(b) If the pH were 4.50, in which direction would electrons flow through the potentiometer?

14-29. The voltage for the cell below is 0.490 V. Find K_b for the organic base RNH_2.

$$Pt(s)|H_2(1.00 \text{ atm})|RNH_2(aq, 0.10 \text{ M}),$$

$$RNH_3^+Cl^-(aq, 0.050 \text{ M})\|S.H.E.$$

14-30. Calculate the standard potential for the half-reaction

$$Pd(OH)_2(s) + 2e^- \rightleftharpoons Pd(s) + 2OH^-$$

given that K_{sp} for $Pd(OH)_2$ is 3×10^{-28} and given that the standard potential for the reaction $Pd^{2+} + 2e^- \rightleftharpoons Pd(s)$ is 0.915 V.

14-31. From the standard potentials for reduction of $Br_2(aq)$ and $Br_2(l)$ in Appendix H, calculate the solubility of Br_2 in water at 25°C. Express your answer as g/L.

14-32. The standard free energy of vaporization of $Cl_2(aq)$ is $\Delta G° = -6.9$ kJ/mol at 298 K. Given that $E°$ for the reaction $Cl_2(g) + 2e^- \rightleftharpoons 2Cl^-(aq)$ is 1.360 V, find $E°$ for the reaction $Cl_2(aq) + 2e^- \rightleftharpoons 2Cl^-(aq)$.

14-33. For the cell below, E (not $E°$) = -0.289 V. Write the net cell reaction and calculate its equilibrium constant. Do not use $E°$ values from Appendix H to answer this question.

$$Pt(s)|VO^{2+}(0.116 \text{ M}), V^{3+}(0.116 \text{ M}), H^+(1.57 \text{ M})$$

$$\|Sn^{2+}(0.031 \text{ 8 M}), Sn^{4+}(0.031 \text{ 8 M})|Pt(s)$$

14-34. The following cell has a voltage of 1.018 V. Find K_a for formic acid, HCO_2H.

$$Pt(s)|UO_2^{2+}(0.050 \text{ M}), U^{4+}(0.050 \text{ M}),$$

$$HCO_2H(0.10 \text{ M}), HCO_2Na(0.30 \text{ M})$$

$$\|Fe^{3+}(0.050 \text{ M}), Fe^{2+}(0.025 \text{ M})|Pt(s)$$

14-35. Using the reaction

$$HPO_4^{2-} + 2H^+ + 2e^- \rightleftharpoons HPO_3^{2-} + H_2O$$

$$E° = -0.234 \text{ V}$$

and any acid dissociation constants from Appendix G, calculate $E°$ for the reaction

$$H_2PO_4^- + H^+ + 2e^- \rightleftharpoons HPO_3^{2-} + H_2O$$

14-36. Given the following information

$$FeY^- + e^- \rightleftharpoons Fe^{2+} + Y^{4-} \quad E° = -0.730 \text{ V}$$

$$FeY^{2-}: \quad K_f = 2.1 \times 10^{14}$$

$$FeY^-: \quad K_f = 1.3 \times 10^{25}$$

calculate the standard potential for the reaction

$$FeY^- + e^- \rightleftharpoons FeY^{2-}$$

where Y is EDTA.

14-37. Write a balanced chemical equation (in acidic solution) for the reaction represented by the question mark on the lower arrow. Calculate $E°$ for the reaction.

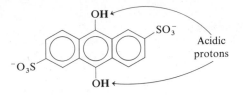

$$BrO_3^- \xrightarrow{1.491} HOBr \xrightarrow{1.584} Br_2(aq) \xrightarrow{1.098} Br^-$$

with overarching 1.441 and lower $?$

14-38. Write a balanced chemical equation (in acid solution) for the reaction represented by the question mark below. Calculate $E°$ for the reaction.

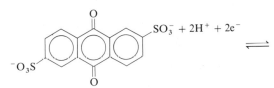

$$NO_3^- \xrightarrow{0.773} NO_2(g) \xrightarrow{1.108} HNO_2 \xrightarrow{?} NO$$

with lower 0.955

14-39. What must be the relationship between $E_1°$ and $E_2°$ if the species X^+ is to disproportionate spontaneously to X^{3+} and $X(s)$? Write a balanced equation for the disproportionation.

$$X^{3+} \xrightarrow{E_1°} X^+ \xrightarrow{E_2°} X(s)$$

14-40. Calculate $E°'$ for the reaction

$$H_2C_2O_4 + 2H^+ + 2e^- \rightleftharpoons 2HCO_2H$$

$$E° = 0.204 \text{ V}$$

14-41. The standard reduction potential of anthraquinone-2,6-disulfonate is 0.229 V.

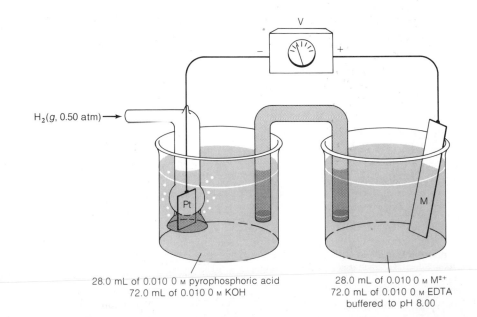

Anthraquinone-2,6-disulfonate

$$SO_3^- + 2H^+ + 2e^- \rightleftharpoons$$

The reduced product is a diprotic acid with $pK_1 = 8.10$ and $pK_2 = 10.52$. Calculate $E°'$ (pH = 7) for anthraquinone-2,6-disulfonate.

14-42. Suppose that HOx is a monoprotic acid with dissociation constant 1.4×10^{-5} and H_2Red^- is a diprotic acid with dissociation constants of 3.6×10^{-4} and 8.1×10^{-8}. Calculate $E°$ for the reaction

$$HOx + e^- \rightleftharpoons H_2Red^- \qquad E°' = 0.062 \text{ V}$$

14-43. Given the information below, find K_a for nitrous acid, HNO_2.

$$NO_3^- + 3H^+ + 2e^- \rightleftharpoons HNO_2 + H_2O$$

$$E° = 0.940 \text{ V}$$

$$E°' = 0.433 \text{ V}$$

14-44. The voltage of the cell below is -0.246 V. The right half-cell contains the metal ion, M^{2+}, whose standard reduction potentials is -0.266 V.

$$M^{2+} + 2e^- \rightleftharpoons M(s) \qquad E° = -0.266 \text{ V}$$

Calculate K_f for the metal–EDTA complex.

14-45. For the cell at the top of the next page, the half-reactions can be written

$H_2(g, 0.50 \text{ atm}) \rightarrow$

Pt

M

V

28.0 mL of 0.010 0 M pyrophosphoric acid
72.0 mL of 0.010 0 M KOH

28.0 mL of 0.010 0 M M^{2+}
72.0 mL of 0.010 0 M EDTA
buffered to pH 8.00

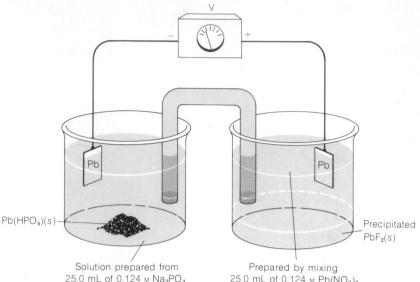

Pb(HPO₄)(s)

Precipitated
PbF₂(s)

Solution prepared from
25.0 mL of 0.124 M Na₃PO₄
plus 25.0 mL of 0.248 M HClO₄

Prepared by mixing
25.0 mL of 0.124 M Pb(NO₃)₂
plus 25.0 mL of 0.248 M KF

Left half-cell: $Pb(s) \rightleftharpoons Pb^{2+}(left) + 2e^-$

Right half-cell: $Pb^{2+}(right) + 2e^- \rightleftharpoons Pb(s)$

(a) Given that K_{sp} for $Pb(HPO_4)(s)$ is 2.0×10^{-10}, find $[HPO_4^{2-}]$ in the left half-cell.

(b) If the measured cell voltage is 0.097 V, calculate K_{sp} for $PbF_2(s)$.

14-46. The following cell was constructed to find the difference in K_{sp} between two naturally occurring forms of $CaCO_3(s)$, called *calcite* and *aragonite*.[†]

$$Pb(s) \,|\, PbCO_3(s) \,|\, CaCO_3(s, \text{calcite}) \,|$$

$$\text{aqueous buffer (pH 7.00)} \,\|\, \text{aqueous buffer (pH 7.00)} \,|$$

$$CaCO_3(s, \text{aragonite}) \,|\, PbCO_3(s) \,|\, Pb(s)$$

Each compartment of the cell contains a mixture of solid $PbCO_3$ ($K_{sp} = 7.4 \times 10^{-14}$) and either calcite or aragonite, both of which have $K_{sp} \approx 5 \times 10^{-9}$. Each solution was buffered to pH 7.00 with an inert buffer, and the cell was completely isolated from atmospheric CO_2. The measured cell voltage was -1.8 mV. Find the ratio of solubility products for calcite and aragonite.

$$\frac{K_{sp} \text{ (for calcite)}}{K_{sp} \text{ (for aragonite)}} = ?$$

14-47. The oxidized form (Ox) of a flavoprotein that functions as a one-electron reducing agent has a molar absorptivity (ε) of 1.12×10^4 $M^{-1} \cdot cm^{-1}$ at 457 nm at pH = 7.00.[†] For the reduced form (Red), $\varepsilon = 3.82 \times 10^3$ at 457 nm at pH 7.00.

$$Ox + e^- \rightleftharpoons Red \qquad E^{\circ\prime} = -0.128 \text{ V}$$

The substrate (S) is the molecule reduced by the protein.

$$Red + S \rightleftharpoons Ox + S^-$$

Both S and S^- are colorless. A solution at pH 7.00 was prepared by mixing enough protein plus substrate (Red + S) so that the initial concentrations of Red and S are each 5.70×10^{-5} M. The absorbance at 457 mm was 0.500 in a 1.00-cm cell.

(a) Calculate the concentrations of Ox and Red from the absorbance data.

(b) Calculate the concentrations of S and S^-.

(c) Calculate the value of $E^{\circ\prime}$ for the reaction $S + e^- \rightleftharpoons S^-$.

14-48. The monstrous cell shown on the facing page was set up. Then 50.0 mL of 0.044 4 M Na_2EDTA was added to the right-hand compartment and 50.0 mL of 0.070 0 M NaOH was added to the left-hand compartment. The cell voltage leveled off at $+0.418$ V. Find the formation constant for CuY^{2-} (where Y = EDTA).

[†] The cell in this problem would not give an accurate result because of the *junction potential* at each liquid junction (Section 15-3). A clever way around this problem, using a cell without any liquid junctions, is described by P. A. Rock, *J. Chem. Ed.*, **52**, 787 (1975).

[†] This problem is an application of Beer's law, which you can read about in Sections 19-1 and 19-2.

14-49. *Without neglecting activities,* calculate the voltage of the cell

$$Ni(s) \mid NiSO_4(0.020 \text{ M}) \parallel CuCl_2(0.030 \text{ M}) \mid Cu(s)$$

14-50. *Do not ignore activity coefficients in this problem.*

If the voltage for the following cell is 0.489 V, find K_{sp} for $Cu(IO_3)_2$.

$$Ni(s) \mid NiSO_4(0.025 \text{ M}) \parallel KIO_3(0.10 \text{ M}) \mid$$
$$Cu(IO_3)_2(s) \mid Cu(s)$$

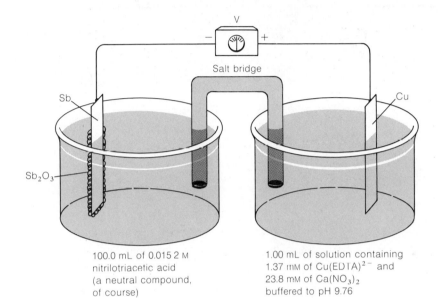

100.0 mL of 0.015 2 M
nitrilotriacetic acid
(a neutral compound,
of course)

1.00 mL of solution containing
1.37 mM of $Cu(EDTA)^{2-}$ and
23.8 mM of $Ca(NO_3)_2$
buffered to pH 9.76

15 Electrodes and Potentiometry

We saw in the last chapter that a galvanic cell can serve as a *probe* for one chemical species when the activities of all other species are known. That is, if we know the activities of all but one species in a cell, the cell voltage tells us the activity of that one species. The measurement of cell voltages to extract chemical information is called **potentiometry.**

Imagine a solution containing an **electroactive species** whose activity (concentration) we wish to measure. An electroactive species is one that can donate or accept electrons from an electrode. We can turn the unknown into a half-cell by inserting an electrode (such as a Pt wire) into the solution to transfer electrons to or from the species of interest. Since this electrode responds directly to the analyte, it is called the **indicator electrode.** We then connect this half-cell to a second half-cell via a salt bridge. The second half-cell should have a known, fixed composition so that it has a known, constant potential. Since the second half-cell has a constant potential, it is called a **reference electrode.** The cell voltage is the difference between a constant potential from the reference electrode and a variable potential that reflects changes in the analyte activity.

Indicator electrode: responds to analyte activity

Reference electrode: maintains a fixed (reference) potential

15-1 REFERENCE ELECTRODES

Suppose that you have a solution with variable amounts of Fe^{2+} and Fe^{3+}. If you are clever, you can make this solution part of a cell in such a way that the voltage will tell you the relative concentrations of these species—that is, the value of $[Fe^{2+}]/[Fe^{3+}]$. Figure 15-1 shows one way to do this. A Pt wire is inserted as an indicator electrode through which Fe^{3+} can receive electrons or Fe^{2+} can lose electrons. The left half-cell serves to complete the galvanic cell and has a known, constant potential.

The two half-reactions can be written as follows:

$$Ag(s) + Cl^- \rightleftharpoons AgCl(s) + e^- \qquad E° = -0.222 \text{ V} \qquad (15\text{-}1)$$

$$Fe^{3+} + e^- \rightleftharpoons Fe^{2+} \qquad E° = 0.771 \text{ V} \qquad (15\text{-}2)$$

The net cell reaction is

$$Fe^{3+} + Ag(s) + Cl^- \rightleftharpoons Fe^{2+} + AgCl(s) \qquad E° = 0.549 \text{ V} \qquad (15\text{-}3)$$

for which the Nernst equation is

$$E = 0.549 - 0.059\ 16 \log \frac{[Fe^{2+}]}{[Fe^{3+}][Cl^-]} \qquad (15\text{-}4)$$

The concentration of Cl^- in the left half-cell is constant, fixed by the solubility of KCl, with which the solution is saturated. Therefore the cell voltage changes only when the quotient $[Fe^{2+}]/[Fe^{3+}]$ changes.

The half-cell on the left in Figure 15-1 can be thought of as a *reference electrode*. We can picture the cell and salt bridge enclosed by the dashed line as a single unit dipped into the analyte solution, as shown in Figure 15-2. The Pt wire is the indicator electrode, whose potential responds to changes in the quotient $[Fe^{2+}]/[Fe^{3+}]$. The reference is necessary to complete the redox reaction and to provide a *constant reference potential* to the left side of the potentiometer. Changes in the cell voltage can be unambiguously assigned to changes in the quotient $[Fe^{2+}]/[Fe^{3+}]$.

Actually, the voltage tells us the quotient of activities, $\mathscr{A}_{Fe^{2+}}/\mathscr{A}_{Fe^{3+}}$. However, we will neglect activity coefficients and continue to write the Nernst equation in terms of concentrations instead of activities.

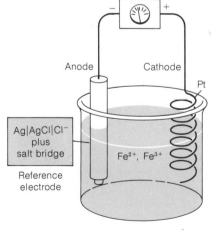

Figure 15-2
Another view of Figure 15-1. The contents of the dashed box in Figure 15-1 are now considered to be a reference electrode dipped into the analyte solution.

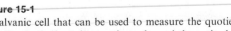

Anode: $Ag + Cl^- \rightleftharpoons AgCl + e^-$ Cathode: $Fe^{3+} + e^- \rightleftharpoons Fe^{2+}$

$$Ag(s)|AgCl(s)|Cl^-(aq)\|Fe^{2+}(aq),\ Fe^{3+}(aq|Pt(s)$$

Figure 15-1
A galvanic cell that can be used to measure the quotient $[Fe^{2+}]/[Fe^{3+}]$ in the right half-cell. The Pt wire is the *indicator electrode*, and the entire left half-cell plus salt bridge (enclosed by the dashed line) can be considered a *reference electrode*.

Another Way to Write the Nernst Equation

We now introduce an alternate way to write the Nernst equation that emphasizes the contribution of each half-cell in Figure 15-2. This equation is used so widely by analytical chemists that we will use it for the remainder of our discussions. The idea is simple: The potentiometer in Figure 15-2 is connected to two electrodes. The measured voltage, E, is the difference between the potentials of the two electrodes:

From now on, we will write the cell voltage as the difference between two electrode potentials.

$$E = E_+ - E_- \qquad (15\text{-}5)$$

where E_+ is the potential of the electrode attached to the positive input terminal of the potentiometer and E_- is the potential of the electrode attached to the negative input terminal of the potentiometer. When we use Equation 15-5, we must *write both half-reactions as reductions* and write the half-cell Nernst equation 14-63 for each.

Let's see how Equation 15-5 works. We begin by writing a reduction half-reaction for each electrode:

Left electrode: $\qquad AgCl(s) + e^- \rightleftharpoons Ag(s) + Cl^- \qquad E° = 0.222 \text{ V} \quad (15\text{-}6)$

Right electrode: $\qquad Fe^{3+} + e^- \rightleftharpoons Fe^{2+} \qquad\qquad E° = 0.771 \text{ V} \quad (15\text{-}7)$

The two electrode potentials are

E_+ is the potential of the electrode attached to the positive input of the potentiometer. E_- is the potential of the electrode attached to the negative input of the potentiometer.

$$E_- = 0.222 - 0.059\,16 \log [Cl^-] \qquad (15\text{-}8)$$

$$E_+ = 0.771 - 0.059\,16 \log \frac{[Fe^{2+}]}{[Fe^{3+}]} \qquad (15\text{-}9)$$

The cell voltage is just

$$E = E_+ - E_- = \left(0.771 - 0.059\,16 \log \frac{[Fe^{2+}]}{[Fe^{3+}]}\right) \qquad (15\text{-}10)$$

$$- (0.222 - 0.059\,16 \log [Cl^-])$$

$$E = 0.549 - 0.059\,16 \log \frac{[Fe^{2+}]}{[Fe^{3+}][Cl^-]} \qquad (15\text{-}11)$$

Equation 15-11 is the same as 15-4. It is just convenient in much of electrochemistry to derive the Nernst equation by subtracting one electrode potential from another.

Let's summarize what we will do in the rest of this chapter. When faced with a cell such as Figure 15-2, we will write *reduction* half-reactions such as 15-6 and 15-7 for both electrodes. Then we will write half-cell Nernst equations such as 15-8 and 15-9 for both electrodes. The net cell voltage, $E = E_+ - E_-$, is the difference between the two electrode potentials.

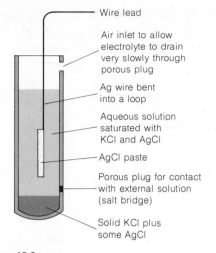

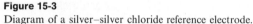

Figure 15-3
Diagram of a silver–silver chloride reference electrode.

Common Reference Electrodes[†]

Silver–silver chloride electrode

The reference electrode enclosed by the dashed line in Figure 15-1 is called a **silver–silver chloride electrode.** Figure 15-3 shows how the half-cell in Figure 15-1 might be reconstructed as a thin, glass-enclosed electrode that can be dipped into the analyte solution in Figure 15-2.

The standard reduction potential for the AgCl|Ag couple is $+0.222$ V at 25°C. This would be the potential of a silver–silver chloride electrode if $\mathscr{A}_{Cl^-}$ were unity. But the activity of Cl^- in a saturated solution of KCl at 25°C is not unity, and the potential of the electrode in Figure 15-3 is found to be $+0.197$ V with respect to a standard hydrogen electrode at 25°C.

The two most common reference electrodes are the silver–silver chloride electrode and the calomel electrode.

$$AgCl(s) + e^- \rightleftharpoons Ag(s) + Cl^- \qquad E° = +0.222 \text{ V}$$
$$E(\text{saturated KCl}) = +0.197 \text{ V} \qquad (15\text{-}12)$$

Calomel electrode

The **calomel electrode** in Figure 15-4 is based on the reaction

$$Hg_2Cl_2(s) + 2e^- \rightleftharpoons 2Hg(l) + 2Cl^- \qquad (15\text{-}13)$$

Mercurous
chloride
(calomel)

$$E = E° - \frac{0.059\ 16}{2} \log [Cl^-]^2 \qquad (15\text{-}14)$$

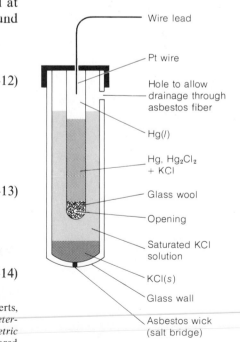

Figure 15-4
A saturated calomel electrode (S.C.E.).

[†] Practical aspects of electrode fabrication are discussed in D. T. Sawyer and J. L. Roberts, Jr., *Experimental Electrochemistry for Chemists* (New York: Wiley, 1974); R. G. Bates, *Determination of pH* (New York: Wiley, 1984); and E. P. Serjeant, *Potentiometry and Potentiometric Titrations* (New York: Wiley, 1984). A versatile junction for reference electrodes can be prepared by attaching a molecular sieve to glass tubing with heat-shrinkable tubing as described by S. Yee and O.-K. Chang, *J. Chem. Ed.,* **65,** 129 (1988).

The standard potential ($E°$) for this reaction is $+0.268$ V. If the cell is saturated with KCl at 25°C, the activity of Cl^- is such that the potential of the electrode is $+0.241$ V. The calomel electrode saturated with KCl is called the **saturated calomel electrode.** It is encountered so frequently that it is abbreviated **S.C.E.** The advantage in using a saturated KCl solution is that the concentration of chloride does not change if some of the liquid evaporates.

Voltage conversions between different reference scales

It is sometimes necessary to convert potentials between different reference scales. If an electrode has a potential of -0.461 V with respect to a calomel electrode, what is the potential with respect to a silver–silver chloride electrode? What would be the potential with respect to the standard hydrogen electrode?

To answer these questions, consider the scale below,[†] in which the positions of the calomel and silver–silver chloride electrodes are labeled with respect to the standard hydrogen electrode:

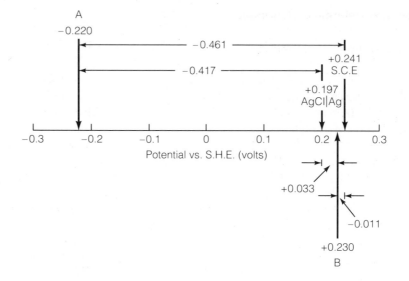

It is obvious that point A, which is -0.461 V from S.C.E., is -0.417 V from the silver–silver chloride electrode and -0.220 V with respect to S.H.E. What about point B, whose potential is $+0.033$ V with respect to silver–silver chloride? Its position is -0.011 V from S.C.E. and $+0.230$ V with respect to S.H.E. By keeping this simple diagram in mind, you can convert potentials from one scale to another.

15-2 INDICATOR ELECTRODES

The most common indicator electrode is the platinum electrode that we have encountered several times already (in Figure 15-2, for example). Platinum is used because it is relatively *inert*—it does not participate in many chemical reactions. When it is used as an electrode, its purpose is simply to transmit electrons to or from a reactive species in solution.

[†] W. -Y. Ng, *J. Chem. Ed.,* **65,** 627 (1988).

In cases where platinum reacts with the electrolyte solution, a gold electrode can usually be used. Gold is more inert than platinum. Various types of carbon are also widely used as indicator electrodes because the rates of many redox reactions on the carbon surface are fast. A metal electrode works best when its surface is large and clean. A brief dip in concentrated nitric acid, followed by rinsing with distilled water, is often effective for cleaning an electrode surface.

Figure 15-5 shows how a silver electrode can be used in conjunction with a calomel reference electrode to measure silver ion concentration (actually activity). The reaction at the silver indicator electrode is

$$Ag^+ + e^- \rightleftharpoons Ag(s) \qquad E° = 0.799 \text{ V} \qquad (15\text{-}15)$$

The reference cell reaction is

$$Hg_2Cl_2(s) + 2e^- \rightleftharpoons 2Hg(l) + 2Cl^- \qquad E° = 0.268 \text{ V} \qquad (15\text{-}16)$$

Using Equation 15-5, we can write a Nernst equation for this cell as follows:

$$E = E_+ - E_-$$

$$= \left(0.799 - 0.059\,16 \log \frac{1}{[Ag^+]}\right) - \left(0.268 - \frac{0.059\,16}{2} \log [Cl^-]^2\right)$$

In this equation E_+ and E_- are each a Nernst equation for a single-electrode reaction written as a reduction.

$[Cl^-]$ refers to concentration of Cl^- in calomel electrode

(15-17)

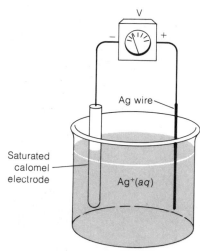

Figure 15-5
Use of silver and calomel electrodes to measure the concentration of Ag^+ in a solution.

But the term E_- in Equation 15-17 is a constant, equal to 0.241 V. It is constant because $[Cl^-]$ is constant inside the reference electrode. The cell voltage can therefore be rewritten in a simpler form:

$$E = \left(0.799 - 0.059\,16 \log \frac{1}{[Ag^+]}\right) - (0.241)$$

$$= 0.558 + 0.059\,16 \log [Ag^+] \qquad (15\text{-}18)$$

That is, the voltage of the cell in Figure 15-5 is a direct measure of the concentration of Ag^+. Ideally, the voltage changes by 59.16 mV (at 25°C) for each factor-of-10 change in $[Ag^+]$.

In Figure 9-8 we used a silver indicator electrode and a *glass* reference electrode. The glass electrode responds to the pH of the solution, as will be discussed in Section 15-4. The cell shown in Figure 9-8 contains a buffer to maintain a constant pH. Therefore, the glass electrode remains at a constant potential throughout the experiment; it is being used in an unconventional way as a reference electrode.

Challenge: In Equation 15-17 we used *one* electron for the silver half-reaction and *two* electrons for the calomel half-reaction. Show that if you had used two electrons for each reaction, Equation 15-18 would still be the same.

EXAMPLE: Potentiometric Precipitation Titration

A 100.0-mL solution containing 0.100 0 M NaCl was titrated with 0.100 0 M AgNO₃, and the voltage of the cell shown in Figure 15-5 was monitored. Calculate the voltage after the addition of 65.0, 100.0, and 103.0 mL of AgNO₃.

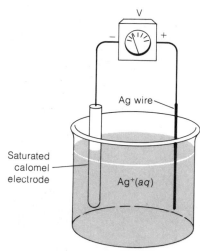

The titration reaction is

$$Ag^+ + Cl^- \rightarrow AgCl(s)$$

for which V_e (the equivalence point) is 100.0 mL.

65.0 mL: 65.0% of the chloride has been precipitated and 35.0% remains in solution:

$$[Cl^-] = \underbrace{(0.350)}_{\substack{\text{Fraction} \\ \text{remaining}}} \underbrace{(0.100\,0)}_{\substack{\text{Initial} \\ \text{concentration} \\ \text{of } Cl^-}} \underbrace{\left(\frac{100.0}{165.0}\right)}_{\substack{\text{Dilution} \\ \text{factor}}} = 0.021\,2 \text{ M}$$

with arrows labeled "Initial volume of Cl^-" pointing to 100.0 and "Total volume of solution" pointing to 165.0.

To find the cell voltage in Equation 15-18, we need to know $[Ag^+]$:

$$[Ag^+] = \frac{K_{sp} \text{ (for AgCl)}}{[Cl^-]}$$

$$= \frac{1.8 \times 10^{-10}}{0.021\,2} = 8.5 \times 10^{-9} \text{ M}$$

The cell voltage is therefore

$$E = 0.558 + 0.059\,16 \log(8.5 \times 10^{-9}) = 0.081 \text{ V}$$

100.0 mL: This is the equivalence point, at which $[Ag^+] = [Cl^-]$:

$$[Ag^+][Cl^-] = [Ag^+]^2 = K_{sp}$$
$$[Ag^+] = \sqrt{K_{sp}} = 1.34 \times 10^{-5} \text{ M}$$
$$E = 0.558 + 0.059\,16 \log(1.34 \times 10^{-5}) = 0.270 \text{ V}$$

103.0 mL: Now there is an excess of 3.0 mL of 0.100 0 M $AgNO_3$ in the solution. Therefore,

$$[Ag^+] = \underbrace{(0.100\,0)}_{\substack{\text{Initial} \\ \text{concentration} \\ \text{of } Ag^+}} \underbrace{\left(\frac{3.0}{203.0}\right)}_{\substack{\text{Dilution} \\ \text{factor}}} = 1.48 \times 10^{-3} \text{ M}$$

with arrows labeled "Volume of excess Ag^+" pointing to 3.0 and "Total volume of solution" pointing to 203.0.

and we can write

$$E = 0.558 + 0.059\,16 \log(1.48 \times 10^{-3}) = 0.391 \text{ V}$$

The cell responds to a change in Cl^- concentration because such a change necessarily changes the Ag^+ concentration, since $[Ag^+][Cl^-] = K_{sp}$.

Using a little imagination, we can say that a *silver electrode is also a halide electrode, if solid silver halide is present in the cell.* For example, if the solution contains $AgCl(s)$, we can write

$$[Ag^+][Cl^-] = K_{sp}$$

$$[Ag^+] = \frac{K_{sp}}{[Cl^-]} \qquad (15\text{-}19)$$

Demonstration 15-1   POTENTIOMETRY WITH AN OSCILLATING REACTION

The principles of potentiometry can be illustrated in a fascinating manner using *oscillating reactions*. An oscillating reaction is one in which the concentrations of various species oscillate between high and low values, instead of monotonically approaching their equilibrium values. In spite of this oscillatory behavior, the free energy of the system decreases throughout the entire reaction.[†]

One interesting example is the Belousov–Zhabotinskii reaction, in which the net reaction is

$$3CH_2(CO_2H)_2 + 2BrO_3^- + 2H^+ \rightarrow 2BrCH(CO_2H)_2 + 3CO_2 + 4H_2O$$
$$\text{Malonic acid} \qquad \text{Bromate} \qquad \qquad \text{Bromomalonic acid}$$

During this cerium-catalyzed oxidation of malonic acid by bromate, the quotient $[Ce^{3+}]/[Ce^{4+}]$ oscillates by a factor of 10 to 100.[‡] When the Ce^{4+} concentration is high, the solution is yellow. When Ce^{3+} predominates, the solution becomes colorless. Using redox indicators (Section 16-3), this reaction can be made to oscillate through a spectacular sequence of color.[§]

The simple oscillation between yellow and colorless is the basis for this demonstration. A 300 mL beaker is loaded with the following solutions:

160 mL of 1.5 M H_2SO_4

40 mL of 2 M malonic acid

30 mL of 0.5 M $NaBrO_3$ (or saturated $KBrO_3$)

4 mL of saturated ceric ammonium sulfate, $(Ce(SO_4)_2 \cdot 2(NH_4)_2SO_4 \cdot 2H_2O)$

After an induction period of 5 to 10 minutes with magnetic stirring, oscillations can be initiated by adding 1 mL of ceric ammonium sulfate solution. The reaction is somewhat temperamental and may need to be treated once or twice more with Ce^{4+} over a five-minute period to initiate oscillations.

A galvanic cell is built around the reaction as shown at the top of the following page. The value of $[Ce^{3+}]/[Ce^{4+}]$ is monitored by a Pt indicator electrode, with an S.C.E. reference electrode. You should be able to write the cell reactions.

In place of a potentiometer (a pH meter), we use a chart recorder to obtain a permanent record of the oscillations. Since the potential oscillates over a range of ~ 100 mV, but is centered near ~ 1.2 V, the cell voltage is offset by ~ 1.2 V with any available power supply.

[†] General discussions of oscillating reactions can be found in I. R. Epstein, K. Kustin, P. De Kepper, and M. Orbán, *Scientific American,* March 1983, p. 112; I. R. Epstein, *Chem. Eng. News,* 30 March 1987, p. 24; and H. Degn, *J. Chem. Ed.,* **49,** 302 (1972).

[‡] The mechanisms of oscillating reactions are discussed by R. J. Field and F. W. Schneider, *J. Chem. Ed.,* **66,** 195 (1989); R. M. Noyes, *J. Chem. Ed.,* **66,** 190 (1989); and P. Ruoff, M. Varga, and E. Körös, *Acc. Chem. Res.,* **21,** 326 (1988).

[§] For some beautiful lecture demonstrations and experiments, see D. Kolb, *J. Chem. Ed.,* **65,** 1004 (1988); R. J. Field, *J. Chem. Ed.,* **49,** 308 (1972); J. N. Demas and D. Diemente, *J. Chem. Ed.,* **50,** 357 (1973); and J. F. Lefelhocz, *J. Chem. Ed.,* **49,** 312 (1972). Computer monitoring of an oscillating reaction is described by P. Aroca, Jr. and R. Aroca, *J. Chem. Ed.,* **64,** 1017 (1987). Other systems of interest include a chemiluminescent oscillator [J. Amrehn, P. Resch, and F. W. Schneider, *J. Phys. Chem.,* **92,** 3318 (1988)], the mercury beating heart [D. Avnir, *J. Chem. Ed.,* **66,** 211 (1989)], a salt-water oscillator [K. Yoshikawa, S. Nakata, M. Yamanaka, and T. Waki, *J. Chem. Ed.,* **66,** 205 (1989)], and a gas flashback oscillator [L. J. Soltzberg, M. M. Boucher, D. M. Crane, and S. S. Pazar, *J. Chem. Ed.,* **64,** 1043 (1987)].

(continued)

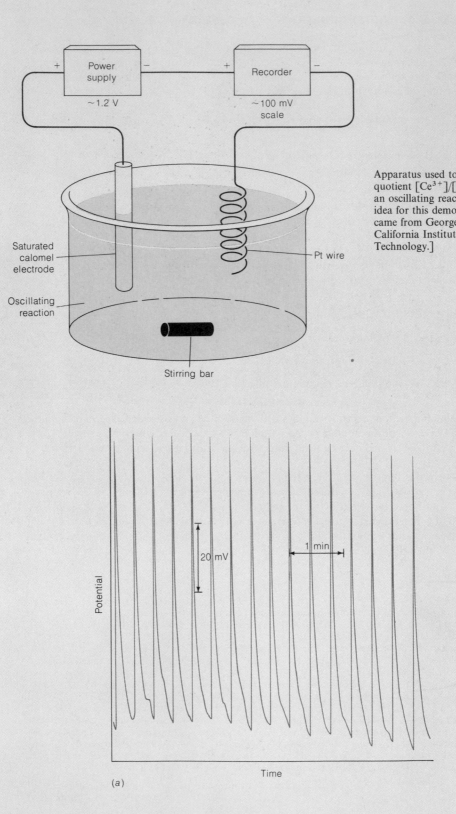

Apparatus used to monitor the quotient $[Ce^{3+}]/[Ce^{4+}]$ for an oscillating reaction. [The idea for this demonstration came from George Rossman, California Institute of Technology.]

Power supply
~1.2 V

Recorder
~100 mV scale

Saturated calomel electrode

Oscillating reaction

Pt wire

Stirring bar

20 mV

1 min

Potential

Time

(a)

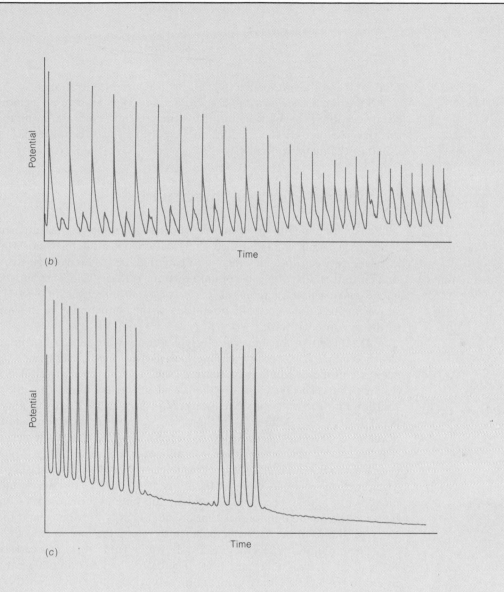

(b)

Potential

Time

(c)

Potential

Time

Using the setup shown, we recorded the traces. Trace a shows what is usually observed. The potential changes rapidly during the abrupt colorless-to-yellow color change and gradually during the gentle yellow-to-colorless change. Trace b shows two different cycles superimposed in the same solution. This unusual event happened spontaneously in a reaction that had been oscillating normally for about 30 minutes. Trace c represents a reaction that had been oscillating for several hours. It stopped for a while, began again spontaneously, and finally died.

This same experiment can be used to demonstrate a bromide ion-selective electrode, since the concentration of Br^- also oscillates in this solution. Simultaneous monitoring of $[Ce^{3+}]/[Ce^{4+}]$ and the Br^- concentration would be worthwhile. An even more striking oscillation, in which the concentration of I^- oscillates over four orders of magnitude, has been described.[†] Monitoring the oscillating I^- concentration with a silver–silver iodide electrode is an informative classroom demonstration.

[†] T. S. Briggs and W. C. Rauscher, *J. Chem. Ed.*, **50**, 496 (1973).

Putting this value of $[Ag^+]$ into Equation 15-18 gives

$$E = 0.558 + 0.059\,16 \log \frac{K_{sp}}{[Cl^-]} \qquad (15\text{-}20)$$

Box 15-1 describes how electrodes can be chemically tailored for specific tasks.

That is, the voltage responds to changes in Cl^- concentration.

Some metals, such as Ag, Cu, Zn, Cd, and Hg, can be used as indicator electrodes for their aqueous ions. Most metals, however, are unsuitable for this purpose because the equilibrium $M \rightleftharpoons M^{n+} + ne^-$ is not readily established at the metal surface.

15-3 WHAT IS A JUNCTION POTENTIAL?

$E(\text{measured}) = E(\text{cell}) + E(\text{junction})$

Since the junction potential is usually unknown, the value of $E(\text{cell})$ is uncertain.

There is a fundamental problem with most potentiometric measurements. It concerns a small voltage, called the **junction potential,** that exists at the interface between the salt bridge and each half-cell. This voltage is usually small, but it is almost always of unknown magnitude. *The junction potential puts a fundamental limitation on the accuracy of direct potentiometric measurements,* because we usually do not know the contribution of the junction to the measured voltage.

Any time two dissimilar electrolyte solutions are placed in contact, an electric voltage develops at their interface. To see why this junction potential occurs, consider a solution containing NaCl in contact with distilled water (Figure 15-6). The Na^+ and Cl^- ions will begin to diffuse from the NaCl solution into the water phase. However, Cl^- ion has a greater **mobility** than Na^+. That is, Cl^- diffuses faster than Na^+. As a result, a region rich in Cl^- develops at the front. This region has excess negative charge. Behind it is a region depleted of Cl^- and thus containing excess positive charge. The result is an electric potential difference at the junction of the NaCl and H_2O phases. The junction potential opposes the movement of Cl^- and accelerates the movement of Na^+. The steady-state junction potential represents a balance between the unequal mobilities that create a charge imbalance and the tendency of the resulting charge imbalance to retard the movement of Cl^-.

The mobilities of several ions are shown in Table 15-1. You can see that K^+ and Cl^- have similar mobilities. Therefore, the junction potential at the interface of a KCl salt bridge with another solution will be slight. Furthermore, the junction potentials at each end of a salt bridge often partially cancel each other. Although a salt bridge necessarily contributes some *unknown* net potential to a galvanic cell, the contribution is fairly small (a few millivolts).

Several liquid junction potentials are listed in Table 15-2. You can see that a high concentration of KCl in one solution reduces the magnitude of the potential. This is why saturated KCl is used in salt bridges. Any time a liquid junction is present in a cell, there will be some (usually unknown)

A junction potential exists at each end of a salt bridge.

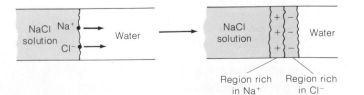

Figure 15-6
Development of the junction potential caused by unequal mobilities of Na^+ and Cl^-.

Table 15-1
Mobilities of ions in water at 25°C

Ion	Mobility $[m^2/(s \cdot V)]^\dagger$
H^+	36.30×10^{-8}
Rb^+	7.92×10^{-8}
K^+	7.62×10^{-8}
NH_4^+	7.61×10^{-8}
La^{3+}	7.21×10^{-8}
Ba^{2+}	6.59×10^{-8}
Ag^+	6.42×10^{-8}
Ca^{2+}	6.12×10^{-8}
Cu^{2+}	5.56×10^{-8}
Na^+	5.19×10^{-8}
Li^+	4.01×10^{-8}
OH^-	20.50×10^{-8}
$Fe(CN)_6^{4-}$	11.45×10^{-8}
$Fe(CN)_6^{3-}$	10.47×10^{-8}
SO_4^{2-}	8.27×10^{-8}
Br^-	8.13×10^{-8}
I^-	7.96×10^{-8}
Cl^-	7.91×10^{-8}
NO_3^-	7.40×10^{-8}
ClO_4^-	7.05×10^{-8}
F^-	5.70×10^{-8}
HCO_3^-	4.61×10^{-8}
$CH_3CO_2^-$	4.24×10^{-8}

† The mobility of an ion is the terminal velocity that the particle achieves in an electric field of 1 V/m. Mobility = velocity/field. The units of mobility are therefore (m/s)/(V/m) = $m^2/(s \cdot V)$.

Table 15-2
Liquid junction potentials at 25°C

Junction	Potential (mV)
0.1 M NaCl|0.1 M KCl	-6.4
0.1 M NaCl|3.5 M KCl	-0.2
1 M NaCl|3.5 M KCl	-1.9
0.1 M HCl|0.1 M KCl	$+27$
0.1 M HCl|3.5 M KCl	$+3.1$

Note: A positive sign means that the right side of the junction becomes positive with respect to the left side.
SOURCE: G. D. Christian, *Analytical Chemistry,* 3rd ed. (New York: Wiley, 1980), p. 294.

junction potential. Therefore, the measured cell voltage cannot be attributed entirely to the two half-reactions.

EXAMPLE: Junction Potential

A 0.1 M NaCl solution was placed in contact with a 0.1 M $NaNO_3$ solution. Which side of the junction will be positive and which will be negative?

Since $[Na^+]$ is equal on both sides, there will be no net diffusion of Na^+ across the junction. However, Cl^- will diffuse into the $NaNO_3$, and NO_3^- will diffuse into the NaCl. Since the mobility of Cl^- is greater than that of NO_3^-, the NaCl region will be depleted of Cl^- faster than the $NaNO_3$ region will be depleted of NO_3^-. The $NaNO_3$ side will become negative, and the NaCl side will become positive.

15-4 HOW ION-SELECTIVE ELECTRODES WORK

Ion-selective electrodes respond selectively to one species in a solution. These electrodes have a thin membrane separating the sample from the inside of

Box 15-1   SURFACE-MODIFIED ELECTRODES[†]

Instead of limiting their horizons to the properties of a few naturally occurring electrode materials, chemists are modifying electrode surfaces to accomplish specific tasks. For example, the reaction

$$HCO_3^- + 2e^- + 2H^+ \rightarrow HCO_2^- + H_2O \qquad \text{(A)}$$

$$\underset{\text{Bicarbonate}}{} \qquad\qquad \underset{\text{Formate}}{}$$

does not occur readily at most electrode surfaces. However, finely dispersed Pd is known to catalyze the hydrogenation reaction

$$H_2 + HCO_3^- \rightarrow HCO_2^- + H_2O \qquad \text{(B)}$$

An ingenious modification[‡] of otherwise unreactive metal electrodes allows reaction (B) to occur near its thermodynamic equilibrium potential.

The surface of a metal can be oxidized electrochemically and the oxide reacts in aqueous solution to create reactive hydroxyl groups chemically bound to the metal surface. Treatment with a silicon compound of the type $RSiX_3$ (X = Cl or OCH_3) in the presence of water gives a layer of silicon polymer covalently attached to the metal surface:

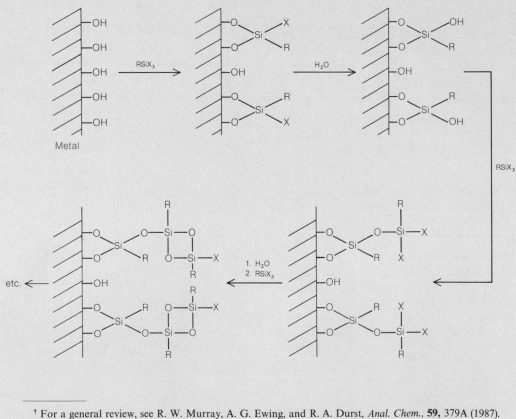

[†] For a general review, see R. W. Murray, A. G. Ewing, and R. A. Durst, *Anal. Chem.,* **59,** 379A (1987).
[‡] C. J. Stalder, S. Chao, and M. S. Wrighton, *J. Amer. Chem. Soc.,* **106,** 3673 (1984).

For the bicarbonate reduction, the silicon compound contained the reducible moiety 4,4'-bipyridine:

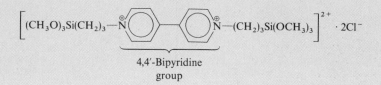

4,4'-Bipyridine
group

Tungsten and platinum electrodes were coated with a polymer formed from this compound, giving a surface coverage of 10^{-7} mol of 4,4'-bipyridine groups per square centimeter of electrode surface. Then $PdCl_4^{2-}$ ions were exchanged for Cl^- in the surface polymer by soaking in K_2PdCl_4 solution. At electrode potentials of -0.3 to -0.75 V (versus S.C.E.), the 4,4'-bipyridine groups were reduced:

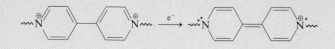

The radical cation thus formed was powerful enough to reduce $PdCl_4^{2-}$ to microscopic particles of metallic Pd.

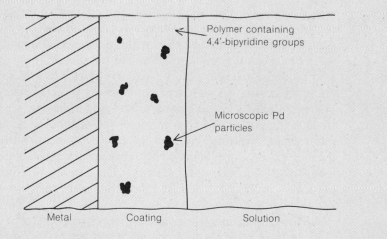

When a potential of -0.7 V (versus S.C.E.) was applied to the electrode immersed in $NaHCO_3$ solution, the 4,4'-bipyridine groups conducted electrons through the polymer into the Pd particles. There H_2O was reduced to H_2, which reacted with HCO_3^- at the Pd surface to give HCO_2^- (Reaction B). Up to 85% of the current went into reduction of HCO_3^-, and the reaction could be run long enough to reduce more than 1 000 HCO_3^- molecules per atom of Pd.

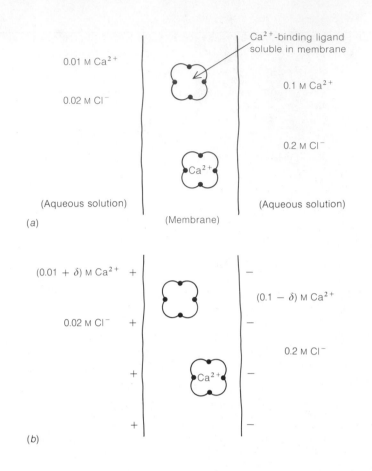

Figure 15-7
Mechanism of ion-selective electrode. (a) Initial conditions prior to Ca^{2+} migration across membrane. (b) After δ moles of Ca^{2+} per liter have crossed the membrane, giving the left side a charge of $+2\delta$ mol/L and the right side a charge of -2δ mol/L.

Mechanism of ion-selective electrode: charge buildup balances tendency to move from high concentration to low concentration.

the electrode. On the inside is a solution containing the ion of interest at a constant activity. On the outside is a variable sample. The potential difference across the membrane depends on the difference in activity of the analyte species between the inside solution and the sample.

To understand how an ion-selective electrode works, imagine a membrane separating two solutions containing $CaCl_2$ (Figure 15-7). The membrane contains a ligand that can bind and transport Ca^{2+}, but *not* Cl^-. Initially the potential difference across the membrane is zero, because both solutions are neutral (Figure 15-7a). However, there will be a tendency for the ions on the side of high activity (concentration) to diffuse to the side with low activity (concentration). Chloride has no way to cross the membrane, but Ca^{2+} can bind to the ligand dissolved in the membrane and cross from one side to the other.

Thermodynamics tells us that the free energy difference between a species in solution at activity $\mathscr{A}_1$ in one place and activity $\mathscr{A}_2$ at another place is

$$\Delta G = -RT \ln \frac{\mathscr{A}_1}{\mathscr{A}_2} \qquad (15\text{-}21)$$

As Ca^{2+} migrates from the region of high activity to the region of low activity, positive charge is built up on the low-activity side (Figure 15-7b). Eventually the charge buildup prevents further migration of Ca^{2+} to the positively charged side. This is the same thing that happens at a liquid junction, and a constant potential difference across the junction or membrane is reached.

The constant potential difference is such that the free energy decrease due to the activity difference is balanced by the free energy increase from repulsion of like charges:

$$\Delta G = -nFE \tag{14-5}$$

$$-RT \ln \frac{\mathscr{A}_1}{\mathscr{A}_2} = -nFE$$

$$E = \frac{RT}{nF} \ln \frac{\mathscr{A}_1}{\mathscr{A}_2}$$

$$E = \frac{0.059\,16}{n} \log \frac{\mathscr{A}_1}{\mathscr{A}_2} \quad \text{(volts, at } 25°\text{C)} \tag{15-22}$$

$$\ln x = (\ln 10)(\log x)$$

A potential difference of $59.16/2 = 29.58$ mV is expected for every factor-of-10 difference in activity of Ca^{2+} across the membrane. No charge buildup results from the Cl^- activity difference because Cl^- has no way to cross the membrane.

The next two sections deal with the most common ion-selective electrodes. The membranes of these electrodes are made of such diverse materials as glass, synthetic polymers, and inorganic crystals.

15-5 pH MEASUREMENT WITH A GLASS ELECTRODE

The most widely employed ion-selective electrode is the **glass electrode** for measuring pH.[†] A diagram of a typical **combination electrode,** incorporating both the glass and the reference electrodes in one body, is shown in Figure 15-8. A line diagram of this cell can be written as follows:

Glass
membrane

$Ag(s)\,|\,AgCl(s)\,|\,Cl^-(aq)\,||\,H^+(aq,\text{ outside})\,|\,H^+(aq,\text{ inside}),\,Cl^-(aq)\,|\,AgCl(s)\,|\,Ag(s)$

| Outer reference electrode | H^+ outside glass electrode (analyte solution) | H^+ inside glass electrode | Inner reference electrode |

The pH-sensitive part of the electrode is the specially constructed thin glass membrane at the bottom of the electrode. Figure 15-9 shows the structure of the silicate lattice of which glass is made. The irregular silicate network contains negatively charged oxygen atoms available for coordination to metal cations of suitable size. The monovalent cations, particularly Na^+, are able to move somewhat through the silicate lattice.

[†] Although the glass electrode is by far the most widely used electrode for pH measurement, many other types of electrodes are also sensitive to pH and can withstand harsher environments than those tolerated by the glass electrode. An electrode of iridium coated with IrO_2 behaves as a pH sensor [S. Bordi, M. Carlà, and G. Papeschi, *Anal. Chem,* **56,** 317 (1984)]. An electrode based on ZrO_2 can be used to measure the pH of solutions up to 300°C [L. W. Niedrach, *Angew. Chem.,* **26,** 161 (1987)]. Platinum and carbon surfaces are also pH-sensitive, but they must be coated to make them selective for H^+ [G. Cheek, C. P. Wales, and R. J. Nowak, *Anal Chem.,* **55,** 380 (1983)]. Ingenious surface-modified wire electrodes for pH measurement have also been described [I. Rubenstein, *Anal. Chem.,* **56,** 1135 (1984)].

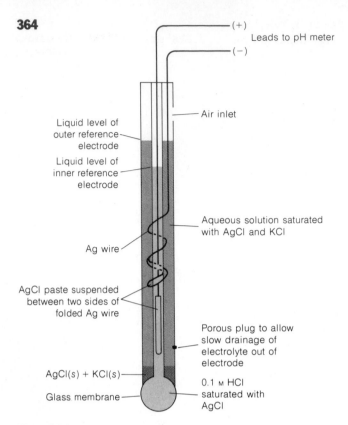

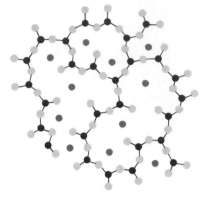

Figure 15-9
Schematic diagram of the structure of glass, which consists of an irregular network of SiO_4 tetrahedra connected through their oxygen atoms. ○ = O, ● = Si, ◉ = Cation. Cations such as Li^+, Na^+, K^+, and Ca^{2+} are coordinated to the oxygen atoms. The silicate network is not planar. This diagram is a projection of each tetrahedron onto the plane of the page. [Adapted from G. A. Perley, *Analy. Chem.*, **21**, 394 (1949).]

Figure 15-8
Diagram of a glass combination electrode having a silver–silver chloride reference electrode. The glass electrode is immersed in a solution of unknown pH to where the porous plug, on the lower light, is below the surface of the liquid. The two silver electrodes measure the voltage across the glass membrane.

A schematic cross section of the glass membrane of the pH electrode is shown in Figure 15-10. The two surfaces exposed to aqueous solution absorb some water and become swollen. Most of the metal cations in these *hydrated gel* regions of the membrane diffuse out of the glass and into solution. Concomitantly, H^+ from solution can diffuse into the swollen silicate lattice and occupy some of the cation binding sites. We refer to the equilibrium in which H^+ replaces metallic cations in the glass as an **ion-exchange equilibrium.** The H^+ ion-exchange equilibrium is depicted schematically in Figure 15-11.

It has been shown by tracer studies using tritium (the radioactive isotope ^3H) that hydrogen ion does not cross the glass membrane of the pH electrode.

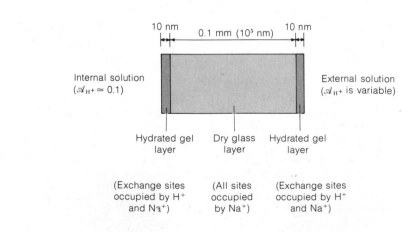

Figure 15-10
Schematic diagram showing a cross section of the glass membrane of a pH electrode.

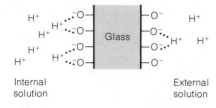

Internal
solution

External
solution

Figure 15-11
Ion-exchange equilibria on the inner and
outer surfaces of the glass membrane.

The membrane is in equilibrium with H^+ on each surface, and Na^+ ions transport charge across the membrane so that one side still "senses" the other side. The mechanism of the glass electrode is therefore similar to that described for other ion-selective electrodes, except that the species that migrates across the membrane is not the same as the species selectively adsorbed at each membrane surface. The resistance of the glass membrane is typically $10^8\ \Omega$, so very little current actually flows across it.

The potential difference between the inner and outer silver–silver chloride electrodes in Figure 15-8 depends on the chloride concentration in each electrode compartment and on the potential difference across the glass membrane. Since the chloride concentration is fixed in each electrode compartment, and since the H^+ concentration is fixed on the inside of the glass membrane, the only factor causing a change in voltage is a change in pH of the analyte solution outside the glass membrane. The thermodynamic argument in the preceding section implies that for every factor-of-10 difference in H^+ activity on each side of the membrane, a potential difference of 59.16 mV (at 25°C) should develop. That is, *the voltage of the ideal pH electrode changes by 59.16 mV for every pH-unit change of the analyte.*

For real glass electrodes, the response to changes in pH is nearly Nernstian and can be described by the equation

$$E = \text{constant} - \beta(0.059\ 16)\log\frac{\mathscr{A}_{H^+}(\text{inside})}{\mathscr{A}_{H^+}(\text{outside})} \quad \text{(at 25°C)} \quad (15\text{-}23)$$

$\uparrow$
Asymmetry
potential

The value of β, the *electromotive efficiency,* is close to 1.00 (typically >0.98). It varies with each type of glass and with each individual electrode. The constant term, called the **asymmetry potential,** would be zero if both sides of the glass were identical. That is, no voltage would be observed if the activities of H^+ were the same inside and outside and the membrane were ideal. However, no two sides of any real object are identical, and a small voltage exists even if $\mathscr{A}_{H^+}$ is the same on both sides of the membrane. We correct for this asymmetry potential by calibrating the electrode in solutions of known pH. Because the asymmetry potential varies with time and conditions (temperature, concentrations, phase of the moon, color of your socks), electrodes must be calibrated frequently.

The pH electrode measures H^+ activity, not H^+ concentration.

Real electrodes come near to obeying the Nernst equation.

A pH electrode **must** be calibrated before it can be used, It should be calibrated about every two hours during sustained use.

Calibrating a Glass Electrode

Many pH standards are available commercially. The pH values of the solutions in Table 15-3 were measured at the U.S. National Institute of Standards and Technology and are considered to be accurate to ±0.01 pH unit.

Table 15-3
pH values of National Institute of Standards and Technology buffers

Temperature (°C)	Saturated (25°C) potassium hydrogen tartrate	0.05 m potassium dihydrogen citrate	0.05 m potassium hydrogen phthalate	0.025 m potassium dihydrogen phosphate 0.025 m disodium hydrogen phosphate	0.008 695 m potassium dihydrogen phosphate 0.030 43 m disodium hydrogen phosphate	0.01 m borax	0.025 m sodium bicarbonate 0.025 m sodium carbonate
0	—	3.863	4.003	6.984	7.534	9.464	10.317
5	—	3.840	3.999	6.951	7.500	9.395	10.245
10	—	3.820	3.998	6.923	7.472	9.332	10.179
15	—	3.802	3.999	6.900	7.448	9.276	10.118
20	—	3.788	4.002	6.881	7.429	9.225	10.062
25	3.557	3.776	4.008	6.865	7.413	9.180	10.012
30	3.552	3.766	4.015	6.853	7.400	9.139	9.966
35	3.549	3.759	4.024	6.844	7.389	9.102	9.925
38	3.548	—	4.030	6.840	7.384	9.081	—
40	3.547	3.753	4.035	6.838	7.380	9.068	9.889
45	3.547	3.750	4.047	6.834	7.373	9.038	9.856
50	3.549	3.749	4.060	6.833	7.367	9.011	9.828
55	3.554	—	4.075	6.834	—	8.985	—
60	3.560	—	4.091	6.836	—	8.962	—
70	3.580	—	4.126	6.845	—	8.921	—
80	3.609	—	4.164	6.859	—	8.885	—
90	3.650	—	4.205	6.877	—	8.850	—
95	3.674	—	4.227	6.886	—	8.833	—

Note: The designation m stands for molality.

In the buffer solution preparations, it is essential to use high-purity materials and to employ freshly distilled or deionized water of specific conductivity not greater than 5 micromho/cm. Solutions having pH 6 or above ahould be stored in plastic containers, and preferably with an NaOH trap to prevent ingress of atmospheric carbon dioxide. They can normally be kept for 2–3 weeks, or slightly longer in a refrigerator.

1. Saturated (25°C) potassium hydrogen tartrate, $KHC_4H_4O_6$. An excess of the salt is shaken with water, and it can be stored in this way. Before use, it should be filtered or decanted at a temperature between 22°C and 28°C.

2. 0.05 m potassium dihydrogen citrate, $KH_2C_6H_5O_7$. Dissolve 11.41 g of the salt in one liter of solution at 25°C.

3. 0.05 m potassium hydrogen phthalate. Although this is not usually essential, the crystals may be dried at 110°C for an hour, then cooled in a desiccator. At 25°C, 10.12 g $C_6H_4(CO_2H)(CO_2K)$ is dissolved in water, and the solution made up to one liter.

4. 0.025 m disodium hydrogen phosphate, 0.025 m potassium dihydrogen phosphate. The anhydrous salts are best used; each should be dried for two hours at 120°C and cooled in a desiccator, since they are slightly hygroscopic. Higher drying temperatures should be avoided to prevent formation of condensed phosphates. Dissolve 3.53 g Na_2HPO_4 and 3.39 g KH_2PO_4 in water to give one liter of solution at 25°C.

5. 0.008 695 m potassium dihydrogen phosphate, 0.030 43 m disodium hydrogen phosphate. Prepare as in Step 4 and dissolve 1.179 g KH_2PO_4 and 4.30 g Na_2HPO_4 in water to give one liter of solution at 25°C.

6. 0.01 m sodium tetraborate decahydrate. Dissolve 3.80 g $Na_2B_4O_7 \cdot 10H_2O$ in water to give one liter of solution. This borax solution is particularly susceptible to pH change from carbon dioxide absorption, and it should be correspondingly protected.

7. 0.025 m sodium bicarbonate, 0.025 m sodium carbonate. Primary standard-grade Na_2CO_3 is dried at 250°C for 90 minutes and stored over $CaCl_2$ and Drierite. Reagent-grade $NaHCO_3$ is dried over molecular sieves and Drierite for two days at room temperature. Do not heat $NaHCO_3$, or it may decompose to Na_2CO_3. Dissolve 2.092 g of $NaHCO_3$ and 2.640 g of Na_2CO_3 in one liter of solution at 25°C.

SOURCE: R. G. Bates, *J. Res. National Bureau of Standards,* **66A,** 179 (1962); and B. R. Staples and R. G. Bates, *J. Res. National Bureau of Standards,* **73A,** 37 (1969). The instructions for preparing these solutions are taken, in part, from G. Mattock in C. N. Reilley, ed., *Advances in Analytical Chemistry and Instrumentation* (New York: Wiley, 1963), Vol. 2, p. 45. See also R. G. Bates, *Determination of pH: Theory and Practice,* 2nd ed. (New York: Wiley, 1973), Chap. 4.

The pH of the calibration standards should bracket the pH of the unknown.

Before using a pH electrode, you must calibrate it, using two (or more) standard buffers selected so that the pH of the unknown lies within the range of the pH values for the buffers. The exact calibration procedure is different for each model of pH meter, so consult the manufacturer's instructions for your meter. Before calibrating the electrode, wash it with distilled water and gently *blot* it dry with a tissue. Do not *wipe* it because this might produce a static electric charge on the glass.

For calibration with most types of pH meters, dip the electrode in a standard buffer whose pH is near 7 and allow the electrode to equilibrate for at least a minute. For best results, all solutions for calibration and measurement should be stirred continuously during the measurement. Following the manufacturer's instructions, adjust the meter reading (usually with a knob labeled "Calibrate") to indicate the pH of the standard buffer. (If the meter has an adjustment of the "isopotential point," it should be set to the pH of this first buffer.) The electrode is then washed, blotted dry, and immersed in a second standard whose pH is further from 7 than the pH of the first standard. If the electrode response were perfectly Nernstian, the voltage would change by 0.059 16 V per pH unit at 25°C. The actual change is slightly less, so these two measurements serve to establish the value of β in Equation 15-23. The pH of the second buffer is set on the meter with a knob that may be labeled "Slope" or "Temperature" on different instruments. This effectively tells the meter the value of β. If your meter does not have an isopotential point adjustment, it may be necessary to repeat the calibration with the two buffers to obtain correct readings for each. Finally, the electrode is dipped in the unknown and the voltage is translated directly into a pH reading by the meter.

A glass electrode must be stored in aqueous solution so that the hydrated gel layer of the glass does not dry out. If the electrode has been allowed to dry out, it should be reconditioned by soaking in water for several hours. If the electrode is to be used at a pH above 9, soak it in a high-pH buffer.

> Do not leave a glass electrode out of water (or in a nonaqueous solvent) any longer than necessary.

Glass electrodes slowly wear out, partly because the composition of the glass changes near the solution interface as ions diffuse in and out. If electrode response becomes sluggish or if the electrode cannot be calibrated properly, try washing it with 6 M HCl, followed by water. As a last resort, the electrode can be soaked in 20% (wt/wt) aqueous ammonium bifluoride, NH_4HF_2, for one minute in a plastic beaker. This reagent dissolves a little of the glass and exposes fresh surface. Wash the electrode with water and try calibrating it again. *Ammonium bifluoride must not contact your skin, because it produces HF burns.*

Errors in pH Measurement

To make intelligent use of a glass electrode, it is important to understand its limitations:

1. Our knowledge of an analyte's pH cannot be any better than our knowledge of the pH of the buffers used to calibrate the meter and the electrode. This error is typically on the order of ± 0.01 pH unit.

2. A *junction potential* exists across the porous plug near the bottom of the electrode in Figure 15-7. The porous plug is the salt bridge connecting the outer silver–silver chloride electrode with the analyte solution. If the ionic composition of the analyte solution is different from that of the standard buffer, the junction potential will change *even if the pH of the two solutions is the same*. This change appears as a change in pH and leads to an uncertainty in the pH of at least ~ 0.01 pH unit. Box 15-2 describes measurements in which junction potential changes lead to significant errors.

> The apparent pH will change if the ionic composition of the analyte changes, even when the actual pH is constant.

3. When the concentration of H^+ is very low and the concentration of Na^+ is high (a typical set of conditions in strongly basic solution), the electrode responds to Na^+ as well as to H^+. This arises because Na^+ can participate in ion exchange with the hydrated gel layer.

Box 15-2 SYSTEMATIC ERROR IN RAINWATER pH MEASUREMENT: THE EFFECT OF JUNCTION POTENTIAL

Combustion products from automobiles and factories include nitrogen oxides and sulfur dioxide, which react with water in the atmosphere to produce acids.

$$SO_2 + H_2O \rightarrow \quad H_2SO_3$$
sulfurous acid

The damage caused by *acid rain* containing these pollutants is a serious threat to lakes and forests around the world. Monitoring the pH of rainwater is a critical component of programs to measure and reduce the production of acid rain by man's activities.

To identify and correct systematic errors in the measurement of pH of rainwater, workers at the U.S. National Institute of Standards and Technology conducted a careful study involving 17 different laboratories.[†] Eight samples were provided to each laboratory, along with explicit instructions for how to conduct the measurements. Each laboratory used two buffers to standardize electrodes and pH meters. Sixteen laboratories successfully measured the pH of Unknown A (within ± 0.02 pH units), which was Standard Reference Material 185f, 0.05 m potassium acid phthalate, whose pH is 4.008 at 25°C (Table 15-3). One lab whose measurement was 0.04 pH units low was later found to have a faulty commercial standardization buffer.

The figure at the upper right shows typical results for the pH of rainwater. The average of the 17 measurements is given by the horizontal line at pH 4.14. Individual results are labeled with the letters s, t, u, v, w, x, y, and z to identify the type of pH electrode used for the measurements. Laboratories using electrode type t, for example, obtained good results. Laboratories using electrode types s and w had relatively large systematic errors in the figure and in measurement of other unknowns in the study. The type s electrode was a combination electrode (Figure 15-8) containing a reference electrode liquid junction with a larger area than those of the other electrodes. Electrode type w was a combination electrode whose reference electrode was filled with a gel.

[†] W. F. Koch, G. Marinenko, and R. C. Paule, *J. Res. National Bureau of Standards,* **91,** 23 (1986).

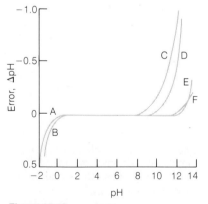

Figure 15-12
Acid and alkaline errors of some glass electrodes. A: Corning 015, H_2SO_4. B: Corning 015, HCl. C: Corning 015, 1 M Na^+. D: Beckman-GP, 1 M Na^+. E: L & N Black Dot, 1 M Na^+. F: Beckman Type E, 1 M Na^+. [R. G. Bates, *Determination of pH: Theory and Practice,* 2nd ed. (New York: Wiley, 1973).]

$$Na^+ \text{ (in aqueous phase)} \rightleftharpoons Na^+ \text{ (in hydrated gel layer)}$$

This reaction is not nearly as important as H^+ ion exchange, but becomes significant when $\mathscr{A}_{H^+}$ is low and $\mathscr{A}_{Na^+}$ is high. Under these conditions, the electrode behaves as if Na^+ were H^+, and the apparent pH is lower than the true pH. This behavior leads to the **alkaline error,** or **sodium error,** and its magnitude varies with the composition of the glass (Figure 15-12). To reduce the alkaline error, lithium has largely replaced sodium in glass electrodes.

4. In very acidic solutions, the measured pH is higher than the actual pH. The reason for this is not clear, but it may be related to the decreased activity of water in concentrated acid solution.

5. Adequate time must be allowed for the glass membrane to equilibrate with a fresh solution. In a well-buffered solution, this takes just seconds with adequate stirring. In a poorly buffered solution (such as near the equivalence point of a titration), it could take many minutes.

6. An electrode that has not been stored in water becomes dehydrated and requires several hours of soaking before it responds to H^+ correctly.

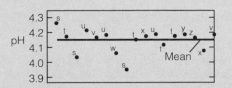

Rainwater pH measured at 17 different labs using standard buffers for calibration. Letters s to z designate different types of electrodes used for pH measurement.

It was hypothesized that variability in the liquid junction potential (Section 15-3) led to variability among the pH measurements. Buffers used for standardization typically have ionic strengths in the neighborhood of 0.05 M, whereas rainwater samples have ionic strengths two or more orders of magnitude lower. To test the hypothesis that junction potential was the cause of the systematic errors, a pure HCl solution with a concentration near 2×10^{-4} M was used as a pH standard in place of high ionic strength buffers. The data below were obtained, with good results from all but the first lab. The standard deviation of all 17 measurements was reduced from 0.077 pH units with the standard buffer to 0.029 pH units with the HCl standard. It was concluded that junction potential is the cause of most of the variability between labs and that a low ionic strength standard is appropriate for rainwater pH measurements.[††]

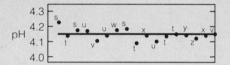

Rainwater pH measured using low ionic strength HCl for calibration.

[††] A commercially available electrode is designed to minimize junction potentials induced by changing ionic strength. The *free diffusion junction* of this electrode is a Teflon capillary tube containing an electrolyte that is periodically renewed by syringe. [A. Kopelove, S. Franklin, and G. M. Miller, *Amer. Lab.,* June 1989, p. 40].

Errors 1 and 2 are unavoidable and effectively limit the accuracy of pH measurement with the glass electrode to ± 0.02 pH unit, at best. Measurement of pH differences *between* solutions can be accurate to about ± 0.002 pH unit with good equipment. But knowledge of the true pH will still be at least an order of magnitude more uncertain. It is instructive to realize that an uncertainty of ± 0.02 pH unit corresponds to an uncertainty of $\pm 5\%$ in $\mathscr{A}_{H^+}$.

One final caution: A pH meter has a temperature knob that allows you to vary the slope (0.059 16) in Equation 15-23 for different temperatures. (That is, a change of 1.00 pH unit corresponds to 54 mV at 0°C, 59 mV at 25°C, and 64 mV at 50°C.) *A pH meter must be calibrated at the same temperature at which the measurement will be made.* You cannot calibrate your equipment at one temperature and then make an accurate measurement at a second temperature simply by changing the temperature adjustment. The meter must be calibrated at the same temperature as that at which the pH of the unknown is measured. Furthermore, the pH of the standard buffer changes with temperature, and this variation must be known in order to use the buffer for calibration purposes.

Challenge: Show that the potential of the glass electrode described by Equation 15-23 changes by 1.3 mV when the analyte H^+ activity changes by 5.0%. Since 59 mV $\approx$ 1 pH unit, 1.3 mV = 0.02 pH unit. *Moral:* A small uncertainty in voltage (1.3 mV) or pH (0.02 units) corresponds to a large uncertainty (5%) in analyte concentration. Similar uncertainties arise in other potentiometric measurements.

Box 15-3 MICROELECTRODES INSIDE LIVING CELLS

To understand how nerve and muscle cells work, an accurate knowledge of the intracellular ionic composition is necessary. The figure below shows the design of a *microelectrode* used to make intracellular measurements of K^+ and Cl^-. It is similar to the liquid-based electrode in Figure 15-13, but the glass tip is drawn out to microscopic dimensions. Inside this tip a droplet of liquid ion exchanger is held by capillary action. To make a measurement, the cell must be impaled with this electrode and also with a reference electrode of similar dimensions. The reference electrode is a silver–silver chloride electrode containing 3 M KCl and no liquid ion exchanger.

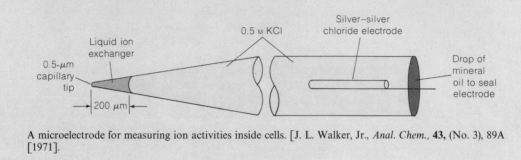

A microelectrode for measuring ion activities inside cells. [J. L. Walker, Jr., *Anal. Chem.*, **43**, (No. 3), 89A [1971].

15-6 ION-SELECTIVE ELECTRODES

There are many imaginative strategies for producing an electrode that is selectively sensitive to one species. Most electrodes fall into one of the following classes:

1. *glass membranes* for H^+ and certain monovalent cations
2. *solid-state electrodes* based on inorganic salt crystals
3. *liquid-based electrodes* using a hydrophobic membrane saturated with a hydrophobic liquid ion exchanger
4. *compound electrodes* involving a species-selective electrode enclosed by a region in which that species is separated from other species or produced by a chemical reaction

Selectivity Coefficient

No electrode responds exclusively to one kind of ion, but the glass electrode is among the most selective. A high-pH glass electrode responds to Na^+ only when $[H^+] \lesssim 10^{-12}$ M and $[Na^+] \gtrsim 10^{-2}$ M (Figure 15-12).

By varying the composition of the glass, it is possible to alter its sensitivity to different ions. Corning 015 glass for pH electrodes contains 22% (wt/wt) Na_2O, 6% CaO, and 72% SiO_2. By contrast, a glass membrane made of 11% Na_2O, 18% Al_2O_3, and 71% SiO_2 is used in a *sodium ion electrode*. Such an electrode is 2800 times more responsive to Na^+ than to K^+ at pH 11. By using different glass compositions, Li^+, K^+, and Ag^+ ion-selective electrodes have been fabricated from glass membranes.

An electrode used for the measurement of the ion X may also respond to the ion Y. The sensitivity of the electrode to different species with the same charge is given by the **selectivity coefficient,** defined as

$$k_{X,Y} = \frac{\text{response to Y}}{\text{response to X}} \qquad (15\text{-}24)$$

Ideally, the selectivity coefficient should be very small ($k \ll 1$), or else significant interference will occur. The sodium ion-selective electrode just described has a selectivity coefficient $k_{Na^+, K^+} = \frac{1}{2\,800}$ at pH 11, and $k_{Na^+, K^+} = \frac{1}{300}$ at pH 7. The selectivity coefficient k_{Na^+, H^+} has a value of 36, which means that the electrode is more sensitive to H^+ than to Na^+, even though the electrode is used for sodium.

In general, the behavior of most ion-selective electrodes can be described by the equation

$$E = \text{constant} \pm \beta \frac{0.059\,16}{n_X} \log\left[\mathscr{A}_X + \sum_Y (k_{X,Y}\mathscr{A}_Y^{n_X/n_Y})\right] \qquad (15\text{-}25)$$

Equation 15-25 describes the response of an electrode to its primary ion, X, and to all interfering ions, Y.

where $\mathscr{A}_X$ is the activity of the ion intended to be measured and $\mathscr{A}_Y$ is the activity of any interfering species (Y). The charge of each species is n_X or n_Y, and $k_{X,Y}$ is the selectivity coefficient defined above. The sign before the log term is positive if X is a cation and negative if X is an anion. The value of β is near 1 for most electrodes.

EXAMPLE: Using the Selectivity Coefficient

A Tl^{3+} ion-selective electrode has a selectivity coefficient $k_{Tl^{3+}, K^+} = 3.4 \times 10^{-4}$. What will be the change in electrode potential when 10^{-1} M K^+ is added to 10^{-5} M Tl^{3+}?

Using Equation 15-25 with $\beta = 1$, $n_X = +3$ and $n_Y = +1$, the potential without K^+ is

$$E = \text{constant} - \frac{0.059\,16}{3} \log[10^{-5}] = \text{constant} - 98.60 \text{ mV}$$

Addition of 10^{-1} M K^+ gives an electrode potential of

$$E = \text{constant} - \frac{0.059\,16}{3} \log[10^{-5} + (3.4 \times 10^{-4})(10^{-1})^{3/1}] = \text{constant} - 98.31 \text{ mV}$$

The change is $98.60 - 98.31 = 0.29$ mV, barely enough to measure.

Solid-State Electrodes

A schematic diagram of a **solid-state ion-selective electrode** based on an inorganic crystal is shown in Figure 15-13. One common electrode of this type is the fluoride electrode, employing a crystal of LaF_3 doped with Eu(II). *Doping* means adding a small amount of Eu(II) in place of La(III). The filling solution contains 0.1 M NaF and 0.1 M NaCl.

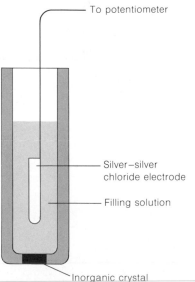

To potentiometer

Silver–silver chloride electrode

Filling solution

Inorganic crystal

Figure 15-13
Schematic diagram of an ion-selective electrode employing an inorganic salt crystal as the ion-sensitive membrane.

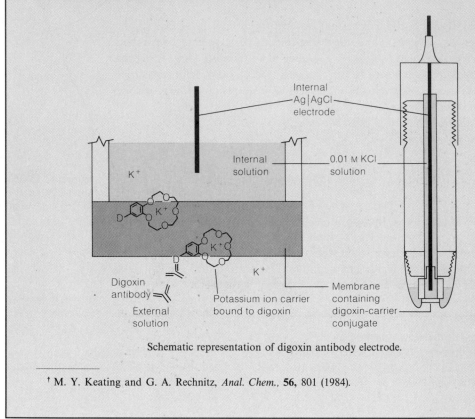
Fluoride ion from solution is selectively adsorbed on each surface of the
crystal. Unlike H^+ ion in the glass pH electrode, F^- ion can actually migrate
through the LaF_3 crystal, as shown in Figure 15-14. By doping the LaF_3
with EuF_2, anion vacancies are created within the crystal. A fluoride ion

Figure 15-14

Migration of F^- through LaF_3 doped with
EuF_2. Since Eu^{2+} has less charge than
La^{3+}, an anion vacancy occurs for every
Eu^{2+}. A neighboring F^- can jump into the
vacancy, moving the vacancy to another
site. Repetition of this process moves F^-
through the lattice.

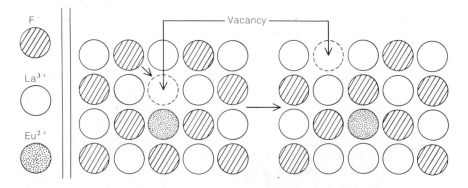

The electrode drawn on the facing page measures the concentration of this antibody. The membrane at the bottom of the electrode contains a K^+–chelator covalently bound to digoxin. (This type of chelator, called a *crown ether*, is discussed in Box 22-1). The electrode acts as a K^+ ion-selective electrode if the concentration of K^+ in the external solution is varied. If the external concentration of K^+ is fixed, but anti-D antibody is added to the external solution, the electrode potential also changes. Apparently, the antibody binds to the digoxin and prevents movement of the carrier across the membrane.

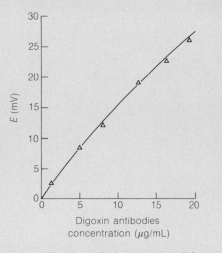

Electrode response to digoxin antibody at constant K^+ concentration.

from a site adjacent to a vacancy can jump into the vacancy, thus leaving a new vacancy behind. In this manner F^- can migrate from one side to the other and establish a potential difference across the crystal necessary for the electrode to work.

By analogy with the pH electrode, we can describe the response of the F^- electrode in the form

$$E = \text{constant} - \beta(0.059\,16)\log\left[\mathscr{A}_{F^-}\,(\text{outside})\right] \qquad (15\text{-}26)$$

where β is close to 1.00. The F^- electrode gives a nearly Nernstian response over a F^- concentration range from about 10^{-6} M to 1 M. The electrode is more responsive to F^- than to other ions by a factor greater than 1 000. The only interfering species is OH^-, for which the selectivity coefficient is $k_{F^-,OH^-} = 0.1$. At low pH, F^- is converted to HF ($pK_a = 3.17$), to which the electrode is insensitive. The F^- electrode is used to monitor continuously and control the fluoridation of municipal water supplies.

EXAMPLE: Response of an Ion-Selective Electrode

When an F^- electrode was immersed in standard F^- solutions (maintained at a constant ionic strength of 0.1 M with $NaNO_3$), the following potentials (versus S.C.E.) were observed:

$[F^-]$(M)	E (mV)	
1.00×10^{-5}	100.0	$\Big\}\ -58.5\ \text{mV}$
1.00×10^{-4}	41.5	
1.00×10^{-3}	-17.0	$\Big\}\ -58.5\ \text{mV}$

Since the ionic strength was held constant, the response should depend upon the logarithm of the F^- *concentration*. What potential is expected if $[F^-] = 5.00 \times 10^{-5}$ M? What concentration of F^- will give a potential of 0.0 V?

We seek to fit the calibration data with an equation in the form of Equation 15-26:

$$E = m \underbrace{\log[F^-]}_{x} + b$$
$$\quad\ \ \overset{y}{}$$

Plotting E versus $\log[F^-]$ gives a straight line with a slope of -58.5 mV and a y-intercept of -192.5 mV. Setting $[F^-] = 5.00 \times 10^{-5}$ M gives

$$E = (-58.5)\log[5.00 \times 10^{-5}] - 192.5 = 59.1 \text{ mV}$$

If $E = 0.0$ mV, we can solve for the concentration of $[F^-]$:

$$0.0 = (-58.5)\log[F^-] - 192.5 \quad \Rightarrow \quad [F^-] = 5.1 \times 10^{-4} \text{ M}$$

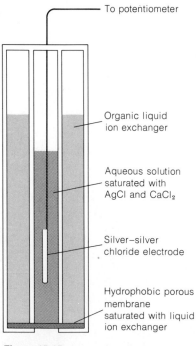

To potentiometer

Organic liquid
ion exchanger

Aqueous solution
saturated with
$AgCl$ and $CaCl_2$

Silver–silver
chloride electrode

Hydrophobic porous
membrane
saturated with liquid
ion exchanger

Figure 15-15
Schematic diagram of a calcium ion-selective electrode based on a liquid ion exchanger.

Another common inorganic crystal electrode uses Ag_2S for the membrane. This electrode responds to Ag^+ and to S^{2-}. By doping the electrode with CuS, CdS, or PbS, it is possible to prepare electrodes sensitive to Cu^{2+}, Cd^{2+}, or Pb^{2+}, respectively. Table 15-4 lists some ion-selective electrodes based on inorganic crystals.[†]

Liquid-Based Ion-Selective Electrodes

The principle of the **liquid-based ion-selective electrode** was described in Section 15-4 (Figure 15-7). Figure 15-15 shows the construction of a calcium ion-selective electrode, which is basically similar to the solid-state electrode in Figure 15-13. The difference is that the solid crystal is replaced by a membrane saturated with a hydrophobic liquid ion exchanger (a calcium chelator). A solution of the exchanger is housed in a reservoir surrounding the internal silver–silver chloride electrode. Calcium ion is selectively trans-

---

[†] The construction of inexpensive solid-state ion-selective electrodes for classroom experiments has been described by A. Palanivel and P. Riyazuddin [*J. Chem. Ed.,* **61,** 920 (1984)] and by W. S. Selig [*J. Chem. Ed.,* **61,** 80 (1984)]. A tremendous number of inorganic anions can be titrated with quaternary ammonium halides using simple home-made indicator electrodes [W. S. Selig, *J. Chem. Ed.,* **64,** 141 (1987)].

Table 15-4
Properties of some solid-state ion-selective electrodes

Ion	Concentration range (M)	Membrane material	pH range	Some interfering species
F^-	$10^{-6}–1$	LaF_3	5–8	OH^-
Cl^-	$10^{-4}–1$	$AgCl$	2–11	$CN^-, S^{2-}, I^-, S_2O_3^{2-}, Br^-$
Br^-	$10^{-5}–1$	$AgBr$	2–12	CN^-, S^{2-}, I^-
I^-	$10^{-6}–1$	AgI	3–12	S^{2-}
SCN^-	$10^{-5}–1$	$AgSCN$	2–12	$S^{2-}, I^-, CN^-, Br^-, S_2O_3^{2-}$
CN^-	$10^{-6}–10^{-2}$	AgI	11–13	S^{2-}, I^-
S^{2-}	$10^{-5}–1$	Ag_2S	13–14	

SOURCE: P. L. Bailey, *Analysis with Ion-Selective Electrodes* (London: Heyden, 1976), pp. 95–99; and *Orion Research Analytical Methods Guide* (Cambridge, Mass.: Orion Research Inc., 1975).

ported across the membrane to establish a voltage related to the difference in activity of Ca^{2+} between the sample and internal solution:

$$E = \text{constant} + \beta \left(\frac{0.059\ 16}{2} \right) \log \mathscr{A}_{Ca^{2+}}(\text{outside}) \qquad (15\text{-}27)$$

where β is close to 1.00. Note that Equations 15-26 and 15-27 have different signs before the log term, because one involves an anion and the other a cation. Note also that the charge of the calcium ion requires a factor of 2 in the denominator before the logarithm.

The ion exchanger in the Ca^{2+} electrode is calcium didecylphosphate dissolved in dioctylphenylphosphonate. The didecylphosphate anion can react with calcium ions at each surface of the membrane and transport Ca^{2+} across the membrane:

$$[(RO)_2PO_2^-]_2Ca \rightleftharpoons 2(RO)_2PO_2^- + Ca^{2+} \qquad (15\text{-}28)$$

The most serious interference with the Ca^{2+} electrode comes from Zn^{2+}, Fe^{2+}, Pb^{2+}, and Cu^{2+}, but high concentrations of Sr^{2+}, Mg^{2+}, Ba^{2+}, and Na^+ also interfere. For one particular Ca^{2+} electrode, $k_{Ca^{2+},Fe^{2+}} = 0.8$ and $k_{Ca^{2+},Mg^{2+}} = 0.01$. Interference from H^+ is substantial below a pH of 4–5. Table 15-5 gives properties of some other liquid-based ion-selective electrodes.

$$[(CH_3(CH_2)_8CH_2O)_2\overset{\overset{\displaystyle O}{\|}}{P}O]_2Ca$$
Calcium didecylphosphate

$$(CH_3(CH_2)_6CH_2O)_2\overset{\overset{\displaystyle O}{\|}}{P}C_6H_5$$
Dioctylphenylphosphonate

Compound Electrodes

Gas-sensing and enzyme-based electrodes are usually of a **compound** design, incorporating a conventional electrode and an additional membrane to isolate (or generate) the species that the electrode detects. A CO_2 gas-sensing electrode is shown in Figure 15-16. It consists of an ordinary glass pH electrode surrounded by an electrolyte solution and enclosed by a semipermeable membrane made of rubber, Teflon, or polyethylene. Immersed in the electrolyte solution is a silver–silver chloride band serving as the reference electrode. When CO_2 diffuses through the semipermeable membrane, it lowers the pH in the electrolyte compartment. The response of the glass electrode

Table 15-5
Properties of some liquid-based ion-selective electrodes

Ion	Concentration range (M)	Carrier	Solvent for carrier	pH range	Some interfering species
Ca^{2+}	$10^{-5}-1$	Calcium didecylphosphate	Dioctylphenyl phosphonate	6–10	Zn^{2+}, Pb^{2+}, Fe^{2+}, Cu^{2+}
NO_3^-	$10^{-5}-1$	Tridodecylhexadecylammonium nitrate	Octyl-2-nitrophenyl ether	3–8	ClO_4^-, I^-, ClO_3^-, Br^-, HS^-, CN^-
ClO_4^-	$10^{-5}-1$	Tris(substituted 1,10-phenanthroline) iron(II) perchlorate	p-Nitrocymene	4–10	I^-, NO_3^-, Br^-
BF_4^-	$10^{-5}-1$	Tris(substituted 1,10-phenanthroline) nickel (II) tetrafluoroborate	p-Nitrocymene	2–12	NO_3^-
Cl^-	$10^{-5}-1$	Dimethyldioctadecylammonium chloride		3–10	ClO_4^-, I^-, NO_3^-, SO_4^{2-}, Br^-, OH^-, HCO_3^-, F^-, acetate

SOURCE: P. L. Bailey, *Analysis with Ion-Selective Electrodes* (London: Heyden, 1976), pp. 127–130; and *Orion Research Analytical Methods Guide* (Cambridge, Mass.: Orion Research Inc., 1975).

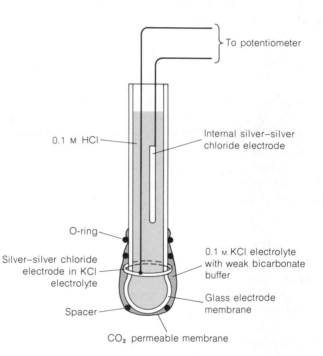

Figure 15-16
Schematic diagram of a CO_2 gas–sensing electrode.

Acidic or basic gases are detected by a pH electrode surrounded by an electrolyte and enclosed in a gas-permeable membrane.

to the change in pH is measured. Other acidic or basic gases, including NH_3, SO_2, H_2S, and NO_x (nitrogen oxides) can be detected in the same manner. These electrodes can be used to measure gases *in the gas phase* or dissolved in solution.

Numerous ingenious compound electrodes using *enzymes* have been built.[†] These devices contain a conventional electrode coated with an enzyme

---

[†] Enzyme-based penicillin and urea electrodes for student experiments have been described by T. E. Mifflin, K. M. Andriano, and W. B. Robbins, *J. Chem. Ed.,* **61,** 638 (1984) and by T. L. Riechel, *J. Chem. Ed.,* **61,** 640 (1984). Interesting compound electrodes with biological applications are described in a review by J. D. Czaban, *Anal. Chem.,* **57,** 345A (1985). For a general review of biosensors, see G. A. Rechnitz, *Chem. Eng. News,* September 5, 1988, p 24.

that catalyzes a reaction of the analyte. The product of the reaction is detected by the electrode.

377
15-6 ION-SELECTIVE ELECTRODES

Use and Abuse of Ion-Selective Electrodes

The advantages of ion-selective electrodes are many. Electrodes respond in a linear manner to most analyte species over a wide range of concentrations (four to six orders of magnitude). They do not destroy the unknown sample, and they introduce negligible contamination. Their response time is usually short (seconds to minutes), so they may be used to monitor flowing samples in industrial or clinical applications. Highly colored or turbid solutions are amenable to measurements with electrodes, but not to spectrophotometry. Finally, specially designed electrodes can be used in otherwise inaccessible environments, such as the interior of living cells.

On the other hand, special care is needed to obtain reliable results. The precision of ion-selective electrode measurements is rarely better than 1%, and is usually worse. Electrodes can be fouled by proteins or other organic solutes, leading to sluggish response and drifting potentials. Certain ionic species interfere with or poison particular electrodes. Some electrodes are fragile and have limited shelf life.

Extreme care in the preparation of samples and standards is critical to obtaining meaningful results. The electrode responds to the *activity* of *uncomplexed* analyte ion. Therefore, potential ligands must be absent or masked. Since we usually wish to know concentrations, not activities, an inert salt is often used to bring all the standards and samples to a high and constant ionic strength. If the activity coefficients remain constant, the electrode potential gives concentrations directly.

Advantages of ion-selective electrodes:

1. wide range of linear response
2. nondestructive
3. noncontaminating
4. short response time
5. unaffected by color or turbidity

A 1-mV error in potential corresponds to a 4% error in monovalent ion activity. A 5-mV error corresponds to a 22% error. The relative error *doubles* for divalent ions and *triples* for trivalent ions.

Electrodes respond to the *activity* of *uncomplexed* ion. If the ionic strength is held constant, the concentration is proportional to the activity and the electrode measures concentrations.

Standard addition method

It is important that the composition of the standard solutions closely approximate the composition of the unknown. The medium in which the analyte exists is called the **matrix.** In cases where the matrix is complex or unknown, the **standard addition method** can be used. In this technique, the electrode is immersed in the unknown and the potential is recorded. Then a known addition of standard solution is made to the unknown. The addition should be of a small volume so as not to change the ionic strength of the sample. The change in potential tells how the electrode responds to analyte and how much analyte was in the original solution.

The matrix (solution composition) of the standards should be the same as the matrix of the unknown.

The *standard addition method* is discussed further is Sections 18-3 and 21-4. A general treatment of the standard addition method can be found in an article by M. Bader, *J. Chem. Ed.,* **57,** 703 (1980).

EXAMPLE: Standard Addition with an Ion-Selective Electrode

A perchlorate ion-selective electrode immersed in 50.0 mL of unknown perchlorate solution gave a potential of 358.7 mV versus S.C.E. When 1.00 mL of 0.050 M $NaClO_4$ was added, the potential changed to 346.1 mV. Assuming that the electrode has a Nernstian response ($\beta = 1.00$), find the concentration of ClO_4^- in the unknown.

The first solution contains x moles of ClO_4^- in 0.050 0 L. The standard addition adds $(0.001\ 00\ L)(0.050\ 0\ M) = 5.00 \times 10^{-5}$ mol of ClO_4^-. Therefore, the second solution contains $x + (5.00 \times 10^{-5})$ mol in 0.051 0 L. We can write a Nernst equation for the first solution:

$$E_1 = \text{constant} - 0.059\ 16 \log [ClO_4^-]_1$$

and another for the second solution:

$$E_2 = \text{constant} - 0.059\ 16 \log[ClO_4^-]_2$$

We set $E_1 = 0.358\ 7$ V, $E_2 = 0.346\ 1$ V, $[ClO_4^-]_1 = x/0.050\ 0$, and $[ClO_4^-]_2 = (x + 5.00 \times 10^{-5})/0.051\ 0$. This allows us to solve for x by subtracting one equation from the other:

$$0.358\ 7 = \text{constant} - 0.059\ 16 \log[x/0.050\ 0]$$

$$\underline{-0.346\ 1 = \text{constant} - 0.059\ 16 \log[(x + 5.00 \times 10^{-5})/0.051\ 0]}$$

$$0.012\ 6 = -0.059\ 16 \log \frac{x/0.050\ 0}{(x + 5.00 \times 10^{-5})/0.051\ 0}$$

$$\frac{x/0.050\ 0}{(x + 5.00 \times 10^{-5})/0.051\ 0} = 0.612 \quad \Rightarrow \quad x = 7.51 \times 10^{-5} \text{ mol}$$

The original perchlorate concentration was therefore

$$\frac{7.51 \times 10^{-5} \text{ mol}}{0.050\ 0 \text{ L}} = 1.50 \text{ mM}$$

The standard addition method works best if the quantity of added analyte is approximately 50–200% of the original analyte. Results are more accurate if the effects of several additions are averaged.

Metal Ion buffers

If you wanted to prepare a 10^{-6} M H^+ standard, you would never dream of doing it by diluting HCl until it was 10^{-6} M. (Why?) Similarly, it is poor form to prepare, say, a 10^{-6} M Ca^{2+} standard by diluting $CaCl_2$ to this low concentration. At such a low concentration, the Ca^{2+} ion could be lost by adsorption on a glass wall or reaction with some impurity.

A better idea is to prepare a **metal ion buffer.** Such a buffer is prepared from the metal and a suitable ligand. For example, consider the reaction of Ca^{2+} with nitrilotriacetic acid (NTA) at a pH high enough for NTA to be in its fully basic form (NTA^{3-}):

$$Ca^{2+} + NTA^{3-} \rightleftharpoons CaNTA^- \tag{15-29}$$

$$K_f = \frac{[CaNTA^-]}{[Ca^{2+}][NTA^{3-}]} = 10^{6.46} \text{ in 0.1 M } KNO_3 \tag{15-30}$$

If equal concentrations of NTA^{3-} and $CaNTA^-$ are present in a solution, the concentration of Ca^{2+} would be

$$[Ca^{2+}] = \frac{[\cancel{CaNTA^-}]}{K_f[\cancel{NTA^{3-}}]} = 10^{-6.46} \text{ M} \tag{15-31}$$

Glass vessels should not be used for low concentration standard solutions, because ions can be adsorbed on the glass surface. Bottles and beakers made of polyethylene or similar materials are much better than glass for handling dilute solutions. Adding strong acid (0.1–1 M) to any solution helps to minimize adsorption of cations on the walls of the container, because H^+ competes with the other cations for ion-exchange sites.

$$HN(CH_2CO_2H)_3^+ = H_4NTA^+$$

$pK_1 = 1.1$	$pK_3 = 2.940$
$pK_2 = 1.650$	$pK_4 = 10.334$

EXAMPLE: Preparing a Metal Ion Buffer

What concentration of NTA^{3-} should be added to 1.00×10^{-2} M $CaNTA^-$ in 0.1 M KNO_3 to give $[Ca^{2+}] = 1.00 \times 10^{-6}$ M?

Using Equation 15-31, we write

$$[NTA^{3-}] = \frac{[CaNTA^-]}{K_f[Ca^{2+}]} = \frac{1.00 \times 10^{-2}}{(10^{6.46})(1.00 \times 10^{-6})} = 3.47 \times 10^{-3}\ \text{M}$$

These are practical concentrations of $CaNTA^-$ and of NTA^{3-}.

15-7 SOLID-STATE CHEMICAL SENSORS

The technology used to fabricate microelectronic chips can also be applied to the construction of solid-state chemical sensors. We encountered one such sensor when we discussed how a piezoelectric quartz crystal coated with a chemical absorbant could be used to detect species that bind to the absorbant (Box 2-1). Analytical chemists are experimenting with chemical coatings that alter the electrical properties of semiconductor devices and make them sensitive to changes in the chemical environment.[†] In this section we will describe one of these devices—the field effect transistor.

Semiconductors

A **semiconductor** is a material whose electrical **resistivity**[‡] (10^{-4}–$10^7\ \Omega \cdot m$) lies between those of conductors ($\sim 10^{-8}\ \Omega \cdot m$) and insulators ($10^{12}$–$10^{20}\ \Omega \cdot m$). Useful semiconductors for electronic components are crystalline solids such as Si (Figure 15-17), Ge, and GaAs. Their common feature is that the valence electrons of the pure materials are all involved in the sigma bonding network, with no pi electrons or nonbonding electrons (Figure 15-18a). When atoms of an impurity such as phosphorus, with five valence electrons, are introduced into the lattice, there is one extra electron beyond

[†] H. Wohltjen, *Anal. Chem.,* **56,** 87A (1984); J. Janata and R. J. Huber, *Solid State Chemical Sensors* (New York: Academic Press, 1985).

[‡] *Resistivity,* ρ, is a measure of how well a substance retards the flow of electric current when an electric field is applied: $J = E/\rho$, where J is the current density (current flowing through a unit cross section of the material, amperes per square meter) and E is the electric field (V/m). The units of resistivity are $V \cdot m/A$ or $\Omega \cdot m$, since $\Omega = V/A$, where $\Omega = ohm$. The reciprocal of resistivity is **conductivity.** Resistivity does not depend on the dimensions of the substance. Resistance, R, does depend on dimensions and is related to resistivity by the equation $R = \rho l/A$, where l is the length and A is the cross-sectional area of the conducting substance.

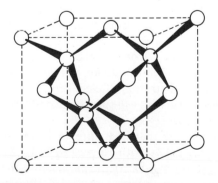

Figure 15-17
Diamondlike structure of silicon. Each atom is tetrahedrally bonded to four neighbors, with an Si—Si distance of 235 pm.

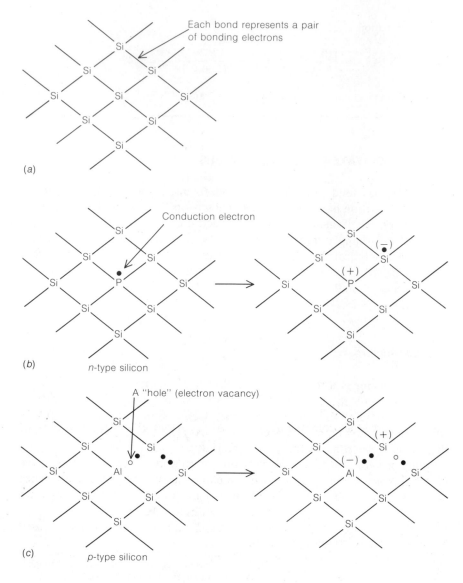

Figure 15-18
(a) The electrons of pure silicon are all involved in the sigma bonding framework. (b) An impurity atom such as phoshorus adds one extra electron (●), which is relatively free to move through the crystal. (c) An aluminum impurity atom lacks one electron needed for the sigma bonding framework. The hole (○) introduced by the Al atom can be occupied by an electron from a neighboring site, effectively moving the hole to the neighboring site.

what is needed for sigma bonding. This extra electron is relatively free to move through the crystal and is called a **conduction electron** (Figure 15-18b). If an atom such as aluminum is present instead of a silicon atom, there is one less bonding electron than necessary to complete the sigma bonds. This vacancy is called a **hole.** An electron from a neighboring bond can move into the hole, effectively moving the hole to an adjacent position (Figure 15-18c). The hole is therefore also a charge carrier, moving through the lattice from one atom to another.

A semiconductor with excess conduction electrons is called *n-type,* and one with excess holes is called *p-type.* An *n*-type semiconductor contains both conduction electrons and holes, but the concentration of electrons is much greater than the concentration of holes. Similarly, a *p*-type semiconductor contains an excess of holes.

A **diode** consists of a *pn* junction, such as that in Figure 15-19, made from doped silicon. If the *n*-Si is made negative with respect to the *p*-Si, electrons flow from the external circuit into the *n*-Si. At the *pn* junction the

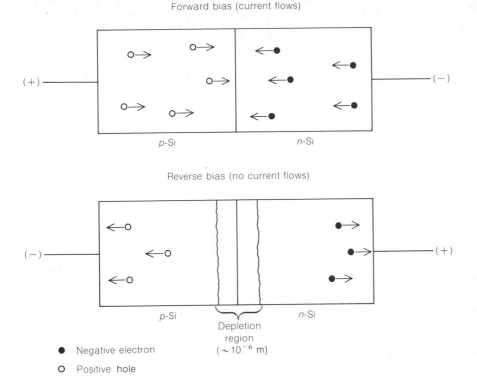

Forward bias (current flows)

p-Si   n-Si

Reverse bias (no current flows)

p-Si   n-Si

Depletion region
($\sim 10^{-6}$ m)

● Negative electron
○ Positive hole

Figure 15-19
Behavior of a *pn* junction, showing that current can flow under forward bias conditions, but is prevented from flowing under reverse bias.

electrons and holes combine. As electrons are withdrawn from the *p*-Si into the circuit, a fresh supply of holes is made available in the *p*-Si. The net result is that current can flow when the *n*-Si is made negative with respect to the *p*-Si. The diode is said to be *forward biased*.

If the opposite polarity is applied to the diode, electrons are drawn out of the *n*-Si into the circuit, and holes are drawn out of the *p*-Si (lower part of Figure 15-19). This leaves a thin *depletion region* devoid of charge carriers near the *pn* junction. Current ceases to flow. Such a diode is said to be in *reverse bias*. The diode does not conduct current in the reverse direction.

An activation energy is needed to get the charge carriers to move across the diode. For Si, ~ 0.6 V of forward bias is required before current will flow. For Ge, the voltage drop is ~ 0.2 V.

For moderate reverse bias voltages, no current flows. If the voltage is sufficiently negative, *breakdown* occurs and current flows in the reverse direction.

Field Effect Transistor

Figure 15-20 is a schematic representation of a **field effect transistor.** The body of this device (the *base*) is *p*-Si with two embedded regions of *n*-Si, called the *source* and *drain*. The top of the base is covered by a thin layer of the insulator, SiO_2. Above this insulator is a metal conductor called the *gate*. The source and the base are held at the same electric potential by a wire connection. When a voltage is applied between the source and drain (Figure 15-20), little current flows. The drain–base interface is a *pn* junction in reverse bias. The drain is surrounded by a depletion region of high resistance.

If the gate is made positive with respect to the base, electrons in the base are attracted toward the gate, and a thin channel rich in electrons forms (Figure 15-20b). Now there is a conducting path between the *n*-Si regions, and current can flow between the source and drain. The current increases as the gate is made more positive. *The potential of the gate therefore regulates the current flow between source and drain.*

The more positive the gate, the more current can flow between source and drain.

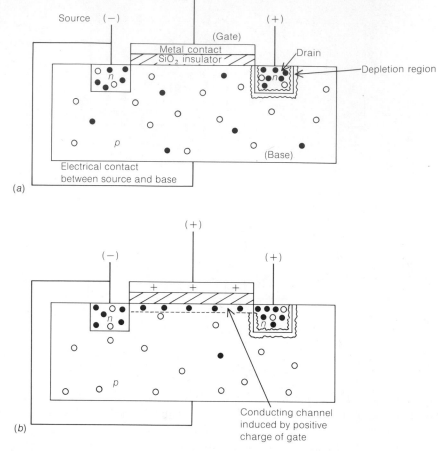

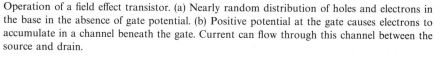

Figure 15-20

Operation of a field effect transistor. (a) Nearly random distribution of holes and electrons in the base in the absence of gate potential. (b) Positive potential at the gate causes electrons to accumulate in a channel beneath the gate. Current can flow through this channel between the source and drain.

Chemical-Sensing Field Effect Transistor

The essential feature of the chemical-sensing field effect transistor in Figure 15-21 is the chemically sensitive layer over the gate. A simple example of such a coating is a layer of AgBr. If this layer is exposed to a solution containing Ag^+ ions, the Ag^+ becomes adsorbed on the surface of the AgBr (Figure 8-2), giving the surface a positive charge. This positive charge attracts electrons from the base of the transistor, inducing a current-carrying channel between the source and drain. As a result of the binding of Ag^+ to the AgBr, more current can pass between the source and drain.

Figure 15-21 shows some additional details of the transistor and outlines one way to make the electrical measurements. The top of the transistor is coated with an insulating SiO_2 layer and a second layer of Si_3N_4 (silicon nitride), which is impervious to ions and improves the electrical stability of the device. The voltage source at the lower right maintains a constant drain–source potential difference. The current meter measures current flowing be-

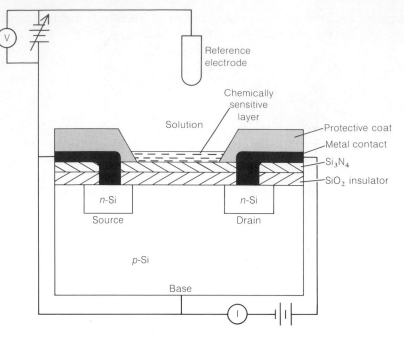

Figure 15-21
Operation of a chemical-sensing field effect transistor.

tween the source and the drain. (No current flows from the base.) The base is held at the same potential as the source, and both are connected to a reference electrode (such as a silver–silver chloride electrode) in contact with the sample solution.

Imagine that the sensing layer is exposed to a solution and that there is a fixed voltage between the source and the drain. A current of, say, 100 μA is observed to flow between source and drain. If Ag^+ is added to the solution, the gate becomes more positive and more current flows. The variable voltage source at the upper left can be adjusted to make the base more positive until the increased potential just cancels the increased gate potential, and 100 μA flows again. The increased voltage needed to bring the current back to its initial value represents the response of the device to Ag^+. Figure 15-22 shows that Ag^+ makes the gates more positive and that adsorption of Br^- makes the gate more negative. The response is close to the theoretical Nernstian value of 59 mV for a 10-fold concentration change.

The coating on the gate is the key to the analytical chemistry that the transistor can perform. Many of the ion exchangers used for ion-selective electrodes can be used on field effect transistors. An advantage of the transistor is that it is much smaller and potentially more rugged than an ion-selective electrode and can therefore be used in a wider variety of applications. The sensing surface of the transistor is typically only 1 mm^2.

Fortuitously, the Si_3N_4 coat on the transistor is itself sensitive to H^+ ions, apparently undergoing an ion exchange reaction with H^+. Many ingenious schemes take advantage of the pH sensitivity of the field effect transistor to make devices sensitive to species other than H^+. For example, the Si_3N_4 can be coated with a polyacrylamide gel containing covalently bound

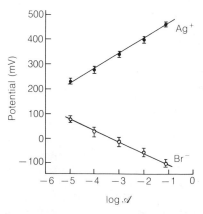

Figure 15-22
Response of a silver bromide-coated field effect transistor. Error bars are 95% confidence intervals for data obtained from 195 sensors prepared from different chips. [R. P. Buck and D. E. Hackleman, *Anal. Chem.*, **49**, 2315 (1977).]

penicillinase.[†] Penicillinase is an enzyme that hydrolyzes the drug penicillin and changes the acid–base properties of the drug.

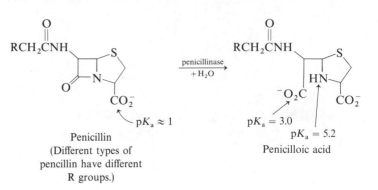

Penicillin
(Different types of
pencillin have different
R groups.)

Penicilloic acid

The altered drug changes the pH within the gel, and the transistor responds to this change in H^+ activity. By this means, the solid-state sensor is used to measure the concentration of penicillin in a solution.

[†] S. D. Caras and J. Janata, *Anal. Chem.*, **57**, 1924 (1985).

Summary

In potentiometric measurements, the indicator electrode responds to changes in the activity of analyte, and the reference electrode is a self-contained half-cell producing a constant reference potential. The most common reference electrodes are the calomel and the silver–silver chloride electrodes. Common indicator electrodes include (1) the inert Pt electrode, (2) a silver electrode responsive to Ag^+, halides, and other ions that react with Ag^+, and (3) ion-selective electrodes. Writing the Nernst equation in the form $E = E_+ - E_-$, you should be able to analyze complex cells, such as those in which complexometric or precipitation titrations are occurring. The existence of small, unknown junction potentials at the interface of two electrolyte solutions sets a fundamental limit on the accuracy of potentiometric measurements. If small voltages must be measured accurately, it is necessary to construct a cell without a liquid junction.

Ion-selective electrodes, including the glass pH electrode, respond selectively to one species in a solution. A gradient of activity in any species produces a gradient of free energy equal to $\Delta G = -RT \ln \mathscr{A}_1/\mathscr{A}_2$. The electric potential difference corresponding to this free energy difference is $E = -\Delta G/nF = (RT/nF) \ln \mathscr{A}_1/\mathscr{A}_2$. Most electrodes are sensitive to many species, and their net response can be described by the equation $E = \text{constant} \pm \beta(0.059\ 16/n_X) \log [\mathscr{A}_X + \sum(k_{X,Y} \mathscr{A}_Y^{n_X/n_Y})]$, where $k_{X,Y}$ is the selectivity coefficient for each species. Common ion-selective electrodes can be classified as solid-state, liquid-based, and compound, depending on their construction. Metal ion buffers are especially appropriate for establishing and maintaining low concentrations of ions. The field effect transistor is an example of a solid-state device that uses a chemically sensitive coating to alter the electrical properties of a semiconductor in response to changes in the chemical environment.

Terms to Understand

alkaline (sodium) error	asymmetry potential	compound electrode
antibody	calomel electrode	conduction electron
antigen	combination electrode	conductivity

diode	junction potential	S.C.E. (saturated calomel electrode)
electroactive species	liquid-based ion-selective electrode	selectivity coefficient
field effect transistor	matrix	semiconductor
glass electrode	metal ion buffer	silver–silver chloride electrode
hole	mobility	solid-state ion-selective electrode
indicator electrode	potentiometry	standard addition method
ion-exchange equilibrium	reference electrode	
ion-selective electrode	resistivity	

Exercises

P. 287

15-A. The apparatus in Figure 9-8 was used to monitor the titration of 100.0 mL of solution containing 50.0 mL of 0.100 M $AgNO_3$ and 50.0 mL of 0.100 M $TlNO_3$. The titrant contained 0.200 M NaBr. Suppose that the glass electrode (used as a *reference electrode* in this experiment) gives a constant potential of +0.200 V. The glass electrode is attached to the *positive* terminal of the pH meter, and the silver wire to the negative terminal. Calculate the cell voltage at each of the following volumes of NaBr, and sketch the titration curve: 1.0, 15.0, 24.0, 24.9, 25.2, 35.0, 50.0, 60.0 mL.

15-B. The apparatus below can be used to follow the course of an EDTA titration and was used to generate the curves in Figure 13-7. The heart of the cell is a pool of liquid Hg in contact with the solution and with a Pt wire. A small amount of HgY^{2-} added to the analyte equilibrates with a very tiny amount of Hg^{2+}:

$$Hg^{2+} + Y^{4-} \rightleftharpoons HgY^{2-}$$

$$K_f = \frac{[HgY^{2-}]}{[Hg^{2+}][Y^{4-}]} = 5 \times 10^{21} \quad (A)$$

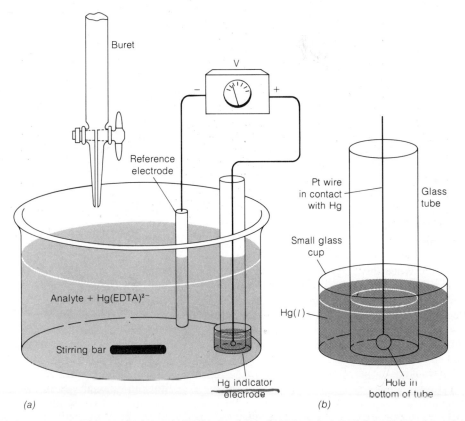

(a) Apparatus used for a potentiometric EDTA titration.   (b) Enlarged view of the mercury electrode.

The redox equilibrium $Hg^{2+} + 2e^- \rightleftharpoons Hg(l)$ is established rapidly at the surface of the Hg electrode, so the Nernst equation for the cell can be written in the form

$$E = E_+ - E_-$$

$$= \left(0.852 - \frac{0.059\ 16}{2} \log \frac{1}{[Hg^{2+}]}\right) - E_- \qquad (B)$$

where E_- is the constant potential of the reference electrode. But from Equation A, we can write $[Hg^{2+}] = [HgY^{2-}]/K_f[Y^{4-}]$, and this can be substituted into Equation B to give

$$E = 0.852 - \frac{0.059\ 16}{2} \log \frac{[Y^{4-}]K_f}{[HgY^{2-}]} - E_-$$

$$= 0.852 - E_- - \frac{0.059\ 16}{2} \log \frac{K_f}{[HgY^{2-}]}$$

$$- \frac{0.059\ 16}{2} \log [Y^{4-}] \qquad (C)$$

where K_f is the formation constant for HgY^{2-}. This apparatus thus responds to the changing EDTA concentration during an EDTA titration.

Suppose that 50.0 mL of 0.010 0 M $MgSO_4$ is titrated with 0.020 0 M EDTA at pH 10.0 using the apparatus above with an S.C.E. reference electrode. Assume that the analyte contains 1.0×10^{-4} M $Hg(EDTA)^{2-}$ added at the beginning of the titration. Calculate the cell voltage at the following volumes of added EDTA, and draw a graph of millivolts versus milliliters: 0, 10.0, 20.0, 24.9, 25.0, and 26.0 mL.

15-C. A solid-state fluoride ion-selective electrode responds to F^-, but not to HF. It also responds to hydroxide ion at high concentration when $[OH^-] \approx [F^-]/10$. Suppose that such an electrode gave a potential of $+100$ mV (versus S.C.E.) in 10^{-5} M NaF and $+41$ mV in 10^{-4} M NaF. Sketch qualitatively how the potential would vary if the electrode were immersed in 10^{-5} M NaF and the pH ranged from 1 to 13.

15-D. One commercial glass-membrane sodium ion-selective electrode has a selectivity coefficient $k_{Na^+,H^+} = 36$. When this electrode was immersed in 1.00 mM NaCl at pH 8.00, a potential of -38 mV (versus S.C.E.) was recorded.
(a) Neglecting activity coefficients and assuming $\beta = 1$ in Equation 15-25, calculate the potential if the electrode were immersed in 5.00 mM NaCl at pH 8.00.
(b) What would be the potential for 1.00 mM NaCl at pH 3.87?
After answering this question, you should realize that pH is a critical variable in the use of a sodium ion-selective electrode.

15-E. An ammonia gas-sensing electrode gave the following calibration points when all solutions contained 1 M NaOH.

NH_3 (M)	E (mV)	NH_3 (M)	E (mV)
1.00×10^{-5}	268.0	5.00×10^{-4}	368.0
5.00×10^{-5}	310.0	1.00×10^{-3}	386.4
1.00×10^{-4}	326.8	5.00×10^{-3}	427.6

A dry food sample weighing 312.4 mg was digested by the Kjeldahl procedure (page 160) to convert all the nitrogen to NH_4^+. The digestion solution was diluted to 1.00 L, and 20.0 mL was transferred to a 100-mL volumetric flask. The 20.0-mL aliquot was treated with 10.0 mL of 10.0 M NaOH plus enough NaI to complex the Hg catalyst from the digestion, and diluted to 100.0 mL. When measured with the ammonia electrode, this solution gave a reading of 339.3 mV. Calculate the percent nitrogen in the food sample.

15-F. Cyanide ion was measured indirectly with a solid-state ion-selective electrode containing a silver sulfide membrane. Assuming that the electrode response is Nernstian and that the ionic strength of all solutions is constant, we can write the electrode response as follows:

$$E = \text{constant} + 0.059\ 16 \log [Ag^+]$$

To an unknown cyanide solution was added $Ag(CN)_2^-$, such that $[Ag(CN)_2^-] = 1.0 \times 10^{-5}$ M in the final solution. This complex ion behaves as a silver ion buffer in the presence of CN^-:

$$Ag^+ + 2CN^- \rightleftharpoons Ag(CN)_2^-$$

$$K = \beta_2 = \frac{[Ag(CN)_2^-]}{[Ag^+][CN^-]^2} = 10^{19.85}$$

(a) Suppose that the unknown contained 8.0×10^{-6} M CN^- and the potential was found to be $+206.3$ mV. Then a *standard addition* of CN^- was made to bring the CN^- concentration to 12.0×10^{-6} M. What will be the new potential?
(b) Now consider a real experiment in which 50.0 mL of unknown gave a potential of 134.8 mV before a standard addition of CN^- was made. After adding 1.00 mL of 2.50×10^{-4} M KCN, the potential dropped to 118.6 mV. What was the concentration of CN^- in the 50.0-mL sample? For this part, we do not know the value of the constant in the equation for E. That is, you cannot use the value of the constant derived from part a of this question.

Problems

A15-1. (a) Write the half-reactions for the silver–silver chloride and calomel reference electrodes.
(b) What is the expected voltage for the cell below.

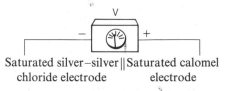

Saturated silver–silver ‖ Saturated calomel
chloride electrode   electrode

A15-2. Suppose that the silver–silver chloride electrode in Figure 15-2 is replaced by a saturated calomel electrode. Calculate the cell voltage if $[Fe^{2+}]/[Fe^{3+}] = 2.5 \times 10^{-3}$.

A15-3. A cell was prepared by dipping a Cu wire and a saturated calomel electrode into 0.10 M $CuSO_4$ solution. The Cu wire was attached to the positive terminal of a potentiometer and the calomel electrode was attached to the negative terminal.
(a) Write a half-reaction for the Cu electrode.
(b) Write the Nernst equation for the Cu electrode.
(c) Calculate the cell voltage.

A15-4. Why is the 0.1 M HCl|0.1 M KCl junction potential of opposite sign and greater magnitude than the 0.1 M NaCl|0.1 M KCl potential in Table 15-2?

A15-5. Which side of the liquid junction 0.1 M KNO_3|0.1 M NaCl will be negative?

A15-6. Describe how you would calibrate a pH electrode and measure the pH of blood (which is ~7.5) *at 37°C*. Use the standard buffers in Table 15-3.

A15-7. List the sources of error associated with pH measurement using the glass electrode.

A15-8. If electrode C in Figure 15-12 is placed in a solution of pH 11.0, what will the pH reading be?

A15-9. Which National Institute of Standards and Technology buffer (s) would you use to calibrate an electrode for pH measurements in the range 3–4?

A15-10. A cyanide ion-selective electrode obeys the equation

$$E = \text{constant} - 0.059\,16\,\log[CN^-]$$

The electrode potential was −0.230 V when the electrode was immersed in 1.00×10^{-3} M NaCN.
(a) Evaluate the constant in the equation above.
(b) Using the result from part a, find the concentration of CN^- if $E = -0.300$ V.
(c) Without using the constant from part a, find the concentration of CN^- if $E = -0.300$ V.

A15-11. By how many volts will the potential of a Mg^{2+} ion-selective electrode change if the electrode is removed from 1.00×10^{-4} M $MgCl_2$ and placed in 1.00×10^{-3} M $MgCl_2$?

A15-12. The selectivities of a lithium ion-selective electrode are indicated on the following diagram:

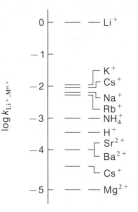

(a) Which alkali metal (Group I) ion causes the most interference?
(b) Do the alkali metal ions cause greater or less interference than the alkaline earth (Group II) ions?

A15-13. A metal ion buffer was prepared from 0.030 0 M ML and 0.020 M L, where ML is a metal–ligand complex and L is free ligand.

$$M + L \rightleftharpoons ML \qquad K_f = 4.0 \times 10^8$$

Calculate the concentration of free metal ion, M, in this buffer.

15-14. For a silver–silver chloride electrode, the following potentials are observed:

$$E° = 0.222 \text{ V} \qquad E(\text{saturated KCl}) = 0.197 \text{ V}$$

Predict the value of E for a calomel electrode saturated with KCl, given that $E°$ for the calomel electrode is 0.268 V. (Your answer will not be exactly the value 0.241 used in this text.)

15-15. Using the potentials given below, calculate the *activity* of Cl^- in 1 M KCl.

$$E°(\text{calomel electrode}) = 0.268 \text{ V}$$

$$E(\text{calomel electrode, 1 M KCl}) = 0.280 \text{ V}$$

15-16. How many seconds will it take for (a) H^+ and (b) NO_3^- to migrate a distance of 12.0 cm in a field of 7.80×10^3 V/m?

15-17. Suppose that the Ag|AgCl outer electrode in Figure 15-8 is filled with 0.1 M NaCl instead of saturated KCl. Suppose that the electrode is calibrated in a dilute buffer containing 0.1 M KCl. at pH 6.54 at 25°C. The electrode is then dipped in a second buffer *at the same pH* and same temperature, but containing 3.5 M KCl. Use Table 15-2 to estimate how much the indicated pH will change.

15-18. Suppose that an ideal hypothetical cell such as that in Figure 14-7 were set up to measure $E°$ for the half-reaction $Ag^+ + e^- \rightleftharpoons Ag(s)$.
(a) Calculate the equilibrium constant for the net cell reaction.
(b) If there were a junction potential of +2 mV (increasing E from 0.799 to 0.801 V), by what percent would the calculated equilibrium constant increase?
(c) Answer parts a and b, using 0.100 V instead of 0.799 V for the value of $E°$ for the silver reaction.

15-19. (a) When the difference in pH across the membrane of a glass electrode at 25°C is 4.63 pH units, how much voltage is generated by the pH gradient?
(b) What would be the voltage for the same pH difference at 37°C?

15-20. A solution prepared by mixing 25.0 mL of 0.200 M KI with 25.0 mL of 0.200 M NaCl was titrated with 0.100 M AgNO₃ in the following cell:

S.C.E.‖titration solution Ag(s)

Call the solubility products of AgI and AgCl K_I and K_{Cl}, respectively. The answers to parts a and b should be expressions containing these constants.
(a) Calculate the concentration of $[Ag^+]$ in the solution when 25.0 mL of titrant has been added.
(b) Calculate the concentration of $[Ag^+]$ in the solution when 75.0 mL of titrant has been added.
(c) Write an expression showing how the cell voltage depends on $[Ag^+]$.

(d) The titration curve is shown below. Calculate the numerical value of the quotient K_{Cl}/K_I.

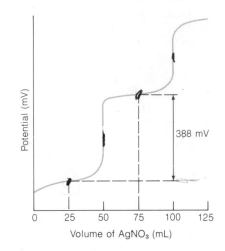

15-21. A solution containing 50.0 mL of 0.100 M EDTA buffered to pH 10.00 was titrated with 50.0 mL of 0.020 0 M Hg(ClO₄)₂ in the cell below.

Hg(*l*)|titration solution‖S.C.E.

From the cell voltage of +0.036 V, calculate the formation constant of Hg(EDTA)²⁻. (*Hint:* See Exercise 15-B.)

15-22. The titration solution in the cell below had a total volume of 50.0 mL and contained 0.100 M Mg^{2+} and 1.00×10^{-5} M Zn(EDTA)²⁻ at a pH of 10.00.

Zn(*s*)|titration solution‖S.C.E.

What will be the cell voltage when 10.0 mL of 0.100 M EDTA has been added? (*Hint:* See Exercise 15-B.)

15-23. Consider the cell Pt(*s*)|cell solution‖S.C.E., whose voltage is +0.126 V. The cell solution contains 2.00 mmol of Fe(NH₄)₂(SO₄)₂, 1.00 mmol of FeCl₃, 4.00 mmol of Na₂EDTA, and lots of buffer, pH 6.78, in a volume of 1.00 L.
(a) Write a half-reaction for the left half-cell.
(b) Find the value of $[Fe^{2+}]/[Fe^{3+}]$ in the cell solution. (This gives the ratio of *uncomplexed* ions.)
(c) Find the quotient of formation constants, $K_f(\text{FeEDTA}^-)/K_f(\text{FeEDTA}^{2-})$.

15-24. Here's a cell you'll really like:

Electrode‖cell solution‖Cu(*s*)

where the electrode is a saturated silver–silver chloride electrode. The cell solution was made by mixing

25.0 mL of 4.00 mM KCN

25.0 mL of 4.00 mM KCu(CN)$_2$

25.0 mL of 0.400 M acid, HA, with pK_a = 9.50

25.0 mL of KOH solution

The measured voltage was −0.440 V. *Calculate the molarity of the KOH solution.* You may assume that essentially all the cuprous ion is present in the form Cu(CN)$_2^-$. For the right half-cell, the reaction is Cu(CN)$_2^-$ + e$^-$ ⇌ Cu(s) + 2CN$^-$ ($E°$ = −0.429 V).

15-25. To determine the *concentration* of a dilute analyte with an ion-selective electrode, why is it desirable to use standards with a constant, high concentration of an inert salt?

15-26. The data below were obtained when a Ca^{2+} ion-selective electrode was immersed in a series of standard solutions whose ionic strength was constant at 2.0 M.

[Ca^{2+}] (M)	E (mV)	[Ca^{2+}] (M)	E (mV)
3.38×10^{-5}	−74.8	3.38×10^{-2}	+10.0
3.38×10^{-4}	−46.4	3.38×10^{-1}	+37.7
3.38×10^{-3}	−18.7		

(a) Make a graph of the data and find the concentration of Ca^{2+} in a sample that gave a reading of −22.5 mV.

(b) Calculate the value of β in Equation 15-27.

15-27. (a) Calculate the slope and the y-intercept (and their standard deviations) of the best straight line through the points in Problem 15-26, using the method of least squares (Section 4-4).

(b) Calculate the concentration (and its associated uncertainty) of a sample that gave a reading of −22.5 (±0.3) mV.

15-28. The selectivity coefficient, $k_{Li^+,Ca^{2+}}$, for a lithium ion-selective electrode is 5×10^{-5}. When this electrode is placed in 3.44×10^{-4} M Li$^+$ solution, the potential is −0.333 V versus S.C.E. What would the potential be if Ca^{2+} were added to give 0.100 M Ca^{2+}?

15-29. A calcium ion-selective electrode obeys Equation 15-25, in which β = 0.970 and $n_{Ca^{2+}}$ = 2. The selectivity coefficients for several ions are listed at the top of the next column.

Interfering ion, Y	$k_{Ca^{2+},Y}$
Mg^{2+}	0.040
Ba^{2+}	0.021
Zn^{2+}	0.081
K$^+$	6.6×10^{-5}
Na$^+$	1.7×10^{-4}

In a pure 1.00×10^{-3} M calcium solution, the reading was +300.0 mV. What would be the voltage if the solution had the same calcium concentration plus [Mg^{2+}] = 1.00×10^{-3} M, [Ba^{2+}] = 1.00×10^{-3} M, [Zn^{2+}] = 5.00×10^{-4} M, [K$^+$] = 0.100 M, and [Na$^+$] = 0.050 0 M? (Use concentrations instead of activities to answer this question.) If they are present at *equal* concentrations, which ion in the table above interferes the most with the Ca^{2+} electrode?

15-30. The following standard addition table assumes that an electrode has a Nernstian response to analyte:

Standard addition table: 10 mL of standard added to 100 mL of sample. (To obtain sample concentration, multiply standard concentration by Q.)

ΔE	Q	ΔE	Q	ΔE	Q
0 mV	1.00	10 mV	0.160	20 mV	0.071 6
1	0.696	11	0.145	21	0.067 1
2	0.529	12	0.133	22	0.062 9
3	0.423	13	0.121	23	0.059 1
4	0.351	14	0.112	24	0.055 6
5	0.297	15	0.103 0	25	0.052 3
6	0.257	16	0.095 2	26	0.049 4
7	0.225	17	0.088 4	27	0.046 6
8	0.199	18	0.082 2	28	0.044 0
9	0.178	19	0.076 7	29	0.041 6

SOURCE: *Orion Research Analytical Methods Guide* (Cambridge, Mass.: Orion Research, Inc., 1975), p. 5.

As an example, suppose chloride ion is measured with an ion-selective electrode and a reference electrode. The electrodes are placed in 100.0 mL of sample, and a reading of 228.0 mV is obtained. Then 10.0 mL of standard containing 100.0 ppm of Cl$^-$ is added, and a new reading of 210.0 mV is observed. Since $|\Delta E|$ = 18.0 mV, Q = 0.082 2 in the table above. Therefore, the original concentration of Cl$^-$ was (0.082 2)(100 ppm) = 8.22 ppm.

(a) What molarity of Cl$^-$ is 8.22 ppm? Assume that the density of unknown is 1.00 g/mL.

(b) Use the original concentration of 8.22 ppm and the original potential of 228.0 mV to show that the second potential should be 210.0 mV if the electrode response obeys the equation

$$E = \text{constant} - 0.059\,16 \log[\text{Cl}^-]$$

(c) How would you change the previous table to use it for Ca^{2+} instead of Cl^-?

15-31. *Derive* the value of 0.696 for the second value of Q in the standard addition of Problem 15-30.

15-32. A calcium ion-selective electrode was calibrated in metal ion buffers whose ionic strength was fixed at 0.50 M. Using the electrode readings below, write an equation for the response of the electrode to Ca^{2+} and Mg^{2+}.

$[\text{Ca}^{2+}]$ (M)	$[\text{Mg}^{2+}]$ (M)	mV
1.00×10^{-6}	0	-52.6
2.43×10^{-4}	0	$+16.1$
1.00×10^{-6}	3.68×10^{-3}	-38.0

15-33. An ion-selective electrode used to measure the cation M^{2+} obeys the equation

$$E = \text{constant}$$
$$+ \frac{0.056\,8\ \text{V}}{2} \log\{[\text{M}^{2+}] + 0.001\,3\,[\text{Na}^+]^2\}$$

When the electrode was immersed in 10.0 mL of unknown containing M^{2+} in 0.200 M NaNO_3, the reading was -163.3 mV (versus S.C.E.). When 1.00 mL of 1.07×10^{-3} M M^{2+} (in 0.200 M NaNO_3) was added to the unknown, the reading increased to -158.8 mV. Find the concentration of M^{2+} in the original unknown.

15-34. A lead ion buffer was prepared by mixing 0.100 mmol of $\text{Pb(NO}_3)_2$ with 2.00 mmol of $\text{Na}_2\text{C}_2\text{O}_4$ in a volume of 10.0 mL.
(a) Given the equilibrium below, find the concentration of free Pb^{2+} in this solution.

$$\text{Pb}^{2+} + 2\text{C}_2\text{O}_4^{2-} \rightleftharpoons \text{Pb(C}_2\text{O}_4)_2^{2-}$$

$$K = \beta_2 = 10^{6.54}$$

(b) How many mmol of $\text{Na}_2\text{C}_2\text{O}_4$ should be used to give $[\text{Pb}^{2+}] = 1.00 \times 10^{-7}$ M?

15-35. A magnesium ion buffer was made by mixing 10.0 mL of 1.00 mM MgSO_4, 10.0 mL of 1.30 mM EDTA, and 5.00 mL of buffer, pH 10.00. What is the concentration of free metal ion in this solution? Answer the same question for MnSO_4 used instead of MgSO_4.

15-36. Do not ignore activities in this problem. Citric acid is a triprotic acid (H_3A) whose anion (A^{3-}) forms stable complexes with many metal ions.

(a) A calcium ion selective electrode gave a calibration curve similar to Figure B-2 in Appendix B. The slope of the curve was 29.58 mV. When the electrode was immersed in a solution having $\mathscr{A}_{\text{Ca}^{2+}} = 1.00 \times 10^{-3}$, the reading was $+2.06$ mV. When the electrode was immersed in the solution to be described in part b of this problem, the reading was -25.90 mV. Calculate the activity of Ca^{2+} in the solution in part b.

(b) Ca^{2+} forms a $1:1$ complex with citrate under the conditions of this problem.

$$\text{Ca}^{2+} + \text{A}^{3-} \xrightleftharpoons{K_f} \text{CaA}^-$$

A solution was prepared by mixing equal volumes of solutions 1 and 2 below.

Solution 1: $[\text{Ca}^{2+}] = 1.00 \times 10^{-3}$ M, pH $= 8.00$, $\mu = 0.10$ M

Solution 2: $[\text{citrate}]_{\text{total}} = 1.00 \times 10^{-3}$ M, pH $= 8.00$, $\mu = 0.10$ M

The activity of the calcium ion in the resulting solution was determined in part a of this problem. Calculate the formation constant, K_f, for CaA^-. For the sake of this calculation, you may assume that the size of CaA^- is 500 pm. At pH 8.00 and $\mu = 0.10$ M, the fraction of free citrate in the form A^{3-} is 0.998.

15-37. Explain how the cell $\text{Ag}(s)|\text{AgCl}(s)|0.1\ \text{M HCl}|$ $0.1\ \text{M KCl}|\text{AgCl}(s)|\text{Ag}(s)$ can be used to measure the $0.1\ \text{M HCl}|0.1\ \text{M KCl}$ junction potential.

15-38. The junction potential, E_j, between solutions α and β can be estimated with the Henderson equation:

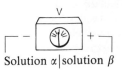

Solution $\alpha|$solution β

$$E_j \approx \frac{\sum\limits_i \dfrac{|z_i|u_i}{z_i}[C_i(\beta) - C_i(\alpha)]}{\sum\limits_i |z_i|u_i[C_i(\beta) - C_i(\alpha)]} \frac{RT}{F} \ln \frac{\sum\limits_i |z_i|u_i\,C_i(\alpha)}{\sum\limits_i |z_i|u_i\,C_i(\beta)}$$

where z_i is the charge of species i, u_i is the *mobility* of species i (Table 15-1), $C_i(\alpha)$ is the concentration of species i in phase α, and $C_i(\beta)$ is the concentration in phase β. (One of the approximations in this equation is the neglect of activity coefficients.) Calculate the junction potential between the following phases at 25°C: (a) 0.1 M HCl$|$0.1 M KCl; (b) 0.1 M HCl$|$3.5 M KCl. Compare your answers with the values in Table 15-2.

16 Redox Titrations

A redox titration is based on an oxidation–reduction reaction between analyte and titrant. For example, hydroquinone can be analyzed by titration with standard dichromate solution:

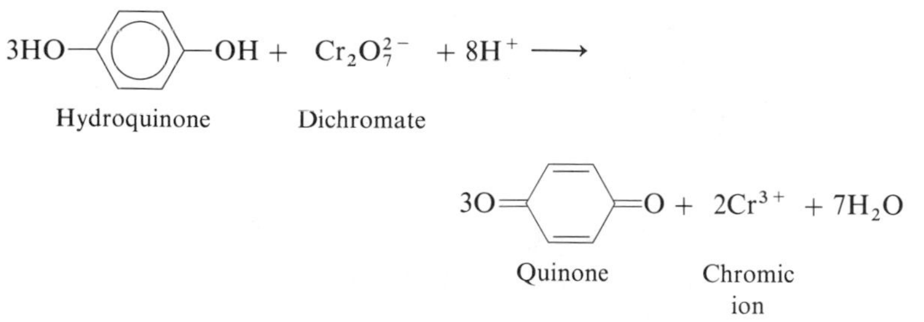

$$3HO-\langle\bigcirc\rangle-OH + \ Cr_2O_7^{2-} \ + 8H^+ \longrightarrow$$

Hydroquinone   Dichromate

$$3O=\langle\bigcirc\rangle=O + \ 2Cr^{3+} \ + 7H_2O \quad (16\text{-}1)$$

Quinone   Chromic
ion

The equilibrium constant for the titration reaction is easy to calculate from its standard reduction potential of $E° = 0.66$ V:

$$K = 10^{nE°/0.059\ 16} = 10^{6(0.66)/0.059\ 16} = 10^{67} \qquad (16\text{-}2)$$

Relation between $E°$ and the equilibrium constant:

$$K = 10^{nE°/0.059\ 16} \qquad \text{at } 25°C.$$

In practice, the reaction is too slow for a titration at room temperature, but it proceeds rapidly enough at 40–60°C to be a useful analytical procedure. The enormous equilibrium constant (typical of many redox reactions) assures us that the reaction will be quantitative (i.e., it will go to completion). The end point is detected with the *redox indicator* diphenylamine, whose color changes from colorless to violet when the titration reaction is complete.

In this chapter we first deal with the theory of redox titrations and then discuss practical aspects of some common reagents. The theory is necessary to understand how redox indicators and potentiometric detection procedures work.

Consider the titration of Fe^{2+} with standard Ce^{4+}, the course of which could be monitored potentiometrically as shown in Figure 16-1. The titration reaction is

The *titration reaction* is

$$Ce^{4+} + Fe^{2+} \rightarrow Ce^{3+} + Fe^{3+}.$$

It goes to completion after each addition of titrant.

$$Ce^{4+} + Fe^{2+} \rightarrow Ce^{3+} + Fe^{3+} \qquad (16\text{-}3)$$

Ceric ion Ferrous ion Cerous ion Ferric ion

for which $K \approx 10^{17}$ in 1 M $HClO_4$. Each mole of ceric ion oxidizes one mole of ferrous ion rapidly and quantitatively. The titration reaction thus creates a mixture of Ce^{4+}, Ce^{3+}, Fe^{2+}, and Fe^{3+} in the beaker in Figure 16-1.

To follow the course of the reaction, a pair of electrodes is inserted into the reaction mixture. At the *calomel reference electrode*, the reaction is

$$2Hg(l) + 2Cl^- \rightleftharpoons Hg_2Cl_2(s) + 2e^- \qquad (16\text{-}4)$$

At the *Pt indicator electrode*, there are *two* reactions that come to equilibrium:

These redox equilibria are established at the Pt electrode.

$$Fe^{3+} + e^- \rightleftharpoons Fe^{2+} \qquad E^\circ = 0.767 \text{ V} \qquad (16\text{-}5)$$

$$Ce^{4+} + e^- \rightleftharpoons Ce^{3+} \qquad E^\circ = 1.70 \text{ V} \qquad (16\text{-}6)$$

The potentials cited here are the formal potentials that apply in 1 M $HClO_4$.

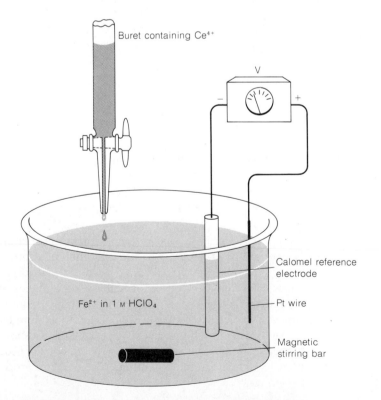

Figure 16-1
Apparatus for potentiometric titration of Fe^{2+} with Ce^{4+}.

The cell reaction can be described in either of two ways:

$$2Fe^{3+} + 2Hg(l) + 2Cl^- \rightleftharpoons 2Fe^{2+} + Hg_2Cl_2(s) \qquad (16\text{-}7)$$

or

$$2Ce^{4+} + 2Hg(l) + 2Cl^- \rightleftharpoons 2Ce^{3+} + Hg_2Cl_2(s) \qquad (16\text{-}8)$$

The cell reactions are not the same as the titration reaction (16-3). The potentiometer has no interest in how the Ce^{4+}, Ce^{3+}, Fe^{3+}, and Fe^{2+} happened to get into the beaker. What it does care about is that electrons want to flow from the anode to the cathode through the meter. If the solution has come to equilibrium, the potential driving both Reaction 16-7 and Reaction 16-8 must be the same. **We may describe the cell voltage using either Reaction 16-7 or Reaction 16-8, or both, however we please.**

The titration reaction goes to completion. The cell reactions proceed to a negligible extent. The potentiometric circuit *measures* the concentrations of species in solution, but does not *change* them.

To reiterate, the physical picture of the titration is this: Ce^{4+} is added from the buret to create a mixture of Ce^{4+}, Ce^{3+}, Fe^{3+}, and Fe^{2+} in the beaker. Since the equilibrium constant for Reaction 16-3 is so large, the titration reaction goes to "completion" after each addition of Ce^{4+}. The potentiometer measures the voltage driving electrons from the reference electrode, through the meter, and out at the Pt electrode. That is, **the circuit measures the potential for reduction of Fe^{3+} or Ce^{4+} at the Pt surface by electrons from the $Hg|Hg_2Cl_2$ couple in the reference electrode.** The titration reaction, on the other hand, is an oxidation of Fe^{2+} and a reduction of Ce^{4+}. The titration reaction produces a certain mixture of Ce^{4+}, Ce^{3+}, Fe^{3+}, and Fe^{2+}. The circuit measures the potential for reduction of Ce^{4+} and Fe^{3+} by Hg. **The titration reaction goes to completion. The cell reaction is negligible. The cell is being used to measure activities, not to change them.**

We now set out to calculate how the cell voltage will change as Fe^{2+} is titrated with Ce^{4+}. The titration curve has three regions:

Region 1: Before the equivalence point

As each aliquot of Ce^{4+} is added, the titration reaction (16-3) goes to "completion," consuming the Ce^{4+} and making an equal number of moles of Ce^{3+} and Fe^{3+}. Prior to the equivalence point, excess unreacted Fe^{2+} remains in the solution. Therefore, we can find the concentrations of Fe^{2+} and Fe^{3+} without any difficult calculations. On the other hand, we cannot find the concentration of Ce^{4+} without solving a fancy little equilibrium problem. Since the amounts of Fe^{2+} and Fe^{3+} are both known, it is *convenient* to calculate the cell voltage using Reaction 16-5 instead of Reaction 16-6.

Either Reaction 16-5 or 16-6 can be used to describe the cell voltage at any time. However, since we know the concentrations of Fe^{2+} and Fe^{3+}, it is more convenient for now to use Reaction 16-5.

$$E = E_+ - E_- \qquad (15\text{-}5)$$

$$E = \left(0.767 - 0.059\,16 \log \frac{[Fe^{2+}]}{[Fe^{3+}]} \right) - (0.241) \qquad (16\text{-}9)$$

Formal potential for Fe^{3+} reduction in 1 M $HClO_4$

Potential of saturated calomel electrode

E_+ is the potential of the half-cell connected to the positive terminal of the potentiometer in Figure 16-1. E_- refers to the half-cell connected to the negative terminal.

$$E = +0.526 - 0.059\,16 \log \frac{[Fe^{2+}]}{[Fe^{3+}]} \qquad (16\text{-}10)$$

One special point is reached before the equivalence point. When the volume of titrant is one-half of the amount required to reach the equivalence point ($V = \frac{1}{2}V_e$), the concentrations of Fe^{3+} and Fe^{2+} are equal. In this case the log term of Equation 16-9 is zero, and $E_+ = E°$ for the $Fe^{3+}|Fe^{2+}$ couple. *The point at which $V = \frac{1}{2}V_e$ is analogous to the point at which $pH = pK_a$ when $V = \frac{1}{2}V_e$ in an acid–base titration.*

For Reaction 16-5,
$E_+ = E°(Fe^{3+}|Fe^{2+})$ when $V = \frac{1}{2}V_e$.

Region 2: At the equivalence point

Exactly enough Ce^{4+} has been added to react with all the Fe^{2+}. Virtually all the cerium is in the form Ce^{3+}, and virtually all the iron is in the form Fe^{3+}. Tiny amounts of Ce^{4+} and Fe^{2+} are present at equilibrium. From the stoichiometry of Reaction 16-3, we can say that

$$[Ce^{3+}] = [Fe^{3+}] \qquad (16\text{-}11)$$

and

$$[Ce^{4+}] = [Fe^{2+}] \qquad (16\text{-}12)$$

To understand why Equations 16-11 and 16-12 are true, imagine that *all* the cerium and iron has been converted to Ce^{3+} and Fe^{3+}. Since we are at the equivalence point, $[Ce^{3+}] = [Fe^{3+}]$. Now let Reaction 16-3 come to equilibrium:

$$Fe^{3+} + Ce^{3+} \rightleftharpoons Fe^{2+} + Ce^{4+} \qquad \text{(reverse of Reaction 16-3)}$$

If a little bit of Fe^{3+} goes back to Fe^{2+}, an equal number of moles of Ce^{4+} must be made. So $[Ce^{4+}] = [Fe^{2+}]$.

At any time, Reactions 16-5 and 16-6 are *both* in equilibrium at the cathode. At the equivalence point, it is *convenient* to use both reactions to describe the cell voltage. The Nernst equations for the reactions are, respectively,

At the equivalence point, it is convenient to use both Reactions 16-5 and 16-6 to calculate the cell voltage. This is strictly a matter of algebraic convenience.

$$E_+ = 0.767 - 0.059\,16 \log \frac{[Fe^{2+}]}{[Fe^{3+}]} \qquad (16\text{-}13)$$

$$E_+ = 1.70 - 0.059\,16 \log \frac{[Ce^{3+}]}{[Ce^{4+}]} \qquad (16\text{-}14)$$

Here is where we stand: Each equation above is a statement of algebraic truth. But neither one alone allows us to find E_+ because we do not know exactly what tiny concentrations of Fe^{2+} and Ce^{4+} are present. It is possible to solve the four simultaneous equations 16-11, 16-12, 16-13, and 16-14, by first *adding* Equations 16-13 and 16-14. This is strictly an algebraic manipulation, the logic and beauty of which will become apparent very soon.

Adding Equations 16-13 and 16-14 gives

$$2E_+ = 0.767 + 1.70 - 0.059\,16 \log \frac{[Fe^{2+}]}{[Fe^{3+}]} - 0.059\,16 \log \frac{[Ce^{3+}]}{[Ce^{4+}]}$$

$$2E_+ = 2.46_7 - 0.059\,16 \log \frac{[Fe^{2+}][Ce^{3+}]}{[Fe^{3+}][Ce^{4+}]} \qquad (16\text{-}15)$$

$\log a + \log b = \log ab.$

But since $[Ce^{3+}] = [Fe^{3+}]$ and $[Ce^{4+}] = [Fe^{2+}]$ at the equivalence point, the ratio of concentrations in the log term is unity. The logarithm is zero. Therefore,

$$2E_+ = 2.46_7\,V$$

$$E_+ = 1.23\,V \qquad (16\text{-}16)$$

and the cell voltage is

$$E = E_+ - E(\text{calomel})$$

$$= 1.23 - 0.241 = 0.99\,V \qquad (16\text{-}17)$$

In this particular titration, the equivalence-point voltage is independent of the concentrations and volumes of the reactants.

Region 3: After the equivalence point

Now virtually all the iron is Fe^{3+}. The moles of Ce^{3+} equal the moles of Fe^{3+}. There is also a known excess of unreacted Ce^{4+}. Since we know both $[Ce^{3+}]$ and $[Ce^{4+}]$, it is *convenient* to use Reaction 16-6 to describe the cathode reaction.

$$E = E_+ - E(\text{calomel})$$

$$= \left(1.70 - 0.059\,16 \log \frac{[Ce^{3+}]}{[Ce^{4+}]}\right) - (0.241) \qquad (16\text{-}18)$$

At the special point when $V = 2V_e$, $[Ce^{3+}] = [Ce^{4+}]$ *and* $E_+ = E°(Ce^{4+}|Ce^{3+}) = 1.70\,V.$

Roughly speaking, the cell voltage will be fairly level before and after the equivalence point, with a rapid rise near the equivalence point. Before the equivalence point, the voltage is roughly in the range $E_+ \approx E°(Fe^{3+}|Fe^{2+})$ or $E = E_+ - E(\text{calomel}) \approx E°(Fe^{3+}|Fe^{2+}) - 0.241\,V$. After the equivalence point, the voltage levels off such that $E_+ \approx E°(Ce^{4+}|Ce^{3+})$ and $E \approx E°(Ce^{4+}|Ce^{3+}) - 0.241\,V.$

After the equivalence point, it is convenient to use Reaction 16-6 because we can easily calculate the concentrations of Ce^{3+} and Ce^{4+}. It is not convenient to use Reaction 16-5 because we do not know the concentration of Fe^{2+}, which has been "used up."

The potential (versus S.H.E.) in a redox titration varies, ranging roughly between the standard potentials of the two couples involved in the titration reaction.

EXAMPLE: Potentiometric Redox Titration

Suppose that 100.0 mL of 0.050 0 M Fe^{2+} is titrated with 0.100 M Ce^{4+}, using the cell in Figure 16-1. The equivalence point occurs when $V_{Ce^{4+}} = 50.0$ mL, since the Ce^{4+} is twice as concentrated as the Fe^{2+}. Calculate the cell voltage at the following points: 36.0, 50.0, and 63.0 mL.

36.0 mL: This is 36.0/50.0 of the way to the equivalence point. Therefore, 36.0/50.0 of the iron is in the form Fe^{3+}, and 14.0/50.0 is in the form Fe^{2+}. Putting the value $[Fe^{2+}]/[Fe^{3+}] = 14.0/36.0$ into Equation 16-10 gives a cell voltage of 0.550 V.

50.0 mL: This is the equivalence point. Equation 16-17 tells us that $V_e = 0.99$ V, regardless of the concentrations of analyte and titrant for this particular redox tiration. For many tirations the voltage at the equivalence point is constant, but, as we will see in the next section, this is not always true.

63.0 mL: The first 50.0 mL of cerium has been converted to Ce^{3+}. Since 13.0 mL of excess Ce^{4+} has been added, the value of $[Ce^{3+}]/[Ce^{4+}]$ in Equation 16-18 is 50.0/13.0, and the cell voltage is 1.424 V.

The calculated titration curve for Reaction 16-3 is shown in Figure 16-2. As in all previous titrations of all types, the end point is marked by a

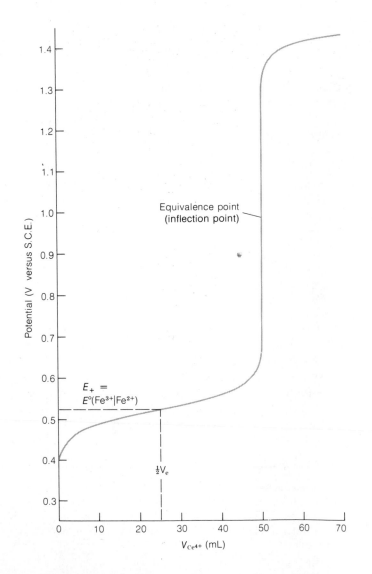

Figure 16-2
Theoretical curve for titration of 100.0 mL of 0.050 0 M Fe^{2+} with 0.100 M Ce^{4+} in 1 M $HClO_4$.

steep rise in the curve. Because both reactants are one-electron redox reagents, the curve is symmetric near the equivalence point. As we shall see in the next section, this is not the case when the stoichiometry of reactants is not 1:1. The calculated value of E_+ when $V_{Ce^{4+}} = \frac{1}{2}V_e$ is the formal potential of the $Fe^{3+}|Fe^{2+}$ couple, since the quotient $[Fe^{2+}]/[Fe^{3+}]$ in Equation 16-9 is unity at this point. The calculated voltage at any point in this titration depends only on the *ratio* of reactants; their *concentrations* do not figure in any calculations in this example. We expect, therefore, that the curve in Figure 16-2 will be independent of dilution. We should observe the same curve even if both reactants were diluted by a factor of 10.

The shape of the curve in Figure 16-2 is essentially independent of the concentrations of analyte and titrant. The curve is symmetric near V_e because the stoichiometry is 1:1.

The voltage before Ce^{4+} is added to the reaction ($V_{Ce^{4+}} = 0$) cannot be calculated because we do not know how much Fe^{3+} is present in solution. If no Fe^{3+} is present, the voltage calculated with Equation 16-10 is $-\infty$. In fact, there must be some Fe^{3+} present in each reagent, either as an impurity or from oxidation of Fe^{2+} by atmospheric oxygen. In any case, the voltage could never be lower than that needed to reduce the solvent ($H_2O + e^- \rightarrow \frac{1}{2}H_2 + OH^-$).

A Slightly More Complicated Redox Calculation

Iodate can be used to titrate Tl^+ in a concentrated solution of HCl:

$$IO_3^- + 2Tl^+ + 2Cl^- + 6H^+ \rightarrow ICl_2^- + 2Tl^{3+} + 3H_2O \quad (16\text{-}19)$$

| Iodate | Thallous ion | | Thallic ion | |

From the standard potentials below

$$IO_3^- + 2Cl^- + 6H^+ + 4e^- \rightarrow ICl_2^- + 3H_2O \qquad E^\circ = 1.24 \text{ V} \quad (16\text{-}20)$$

$$Tl^{3+} + 2e^- \rightarrow Tl^+ \qquad E^\circ = 0.77 \text{ V} \quad (16\text{-}21)$$

we expect E° for Reaction 16-19 to be $1.24 - 0.77 = 0.47$ V and $K = 10^{4E^\circ/0.059\,16} = 10^{32}$. The equilibrium constant is very large, so the reaction goes to completion after each addition of IO_3^-.

Let's calculate the theoretical titration curve that results when 100.0 mL of 0.010 0 M Tl^+ is titrated with 0.010 0 M IO_3^-. We will assume that both solutions have a constant concentration of HCl equal to 1.00 M. Beause one mole of IO_3^- consumes two moles of Tl^+, the equivalence point occurs at $V_{IO_3^-} = 50.0$ mL. Suppose that the reference electrode is a saturated calomel electrode, and the apparatus shown in Figure 16-1 is used for the titration. We will calculate the voltage at one point in each region of the titration.

Region 1: Before the equivalence point

Suppose $V_{IO_3^-} = 10.0$ mL. Since the entire reaction requires 50.0 mL, the reaction is 10.0/50.0 complete. That is, one-fifth of the thallium is in the form Tl^{3+}, and four-fifths is in the form Tl^+. We may choose Reaction 16-21 to calculate E_+ because we know the concentrations of both Tl^+ and Tl^{3+}:

Before the equivalence point, we know the concentrations of Tl^+ and Tl^{3+}, so it is convenient to use Reaction 16-21 to calculate E_+.

$$E = E_+ - E(\text{calomel})$$

$$= \underbrace{\left(0.77 - \frac{0.059\,16}{2} \log \frac{[\text{Tl}^+]}{[\text{Tl}^{3+}]}\right)}_{\substack{\text{Nernst equation for} \\ \text{Reaction 17-21}}} - \underbrace{(0.241)}_{\substack{\text{Potential of} \\ \text{calomel electrode}}} \qquad (16\text{-}22)$$

$$= \left(0.77 - \frac{0.059\,16}{2} \log \frac{\frac{4}{5}}{\frac{1}{5}}\right) - 0.241 = 0.511 \text{ V}$$

Region 2: At the equivalence point

Now virtually all the reactants have been converted to products. We can say that

$$[\text{Tl}^{3+}] = 2[\text{ICl}_2^-] \qquad (16\text{-}23)$$

since two moles of Tl^{3+} are created for each mole of ICl_2^-. Although only tiny amounts of the reactants are present, we can say that

$$[\text{Tl}^+] = 2[\text{IO}_3^-] \qquad (16\text{-}24)$$

You can justify this by imagining that only products are present. When they react, they produce a tiny amount of Tl^+ and IO_3^- in a 2:1 mole ratio.

It is convenient to use Reactions 16-20 and 16-21 *together* to find E_+ at the equivalence point. We do not have enough information to use either reaction by itself.

Once again it is convenient to use both half-reactions, 16-20 and 16-21, to calculate E_+ at the equivalence point. From Reaction 16-20, we can say that

$$E_+ = 1.24 - \frac{0.059\,16}{4} \log \frac{[\text{ICl}_2^-]}{[\text{IO}_3^-][\text{Cl}^-]^2[\text{H}^+]^6} \qquad (16\text{-}25)$$

From Reaction 16-21, we may write

$$E_+ = 0.77 - \frac{0.059\,16}{2} \log \frac{[\text{Tl}^+]}{[\text{Tl}^{3+}]} \qquad (16\text{-}26)$$

To solve the four simultaneous equations (16-23 to 16-26), it is algebraically expedient to add Equations 16-25 and 16-26 together,. However, Equation 16-25 should first be multiplied by 2 so that the coefficients of the log terms are equal. Multiplying Equation 16-25 by 2 gives

$$2E_+ = 2.48 - \frac{0.059\,16}{2} \log \frac{[\text{ICl}_2^-]}{[\text{IO}_3^-][\text{Cl}^-]^2[\text{H}^+]^6} \qquad (16\text{-}27)$$

Adding Equations 16-26 and 16-27 gives

$$3E_+ = 3.25 - \frac{0.059\,16}{2} \log \frac{[\text{ICl}_2^-][\text{Tl}^+]}{[\text{IO}_3^-][\text{Cl}^-]^2[\text{H}^+]^6[\text{Tl}^{3+}]} \qquad (16\text{-}28)$$

Substituting $[\text{Tl}^{3+}] = 2[\text{ICl}_2^-]$ and $[\text{Tl}^+] = 2[\text{IO}_3^-]$ into Equation 16-28 simplifies the log term:

$$3E_+ = 3.25 - \frac{0.059\ 16}{2} \log \frac{[ICl_2^-]^2[IO_3^-]}{[IO_3^-][Cl^-]^2[H^+]^6{}_2^2[ICl_2^-]} \quad (16\text{-}29)$$

Since we have stipulated that $[Cl^-] = [H^+] = 1.00$ M in this example, the log term is again zero. If the concentrations were not 1.00 M, we would simply put their values into Equation 16-29 and evaluate the log term. Putting $[H^+] = [Cl^-] = 1.00$ M into Equation 16-29 gives

$$3E_+ = 3.25\ \text{V} \quad \Rightarrow \quad E_+ = 1.08_3\ \text{V} \quad (16\text{-}30)$$

$$E = E_+ - E(\text{calomel}) = 1.08_3 - 0.241 = 0.84\ \text{V} \quad (16\text{-}31)$$

In some other titrations, the concentrations of products do not entirely cancel in the log term. In such cases, we simply calculate the molarity of product in the solution at the equivalence point, assuming that all reactant has been converted to product.

Region 3: After the equivalence point

If $V_{IO_3^-} = 57.6$ mL, we can say that

$$\frac{[ICl_2^-]}{[IO_3^-]} = \frac{50.0}{7.6} \begin{array}{l} \leftarrow \text{ Volume at equivalence point} \\ \leftarrow \text{ Volume past equivalence point} \end{array} \quad (16\text{-}32)$$

Using Reaction 16-20 to calculate E_+, we may write

$$E = E_+ - E(\text{calomel})$$

$$= \left(1.24 - \frac{0.059\ 16}{4} \log \frac{[ICl_2^-]}{[IO_3^-][Cl^-]^2[H^+]^6} \right) - E(\text{calomel}) \quad (16\text{-}33)$$

Now we know both $[ICl_2^-]$ and $[IO_3^-]$, so we can use Reaction 16-20 to find E_+.

Substituting $[ICl_2^-]/[IO_3^-] = 50.0/7.6$ and $[H^+] = [Cl^-] = 1.00$ M, we obtain

$$E = 1.24 - \frac{0.059\ 16}{4} \log \frac{50.0}{(7.6)(1.00)^2(1.00)^6} - 0.241 \quad (16\text{-}34)$$

$$E = 0.987\ \text{V} \quad (16\text{-}35)$$

The theoretical titration curve is shown in Figure 16-3. *Note that the curve is not symmetric about the equivalence point.* This is true whenever the stoichiometry of reactants is not 1:1. Still, the curve is so steep near the equivalence point that negligible error is introduced if the center of the steepest portion is taken as the end point. Demonstration 16-1 illustrates a titration curve with an asymmetric end point.

The magnitude of the change in potential near the equivalence point is smaller in Figure 16-3 than in Figure 16-2 because $E°$ for Reaction 16-3 is greater than $E°$ for Reaction 16-19. In general, the clearest results are achieved with the strongest oxidizing and reducing agents. This is analogous to acid–base titrations, where we normally use a strong acid or strong base as titrant, to get the sharpest break at the equivalence point.

When the stoichiometry of the titration reaction is not 1:1, the curve is not symmetric around the equivalence point.

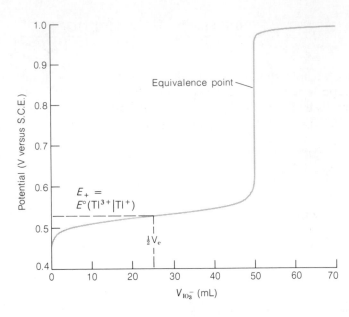

Figure 16-3
Theoretical curve for titration of 100.0 mL of 0.010 0 M Tl^+ with 0.010 0 M IO_3^- in 1.00 M HCl.

16-2 TITRATION OF A MIXTURE

The titration of two species will exhibit two breaks if the standard potentials of the redox couples are sufficiently different. Figure 16-4 shows the theoretical titration curve for an equimolar mixture of Tl^+ and Sn^{2+} titrated with IO_3^-. The two *titration reactions* are

Sn^{2+} reacts before Tl^+ because Sn^{2+} is a stronger reducing agent than Tl^+. The strongest reagents react first, just as the strongest acids or bases react first in acid–base reactions.

First: $IO_3^- + 2Sn^{2+} + 2Cl^- + 6H^+ \rightarrow$

$$ICl_2^- + 2Sn^{4+} + 3H_2O \quad (16\text{-}36)$$

Second: $IO_3^- + 2Tl^+ + 2Cl^- + 6H^+ \rightarrow$

$$ICl_2^- + 2Tl^{3+} + 3H_2O \quad (16\text{-}37)$$

The relevant half-reactions are Reactions 16-20 and 16-21 and the following:

$$Sn^{4+} + 2e^- \rightleftharpoons Sn^{2+} \quad E° = 0.139 \text{ V} \quad (16\text{-}38)$$

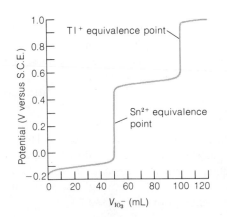

Figure 16-4
Theoretical curve for 0.010 0 M Tl^+ plus 0.010 0 M Sn^{2+} titrated with 0.010 0 M IO_3^-. The initial volume of analyte is 100.0 mL, and all solutions contain 1.00 M HCl.

Because the $Sn^{4+}|Sn^{2+}$ couple has a lower reduction potential, Sn^{2+} will be oxidized before Tl^+. That is, the equilibrium constant for Reaction 16-36 is larger than for Reaction 16-37. This is another way of saying that Sn^{2+} is a stronger reducing agent than Tl^+.

The first part of the titration curve in Figure 16-4 is calculated using Reaction 16-36, as if no Tl^+ were present. After the Sn^{2+} equivalence point, the remainder is calculated using Reaction 16-37, as if no Sn^{2+} were present.

Challenge: Calculate the value of E_+ at $V = 25.0$ and $V = 75.0$ mL in the titration of Figure 16-4.

16-3 REDOX INDICATORS

A chemical indicator may be used to detect the end point of a redox titration, just as an indicator may be used in an acid–base titration. A **redox indicator** is a compound that changes color when it goes from its oxidized to its reduced state. One common indicator is ferroin, whose color change is from pale blue (almost colorless) to red.

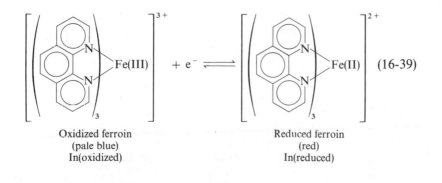

Oxidized ferroin   Reduced ferroin
(pale blue)   (red)
In(oxidized)   In(reduced)

$$+ \; e^- \rightleftharpoons \qquad\qquad\qquad\qquad\qquad\qquad (16\text{-}39)$$

To predict the potential range over which the indicator color will change, we first write a Nernst equation for the indicator.

$$In(oxidized) + ne^- \rightleftharpoons In(reduced) \tag{16-40}$$

$$E = E° - \frac{0.059\ 16}{n} \log \frac{[In(reduced)]}{[In(oxidized)]} \tag{16-41}$$

As with acid–base indicators, the color of In(reduced) will be observed when

$$\frac{[In(reduced)]}{[In(oxidized)]} \gtrsim \frac{10}{1} \tag{16-42}$$

and the color of In(oxidized) will be observed when

$$\frac{In(reduced)}{In(oxidized)} \lesssim \frac{1}{10} \tag{16-43}$$

Putting these quotients into Equation 16-41 tells us that the color change will occur over the range

$$E = \left(E° \pm \frac{0.059}{n} \right) \text{volts} \tag{16-44}$$

A redox indicator changes color over a range of $\pm(59/n)$ mV, centered at $E°$ for the indicator.

For ferroin, with $E° = 1.147$ V (Table 16-1), we expect the color change to occur in the approximate range 1.088 V to 1.206 V with respect to the standard hydrogen electrode. If a saturated calomel electrode is used as the reference instead, the indicator transition range will be

See the diagram on page 352 for a better understanding of Equation 16-45.

$$\begin{pmatrix} \text{Indicator transition} \\ \text{range versus calomel} \\ \text{electrode} \end{pmatrix} = \begin{pmatrix} \text{indicator transition} \\ \text{range versus standard} \\ \text{hydrogen electrode} \end{pmatrix} - E(\text{calomel}) \tag{16-45}$$

$$= (1.088 \text{ to } 1.206) - (0.241)$$

$$= 0.847 \text{ V to } 0.965 \text{ V (versus S.C.E.)}$$

The indicator's transition range should match the steep part of the titration curve.

Ferroin would therefore be a useful indicator for the titrations in Figures 16-2 and 16-3.

We have seen that the larger the difference in standard potential (or formal potential) between titrant and analyte, the sharper the break in the titration curve at the equivalence point. A redox titration is usually feasible if the difference in potentials between analyte and titrant is $\gtrsim 0.2$ V. However, the end point of such a titration is not very sharp, and it is probably best detected potentiometrically. If the difference in formal potentials is $\gtrsim 0.4$ V, then a redox indicator will usually give a satisfactorily sharp end point.

Challenge: Select suitable indicators from Table 16-1 for the two end points of the titration in Figure 16-4. What color change might you expect at each of the two end points? Remember that *both* indicators will be in the solution together.

The Starch—Iodine Complex

Numerous analytical procedures are based on titrations involving iodine. Starch is the indicator of choice for these procedures because it forms an

Table 16-1
Some redox indicators

Indicator	Color		$E°$
	Oxidized	Reduced	
Phenosafranine	Red	Colorless	0.28
Indigo tetrasulfonate	Blue	Colorless	0.36
Methylene blue	Blue	Colorless	0.53
Diphenylamine	Violet	Colorless	0.75
4'-Ethoxy-2,4-diaminoazobenzene	Yellow	Red	0.76
Diphenylamine sulfonic acid	Red-violet	Colorless	0.85
Diphenylbenzidine sulfonic acid	Violet	Colorless	0.87
Tris(2,2'-bipyridine)iron	Pale blue	Red	1.120
Tris(1,10-phenanthroline)iron (ferroin)	Pale blue	Red	1.147
Tris(5-nitro-1,10-phenanthroline)iron	Pale blue	Red-violet	1.25
Tris(2,2'-bipyridine)ruthenium	Pale blue	Yellow	1.29

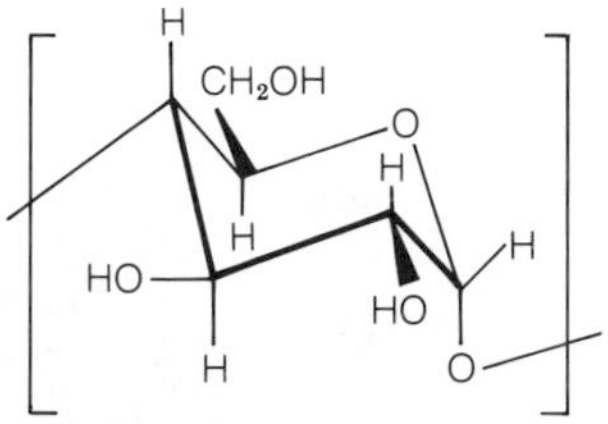

intense blue complex with iodine. Starch is not a redox indicator because it responds specifically to the presence of I_2, not to a change in redox potential.

The active fraction of starch is amylose, a polymer of the sugar α-D-glucose, with the repeating unit shown in Figure 16-5. The polymer exists as a coiled helix into which small molecules can fit. In the presence of starch and I^-, iodine molecules form long chains of I_5^- ions that occupy the center of the amylose helix (Figure 16-6).

$$\cdots[I—I—I—I—I]^- \cdots [I—I—I—I—I]^- \cdots$$

It is a visible absorption band of this I_5^- chain bound within the helix that gives rise to the characteristic starch–iodine color.

Starch is readily biodegraded, so either it should be freshly dissolved, or the solution should contain a preservative, such as HgI_2 or thymol. One of the hydrolysis products of starch is glucose, which is a reducing agent. A partially hydrolyzed solution of starch could thus be a source of error in a redox titration.

16-4 COMMON REDOX REAGENTS

In this section we will discuss some of the more common redox agents used in volumetric analysis.[†] Table 16-2 lists some frequently encountered oxidizing and reducing agents. Most of the oxidizing agents can be used as titrants. However, most of the reducing agents react with oxygen and are therefore less suitable as titrants. Only a few of the reagents in Table 16-2 can be used as primary standards.

[†] Some sources of information on redox titrations include J. Bassett, R. C. Denney, G. H. Jeffery, and J. Mendham, *Vogel's Textbook of Inorganic Analysis,* 4th ed. (Essex, England: Longman, 1978); H. A. Laitinen and W. E. Harris, *Chemical Analysis,* 2nd ed. (New York: McGraw-Hill, 1975); I. M. Kolthoff, R. Belcher, V. A. Stenger, and G. Matsuyama, *Volumetric Analysis,* Vol. 3 (New York: Wiley, 1957); A. Berka, J. Vulterin, and J. Zýka, *Newer Redox Titrants* (H. Weisz, trans., Oxford: Pergamon, 1965).

Figure 16-5
Structure of the repeating unit of the sugar amylose.

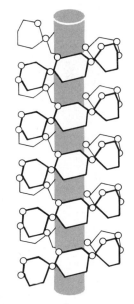

Figure 16-6
Schematic structure of the starch–iodine complex. The amylose sugar chain forms a helix around the nearly linear iodine chain. [R. C. Teitelbaum, S. L. Ruby, and T. J. Marks, *J. Amer. Chem. Soc.,* **102,** 3322 (1980).]

Table 16-2
Some common oxidizing and reducing agents

Oxidants		Reductants	
BiO_3^-	bismuthate		
BrO_3^-	bromate	AsO_3^{3-}	arsenite
Br_2	bromine		ascorbic acid
Ce^{4+}	ceric	Cr^{2+}	chromous
CH_3—⟨⟩—SO_2NCl^-	chloramine T	$S_2O_4^{2-}$	dithionite
$Cr_2O_7^{2-}$	dichromate	$Fe(CN)_6^{4-}$	ferrocyanide
H_2O_2	hydrogen peroxide	Fe^{2+}	ferrous
OCl^-	hypochlorite	N_2H_4	hydrazine
IO_3^-	iodate	HO—⟨⟩—OH	hydroquinone
I_2	iodine	NH_2OH	hydroxylamine
$Pb(acetate)_4$	lead(IV) acetate	Hg_2^{2+}	mercurous
$HClO_4$	perchloric acid	Sn^{2+}	stannous
IO_4^-	periodate	SO_3^{2-}	sulfite
MnO_4^-	permanganate	SO_2	sulfur dioxide
$S_2O_8^{2-}$	peroxydisulfate	$S_2O_3^{2-}$	thiosulfate
		H_3PO_2	hypophosphorous acid

Adjustment of Analyte Oxidation State

It is necessary to remove the excess preadjustment reagent so that it will not interfere in the subsequent titration.

For many analyses it is necessary to adjust the oxidation state of the analyte to one that can be titrated with an oxidizing or a reducing agent. For example, Mn(II) might be quantitatively **preoxidized** to MnO_4^- and then titrated with standard Fe^{2+}. The preadjustment reaction must be quantitative, and it must be possible to remove or destroy the excess preadjustment reagent.

Preoxidation

There are several powerful oxidants available that can be easily removed after preoxidation. *Peroxydisulfate* ($S_2O_8^{2-}$, also called *persulfate*) is a strong oxidant that requires Ag^+ as a catalyst.

$$S_2O_8^{2-} + Ag^+ \rightarrow SO_4^{2-} + \underbrace{SO_4^- + Ag^{2+}}_{\text{Two powerful oxidants}} \qquad (16\text{-}46)$$

The excess reagent is destroyed by boiling the solution after the oxidation of analyte is complete.

$$2S_2O_8^{2-} + 2H_2O \xrightarrow{\text{boiling}} 4SO_4^{2-} + O_2 + 4H^+ \qquad (16\text{-}47)$$

The $S_2O_8^{2-}/Ag^+$ mixture is able to oxidize Mn(II) to MnO_4^-, Ce(III) to Ce(IV), Cr(III) to $Cr_2O_7^{2-}$, and V(IV) to V(V).

Silver (II) oxide (AgO) dissolves in concentrated mineral acids to give Ag^{2+}, with oxidizing power similar to the $S_2O_8^{2-}/Ag^+$ combination. Excess Ag^{2+} can be removed by boiling:

$$4Ag^{2+} + 2H_2O \xrightarrow{\text{boiling}} 4Ag^+ + O_2 + 4H^+ \qquad (16\text{-}48)$$

Solid *sodium bismuthate* ($NaBiO_3$) is of similar oxidizing strength to Ag^{2+} and $S_2O_8^{2-}$. The excess solid oxidant is removed by filtration.

Hydrogen peroxide is a good oxidant in basic solution. It can transform Co(II) to Co(III), Fe(II) to Fe(III), and Mn(II) to Mn(IV). In acidic solution it can *reduce* $Cr_2O_7^{2-}$ to Cr^{3+} and MnO_4^- to Mn^{2+}. Excess H_2O_2 spontaneously **disproportionates** in boiling water.

$$2H_2O_2 \xrightarrow{\text{boiling}} O_2 + 2H_2O \qquad (16\text{-}49)$$

Do you remember what "disproportionation" means? Look in the Glossary if you don't.

Challenge: Write one half-reaction in which H_2O_2 behaves as an oxidant and one half-reaction in which it behaves as a reductant.

Prereduction

Stannous chloride ($SnCl_2$) has been used to prereduce Fe(III) to Fe(II) in hot HCl. The excess reducing agent is destroyed by adding excess $HgCl_2$:

$$\text{Sn(II)} + 2HgCl_2 \rightarrow \text{Sn(IV)} + Hg_2Cl_2 + 2Cl^- \qquad (16\text{-}50)$$

The Fe(II) is then titrated with an oxidant.

Chromous chloride is a very powerful reductant sometimes used for prereduction. Any excess Cr^{2+} is oxidized by atmospheric oxygen. *Sulfur dioxide* and *hydrogen sulfide* are mild reducing agents that can be expelled by boiling an acidic solution after the reduction is complete.

An important prereduction technique uses a column packed with a solid reducing agent. Figure 16-7 shows the *Jones reductor*, which is packed with zinc coated with zinc **amalgam.** The amalgam is prepared by mixing granular zinc with 2% (wt/wt) aqueous $HgCl_2$ for 10 minutes, then washing with water. For example, a sample of Fe(III) can be reduced to Fe(II) by being passed through a Jones reductor using 1 M H_2SO_4 as solvent. The column is washed well with water, and the combined washings can be titrated with standard MnO_4^-, Ce(IV), or $Cr_2O_7^{2-}$. It is necessary to do a blank determination on a solution passed through the reductor in the same manner as for the unknown.

Most reduced analytes are easily reoxidized by atmospheric oxygen. To avoid air oxidation, the reduced analyte may be collected in a solution containing excess Fe(III). The ferric ion is immediately reduced to Fe(II), which is stable in acid. The Fe(II) is then titrated with an oxidant. By this means, such elements as Cr, Ti, V, and Mo can be analyzed.

Zinc is such a powerful reducing agent that the Jones reductor is not very selective.

$$Zn^{2+} + 2e^- \rightleftharpoons Zn(s) \qquad E° = -0.764 \text{ V} \qquad (16\text{-}51)$$

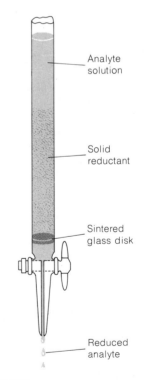

Figure 16-7
A column filled with a solid reagent used for prereduction of analyte is called a *reductor*. Often the analyte is drawn through by suction.

Analyte solution

Solid reductant

Sintered glass disk

Reduced analyte

More selective is the *Walden reductor*, filled with solid Ag and 1 M HCl. The reduction potential for the silver–silver chloride couple (0.222 V) is high enough that such species as Cr^{3+} and TiO^{2+} are not reduced and therefore do not interfere in the analysis of a metal such as Fe^{3+}. Another more selective reductor uses a filling of granular Cd metal. In determining levels of nitrogen oxides for air-pollution monitoring,[†] the gases are first converted to NO_3^-, which is not easy to analyze. Passing the nitrate through a Cd-filled column reduces NO_3^- to NO_2^-, for which a convenient spectrophotometric analysis is available. Significant disadvantages of all reductors are the time, labor, and expense associated with their use.

Once the analyte has been adjusted to the desired oxidation state, it may be titrated with an appropriate oxidant or reductant. We will now examine a few of the most common titrants.

Oxidation with Potassium Permanganate

Potassium permanganate ($KMnO_4$) is an oxidizing agent of an intense violet color. In strongly acidic solutions (pH $\lesssim$ 1), it is reduced to colorless Mn^{2+}.

$$\underset{\text{Permanganate}}{MnO_4^-} + 8H^+ + 5e^- \rightleftharpoons \underset{\text{Manganous}}{Mn^{2+}} + 4H_2O \qquad E° = 1.507 \text{ V}$$

$$(16\text{-}52)$$

In neutral or alkaline solution, the product is the brown solid, MnO_2.

$$MnO_4^- + 4H^+ + 3e^- \rightleftharpoons \underset{\substack{\text{Manganese} \\ \text{dioxide}}}{MnO_2(s)} + 2H_2O \qquad E° = 1.692 \text{ V} \quad (16\text{-}53)$$

In very strongly alkaline solution (2 M NaOH), green manganate ion is produced.

$$MnO_4^- + e^- \rightleftharpoons \underset{\text{Manganate}}{MnO_4^{2-}} \qquad E° = 0.56 \text{ V} \qquad (16\text{-}54)$$

Some representative permanganate titrations are listed in Table 16-3. For titrations in strongly acidic solution, $KMnO_4$ serves as its own indicator (Color Plate 9). The product, Mn^{2+}, is colorless. The end point is taken as the first persistent appearance of pale pink MnO_4^-. If the titrant is too dilute to be seen, an indicator such as ferroin may be used.

Preparation and standardization

Potassium permanganate is not pure enough to be a primary standard, since traces of MnO_2 are invariably present. In addition, distilled water usually contains enough organic impurities to reduce some freshly dissolved MnO_4^- to MnO_2. To prepare a stable solution, $KMnO_4$ is dissolved in distilled water, boiled for an hour to hasten the reaction between MnO_4^- and organic impurities, and filtered through a clear, sintered glass filter to remove precipitated MnO_2. Filter paper (organic matter!) should never be used. The

Margin notes:

$KMnO_4$ serves as its own indicator in acidic solution.

$KMnO_4$ is not a primary standard.

A properly stored solution of 0.02 M $KMnO_4$ decomposes $\leq 0.1\%$ in several months. More dilute solutions should be prepared fresh from this stock solution and standardized.

[†] J. H. Margeson, J. C. Suggs, and M. R. Midgett, *Anal. Chem.*, **52**, 1955 (1980).

Table 16-3 **407**
Some analytical applications of permanganate titrations

Species analyzed	Oxidation reaction	Notes
Fe^{2+}	$Fe^{2+} \rightleftharpoons Fe^{3+} + e^-$	Fe^{3+} is reduced to Fe^{2+} with Sn^{2+} or a Jones reductor. Titration is carried out in 1 M H_2SO_4 or 1 M HCl containing Mn(II), H_3PO_4, and H_2SO_4. Mn(II) inhibits oxidation of Cl^- by MnO_4^-. H_3PO_4 complexes Fe(III) to prevent formation of yellow Fe(III)–chloride complexes.
$H_2C_2O_4$	$H_2C_2O_4 \rightleftharpoons 2CO_2 + 2H^+ + 2e^-$	Add 95% of titrant at 25°C, then complete titration at 55–60°C.
Br^-	$Br^- \rightleftharpoons \frac{1}{2}Br_2(g) + e^-$	Titrate in boiling 2 M H_2SO_4 to remove $Br_2(g)$.
H_2O_2	$H_2O_2 \rightleftharpoons O_2(g) + 2H^+ + 2e^-$	Titrate in 1 M H_2SO_4.
HNO_2	$HNO_2 + H_2O \rightleftharpoons NO_3^- + 3H^+ + 2e^-$	Add excess standard $KMnO_4$ and back-titrate after 15 minutes at 40°C with Fe(II).
As(III)	$H_3AsO_3 + H_2O \rightleftharpoons H_3AsO_4 + 2H^+ + 2e^-$	Titrate in 1 M HCl with KI or ICl catalyst.
Sb(III)	$H_3SbO_3 + H_2O \rightleftharpoons H_3SbO_4 + 2H^+ + 2e^-$	Titrate in 2 M HCl.
Mo(III)	$Mo^{3+} + 2H_2O \rightleftharpoons MoO_2^{2+} + 4H^+ + 3e^-$	Reduce Mo in a Jones reductor, and run the Mo^{3+} into excess Fe^{3+} in 1 M H_2SO_4. Titrate the Fe^{2+} formed.
W(III)	$W^{3+} + 2H_2O \rightleftharpoons WO_2^{2+} + 4H^+ + 3e^-$	Reduce W with Pb(Hg) at 50°C and titrate in 1 M HCl.
U(IV)	$U^{4+} + 2H_2O \rightleftharpoons UO_2^{2+} + 4H^+ + 3e^-$	Reduce U to U^{3+} with a Jones reductor. Expose to air to produce U^{4+}, which is titrated in 1 M H_2SO_4.
Ti(III)	$Ti^{3+} + H_2O \rightleftharpoons TiO^{2+} + 2H^+ + e^-$	Reduce Ti to Ti^{3+} with a Jones reductor, and run the Ti^{3+} into excess Fe^{3+} in 1 M H_2SO_4. Titrate the Fe^{2+} that is formed.
$Mg^{2+}, Ca^{2+}, Sr^{2+}, Ba^{2+}, Zn^{2+}, Co^{2+}, La^{3+}, Th^{4+}, Pb^{2+}, Ce^{3+}, BiO^+, Ag^+$	$H_2C_2O_4 \rightleftharpoons 2CO_2 + 2H^+ + 2e^-$	Precipitate the metal oxalate. Dissolve in acid and titrate the $H_2C_2O_4$.
K^+	$K_2NaCo(NO_2)_6 + 6H_2O \rightleftharpoons Co^{2+} + 6NO_3^- + 12H^+ + 2K^+ + Na^+ + 11e^-$	Precipitate potassium sodium cobaltinitrite. Dissolve in acid and titrate. Co^{3+} is *reduced* to Co^{2+}, and HNO_2 is oxidized to NO_3^-.
Na^+	$U^{4+} + 2H_2O \rightleftharpoons UO_2^{2+} + 4H^+ + 2e^-$	Precipitate $NaZn(UO_2)_3(acetate)_9$. Dissolve in acid, reduce the UO_2^{2+} (as above), and titrate.
$S_2O_8^{2-}$	$S_2O_8^{2-} + 2Fe^{2+} + 2H^+ \rightleftharpoons 2Fe^{3+} + 2HSO_4^-$	Peroxydisulfate is added to excess standard Fe^{2+} containing H_3PO_4. Unreacted Fe^{2+} is titrated with MnO_4^-.
PO_4^{3-}	$Mo^{3+} + 2H_2O \rightleftharpoons MoO_2^{2+} + 4H^+ + 3e^-$	$(NH_4)_3PO_4 \cdot 12MoO_3$ is precipitated and dissolved in H_2SO_4. The Mo(VI) is reduced (as above) and titrated.

reagent is stored in a dark glass bottle. Thermodynamically, aqueous $KMnO_4$ is unstable by virtue of the reaction

$$4MnO_4^- + 2H_2O \rightarrow 4MnO_2(s) + 3O_2 + 4OH^- \qquad (16\text{-}55)$$

But the oxidation of water by MnO_4^- is very slow in the absence of such agents as MnO_2, Mn^{2+}, heat, light, acids, and bases.

Potassium permanganate can be standardized by titration of sodium oxalate ($Na_2C_2O_4$), pure electrolytic iron wire, or arsenious oxide (As_4O_6). Dry (105°C, two hours) sodium oxalate (available in a 99.9–99.95% pure form) is dissolved in 1 M H_2SO_4 and treated with 90–95% of the required $KMnO_4$ solution at room temperature. The solution is then warmed to 55–60°C, and the titration is completed by slow addition of $KMnO_4$. A blank value is subtracted to account for the quantity of titrant (usually one drop) needed to impart a pink color to the solution.

$$2MnO_4^- + 5H_2C_2O_4 + 6H^+ \rightarrow 2Mn^{2+} + 10CO_2 + 8H_2O \quad (16\text{-}56)$$

If pure Fe wire is used as a standard, it is dissolved in warm 1.5 M H_2SO_4 under nitrogen. The product is Fe(II), and the cooled solution can be used to standardize $KMnO_4$ (or other oxidants) with no special precautions. Addition of 5 mL of 86% (wt/wt) phosphoric acid per 100 mL of solution masks the yellow color of Fe^{3+} and makes the end point easier to see. Alternatively, metallic iron can be dissolved less cautiously, and Fe^{3+} reduced to Fe^{2+} with $SnCl_2$. Ferrous ammonium sulfate—$Fe(NH_4)_2(SO_4)_2 \cdot 6H_2O$—and ferrous ethylenediammonium sulfate—$Fe(H_3NCH_2CH_2NH_3)(SO_4)_2 \cdot 2H_2O$—are available in sufficiently pure forms to be used as primary standards for most purposes.

Oxidation with Cerium(IV)

The reduction of Ce(IV) to Ce(III) proceeds cleanly in acidic solution. The aquo ion—$Ce(H_2O)_n^{4+}$—probably does not exist in any of these solutions, as the cerium ion binds the acid counterion (ClO_4^-, SO_4^{2-}, NO_3^-, Cl^-) very strongly to give a variety of complexes. The variation of the $Ce^{4+}|Ce^{3+}$ formal potential with the medium is indicative of these interactions:

The variation of potential in each solvent implies that different species of cerium are present in each solvent.

$$Ce(IV) + e^- \rightleftharpoons Ce^{3+} \quad (16\text{-}57)$$

$$\text{Formal potential} = 1.70 \text{ V in 1 F } HClO_4$$

$$= 1.61 \text{ V 1 F } HNO_3$$

$$= 1.44 \text{ V in 1 F } H_2SO_4$$

$$= 1.28 \text{ V in 1 F } HCl$$

Ce(IV) is yellow and Ce(III) is colorless, but the color change is not distinct enough for cerium to be its own indicator. Ferroin and other substituted phenanthroline redox indicators (Table 16-1) are well suited to titrations with Ce(IV).

Preparation and standardization

$(NH_4)_2Ce(NO_3)_6$ is a primary standard.

For preparation of solutions of Ce(IV), primary standard-grade ammonium hexanitratocerate(IV)—$(NH_4)_2Ce(NO_3)_6$—can be dissolved in 1 M H_2SO_4 and used directly. Although the oxidizing strength of Ce(IV) is greater in

$HClO_4$ or HNO_3, solutions in these acids undergo slow photochemical decomposition with concomitant oxidation of water. Solutions of Ce(IV) in H_2SO_4 are stable indefinitely. Solution in HCl are unstable because Cl^- is oxidized to Cl_2. Sulfuric acid solutions of Ce(IV) can be used to titrate unknowns (such as Fe^{2+}) in HCl solution because the reaction with analyte is favored over the slow reaction with Cl^-.

Less expensive, less pure salts of Ce(IV), including $Ce(HSO_4)_4$, $(NH_4)_4Ce(SO_4)_4 \cdot 2H_2O$, and $CeO_2 \cdot xH_2O$—also called $Ce(OH)_4$—are perfectly adequate for preparing titrants that are subsequently standardized. The procedures for Ce(IV) standardization are similar to those for MnO_4^-, with As_4O_6, $Na_2C_2O_4$, and Fe being useful primary standards.

Analytical applications

Ce(IV) can be used in place of $KMnO_4$ in most of the procedures mentioned in the previous section. In addition, ceric ion finds applications in analysis of many organic compounds. In the oscillating reaction in Demonstration 15-1, Ce(IV) oxidizes malonic acid to CO_2 and formic acid:

$$CH_2(CO_2H)_2 + 2H_2O + 6Ce(IV) \rightarrow 2CO_2 + HCO_2H + 6Ce(III) + 6H^+$$

Malonic acid　　　　　　　　　　　　　　　Formic acid

$$(16\text{-}58)$$

This reaction can be used for quantitative analysis of malonic acid by heating a sample in 4 M $HClO_4$ with excess standard Ce(IV) and back-titrating the unreacted Ce(IV) with Fe^{2+}. Analogous procedures are available for many alcohols, aldehydes, ketones, and carboxylic acids.

Oxidation with Potassium Dichromate

In acidic solution, the orange dichromate ion is a powerful oxidant that is reduced to chromic ion, Cr^{3+}:

$$Cr_2O_7^{2-} + 14H^+ + 6e^- \rightleftharpoons 2Cr^{3+} + 7H_2O \qquad E° = 1.36 \text{ V} \quad (16\text{-}59)$$

Chromium waste is toxic and should not be poured down the drain. See Box 2-2.

In 1 M HCl, the formal potential is just 1.00 V, and in 2 M H_2SO_4, it is 1.11 V; so dichromate is a less powerful oxidizing agent than MnO_4^- or Ce(IV). In basic solution, $Cr_2O_7^{2-}$ is converted to the yellow chromate ion (CrO_4^{2-}), whose oxidizing ability is nil:

$$CrO_4^{2-} + 4H_2O + 3e^- \rightleftharpoons Cr(OH)_3(s, \text{ hydrated}) + 5OH^- \quad (16\text{-}60)$$

$$E° = -0.12 \text{ V}$$

The advantages of $K_2Cr_2O_7$ are that it is pure enough to be a primary standard, its solutions are stable, and it is cheap. Because $Cr_2O_7^{2-}$ is orange and complexes of Cr^{3+} range from green to violet, indicators with very distinctive color changes must be used to find a dichromate end point. Indicators such as diphenylamine sulfonic acid or diphenylbenzidine sulfonic acid are suitable. Alternatively, the reaction can be monitored with Pt and calomel electrodes.

Diphenylbenzidine sulfonate (reduced) (colorless)

Diphenylbenzidine sulfonate (oxidized) (violet)
$$+ 2H^+ + 2e^-$$

Since potassium dichromate is not as strong an oxidant as $KMnO_4$ or Ce(IV), it is not used as widely. It is employed chiefly for the determination of Fe^{2+} and, indirectly, for many species that will oxidize Fe^{2+} to Fe^{3+}. For indirect analyses, the unknown is treated with a measured excess of Fe^{2+}, and the unreacted Fe^{2+} then titrated with $K_2Cr_2O_7$. Among the species that can be analyzed in this way are ClO_3^-, NO_3^-, MnO_4^-, and organic peroxides.

Methods Involving Iodine

Iodimetry: titration *with* iodine

Iodometry: titration *of* iodine produced by a chemical reaction

A great many analytical procedures are based on reactions of iodine. When a reducing analyte is titrated directly with iodine (to produce I^-), the method is called **iodimetry.** In **iodometry,** an oxidizing analyte is added to excess I^- to produce iodine, which is then titrated with standard thiosulfate solution.

Molecular iodine is only slightly soluble in water (1.3×10^{-3} M at 20°C), but its solubility is greatly enhanced by complexation with iodide.

The form of iodine used as a titrant or produced by reaction with analyte is almost always I_3^-, not I_2.

$$I_2(aq) + \quad I^- \quad \rightleftharpoons \quad I_3^- \qquad K = 7 \times 10^2$$

$$\text{Iodine} \quad \text{Iodide} \quad \text{Triiodide} \tag{16-61}$$

A typical 0.05 M solution of I_3^- for titrations is prepared by dissolving 0.12 mol of KI plus 0.05 mol of I_2 in one liter of water. When we speak of using iodine as a titrant, we almost always mean that we are using a solution of I_2 plus excess I^-.

Use of starch indicator

As described in Section 16-3, starch is used as an indicator for iodine. In a solution with no other colored species, it is possible to see the color of $\sim 5 \times 10^{-6}$ M I_3^-. With a starch indicator, the limit of detection is extended by about a factor of ten.

In iodimetry (titration *with* I_3^-), starch can be added at the beginning of the titration. The first drop of excess I_3^- after the equivalence point causes the solution to turn dark blue. In iodometry (titration *of* I_3^-), I_3^- is present throughout the reaction up to the equivalence point. *Starch should not be added to such a reaction until immediately before the equivalence point* [as detected visually, by fading of the I_3^- (Color Plate 10)]. Otherwise some iodine tends to remain bound to starch particles after the equivalence point is reached.

An alternative to using starch is to add a few milliliters of chloroform to the vigorously stirred titration vessel. After each addition of reagent near the end point, stirring is stopped long enough to examine the color of the chloroform phase at the bottom of the flask. I_2 is much more soluble in $CHCl_3$ than in water, and its color is readily detected in the organic phase.

The starch–iodine complex is very temperature-sensitive. At 50°C, the color is only one-tenth as intense as at 25°C.[†] Organic solvents also decrease the affinity of iodine for starch and markedly reduce the utility of the indicator.

Preparation and standardization of I_3^- solutions

There is a significant vapor pressure of toxic I_2 above solid I_2 and aqueous I_3^-. Vessels containing I_2 or I_3^- should be sealed or, better, kept in a fume hood. Waste solutions of I_3^- should not be dumped into a sink in the open lab.

As described previously, I_3^- is prepared by dissolving solid I_2 in excess KI. Sublimed I_2 is pure enough to be a primary standard, but it is seldom used as a standard due to significant vaporization of the solid during the weighing procedure. Instead, the approximate amount is rapidly weighed, and the

---

[†] If maximum sensitivity is required, it is best to cool the reaction in ice water [G. L. Hatch, *Anal. Chem.*, **54**, 2002 (1984)].

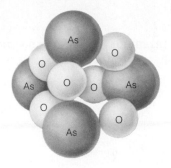

Figure 16-8
The As_4O_6 molecule consists of an As_4 pyramid with a bridging oxygen atom on each edge.

solution of I_3^- is standardized with a pure sample of the intended analyte or with As_4O_6 or $Na_2S_2O_3$ as described below.

Acidic solutions of I_3^- are unstable because the excess I^- is slowly oxidized by air:

$$6I^- + O_2 + 4H^+ \rightarrow 2I_3^- + 2H_2O \qquad (16-62)$$

In neutral solutions, Reaction 16-62 is insignificant in the absence of heat, light, and metal ions.

Triiodide can be standardized by reaction with primary standard-grade arsenious oxide, As_4O_6, the structure of which is shown in Figure 16-8. When As_4O_6 is dissolved in acidic solution, arsenious acid is formed:

$$As_4O_6(s) \; + 6H_2O \rightleftharpoons \; 4H_3AsO_3 \qquad (16-63)$$

Arsenious oxide   Arsenious acid

The latter reacts with I_3^- as follows:

$$H_3AsO_3 + I_3^- + H_2O \rightleftharpoons H_3AsO_4 + 3I^- + 2H^+ \qquad (16-64)$$

Because the equilibrium constant for Reaction 16-64 is small, the concentration of H^+ must be kept low to ensure complete reaction. If the pH is *too* high (pH $\gtrsim$ 11), triiodide disproportionates to hypoiodous acid, iodate, and iodide. For best results, the standardization is carried out at pH 7–8 with bicarbonate as a buffer.

An alternative and excellent way to make a standard solution of I_3^- is to add a weighed quantity of pure potassium iodate to a small excess of KI. Addition of excess strong acid (to give pH $\approx$ 1) gives a quantitative reverse disproportionation reaction in which I_3^- is formed:

$$IO_3^- + 8I^- + 6H^+ \rightleftharpoons 3I_3^- + 3H_2O \qquad (16-65)$$

Iodate

A freshly acidified solution of iodate plus iodide can be used to standardize thiosulfate or any other reagent that reacts with I_3^-. The reagent must be used immediately, or else air oxidation of I^- takes place. The only disadvantage of KIO_3 is its low molecular weight relative to the number of electrons it accepts. This leads to a larger-than-desirable relative weighing error in preparing solutions.

Challenge: Use standard potentials from Appendix H to show that the equilibrium constant for Reaction 16-64 is 0.04.

HOI: hypoiodous acid
IO_3^-: iodate

KIO_3 is a good primary standard for the generation of I_3^-.

Use of sodium thiosulfate

Reaction of iodine with thiosulfate.

Sodium thiosulfate is the almost universal titrant for triiodide. In neutral or acidic solution, triiodide oxidizes thiosulfate to tetrathionate:

$$I_3^- + 2S_2O_3^{2-} \rightleftharpoons 3I^- + O=\overset{\overset{\displaystyle O}{\|}}{\underset{\underset{\displaystyle O^-}{|}}{S}}-S-S-\overset{\overset{\displaystyle O}{\|}}{\underset{\underset{\displaystyle O^-}{|}}{S}}=O \qquad (16\text{-}66)$$

Thiosulfate   Tetrathionate

(In basic solution, I_3^- disproportionates to I^- and HOI. Because hypoiodite oxidizes thiosulfate to sulfate, the stoichiometry of Reaction 16-66 changes and the titration of I_3^- with thiosulfate is carried out below pH 9.) The common form of thiosulfate, $Na_2S_2O_3 \cdot 5H_2O$, is not pure enough to be a primary standard. Instead, thiosulfate is usually standardized by reaction with a fresh solution of I_3^- prepared from KIO_3 plus KI or a solution of I_3^- standardized with As_4O_6.

A stable solution of $Na_2S_2O_3$ can be prepared by dissolving the reagent in high-quality, freshly boiled distilled water. The quality of the water is important because dissolved CO_2 promotes disproportionation of $S_2O_3^{2-}$:

$$S_2O_3^{2-} + H^+ \rightleftharpoons HSO_3^- + S(s) \qquad (16\text{-}67)$$

Bisulfite   Sulfur

and metal ions catalyze the atmospheric oxidation of thiosulfate:

$$2Cu^{2+} + 2S_2O_3^{2-} \rightarrow 2Cu^+ + S_4O_6^{2-} \qquad (16\text{-}68)$$

$$2Cu^+ + \tfrac{1}{2}O_2 + 2H^+ \rightarrow 2Cu^{2+} + H_2O \qquad (16\text{-}69)$$

A solution of thiosulfate should be stored in the dark. Addition of 0.1 g of sodium carbonate per liter maintains the pH in an optimum range for stability of the solution. Three drops of chloroform should also be added to each bottle of thiosulfate solution to help prevent bacterial growth. An acidic solution of thiosulfate is unstable, but the reagent can be used to titrate I_3^- in acidic solution because the reaction with triiodide is faster than Reaction 16-67.

Anhydrous sodium thiosulfate can be prepared from the pentahydrate and is suitable as a primary standard.[†] The standard is prepared by refluxing 21 g of $Na_2S_2O_3 \cdot 5H_2O$ with 100 mL of methanol for 20 minutes. The anhydrous salt is then filtered, washed with 20 mL of methanol, and dried at 70°C for 30 minutes.

Analytical applications of iodine

Reducing agent + $I_3^- \rightarrow 3I^-$

Reducing agents can be titrated directly with standard I_3^- in the presence of starch. The end point is marked by the appearance of the intense blue starch–iodine complex. An example is the iodimetric determination of vitamin C:

[†] A. A. Woolf, *Anal. Chem.*, **54**, 2134 (1982).

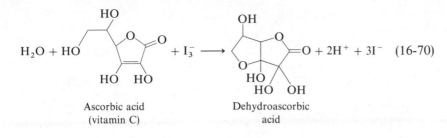

Ascorbic acid
(vitamin C)

Dehydroascorbic
acid

In some cases, excess standard I_3^- is used to drive the reaction to completion, as in the analysis of glucose or other reducing sugars (sugars with an aldehyde group).

Aldehyde ⟶
group

$$
\begin{array}{ccc}
\mathrm{HC{=}O} & & \mathrm{CO_2^-} \\
| & & | \\
\mathrm{HCOH} & & \mathrm{HCOH} \\
| & & | \\
\mathrm{HOCH} + 3OH^- + I_3^- \longrightarrow & \mathrm{HOCH} & + 3I^- + 2H_2O \\
| & & | \\
\mathrm{HCOH} & & \mathrm{HCOH} \\
| & & | \\
\mathrm{HCOH} & & \mathrm{HCOH} \\
| & & | \\
\mathrm{CH_2OH} & & \mathrm{CH_2OH}
\end{array}
$$

Glucose

Gluconate

(16-71)

Reaction 16-71 is carried out in a basic solution, which is then acidified and the excess I_3^- back-titrated with standard thiosulfate. Examples of iodimetric analyses are given in Table 16-4.

Oxidizing agents can be treated with excess I^- to produce I_3^- (Table 16-5). The iodometric analysis is completed by titrating the liberated I_3^- with standard thiosulfate. Starch is not added until just before the end point. Box 16-1 describes an iodometric titration used to establish the composition of high-temperature *superconductors*.

Oxidizing agent $+ 3I^- \rightarrow I_3^-$

Analysis of Organic Compounds with Periodic Acid

Periodic acid is a powerful oxidizing agent that exists under different conditions as *paraperiodic acid* (H_5IO_6), *metaperiodic acid* (HIO_4), and various deprotonated forms of the acids. Sodium metaperiodate ($NaIO_4$) is usually used to prepare a solution that is standardized by addition to excess KI in bicarbonate solution at pH 8—9:

H_5IO_6: paraperiodic acid

IO_4^-: metaperiodate

$$IO_4^- + 3I^- + H_2O \rightleftharpoons IO_3^- + I_3^- + 2OH^- \qquad (16\text{-}72)$$

The I_3^- released is titrated with thiosulfate to complete the standardization.

Periodate solutions are especially useful for the analysis of organic compounds (such as carbohydrates) containing the following functional groups:

In acidic solution, the reaction goes further:

$H_5IO_6 + 11I^- + 7H^+ \rightleftharpoons$

$4I_3^- + 6H_2O$

hydroxyl adjacent to hydroxyl   amine adjacent to hydroxyl

carbonyl adjacent to hydroxyl   amine adjacent to carbonyl

carbonyl adjacent to carbonyl

Table 16-4
Titrations with standard triiodide (iodimetric titrations)

Species analyzed	Oxidation reaction	Notes
As(III)	$H_3AsO_3 + H_2O \rightleftharpoons H_3AsO_4 + 2H^+ + 2e^-$	Direct titration in $NaHCO_3$ solution with I_3^-.
Sb(III)	$SbO(O_2CCHOHCHOHCO_2)^- + H_2O \rightleftharpoons$ $SbO_2(O_2CCHOHCHOHCO_2)^- + 2H^+ + 2e^-$	Direct titration in $NaHCO_3$ solution, using tartrate to mask As(III).
Sn(II)	$SnCl_4^{2-} + 2Cl^- \rightleftharpoons SnCl_6^{2-} + 2e^-$	Sn(IV) is reduced to Sn(II) with granular Pb or Ni in 1 M HCl and titrated in the absence of oxygen.
N_2H_4	$N_2H_4 \rightleftharpoons N_2 + 4H^+ + 4e^-$	Titrate in $NaHCO_3$ solution.
SO_2	$SO_2 + H_2O \rightleftharpoons H_2SO_3$ $H_2SO_3 + H_2O \rightleftharpoons SO_4^{2-} + 4H^+ + 2e^-$	Add SO_2 (or H_2SO_3 or HSO_3^- or SO_3^{2-}) to excess standard I_3^- in dilute acid and back-titrate unreacted I_3^- with standard thiosulfate.
H_2S	$H_2S \rightleftharpoons S(s) + 2H^+ + 2e^-$	Add H_2S to excess I_3^- in 1 M HCl and back-titrate with thiosulfate.
Zn^{2+}, Cd^{2+}, Hg^{2+}, Pb^{2+}, etc.	$M^{2+} + H_2S \rightarrow MS(s) + 2H^+$ $MS(s) \rightleftharpoons M^{2+} + S + 2e^-$	Precipitate and wash metal sulfide. Dissolve in 3 M HCl with excess standard I_3^- and back-titrate with thiosulfate.
Cysteine, glutathione, thioglycolic acid, mercaptoethanol	$2\,RSH \rightleftharpoons RSSR + 2H^+ + 2e^-$	Titrate the sulfhydryl compound at pH 4–5 with I_3^-.
HCN	$I_2 + HCN \rightleftharpoons ICN + I^- + H^+$	Titrate in carbonate–bicarbonate buffer, using $CHCl_3$ as an extraction indicator.
$H_2C{=}O$	$H_2CO + 3OH^- \rightleftharpoons HCO_2^- + 2H_2O + 2e^-$	Add excess I_3^- plus NaOH to the unknown. After five minutes, add HCl and back-titrate with thiosulfate.
Glucose (and other reducing sugars)	$RCH{=}O + 3OH^- \rightleftharpoons RCO_2^- + 2H_2O + 2e^-$	Add excess I_3^- plus NaOH to the sample. After five minutes add HCl and back-titrate with thiosulfate.
Ascorbic acid (vitamin C)		Direct titration with I_3^-.
H_3PO_3.	$H_3PO_3 + H_2O \rightleftharpoons H_3PO_4 + 2H^+ + 2e^-$	Titrate in $NaHCO_3$ solution,

In this oxidation, known as the *Malaprade reaction,* the carbon–carbon bond
between the two functional groups is broken and the following changes occur:

(a) (b) (c)

COLOR PLATE 1 HCl Fountain (Demonstration 5-1) (a) Basic indicator solution in beaker. (b) Indicator is drawn into flask and changes to acidic color. (c) Solution levels at end of experiment.

(a) (b)

COLOR PLATE 2 Effect of Ionic Strength on Ionic Dissociation (Demonstration 6-1) (a) Two beakers containing identical solutions with $FeSCN^{2+}$, Fe^{3+} and SCN^-. (b) Change when KNO_3 is added to right beaker.

(a) (b) (c)

COLOR PLATE 3 Colloids and Dialysis (Demonstration 8-1) (a) Ordinary aqueous Fe(III) (right) and colloidal Fe(III) (left). (b) Dialysis bags containing colloidal Fe(III) (left) and a solution of Cu(II) (right) immediately after placement in flasks of water. (c) After 24 hours of dialysis, the Cu(II) has diffused uniformly between the bag and flask, but the colloidal Fe(III) remains inside the bag.

(a)   (b)   (c)

COLOR PLATE 4 Fajans Titration of Cl⁻ with AgNO₃ Using Dichlorofluorescein (Demonstration 9-1)
(*a*) Indicator before beginning titration. (*b*) AgCl precipitate before endpoint. (*c*) Indicator adsorbed on precipitate after endpoint.

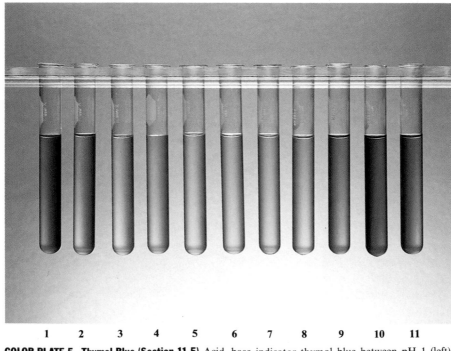

1   2   3   4   5   6   7   8   9   10   11

COLOR PLATE 5 Thymol Blue (Section 11-5) Acid–base indicator thymol blue between pH 1 (left) and 11 (right). The pK values are 1.7 and 8.9.

(a)   (b)   (c)

(d)   (e)

COLOR PLATE 6 Indicators and Acidity of CO_2 (Demonstration 11-1) (a) Cylinders before adding dry
ice. Ethanol indicator solutions of phenolphthalcin (left) and bromothymol blue (right) have not
yet mixed with entire cylinder. (b) Adding dry ice causes bubbling and mixing. (c) Further mixing.
(d) Phenolphthalein changes to its colorless acidic form. Color of bromothymol blue is due to
mixture of acidic and basic forms. (e) After addition of HCl and stirring of right-hand cylinder,
bubbles of CO_2 can be seen leaving solution, and indicator changes completely to its acidic color.

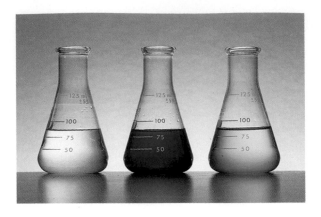

COLOR PLATE 7 Titration of Cu(II) with EDTA Using Auxiliary Complexing Agent (Section 13-4) 0.02 M $CuSO_4$ before titration (left). Color of Cu(II)-ammonia complex after adding ammonia buffer, pH 10 (center). End point color when all ammonia ligands have been displaced by EDTA (right).

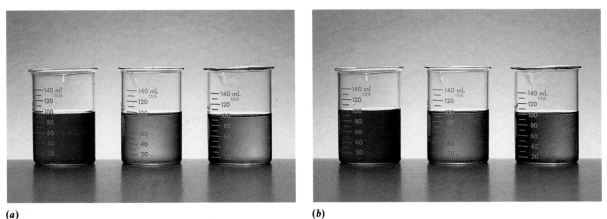

(a)   *(b)*

COLOR PLATE 8 Titration of Mg^{2+} by EDTA Using Eriochrome Black T Indicator (Demonstration 13-1)
(*a*) Before (left), near (center), and after (right) equivalence point. (*b*) Same titration with methyl red added as inert dye to alter colors.

COLOR PLATE 9 Titration of VO^{2+} with Potassium Permanganate (Section 16-4; Experiment 25-9) Blue VO^{2+} solution prior to titration (left). Mixture of blue VO^{2+} and yellow VO_2^+ observed during titration (center). Dark color of MnO_4^- at end point (right).

COLOR PLATE 10 Iodometric Titration (Section 16-4; Experiment 25-10) I_3^- solution (left). I_3^- solution before end point in titration with $S_2O_3^{2-}$ (left center). I_3^- solution immediately before end point with starch indicator present (right center). At the end point (right).

(a)

(b)

(c)

COLOR PLATE 11 Electrochemical Writing (Demonstration 17-1) (a) Stylus used as cathode. (b) Stylus used as anode. (c) Foil backing has opposite polarity from stylus and produces reverse color on bottom sheet of paper

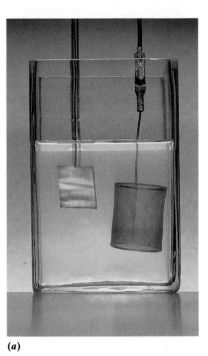

(a)

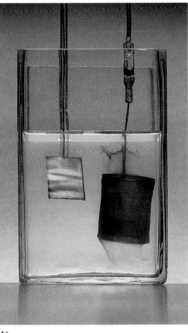

(b)

COLOR PLATE 12 Formation of Diffusion Layer during Electrolysis (Box 18-1) (a) Cu electrode (flat plate, left) and Pt electrode (mesh basket, right) immersed in solution containing KI and starch, with no electric current. (b) I_3^--starch complex forms at surface of Pt anode when current flows.

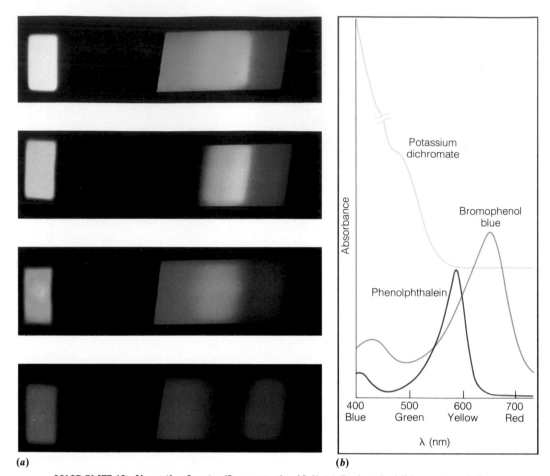

COLOR PLATE 13 Absorption Spectra (Demonstration 19-1) (*a*) Projected visible spectra of (from top to bottom) white light, potassium dichromate, bromophenol blue, and phenolphthalein. (*b*) Spectrophotometric visible spectra of colored compounds whose projected spectra are shown in Part A. Bromophenol blue and potassium dichromate spectra are displaced upward for clarity.

COLOR PLATE 14 Grating Dispersion (Section 19-4) Visible spectrum produced by grating inside spectrophotometer.

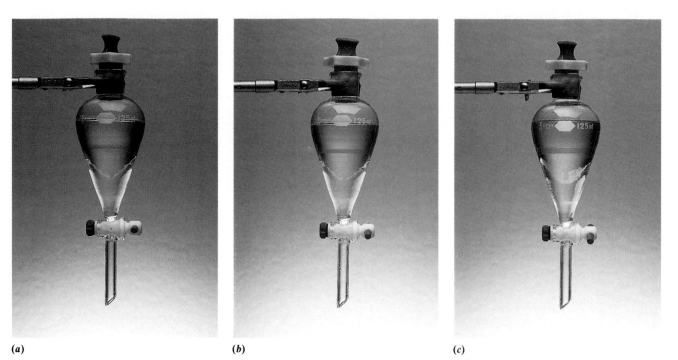

(a)   (b)   (c)

COLOR PLATE 15 Extraction of Uranyl Nitrate from Water into Ether (Section 22-1) (a) Separatory funnel with lower aqueous layer containing yellow 1 M $UO_2(NO_3)_2$ (plus 3 M HNO_3 and 4 M $Ca(NO_3)_2$) beneath colorless diethyl ether layer prior to mixing. (b) Yellow uranyl nitrate is distributed in both layers after shaking. (c) After eight extractions with ether, almost all yellow uranyl nitrate has been removed from the aqueous phase.

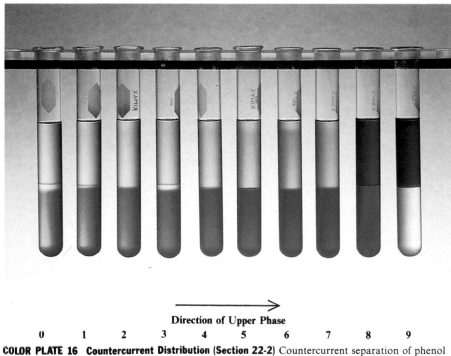

$\longrightarrow$
Direction of Upper Phase

0   1   2   3   4   5   6   7   8   9

COLOR PLATE 16 Countercurrent Distribution (Section 22-2) Countercurrent separation of phenol red and bromocresol green. Upper phase (far right) is 1-butanol; lower phase (far left) is 0.1 M aqueous Na_2CO_3.

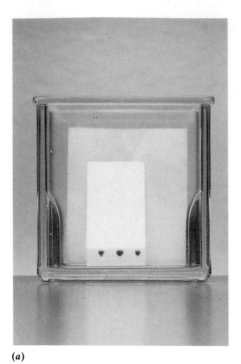

(a)

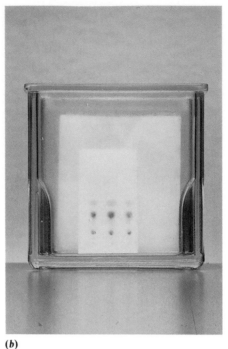

(b)

COLOR PLATE 17   Thin Layer Chromatography (Section 23-2) (*a*) Solvent ascends past mixture of dyes near bottom of plate. (*b*) Separation achieved after solvent has ascended most of the way up the plate.

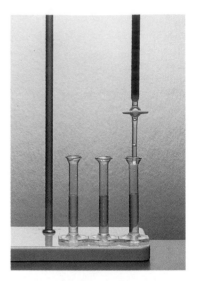

COLOR PLATE 18   Cation Exchange Separation of Co(II) and Ni(II) (Section 23-4) A solution containing Co(II) and Ni(II) in 6 M HCl was applied to the top of the cation exchange chromatography column. The blue Ni(II) was eluted first and collected in the graduated cylinders. Green Co(II) is retained more strongly and has not yet been eluted from the column.

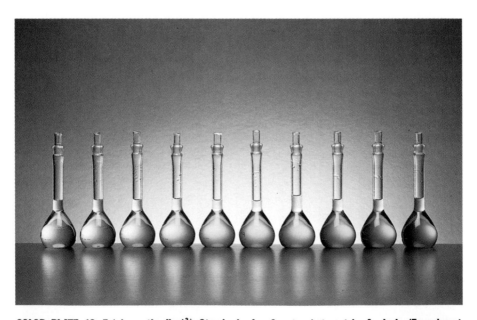

COLOR PLATE 19   Fe(phenanthroline)$_3^{2+}$ Standards for Spectrophotometric Analysis (Experiment 25-16) Volumetric flasks containing Fe(phenanthroline)$_3^{2+}$ solutions with iron concentrations ranging from 1 mg/L (left) to 10 mg/L (right).

Species analyzed	Reaction	Notes
Cl_2	$Cl_2 + 3I^- \rightleftharpoons 2Cl^- + I_3^-$	Reaction in dilute acid.
HOCl	$HOCl + H^+ + 3I^- \rightleftharpoons Cl^- + I_3^- + H_2O$	Reaction in 0.5 M H_2SO_4.
Br_2	$Br_2 + 3I^- \rightleftharpoons 2Br^- + I_3^-$	Reaction in dilute acid.
BrO_3^-	$BrO_3^- + 6H^+ + 9I^- \rightleftharpoons Br^- + 3I_3^- + 3H_2O$	Reaction in 0.5 M H_2SO_4.
IO_3^-	$2IO_3^- + 16I^- + 12H^+ \rightleftharpoons 6I_3^- + 6H_2O$	Reaction in 0.5 M HCl.
IO_4^-	$2IO_4^- + 22I^- + 16H^+ \rightleftharpoons 8I_3^- + 8H_2O$	Reaction in 0.5 M HCl.
O_2	$O_2 + 4Mn(OH)_2 + 2H_2O \rightleftharpoons 4Mn(OH)_3$ $2Mn(OH)_3 + 6H^+ + 6I^- \rightleftharpoons 2Mn^{2+} + 2I_3^- + 6H_2O$	The sample is treated with Mn^{2+}, NaOH, and KI. After one minute, it is acidified with H_2SO_4, and the I_3^- is titrated.
H_2O_2	$H_2O_2 + 3I^- + 2H^+ \rightleftharpoons I_3^- + 2H_2O$	Reaction in 1 M H_2SO_4 with NH_4MoO_3 catalyst.
O_3	$O_3 + 3I^- + 2H^+ \rightleftharpoons O_2 + I_3^- + H_2O$	O_3 is passed through neutral 2% (wt/wt) KI solution. Add H_2SO_4 and titrate.
NO_2^-	$2HNO_2 + 2H^+ + 3I^- \rightleftharpoons 2NO + I_3^- + 2H_2O$	The nitric oxide is removed (by bubbling CO_2 generated *in situ*) prior to titration of I_3^-.
As(V)	$H_3AsO_4 + 2H^+ + 3I^- \rightleftharpoons H_3AsO_3 + I_3^- + H_2O$	Reaction in 5 M HCl.
Sb(V)	$SbCl_6^- + 3I^- \rightleftharpoons SbCl_4^- + I_3^- + 2Cl^-$	Reaction in 5 M HCl
$S_2O_8^{2-}$	$S_2O_8^{2-} + 3I^- \rightleftharpoons 2SO_4^{2-} + I_3^-$	Reaction in neutral solution. Then acidify and titrate.
Cu^{2+}	$2Cu^{2+} + 5I^- \rightleftharpoons 2CuI(s) + I_3^-$	NH_4HF_2 is used as a buffer.
$Fe(CN)_6^{3-}$	$2Fe(CN)_6^{3-} + 3I^- \rightleftharpoons 2Fe(CN)_6^{4-} + I_3^-$	Reaction in 1 M HCl.
MnO_4^-	$2MnO_4^- + 8H^+ + 15I^- \rightleftharpoons 2Mn^{2+} + 5I_3^- + 8H_2O$	Reaction in 0.1 M HCl.
MnO_2	$MnO_2(s) + 4H^+ + 3I^- \rightleftharpoons Mn^{2+} + I_3^- + 2H_2O$	Reaction in 0.5 M H_3PO_4 or HCl.
$Cr_2O_7^{2-}$	$Cr_2O_7^{2-} + 14H^+ + 9I^- \rightleftharpoons 2Cr^{3+} + 3I_3^- + 7H_2O$	Reaction in 0.4 M HCl requires five minutes for completion and is particularly sensitive to air oxidation.
Ce(IV)	$2Ce(IV) + 3I^- \rightleftharpoons 2Ce(III) + I_3^-$	Reaction in 1 M H_2SO_4.

1. A hydroxyl group is oxidized to an aldehyde or a ketone.

2. A carbonyl group is oxidized to a carboxylic acid.

3. An amine is converted to an aldehyde plus ammonia (or a substituted amine if the original compound was a secondary amine).

When there are three or more adjacent functional groups, oxidation begins near one end of the molecule.

    The reactions are performed at room temperature for about one hour with a known excess of periodate. At higher temperatures, further nonspecific oxidations occur. Solvents such as methanol, ethanol, dioxane, or acetic acid may be added to the aqueous solution to increase the solubility of the organic reactant. After the reaction is complete, the unreacted periodate is analyzed by using Reaction 16-72, followed by thiosulfate titration of the liberated I_3^-.

    Some examples of the Malaprade reaction follow:

$$(1) \quad \underset{\text{2,3-Dihydroxybutane}}{CH_3\overset{\overset{\displaystyle OH}{|}}{CH} \text{---} \overset{\overset{\displaystyle OH}{|}}{CH}CH_3} + IO_4^- \longrightarrow CH_3\overset{\overset{\displaystyle O}{\|}}{CH} + H\overset{\overset{\displaystyle O}{\|}}{C}CH_3 + IO_3^- + H_2O$$

Box 16-1 IODOMETRIC ANALYSIS OF HIGH-TEMPERATURE SUPERCONDUCTORS

A **superconductor** is a material that loses all electric resistance when cooled below a critical temperature. Prior to 1987, all known superconductors required cooling to temperatures near that of liquid helium (4 K), which is costly and impractical for all but a few applications. One of the important applications is the huge electromagnets needed for medical magnetic resonance imaging. Ordinary conductors in such electromagnets would consume a huge amount of electric power. Since electricity moves through a superconductor with no resistance, the voltage can be removed from the electromagnetic coil once the current has started. The current then continues to flow virtually forever. The power consumption (Equation 14-8) is *zero* because the resistance is zero.

If superconductors could operate at room temperature, revolutionary electrical applications would be possible. A giant step toward this goal was taken when *high-temperature superconductors* that operate above the temperature of liquid nitrogen (77 K) were discovered in 1987. Liquid nitrogen is a far more economical means of cooling than is liquid helium, so many more applications of superconductors can now be contemplated.

The breakthrough came with the discovery[†] of yttrium barium copper oxide, $YBa_2Cu_3O_7$, whose crystal structure is shown below. When heated, the material readily loses oxygen atoms from the Cu–O, chains and any composition between $YBa_2Cu_3O_7$ and $YBa_2Cu_3O_6$ is observable.

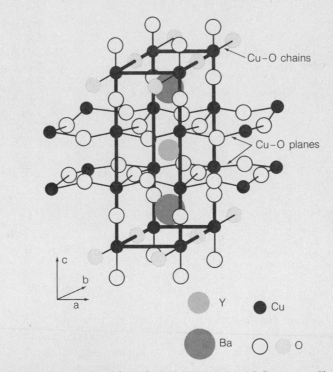

Structure of $YBa_2Cu_3O_7$ reproduced from G. F. Holland and A. M. Stacy, *Acc. Chem. Res.,* **21,** 8 (1988). One-dimensional Cu–O chains run along the crystallographic *b* axis and two-dimensional Cu–O sheets lie in the *a–b* plane. Loss of shaded oxygen atoms from the chains at elevated temperature results in $YBa_2Cu_3O_6$.

[†] Accounts of the discovery of high-temperature superconductors are given by K. A. Müller and J. G. Bednorz, *Science,* **237,** 1133 (1987), and by R. Pool, *Science,* **241,** 655 (1988).

When high-temperature superconductors were first discovered, the oxygen composition in the formula $YBa_2Cu_3O_x$ was unknown. There is nothing at all obvious about the formula $YBa_2Cu_3O_7$, which represents an unusual set of oxidation states. The only common oxidation states of yttrium and barium are Y^{3+} and Ba^{2+}, while common states of copper are Cu^{2+} and Cu^+. If all copper were Cu^{2+}, the formula of the superconductor would be $(Y^{3+})(Ba^{2+})_2(Cu^{2+})_3(O^{2-})_{6.5}$, with a cation charge of $+13$ and an anion charge of -13. If Cu^+ is present, the oxygen content would be less than 6.5 per formula unit. In order to produce the composition $YBa_2Cu_3O_7$, one of the three copper atoms must be described as Cu^{3+}, which is rather rare. Formally, $YBa_2Cu_3O_7$ can be thought of as $(Y^{3+})(Ba^{2+})_2(Cu^{2+})_2(Cu^{3+})(O^{2-})_7$ with a cation charge of $+14$ and an anion charge of -14.

To measure the oxidaton state of copper, and thereby deduce the oxygen content of $YBa_2Cu_3O_x$, analytical chemists rushed to the rescue![†] A simple iodometric method involves two experiments. In Experiment A, $YBa_2Cu_3O_x$ is dissolved in dilute acid, in which Cu^{3+} is converted to Cu^{2+}. For simplicity, we write the equations for the formula $YBa_2Cu_3O_7$, but you could balance these equations for $x \neq 7$.[‡]

$$YBa_2Cu_3O_7 + 12H^+ \rightarrow Y^{3+} + 2Ba^{2+} + 3Cu^{2+} + 6H_2O + \tfrac{1}{2}O_2 \qquad (1)$$

The total copper content can then be measured by treatment with iodide:

$$3Cu^{2+} + \tfrac{15}{2}I^- \rightarrow 3CuI(s) + + \tfrac{3}{2}I_3^- \qquad (2)$$

and titration of the liberated triiodide with standard thiosulfate by Reaction 16-66. Each mole of Cu in $YBa_2Cu_3O_7$ is equivalent to one mole of $S_2O_3^{2-}$ in Experiment A.

In Experiment B, $YBa_2Cu_3O_x$ is dissolved in dilute acid containing I^-. Each mole of Cu^{3+} produces one mole of I_3^- and each mole of Cu^{2+} produces half a mole of I_3^-.

$$Cu^{3+} + 4I^- \rightarrow CuI(s) + I_3^- \qquad (3)$$

$$2Cu^{2+} + 5I^- \rightarrow 2CuI(s) + I_3^- \qquad (4)$$

The moles of thiosulfate required in Experiment A equal the total moles of Cu in the superconductor. The difference in thiosulfate required between Experiments B and A gives the Cu^{3+} content. From this difference, it is possible to calculate the value of x in the formula $YBa_2Cu_3O_x$. A procedure for this analysis is described in Experiment 25-11.

Although we can balance cation and anion charge in the formula $YBa_2Cu_3O_7$ by including Cu^{3+} in the formula, there is no evidence for discrete Cu^{3+} ions in the crystal. There is also no evidence that some of the oxygen is in the form of peroxide, O_2^{2-}, which would also balance the cation and anion charges. The best description of the valence state in the solid crystal involves electrons and holes delocalized in the Cu–O planes and chains. Nonetheless, the formal designation of Cu^{3+} and the chemistry in Equations 1–4 accurately describe the redox chemistry of $YBa_2Cu_3O_7$.[§]

[†] D. C. Harris, M. E. Hills, and T. A. Hewston, *J. Chem. Ed.*, **64,** 847 (1987). This article includes a lecture demonstration of magnetic levitation by superconductors. Demonstration kits can be purchased from several vendors, including Sargent-Welch, 7400 N. Linder Ave., Skokie, IL 60077-1026.

[‡] Experiments with an ^{18}O-enriched superconductor show that the O_2 evolved in Reaction (1) is all derived from the solid, not from the solvent.

[§] A more sensitive and elegant iodometric procedure is described by E. H. Appelman, L. R. Morss, A. M. Kini, U. Geiser, A. Umezawa, G. W. Crabtree, and K. D. Carlson, *Inorg. Chem.*, **26,** 3237 (1987). The method of Appelman et al. can be modified by adding standard Br_2 to analyze superconductors with oxygen in the range 6.0–6.5, in which there is formally no Cu^{3+}, but there is Cu^+.

(2)
$$\underset{\text{Glycerol}}{CH_2 \overset{OH}{\underset{|}{\overset{|}{C}}} \overset{OH}{\underset{|}{CH}} \overset{OH}{\underset{|}{CH_2}}} + IO_4^- \longrightarrow CH_2 + HC - CH_2 + IO_3^- + H_2O$$

$$HC \overset{O}{\underset{||}{C}} \overset{OH}{\underset{|}{CH_2}} + IO_4^- \longrightarrow HCOH + CH_2 + IO_3^-$$

(3)
$$\underset{\text{Serine}}{CH_2 \overset{OH}{\underset{|}{\overset{|}{C}}} \overset{NH_3^+}{\underset{|}{CH}} - CO_2^-} + IO_4^- \longrightarrow CH_2 + CH - CO_2^- + NH_4^+ + IO_3^-$$

Titrations with Reducing Agents

Most analytical redox titrations involve an oxidizing titrant. Reducing titrants are less common because they are generally unstable in the presence of oxygen and must therefore be stored and used under an inert atmosphere. Fewer indicators are suitable for reductive titrations, so potentiometry is usually used to find the end point. In Section 17-4, we will see how reducing agents can be conveniently generated *in situ* by electrolytic reduction of an appropriate precursor. Some reagents used for reductive titrations include Fe(II), Cr(II), Ti(III), $Hg_2(NO_3)_2$, and ascorbic acid. Solutions of Fe(II) in 1 M H_2SO_4 are stable to oxygen and are used to standardize strong oxidants such as MnO_4^-, $Cr_2O_7^{2-}$, Ce(IV), Au(III), and V(V).

Chromous ion, prepared by reducing $K_2Cr_2O_7$ with H_2O_2, followed by zinc amalgam, is the most powerful reducing agent commonly used in volumetric analysis:

$$Cr_2O_7^{2-} + 3H_2O_2 + 8H^+ \rightarrow 2Cr^{3+} + 3O_2(g) + 7H_2O \qquad (16\text{-}73)$$

$$2Cr^{3+} + Zn(Hg) \rightarrow 2Cr^{2+} + Zn^{2+} \qquad (16\text{-}74)$$

The chromous solution must be completely protected from oxygen in storage and use, which is difficult. It can be standardized by titrating Cu(II) in 6 M HCl, giving Cu(I) and Cr^{3+}.

Species that can be analyzed with Cr(II) titrant include $Fe(CN)_6^{3-}$, NO_3^-, CN^-, Fe(III), Ti(IV), V(V), Cr(VI), Mo(VI), W(VI), Ag(I), Au(III), Hg(II), Sn(IV), Sb(V), Bi(III), and Se(IV). Organic nitro, nitroso, and azo compounds are reduced rapidly at room temperature to their corresponding amines by excess Cr(II).

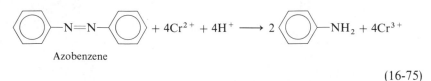

Azobenzene

$$(16\text{-}75)$$

The excess Cr(II) is back-titrated with standard Fe(III). Ti(III) is not as strong a reducing agent as Cr(II), but has similar analytical applications, including the reactions with nitro, nitroso, and azo compounds.

Generating a reagent *in situ* (literally, "in place") means that it is produced in the same solution in which it will be used.

$Fe(NH_4)_2(SO_4)_2 \cdot 6H_2O$ is usually used as a primary standard for preparing solutions of Fe(II).

$R-NO_2$: a nitro compound
$R-NO$: a nitroso compound
$R-N=N-R$: an azo compound

Summary

Redox titrations are based on an oxidation–reduction reaction between analyte and titrant. Sometimes a quantitative chemical preoxidation (with such reagents as $S_2O_8^{2-}$, AgO, $NaBiO_3$, H_2O_2, or $HClO_4$) or prereduction (with such reagents as $SnCl_2$, $CrCl_2$, SO_2, H_2S, or a metallic reductor column) is necessary to adjust the oxidation state of the analyte prior to analysis. The end point of a redox titration is commonly detected with a redox indicator or by potentiometry. A useful indicator must have a transition range ($= E°$(indicator) $\pm$ $0.059/n$ V) that overlaps the abrupt change in potential of the titration curve.

The greater the difference in reduction potential between the analyte and titrant, the sharper will be the end point of a titration. The plateaus before and after the equivalence point are centered near $E°$(analyte) and $E°$(titrant). Calculations regarding the shape of the titration curve fall into three natural categories. Prior to the equivalence point, the half-reaction involving analyte is used because the concentration of both the oxidized and the reduced forms of analyte can be readily calculated. After the equivalence point, the half-reaction involving titrant is employed for the same reason. At the equivalence point, both half-reactions must be used simultaneously to find the voltage.

Common oxidizing titrants include $KMnO_4$, Ce(IV), and $K_2Cr_2O_7$. Periodic acid is an oxidant used specifically to analyze organic reagents with certain adjacent functional groups (Malaprade reaction). A very large number of procedures is based on oxidation with I_3^- or titration of I_3^- liberated in a chemical reaction. Titrations with reducing agents such as Fe(II), Cr(II), or Ti(III) are not so common because the reductant must be protected from the air.

Terms to Understand

amalgam	preoxidation
disproportionation	prereduction
iodimetry	redox indicator
iodometry	superconductor

Exercises

16-A. Consider the titration of 120.0 mL of 0.010 0 M Fe^{2+} (buffered to pH 1.00) with 0.020 0 M $Cr_2O_7^{2-}$. Calculate the potential (versus a silver–silver chloride anode saturated with KCl) at the following volumes of $Cr_2O_7^{2-}$: 0.100, 2.00, 4.00, 6.00, 8.00, 9.00, 9.90, 10.00, 10.10, 11.00, 12.00 mL. Sketch the titration curve.

16-B. Vanadium (II) undergoes three stepwise oxidation reactions:

$$V^{2+} \rightarrow V^{3+} \rightarrow VO^{2+} \rightarrow VO_2^+$$

Calculate the potential (versus S.C.E. anode) at each of the following volumes when 10.0 mL of 0.010 0 M V^{2+} is titrated with 0.010 0 M Ce(IV) in 1 M $HClO_4$: 5.0, 15.0, 25.0, and 35.0 mL. Sketch the titration curve.

16-C. Select an indicator from Table 16-1 that would be useful for finding (a) the second end point and (b) the third end point of the titration in Problem 16-B. What color change would you see at each point?

16-D. A 128.6-mg sample of a protein (M.W. 58 600) was treated with 2.000 mL of 0.048 7 M $NaIO_4$ to react with all the serine and threonine residues.

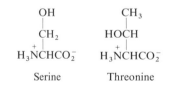

Serine   Threonine

The solution was then treated with excess iodide to convert the unreacted periodate to iodine (Equation 16-72). Titration of the iodine required 823 μL of 0.098 8 M thiosulfate.

(a) Write balanced equations for the reaction of IO_4^- with serine and threonine.

(b) Calculate the number of serine plus threonine residues per molecule of protein. Answer to the nearest integer.

(c) How many milligrams of As_4O_6 (F.W. 395.68) would be required to react with the I_3^- liberated in this experiment?

Problems

A16-1. What is the difference between the *titration reaction* and the *cell reaction(s)* in a potentiometric titration?

A16-2. Consider the titration in Figure 16-2.

(a) Write a balanced titration reaction.
(b) Write two different half-reactions for the indicator electrode.
(c) Write two different Nernst equations for the net cell reaction.
(d) Calculate E at the following volumes of Ce^{4+}: 10.0, 20.0, 25.0, 30.0, 40.0, 49.0, 50.0, 51.0, and 60.0 mL. Compare your results with Figure 16-2.

A16-3. The part of the titration curve in Figure 16-4 prior to the Sn^{2+} equivalence point (between 0 and 50 mL) can be calculated using the Nernst equation for Reaction 16-38. The part of the titration curve between 50 and 100 mL depends only on Reaction 16-21. Calculate E at the following volumes of IO_3^- and compare your results with the figure: 20.0, 25.0, 60.0, 75.0, 100.0, and 110.0 mL. To see how to handle the first equivalence point (50 mL), refer to Problem 16-16.

A16-4. Select indicators from Table 16-1 that would be suitable for finding the end point in Figure 16-3 state what color changes would be observed.

A16-5. A solution of I_3^- was standardized by titrating freshly dissolved arsenious oxide (As_4O_6, F.W. 395.683). The titration of 25.00 mL of a solution prepared by dissolving 0.366 3 g of As_4O_6 in a volume of 100.0 mL required 31.77 mL of I_3^-.
(a) Calculate the molarity of the I_3^- solution.
(b) Does it matter whether starch indicator is added at the beginning or near the end point in this titration?

A16-6. Explain the terms *preoxidation* and *prereduction*. Why is it important to be able to destroy the reagents used for these purposes?

A16-7. What is a Jones reductor and what is it used for?

A16-8. Write balanced half-reactions in which MnO_4^- acts as an oxidant at (a) pH = 0, (b) pH = 10, and (c) pH = 15.

A16-9. List three ways to standardize triiodide solution.

16-10. Calculate the potential (versus S.C.E. anode) at each point listed for the titration of 25.00 mL of 0.020 00 M Cr^{2+} with 0.010 00 M Fe^{3+}: 5.00, 25.00, 50.00, and 100.0 mL.

16-11. Suppose 50 mL of 0.020 8 M Fe^{3+} was titrated with 0.017 3 M ascorbic acid at pH 1.00.

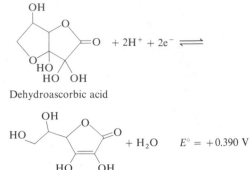

Dehydroascorbic acid

Ascorbic acid

The potential was measured with a Pt cathode and an S.C.E anode.
(a) Write a balanced equation for the titration reaction
(b) Write two balanced equations to describe the cell reactions. (This question is asking for two complete reactions, not two half-reactions.)
(c) Calculate the potential after 50.0 mL of titrant has been added.
(d) Calculate the potential at the equivalence point.

16-12. Suppose that 50.0 mL of 0.050 0 M Fe^{2+} was titrated with 0.050 0 M MnO_4^- in a solution buffered to pH 1.00. Calculate the potential (versus S.C.E. anode) at each volume of MnO_4^-: 2.00, 4.00, 5.00, 6.00, 8.00, 9.00, 9.90, 10.00, 10.10, 11.00, 13.00 mL. Sketch the titration curve.

16-13. A solution containing 50.0 mL of 0.050 M UO_2^{2+} in 1 M HCl was titrated with 0.100 M Sn^{2+}, and the potential was measured with a Pt electrode relative to a saturated silver–silver chloride anode. The *unbalanced* reaction is

$$UO_2^{2+} + Sn^{2+} \rightarrow U^{4+} + Sn^{4+}$$

(a) Calculate the potential when 35.0 mL of Sn^{2+} has been added.
(b) Calculate the potential at the equivalence point.

16-14. Suppose that the 0.010 0 M IO_3^- reagent in Figure 16-4 were replaced by 0.010 0 M $Cr_2O_7^{2-}$. Calculate the potential at the following volumes: $\frac{1}{2}V_e$, $\frac{3}{2}V_e$, $2V_e$, and $3V_e$. (V_e is the volume at the first equivalence point.)

16-15. A 25.0-mL solution containing a mixture of UO_2^+ and Fe^{2+} in 1 M $HClO_4$ was titrated with 0.009 87 M $KMnO_4$.
(a) Write balanced equations for the two titration reactions in the order in which they occur.

(b) The two potentiometric end points were observed at 12.73 and 31.21 mL. Calculate the molarities of UO_2^+ and Fe^{2+} in the unknown.

(c) Calculate the potential (versus S.C.E anode) at $\frac{1}{2}V_1$, $V_1 + \frac{1}{2}V_2$, V_2, and $V_2 + 1.00$ mL, where V_1 and V_2 are the two equivalence points. Assume that $[H^+]$ is constant at 1.00 M.

16-16. Problem A16-3 deals with the potentiometric titration of a mixture of Sn^{2+} and Tl^+ by IO_3^-. All of the points except the first equivalence point at 50 mL are readily calculated by the individual titrations of Sn^{2+} by IO_3^- and Tl^+ by IO_3^-. To find the potential at the first equivalence point, we use the algebraic trick of adding the Nernst equations for the thallium and tin half-reactions:

$$Tl^{3+} + 2e^- \rightleftharpoons Tl^+ \qquad E° = 0.77 \text{ V}$$

$$Sn^{4+} + 2e^- \rightleftharpoons Sn^{2+} \qquad E° = 0.139 \text{ V}$$

$$E_+ = 0.77 - \frac{0.059\,16}{2} \log \frac{[Tl^+]}{[Tl^{3+}]}$$

$$E_+ = 0.139 - \frac{0.059\,16}{2} \log \frac{[Sn^{2+}]}{[Sn^{4+}]}$$

$$\overline{2E_+ = 0.909 - \frac{0.059\,16}{2} \log \frac{[Sn^{2+}][Tl^+]}{[Sn^{4+}][Tl^{3+}]}} \quad \text{(a)}$$

At the first equivalence point, essentially all of the Sn^{2+} has been oxidized to Sn^{4+} by iodate and virtually none of the Tl^+ has reacted. The Sn^{4+} and Tl^+ will have the following equilibrium, in which small and equal amounts of Sn^{2+} and Tl^{3+} are created:

$$Sn^{4+} + Tl^+ \rightleftharpoons Sn^{2+} + Tl^{3+}$$

The small, unknown concentrations of Sn^{2+} and Tl^{3+} are equal to each other at the first equivalence point and can be canceled in the combined Nernst equation (a) above. Since we know the concentrations of all remaining species in the equation, we can calculate the potential at the 50-mL point. Calculate this potential and compare your answer with Figure 16-4.

16-17. Suppose 25 mL of $0.010\,0$ M U^{4+} at pH 1.00 was oxidized to UO_2^{2+} with $0.005\,00$ M $Br_2(aq)$. Calculate the potential (versus S.C.E. anode) after addition of 10.0, 50.0, and 51.0 mL of Br_2.

16-18. Would indigo tetrasulfonate be a suitable redox indicator for the titration of $Fe(CN)_6^{4-}$ with Tl^{3+} in 1 F HCl?

16-19. Would tris(2,2′-bipyridine) iron be a useful indicator for the titration of Sn^{2+} with $Mn(EDTA)^-$?

16-20. Suppose that 100.0 mL of solution containing 0.100 M Fe^{2+} and $0.005\,00$ M tris(1,10-phenanthroline)Fe(II) (ferroin) is titrated with $0.050\,0$ M

Ce^{4+}. Calculate the potential (versus S.C.E. anode) at the following volumes of Ce^{4+}: 1.0, 10.0, 100.0, 190.0, 199.0, 201.0, 205.0, 209.0, 210.0, 211.0, and 220.0 mL. Assume that the reaction contains 1 M $HClO_4$. Sketch the titration curve. Is ferroin a suitable indicator for the titration of Fe^{2+} by Ce^{4+}?

16-21. *Iodometric analysis of high-temperature superconductor.* The procedure in Box 16-1 was carried out to find the effective copper oxidation state, and therefore the number of oxygen atoms, in the formula $YBa_2Cu_3O_{7-z}$ where z ranges from 0 to 0.5.

(a) In Experiment A of Box 16-1, 1.00 g of superconductor required 4.55 mmol of $S_2O_3^{2-}$. In Experiment B, 1.00 g of superconductor required 5.68 mmol of $S_2O_3^{2-}$. Calculate the value of z in the formula $YBa_2Cu_3O_{7-z}$ (F.W. $666.246 - 15.999\,4\ z$).

(b) *Propagation of Uncertainty.* In several replications of Experiment A, the thiosulfate required was 4.55 (± 0.10) mmol of $S_2O_3^{2-}$ per gram of $YBa_2Cu_3O_{7-z}$. In Experiment B, the thiosulfate required was 5.68 (± 0.05) mmol of $S_2O_3^{2-}$ per gram. Calculate the uncertainty of x in the formula $YBa_2Cu_3O_x$.

16-22. An enzyme involved in the catalysis of redox reactions has an oxidized form and a reduced form differing by two electrons. The oxidized form of this enzyme was mixed at pH 7 with the oxidized form of a redox indicator whose two forms differ by one electron. The mixture was protected by an inert atmosphere and partially reduced with sodium dithionite. The equilibrium composition of the mixture was determined by spectroscopic analysis:

Enzyme(oxidized)	4.2×10^{-5} M
Indicator(oxidized)	3.9×10^{-5} M
Enzyme(reduced)	1.8×10^{-5} M
Indicator(reduced)	5.5×10^{-5} M

Given that $E°' = -0.187$ V for the indicator, find $E°'$ for the enzyme.

16-23. From the reduction potentials below

$$I_2(s) + 2e^- \rightleftharpoons 2I^- \qquad E° = 0.535 \text{ V}$$

$$I_2(aq) + 2e^- \rightleftharpoons 2I^- \qquad E° = 0.620 \text{ V}$$

$$I_3^- + 2e^- \rightleftharpoons 3I^- \qquad E° = 0.535 \text{ V}$$

(a) calculate the equilibrium constant for the reaction $I_2(aq) + I^- \rightleftharpoons I_3^-$.

(b) calculate the equilibrium constant for the reaction $I_2(s) + I^- \rightleftharpoons I_3^-$.

(c) calculate the solubility (g/L) of $I_2(s)$ in water.

16-24. When 25.00 mL of unknown was passed through a Jones reductor, molybdate ion (MoO_4^{2-}) was converted to Mo(III). The filtrate required 16.43 mL of 0.010 33 M $KMnO_4$ to reach the purple end point.

$$MnO_4^- + Mo^{3+} \rightarrow Mn^{2+} + MoO_2^{2+}$$

A blank required 0.04 mL. Find the molarity of molybdate in the unknown.

16-25. An aqueous glycerol solution weighing 100.0 mg was treated with 50.0 mL of 0.083 7 M Ce(IV) in 4 M $HClO_4$ at 60°C for 15 minutes to oxidize the glycerol to formic acid:

$$\begin{array}{cc}
CH_2-CH-CH_2 & 3HCO_2H \\
| \quad | \quad | & \\
HO \quad HO \quad HO & \\
\text{Glycerol} & \text{Formic acid}
\end{array}$$

The excess Ce(IV) required 12.11 mL of 0.044 8 M Fe(II) to reach a ferroin end point. What is the weight percent of glycerol in the unknown?

16-26. A mixture of nitrobenzene and nitrosobenzene weighing 24.43 mg was titrated with Cr^{2+} to give aniline:

Nitrosobenzene	Nitrobenzene
F.W. 107.112	F.W. 123.111

Aniline

The titration required 21.57 mL of 0.050 00 M Cr^{2+} to reach a potentiometric end point. Find the weight percent of nitrosobenzene in the mixture.

16-27. Potassium bromate, $KBrO_3$, is a good primary standard for the generation of Br_2 in acidic solution:

$$BrO_3^- + 5Br^- + 6H^+ \rightleftharpoons 3Br_2(aq) + 3H_2O$$

The Br_2 can be used to analyze many unsaturated organic compounds. Al^{3+} was analyzed as follows: An unknown was treated with 8-hydroxyquinoline (oxine) at pH 5 to precipitate aluminum oxinate, $Al(C_9H_6ON)_3$. The precipitate was washed, dissolved in warm HCl containing excess KBr, and treated with 25.00 mL of 0.020 00 M $KBrO_3$.

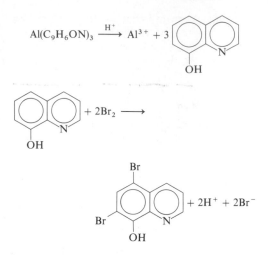

The excess Br_2 was reduced with KI, which was converted to I_3^-. The I_3^- required 8.83 mL of 0.051 13 M $Na_2S_2O_3$ to reach a starch end point. How many milligrams of Al were in the unknown?

16-28. Nitrite (NO_2^-) can be determined by oxidation with excess Ce^{4+}, followed by back titration of the unreacted Ce^{4+}. A 4.030-g sample of solid containing only $NaNO_2$ and $NaNO_3$ was dissolved in 500.0 mL. A 25.00-mL sample of this solution was treated with 50.00 mL of 0.118 6 M Ce^{4+} in strong acid for five minutes, and the excess Ce^{4+} was back-titrated with 31.13 mL of 0.042 89 M ferrous ammonium sulfate.

$$2Ce^{4+} + NO_2^- + H_2O \rightarrow 2Ce^{3+} + NO_3^- + 2H^+$$
$$Ce^{4+} + Fe^{2+} \rightarrow Ce^{3+} + Fe^{3+}$$

Calculate the weight percent of $NaNO_2$ in the solid.

16-29. A 50.00-mL sample containing La^{3+} was treated with sodium oxalate to precipitate $La_2(C_2O_4)_3$, which was washed, dissolved in acid, and titrated with 18.04 mL of 0.006 363 M $KMnO_4$. Calculate the molarity of La^{3+} in the unknown.

16-30. Write a balanced equation for the reaction of periodate (IO_4^-) with

(a)
$$\begin{array}{c}
CHO \\
H\!-\!\!-\!OH \\
OH\!-\!\!-\!H \\
H\!-\!\!-\!OH \\
H\!-\!\!-\!OH \\
CH_2OH \\
\text{Glucose}
\end{array}$$

(b)
$$\begin{array}{c}
CHO \\
H\!-\!\!-\!NH_2 \\
HO\!-\!\!-\!H \\
H\!-\!\!-\!OH \\
H\!-\!\!-\!OH \\
CH_2OH \\
\text{2-Amino-2-deoxyglucose}
\end{array}$$

(c)

Dihydroxyacetone

16-31. A solution containing Be and several other metals was treated with excess EDTA to *mask* the other metals. Then excess acetoacetanilide was added at 50°C at pH 7.5 to precipitate beryllium ion:

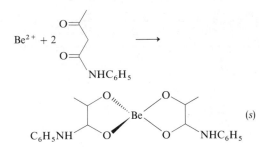

The precipitate was dissolved in 6 M HCl and treated with 50.0 mL of solution containing 0.139 2 g of KBrO$_3$ (F.W. 167.00) plus 0.5 g of KBr. Bromine produced by these reagents reacts with acetoacetanilide as follows:

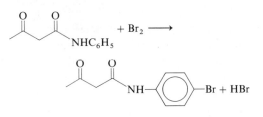

After five minutes, the excess Br$_2$ was destroyed by adding 2 g of KI (F.W. 166.00). The I$_2$ released by the Br$_2$ required 19.18 mL of 0.050 00 M Na$_2$S$_2$O$_3$. Calculate the number of milligrams of Be in the original solution.

16-32. A sensitive titration of Bi^{3+} is based on the following sequence:[†]

1. Bi^{3+} is precipitated by Cr(SCN)$_6^{3-}$:

$$Bi^{3+} + Cr(SCN)_6^{3-} \rightarrow$$
$$Bi[Cr(SCN)_6](s) \quad (a)$$

2. The precipitate is filtered, washed, and treated with bicarbonate to release Cr(SCN)$_6^{3-}$:

$$Bi[Cr(SCN)_6] + HCO_3^- + H_2O \rightarrow$$
$$(BiO)_2CO_3(s) + Cr(SCN)_6^{3-} + H^+ \quad (b)$$

3. I$_2$ is added to the filtrate after removal of the (BiO)$_2$CO$_3$(s):

$$Cr(SCN)_6^{3-} + I_2 + H_2O \rightarrow$$
$$SO_4^{2-} + ICN + I^- + H^+ + Cr^{3+} \quad (c)$$

[†] G. A. Parker, *J. Chem. Ed.*, **57**, 721 (1980).

4. Upon acidification to pH 2.5, HCN is created:

$$ICN + I^- + H^+ \rightarrow I_2 + HCN \quad (d)$$

5. The I$_2$ in this mixture is removed by extraction with chloroform. Excess bromine water is then added to the aqueous phase to convert iodide to iodate and HCN to BrCN:

$$Br_2 + I^- + H_2O \rightarrow IO_3^- + H^+ + Br^- \quad (e)$$
$$Br_2 + HCN \rightarrow BrCN + H^+ + Br^- \quad (f)$$

6. Excess Br$_2$ is destroyed with formic acid:

$$Br_2 + HCO_2H \rightarrow Br^- + CO_2 + H^+ \quad (g)$$

7. Addition of excess I$^-$ produces I$_2$:

$$IO_3^- + I^- + H^+ \rightarrow I_2 + H_2O \quad (h)$$
$$BrCN + I^- + H^+ \rightarrow I_2 + HCN + Br^-$$
$$(i)$$

8. Finally, the iodine is titrated with a standard solution of sodium thiosulfate.

Show that 228 mol of thiosulfate will be required for each mole of Bi^{3+} that is analyzed.

16-33. Chromous ion (Cr^{2+}) was titrated with chlorate, ClO$_3^-$, at pH = −0.30 using Pt and a saturated Ag|AgCl electrode. The unbalanced titration reaction is

$$Cr^{2+} + ClO_3^- \rightarrow Cr^{3+} + Cl^-$$

(a) Using just the information below, find $E°$ for the chlorate half-reaction and write a balanced equation for this half-reaction.

$$ClO_3^- + 6H^+ + 5e^- \rightleftharpoons$$
$$\tfrac{1}{2}Cl_2(g) + 3H_2O \quad E° = 1.458 \text{ V}$$
$$Cl_2(g) + 2e^- \rightleftharpoons 2Cl^- \quad E° = 1.360 \text{ V}$$

(b) Calculate E at the equivalence point.

16-34. Ten mL of 3.00 mM Fe(bipyridyl)$_3^{3+}$ was titrated with 1.00 mM sodium dithionite Na$_2$S$_2$O$_4$, at pH 9.00 using Pt and a saturated Ag|AgCl electrode. Calculate the voltage expected when the volume of titrant is 10.0% greater than the equivalence volume.

Bipyridyl
(also called 2,2′-bipyridine)

17 Electrogravimetric and Coulometric Analysis

Charles Martin Hall. [Photo courtesy of Alcoa.]

So far we have seen how *spontaneous* electrochemical reactions can be adapted for analytical purposes. Potentiometric measurements with galvanic cells (in which negligible current is permitted to flow) can be used to measure concentrations of species in the cell. In redox titrations a spontaneous reaction between analyte and titrant is employed. In this chapter and Chapter 18 we will see how *nonspontaneous* redox reactions, driven by an external source of electricity, are used in analytical chemistry.

This chapter deals with electrogravimetric and coulometric techniques. In **electrogravimetric analysis** the analyte is quantitatively deposited as a solid on the cathode or anode. The increase in mass of the electrode is a direct measure of the amount of analyte. **Coulometric analyses** are based on the measurement of current and time needed to complete a chemical reaction. Before discussing these methods, we must examine how the voltage of an electrochemical cell changes when a significant current flows.

17-1 ELECTROLYSIS: PUTTING ELECTRONS TO WORK

Electrolysis is the process in which a reaction is driven in its nonspontaneous direction by the application of an electric current. For example, the following reaction has a standard potential of -1.458 V, which means that it is spontaneous to the left, not the right.

$$Pb^{2+} + 2H_2O \rightleftharpoons PbO_2(s) + H_2(g) + 2H^+ \qquad E^\circ = -1.458 \text{ V} \qquad (17\text{-}1)$$

Suppose you wish to drive this reaction to the *right* in a solution containing 5.00 mM Pb^{2+}, 2.00 M HNO_3, and solid PbO_2 under 1.00 atm of $H_2(g)$. Under these conditions, we calculate the potential to be

$$E = E^\circ - \frac{0.059\ 16}{2} \log \frac{P_{H_2}[H^+]^2}{[Pb^{2+}]}$$

$$= -1.458 - \frac{0.059\ 16}{2} \log \frac{(1.00)(2.00)^2}{(5.00 \times 10^{-3})} = -1.544\ \text{V} \qquad (17\text{-}2)$$

For Reaction 17-1,

$$\Delta G = -nFE = -nF(-1.544\ \text{V})$$
$$= +298\ \text{kJ/mol}$$

We predict that a voltage greater than 1.544 V must be *applied* to the solution to force Reaction 17-1 to proceed as written.

Figure 17-1 shows how this might be done. A pair of Pt electrodes is immersed in the solution, and a voltage more positive than 1.544 V is applied with an external power supply. At the cathode (where reduction takes place), the reaction is

$$2H^+ + 2e^- \rightarrow H_2(g) \qquad (17\text{-}3)$$

and at the anode (where oxidation occurs), the reaction is

$$Pb^{2+} + 2H_2O \rightarrow PbO_2(s) + 4H^+ + 2e^- \qquad (17\text{-}4)$$

The net reaction is Reaction 17-1.

If a current I flows for a time t, the amount of charge q that has passed any point in the circuit is

$$\underset{\text{Coulombs}}{q} = \underset{\text{Amperes}}{I} \cdot \underset{\text{Seconds}}{t} \qquad (17\text{-}5)$$

Amperes = coulombs/second

$$F = 9.648\ 530\ 9 \times 10^4\ \text{C/mol}$$

The number of moles of electrons is

$$\text{Moles of } e^- = \frac{\text{coulombs}}{\text{coulombs/mole}} = \frac{I \cdot t}{F} \qquad (17\text{-}6)$$

Moles of electrons $= \dfrac{I \cdot t}{F}$

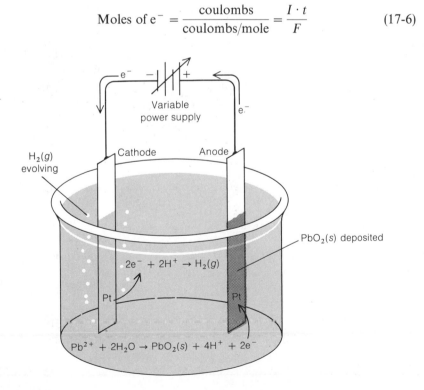

Figure 17-1
Electrolysis of Pb^{2+} in acidic solution. The symbol $\dashv\!\!\!\mid\!\!\!\mid\!\!\!\vdash$ will be used to represent a variable source of direct current.

Demonstration 17-1   ELECTROCHEMICAL WRITING[†]

Approximately 7% of the electric power output of the United States goes into electrolytic chemical production. Electrolysis is dramatically demonstrated in this experiment. The apparatus consists of a sheet of aluminum foil taped or cemented to a glass or wood surface. Any size will work, but an area about 15 cm on a side is convenient for a classroom demonstration. On the metal foil is taped (at one edge only) a sandwich consisting of filter paper, typing paper, and another sheet of filter paper. A stylus is prepared from a length of copper wire (18 gauge or thicker) looped at the end and passed through a length of glass tubing.

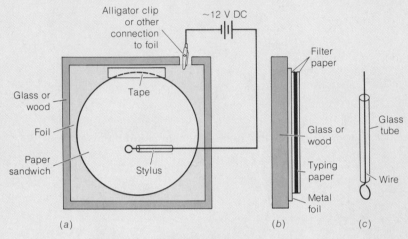

(a) Front view. (b) Side view. (c) Stylus.

A fresh solution is prepared from 1.6 g of KI, 20 mL of water, 5 mL of 1% (wt/wt) starch solution, and 5 mL of phenolphthalein indicator solution. (If the solution darkens after standing for several days, it can be decolorized by adding drops of dilute $Na_2S_2O_3$.) The three layers of paper are soaked with the KI–starch–phenolphthalein solution. Connect the stylus and foil to a 12 V DC power source, and write on the paper with the stylus.

When the stylus is the cathode, pink color appears from the reaction of OH^- with phenolphthalein:

$$\text{Cathode:} \qquad H_2O + e^- \rightarrow \tfrac{1}{2}H_2(g) + OH^-$$

When the polarity is reversed and the stylus is the anode, a black (very dark blue) color appears from the reaction of I_2 with starch:

$$\text{Anode:} \qquad I^- \rightarrow \tfrac{1}{2}I_2 + e^-$$

Pick up the top sheet of filter paper and the typing paper, and you will discover that the writing appears in the opposite color on the bottom sheet of filter paper. This sequence is shown in Color Plate 11.

[†] E. C. Gilbert in H. N. Alyea and F. B. Dutton, eds., *Tested Demonstrations in Chemistry* (Easton, Penn.: Journal of Chemical Education, 1965), p. 145.

If a species requires n electrons per molecule in its redox half-reaction, the moles of that species that will have reacted in a time t is simply

$$\text{Moles reacted} = \frac{I \cdot t}{nF} \qquad (17\text{-}7)$$

EXAMPLE: Relating Current, Time, and Amount of Reaction

If a current of 0.17 A flows for 16 minutes through the cell in Figure 17-1, how many grams of PbO_2 will have been deposited?

To anwer this, we first calculate the moles of e^- that have flowed through the cell:

$$\text{Moles of } e^- = \frac{I \cdot t}{F} = \frac{\left(0.17 \frac{\cancel{C}}{\cancel{s}}\right)(16 \cancel{\min})\left(60 \frac{\cancel{s}}{\cancel{\min}}\right)}{96\,485\left(\frac{\cancel{C}}{\text{mol}}\right)} = 1.69 \times 10^{-3} \text{ mol}$$

For the half-reaction

$$Pb^{2+} + 2H_2O \rightleftharpoons PbO_2(s) + 4H^+ + 2e^-$$

two moles of electrons are required for each mole of PbO_2 deposited. Therefore,

$$\text{Moles of } PbO_2 = \tfrac{1}{2}(\text{moles of } e^-) = 8.45 \times 10^{-4} \text{ mol}$$

The mass of PbO_2 will be

$$\text{Grams of } PbO_2 = (8.45 \times 10^{-4} \text{ mol})(239.2 \text{ g/mol}) = 0.20 \text{ g}$$

Demonstration 17-1 is an impressive classroom illustration of electrolysis. Box 17-1 describes one approach to the conversion of solar energy into storable fuel by means of photoelectrolysis.

17-2 WHY THE VOLTAGE CHANGES WHEN CURRENT FLOWS THROUGH A CELL

In previous chapters we considered the voltage of galvanic cells only under conditions of negligible current flow. For a cell to do any useful work or for an electrolysis to occur, a significant current must flow. Whenever current flows, three factors act to decrease the output voltage of a galvanic cell or to increase the applied voltage needed for electrolysis. These factors are called the *ohmic potential, concentration polarization,* and *overpotential.*

Ohmic Potential

Any cell possesses some electric resistance. The voltage needed to force current (ions) to flow through the cell is called the **ohmic potential** and is given by Ohm's law:

$$E_{ohmic} = IR \qquad (17\text{-}8)$$

where I is the current and R is the resistance of the cell.

When current flows through a galvanic cell, the observed voltage is *lower* than the equilibrium voltage.

Performing an electrolysis requires a voltage *higher* than the calculated equilibrium voltage for the reaction.

Box 17-1 PHOTOELECTROLYSIS

The most intense solar radiation reaching the earth's surface is visible light (400–750 nm in the spectrum below).

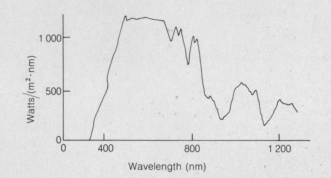

Spectral distribution of solar energy with the sun at 60° from the zenith. Absorption bands are due to atmospheric species. [Reproduced from A. W. Adamson, *A Textbook of Physical Chemistry,* 2nd ed. (New York: Academic Press, 1979).]

If solar energy could be used with 10% efficiency, just 3% of the sunlight falling on the earth's deserts would provide all the energy used in the world around 1980. A major goal of solar energy research is to devise a means of converting solar energy into a fuel that can be stored and used later. Mother Nature accomplishes this in photosynthesis. The green leaves of plants use the energy of sunlight to reduce atmospheric CO_2 to carbohydrates, which can be "burned" by living organisms to provide chemical energy.

Procedures for converting solar energy (visible light) to fuels are not yet economically practical, but are the subject of much research. The experimental system described below[†] produces hydrogen and oxygen from sunlight with approximately 0.3% efficiency, but efficiencies up to 10% are claimed for related schemes[‡]. Practical systems would store the H_2 for later use in an engine or fuel cell.

The cell shown at right contains thin semiconducting layers of CdSe and CoS. (Semiconductors were described in Section 15-7.) The valence electrons of a semiconductor lie in a band of energy levels called the *valence band*. At higher energy lies the *conduction band,* whose electrons are free to roam through the bulk material. In a conductor, the conduction band lies directly above the valence band and is populated by electrons with thermal energy. In an insulator, the *band gap* (energy difference) between the valence and conduction bands is large, and the conduction band is unpopulated. A semiconductor has a band gap intermediate between those of a conductor and an insulator, and has intermediate electrical conductivity.

When the CdSe is irradiated with sunlight whose energy is greater than that of the band gap, electrons are promoted from the valence band to the conduction band, leaving positively charged, mobile holes (designated h^+) behind in the valence band.[§] The internal electric fields generated by contact between the solid layers and electrolyte solutions are such that the electrons are attracted to the CoS layer and the holes move to the CdSe|electrolyte interface.

[†] E. S. Smotkin, S. Cervera-March, A. J. Bard, A. Campion, M. A. Fox, T. Mallouk, S. E. Webber, and J. M. White, *J. Phys. Chem.,* **91,** 6 (1987).

[‡] *Chem. Eng. News,* June 26, 1989, p. 37.

[§] Readable and excellent articles dealing with photochemistry at semiconductor surfaces can be found in B. Parkinson, *J. Chem. Ed.,* **60,** 338 (1983); M. S. Wrighton, *J. Chem. Ed.,* **60,** 335 (1983); H. O. Finklea, *J. Chem. Ed.,* **60,** 325 (1983); J. A. Turner, *J. Chem. Ed.,* **60,** 327 (1983); M. T. Spitler, *J. Chem. Ed.,* **60,** 330 (1983); and A. B. Ellis, *J. Chem. Ed.,* **60,** 332 (1983).

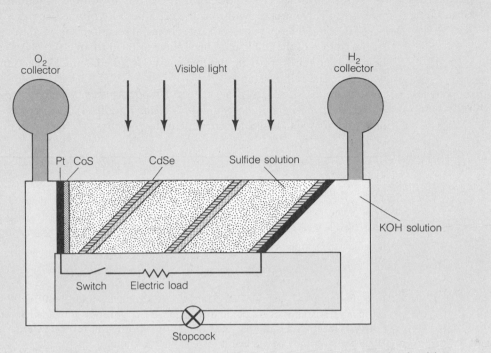

Schematic diagram of photolysis cell using sunlight to make H_2 and O_2, or to make electricity.

The sulfide electrolyte solution contains 1 M KOH, 1 M Na_2S, and 1 M S, which generates Na_2S_2. At the CdSe|electrolyte interface, the holes oxidize S^{2-}:

$$S^{2-} + h^+ \rightarrow \tfrac{1}{2}S_2^{2-}$$

and at the CoS|electrolyte interface electrons reduce S_2^{2-}:

$$\tfrac{1}{2}S_2^{2-} + e^- \rightarrow S^{2-}$$

If the electrical switch connecting the two Pt electrodes is closed, and the stopcock at the bottom of the cell is closed, electricity flows across the electric load when the cell is irradiated with sunlight. More CdSe|CoS|electrolyte layers in series generate a higher voltage. With an array of six CdSe|CoS layers, the fraction of solar energy converted to electrical energy is 1.9%. If the electric circuit is open, and the stopcock is open, there is a conducting path of KOH solution between the two Pt electrodes. In this case, electrons are available to reduce water on the right side and holes oxidize water at the left electrode:

$$\text{Right-hand Pt electrode:} \quad H_2O + e^- \rightarrow \tfrac{1}{2}H_2 + OH^-$$

$$\text{Left-hand Pt electrode:} \quad OH^- + h^+ \rightarrow \tfrac{1}{4}O_2 + \tfrac{1}{2}H_2O$$

The efficiency of conversion of solar energy to chemical energy is only 0.32%, but it can be improved by using an anode material other than Pt, whose *overpotential* (Section 17-2) for O_2 production is high. Using semiconductors with smaller band gaps also improves efficiency.

Major challenges in the conversion of sunlight to chemical fuels include increasing efficiency and preventing deterioration of the cell during operation. The sulfide electrolyte in the cell illustrated is chosen to prolong the lifetime of the semiconductor materials.

The output voltage of a galvanic cell is decreased by $I \cdot R$.

In a galvanic cell at equilibrium, there is no ohmic potential because $I = 0$. If a current is drawn from the cell, the cell voltage decreases because part of the free energy released by the chemical reaction is needed to overcome the resistance of the cell itself. The voltage applied to an electrolysis cell must be great enough to provide the free energy for the chemical reaction and to overcome the cell resistance.

The magnitude of the input voltage for an electrolysis cell must be increased by $I \cdot R$.

In the absence of any other effects, *the voltage of a galvanic cell is decreased by IR, and the magnitude of the applied voltage in an electrolysis must be increased by IR* in order for current to flow.

EXAMPLE: Effect of Ohmic Pontential
Consider the cell

$$Cd(s)\,|\,CdCl_2(aq, 0.167\ \text{м})\,|\,AgCl(s)\,|\,Ag(s)$$

in which the spontaneous chemical reaction is

$$Cd(s) + 2AgCl(s) \rightarrow Cd^{2+} + 2Ag(s) + 2Cl^-$$

In Equation 14-47, we calculated that the cell voltage should be 0.764 V. (a) If the cell has a resistance of 6.42 Ω and a current of 28.3 mA is drawn, what will be the cell voltage? (b) Suppose that the same cell is operated in reverse as an electrolysis. What voltage must be applied to reverse the reaction?

(a) In the absence of electron flow, the voltage (E_{eq}, where the subscript eq means equilibrium) is 0.764 V. With a current of 28.3 mA, the voltage will *decrease* to

Volts = amperes · ohms

$$E = E_{eq} - IR = 0.764 - (0.028\ 3\ \text{A})(6.42\ \Omega) = 0.582\ \text{V}$$

(b) By convention, the voltage applied to an electrolysis cell is given a negative sign. The voltage needed to reverse the spontaneous reaction will be

$$E = -E_{eq} - IR = -0.764 - (0.028\ 3\ \text{A})(6.42\ \Omega) = -0.946\ \text{V}$$

Notice that the magnitude of any galvanic cell voltage is *decreased* by the ohmic potential. The magnitude of the voltage that must be applied to any electrolysis cell is *increased* by the ohmic potential.

Concentration Polarization

Consider the cadmium anode in Figure 17-2, for which the reaction is

$$Cd(s) \rightarrow Cd^{2+} + 2e^- \tag{17-9}$$

This reaction creates Cd^{2+} ions in the layer of solution immediately surrounding the Cd electrode. If these ions diffuse rapidly or are transported rapidly away from the electrode by stirring, the concentration of Cd^{2+} will be essentially constant in the entire solution. We will call the concentration of Cd^{2+} in the bulk solution $[Cd^{2+}]_0$. We will call the concentration of Cd^{2+} in the immediate vicinity of the electrode surface $[Cd^{2+}]_s$. This concentration will equal $[Cd^{2+}]_0$ if diffusion or stirring is sufficiently rapid. Otherwise, $[Cd^{2+}]_s$ will be greater than $[Cd^{2+}]_0$, since Cd^{2+} ions are created at the electrode.

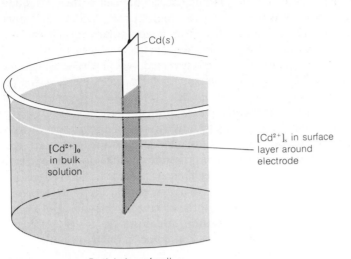

Partial view of cell

Figure 17-2
The potential of the $Cd^{2+}|Cd$ couple depends on the concentration of Cd^{2+} in the layer surrounding the electrode. The layer is greatly exaggerated in this drawing; it is only a few molecules thick.

The anode potential depends on $[Cd^{2+}]_s$, not $[Cd^{2+}]_0$, because $[Cd^{2+}]_s$ is the actual concentration at the electrode surface. Reversing Reaction 17-9 to write it as a reduction, the anode potential is given by the equation

$$E(\text{anode}) = E°(\text{anode}) - \frac{0.059\,16}{2} \log \frac{1}{[Cd^{2+}]_s} \qquad (17\text{-}10)$$

If $[Cd^{2+}]_s = [Cd^{2+}]_0$, the anode potential will be that expected from the bulk Cd^{2+} concentration.

If the current is flowing so fast that Cd^{2+} cannot escape from the region around the electrode as fast as it is made, $[Cd^{2+}]_s$ will be greater than $[Cd^{2+}]_0$. When $[Cd^{2+}]_s \neq [Cd^{2+}]_0$, we say that **concentration polarization** exists. The anode potential in Equation 17-10 will become more positive and the cell voltage ($= E(\text{cathode}) - E(\text{anode})$) will decrease. This behavior is shown in Figure 17-3, where the straight line shows the behavior expected

The electrode potential depends on the concentration of species in the region immediately surrounding the electrode.

When ions are not transported to or from an electrode as rapidly as they are consumed or created, we say that *concentration polarization* exists. That is, concentration polarization means that $[X]_s \neq [X]_0$.

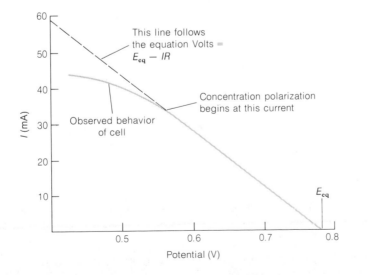

Figure 17-3
The hypothetical behavior of a galvanic cell, illustrating concentration polarization that occurs when $[Cd^{2+}]_s > [Cd^{2+}]_0$. The resistance of the cell is 6.42 Ω.

Concentration polarization *decreases* the voltage of a galvanic cell and *increases* the magnitude of the voltage necessary for electrolysis.

To decrease concentration polarization:

1. Raise the temperature.
2. Increase stirring.
3. Increase electrode surface area.
4. Change ionic strength to increase or decrease electrostatic interaction between the electrode and the reactive ion.

if only the ohmic potential (IR) affects the net cell voltage. The deviation of the curve from the straight line at high currents is due to concentration polarization. In a galvanic cell, concentration polarization *decreases* the voltage below the value expected in the absence of polarization.

In electrolysis the situation is reversed; reactant is depleted and product accumulates. Therefore, *concentration polarization requires us to apply a voltage of greater magnitude (more negative) than that expected in the absence of polarization.* By custom, the applied voltage is given a negative sign.

Among the factors causing ions to move toward or away from the electrode are diffusion, convection, and electrostatic attraction or repulsion. Raising the temperature increases the rate of diffusion and thereby decreases concentration polarization. Mechanical stirring is very effective in transporting species through the cell. Increasing ionic strength decreases the electrostatic forces between ions and the electrode. These factors can all be used to affect the degree of polarization. Also, the greater the electrode surface area, the more current can be passed without polarization.

Overpotential

Even when concentration polarization is absent and ohmic potential is taken into account, the voltages of some electrochemical cells still show unexpected values. Some electrolyses require a greater-than-expected applied voltage, and some galvanic cells produce less voltage than we anticipate. The difference between the expected voltage (after accounting for IR drop and concentration polarization) and the applied voltage is called the **overpotential.** *The faster you wish to drive an electrode reaction, the greater the overpotential that must be applied.*

The overpotential can be traced to the activation energy barrier for the electrode reaction.[†] Figure 17-4 shows a schematic "reaction coordinate" diagram for a chemical reaction. The activation energy for any chemical reaction is the barrier that must be overcome before reactants can be converted to products. The higher the temperature, the more molecules have sufficient energy to overcome the barrier, and the faster the reaction proceeds.

Now consider the reaction in which an electron from a metal electrode is transferred to H_3O^+ to initiate the reduction to H_2:

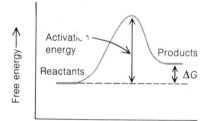

Figure 17-4
Schematic energy profile of a chemical reaction

$$H_3O^+ + e^- \rightarrow \tfrac{1}{2}H_2 + H_2O \qquad (17\text{-}11)$$

Figure 17-5a shows a schematic energy-level diagram for this process. A certain energy barrier must be overcome for the electron transfer to occur, and the rate of reduction may be very slow. If an electric potential (the overpotential) is applied to the electrode, the energy of the electrons *in the metal electrode* is increased (Figure 17-5b). This decreases the height of the activation energy barrier and increases the rate of electron transfer. *The overpotential is that voltage needed to sustain a particular rate of electron transfer.* The greater the desired rate, the higher must be the overpotential. Thus, the current density (A/m^2) increases as the overpotential increases (Table 17-1). Because overpotential is a kinetic phenomenon, it is also called *kinetic polarization.*

Overpotential is needed to overcome the activation energy barrier for a reaction.

[†] J. O'M. Bockris, *J. Chem. Ed.,* **48,** 352 (1971).

Table 17-1

Overpotential (V) for gas evolution at 25°C

Electrode	10 A/m²		100 A/m²		1 000 A/m²		10 000 A/m²	
	H_2	O_2	H_2	O_2	H_2	O_2	H_2	O_2
Platinized Pt	0.015 4	0.398	0.030 0	0.521	0.040 5	0.638	0.048 3	0.766
Smooth Pt	0.024	0.721	0.068	0.85	0.288	1.28	0.676	1.49
Cu	0.479	0.422	0.584	0.580	0.801	0.660	1.254	0.793
Ag	0.475 1	0.580	0.761 8	0.729	0.874 9	0.984	1.089 0	1.131
Au	0.241	0.673	0.390	0.963	0.588	1.244	0.798	1.63
Graphite	0.599 5		0.778 8		0.977 4		1.220 0	
Sn	0.856 1		1.076 7		1.223 0		1.230 6	
Pb	0.52		1.090		1.179		1.262	
Zn	0.716		0.746		1.064		1.229	
Cd	0.981		1.134		1.216		1.254	
Hg	0.9		1.0		1.1		1.1	
Fe	0.403 6		0.557 1		0.818 4		1.291 5	
Ni	0.563	0.353	0.747	0.519	1.048	0.726	1.241	0.853

SOURCE: *International Critical Tables*, **6**, 339 (1929). This reference also gives overpotentials for Cl_2, Br_2, and I_2.

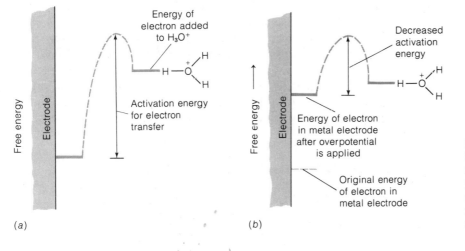

(a) (b)

Figure 17-5

Schematic energy profile for electron transfer from a metal to H_3O^+ (a) with no applied potential; (b) after a potential is applied to the electrode. The overpotential increases the energy of the electrons in the electrode.

The overpotential depends on the composition of the electrode, since the potential energy of an electron depends on what metal it is in. The overpotential also depends on how a given reactant interacts with the metal surface. Overpotential tends to decrease with increasing temperature.

17-3 ELECTROGRAVIMETRIC ANALYSIS

In *electrogravimetric analysis* the analyte is electrolytically deposited as a solid on an electrode. The increase in the mass of the electrode tells us how much analyte was present.

Electrogravimetric analysis is a combination of electrolysis and gravimetric analysis.

EXAMPLE: Electrogravimetric Deposition of Cobalt

A solution containing 0.402 49 g of $CoCl_2 \cdot xH_2O$ was exhaustively electrolyzed to deposit 0.099 37 g of metallic cobalt on a platinum cathode.

$$Co^{2+} + 2e^- \rightarrow Co(s)$$

Calculate the number of moles of water per mole of cobalt in the reagent.

Assuming that the reagent contains only $CoCl_2$ and H_2O, we can write

$$\text{Grams of } CoCl_2 = \left(\frac{\text{grams of Co}}{\text{atomic weight of Co}} \right)(\text{F.W. of } CoCl_2)$$

$$= 0.218\ 93 \text{ g}$$

$$\text{Grams of } H_2O = 0.402\ 49 - 0.218\ 93 = 0.183\ 56 \text{ g}$$

$$\frac{\text{Moles of } H_2O}{\text{Moles of Co}} = \frac{0.183\ 56/\text{F.W. of } H_2O}{0.099\ 37/\text{atomic weight of Co}} = 6.043$$

The reagent has a composition apparently close to $CoCl_2 \cdot 6H_2O$.

The basic requirements of an apparatus for electrogravimetric analysis are shown in Figure 17-6. The analyte is typically deposited on a carefully cleaned Pt gauze cathode, used because of its large surface area and chemical inertness. A simple power supply is adequate, but more elaborate apparatus, such as that in Figure 17-7, is available.

How do you know when an electrolysis is complete? One way is to observe the disappearance of color in a solution from which a colored ion (such as Co^{2+} or Cu^{2+}) is removed. Another way is to expose most, but not all, of the surface of the cathode to the solution during electrolysis. When you believe the reaction is complete, raise the level of the beaker or add some water so that some fresh surface of the cathode is exposed to the solution.

Tests for completion of the deposition:

1. disappearance of color
2. deposition on freshly exposed electrode surface
3. qualitative test for analyte in solution

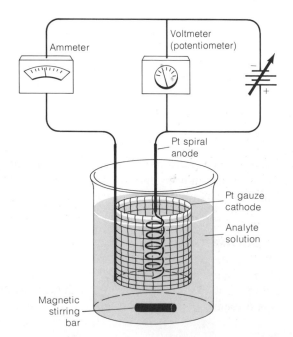

Figure 17-6

Simple apparatus for electrogravimetric analysis. The analyte is deposited on the large Pt gauze electrode. If the electrolytic deposition involves oxidation of the analyte, rather than reduction, the polarity of the power supply is simply reversed so that the deposition always occurs on the large electrode.

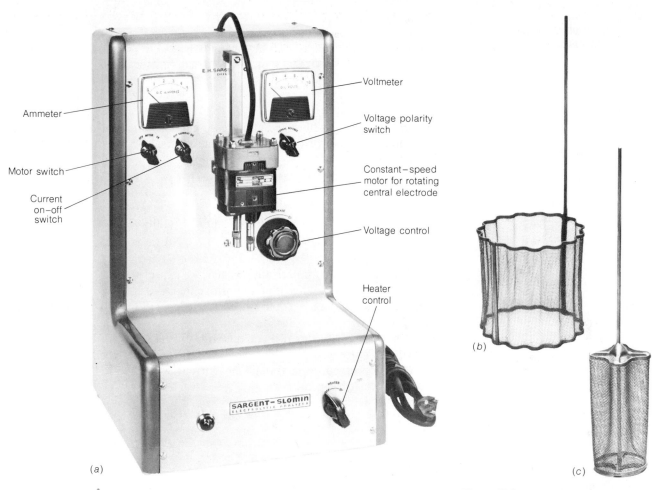

Figure 17-7
(a) Apparatus for electrogravimetric analysis. Spinning of the inner electrode serves to stir the solution. Heating is available to facilitate some electrolyses. (b) Outer electrode, usually the cathode. (c) Inner, rotating electrode. [Courtesy Sargent–Welch Co., Skokie, Ill.]

Ammeter

Motor switch

Current on–off switch

Voltmeter

Voltage polarity switch

Constant–speed motor for rotating central electrode

Voltage control

Heater control

(a)

(b)

(c)

After an additional period of electrolysis (15 minutes, say), see if the newly exposed electrode surface has a deposit. If it does, repeat the procedure. If not, the electrolysis is done. A third method is to remove a small sample of solution and perform a qualitative test for the analyte species.

Electrogravimetric analysis would be a simple matter if it merely involved electrolysis of a single analyte species from an otherwise inert solution. In practice, there may be other **electroactive species** that interfere by codeposition with the desired analyte. Even the solvent (water) is electroactive, since it decomposes to $H_2 + \frac{1}{2}O_2$ at a sufficiently high voltage. Although these gases are liberated from the solution, their presence at the electrode surface interferes with deposition of solids. Because of these complications, control of the electrode potential is an important feature of a successful gravimetric analysis.

Electroactive species are those that can be oxidized or reduced at an electrode.

Current–Voltage Behavior during Electrolysis

One of the most common applications of electrogravimetry is the analysis of copper. Suppose that a solution containing 0.20 M Cu^{2+} and 1.0 M H^+ is electrolyzed to deposit Cu(s) on a Pt cathode and to liberate O_2 at a Pt anode.

435

Cathode:	$Cu^{2+} + 2e^- \rightleftharpoons Cu(s)$	$E° =$	0.339
Anode:	$H_2O \rightleftharpoons \frac{1}{2}O_2(g) + 2H^+ + 2e^-$	$E° =$	-1.229
Net reaction:	$H_2O + Cu^{2+} \rightleftharpoons Cu(s) + \frac{1}{2}O_2(g) + 2H^+$	$E° =$	-0.890

$$(17\text{-}12)$$

Assuming that O_2 is liberated at a pressure of 0.20 atm, we naively calculate the voltage needed for the electrolysis as follows:

$$E = E° - \frac{0.059\ 16}{n} \log \frac{P_{O_2}^{1/2}[H^+]^2}{[Cu^{2+}]}$$

$$= -0.890 - \frac{0.059\ 16}{2} \log \frac{(0.20)^{1/2}(1.0)^2}{(0.20)} = -0.900\ \text{V} \qquad (17\text{-}13)$$

In the absence of any polarization effects, we expect that no reaction should occur if the applied voltage is more positive than -0.900 V. When the voltage is more negative than -0.900 V, we expect deposition of Cu and liberation of O_2 to occur.

The actual behavior of the electrolysis (using a pair of Pt electrodes) is shown in Figure 17-8. At low voltage, a small current called the **residual current** is observed, even though no current was expected. At -0.900 V, nothing special happens. Near -2 V the reaction begins in earnest.

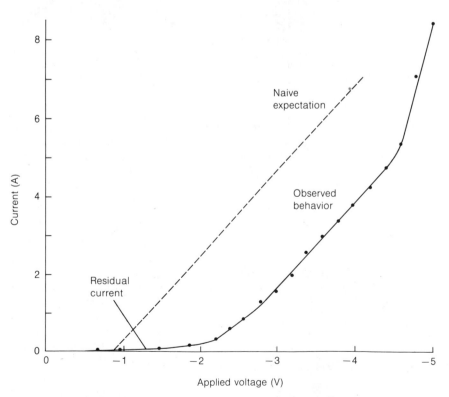

Figure 17-8
Current–voltage relationship for electrolysis of 0.2 M $CuSO_4$ in 1 M $HClO_4$ under N_2, using the apparatus shown in Figure 17-7.

Why is it necessary to apply a voltage considerably more negative than -0.900 V to drive the reaction at an appreciable rate? The principal reason is the overpotential for oxygen formation at a Pt surface. In Table 17-1, we see that an extra potential of around one volt is needed to overcome the activation energy for oxidation of water at a smooth platinum anode, giving $O_2(g)$ as the product.

It requires about 1 V of extra potential to overcome the barrier to O_2 formation at the anode.

Why is a residual current observed at a small voltage? There must be some oxidation occurring at the anode and some equal amount of reduction occurring at the cathode to support this residual current. Probably, these reactions involve mainly impurities. For example, dissolved oxygen can be reduced to H_2O_2, and Fe^{3+} could be reduced to Fe^{2+} at the cathode. Electrode surface oxide impurities could be reduced as well. At the anode, some small amount of water can be oxidized, or some oxidizable impurity in the solution might react.

Another reason why the calculation in Equation 17-13 is naive is that when copper is deposited on a platinum electrode, the initial activity of $Cu(s)$ is infinitely small. The calculation in Equation 17-13 assumes that copper is being deposited on a copper surface, for which the activity of copper is unity. Suppose that at some time early in the deposition of $Cu(s)$ on Pt, the activity of $Cu(s)$ on the Pt surface is 10^{-6}. Then a more realistic estimate of the potential for Reaction 17-12 is the following:

$$E = E° - \frac{0.059\ 16}{2} \log \frac{P_{O_2}^{1/2}[H^+]^2[Cu(s)]}{[Cu^{2+}]}$$

Note the value of $[Cu(s)]$ in this equation.

$$= -0.890 - \frac{0.059\ 16}{2} \log \frac{(0.20)^{1/2}(1.0)^2(10^{-6})}{(0.20)} = -0.722 \text{ V} \quad (17\text{-}14)$$

That is, when very little $Cu(s)$ has been deposited, the activity of $Cu(s)$ is much less than unity. It is not necessary to apply as much voltage to electrolyze Cu^{2+} in these circumstances.

The curve in Figure 17-8 can therefore be explained in the following way:

1. Initially, a small residual current is present, owing both to oxidation and reduction of impurities and to a small amount of the intended electrolysis.

   Residual current

2. At a sufficiently negative voltage, the desired electrolysis is the main reaction. The voltage is shifted from E_{eq} by the overpotential for oxygen formation.

3. At more negative voltages, the current is essentially linear with respect to voltage, according to Ohm's law. The ratio between current and voltage tells us the resistance of the cell.

   Ohmic potential

4. At -4.6 V, the curve deviates from Ohm's law. Reduction of water to H_2 begins to occur at the cathode, and the current shoots upward.

   Electrolysis of solvent

EXAMPLE: Current–Voltage Relation for Electrolysis

The resistance of the cell in Figure 17-8 is $0.44\ \Omega$. Estimate the voltage needed to maintain a current of 2.0 A. Assume that the smooth Pt anode is supporting a current density of 1 000 A/m^2 and that there is no concentration polarization.

Our estimate includes contributions from E_{eq}, the ohmic potential, and the overpotential.

$$E(\text{applied}) = \underbrace{E(\text{cathode}) - E(\text{anode})}_{\substack{E_{eq} \\ = 0.900 \text{ V} \\ \text{(from Equation 17-13)}}} - IR - \underbrace{\text{overpotential}}_{\substack{1.28 \text{ V} \\ \text{(from Table 17-1)}}}$$

$$= -0.900 - (2.0 \text{ A})(0.44 \ \Omega) - 1.28 \text{ V}$$

$$= -0.900 - 0.88 - 1.28 = -3.04 \text{ V}$$

Question: If the cell in Figure 17-8 has a resistance of 0.44 Ω, what will be the slope (A/V) of the linear portion of the curve between about -3.0 and -4.5 V? Measure this slope in the figure.

Constant-Voltage Electrolysis

We now know that the voltage calculated for a simple reversible cell must be modified to include the contributions of the ohmic potential and overpotential.

$$E(\text{applied}) = \underbrace{E(\text{cathode}) - E(\text{anode})}_{\substack{E_{eq} \text{ calculated for} \\ \text{reversible cell with} \\ \text{negligible current flow}}} - IR - \text{overpotential} \quad (17\text{-}15)$$

Question: Why do the ohmic potential and overpotential decrease as the current decreases?

Suppose that the applied voltage for an electrolysis is held at a constant value. For example, we might electrolyze 0.10 M Cu^{2+} in 1.0 M HNO_3 at -2.0 V. As Cu(s) is deposited, the concentration of Cu^{2+} in solution decreases. Eventually, there is too little Cu^{2+} in solution, and Cu^{2+} cannot be transported to the cathode rapidly enough to maintain the initial current of the cell. Since the current decreases, the ohmic potential and the overpotential in Equation 17-15 decrease. The value of $E(\text{anode})$ is fairly constant owing to the high concentration of solvent being oxidized at the anode. (Recall that the anode oxidation reaction is $H_2O \rightarrow \frac{1}{2}O_2 + 2H^+ + 2e^-$.)

$E(\text{cathode})$ becomes more negative as Cu^{2+} is consumed because of concentration polarization.

Now think about Equation 17-15: If $E(\text{applied})$ and $E(\text{anode})$ are constant and if IR and overpotential decrease in magnitude, $E(\text{cathode})$ must become more negative to maintain the algebraic equality. This is shown schematically in Figure 17-9. The value of $E(\text{cathode})$ becomes more negative with time because of the concentration polarization that occurs as Cu^{2+} is consumed.

The value of $E(\text{cathode})$ continues to drop until it is negative enough to reduce H^+ to H_2:

$$H^+ + e^- \rightarrow \tfrac{1}{2}H_2(g) \quad (17\text{-}16)$$

The overpotential of Reaction 17-16 determines where the curve in Figure 17-9 will level off. At a potential near -0.4 V, steady reduction of H^+ ensues. As $E(\text{cathode})$ falls from its initial value of $+0.3$ V to its steady value near

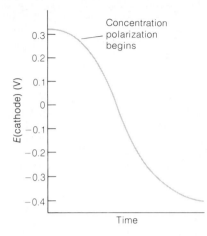

Figure 17-9
Schematic drawing showing the change of E(cathode) with time for an electrolysis conducted at a constant applied potential.

-0.4 V, other ions in the solution might react. For example, Co^{2+}, Sn^{2+} and Ni^{2+}, if present, would be reduced. In general, then, **constant-voltage electrolysis** is not selective. Any solute more easily reduced than H^+ will be electrolyzed.

Formation of H_2 sometimes weakens the cathode deposit, causing it to crumble and fall off the electrode. To avoid this and to prevent the cathode potential from becoming so negative that unintended ions are reduced, a **cathodic depolarizer** such as NO_3^-, is added to the solution. The cathodic depolarizer is more easily reduced than H^+ and yields a harmless product.

$$NO_3^- + 10H^+ + 8e^- \rightarrow NH_4^+ + 3H_2O \qquad (17\text{-}17)$$

From inspection of reduction potentials, we would expect that NO_3^- would be reduced *before* Cu^{2+}, since $E°$ for Reaction 17-17 is 0.880 V while $E°$ for the reduction of Cu^{2+} to $Cu(s)$ is 0.339 V. Apparently, the overpotential for reduction of NO_3^- is high enough that reduction of Cu^{2+} happens first.

Electrolysis can be made somewhat selective through control of pH. In strongly acidic solution, Cu^{2+} can be reduced without concomitant reduction of such ions as Zn^{2+}, Ni^{2+}, or Cd^{2+}, because H^+ is more readily reduced than Zn^{2+}, Ni^{2+}, or Cd^{2+}. If the concentration of H^+ is not great enough, these metal ions will be reduced ahead of H^+. The redox behavior of various analytes is also affected in a predictable way by chelating agents, which tend to stabilize the cations and thereby make them harder to reduce. Analytes that are not separable in one electrolyte may be separable in a different electrolyte with different coordinating abilities.

A *cathodic depolarizer* is reduced in preference to solvent. It prevents E(cathode) from becoming so negative that water and impurities are reduced. For *oxidation* reactions, common *anodic depolarizers* include hydrazine, N_2H_4, and hydroxylamine, NH_2OH.

The selectivity of electrolysis can be affected by pH and by chelate concentrations.

Constant-Current Electrolysis

Constant-current electrolysis is the least selective mode of electrolysis. Concentration polarization reduces the current if the applied voltage is constant.

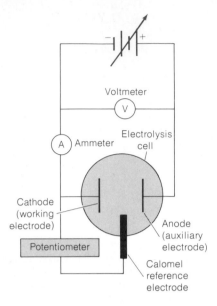

Figure 17-10
Circuit used for electrolysis using constant cathode potential. Commonly used schematic symbols for electrodes are

———○ working electrode
———┤ auxiliary electrode
———→ reference electrode

To maintain constant current, the magnitude of the applied voltage must be progressively increased. The cathode potential rapidly becomes more negative until it levels off at a value fixed by the reduction of H^+ or a cathodic depolarizer. Reduction of the desired analyte continues as well.

Controlled-Potential Electrolysis

We have seen that electrolysis using a constant applied voltage or a constant current is rather unselective. In both techniques the cathode potential becomes more and more negative, leading to reduction of species other than the intended analyte. A *three-electrode cell* (as shown in Figure 17-10) can be used to maintain a *constant cathode potential* and thereby greatly increase the selectivity of the electrolysis.

The electrode at which the reaction of interest occurs is called the **working electrode.** The calomel electrode serves as a **reference electrode** against which the potential of the working electrode can be measured. The third electrode (the current-supporting partner of the working electrode) is called the **auxiliary electrode.** A significant current flows at the working and auxiliary electrodes. Negligible current flows at the reference electrode.

In a **controlled-potential electrolysis,** the voltage applied between the working and auxiliary electrodes is varied in such a way that the voltage between the working and reference electrodes is *constant.* The cathode potential can be maintained at a constant value by using an electronic device called a **potentiostat.**

To see why a constant cathode potential permits selective deposition of analyte, consider a solution containing 0.1 M Cu^{2+} and 0.1 M Sn^{2+}. From the standard potentials below, we expect Cu^{2+} to be reduced more easily than Sn^{2+}:

$$Cu^{2+} + 2e^- \rightleftharpoons Cu(s) \qquad E° = 0.339 \text{ V}$$

Question: How would the curve in Figure 17-9 be different if the electrolysis were conducted at a constant current?

Working electrode: where the analytical reaction occurs

Auxiliary electrode: the other electrode needed for current flow

Reference electrode: the third electrode, used to measure the potential of the working electrode.

In *controlled-potential* electrolysis, there is a constant voltage between the working and reference electrodes. The voltage between the working and auxiliary electrodes does change. In *constant-voltage* electrolysis there is a constant voltage between the working and auxiliary electrodes.

$$Sn^{2+} + 2e^- \rightleftharpoons Sn(s) \qquad E° = -0.141 \text{ V}$$

The cathode potential at which Cu^{2+} ought to be reduced is calculated as

$$E(\text{cathode}) = 0.339 - \frac{0.059\ 16}{2} \log \frac{1}{0.1} = 0.31 \text{ V} \qquad (17\text{-}18)$$

If 99.99% of the Cu^{2+} were deposited, the concentration of Cu^{2+} remaining in solution would be 10^{-5} M, and the cathode potential required to continue reduction would be

$$E(\text{cathode}) = 0.339 - \frac{0.059\ 16}{2} \log \frac{1}{10^{-5}} = 0.19 \text{ V} \qquad (17\text{-}19)$$

At a cathode potential of 0.19 V, then, rather complete deposition of copper is expected. Would Sn^{2+} be reduced at this potential? To deposit $Sn(s)$ from a solution containing 0.1 M Sn^{2+}, a cathode potential of -0.17 V is required:

$$E(\text{cathode, for reduction of } Sn^{2+}) = -0.141 - \frac{0.059\ 16}{2} \log \frac{1}{[Sn^{2+}]}$$

$$= -0.141 - \frac{0.059\ 16}{2} \log \frac{1}{0.1} = -0.17 \text{ V} \qquad (17\text{-}20)$$

We do not expect significant reduction of Sn^{2+} at a cathode potential more positive than -0.17 V.

If the *cathode* potential is kept constant at a value near 0.19 V, we predict that 99.99% of the Cu^{2+} will react without deposition of Sn^{2+}. On the other hand, if the cell were run with a constant voltage between the working and auxiliary electrodes, the cathode potential would behave as in Figure 17-9 and Sn^{2+} would be reduced. The price of achieving selective reduction with a constant cathode potential is that the current decreases and the reaction is slower than for a constant-voltage electrolysis.

The cathode potentials calculated in Equations 17-18 through 17-20 are implicitly stated with respect to the standard hydrogen electrode (since $E°$ is taken from a table of standard potentials). To calculate the potential that would be measured with respect to a saturated calomel electrode in Figure 17-10, we use the relation

$$E(\text{versus S.C.E.}) = E(\text{versus S.H.E.}) - E(\text{calomel electrode})$$

$$= E(\text{versus S.H.E.}) - 0.241 \qquad (17\text{-}21)$$

To obtain a cathode potential of 0.19 V (versus S.H.E.), we must maintain a cathode potential of $0.19 - 0.241 = -0.05$ V with respect to the saturated calomel reference electrode in the three-electrode cell of Figure 17-10.

Reduction reactions occur at working electrode potentials (measured with respect to the reference electrode) that are more *negative* than that required to start the reaction. *Oxidations* occur when the working electrode is more *positive* than necessary to start the reaction (Figure 17-11).

Controlled potential means that a constant potential difference is maintained between the working and *reference* electrodes. *Constant voltage* means that a constant potential difference is maintained between the working and *auxiliary* electrodes. Controlled potential affords high selectivity, but the procedure is slower than constant-voltage electrolysis.

The diagram on page 352 should remind you why Equation 17-21 is true.

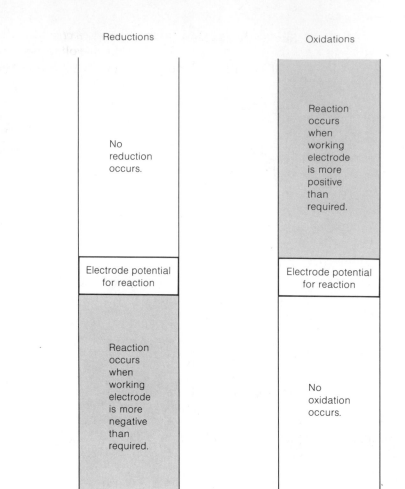

Figure 17-11

Reduction occurs at working electrode potentials more *negative* than that required to start the reaction. *Oxidation* occurs at working electrode potentials more *positive* than that required to start the reaction.

17-4 COULOMETRIC ANALYSIS

Coulometric methods are based on measuring the number of electrons that participate in a chemical reaction.

Coulometry is based on counting the number of electrons used in a chemical reaction. For example, cyclohexene may be titrated with Br_2 generated by electrolytic oxidation of Br^-:

$$2Br^- \rightarrow Br_2 + 2e^- \tag{17-22}$$

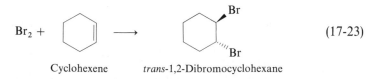

$$\text{Br}_2 + \quad \longrightarrow \tag{17-23}$$

Cyclohexene   *trans*-1,2-Dibromocyclohexane

In this sequence of reactions, just enough Br_2 is generated to react with all of the cyclohexene. The moles of electrons liberated in Reaction 17-22 will be equal to twice the moles of Br_2 and therefore twice the moles of cyclohexene.

The reaction is conveniently carried out using the apparatus in Figure 17-12. Br_2 is generated by the Pt anode at the left. As the Br_2 is formed, it reacts with cyclohexene. When all of the cyclohexene has been consumed, the concentration of Br_2 in the solution suddenly rises, signaling the end of the reaction.

The rise in Br_2 concentration is detected *amperometrically*, using the circuit shown on the right in Figure 17-12. A small voltage (~ 0.25 V) is applied between the two electrodes on the right. This voltage is not great enough to electrolyze any of the solutes, so only a small residual current of < 1 μA flows through the sensitive ammeter (a device that measures electric current). When $[Br_2]$ suddenly increases, a current can flow by virtue of the following reactions:

Amperometry is based on the measurement of electric current. Amperometric methods are discussed further in Chapter 18.

$$\text{Detector anode:} \qquad 2Br^- \rightarrow Br_2 + 2e^- \qquad (17\text{-}24)$$

$$\text{Detector cathode:} \qquad Br_2 + 2e^- \rightarrow 2Br^- \qquad (17\text{-}25)$$

The sudden increase of current is taken as the end point of the cyclohexene titration.

In practice, enough Br_2 is first generated in the absence of cyclohexene to give a detector current of 20.0 μA. When cyclohexene is added, the current decreases to a very small value because bromine is consumed. Bromine is then generated by the coulometric circuit, and the end point is taken when the detector again indicates 20.0 μA. This end point is easier to reproduce than "a sudden increase of current." Furthermore, since the reaction is begun with enough Br_2 to carry 20.0 μA, any impurities that can react with Br_2 are eliminated.

The current for the bromine-generating electrodes can be controlled by a hand-operated switch. As the detector current approaches 20.0 μA, the operator closes the switch for shorter and shorter intervals. This is closely analogous to adding titrant dropwise from a buret near the end of a titration. *The switch in the coulometer circuit serves as a "stopcock" for addition of Br_2 to the reaction.*

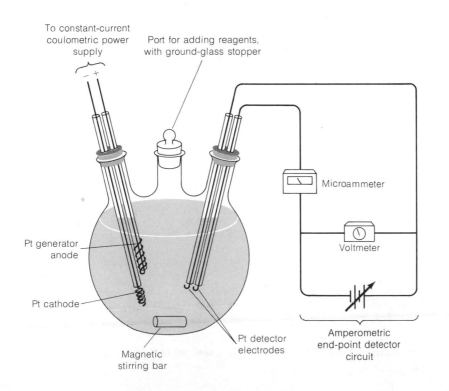

To constant-current coulometric power supply

Port for adding reagents, with ground-glass stopper

Microammeter

Voltmeter

Pt generator anode

Pt cathode

Magnetic stirring bar

Pt detector electrodes

Amperometric end-point detector circuit

Figure 17-12
Apparatus for coulometric titration of cyclohexene with Br_2. The solution contains cyclohexene, 0.15 M KBr, and 3 mM mercuric acetate in a mixed solvent of acetic acid, methanol, and water. The mercuric acetate catalyzes the addition of Br_2 to the olefin. [Adapted from D. H. Evans, *J. Chem. Ed.*, **45**, 88 (1968).]

EXAMPLE: Coulometry Calculation

A 2.000-mL volume of solution containing 0.611 3 mg of cyclohexene/mL is to be titrated in the apparatus shown in Figure 17-12. If the coulometer is operated at a constant current of 4.825 mA, how much time will be required for complete titration?

The quantity of cyclohexene is

$$\frac{(2.000 \text{ mL})(0.611 3 \text{ mg/mL})}{(82.146 \text{ mg/mmol})} = 0.014 88 \text{ mmol}$$

In Reactions 17-22 and 17-23, each mole of cyclohexene requires one mole of Br_2, which in turn requires two moles of electrons to flow through the circuit. For 0.014 88 mmol of cyclohexene to react, 0.029 76 mmol of electrons must flow. From Equation 17-6, we can write

$$\text{Moles of } e^- = \frac{I \cdot t}{F} \quad \Rightarrow \quad t = \frac{(\text{moles of } e^-)F}{I}$$

$$t = \frac{(0.029 76 \times 10^{-3} \text{ mol})(96 485 \text{ C/mol})}{(4.825 \times 10^{-3} \text{ C/s})} = 595.1 \text{ s}$$

It will require just under 10 minutes to complete the reaction if the current is constant.

Advantages of coulometry:

1. accuracy
2. sensitivity
3. generation of unstable reagents *in situ*

This example illustrates the accuracy and sensitivity afforded by coulometric titrations. A common commercial coulometric power supply delivers current with an accuracy of $\sim 0.1\%$ (Figure 17-13). With extreme care, the value of the Faraday constant has been measured with an accuracy of several parts per million by a coulometric procedure (see Box 17-2). With careful attention to electronic circuitry, coulometric titration of Cl^- at a concentration of 10^{-5} M (0.3 ppm) is possible.[†] Another advantage of coulometric titrations is that such unstable reagents as Ag(II), Cu(I), Mn(III), and Ti(III) are generated and used in the same vessel. There is no need to handle reagents in the air or transfer them between vessels.

Some Details of Coulometry

Type of coulometry

The two common coulometric methods employ either a *constant current* or a *controlled potential*. Constant-current methods, as in the preceding Br_2/cyclohexene example, are called **coulometric titrations.** If we know the current, it is only necessary to measure the time needed for complete reaction, in order to count the coulombs (Equation 17-5):

$$q = I \cdot t \qquad (17-5)$$

Controlled-potential coulometry is inherently more selective than constant-current coulometry. This is for the same reason that controlled-potential electrogravimetric analysis is more selective than constant-voltage electrolysis.

Question: What was that reason?

[†] M. J. Zetlmeisl and D. F. Lawrence, *Anal. Chem.,* **49,** 1557 (1977).

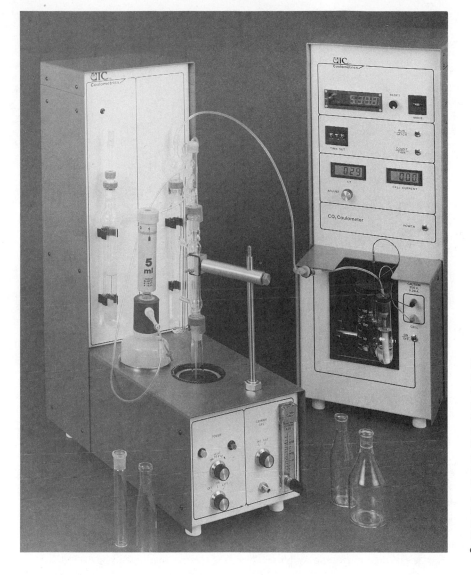

Figure 17-13
Commercial coulometer (right) for CO_2 detection. Unit at left is used to acidify inorganic carbonates and pass the released CO_2 into the coulometer. The coulometer can also be attached to a combustion train for analysis of carbon in organic compounds (Figure 8-1). A modification of the coulometer is used to measure sulfites or hydrogen sulfide [J. Greyson and S. Zeller, *Am. Lab.*, July 1987, p. 44]. Sulfite measurement is important because sulfite preservatives in foods cause allergic reactions in 5–10% of people with asthma. [Photo courtesy UIC, Inc., Joliet, Illinois.]

In controlled-potential coulometry, the initial current is high, but decreases exponentially as the analyte concentration decreases. Since the current is not constant, the coulombs must be measured by integrating the current over the time of the reaction:

$$q = \int_0^t I \, dt \qquad (17\text{-}26)$$

In coulometric titrations, a constant current is delivered whenever the power supply is connected to the electrodes. The power supply is built to automatically measure the time of operation. Multiplying current by the length of time gives the number of coulombs used in the titration. In controlled-potential coulometry, the current decreases with time. The power supply contains a circuit that automatically integrates Equation 17-26 to calculate the number of coulombs delivered.

Box 17-2 MEASURING THE FARADAY CONSTANT

The most accurate measurement of the Faraday constant comes from a very careful coulometric experiment conducted at the U.S. National Institute of Standards and Technology.[†] A schematic diagram of the experiment is shown in part a of the figure on the facing page. Part b is a photograph of the actual equipment. The method consists of the electrolytic dissolution of a highly purified metallic Ag anode in an aqueous solution of 20% (wt/wt) $HClO_4$ containing 0.5% (wt/wt) $AgClO_4$.

Ag anode: $\qquad Ag(s) \rightarrow Ag^+ + e^-$

Pt cathode: $\quad Ag^+ + e^- \rightarrow Ag(s)$

This electrolyte was chosen because metallic Ag is very stable in this solution and does not spontaneously dissolve.

In a typical experiment, electrolysis was conducted using a voltage of 1.018 209 8 V and a current of 0.203 639 0 A for 18 000.075 s. The loss of mass at the anode amounted to 4.097 900 g. The number of coulombs passed through the cell is therefore

$$q = I \cdot t = (0.023\ 639\ 0\ A)(18\ 000.075\ s) = 3\ 665.517\ 3\ C$$

The silver lost from the anode amounts to

$$\text{Moles of Ag} = \frac{4.097\ 900\ g}{107.868\ g/mol} = 3.798\ 995 \times 10^{-2}\ mol$$

The coulombs and moles can be combined to find the Faraday constant:

$$F = \frac{\text{coulombs}}{\text{moles}} = \frac{3\ 665.517\ 3}{3.798\ 995 \times 10^{-2}} = 96\ 486.5\ C/mol$$

Not all the silver lost from the anode was oxidized. A fraction (ranging from 0.01% to 15%) of the Ag simply fell off as contiguous parts of the electrode were electrolyzed. This sediment was collected and weighed at the end of the experiment, so that the true mass of oxidized silver could be calculated. The intermediate beakers and siphons in the apparatus shown in the figure are used to physically separate the anode and cathode compartments. This prevents Ag deposited on the Pt cathode from falling off and being weighed with the anode sediment.

Very great care was taken to purify the Ag anode. Needless to say, dust had to be excluded from all phases of the electrolysis and from the purification of the metallic silver. The electrode was prepared from "pure" electrolytic silver, obtained from the U.S. Mint, dissolved in HNO_3, and crystallized as $AgNO_3$. The silver was then deposited in metallic form on the purest available Ag electrode. The fresh deposit was scraped off the electrode, and the scrapings were washed for two weeks with HF (to dissolve any silica from the glass apparatus) and for two weeks with very pure water. The dry scrapings were fused in silica tubes from which impurities had been leached with hot, concentrated HNO_3. The fused Ag was again etched with HF until a constant mass was achieved. Under vacuum, the metal was melted to remove any oxide, then etched with 10% aqueous NH_3, and washed with very pure water. Spectrochemical analysis indicated less than 1 ppm of impurities. The Faraday constant was corrected for this tiny quantity of impurities. Also, the quotient $^{107}Ag/^{109}Ag$ was monitored throughout the purification and electrolysis procedures. No variation was found.

[†] D. N. Craig, J. I. Hoffman, C. A. Law, and W. J. Hamer, *J. Res. National Bureau of Standards,* **64A,** 381 (1960). A fascinating account of another determination of the Faraday constant using coulometric titration of 4-aminopyridine can be found in H. Diehl, *Anal. Chem.,* **51,** 318A (1979).

447

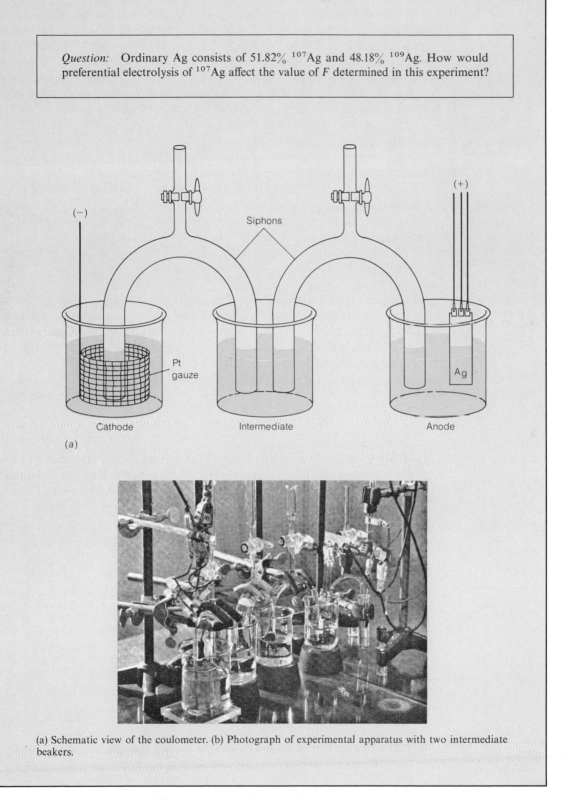

Question: Ordinary Ag consists of 51.82% ^{107}Ag and 48.18% ^{109}Ag. How would preferential electrolysis of ^{107}Ag affect the value of F determined in this experiment?

(a) Schematic view of the coulometer. (b) Photograph of experimental apparatus with two intermediate beakers.

End-point detection

Methods of end-point detection used for other types of titrations apply to coulometric titrations as well. These include the use of indicators and spectrophotometric methods. Potentiometric methods are useful for acid–base reactions (glass electrode), redox reactions (Pt electrode), EDTA titrations (mercury electrode, as in Exercise 15-B), and reactions for which ion-selective electrodes are available. For halide precipitations, a silver electrode might be used. The amperometric end-point detector in Figure 17-12 is useful for a wide variety of coulometric titrations.

In controlled-potential coulometry, the equivalence point is never reached because the current decays exponentially. However, you can approach the equivalence point by letting the current decay to an arbitrarily set value. For example, the current will ideally be 1% of its initial value when 99% of the analyte has been consumed, and it will be 0.1% of its initial value when 99.9% of the analyte has been consumed. (These figures refer to the current *above* the residual current.)

Mediators

A necessary condition for coulometric analysis is that the analytical reaction(s) proceed with 100% electrochemical efficiency. Electrons cannot be siphoned off into side reactions, or the measurement of total coulombs becomes meaningless. While a few reactions meet this requirement directly, most analyses incorporate a **mediator** to improve the efficiency.

For example, Fe^{2+} might be analyzed coulometrically by titration to Fe^{3+}:

$$Fe^{2+} \rightarrow Fe^{3+} + e^- \qquad (17\text{-}27)$$

Initially, this reaction accounts for all of the current. When the concentration of Fe^{2+} decreases sufficiently, concentration polarization develops. The anode potential might increase to the point where water could be oxidized to O_2 and current would be carried mainly by the side reaction

$$H_2O \rightarrow \tfrac{1}{2}O_2 + 2H^+ + 2e^- \qquad (17\text{-}28)$$

Question: Why is Reaction 17-28 undesirable? Will it lead to an estimate of Fe^{2+} concentration that is higher or lower than the true value?

To avoid this side reaction, excess Ce^{3+} can be added to the solution. As the electrode becomes polarized, the reaction

$$Ce^{3+} \rightarrow Ce^{4+} + e^- \qquad (17\text{-}29)$$

begins at a lower potential than that needed for oxidation of water. The Ce^{4+} diffuses into the solution and oxidizes any Fe^{2+} it meets:

$$Ce^{4+} + Fe^{2+} \rightarrow Ce^{3+} + Fe^{3+} \qquad (17\text{-}30)$$

A *mediator* transports electrons quantitatively between the analyte and the working electrode. The mediator undergoes no net reaction itself.

As long as Reaction 17-30 is rapid, the net reaction is the oxidation of Fe^{2+} to Fe^{3+}. The Ce^{3+} ion, which undergoes no net change, acts as a *mediator*.

An important application of mediators is for determination of glucose in the blood of diabetic patients. The management of blood sugar (glucose) levels with insulin is a critical part of the treatment of diabetes. An electrode used to detect glucose contains the *enzyme* glucose oxidase covalently attached to a graphite surface.[†] The enzyme catalyzes oxidation of glucose inside living cells:

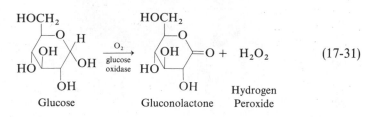

$$\text{Glucose} \qquad \qquad \text{Gluconolactone} \qquad \begin{array}{c}\text{Hydrogen}\\\text{Peroxide}\end{array} \tag{17-31}$$

Hydrogen peroxide produced by the reaction is measured with an amperometric electrode system. The moles of hydrogen peroxide produced by the enzyme reaction equal the moles of glucose present in the blood sample.

The O_2 in Reaction 17-31 oxidizes the enzyme, which in turn oxidizes the glucose. Unfortunately, the reaction requires more O_2 than is generally present in blood. Therefore the analytical procedure makes use of the mediator 1,1'-dimethylferrocene to oxidize the enzyme.

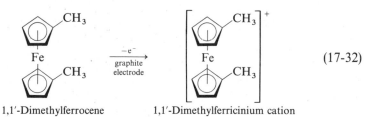

1,1'-Dimethylferrocene   1,1'-Dimethylferricinium cation   (17-32)

The enzyme itself cannot be oxidized at the graphite electrode because the active redox site is buried too deeply within the enzyme. The ferrocene is oxidized to the ferricinium ion at the graphite surface. The ferricinium ion rapidly diffuses and oxidizes the enzyme, which can then oxidize glucose by Reaction 17-31. The ferricinium ion is reduced back to ferrocene, which can be oxidized again at the electrode.

Ferrocene contains flat five-membered aromatic carbon rings, similar to benzene. Each ring formally carries one negative charge, so the oxidation state of iron is $+2$ (ferrous). The iron atom sits in the middle of the face of each ring. Because of its shape, this type of molecule is called a *sandwich complex.*

Separation of anode and cathode reactions

In Figure 17-12, the reactive species (Br_2) is generated at the anode (the working electrode). The cathode products (H_2 from solvent and Hg from the catalyst) do not interfere with the reaction of Br_2 and cyclohexene. Therefore, the cathode can be in the same compartment with the analyte. In some cases, the H_2 or Hg might react with the analyte. Frequently, then, it is desirable to separate the electrolysis product of the auxiliary electrode from the bulk solution. This can be done with the cell in Figure 17-14. An oxidizing agent, for instance, might be generated at the left-hand electrode operating as an anode. Gaseous H_2 produced at the cathode bubbles innocuously out of the cathode chamber without mixing with the bulk solution.

[†] A. E. G. Cass, G. Davis, G. D. Francis, H. A. O. Hill, W. J. Aston, I. J. Higgins, E. V. Plotkin, I. D. L. Scott, and A. P. F. Turner, *Anal. Chem.,* **56,** 667 (1984). For a general review of electrochemical biosensors, see J. E. Frew and H. A. O. Hill, *Anal. Chem., **59,** 933A (1987).

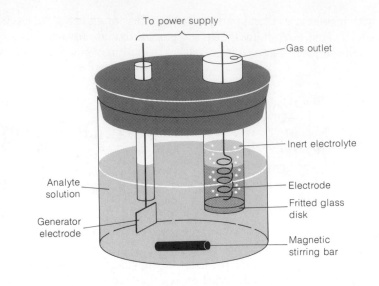

Figure 17-14
Cell showing how one electrode can be isolated from the analyte. Electrical contact
is made through the porous fritted glass
disk.

Summary

In electrolysis, a chemical reaction is forced to occur by
the flow of electricity through a cell. The moles of electrons
flowing through the cell are simply It/F, where I is current,
t is time, and F is the Faraday constant. The voltage that
must be applied to an electrolysis cell is greater than the
equilibrium voltage predicted by the Nernst equation because of three factors:

1. Ohmic potential ($=IR$) is that voltage needed to
   overcome the resistance of the cell itself.
2. Concentration polarization describes the condition
   in which the concentration of electroactive species
   near an electrode is not the same as the concentration
   in bulk solution. It always opposes the desired reaction and requires that a greater voltage be applied.
3. Overpotential describes that voltage required to
   overcome the activation energy of an electrode reaction. A greater overpotential is required to drive
   a reaction at a faster rate.

In electrogravimetric analysis, the analyte is electrolytically deposited on an electrode, whose increase in mass
is then measured. With a constant voltage between the
working and auxiliary electrodes, electrolysis is not very
selective. Greater selectivity results from maintaining a
constant voltage between the working and reference electrodes. The most rapid and least selective electrolysis
results from maintaining a constant current between the
working and auxiliary electrodes.

Coulometry comprises a series of analytical techniques
in which the quantity of electricity needed to carry out
a chemical reaction is used to measured the quantity of
analyte present. These techniques are particularly well
suited to automation. Coulometric titrations are performed with a constant current. Measuring the time
needed for complete reaction thus directly measures the
number of electrons consumed. The end point can be
found by any conventional means, but amperometry is
especially appropriate in many cases. Reactive mediators
are often required to transfer electrons between the electrode and the analyte. Controlled-potential coulometry
is inherently more selective than constant-current coulometry, but slower. Measuring the electrons consumed
in the reaction is accomplished by electronic integration
of the current-versus-time curve.

Terms to Understand

ampere	coulomb	mediator
amperometry	coulometric titration	ohmic potential
auxiliary electrode	coulometry	overpotential
concentration polarization	depolarizer	potentiostat
constant-current electrolysis	electroactive species	reference electrode
constant-voltage electrolysis	electrogravimetric analysis	residual current
controlled-potential electrolysis	electrolysis	working electrode

Exercises

17-A. Suppose that the cell below has a constant voltage of 1.02 V and is used to operate a light bulb with a resistance of 2.8 Ω

$$Zn(s)|Zn^{2+}(aq)\|Cu^{2+}(aq)|Cu(s)$$

How many hours are required for 5.0 g of Zn to be consumed?

17-B. A dilute Na_2SO_4 solution is to be electrolyzed with a pair of smooth Pt electrodes at a current density of 100 A/m² and a current of 0.100 A. The products of the electrolysis are $H_2(g)$ and $O_2(g)$. Calculate the voltage necessary to cause electrolysis if the cell resistance is 2.00 Ω and there is no concentration polarization. Assume that H_2 and O_2 are both produced at 1.00 atm. What would your answer be if the Pt electrodes were replaced by Au electrodes?

17-C. (a) What percent of 0.10 M Cu^{2+} could be reduced electrolytically before 0.010 M SbO^+ in the same solution begins to be reduced at pH 0.00? Consider the reaction

$$SbO^+ + 2H^+ + 3e^- \rightleftharpoons Sb(s) + H_2O$$

$$E° = 0.208\ V$$

(b) What will be the cathode potential (measured with respect to a silver–silver chloride electrode saturated with KCl) when Sb(s) deposition commences?

17-D. Calculate the cathode potential (versus S.C.E.) needed to reduce the total concentration of cobalt(II) to 1.0 μM in each solution below. In each case, Co(s) is the product of the reaction.
(a) A solution containing 0.10 M $HClO_4$.
(b) A solution containing 0.10 M $C_2O_4^{2-}$. Use the reaction

$$Co(C_2O_4)_2^{2-} + 2e^- \rightleftharpoons Co(s) + 2C_2O_4^{2-}$$

$$E° = -0.474\ V$$

This question is asking you to find the potential at which $[Co(C_2O_4)_2^{2-}]$ will be reduced to 1.0 μM.
(c) A solution containing 0.10 M EDTA at pH 7.00.

17-E. Ions that react with Ag^+ can be determined electrogravimetrically by deposition on a silver anode:

$$Ag(s) + X^- \rightarrow AgX(s) + e^-$$

(a) What will be the final mass of a silver anode used to electrolyze 75.00 mL of 0.023 80 M KSCN if the initial mass of the anode is 12.463 8 g?

(b) At what anode potential (versus S.C.E. cathode) will 0.10 M Br^- be deposited as AgBr(s)?

(c) Is it theoretically possible to separate 99.99% of 0.10 M KI from 0.10 M KBr by controlled anode-potential electrolysis?

17-F. The galvanic cell below can be used to measure the concentration of O_2 in gases.[†] Oxygen is quantitatively reduced as it passes through the porous Ag bubbler, and Cd is oxidized to Cd^{2+} to complete the cell. Suppose that gas at 293 K and 1.00 atm is bubbled through the cell at a constant rate of 30.0 mL/min. Calculate the current that would be measured if the gas contains 1.00 ppt or 1.00 ppm (vol/vol) O_2.

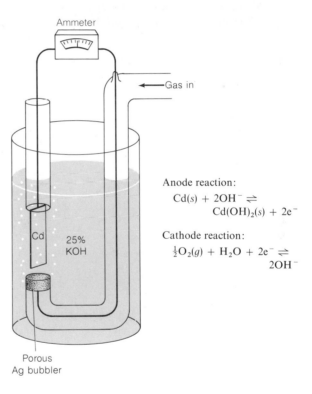

Anode reaction:
$$Cd(s) + 2OH^- \rightleftharpoons$$
$$Cd(OH)_2(s) + 2e^-$$

Cathode reaction:
$$\tfrac{1}{2}O_2(g) + H_2O + 2e^- \rightleftharpoons$$
$$2OH^-$$

17-G. An interesting ligand for coulometric complexometric titrations is R-(−)-1,2-propylenediaminetetraacetic acid (PDTA), an optically active relative of EDTA.[‡]

N(CH₂CO₂H)₂
|
C·····H
H₃C   CH₂N(CH₂CO₂H)₂

R-(−)-1,2-Propylene-diaminetetraacetic acid (PDTA)

[†] F. A. Keidel, *Ind. Eng. Chem.*, **52**, 491 (1960).
[‡] R. A. Gibbs and R. J. Palma, Sr., *Anal. Chem.*, **48**, 1983 (1976).

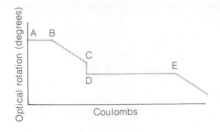

This chelate has a negative optical rotation at 365 nm and gives metal complexes with positive optical rotations. The reaction of PDTA with a metal can be studied by observing changes in the optical rotation of the solution. A schematic coulometric titration curve is shown above.

The initial solution contains PDTA and excess Hg^{2+}. Between points A and B, the cathode current reduces excess Hg^{2+} to Hg(l), which has no effect on the optical rotation. Between points B and C, $Hg(PDTA)^{2-}$ is reduced to Hg(l) plus $PDTA^{4-}$. This decreases the optical rotation. A large amount of $Hg(PDTA)^{2-}$ is still in solution at point C. When analyte solution containing, say, Zn^{2+} is added, the Zn^{2+} displaces Hg^{2+} from PDTA, changing the optical rotation from C to D. Then coulometric reduction of the liberated Hg^{2+} is continued beyond point D. At point E, the free Hg^{2+} is used up and $Hg(PDTA)^{2-}$ begins to be reduced, lowering the optical rotation once again. In a typical experiment, 2.000 mL of Zn^{2+} was added to the cell at point C. The coulombs measured in each region are

AB	2.60 C
BC	3.89 C
DE	14.47 C

Calculate the molarity of Zn^{2+} in the unknown.

Problems

A17-1. What is the difference between a galvanic cell and an electrolysis cell?

A17-2. Explain how the amperometric end-point detector operates in Figure 17-12.

A17-3. Explain why *controlled-potential* electrolysis (constant voltage between working and reference electrodes) is more selective than *constant-voltage* electrolysis (constant voltage between working and auxiliary electrodes).

A17-4. Which voltage, V_1 or V_2 in the diagram below, is constant in *constant-voltage* electrolysis? in *controlled-potential* electrolysis?

A17-5. Would you use an anodic or a cathodic depolarizer to prevent the potential of the working electrode from becoming too negative during reduction of Cu^{2+} to Cu(s)?

A17-6. Why does overpotential increase with current density?

A17-7. How many hours are required for 0.100 mol of electrons to flow through a circuit if the current is 1.00 A?

A17-8. The sensitivity of a coulometer is governed by the delivery of its minimum current for its minimum time. Suppose that 5 mA can be delivered for 0.1 s.
(a) To how many moles of electrons does this correspond?
(b) How many milliliters of a 0.01 M solution of a two-electron reducing agent are required to deliver the same number of electrons?

A17-9. Consider the electrolysis reactions

Cathode:

$$H_2O(l) + e^- \rightleftharpoons$$

$$\tfrac{1}{2}H_2(g, 1.0 \text{ atm}) + OH^-(aq, 0.10 \text{ M})$$

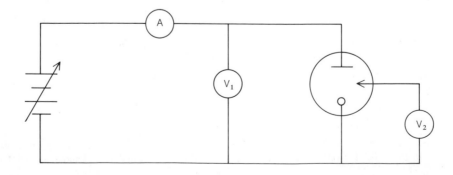

Anode:

$$Br^-(aq, 0.10 \text{ M}) \rightleftharpoons \tfrac{1}{2}Br_2(l) + e^-$$

(a) Calculate the equilibrium voltage needed to drive the net reaction.

(b) Suppose the cell has a resistance of 2.0 Ω and a current of 100 mA is flowing. How much voltage is needed to overcome the cell resistance? This is the ohmic potential.

(c) Suppose that the anode reaction has an overpotential (activation energy) of 0.20 V and that the cathode overpotential is 0.40 V. What voltage is necessary to overcome these effects combined with those of parts a and b?

(d) Suppose that concentration polarization occurs. The concentration of OH^- at the cathode surface increases to 1.0 M and the concentration of Br^- at the anode surface decreases to 0.010 M. What voltage is necessary to overcome these effects combined with those of parts b and c?

A17-10. An unknown weighing 0.326 8 g and containing lead lactate, $Pb(CH_3CHOHCO_2)_2$, plus inert material was electrolyzed to produce 0.111 1 g of PbO_2. Was the PbO_2 deposited at the anode or at the cathode? Find the weight percent of lead lactate in the unknown.

A17-11. A solution of Sn^{2+} is to be electrolyzed to reduce the Sn^{2+} to $Sn(s)$. Calculate the cathode potential (versus S.H.E.) needed to reduce the Sn^{2+} concentration to 1.0×10^{-8} M if no concentration polarization occurs. What would be the potential versus S.C.E. instead of S.H.E.? Would the potential be more positive or more negative if concentration polarization occurred?

A17-12. The experiment in Figure 17-12 required electrolysis at 5.32 mA for 964 s for complete reaction of a 5.00-mL aliquot of unknown cyclohexene solution.

(a) How many moles of electrons passed through the cell?

(b) How many moles of cyclohexene reacted?

(c) What was the molarity of cyclohexene in the unknown?

17-13. The two reactions occurring in the electrolysis cell at right are

$$Mn(s) \rightarrow Mn^{2+} + 2e^-$$

$$M^{3+} + 3e^- \rightarrow M(s)$$

The initial volume of the cell is 1.00 L, and the initial concentration of Mn^{2+} is 0.025 0 M.

(a) Is the Mn electrode the anode or the cathode?

(b) A constant current of 2.60 A was passed through the cell for 18.0 minutes, causing 0.504 g of the metal M to plate out on the Pt electrode. What is the atomic weight of M?

(c) What will be the concentration of Mn^{2+} in the cell at the end of the experiment?

17-14. The Weston cell shown at the top of the next page is a very stable source of potential used as a voltage standard in potentiometers. (The potentiometer compares the unknown input potential to the potential of the standard. In contrast to the conditions of this problem, very little current may be drawn from the cell if it is to be an accurate voltage standard.)

(a) How much work (J) can be done by the Weston cell if the voltage is 1.02 V and 1.00 mL of Hg (density = 13.53 g/mL) is deposited?

(b) Suppose that the cell is used to pass current through a 100-Ω resistor. If the heat dissipated by the resistor is 0.209 J/min, how many grams of cadmium are oxidized each hour? This part of the problem is not meant to be consistent with part a. That is, the voltage is no longer 1.02 volts, and you do not know what the voltage is.

17-15. The electrolysis cell at the bottom of the next page was run at a constant current of 0.021 96 A. On one side, 49.22 mL of H_2 was produced (at 303 K and 0.983 atm); on the other side, Cu metal was oxidized to Cu^{2+}.

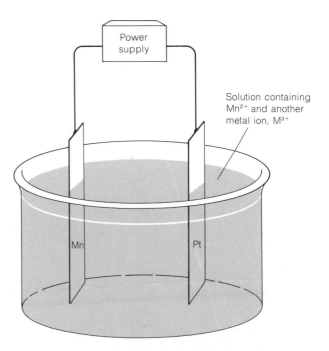

Power supply

Solution containing Mn^{2+} and another metal ion, M^{3+}

Mn   Pt

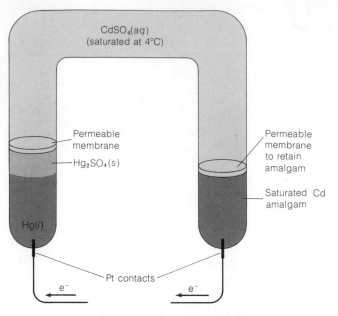

CdSO₄(aq)
(saturated at 4°C)

Permeable
membrane
—Hg₂SO₄(s)

Permeable
membrane
to retain
amalgam

Saturated Cd
amalgam

Hg(l)

Pt contacts

e^- e^-

A saturated Weston cell contains solid $CdSO_4$ and is therefore saturated with this salt at room temperature. The saturated cell is a more precise voltage standard than the unsaturated cell, but it is more sensitive to temperature and mechanical shock and cannot be easily incorporated into portable electronic equipment.

(a) How many moles of H_2 were produced?
(b) If 47.36 mL of EDTA were required to titrate the Cu^{2+} produced by the electrolysis, what was the molarity of the EDTA?
(c) For how many hours was the electrolysis run?

17-16. Consider the cell at the top of the next column, whose resistance is 3.50 Ω.

$Pt(s)|Fe^{2+}(0.10\ \text{M}), Fe^{3+}(0.10\ \text{M}), HClO_4(1\ \text{M})\|$
$Ce^{3+}(0.050\ \text{M}), Ce^{4+}(0.10\ \text{M}), HClO_4(1\ \text{M})|Pt(s)$

Suppose that there is no concentration polarization or overpotential.
(a) Calculate the voltage of the galvanic cell if it produces 30.0 mA.
(b) Calculate the voltage that must be applied to run the reaction in reverse, as an electrolysis, at 30.0 mA.

17-17. Suppose that the galvanic cell in Problem 17-16 delivers 100 mA under the following conditions: $[Fe^{2+}]_s = 0.050$ M, $[Fe^{3+}]_s = 0.160$ M, $[Ce^{3+}]_s = 0.180$ M, and $[Ce^{4+}]_s = 0.070$ M. Considering the ohmic potential and concentration polarization, calculate the cell voltage.

17-18. Calculate the initial voltage that should be applied to electrolyze 0.010 M $Zn(OH)_4^{2-}$ in 0.10 M NaOH using Ni electrodes. Assume that the current is 0.20 A, the anode current density is 100 A/m², the cell resistance is 0.35 Ω, and O_2 is evolved at 0.20 atm. The reactions are

Cathode: $Zn(OH)_4^{2-} + 2e^- \rightleftharpoons Zn(s) + 4OH^-$
$$E° = -1.214\ \text{V}$$

Anode: $H_2O \rightleftharpoons \frac{1}{2}O_2 + 2H^+ + 2e^-$

17-19. The free energy change for the formation of $H_2(g) + \frac{1}{2}O_2(g)$ from $H_2O(l)$ is $\Delta G° = +237.19$ kJ. Calculate the standard voltage needed to decompose water into its elements by electrolysis.

17-20. A mixture of trichloroacetate and dichloroacetate can be analyzed by selective reduction in a solution containing 2 M KCl, 2.5 M NH_3, and

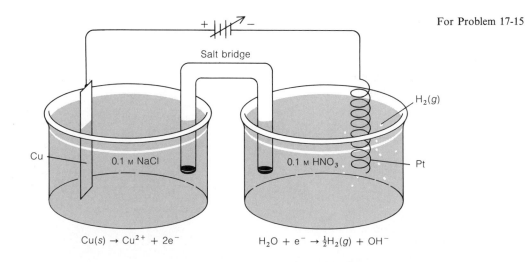

Salt bridge

For Problem 17-15

+ ⊣⊢⊢⊢ ↗ −

H₂(g)

Cu

0.1 M NaCl

0.1 M HNO₃

Pt

$Cu(s) \rightarrow Cu^{2+} + 2e^-$ $H_2O + e^- \rightarrow \frac{1}{2}H_2(g) + OH^-$

1 M NH_4Cl. At a mercury cathode potential of -0.90 V (versus S.C.E.), only trichloroacetate is reduced:

$$Cl_3CCO_2^- + H_2O + 2e^- \rightarrow$$
$$Cl_2CHCO_2^- + OH^- + Cl^-$$

At a potential of -1.65 V, dichloroacetate will react:

$$Cl_2CHCO_2^- + H_2O + 2e^- \rightarrow$$
$$ClCH_2CO_2^- + OH^- + Cl^-$$

A hygroscopic mixture of trichloro- and dichloroacetic acid containing an unknown quantity of water weighed 0.721 g. Upon controlled potential electrolysis, 224 C passed at -0.90 V, and 758 C was required to complete the electrolysis at -1.65 V. Calculate the weight percent of each acid in the mixture.

17-21. H_2S in aqueous solution can be analyzed by titration with coulometrically generated I_2:

$$H_2S + I_2 \rightarrow S(s) + 2H^+ + 2I^-$$

To 50.00 mL of sample was added 4 g of KI. Electrolysis required 812 s at a constant current of 52.6 mA. Calculate the concentration of H_2S ($\mu g/mL$) in the sample.

17-22. What equilibrium cathode potential (versus S.H.E.) is required to reduce 99.99% of Cd(II) from a solution containing 0.10 M Cd(II) in 1.0 M ammonia? Consider the reactions

$$Cd^{2+} + 4NH_3 \rightleftharpoons Cd(NH_3)_4^{2+} \quad \beta_4 = 3.6 \times 10^6$$
$$Cd^{2+} + 2e^- \rightleftharpoons Cd(s) \quad E° = -0.402 \text{ V}$$

17-23. Is it possible to remove 99% of a 1.0 μM CuY^{2-} impurity from a 10 mM CoY^{2-} solution at pH 4 without reducing any cobalt? Here, Y is EDTA, and the total concentration of free EDTA is 10 mM.

17-24. Ti^{3+} is to be generated in 0.10 M $HClO_4$ solution for coulometric reduction of azobenzene.

$$TiO^{2+} + 2H^+ + e^- \rightleftharpoons$$
$$Ti^{3+} + H_2O \quad E° = 0.100 \text{ V}$$

$$4Ti^{3+} + C_6H_5N{=}NC_6H_5 + 4H_2O \rightarrow$$
$$\text{Azobenzene}$$
$$2C_6H_5NH_2 + 4TiO^{2+} + 4H^+$$

At the counterelectrode, water is oxidized, and O_2 is liberated at a pressure of 0.20 atm. Both electrodes are made of smooth Pt, and each has a total surface area of 1.00 cm². The rate of reduction of the azobenzene is 25.9 nmol/s, and the resistance of the solution between the generator electrodes is 52.4 Ω.

(a) Calculate the current density (A/m²) at the electrode surface. Use Table 17-1 to estimate the overpotential for O_2 liberation.
(b) Calculate the equilibrium cathode potential (versus S.H.E.) assuming that $[TiO^{2+}]_{surface} = [TiO^{2+}]_{bulk} = 0.050$ M and $[Ti^{3+}]_{surface} = 0.10$ M.
(c) Calculate the equilibrium anode potential (versus S.H.E.).
(d) What should be the total applied voltage?

17-25. The chlor-alkali process,[†] in which seawater is electrolyzed to produce Cl_2 and NaOH, is the second most important commercial electrolysis, behind production of aluminum.

Anode: $\qquad Cl^- \rightarrow \frac{1}{2}Cl_2 + e^-$

Cathode: $\quad Na^+ + H_2O + e^- \rightarrow NaOH + \frac{1}{2}H_2$

The semipermeable Nafion membrane used to separate the anode and cathode compartments is resistant to chemical attack. Its many anionic side chains permit conduction of Na^+, but not anions. The cathode compartment is loaded with pure water, and the anode compartment contains seawater from which Ca^{2+} and Mg^{2+} have been removed. Explain how the membrane allows NaOH free of NaCl to be formed.

$$-[(CF_2CF_2)_n{-}CFCF_2]_x-$$
$$|$$
$$O$$
$$|$$
$$CF_2$$
$$|$$
$$CF{-}CF_3$$
$$|$$
$$O$$
$$|$$
$$CF_2CF_2SO_3^- Na^+$$
$$\text{Nafion}$$

† S. Venkatesh and B. V. Tilak, *J. Chem. Ed.*, **60**, 276 (1983); S. C. Stinson, *Chem. Eng. News*, March 15, 1982, p. 22; E. J. Taylor, R. Waterhouse, and A. Gelb, *Chemtech*, **17**, 316 (1987). Nafion is a trademark of the Du Pont Company.

18 Voltammetry

Voltammetry encompasses a sophisticated collection of analytical techniques in which the relationship between voltage and current is observed during electrochemical processes. The major subdivision of voltammetry is *polarography,* a sensitive electroanalytical technique that is especially useful for trace analysis. A second major classification within voltammetry is *amperometry,* which was introduced in conjunction with coulometric end-point detection in Chapter 17.

18-1 POLAROGRAPHY

In *polarography,* we measure the current as a function of the potential of the working electrode.

Consider an experiment in which analyte is reduced or oxidized at the working electrode of an electrolysis cell. In **polarography** the current flowing through the cell is measured as a function of the potential of the working electrode. Usually, this current is proportional to the concentration of analyte. The most sensitive polarographic procedures have a detection limit near 10^{-9} M and a precision around 5%. Less sensitive polarographic methods operating with $\sim 10^{-3}$ M analyte are capable of a precision of a few tenths of a percent, though 2–3% is most common.

Apparatus for a direct current polarographic experiment is shown in Figure 18-1. The working electrode is a mercury droplet suspended from the bottom of a glass capillary tube. Analyte is either reduced or oxidized at the surface of the mercury drop. The current-carrying auxiliary electrode is a platinum wire, and for reference a saturated calomel electrode is used. The potential of the mercury drop is measured with respect to the calomel electrode, through which negligible current flows.

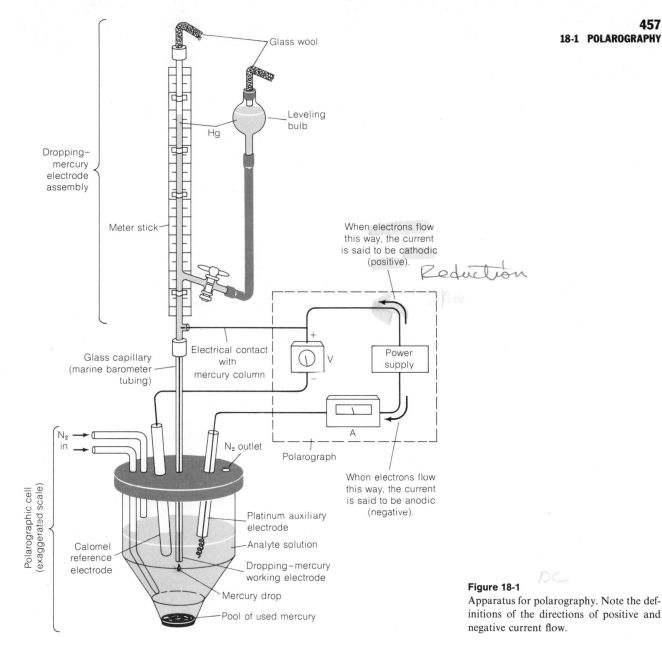

Reduction

Figure 18-1
Apparatus for polarography. Note the definitions of the directions of positive and negative current flow.

Why We Use the Dropping-Mercury Electrode

At first sight, the **dropping-mercury electrode** in Figure 18-1 seems a little strange. It consists of a very-small-diameter capillary through which mercury drips from an adjustable reservoir. The height of mercury, measured from the outlet of the capillary tube, is typically ~30 cm. With a capillary length of 10–20 cm and a capillary diameter of 0.05 mm, drops of mercury with a diameter of 0.5 mm are formed at a rate of 10–20 drops per minute. The drop interval is therefore 3–6 seconds. The drop rate is controlled by raising or lowering the leveling bulb at the top right of Figure 18-1.

The experiment in Figure 18-1 was first done by J. Heyrovsky in 1922. For his pioneering work in polarography, Heyrovsky received the Nobel Prize in 1959.

A *dropping-mercury electrode* is used because fresh Hg is continuously exposed to the analyte. This gives more reproducible behavior than does a static surface, whose characteristics change with use.

The reason for using the dropping-mercury electrode is its ability to yield reproducible current–potential data. This reproducibility can be attributed to the continuous exposure of fresh surface on the growing mercury drop. With any other electrode (such as Pt in various forms), the potential depends on its surface condition and therefore on its previous treatment.

The vast majority of reactions studied with the mercury electrode are reductions. At a Pt surface, reduction of solvent is expected to compete with reduction of many analyte species, especially in acidic solutions:

$$2H^+ + 2e^- \rightarrow H_2(g) \qquad E^\circ = 0 \qquad (18\text{-}1)$$

Another reason for using a mercury electrode is that there is a high overpotential for H^+ reduction at the mercury surface. Therefore, H^+ reduction does not interfere with many reductions.

But Table 17-1 showed that there is a rather large *overpotential* for reduction of H^+ at the Hg surface. Therefore, reactions thermodynamically more difficult than Reaction 18-1 can be carried out without competitive reduction of H^+. In neutral or basic solutions, even alkali metal cations can be reduced more easily than H^+, despite their lower standard potentials. This is partly because the potential for reduction of a metal forming a mercury amalgam is more positive than its potential for reduction to the solid state:

It is easier to reduce most metals to their amalgam than to the solid metal.

$$K^+ + e^- \rightarrow K(s) \qquad E^\circ = -2.936 \text{ V} \qquad (18\text{-}2)$$

$$K^+ + e^- + Hg \rightarrow K(\text{in Hg}) \qquad E^\circ = -1.975 \text{ V} \qquad (18\text{-}3)$$

Question: What do these potentials imply about the relative stabilities of K(s) and K(in Hg)?

A mercury electrode is not very useful for performing oxidations, because Hg is too easily oxidized. For most oxidations, some other working electrode must be employed. In a noncomplexing medium, Hg is oxidized near +0.25 V (versus S.C.E.). If the concentration of the complexing ion, Cl^-, is 1 M, the Hg oxidation potential is near 0.0 V. The oxidation is made easier by the stabilization of the Hg(II) product:

$$Hg(l) + 4Cl^- \rightleftharpoons HgCl_4^{2-} + 2e^- \qquad (18\text{-}4)$$

With inert electrodes and appropriate solvents, a wide range of redox potentials is accessible (Table 18-1).

The capillary of the dropping-mercury electrode will function for years if it is not allowed to clog. A flow of mercury should always be started prior to immersing the capillary in any solution. At the conclusion of an experiment, the electrode is raised and rinsed with distilled water with mercury still flowing. The clean electrode is stored either dry or in a pool of liquid mercury.

Respect your local electrode!

Mercury is toxic and slightly volatile, and spills are almost inevitable. To clean a spill, consolidate the droplets as much as possible with a piece of cardboard. Then suck the mercury into a filter flask using an aspirator or other source of vacuum. A disposable Pasteur pipet attached to a hose makes a good vacuum cleaner. To remove residual mercury, sprinkle elemental zinc powder on the surface and dampen the powder with 5% aqueous H_2SO_4 to give a paste consistency. The mercury dissolves in the zinc. After working the paste into contaminated areas with a sponge or brush, allow the paste to dry and then sweep it up. Discard the powder appropriately as contaminated mercury waste.[†]

[†] This procedure is better than sprinkling sulfur on the spill. Sulfur coats the mercury, but does not react with the bulk of the droplet. [D. N. Easton, *Am. Lab.,* July 1988, p. 66.]

Table 18-1
Potential ranges (versus S.C.E.) of various solvents[†]

Solvent	Supporting electrolyte	Cathodic limit (V)	Anodic limit (V)
Acetic acid	CH_3CO_2Na	−1.0	+2.0
Acetonitrile	$LiClO_4$	−3.0	+2.5
Ammonia	TBAI	−2.8(Hg)	+0.3
Dichloromethane	TBAP	−1.7	+1.8
Dimethylformamide	TBAP	−2.8	+1.9
Dimethylsulfoxide	$LiClO_4$	−3.4	+1.3
Hydrogen fluoride	NaF	−1.0	+3.5
Methanol	KOH	−1.0	+0.6
Methanol	KCN	−1.0	+1.7
Propylene carbonate	TEAP	−1.9	+1.7
Pyridine	TEAP	−2.2	+3.3
Sulfolane	TEAP	−2.9	+3.0
Tetrahydrofuran	$LiClO_4$	−3.2	+1.6
Trifluoroacetic acid	CF_3CO_2Na	−0.6	+2.4
Water	TBAP	−2.7(Hg)	+1.3

(handwritten annotation: reduce the analyte)

[†] Electrodes are Pt unless Hg is indicated. Abbreviations: TBAI = tetrabutylammonium iodide; TBAP = tetrabutylammonium perchlorate; TEAP = tetraethylammonium perchlorate.
SOURCE: N. L. Weinberg, *J. Chem. Ed.* **60**, 268 (1983).

18-2 SHAPE OF THE POLAROGRAM

A graph of current versus potential in a polarographic experiment is called a **polarogram.** In Figure 18-2, we see the result of the polarographic reduction of Cd^{2+} in HCl solution.

$$Cd^{2+} + 2e^- \rightleftharpoons Cd(\text{in Hg}) \qquad (18\text{-}5)$$

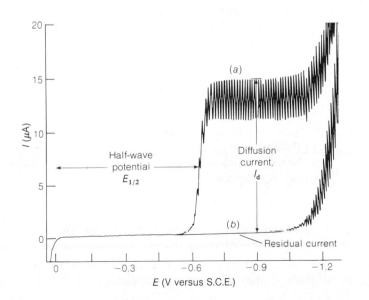

(handwritten annotations: v = IR; Reduction $Cd^{2+} \rightarrow Cd^0$ more electron flow higher I.)

Figure 18-2
Polarograms. (a) 5×10^{-4} M Cd^{2+} in 1 M HCl. (b) 1 M HCl alone. Note that the scale of current is μA. [D. T. Sawyer and J. L. Roberts, Jr., *Experimental Electrochemistry for Chemists* (New York: Wiley, 1974).]

The magnitude of the diffusion current is proportional to the concentration of analyte. This is why polarography is used for quantitative analysis.

Speaking of "reduction of H^+" is equivalent to speaking of "reduction of H_2O."

When the potential is only slightly negative with respect to the calomel electrode, essentially no reduction of Cd^{2+} occurs. Only a small **residual current** flows. At a sufficiently negative potential, reduction of Cd^{2+} commences and the current increases. The reduced Cd dissolves in the Hg to form an amalgam. After a steep increase in current, **concentration polarization** sets in: The rate of electron transfer becomes limited by the rate at which Cd^{2+} can diffuse from bulk solution to the surface of the electrode. The magnitude of this **diffusion current** (I_d) is proportional to Cd^{2+} concentration and is used for quantitative analysis. The upper trace in Figure 18-2 is called a **polarographic wave.**

When the potential is sufficiently negative, around -1.2 V, reduction of H^+ begins and the curve rises steeply. This can be seen at the right-hand edge of Figure 18-2. At positive potentials (near the left side of the polarogram), oxidation of the Hg electrode produces a negative current. *By convention, a negative current means that the working electrode is behaving as the anode with respect to the auxiliary electrode. A positive current means that the working electrode is behaving as the cathode.*

The oscillating current in Figure 18-2 is due to the growth and fall of the Hg drops. As each drop begins to form, there is very little Hg surface, and correspondingly little current can flow. As the drop grows, its area increases, more solute can reach the surface in a given time, and more current flows. The current increases as the drop grows until, finally, the drop falls off and the current decreases sharply.

Diffusion Current

The diffusion current is the limiting current when the rate of electrolysis is controlled by the rate of diffusion of species to the electrode.

When the potential of the working electrode is sufficiently negative, the rate of reduction of Cd^{2+} ions in Equation 18-5 is governed by the rate at which Cd^{2+} can reach the electrode. In Figure 18-2, this occurs at potentials more negative than -0.7 V. In an unstirred solution, the rate of reduction is controlled by the rate of diffusion of analyte to the electrode. In this case, the limiting current is called the *diffusion current*. The solution must be perfectly quiet to reach the diffusion limit in polarography.

The rate of diffusion of a solute from bulk solution to the surface of the electrode is proportional to the concentration difference between the two regions:

The symbol $\propto$ is read "is proportional to."

$$\text{Current} \propto \text{rate of diffusion} \propto [C]_0 - [C]_s \qquad (18\text{-}6)$$

where $[C]_0$ is the concentration in bulk solution and $[C]_s$ is the concentration at the surface of the electrode (see Figure 17-2). The greater the difference in concentrations, the more rapid will be the diffusion. At a sufficiently negative potential, the reduction is so fast that $[C]_s \ll [C]_0$ and Equation 18-6 reduces to the form

Equations 18-6 and 18-7 should be formulated in terms of activities instead of concentrations. However, in the presence of a large quantity of supporting electrolyte, the ionic strength is constant, so the activity coefficients are also constant. Under these conditions, both equations are valid in terms of concentrations.

$$\text{Limiting current} = \text{diffusion current} \propto [C]_0 \qquad (18\text{-}7)$$

The ratio of the diffusion current to the bulk-solute concentration is the basis for the use of polarography in analytical chemistry.

The magnitude of the diffusion current, *measured at the top of each oscillation* in Figure 18-2, is given (to an accuracy of a few percent) by the Ilkovič equation:[†]

$$I_d = (7.08 \times 10^4)nCD^{1/2}m^{2/3}t^{1/6} \qquad (18\text{-}8)$$

where I_d = diffusion current, measured at the top of the oscillations in Figure 18-2, with the units μA

n = number of electrons per molecule involved in the oxidation or reduction of the electroactive species.

C = concentration of electroactive species, with the units mmol/L

D = diffusion coefficient of electroactive species, with the units m^2/s[‡]

m = rate of flow of Hg, in mg/s

t = drop interval, in s

The number 7.08×10^4 is a combination of several constants whose dimensions are such that I_d will be given in μA.

Clearly, the magnitude of the diffusion current depends on several factors in addition to analyte concentration. The quantity $m^{2/3}t^{1/6}$ in Equation 18-8 is called the **capillary constant,** and is very nearly proportional to the square root of the Hg column height (h), measured from the top of the Hg meniscus to the bottom of the Hg electrode in Figure 18-1. To demonstrate that the limiting current is indeed *diffusion*-controlled, you can measure the current at various heights of the Hg column and see if the current is proportional to $\sqrt{h}$.

In quantitative polarography, it is important to control the temperature within a few tenths of a degree. This is because the diffusion coefficient for most species increases by about 2% per degree. If the capillary constant is to be used, it should be measured at the same electrode potential used to measure I_d, since t depends on the potential.

The transport of solute to the electrode depends on three factors: diffusion, mechanical transport (stirring and convection), and electrostatic attraction. In polarography, we try to minimize the latter two mechanisms. We want the limiting current to be controlled solely by the rate of diffusion of Cd^{2+} to the electrode.

To minimize mechanical transport, the solution is *not* stirred, and effort is made to reduce vibrations. Setting the apparatus on a heavy base (such as the marble slab used for sensitive analytical balances) helps to reduce the effect of vibrations of the building.

Under proper, analytically useful conditions, the limiting current should be proportional to $\sqrt{h}$. Otherwise, effects other than diffusion are controlling the rate of reaction.

Three ways that electrolytes get to the electrode:

1. diffusion (desired in polarography)
2. mechanical transport (not desired)
3. electrostatic attraction (not desired)

[†] In the older literature, current was measured at the center of each oscillation, using a damped ballistic galvanometer. In that case, the constant in Equation 18-8 should be 6.07×10^4. [A. J. Bard and L. R. Faulkner, *Electrochemical Methods* (New York: Wiley, 1980), pp. 147–150.]

[‡] The diffusion coefficient is defined from Fick's first law of diffusion: The rate (J) at which molecules diffuse across a plane of unit area is given by

$$J = -D\frac{dc}{dx}$$

where D is the diffusion coefficient and dc/dx is the gradient of concentration in the direction of diffusion. The larger the diffusion coefficient, the more rapidly the molecules diffuse.

Table 18-2

Influence of supporting electrolyte on the limiting current for reduction of Pb^{2+}

Electrolyte concentration (M)	Limiting current (μA) KCl	KNO$_3$
0	17.6	17.6
0.000 1	16.3	16.2
0.000 2	15.9	15.0
0.000 5	13.3	13.4
0.001	11.8	12.0
0.005	9.8	9.8
0.1	8.35	8.45
1.0	8.00	8.45

Note: 50 mL of 9.5×10^{-4} M PbCl$_2$ was analyzed at 25°C with 0.2 mL of 0.1% (wt/wt) sodium methyl red present as a maximum suppressor.

SOURCE: I. M. Kolthoff and J. J. Lingane, *Polarography*, Vol. I (New York: Wiley, 1952), p. 123.

Electrostatic attraction of analyte ions is minimized by using a high concentration of supporting electrolyte.

Challenge: See if you can understand why the presence of supporting electrolyte will *increase* the limiting current for reduction of an *anion*, such as IO$_3^-$.

I_d = limiting current − residual current

Current flow due to electrostatic attraction (or repulsion) of analyte ions by the electrode is reduced to a negligible level by the presence of a high concentration of **supporting electrolyte** (1 M HCl in Figure 18-2). As shown in Table 18-2, increasing concentrations of electrolyte reduce the net current, since the rate of arrival of cationic analyte at the negative Hg surface is decreased. Typically, a supporting electrolyte concentration 50–100 times greater than the analyte concentration will reduce electrostatic transport of the analyte to a negligible level.

Residual Current

To measure the value of the diffusion current in a polarogram, we subtract the value of the *residual current* from the observed limiting current in the diffusion-controlled region. This difference is labeled I_d in Figure 18-2, where the magnitude of the residual current is quite small. At lower analyte concentration, the proportion of residual current will be greater, since residual current will be the same, while diffusion current decreases. For accurate work, you should always measure the residual current of a solution containing the same supporting electrolyte as your sample. The reagents should come from the same stock solutions, since different batches will have different impurities contributing to the residual current.

An enlarged plot of the residual current of a 0.1 M HCl solution is shown in Figure 18-3. The dashed lines show that the residual current has an inflection point near −0.5 V in this case. At this point, called the **electrocapillary maximum,** the charge on the drop of mercury is zero. At more positive potentials, the charge on the drop is positive with respect to the solution. At more negative potentials, the charge is negative. Because the mercury drop

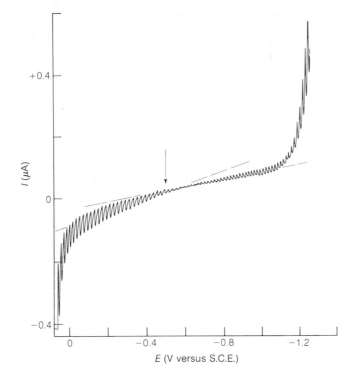

Figure 18-3
Residual current of 0.1 M HCl. The arrow marks the inflection point that occurs at the electrocapillary maximum. This occurs when the charge of the Hg drop is zero with respect to the solution. If there were no faradaic current, the net current would be zero at the electrocapillary maximum. [L. Meites, *Polarographic Techniques,* 2nd ed. (New York: Wiley, 1965).]

and the surrounding solution can have different charges, the mercury–solution interface behaves as a capacitor. Box 18-1 describes the structure of the electrode–solution interface.

The residual current has two components. One is the current needed to charge or discharge the capacitor formed by the mercury–solution interface. This is called the **condenser current** or **charging current.** It is present in all polarographic experiments, regardless of the purity of reagents. As each drop of mercury falls, it carries its charge with it to the bottom of the cell. The new drop requires more current for charging.

Any current that flows as a result of reduction or oxidation of a species in solution is called a **faradaic current.** The second component of the residual current is a small faradaic current due mainly to the reduction (or oxidation) of impurities in the supporting electrolyte. Since the concentration of supporting electrolyte is very high, the concentration of trace impurities can be significant.

Residual current has two components:

1. condenser current (due to charging of Hg drops)
2. faradaic current (due to redox reactions)

Shape of the Polarographic Wave

The two most common reactions at the dropping Hg electrode involve reduction of an ion to an amalgam or reduction of a soluble ion to another soluble ion:

$$M^{n+} + ne^- + Hg \rightleftharpoons M(\text{in Hg}) \qquad (18\text{-}9)$$

$$X^{a+} + ne^- \rightleftharpoons X^{(a-n)+} \qquad (18\text{-}10)$$

Box 18-1 THE ELECTRIC DOUBLE LAYER

A generally accepted model for the structure of the electrode–solution interface is shown below.

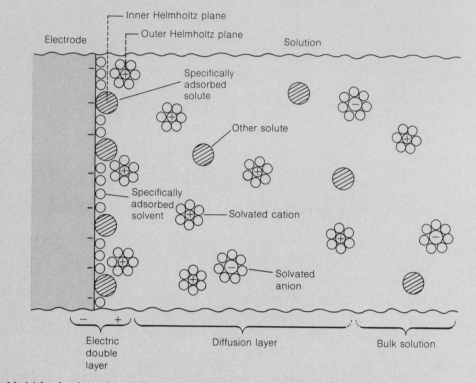

Model for the electrode–solution interface. The plane going through the center of charge of the adsorbed species is called the *inner Helmholtz plane*. The plane through the center of charge of the next layer of nonspecifically adsorbed ions is the *outer Helmholtz plane*. The *diffusion layer* contains an excess of cations, attracted by the negatively charged electrode.

The external power supply pumps electrons into or out of the electrode. In general, the electrode is either positively or negatively charged with respect to the solution. For a given solution composition, there is one *potential of zero charge* at which there is no excess charge on the electrode. This potential is -0.58 V (versus a calomel electrode containing 1 M KCl) for a mercury electrode immersed in 0.1 M KBr. The potential of zero charge is shifted to -0.72 V for the same electrode in 0.1 M KI.

The first layer of molecules at the surface of the electrode is *specifically adsorbed:* The molecules are held tightly by van der Waals and electrostatic forces. The figure above shows solvent and solute molecules on the electrode surface. The adsorbed solute could be neutral molecules, anions, or cations. Anions tend to be more *polarizable* than cations, and therefore subject to stronger van der Waals forces. This is why the potential of zero charge for KI is more negative than for KBr. Iodide is more strongly adsorbed than is bromide, and a more negative electrode potential is necessary to expel the iodide from the electrode surface. The figure to the above right shows that the quantity of adsorbed solute increases as the solution concentration of solute increases.

The next layer of molecules beyond the specifically adsorbed layer is rich in cations attracted to the negatively charged electrode. These cations are fully solvated and held near the electrode by coulombic attraction. The charged electrode and the oppositely charged ions attracted to it constitute the **electric double layer.** The excess population of cations over anions in the solution decreases as the distance from the electrode increases. This region, whose composition

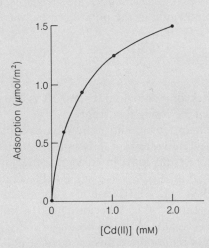

Concentration dependence of adsorption of Cd(II) (probably $Cd(SCN)_4^{2-}$) from 0.5 M NaSCN + 0.5 M $NaNO_3$ on a mercury electrode at -0.2 V (versus S.C.E.). This figure is from a very readable article by F. C. Anson and R. A. Osteryoung [*J. Chem. Ed.,* **60,** 293 (1983)], which explains how *chronocoulometry* is used to measure adsorption. A quantitative electrode adsorption experiment is described by D. Marin and F. Mendicuti, *J. Chem. Ed.,* **65,** 712 (1988).

is different from that of bulk solution, is called the **diffusion layer.** The thickness of this layer, typically in the range 0.3-10 nm, decreases as the concentration of ions increases, as the charge of the ions increases, and as the temperature decreases. In the diffusion layer there is a balance between attraction toward the electrode and randomization by thermal motion.

When a species is created or destroyed by an electrochemical reaction, its concentration near the electrode is generally different from the concentration in bulk solution (Color Plate 12). The figure below shows the measured concentration of tri(*p*-methoxyphenyl)amine cation at

$$\left(CH_3O-\!\!\!\bigcirc\!\!\!-\right)_3\!\!N^{\cdot+} \quad \text{Tri(}p\text{-methoxyphenyl)amine cation}$$

various time intervals after stepping up the potential of a platinum electrode from 0 to 0.8 V (versus S.C.E.). At 0 V, no cation is present. When the potential is suddenly changed to 0.8 V, the amine is oxidized to a cation present only at the electrode surface. As time goes on, more cation is generated and it diffuses away from the electrode. The thickness of this layer of diffusing product is proportional to $\sqrt{Dt}$, where D is the diffusion coefficient of the cation and t is time.

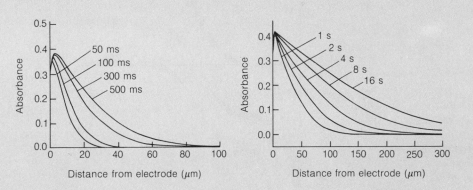

Diffusion profile of cation created by sudden oxidation of tri(*p*-methoxyphenyl)amine at a platinum electrode. The ordinate scale (absorbance of 633-nm visible light) is proportional to the concentration of the cation. Times are measured from the start of the oxidation. These elegant measurements were made with a photodiode array (see Figure 20-22) by C.-C. Jan, R. L. McCreery and F. T. Gamble. [*Anal. Chem.,* **57,** 1763 (1985); *Anal. Chem.,* **58,** 2771 (1986).] Microelectrodes can also be used to probe the nature of the diffusion layer. See R. C. Engstrom, T. Meaney, R. Tople, and R. M. Wightman, *Anal. Chem.,* **59,** 2005 (1987).

It can be shown for both cases that if the reactions are reversible, the equation relating the current and the potential in the polarographic wave is

$$E = E_{1/2} - \frac{0.059\,16}{n} \log \frac{I}{I_d - I} \tag{18-11}$$

$E_{1/2}$ is the **half-wave potential,** drawn in Figure 18-2. It is the potential at which $I = \frac{1}{2}I_d$ (both corrected for residual current).

Reversible anodic and composite anodic–cathodic waves have the same shape defined by Equation 18-11. Figure 18-4 shows polarograms of (a) ferric, (b) ferric plus ferrous, and (c) ferrous ions. In curve a, Fe(III) is reduced to Fe(II) at a half-wave potential of +0.05 V (versus S.C.E.). Only a residual current is observed at +0.15 V, since Fe(III) is not reduced at this potential. Curve c shows an anodic current at +0.15 V because Fe(II) is being oxidized. At −0.05 V, only a residual current is observed because no Fe(III) is present to be reduced. Curve b shows an anodic diffusion current at +0.15 V due to oxidation of Fe(II) and a cathodic current at −0.05 V due to reduction of Fe(III). All three curves have the same value of $E_{1/2}$ and the same shape.

The graph of Equation 18-11 in Figure 18-5 shows how the shape of a reversible polarographic wave depends on n, the number of electrons in the half-reaction. The larger the value of n, the steeper the polarographic wave. The function in Equation 18-11 predicts that a graph of E versus $\log(I/I_d - I)$ should be linear, with a slope of $-0.059\,16/n$. Such a graph allows us to verify that the reaction is reversible and to find n, the number of electrons involved. The best results are obtained when the effect of cell resistance on the current is also considered.

A *cathodic current* represents a flow of electrons from the Hg electrode to the analyte (reduction). An *anodic current* represents a flow of electrons from the analyte to the electrode (oxidation).

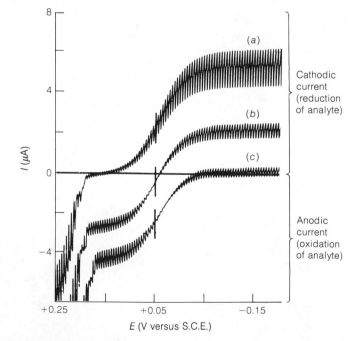

Figure 18-4
Polarograms. (a) 1.4 mM Fe(III). (b) 0.7 mM Fe(III) plus 0.7 mM Fe(II). (c) 1.4 mM Fe(II). Each solution contains saturated oxalic acid as supporting electrolyte and 0.000 2% methyl red as maximum suppressor. Vertical lines are drawn at the observed half-wave potentials. [L. Meites, *Polarographic Techniques,* 2nd ed. (New York: Wiley, 1965).]

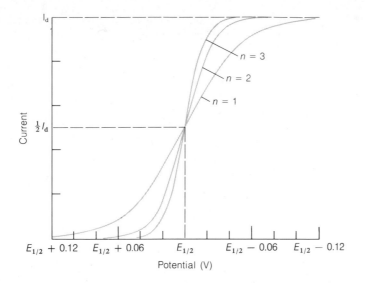

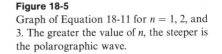

Figure 18-5
Graph of Equation 18-11 for $n = 1, 2$, and 3. The greater the value of n, the steeper is the polarographic wave.

Relation Between $E_{1/2}$ and $E°$

For reversible electrochemical reactions (those with negligible overpotential), there is a relatively straightforward thermodynamic interpretation of the half-wave potential, $E_{1/2}$. Consider Reaction 18-9, in which the product dissolves in mercury. If the reduction product were not stabilized by dissolution in mercury, then we would expect $E_{1/2}$ to be equal to $E°$. In fact, the product is stabilized when it dissolves in the mercury, so $E_{1/2} \neq E°$.

The relation between the half-wave potential and $E°$ is[†]

$$E_{1/2} \approx E° + E_s + \frac{0.059\ 16}{n} \log C\gamma \qquad (18\text{-}12)$$

The symbols in Equation 18-12 have the following meanings: $E°$ is the ordinary standard reduction potential for M^{n+} with respect to the reference electrode used in the polarographic cell. For a calomel reference electrode, $E° = E°$ (versus S.H.E.) $- 0.241$ V. E_s is the standard potential for the cell $M(s)|M^{n+}|M(\text{in Hg})_{\text{saturated}}$, in which the left electrode is solid metal and the right electrode is a saturated amalgam of the same metal. C is the concentration of M in the saturated amalgam, and γ is the activity coefficient of M in the saturated amalgam; therefore, $C\gamma$ is the activity of metal M in the saturated amalgam.

The term E_s in Equation 18-12 measures the tendency of electrons to flow between solid analyte and analyte amalgam. A positive value means that electrons flow spontaneously from pure reduced analyte to reduced analyte dissolved in mercury. The more positive the value of E_s, the easier it will be to reduce the species in the presence of mercury (compared with reduction in the absence of mercury). The log term in Equation 18-12 simply gives the free energy change for dissolving reduced analyte in mercury. The more soluble the reduced product is in mercury, the easier it will be to carry out the reduction.

Equation 18-12 tells us that the reaction $M^{n+} + ne^- \rightleftharpoons M$ (in Hg) is facilitated if

1. Electrons flow spontaneously from $M(s)$ to $M(\text{in Hg})$.
2. M is very soluble in liquid mercury.

[†] I. M. Kolthoff and J. L. Lingane, *Polarography* (New York: Wiley, 1952), p. 201. Equation 18-12 ignores small terms involving diffusion and activity coefficients.

Table 18-3
Test of Equation 18-12 for several metals at 25°C

Ion	$E°$(V)	E_s(V)	C(M)	γ	$E_{1/2}$[Eq. 18-12]	$E_{1/2}$(observed)
Pb^{2+}	−0.372	0.006	0.96	0.72	−0.371	−0.388
Tl^+	−0.582	0.003	27.4	8.3	−0.440	−0.459
Cd^{2+}	−0.647	0.051	6.40	1.15	−0.570	−0.578
Zn^{2+}	−1.008	0	4.37	0.74	−0.993	−0.997
Na^+	−2.961	0.780	3.52	1.3	−2.14	−2.12
K^+	−3.170	1.001	1.69	5.6	−2.11	−2.14

Note: $E°$ and E_s are with respect to S.C.E.
SOURCE: I. M. Kolthoff and J. L. Lingane, *Polarography* (New York: Wiley, 1952), p. 198.

Table 18-3 gives some data that demonstrate these effects. In the case of Pb^{2+}, both E_s and the log term are small, and $E_{1/2}$ is close to $E°$. For the other ions in the table, either E_s or the log term is sufficiently large so that $E_{1/2}$ is not very close to $E°$.

Now consider the reversible Reaction 18-10, in which the oxidized and reduced species remain in solution. In this case the relation between $E_{1/2}$ and $E°$ is given by

> For the reaction
> $X^{a+} + ne^- \rightleftharpoons X^{(a-n)+}$,
> $E_{1/2}$ is usually close to $E°$.

$$E_{1/2} = E° - \frac{0.059\ 16}{n} \log \frac{\gamma_{red}D_{ox}^{1/2}}{\gamma_{ox}D_{red}^{1/2}} \tag{18-13}$$

where $E°$ is the standard reduction potential for Reaction 18-10, with respect to S.C.E. The activity coefficients of the reduced and oxidized species are γ_{red} and γ_{ox}. The corresponding diffusion coefficients are D_{red} and D_{ox}. The quotient in the log term is usually close to unity, so the logarithm is usually small. Therefore, $E_{1/2}$ for reaction of soluble reactants and products is usually close to $E°$ for the redox couple.

Other Factors Affecting the Shape of the Polarogram

Current maxima

Figure 18-6 shows the polarogram of a mixture of Pb(II) and Zn(II) in 2 M NaOH. The wave near −0.8 V is due to reduction of Pb(II), and the smaller wave near −1.5 V is due to reduction of Zn(II). In trace a, the current "overshoots" the value of I_d, then settles back down. These current maxima are common and have been attributed to convection currents near the surface of the electrode. Adding traces of certain agents, such as gelatin, Triton X-100, or methyl red, generally eliminates this behavior, as shown in trace b of Figure 18-6. These agents are called **maximum suppressors** and apparently alter the convective behavior of the solution. Too much suppressor will distort the polarogram.

> Small quantities of certain surface-active agents are routinely used to suppress current maxima.

Oxygen

Figure 18-7 shows that oxygen gives rise to a pair of intense polarographic waves. The first wave is due to its reduction to H_2O_2, and the second is from further reduction to H_2O.

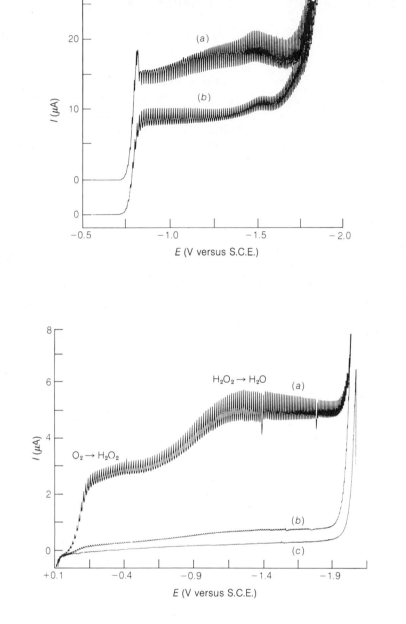

Figure 18-6
Polarograms of 3 mM Pb(II) and 0.25 mM Zn(II) in 2 M NaOH. (a) In the absence of a suppressor. (b) In the presence of 0.002% Triton X-100. [L. Meites, *Polarographic Techniques*, 2nd ed. (New York: Wiley, 1965).]

Figure 18-7
Polarogram of 0.1 M KCl. (a) Saturated with air. (b) After partial deaeration. (c) After further deaeration. [L. Meites, *Polarographic Techniques*, 2nd ed. (New York: Wiley, 1965).]

$$O_2 + 2H^+ + 2e^- \rightleftharpoons H_2O_2 \qquad E_{1/2} \approx -0.1 \text{ V (versus S.C.E.)} \quad (18\text{-}14)$$

$$H_2O_2 + 2H^+ + 2e^- \rightleftharpoons 2H_2O \qquad E_{1/2} \approx -0.9 \text{ V (versus S.C.E)} \quad (18\text{-}15)$$

Since oxygen is dissolved in any solution exposed to air, these waves would be superimposed on the polarogram of the analyte if the oxygen were not removed. Bubbling nitrogen gas through the solution for 3–15 minutes (depending on the apparatus) removes oxygen and eliminates these waves. The cell in Figure 18-1 has ports for purging the solution with nitrogen and for maintaining a blanket of nitrogen over the solution during the experiment.

18-3 APPLICATIONS OF POLAROGRAPHY

Polarography serves many purposes in analytical chemistry. One of the most straightforward is in the qualitative identification of an unknown. For a reversible redox process, the half-wave potential is characteristic of the electroactive species (the analyte) and the medium in which it is analyzed. Such factors as analyte concentration and the electrode capillary constant have no effect on $E_{1/2}$. Extensive tables of half-wave potentials exist,[†] so $E_{1/2}$ for an unknown can be compared with known values to try to identify the species by polarography. For irreversible processes, the waves are broader and $E_{1/2}$ depends on concentration and capillary characteristics. Nonetheless, the variation in half-wave potential is small enough so that polarography is useful in qualitative analysis.

A unique identification of an unknown cannot be made with a single half-wave potential. However, once a list of suspect ions is compiled, varying the medium usually permits further shortening of the list of possibilities. For example, two ions that have similar reduction potentials in 0.1 M tartrate solution are not likely to have similar potentials in ammonia solution. If a species undergoes more than one redox process, the positions of successive waves (as well as their relative heights) can be very diagnostic.

Many organic functional groups give rise to polarographic waves (Table 18-4). Once again, the half-wave potentials can help to distinguish one possible functional group from another. Table 18-4 might lead you to ask, "Why not do synthetic redox chemistry by controlled-potential electrolysis?" In fact, many reactions can be conducted electrochemically, and a large body of literature based on this technique does exist.[‡]

> For qualitative analysis, the half-wave potential of an unknown is measured in several different complexing media. Comparison with a table of half-wave potentials allows the unknown to be identified.

> Oxidations and reductions can be carried out on a synthetically useful scale by controlled-potential electrolysis, as well as by using oxidizing or reducing agents.

Quantitative Analysis

The principal use of polarography is in quantitative analysis. Since the magnitude of the diffusion current is proportional to the concentration of analyte, the height of a polarographic wave tells how much analyte is present. In the following sections, we will describe the use of *standard curves, standard additions,* and *internal standards* for quantitative analysis. These methods are completely general and by no means restricted to polarography; they can be used in conjunction with any quantitative technique, such as potentiometry, spectrophotometry, or chromatography.

Standard curves

The most reliable, but tedious, method of quantitative analysis is to prepare a series of known concentrations of analyte in otherwise identical solutions. A polarogram of each solution is recorded, and a graph of the diffusion current versus analyte concentration is prepared. Finally, a polarogram of the unknown is recorded, using the same conditions. From the measured

---

[†] L. Meites, *Handbook of Analytical Chemistry* (New York: McGraw-Hill, 1963), pp. 5-53 to 5-103.

[‡] N. L. Weinberg, ed., *Technique of Electroorganic Synthesis* (New York: Wiley, 1974); J. Chang, R. F. Large, and G. Popp in A. Weissberger and B. W. Rossiter, eds., *Physical Methods of Chemistry,* Vol. I, Part IIB (New York: Wiley, 1971); Z. Nagy, *Electrochemical Synthesis of Inorganic Compounds: A Bibliography* (New York: Plenum Press, 1985); and J. H. Wagenknecht, *J. Chem. Ed.,* **60,** 271 (1983).

Table 18-4
Polarographic behavior of some functional groups

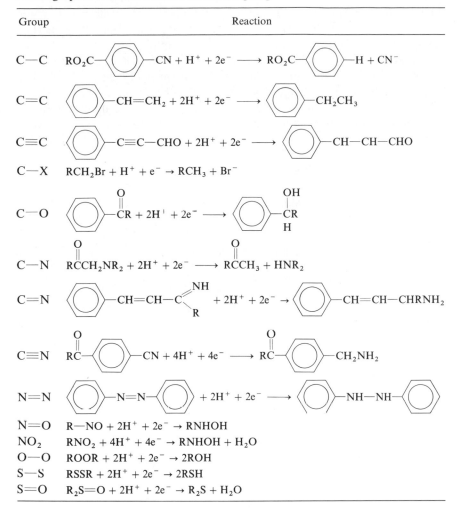

Group	Reaction

C—C RO_2C—⟨◯⟩—$CN + H^+ + 2e^- \longrightarrow RO_2C$—⟨◯⟩—$H + CN^-$

C=C   ⟨◯⟩—$CH=CH_2 + 2H^+ + 2e^- \longrightarrow$ ⟨◯⟩—CH_2CH_3

C≡C   ⟨◯⟩—$C≡C—CHO + 2H^+ + 2e^- \longrightarrow$ ⟨◯⟩—$CH—CH—CHO$

C—X $RCH_2Br + H^+ + e^- \rightarrow RCH_3 + Br^-$

C—O   ⟨◯⟩—$\overset{O}{\overset{\|}{C}}R + 2H^+ + 2e^- \longrightarrow$ ⟨◯⟩—$\overset{OH}{\underset{H}{\overset{|}{C}R}}$

C—N $R\overset{O}{\overset{\|}{C}}CH_2NR_2 + 2H^+ + 2e^- \longrightarrow R\overset{O}{\overset{\|}{C}}CH_3 + HNR_2$

C=N   ⟨◯⟩—$CH=CH—\overset{NH}{\underset{R}{\overset{\diagup}{C}\diagdown}} + 2H^+ + 2e^- \rightarrow$ ⟨◯⟩—$CH=CH—CHRNH_2$

C≡N $R\overset{O}{\overset{\|}{C}}$—⟨◯⟩—$CN + 4H^+ + 4e^- \longrightarrow R\overset{O}{\overset{\|}{C}}$—⟨◯⟩—$CH_2NH_2$

N=N   ⟨◯⟩—$N=N$—⟨◯⟩$ + 2H^+ + 2e^- \longrightarrow$ ⟨◯⟩—$NH—NH$—⟨◯⟩

N=O $R—NO + 2H^+ + 2e^- \rightarrow RNHOH$

NO_2 $RNO_2 + 4H^+ + 4e^- \rightarrow RNHOH + H_2O$

O—O $ROOR + 2H^+ + 2e^- \rightarrow 2ROH$

S—S $RSSR + 2H^+ + 2e^- \rightarrow 2RSH$

S=O $R_2S=O + 2H^+ + 2e^- \rightarrow R_2S + H_2O$

diffusion current and the **standard curve,** the concentration of analyte can be determined. Figure 18-8 shows an example of the linear relationship between diffusion current and concentration.

EXAMPLE: Using a Standard Curve

Suppose that 5.00 mL of an unknown sample of Al(III) was placed in a 100-mL volumetric flask containing 25.00 mL of 0.8 M sodium acetate (pH 4.7) and 2.4 mM pontachrome violet SW (a maximum suppressor). After dilution to 100 mL, an aliquot of the solution was analyzed by polarography. The height of the polarographic wave was 1.53 μA, and the residual current—measured at the same potential with a similar solution containing no Al(III)—was 0.12 μA. Find the concentration of Al(III) in the unknown.

The corrected diffusion current is $1.53 - 0.12 = 1.41$ μA. In Figure 18-8, 1.41 μA corresponds to [Al(III)] = 0.126 mM. Since the unknown was diluted by a factor of 20.0 (from 5.00 mL to 100 mL) for analysis, the original concentration of unknown must have been $(20.0)(0.126) = 2.46$ mM.

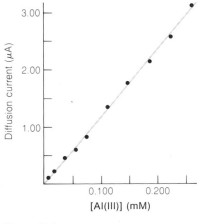

Figure 18-8

Standard curve for polarographic analysis of Al(III) in 0.2 M sodium acetate, pH 4.7, with 0.6 mM pontachrome violet SW used as a maximum suppressor. I_d is corrected for the residual current and for the diffusion current due to reaction of the maximum suppressor. [Data from H. H. Willard and J. A. Dean, *Anal. Chem.,* **22,** 1264 (1950).]

In the method of *standard additions,* a known amount of analyte is added to the unknown. The increase in signal intensity tells us how much analyte was present prior to the standard addition.

The method of standard additions relies on a linear relation between signal and concentration. If the polarographic response is not proportional to analyte concentration over the whole range of concentrations employed, there will be an error in the calculated concentration of unknown. It is impossible to detect this with a single standard addition, but several standard additions *might* indicate that the relation is not linear.

To apply the method of standard additions to any other technique, substitute the ratio of whatever signals are measured (such as absorbance or the height of chromatographic peaks) for the ratio I_d (unknown + standard)/ I_d(unknown).

The **standard addition method** is most useful when the sample matrix is unknown or difficult to duplicate in synthetic standard solutions. This method is faster but usually not as reliable as the method employing a standard curve.

First, a polarogram of the unknown is recorded. Then, a small volume of concentrated solution containing a known quantity of the analyte is added to the sample. *With the assumption that the response is linear,* the increase in diffusion current of this new solution (Figure 18-9) can be used to estimate the amount of unknown in the original solution. For greatest accuracy, several standard additions are made. Section 21-4 describes how a sequence of several consecutive standard additions can be used.

To see how the wave height is affected by the standard addition, consider the first sample containing just the unknown in supporting electrolyte. The diffusion current of the unknown will be proportional to the concentration of unknown, C_x:

$$I_d(\text{unknown}) = kC_x \tag{18-16}$$

where k is a constant of proportionality. Let the concentration of standard solution be C_s. When V_s mL of standard solution is added to V_x mL of unknown, the diffusion current is the sum of diffusion currents due to the unknown and the standard.

$$I_d(\text{unknown} + \text{standard}) = kC_x\left(\frac{V_x}{V_x + V_s}\right) + kC_s\left(\frac{V_s}{V_x + V_s}\right) \tag{18-17}$$

where the quantities in parentheses are dilution factors. Dividing the left side of Equation 18-17 by I_d(unknown), and dividing the right side by kC_x, allows us to rearrange and solve for C_x:

$$C_x = \frac{C_s V_s}{R(V_x + V_s) - V_x} \tag{18-18}$$

where R is the quotient I_d(unknown + standard)/I_d(unknown). All the quantities on the right side of Equation 18-18 are known, so C_x can be calculated. The standard addition method is most accurate when the wave height of the combined solution is about twice that of the unknown solution.

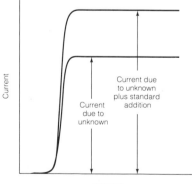

Current due to unknown plus standard addition

Current due to unknown

Current

Potential

Figure 18-9
Illustration of standard addition. The increased signal (current) when standard is added tells us how much unknown was present initially.

EXAMPLE: Standard Addition Calculation
A 25.0-mL sample of Ni^{2+} gave a wave height of 2.36 μA (corrected for residual current) in a polarographic analysis. When 0.500 mL of solution containing 28.7 mM Ni^{2+} was added, the wave height increased to 3.79 μA. Find the concentration of Ni^{2+} in the unknown.

Using Equation 18-18, we can write

$$C_x = \frac{(28.7 \text{ mM})(0.500 \text{ mL})}{\left(\dfrac{3.79 \ \mu A}{2.36 \ \mu A}\right)(25.0 + 0.500 \text{ mL}) - 25.0 \text{ mL}} = 0.900 \text{ mM}$$

The use of an **internal standard** is based on the fact that in a particular medium the diffusion currents due to two different species will have a ratio in direct proportion to their concentration ratio. The half-wave potentials of the two species must be sufficiently far apart ($\gtrsim 0.2$ V) that the limiting current of the first can be measured before the onset of the second wave.

To use this method, a polarogram of a known mixture of analyte (say, Tl^+) plus internal standard (say, Cd^{2+}) must be recorded (Figure 18-10). This polarogram establishes the relative response to the two species. Next, the unknown (containing an unknown concentration of Tl^+) is mixed with a known quantity of internal standard (Cd^{2+}), and the polarogram is recorded again. Comparing the ratios of diffusion currents in the two experiments tells us the concentration of unknown.

An *internal standard* is a known amount of a second compound added to the analyte. The ratio of signals due to known and unknown is compared with the ratio of signals from a solution in which both concentrations are known.

EXAMPLE: Using an Internal Standard

Chloroform and the pesticide DDT exhibit the following half-wave potentials in a medium consisting of 0.05 M $(CH_3)_4NBr$ in 3:1 (vol/vol) dioxane/water:

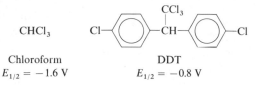

$CHCl_3$

Chloroform
$E_{1/2} = -1.6$ V

DDT
$E_{1/2} = -0.8$ V

Suppose that a polarogram of a mixture containing 0.500 mM chloroform and 0.800 mM DDT gives the following relative wave heights:

$$\frac{\text{Wave height of } CHCl_3}{\text{Wave height of } DDT} = 1.53 \quad \text{when} \quad \frac{[CHCl_3]}{[DDT]} = \frac{0.500 \text{ mM}}{0.800 \text{ mM}} = 0.625$$

To use chloroform as an *internal standard* for the analysis of DDT, a known concentration of chloroform is added to an unknown solution of DDT. Suppose that when the concentration of internal standard is 0.462 mM, the relative wave heights are

$$\frac{\text{Wave height of } CHCl_3}{\text{Wave height of } DDT} = 1.11 \quad \text{when} \quad \frac{[CHCl_3]}{[DDT]} = \frac{0.462 \text{ mM}}{\text{unknown}} = x$$

Find the concentration of DDT in the unknown.

To do this, we can set up a ratio. Let x be the quotient $[CHCl_3]/[DDT]$ in the mixture of unknown plus internal standard. When the relative wave heights were 1.53, the relative concentrations were 0.625. When the relative wave heights are 1.11, the relative concentrations, x, must be given by

$$\frac{\text{Relative concentrations in unknown}}{\text{Relative concentrations in known}} = \frac{\text{relative wave heights in unknown}}{\text{relative wave heights in known}}$$

$$\frac{x}{0.625} = \frac{1.11}{1.53} \quad \Rightarrow \quad x = 0.453 = \frac{[CHCl_3]}{[DDT]}$$

Putting in $[CHCl_3] = 0.462$ mM, we find

$$[DDT] = \frac{[CHCl_3]}{x} = \frac{0.462}{0.453} = 1.02 \text{ mM}$$

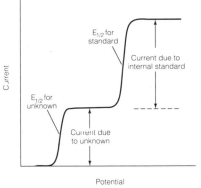

Figure 18-10
Use of an internal standard. The relative size of the two signals tells us how much unknown is present.

In the older literature, an intenal standard in polarography was called a "pilot ion."

Internal standards are most useful when loss of sample is unavoidable in the course of a chemical procedure. Suppose that a known amount of internal standard is added to an unknown. During subsequent operations, much of the sample might be lost or diluted. However, the *ratio* of concentrations of internal standard and unknown remains constant throughout. An analysis performed at the end of the operations will give the correct ratio of unknown to standard, regardless of any losses that occurred. As with the method of standard additions, the use of internal standard relies on a linear response to both analyte and standard over the entire range of concentrations employed.

Polarographic Study of Chemical Equilibrium

Polarography is one of the common techniques used in measuring equilibrium constants. To see why, consider a solution containing 1.00 mM Fe^{3+} and 1.00 mM Fe^{2+}. The potential for reduction of Fe^{3+} is given by

Any equilibrium that affects either $[Fe^{3+}]$ or $[Fe^{2+}]$ will alter the reduction potential of the solution.

$$E = 0.771 - 0.059\,16 \log \frac{[Fe^{2+}]}{[Fe^{3+}]} = 0.770 \qquad (18\text{-}19)$$

where 0.771 V is the standard reduction potential of Fe^{3+}. Suppose that a ligand that binds only to Fe^{3+} is added to the solution. The potential needed for reduction of Fe^{3+} will no longer be 0.771 V, because the quotient $[Fe^{2+}]/[Fe^{3+}]$ is no longer 1.00. Since the ligand decreases the concentration of Fe^{3+}, E will become more negative. It will be harder to reduce Fe^{3+} after the ligand has been added than in its absence.

The change in reduction potential can be predicted if we know the formation constant of the metal-ligand complex. Alternatively, measurement of the reduction potential tells us the magnitude of the formation constant. In polarography, the half-wave potential is related to the reduction potential of analyte and will therefore be sensitive to any equilibria involving the analyte.

As an example, consider the reduction of a complex ion to yield an amalgam plus free ligand:

$$ML_p^{n-pb} + ne^- + Hg \rightleftharpoons M(\text{in Hg}) + pL^{-b} \qquad (18\text{-}20)$$

where M = metal

    L = ligand

    n = charge of free metal

    b = charge of free ligand

    p = stoichiometry coefficient

For such a reaction, it can be shown that $E_{1/2}$ is given by

For Reaction 18-20, a graph of $E_{1/2}$ versus log $[L^{-b}]$ will give the values of p and the equilibrium constant.

$$E_{1/2} = E_{1/2}(\text{for free } M^{n+}) - \frac{0.059\,16}{n} \log \beta_p - \frac{0.059\,16p}{n} \log [L^{-b}] \quad (18\text{-}21)$$

where β_p is the equilibrium constant for the reaction $M^{n+} + pL^{-b} \rightleftharpoons ML_p^{n-pb}$. The value of $E_{1/2}$ (for free M^{n+}) refers to $E_{1/2}$ in a noncomplexing medium.

A graph of $E_{1/2}$ versus $\log[L^{-b}]$ should have a slope of $-0.059\ 16p/n$ and a y-intercept of $[E_{1/2}$ (for free M^{n+}) $-(0.059\ 16/n)\log\beta_p]$.

18-4 PULSE POLAROGRAPHY

Clever improvements have replaced the **direct current polarography** that we have been discussing. Our efforts have not been wasted, however, because it is necessary to understand classical polarography before studying the improvements. In favorable cases, direct current polarography has a detection limit around 10^{-5} M and can resolve species with half-wave potentials differing by $\gtrsim 0.2$ V. **Differential pulse polarography** is now the routine polarographic method, and it is likely to be replaced by **square wave polarography.** These techniques provide detection limits near 10^{-7} M and resolutions of 0.05 V. The arrangement of electrodes and sample is the same for direct current and pulse methods. It is the electronics that makes the difference.

Differential Pulse Polarography

In direct current polarography, the voltage applied to the working electrode increases linearly with time, as shown in Figure 18-11a. The current is recorded

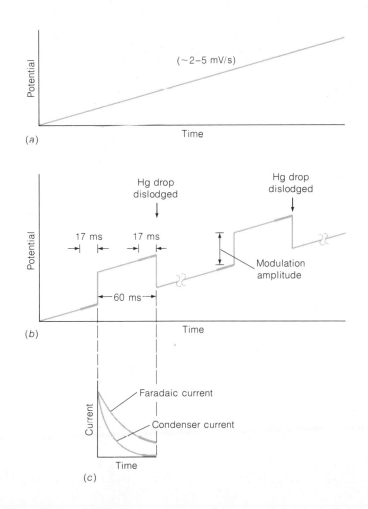

Figure 18-11

(a) Linear voltage ramp used in direct current polarography. (b) Pulsed ramp of differential pulse polarography. Current is measured only during the intervals shown by heavy lines. (c) Behavior of faradaic and condenser currents during each pulse.

Differential pulse polarography produces a curve that is nearly the derivative of an ordinary direct current polarogram.

continuously, and a polarogram such as that in Figure 18-2 results. The shape of the plot in Figure 18-11a is called a **linear voltage ramp.** In differential pulse polarography, small voltage pulses are superimposed on the linear voltage ramp, as in Figure 18-11b. The height of a pulse is called its **modulation amplitude.** Each pulse of magnitude 5–100 mV is applied during the last 60 ms of the life of each mercury drop. The drop is then mechanically dislodged. The current is not measured continuously. Rather, it is measured once before the pulse and again for the last 17 ms of the pulse. The polarograph subtracts the first current from the second and plots this difference versus the applied potential (measured just before the voltage pulse). The resulting differential pulse polarogram is nearly the *derivative* of a direct current polarogram, as shown in Figure 18-12.

To see why a derivative is produced, consider the idealized polarographic wave in Figure 18-13. Periodically, the polarograph steps up the voltage and

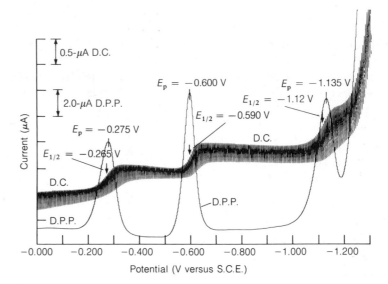

Figure 18-12

Comparison of direct current (D.C.) and differential pulsed polarography (D.P.P.) of 1.2×10^{-4} M chlordiazepoxide (the drug Librium) in 3 mL of 0.05 M H_2SO_4. Modulation amplitude = 50 mV. Note the different current scales for each curve. [M. R. Hackman, M. A. Brooks, J. A. F. de Silva, and T. S. Ma, *Anal. Chem.,* **46,** 1075 (1974).] The chemistry responsible for each wave is believed to be as shown below. [E. Jacobsen and T. V. Jacobsen, *Anal. Chim. Acta,* **55,** 293 (1971).]

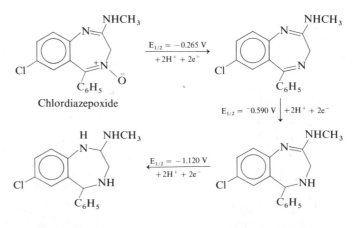

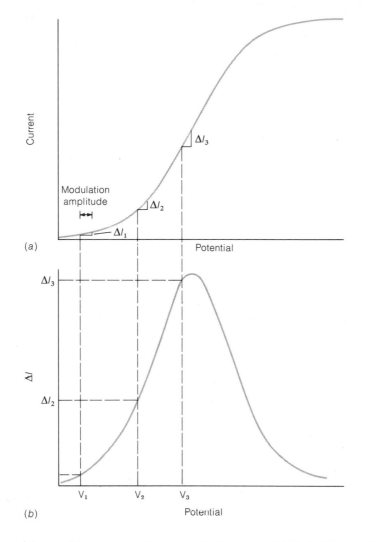

Figure 18-13
Illustration showing why the differential pulse polarogram, (b), is very nearly the derivative of the direct current polarogram, (a).

measures the resulting current increase. At the potential V_1 in Figure 18-13, the current increase (ΔI_1) is rather small. At the potential V_2, ΔI_2 is larger, and at the potential V_3, ΔI_3 is still larger. By plotting ΔI versus V, the differential pulse polarograph produces a curve that is nearly the derivative of the direct current polarogram.

Figure 18-14 shows the effect of increasing modulation amplitude in the analysis of a mixture of Fe(III) and Mn(II). As the modulation amplitude is increased, the signal increases, but the resolution (separation of neighboring peaks) decreases. If the modulation amplitude is made too great, the differential pulse polarogram no longer gives a faithful derivative shape.

Increasing the modulation amplitude increases the signal height, but decreases resolution of neighboring peaks.

The enhanced sensitivity of pulsed polarography compared with direct current polarography is due mainly to an increase in the *faradaic current* and a decrease in the *condenser current*. Consider the sample solution when the cathode potential is -0.200 V. The surface concentration of electroactive species is that which is consistent with -0.200 V. Suddenly, a pulse is applied and the voltage is changed to -0.250 V. If the voltage had been *slowly* changed from -0.200 to -0.250 V, the concentration of analyte near the electrode surface would have decreased. Instead, when the pulse is applied, the concentration of analyte is at the higher level consistent with -0.200 V.

Faradaic current is due to redox reactions. *Condenser current* is due to charging of the Hg drop.

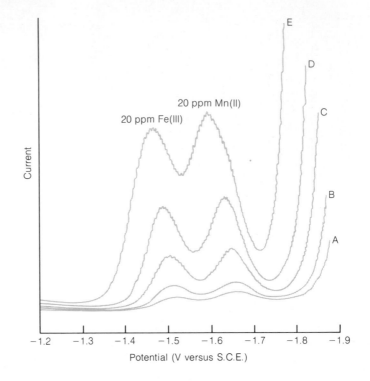

Figure 18-14

Effect of modulation amplitude on peak height and resolution in differential pulse polarography. Modulation amplitude: A = 5 mV, B = 10 mV, C = 25 mV, D = 50 mV, E = 100 mV. [Courtesy Princeton Applied Research Corp., Application Note 151.]

Condenser current decays exponentially with time, whereas faradaic current decay is proportional to $1/\sqrt{time}$. The exponential decay is faster.

Differential pulse polarography provides greater sensitivity and better resolution than does direct current polarography.

At the instant that each pulse is applied, there is a surge of electroactive species toward the working electrode. The faradaic current suddenly increases as the species reacts, and its concentration approaches a new steady-state value consistent with the new applied voltage. As the concentration of electroactive species is reduced, the faradaic current decays, as shown schematically in Figure 18-11c. At the instant the pulse is applied, the condenser current also increases to charge the drop up to its new equilibrium charge. As the drop becomes charged, the condenser current also decays, as shown in Figure 18-11c. The condenser current decays faster than the faradaic current. By the time the total current is measured (~40 ms after applying the pulse), the condenser current has decayed to near zero, but the faradaic current is substantial.

As compared with direct current polarography, pulsed measurement increases the faradaic current and almost eliminates the condenser current. Both of these effects enhance the sensitivity of the pulse technique. Another reason for increased sensitivity is that the current is measured only during the last 17 ms of the drop life, so the maximum area of the drop is available for electron transfer. Differential pulse polarography also provides better resolution of adjacent waves because it is easier to distinguish partially overlapping derivative maxima than partially overlapping polarographic waves.

The classical polarographic apparatus of Figure 18-1 is being replaced by more sophisticated equipment such as that in Figure 18-15. The cell and electrode are contained in the unit at the left. This device uses an electrically controlled, mechanical mechanism to suspend a static drop of mercury at the base of the electrode. After recording the current and the potential, the drop is mechanically dislodged and a fresh, identical drop is created. A new current is recorded at the new potential of the new drop. Because the mercury drop does not change its size during the polarographic measurements, no

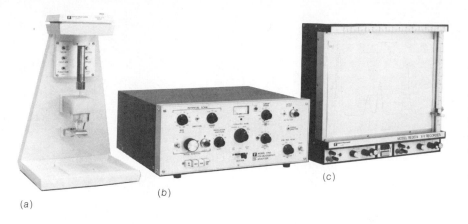

Figure 18-15
Modern polarographic apparatus. (a) This unit includes the dropping mercury electrode and polarographic cell. (b) The analyzer can perform direct current or pulsed experiments. (c) The results of the experiments are displayed on the $X-Y$ recorder. [Courtesy EG&G Princeton Applied Research Corp., Princeton, N.J.]

oscillations are observed in the polarogram. Furthermore, by waiting for a short time after the creation of each new drop, the condenser current decays to near zero (Figure 18-11c), and the signal-to-noise ratio is increased.

Square Wave Polarography

The advent of microprocessor-controlled electrochemical instrumentation allows chemists to create arbitrary waveforms that were not easily attained in the past. The *square wave polarography* waveform in Figure 18-16 offers advantages that may lead to the replacement of differential pulse as the standard form of polarography. The waveform in Figure 18-16 consists of a square wave superimposed on a voltage staircase. The current is measured for short times at the top and bottom of every square wave (regions 1 and 2 in the figure) and the difference $(I_1 - I_2)$ is plotted versus staircase potential. The resulting square wave polarogram is a symmetric peak similar to the differential pulse polarogram (Figure 18-13b).

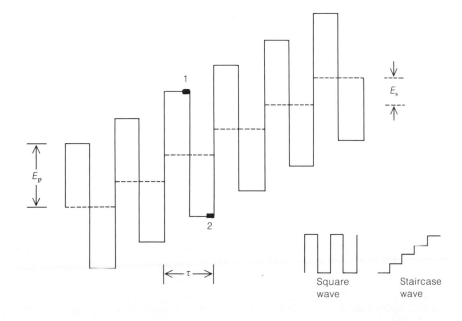

Figure 18-16
Waveform for square wave polarography. Typical parameters are pulse potential $(E_p) = 25$ mV, step height $(E_s) = 10$ mV, and pulse period $(\tau) = 5$ ms.

Figure 18-17
Three-dimensional square wave polarograms used to monitor elution of (a) *N*-nitrosoproline and (b) *N*-nitrosodiethanolamine from a liquid chromatography column. Polarographic peaks occur when the electroactive compounds reach the detector at the end of the column. [J. G. Osteryoung and R. A. Osteryoung, *Anal. Chem.*, **57**, 101A (1985).]

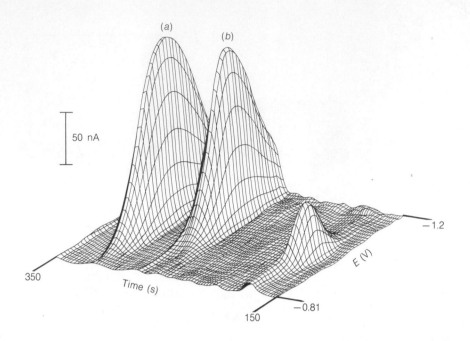

N-Nitrosoproline

$(HOCH_2CH_2)_2N—N=O$
N-Nitrosodiethanolamine

Square wave polarography is more sensitive and much faster than differential pulse polarography. The square wave is also better at rejecting background signals such as those generated by reduction of oxygen.

Square wave polarography offers advantages of sensitivity and speed over differential pulse polarography. The reverse (anodic) square wave pulse in Figure 18-16 causes reoxidation of the product of each forward (cathodic) pulse. The polarographic signal is the difference between the two currents. It is larger than in differential pulse polarography because the reverse current is not used in the differential pulse polarogram. A given concentration of analyte gives a larger signal in square wave polarography than in differential pulse polarography. The optimum pulse potential is about $50/n$ mV, where n is the number of electrons in the redox reaction. For reversible (i.e., rapid) electrode reactions, square wave polarography is about five times more sensitive than differential pulse polarography.

The square wave experiment is necessarily carried out faster than the differential pulse experiment. With a typical pulse period of 5 ms and step height of 10 mV, an entire polarogram with a 1-V width is obtained with *one drop of* Hg *in* 0.5 s. The corresponding differential pulse experiment takes about 100 times longer.

The increased speed opens applications not heretofore practical. Figure 18-17 shows how square wave polarography can be used for detection of compounds emerging from a liquid chromatograph. At each time interval, a 0.4-V scan is made and a complete polarogram is recorded. The compounds are distinguished by the times at which they emerge from the chromatography column and by their half-wave potentials.

18-5 STRIPPING ANALYSIS

In *stripping analysis,* analyte is first concentrated into a drop of Hg by reduction. The concentrated analyte is then oxidized by making the potential more positive. The polarographic signal is recorded during the oxidation process.

In **stripping analysis,** the analyte from a dilute solution is first concentrated in a single drop of Hg by electroreduction. The electroactive species is then *stripped* from the electrode by reversing the direction of the voltage sweep. The potential becomes more *positive, oxidizing* the species back into solution. The current measured during the oxidation is related to the quantity of analyte

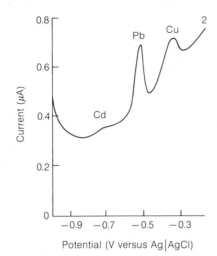

Figure 18-18
Anodic stripping voltammogram (differential pulse mode) of Sargasso seawater acidified to pH 2. Peaks for Cd and Cu correspond to 0.02 and 1.3 nmol/kg of seawater, respectively. Extreme precautions are necessary to avoid contamination when analyte concentrations are so low. [S. R. Piotrowicz, M. Springer-Young, J. A. Puig, and M. J. Spencer, *Anal. Chem.*, **54**, 1367 (1982).]

that was initially deposited. Stripping analysis can be done by conventional direct current polarography or by any variation, such as differential pulse polarography (Figure 18-18).

The customary setup for stripping analysis involves a **hanging-drop electrode.** Apparatus such as that in Figure 18-15 can create and suspend a single drop of Hg. Analyte is reduced and dissolved in the hanging drop by applying a constant potential more negative than the half-wave potential for the analyte. Because only a fraction of analyte from the solution is deposited, the deposition must be done for a reproducible time (such as 5 minutes) and with reproducible stirring. After concentrating analyte in the drop for a certain time, stirring is stopped and the potential is made more positive at a constant rate to reoxidize the analyte from the drop. The current measured during this oxidation reaches a maximum value that is proportional to the quantity of analyte that was deposited. A fresh drop of mercury is created for each analysis.

Stripping analysis is the most sensitive of polarographic techniques, because analyte is first concentrated from a dilute solution onto the electrode. The longer the period of concentration, the more sensitive is the analysis. With a 30-minute period of concentration, Ag^+ can be detected at a concentration of 2×10^{-12} M.[†]

Box 18-2 describes the measurement of Pb in blood by stripping analysis.

18-6 CYCLIC VOLTAMMETRY

Cyclic voltammetry is used principally to characterize the redox properties of compounds and to study the mechanisms of redox reactions. In this technique, the triangular waveform in Figure 18-19 is applied to the working electrode. The portion between times t_0 and t_1 is a linear voltage ramp. Unlike ordinary polarography, in which the ramp is applied over a period of a few minutes, in cyclic voltammetry the time is on the order of seconds. Furthermore, in cyclic voltammetry the ramp is then reversed to bring the potential back to its initial value at time t_2. The cycle may be repeated many more times.

Cyclic voltammetry uses a periodic, triangular waveform.

[†] S. Dong and Y. Wang, *Anal. Chim. Acta*, **212**, 341 (1988). For a general discussion of stripping analysis, see J. Wang, *Stripping Analysis: Principles, Instrumentation and Applications*, (Deerfield Beach, Florida: VCH Publishers, 1984).

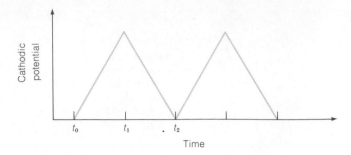

The current decreases after the cathodic peak because of *concentration polarization.*

The reaction in the upper curve of Figure 18-20 is

$$O_2 + e^- \rightleftharpoons O_2^-$$
Superoxide

Cyclic voltammograms are recorded with a computer, an oscilloscope, or a fast $X-Y$ recorder. The initial portions of the current–potential curves in Figure 18-20 look like ordinary polarograms, with a residual current followed by a **cathodic wave.** Instead of leveling off at the top of the wave, the current decreases as the potential is increased further. This happens because the electroactive species becomes depleted in the region around the electrode surface, and diffusion from bulk solution is too slow to replenish the concentration near the electrode. In the upper curve, at the time of the

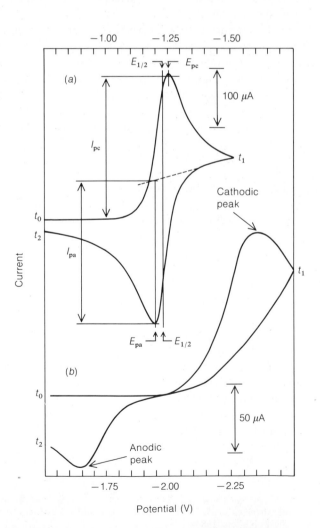

Box 18-2 THIN-LAYER DIFFERENTIAL PULSE POLAROGRAPHY[†]

The high sensitivity of differential pulse polarographic stripping analysis can be combined with the very small sample volume required by a *thin-layer cell* as shown in the figure below. Analyte is first reduced into a thin layer of Hg on the surface of the graphite working electrode at the bottom of the cell. Salt bridges leading to reference and auxiliary electrodes are at the top of the cell. Although the entire volume of sample in the cell is 70 μL, only 6 μL of solution in a 0.30-mm layer above the graphite surface is electrolyzed.

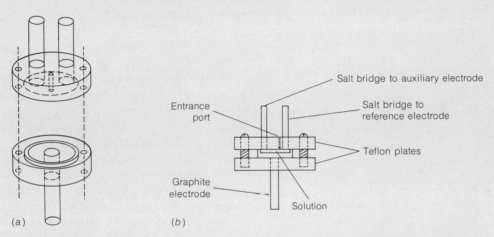

Thin-layer cell used for stripping analysis of very-small-volume samples. (a) Exploded view of cell. (b) Side view of assembled cell. [T. P. DeAngelis and W. R. Heineman, *Anal. Chem.,* **48,** 2262 (1976).]

The thin layer of Hg on the working electrode is actually deposited along with analyte by electrolysis of the sample solution, which contains 10^{-4} M Hg^{2+}. The electrolysis also reduces O_2 to H_2O, eliminating the need to degas the cell.

In one application, the cell was used to measure the Pb content of human blood, using the standard addition method. Standard Pb^{2+} at a concentration of 10^{-7} M (25 ppb) gave a signal height of 3 μA in the polarogram. Deposition and analysis each required just 60 seconds. Replicate 100-μL samples of blood gave a precision of 4% and an accuracy estimated at 9%.

[†] T. P. DeAngelis, R. E. Bond, E. E. Brooks, and W. R. Heineman, *Anal. Chem.,* **49,** 1792 (1977); T. P. DeAngelis and W. R. Heineman, *Anal. Chem.,* **48,** 2262 (1976).

maximum voltage (t_1), the cathodic current has decayed to a fairly small value. After t_1, the potential change is reversed, but a cathodic current continues to flow because the potential is still negative enough to reduce the analyte. When the potential becomes sufficiently less negative, the reduced analyte in the layer around the electrode surface begins to be oxidized. This gives rise to an **anodic wave.** Finally, as the reduced species is depleted, the anodic current decays back toward the initial value of the residual current.

Figure 18-20a illustrates the behavior of a *reversible* electrode reaction. By reversible, we mean that the redox reaction is fast enough to maintain equilibrium concentrations of reactant and product *at the electrode surface* as the electrode potential is varied. The peak anodic and peak cathodic currents have equal magnitudes in a reversible process, and the difference in potential between the peaks is

$$E_{pa} - E_{pc} = \frac{2.22\ RT}{nF} = \frac{57.0}{n}\ (mV) \tag{18-22}$$

where E_{pa} and E_{pc} are the potentials at which the *peak a*nodic and *peak c*athodic currents are observed and n is the number of electrons in the half-reaction. The half-wave potential, $E_{1/2}$, lies midway between the two peak potentials. With an irreversible reaction, the cathodic and anodic peaks become more drawn out and more separated (Figure 18-20b). At the limit of irreversibility, where the oxidation is very slow, no anodic peak would be seen.

> For a reversible reaction, $E_{1/2}$ lies midway between the cathodic and anodic peaks.

A study of the peak current as a function of the rate of change of applied potential permits an evaluation of the rate constant for the electrochemical reaction. If there are chemical reactions competing for the electrochemical reactants or products, the shape of the voltammogram will reflect the rates of these competing reactions.

Cyclic voltammetry is widely used to characterize the redox behavior of compounds and to elucidate the kinetics of electrode reactions and competing chemical reactions.[†] Common electrodes of Pt, C, Au, or Hg permit the study of either oxidations or reductions. Nonaqueous solvents such as alcohols, dioxane, acetonitrile, dimethylsulfoxide, and dimethylformamide are routinely employed with supporting electrolytes such as LiCl, LiClO$_4$, or tetraalkylammonium salts.

Identifying the species in a complex sequence of redox reactions is not a trivial matter. Cleverly designed cells permit us to measure optical and magnetic resonance spectra of some intermediates, thereby helping to establish structures. Box 18-3 describes how the optical absorbance spectrum can be measured concurrently with a voltammetric experiment.

Microelectrodes

> Advantages of microelectrodes:
>
> 1. They fit into small places.
> 2. They are useful in resistive, nonaqueous media (because of small ohmic losses).
> 3. Rapid voltage scans (possible because of small double-layer capacitance) allow short-lived species to be studied.

Extremely small electrodes with diameters of approximately 10 μm (one-tenth the diameter of a human hair) offer significant advantages in electrochemical analysis. A microelectrode can fit into very small places, such as the inside of a living cell. Since the surface area of the electrode is small, the rate at which molecules can reach the surface to be oxidized or reduced is low, and the current is tiny. With a small current, the ohmic drop ($= IR$) in a highly resistive medium is small. This allows microelectrodes to be used in poorly conducting nonaqueous solutions and even inside polymer membranes. The electrical capacitance of the double layer (Box 18-1) of a microelectrode is very small. Low capacitance allows the potential of the electrode to be varied much more rapidly than the potential of a conventional electrode. (Rates of 500 kV/s are possible.) Rapid scanning of the potential in cyclic voltammetry allows short-lived species with lifetimes less than 1 μs to be studied.

[†] The application of cyclic voltammetry to the study of chemical reaction mechanisms is nicely described by G. A. Mabbott, *J. Chem. Ed.,* **60,** 697 (1983): P. T. Kissinger and W. R. Heineman, *J. Chem. Ed.,* **60,** 702 (1983); and D. H. Evans, K. M. O'Connell, R. A. Petersen, and M. J. Kelly, *J. Chem. Ed.,* **60,** 290 (1983).

Box 18-3   AN OPTICALLY TRANSPARENT THIN-LAYER ELECTRODE

A very clever cell design permits the simultaneous measurement of the optical and electro-chemical properties of a solution.[†] The key element is a thin gold screen, which is the working electrode. As shown below, the screen is sandwiched between two microscope slides, which form a thin-layer cell. The screen has an optical transmittance of 82%, which means that a visible spectrum of the solution can be recorded by placing the cell in a spectrophotometer. The working volume of the cell is only 30–50 μL, and complete electrolysis of the solute requires only 30–60 s.

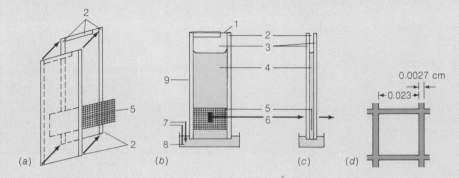

Diagram of an optically transparent thin-layer electrode. (a) Assembly of the cell. (b) Front view. (c) Side view. (d) Dimensions of gold grid. *Key:* (1) Point where suction is applied to change solution. (2) Teflon tape spacers. (3) Microscope slides. (4) Solution. (5) Gold grid electrode. (6) Optical path of spectrometer. (7) Reference and auxiliary electrodes. (8) Solution cup. (9) Epoxy cement. [T. P. DeAngelis and W. R. Heineman, *J. Chem. Ed.,* **53,** 594 (1976).]

The cell can be used for such purposes as characterization of polarographic products. It has also been used to measure the reduction potentials of colored redox enzymes[‡] and the lifetime of reduced chlorophyll.[§] The figure below shows a series of spectra recorded during the reduction of *o*-tolidine. From these spectra the fraction of *o*-tolidine reduced at each potential can be measured.

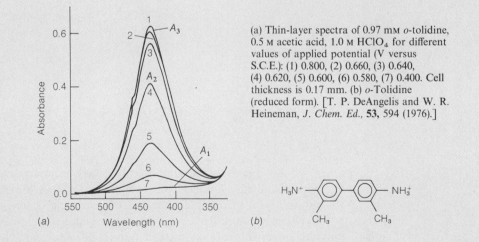

(a) Thin-layer spectra of 0.97 mM *o*-tolidine, 0.5 M acetic acid, 1.0 M $HClO_4$ for different values of applied potential (V versus S.C.E.): (1) 0.800, (2) 0.660, (3) 0.640, (4) 0.620, (5) 0.600, (6) 0.580, (7) 0.400. Cell thickness is 0.17 mm. (b) *o*-Tolidine (reduced form). [T. P. DeAngelis and W. R. Heineman, *J. Chem. Ed.,* **53,** 594 (1976).]

[†] T. P. DeAngelis and W. R. Heineman, *J. Chem. Ed.,* **53,** 594 (1976). A cell that uses an optical fiber and eliminates the gold screen is described by C. Zhang and S.-M. Park, *Anal. Chem.,* **60,** 1639 (1988).
[‡] W. R. Heineman, B. J. Norris, and J. F. Goelz, *Anal. Chem.,* **47,** 79 (1975).
[§] T. Watanabe and K. Honda, *J. Amer. Chem. Soc.,* **102,** 370 (1980).

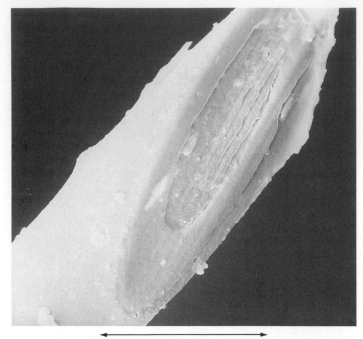

Figure 18-21
Electron micrograph of the tip of a Nafion-coated carbon fiber electrode. The carbon inside the electrode has a diameter of 10 μm. Nafion permits cations to pass but excludes anions. [Photo courtesy R. M. Wightman. Reproduced from R. M. Wightman, L. J. May and A. C. Michael, *Anal. Chem.*, **60**, 769A (1988).]

Figure 18-22
Cyclic voltammogram of dopamine using Nafion-coated carbon fiber electrode at pH 7.4. The scan begins and ends at -0.2 V. [Reproduced from R. M. Wightman, L. J. May, and A. C. Michael, *Anal. Chem.*, **60**, 769A (1988).] The oxidation half-reaction is

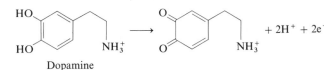

Dopamine

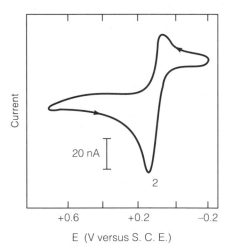

Figure 18-21 shows a carbon fiber microelectrode coated with a cation-exchange membrane called Nafion. This membrane, whose molecular structure was shown in Problem 17-25, resists attack in most chemical environments. Its key features are fixed negative charges and mobile cations. The Na^+ ions exchange with cations in the solution contacting the Nafion. Cations are transported rapidly through the membrane, but anions are excluded by the immobile negatively charged groups of the membrane. Figure 18-22 shows the cyclic voltammogram of the neurotransmitter dopamine, which is oxidized near $+0.12$ V (versus S.C.E.). Dopamine concentrations inside the brain of a rat can be measured with a Nafion-coated carbon fiber electrode. Ascorbate is present in high concentrations and would ordinarily interfere with the dopamine analysis. However, the Nafion membrane excludes the ascorbate anion but allows the dopamine cation to pass easily. The response to dopamine is 1000 times higher than the respone to ascorbate at the same concentration. By this means, the response of neurons to chemical and electrical stimuli can be studied.[†]

18-7 AMPEROMETRIC TITRATIONS

Amperometry refers to the measurement of electric current. An **amperometric titration** employs a current measurement to detect the end point of a titration. In Figure 17-12, an amperometric circuit was used to detect excess Br_2 at the end of a coulometric titration of cyclohexene. The detection system was simply a pair of platinum electrodes with a voltage of 0.2 V applied between them. Only a small residual current was observed until the end point was reached; beyond the end point, the current increased markedly. The reason for this behavior is that prior to the end point, only Br^- was present; beyond it, both Br_2 and Br^- were present. In the presence of both species, the two reactions below can carry substantial current between the electrodes.

$$\text{Cathode:} \quad Br_2 + 2e^- \rightarrow 2Br^- \quad (18\text{-}23)$$

$$\text{Anode:} \quad 2Br^- \rightarrow Br_2 + 2e^- \quad (18\text{-}24)$$

A platinum electrode is said to be **polarizable** because its potential is easily changed when only a small current flows. In contrast, a calomel electrode is said to be **nonpolarizable** because its potential remains very nearly constant unless a large current is flowing. Amperometric titrations utilize either one or two polarizable electrodes.

Systems with One Polarizable Electrode

Conventional polarography, utilizing a polarizable working electrode and a nonpolarizable reference electrode, can be used to follow the progress of a titration. Figure 18-23 shows polarographic results for the titration of 10^{-5} M Br^- with Ag^+. For each drop, before the equivalence point, almost all the Ag^+ precipitates, so only a tiny residual current is observed for the polarographic reduction of Ag^+.

The end point in Figure 18-23 is taken as the intersection of the two linear portions. Note that it is necessary to record points out as far as twice the equivalence volume in order to define the truly linear portion of the curve beyond the end point. Premature extrapolation of the data just beyond the equivalence volume gives an end point that is 8% low.

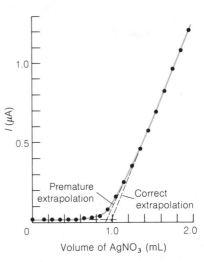

Figure 18-23
Argentometric titration of 100 mL of 10^{-5} M Br^- in 0.01 M HNO_3 with 0.005% gelatin. The potential of the rotating Pt working electrode was +0.15 V versus S.C.E. [J. T. Stock, *Amperometric Titrations* (New York: Wiley, 1965).]

The potential of a *polarizable electrode* is easily changed. The potential of a *nonpolarizable electrode* is hard to change.

Challenge: Write the titration reaction and the working electrode half-reaction for the titration in Figure 18-23.

[†] R. M. Wightman, L. J. May, and A. C. Michael, *Anal. Chem.*, **60**, 769A (1988); A. J. Cunningham and J. B. Justice, Jr., *J. Chem. Ed.*, **64**, A34 (1987). For a review of microelectrodes, see R. M. Wightman, *Science*, **240**, 415 (1988).

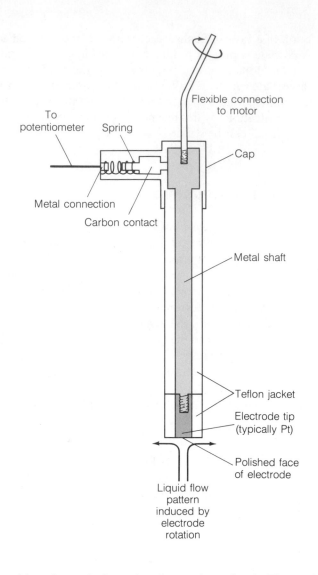

Figure 18-24
Rotating disk electrode.

Box 18-4 describes one of the most important applications of amperometry: the measurement of dissolved oxygen.

The working electrode for a titration such as that in Figure 18-23 is the **rotating disk electrode** in Figure 18-24. This electrode spins at a constant rate of ~ 600 rpm, bringing analyte to the surface by convection as shown in Figure 18-24. The resulting current is much greater than that arising from diffusion alone. The polished bottom surface of the electrode is the only part in electrical contact with the analyte. While a typical diameter for a rotating disk electrode is 5 mm, rotating microelectrodes with dimensions measured in micrometers are also used.

The rotating Pt electrode is preferred over the dropping Hg electrode for very easily reduced species—such as Ag^+, Br_2, and Fe(III)—and for anodic reactions. The Hg electrode cannot be used for many anodic reactions because the electrode itself is easily oxidized. The Pt electrode has a less useful cathodic range because of the low overpotential for H^+ reduction at the Pt surface. For anodic reactions with the Pt electrode, the solution need not be deoxygenated, since the potential is too positive to reduce O_2. The current produced by the rotating Pt electrode is not as reproducible as that of the Hg electrode, but this is not critical for finding the end point of a titration.

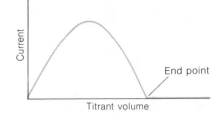

Figure 18-25
Schematic biamperometric titration curve for the addition of $S_2O_3^{2-}$ to I_2.

Systems with Two Polarizable Electrodes

In ordinary polarography, we use one polarizable (working) electrode and one nonpolarizable (reference) electrode. For the amperometric detection of a titration end point, it is often desirable to have two polarizable electrodes. An amperometric titration employing two polarizable electrodes is called a **biamperometric titration.**

Figure 18-25 illustrates the shape of the curve obtained for the biamperometric titration of I_2 with $S_2O_3^{2-}$. The titration reaction is

$$2S_2O_3^{2-} + I_2 \rightarrow 2I^- + S_4O_6^{2-} \qquad (18\text{-}25)$$

The $I_2 | I^-$ couple reacts reversibly at a Pt electrode, but the $S_2O_3^{2-} | S_4O_6^{2-}$ couple does not. That is, of the following reactions, the first occurs but the second does not.

$$I_2 + 2e^- \rightleftharpoons 2I^- \qquad (18\text{-}26)$$

$$S_4O_6^{2-} + 2e^- \rightleftharpoons 2S_2O_3^{2-} \qquad (18\text{-}27)$$

Now consider what happens when $S_2O_3^{2-}$ is added to I_2. At the beginning of the titration, no I^- is present and only residual current is observed. As the reaction proceeds, I^- is created and current is conducted between the two polarizable electrodes by means of the following reactions:

$$\text{Cathode:} \qquad I_2 + 2e^- \rightarrow 2I^- \qquad (18\text{-}28)$$

$$\text{Anode:} \qquad 2I^- \rightarrow I_2 + 2e^- \qquad (18\text{-}29)$$

The current reaches a maximum near the middle of the titration, when both I_2 and I^- are present. The current then decreases as more I_2 is consumed by Reaction 18-25. Beyond the end point, there is no I_2 present and only residual current is observed.

In a *biamperometric* titration, the current between two electrodes that are held at a constant potential difference is measured.

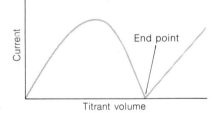

Figure 18-26
Schematic biamperometric titration curve for the addition of Ce^{4+} to Fe^{2+}.

The end point of a biamperometric titration is sometimes called a *dead stop* end point.

Challenge: A schematic biamperometric titration curve for the addition of Ce^{4+} to Fe^{2+} is shown in Figure 18-26. The titration reaction is

$$Ce^{4+} + Fe^{2+} \rightarrow Ce^{3+} + Fe^{3+}$$

and both couples ($Ce^{4+} | Ce^{3+}$ and $Fe^{3+} | Fe^{2+}$) react reversibly at the Pt electrodes. Explain the shape of the titration curve.

Box 18-4 OXYGEN SENSORS

The simplest means of measuring dissolved O_2 is with a **Clark electrode.** This combination electrode consists of a Pt cathode held at a potential of -0.6 V with respect to a silver–silver chloride anode. The cell is covered by a semipermeable Teflon membrane, across which oxygen can diffuse in a few seconds. The electrode is dipped into a sample solution, and a short time is allowed for oxygen to equilibrate between the sample and the electrode solution. The current that flows between the two electrodes is proportional to the dissolved oxygen concentration:

Cathode: $\qquad O_2 + 2H^+ + 2e^- \rightleftharpoons H_2O_2$

Anode: $\qquad 2Ag + 2Cl^- \rightleftharpoons 2AgCl + 2e^-$

The electrode must first be calibrated in solutions of known oxygen concentration.[†]

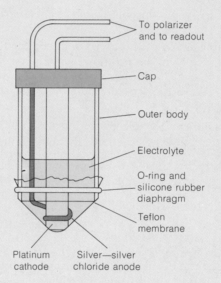

Clark oxygen electrode. [D. T. Sawyer and J. L. Roberts, Jr., *Experimental Electrochemistry for Chemists* (New York: Wiley, 1974).]

[†] Construction of an inexpensive oxygen electrode has been described by J. E. Brunet, J. I. Gardiazabal, and R. Schrebler, *J. Chem. Ed.,* **60,** 677 (1983).

Karl Fischer titration of H$_2$O

A very important technique in analytical chemistry is the analysis of water by the **Karl Fischer titration.** This sensitive method can be used to measure residual water in purified solvents and water of hydration in crystals. The Karl Fischer reagent consists of I_2, pyridine, and SO_2 in a $1:10:3$ mole ratio, dissolved in methanol or in ethylene glycol monomethyl ether ($CH_3OCH_2CH_2OH$). Addition of water from the analyte begins the following sequence of reactions:

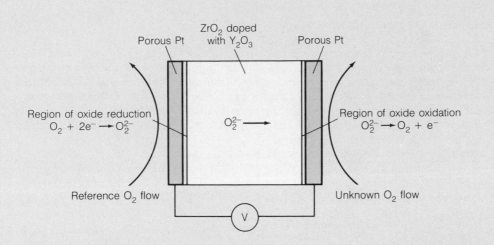

The Clark electrode can be miniaturized to fit into the tip of a 1.5-mm-diameter surgical catheter that is stored in a dry, sterile state.[‡] The catheter is inserted into the aorta of a newborn infant via the umbilical artery. Upon contact with blood, water diffuses into the KCl electrolyte and activates the electrode, which monitors blood oxygen concentration during respiratory distress. The sensor responds within 20–50 s to changes to P_{O_2} associated with mechanical ventilation of the child, or breathing pure O_2.

A completely different type of sensor was developed by the National Aeronautics and Space Administration to monitor the oxygen concentration in hot combustion gases.[§] The heart of the sensor is a disk of Y_2O_3-doped ZrO_2 coated with porous Pt electrodes. Just as LaF_3 doped with EuF_2 contains mobile fluoride ions (Figure 15-14), ZrO_2 doped with Y_2O_3 contains oxygen vacancies that allow oxide ions to diffuse readily through the solid at elevated temperature. In the diagram above, one side of the sensor is exposed to unknown exhaust gas and the other side is exposed to a reference gas (such as the atmosphere). Suppose that the O_2 concentration is higher on the left side than on the right. Oxygen gas diffuses through the porous Pt electrode at the left and is reduced to oxide ion at the surface of the Y_2O_3-doped ZrO_2. The oxide ions diffuse through the solid and are oxidized at the right-hand electrode. The driving force is the difference in concentration of O_2 on the two sides of the sensor. The redox chemistry gives a voltage difference between the two electrodes that is measured by a potentiometer. The sensor temperature is maintained at 843°C for this application and is calibrated at known O_2 concentrations.

[‡] D. Parker, *J. Phys. E: Sci. Instrum.,* **20,** 1103 (1987).
[§] *NASA Technical Briefs,* **9,** 105 (1985).

$$\langle\bigcirc\rangle N\cdot I_2 + \langle\bigcirc\rangle N\cdot SO_2 + \langle\bigcirc\rangle N + H_2O \longrightarrow$$

$$2\langle\bigcirc\rangle NH^+I^- + \langle\bigcirc\rangle \overset{+}{N}-SO_3^- \quad (18\text{-}30)$$

$$\langle\bigcirc\rangle \overset{-}{N}{}^+-SO_3^- + CH_3OH \longrightarrow \langle\bigcirc\rangle NH^+CH_3OSO_3^- \quad (18\text{-}31)$$

Demonstration 18-1 THE KARL FISCHER JACKS OF A pH METER

Most pH meters contain a pair of sockets at the back that are labeled "K–F" or "Karl Fischer." When the manufacturer's instructions are followed, a constant current (usually around 10 μA) is applied across these terminals. To perform a bipotentiometric titration, a pair of Pt electrodes is connected to the K–F sockets. The meter is set to the millivolt scale, which indicates the voltage needed to maintain the constant current between the electrodes immersed in a solution.

The figure below shows the results of a bipotentiometric titration of ascorbic acid with I_3^-. Ascorbic acid (146 mg) was dissolved in 200 mL of water in a 400-mL beaker. Two Pt electrodes spaced about 4 cm apart were immersed in the solution, which was magnetically stirred. Each electrode was attached to one of the K-F outlets of the pH meter. The solution was titrated with 0.04 M I_3^- (prepared by dissolving 2.4 g of KI plus 1.2 g of I_2 in 100 mL of water), and the voltage was recorded after each addition.

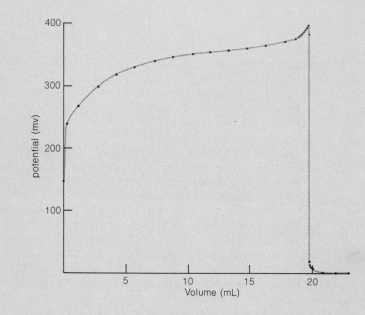

Prior to the equivalence point, all the I_3^- is reduced to I^- by the excess ascorbic acid. Reaction 18-29 can take place at the anode, but Reaction 18-28 cannot occur at the cathode. A voltage of about 300 mV is required to support a constant current of 10 μA. (The ascorbate|dehydroascorbate couple does not react at a Pt electrode and cannot carry current.) After the equivalence point, excess I_3^- is present, so Reactions 18-28 and 18-29 can both occur, and the voltage drops precipitously.

The alcoholic solvent is needed to drive Reaction 18-31 (and therefore Reaction 18-30) to the right. The most common procedure is to titrate a known volume of standardized Karl Fischer reagent with a solution of the unknown in methanol. The reagent is first standardized with a methanol solution containing a known amount of water.

A *bipotentiometric* measurement is the most common way to detect the end point of a Karl Fischer titration. Most pH meters have a pair of electrode receptacles for a Karl Fischer (or other bipotentiometric) titration. The meter

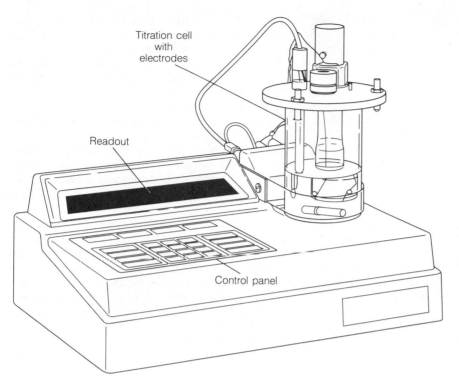

Titration cell
with
electrodes

Readout

Control panel

Figure 18-27
Automatic coulometric Karl Fischer titrator. [Courtesy Photovolt Corp.]

maintains a *constant current* (usually 5 or 10 μA) between electrodes plugged
into these sockets. When these electrodes are immersed in a solution containing I_2 and I^-, this current flows at a very low applied voltage. Prior to
the equivalence point, the solution contains I_2 from the Karl Fischer reagent
and I^- produced by Reaction 18-30. At the end point, the I_2 has been
consumed. In order to maintain a current of 10 μA, the cathode potential
must change dramatically to carry current by means of the reduction of
solvent:

$$CH_3OH + e^- \rightleftharpoons CH_3O^- + \tfrac{1}{2}H_2 \qquad (18\text{-}32)$$

The sudden change of voltage (read directly on the pH meter) marks the end
point. The entire operation can be carried out automatically (Figure 18-27).
Demonstration 18-1 shows how the Karl Fischer jacks of a pH meter are
used.

Summary

In direct current polarography, we observe the current
flowing through an electrochemical cell as a function of
the potential applied to a dropping-mercury working
electrode. This electrode gives reproducible polarograms
because fresh surface is continuously exposed. The overpotential for H^+ reduction allows the observation of cathodic processes that are inherently less favorable than H^+
reduction. Most anodic processes must be observed with
other electrodes because Hg is too readily oxidized. Polarography is useful in quantitative analysis because the
diffusion current ($=$ limiting current $-$ residual current)
is generally proportional to analyte concentration. Methods employing standard curves, standard additions, or
internal standards are most common in quantitative analysis. Polarography is also useful in qualitative analysis,
since $E_{1/2}$ varies characteristically with each electroactive

species. The shape of a polarographic wave can be used to find the number of electrons participating in a reversible electrode reaction. The dependence of $E_{1/2}$ on ligand concentration can be used to measure metal–ligand stability constants.

In differential pulse polarography, the voltage ramp is periodically stepped up, and current is measured during the last part of the life of each mercury drop. This process produces a derivative-shaped polarogram. Residual current is diminished because the condenser current for each mercury drop decreases more rapidly than does the faradaic current. The signal height can be increased (up to a point) by increasing the modulation amplitude, and the resolution of neighboring waves is better than in direct current polarography. Square wave polarography provides a further improvement in sensitivity because both reduction and oxidation reactions contribute to the observed signal. The greatly increased speed of square wave polarography allows real-time measurements not possible with other electrochemical methods.

In anodic stripping voltammetry, the most sensitive form of polarography, analyte is concentrated into a single drop of mercury by reduction at a fixed voltage for a fixed time. The potential is then made more positive, and current is measured as the analyte is spontaneously oxidized and leaves the mercury drop. In cyclic voltammetry, a triangular waveform is applied, and cathodic and anodic processes are observed in succession. Microelectrodes are particularly useful because they fit into small places and their low current allows them to be used in resistive, nonaqueous media. Their low capacitance permits very rapid voltage scanning, which allows very short-lived species to be studied.

In amperometric titrations, the current flowing between a pair of electrodes is used to locate the end point. Systems with one polarizable electrode typically use ordinary polarographic apparatus or a rotating Pt electrode to measure the concentration of one species in the reaction. A system with two polarizable electrodes can be used for two types of measurements. In a biamperometric titration, there is a fixed potential difference between the electrodes, and the current needed to sustain it is recorded. In a bipotentiometric titration, the voltage needed to maintain a constant current is measured. The current or the voltage changes abruptly at the end point in these titrations, because at the end point one member of an electroactive redox couple is either created or destroyed.

Terms to Understand

amperometric titration
amperometry
anodic wave
biamperometric titration
bipotentiometric titration
capillary constant
cathodic wave
charging current
Clark electrode
concentration polarization
condenser current
current maximum
cyclic voltammetry
differential pulse polarography

diffusion current
diffusion layer
direct current polarography
dropping-mercury electrode
electric double layer
electrocapillary maximum
faradaic current
half-wave potential
hanging-drop electrode
internal standard
Karl Fischer titration
linear voltage ramp
maximum suppressor
microelectrode

modulation amplitude
nonpolarizable electrode
polarizable electrode
polarogram
polarographic wave
polarography
residual current
rotating disk electrode
square wave polarography
standard addition method
standard curve
stripping analysis
supporting electrolyte
voltammetry

Exercises

18-A. The analysis of Ni(II) at the nanogram level is possible using differential pulse polarography.[†]

[†] C. J. Flora and E. Neiboer, *Anal. Chem.,* **52,** 1013 (1980).

Addition of dimethylglyoxime to an ammonium citrate buffer enhances the response to Ni(II) by a factor of 15. Some representative data are given at the top of the next column.

Ni(II)(ppb)	Peak current (μA)
19.1	0.095
38.2	0.173
57.2	0.258
76.1	0.346
95.0	0.429
114	0.500
132	0.581
151	0.650
170	0.721

Construct a standard curve from these data and use it to answer the following question: What current is expected if 54.0 μL of solution containing 10.0 ppm Ni(II) is added to 5.00 mL of buffer for polarographic analysis?

18-B. Cd^{2+} was used as an internal standard in the analysis of Pb^{2+} by differential pulse polarography. Cd^{2+} gives a reduction wave at $-0.60\,(\pm0.02)$ V and Pb^{2+} gives a reduction wave at $-0.40\,(\pm0.02)$ V. It was first verified that the ratio of peak heights is proportional to the ratio of concentrations over the whole range employed in the experiment. Results for known and unknown mixtures are given below.

	Concentration (M)	Current (μA)
KNOWN		
Cd^{2+}	$3.23\,(\pm0.01)\times10^{-5}$	$1.64\,(\pm0.03)$
Pb^{2+}	$4.18\,(\pm0.01)\times10^{-5}$	$1.58\,(\pm0.03)$
UNKNOWN		
Cd^{2+}	?	$2.00\,(\pm0.03)$
Pb^{2+}	?	$3.00\,(\pm0.03)$

The second sample in the table was prepared by mixing 25.00 (±0.05) mL of unknown plus 10.00 (±0.05) mL of $3.23\,(\pm0.01)\times10^{-4}$ M Cd^{2+} and diluting to 50.00 (±0.05) mL.
(a) Without regard to uncertainties, find the concentration of Pb^{2+} in the undiluted unknown.
(b) Find the absolute uncertainty associated with the answer to part a.

18-C. A large quantity of Fe(III) interferes with the polarographic analysis of Cu(II) because the Fe(III) is reduced at less negative potentials than Cu(II) in most supporting electrolytes. The Fe(III) reduction wave can be eliminated by addition of hydroxylamine (NH_2OH), which reduces Fe(III), but not Cu(II). The figure in the next column shows the differential pulse polarogram of Cu(II) in the presence of 1000 ppm of Fe(III) with saturated $NH_3OH^+Cl^-$ as supporting electrolyte. Each sample in this analysis was made up to the *same final volume*. Averaging the response for the two standard additions, calculate the concentration of the sample solution.

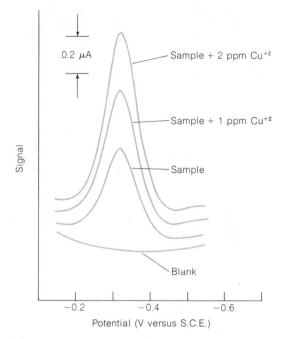

Differential pulse polarograms. Modulation amplitude: 25 mV. Drop interval: 1 second. Scan rate: 2 mV/s. Note that each scan is offset (displaced vertically) from the previous one. When you measure the peak heights relative to the blank, subtract the vertical displacement at the left edge of the scan. [Courtesy EG&G Princeton Applied Research Corp., Application Note 151.]

18-D. Polarographic data for the reaction

$$HPbO_2^- + H_2O + 2e^- + Hg \rightleftharpoons$$
$$Pb(Hg) + 3OH^-$$

are given below.[†]

$E_{1/2}$(V, versus S.C.E.)	$[OH^-]$ (M, free OH^-)
-0.603	0.011
-0.649	0.038
-0.666	0.060
-0.681	0.099
-0.708	0.201
-0.734	0.448
-0.747	0.702
-0.755	1.09

[†] J. J. Lingane, *Chem. Rev.*, **29**, 1 (1941).

The species $HPbO_2^-$ can be treated as $Pb(OH)_3^-$ by virtue of the equilibrium

$$Pb(OH)_3^- \rightleftharpoons HPbO_2^- + H_2O$$

Use the data above to show that $p = 3$ in Equation 18-20, and find the value of β_3 for $Pb(OH)_3^-$. The value of $E_{1/2}$ (for free Pb^{2+}) in Equation 18-21 is -0.41 V. This is the half-wave potential for the reaction $Pb^{2+} + 2e^- \rightleftharpoons Pb(Hg)$ in 1 M HNO_3.

18-E. Shown at the right is a cyclic voltammogram of the compound $Co(III)(B_9C_2H_{11})_2^-$ in 1,2-dimethoxyethane solution.

$E_{1/2}$(V, versus S.C.E.)	I_{pa}/I_{pc}	$E_{pa} - E_{pc}$(mV)
-1.38	1.01	60
-2.38	1.00	60

Suggest a chemical reaction to account for each wave. Are the reactions reversible? How many electrons are involved in each step? Sketch the direct current and differential pulse polarograms expected for this compound.

18-F. Amperometric titration curves for the precipitation of Pb(II) with $Cr_2O_7^{2-}$ are shown below. The potential of the dropping-mercury electrode was -0.8 V versus S.C.E. for curve a and 0.0 V versus S.C.E. for curve b. Explain the shapes of the two curves.

18-G. Sketch the shape of a bipotentiometric titration curve (E versus volume added) for the addition of Ce(IV) to Fe(II). Both the Ce(IV)|Ce(III) and Fe(III)|Fe(II) couples react reversibly at Pt electrodes.

18-H. The Karl Fischer reagent is usually standardized by titration with H_2O dissolved in methanol. A 25.00-mL aliquot of Karl Fischer reagent reacted with 34.61 mL of methanol to which was added

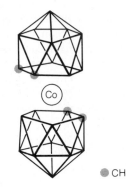

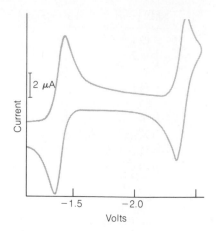

Cyclic voltammogram of $Co(III)(B_9C_2H_{11})_2^-$. [W. E. Geiger, Jr., W. L. Bowden, and N. El Murr, *Inorg. Chem.*, **18**, 2358 (1979).]

4.163 mg of H_2O per mL. When pure "dry" methanol was titrated, 25.00 mL of methanol reacted with 3.18 mL of the same Karl Fischer reagent. A suspension of 1.000 g of a hydrated crystalline salt in 25.00 mL of methanol consumed a total of 38.12 mL of Karl Fischer reagent. Calculate the weight percent of water in the crystal.

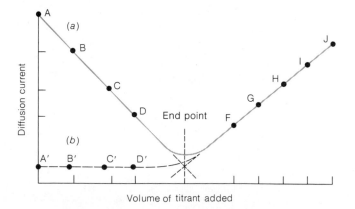

Amperometric titration curves. [D. T. Sawyer and J. L. Roberts, Jr., *Experimental Electrochemistry for Chemists* (New York: Wiley, 1974).]

Problems

A18-1. Why does the reaction $Cu(I) \rightarrow Cu(0)$ have different half-wave potentials with Hg and Pt electrodes? Will the half-wave potential for Reaction 18-10 be dependent on the composition of the electrode?

A18-2. Why is a dropping-mercury electrode preferred for cathodic reactions in amperometric titrations, while a rotating platinum disk electrode is preferred for anodic reactions?

A18-3. In 1 M NH_3–1 M NH_4Cl solution, Cu(II) is reduced to Cu(I) near -0.3 V (vs S.C.E.), and Cu(I) is reduced to Cu(Hg) near -0.6 V.
  (a) Sketch a qualitative polarogram for a solution of Cu(I).
  (b) Sketch a qualitative polarogram for a solution of Cu(II).
  (c) Suppose that Pt, instead of Hg, were used as the working electrode. Which, if any, reduction potential would you expect to change?

A18-4. For the data in Table 18-2, $m^{2/3}t^{1/6} = 2.28$ $mg^{2/3}/s^{1/2}$. Calculate the diffusion coefficient for Pb^{2+} in 1.0 M KNO_3. The limiting current in Table 18-2 is given by Equation 18-8, using the constant 6.07×10^4 for the reason described in the footnote on page 461.

A18-5. What is a Clark electrode and how does it work?

A18-6. Draw graphs showing the voltage ramps used in direct current and differential pulse polarography. Label the axes. Label the modulation amplitude and show the two regions between which ΔI is measured on the differential pulse ramp.

A18-7. Why is differential pulse polarography more sensitive than direct current polarography?

A18-8. Explain how the amperometric end-point detection in the experiment in Figure 17-12 works.

A18-9. A solution for polarographic analysis was prepared by mixing 10.0 mL of unknown containing Al(III) with buffer and diluting to 100.0 mL. The final composition was the same as in Figure 18-8. The measured polarographic diffusion current was 2.0 μA. Use the standard curve in Figure 18-8 to calculate the concentration of Al(III) in the unknown. Remember that the sample was diluted before making the measurement.

A18-10. An unknown gave a polarographic signal of 10.0 μA. When 1.00 mL of 0.050 0 M standard was added to 100.0 mL of unknown, the signal increased to 14.0 μA. Use the standard addition formula given by Equation 18-18 to find the concentration of the original unknown.

A18-11. Consider the example of internal standards involving chloroform and DDT on page 473. Suppose that a mixture containing 1.00 mM $CHCl_3$ and 1.00 mM DDT gave polarographic signals in the proportion

$$\frac{\text{Wave height of } CHCl_3}{\text{Wave height of DDT}} = 1.40$$

An unknown solution of DDT was treated with a tiny amount of pure $CHCl_3$ to give a concentration of 0.500 mM $CHCl_3$, without significantly changing the concentration of unknown. Now the relative signals are found to be

$$\frac{\text{Wave height of } CHCl_3}{\text{Wave height of DDT}} = 0.86$$

Find the concentration of DDT in the unknown.

18-12. For the polarogram in Figure 18-2, the diffusion current is 14 μA.
  (a) If the solution contains 25 mL of 0.50 mM Cd^{2+}, calculate the fraction of Cd^{2+} reduced per minute of current passage.
  (b) If the drop interval is 4 s, how many minutes were required to scan from -0.6 V to -1.2 V?
  (c) By what percent has the concentration of Cd^{2+} changed during the scan from -0.6 V to -1.2 V?

18-13. Use Equation 18-11 to derive an expression for $E_{3/4} - E_{1/4}$, where $E_{3/4}$ is the potential when $I = \frac{3}{4}I_d$ and $E_{1/4}$ applies when $I = \frac{1}{4}I_d$. The value of $E_{3/4} - E_{1/4}$ is used as a criterion for reversibility of a polarographic wave. If $E_{3/4} - E_{1/4}$ is close to the value you have calculated, the reaction is probably reversible.

18-14. Polarographic data for the reduction of Al(III) in 0.2 M sodium acetate, pH 4.7, are given on the next page.[†] Construct a standard curve and determine the best straight line by the method of least squares. Calculate the standard deviation for the slope and the intercept. If an unknown solution gives $I_d = 0.904$ μA, calculate the concentration of Al(III) and estimate the uncertainty in concentration, using the procedure of Section 4-4.

---

† H. H. Willard and J. A. Dean, *Anal. Chem.*, **22**, 1264 (1950).

[Al(III)] (mM)	I_d (corrected for residual current) (µA)
0.009 25	0.115
0.018 5	0.216
0.037 0	0.445
0.055 0	0.610
0.074 0	0.842
0.111	1.34
0.148	1.77
0.185	2.16
0.222	2.59
0.259	3.12

18-15. The diffusion currents below were measured at -0.6 V for $CuSO_4$ in 2 M $NH_4Cl/2$ M NH_3.[†] Use the method of least squares in Section 4-4 to estimate the molarity and uncertainty in molarity of an unknown solution giving $I_d = 15.6$ µA.

[Cu(II)] (mM)	I_d (µA)	[Cu(II)] (mM)	I_d (µA)
0.039 3	0.256	0.990	6.37
0.078 0	0.520	1.97	13.00
0.158 5	1.058	3.83	25.0
0.489	3.06	8.43	55.8

18-16. The differential pulse polarogram of 3.00 mL of solution containing the antibiotic tetracycline in 0.1 M acetate, pH 4, gives a maximum current of 152 nA at a half-wave potential of -1.05 V (versus S.C.E.). When 0.500 mL containing 2.65 ppm of tetracycline was added, the current increased to 206 nA. Calculate the parts per million of tetracycline in the original solution.

18-17. Br_2 can be generated for quantitative analysis by addition of standard BrO_3^- to excess Br^- in acidic solution:

$$BrO_3^- + 5Br^- + 6H^+ \rightleftharpoons 3H_2O + 3Br_2$$

Consider the titration of As(III) with Br_2, using biamperometric end-point detection:

$$Br_2 + H_3AsO_3 + H_2O \rightleftharpoons$$
$$2Br^- + H_3AsO_4 + 2H^+$$

A solution containing As(III) and Br^- is titrated with standard BrO_3^-. Given that the $H_3AsO_4|$

[†] I. M. Kolthoff and J. J. Lingane, *Polarography,* Vol. I (New York: Wiley, 1952), p. 378.

H_3AsO_3 couple does not react at a Pt electrode, predict the shape of the titration curve. The titration curve is a graph of I versus volume of titrant.

18-18. Predict the shape of a bipotentiometric titration curve for the titration of H_3AsO_3 with I_2. You should sketch a graph of E versus volume of I_2 added.

18-19. Ammonia can be titrated with hypobromite, but the reaction is somewhat slow:

$$2NH_3 + 3OBr^- \rightleftharpoons N_2 + 3Br^- + 3H_2O$$

The titration can be performed with a rotating Pt electrode held at $+0.20$ V (versus S.C.E.) to monitor the concentration of OBr^-:

Cathode: $OBr^- + 2H^+ + 2e^- \rightleftharpoons Br^- + H_2O$

The current would be near zero before the equivalence point if the OBr^- were consumed quickly in the titration. However, the sluggish titration reaction does not consume all the OBr^- after each addition, and some current is observed. The increasing current beyond the equivalence point can be extrapolated back to the residual current to find the endpoint. A solution containing 30.0 mL of 4.43×10^{-5} M NH_4Cl was titrated with NaOBr in 0.2 M $NaHCO_3$, with the following results:

OBr^- (mL)	I (µA)	OBr^- (mL)	I (µA)
0.000	0.03	0.700	1.63
0.100	1.42	0.720	2.89
0.200	2.61	0.740	4.17
0.300	3.26	0.760	5.53
0.400	3.74	0.780	6.84
0.500	3.79	0.800	8.08
0.600	3.20	0.820	9.36
0.650	2.09	0.840	10.75

Prepare a graph of current versus volume of OBr^-, and find the molarity of the NaOBr solution.

18-20. The drug Librium gives a polarographic wave with $E_{1/2} = -0.265$ V (versus S.C.E.) in 0.05 M H_2SO_4. A 50.0-mL sample containing librium gave a wave height of 0.37 µA. When 2.00 mL of 3.00 mM librium in 0.05 M H_2SO_4 was added to the sample, the wave height increased to 0.80 µA. Find the molarity of librium in the unknown.

18-21. By now you are probably familiar with the electrochemical system in which a copper wire is immersed in a solution of cupric chloride. But what about the case in which a copper wire is immersed in ferric chloride solution? This *mixed* redox couple system can be described with the experimental voltammograms at right[†] Curve A represents the response of a rotating copper disk anode immersed in a solution containing 3 M Cl^-. At potentials near 0 V, the copper is oxidized and a large current is observed. Curve B was generated by a rotating platinum disk electrode in a 3 M Cl^- solution containing Fe(III). The experimentally observed current–voltage relation when the copper electrode is immersed in the Fe(III) solution is the same as the sum of curves A and B. Sketch this curve and estimate the *mixed potential* at which the current is zero. This is the potential at which the reaction $Fe^{3+} + Cu(s) \rightleftharpoons Fe^{2+} + Cu^+$ comes to equilibrium.

18-22. The cyclic voltammogram of the antibiotic chloramphenicol (abbreviated RNO_2) is shown at right. The scan was started at 0 V, and potential was swept toward negative voltage. The first cathodic wave, A, is from the reaction $RNO_2 + 4e^- + 4H^+ \rightarrow RNHOH + H_2O$. Explain what happens at peaks B and C using the reaction $RNO + 2e^- + 2H^+ \rightleftharpoons RNHOH$. Why was peak C not seen in the initial scan?

18-23. The cyclic voltammograms below are due to the irreversible reduction of *trans*-1,2-dibromocyclohexane. At room temperature, just one peak is seen. At low temperature, two peaks are seen. At $-60°C$, the relative size of the peak near -3.1 V increases if the scan rate is increased.

[†] G. P. Power and I. M. Ritchie, *J. Chem. Ed.,* **60,** 1022 (1983).

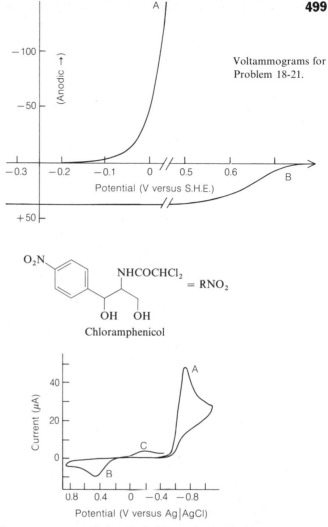

Voltammograms for Problem 18-21.

Chloramphenicol

Cyclic voltammogram of 3.7×10^{-4} M chloramphenicol in 0.1 M acetate buffer, pH 4.62. The voltage of the carbon paste working electrode was scanned at a rate of 350 mV/s. [P. T. Kissinger and W. R. Heineman, *J. Chem. Ed.,* **60,** 702 (1983).]

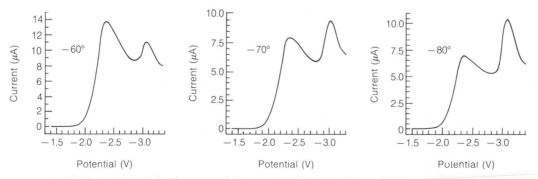

Voltammograms of 2.0 mM *trans*-1,2-dibromocyclohexane at a frozen mercury-drop electrode in butyronitrile with 0.10 M (n-C_4H_9)$_4$NClO$_4$ electrolyte. Scan rate = 1.00 V/s. [D. H. Evans, K. M. O'Connell, R. A. Petersen, and M. J. Kelly, *J. Chem. Ed.,* **60,** 290 (1983).]

Explain these observations with the scheme below:

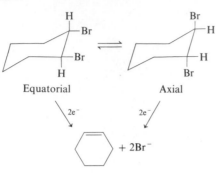

Equatorial   Axial

$+ 2Br^-$

18-24. A mixture containing Tl^+, Cd^{2+}, and Zn^{2+} exhibited the following diffusion currents in two different experiments, A and B, run with the same electrolyte on different occasions:

	Concentration (mM)	I_d (μA)
Tl^+:		
A	1.15	6.38
B	1.21	6.11
Cd^{2+}:		
A	1.02	6.48
B	?	4.76
Zn^{2+}:		
A	1.23	6.93
B	?	8.54

Calculate the Cd^{2+} and Zn^{2+} concentrations in experiment B.

18-25. A series of spectra for o-tolidine in an optically transparent, thin-layer electrode was shown in Box 18-3. The Nernst equation for the reaction

o-Tolidine(oxidized) $+ ne^- \rightleftharpoons$

o-tolidine(reduced)

can be written

$$E_{applied} = E' - \frac{0.059\ 16}{n} \log \frac{[tolidine(reduced)]}{[tolidine(oxidized)]}$$

where E' is the formal reduction potential in the medium used for the experiment. The spectrum labeled 1 is that of the oxidized form of o-tolidine. The spectrum labeled 7 is that of the reduced form. Spectra 2 through 6 represent mixtures of both forms. The ratio of species can be calculated from Beer's law.[†] For example, for spectrum 4,

$$\frac{[Tolidine(reduced)]}{[Tolidine(oxidized)]} = \frac{A_3 - A_2}{A_2 - A_1}$$

(a) Measure the quotient [tolidine(reduced)]/[tolidine(oxidized)] using the absorption maxima for curves 2 through 6.
(b) Prepare a graph of $E_{applied}$ versus log ([tolidine(reduced)]/[tolidine(oxidized)]).
(c) Use this graph to find E' (versus S.C.E.). Convert your value to E'(versus S.H.E.), which is given by

$$E'(\text{versus S.H.E.}) = E'(\text{versus S.C.E.}) + 0.241\ V$$

(d) Use the slope of the graph to calculate the number of electrons in the half-reaction. Propose a structure for the oxidized form of o-tolidine.

_____

[†] If you are not familiar with Beer's law, you can read about it in Sections 19-1 and 19-2.

19 Spectrophotometry

Spectrophotometry refers to the use of light to measure chemical concentrations. In this chapter, we discuss the fundamental principles of absorption and emission of radiation by molecules and how these processes are used in quantitative analysis.

19-1 PROPERTIES OF LIGHT

It is convenient to describe light in terms of both particles and waves. Light waves consist of perpendicular, oscillating electric and magnetic fields. For simplicity, a *plane-polarized* wave is shown in Figure 19-1. In this figure, the electric field is confined to the *xy* plane, and the magnetic field is confined to the *xz* plane. The **wavelength,** λ, is the crest-to-crest distance between waves. The **frequency,** v, is the number of complete oscillations that the wave makes each second. The unit of frequency is s^{-1}. One oscillation per second is also called one **hertz** (Hz). A frequency of $10^6 \ s^{-1}$ is therefore said to be 10^6 Hz, or one *megahertz* (MHz).

The electric field of *plane-polarized* light is confined to a single plane. Ordinary, unpolarized light has electric field components in all planes.

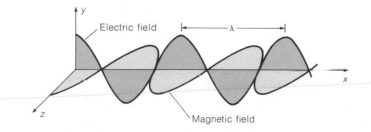

Figure 19-1
Plane-polarized electromagnetic radiation of wavelength λ, propagating along the *x* axis.

The relation between frequency and wavelength is

$$\lambda v = c \tag{19-1}$$

where c is the speed of light ($2.997\,924\,58 \times 10^8$ m/s in vacuum). In any medium other than vacuum, the speed of light is c/n, where n is the **refractive index** of that medium. Since n is always ≥ 1, light travels more slowly through any medium than through vacuum. When light enters a medium of higher refractive index, its frequency remains unchanged but its wavelength decreases.

With regard to its energy, it is more convenient to think of light as particles called **photons.** Each photon carries the energy, E, which is given by

Fundamental equations:

$$\lambda v = c$$

$$E = hv$$

$$E = hv \tag{19-2}$$

where h is **Planck's constant** ($= 6.626\,075\,5 \times 10^{-34}$ J·s). One mole of photons is called one *einstein.*

Equation 19-2 states that energy is proportional to frequency. Combining Equations 19-1 and 19-2, we can write

$$E = \frac{hc}{\lambda} = hc\tilde{v} \tag{19-3}$$

Ultraviolet light, ozone, and chlorofluorocarbons: Ozone (O_3) present in the stratosphere absorbs ultraviolet (uv) light and thereby protects us from harmful effects, including skin cancer. Chlorofluorocarbons such as Freon-12 from refrigerators, air conditioners, and spray cans diffuse to the stratosphere, where they catalyze ozone decomposition:

$$CCl_2F_2 \xrightarrow{\text{light}} CClF_2 + Cl$$
Freon-12

$$Cl + O_3 \rightarrow ClO + O_2$$

$$ClO \xrightarrow{\text{light}} Cl + O$$

$$O + O_3 \rightarrow 2O_2$$

A single Cl atom in this chain reaction can destroy 10^5 O_3 molecules. There is now a worldwide effort to find substitutes for chlorofluorocarbons to protect the earth's ozone layer.

where $\tilde{v}$, equal to $1/\lambda$, is called the **wavenumber.** We see that energy is inversely proportional to wavelength and directly proportional to wavenumber. Red light, with a longer wavelength than blue light, is thus less energetic than blue light. The SI unit for wavenumber is m^{-1}. However, the most common unit of wavenumber in the chemical literature is cm^{-1}, read "reciprocal centimeter" or "wavenumber."

The major regions of the **electromagnetic spectrum** are labeled in Figure 19-2. The various names reflect the history of physical science. There are no discontinuities in the properties of radiation as we pass from one region of the spectrum to another. Note that visible light, which is the kind our eyes detect, represents only a very small fraction of the electromagnetic spectrum.

19-2 ABSORPTION OF LIGHT

When a molecule absorbs a photon, the energy of the molecule is increased. We say that the molecule is promoted to an **excited state** (Figure 19-3). If a molecule emits a photon, its energy is lowered. The lowest energy state of a molecule is called the **ground state.**

When light is absorbed by a sample, the **radiant power** of the beam of light is decreased. Radiant power, P, refers to the energy per second per unit area of the light beam. A rudimentary spectrophotometric experiment is

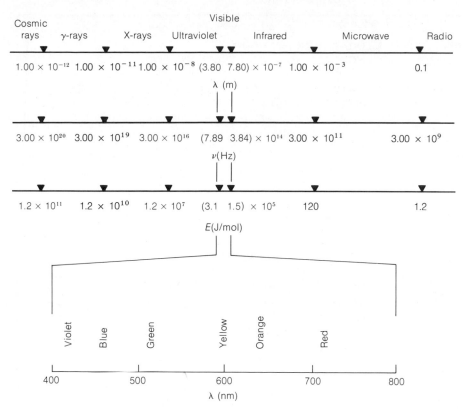

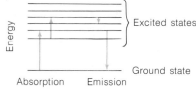

Figure 19-3
Absorption of light increases the energy of a molecule. Emission of light decreases the energy.

Figure 19-2
The electromagnetic spectrum. The units of wavelength in the visible portion of the spectrum are nanometers (1 nm = 10^{-9} m). Definitions of the different regions are those recommended by the International Union of Pure and Applied Chemistry [H. A. Willis and J. C. Rigg, *Pure Appl. Chem.*, **57**, 105 (1985)].

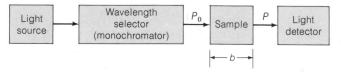

Figure 19-4
*Schematic diagram of a spectrophotometric experiment.

illustrated in Figure 19-4. Light is passed through a **monochromator** (a prism, grating, or even a filter) to select one wavelength. Light of this wavelength, with radiant power P_0, strikes a sample of length b. The radiant power of the beam emerging from the other side of the sample is P. Some of the light may be absorbed by the sample, so $P \leq P_0$.

The **transmittance,** T, is defined as the fraction of the original light that passes through the sample.

Monochromatic light consists of a single color (wavelength).

$$T = \frac{P}{P_0} \qquad (19\text{-}4)$$

504

Box 19-1   WHY IS THERE A LOGARITHMIC RELATION BETWEEN TRANSMITTANCE AND CONCENTRATION?

Beer's law, Equation 19-6, states that the *absorbance* of a sample is directly proportional to the concentration of the absorbing species. The fraction of light passing through the sample (the *transmittance*) is related logarithmically, not linearly, to the sample concentration. Why should this be?

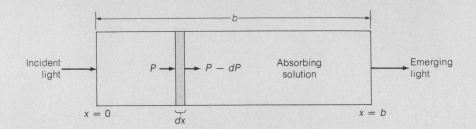

Imagine light of radiant power P passing through an infinitesimally thin layer of solution whose thickness is dx. The decrease in power (dP) is proportional to the incident power (P), to the concentration of absorbing species (c), and to the thickness of the section (dx):

$$dP = -\beta P c \, dx$$

where β is a constant of proportionality. The equation above can be rearranged and integrated quite simply:

$$-\frac{dP}{P} = \beta c \, dx$$

$$-\int_{P_0}^{P} \frac{dP}{P} = \beta c \int_0^b dx$$

The limits of integration are $P = P_0$ at $x = 0$ and $P = P$ at $x = b$. Evaluating the integrals gives us

$$-\ln P - (-\ln P_0) = \beta c b$$

$$\ln\left(\frac{P_0}{P}\right) = \beta c b$$

Finally, converting ln to log, using the relation $\ln z = (\ln 10)(\log z)$, gives

$$\log\left(\frac{P_0}{P}\right) = \underbrace{\left(\frac{\beta}{\ln 10}\right)}_{\text{A constant}} c b$$

or

$$A = \varepsilon c b$$

which is Beer's law!

The logarithmic relation of P_0/P and concentration arises from the fact that in each infinitesimal portion of the total volume, *the decrease in power is proportional to the power incident upon that section.* As light travels through the sample, the drop in power in each succeeding layer decreases, because the magnitude of the incident power that reaches each layer is decreasing.

Therefore, T has the range zero to one. The *percent transmittance* is simply $100 \cdot T$ and ranges between zero and 100 percent. A more useful quantity is the **absorbance,** defined as

$$A = \log_{10}\left(\frac{P_0}{P}\right) = -\log T \qquad (19\text{-}5)$$

When no light is absorbed, $P = P_0$ and $A = 0$. If 90% of the light is absorbed, 10% is transmitted and $P = P_0/10$. This gives $A = 1$. If only 1% of the light is transmitted, $A = 2$. The absorbance is sometimes called *optical density*, abbreviated OD or, occasionally, E.

The reason absorbance is so important is that it is directly proportional to the concentration of light-absorbing species in the sample:

$$\boxed{A = \varepsilon bc} \qquad (19\text{-}6)$$

Relation between transmittance and absorbance:

P/P_0	$\%T$	A
1	100	0
0.1	10	1
0.01	1	2

Beer's law: $A = \varepsilon bc$.

Equation 19-6, which is the heart of spectrophotometry as applied to analytical chemistry, is called the Beer–Lambert law, or simply **Beer's law.** The absorbance, A, is dimensionless. The concentration of the sample, c, is usually given in units of moles per liter (M). The pathlength, b, is commonly expressed in centimeters. The quantity ε (epsilon) is called the **molar absorptivity** (or *extinction coefficient,* in the older literature) and has the units $\text{M}^{-1} \cdot \text{cm}^{-1}$ (since the product εbc must be dimensionless). Molar absorptivity is the characteristic of the substance that tells how much light is absorbed at a particular wavelength.

Box 19-1 explains why absorbance, not transmittance, is proportional to concentration.

Equation 19-6 could be written

$$A_\lambda = \varepsilon_\lambda bc \qquad (19\text{-}7)$$

because the values of A and ε depend on the wavelength of light. The quantity ε is simply a coefficient of proportionality between absorbance and the product bc. The larger the value of ε, the greater is A. An **absorption spectrum** is a graph showing how A (or ε) varies with wavelength. Demonstration 19-1 illustrates the meaning of an absorption spectrum and shows several examples of spectra.

The part of a molecule responsible for light absorption is called a **chromophore.** Any substance that absorbs visible light will appear colored when white light is transmitted through it or reflected from it. The substance absorbs certain wavelengths of the white light, and our eyes detect the wavelengths that are not absorbed. A rough guide to colors is given in Table 19-1. The observed color is said to be the *complement* of the absorbed color. As an example, bromophenol blue has a visible absorbance maximum at 614 nm, and its observed color is blue.

The color of a solution is the complement of the color of light that it absorbs.

When Beer's Law Fails

Beer's law states that absorbance is proportional to the concentration of the absorbing species. It works very well for dilute solutions ($\lesssim 0.01$ M) of most

Beer's law works for dilute solutions in which the absorbing species is not participating in a concentration-dependent equilibrium.

TABLE 19-1
Colors of visible light

Wavelength of maximum absorption (nm)	Color absorbed	Color observed
380–420	Violet	Green-yellow
420–440	Violet-blue	Yellow
440–470	Blue	Orange
470–500	Blue-green	Red
500–520	Green	Purple
520–550	Yellow-green	Violet
550–580	Yellow	Violet-blue
580–620	Orange	Blue
620–680	Red	Blue-green
680–780	Purple	Green

substances. Apparent deviations from Beer's law at higher concentrations can be traced to changes in the absorbing species or in the properties of the bulk solution.

As a solution becomes more concentrated, solute molecules begin to influence each other due to their proximity. When one solute molecule interacts with another, the electrical properties of each (including absorption of light) are likely to change. The apparent result is that a graph of absorbance versus concentration is no longer a straight line. In the extreme case, at very high concentration, the solute *becomes* the solvent. Clearly, you cannot expect the electrical properties of a molecule to be the same in different solvents.[†] Sometimes, nonabsorbing solutes in a solution may interact with the absorbing species and alter the apparent absorptivity.

The physical interaction of two solutes is just one example of a chemical equilibrium (the association of two solutes) affecting the apparent absorbance. An even simpler example is that of a weak electrolyte, such as a weak acid. In concentrated solution, the predominant form of the acid will be the undissociated form, HA. As the solution is diluted, more dissociation occurs. If the absorptivity of A^- is not the same as that of HA, the solution will appear not to obey Beer's law as it is diluted.

19-3   WHAT HAPPENS WHEN A MOLECULE ABSORBS LIGHT?

When a molecule absorbs a photon, the molecule is necessarily promoted to a more energetic *excited state* (Figure 19-3). Conversely, when a molecule emits a photon, the energy of the molecule necessarily drops by an amount equal to the energy of the photon. We will now consider the physical processes associated with absorption and emission of radiation.

[†] The apparent absorbance also depends on the refractive index, n, of the solution. At sufficiently high concentrations of solute, the refractive index will change and the absorbance will appear to deviate from Beer's law. The dependence on refractive index is given by $A = \varepsilon bcn/(n^2 + 2)^2$.

Demonstration 19-1   ABSORPTION SPECTRA[†]

The spectrum of visible light can be projected on a screen in a darkened room in the following manner: Four layers of plastic diffraction grating[‡] are mounted on a cardboard frame having a square hole large enough to cover the lens of an overhead projector. This assembly is taped over the projector lens facing the screen. An opaque cardboard surface with two 1 × 3 cm slits is placed on the working surface of the projector.

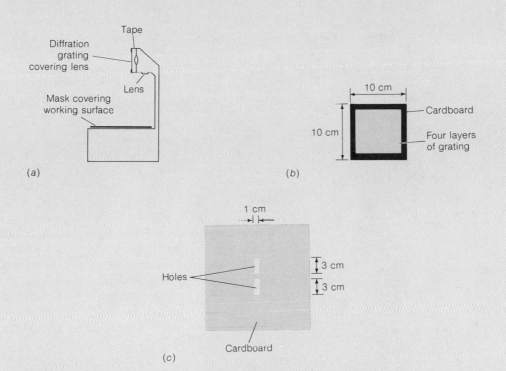

(a) Overhead projector. (b) Diffraction grating mounted on cardboard. (c) Mask for working surface.

When the lamp is turned on, the white image of each slit is projected on the center of the screen. A visible spectrum appears on either side of each image. By placing a beaker of colored solution over one slit, you can see its color projected on the screen where the white image previously appeared. The spectrum beside the colored image loses its intensity in regions where the colored species absorbs light.

Shown in Color Plate 13A are the spectrum of white light and the absorption spectra of three different colored solutions. You can see that potassium dichromate, which appears orange or yellow, absorbs blue wavelengths. Bromophenol blue absorbs red wavelengths and appears blue to our eyes. The absorption of phenolphthalein is located near the center of the visible spectrum. For comparison, the spectra of these three solutions as recorded with a spectrophotometer are shown in Color Plate 13B.

This same setup can be used to demonstrate fluorescence and the properties of colors.[†]

[†] D. H. Alman and F. W. Billmeyer, Jr., *J. Chem. Ed.,* **53,** 166 (1976). For another approach to spectroscopy in lecture halls, see F. H. Juergens, *J. Chem. Ed.,* **65,** 266, 1006 (1988).

[‡] One inexpensive 8½ × 11 inch sheet of plastic diffraction grating is all that is required. It is available from Edmund Scientific Co., 5975 Edscorp Building, Barrington, N.J. 08007, catalog no. 40, 267.

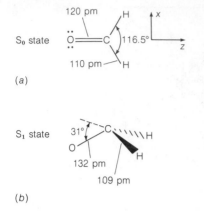

S_0 state

120 pm

116.5°

110 pm

(a)

S_1 state

31°

132 pm

109 pm

(b)

Figure 19-5
Geometry of formaldehyde. (a) Ground state. (b) Lowest excited singlet state.

Excited States of Molecules

As a concrete example, let's consider the molecule formaldehyde, whose structure is shown in Figure 19-5a. In its ground state, the molecule is planar with a double bond between carbon and oxygen. From the simplest electron-dot description of formaldehyde, we expect two pairs of nonbonding electrons to be localized on the oxygen atom. The double bond consists of a sigma bond between carbon and oxygen and a pi bond made from the $2p_y$ (out-of-plane) atomic orbitals of carbon and oxygen.

Electronic states of formaldehyde

A molecular orbital description of the valence shell of formaldehyde is given in Figure 19-6. The contours denote the electron density in the various molecular orbitals. The **molecular orbitals** describe the distribution of electrons in a molecule, just as *atomic orbitals* describe the distribution of electrons

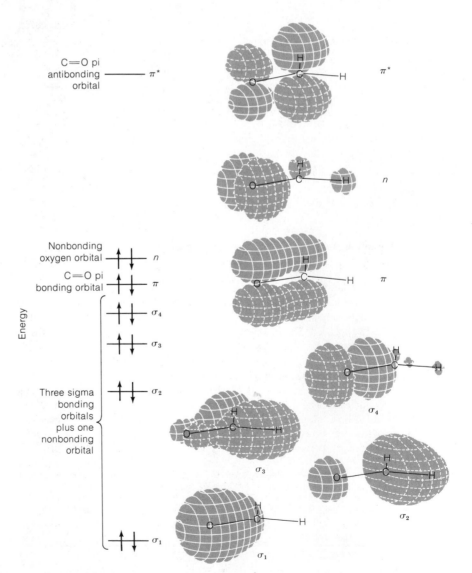

C=O pi antibonding orbital —— π^*

π^*

n

Nonbonding oxygen orbital —— n

C=O pi bonding orbital —— π

π

Energy

σ_4

σ_3

Three sigma bonding orbitals plus one nonbonding orbital

σ_2

σ_4

σ_3

σ_2

σ_1

σ_1

Figure 19-6
Molecular orbital diagram of formaldehyde, showing energy levels and orbital drawings. [Orbital drawings from W. L. Jorgensen and L. Salem, *The Organic Chemist's Book of Orbitals* (New York: Academic Press, 1973).]

in an atom. In the molecular orbital description of formaldehyde, one of the nonbonding orbitals of oxygen is thoroughly mixed with the three sigma-bonding orbitals. These four orbitals are labeled σ_1 through σ_4 in Figure 19-6, and each is occupied by a pair of electrons with opposite spin (spin quantum numbers $= +\frac{1}{2}$ and $-\frac{1}{2}$). At higher energy is an occupied pi-bonding orbital (π), made of the p_y atomic orbitals of carbon and oxygen. The highest-energy occupied orbital is the nonbonding orbital (n), composed principally of the oxygen $2p_x$ atomic orbital. The lowest-energy unoccupied orbital is the pi-antibonding orbital (π^*). An electron in this orbital produces a repulsion, rather than an attraction, between the carbon and oxygen atoms.

In an **electronic transition,** an electron from one molecular orbital moves to another orbital, with a concomitant increase or decrease in the energy of the molecule. The lowest-energy electronic transition of formaldehyde involves the promotion of a nonbonding (n) electron to the antibonding pi-orbital (π^*). There are actually two possible transitions, depending on the spin quantum numbers in the excited state (Figure 19-7). The state in which the spins are opposed ($+\frac{1}{2}$, $-\frac{1}{2}$) is called a **single state.** If both electrons have the same spin quantum number ($+\frac{1}{2}$, $+\frac{1}{2}$), we call the excited state a **triplet state.**

The lowest-energy excited singlet and triplet states are called S_1 and T_1, respectively. In general, T_1 is of lower energy than S_1. In formaldehyde, the transition $n \rightarrow \pi^*(T_1)$ requires the absorption of visible light with a wavelength of 397 nm. The $n \rightarrow \pi^*(S_1)$ transition occurs when ultraviolet radiation with a wavelength of 355 nm is absorbed. In general, higher excited states of the molecule are designated $S_2, S_3, \ldots$, and $T_2, T_3, \ldots$.

With an electronic transition near 397 nm, you might expect solutions of formaldehyde to be green-yellow (Table 19-1) in appearance. In fact, formaldehyde is colorless, because the probability of undergoing the $n \rightarrow \pi^*(T_1)$ transition is exceedingly small. The solution absorbs so little light at 397 nm that our eyes do not detect any absorbance at all. The reason for the low probability is that in the ground state the two spin quantum numbers are $+\frac{1}{2}$ and $-\frac{1}{2}$. In the T_1 excited state, the quantum numbers are $+\frac{1}{2}$ and $+\frac{1}{2}$. The probability of simultaneously changing orbitals and spin quantum numbers is very low. Therefore, very little visible light is absorbed by formaldehyde. The $n \rightarrow \pi^*(S_1)$ transition is much more probable, and the ultraviolet absorption is more intense.

Although formaldehyde is planar in its ground state, it has a pyramidal structure in both the T_1 and the S_1 excited states (Figure 19-5). The excited-state electron distribution actually leads to a change in the geometry of the molecule. Promotion of a nonbonding electron to an antibonding C—O orbital also leads to considerable lengthening of the C—O bond.

Vibrational and rotational states of formaldehyde

We have seen that absorption of visible or ultraviolet radiation can promote electrons to higher-energy orbitals in formaldehye. Infrared and microwave radiation are not energetic enough to induce electronic transitions, but they can cause changes in the vibrational or rotational motion of the molecule.

Consider formaldehyde as a collection of four atoms. Each atom can move along three axes in space, so the entire molecule can move in $4 \times 3 = 12$ different ways. We say that formaldehyde has 12 *degrees of freedom.* Three degrees of freedom correspond simply to translation of the entire molecule

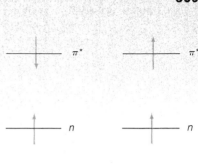

Figure 19-7
Diagram showing the two possible electronic states arising from an $n \rightarrow \pi^*$ transition. (a) Excited singlet state, S_1. (b) Excited triplet state, T_1.

The terms *singlet* and *triplet* are used because a triplet state is split into three slightly different energy levels in the presence of a magnetic field, but a singlet state is not split.

The shorter the wavelength of light, the greater the energy.

A nonlinear molecular with n atoms has $3n - 6$ vibrational modes and three possible rotations. A linear molecule can rotate about only two axes; it therefore has $3n - 5$ vibrational modes and two rotations.

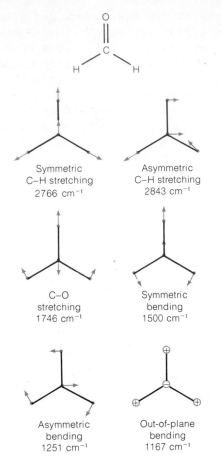

Figure 19-8
The six kinds of vibrations of formaldehyde. The wavenumber of each vibration is given in units of cm^{-1}. These are the wavenumbers of infrared radiation needed to stimulate each kind of motion.

Just as the geometry of the excited electronic state is not the same as the geometry of the ground state, the vibrational and rotational energy levels of the excited states are not the same as in the ground state. For example, the C—O stretching vibration of formaldehyde is reduced from 1 746 cm^{-1} in the S_0 state to 1 183 cm$^-$ in the S_1 state. This is reasonable because the strength of the C—O bond is lessened when the antibonding π^* orbital is populated.

in the x, y, and z directions. Another three degrees of freedom correspond to rotation about the x, y, and z axes of the molecule. The remaining six degrees of freedom represent vibrations of the molecule. The six possible kinds of vibrations of formaldehyde are shown in Figure 19-8.

When a molecule of formaldehyde absorbs an infrared photon with a wavenumber of, say, 1 251 cm^{-1}, the asymmetric bending vibration depicted in Figure 19-8 is stimulated. This means that the oscillations of the atoms are increased in amplitude, and the energy of the molecule increases.

EXAMPLE: Relation of Energy and Wavenumber
By how many kilojoules per mole is the energy of formaldehyde increased when 1 251 cm^{-1} radiation is absorbed?
   The energy is increased by

$$\Delta E = h\nu = h\frac{c}{\lambda} = hc\tilde{\nu} \qquad \left(\text{since } \tilde{\nu} = \frac{1}{\lambda}\right)$$

$$= (6.626\ 1 \times 10^{-34}\ \text{J·s})(2.997\ 9 \times 10^8\ \text{m/s})(1\ 251\ \text{cm}^{-1})(100\ \text{cm/m})$$

$$= 2.485 \times 10^{-20}\ \text{J/molecule} = 14.97\ \text{kJ/mol}$$

The rotational energy levels of a molecule lie at even lower energy than do the vibrational energy levels. The three lowest rotational energy levels of formaldehyde lie at 0, 0.029 07, and 0.087 16 kJ/mol. A molecule in the rotational ground state could absorb microwave photons with energies of 0.029 07 or 0.087 16 kJ/mol (wavelengths of 4.115 or 1.372 mm) to be promoted to the two lowest excited states. A molecule in the lowest excited state could absorb a photon whose energy is 0.058 09 (= 0.087 16 − 0.029 07) kJ/mol and be promoted to the second excited state. In general, absorption of microwave radiation leads to rotational excitation of molecules. In a rotationally excited state, the molecule rotates faster than it does in its ground state.

Combined electronic, vibrational, and rotational transitions

In general, when a molecule absorbs light having sufficient energy to cause an electronic transition, **vibrational** and **rotational transitions**—that is, changes in the vibrational and rotational states—can occur as well. Thus, for example, formaldehyde can absorb one photon with just the right energy to cause a transition from the S_0 to the S_1 electronic state, from the ground vibrational state of S_0 to an excited vibrational state of S_1, and from one rotational state of S_0 to a different rotational state of S_1.

The reason why electronic absorption bands are usually very broad (as in Color Plate 13) is that many different vibrational and rotational levels are available at slightly different energies. Therefore, a molecule could absorb photons with a fairly wide range of energies and still be promoted from the ground electronic state to one particular excited electronic state.

Vibrational transitions usually involve simultaneous rotational transitions. Electronic transitions usually involve simultaneous vibrational and rotational transitions.

What Happens to Absorbed Energy?

Suppose that the absorption promotes the molecule from the ground electronic state, S_0, to a vibrationally and rotationally excited level of the excited electronic state S_1 (Figure 19-9). Usually, the first process following this absorption is *vibrational relaxation* to the ground vibrational level of S_1. This

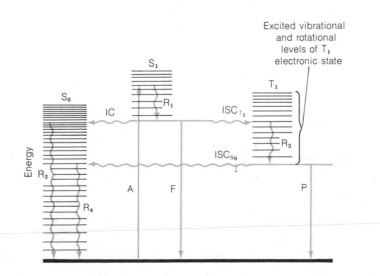

Figure 19-9
Diagram illustrating some of the physical processes that can occur after a molecule absorbs a photon. S_0 is the ground electronic state of the molecule. S_1 and T_1 are the lowest excited singlet and triplet states, respectively. Straight arrows represent processes involving photons, and wavy arrows represent radiationless transitions. A = absorption, F = fluorescence, P = phosphorescence, IC = internal conversion, ISC = intersystem crossing, R = vibrational relaxation.

Internal conversion is a radiationless transition between states with the same spin quantum numbers (e.g., $S_1 \rightarrow S_0$).

Intersystem crossing is a radiationless transition between states with different spin quantum numbers (e.g., $T_1 \rightarrow S_0$).

Flurorescence is a radiational transition between states with the same spin quantum numbers (e.g., $S_1 \rightarrow S_0$). *Phosphorescence* is a radiational transition between states with different spin quantum numbers (e.g., $T_1 \rightarrow S_0$).

The *lifetime* of a state is the time needed for the population of that state to decay to $1/e$ of its initial value, where e is the base of natural logarithms.

For a report on chemiluminescence in chemical analysis, see M. L. Grayeski, *Anal. Chem.*, **59**, 1243A (1987).

Monochromatic radiation is radiation of a single wavelength. Of course, it is impossible to produce truly monochromatic light. However, the better the monochromator, the narrower is the range of wavelengths.

radiationless transition is labeled R_1 in Figure 19-9. The vibrational energy lost in this relaxation is transferred to other molecules (solvent, for example) through collisions. The net effect is to convert part of the energy of the absorbed photon into heat spread throughout the entire medium.

From the S_1 level, many things can happen. The molecule could enter a very highly excited vibrational state of S_0 having the same energy as S_1. This is called **internal conversion.** From this excited state, the molecule can relax back to the ground vibrational state and transfer its energy to neighboring molecules through collisions. This radiationless process is labeled R_2. If a molecule follows the path $A–R_1–IC–R_2$ in Figure 19-9, the entire energy of the photon will have been converted to heat.

Alternatively, the molecule could cross from S_1 into an excited vibrational level of T_1. Such an event is known as **intersystem crossing.** Following the radiationless vibrational relaxation R_3, the molecule finds itself at the lowest vibrational level of T_1. From here, the molecule might undergo a second intersystem crossing to S_0, followed by the radiationless relaxation R_4. All of the processes mentioned so far have the net effect of converting light to heat.

In contrast, from S_1 or T_1, a molecule could relax to S_0 by emitting a photon. The transition $S_1 \rightarrow S_0$ is called **fluorescence,** and the transition $T_1 \rightarrow S_0$ is called **phosphorescence.** The relative rates of internal conversion, intersystem crossing, fluorescence, and phosphorescence depend on the molecule, the solvent, and physical conditions such as temperature and pressure. Box 19-2 describes some occasions when light is emitted by excited molecules.

Most molecules under ordinary conditions return to the ground state through radiationless processes. Fluorescence and phosphorescence are relatively rare. The *lifetime* of fluorescence is always very short (10^{-4}–10^{-8} s). If a molecule is *not* to fluoresce, internal conversion or intersystem crossing must be even more rapid. The lifetime of phosphorescence is much longer, being in the range 10^{-4}–10^2 s. This means that phosphorescence is even rarer than fluorescence, since a molecule in the T_1 state has a good chance of undergoing intersystem crossing to S_1 before phosphorescence can occur. In molecules containing transition metals, electronic states other than singlets and triplets are possible. Emission from a transition metal complex is usually called simply **luminescence,** which makes no distinction between fluorescence and phosphorescence.

A molecule can dissipate the energy of an absorbed photon by emitting a photon or by creating heat throughout the medium. Alternatively, a bond may break, and *photochemistry* may occur. That is, the energy of the excited molecule could overcome the activation energy of a chemical reaction. In some chemical reactions (not necessarily stimulated by light), part of the energy released appears in the form of light. Light emitted during a chemical reaction is called **chemiluminescence.**

19-4 THE SPECTROPHOTOMETER

The minimum requirements for a spectrophotometer—a device to measure absorbance of light—were shown in Figure 19-4. Light from a continuous source is passed through a monochromator, which selects a narrow band of wavelengths from the incident beam. This **"monochromatic" light** travels through a sample of pathlength b, and the radiant power of the emergent light is measured.

Box 19-2 FLUORESCENT LAMPS AND LITTLE-KNOWN FLUORESCENT OBJECTS

A fluorescent lamp is a glass tube filled with mercury vapor; the inner walls are coated with a *phosphor* (luminescent substance) consisting of a calcium halophosphate ($Ca_5(PO_4)_3F_{1-x}Cl_x$) doped with Mn^{2+} and Sb^{3+}. The mercury atoms, promoted to an excited state by electric current passing through the lamp, emit mostly ultraviolet radiation at 254 and 185 nm. This radiation is absorbed by the Sb^{3+} dopant, and some of the energy is passed on to Mn^{2+}. The Sb^{3+} emits blue light, and the Mn^{2+} emits yellow light, with the combined emission appearing white. The emission spectrum is shown below. Fluorescent lamps are important energy-saving devices because they are more efficient than incandescent lamps in the conversion of electricity to light.

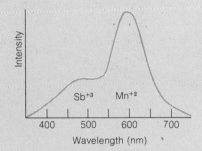

Emission spectrum of the phosphor used in fluorescent lamps. [J. A. DeLuca, *J. Chem. Ed.*, **57**, 541 (1980).]

A little-known occurrence of photoemission is that from most white fabrics. Just for fun, turn on an ultraviolet lamp in a darkened room containing several people. You will discover a surprising amount of emission from white fabrics (shirts, pants, shoelaces, and lots more) that have been treated with fluorescent compounds to enhance their whiteness. You may also be surprised to see fluorescence from teeth and from recently bruised areas of skin that show no surface damage.

One of the most popular ways to measure absorbance of visible light in the undergraduate laboratory is with the Bausch and Lomb Spectronic 20 spectrophotometer, shown in Figure 19-10. In this instrument, the light source is an ordinary tungsten lamp whose emission covers the entire visible spectrum, extending somewhat into the ultraviolet and infrared regions. The light is dispersed into its component wavelengths by a grating (see Color Plate 14), and only one small band of wavelengths is passed through the sample. The detector is a phototube that creates an electric current proportional to the radiant power of light striking the tube. The output is expressed on a meter that reports both transmittance and absorbance (Figure 3-1). It is important to be aware of which scale you are reading on such an instrument.

The sample is introduced in a cell called a **cuvet.** We do not measure the incident radiant power, P_0, directly. Rather, the radiant power of light passing through a cuvet containing pure solvent is *defined* as P_0. This cuvet is then removed and replaced by an identical one containing sample. The radiant power of light striking the detector is then taken as P, permitting T or A to be determined. The *reference cuvet,* containing pure solvent, compensates for reflection, scattering, or absorption of light by the cuvet and solvent. The radiant power of light striking the detector would not be the same if the reference cuvet were removed from the beam.

If you record transmittance when you really want absorbance, you can use Equation 19-5 to convert one to the other.

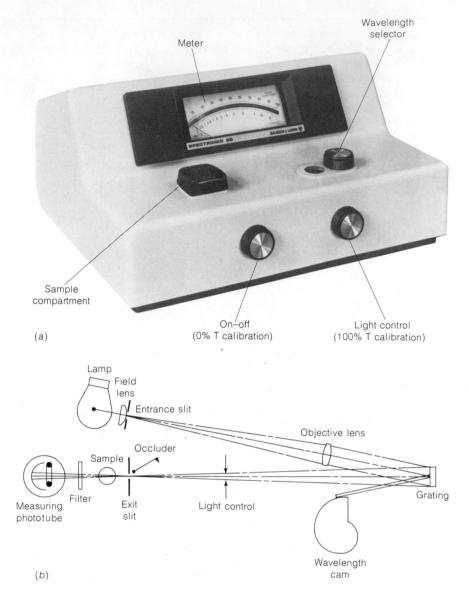

Figure 19-10

(a) Bausch and Lomb Spectronic 20 spectrophotometer. (b) Schematic diagram of optical train. [Courtesy Bausch and Lomb, Analytical Systems Division, Rochester, N.Y.]

We will use the symbol P_r for the radiant power measured when the reference is placed in the spectrophotometer. This is effectively the value of P_0 in Equation 19-5, because it is the maximum power that can reach the detector in the absence of an absorbing species in the solution. We denote by the symbol P_s the power measured at the detector with the sample in the beam. The absorbance is therefore

$$A = \log \frac{P_r}{P_s} \tag{19-8}$$

Double-Beam Strategy

The Spectronic 20 is an example of a *single-beam spectrophotometer,* so called because the beam of light follows a single path through one sample at a time.

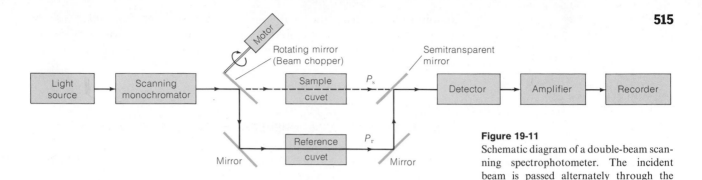

Figure 19-11
Schematic diagram of a double-beam scanning spectrophotometer. The incident beam is passed alternately through the sample and reference cuvets by the rotating beam chopper.

To measure the absorbance of a sample, we must first measure the value of P_0 with a separate reference sample. This process is inconvenient because two different samples must be placed alternately in the beam. It is inaccurate because both the output of the source and the response of the detector fluctuate. If there is a change in either one between the measurement of the reference solution and that of the sample solution, the apparent absorbance will be in error. A single-beam instrument is poorly suited to continuous measurements of absorbance, as in a kinetics experiment, because both the source intensity and the detector response drift.

In a *double-beam spectrophotometer* (Figure 19-11), light passes alternately through the sample and the reference cuvets. This is accomplished by a motor that rotates a mirror into and out of the light path. When the *chopper* is not diverting the beam, the light passes through the sample, and the detector measures the radiant power P_s. When the chopper diverts the beam through the reference cuvet, the detector measures P_r. The beam is chopped several times per second, and the circuitry automatically compares P_r and P_s to obtain absorbance (using Equation 19-8). This procedure provides automatic correction for drift of the source intensity or detector response, since the power emerging from the two samples is compared so frequently. Most research-quality spectrophotometers also provide for automatic wavelength scanning and continuous recording of the absorbance. The next chapter describes components of a spectrophotometer in greater detail.

The double-beam instrument makes an essentially continuous measurement of the light emerging from the sample and the reference cells.

In recording an absorbance spectrum, it is a routine procedure to first record the baseline spectrum with reference solutions (pure solvent or a reagent blank) in both cuvets. The absorbance of the reference is then subtracted from the absorbance of the sample to obtain the true absorbance of the sample at any wavelength.

Precautions

Most spectrophotometers exhibit their minimum error at intermediate levels of absorbance (say, $A \approx 0.4-0.9$). If too little light gets through the sample (high absorbance), the intensity is hard to measure. If too much light gets through (low absorbance), then it is hard to detect the difference between the sample and the reference cuvets. It is therefore desirable to adjust the concentration of the sample so that its absorbance falls in this intermediate range.

Errors in sample preparation and handling can be minimized by suitable care and common sense. For example, samples must be dust-free because small particles scatter light and increase the apparent "absorbance" of the sample. Keeping all containers covered will lower the concentration of dust

Keep your fingers off the faces of the cuvet!

in solutions. Filtering the final solution through a very fine filter may be necessary in critical work. Cuvets should not be handled with fingers and must be kept scrupulously clean to avoid surface contamination, which leads to scattering.

19-5   A TYPICAL PROCEDURE: SERUM IRON DETERMINATION

Spectrophotometric analyses employing visible radiation are called *colorimetric* analyses.

For a compound to be analyzed by spectrophotometry, it must absorb light, and this absorption should be distinguishable from that due to other species in the sample. Since most compounds absorb ultraviolet radiation, those results tend to be inconclusive, and analysis is usually restricted to the visible spectrum. If there are no interfering species, however, ultraviolet absorbance can be used as well. Solutions of proteins are normally assayed in the ultraviolet region at 280 nm because the aromatic groups present in virtually every protein have an absorbance maximum at 280 nm. In this section, we will describe a typical procedure utilizing absorption spectrophotometry for the quantitative analysis of iron in blood serum.

Iron for biosythesis is transported through the bloodstream attached to the protein transferrin. The procedure described below is used to measure the iron content of transferrin.[†] This analysis is quite sensitive, with only about 1 μg (microgram; 1 μg $= 10^{-6}$ g) of iron needed to provide an accuracy of $\sim 2–5\%$. Human blood usually contains about 45% (vol/vol) cells and 55% plasma (liquid). If blood is collected without an anticoagulant, the blood clots, and the liquid that remains is called *serum*. Serum normally contains about 1 μg of Fe/mL attached to transferrin.

To measure the serum iron content three steps are needed:

$HSCH_2CO_2H$

Thioglycolic acid

Ascorbic acid (vitamin C)

1. Fe(III) in transferrin is reduced to Fe(II) and thereby released from the protein. Commonly employed reducing agents are hydroxylamine hydrochloride ($NH_3OH^+Cl^-$), thioglycolic acid, or ascorbic acid.

2. Trichloroacetic acid (Cl_3CCO_2H) is added to precipitate all of the proteins, leaving Fe(II) in solution. The proteins are removed by centrifugation. If protein is left in the solution, it will partially precipitate in the final solution. Light scattering by particles of precipitate would be mistaken for absorbance.

3. A measured volume of supernatant liquid from Step 2 is transferred to a fresh vessel and treated with excess ferrozine to form a purple complex, whose absorbance is measured (Figure 19-12). A buffer is also added to keep the pH in a range in which the formation of the ferrozine-iron complex is complete.

The blank should contain all sources of absorbance other than the analyte.

In most spectrophotometric analyses, it is important to prepare a **reagent blank** containing all reagents, but with analyte replaced by distilled water. Any absorbance of the blank is due to the color of uncomplexed ferrozine plus the color caused by the iron impurities in the reagents and glassware. *The absorbance of the blank is subtracted from all other absorbances before any calculations are done.*

It is also important to use a series of iron standards to establish a calibration curve. Figure 19-13 is a typical calibration curve for this analysis.

[†] D. C. Harris, *J. Chem. Ed., 55,* 539 (1978).

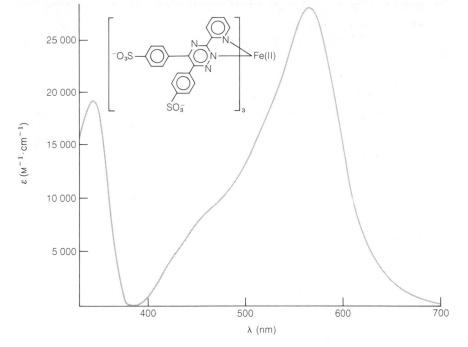

Figure 19-12
Visible absorption spectrum of the complex (ferrozine)$_3$Fe(II) used in the colorimetric analysis
of iron.

Beer's law is clearly obeyed over the concentration range in this illustration.
The standards should always be prepared using the same procedure as for
the unknowns. The absorbance of the unknown should always fall within the
region covered by the standards, so that there is no question about the validity
of the calibration curve.

 If all samples and standards are prepared in the same way and with
identical volumes, then the quantity of iron in the unknown can be read
directly from the calibration curve. For example, if the unknown has an
absorbance of 0.357 (after subtracting the absorbance of the blank), Figure
19-13 tells us that it contains 3.59 μg of iron. To appreciate the uncertainty
of the result, the method of least squares in Section 4-4 should be used.

 In the serum iron determination just described, the values obtained would
be about 10% high due to reaction of serum copper with the ferrozine. This
interference can be eliminated if neocuproine or thiourea is added.[†] These
reagents form strong complexes with copper, thereby **masking** it.

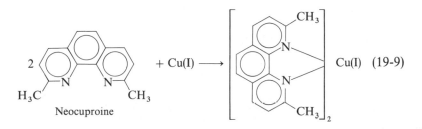

$$2 \quad \text{Neocuproine} \quad + \text{Cu(I)} \longrightarrow \quad \left[\quad \right]_2 \text{Cu(I)} \quad (19\text{-}9)$$

[†] J. R. Duffy and J. Gaudin, *Clin. Biochem.*, **10**, 122 (1977).

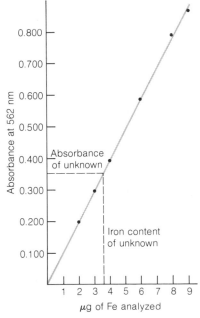

Figure 19-13
Calibration curve showing the validity of
Beer's law for the (ferrozine)$_3$Fe(II) complex used in the serum iron determination.
Each sample was diluted to a final volume
of 5.00 mL. Therefore, 1.00 μg of iron gives
a concentration of 3.58×10^{-6} M.

EXAMPLE: Serum Iron Analysis

Serum iron and standard iron solutions were analyzed according to the following procedure.

1. To 1.00 mL of sample are added 2.00 mL of reducing agent and 2.00 mL of acid to reduce and release Fe from transferrin.

2. The serum proteins are precipitated with 1.00 mL of 30% (wt/wt) trichloroacetic acid. The volume change of the solution is negligible when the protein precipitates and can be said to remain $1.00 + 2.00 + 2.00 + 1.00 = 6.00$ mL (assuming no changes in volume due to mixing). The mixture is centrifuged to remove protein.

3. A 4.00-mL aliquot of solution is transferred to a fresh test tube and treated with 1.00-mL of solution containing ferrozine and buffer. The absorbance of this solution is measured after a 10-minute waiting period.

The following data were obtained:

Sample	A (at 562 nm in 1.000-cm cell)
Blank	0.038
3.00 μg Fe standard	0.239
Serum sample	0.129

Assuming that Beer's law has been shown to be valid in control experiments, use the data above to find the concentration of Fe in the serum. Also, calculate the molar absorptivity of (ferrozine)$_3$Fe(II).

The calculation of the serum iron content is trivial. Since the sample and standard were prepared in an identical manner, their Fe ratio must be equal to their absorbance ratio (corrected for the blank absorbance).

$$\frac{\text{Fe in sample}}{\text{Fe in standard}} = \frac{\text{corrected absorbance of sample}}{\text{corrected absorbance of standard}} = \frac{0.129 - 0.038}{0.239 - 0.038} = 0.453$$

Since the standard contained 3.00 μg of Fe, the sample must have contained $(0.453)(3.00 \ \mu\text{g}) = 1.359 \ \mu\text{g}$ of Fe. The concentration of Fe in the serum is

$$[\text{Fe}] = \text{moles of Fe/liters of serum}$$

$$= \left(\frac{1.359 \times 10^{-6} \text{ g Fe}}{55.847 \text{ g Fe/mol Fe}}\right)\bigg/ (1.00 \times 10^{-3} \text{ L}) = 2.43 \times 10^{-5} \text{ M}$$

To find ε for (ferrozine)$_3$Fe(II), we can use the absorbance of the standard. In the procedure above, a volume of 1.00 mL of standard containing 3.00 μg of Fe was diluted to 6.00 mL with other reagents. Then 4.00 mL (containing $4.00/6.00 \times 3.00 \ \mu\text{g} = 2.00 \ \mu\text{g}$ of Fe) was transferred to a new vessel and diluted with 1.00 mL of reagent. The final concentration of Fe is

$$[\text{Fe}] = \left(\frac{2.00 \times 10^{-6} \text{ g Fe}}{55.847 \text{ g Fe/mol Fe}}\right)\bigg/ (5.00 \times 10^{-3} \text{ L}) = 7.16 \times 10^{-6} \text{ M}$$

All this Fe is in the form (ferrozine)$_3$Fe(II). The molar absorptivity is

$$\varepsilon = \frac{A}{bc} = \frac{0.239 - 0.038}{(1.000 \text{ cm})(7.16 \times 10^{-6} \text{ M})} = 2.81 \times 10^4 \text{ M}^{-1} \cdot \text{cm}^{-1}$$

The absorbance of a solution at a particular wavelength is the sum of the absorbances at that wavelength of each species in the solution:

$$A = \varepsilon_X b[X] + \varepsilon_Y b[Y] + \varepsilon_Z b[Z] + \cdots \qquad (19\text{-}10)$$

Absorbance is additive.

where ε is the molar absorptivity of each species at the wavelength in question.

What to Do When the Individual Spectra Overlap

Let's apply Equation 19-10 to the analysis of a mixture containing two components whose spectra overlap each other a great deal. The problem is illustrated for a mixture of H_2O_2 complexes of Ti^{4+} and V^{5+} in H_2SO_4 solution in Figure 19-14. The absorbance of the mixture (A_m) at any chosen wavelength is

$$A_m = \varepsilon_X b[X] + \varepsilon_Y b[Y] \qquad (19\text{-}11)$$

where X and Y refer to Ti^{4+} and V^{5+}, respectively. If a standard solution of species X with concentration $[X]_s$ is prepared, its absorbance will be

$$A_{X_s} = \varepsilon_X b[X]_s \qquad (19\text{-}12)$$

Similarly, a standard solution of species Y with concentration $[Y]_s$ will have absorbance

$$A_{Y_s} = \varepsilon_Y b[Y]_s \qquad (19\text{-}13)$$

Solving Equations 19-12 and 19-13 for ε_X and ε_Y and substituting these values into Equation 19-11 gives

$$\frac{A_m}{A_{X_s}} = \frac{[X]}{[X]_s} + \frac{A_{Y_s}}{A_{X_s}}\frac{[Y]}{[Y]_s} \qquad (19\text{-}14)$$

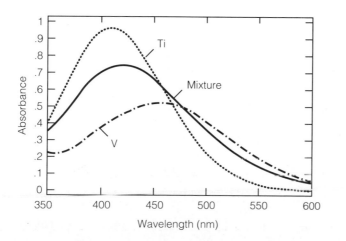

Figure 19-14

Visible spectrum of 1.32 mM Ti^{4+}, 1.89 mM V^{5+}, and an unknown mixture containing both ions. All solutions contain 0.5% (wt/wt) H_2O_2 and ~0.01 M H_2SO_4. [From M. Blanco, H. Iturriaga, S. Maspoch, and P. Tarín, *J. Chem. Ed.*, **66**, 178 (1989).]

Table 19-2
Absorbance values for Figure 19-14

Wavelength (nm)	A_{X_s} titanium standard	A_{Y_s} vanadium standard	A_m mixture	A_m/A_{X_s}	A_{Y_s}/A_{X_s}
390	0.895	0.326	0.651	0.727_3	0.364_2
430	0.884	0.497	0.743	0.840_5	0.562_2
450	0.694	0.528	0.665	0.958_2	0.760_8
470	0.481	0.512	0.547	1.137_2	1.064_4
510	0.173	0.374	0.314	1.815_0	2.161_8

SOURCE: M. Blanco, H. Iturriaga, S. Maspoch, and P. Tarín, *J. Chem. Ed.,* **66**, 178 (1989).

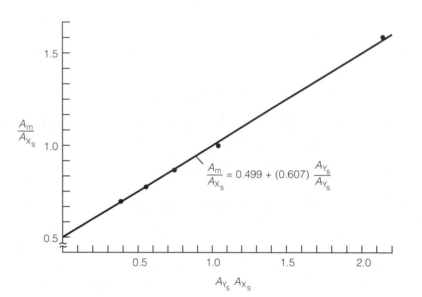

Figure 19-15
Graph of A_m/A_{X_s} versus A_{Y_s}/A_{X_s} using data from Table 19-2.

Analysis of a mixture:

Plot A_m/A_{X_s} versus A_{Y_s}/A_{X_s}

Slope = $[Y]/[Y]_s$

Intercept = $[X]/[X]_s$

That is, a graph of A_m/A_{X_s} versus A_{Y_s}/A_{X_s} at various wavelengths will have a slope of $[Y]/[Y]_s$ and an intercept of $[X]/[X]_s$. Since we know the standard concentrations $[X]_s$ and $[Y]_s$, we can find the concentrations $[X]$ and $[Y]$ in the unknown mixture.

Table 19-2 gives experimental absorbance values for the spectra in Figure 19-14. Figure 19-15 is a graph of A_m/A_{X_s} versus A_{Y_s}/A_{X_s}. The equation of the least-squares straight line through the points is

$$\frac{A_m}{A_{X_s}} = 0.499 + (0.607)\frac{A_{Y_s}}{A_{X_s}} \qquad (19\text{-}15)$$

from which we can say

$$\text{Slope} = 0.607 = [Y]/[Y]_s$$

$$[Y] = [V^{5+}] = (0.607)(1.89\ \text{mM}) = 1.15\ \text{mM} \qquad (19\text{-}16)$$

$$\text{Intercept} = 0.499 = [X]/[X]_s$$

$$\Rightarrow\quad [X] = [Ti^{4+}] = (0.499)(1.32\ \text{mM}) = 0.659\ \text{mM} \qquad (19\text{-}17)$$

In this method we used absorbance at many wavelengths to decompose the spectrum of the mixture into its two components. The best wavelength region to use is the one where overlap of the two individual spectra occurs, and where the errors in absorbance of the mixture are least.

What to Do When the Individual Spectra Are Well Resolved

Now we apply Equation 19-10 to the analysis of a mixture of two components whose individual spectra do not overlap very much. Suppose that species X has an absorbance maximum at wavelength λ' that is fairly well separated from the absorbance maximum of species Y at wavelength λ''. The absorbance at any wavelength is the sum of absorbances of each component at that wavelength. For the absorbance at wavelengths λ' and λ'' we can write

$$A' = \varepsilon'_X b[X] + \varepsilon'_Y b[Y] \qquad A'' = \varepsilon''_X b[X] + \varepsilon''_Y b[Y] \qquad (19\text{-}18)$$

where the ε values apply to each species at each wavelength. The absorptivities of X and Y at each wavelength must be measured in a separate experiment.

We can solve the two Equations 19-18 for the two unknowns [X] and [Y]. The result is

$$[X] = \frac{\begin{vmatrix} A' & \varepsilon'_Y b \\ A'' & \varepsilon''_Y b \end{vmatrix}}{\begin{vmatrix} \varepsilon'_X b & \varepsilon'_Y b \\ \varepsilon''_X b & \varepsilon''_Y b \end{vmatrix}} \qquad [Y] = \frac{\begin{vmatrix} \varepsilon'_X b & A' \\ \varepsilon''_X b & A'' \end{vmatrix}}{\begin{vmatrix} \varepsilon'_X b & \varepsilon'_Y b \\ \varepsilon''_X b & \varepsilon''_Y b \end{vmatrix}} \qquad (19\text{-}19)$$

The determinant $\begin{vmatrix} a & b \\ c & d \end{vmatrix}$ means $ad - bc$.

In Equation 19-19, each symbol $\begin{vmatrix} a & b \\ c & d \end{vmatrix}$ is called a *determinant*. It is a shorthand way of writing the product $a \cdot d$ minus the product $b \cdot c$. Thus the determinant $\begin{vmatrix} 1 & 2 \\ 3 & 4 \end{vmatrix}$ means $1 \cdot 4 - 2 \cdot 3 = -2$.

To analyze a mixture of two compounds, therefore, it is necessary to measure the absorbances at two wavelengths and to know ε at each wavelength for each compound. Similarly, a mixture of n components may be analyzed by making n absorbance measurements at n wavelengths.

EXAMPLE: Analysis of a Mixture Using Equations 19-19

The molar absorptivities of compounds X and Y were measured with pure samples of each:

	$\varepsilon[\text{M}^{-1} \cdot \text{cm}^{-1}]$	
	At 272 nm	At 327 nm
X	16 440	3 990
Y	3 870	6 420

A mixture of compounds X and Y in a 1.00-cm cell had an absorbance of 0.957 at 272 nm and 0.559 at 327 nm. Find the concentrations of X and Y in the mixture.

Using equations 19-19 and setting $b = 1.00$, we find

$$[X] = \frac{\begin{vmatrix} 0.957 & 3\,870 \\ 0.559 & 6\,420 \end{vmatrix}}{\begin{vmatrix} 16\,400 & 3\,870 \\ 3\,990 & 6\,420 \end{vmatrix}} = \frac{(0.957)(6\,420) - (3\,870)(0.559)}{(16\,400)(6\,420) - (3\,870)(3\,990)} = 4.43 \times 10^{-5} \text{ M}$$

$$[Y] = \frac{\begin{vmatrix} 16\,400 & 0.957 \\ 3\,990 & 0.559 \end{vmatrix}}{\begin{vmatrix} 16\,400 & 3\,870 \\ 3\,990 & 6\,420 \end{vmatrix}} = 5.95 \times 10^{-5} \text{ M}$$

Isosbestic Points

Often one absorbing species, X, is converted to another absorbing species, Y, during the course of a chemical reaction. This sort of transformation leads to a very obvious and characteristic behavior, shown in Figure 19-16. If the spectra of pure X and pure Y cross each other at any wavelength, then every spectrum recorded during this chemical reaction will cross at that same point, called an **isosbestic point.**

Methyl red is an acid–base indicator that has two pK_a values:

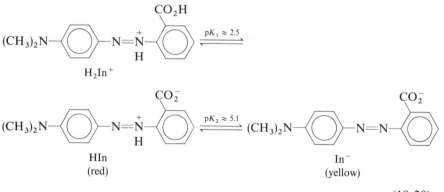

(19-20)

In a solution of methyl red at pH 4.5, the predominant species is the red compound HIn. As the pH is raised, yellow In$^-$ is formed. The spectra of these two species (at the same concentration) happen to cross at 465 nm.

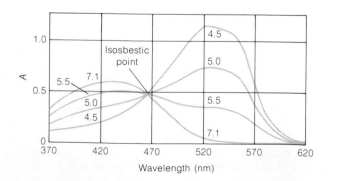

Figure 19-16
Absorption spectrum of 3.7×10^{-4} M methyl red as a function of pH between pH 4.5 and 7.1. [E. J. King, *Acid–Base Equilibria* (Oxford: Pergamon Press, 1965).]

Since the solution contains essentially HIn and In$^-$ in the pH range 4.5–7.1, all spectra in Figure 19-16 cross at one point.

To see why this is so, we write an equation for the absorbance of the solution at 465 nm:

$$A^{465} = \varepsilon_{HIn}^{465}b[HIn] + \varepsilon_{In^-}^{465}b[In^-] \tag{19-21}$$

But since the spectra of pure HIn and pure In$^-$ (at the same concentration) cross at 465 nm, ε_{HIn}^{465} must be equal to $\varepsilon_{In^-}^{465}$. Setting $\varepsilon_{HIn}^{465} = \varepsilon_{In^-}^{465} = \varepsilon^{465}$, Equation 19-21 can be factored as follows:

$$A^{465} = \varepsilon^{465}b([HIn] + [In^-]) \tag{19-22}$$

In Figure 19-16, all solutions contain the same total concentration of methyl red ($=[HIn] + [In^-]$). Only the pH varies. Therefore, the sum of concentrations in Equation 19-20 is constant, and A^{465} is constant. *The existence of an isosbestic point during a chemical reaction is good evidence that only two principal species are present.*[†]

19-7 SPECTROPHOTOMETRIC TITRATIONS

Absorption of light is one of many common physical properties whose change may be used to monitor the progress of a titration. For example, a solution of the iron-transport protein, transferrin, may be titrated with iron to measure the transferrin content. Transferrin without iron, called apotransferrin, is colorless. Each molecule binds two atoms of Fe(III) and has a molecular weight of 81 000. When the iron binds to the protein, a red color ($\lambda_{max} = 465$ nm) develops. The appearance of the red color may be used to follow the course of a titration of an unknown amount of transferrin with a standard solution of Fe(III).

$$\underset{\text{(Colorless)}}{\text{Apotransferrin}} + 2\text{Fe(III)} \rightarrow \underset{\text{(Red)}}{[\text{Fe(III)}]_2\text{transferrin}} \tag{19-23}$$

Figure 19-17 shows the results of a titration of 2.000 mL of a solution of apotransferrin with 1.79×10^{-3} M ferric nitrilotriacetate solution. As Fe(III) is added to the protein, the red color develops and the absorbance increases. When the protein is saturated with iron, no further color-forming reaction can occur, and the curve levels off abruptly. The extrapolated intersection of the two straight portions of the titration curve at 203 μL of Fe(III) in Figure 19-17 is taken as the end point. The absorbance continues to rise slowly after the equivalence point because the ferric nitrilotriacetate solution has some absorbance at 465 nm.

In constructing the graph in Figure 19-17, the effect of dilution must be considered, because the volume is different at each point. Each point plotted on the graph represents the absorbance that would be observed *if the solution had not been diluted from its original volume of 2.000 mL.* For example, the observed absorbance after adding 125 μL ($=0.125$ mL) of Fe(III) was 0.260.

An isosbestic point occurs when $\varepsilon_X = \varepsilon_Y$ and $[X] + [Y]$ is constant.

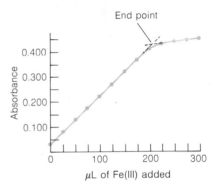

Figure 19-17
Spectrophotometric titration of transferrin with Fe(III). The initial absorbance of the solution, before iron is added, is due to a colored impurity.

Ferric nitrilotriacetate is used because uncomplexed ferric ion precipitates in neutral solution. Nitrilotriacetate is $N(CH_2CO_2^-)_3$.

[†] Under certain conditions it is possible for a solution with more than two principal species to exhibit an isosbestic point. See D.V. Stynes, *Inorg. Chem.,* **14,** 453 (1975).

The solution volume was 2.000 + 0.125 = 2.125 mL. If the volume had been 2.000 mL, the absorbance would have been greater than 0.260 by a factor of (2.125)/(2.000).

$$\text{Corrected absorbance} = \left(\frac{\text{total volume}}{\text{initial volume}}\right)(\text{observed absorbance})$$

$$= \left(\frac{2.125}{2.000}\right)(0.260) = 0.276 \qquad (19\text{-}24)$$

The absorbance plotted in the figure is 0.276, the corrected absorbance.

19-8 MEASURING AN EQUILIBRIUM CONSTANT: THE SCATCHARD PLOT

Since absorbance is proportional to *concentration* (not activity), concentrations must be converted to activities to get true equilibrium constants. Sometimes this is easy to do, and often it is not. We will neglect activity coefficients in this section.

Measuring an equilibrium constant requires that we measure the concentration (actually activity) of the species involved in the equilibrium. It is usually not necessary to measure all concentrations, because some are related to others through various mass balance equations. In general, any physical property related to concentration or activity can be useful in measuring the equilibrium constant. We have seen how pH, other potentiometric measurements, and polarographic measurements can be used to find equilibrium constants. In this section, we will see how absorbance can be used to measure an equilibrium constant.[†]

We confine our attention to the simplest equilibrium, in which the species P and X react to form PX.

$$P + X \rightleftharpoons PX \qquad (19\text{-}25)$$

Neglecting activity coefficients, we can write

$$K = \frac{[PX]}{[P][X]} \qquad (19\text{-}26)$$

Consider a series of solutions in which increments of X are added to a constant amount of P. Letting the total concentration of P (in the form P or PX) be called P_0, we can write

$$P_0 = [P] + [PX] \qquad (19\text{-}27)$$

Clearing the cobwebs from your brain, you should realize Equation 19-27 is a mass balance equation.

or

$$[P] = P_0 - [PX] \qquad (19\text{-}28)$$

Now the equilibrium expression, Equation 19-26, can be rearranged as follows:

$$\frac{[PX]}{[X]} = K[P]$$

$$= K(P_0 - [PX]) \qquad (19\text{-}29)$$

[†] An excellent article dealing with practical aspects of this subject has been written by R. W. Ramette, *J. Chem. Ed.*, **44**, 647 (1967).

A graph of $[PX]/[X]$ versus $[PX]$ will have a slope of $-K$ and is called a **Scatchard plot.**[†] It is widely used (in various forms) to measure equilibrium constants, especially in biochemistry.

A *Scatchard plot* is a graph of $[PX]/[X]$ versus $[PX]$. The slope is $-K$.

If we can find $[PX]$, we can find $[X]$ with the mass balance:

$$X_0 = [\text{total X}] = [PX] + [X] \tag{19-30}$$

To measure $[PX]$, we might use the absorbance of the solution. Suppose that P and PX each have some absorbance at a certain wavelength, but X has no absorbance at this wavelength. Suppose, for simplicity, that all measurements are made in a cell of pathlength 1.00 cm. This will allow us to omit b ($=1.00$ cm) when writing Beer's law.

The absorbance of the solution at some wavelength is the sum of absorbances of PX and P:

$$A = \varepsilon_{PX}[PX] + \varepsilon_P[P] \tag{19-31}$$

Substituting $[P] = P_0 - [PX]$, we can write

$$A = \varepsilon_{PX}[PX] + \underbrace{\varepsilon_P P_0}_{A_0} - \varepsilon_P[PX] \tag{19-32}$$

But $\varepsilon_P P_0$ is A_0, the initial absorbance before any X is added. Regrouping Equation 19-32 gives

$$A = [PX](\varepsilon_{PX} - \varepsilon_P) + \varepsilon_P P_0$$
$$= [PX]\Delta\varepsilon + A_0 \tag{19-33}$$

where $\Delta\varepsilon = \varepsilon_{PX} - \varepsilon_P$. Solving Equation 19-33 for $[PX]$ gives

$$[PX] = \frac{A - A_0}{\Delta\varepsilon} = \frac{\Delta A}{\Delta\varepsilon} \tag{19-34}$$

where ΔA is the observed absorbance minus the initial absorbance for each point in the titration.

Substituting the value of $[PX]$ from Equation 19-34 into Equation 19-29 gives a useful result:

$$\frac{\Delta A}{\Delta\varepsilon[X]} = K\left(P_0 - \frac{\Delta A}{\Delta\varepsilon}\right)$$

$$\frac{\Delta A}{[X]} = K\,\Delta\varepsilon P_0 - K\,\Delta A \tag{19-35}$$

A useful result.

That is, a graph of $\Delta A/[X]$ versus ΔA should be a straight line with a slope of $-K$. In this way, absorbances measured while P is titrated with X can be used to find the equilibrium constant for the reaction of X with P.

Two cases commonly arise in the application of Equation 19-35. If the binding constant is small, then large concentrations of X are needed to observe

[†] G. Scatchard, *Ann. N. Y. Acad. Sci.,* **51,** 660 (1949).

the formation of PX. Therefore, $X_0 \gg P_0$, and the concentration of unbound X in Equation 19-35 can be set equal to the total concentration, X_0. Alternatively, if K is not small, then $[X]$ is not equal to X_0, and $[X]$ must be measured somehow. The best approach is to have an independent measurement of $[X]$, either by measurement at another wavelength or by measurement of a different physical property.

Challenge: Use the substitution $[X] = X_0 - [PX]$ to show that Equation 19-35 can be written in the form

$$\frac{\Delta A}{X_0 - \dfrac{\Delta A}{\Delta \varepsilon}} = K P_0 \, \Delta \varepsilon - K \, \Delta A \qquad (19\text{-}36)$$

If $\Delta \varepsilon \; (= \varepsilon_{PX} - \varepsilon_P)$ is known, Equation 19-36 allows us to use only the measured absorbance to make the Scatchard plot: $\Delta A / (X_0 - (\Delta A / \Delta \varepsilon))$ versus ΔA.

In practice, the errors inherent in a Scatchard plot are often substantial and sometimes overlooked. Defining the fraction of saturation of P as

$$\text{Fraction of saturation} = S = \frac{[PX]}{P_0} \qquad (19\text{-}37)$$

it can be shown that the most accurate data are obtained for $0.2 \lesssim S \lesssim 0.8$.[†] Furthermore, data should be obtained throughout a range representing about 75% of the total saturation curve before it can be verified that the equilibrium (Equation 19-25) is obeyed. Many people have made mistakes by exploring too little of the binding curve and by not including the region $0.2 \lesssim S \lesssim 0.8$.

19-9 METHOD OF CONTINUOUS VARIATION

In the preceding section we considered the equilibrium

$$P + X \rightleftharpoons PX \qquad (19\text{-}25)$$

Suppose that several complexes can form

$$P + 2X \rightleftharpoons PX_2 \qquad (19\text{-}38)$$

$$P + 3X \rightleftharpoons PX_3 \qquad (19\text{-}39)$$

but that under a given set of conditions, one complex (say, PX_2) predominates. The **method of continuous variation** (also called *Job's method*) is a technique for identifying the stoichiometry of the predominant complex.

The classical procedure calls for mixing aliquots of equimolar solutions of P and X (perhaps followed by dilution to a constant volume) such that the total (formal) concentration of P + X remains constant. For example, stock solutions containing 2.50 mM P and 2.50 mM X could be mixed as

---

[†] D. A. Deranleau, *J. Am. Chem. Soc.,* **91,** 4044 (1969).

Table 19-3
Solutions for the method of continuous variation [†]

mL of 2.50 mM P	mL of 2.50 mM X	Mole ratio (X:P)	Mole fraction of X $\left(\dfrac{\text{mol X}}{\text{mol X + mol P}}\right)$
1.00	9.00	9.00:1	0.900
2.00	8.00	4.00:1	0.800
2.50	7.50	3.00:1	0.750
3.33	6.67	2.00:1	0.667
4.00	6.00	1.50:1	0.600
5.00	5.00	1.00:1	0.500
6.00	4.00	1:1.50	0.400
6.67	3.33	1:2.00	0.333
7.50	2.50	1:3.00	0.250
8.00	2.00	1:4.00	0.200
9.00	1.00	1:9.00	0.100

[†] All solutions are diluted to a total volume of 25.0 mL with a buffer.

shown in Table 19-3 to give a variable X:P ratio, but a constant total concentration of 1.00 mM. The absorbance of each solution is measured at a suitable wavelength, and a graph is made showing *corrected* absorbance (defined below) versus mole fraction of X. *A maximum absorbance is reached at the composition corresponding to the stoichiometry of the predominant complex.*

The corrected absorbance is defined as the measured absorbance minus the absorbance that would be produced by free P and free X alone:

$$\text{Corrected absorbance} = \text{measured absorbance} - \varepsilon_P \cdot b \cdot P_T - \varepsilon_X \cdot b \cdot X_T$$
$$(19\text{-}40)$$

where ε_P and ε_X are the molar absorptivities of pure P and pure X, b is the sample pathlength, and P_T and X_T are the total (formal) concentrations of P and X in the solution. For the first solution in Table 19-3, $P_T = (1.00/25.0)(2.50 \text{ mM}) = 0.100 \text{ mM}$ and $X_T = (9.00/25.0)(2.50 \text{ mM}) = 0.900 \text{ mM}$. In many experiments P and/or X do not absorb at the wavelength of interest, so no absorbance correction is needed.

The maximum absorbance occurs at the mole fraction of X corresponding to the stoichiometry of the complex. If the predominant complex is PX_2, the maximum occurs at (mole fraction of X) $= 2/(2 + 1) = 0.667$. If the predominant complex were P_3X, the maximum would occur at (mole fraction of X) $= 1/(1 + 3) = 0.250$.

Some precautions to take with this procedure include the following:

1. Verify that the complex follows Beer's law.

2. Use a constant ionic strength and pH, if applicable.

3. Take readings at more than one wavelength; the maximum should occur at the same mole fraction for each wavelength.

4. Do experiments at different total concentrations of P + X. If a second set of solutions were prepared as in Table 19-3, but the stock concentrations were 5.00 mM, the maximum should still occur at the same mole fraction.

Figure 19-18

(a) Spectrophotometric titration of 30.0 mL of EDTA in acetate buffer with $CuSO_4$ in the same buffer. *Upper curve:* $C_{EDTA} = C_{Cu^{2+}} = 5.00$ mM. *Lower curve:* $C_{EDTA} = C_{Cu^{2+}} = 2.50$ mM. The absorbance has not been "corrected" in any way. (b) Transformation of data from part a into mole fraction format. Absorbance of free $CuSO_4$ at the same formal concentration has been subtracted from each point in part a. [From Z. D. Hill and P. MacCarthy, *J. Chem. Ed.,* **63,** 162 (1986).]

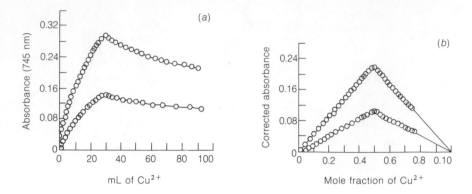

Method of continuous variation:

$$P + nX \rightleftharpoons PX_n$$

Maximum absorbance occurs when (mole fraction of X) = $n/(n + 1)$.

Luminescence refers to any emission of radiation, including both fluorescence (singlet → singlet emission) and phosphorescence (triplet → singlet emission).

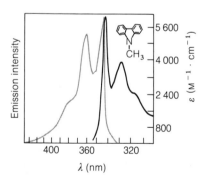

Figure 19-19

Absorption (*black line*) and emission (*gray line*) spectra of *N*-methylcarbazole in cyclohexane solution, illustrating the approximate mirror-image relationship between absorption and emission. [I. B. Berlman, *Handbook of Fluorescence Spectra of Aromatic Molecules* (New York: Academic Press, 1971).]

Although the method of continuous variation can be carried out with many separate solutions, as in Table 19-3, it is more sensible to do one titration and plot the data in the format of Figure 19-18b.[†] Figure 19-18a shows results for the titration of EDTA with Cu^{2+}. In Figure 19-18b the abscissa has been transformed into mole fraction of Cu^{2+} (=moles of Cu^{2+}/[moles of Cu^{2+} + moles of EDTA]) instead of mL of Cu^{2+}. Each absorbance in the second curve is corrected for the absorbance of an equal amount of pure Cu^{2+} solution, which absorbs the wavelength of interest. EDTA is transparent at this wavelength.

The sharp maximum at a mole fraction of 0.5 indicates formation of a 1:1 complex. If the equilibrium constant is not large, the maximum is more curved than in Figure 19-18b.

19-10   LUMINESCENCE

Although absorption measurements account for the majority of analytical spectrophotometric methods at present, luminescence measurements will probably find increasing use in the future. This is because luminescence measurements are inherently more sensitive than absorption measurements and because instruments designed to exploit this advantage are becoming more common. However, luminescence measurements are not universally applicable because many molecules produce weak or negligible emission when they are irradiated.

Relation between Absorption and Emission Spectra

In general, molecular fluorescence or phosphorescence is observed at a *lower* energy than that of the absorbed radiation (the *excitation* energy). That is, the radiation emitted by molecules is of longer wavelength than that of the radiation they absorb. A typical example is shown in Figure 19-19. Let's try to understand why emission occurs at lower energy, why there is so much structure in Figure 19-19, and why the emission spectrum is the approximate mirror image of the absorption spectrum.

[†] Z. D. Hill and P. MacCarthy, *J. Chem. Ed.,* **63,** 162 (1986).

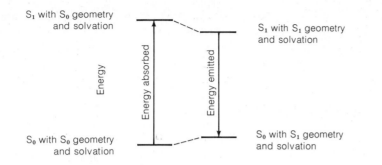

S₁ with S₀ geometry and solvation

S_1 with S_0 geometry and solvation

S_1 with S_1 geometry and solvation

Energy

Energy absorbed

Energy emitted

S_0 with S_0 geometry and solvation

S_0 with S_1 geometry and solvation

Figure 19-20
Diagram showing why the absorbed energy is greater than the emitted energy.

A molecule absorbing radiation is initially in its electronic ground state, S_0. This molecule possesses a certain geometry and solvation. Suppose that the excited state is S_1. When radiation is first absorbed the excited molecule still possesses its S_0 geometry and solvation (Figure 19-20). Very shortly after the excitation, the geometry and solvation revert to their most favorable values for the S_1 state. This must lower the energy of the excited molecule. When an S_1 molecule fluoresces, it returns to the S_0 state but retains the S_1 geometry and solvation. This unstable configuration must have a higher energy than that of an S_0 molecule with S_0 geometry and solvation. As shown in Figure 19-20, the net effect is that the emission energy is less than the excitation energy.

Figure 19-21 explains the structure in the spectra and shows why the emission spectrum is roughly the mirror image of the absorption spectrum. The structure in the absorption spectrum is due to absorption of zero or more quanta of vibrational energy in addition to one quantum of electronic energy. In polar solvents, the vibrational structure is often broadened beyond recognition, and only a broad envelope of absorption is observed. In Figure 19-19, the solvent is cyclohexane, and the vibrational structure is very evident. Following absorption, the vibrationally excited S_1 molecule relaxes back to the ground vibrational level of S_1 prior to emitting any radiation. As shown in Figure 19-21, emission from S_1 can occur to any of the vibrational levels

Electronic transitions are so fast, relative to nuclear motion, that each atom has nearly the same position and momentum before and after a transition. This is called the *Franck–Condon principle.*

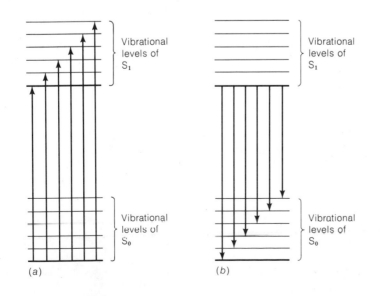

Vibrational levels of S_1

Vibrational levels of S_1

Vibrational levels of S_0

Vibrational levels of S_0

(a)

(b)

Figure 19-21
Energy-level diagram showing why structure is seen in the (a) absorption and (b) emission spectra, and why the spectra seem roughly mirror images of each other.

of S_0. This gives rise to a series of peaks in the emission spectrum. The absorption and emission spectra will have an approximate mirror-image relationship if the spacings between vibrational levels are roughly equal and if the transition probabilities are similar.[†]

Emission Intensity

The general outline of an emission experiment is shown in Figure 19-22. An excitation wavelength is selected by one monochromator, and the luminescence is examined with a second monochromator, usually positioned at a 90° angle to the incident light. By holding the excitation wavelength (λ_{ex}) fixed and scanning through the emitted radiation, an **emission spectrum** is produced. An emission spectrum is a graph of emission intensity versus emission wavelength. If the emission wavelength (λ_{em}) is held constant and the excitation wavelength is varied, an **excitation spectrum** is produced. An excitation spectrum is a graph of emission intensity versus excitation wavelength.

In emission spectroscopy, we are measuring the absolute intensity of the emission, rather than the fraction of radiant power striking the detector. Since all photomultipliers (or other detectors) have different responses for different wavelengths, the recorded emission spectrum is usually not a true profile of emission intensity versus emission wavelength. For analytical measurements employing a single emission wavelength, this effect is inconsequential. If a true profile is required (which is rare), it is necessary to calibrate the detector.

To derive a relation between the incident radiant power and the emission intensity, consider the sample cell in Figure 19-22. We expect the emission intensity to be proportional to the radiant power absorbed by the sample. That is, a certain proportion of the absorbed radiation will appear as emission

Emission spectrum: constant λ_{ex} and variable λ_{em}

Excitation spectrum: variable λ_{ex} and constant λ_{em}

Challenge: Explain why an excitation spectrum will bear a very close resemblance to the absorption spectrum.

[†] The pH dependence of emission intensity is a reflection of the acid-base properties of the excited state. An experiment to measure pK_a of the excited state of 2-naphthol has been described by J. van Stam and J.-E. Löfroth, *J. Chem. Ed.,* **63,** 181 (1986).

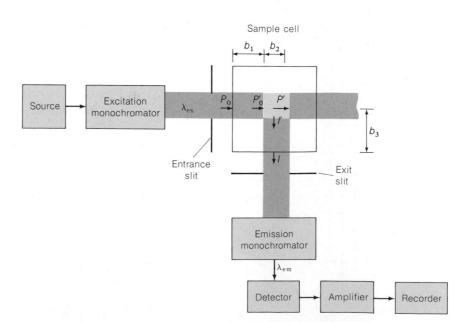

Figure 19-22
Block diagram of a fluorescence spectrophotometer, with the sample cell magnified to show various pathlengths.

under a given set of conditions (solvent, temperature, etc.). The exit slit in Figure 19-22 is set to observe emission from a region whose width is b_2.

Let the incident radiant power striking the cell be called P_0. Some of this is absorbed by the sample over the pathlength b_1 in Figure 19-22. The radiant power striking the central region of the cell is diminished by absorbance over the pathlength b_1:

$$\text{Power striking central region} = P_0' = P_0 \cdot 10^{-\varepsilon_{ex}b_1c} \qquad (19\text{-}41)$$

where ε_{ex} is the molar absorptivity for the wavelength λ_{ex}. The radiant power of the beam when it has traveled the additional distance b_2 is

$$P' = P_0' \cdot 10^{-\varepsilon_{ex}b_2c} \qquad (19\text{-}42)$$

Equation 19-41 follows from Beer's law:

$$\varepsilon_{ex}b_1c = A = \log\left(\frac{P_0}{P_0'}\right)$$

$$\Rightarrow \quad P_0' = P_0 10^{-\varepsilon_{ex}b_1c}$$

The emission intensity, I, is proportional to the radiant power absorbed in the central region of the cell:

$$\text{Emission intensity} = I' = k'(P_0' - P') \qquad (19\text{-}43)$$

where k' is a constant of proportionality dependent on the emitting molecule and the conditions. Not all the radiation emitted from the center of the cell and directed at the exit slit is observed. Part of it is absorbed by the solution between the center and the edge of the cell. The emission intensity, I, emerging from the cell is given by Beer's law:

$$I = I' \cdot 10^{-\varepsilon_{em}b_3c} \qquad (19\text{-}44)$$

where ε_{em} is the molar absorptivity at the emission wavelength and b_3 is the distance from the center to the side of the cell (Figure 19-22).

Combining Equations 19-43 and 19-44, we obtain an expression for the observed emission intensity:

$$I = k'(P_0' - P')10^{-\varepsilon_{em}b_3c} \qquad (19\text{-}45)$$

Substituting values of P_0' and P' from Equations 19-41 and 19-42, we obtain a relation between the incident radiant power and the emission intensity:

$$I = k'(P_0 \cdot 10^{-\varepsilon_{ex}b_1c} - P_0 \cdot 10^{-\varepsilon_{ex}b_1c} \cdot 10^{-\varepsilon_{ex}b_2c})10^{-\varepsilon_{em}b_3c}$$

$$= k'P_0 \cdot 10^{-\varepsilon_{ex}b_1c}(1 - 10^{-\varepsilon_{ex}b_2c})10^{-\varepsilon_{em}b_3c} \qquad (19\text{-}46)$$

Equation 19-46 allows us to calculate the emission intensity as a function of solute concentration. At low concentrations, the emission intensity increases with increasing concentration of analyte, because absorption is small and emission is proportional to emitter concentration. At high concentration, the emission intensity actually decreases, because the absorption increases more rapidly than the emission (Figure 19-23). We say the emission is *quenched* by self-absorption. At high concentration, even the *shape* of the emission spectrum can change, because absorption and emission both depend on wavelength.

For quantitative analysis, it is helpful to have a simple, monotonic, preferably linear relation between emission intensity and solute concentration.

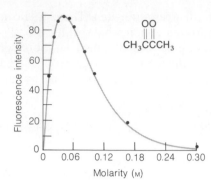

Figure 19-23
Concentration dependence of fluorescence intensity of biacetyl in CCl_4 with $\lambda_{ex} = 422$ nm and $\lambda_{em} = 464$ nm. [From G. Henderson, *J. Chem. Ed.*, **54**, 57 (1977).]

When the concentration c in Equation 19-46 is sufficiently low, the equation can be greatly simplified. Analytical emission experiments are ordinarily performed with solutions so dilute that their absorbance is negligible. This means that the exponents $\varepsilon_{ex}b_1c$, $\varepsilon_{ex}b_2c$, and $\varepsilon_{em}b_3c$ are all very small, and the terms $10^{-\varepsilon_{ex}b_1c}$, $10^{-\varepsilon_{ex}b_2c}$, and $10^{-\varepsilon_{em}b_3c}$ are all very close to unity. We can replace $10^{-\varepsilon_{ex}b_1c}$ and $10^{-\varepsilon_{em}b_3c}$ by unity in Equation 19-46, whenever the absorbance is negligible. We cannot replace $10^{-\varepsilon_{ex}b_2c}$ by unity because it appears in the term $1 - 10^{-\varepsilon_{ex}b_2c}$, which would become zero.

To find the value of $1 - 10^{-\varepsilon_{ex}b_2c}$ when $10^{-\varepsilon_{ex}b_2c}$ is close to unity, we can expand $10^{-\varepsilon_{ex}b_2c}$ in the power series:

The series in Equation 19-47 follows from the relation $10^{-A} = (e^{\ln 10})^{-A} = e^{-A \ln 10}$ and the power series expansion of e^x:

$$e^x = 1 + \frac{x}{1!} + \frac{x^2}{2!} + \frac{x^3}{3!} + \cdots$$

$$10^{-\varepsilon_{ex}b_2c} = 1 - \varepsilon_{ex}b_2c \ln 10 + \frac{(\varepsilon_{ex}b_2c \ln 10)^2}{2!} - \frac{(\varepsilon_{ex}b_2c \ln 10)^3}{3!} + \cdots \quad (19\text{-}47)$$

The term $1 - 10^{-\varepsilon_{ex}b_2c}$ in Equation 19-46 becomes

$$1 - 10^{-\varepsilon_{ex}b_2c} = \varepsilon_{ex}b_2c \ln 10 - \frac{(\varepsilon_{ex}b_2c \ln 10)^2}{2!} + \frac{(\varepsilon_{ex}b_2c \ln 10)^3}{3!} - \cdots \quad (19\text{-}48)$$

But when the absorbance of the solution is very small, $\varepsilon_{ex}b_2c$ is very small, and it is a good approximation to neglect all but the first term in the power series:

$$1 - 10^{-\varepsilon_{ex}b_2c} \approx \varepsilon_{ex}b_2c \ln 10 \qquad \text{(when } \varepsilon_{ex}b_2c \text{ is small)} \quad (19\text{-}49)$$

Substituting Equation 19-49 into Equation 19-46, and setting the other exponential terms equal to unity, gives

$$I = k'P_0\varepsilon_{ex}b_2c \ln 10$$

or

In analytical experiments, the absorbance is low, and the emission intensity is given simply by $I = kP_0c$.

$$\boxed{I = kP_0c} \quad (19\text{-}50)$$

where $k = k'\varepsilon_{ex}b_2 \ln 10$. That is, *when the absorbance is small, the emission intensity is directly proportional to the sample concentration, c, and to the incident radiant power, P_0.*

For most analytical applications, the concentration of analyte is sufficiently small that Equation 19-50 is obeyed and the emission intensity is directly proportional to concentration. The linear relationship between I and P_0 does not extend to arbitrarily high power levels, but it provides substantially increased sensitivity in emission measurements as compared with absorbance measurements. That is, doubling the incident radiant power will double the emission intensity, whereas doubling P_0 has no effect whatsoever on the absorbance. Another advantage of fluorescence is that sensitivity can be increased simply by using a more sensitive detector.

Luminescence in Analytical Chemistry

Some analytes are naturally fluorescent and can be analyzed directly. A typical procedure involves establishing a working curve of luminescence intensity versus analyte concentration. (Blank samples invariably scatter light and must be run in every analysis.) Among the more important naturally fluorescent compounds are riboflavin (vitamin B$_2$), many drugs, polycyclic aromatic compounds (an important class of carcinogens), and proteins.

Most compounds are not naturally luminescent enough to be analyzed directly. However, coupling to a fluorescent moiety provides an easy route to sensitive fluorimetric analyses. For example, airborne aliphatic isocyanates (RNCO) found in workplaces employing polyurethane foam are a significant health hazard. In one sensitive analytical procedure, a mixture of ioscyanates collected from an air sample is treated with 1-naphthyl-methylamine to form fluorescent derivatives that can be separated by liquid chromatography.[†]

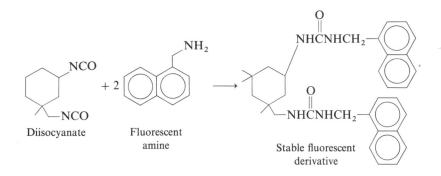

Diisocyanate   Fluorescent amine   Stable fluorescent derivative

The fluorescence intensities of fractions eluted from the chromatography column are used to quantify the various isocyanates in the air. Box 19-3 describes how fluorescent labels are used for sequencing DNA.

Metal ions can be analyzed following their reaction with a fluorescent chelating agent. For example, calcein forms a fluorescent complex with calcium. Fluoride ion can be analyzed because of its ability to *quench* (decrease) the fluorescence of the Al^{3+} complex of alizarin garnet R. A working curve of fluorescence intensity versus F$^-$ concentration decreases as [F$^-$] increases.

With sophisticated laser fluorometers, it is possible to measure the time dependence of fluorescence. Each member of a multicomponent mixture is likely to have a different fluorescence lifetime, even if the emission wavelengths are similar. By measuring intensity as a function of time and wavelength, complex mixtures with overlapping spectra may be analyzed.

[†] S. P. Levine, J. H. Hoggott, E. Chladek, G. Jungclaus, and J. L. Gerlock, *Anal. Chem.,* **51**, 1106 (1979).

This is the same reason nephelometry is more sensitive than turbidimetry. See Box 9-1 if you have forgotten about these techniques.

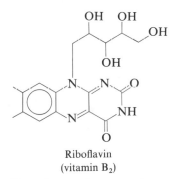

Riboflavin
(vitamin B$_2$)

For a lecture demonstration of riboflavin fluorescence, see D. S. Chatellier and H. B. White, III, *J. Chem Ed.,* **65**, 814 (1988).

If the analyte is not fluorescent, it may be coupled to something that is, and then analyzed. Fluorescent labels of fingerprints are a powerful tool in forensic analysis. For a report on this application, see E. R. Menzel, *Anal. Chem.,* **61**, 557A (1989).

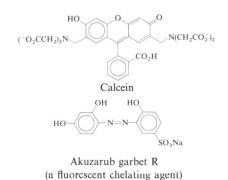

Calcein

Akuzarub garbet R
(a fluorescent chelating agent)

Box 19-3 DNA SEQUENCING WITH FLUORESCENT LABELS

Genetic information is coded in the sequence of nucleotides in deoxyribonucleic acid (DNA). Biologists are beginning the arduous task of mapping human genes by learning the sequence of nucleotides in our entire set of chromosomes. This, in turn, will shed light at a molecular level on the how living organisms operate. Fluorescent labels are central to automated DNA sequencing.

DNA is made of two very long strands wrapped around each other to form a helix. The two strands are connected by hydrogen bonds as shown below.

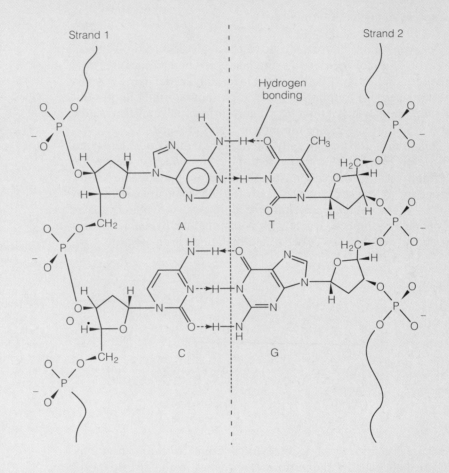

The nucleotides are designated A, T, C, and G, the letters standing for adenine, thymine, cytosine, and guanosine. A and T are always hydrogen bonded to each other, as are C and G. Sequencing begins by separating the two strands and selecting one.

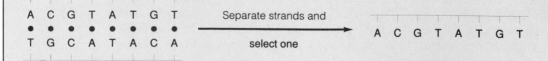

The single strand is then used as a *template* for constructing a new strand with a complementary sequence. For example, A on the original strand will give T on the complementary strand, and

G on the original strand will give C on the complementary strand. The enzyme DNA polymerase carries out the synthesis of the new strand, using the template strand and a supply of the four building blocks, designated dATP, dTTP, dCTP, and dGTP.

```
 ┌ ┬ ┬ ┬ ┬ ┬ ┬ ┬ ┐                              A  C  G  T  A  T  G  T
                          DNA polymerase         •  •  •  •  •  •  •  •
 A  C  G  T  A  T  G  T   ──────────────────→    T  G  C  A  T  A  C  A
                          dATP, dTTP, dCTP, dGTP
 └ ┴ ┴ ┴ ┴ ┴ ┴ ┴ ┘                              └ ┴ ┴ ┴ ┴ ┴ ┴ ┴ ┘
```

For DNA sequencing, *chain-terminating building blocks* are mixed with the supply of normal building blocks. The chain-terminating building blocks are similar to dATP, dTTP, dCTP, and dGTP, but they lack a hydroxyl group necessary to make the next bond in the chain.

Chain-terminating G building block lacks OH group necessary to continue chain beyond this point

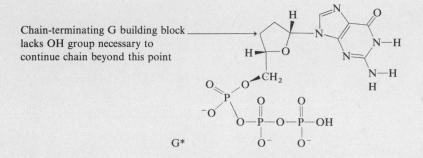

G*

When one of these chain-terminating building blocks, designated A*, T*, C*, and G*, is incorporated into the growing strand of DNA, no further growth can occur. Since this happens at random, complementary strands of DNA of every possible length are synthesized:

```
 ┌ ┬ ┬ ┬ ┬ ┬ ┬ ┬ ┐                              DNA polymerase
                                                ──────────────────────→
 A  C  G  T  A  T  G  T                          dATP, dTTP, dCTP, dGTP
 └ ┴ ┴ ┴ ┴ ┴ ┴ ┴ ┘                                  A*, T*, C*, G*
```

```
A  C  G  T  A  T  G  T        A  C  G  T  A  T  G  T        A  C  G  T  A  T  G  T
•  •  •                       •  •  •  •  •                 •  •  •  •  •  •  •  •
T  G  C*                      T  G  C  A  T*                 T  G  C  A  T  A  C  A*
```

```
A  C  G  T  A  T  G  T        A  C  G  T  A  T  G  T        A  C  G  T  A  T  G  T
•  •  •  •  •  •               •  •                          •
T  G  C  A  T  A*             T  G*                          T*
```

```
        A  C  G  T  A  T  G  T        A  C  G  T  A  T  G  T
        •  •  •  •  •  •  •            •  •  •  •
        T  G  C  A  T  A  C*          T  G  C  A*
```

Box 19-3 (Cont.)

When the strands are separated again, pieces of DNA of every possible length (from one to eight nucleotides in the illustration above) are present, and each ends with one chain-terminating building block.

For sequencing DNA, the chain-terminating building blocks are labeled with fluorescent side groups. An example is G-505, whose side chain has a fluorescence maximum at 505 nm.

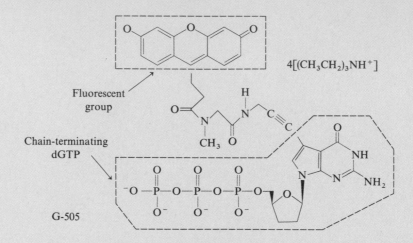

The side groups for the four terminators are selected so that they fluoresce at slightly different wavelengths.

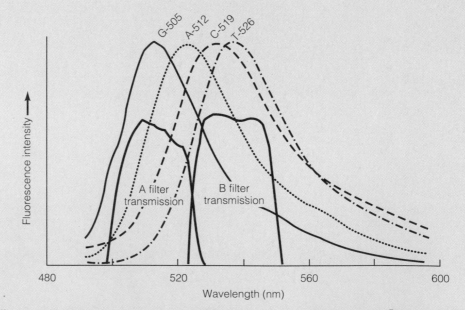

Fluorescence spectra of chain-terminating building blocks used for DNA sequencing. [Reproduced from J. M. Prober, G. L. Trainor, R. J. Dam, F. W. Hobbs, C. W. Robertson, R. J. Zagursky, A. J. Cocuzza, M. A. Jensen, and K. Baumeister, *Science,* **238,** 336 (1987).]

The mixture of terminated DNA strands now contains pieces of every possible length. Each one ending with G fluoresces at 505 nm. Each one terminated by A fluoresces at 512 nm. C- and T-terminated fragments fluoresce at 519 and 526 nm, respectively. By viewing the fluorescence through the two filters whose transmission is shown at the lower left, it is possible to determine which terminator is being observed. For example, a fragment ending in T* will have maximum fluorescence with the B filter and minimum transmission with the A filter. A fragment ending with A* will have about equal transmission through each filter.

The mixture of fragments is separated by *electrophoresis,* in which a strong electric field causes the negatively charged strands to migrate toward the positive pole. The longest strands are the slowest, giving a pattern that looks like this:

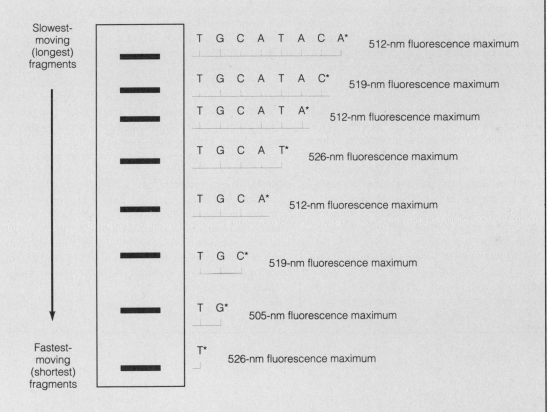

The bands in the electrophoretic experiment are invisible. However, when excited by a 488-nm laser, each fluoresces a color determined by its terminator. By scanning the gel from one end to the other with the laser, and observing the fluorescence through the two filters, it is possible to read off the exact sequence of nucleotides in the DNA fragment. By this means we hope to learn the sequence of the entire human genome, which consists of 46 chromosomes divided into 10^5 genes comprising $10^{9.5}$ nucleotides!

Summary

Light can be thought of as waves whose frequency (λ) and wavelength (v) have the important relation $\lambda v = c$, where c is the speed of light. Alternatively, light may be viewed as consisting of photons whose energy (E) is given by $E = hv = hc/\lambda = hc\tilde{v}$, where h is Planck's constant and $\tilde{v}\ (= 1/\lambda)$ is the wavenumber. Absorption of light is commonly measured by absorbance (A) or transmittance (T), defined as $A = \log(P_0/P)$ and $T = P/P_0$, where P_0 is the radiant power of light incident on a sample and P is the power emerging from the other side. The major analytical utility of absorption spectroscopy is derived from the fact that absorbance is proportional to the concentration of the absorbing species in dilute solution (Beer's law): $A = \varepsilon bc$. In this equation, b is pathlength, c is concentration, and the constant of proportionality, ε, is the molar absorptivity. If a sample does not obey Beer's law, it is likely that a chemical reaction is altering the concentration of chromophore when the sample concentration is changed.

The basic components of a spectrophotometer include a radiation source, a monochromator, a sample cell, and a detector. In double-beam spectrophotometry, light is passed alternately through the sample and the reference cuvets by a rotating beam chopper, and the light beams emerging from each are continually compared to measure absorbance. To minimize errors in spectrophotometry, samples should be free of particles, and cuvets should be clean. Measurements should be made at a wavelength of maximum absorbance. Instrument errors tend to be minimized if the absorbance falls in the approximate range $A \approx 0.4{-}0.9$.

The most common analytical application of spectrophotometry makes use of the proportionality between absorbance and concentration. If the absorbance of a series of standards is measured, the concentration of an unknown treated in the same way can be obtained by direct comparison to the standards. In such analyses, a suitable reagent blank should be prepared, and interfering species should be removed, masked, or otherwise accounted for. The absorbance spectrum of a mixture can be used to find the amount of each component in the mixture. If the spectra of individual components overlap a great deal, a graphical method of analysis is preferred. If the spectra overlap only partially, then n measurements of absorbance at n wavelengths of maximum absorption are, in principle, sufficient to find the concentrations of the n absorbing components. Spectrophotometry can also be used to follow the course of a titration reaction, to measure equilibrium constant, and to determine the stoichiometry of a complex.

When a molecule absorbs light, it is promoted to an excited state from which it may return to the ground state by radiationless processes or by fluorescence (singlet $\rightarrow$ singlet emission) or phosphorescence (triple $\rightarrow$ singlet emission). Any form of luminescence is potentially useful for quantitative analysis, because emission intensity is proportional to sample concentration at low concentration. At higher concentration, self-absorption and self-quenching distort the emission. An excitation spectrum (a graph of emission intensity versus excitation wavelength) is very similar to an absorption spectrum (a graph of absorbance versus wavelength). An emission spectrum (a graph of emission intensity versus emission wavelength) comes at lower energy and tends to be the mirror image of the absorption spectrum.

Terms to Understand

absorbance	hertz	radiant power
absorption spectrum	internal conversion	reagent blank
Beer's law	intersystem crossing	refractive index
chemiluminescence	isosbestic point	rotational transition
chromophore	luminescence	Scatchard plot
cuvet	masking	singlet state
electromagnetic spectrum	method of continuous variation	spectrophotometry
electronic transition	molar absorptivity	transmittance
emission spectrum	molecular orbital	triplet state
excitation spectrum	monochromatic light	vibrational transition
excited state	monochromator	wavelength
fluorescence	phosphorescence	wavenumber
frequency	photon	
ground state	Planck's constant	

Exercises

19-A. (a) What value of absorbance corresponds to 45.0% T?

(b) If a 0.010 0 M solution exhibits 45.0% T at some wavelength, what will be the percent transmittance for a 0.020 0 M solution of the same substance?

19-B. Ammonia can be determined spectrophotometrically by reaction with phenol in the presence of hypochlorite (OCl^-):

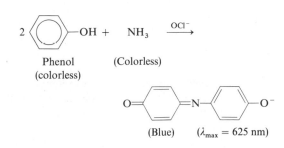

Phenol   (Colorless)
(colorless)

(Blue)   (λ_{max} = 625 nm)

A 4.37-mg sample of protein was chemically digested to convert all of its nitrogen to ammonia. After this treatment, the volume of the sample was 100.0 mL. Then 10.0 mL of the solution was placed in a 50-mL volumetric flask and treated with 5 mL of phenol solution plus 2 mL of sodium hypochlorite solution. The sample was diluted to 50.0 mL, and the absorbance at 625 nm was measured in a 1.00-cm cuvet after 30 minutes. For reference, a standard solution was prepared from 1.00×10^{-2} g of NH_4Cl dissolved in 1.00 L of water. Then 10.0 mL of this standard was placed in a 50-mL volumetric flask and analyzed in the same manner as the unknown. A reagent blank was prepared using distilled water in place of unknown.

Sample	Observed absorbance at 625 nm
Blank	0.140
Reference	0.308
Unknown	0.592

(a) Calculate the molar absorptivity of the blue product.

(b) Calculate the weight percent of nitrogen in the protein.

19-C. Cu(I) reacts with neocuproine to form a colored complex with an absorption maximum at 454 nm (Equation 19-9). Neocuproine is particularly useful because it reacts with few other metals. The copper complex is soluble in 3-methyl-1-butanol (isoamyl alcohol), an organic solvent that does not dissolve appreciably in water. This means that if isoamyl alcohol is added to water, a two-layered mixture results, with the denser water layer at the bottom. If Cu(I)–neocuproine is present, virtually all of it goes into the organic phase. For the purpose of this problem, assume that the isoamyl alcohol does not dissolve in the water at all and that all of the colored complex will be in the organic phase. Suppose that the following procedure is carried out:

1. A rock containing copper is pulverized, and all metals are extracted from it with strong acid. The acidic solution is neutralized with base and made up to 250.0 mL in flask A.

2. Next 10.00 mL of the solution is transferred to flask B and treated with 10.00 mL of a reducing agent to reduce all Cu to Cu(I). Then 10.00 mL of buffer is added to bring the pH to a value suitable for complex formation with neocuproine.

3. After that, 15.00 mL of this solution is withdrawn and placed in flask C. To the flask is added 10.00 mL of an aqueous solution containing neocuproine and 20.00 mL of isoamyl alcohol. After shaking well and allowing the phases to separate, all Cu(I)–neocuproine is in the organic phase.

4. A few milliliters of the upper layer are withdrawn, and the absorbance at 454 nm is measured in a 1.00-cm tube. A blank carried through the same procedure gave an absorbance of 0.056.

(a) Suppose that the rock contained 1.00 mg of Cu. What will be the concentration of Cu (moles per liter) in the isoamyl alcohol phase?

(b) If the molar absorptivity of Cu(I)–neocuproine is 7.90×10^3 $M^{-1} \cdot cm^{-1}$, what will be the observed absorbance? Remember that a blank carried through the same procedure gave an absorbance of 0.056.

(c) A rock is analyzed and found to give a final absorbance of 0.874 (uncorrected for the blank). How many milligrams of Cu are in the rock?

19-D. Transferrin is the iron-transport protein found in blood. It has a molecular weight of 81 000 and carries two Fe(III) ions. Desferrioxamine B is a potent iron chelator used to treat patients with iron overload. It has a molecular weight of about 650 and can bind one iron atom as Fe(III). Desferrioxamine can take iron from many sites within the body and is excreted (with its iron) through the kidneys. The molar absorptivities of these compounds (saturated with iron) at two wavelengths are given below. Both compounds are colorless (no visible absorption) in the absence of iron.

	$\varepsilon[M^{-1} \cdot cm^{-1}]$		
λ (nm)	Transferrin	Desferrioxamine	λ_{max} (nm)
428	3 540	2 730	Transferrin: 470
470	4 170	2 290	Desferrioxamine: 428

(a) A solution of transferrin exhibits an absorbance of 0.463 at 470 nm in a 1.000-cm cell. Calculate the concentration of transferrin in milligrams per milliliter and the concentration of iron in micrograms per milliliter.

(b) A short time after adding some desferrioxamine (which dilutes the sample), the absorbance at 470 nm was 0.424, and the absorbance at 428 nm was 0.401. Calculate the fraction of iron in transferrin and the fraction in desferrioxamine. Remember that transferrin binds two iron atoms and desferrioxamine binds only one.

19-E. The metal chelator semi-xylenol orange is yellow at pH 5.9, but turns red ($\lambda_{max} = 490$ nm) when it reacts with Pb^{2+}. A 2.025 mL sample of semi-xylenol orange at pH 5.9 was titrated with 7.515×10^{-4} M $Pb(NO_3)_2$, with the results shown at the top of the next column.

Make a graph of A versus μL of Pb^{2+} added. Be sure to correct the absorbances in the table for the effect of dilution. That is, the corrected absorbance is what would be observed if the volume were not changed from its initial value of 2.025 mL. Assuming that the reaction of semi-xylenol orange with Pb^{2+} has a 1:1 stoichiometry, find the molarity of semixylenol orange in the original solution.

Total μL Pb^{2+} added	$A^{1\,cm}_{490\,nm}$
0.0	0.227
6.0	0.256
12.0	0.286
18.0	0.316
24.0	0.345
30.0	0.370
36.0	0.399
42.0	0.425
48.0	0.445
54.0	0.448
60.0	0.449
70.0	0.450
80.0	0.447

19-F. The compound P, which absorbs light at 305 nm, was titrated with X, which does not absorb at this wavelength. The product, PX, also absorbs at 305 nm. The absorbance of each solution was measured in a 1.000-cm cell, and the concentration of free X was determined by an independent method. The results are shown below.

Experiment	P_0 (M)	X_0 (M)	A	[X] (M)
0	0.010 0	0	0.213	0
1	0.010 0	0.001 00	0.303	4.42×10^{-6}
2	0.010 0	0.002 00	0.394	9.10×10^{-6}
3	0.010 0	0.003 00	0.484	1.60×10^{-5}
4	0.010 0	0.004 00	0.574	2.47×10^{-5}
5	0.010 0	0.005 00	0.663	3.57×10^{-5}
6	0.010 0	0.006 00	0.752	5.52×10^{-5}
7	0.010 0	0.007 00	0.840	8.20×10^{-5}
8	0.010 0	0.008 00	0.926	1.42×10^{-4}
9	0.010 0	0.009 00	1.006	2.69×10^{-4}
10	0.010 0	0.010 0	1.066	5.87×10^{-4}
11	0.010 0	0.020 0	1.117	9.66×10^{-3}

Prepare a Scatchard plot and find the equilibrium constant for the reaction $X + P \rightleftharpoons PX$.

19-G. Consider a fluorescence experiment in which the cell in Figure 19-22 is arranged so that b_1 and b_3 are negligible and, therefore, self-absorption can be neglected. To a first approximation, the emission intensity will simply be proportional to solute concentration. At what absorbance ($= \varepsilon_{ex}b_2c$) will the emission be 5% below the value expected if emission is proportional to concentration?

Problems

A19-1. Fill in the blanks:
  (a) If you double the frequency of electromagnetic radiation, you ____ the energy.
  (b) If you double the wavelength, you ____ the energy.
  (c) If you double the wavenumber, you ____ the energy.

A19-2. (a) How much energy (in kilojoules) is carried by one einstein of red light with $\lambda = 650$ nm?
  (b) How much is carried by one einstein of blue light with $\lambda = 400$ nm?

A19-3. Calculate the frequency (in hertz), wavenumber (in cm^{-1}) and energy (in joules per photon and joules per einstein) of visible light with a wavelength of 562 nm.

A19-4. The absorbance of a 2.31×10^{-5} M solution of a compound is 0.822 at a wavelength of 266 nm in a 1.00-cm cell. Calculate the molar absorptivity at 266 nm.

A19-5. What color would you expect to observe for a solution of the ion Fe(ferrozine)$_3^{4-}$, which has a visible absorbance maximum at 562 nm?

A19-6. What is the difference between fluorescence and phosphorescence?

A19-7. Equation 19-50 applies to fluorescence analysis of a dilute solution. Fill in the blanks:
  (a) If you double the sample concentration, you ____ the fluorescence intensity.
  (b) If you double the incident power, you ____ the fluorescence intensity.

A19-8. A 15.0-mg sample of a compound with a molecular weight of 384.63 was dissolved in a 5-mL volumetric flask. A 1.00-mL aliquot was withdrawn, placed in a 10-mL volumetric flask, and diluted to the mark.
  (a) Find the concentration of sample in the 5-mL flask.
  (b) Find the concentration in the 10-mL flask.
  (c) The 10-mL sample was placed in a 0.500-cm cuvet and gave an absorbance of 0.634 at 495 nm. Find the molar absorptivity (ε_{495}, with units of M$^{-1} \cdot$cm^{-1}) at this wavelength.

A19-9. A compound with a molecular weight of 292.16 was dissolved in a 5-mL volumetric flask. A 1.00-mL aliquot was withdrawn, placed in a 10-mL volumetric flask, and diluted to the mark. The absorbance measured at 340 nm was 0.427 in a 1.000-cm cuvet. The molar absorptivity for this compound at 340 nm is $\varepsilon_{340} = 6\,130$ M$^{-1} \cdot$cm^{-1}.
  (a) Calculate the concentration of compound in the cuvet.

(b) What was the concentration of compound in the 5-mL flask?
(c) How many milligrams of compound were used to make the 5-mL solution?

A19-10. Consider the compounds X and Y in the example in Section 19-6. A mixture of X and Y in a 0.100-cm cell had an absorbance of 0.233 at 272 nm and 0.200 at 327 nm. Find the concentrations of X and Y in the mixture.

A19-11. Explain the significance of the isosbestic point in Figure 19-16.

A19-12. A 2.00-mL solution of transferrin (Section 19-7) was titrated as in Figure 19-17. It required 163 μL of 1.43 mM ferric nitrilotriacetate to reach the end point.
  (a) How many moles of Fe(III) (=ferric nitrilotriacetate) were required to reach the end point?
  (b) Each transferrin molecule binds two ferric ions. Find the concentration of transferrin in the 2.00-mL solution.

A19-13. Compound P was titrated with X to form the complex PX. A series of solutions was prepared with the total concentration of P remaining fixed at 1.00×10^{-5} M. Both P and X have no visible absorbance, but PX has an absorption maximum at 437 nm. The following table shows how the absorbance at 437 nm in a 5.00-cm cell depends on the total concentration of added X ($X_T = [X] + [PX]$).

X_T (M)	A
0	0.000
0.002 00	0.125
0.004 00	0.213
0.006 00	0.286
0.008 00	0.342
0.010 0	0.406
0.020 0	0.535
0.040 0	0.631
0.060 0	0.700
0.080 0	0.708
0.100	0.765

(a) Make a Scatchard plot of $\Delta A/[X]$ versus ΔA. In this plot, [X] refers to the species X, not to X_T. However, since $X_T \gg [P]$, we can safely say that $[X] \approx X_T$ in this experiment.
(b) Refer to Equation 19-35. From the slope of the graph, find the equilibrium constant, K.

19-14. The characteristic orange light produced by sodium in a flame is due to an intense emission called the sodium "D" line. This "line" is actually a doublet, with wavelengths (measured in vacuum) of 589.157 88 and 589.755 37 nm. The index of refraction of air at a wavelength near 589 nm is 1.000 292 6. Calculate the frequency, wavelength, and wavenumber of each component of the "D" line, measured in air.

19-15. During an assay of the thiamine (vitamin B_1) content of a pharmaceutical preparation, the percent transmittance scale was accidentally read, instead of the absorbance scale of the spectrophotometer. One sample gave a reading of 82.2% T, and a second sample gave a reading of 50.7% T at a wavelength of maximum absorbance. What is the ratio of concentrations of thiamine in the two samples?

19-16. Nitrite ion, NO_2^-, is used as a preservative for bacon and other foods. It has been the center of controversy because it is potentially carcinogenic. A spectrophotometric determination of NO_2^- makes use of the following reactions:

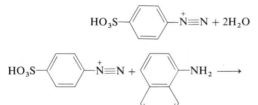

Sulfanilic acid

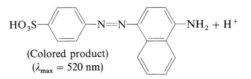

1-Aminonaphthalene

(Colored product)
(λ_{max} = 520 nm)

An abbreviated procedure for the determination is given below:

1. To 50.0 mL of unknown solution containing nitrite is added 1.00 mL of sulfanilic acid solution.

2. After 10 minutes, 2.00 mL of 1-aminonaphthalene solution and 1.00 mL of buffer are added.

3. After 15 minutes, the absorbance is read at 520 nm in a 5.00 cm cell.

The following solutions were analyzed:

A. 50.0 mL of food extract known to contain no nitrite (that is, a negligible amount); final absorbance = 0.153.

B. 50.0 mL of food extract suspected of containing nitrite; final absorbance = 0.622.

C. same as B, but with 10.0 μL of 7.50 × 10^{-3} M $NaNO_2$ added to the 50.0-mL sample; final absorbance = 0.967.

(a) Calculate the molar absorptivity, ε, of the colored product. Remember that a 5.00-cm cell was used.

(b) How many micrograms of NO_2^- were present in 50.0 mL of food extract?

19-17. Spectrophotometric analysis of phosphate can be performed by the following procedure:

Standard solutions

A. KH_2PO_4 (F.W. 136.09)—81.37 mg dissolved in 500.0 mL of water

B. $Na_2MoO_4 \cdot 2H_2O$ (sodium molybdate)— 1.25 g in 50 mL of 5 M H_2SO_4

C. $H_3NNH_3^{2+}SO_4^{2-}$ (hydrazine sulfate)— 0.15 g in 100 mL of H_2O

Procedure
Place the sample (either an unknown or the standard phosphate solution, A) in a 5-mL volumetric flask, and add 0.500 mL of B plus 0.200 mL of C. Dilute to almost 5 mL with water, and heat at 100°C for 10 minutes to form a blue product ($H_3PO_4(MoO_3)_{12}$, 12-molybdophosphoric acid). Cool the flask to room temperature, dilute to the mark with water, mix well, and measure the absorbance at 830 nm in a 1.00-cm cell.

(a) When 0.140 mL of solution A was analyzed, an absorbance of 0.829 was recorded. A blank carried through the same procedure gave an absorbance of 0.017. Find the molar absorptivity of blue product.

(b) A solution of the phosphate-containing iron-storage protein ferritin was analyzed by this procedure. The unknown contained 1.35 mg of ferritin, which was digested in a total volume of 1.00 mL to release phosphate from the protein. Then 0.300 mL of this solution was analyzed by the procedure above and found to give an absorbance of 0.836. A blank carried through this procedure gave an absorbance of 0.038. Find the weight percent of phosphorus in the ferritin.

19-18. If a sample for spectrophotometric analysis is placed in a 10-cm cell, the absorbance will be 10 times greater than the absorbance in a 1-cm cell. Will the absorbance of the reagent-blank solution also be increased by a factor of 10?

19-19. When I was a boy, Uncle Wilbur let me watch as he analyzed the iron content of runoff from his banana ranch. A 25.0-mL sample was acidified with nitric acid and treated with excess KSCN to form a red complex. (KSCN itself is colorless.) The solution was then diluted to 100.0 mL and put in a variable-pathlength cell. For comparison, a 10.0-mL reference sample of 6.80×10^{-4} M Fe^{3+} was treated with HNO_3 and KSCN and diluted to 50.0 mL. The reference was placed in a cell with a 1.00-cm light path. The runoff sample exhibited the same absorbance as the reference when the pathlength of the runoff cell was 2.48 cm. What was the concentration of iron in Uncle Wilbur's runoff?

19-20. The figure below shows the spectra of 1.00×10^{-4} M MnO_4^-, 1.00×10^{-4} $Cr_2O_7^{2-}$ and an unknown mixture of both. Absorbances at several wavelengths are given in the table. Use Equation 19-14 to find the concentration of each species in the mixture.

Wavelength (nm)	MnO_4^- Standard	$Cr_2O_7^{2-}$ Standard	Mixture
266	0.042	0.410	0.766
288	0.082	0.283	0.571
320	0.168	0.158	0.422
350	0.125	0.318	0.672
360	0.056	0.181	0.366

19-21. Infrared spectra are customarily recorded on a % transmittance scale so that weak and strong bands can be displayed on the same scale. The region near $2\,000$ cm^{-1} in the infrared spectra of compounds A and B is shown below. Note that absorption corresponds to a downward peak on this scale. The spectra were recorded using a 0.010 0 M solution of each, in 0.005 00-cm pathlength cells. A mixture of A and B in a 0.005 00-cm cell gave a transmittance of 34.0% at $2\,022$ cm^{-1} and 38.3% at $1\,993$ cm^{-1}. Find the concentrations of A and B.

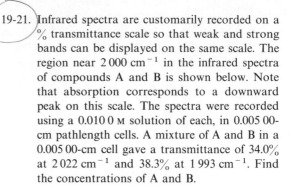

	Wavenumber (cm^{-1})	
	2 022	1 993
pure A	31.0% T	79.7% T
pure B	97.4% T	20.0% T

50,83

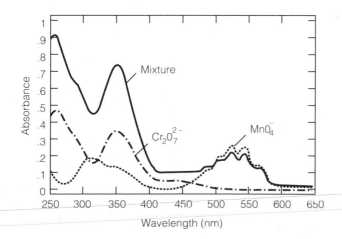

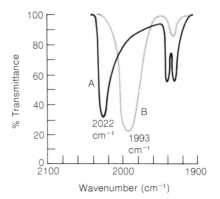

Visible spectrum of MnO_4^-, $Cr_2O_7^{2-}$, and an unknown mixture containing both ions. [From M. Blanco, H. Iturriaga, S. Maspoch, and P. Tarín, *J. Chem. Ed.,* **66,** 178 (1989).]

33.94

19-22. The coenzyme NADP$^+$ can be assayed by a titration in which it is converted to the fluorescent product NADPH by the action of adenosine triphosphate (ATP) plus several enzymes. A hypothetical titration curve is shown below.

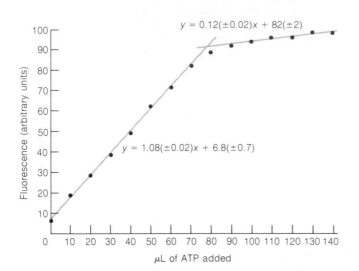

In this titration, the fluorescence intensity is plotted versus microliters of added ATP. The first eight points lie on one line, and the points from 90 to 140 μL lie on a second line. The uncertainties in slope and intercept of these lines are one standard deviation, as determined by the method of least squares. The end point of the titration lies at the intersection of the two lines. Using the equations for the two lines, *determine the volume of ATP (in microliters) at the equivalence point.* Also, use the standard deviations of the slopes and intercepts to *estimate the standard deviation of the volume of ATP at the equivalence point.* Express your answer (μL $\pm$ standard deviation) with an appropriate number of significant figures.

19-23. The metal ion indicator xylenol orange (Table 13-3) is yellow at pH 6 ($\lambda_{max} = 439$ nm). The spectral changes that occur as the indicator is titrated with VO^{2+} ion at pH 6 are shown below. The molar ratio VO^{2+}/xylenol orange at each point is

Trace	Ratio	Trace	Ratio
0	0	9	0.90
1	0.10	10	1.0
2	0.20	11	1.1
3	0.30	12	1.3
4	0.40	13	1.5
5	0.50	14	2.0
6	0.60	15	3.1
7	0.70	16	4.1
8	0.80		

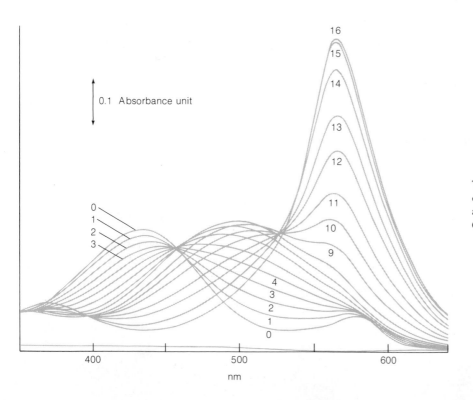

Titration curve for the reaction of xylenol orange with VO^{2+} at pH 6.0 [D. C. Harris and M. H. Gelb, *Biochim. Biophys. Acta,* **623,** 1 (1980).]

Suggest a sequence of chemical reactions to explain the spectral changes, especially the isosbestic points at 457 and 528 nm.

19-24. *Method of continuous variation.* Make a graph of absorbance versus mole fraction of thiocyanate for the data in the following table.

mL Fe^{3+} solution	mL SCN^- solution	Absorbance
30.00	0	0.001
27.00	3.00	0.122
24.00	6.00	0.226
21.00	9.00	0.293
18.00	12.00	0.331
15.00	15.00	0.346
12.00	18.00	0.327
9.00	21.00	0.286
6.00	24.00	0.214
3.00	27.00	0.109
0	30.00	0.002

Note: Fe^{3+} solution: 1.00 mM $Fe(NO_3)_3$ + 10.0 mM HNO_3
SCN^- solution: 1.00 mM KSCN + 15.0 mM HCl
Source: Adapted from Z. D. Hill and P. MacCarthy, *J. Chem. Ed.,* **63,** 162 (1986).

(a) What is the stoichiometry of the predominant $Fe(SCN)_n^{3-n}$ species?
(b) Why is the peak not as sharp as the ones in Figure 19-18?
(c) Why does one solution contain 10.0 mM acid while the other contains 15.0 mM acid?

19-25. What is the difference between a fluorescence excitation spectrum and a fluorescence emission spectrum? Which one resembles an absorption spectrum?

19-26. How can the sample cell in Figure 19-22 be positioned to minimize the self-absorption expressed in the terms $10^{-\varepsilon_{ex}b_1c}$ and $10^{-\varepsilon_{em}b_3c}$ in Equation 19-46?

19-27. Consider a fluorescence experiment in which $\varepsilon_{ex} = 1\,530$ $M^{-1}\cdot cm^{-1}$, $\varepsilon_{em} = 495$ $M^{-1}\cdot cm^{-1}$, $b_1 = 0.400$ cm, $b_2 = 0.200$ cm, and $b_3 = 0.500$ cm in Equation 19-46. Make a graph of relative fluorescence intensity versus concentration for the following concentrations of solute: 1.00×10^{-7}, 1.00×10^{-6}, 1.00×10^{-5}, 1.00×10^{-4}, 1.00×10^{-3}, and 1.00×10^{-2} M.

19-28. A sensitive assay for ATP is based on its participation in the light-producing reaction of the firefly.[†] The reaction catalyzed by the enzyme luciferase is

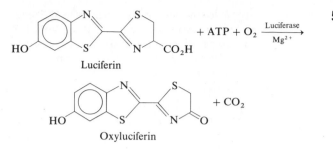

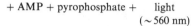

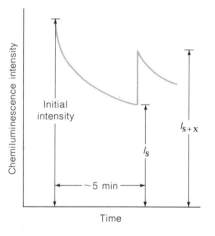

+ AMP + pyrophosphate + light
(~ 560 nm)

When the reactants are mixed, the solution gives off light. The light intensity decays slowly due to product inhibition of the reaction. Otherwise, the light would have a steady intensity because *the rate at which reactants are consumed is negligible.* That is, ATP and luciferin maintain their original concentrations throughout the few minutes that the reaction might be monitored. Some typical experimental results are shown below.

Let the initial concentration of ATP in the reaction be [S]. Suppose that additional ATP is added, increasing the concentration in the reaction to [S] + [X]. The kinetic description of the reaction predicts that the increase in light intensity after the addition will be given by

$$\frac{I_S}{I_{S+X}} = \frac{1}{[S]+[X]}\left(\frac{K[S]}{K+[S]}\right) + \frac{[S]}{K+[S]}$$

where K is constant.
(a) Suppose that [S] = 250 μM and after 5 minutes $I_S = 58.7$ arbitrary intensity units. Then a standard addition of [X] = 200 μM is made, and I_{S+X} is found to be 74.5 units. Use these data to find the value of K in the equation above.

[†] J. J. Lemasters and C. R. Hackenbrock, *Methods of Enzymology,* **57,** 36 (1978). A student experiment using luciferase for the assay of ATP or reduced nicotine adenine dinucleotide (NADH) is described by T. C. Selig, K. A. Drozda, and J. A. Evans, *J. Chem. Ed.,* **61,** 918 (1984).

(b) When the intensity had decayed to 63.5 units, an unknown aliquot of ATP was added to the reaction, and the intensity increased to 74.6 units. How much was the increase in concentration caused by the unknown aliquot?

19-29. *Kinetic method for trace determination of iron.* Very sensitive spectrophotometric analyses are based on the ability of traces of analyte to catalyze a reaction of a species present at much higher concentration than the analyte. An example is the Fe^{3+} catalysis of oxidation of N,N-dimethyl-p-phenylenediamine (DPD) by H_2O_2.

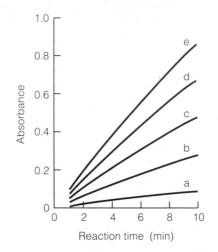

DPD → DPD⁺

The reactant DPD has no visible absorbance, but the product has an absorbance maximum at 514 nm. The analytical procedure follows: To a 25-mL volumetric flask is added a measured volume of unknown containing Fe^{3+}. After addition of 1 mL of acetate buffer (pH 5.7, treated to remove Fe^{3+} impurity) and 1 mL of 3.5% H_2O_2, the solution is dilute to the mark. The solution is kept at $25.0 \pm 0.1°C$. At time $t = 0$ 1.00 mL of 0.024 M DPD (also treated to remove Fe^{3+}) is added. After rapid mixing, a portion of the solution is transferred to a spectrophotometer cell kept at 25.0°C. The absorbance at 514 min is recorded at $t = 10.00$ min. The concentrations of all reactants are essentially constant throughout the 10-min period. The appearance of product is a linear function of time and Fe^{3+} concentration, as shown at right.

(a) From the graph below, find the constants k and b in the equation $A_t = k[Fe^{3+}] + b$, where A_t is the absorbance at time t, $[Fe^{3+}]$ is the iron concentration in ng/mL, and b is the reagent blank value for a solution with no deliberate addition of Fe^{3+}. Use $t = 10$ min.

Absorbance versus time for different Fe^{3+} concentrations (ng/mL) in 26.00-mL reaction solution: (a) 0, (b) 0.40, (c) 0.80, (d) 1.20, (e) 1.60. [From K. Hirayama and N. Unohara, *Anal. Chem.*, **60**, 2573 (1988).]

(b) The analytical procedure was carried out starting with 5.00 mL of unknown. The absorbance at 10 min was 0.515. Find the concentration of Fe^{3+} in the unknown, expressed as ng/mL and mol/L.

20 Instrumental Aspects of Spectrophotometry

The previous chapter emphasized how absorption and emission of light are used in chemical analysis. Now we examine the workings of the spectrophotometer and introduce Fourier transform spectroscopy.

20-1 ABSORPTION, REFLECTION, REFRACTION, AND EMISSION OF LIGHT

What happens to light when it strikes a sample? Consider the material in Figure 20-1 to be a uniform solid that absorbs some of the light striking it. The radiant power of the light incident on the first surface is called P_0. Radiant power refers to the energy per second striking each unit area and has the

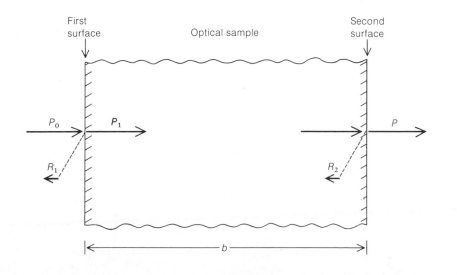

Figure 20-1
Light of radiant power P_0 strikes first surface of sample whose thickness is b. Radiant power R_1 is reflected and radiant power P_1 enters the sample. Some of P_1 is absorbed as it travels through the sample and power P_2 arrives at the second surface. There, power P is transmitted and power R_2 is reflected.

Light striking a sample can be

1. transmitted
2. absorbed
3. reflected
4. scattered

An example of a compressed poly-crystalline solid is a KBr pellet used for infrared spectroscopy, described in Section 20-2.

Three confusing terms:

a = absorptance = fraction of light absorbed by sample

α = absorption coefficient (in Equation 20-2)

A = absorbance (Equation 19-5)

units W/m^2. At the first surface, some of the radiant power (R_1) is reflected back toward the source, and some (P_1) enters the sample. After this fraction of radiant power travels through the sample, some energy is absorbed, and radiant power P_2 ($< P_1$) strikes the second surface. At this point, some power (R_2) is reflected, and some (P) is transmitted out of the sample to emerge on the other side. The internal reflections continue indefinitely. Some of R_2 is absorbed as it travels back to the first surface. Here, some of this return ray is transmitted back out of the sample (toward the left in the figure) and some is reflected back into the sample. The power involved gets smaller with successive passes, becoming negligible after several cycles.

Four things can happen to light impinging on a sample:

1. Some light is transmitted through the sample. The *transmittance* measured in the lab is the fraction of incident light emerging from the other side:

$$T = \frac{P}{P_0} \qquad (20\text{-}1)$$

2. Some light is absorbed by the material. Box 19-1 demonstrated that there is an exponential relationship between power absorbed within the sample and the pathlength, b, of the sample:

$$\frac{P_2}{P_1} = e^{-\alpha b} \qquad (20\text{-}2)$$

where α in Equation 20-2 is the product βc in Box 19-1. The radiant powers P_1 and P_2 defined in Figure 20-1, represent light *inside* the sample. The quantity α in Equation 20-2 is a property of the sample called the **absorption coefficient** and customarily has units of cm^{-1}. If $\alpha = 1\ cm^{-1}$, the radiant power inside the sample decays to $1/e$ times its initial value after traversing a pathlength of 1 cm. The greater the value of α, the more light is absorbed.

3. Some light is reflected at each surface and eventually leaves the sample in the direction of the light source.

4. Some light is scattered to the side. If the sample is a liquid, light can be scattered by dust particles in the liquid or by very large solute molecules. If the sample is a solid made of many microcrystalline particles compressed together, scattering occurs at boundaries between the particles (called *grain boundaries*) or at microscopic voids between particles. If scattering is significant, another term must be added to Equation 20-2, since radiant power is lost to both absorption and scattering:

$$\frac{P_2}{P_1} = e^{-(\alpha_a + \alpha_s)b} \qquad (20\text{-}3)$$

where α_a is the true absorption coefficient and α_s describes the loss of transmitted light due to scattering.

The quantity P_2/P_1 in Equation 20-2 is the *internal transmittance* of the sample. It is the fraction of radiant power inside the sample that is transmitted through the material. The measured (*external*) transmittance is P/P_0. The **absorptance** is defined as the fraction of incident radiation that is absorbed by the sample. In the absence of scattering, absorptance is given by

$$\text{Absorptance} = a = \frac{P_1 - P_2}{P_0} \qquad (20\text{-}4)$$

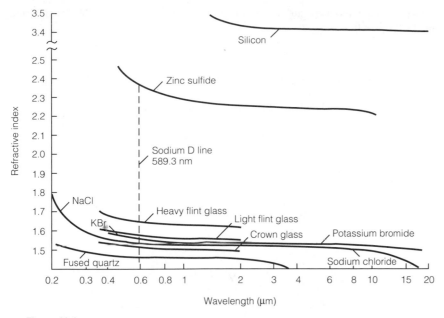

Figure 20-2
Dependence of refractive index on wavelength. Wavelength scale is logarithmic.

where $P_1 - P_2$ is the radiant power absorbed by the sample. In the absence of scattering, the quantities transmittance, absorptance, and **reflectance** (R, the fraction of P_0 that is reflected) must sum to unity:

$$T + a + R = 1 \qquad (20\text{-}5)$$

Refractive Index and Reflection

The speed of light in a medium of **refractive index** n is c/n, where c is the speed of light in vacuum. That is, for vacuum, $n = 1$. The refractive index of a material is a function of the wavelength of light (Figure 20-2) and the temperature of the material. The index is most often measured at 20°C at the wavelength of the sodium D line ($\lambda = 589.3$ nm). Representative values of refractive index are shown in Table 20-1.

When light passes from one medium to another, its path is bent, as shown in Figure 20-2. This phenomenon is called **refraction,** and the degree of bending is described by *Snell's law:*

$$n_1 \sin \theta_1 = n_2 \sin \theta_2 \qquad (20\text{-}6)$$

where n_1 and n_2 are the refractive indexes of the two media and θ_1 and θ_2 are angles defined in Figure 20-3.

Table 20-1
Refractive index at 589.3 nm
(sodium D line)

Substance	Refractive index
Vacuum	1
Air (0°C, 1 atm)	1.000 29
Water	1.33
Ethanol	1.36
Pentane	1.36
Magnesium fluoride	1.38
Fused quartz	1.46
Carbon tetrachloride	1.47
Benzene	1.50
Sodium chloride	1.54
Potassium bromide	1.56
Carbon disulfide	1.63
Bromine	1.66
Silver chloride	2.07
Zinc sulfide	2.36
Iodine	3.34

Refraction: bending of light when it passes between media with different refractive indexes.

EXAMPLE: Refraction of Light by Water
Visible light travels from air (medium 1) into water (medium 2) at a 45° angle (θ_1 in Figure 20-3). At what angle, θ_2, does the light ray pass through the water?

The refractive index for air is close to 1, and for water, ~ 1.33 (Table 20-1). Using Snell's law, we find that

$$(1.00)(\sin 45°) = (1.33)(\sin \theta_2) \quad \Rightarrow \quad \theta_2 = 32°$$

Now find θ_2 if the ray is perpendicular to the surface (i.e., $\theta_1 = 0°$). In this case,

$$(1.00)(\sin 0°) = (1.33)(\sin \theta_2) \quad \Rightarrow \quad \theta_2 = 0°$$

A perpendicular ray is not refracted.

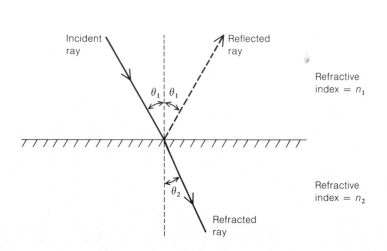

Figure 20-3
Illustration of Snell's law: $n_1 \sin \theta_1 = n_2 \sin \theta_2$.

A perfectly smooth surface will reflect light at the same angle as the angle of incidence, as shown in Figure 20-3. The reflection of light at this angle is called **specular reflection.** Real surfaces are somewhat rough and reflect some light in all directions. Reflection at other than the angle of incidence is called **diffuse reflection.** The remainder of our discussion applies to specular reflection from a perfectly smooth surface.

Light traveling from medium 1 to medium 2 with $\theta_1 = \theta_2 = 0°$ in Figure 20-3 is not refracted, but is partially reflected. For such a perpendicular ray, the ratio of radiant power reflected (P_r) to incident radiant power (P_0) is

$$R = \frac{P_r}{P_0} = \left(\frac{n_1 - n_2}{n_1 + n_2}\right)^2 \tag{20-7}$$

Reflection of a perpendicular ray from one surface of a transparent material.

EXAMPLE: Reflection of Light by Water
What fraction of visible light traveling from air to water at normal incidence ($\theta_1 = \theta_2 = 0°$ in Figure 20-3) is reflected?
We use Equation 20-7 with $n_1 = 1.00$ and $n_2 = 1.33$:

$$R = \frac{P_r}{P_0} = \left(\frac{1.00 - 1.33}{1.00 + 1.33}\right)^2 = 0.020$$

The reflected radiant power is 2.0% of the incident power.

In Figure 20-1 we see that reflection occurs at each surface. The reflected ray bounces back and forth, undergoing partial transmission and partial reflection at each surface until its power is negligible. If the sample absorbs no light and the only processes that occur are reflection and transmission, then the theoretical transmittance of a ray striking perpendicular to the surface will be

$$T = \frac{P}{P_0} = \frac{1 - R}{1 + R} \qquad (20\text{-}8)$$

Transmittance of a plate that absorbs no light.

where R is defined in Equation 20-7. The theoretical reflection is $1 - T$, since we stipulate that all light is either reflected or transmitted. Therefore,

$$\text{Total reflection} = 1 - T = 1 - \frac{1 - R}{1 + R} = \frac{2R}{1 + R} \qquad (20\text{-}9)$$

Reflection of a perpendicular ray from a plate with two surfaces.

EXAMPLE: Transmission of a Quartz Plate

What fraction of light striking a quartz plate in air at normal incidence. (i.e., perpendicular to the surface) is transmitted?

The refractive index of quartz is 1.46. Putting $n_1 = 1.00$ and $n_2 = 1.46$ into Equation 20-7 gives $R = 0.035\,0$. Putting this value of R in Equation 20-8 gives

$$T = \frac{1 - 0.035\,0}{1 + 0.035\,0} = 0.93$$

The quartz transmits 93% of the incident power and reflects 7%. This has a practical significance in spectrophotometry because a liquid sample must be contained in a cuvet that necessarily transmits less than 100% of the light that strikes the cuvet. If you want to measure the absorbance of a sample, you must have a reference that compensates for the reflectance of the cuvet.

The discussion of reflection by transparent materials should make you realize that a significant fraction of the light in a spectrophotometer can be lost by reflection from the optical components. Fortunately, **antireflection coatings** can be applied to greatly decrease reflection. Ideally the coating should have a refractive index given by $\sqrt{n_1 n_2}$, where n_1 is the refractive index of the external medium (usually air) and n_2 is the refractive index of the optical component (such as a lens). If the coating thickness is one-fourth of the wavelength of incident light (wavelength in the coating) and the refractive index is $\sqrt{n_1 n_2}$, reflection is theoretically reduced to zero. However, it is not possible to achieve zero reflection over a range of wavelengths because λ and n both vary. Antireflection coatings can also be made from layers of materials with the refractive index steadily changing from n_1 to n_2. Figure 20-4 shows what can be achieved with a multilayer antireflection coating.

Recall that $\lambda v = c$. The speed of light in a medium depends on the refractive index, n, of the medium. In going from one medium to another, the frequency is the same, but λ changes if n changes.

Optical Fibers

An **optical fiber** carries light from one place to another. It is made of a high-refractive-index optically transmitting core enclosed in a lower-refractive-index cladding, each typically 0.1-mm thick (Figure 20-5a). The cladding is enclosed in a protective plastic jacket.

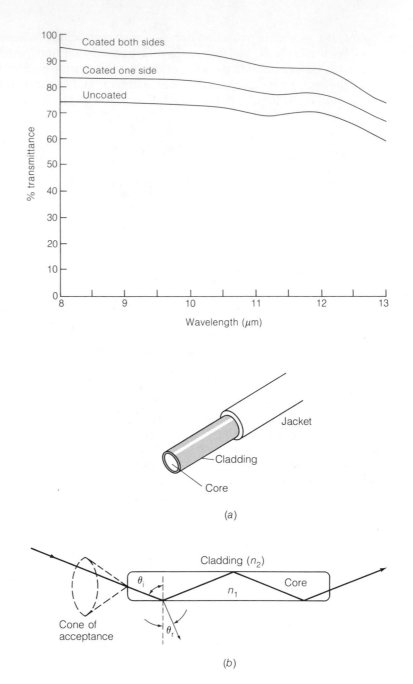

Figure 20-4
Effect of antireflection coating on a 2-mm-thick zinc sulfide plate used as an infrared window. [Reproduced from "Kodak Irtran Infrared Optical Materials" (Eastman Kodak Co., Publication U-72, 1981). Reprinted courtesy of Eastman Kodak Company.]

Figure 20-5
Optical fiber construction and principle of operation.

The principle of operation is shown in Figure 20-5b. Consider the ray striking the wall of the core at the angle of incidence θ_i. Part of the ray is reflected inside the core, and part might be transmitted into the cladding at the angle of refraction, θ_r. If the index of refraction of the core is n_1 and the index of the cladding is n_2, we can say from Snell's law (Equation 20-6) that

$$n_1 \sin \theta_i = n_2 \sin \theta_r \quad \Rightarrow \quad \sin \theta_r = \frac{n_1}{n_2} \sin \theta_i \qquad (20\text{-}10)$$

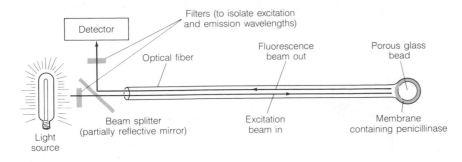

Filters (to isolate excitation and emission wavelengths)

Detector

Optical fiber

Fluorescence beam out

Porous glass bead

Light source

Beam splitter (partially reflective mirror)

Excitation beam in

Membrane containing penicillinase

Figure 20-6
Optical fiber sensor used to measure penicillin concentration in the bloodstream.

If the product $(n_1/n_2) \sin \theta_i$ is greater than 1, then no light is transmitted into the cladding because $\sin \theta_r$ cannot be greater than 1. In such a case, we say that θ_i exceeds the *critical angle* for total internal reflection. *If $n_1/n_2 > 1$, there is a range of angles θ_i in which all light will be reflected at the walls of the core, and none will enter the cladding.* A ray entering one end of the fiber within the cone of acceptance will emerge from the other end of the fiber with very little loss. Optical fibers are flexible and can be bent (within reason) to transmit light from one place to another.

Light traveling from a region of high refractive index to a region of low refractive index is totally reflected if the angle of incidence exceeds the *critical angle*.

Figure 20-6 illustrates a sensor based on an optical fiber that is used to measure penicillin concentration inside a human vein.[†] An optical fiber carries light to a porous glass bead coated with a membrane containing the enzyme penicillinase. Penicillin in the blood is converted to penicilloic acid (page 384), which is fluorescent. The fluorescence signal returning to the detector is proportional to the penicillin concentration in the blood.

Optical fibers are replacing electric wires for telephone communication. Fibers are immune to electrical noise, can transmit data at a higher rate, and can handle more signals simultaneously than can wires.

Blackbody Radiation

When an object is heated, it emits radiation—it glows. Even at room temperature, all objects are glowing with infrared radiation. Imagine a hollow sphere whose inside is perfectly black. That is, the surface absorbs all radiation striking it. The surface also emits radiation. As photons are absorbed and emitted by the inside surface of this sphere, they come to thermal equilibrium with the sphere. If a small hole is made in the wall, we would observe escaping radiation with a continuous spectral distribution. The object is called a *blackbody,* and the radiation is called **blackbody radiation.** Emission from real objects such as the tungsten filament of a light bulb closely resembles that from an ideal blackbody.

The power per unit area radiating from the surface of an object is called the **exitance,** M. For a blackbody, M is given by

$$M = \sigma T^4 \tag{20-11}$$

Exitance is the power emitted per unit surface area. It was formerly called *emittance*.

where

$$\sigma = \frac{2\pi^5 k^4}{15h^3 c^2} = 5.669\ 8 \times 10^{-8}\ \text{W/(m}^2\text{K}^4) \tag{20-12}$$

[†] K. J. Skogerboe, *Anal. Chem.,* **60,** 1271A (1988); R. E. Dessy, *Anal. Chem.,* **61,** 1079A (1989).

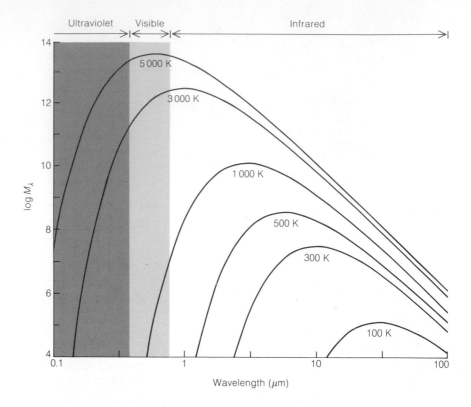

Figure 20-7
Spectral distribution of blackbody emission. This is a graph of Equation 20-13. The units of M_λ are W/m³. Note that both axes are logarithmic.

A *two-color pyrometer* (infrared thermometer) indicates the temperature of an object by measuring the ratio of infrared energy emitted by the object at two wavelengths. Using Equation 20-13, this ratio gives the temperature of the object.

Challenge: Explain how an infrared camera can "see" at night. Why is nighttime aerial infrared photography used in searching for a lost person in the wilderness?

Here T is temperature, k is Boltzmann's constant, h is Planck's constant, and c is the speed of light. A blackbody whose temperature is 1 000 K radiates 5.67×10^4 watts per square meter of surface area. If the temperature is doubled, the exitance increases by a factor of $2^4 = 16$.

The emission spectrum changes with temperature (Figure 20-7). At higher temperatures the maximum exitance shifts to shorter wavelengths (higher energies). The wavelength dependence of the exitance is given by the *Planck distribution:*

$$M_\lambda = \frac{2\pi hc^2}{\lambda^5}\left(\frac{1}{e^{hc/\lambda kT} - 1}\right) \tag{20-13}$$

where λ is wavelength, T is kelvins, and h, c, and k are the same as in Equation 20-12. The units of M_λ are W/m³, which can be thought of as watts per square meter of surface per meter of wavelength. The area under each curve between two wavelengths in Figure 20-7 is equal to the energy (W/m²) emitted between those two wavelengths.

At low temperature the maximum emission in Figure 20-7 occurs at infrared wavelengths. At temperatures above several thousand degrees (such as the temperature of the sun's surface), emission is greatest at visible wavelengths. For temperatures of 100 K and above, an excellent approximation for the wavelength of maximum emission (λ_{max}) in Figure 20-7 is given by the *Wien displacement law:*

$$\lambda_{max} \cdot T = 2.878 \times 10^{-3}\ \text{m} \cdot \text{K} \tag{20-14}$$

where T is kelvins.

The **emissivity** of an object is defined as the radiant power emitted by the object divided by the radiant power emitted by a blackbody at the same temperature:

$$\text{Emissivity} = \frac{\text{radiant power emitted by object}}{\text{radiant power emitted by blackbody}} \qquad (20\text{-}15)$$

Emissivity is a fraction between zero and one.

20-2 COMPONENTS OF A SPECTROPHOTOMETER

Figure 20-8 shows the optical components of a representative infrared spectrometer. Light from the source at the right is directed by two mirrors into the sample and reference. A rotating mirror alternately sends light from sample and reference into the grating monochromator, which transmits one narrow band of wavelengths to the detector. The spectrum produced by the instrument is a graph of transmission or absorbance versus wavelength. We now describe in some detail the various components of spectrophotometers.

Sources

A *tungsten lamp* is an excellent source of visible and near-infrared light. A typical tungsten filament operates at a temperature near 3 000 K and produces useful radiation in the range 320–2 500 nm (Figure 20-9). This covers the entire visible region and parts of the infrared and ultraviolet regions as well. Ultraviolet spectroscopy normally employs a *deuterium arc lamp* in which an electric discharge (a spark) causes D_2 to dissociate and emit ultraviolet light in the approximate range 200–400 nm (Figure 20-9). In a typical ultraviolet–visible spectrophotometer, a change is made between deuterium and tungsten lamps when passing through 360 nm, so that the more powerful source is always employed.

Mid-infrared radiation (4 000–200 cm^{-1}) is commonly obtained from a silicon carbide rod called a *globar,* heated to near 1 500 K by passage of an

Ultraviolet light is harmful to the naked eye and should not be viewed without protection.

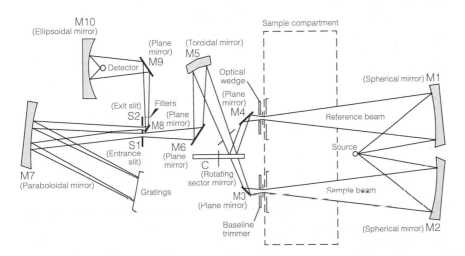

Figure 20-8
Optical schematic of the Perkin-Elmer 1320 infrared spectrophotometer. [Courtesy Perkin-Elmer Corp.]

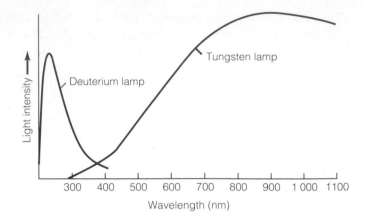

Figure 20-9
Intensity of a tungsten filament at 3 200 K
and a deuterium arc lamp.

Monochromatic—one wavelength
Polychromatic—many wavelengths

Typical properties of laser light:

Monochromatic	one wavelength
Extremely bright	high power at one wavelength
Collimated	parallel rays
Polarized	electric field of waves oscillating in one plane
Coherent	all waves in phase

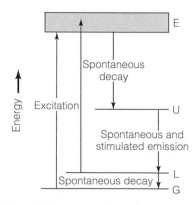

Figure 20-10
Principle of operation of a laser.

electric current through the rod. The hot globar emits radiation with approximately the same spectrum as a blackbody at 1 000 K (Figure 20-7).

Lasers are important sources of **monochromatic** radiation for many laboratory experiments. How monochromatic is a laser? A laser with a wavelength of 3 μm might have a **bandwidth** (range of wavelengths) of 3×10^{-14} to 3×10^{-8} μm. The bandwidth is measured at the frequencies at which the power falls to half of its maximum value. The brightness of a low-power laser at its frequency of operation is 10^{13} times as great as that of the sun at its brightest (yellow) wavelength. (Of course, the sun emits all wavelengths, whereas the laser emits only one. The total brightness of the sun is greater than that of the laser.) The angular divergence of the laser beam from its direction of travel is typically less than 0.05°. Laser light is typically *plane polarized,* with the electric field oscillating in one plane perpendicular to the direction of travel (Figure 19-1). Another characteristic of laser light is *coherence,* which means that all light waves emerging from the laser oscillate in phase with each other.

Figure 20-10 illustrates the principle of operation of a laser. A necessary condition for production of laser light is a *population inversion,* in which a higher energy state has a greater population than a lower energy state. In Figure 20-10 molecules in the ground state (G) are pumped to an excited state (E) by broadband irradiation from a powerful lamp or by an electric discharge. Molecules in state E rapidly decay to an upper excited state (U), which has a relatively long lifetime. State U can decay to a lower state (L) that rapidly decays to the ground state. A photon with an energy that exactly spans two molecular states can be absorbed to raise the molecule to an excited state. A photon with the correct energy can also stimulate the excited molecule to emit a photon and return to the lower state. This is called *stimulated emission.* A laser contains a high population of molecules in state U. A molecule spontaneously dropping from U to L emits a photon. When this photon strikes another molecule in state U, a second photon is emitted by the second molecule as it drops to state L. If there are many molecules in state U, the first photon stimulates the emission of many photons as it travels through the laser. Each new photon has the same phase and polarization as the incident photon.

Figure 20-11 shows a helium–neon laser, which is a common source of red light with a wavelength of 632.8 nm and an output power of 0.1–25 mW. An electric discharge pumps helium atoms to state E in Figure 20-10. The

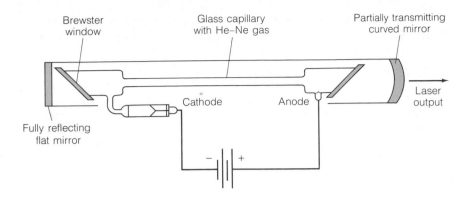

Brewster window

Glass capillary with He–Ne gas

Partially transmitting curved mirror

Laser output

Fully reflecting flat mirror

Cathode

Anode

– +

Figure 20-11
Construction of a helium–neon laser. Helium pressure is typically 1 torr and neon pressure is 0.1 torr.

excited helium transfers energy by colliding with a neon atom, raising the neon to state U. A high concentration of helium and intense electric pumping create a greater population of neon in state U than in the lower state L (i.e., a *population inversion*). Neon in state U can emit a photon and drop to state L. A few of these photons are emitted parallel to the axis of the laser tube. As they travel the length of the tube, they stimulate more emission. The mirrors on both ends of the tube reflect most of the photons back and forth through the tube, stimulating more emission with each pass. One mirror is fully reflecting, but the other is only partially silvered, allowing some fraction of the emission to pass out of the end of the laser, where we can use the light. The *Brewster windows* are tilted at an angle that ensures that light emitted from the laser cavity is polarized perpendicular to the direction of travel.

If there are more molecules in state U than in state L (a population inversion), passing photons are more likely to stimulate emission from U to L than to be absorbed from L to U.

Liquid Sample Cells

Cells come in all sizes and shapes appropriate for various experiments (Figure 20-12). The most common cuvets for measuring visible and ultraviolet spectra are made of quartz, have a 1.000-cm pathlength, and are sold in matched sets (for sample and reference). Glass cells are suitable for visible-light measurements, but not for ultraviolet spectroscopy (because they absorb the ultraviolet light). Quartz is transparent through the normally accessible visible and ultraviolet regions.

Cells for infrared measurements on liquids are commonly constructed of NaCl, KBr, or AgCl, which transmit infrared radiation. For measurements in the $400-50 \text{ cm}^{-1}$ region, polyethylene is a suitable transparent window material. Solids commonly are ground to a fine powder, which can be added to mineral oil (a viscous hydrocarbon also called Nujol) to give a dispersion called a *mull*. The mull is pressed between two infrared windows (such as KBr) and placed in the sample compartment. The spectrum of the analyte is obscured in a few regions in which the mineral oil absorbs infrared radiation. Alternatively the fine powder can be combined with excess KBr (KBr/solid $\approx 100 \text{ g/g}$), and the resulting mixture pressed into a translucent pellet at a pressure of $\sim 60 \text{ MPa}$ (600 atm). Solids and powders also can be examined by *diffuse reflectance,* in which reflected infrared light, instead of transmitted infrared light, provides signal to the detector. Wavelengths from the source that are absorbed by the sample will not be reflected as well as other wavelengths. This technique is sensitive to only the surface of the sample.

Box 20-1 (pages 560–561) describes a continuous analytical method that makes use of flow cells.

Approximate low-energy cutoff for common infrared windows:

Sapphire (Al_2O_3)	$1\,500 \text{ cm}^{-1}$
NaCl	650 cm^{-1}
KBr	350 cm^{-1}
AgCl	350 cm^{-1}
CsBr	250 cm^{-1}
CsI	200 cm^{-1}

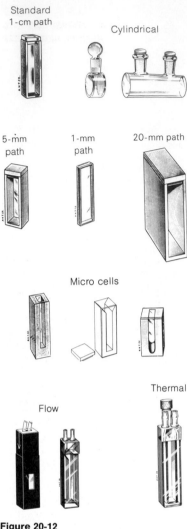

Cylindrical

5-mm
path

1-mm
path

20-mm path

Micro cells

Flow

Thermal

Figure 20-12

Common cuvets for visible and ultraviolet spectroscopy. The flow cells allow for continuous flow of solution through the cell. They are especially useful for measuring the absorbance of a solution flowing out of a chromatography column. The thermal cell permits liquid from a constant-temperature bath to flow through the cell jacket and thereby maintains the contents of the cell at a desired temperature. [Courtesy A. H. Thomas Co., Philadelphia, Pa.]

Figure 20-13

Diagram of a Czerney–Turner grating monochromator.

After our discussion of reflection and scattering of light in Section 20-1, it should be clear that the sample cell contributes significantly to the observed transmittance. Quantitative spectrophotometry requires that a suitable reference sample be available. The reference might be solvent or a reagent blank (containing all reagents except analyte) in a cell that is identical to the sample cell. Reflection, scattering, and absorption by the cell, solvent, and other reagents should be virtually the same in both samples. The difference in transmittance between the sample and the reference, then, will be due to the analyte. Only under these conditions can the absorbance of the sample be equated to log P_0/P.

Monochromators

Light of one wavelength (or color) is said to be *monochromatic*. Gratings and prisms are the two most common devices used to disperse light into its component wavelengths.

Gratings

A typical design of a grating monochromator is shown in Figure 20-13. *Polychromatic* radiation from the entrance slit is collimated into a beam of parallel rays by a concave mirror. These rays fall on a reflection **grating,** whereupon different wavelengths are *diffracted* at different angles. The light strikes a second concave mirror, which focuses each wavelength at a different point on the focal plane. The orientation of the reflection grating directs only one narrow band of wavelengths to the exit slit of the monochromator. By rotation of the grating, different wavelengths are allowed to pass through the exit slit.

The principle of the **diffraction** grating is shown in Figure 20-14. The grating is ruled with a series of closely spaced parallel grooves. The grating is coated with aluminum to make it reflective. On top of the aluminum is a thin protective layer of silica, SiO_2, to prevent the metal surface from tarnishing (oxidizing), which would reduce its reflectivity. When light is reflected

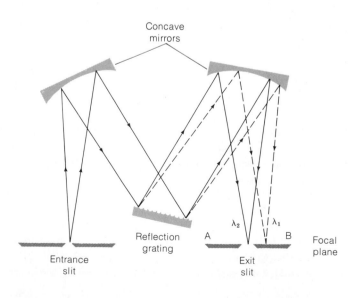

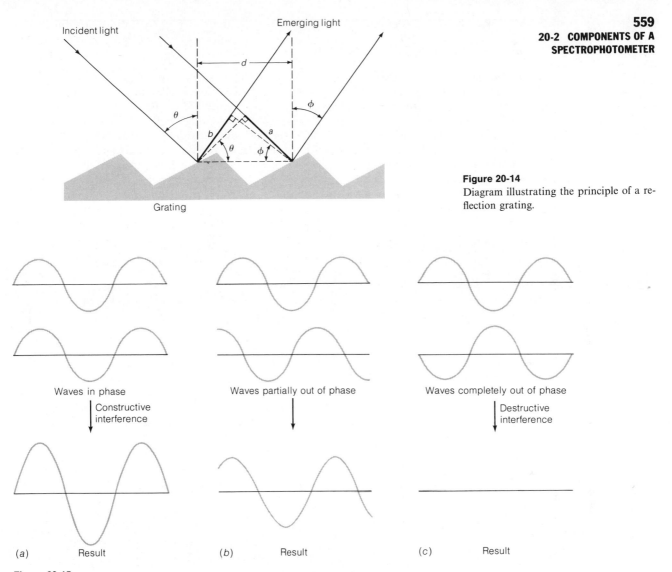

Figure 20-14
Diagram illustrating the principle of a reflection grating.

Figure 20-15
Interference of adjacent waves that are (a) 0°, (b) 90°, and (c) 180° out of phase.

from the grating, each groove behaves as a source of radiation. When adjacent light rays are in phase, they reinforce each other. When they are not in phase, they partially or completely cancel (Figure 20-15).

Consider the two rays shown in Figure 20-14. Fully constructive interference will occur only if the difference in length of the two paths is exactly equal to the wavelength of light. The difference in path is equal to the distance $a - b$ in Figure 20-14. Constructive interference occurs if

$$n\lambda = a - b \qquad (20\text{-}16)$$

where $n = 1, 2, 3, 4, \ldots$. The interference maximum for which $n - 1$ is called *first-order diffraction*. When $n = 2$, we have *second-order diffraction,* and so on. The most intense component of diffracted light occurs when $n = 1$.

Box 20-1   FLOW INJECTION ANALYSIS

In **flow injection analysis** a sample is injected into a liquid carrier stream to which various reagents can be added. After a suitable time the reacted sample reaches a detector, which is usually a spectrophotometric cell through which the stream flows. In the schematic diagram below, a solvent carrier stream (usually an aqueous solution) is pumped continuously through the sample injector, in which 40- to 200-μL volumes of sample can be added to the stream. A reagent stream is then combined with the carrier stream, and the resulting solution passes through a coil to give the reagent and sample time to react. By measuring the absorbance of the stream in a detector cell such as that in Figure 23-12, the concentration of analyte in the sample is determined. Typical total flow rates are 0.5–2.5 mL/min and the diameter of the inert Teflon tubing from which the system is constructed is about 0.5 mm. Coils are 10–200 cm in length to allow suitable reaction times. Replicate analyses can be completed in 20–60 s.

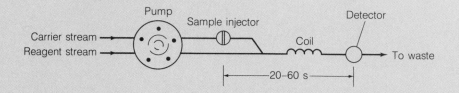

The following schematic gives an idea of variations possible with flow injection analysis. In this example reagents 1 and 2 are mixed and added to the sample in the carrier before reagent 3 is added. Commercial equipment allows different flow paths to be readily assembled. Streams can be passed through reactive columns, ion exchangers, dialysis tubes, gas diffusers, and solvent extractors. Detectors might measure absorbance, fluorescence, or luminescence, or could utilize potentiometry or amperometry.

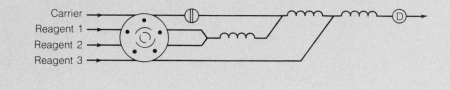

From the geometry in Figure 20-14, we see that $a = d \sin \theta$ and $b = d \sin \phi$. Therefore, the condition for constructive interference is

$$n\lambda = d(\sin \theta - \sin \phi) \tag{20-17}$$

For each angle, θ, there is a series of angles, ϕ, at which a given wavelength will produce maximum constructive interference. Many sophisticated spectrophotometers contain two monochromators in series (a *double monochromator*) to produce radiation of an even narrower bandwidth.

Prisms

Figure 20-16 shows a sodium chloride *Littrow prism,* which has a mirrored back surface. This type of prism conserves space within a spectrophotometer

A key feature of flow injection is rapid, repetitive analysis. The chart below is an example in which standards and samples were run in triplicate to measure the ethanol content of beverages. The method makes use of an enzymatic reaction that leads to color formation. Flow injection is widely used in medical and pharmaceutical analysis, water analysis, and industrial process control.

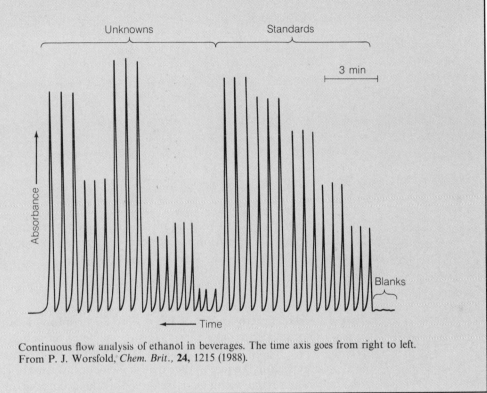

Continuous flow analysis of ethanol in beverages. The time axis goes from right to left. From P. J. Worsfold, *Chem. Brit.*, **24**, 1215 (1988).

because the incident and reflected rays are both on the same side of the prism. Sodium chloride is useful for infrared light that is absorbed by other common prism materials, such as glass. Sodium chloride is transparent to light with wavelengths from approximately 0.2 (ultraviolet) to 15 μm (infrared). Unfortunately, sodium chloride is soluble in water and must be protected from atmospheric moisture in order to retain its optical quality.

A monochromator based on a Littrow prism is shown in Figure 20-17. Polychromatic radiation from a light bulb enters the monochromator through an entrance slit. After reflection from the prism, one narrow band of wavelengths emerges from the exit slit. The exiting wavelength band is selected by rotating the prism.

To see how the prism chooses one wavelength, consider the behavior of two coincident light rays with wavelengths λ_1 and λ_2 in Figure 20-16. Ray 1 is accurately drawn for $\lambda_1 = 10.0$ μm, for which the refractive index of the

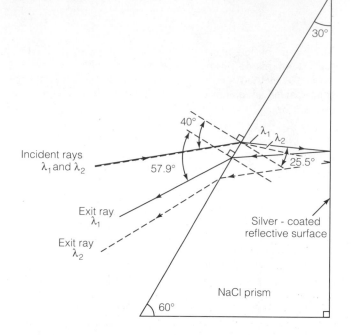

Figure 20-16
Littrow prism with mirrored back surface.

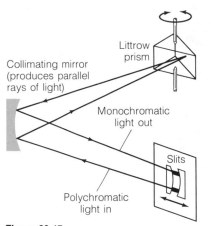

Figure 20-17
Monochromator design based on Littrow prism.

NaCl prism is $n_{NaCl} = 1.494\ 73$. The angle of incidence is arbitrarily taken as $40°$. The first angle of refraction can be computed from Snell's law (Equation 20-6):

$$n_{air} \sin \theta_{air} = n_{NaCl} \sin \theta_{NaCl} \qquad (20\text{-}18)$$

where n_{air} is the refractive index of air ($= 1.000\ 26$ for infrared radiation), θ_{air} is the angle of incidence ($40°$) and θ_{NaCl} is the angle of refraction. The angle of refraction computed from Equation 20-18 is $25.476\ 9°$. At the mirror surface the angle of reflection is equal to the angle of incidence. When the ray emerges from the front of the prism, it is refracted again according to Equation 20-18. The ray that entered the prism at an angle of $40°$ emerges at an angle of $57.876\ 2°$.

Now consider wavelength $\lambda_2 = 9.9\ \mu m$. The deviation of the λ_2 ray from the λ_1 ray in Figure 20-16 is greatly exaggerated for the purpose of illustration. At a wavelength of $9.9\ \mu m$, $n_{NaCl} = 1.495\ 38$ and n_{air} is unchanged at $1.000\ 26$. Using Equation 20-18 we find that an angle of incidence of $40°$ gives an angle of refraction of $25.465\ 1°$. This ray is then reflected from the mirror and refracted at the front surface again, emerging at an angle of $57.943\ 3°$. The difference in exit angle for the two wavelengths is $57.943\ 3 - 57.876\ 2 = 0.067\ 1°$. If the exit slit is positioned to separate rays differing by $0.067\ 1°$, then these two wavelengths can be resolved from each other. In practice, infrared spectrometers can generally resolve wavelengths that are from 2 to 100 times closer together than λ_1 and λ_2 in this example.

Resolution and Dispersion of Prisms and Gratings

The ability of a prism to separate light of different wavelengths arises because the refractive index of the prism material varies with wavelength, as was shown in Figure 20-2. If the refractive indexes for λ_1 and λ_2 were the same in Figure 20-16, then both rays would emerge at the same angle. The greater the difference in refractive index for the two wavelengths, the greater the difference in angle between the emerging rays. The rate of change of refractive index with wavelength, $dn/d\lambda$, is called the **dispersion** of the prism material. The larger the dispersion, the greater the separation of adjacent wavelengths. The resolving ability of a prism is also proportional to its size. If you double the thickness of a prism, you will increase the **resolution** (separation) of adjacent wavelengths by a factor of 2.

The resolution of a grating monochromator is proportional to the number of grating lines illuminated by radiation from the entrance slit. Therefore a grating with more lines per centimeter will have a greater resolution for a given entrance slit width. Using a higher diffraction order (greater value of n in Equation 20-17) also increases resolution. Usually, gratings can be designed to give greater resolution than a prism of the same size. Another advantage of reflection gratings is that they can be used for ultraviolet and infrared wavelengths that are absorbed by many prism materials.

For any type of monochromator, the width of the exit slit determines what range of wavelengths is passed on to the sample. The narrower the slit width, the narrower the bandwidth emerging from the monochromator. A narrow slit is needed to resolve closely spaced peaks. However, the narrower the slit, the less light is passed through the sample, and the smaller is the signal measured by the detector. Thus resolution is achieved at the expense of increased detector noise. For quantitative analysis involving broad peaks, it is best to use a broad bandwidth (wide monochromator slits), so that there is more signal to measure. The bandwidth should still be significantly smaller than the width of the absorption peak being measured.

Figure 20-18 shows an important difference between the dispersion produced by prisms and gratings. A properly designed grating monochromator gives a linear variation of wavelength with distance along the focal plane of the exit slit. That is, as the exit slit is moved left or right in Figure 20-13, the wavelength varies linearly. Moving the slit 2 cm will change the wavelength twice as much as moving it 1 cm. A prism monochromator gives a nonlinear dispersion, because the variation of refractive index with wavelength is not linear (Figure 20-2). Therefore a spectrum produced by a prism has a nonlinear wavelength scale, whereas the spectrum from a grating instrument is linear in wavelength.

Resolution is a measure of the ability to separate two closely spaced wavelengths. The greater the resolution, the smaller is the difference between two wavelengths that can be distinguished from each other.

Resolution of a prism increases with

Increased size of the prism

Increased dispersion ($dn/d\lambda$)

Resolution of a grating increases with

Increased number of grating lines illuminated by entrance slit

Increased diffraction order (higher value of n in Equation 20-16)

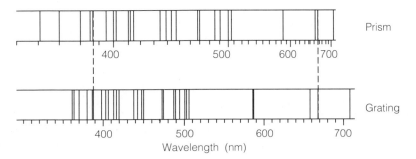

Figure 20-18

Comparison of the spectrum of helium produced with prism and grating monochromators. Each line represents emission at that wavelength by He atoms. Note that the wavelength scale is linear for the grating and nonlinear for the prism.

Filters permit certain bands of wavelength to pass through.

It is frequently necessary to filter (remove) wide bands of radiation from a signal. For example, the grating monochromator in Figure 20-13 directs first-order diffraction of a small wavelength band to the exit slit. (By "first order" we mean diffraction for which $n = 1$ in Equation 20-17.) Let λ_1 be the wavelength whose first-order diffraction reaches the exit slit. Inspection of Equation 20-17 shows that if $n = 2$, the wavelength $\frac{1}{2}\lambda_1$ also reaches the same exit slit (because the wavelength $\frac{1}{2}\lambda_1$ also gives constructive interference at the same angle as λ_1). For $n = 3$, the wavelength $\frac{1}{3}\lambda_1$ also reaches the slit. One solution for selecting just λ_1 is to place a broad **band pass filter** in the beam, so that only some range of wavelengths around λ_1 can pass through the filter. The very different wavelengths $\frac{1}{2}\lambda_1$ and $\frac{1}{3}\lambda_1$ will be blocked by the filter. To cover a wide range of wavelengths, it may be necessary to use several filters and to change them as the desired wavelength region changes.

The simplest kind of filter is just colored glass, in which the coloring species absorbs a broad portion of the spectrum and transmits other portions. Table 19-1 shows the relation between absorbed and transmitted colors. For much finer control of filtering behavior, specially designed *interference filters* are constructed to pass light in the region of interest and reflect other wavelengths. Figure 20-19 shows examples of filters designed to pass a broad band and a narrow band of wavelengths. These devices derive their behavior from contructive or destructive interference of light waves within the filter.

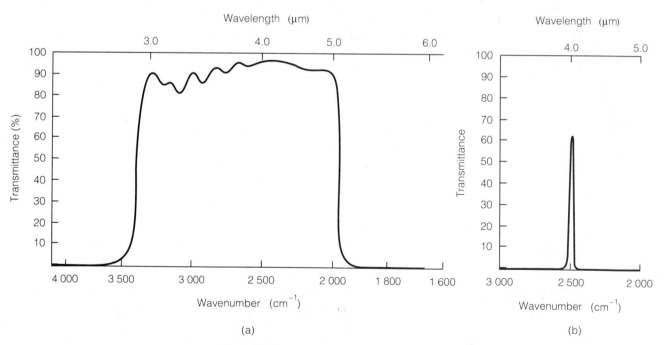

(a)   (b)

Figure 20-19
Transmission spectra of infrared interference filters. (*a*) Wide band pass filter has $\sim 90\%$ transmission in the 3- to 5-μm-wavelength range, but $<0.01\%$ transmittance outside this range. (*b*) Narrow band pass filter has a width of 0.1 μm centered around 4 μm. [Courtesy Barr Associates, Inc.]

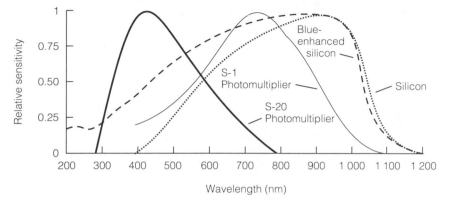

Figure 20-20
Response of several different detectors. The greater the sensitivity, the greater the output (current or voltage) of the detector for a given incident power (watts per unit area) of photons. Each curve is normalized to a maximum value of 1. [Courtesy Barr Associates, Inc.]

Detectors

The general property of a detector is the ability to produce an electric signal when it is struck by photons. As Figure 20-20 shows, the response of most detectors depends on the wavelength of the incident photons. For example, the S-20 photomultiplier produces about four times as great a current for a given energy flux (W/m^2) of 420-nm light than for the same energy flux of 300-nm light. The response below 280 nm and above 800 nm is essentially zero. When using a single beam spectrophotometer such as the Spectronic 20 (Figure 19-10), the 100% transmittance control must be readjusted each time the wavelength is changed because the detector response changes with wavelength. This adjusts the spectrophotometer to the maximum detector output that can be obtained at each wavelength. Subsequent readings are scaled to the 100% reading.

Detector response is a function of wavelength of incident light.

Photomultiplier Tube

A **phototube** emits electrons from a photosensitive, negatively charged surface when struck by visible or ultraviolet light. The electrons flow through a vacuum to a positively charged collector whose current is essentially proportional to the radiation intensity.

A more sophisticated and very sensitive device is the **photomultiplier tube** (Figure 20-21). In this device, the electrons emitted from the photosensitive surface strike a second surface, called a *dynode,* which is kept positive with respect to the photosensitive emitter. The electrons from the emitter are therefore accelerated and strike the dynode with more than their original kinetic energy. These energetic electrons cause more electrons to be emitted from the dynode than those striking the dynode. These new electrons are accelerated toward a second dynode, which is more positive than the first dynode. Upon striking the second dynode, even more electrons are knocked off and accelerated toward a third dynode. This process is repeated several times, with the result that more than 10^6 electrons are finally collected for each photon striking the first surface. By this means, extremely low light intensities may be translated into measurable electric signals.

The human eye is a better amplifier than a photomultiplier. See Box 20-2.

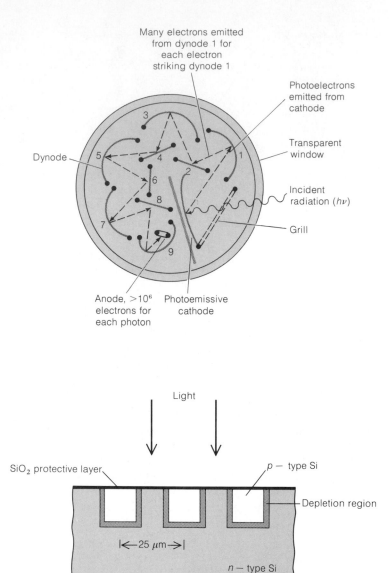

Figure 20-21
Diagram of a photomultiplier tube with nine dynodes. Amplifiation of the signal occurs at each dynode, which is approximately 90 volts more positive than the previous dynode.

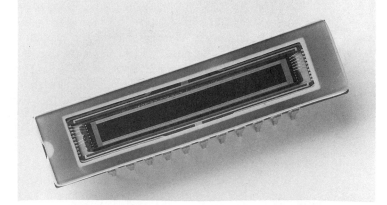

Figure 20-22
(a) Schematic cross-sectional view of photodiode array. (b) Photograph of array with 1 024 elements, each 25 μm wide and 2.5 mm high. The central black rectangle is the photosensitive area. The entire chip is 5 cm in length. [Courtesy Oriel Corporation.]

(a)

(b)

Conventional spectrophotometers scan slowly through a spectrum, one wavelength at a time. Newer systems record the entire spectrum at once in a fraction of a second. These instruments will find increased use in such areas as chemical kinetics (looking at intermediates as they are formed) and chromatography (recording the full spectrum of a compound as it comes off the chromatography column).

At the heart of new developments is the **photodiode array**[†] shown in Figure 20-22. In this device, bars of p-type silicon are formed on a substrate (the underlying body) of n-type silicon, creating a series of p-n junction diodes. A reverse bias is applied to each diode, drawing electrons and holes away from the junction. There is a depletion region at each junction, in which there are few electrons and holes. The junction behaves as a capacitor, with charge stored on either side of the depletion region. At the beginning of each measurement cycle, each diode is fully charged.

When light strikes the semiconductor, free electrons and holes are created. These migrate into regions of opposite charge and partially discharge the capacitor. The more light that strikes each diode, the less charge is left at the end of the measurement cycle. The longer the array is exposed to light between readings, the more each capacitor is discharged. The state of each capacitor is determined at the end of the cycle by measuring the current needed to recharge the capacitor.

The key feature of a spectrophotometer utilizing a photodiode array is that *white light* (with all wavelengths) is passed through the sample, as shown in Figure 20-23. The light then enters a **polychromator**, which disperses the light into its component wavelengths and directs the light at the diode array. *A different wavelength band strikes each diode*, and the resolution depends on how closely spaced the diodes are and how much dispersion is produced by the polychromator. The spectrum in Figure 21-24 in the next chapter was produced with a photodiode array detector.

The entire spectrum is observed simultaneously in such a spectrophotometer. An array of several thousand diodes can be read in a fraction of a second. The information from each detector element is digitized and available for computer storage and manipulation.

For a refresher on semiconductors, see Section 15-7 and Box 17-1.

A typical photodiode array has a response curve similar to that of the blue-enhanced silicon in Figure 20-20.

[†] D. G. Jones, *Anal. Chem.,* **57,** 1057A, 1207A (1985).

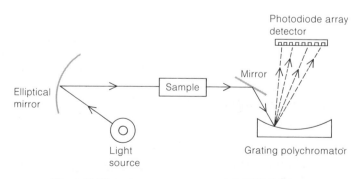

Figure 20-23
Schematic design of photodiode array spectrophotometer.

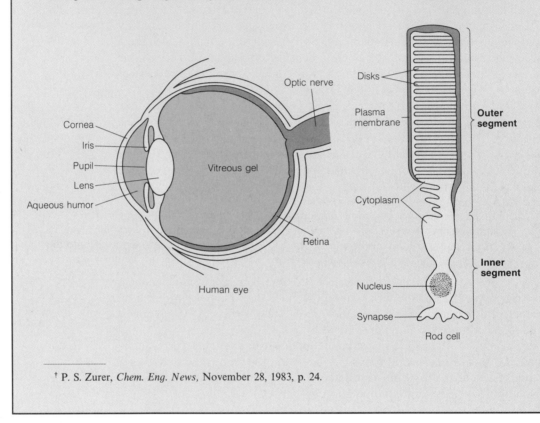
Charge Coupled Device

Very sensitive television cameras use charge coupled devices to record the video image.

A **charge coupled device** is an extremely sensitive silicon detector that stores photogenerated charge in a two-dimensional array. As shown schematically in Figure 20-24a, the array is constructed of *p*-doped Si on an *n*-doped substrate. This diode structure is capped with an insulating layer of SiO_2, on top of which is placed a pattern of conducting Si electrodes. When light is absorbed in the *p*-doped region, an electron is introduced into the conduction band and a hole is left in the valence band. The electron is attracted to the region under the positive electrode, where it is stored. The hole migrates into the *n*-doped substrate, where it combines with an electron. The potential well beneath each electrode can typically hold up to 10^5 electrons before it is filled and electrons spill out into adjacent elements.

A stack of about 1 000 *disks* in the schematic rod cell at the left contains the light-sensing protein *rhodopsin*, in which the chromophore 11-*cis*-retinal (derived from Vitamin A) is covalently attached to the protein called *opsin*. When visible light is absorbed by rhodopsin, a series of rapid transformations occurs, resulting in the release from the protein of all-*trans*-retinal.

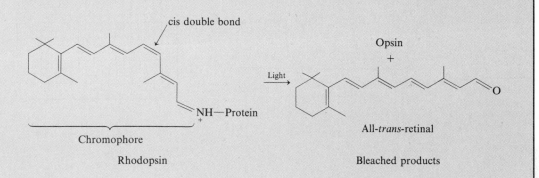

At this stage the pigment is *bleached* (losing all color), and cannot respond to more light until the retinal isomerizes back to the 11-*cis* form and recombines with the protein.

Within a second of absorbing a photon, an electric signal is generated in the optic nerve. In the dark, there is a continuous flow of 10^9 Na$^+$ ions per second from outside the cell into the outer segment of the rod cell. The Na$^+$ moves from the outer segment to the inner segment, where an energy-dependent process using adenosine triphosphate (ATP) and oxygen pumps the ions back out of the cell. When light is absorbed and rhodopsin is bleached, the channels through which Na$^+$ flows into the cell are somehow shut down, and the electric current moving along the length of the cell is changed. A single photon reduces the ion current by 3%— corresponding to a decreased current of 3×10^7 ions per second. The *amplification* is greater *than that of a photomultiplier tube!* The electric current returns to its dark value as the protein and retinal recombine. A great deal of work remains to be done before the process of transduction of light to nerve impulses in the human eye is fully understood.

The charge coupled device is divided into a two-dimensional array, as shown in Figure 20-24b. After the desired observation time, the electrons stored in each *pixel* (picture element) of the top row are moved into the serial register at the top, and then moved, one pixel at a time, to the top right position, where the stored charge is read out. Then the next row is moved up and read out, and the sequence is repeated until the entire array has been read. The transfer of stored charges is carried out by an array of electrodes considerably more complex than we have indicated in Figure 20-24a. The transfer of charge from one pixel to the next is extremely efficient, with a loss of approximately five of every million electrons.

The charge coupled device is extremely sensitive. The minimum detectable signal for visible light in Table 20-2 is 0.2 photons per second per pixel for a 100-second integration time. The sensitivity of the charge coupled device

Electrons from several adjacent pixels can be combined to create a single, larger picture element. This process, called *binning*, increases the sensitivity of the charge coupled device at the expense of resolution.

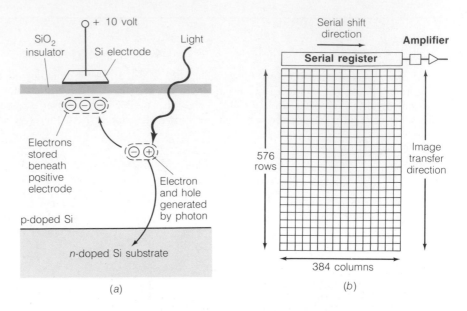

Figure 20-24
Schematic representation of a charge coupled device. (*a*) Cross-sectional view indicating charge generation and storage in each pixel. (*b*) Top view showing two-dimensional nature of one particular array. An actual array is about the size of a postage stamp.

Table 20-2
Minimum detectable signal (photons/s/detector element) of ultraviolet/visible detectors

Signal acquisition time (s)	Photodiode array		Photomultiplier tube		Charge coupled device	
	Ultraviolet	Visible	Ultraviolet	Visible	Ultraviolet	Visible
1	6 000	3 300	30	122	31	17
10	671	363	6.3	26	3.1	1.7
100	112	62	1.8	7.3	0.3	0.2

SOURCE: R. B. Bilhorn, J. V. Sweedler, P. M. Epperson, and M. B. Denton, *Appl. Spec.*, **41**, 1114 (1987).

is derived from its high *quantum efficiency* (electrons generated per incident photon), low background electrical noise (thermally generated free electrons), and low noise associated with readout. Figure 20-25 compares spectra recorded under the same conditions with a photomultiplier tube and a charge coupled device. Although a photomultiplier is very sensitive, the charge coupled device is even better.

Figure 20-26 shows one example of how a charge coupled device can be used to record a rapidly changing event. The grating polychromator illuminates one masked row of the charge coupled device. Each pixel of that row records the light intensity at a different wavelength. After a preselected time, the second row is unmasked and the first row is masked. The second row then records the next spectrum. Each row of the device records a spectrum at a different time. When the process is finished, the rows are read sequentially to reconstruct the event.

Infrared Detectors

For infrared detection, several kinds of devices are used. A **thermocouple** is a junction between two different electrical conductors. Since electrons have

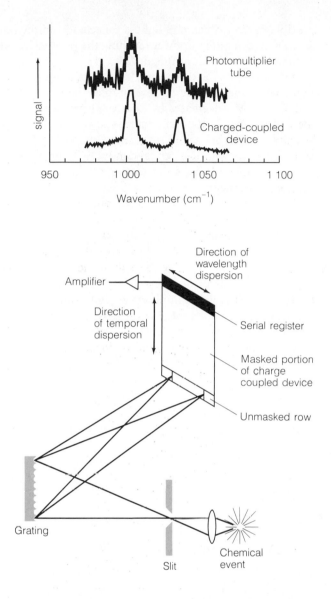

Figure 20-25
Comparison of spectra recorded in 5 min using a photomultiplier tube or a charge coupled device. [From P. M. Epperson, J. V. Sweedler, R. B. Bilhorn, G. R. Sims, and M. B. Denton, *Anal. Chem.,* **60,** 327A (1988).]

Figure 20-26
Use of a charge coupled device for time-resolved spectroscopy. The mask is moved periodically, so that, each row of the detector records a spectrum at a different time. [From P. M. Epperson, J. V. Sweedler, R. B. Bilhorn, G. R. Sims, and M. B. Denton, *Anal. Chem.,* **60,** 327A (1988).]

a lower free energy in one conductor than the other, they flow from one conductor to the other until the resulting small voltage difference prevents further flow. This solid-solid junction potential is temperature-dependent, since electrons can flow back to the high-energy conductor at higher temperature. When the thermocouple is blackened so that it absorbs light, its temperature (and hence the voltage) becomes sensitive to light. A typical sensitivity is 6 V per watt of light absorbed.

A **ferroelectric material** such as triglycine sulfate has a permanent electric polarization in the absence of an applied field because of alignment of the molecules in the crystal. One face of the crystal is positively charged and the opposite face is negative. The polarization is temperature-dependent, and the variation of polarization with temperature is called the **pyroelectric effect.** The crystal absorbs infrared light, which changes the temperature and the electric polarization. The change in polarization is the signal in a pyroelectric detector. A **Golay cell** uses thermal expansion of a gas in a blackened chamber

In a *ferroelectric material,* the dipole moments of the molecules remain aligned in the absence of an external field. This gives the material a permanent electric polarization.

Semiconductors, valence bands, conduction bands, *p-n* junctions, electrons, and holes were discussed in Section 15-7 and Box 17-1.

Box 20-3 describes types of environmental analysis, some of which use infrared detection.

to deform a flexible mirror. A beam of light is reflected from the mirror onto a photocell. The change in radiant power reaching the photocell as the mirror is deflected is the detector signal in this device.

A **bolometer,** also called a **thermistor,** is a device whose electrical resistance changes with temperature. For example, the resistance of flakes of oxides of Ni, Co, or Mn decreases by about 4% per degree Celsius. In a bolometer, the temperature of the flake increases when infrared light is absorbed by the flake, and the change in resistance is the detector signal. These oxides are semiconductors whose electrical conductivity increases when heat excites electrons from the valence band to the conduction band. A **photoconductive detector** is a semiconductor whose electrical conductivity increases when infrared light directly excites electrons from the valence band to the conduction band. **Photovoltaic detectors** contain *p–n* semiconductor junctions, across which an electric field normally exists. Absorption of infrared light creates more electrons and holes, which are attracted to opposite sides of the *p–n* junction, and which change the voltage across the junction. This voltage change is the detector signal. Mercury cadmium telluride ($Hg_{1-x}Cd_xTe$, $0 < x < 1$) is an example of an important infrared detector material whose sensitivity to different wavelengths of infrared light is affected by the metal atom stoichiometry coefficient, x. Photoconductive and photovoltaic devices are often cooled to liquid nitrogen temperature to reduce thermal electric noise.

Box 20-3 ENVIRONMENTAL CARBON ANALYSIS AND OXYGEN DEMAND

Industrial waste streams are frequently characterized and regulated based on their carbon content or oxygen demand. *Total carbon* (TC) is defined by the amount of CO_2 produced when a sample is completely oxidized at high temperature:

$$\text{Total carbon analysis:} \qquad \text{All carbon} \xrightarrow[\text{catalyst}]{O_2,\ 900°C} CO_2$$

Commercial on-line analyzers subject 20-μL water samples to this procedure and measure the CO_2 by infrared absorption.[†] The entire analysis requires 3 minutes. Total carbon includes dissolved organic material (called *total organic carbon*, TOC) and dissolved carbonate and bicarbonate (called *inorganic carbon*, IC). By definition, TC = TOC + IC.

To discriminate between TOC and IC, the pH of a fresh sample is lowered below 2 to convert carbonate and bicarbonate to CO_2. Purging with N_2 then removes all of the IC from the sample, which is then subjected to the total carbon analysis above. Since IC has been removed, the CO_2 produced in this experiment corresponds to TOC only. The IC content is calculated from the difference between the two experiments. TOC measurement is widely used to determine compliance with discharge laws.

Total oxygen demand (TOD) tells us how much oxygen is required for complete combustion of the pollutants in the waste stream. A volume of nitrogen containing a known quantity of oxygen is mixed with sample and complete combustion is carried out as in TC analysis. The oxygen remaining in the resulting mixture is measured by the potentiometric ZrO_2 sensor in

[†] K. M. Queeney and F. B. Hoek, *Am. Lab.,* October 1989, p. 26.

For a given spectrophotometer, the manufacturer normally provides some specifications for the instrument errors to be expected. Additional errors associated with sample preparation and cell positioning will increase the uncertainty in any measurement.

Choosing the Wavelength and Bandwidth

Consider the absorption spectrum of the analytically important complex (ferrozine)$_3$Fe(II) in Figure 19-12. If this species is to be used for the spectrophotometric analysis of iron, it would be sensible to choose the wavelength of maximum absorbance (562 nm) because this gives the greatest absorbance for a given concentration of iron.

Beer's law, which states that the absorbance is proportional to concentration, applies for monochromatic radiation. At the wavelength of maximum absorption, the variation of absorbance with wavelength ($dA/d\lambda$) is a minimum. Therefore, the effect of using light that is not perfectly monochromatic will be insignificant, since the molar absorptivity (ε in Beer's law) is fairly constant over a small range of wavelengths.

Box 18-4. This measurement is sensitive to the oxidation states of species in the waste stream. For example, urea consumes five times as much oxygen as formic acid. The method responds to such noncarbon species as ammonia and sulfides.

Pollutants in water can be oxidized by refluxing with dichromate ($Cr_2O_7^{2-}$). The quantity of dichromate consumed in this process is a measure of the pollutant level in the water. *Chemical oxygen demand* (COD) is defined as the O_2 that is chemically equivalent to the dichromate consumed in this process. Since each mole of $Cr_2O_7^{2-}$ consumes 6 moles of electrons (to make two moles of Cr^{3+}), and since each mole of O_2 can consume 4 moles of electrons (to make H_2O), one mole of $Cr_2O_7^{2-}$ is equivalent to 1.5 moles of O_2 in this computation. The COD analysis is carried out by refluxing the polluted water for two hours with excess standard dichromate in a sulfuric acid solution containing Ag^+ catalyst. Unreacted dichromate is then measured by titration with Fe^{2+} or by spectrophotometry. Many permits for industrial operations are defined in terms of COD analysis of waste streams.

Biochemical oxygen demand (BOD) measures the oxygen required for biochemical degradation of organic material by microorganisms. BOD also includes oxygen required by such species as sulfide and Fe^{2+} that may be in the water. Specific inhibitors are added to prevent oxidation of nitrogen species such as NH_3. The measurement requires five days of incubation at 20°C in the dark in a sealed container with no air space. The oxygen dissolved in the solution is measured before and after the incubation. The difference is the BOD.[‡]

[‡] Procedures for measuring BOD and COD are described in *Standard Methods for the Examination of Wastewater*, 16th ed. (Washington, D. C. American Public Health Association, 1985), p. 525. This book is the standard reference for numerous water analyses.

For quantitative analysis:

1. Absorbance should not be too large or too small.
2. Absorbance should be measured at a peak or shoulder where $dA/d\lambda$ is small.
3. Monochromator bandwidth should be as large as possible, but small compared with the band being measured.

The most accurate spectrophotometric measurements generally occur when the absorbance lies between 0.4 and 0.9.

The breadth of the peak at 562 nm determines how great a monochromator bandwidth can be used. A bandwidth of 4 nm would lead to negligble deviation from Beer's law due to polychromaticity. In solutions, vibrational structure of an electronic absorption band is usually broadened beyond recognition. Under the broad band in Figure 19-12 are myriad vibrational and rotational transitions. They are so broadened that only a single, nearly featureless envelope is seen. Samples with sharper peaks require a narrower monochromator bandwidth to give a faithful reproduction of the spectrum and accurate quantitative analysis.

The bandwidth should generally be kept as wide as the spectrum permits, since the narrower the bandwidth, the less light will reach the detector. Less light means a smaller signal-to-noise ratio, so that the precision of the measurement will be reduced.

Instrument Errors

In general, spectrophotometric measurements are most reliable at intermediate absorbance values ($A \approx 0.4-0.9$). At low absorbance, the radiant power coming through the sample is similar to the power coming through the reference, and there is a large relative uncertainty in the measured difference between these two numbers. At high absorbance, too little light reaches the detector, so the precision of the measurement is reduced.

Figure 20-27 shows some measured errors for a research-quality spectrophotometer. The curve labeled "Dark current noise" gives the imprecision

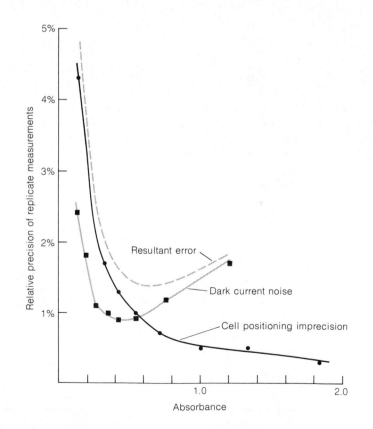

Figure 20-27
Errors in spectrophotometric measurements due to dark current noise and cell positioning imprecision. [Data from L. D. Rothman, S. R. Crouch, and J. D. Ingle, Jr., *Anal. Chem., **47,*** 1226 (1975).]

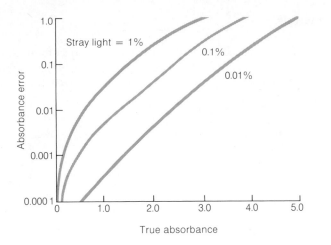

Figure 20-28
Absorbance error introduced by different levels of stray light. The stray light is expressed as a percentage of the radiant power incident on the sample. This illustration is from M. R. Sharp, *Anal. Chem.,* **53**, 339A (1984), which provides an excellent discussion of stray light.

due to instrument noise that was essentially independent of the sample absorbance. This noise is caused by such factors as thermal motion of electrons in electronic components and noise in the readout device. By comparison, noise due to flicker of the light source and photomultiplier tube was negligible.

For the data in Figure 20-27, *the largest source of imprecision for A < 0.6 was nonreproducible positioning of the cuvet in the sample holder.* This error was large despite great care in placement and cleaning of the cuvet. The resulting error curve in the figure reaches a minimum near $A = 0.6$. At lower absorbances, the relative error due to cell positioning is limiting, and at higher absorbances, instrumental noise is most important.

Another source of instrument error in visible and ultraviolet spectrophotometry is *stray light,* defined as light with wavelengths outside the narrow bandwidth expected from the monochromator. Most stray light comes through the monochromator from the spectrophotometer light source. It arises from unintended scattering from the optical components and walls of the monochromator. Stray light can also come from outside the instrument if the sample compartment is not closed tightly. Openings into the sample chamber for tubing or wires connecting accessories may allow room light to enter. Figure 20-28 shows the error introduced in quantitative analysis by stray light.

For greatest precision, samples should be free of dust, and cuvets must be free of fingerprints or other contamination.

20-4 FOURIER TRANSFORM INFRARED SPECTROSCOPY

We have seen that a photodiode array or charge coupled device can be used to measure an entire spectrum at once. The spectrum is spread into its component wavelengths, and each small band of wavelengths is directed onto one detector element. For the infrared region, a more important and widely used method for observing the entire spectrum at once is Fourier transform spectroscopy.[†]

[†] Good references for this topic are P. R. Griffiths, *Chemical Infrared Fourier Transform Spectroscopy* (New York: Wiley, 1975) and W. D. Perkins, *J. Chem. Ed.,* **63**, A5 (1986); **64**, A269, A296 (1987).

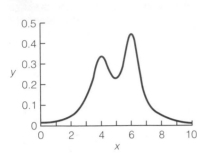

Figure 20-29
A curve to be decomposed into a sum of sine and cosine terms by Fourier analysis.

Table 20-3
Fourier coefficients for Figure 20-30

n	a_n	b_n
0	0	0.136 912
1	−0.006 906	−0.160 994
2	0.015 185	0.037 705
3	−0.014 397	0.024 718
4	0.007 860	−0.043 718
5	0.000 089	0.034 864
6	−0.004 813	−0.018 858
7	0.006 059	0.004 580
8	−0.004 399	0.003 019

Recall that any transparent plate with refractive index $\neq 1$ will reflect some light and transmit the rest.

Fourier Analysis

Fourier analysis is a procedure in which any curve is decomposed into a sum of sine and cosine terms, called a *Fourier series*. To analyze the curve in Figure 20-29, which spans the interval $x_1 = 0$ to $x_2 = 10$, the Fourier series has the form

$$y = a_0 \sin(0\omega x) + b_0 \cos(0\omega x) + a_1 \sin(1\omega x) + b_1 \cos(1\omega x)$$

$$+ a_2 \sin(2\omega x) + b_2 \cos(2\omega x) + \cdots$$

$$= \sum_{n=0}^{\infty} \left[a_n \sin(n\omega x) + b_n \cos(n\omega x) \right] \qquad (20\text{-}19)$$

where

$$\omega = \frac{2\pi}{x_2 - x_1} = \frac{2\pi}{10 - 0} = \frac{\pi}{5} \qquad (20\text{-}20)$$

Equation 20-19 says that the value of y for any value of x can be expressed by an infinite sum of sine and cosine waves. Successive terms correspond to sine or cosine waves with increasing frequency.

Figure 20-30 shows how sequences of 2, 4, or 8 sine and cosine waves give better and better approximations to the curve in Figure 20-29. The coefficients a_n and b_n required to construct the curves in Figure 20-30 are given in Table 20-3. The larger the coefficient, the more that term contributes to the sum. The table tells us that the functions $\sin(2\omega x)$ and $\sin(3\omega x)$ make large contributions, but $\sin(5\omega x)$ makes a very small contribution.

Interferometry

The heart of a Fourier transform infrared spectrophotometer is the **interferometer** designed by Albert Michelson in 1891. In this device, light is shined on a **beamsplitter** that transmits some light and reflects some light (Figure 20-31). For the sake of this discussion, let us assume that the source produces a narrow beam of monochromatic radiation. Let's also suppose that the beamsplitter reflects half the light and transmits half. When light strikes the beamsplitter at point O, some is reflected to a stationary mirror at a distance OS and some is transmitted to a movable mirror at a distance OM. The rays reflected by the mirrors travel back to the beamsplitter, where half of each ray is transmitted and half is reflected. One recombined ray travels in the direction of the detector, and another heads back to the source.

In general, the paths OM and OS are not equal, so the two waves reaching the detector are not in phase. As shown in Figure 20-15, if the two waves are in phase, they interfere constructively to give a wave with twice the amplitude. If the waves are one-half wavelength (180°) out of phase, they interfere destructively and cancel. For any intermediate-phase difference, there is partial cancellation.

The difference in pathlength followed by the two waves in the interferometer in Figure 20-31 is 2(OM − OS). This difference is called the **retardation,** δ. Constructive interference occurs whenever δ is an integral multiple

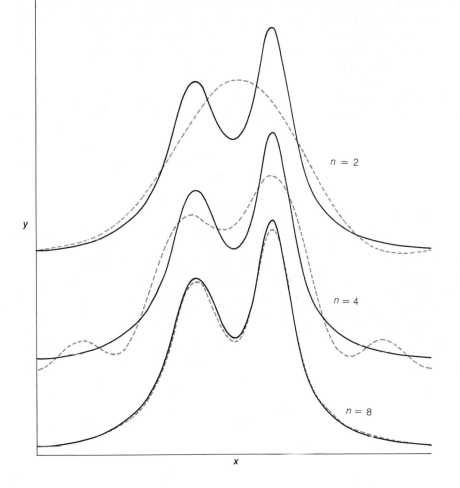

$n = 2$

$n = 4$

$n = 8$

Figure 20-30
Fourier series reconstruction of the curve
in Figure 20-29. Solid line is the original
curve and dashed lines are made from a
series of $n = 0$ to $n = 2$, 4, or 8 in Equation
20-19. Coefficients a_n and b_n are given in
Table 20-3.

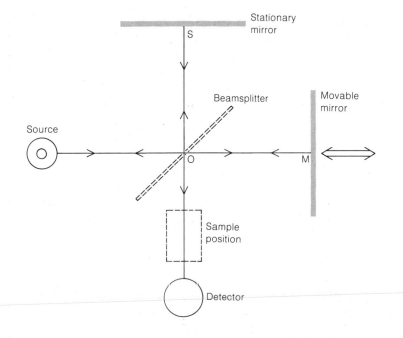

Figure 20-31
Schematic diagram of Michelson interferometer.

of the wavelength (λ) of the light. A minimum appears when δ is a half-integral multiple of λ. If one mirror moves away from the beamsplitter at a constant speed, the light emerging toward the detector goes through a steady sequence of maxima and minima as the interference alternates between constructive and destructive phases.

A graph of output light intensity versus retardation, δ, is called an **interferogram.** If the light from the source is monochromatic, the interferogram is a simple cosine wave:

$$I(\delta) = B(\tilde{v}) \cos \left(\frac{2\pi\delta}{\lambda} \right) = B(\tilde{v}) \cos (2\pi\tilde{v}\delta) \qquad (20\text{-}21)$$

where $I(\delta)$ is the intensity of light reaching the detector and $\tilde{v}$ is the wavenumber ($= 1/\lambda$) of the light. Clearly, I is a function of the retardation, δ. $B(\tilde{v})$ is a constant that accounts for the intensity of the light source, efficiency of the beamsplitter (which never gives exactly 50% reflection and 50% transmission), and response of the detector. All these factors depend on $\tilde{v}$. In the case of monochromatic light, there is only one value of $\tilde{v}$.

The top row of Figure 20-32 shows the interferogram produced by monochromatic radiation of wavenumber $\tilde{v}_0 = 2$ cm^{-1}. The wavelength (repeat distance) of the interferogram can be seen in the figure to be $\lambda = 0.5$ cm, which is equal to $1/\tilde{v}_0 = 1/(2$ cm$^{-1})$. The second row of Figure 20-32 gives the interferogram that results from a source with two monochromatic waves ($\tilde{v}_0 = 2$ and $\tilde{v}_0 = 8$ cm^{-1}) with relative intensities 1:2. The interferogram contains a short wave oscillation ($\lambda = \frac{1}{8}$ cm) superimposed on a long wave oscillation ($\lambda = \frac{1}{2}$ cm). The interferogram is a sum of two terms in this case:

$$I(\delta) = B_1 \cos (2\pi\tilde{v}_1\delta) + B_2 \cos (2\pi\tilde{v}_2\delta) \qquad (20\text{-}22)$$

where $B_1 = 1$, $\tilde{v}_1 = 2$ cm^{-1}, $B_2 = 2$, and $\tilde{v}_2 = 8$ cm^{-1}.

Fourier analysis is a way to decompose a curve into its component wavelengths. Fourier analysis of the top interferogram in Figure 20-32 gives the (trivial) result that the interferogram is made from a single wavelength function, with $\lambda = \frac{1}{2}$ cm. Fourier analysis of the second interferogram in Figure 20-32 gives the slightly more interesting result that the interferogram is composed of two wavelengths ($\lambda = \frac{1}{2}$ and $\lambda = \frac{1}{8}$ cm) with relative contributions 1:2. We say that the spectrum is the *Fourier transform* of the interferogram.

The third interferogram in Figure 20-32 is a less trivial case in which the spectrum consists of a packet of wavelengths centered around $\tilde{v}_0 = 4$ cm^{-1}. The interferogram is the sum of contributions from all source wavelengths. The Fourier transform of the third interferogram in Figure 20-32 is the third spectrum in Figure 20-32. That is, decomposition of the interferogram into its component wavelengths gives the packet of wavelengths centered around $\tilde{v}_0 = 4$ cm^{-1}. *Fourier analysis of the interferogram gives back its component wavelengths.*

The bottom interferogram in Figure 20-32 is obtained from the two wavelength packets at the bottom left. The Fourier transform of the bottom interferogram gives back the bottom spectrum.

Fourier analysis of the interferogram gives back the spectrum from which the interferogram is made. *The spectrum is the Fourier transform of the interferogram.*

Spectrum

Interferogram

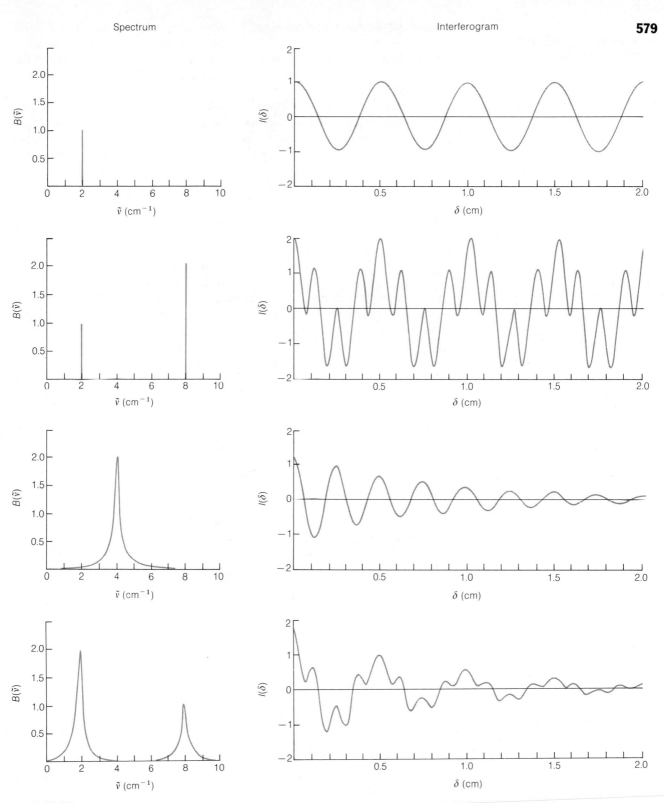

Figure 20-32
Interferograms produced by different spectra.

The interferogram loses those wavelengths absorbed by the sample.

Resolution $\approx 1/\Delta \, \text{cm}^{-1}$
Δ = maximum retardation

For a spectral width of $\Delta\tilde{v} \, \text{cm}^{-1}$, points must be taken at retardation intervals of $1/(2 \, \Delta\tilde{v})$.

Fourier Transform Spectroscopy

In a Fourier transform spectrophotometer, the sample is usually placed between the output of the interferometer and the detector, as shown in Figure 20-31. Since the sample absorbs certain wavelengths of light, *the interferogram contains the spectrum of the source minus the spectrum of the sample.* An interferogram of a reference sample containing the cell and solvent is first recorded and transformed into a spectrum. Then the interferogram of a sample in the same kind of solvent and cell is recorded and transformed into a spectrum. The quotient of the second spectrum divided by the first is the infrared spectrum of the sample (Figure 20-33). Expressing the quotient of the two spectra is the same as computing P/P_0 to find transmittance. P_0 is the radiant power received at the detector through the reference and P is the radiant power received after passage through the sample.

The interferogram is recorded not continuously but at discrete intervals. The greater the number of data collected, the more time and memory is consumed by the Fourier transform computation. The resolution of the spectrum (ability to discern closely spaced peaks) is approximately equal to $1/\Delta \, \text{cm}^{-1}$, where Δ is the maximum retardation. If the mirror travel is ± 2 cm, the retardation is ± 4 cm and the resolution is 0.25 cm^{-1}.

The mathematics of the Fourier transform dictate that the wavelength range of the spectrum is determined by how the interferogram is sampled. The closer the spacing between data points, the greater the wavelength range of the spectrum. To cover a width of $\Delta\tilde{v}$ wavenumbers requires sampling the interferogram at retardation intervals of $\delta = 1/(2 \, \Delta\tilde{v})$. If $\Delta\tilde{v}$ is 4 000 cm^{-1}, sampling must occur at intervals of $\delta = 1/(2 \cdot 4\,000 \, \text{cm}^{-1}) = 1.25 \times 10^{-4}$ cm = 1.25 μm. This corresponds to a mirror motion of 0.625 μm. For every centimeter of mirror travel, 1.6×10^4 data points must be collected. If the mirror moves at a rate of 0.2 cm per second, the data collection rate would be 3.2×10^3 points per second.

The source, beamsplitter, and detector place a physical limitation on the wavelength range, as well. Clearly, the instrument cannot respond to a wavelength that is absorbed by the beamsplitter or to which the detector does not respond. The beamsplitter for the mid-infrared region ($\sim 4\,000\text{--}400$ cm^{-1}) is typically a layer of germanium evaporated onto a KBr plate. For longer wavelengths ($\tilde{v} < 400$ cm^{-1}), a film of the organic polymer Mylar is a suitable beamsplitter.

To control the sampling interval for the interferogram, a monochromatic visible laser beam is passed through the interferometer along with the polychromatic infrared light. The laser beam gives destructive interference whenever the retardation is a half-integral multiple of the laser wavelength. These zeros in the laser signal, observed with a detector of visible light, are used to control sampling of the infrared interferogram. For example, an infrared data point might be taken at every second zero point of the visible-light interferogram. The precision with which the laser frequency is known gives an accuracy of 0.01 cm^{-1} in the infrared spectrum. This accuracy is an order of magnitude greater than that available with dispersive instruments and permits more reliable computer manipulation of spectra that is not possible with a dispersive instrument.

The Fourier transform infrared spectrophotometer is a very sophisticated instrument. However, it is rapidly replacing traditional dispersive instruments

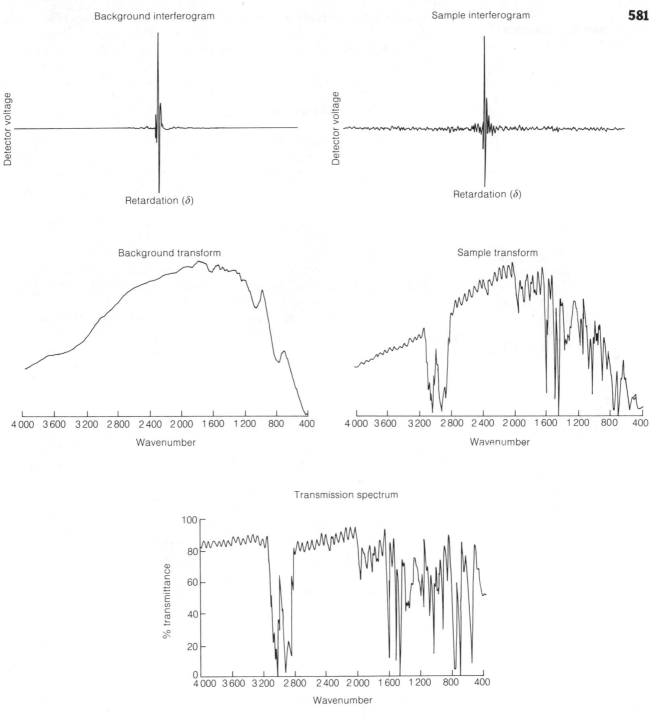

Figure 20-33
Fourier transform infrared spectrum of polystyrene film. The Fourier transform of the background interferogram gives a spectrum determined by the source intensity, beamsplitter efficiency, and detector response. The transform of the sample interferogram is a measure of all the instrumental factors, plus absorption by the sample. The transmission spectrum is obtained by dividing the sample transform by the background transform. Each interferogram is an average of 32 scans and contains 4 096 data points, giving a resolution of 4 cm^{-1}. The mirror velocity was 0.693 cm/s. [Figure courtesy Dr. M. Nadler.]

because it offers advantages of speed, frequency accuracy, more efficient use of radiation by the interferometer compared with a monochromator, improved signal-to-noise ratio at a given resolution, and built-in data-handling capabilities.

20-5 SIGNAL AVERAGING

An advantage of Fourier transform spectroscopy is that the entire interferogram is recorded in a few seconds and stored in a computer. The quality of the spectrum can be improved by collecting many interferograms and averaging them. A typical Fourier transform spectrum is obtained by averaging tens or hundreds of interferograms before computing the transform.

Signal averaging is used in many kinds of experiments to improve the quality of the data, as illustrated in Figure 20-34. The lowest trace is a simulated spectrum containing a great deal of noise. One simple way to estimate the noise level is to measure the maximum amplitude of the noise in a region free of signal. Such a measurement in the lowest trace of Figure 20-34 gives a signal-to-noise ratio of 14/9 = 1.6.

A more precise measurement of noise is the **root-mean-square (rms) noise**, defined as

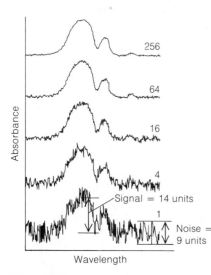

Figure 20-34
Effect of signal averaging on a simulated noisy spectrum. Labels refer to number of scans averaged. [R. Q. Thompson, *J. Chem. Ed.*, **62**, 866 (1985).]

$$\text{rms noise} = \left(\frac{1}{|\lambda_2 - \lambda_1|} \int_{\lambda_1}^{\lambda_2} [N(\lambda) - \overline{N}]^2 \, d\lambda \right)^{1/2} \qquad (20\text{-}23)$$

where $N(\lambda)$ is the noise at wavelength λ, $\overline{N}$ is the average noise, and the integration is carried out between wavelengths λ_1 and λ_2. To measure rms noise it is usually necessary to digitize the spectral data so that calculations can be done by computer. The integration is normally carried out in a region free of signal.

Now consider recording the spectrum twice in a row and adding the two results. The signal is the same in both spectra and adds to give twice the value of each spectrum. If n spectra are added, the signal will be n times as large as in the first spectrum. Noise is random, so it may be positive or negative at any point in any spectrum. It can be shown that if n spectra are added, the amplitude of the noise increases in proportion to $\sqrt{n}$. Since the signal increases in proportion to n, the signal-to-noise ratio increases in proportion to $n/\sqrt{n} = \sqrt{n}$.

By averaging n spectra, the signal-to-noise ratio is improved by $\sqrt{n}$. To improve the signal-to-noise ratio by a factor of 2 requires averaging four spectra. To improve the signal-to-noise ratio by a factor of 10 requires averaging 100 spectra. Figure 20-34 shows this effect. Common spectroscopic techniques use signal averaging to improve the signal-to-noise ratio by recording as many as 10^4 to 10^5 scans. It is rarely possible to do better than this (or even this well) because of instrumental instabilities that degrade the summation of signals after a long period of time.[†]

To improve the signal-to-noise ratio by a factor of n requires averaging n^2 spectra. See Problem 20-30.

Question: By what factor should the signal-to-noise ratio be improved when 16 spectra are averaged? Measure the noise level in Figure 20-34 to see whether your prediction is correct.

[†] There are many excellent digital and electronic techniques for improving the quality of a spectrum without recording numerous scans. For leading references, see R. Q. Thompson, *J. Chem. Ed.*, **62**, 866 (1985); M. P. Eastman, G. Kostal, and T. Mayhew, *J. Chem. Ed.*, **63**, 453 (1986); and B. H. Vassos and L. López, *J. Chem. Ed.*, **62**, 543 (1985). A signal averaging experiment including a circuit to generate noise is described by D. C. Tardy, *J. Chem. Ed.*, **63**, 648 (1986).

Summary

Light striking a sample can be reflected, absorbed, scattered, or transmitted. Light whose angle of reflection is equal to the angle of incidence is called specular reflection. All real surfaces are rough and give diffuse reflection in all directions, as well. Reflection can be decreased by an antireflection coating. Once light enters the sample, the radiant power decreases exponentially: $P_2/P_1 = e^{-\alpha b}$, where P_2 is the radiant power reaching a depth b, P_1 is the radiant power that penetrated the first surface, and α is the absorption coefficient. This exponential relationship is the basis for Beer's law, which states that absorbance is proportional to the concentration of absorbing chromophore and to sample path-length. The absorptance is defined as the fraction of incident radiant power that is absorbed by the sample. If the sample does not scatter light, the sum of absorptance, transmittance, and reflectance is unity. Mathematically, scattering is treated in the same manner as absorption, with an exponential dependence on path-length. When light passes from a region of refractive index n_1 to a region of refractive index n_2, the angle of refraction (θ_2) is related to the angle of incidence (θ_1) by Snell's law: $n_1 \sin \theta_1 = n_2 \sin \theta_2$. The greater the difference in refractive index between two media, the more light is reflected at their interface. Optical fibers transmit light by a series of total internal reflections.

A blackbody is an object that absorbs all light striking it. The total emission of radiation from the surface of the blackbody is proportional to the fourth power of temperature. The spectral distribution of radiant emission follows the Planck distribution. The emission maximum shifts to shorter wavelengths as the temperature increases, as given by the Wien displacement law. Emissivity is the quotient of radiant power emitted by an object to the radiant power emitted by a blackbody at the same temperature.

Components of spectrophotometers include the source, sample cell, monochromator, and detector. Tungsten and deuterium lamps provide visible and ultraviolet radiation, while a silicon carbide globar is a good infrared source. Lasers are important sources of high-intensity monochromatic radiation. Lasers produce light by stimulated emission of radiation from a medium in which an excited state has been pumped to a higher population than that of a lower state. Sample cells must be transparent to the radiation of interest. A reference sample is a necessity in quantitative analysis to compensate for reflection and scattering by the cell and solvent. In flow injection analysis, sample injected into a carrier stream is mixed with reagent and passes into a flow-through detector. Monochromators containing gratings or prisms disperse light into its component wavelengths. Narrow monochromator slits provide a narrow bandwidth and greater resolution of neighboring peaks in the spectrum, but increase the noise level because less light reaches the detector. The dispersion of a grating is a linear function of wavelength, whereas prism dispersion is nonlinear. Filters pass entire bands of wavelength and reject other bands.

A photomultiplier tube is a sensitive detector of visible and ultraviolet radiation in which photons cause electrons to be ejected from a metallic cathode. The signal is amplified at each successive dynode on which the photoelectrons impinge. Photodiode arrays and charge coupled devices are solid-state detectors in which photons create electrons and holes in semiconductor materials. Coupled to a polychromator, these devices can record all wavelengths of a spectrum simultaneously, with resolution limited by the number and spacing of detector elements. Infrared detectors include thermocouples, ferroelectric materials, Golay cells, thermistors, and photoconductive and photovoltaic devices. The absorbance maximum is usually chosen for a spectrophotometric analysis to maximize sensitivity and minimize the effects of imperfectly monochromatic light. Reproducible cell positioning and stray light are important factors in quantitative analysis.

Fourier analysis is a mathematical way to decompose a signal into its component wavelengths. An interferometer contains a beamsplitter, a stationary mirror, and a moveable mirror. Reflection of light from the two mirrors creates an interferogram. Fourier analysis of the interferogram tells us what frequencies went into the interferogram. In a Fourier transform spectrophotometer, the interferogram of the source is first measured without a sample present. Then the sample is placed in the beam and a second interferogram is recorded. The transforms of the interferograms tell what frequencies of light reach the detector with and without the sample present. The quotient of the two transforms is the transmission spectrum. The resolution of a Fourier transform spectrum is approximately $1/\Delta$, where Δ is the maximum retardation. To cover a wavenumber range of $\Delta\tilde{\nu}$ requires sampling the interferogram at intervals of $\delta = 1/(2\,\Delta\tilde{\nu})$. Signal-to-noise ratio can be improved by averaging many spectra. The theoretical signal-to-noise improvement equals $\sqrt{n}$, where n is the number of scans that are averaged.

Terms to Understand

absorptance

absorption coefficient

antireflection coating

band pass filter

bandwidth

beamsplitter

blackbody radiation

bolometer

charge coupled device

diffraction

diffuse reflection

dispersion

emissivity

exitance

ferroelectric material

flow injection analysis

Fourier analysis

Golay cell

grating

interferogram

interferometer

laser

optical fiber

photoconductive detector

photodiode array

photomultiplier tube

phototube

photovoltaic detector

polychromator

prism

pyroelectric effect

reflectance

refraction

refractive index

resolution

retardation

root-mean-square (rms) noise

Snell's law

specular reflection

thermistor

thermocouple

Exercises

20-A. A 4.00-mm-thick plate that does not absorb or scatter light transmits 82.0% of monochromatic light striking the plate at normal incidence $(\theta_1 = 0°$ in Figure 20-3) in air. Calculate the refractive index of the plate. You will be dealing with a quadratic equation that has two roots. Choose the root that is greater than 1.

20-B. The transmittance of a solid that both absorbs and reflects light is given by

$$T = \frac{(1 - R)^2 e^{-\alpha b}}{1 - R^2 e^{-2\alpha b}}$$

where the single-surface reflection, R, was given in Equation 20-7. The pathlength is b and the absorption coefficient is α

(a) Estimate the transmittance of a 1.20-cm-thick plate of ZnS in the air at a wavelength of 12.0 μm, at which $\alpha = 0.47$ cm^{-1} and the refractive index is 2.17.

(b) What would be the transmittance of a 1.20-mm-thick sample?

(c) If the material scatters light, as well as absorbing light, then the coefficient α should be modified to the form $\alpha = \alpha_a + \alpha_s$, where α_a is the contribution of absorption and α_s is the contribution of scattering. Find the transmittance of a 1.20-cm-thick plate of ZnS if $\alpha_a = 0.47$ cm^{-1} and $\alpha_s = 0.20$ cm^{-1}.

20-C. *Derivation of Wien displacement law.*

(a) The term $e^{hc/\lambda kT}$ in the Planck distribution for blackbody radiation (Equation 20-13) is much larger than 1 at the wavelength of maximum emission at most temperatures. Under this condition the denominator is approximately $e^{hc/\lambda kT}$ and the equation can be written

$$M_\lambda \approx 2\pi hc^2 \lambda^{-5} e^{-hc/\lambda kT}$$

Find the wavelength of maximum emission, λ_{max}, by setting the derivative $dM_\lambda/d\lambda$ equal to zero and solving for $\lambda_{max} \cdot T$. When you evaluate the constants in this expression, your answer should reproduce Equation 20-14.

(b) Calculate λ_{max} (in μm) at 100, 500, and 5 000 K and compare your results to Figure 20-7.

20-D. *Prism refraction.* The ray λ_1 in Figure 20-16 is drawn for a wavelength of 10.0 μm, at which the index of refraction of the prism is 1.494 73 and the index of refraction of air is 1.000 26. For the incident angle of 40°, show that the exit angle is 57.876 2°.

20-E. Refer to Figure 20-28. The true absorbance of a sample is 1.0, but the monochromator passes 1% stray light. From the graph, estimate the relative error in the calculated concentration of the sample. Note that the ordinate of Figure 20-28 is logarithmic. The absorbance error is such that the measured concentration is less than the true concentration.

20-F. Refer to the Fourier transform infrared spectrum in Figure 20-33.

(a) The interferogram was sampled at retardation intervals of $1.266\,0 \times 10^{-4}$ cm. What is the theoretical wavenumber range (0 to ?) of the spectrum?

(b) A total of 4 096 data points was collected from

$\delta = -\Delta$ to $\delta = +\Delta$. Compute the value of Δ, the maximum retardation.

(c) Calculate the approximate resolution of the spectrum.

(d) How many microseconds elapse between each datum?

(e) How many seconds were required to record each interferogram once?

(f) What kind of beamsplitter is typically used for the region $400–4\,000$ cm^{-1} covered in Figure 20-33? Explain why the region below 400 cm^{-1} was not observed. Does the background transform agree with your explanation?

20-G. Shown in Table 20-4 are real signal-to-noise ratios recorded in a nuclear magnetic resonance experiment. Construct graphs of (a) signal-to-noise ratio versus n and (b) signal-to-noise ratio versus $\sqrt{n}$, where n is the number of scans. Draw error bars corresponding to the standard deviation at each point. Is the signal-to-noise ratio proportional to $\sqrt{n}$?

Table 20-4
Signal-to-noise ratio at the aromatic protons of 1% ethylbenzene in CCl$_4$

Number of experiments	Number of accumulations (n)	Signal-to-noise ratio	Standard deviation
8	1	18.9	1.9
6	4	36.4	3.7
6	9	47.3	4.9
8	16	66.7	7.0
6	25	84.6	8.6
6	36	107.2	10.7
6	49	130.3	13.3
4	64	143.4	15.1
4	81	146.2	15.0
4	100	159.4	17.1

Note: Data from M. Henner, P. Levoir, and B. Ancian, *J. Chem. Ed.*, **56**, 685 (1979).

Problems

A20-1. Distinguish the terms absorbance, absorption coefficient, absorptance, and molar absorptivity.

A20-2. What is the difference between specular and diffuse reflection?

A20-3. A sample that does not scatter light has an absorptance of 6% and a reflectance of 16%. Calculate the transmittance.

A20-4. Calculate the fraction of radiant power transmitted (P_2/P_1 in Figure 20-1) by a 3.00-mm-thick plate with absorption coefficient $\alpha = 0.100$ cm^{-1}.

A20-5. Light passes from benzene (medium 1) to water (medium 2) in Figure 20-3 at (a) $\theta_1 = 30°$ or (b) $\theta_1 = 0°$. Find the angle θ_2 in each case.

A20-6. Calculate the power per unit area (the exitance, W/m^2) radiating from a blackbody at 77 K (liquid nitrogen temperature) and at 298 K (room temperature).

A20-7. The prism at right is used to totally reflect light at a 90° angle. No surface of this prism is silvered. Use Snell's law to explain why total reflection occurs. What is the minimum refractive index of the prism for total reflection?

A20-8. Use Figure 20-2 to decide whether NaCl or KBr has greater dispersion at a wavelength of 13 μm.

A20-9. Explain how an optical fiber works. Why does the fiber still work when it is bent?

A20-10. Explain how a laser generates light. List important properties of laser light.

A20-11. Would you use a tungsten or a deuterium lamp as a source of 300-nm radiation?

A20-12. What variables increase the resolution of a prism and a grating?

A20-13. Which monochromator (prism or grating) gives constant dispersion ($dn/d\lambda$)?

A20-14. Why are filters generally required for a grating monochromator?

A20-15. What are the advantages and disadvantages of decreasing monochromator slit width?

A20-16. The interferometer mirror of a Fourier transform infrared spectrophotometer travels ± 1 cm.

(a) How many centimeters is the maximum retardation, Δ?

(b) State what is meant by resolution.

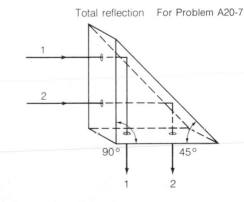
Total reflection   For Problem A20-7

(c) What is the approximate resolution (cm^{-1}) of the instrument?

(d) At what retardation interval, δ, must the interferogram be sampled (converted to digital form) to cover a spectral range of 0–2 000 cm^{-1}?

A20-17. Explain why the transmission spectrum in Figure 20-33 is calculated from the quotient (sample transform)/(background transform) instead the difference (sample transform) − (background transform).

A20-18. A spectrum has a signal-to-noise ratio of 8/1. How many spectra must be averaged to increase the signal-to-noise ratio to 20/1?

20-19. At a wavelength of 24 μm the absorption coefficient (α) for KBr is 0.25 cm^{-1} and the refractive index (n) is 1.47.

(a) What wavenumber ($\tilde{\nu}$) corresponds to $\lambda = 24\mu$m?

(b) What fraction of radiant flux is reflected when 24-μm infrared light strikes a KBr surface from air at normal incidence? Consider reflection just from this one surface.

(c) The fraction of radiant power that is transmitted after the light enters the sample is $e^{-\alpha b}$, where b is the pathlength. Calculate this internal transmittance for a 6.0-mm-thick KBr plate.

(d) Use the equation in Exercise 20-B to calculate the transmittance of the 6.0-mm-thick KBr plate.

20-20. A 1.00-cm-thick plate absorbs no light, but scatters 5.0% of the light passing through it. Find the value of the scattering coefficient, α_s, in Equation 20-4.

20-21. When light from the sodium D line passes through the interface between benzene and sodium chloride at normal incidence, what fraction of the radiant power is reflected?

20-22. The variation of refractive index (n) with wavelength for fused quartz is given by

$$n^2 - 1 = \frac{(0.696\,166\,3)\lambda^2}{\lambda^2 - (0.068\,404\,3)^2} + \frac{(0.407\,942\,6)\lambda^2}{\lambda^2 - (0.116\,241\,4)^2}$$
$$+ \frac{(0.897\,479\,4)\lambda^2}{\lambda^2 - (9.896\,161)^2}$$

where λ is expressed in μm.

(a) Make a graph of n versus λ with points at the following wavelengths: 0.2, 0.4, 0.6, 0.8, 1, 2, 3, 4, 5, and 6 μm.

(b) Is the dispersion of fused quartz greater for blue light or red light?

20-23. (a) For the 60° prism above right, show that light traveling through the prism parallel to the base enters and exits at the same angle, θ.

(b) The index of refraction of the prism is 1.500 and the index of refraction of air is 1.000. Find the angle θ in the diagram.

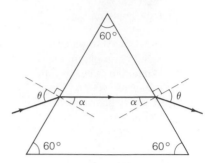

20-24. Find the minimum angle θ_i for total reflection in the optical fiber in Figure 20-5b if the index of refraction of the cladding is 1.400 and the index of refraction of the core is (a) 1.600 or (b) 1.800.

20-25. Which material(s) in Table 20-1 might be useful for making an antireflection coating for silver chloride?

20-26. The power radiating from one square meter of a blackbody surface in the wavelength range λ_1 to λ_2 is obtained by integrating the Planck distribution function in Equation 20-13:

$$\text{Power emitted} = \int_{\lambda_1}^{\lambda_2} M_\lambda\, d\lambda$$

For a narrow wavelength range, $\Delta\lambda$, the value of M_λ is nearly constant and the power emitted is simply the product $M_\lambda \Delta\lambda$.

(a) Evaluate M_λ at $\lambda = 2.00$ μm and at $\lambda = 10.00$ μm at $T = 1\,000$ K.

(b) Calculate the power emitted per square meter at 1 000 K in the interval $\lambda = 1.99$ μm to $\lambda = 2.01$ μm by evaluating the product $M_\lambda \Delta\lambda$, where $\Delta\lambda = 0.02$ μm.

(c) Repeat part b for the interval 9.99–10.01 μm.

(d) The quantity $[M_\lambda\ (\lambda = 2\ \mu\text{m})]/[M_\lambda\ (\lambda = 10\ \mu\text{m})]$ is the relative exitance at the two wavelengths. Compare the relative exitance at these two wavelengths at 1 000 K to the relative exitance at 100 K. What does your answer mean?

20-27. Consider the spectral distribution of solar energy in Box 17-1. The power incident on each square meter of the earth between wavelengths λ_1 and λ_2 is the area under the curve between λ_1 and λ_2. Use the figure in Box 17-1 to estimate the incident power per square meter between (a) 400 and 500 nm and (b) 500 and 600 nm.

20-28. Consider a reflection grating operating with an incident angle of 40° in Figure 20-14.

(a) How many lines per centimeter should be

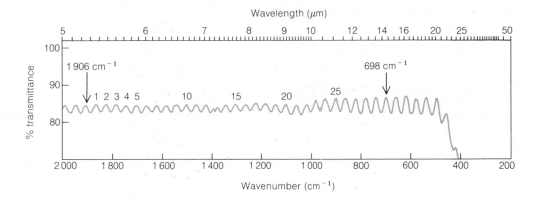

Wavelength (μm)

% transmittance

1 906 cm⁻¹ → $1\ 906\ \text{cm}^{-1}$

698 cm⁻¹ → $698\ \text{cm}^{-1}$

Wavenumber (cm⁻¹)

etched in the grating if the first-order diffraction angle for 600 nm (visible) light is to be 30°?

(b) Answer the same question for $1\ 000\ \text{cm}^{-1}$ (infrared) light.

20-29. The pathlength of a cell for infrared spectroscopy can be measured by counting *interference fringes* (ripples in the transmission spectrum). The spectrum above shows 30 interference maxima between 1 906 and 698 cm^{-1} obtained by placing an empty KBr cell in a spectrophotometer.

The fringes arise because light reflected from the cell compartment interferes constructively or destructively with the unreflected beam (as shown below).

If the reflected beam travels an extra distance λ, it will interfere constructively with the unreflected beam. If the reflection pathlength is $\lambda/2$, destructive interference occurs. Peaks therefore arise when $m\lambda = 2b$ and troughs occur when $m\lambda/2 = 2b$, where m is an integer. If the medium between the KBr plates has refractive index n, the wavelength in the medium is λ/n, so the equations become $m\lambda/n = 2b$ and $m\lambda/2n = 2b$. It can be shown that the cell pathlength is given by

$$b = \frac{N}{2n} \cdot \frac{\lambda_1 \lambda_2}{\lambda_2 - \lambda_1} = \frac{N}{2n} \cdot \frac{1}{\tilde{\nu}_2 - \tilde{\nu}_1}$$

where N maxima occur between wavelengths λ_1 and λ_2. Calculate the pathlength of the cell that gave the interference fringes shown above.

20-30. A measurement with a signal-to-noise ratio of 100/1 can be thought of as a signal (S) with 1% uncertainty (e). That is, the measurement is $S \pm e = 100 \pm 1$.

(a) Use the rules for propagation of uncertainty to show that if you add two such signals the result is: total signal $= 200 \pm \sqrt{2}$, giving a signal-to-noise ratio of $200/\sqrt{2} = 141/1$.

(b) Show that if you add four such measurements, the signal-to-noise ratio increases to 200/1.

(c) Show that averaging n measurements increases the signal-to-noise ratio by a factor of $\sqrt{n}$ compared to the value for one measurement.

20-31. Suppose you have a noisy spectrum that is digitized every 0.01 cm^{-1} in the interval 1 800–1 850 cm^{-1} and there is no sample absorption in this region. Write the steps that would be needed to compute the root-mean-square noise. For this purpose, average over wavenumber instead of wavelength. That is, replace λ by $\tilde{\nu}$ (not $1/\tilde{\nu}$) in Equation 20-23.

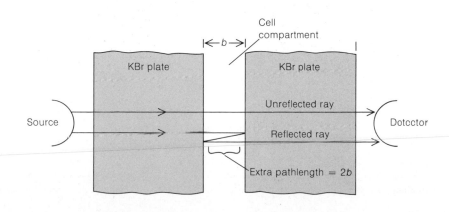

Cell compartment

KBr plate

$\leftarrow b \rightarrow$

KBr plate

Unreflected ray

Source

Reflected ray

Detector

Extra pathlength = 2b

21 Atomic Spectroscopy

When heated to a sufficiently high temperature, most compounds break apart into atoms in the gaseous phase. Unlike the optical spectra of condensed phases, the spectra of these atoms consist of very sharp lines. For example, the spectrum of an iron complex in solution typically has broad bands, each 100 nm in width, but the spectrum of gaseous Fe is a series of sharp lines, whose natural width is < 0.01 nm (Figure 21-1). This spectrum arises from

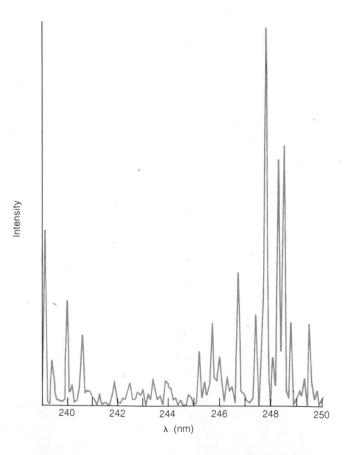

Figure 21-1
A small portion of the spectrum of a hollow-cathode Fe lamp, showing the series of sharp lines characteristic of gaseous Fe atoms. The linewidths in this spectrum are artificially broadened by the monochromator, whose bandwidth is 0.08 nm.

transitions between the electronic states of the Fe atom. Each element has its own characteristic spectrum. Because the lines are so sharp, there is usually little overlap between the spectra of different elements in the same sample.

In *atomic spectroscopy,* samples are vaporized at very high temperatures and the concentrations of selected atoms are determined by measuring absorption or emission at their characteristic wavelengths. Because of its high sensitivity and the ease with which many samples can be examined, atomic spectroscopy has become one of the principal tools of analytical chemistry, especially in industrial settings. Analyte concentrations at the parts-per-million level are routine, and parts-per-trillion levels are amenable to analysis in some cases. In analyzing the major constituents of an unknown, the sample is usually diluted to reduce concentrations to the parts-per-million level. Atomic spectroscopy is not as accurate as some wet chemical methods, since its precision is rarely better than 1–2%. The equipment is expensive, but widely available.

The unit ppm (parts per million) refers to micrograms of solute per gram of solution. Since the density of dilute aqueous solutions is close to 1 g/mL, ppm is often used to mean μg/mL. A concentration of 1.00 ppm of Fe corresponds to 1.00×10^{-6} g Fe/mL $= 1.79 \times 10^{-5}$ M.

21-1 ABSORPTION, EMISSION, AND FLUORESCENCE

In conventional molecular spectroscopy, the absorbance of a sample placed in the beam of light is measured. Alternatively, the sample is irradiated, and its luminescence (fluorescence or phosphorescence) is measured in a direction perpendicular to the incident beam. Both of these experiments can also be done with an atomic vapor (Figure 21-2). In addition, at the high temperature

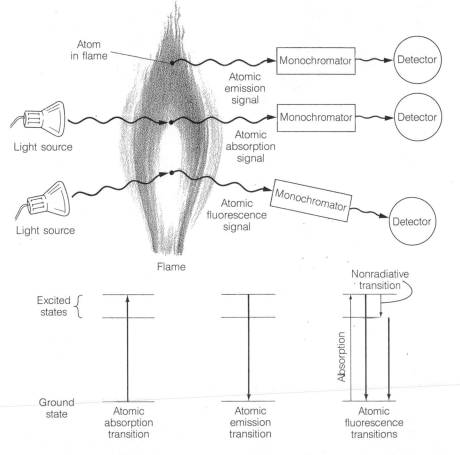

Figure 21-2
Schematic representation of atomic absorption, emission, and fluorescence, using atoms in a flame. In atomic absorption the atoms absorb part of the light from the source and the unabsorbed light reaches the detector. Atomic emission comes from atoms that are in an excited state because of the high thermal energy of the flame. For atomic fluorescence the atom must first be excited by absorption of radiation. The atom can emit the same wavelength that was absorbed, or it can fall to other states and emit other wavelengths.

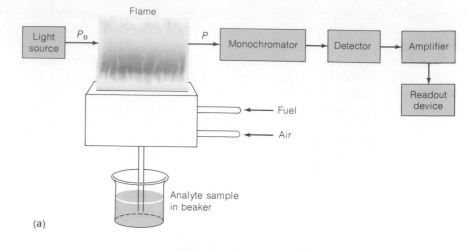

(a)

Figure 21-3

(*a*) Outline of an atomic absorption spectrometer. (*b*) Photograph of a research-quality instrument for atomic absorption and emission. The sample in the flask is being aspirated into the burner, which is behind the metal cage. Valves on the left control the flow rate of fuel and oxidizer. Dials on the right are used to select wavelengths, monochromator bandwidth, and observation modes. Results are displayed on the video tube or a printer. [Courtesy Instrumentation Laboratory, Wilmington, Mass.]

(*b*)

Atomic spectroscopy:

1. absorption
2. emission (luminescence from a thermally populated excited state)
3. fluorescence (luminescence following absorption of radiation)

of the vapor, many atoms are already in thermally populated, excited electronic states. They can spontaneously emit photons and return to a lower state. Therefore, atomic spectroscopy falls into three classes commonly designated as *absorption, fluorescence,* and *emission.* Routinely available instruments can perform absorption and emission experiments with equal ease. Equipment for **atomic fluorescence spectroscopy,** a technique potentially a thousand times more sensitive than absorption and emission techniques, is not yet in general use.

Apparatus for **atomic absorption spectroscopy** is shown in Figure 21-3. The liquid sample is aspirated (sucked) into a flame whose temperature is 2 000–3 000 K. The sample is **atomized** (broken into atoms) in the flame, which replaces the cuvet in conventional spectrophotometry. The pathlength of the flame is typically 10 cm. To measure the light absorbance of Fe atoms in the flame, the light source uses a cathode made of Fe. This source emits light with the characteristic frequencies of Fe atoms. The remainder of the apparatus is not very different from an ordinary spectrophotometer.

Atomic emission spectroscopy is similar to atomic absorption spectroscopy, but no light source is needed. Some of the atoms in the flame are promoted to excited electronic states by collision with other atoms. The excited atoms emit their characteristic radiation as they return to their ground state. In atomic emission spectroscopy, the emission intensity at a characteristic wavelength of an element is nearly proportional to the concentration of the element in the sample. For both absorption and emission, standard curves are usually used to establish the relation between signal and concentration.

Atomic emission requires equipment similar to that for atomic absorption, but the lamp is not used.

21-2 ATOMIZATION: FLAMES, FURNACES, AND PLASMAS

The essential feature that distinguishes atomic spectroscopy from ordinary spectroscopy is that the sample must be atomized. This is usually accomplished with a flame, an electrically heated oven, or a radio-frequency plasma. The sensitivity and interfering effects observed in atomic spectroscopy depend on the details of the heating process.

Premix Burner

Nebulization

Most flame atomic spectrometers use a **premix burner,** such as that in Figure 21-4, in which the sample, oxidant, and fuel are mixed before introduction into the flame. The sample solution (which need not be aqueous) is drawn into the *nebulizer* by the rapid flow of oxidant (usually air) past the tip of the sample capillary. The liquid breaks into a fine mist as it leaves the tip of the nebulizer. The spray is directed at high speed against a glass bead, upon which the droplets are broken into even smaller particles. The formation of particles in this manner is termed **nebulization.** Then the mist, oxidant, and fuel flow past a series of baffles that promotes further mixing and blocks large droplets of liquid. Liquid that collects at the bottom of the spray chamber flows out to a drain. Only a very fine mist containing about 5% of the initial sample reaches the flame. Figure 21-5 shows the distribution of droplet sizes from three different kinds of nebulizers. In general, the narrowest size distribution and smallest size are most desirable.

The flame

The most common fuel–oxidizer combination is acetylene and air, which produces a flame temperature of $\sim 2\,400$–$2\,700$ K. Other fuels and oxidizers are listed in Table 21-1. When a hotter flame is required, the acetylene–nitrous oxide combination is usually used. A flame profile is shown in Figure 21-4. Gas that enters the preheating region from the burner head is heated by downward conduction and radiation from the primary reaction zone (the blue cone). Combustion is completed in the outer cone, where surrounding air is drawn into the flame. Each flame has its own emission spectrum and therefore obscures the spectrum of analyte in certain regions.

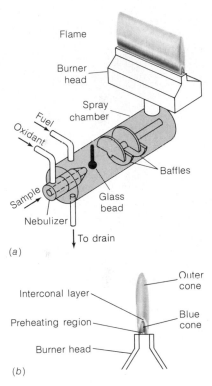

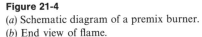

Figure 21-4
(*a*) Schematic diagram of a premix burner. (*b*) End view of flame.

Organic solvents having less surface tension than water are excellent for atomic spectroscopy because they form smaller droplets, leading to more efficient sample atomization.

Table 21-1

Maximum flame temperatures

Fuel	Oxidant	Temperature (K)
Acetylene	Air	2 400–2 700
Acetylene	Nitrous oxide	2 900–3 100
Acetylene	Oxygen	3 300–3 400
Hydrogen	Air	2 300–2 400
Hydrogen	Oxygen	2 800–3 000
Cyanogen	Oxygen	4 800

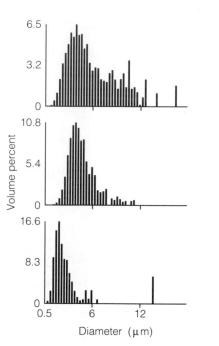

Figure 21-5
Distribution of droplet sizes produced by different types of nebulizers. [From R. H. Clifford, I. Ishii, A. Montaser, and G. A. Meyer, *Anal. Chem.,* **62**, 390 (1990).]

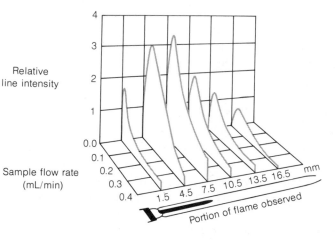

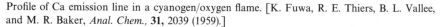

Figure 21-6
Profile of Ca emission line in a cyanogen/oxygen flame. [K. Fuwa, R. E. Thiers, B. L. Vallee, and M. R. Baker, *Anal. Chem.,* **31**, 2039 (1959).]

Hotter flames are needed for *refractory* (high vaporization temperature) elements or to decompose such species as metal oxides formed during passage through the flame.

Droplets entering the flame first lose their water through evaporation; then the remaining sample must vaporize and decompose into atoms. Many elements form oxides as they rise through the outer cone. Oxides do not have the same spectra as their free elements, so the resulting atomic signal is lower. If the flame is kept relatively rich in fuel (a "rich" flame), there is an excess of carbon species, which might reduce the metal oxides and thereby increase sensitivity. The opposite of a rich flame is a "lean" flame, which has excess oxidant and is hotter than a rich flame. Lean or rich flames are recommended in the analysis of different elements.

The position in the flame at which maximum atomic absorption or emission is observed depends on the element being measured as well as the flow rate of sample, fuel, and oxidizer. A profile of emission from Ca atoms in a cyanogen (N≡C—C≡N)/oxygen flame is shown in Figure 21-6. The decreasing intensity at higher flow rates is attributed to cooling of the flame by water from the sample. Different elements have different absorption and emission profiles. The sample flow rate, fuel and oxidant flow rates, and the level at which the flame is observed can be optimized for each element.

The electrically heated furnace offers greater sensitivity than that afforded by flames and requires a smaller volume of sample. Figure 21-7 shows a **graphite furnace** mounted in the beam of a spectrometer. From 1 μL to 100 μL of sample is injected into the oven through the hole at the center. Each end of the oven is a window through which the light beam travels. The maximum temperature for a graphite furnace is around 3 000°C.

A graphite furnace provides higher sensitivity because the entire sample is confined in the light path for a few seconds. In flame spectroscopy, the sample is highly diluted by the time it has been nebulized, and it spends only a fraction of a second in the light path as it rises through the flame. Flames also require a much larger volume of sample, since sample is constantly flowing into the flame. Whereas 1–2 mL is the minimum necessary for flame analysis, as little as 1 μL is adequate for a furnace. In one extreme example where only nanoliter volumes of kidney tubular fluid were available from micropuncture, a method was devised to reproducibly deliver 0.1-nL volumes to a graphite furnace for analysis of Na and K.[†]

Manually operated furnaces yield less precision than that possible with flames. Reproducibility is rarely better than 5–10% with manual sample introduction. Automated sample injection gives greatly improved precision. Furnaces require more operator skill and greater effort to determine the proper conditions for each type of sample. The reason for the increased effort is that the furnace must be heated in three or more steps to properly atomize the sample. As an example, to analyze iron in the iron-storage protein ferritin, 10 μL of sample containing $\sim$0.1 ppm Fe was injected into the cold oven. The furnace was programmed to dry the sample at 125°C for 20 s to remove solvent. This was followed by 60 s of charring at 1 200°C to destroy organic matter, which would otherwise create a great deal of smoke and interfere with the Fe determination. Finally, atomization was accomplished by heating to 2 700°C for 10 s. During this period of heating, the absorbance reached a maximum and then decreased as the Fe evaporated from the oven. Either the maximum absorbance measured on a recorder or the time-integrated absorbance can be taken as the analytical signal. It is important to record these events with a recorder or an oscilloscope, because signals are also observed from smoke produced during charring and from the glow of the red-hot oven in the latter part of atomization. A skilled operator must interpret which signal was due to sample and which to other effects.

Furnaces offer increased sensitivity and require less sample than a flame, but usually give poorer precision.

The operator must determine reasonable time and temperature for each stage of the analysis. Once a program is established, it can be applied to a large number of similar samples.

A cool-down to room temperature after charring and before atomization has been recommended as a means to increase analyte signal slightly. In some instances the cool-down reduces or eliminates interferences that cannot be controlled otherwise.

[†] L. A. Nash, L. N. Peterson, S. P. Nadler, and D. Z. Levine, *Anal. Chem.,* **60,** 2413 (1988).

Figure 21-7
Photograph of an electrically heated graphite-rod furnace used for flameless atomic spectroscopy. Light travels the length of the furnace ($\sim$38 mm in this case), and sample is injected through the hole at the top. [Courtesy Instrumentation Laboratory, Wilmington, Mass.]

Improved performance of the graphite furnace is obtained when a small platform called a *L'vov platform* is placed in the furnace (Figure 21-8a). Since the temperature of the sample on the platform lags behind the rising temperature of the walls of the furnace, the analyte does not vaporize until the walls have reached their constant temperature (Figure 21-8b). Under conditions of constant furnace temperature the area of the signal-versus-time curve is a reliable measure of the total analyte vaporized from the sample. Heating rates of 2 000°C/s are used with this technique.

Sometimes the temperature needed to char the sample **matrix** (the medium containing the analyte) also causes analyte evaporation. Addition of appropriate **matrix modifiers** can retard evaporation of the analyte until the matrix has charred away. The effect of a $Mg(NO_3)_2$ matrix modifer in the analysis of Mn is shown in Figure 21-9. A modifier consisting of a mixture of $Mg(NO_3)_2$ and $Pd(NO_3)_2$ allows the charring temperature for a sample containing Sn to be raised from 800°C to 1 400°C without Sn analyte evaporation.

A careful study of the effect of $Mg(NO_3)_2$ on Al determination has been conducted.[†] At high temperature, $Mg(NO_3)_2$ is converted to $MgO(s)$, which steadily evaporates and maintains some vapor pressure of $MgO(g)$. Aluminum in the unknown sample is converted to Al_2O_3 during heating. At a sufficiently

---

[†] D. L. Styris and D. A. Redfield, *Anal. Chem.,* **59,** 2891 (1987).

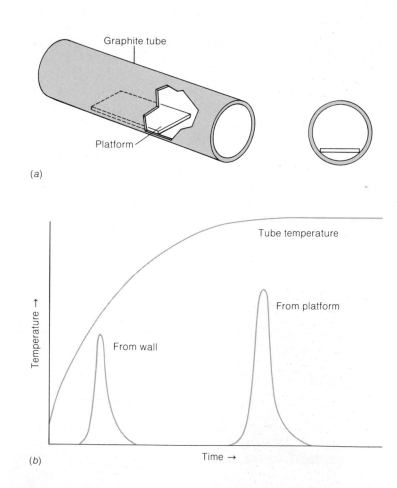

Figure 21-8
(*a*) L'vov platform in graphite furnace. (*b*) Heating profile comparing analyte evaporation from wall and from platform. [W. Slavin, *Anal. Chem.,* **54,** 685A (1982).]

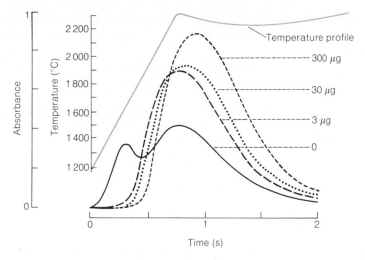

Figure 21-9
Effect of adding $Mg(NO_3)_2$ matrix modifier (0–300 µg) on signal from 0.4 ng of Mn. [W. Slavin, *Anal. Chem.*, **54**, 685A (1982).]

high temperature, Al_2O_3 decomposes to Al and O, and the Al evaporates. However, the evaporation of Al is retarded as long as MgO is present, by virtue of the reaction

$$3MgO(g) + 2Al(s) \rightleftharpoons 3Mg(g) + Al_2O_3(s) \qquad (21\text{-}1)$$

When all of the MgO has evaporated, Reaction 21-1 no longer occurs and Al_2O_3 finally decomposes and evaporates. By this means the $Mg(NO_3)_2$ matrix modifier prevents Al from evaporating until a high temperature is reached.

Inductively Coupled Plasma

The **inductively coupled plasma** is a type of flame that reaches a much higher temperature than that reached in ordinary combustion flames and is useful for emission spectroscopy. Its high temperature and stability eliminate many of the interferences and sources of error encountered with conventional flames. Because of these desirable features, the inductively coupled plasma is replacing conventional flame burners. The plasma's principal disadvantage is its expense to purchase and operate.

A cross-sectional view of an inductively coupled plasma burner head is shown in Figure 21-10. Two turns of a radio-frequency induction coil are wrapped around the upper opening of the quartz apparatus. High-purity argon gas is fed through the plasma gas inlet. A spark from a Tesla coil is used to ionize the Ar gas. The Ar^+ ions are immediately accelerated by the powerful radio-frequency field that oscillates about the load coil at a frequency of 27 MHz. The accelerated ions transfer energy to the entire gas by collisions between atoms. Once the process is begun, the ions absorb enough energy from the electric field to maintain a temperature of 6 000–10 000 K in the plasma (Figure 21-11). It is so hot (especially near the coils) that the quartz burner must be protected by argon coolant gas flowing around the outer edge of the apparatus.

Most elements are ionized to the +1 state in the plasma, and emission from the excited ions is the analytical signal. The predominant ions observed in the flame when an aqueous sample is used are Ar^+, O^+, ArH^+, and H_2O^+.

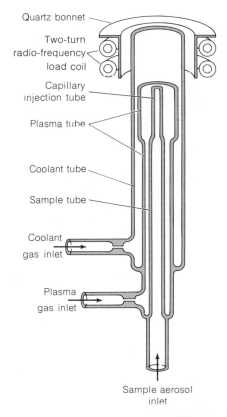

Quartz bonnet

Two-turn radio-frequency load coil

Capillary injection tube

Plasma tube

Coolant tube

Sample tube

Coolant gas inlet

Plasma gas inlet

Sample aerosol inlet

Figure 21-10
Diagram of an inductively coupled plasma burner head. [R. N. Savage and G. M. Hieftje, *Anal. Chem.*, **51**, 408 (1979).]

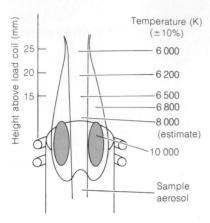

Figure 21-11
Temperature profile of a typical inductively coupled plasma used in analytic spectroscopy. [V. A. Fassel, *Anal. Chem.*, **51**, 1290A (1979).]

The **Boltzmann distribution** applies to a system at thermal equilibrium.

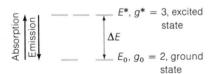

Figure 21-12
Two energy levels with different degeneracies. Ground-state atoms can absorb light to be promoted to the excited state. Excited-state atoms can emit light to return to the ground state.

It is also possible to sample the plasma by mass spectrometry, and such measurements are potentially more sensitive than those obtained with atomic emission.[†]

Sample can be introduced into the plasma by a conventional nebulizer. However, the concentration of analyte needed for adequate signal can be reduced by a factor of 10–20 with an *ultrasonic nebulizer*, in which sample solution is delivered onto a piezoelectric crystal (Box 2-1) oscillating at a frequency near 1 MHz. This creates a very fine **aerosol** (a suspension of solid or liquid particles in a gas) that is carried by a stream of argon through a heated tube where solvent is vaporized. The stream then passes through a refrigerated zone in which solvent condenses and is removed. The analyte reaches the plasma flame as a fine dry cloud. Flame energy is not needed to evaporate solvent and more energy is available for atomization.

Effect of Temperature in Atomic Spectroscopy

Temperature is a critical factor in determining the degree to which a given sample breaks down to atoms. In addition, temperature determines the extent to which a given atom is found in its ground, excited, or ionized states.

Boltzmann distribution

Consider a molecule with two available energy levels separated by energy ΔE (Figure 21-12). Call the lower level E_0 and the upper level E^*. In general, an atom (or molecule) may have more than one state available at a given energy level. In Figure 21-12, we show three states available at E^* and two available at E_0. The number of states available at each energy level is called the *degeneracy* of the level. We will call the degeneracies g_0 and g^*.

If thermal equilibrium exists (which is not true in the blue cone of a flame, but is probably true above the blue cone), the relative populations of any two energy levels are given by

$$\frac{N^*}{N_0} = \frac{g^*}{g_0} e^{-\Delta E/kT} \qquad (21\text{-}2)$$

where N is the population of each state, T is kelvins, and k is Boltzmann's constant ($1.380\,658 \times 10^{-23}$ J/K).

Effect of temperature on excited-state population

The lowest excited state of a sodium atom lies 3.371×10^{-19} J/atom above the ground state. The degeneracy of the excited state is 2, while that of the ground state is 1. Let's calculate the fraction of sodium atoms in the excited state in an acetylene–air flame at 2 600 K. Using Equation 21-2, we find

$$\frac{N^*}{N_0} = \left(\frac{2}{1}\right) e^{-3.371 \times 10^{-19}/(1.381 \times 10^{-23} \cdot 2\,600)} = 1.67 \times 10^{-4} \qquad (21\text{-}3)$$

That is, less than 0.02% of the atoms are in the excited state.

[†] R. S. Houk, Anal. Chem., **58**, 97A (1986).

Table 21-2
Effect of energy separation and temperature on population of excited states[†]

Wavelength separation of states (nm)	Energy separation of states (J/atom)	Excited-state fraction (N^*/N_0)	
		2 500 K	6 000 K
250	7.95×10^{-19}	1.0×10^{-10}	6.8×10^{-5}
500	3.97×10^{-19}	1.0×10^{-5}	8.3×10^{-3}
750	2.65×10^{-19}	4.6×10^{-4}	4.1×10^{-2}

[†] Based on the equation $N^*/N_0 = (g^*/g_0)e^{-\Delta E/kT}$ in which $g^* = g_0 = 1$.

How would the fraction of atoms in the excited state change if the temperature were 2 610 K instead?

$$\frac{N^*}{N_0} = \left(\frac{2}{1}\right)e^{-3.371 \times 10^{-19}/(1.381 \times 10^{-23} \cdot 2\,610)} = 1.74 \times 10^{-4} \quad (21\text{-}4)$$

A 10 K temperature rise changes the excited-state population by 4% in this example.

The fraction of atoms in the excited state is still less than 0.02%, but that fraction has changed by $100(1.74 - 1.67)/1.67 = 4\%$.

Effect of temperature on absorption and emission

In the preceding section, we saw that more than 99.98% of the sodium atoms are in their ground state at 2 600 K. *Varying the temperature by* 10 K *hardly affects the ground-state population and would not noticeably affect the signal in an atomic absorption experiment.*

It turns out that the emission spectrum of sodium is much more intense than the absorption spectrum, because the efficiency of emission is very much greater than the efficiency of absorption for this element. How would the emission intensity be affected by a 10 K rise in temperature?

In Figure 21-12, it can be seen that absorption arises from ground-state atoms, but emission arises from excited-state atoms. The emission intensity should be proportional to the population of the excited state. *Since the excited-state population changes by* 4% *when the temperature rises* 10 K, *the emission intensity will also rise* 4%. It is critical in atomic *emission* spectroscopy that flame conditions be very stable, or the emission intensity will vary significantly. In atomic *absorption* spectroscopy, flame temperature variation is not as critical.

The inductively coupled plasma is so hot that a substantial population of excited-state atoms and ions exists. The plasma is therefore almost always used for emission, not absorption, measurements. Relative to a flame, the plasma has a more uniform temperature profile and therefore gives more reproducible emission intensities. Table 21-2 compares excited-state populations for a flame at 2 500 K and a plasma at 6 000 K.

The intensity of atomic absorption is not very sensitive to temperature. The intensity of atomic emission is very sensitive to temperature.

21-3 INSTRUMENTATION

The fundamental requirements for an atomic absorption experiment are shown in Figure 21-3. The principal differences between atomic spectroscopy and ordinary molecular spectroscopy lie in the light source, the sample container (the flame), and the need to subtract the flame emission spectrum from the observed signal.

Source of Radiation

The linewidth problem

The bandwidth of the source must be narrower than the bandwidth of the atomic vapor for Beer's law to be obeyed.

Beer's law applies to monochromatic radiation. In practical terms, this means that the linewidth of the radiation being measured should be substantially narrower than the bandwidth of the absorbing sample. Otherwise, the measured absorbance will not be proportional to the sample concentration.

Atomic absorption lines are very sharp, with an inherent width of only $\sim 10^{-4}$ nm. Two mechanisms serve to broaden the lines in atomic spectroscopy. One is the **Doppler effect.** An atom moving toward the radiation source samples the oscillating electromagnetic wave more frequently than one moving away from the source (Figure 21-13). That is, an atom moving toward the source "sees" higher-frequency light than that encountered by one moving away. In the laboratory frame of reference, the atom moving toward the source absorbs lower-frequency light than that absorbed by the one moving away. The linewidth, Δv, due to the Doppler effect, is given approximately by

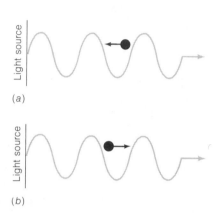

(a)

(b)

Figure 21-13
The Doppler effect. A molecule moving (a) toward the radiation source "feels" the electromagnetic field oscillate more often than one moving (b) away from the source.

$$\Delta v \approx v(7 \times 10^{-7}) \sqrt{\frac{T}{M}} \qquad (21\text{-}5)$$

where v is the frequency (in Hz) of the peak maximum, T is the temperature in kelvins, and M is the mass of the atom in atomic mass units; Δv is the width of the absorption line measured at half the height of the peak.

Another factor that affects atomic absorption linewidths is called **pressure broadening.** This arises because an atom does not behave as an isolated system during a collision with another atom. Its energy levels are perturbed, and it will not absorb the same frequency of radiation as an isolated atom. The pressure broadening, Δv_p, is roughly equal to the collision frequency and is proportional to pressure.

Doppler and pressure effects broaden the atomic lines by 1–2 orders of magnitude as compared with their inherent linewidths.

The Doppler effect and pressure broadening are of a similar order of magnitude. Together they yield linewidths of 10^{-3}–10^{-2} nm in atomic spectroscopy.

Hollow-cathode lamp

There is no monochromator that can isolate linewidths smaller than 10^{-3}–10^{-2} nm. To produce narrow lines of the correct frequency, we use a **hollow cathode lamp** containing the same element as that being analyzed.

A hollow-cathode lamp, such as that shown in Figure 21-14, is filled with Ne or Ar at a pressure of ~ 130–700 Pa (1–5 torr). When a high-enough voltage is applied between the anode and cathode, the filler gas becomes ionized and positive ions are accelerated toward the cathode. They strike the cathode with enough energy to "sputter" metal atoms from the cathode into the gas phase. Many of the sputtered atoms are in excited states; they emit photons and then return to the ground state. This atomic radiation is of exactly the same frequency as that absorbed by atoms of analyte in the flame or furnance. The linewidth is sufficiently sharp (narrow), with respect to that of the high-temperature analyte, to be nearly "monochromatic" (Figure 21-15).

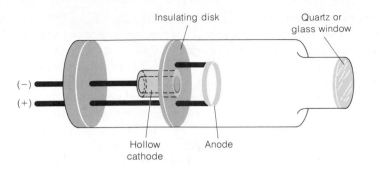

Insulating disk

Quartz or glass window

(−)
(+)

Hollow cathode

Anode

Figure 21-14
A hollow-cathode lamp.

Most of the sputtered atoms condense back on the cathode, but some are deposited on the glass walls of the lamp, eventually ruining the lamp. A different lamp is usually required for each element, although some lamps are made with more than one element in the cathode.

The emission from a lamp having a hollow iron cathode is shown in Figure 21-1. The linewidths are actually much narrower than the figure indicates, because the monochromator used to record the spectrum had a bandwidth of 0.08 nm. Some lines produced by the lamp arise from gaseous ions, such as Fe^+, Ne^+, or Ar^+.

The Spectrophotometer

We have already discussed the two most unusual features of an atomic absorption spectrophotometer. One is that the sample "container" is a flame or furnace. The other is that the lamp emits only a few sharp lines at precisely those frequencies absorbed by the analyte. The remainder of the spectrophotometer is not very different from one used for ordinary absorption spectroscopy.

An important difference between ordinary spectrophotometers and those used in atomic spectroscopy is that some means must be provided in atomic spectroscopy to distinguish the analyte signal from the background spectrum of the flame or red-hot graphite furnace. For example, Figure 21-16 shows

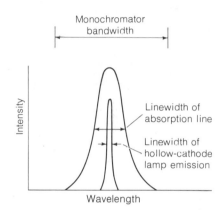

Figure 21-15
Relative linewidths of hollow-cathode emission, atomic absorption, and monochromator. Linewidths are measured at half the signal height.

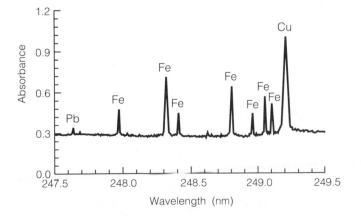

Figure 21-16
Graphite furnace absorption spectrum of bronze dissolved in HNO_3. [Reproduced from B. T. Jones, B. W. Smith, and J. D. Winefordner, *Anal. Chem.,* **61**, 1670 (1989).]

the absorption spectrum of a sample containing Fe, Cu, and Pb in a graphite furnace. Note that the sharp atomic signals are superimposed on a broad background absorbance. The Cu signal at 249.2 nm has a peak absorbance of 1.0. However, the baseline absorbance is 0.3. The true absorbance of Cu is $1.0 - 0.3 = 0.7$. Some means must be available to measure the background absorbance in atomic spectroscopy, or significant errors result.

Background correction

A **beam chopper** is usually used to distinguish the signal due to the flame from the desired atomic line at the same wavelength. As shown in Figure 21-17, the beam from the lamp is periodically blocked by the rotating chopper. The signal that reaches the detector while the beam is blocked must be due to flame emission. The signal reaching the detector when the beam is not blocked is the sum of the signals from the lamp and the flame. The difference between these two signals is the desired analytical signal, which is displayed by the spectrometer.

An alternative means of measuring the flame background involves modulating the lamp with an alternating current. The signal reaching the detector will therefore contain a constant contribution from the flame and a modulated contribution from the lamp. By electronically decomposing the detected signal into these two components, a flame emission correction is provided.

Some samples produce particulate matter (smoke or unvaporized particles), which scatters a significant fraction of light from the hollow-cathode lamp. Smoke is particularly troublesome with graphite furnaces. The detector cannot distinguish scattering from absorption, so this can lead to a systematic error in the measurement of analyte.

Many spectrometers provide an additional means to correct for background scattering and broad-band background absorption. Emission from a continuous radiation source (usually a D_2 lamp) is passed through the flame in alternation with that from the hollow cathode. The monochromator bandwidth is sufficiently wide that a negligible fraction of the D_2 lamp radiation is absorbed by the analyte's atomic absorption line. The attenuation of this continuous radiation is due to background scattering or broad-band absorption. The spectrometer corrects the analytical signal for the attenuation suffered by the background correction beam.

A better, but more expensive, background correction technique involves the *Zeeman effect* (pronounced ZAY-mon). This refers to the shifting of energy levels of atoms and molecules in the presence of a magnetic field. The atomic signals are very sharp, while the background is very broad. When a strong magnetic field is applied, the atomic energy levels shift enough that the hollow-cathode lamp frequency is not absorbed by analyte atoms. To use the Zeeman

Beam chopping and source modulation can correct for flame emission, but not for scattering

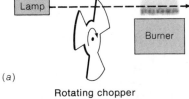

(a)

Rotating chopper

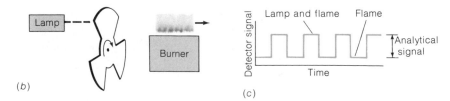

(b)

(c)

Figure 21-17
Operation of a beam chopper for subtracting the signal due to flame background emission. (*a*) Lamp and flame emission reach detector. (*b*) Only flame emission reaches detector. (*c*) Resulting signal.

effect for background correction, a strong magnetic field is pulsed on and off. The analytical signal (sample plus background) is observed when the field is off, and the background signal is observed when the field is on. Figure 21-18 shows the effect of a 1.2 tesla magnetic field on the atomic fluorescence spectrum of Co in a graphite furnace. When the field is turned on, the atomic energy levels split and only a small fraction of the original signal is left at the central wavelength of fluorescence in zero field.

Sensitivity and detection limit

The **sensitivity** of an atomic absorption spectrometer for a given element is defined as the concentration of that element needed to produce 99% transmittance (which corresponds to an absorbance of 0.004 36). The **detection limit** is the concentration of an element that gives a signal equal to twice the peak-to-peak noise level of the baseline (Figure 21-19).[†] The baseline noise level should be measured while a blank sample is being aspirated into the flame.

Figure 21-20 compares the detection limits of flame and flameless (furnace) operation for a particular instrument. You can see that most elements can be determined by atomic absorption and that there are wide variations in the detection limits for different elements. This is attributable to the varying efficiencies of atomization and the differing absorptivities (ε) of different elements. The detection limit for flameless operation is typically two orders of magnitude lower than that observed with a flame. The main reason for this is that the sample is confined in a small volume for a relatively long time in the furnace, compared with its fleeting moment in a flame. Detection limits for inductively coupled plasmas are similar to those of flames.

[†] Detection limit is really the lowest concentration that can be "reliably" detected. Numerous statistical definitions of this ill-defined quantity are possible. See, for example, G. L. Long and J. D. Winefordner, *Anal. Chem.*, **55**, 712A (1983); W. R. Porter, *Anal. Chem.*, **55**, 1290A (1983); and J. E. Knoll, *J. Chromatographic Sci.*, **23**, 422 (1985).

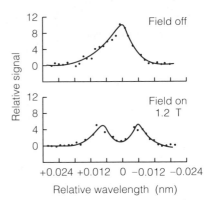

Figure 21-18
Zeeman effect on Co fluorescence in a graphite furnace with excitation at 301 nm and detection at 341 nm. [Reproduced from J. P. Dougherty, F. R. Preli, Jr., J. T. McCaffrey, M. D. Seltzer, and R. G. Michel, *Anal. Chem.*, **59**, 1112 (1987).]

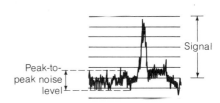

Figure 21-19
Illustration of the measurement of peak-to-peak noise level and signal level. The signal is measured from its base at the midpoint of the noise component along the slightly slanted baseline. This sample exhibits a signal-to-noise ratio of 3.2.

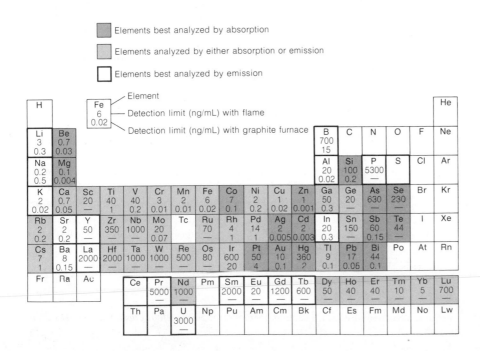

☐ Elements best analyzed by absorption

☐ Elements analyzed by either absorption or emission

☐ Elements best analyzed by emission

Figure 21-20
Atomic absorption detection limits (ng/mL) with the Perkin–Elmer 703 spectrophotometer. Accurate quantitative analysis usually requires a concentration 10–100 times greater than the detection limit.

21-4 ANALYTICAL METHODS

Most elements of the periodic table can be analyzed by either atomic absorption or atomic emission, or both. Figure 21-20 shows the most suitable method for each element. The equipment for both techniques is the same, except that a lamp is required for atomic absorption measurements. The elements most commonly analyzed by atomic emission are Li, Na, and K (Box 21-1). The choice of which method to use for other elements is often dictated by the availability of a lamp.

Standard Curve

The most common technique for quantitative analysis is to construct a standard curve, such as that in Figure 21-21, using known amounts of the desired element in a solution with a similar composition to that of the unknown. The standard curve is then used to find the concentration of unknown from its absorbance. It is *critical* that the composition of the standards be as close as possible to that of the unknown, because different solutions can have different types of interferences that affect the signal.

> The composition of the standards should be as close as possible to the composition of the unknown.

Standard Addition Method

In the **standard addition method,** known quantities of the desired element are added to the analyte, and the increase in signal is measured.[†] Each solution is diluted to the same total volume and should have the same final composition (except for analyte concentration). If the concentration of unknown is [X] and the concentration of added standard is [S], we can say that

$$\frac{[X]}{[X] + [S]} = \frac{A_X}{A_{S+X}} \qquad (21\text{-}6)$$

where A_X is the absorbance (or emission intensity) of unknown and A_{S+X} is the absorbance (or emission intensity) of unknown plus standard. Equation 21-6 applies only if the absorbance or emission is linearly related to concentration. Most elements exhibit some concentration range where this is true.

Equation 21-6 can be solved directly for [X]. Alternatively, a series of standard additions can be made, and the results can be plotted on a graph, as in Figure 21-22 to find the concentration of unknown. In this graph, the *x* axis is the concentration of added analyte *after* it has been mixed with sample. The *x* intercept of the extrapolated line is equal to the concentration of unknown *after* the unknown has been diluted to the final volume. In Figure 21-22, this value is near 4.2 μg/mL. Statistically, the most useful range of standard additions should increase the analyte concentration to between 1.5 and 3 times its original value. The main advantage of the standard addition method is that the matrix remains constant for all samples.

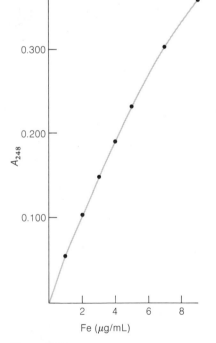

Figure 21-21
Atomic absorption calibration curve for Fe.

> The graphic treatment in Figure 21-22 requires that the response be linear.

[†]For a discussion of standard addition methods, see M. Bader, *J. Chem. Ed.,* **57,** 703 (1980).

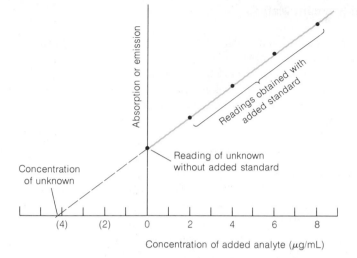

Absorption or emission

Readings obtained with added standard

Reading of unknown without added standard

Concentration of unknown

(4)   (2)   0   2   4   6   8

Concentration of added analyte (μg/mL)

Figure 21-22
Graphic treatment of the method of standard additions.

Box 21-1   THE FLAME PHOTOMETER IN CLINICAL CHEMISTRY

Measurement of sodium and potassium in serum and urine is a routine procedure in diagnostic clinical analysis. This was not true before the advent of atomic spectroscopy because laborious gravimetric techniques were required.

The flame photometer shown below is typical of clinical instruments. It operates on oxygen and natural gas. As many as 18 samples containing 20 μL of serum plus 2 mL of water are placed in the rack on the right. Each sample is automatically aspirated into the flame in the housing above the sample rack. Simple optical filters are sufficient to isolate the strong emission lines of sodium, potassium, and lithium. The concentrations are read out directly on the meter, which is calibrated by aspirating standard samples. Lithium is often used as an internal standard for sodium and potassium.

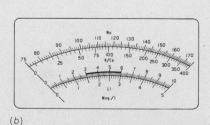

(a)

(b)

(a) The Coleman Model 51 Flame Photometer used in clinical analysis.

(b) Meter scale. [Courtesy A. H. Thomas Co., Philadelphia, Pa.]

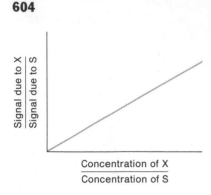

Signal due to X / Signal due to S

Concentration of X / Concentration of S

Figure 21-23
Calibration curve for an internal standard. To construct such a curve, the concentration of X would be varied while the concentration of S is held fixed.

Internal Standard Method

An **internal standard** is a known amount of an element, not already present, which is added to a sample. To use an internal standard, *known* mixtures of standard (S) and analyte (X) are used to construct a standard curve, such as that in Figure 21-23. When a known amount of standard is added to an unknown sample, the calibration curve can be used to find the concentration of unknown.

EXAMPLE: Using an Internal Standard

A solution was prepared by mixing 5.00 mL of unknown (X) with 2.00 mL of solution containing 4.13 μg of standard (S) per millimeter, and diluting to 10.0 mL. The measured ratio of signals was

$$\frac{\text{Signal due to X}}{\text{Signal due to S}} = 0.808$$

In a separate experiment, it was found that for equal concentrations of X and S, the signal due to X was 1.31 times as intense as the signal due to S. Find the concentration of X in the unknown.

The concentration of internal standard in the unknown mixture was

$$[S] = (4.13 \ \mu g/mL)\underbrace{\left(\frac{2.00}{10.0}\right)}_{\substack{\text{Dilution} \\ \text{factor}}} = 0.826 \ \mu g/mL$$

Since the measured signal ratio is 1.31 when the concentration ratio is unity, we can say that

$$\frac{\text{Concentration ratio in unknown}}{\text{Concentration ratio in standard mixture}} = \frac{\text{absorbance ratio in unknown}}{\text{absorbance ratio in standard mixture}}$$

$$\frac{[X]/[S]}{1} = \frac{0.808}{1.31}$$

$$\frac{[X]}{[S]} = 0.617$$

Since $[X]/[S] = 0.617$ and since $[S] = 0.826 \ \mu g/mL$, $[X] = (0.617)(0.826) = 0.510 \ \mu g/mL$. But X was diluted by a factor of 2.00 when it was mixed with the standard. Therefore, the original concentration X was 1.02 μg/mL.

Internal standards are most useful when unavoidable sample losses are expected. If a standard is added to a sample prior to any losses, then the fraction of standard lost is the same as the fraction of sample lost, and the quotient (concentration of unknown)/(concentration of standard) remains constant. Internal standards are also useful when the sample matrix cannot be controlled. For example, the apparent concentration of Fe goes down by 20% as the NaCl concentration of an aqueous solution is increased from 0 to 10%. The reason is probably that the solution becomes more viscous and flows into the nebulizer more slowly. Decreased flow means that fewer atoms of iron reach the flame. The absorbance decreases by 20%. A way to circumvent this problem is to inject Mn as an internal standard. If the viscosity of the sample increased such that the flow rate decreased by 6%, *both* absorbances would decrease by 6%, but their *ratio* would remain the same. Some spectrometers detect absorption and emission at two wavelengths simultaneously. Such instruments are especially convenient for use with an internal standard.

We assume that changing the sample composition affects the signal from both elements equally.

21-5 INTERFERENCE

By *interference,* we mean any effect that changes the signal when analyte concentration remains unchanged. In the measurement of atomic absorption or emission signals, interference is widespread and easy to overlook. If you are clever enough to discern that interference is occurring, it may be corrected by counteracting the source of interference or by preparing standards that exhibit the same interference.

Spectral interference refers to the overlap of analyte signal with signals due to other elements or molecules in the sample or with signals due to the flame or furnace. Interference from the flame can be subtracted by using D_2 lamp or Zeeman background correction. The best means of dealing with overlap between lines of different elements in the sample is to choose another wavelength for analysis.

Elements that form very stable diatomic oxides are said to be *refractory* because they are incompletely atomized at the temperature of the flame or furnace. The spectrum of a molecule is much broader and more complex that that of an atom, because vibrational and rotational transitions are combined with electronic transitions (as described in Section 19-3). The broad spectrum leads to spectral interference at many wavelengths. Figure 21-24 shows an example of a plasma containing Y and Ba atoms, as well as YO molecules. Note how broad the molecular emission is in comparison to the atomic emission.

Chemical interference is caused by any component of the sample that decreases the extent of atomization of analyte. For example, SO_4^{2-} and PO_4^{3-} hinder the atomization of Ca^{2+}, perhaps by forming involatile salts. **Releasing agents** are chemicals that can be added to a sample to decrease chemical interference. EDTA and 8-hydroxyquinoline protect Ca^{2+} from the interfering effects of SO_4^{2-} and PO_4^{3-}. La^{3+} can also be used as a releasing agent, apparently because it preferentially reacts with PO_4^{3-} and frees the Ca^{2+}. Use of a fuel-rich flame is recommended to reduce certain oxidized analyte species that would otherwise hinder atomization. Higher flame temperatures eliminate many kinds of chemical interference.

Ionization interference can be a problem in the analysis of alkali metals at relatively low flame temperature, and for other elements at higher temperature. For any element, we can write a gas-phase ionization reaction:

$$M(g) \rightarrow M^+(g) + e^-(g) \tag{21-7}$$

$$K = \frac{[M^+][e^-]}{[M]} \tag{21-8}$$

Types of interference:

1. spectral: unwanted signals overlapping analyte signal
2. chemical: chemical reactions decreasing the concentration of analyte atoms
3. ionization: ionization of analyte atoms decreases the concentration of neutral analyte atoms in the flame

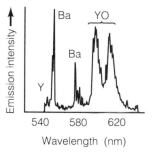

Figure 21-24

Atomic and molecular emission from a plasma produced by striking the high-temperature superconductor $YBa_2Cu_3O_7$ with high-intensity laser light. The solid sample is vaporized by the laser, and excited atoms and molecules in the gas phase emit light at their characteristic wavelengths. [Reproduced from W. A. Weimer, *Appl. Phys. Lett.,* **52,** 2171 (1988).]

Since the alkali metals have the lowest ionization potentials, they are most extensively ionized in a flame. At 2 450 K and a pressure of 0.1 Pa, sodium is expected to be 5% ionized. Potassium, with a lower ionization potential, is expected to be 33% ionized under the same conditions. Since ionized atoms have different energy levels from neutral atoms, the desired atomic signal is decreased.

An **ionization suppressor** is an element added to a sample to decrease the extent of ionization of analyte. For example, in the analysis of potassium, it is recommended that solutions contain 1 000 ppm of CsCl, since cesium is more easily ionized than potassium. By producing a high concentration of electrons in the flame, ionization of Cs suppresses ionization of K. Ionization suppression is desirable in a low-temperature flame in which we want to observe neutral atoms. In a high-temperature plasma we want to observe ions, so ionization suppression is, of course, not desired.

This is an application of Le Châtelier's principle to Reaction 21-7.

Virtues of the inductively coupled plasma

Many common interferences in atomic spectroscopy are eliminated by using an inductively coupled argon plasma for emission measurements. The plasma is twice as hot as a conventional flame, and the residence time of analyte in the flame is about twice as great. Therefore, atomization is more complete than in a flame, and the signal is correspondingly enhanced. Reaction of the analyte to form metal oxide molecules is eliminated. The plasma is also remarkably free of background radiation in the region where sample emission is observed (15–35 mm above the load coil). The background concentration of electrons due to plasma formation is fairly high and uniform. The temperature is so high that most elements are observed as ions, and the concentrations of these ions appear to be fairly insensitive to the presence of potential suppressors.

A common problem in flame emission spectroscopy is that there is a lower concentration of electronically excited atoms in the cooler, outer part of the flame than in the warmer, central part of the flame. As a result, emission from the central region is absorbed in the outer region. This **self-absorption** increases with increasing concentration of analyte and leads to nonlinear calibration curves. In a plasma, the flame temperature is more uniform, and self-absorption is not nearly so important. Figure 21-25 shows calibration curves for plasma emission that are linear over nearly five orders of magnitude. In conventional flames and furnaces, the linear range covers about two orders of magnitude.

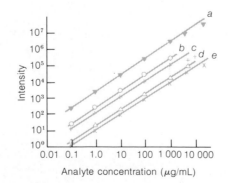

Figure 21-25
Analytical calibration curves for emission from (a) Ba^{2+}, (b) Cu^+, (c) Na^+, (d) Fe^+, and (e) Ba^+ in an inductively coupled plasma. [R. N. Savage and G. M. Hieftje, *Anal. Chem.*, **51**, 408 (1979).]

Summary

In atomic spectroscopy, the absorption, emission, or fluorescence from gaseous atoms is measured. Liquid samples may be atomized by a flame, furnace, or plasma. In a premix burner, the sample is mixed with fuel and oxidant before it flows into the flame, whose temperature is usually in the range 2 300– 3 400 K. The fuel and oxidant chosen determine the temperature of the flame and affect the extent of spectral, chemical, or ionization interference that will be encountered. Flame temperature instability has little effect on atomic absorption, but a large effect on atomic emission, because the excited-state population is very temperature-dependent. An electrically heated furnace requires much less sample than a flame and has a lower detection limit. In an inductively coupled plasma, a radio-frequency induction coil is used to heat Ar^+ ions in an argon gas stream to 6 000—10 000 K. At this high temperature, emission from excited atomic ions is the dominant signal. There is little chemical interference in an inductively coupled plasma, the temperature is very stable, and little self-absorption is observed.

Instrumentation for atomic spectroscopy is similar to that for molecular spectroscopy, but the sample must be atomized, the radiation source must be highly monochromatic, and the background signal must be subtracted. Lamps with a hollow cathode made of the analyte element are usually used to obtain atomic lines sharper than those of the atomic vapor (whose lines are broadened by collisions and by the Doppler effect). Correction for background emission due to the flame usually involves a mechanical beam chopper. Correction for inadvertent light scattering and spectral background can be made by measuring absorption using a broad-band deuterium lamp or by Zeeman background correction, in which the atomic energy levels are alternately shifted in and out of resonance with the lamp frequency by a magnetic field. Chemical interference can sometimes be reduced by appropriate releasing agents, which protect the analyte from reacting with interfering species. Ionization interference in flames often can be suppressed by adding to the sample such easily ionized elements as Cs. Analytical methods employing standard curves, standard additions, or internal standards can be applied to quantitative analysis by atomic spectroscopy.

Terms to Understand

aerosol	chemical interference	ionization interference	releasing agent
atomic absorption spectroscopy	detection limit	ionization suppressor	self-absorption
atomic emission spectroscopy	Doppler effect	matrix	sensitivity
atomic fluorescence spectroscopy	graphite furnace	matrix modifier	spectral interference
atomization	hollow-cathode lamp	nebulization	standard addition method
beam chopper	inductively coupled plasma	premix burner	
Boltzmann distribution	internal standard	pressure broadening	

Exercises

21-A. Li was determined by the method of standard additions using atomic emission. From the data in the following table, calculate the concentration of Li in pure unknown. The Li standard contained 1.62 μg Li/mL.

Unknown (mL)	Standard (mL)	Final volume (mL)	Emission intensity (arbitrary units)
10.00	0.00	100.0	309
10.00	5.00	100.0	452
10.00	10.00	100.0	600
10.00	15.00	100.0	765
10.00	20.00	100.0	906

21-B. Mn was used as an internal standard for measuring Fe by atomic absorption. A standard mixture containing 2.00 μg Mn/mL and 2.50 μg Fe/mL gave a signal quotient (Fe signal/Mn signal) = 1.05. A mixture of volume 6.00 mL was prepared by mixing 5.00 mL of unknown Fe solution with 1.00 mL containing 13.5 μg Mn/mL solution. The absorbance of this mixture at the Mn wavelength was 0.128, and the absorbance at the Fe wavelength was 0.185. Calculate the molarity of the unknown Fe solution.

21-C. The atomic absorption signal on page 608 (left) was obtained with 0.048 5 μg Fe/mL in a graphite furnace. Estimate the detection limit for Fe.

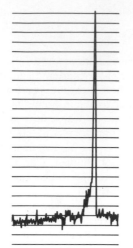

21-D. The laser atomic fluorescence excitation and emission spectra of sodium in an air–acetylene flame are shown at the right. In the *excitation* spectrum, the laser (bandwidth = 0.03 nm) was scanned through various wavelengths, while the detector monochromator (bandwidth = 1.6 nm) was held fixed near 589 nm. In the *emission* spectrum, the laser was fixed at 589.0 nm, and the detector monochromator wavelength was varied. Explain why the emission spectrum gives one broad band, while the excitation spectrum gives two sharp lines. How can the excitation linewidths be much narrower than the detector monochromator bandwidth?

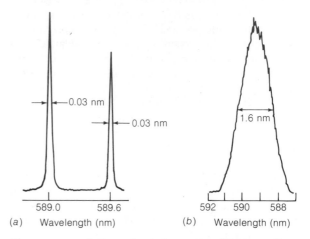

Fluorescence excitation and emission spectra of the two sodium D lines in an air–acetylene flame. (a) In the excitation spectrum the laser was scanned. (b) In the emission spectrum the monochromator was scanned. The monochromator slit width was the same for both spectra. [S. J. Weeks, H. Haraguchi, and J. D. Winefordner, *Anal. Chem.*, **50**, 360 (1978).]

Problems

A21-1. In which technique, atomic absorption or atomic emission, is flame-temperature stability more critical? Why?

A21-2. State the advantages and disadvantages of furnaces compared with flames in atomic absorption spectroscopy.

A21-3. State the advantages and disadvantages of the inductively coupled plasma compared with conventional flames in atomic spectroscopy.

A21-4. Explain what is meant by the Doppler effect. Rationalize why Doppler broadening increases with increasing temperature and decreasing mass in Equation 21-5.

A21-5. Explain how the following background correction techniques work:
(a) beam chopping
(b) deuterium lamp
(c) Zeeman effect

A21-6. Explain what is meant by spectral, chemical, and ionization interference.

A21-7. Why is an internal standard most appropriate for quantitative analysis when unavoidable sample losses are expected during sample preparation?

A21-8. *Standard addition.* An unknown containing element X was mixed with aliquots of a standard solution of element X for atomic absorption spectroscopy. The standard solution contained 1 000.0 μg of X per milliliter.

Volume of unknown (mL)	Volume of standard (mL)	Total volume (mL)	Absorbance
10.00	0	100.0	0.163
10.00	1.00	100.0	0.240
10.00	2.00	100.0	0.319
10.00	3.00	100.0	0.402
10.00	4.00	100.0	0.478

(a) Calculate the concentration (μg X/mL) of added standard in each solution.
(b) Prepare a graph similar to the one in Figure 21-22. From the graph determine the concentration of X in the unknown.

A21-9. *Internal standard.* Follow the steps of the Example dealing with internal standards in Section 21-4 when working this problem. A solution was prepared by mixing 10.00 mL of unknown (X) with 5.00 mL of standard (S) containing 8.24 μg S/mL, and diluting to 50.0 mL. The measured signal quotient was (signal due to X/signal due to S) = 1.69.

   (a) In a separate experiment it was found that for equal concentrations of X and S, the signal due to X was 0.93 times as intense as the signal due to S. Find the concentration of X in the unknown.

   (b) Answer the same question if in a separate experiment it was found that for the concentration of X equal to 3.42 times the concentration of S, the signal due to X was 0.93 times as intense as the signal due to S.

A21-10. Referring to Equation 19-2, calculate the wavelength (nanometers) of emission of excited atoms that lie 3.371×10^{-19} J per molecule above the ground state.

A21-11. Derive the entries for 500 nm in Table 21-2. What would be the value of N^*/N_0 at 6 000 K if $g^* = 3$ and $g_0 = 1$?

A21-12. A series of potassium standards gave the following emission intensities at 404.3 nm. Find the concentration of potassium in the unknown.

Sample (μg K/mL)	Relative emission
Blank	0
5.00	124
10.00	243
20.0	486
30.0	712
Unknown	417

21-13. Calculate the Doppler linewidth in Hz ($= s^{-1}$) for the 589-nm line of Na and for the 254-nm line of Hg, both at 2 000 K.

21-14. For Ca atoms, the first excited state is reached by absorption of light with $\lambda = 422.7$ nm.

   (a) What is the energy difference (kilojoules per mole) between the ground state and the excited state?

   (b) The relative degeneracies are $g^*/g_0 = 3$ for Ca. What is the ratio N^*/N_0 at 2 500 K?

   (c) By what percentage will the fraction in part b be changed by a 15 K rise in temperature?

   (d) What will be the ratio N^*/N_0 at 6 000 K?

21-15. A series of Ca and Cu samples was run to determine the atomic absorbance of each element.

Ca (μg/mL)	$A_{422.7}$	Cu (μg/mL)	$A_{324.7}$
1.00	0.086	1.00	0.142
2.00	0.177	2.00	0.292
3.00	0.259	3.00	0.438
4.00	0.350	4.00	0.576

   (a) Determine the average relative absorbance ($A_{324.7}/A_{422.7}$) produced by equal concentrations (μg/mL) of Ca and Cu.

   (b) Copper was used as an internal standard in a Ca determination. A sample known to contain 2.47 μg Cu/mL gave $A_{324.7} = 0.269$ and $A_{422.7} = 0.218$. Calculate the concentration of Ca in micrograms per milliliter.

21-16. In Figure 21-21, a sample containing 1.00 μg Fe/mL gives an absorbance of 0.055. Estimate the sensitivity of this spectrometer for Fe.

21-17. Fluorescence excitation spectra of Mn and Mn + Ga solutions are shown below. The Ga line is within 0.2 nm of all three Mn lines. Explain why there is no spectral interference by Ga in the laser atomic fluorescence excitation analysis of Mn, even though the detector monochromator bandwidth is set to 1.0 nm. You may wish to refer to Exercise 21-D.

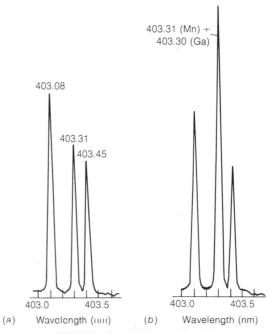

Fluorescence excitation of solution containing (a) 1 μg Mn/mL and (b) 1 μg Mn/mL plus 5 μg Ga/mL. [S. J. Weeks, H. Haraguchi, and J. D. Winefordner, *Anal. Chem.*, **50**, 360 (1978).]

22 Introduction to Analytical Separations

The vast majority of real analytical problems begins with a complex mixture from which we must isolate, identify and quantitate one or more components. The question to be answered may be qualitative ("What is that component?") or quantitative ("How much of the component is present?"). For example, a food processing plant received a shipment of plastic packaging that had a musty odor, making it unsuitable for food wrapping. The questions to be answered were, "What is causing the odor and how do we prevent it from occurring again?" The first step was to isolate the musty-smelling component from the complex mixture of volatile organic compounds that evaporate from the plastic. To do this, the plastic was heated to 100°C in a closed jar equipped with a rubber septum (a soft disk) through which gas could be removed by syringe. The gas was then injected into a gas chromatograph like the ones that we will study in the next chapter. Figure 22-1 shows 10 major components and many minor ones released from the plastic. It was found that the musty odor was associated only with the compound that emerged at 19.77 min. In subsequent experiments, this component was transferred from the chromatograph into a mass spectrometer or an infrared spectrophotometer, both of which helped suggest chemical structures. Finally, an authentic sample of the suspected compound was synthesized and was found to have the same chromatographic and spectroscopic properties—and odor—as the unknown.

It turned out that the musty unknown was 4,4,6-trimethyl-1,3-dioxane, which is produced by the reaction of 2-methyl-2,4-pentanediol with formaldehyde:

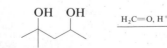

H₂C=O, H⁺ →

2-Methyl-2,4-pentanediol   4,4,6-Trimethyl-1,3-dioxane
(musty-smelling unknown)

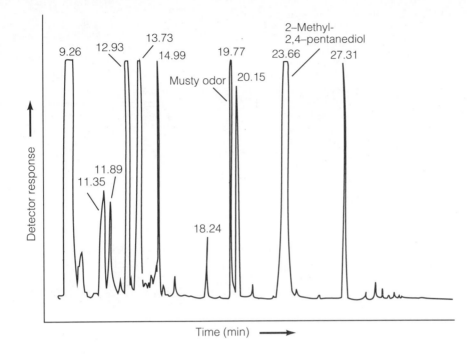

Figure 22-1
Volatile compounds released from plastic food wrapping. Each peak corresponds to a component emerging from the gas chromatography column at a different time. [From R. J. McGorrin, T. R. Pofahl, and W. R. Croasmun, *Anal. Chem.*, **59**, 1109A (1987).]

The 2-methyl-2,4-pentanediol is used as a coating to help ink adhere to the plastic film. In this particular lot, excess coating remained and apparently reacted with formaldehyde (from unknown sources) during storage. Knowing how the musty odor developed, it was possible for the plastic manufacturer to prevent its occurrence in the future.

Isolating the desired unknown from a mixture is a challenging and important step in the analytical process. In this chapter we will discuss fundamentals of separations and in the next chapter we will describe important separation methods.

22-1 SOLVENT EXTRACTION

In its simplest form, **extraction** refers to the transfer of a solute from one liquid phase to another. The most common case is the extraction of an aqueous solution with an organic solvent. Diethyl ether, benzene, and other hydrocarbons are common solvents that are less dense than water and form a phase that sits on top of the aqueous phase (Color Plate 15). Chloroform, dichloromethane, and carbon tetrachloride are common solvents that are *immiscible* with and denser than water.[†] In a two-phase mixture, some of each solvent is found in *both* phases, but one phase is predominantly water and the other phase is predominantly organic. The volumes of each phase after mixing are not exactly equal to the volumes that were mixed. For simplicity, however, we will assume that the volumes of each phase are not changed by mixing.

To be **miscible** means that the two liquids form a single phase when they are mixed in any ratio. To be *immiscible* means that each liquid remains in a separate phase (Color Plate 15).

[†] Whenever a choice exists between $CHCl_3$ and CCl_4, the less toxic $CHCl_3$ should be tried. Also, toluene is greatly preferred over benzene (a carcinogen), if that choice must be made.

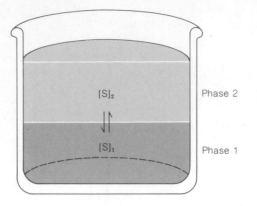

Figure 22-2
Partitioning of a solute between two liquid phases.

Suppose that solute S is partitioned between phases 1 and 2 (Figure 22-2). The **partition coefficient**, K, is the equilibrium constant for the reaction

$$S \text{ (in phase 1)} \rightleftharpoons S \text{ (in phase 2)} \tag{22-1}$$

$$K = \frac{\mathscr{A}_{S_2}}{\mathscr{A}_{S_1}} \approx \frac{[S]_2}{[S]_1} \tag{22-2}$$

where $\mathscr{A}_{S_1}$ refers to the activity of solute in phase 1. In dilute solution, the ratio of activities can be replaced with a ratio of concentrations.

In more concentrated solutions, which are often encountered, the concentration could deviate markedly from the activity. Nonetheless, lacking knowledge of the activity coefficients, we will write the partition coefficient in terms of concentrations.

Suppose that a volume V_1 of solvent 1 containing m moles of solute S is extracted with volume V_2 of solvent 2. Let q be the fraction of S remaining in phase 1 when equilibrium is achieved. The molarity in phase 1 will therefore be qm/V_1. The fraction of total solute transferred to phase 2 will be $(1 - q)$, and the molarity in phase 2 is $(1 - q)m/V_2$. Putting these values of molarity into Equation 22-2 gives

$$K = \frac{(1 - q)m/V_2}{qm/V_1} \tag{22-3}$$

from which we can solve for q:

$$q = \frac{V_1}{V_1 + KV_2} \tag{22-4}$$

Equation 22-4 says that the fraction of solute remaining in phase 1 depends on the partition coefficient and the two volumes. If the phases are separated and a new quantity of fresh solvent 2 is mixed with phase 1, the fraction of solute remaining in phase 1 at equilibrium will be

Example of Equation 22-5: If $q = \frac{1}{4}$, then $\frac{1}{4}$ of the solute remains in phase 1 after one extraction. A second extraction will reduce the concentration to $\frac{1}{4}$ of the value after the first extraction $= (\frac{1}{4})(\frac{1}{4}) = \frac{1}{16}$ of the initial concentration.

$$\begin{array}{c} \text{Fraction remaining after} \\ \text{two extractions} \end{array} = q \cdot q = \left(\frac{V_1}{V_1 + KV_2}\right)^2 \tag{22-5}$$

After n extractions with volume V_2, the fraction remaining in phase 1 is

$$\begin{array}{c} \text{Fraction remaining after} \\ n \text{ extractions} \end{array} = q^n = \left(\frac{V_1}{V_1 + KV_2}\right)^n \tag{22-6}$$

EXAMPLE: Extraction Efficiency

Solute A has a partition coefficient of 3 between benzene and water (with 3 times as much in the benzene phase.) Suppose that 100 mL of a 0.01 M aqueous solution of A is extracted with benzene. What fraction of A remains in the aqueous phase (a) if one extraction with 500 mL is performed and (b) if five extractions with 100 mL are performed?

(a) Taking water as phase 1 and benzene as phase 2, Equation 22-4 says that after a 500 mL extraction, the fraction remaining in the aqueous phase is

$$q = \frac{100}{100 + (3)(500)} = 0.062 \approx 6\%$$

(b) With five 100-mL extractions, the fraction remaining is given by Equation 22-6:

$$\text{Fraction remaining} = \left(\frac{100}{100 + (3)(100)} \right)^5 = 0.000\,98 \approx 0.1\%$$

Many small extractions are much more effective than a few large extractions.

It is much more efficient to do several small extractions than one big extraction.

pH Effects

Suppose that the solute being partitioned between phases 1 and 2 is an amine with a base constant K_b. Let's also assume that BH^+ is soluble *only* in the aqueous phase (1). Suppose that the neutral form, B, has partition coefficient, K, between the phases. The **distribution coefficient,** D, is defined as

$$D = \frac{\text{total concentration in phase 2}}{\text{total concentration in phase 1}} \qquad (22\text{-}7)$$

which becomes

$$D = \frac{[B]_2}{[B]_1 + [BH^+]_1} \qquad (22\text{-}8)$$

Combining the relationships $K = [B_2]/[B_1]$ and $K_a = [H^+][B]/[BH^+] = K_w/K_b$ with Equation 22-8 leads to

$$D = \frac{K \cdot K_a}{K_a + [H^+]} = K \cdot \alpha_{\text{neutral}} \qquad (22\text{-}9)$$

where α_{neutral} is the fraction of weak base in the neutral form, B. (The fraction of an acid or base in each protonated form was given in Section 12-2.) Equation 22-9 tells us that the distribution of solute between the two phases will be pH-dependent.

EXAMPLE: Effect of pH on Extraction

Suppose that $K = 3.0$ and $K_a = 1.0 \times 10^{-9}$. If 50 mL of 0.010 M aqueous amine is extracted with 100 mL of solvent 2, what will be the formal concentration remaining in the aqueous phase (a) at pH 10.00 and (b) at pH 8.00?

(a) At pH 10.00, $D = (3.0)(1.0 \times 10^{-9})/(1.0 \times 10^{-9} + 1.0 \times 10^{-10}) = 2.73$ (from Equation 22-9). Equation 22-4 says that the fraction remaining in the aqueous phase will be

$$q = \frac{50}{50 + (2.73)(100)} = 0.15 \quad \Rightarrow \quad 15\% \text{ left in water}$$

The concentration of amine in the aqueous phase will be 15% of 0.010 M = 0.001 5 M. *In the preceding equation, we have used the distribution coefficient, D, in place of the partition coefficient, K, in Equation 22-4.*

(b) At pH 8.00, $D = (3.0)(1.0 \times 10^{-9})/(1.0 \times 10^{-9} + 1.0 \times 10^{-8}) = 0.273$. This time we find

$$q = \frac{50}{50 + (0.273)(100)} = 0.65 \quad \Rightarrow \quad 65\% \text{ left in water}$$

The concentration in the aqueous phase is 0.006 5 M. At pH 10, the base is predominantly in the form B and is extracted into the organic solvent. At pH 8, it is in the form BH^+ and remains in the water.

Figure 22-3 shows the effect of pH on the distribution ratio in the above example. When you want to extract a base into water, it is clearly desirable to use a low enough pH to convert it to BH^+. By the same reasoning, to extract an acid (HA) into water, you should use a high enough pH to convert the acid to A^-.

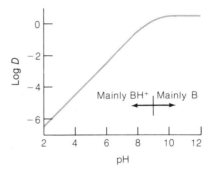

Figure 22-3
Effect of pH on the distribution coefficient for the extraction of a base into an organic solvent. In this example, $K = 3.0$ and $K_b = 1.0 \times 10^{-5}$.

Equation for the distribution of HA between two phases.

Challenge: Suppose that the acid HA (with dissociation constant K_a) is partitioned between aqueous phase 1 and organic phase 2. Calling the partition coefficient K for HA and assuming that A^- is not soluble in the organic phase, show that the distribution coefficient is given by

$$D = \frac{K \cdot [H^+]}{[H^+] + K_a} = K \cdot \alpha_{neutral} \tag{22-10}$$

where $\alpha_{neutral}$ is the fraction of weak acid in the neutral form, HA.

Extraction with a Metal Chelator

One scheme for separating metal ions from each other is to selectively complex one ion using an organic ligand and extract it into an organic solvent. Three ligands commonly employed for this purpose are shown below.

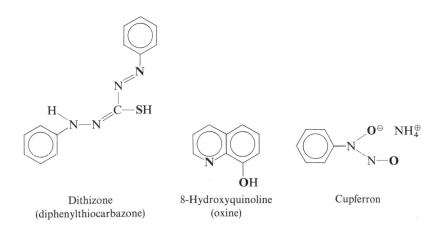

Dithizone
(diphenylthiocarbazone)

8-Hydroxyquinoline
(oxine)

Cupferron

Each ligand can be represented as a weak acid, HL, which loses one proton when it binds to a metal ion through the atoms shown in bold type.

$$HL(aq) \rightleftharpoons H^+(aq) + L^-(aq) \qquad K_a = \frac{[H^+]_{aq}[L^-]_{aq}}{[HL]_{aq}} \qquad (22\text{-}11)$$

$$nL^-(aq) + M^{n+}(aq) \rightleftharpoons ML_n(aq) \qquad \beta = \frac{[ML_n]_{aq}}{[M^{n+}]_{aq}[L^-]_{aq}^n} \qquad (22\text{-}12)$$

Each of these ligands can react with many different metal ions, but some selectivity is achieved by controlling the pH. Most complexes that can be extracted into organic solvents must be neutral. Charged complexes, such as $Fe(EDTA)^-$ or $Fe(1,10\text{-phenanthroline})_3^{2+}$, are not very soluble in organic solvents.

Let's derive an equation for the distribution coefficient of a metal between the two phases under a particular (but common) set of circumstances. We will assume that essentially all of the metal in the aqueous phase is in the form M^{n+} and that all of the metal in the organic phase is in the form ML_n (Figure 22-4). We define the partition coefficients for ligand and complex as follows:

$$HL(aq) \rightleftharpoons HL(org) \qquad K_L = \frac{[HL]_{org}}{[HL]_{aq}} \qquad (22\text{-}13)$$

$$ML_n(aq) \rightleftharpoons ML_n(org) \qquad K_M = \frac{[ML_n]_{org}}{[ML_n]_{aq}} \qquad (22\text{-}14)$$

where *org* refers to the organic phase.

The distribution coefficient we are seeking is

$$D = \frac{[\text{total metal}]_{org}}{[\text{total metal}]_{aq}} \approx \frac{[ML_n]_{org}}{[M^{n+}]_{aq}} \qquad (22\text{-}15)$$

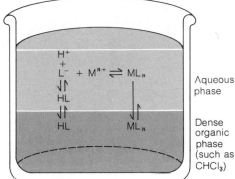

Figure 22-4
Extraction of a metal ion with a chelator. It is assumed that the predominant form of metal in the aqueous phase is M^{n+}, and the predominant form in the organic phase is ML_n.

We assume that M^{n+} is in the aqueous phase and that ML_n is in the organic phase.

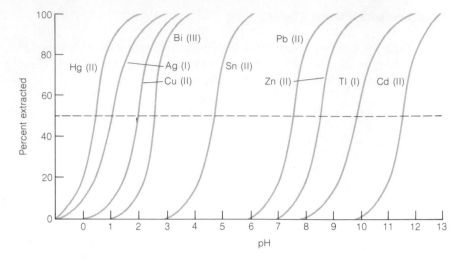

Figure 22-5
Extraction of metal ions by dithizone into CCl_4. [Adapted from G. H. Morrison and H. Freiser in C. L. Wilson and D. Wilson, eds., *Comprehensive Analytical Chemistry,* Vol. IA (New York: Elsevier, 1959).]

From Equations 22-14 and 22-12, we can write

$$[ML_n]_{org} = K_M[ML_n]_{aq} = K_M\beta[M^{n+}]_{aq}[L^-]^n_{aq} \qquad (22\text{-}16)$$

Using the value of $[L^-]_{aq}$ from Equation 22-11 gives

$$[ML_n]_{org} = \frac{K_M\beta[M^{n+}]_{aq}K_a^n[HL]^n_{aq}}{[H^+]^n_{aq}} \qquad (22\text{-}17)$$

Putting this value of $[ML_n]_{org}$ into Equation 22-15 gives

$$D \approx \frac{K_M\beta K_a^n[HL]^n_{aq}}{[H^+]^n_{aq}} \qquad (22\text{-}18)$$

Since most of the HL is usually in the organic phase, we can use Equation 22-13 to rearrange Equation 22-18 to its most useful form:

$$D \approx \frac{K_M\beta K_a^n}{K_L^n}\frac{[HL]^n_{org}}{[H^+]^n_{aq}} \qquad (22\text{-}19)$$

See Problem 22-18 for an alternative view of Equation 22-19.

Distribution of the metal between the two phases is pH-dependent. You can select a pH to bring the metal into either phase.

Equation 22-19 says that the distribution coefficient for metal ion extraction depends on the pH and the ligand concentration. Since the various equilibrium constants are different for each metal, it is often possible to select a pH where D is large for one metal and small for another. For example, Figure 22-5 shows that Cu^{2+} could be separated from Pb^{2+} and Zn^{2+} by extraction with dithizone at pH 5. Demonstration 22-1 illustrates the pH-dependence of an extraction with dithizone.

Some Extraction Strategies

Separation by extraction is very much an art, with clever ideas and trial-and-error discovery of conditions (solvents, pH, chelators, etc.) being necessary. We will simply mention a few strategies that have been used in successful separations.

Demonstration 22-1 EXTRACTION WITH DITHIZONE

Dithizone (diphenylthiocarbazone) is a green compound, soluble in nonpolar organic solvents and insoluble in water below pH 7. It forms red, hydrophobic complexes with most di- and trivalent metal ions.

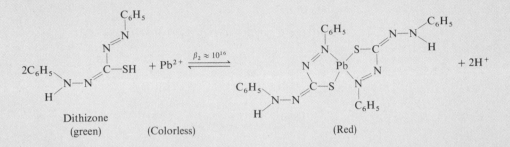

Dithizone
(green)
(Colorless)
(Red)

Dithizone is widely used for analytical extractions, for colorimetric determinations of metal ions, and for removing traces of metals from aqueous buffers.

In the latter application, an aqueous buffer is extracted repeatedly with a green solution of dithizone in $CHCl_3$ or CCl_4. As long as the organic phase turns red, metal ions are being extracted from the buffer. When the extracts are green, the last traces of metal ions have been removed. Elements that react with dithizone include Mn, Fe, Co, Ni, Pd, Pt, Cu, Ag, Au, Zn, Cd, Hg, Ga, In, Tl, Sn, Pb, Bi, Se, Te, and Po. In Figure 22-5 we see that only certain metal ions can be extracted at a given pH.[†]

The equilibrium between the green ligand and red complex is nicely demonstrated using three large test tubes sealed with tightly fitting rubber stoppers. In each tube is placed some hexane plus a few milliliters of dithizone solution (prepared by dissolving 1 mg of dithizone in 100 mL of $CHCl_3$). To tube A is added distilled water; to tube B is added tap water; and to tube C is added 2 mM $Pb(NO_3)_2$. After shaking and settling, tubes B and C contain a red upper phase, while A remains green.

The proton equilibrium indicated by the reaction above is shown by adding a few drops of 1 M HCl to tube C. After shaking, the dithizone turns green again. Competition with a stronger ligand is shown by adding a few drops of 0.05 M EDTA solution to tube B. Again, shaking causes a reversion to the green color.

[†] An experiment using a pH-dependent separation of radioactive $^{212}Bi^{3+}$ from $^{212}Pb^{2+}$ with dithizone at pH $\approx$ 3 has been described by D. M. Downey, D. D. Farnsworth, and P. G. Lee [*J. Chem. Ed.,* **61,** 259 (1984)].

Sometimes **ion pairs** can be extracted from an aqueous phase into an organic phase. For example, $FeCl_4^-$ ion can be extracted from 6 M HCl into diethyl ether as a tightly associated unit that can be written $FeCl_4^- \cdot H^+$, with no net charge. The distribution coefficient is a delicate function of conditions, with too much or too little HCl leading to incomplete extraction of $FeCl_4^- \cdot H^+$. The optimum conditions can be found only by trial and error.

An ion that forms an extractable complex with a nonselective chelator might be isolated by masking other ions in the sample. For example, citrate and tartrate form strong polar complexes with many ions. These complexes would remain in the aqueous phase, while the desired ion might be extracted with another chelator. Clearly, the relative stability constants are all-important in this scheme.

An ion pair is a pair of ions held closely together by electrostatic attraction. In solvents with a low dielectric constant, ions tend to aggregate in pairs.

Some extraction techniques:

1. extraction of ion pairs
2. masking interfering ions
3. use of hydrophobic counterion
4. use of hydrophobic ligand

Sometimes a relatively polar ion can be extracted into an organic phase in the presence of a hydrophobic counterion. The tetrabutylammonium cation $[(C_4H_9)_4N^+]$ is commonly used for this purpose. Another way to extract a metal ion is to use a nonpolar molecule with one good ligand atom that may occupy one coordination site of a metal ion, thus pulling it into the organic phase. Trioctylphosphine oxide and dioctylsulfoxide are examples of hydrophobic ligands that can interact through their oxygen atoms with cations:

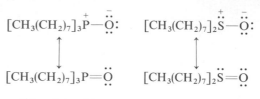

Trioctylphosphine   Dioctylsulfoxide
oxide

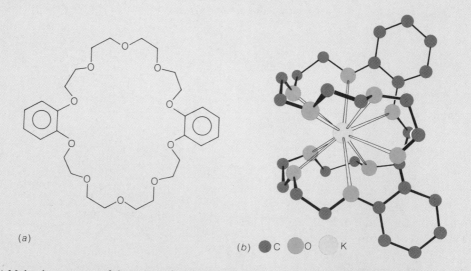

Box 22-1 CROWN ETHERS AND IONOPHORES

The problem of extracting metal ions into a nonpolar phase is important to living cells and synthetic chemists, as well as to analytical chemists. A class of synthetic compounds called *crown ethers* is able to envelop metal ions (especially alkali metal cations) in a pocket of oxygen ligands. The outside of the crown ether is very hydrophobic and soluble in nonpolar solvents. Crown ethers are used as **phase transfer catalysts.** They can extract a water-soluble ionic reagent into a nonpolar solvent, where reaction with a hydrophobic compound can occur. The structure of the potassium complex of dibenzo-30-crown-10 below shows that the K^+ ion is engulfed by ten oxygen atoms with K—O distances averaging 288 pm. Only the hydrophobic outside of the complex is exposed to the solvent.

(a)

(b) ● C   ● O   ○ K

(a) Molecular structure of the crown ether designated dibenzo-30-crown-10. (b) The three-dimensional structure of its K^+ complex. [Adapted from M. A. Bush and M. R. Truter, *J. Chem. Soc. Chem. Commun.,* 1439 (1970).]

Box 22-1 *(continued)*

Living cells must transport hydrophilic molecules through very hydrophobic membranes. For example, a nerve impulse requires that K^+ and Na^+ ions flow across the cell membrane of a neuron. The carriers of these ions appear to be an integral part of the membrane.

A class of antibiotics called **ionophores** works the same way as the crown ethers. Such compounds as nonactin, valinomycin, gramicidin, and nigericin alter the permeability of bacterial cells to metal ions and thus disrupt their metabolism. The structure of nonactin is shown below.

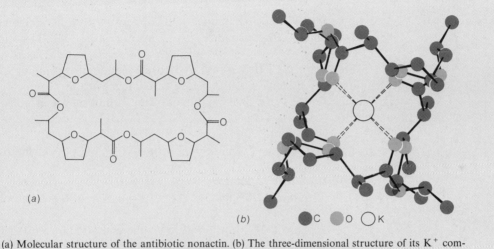

(a) Molecular structure of the antibiotic nonactin. (b) The three-dimensional structure of its K^+ complex, which can penetrate hydrophobic barriers. [Adapted from *Biochemistry* by Lubert Stryer. W. H. Freeman and Company. Copyright © 1975.]

Crown ethers and ionophores, described in Box 22-1, are specifically designed to extract metal ions into nonpolar solvents.

22-2 COUNTERCURRENT DISTRIBUTION

Countercurrent distribution is a serial extraction process devised by L. C. Craig in 1949. The process is a powerful improvement of liquid–liquid extraction. Although countercurrent distribution has been almost totally supplanted by chromatographic methods of separation, its theory is worth studying because it provides a basis for understanding chromatography. Automated countercurrent separation apparatus using a centrifugal field to help separate phases is still the method of choice for certain large-scale separations of delicate drugs and biological molecules.[†]

The object of countercurrent distribution is to separate two or more solutes from each other by a series of partitions between two liquid phases. The scheme

---

[†] J. Cazes and K. Nunogaki, *Am. Lab.,* February 1987, p. 126.

is shown in Figure 22-6. In Step 0, a simple extraction is performed. Some of each solute will appear in each phase. A necessary condition for separation is that the distribution coefficients for the two solutes be different. For example, suppose that the two solutes distribute themselves in the following fashion:

$$D_A = \frac{[A]_{\text{upper phase}}}{[A]_{\text{lower phase}}} = 4 \qquad D_B = \frac{[B]_{\text{upper phase}}}{[B]_{\text{lower phase}}} = 1 \qquad (22\text{-}20)$$

If 1 mmol of each solute is present, the quantities present in each phase upon equilibration in Step 0 of Figure 22-6 will be

A: upper phase (U0)—0.8 mmol   B: upper phase (U0)—0.5 mmol
     lower phase (L0)—0.2 mmol   lower phase (L0)—0.5 mmol

In Step 1, phase U0 is transferred to a second tube containing *fresh* lower phase, L1. Likewise, L0 is placed in contact with *fresh* upper phase, U1. When both tubes are shaken, the following equilibria are established:

A: U1—0.16 mmol ⎱ total =   B: U1—0.25 mmol ⎱ total =
    L0—0.04 mmol ⎰ 0.2 mmol   L0—0.25 mmol ⎰ 0.5 mmol

    U0—0.64 mmol ⎱ total =   U0—0.25 mmol ⎱ total =
    L1—0.16 mmol ⎰ 0.8 mmol   L1—0.25 mmol ⎰ 0.5 mmol

In Step 2, phase U0 is transferred to a tube containing *fresh* L2. Phase U1 is transferred to the tube containing *old* L1. *Fresh* phase U2 is placed in the tube containing *old* L0. After equilibration, solute A will have a ratio 4:1 (upper:lower) in each tube, and solute B will have a ratio 1:1 (upper:lower) in each tube. The process is followed for two steps in Table 22-1. The percent

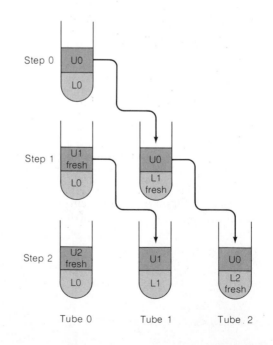

Figure 22-6
The scheme of extractions in countercurrent distribution.

Table 22-1
Countercurrent distribution of two solutes with distribution coefficients $D_A = 4$ and $D_B = 1$

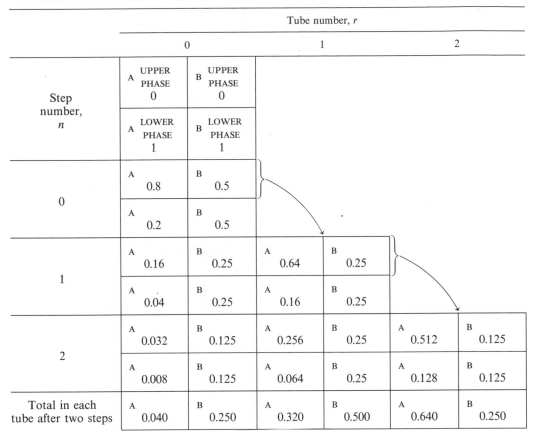

	Tube number, r					
	0		**1**		**2**	
Step number, n	A UPPER PHASE 0	B UPPER PHASE 0				
	A LOWER PHASE 1	B LOWER PHASE 1				
0	A 0.8	B 0.5				
	A 0.2	B 0.5				
1	A 0.16	B 0.25	A 0.64	B 0.25		
	A 0.04	B 0.25	A 0.16	B 0.25		
2	A 0.032	B 0.125	A 0.256	B 0.25	A 0.512	B 0.125
	A 0.008	B 0.125	A 0.064	B 0.25	A 0.128	B 0.125
Total in each tube after two steps	A 0.040	B 0.250	A 0.320	B 0.500	A 0.640	B 0.250

of each solute in each tube (total of upper + lower phase) after one, three, and five steps is shown in Figure 22-7.

What Figure 22-7 shows is that both solutes A and B are carried to the right in successive steps. However, solute A moves faster than B because of the larger distribution coefficient of A. The net result is that after several steps A and B begin to separate from each other. For example, the mixture initially contained 1 mmol each of A and B. After five steps of countercurrent distribution, we might pool tubes 0, 1, and 2 in one flask, and tubes 4 and 5 in a second flask. The first flask would contain 0.5 mmol of B and 0.057 92 mmol of A. We would have a 50% recovery of 89% pure B. Tubes 4 and 5 would contain 0.737 28 mmol of A plus 0.187 5 mmol of B. This fraction gives a 74% recovery of 80% pure A. Instead of stopping the procedure after five steps, it could be continued to obtain further separation of A from B.

Color Plate 16 shows an experiment very much like that in Figure 22-7. A countercurrent extraction was performed beginning with 2 mL of 0.1 M aqueous Na_2CO_3 (saturated with 1-butanol) containing 40 μg of phenol red and 50 μg of bromocresol green. This lower phase was extracted with an upper phase of butanol (saturated with 0.1 M Na_2CO_3) to produce the 9-tube distribution shown. The bromocresol green is extracted readily by the butanol

Challenge: Verify the numbers for Step 2 in Table 22-1.

Purity is expressed as $\dfrac{[B]}{[A] + [B]}$.

The solute with the larger distribution coefficient moves faster.

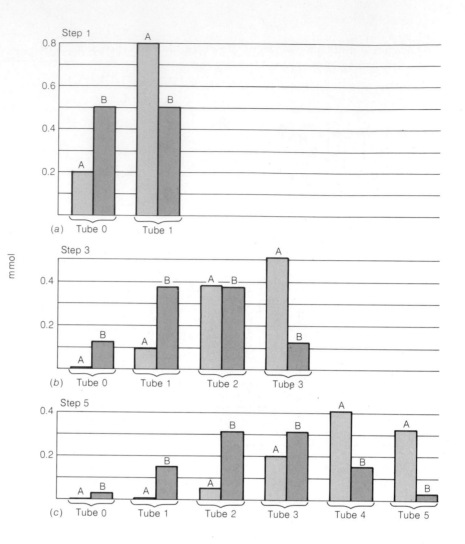

Figure 22-7
Total of each solute in each tube after (a) one, (b) three, and (c) five steps of countercurrent distribution.

and moves rapidly to the right in the upper phase. The phenol red is more soluble in the aqueous phase and moves very slowly to the right.[†]

Theoretical Distribution

Let's now develop a few equations describing the distribution of solute in the fractions of a countercurrent distribution. The pattern of countercurrent extractions is shown below.

U9 fresh	U8 fresh	U7 fresh	U6 fresh	U5	U4	U3	U2	U1	U0	→				Mobile phase
			L0	L1	L2	L3	L4	L5	L6 fresh	L7 fresh	L8 fresh	L9 fresh		Stationary phase

[†] This extraction can be used for a nice experiment on countercurrent distribution. See B. Arreguín, J. Padilla, and J. Herrán, *J. Chem. Ed.,* **39,** 539 (1962). A countercurrent separation of ink pigments is the subject of another experiment by C. E. Bricker and G. T. Sloop, *J. Chem. Ed.,* **62,** 1109 (1985).

The upper phase, which can be thought of as sliding along a fixed lower phase, will be called the **mobile phase.** The lower phase is called the **stationary phase.**

Consider a *single solute* in phase L0. Before equilibration with U0, the concentration in L0 will be called 1, and the concentration in U0 is 0.

Before zeroth equilibration

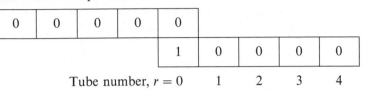

Tube number, $r = 0$ 1   2   3   4

After the zeroth equilibration, we will call the *fraction of solute in the upper phase p* and the *fraction in the lower phase q.* Obviously, $p + q = 1$. The situation after the zeroth equilibration looks like this:

After zeroth equilibration

From Equation 22-20, we can say that $p = D/(D + 1)$.

Tube number, $r = 0$ 1   2   3   4

Now we advance the mobile phase by one tube. Calling this first advance $n = 1$, we find p in tube 1 and q in tube 0:

Before first equilibration

($n = 1$ in Table 22-2)

Tube number, $r = 0$ 1   2   3   4

After equilibration of the phases, a fraction p must be in each upper phase, and a fraction q must be in each lower phase. Since tube 0 has a total quantity q, the quantities in the upper and lower phases are $p \cdot q$ and $q \cdot q$, respectively. The situation looks like this:

After first equilibration

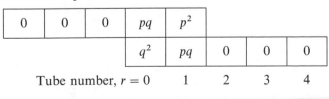

Tube number, $r = 0$ 1   2   3   4

Upon making a second advance (defined as $n = 2$), the distributions before and after equilibration are as follows:

Before second equilibration

0	0	0	pq	p^2	($n = 2$ in Table 22-2)

	q^2	pq	0	0	0

Tube number, $r = 0 \quad 1 \quad 2 \quad 3 \quad 4$

After second equilibration

0	0	pq^2	$2p^2q$	p^3

	q^3	$2pq^2$	p^2q	0	0

Tube number, $r = 0 \quad 1 \quad 2 \quad 3 \quad 4$

Consider tube $r = 1$. After the second advance, it has a total concentration of $2pq$. The upper phase will have concentration $p \cdot 2pq$ and the lower phase will have concentration $q \cdot 2pq$.

The fraction of solute in each tube upon each advance is shown in Table 22-2. You may recognize the coefficients in Table 22-2 as the binomial expansion $(q + p)^n$. The fraction of solute, f, in each tube, r, at each step, n, is given by

$$f = \frac{n!}{(n - r)!r!} p^r q^{n-r} \tag{22-21}$$

The symbol $n!$ is read "n factorial" and means the product $n(n - 1)(n - 2) \cdots (1)$. For example, $6! = 6 \cdot 5 \cdot 4 \cdot 3 \cdot 2 \cdot 1 = 720$. The quantity $0!$ is defined as 1.

EXAMPLE: Fraction of Solute in Each Tube

Use Equation 22-21 to verify the quantity of each solute in tube 1 ($r = 1$) after two steps ($n = 2$) in Table 22-1.

For solute A, the distribution coefficient is 4 (Equation 22-20). This means that $p = \frac{4}{5}$ and $q = \frac{1}{5}$. Putting these values into Equation 22-21 gives:

$$f_A = \frac{2!}{(2 - 1)!1!} \left(\frac{4}{5}\right)^1 \left(\frac{1}{5}\right)^{2-1} = \frac{(2 \cdot 1)}{(1)(1)} \left(\frac{4}{5}\right)^1 \left(\frac{1}{5}\right)^1 = 0.320$$

which is the value in Table 22-1.

For solute B, the distribution coefficient is 1 (Equation 22-20). Setting $p = q = \frac{1}{2}$ gives

$$f_B = \frac{2!}{(2 - 1)!(1)} \left(\frac{1}{2}\right)^1 \left(\frac{1}{2}\right)^{2-1} = 0.500$$

Table 22-2
Fraction of solute in each tube after each advance

Step number, n	Tube number, r				
	0	1	2	3	4
1	q	p			
2	q^2	$2pq$	p^2		
3	q^3	$3pq^2$	$3p^2q$	p^3	
4	q^4	$4pq^3$	$6p^2q^2$	$4p^3q$	p^4

Peak Position, Bandwidth, and Resolution

-2 COUNTERCURRENT DISTRIBUTION

The binomial distribution in Equation 22-21 closely approximates a Gaussian distribution when n and r are large:

$$f \approx \frac{1}{\sqrt{npq}\sqrt{2\pi}} e^{-(r-np)^2/2npq} \qquad (22\text{-}22)$$

At this point, it is worth recalling the Gaussian distribution, Equation 4-4:

$$y = \frac{1}{\sigma\sqrt{2\pi}} e^{-(x-\mu)^2/2\sigma^2}$$

The standard deviation is σ, and the mean is μ. By analogy between Equations 22-22 and 4-4, we see that the "standard deviation" of a solute band in countercurrent distribution is

$$\text{``}\sigma\text{''} \approx \sqrt{npq} \qquad (22\text{-}23)$$

For the solute with fraction p in the mobile phase, the tube (r_{max}) with maximum solute concentration is

$$r_{max} \approx np \qquad (22\text{-}24)$$

We are now in a position to calculate the degree of overlap of two bands. If the distribution coefficients are $D_A = 4$ and $D_B = 1$, the following parameters apply when $n = 100$ steps:

Solute A: $p = \frac{4}{5}$ Solute B: $p = \frac{1}{2}$

$q = \frac{1}{5}$ $q = \frac{1}{2}$

$\sigma_A = \sqrt{npq} = 4$ $\sigma_B = \sqrt{npq} = 5$

$r_{max} = np = 80$ $r_{max} = np = 50$

A graph of the Gaussian curves describing these two bands is shown in the upper part of Figure 22-8. In a real experiment, each band would undoubtedly be less pure than we have calculated because we are likely to leave some of each phase behind every time a phase is transferred from one tube to the next.

The greater the number of separation steps, the broader is the band of solute.

$r_{max} \approx np$ means that the solute peak has always traveled a constant fraction of the distance that the mobile phase has traveled.

In real separations, bands tend to be more spread at the extremes (especially at the rear of the band) than an ideal calculation would predict.

EXAMPLE: Countercurrent Separation

The top portion of Figure 22-8 shows the separation of solutes A and B after 100 steps when $D_A = 4$ and $D_B = \frac{7}{3}$. If fractions 59–72 and 81–90 are each pooled, find the percent recovery and purity of each component.

For these parameters, we calculate the following:

Solute A: $p = 0.80$ Solute B: $p = 0.70$

$q = 0.20$ $q = 0.30$

$\sigma_A = 4$ $\sigma_B = 4.58$

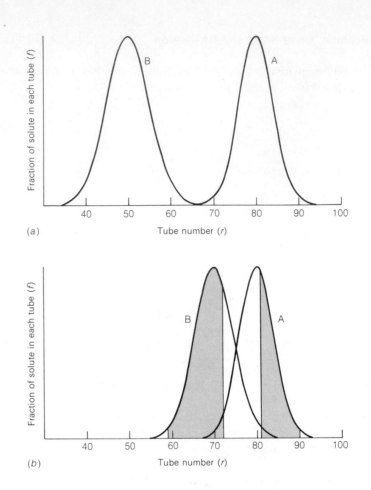

Figure 22-8
Countercurrent distribution separation of solutes A and B using 100 steps. (a) The distribution coefficients are $D_A = 4$ and $D_B = 1$. (b) $D_A = 4$ and $D_B = \frac{7}{3}$. Each Gaussian curve is normalized to the same maximum height.

Tube 70 is r_{max} for solute B. To find the fraction in tubes 59–72, we divide the curve into two portions expressed as multiples of the standard deviation (4.58), and add their areas:

$$z = \frac{|r - r_{max}|}{\sigma} \quad \text{(Equation 4-5)}$$

Tubes:	59–70	70–72	
Width:	$z = \dfrac{11}{4.58} = 2.40$	$z = \dfrac{2}{4.58} = 0.437$	
Area:	$\underbrace{\qquad 0.491\,8 \qquad}$	$0.168\,8$	$\begin{cases} \text{From interpolation} \\ \text{in Table 4-1} \end{cases}$

Total area $= 0.491\,8 + 0.168\,8 = 0.660\,6$

That is, tubes 59–72 contain 66% of B. The fraction of B to the right of tube 81 is $0.500\,0 - 0.491\,8 = 0.008\,2$, because for tube 81 $z = 11/4.58 = 2.40$.

The calculations for band A look like this:

Tubes:	81–90	80–81	
Width:	$z = \dfrac{9}{4} = 2.25$	$z = \dfrac{1}{4} = 0.25\sigma_A$	
Area:	$\underbrace{\qquad 0.487\,6 \qquad}$	$0.098\,6$	$\begin{cases} \text{From interpolation} \\ \text{in Table 4-1} \end{cases}$

Total area $= 0.487\,6 - 0.098\,6 = 0.389\,0$

↑

Note the subtraction

For tube 72 $z = \frac{8}{4} = 2$. Therefore the fraction of A lying on the left of tube 72 is 0.022 7.

The final results for this separation are

Tubes 59–72		Tubes 81–90	
Fraction of B	66.06%	Fraction of A	38.90%
Fraction of A	2.27%	Fraction of B	0.82%
Purity	96.68%	Purity	97.94%

Band spreading

Important insight concerning countercurrent separations can be gained from Equations 22-23 and 22-24:

1. Since $r_{max} \approx np$, each solute migrates a distance equal to a constant fraction of the solvent "front" (the position to which fresh mobile phase moves).
2. The bandwidth increases with the square root of the number of separation steps, because $\sigma \approx \sqrt{npq}$. The farther the front has moved, the larger is n and the broader will be each solute band.
3. Increased separation is afforded when n is increased, because the distance that each band travels is proportional to n, but the band spreading is only proportional to $\sqrt{n}$.

Purification is possible because separation is proportional to n but spreading is only proportional to $\sqrt{n}$.

These physical properties apply to all forms of partition chromatography, as well as to countercurrent distribution.

22-3 CHROMATOGRAPHY

Chromatography is a logical extension of countercurrent distribution. Instead of a series of discrete extractions being performed, a continuous equilibration of solute between two phases occurs. In chromatography, the mobile phase is either a liquid or a gas. The stationary phase is most commonly a liquid coated on the inside of a capillary tube or on the surface of solid particles used to pack a column. Alternatively, the solid particles themselves may serve as a stationary phase. In any case, the partitioning of solutes between the mobile and stationary phases accounts for the separation of solutes, just as it does in countercurrent distribution. This is shown schematically in Figure 22-9. The solute that has a greater affinity for the stationary phase will move through the column more slowly.

As it is currently practiced, chromatography is divided into several classes (Figure 22-10):

1. **Adsorption chromatography.** The oldest form of chromatography, this makes use of a solid stationary phase and a liquid or gaseous mobile phase. Solute can be adsorbed on the surface of the solid particles. Equilibration between the adsorbed state and the solution accounts for the separation of solute molecules.
2. **Partition chromatography.** In this technique, a stationary phase forms a thin film on the surface of a solid support. Solute equilibrates between this stationary liquid and a liquid or gaseous mobile phase.

In 1903, M. Tswett first applied adsorption chromatography to the separation of plant pigments, using a hydrocarbon solvent and $CaCO_3$ stationary phase. The separation of colored bands led to the name *chromatography*, from the Greek word *chromatos*, meaning color.

For their pioneering work on liquid–liquid partition chromatography in 1941, A. J. P. Martin and R. L. M. Synge received the Nobel Prize in 1954.

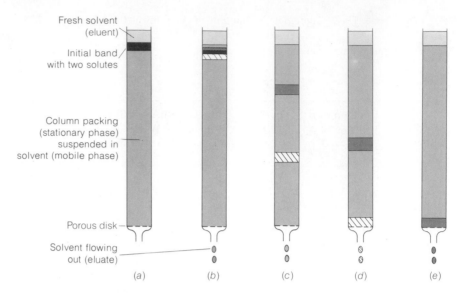

Fresh solvent (eluent)

Initial band with two solutes

Column packing (stationary phase) suspended in solvent (mobile phase)

Porous disk

Solvent flowing out (eluate)

(a) (b) (c) (d) (e)

Figure 22-9
Schematic representation of a chromatographic separation. The solute with a greater affinity for the stationary phase remains on the column longer.

B. A. Adams and E. L. Holmes developed the first synthetic ion-exchange resins in 1935. *Resins* are relatively hard, amorphous organic solids. *Gels* are relatively soft.

Large molecules pass through the column *faster* than small molecules

3. **Ion-exchange chromatography.** Anions (such as $—SO_3^-$) or cations (such as $—N(CH_3)_3^+$) are covalently attached to the stationary solid phase, usually a *resin*, in this type of chromatography. Solute ions of the opposite charge are attracted to the stationary phase by electrostatic force. The mobile phase is a liquid.

4. **Molecular exclusion chromatography.** Unlike other forms of chromatography, there is no attractive interaction between the "stationary phase" and the solute in the ideal case of molecular exclusion chromatography. Rather, the liquid or gaseous mobile phase passes through a porous gel. The pore sizes are small enough to exclude large solute molecules, but not small ones. The large molecules stream past without entering the gel. The small molecules take longer to pass through the column because they enter the gel and therefore must flow through a larger volume before leaving the column. Also called **gel filtration** or **gel permeation** chromatography, this technique separates molecules by size, with the larger solutes passing through most quickly.

5. **Affinity chromatography.** This newest and most selective kind of chromatography utilizes highly specific interactions between one kind of solute molecule and a second molecule covalently attached (immobilized) to the stationary phase. For example, the immobilized molecule might be an antibody to a particular protein. When a crude mixture containing a thousand proteins is passed through the column, only the one protein that reacts with the antibody is bound to the column. After washing all the other solutes off the column, the desired protein is dislodged from the antibody by changing the pH or ionic strength.

Some Terminology

The mobile phase is called the **eluent.** When it emerges from the end of the column it is called the **eluate:**

Elu*ent*—in.
Elu*ate*—out.

$$\text{Eluent} \atop \text{in} \longrightarrow \overline{\underline{\text{COLUMN}}} \longrightarrow \text{eluate} \atop \text{out}$$

The process of passing liquid (or gas) through a chromatography column is called **elution.**

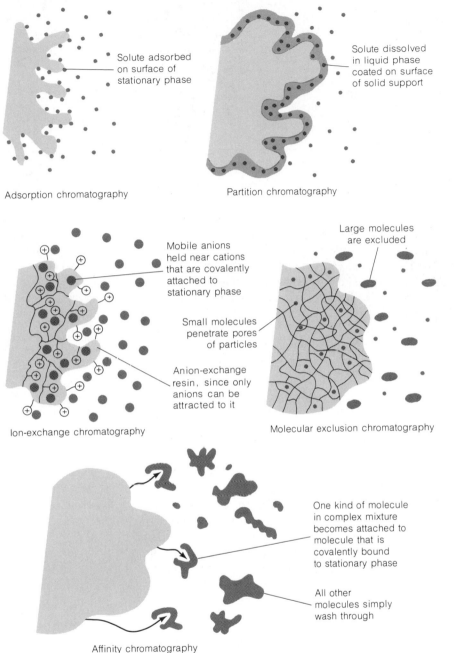

Adsorption chromatography

Solute adsorbed on surface of stationary phase

Solute dissolved in liquid phase coated on surface of solid support

Partition chromatography

Mobile anions held near cations that are covalently attached to stationary phase

Anion-exchange resin, since only anions can be attracted to it

Ion-exchange chromatography

Large molecules are excluded

Small molecules penetrate pores of particles

Molecular exclusion chromatography

One kind of molecule in complex mixture becomes attached to molecule that is covalently bound to stationary phase

All other molecules simply wash through

Affinity chromatography

Figure 22-10
Major types of chromatography.

The speed of the mobile phase through a chromatography column is commonly expressed as a volume or linear flow rate. Consider a liquid chromatography experiment in which the column has an inner diameter of 0.60 cm and the mobile phase occupies 20% of the column volume. Each centimeter of column length has a volume of $(\pi r^2 \times \text{length}) = \pi(0.30 \text{ cm})^2(1 \text{ cm}) = 0.283 \text{ mL}$, of which 20% $(= 0.056\ 5 \text{ mL})$ is mobile liquid phase (solvent). The **volume flow rate** simply tells how many milliliters of solvent per minute travel through the column. A typical rate might be 0.3 mL

Volume flow rate
= volume of solvent per minute traveling through column

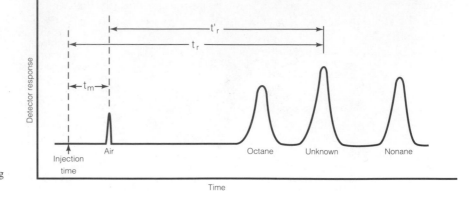

Figure 22-11
Schematic gas chromatogram showing measurement of retention times.

Linear flow rate
= length of column traveled by solvent in 1 minute

per minute. The **linear flow rate** tells how many centimeters of column length are traveled in one minute by the solvent. Since 1 cm of column length contains 0.056 5 mL, 0.3 mL would occupy $(0.3 \text{ mL})/(0.056 5 \text{ mL/cm}) = 5.3$ cm of column length. The linear flow rate corresponding to 0.3 mL/min is 5.3 cm/min.

Solutes eluted from a chromatography column are observed by various detectors described in the next chapter. A **chromatogram** is a graph showing the detector response as a function of elution time. Figure 22-11 is a schematic chromatogram that might be observed when a mixture of octane, nonane, and an unknown are separated by gas chromatography, which is described in the next chapter. The **retention time,** t_r, for each component is the time needed after injection of the mixture onto the column until that component reaches the detector. The **adjusted retention time** is defined as

$$\text{adjusted retention time} = t_r' = t_r - t_m \qquad (22\text{-}25)$$

where t_m is the time needed for the mobile phase to travel the length of the column. An unretained solute would be eluted in time t_m. When a thermal conductivity detector is used in gas chromatography, t_m is taken as the time needed for air to travel through the column (Figure 22-11). This can be determined by injecting a few microliters of air along with the sample (which is typically 0.1–10 μL). Even chromatography columns that retain the components of air usually display a discontinuity in the baseline of the chromatogram at time t_m because of the pressure burst associated with sample injection. For any two components 1 and 2, the **relative retention,** α, is

$$\text{Relative retention} = \alpha = \frac{t_{r2}'}{t_{r1}'} \qquad (22\text{-}26)$$

where $t_{r2}' > t_{r1}'$, so $\alpha > 1$.

For each peak in the chromatogram, the **capacity factor,** k, is defined as

$$\text{Capacity factor} = k = \frac{t_r - t_m}{t_m} \qquad (22\text{-}27)$$

Capacity factor is also called *retention factor, capacity ratio,* or *partition ratio.* The symbol k' is frequently used instead of k.

The longer a component is retained by the column, the greater is the capacity factor. Large capacity factors favor good separation but also increase elution time and bandwidth. To monitor the performance of a particular chromatography column, it is a good practice to measure periodically the capacity factor of a standard and the efficiency (number of plates, Equation 22-36) of the column. Changes reflect degradation of the column.

The definition of capacity factor in Equation 22-27 is equivalent to saying[†]

$$k = \frac{\text{time solute spends in stationary phase}}{\text{time solute spends in mobile phase}} \qquad (22\text{-}28)$$

Let's see why this is true. If the solute spends all of its time in the mobile phase and none in the stationary phase, it would be eluted in time t_m, by definition. Putting $t_r = t_m$ into Equation 22-27 gives $k = 0$, because solute spends no time in the stationary phase. Suppose that solute spends equal time in the stationary and mobile phases. The retention time would then be $t_r = 2t_m$ and $k = (2t_m - t_m)/t_m = 1$. If solute spends three times as much time in the stationary phase as in the mobile phase, $t_r = 4t_m$ and $k = (4t_m - t_m)/t_m = 3$.

The quotient in Equation 22-28 is equivalent to

$$\frac{\text{Time solute spends in stationary phase}}{\text{Time solute spends in mobile phase}} = \frac{C_s V_s}{C_m V_m} \qquad (22\text{-}29)$$

where V_s is the volume of stationary phase and V_m is the volume of mobile phase. The quotient C_s/C_m is the ratio of concentrations of solute in the stationary and mobile phases when solute is allowed to equilibrate with a mixture of the two phases. The quotient C_s/C_m is the *partition coefficient*, K, introduced in connection with solvent extraction. Why is Equation 22-29 true? The product $C_s V_s$ is (moles of solute per liter of stationary phase)(liters of stationary phase) = moles of solute in the stationary phase. Similarly, the product $C_m V_m$ is moles of solute in the mobile phase. The ratio of moles of solute in each phase is just the ratio of time spent by the solute in each phase. If solute spends three times as much time in the stationary phase as in the mobile phase, there will be three times as many moles of solute in the stationary phase as in the mobile phase at any time. Equation 22-27, 22-28, and 22-29 can now be combined to give

$$k = \frac{t_r - t_m}{t_m} = K\frac{V_s}{V_m} \qquad (22\text{-}30)$$

Retention volume, V_r, is the volume of mobile phase required to elute a particular solute from the chromatography column. The retention volume is just

$$V_r = t_r \cdot F \qquad (22\text{-}31)$$

where F is the volume flow rate of the mobile phase. The retention volume of a particular solute is constant over a wide range of flow rates.

Parition coefficient $= K = \dfrac{C_s}{C_m}$.
Unfortunately, this definition is the inverse of the partition coefficient (actually D) defined in countercurrent distribution.

Because $t_r' \propto k \propto K$, relative retention can also be expressed as

$$\alpha = \frac{t_{r2}'}{t_{r1}'} = \frac{k_2}{k_1} = \frac{K_2}{K_1}$$

Because volume is proportional to time, any relation expressed as a ratio of times can be written as the corresponding ratio of volumes. For example

$$k = \frac{t_r - t_m}{t_m} = \frac{V_r - V_m}{V_m}$$

[†] A related quantity is the *retention ratio*, R, defined as

$$R = \frac{\text{time for solvent to pass through column}}{\text{time for solute to pass through column}} = \frac{t_m}{t_r}$$

The relation between R and k is $R = 1/(k + 1)$.

With Equations 22-30 and 22-31 we can derive one more equation that will be useful in the next chapter. First rearrange Equation 22-30 to solve for t_r:

$$t_r = t_m \left(K \frac{V_s}{V_m} + 1 \right) \tag{22-32}$$

Multiplying both sides of Equation 22-32 by F gives

$$V_r = t_r \cdot F = \underbrace{t_m \cdot F}_{V_m} \left(K \frac{V_s}{V_m} + 1 \right) \tag{22-33}$$

But the quantity $t_m \cdot F$ is just V_m, the volume of mobile phase in the column. Replacing $t_m \cdot F$ by V_m in Equation 22-33 gives the equation we are looking for:

$$V_r = K V_s + V_m \tag{22-34}$$

Equation 22-34 will be used in the next chapter in connection with molecular exclusion chromatography.

Plate Theory of Chromatography

The simplest way to think about chromatography is to imagine that a very large number of countercurrent distribution equilibrations occur between the mobile and stationary phases as the solute travels through the column. Although chromatography is a continuous process, it can be imagined that the column is divided into N segments, in each of which one equilibration occurs. Each of these imaginary segments is called a **theoretical plate.** If the total length of the column is L, the **height equivalent to a theoretical plate** (H.E.T.P.) is

$$\text{H.E.T.P.} = \frac{L}{N} \tag{22-35}$$

It is not unusual for a chromatography column to possess several thousand theoretical plates. If a 1-m column possesses 10^4 theoretical plates, H.E.T.P. $= 1 \text{ m}/10^4 = 10^{-4}$ m.

Figure 22-12 shows an ideal chromatogram with a Gaussian band shape. By measuring the retention time (or retention volume) and the bandwidth, the number of theoretical plates can be calculated from the equation

If you want to measure the number of theoretical plates from a real chromatogram, choose a peak with a capacity factor greater than 5.

$$N = \frac{t_r^2}{\sigma^2} = \frac{16 t_r^2}{w^2} \tag{22-36}$$

where t_r is the retention time of the peak, σ is its standard deviation, and w is the width of the band (in units of time) measured at the base of the peak (as shown in Figure 22-12). This width is equal to four standard deviations for a Gaussian peak. Alternatively, if the width of the band is measured at a height equal to half of the peak height, then

$$N = \frac{5.55 t_r^2}{w_{1/2}^2} \tag{22-37}$$

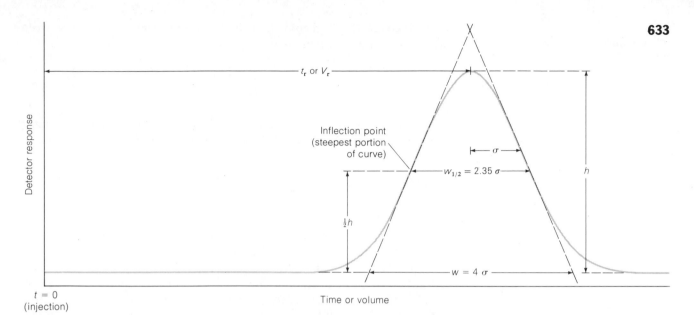

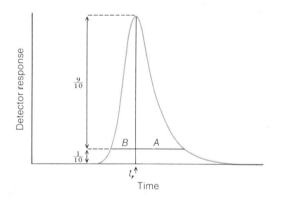

Figure 22-12

Idealized Gaussian chromatogram showing how w and $w_{1/2}$ are measured. The value w is obtained by extrapolating the tangents to the inflection points down to the baseline.

Figure 22-13

Asymmetric peak showing parameters used to estimate the number of theoretical plates.

where $w_{1/2}$ is the width at half-height. In Equations 22-36 and 22-37, retention volume may be used in place of retention time. In this case, the width is measured in units of volume instead of time.

    Equation 22-36 can be derived from the Gaussian distribution function for countercurrent distribution, using several approximations. Clearly, it is a simplification to consider a chromatography column as divided into N theoretical plates. Since a theoretical plate is an imaginary construct, Equation 22-36 can be considered to *define* a theoretical plate. Columns often behave as if they have different numbers of plates for different solutes in a mixture.

    Most peaks eluted from chromatography columns display some asymmetry, as indicated in Figure 22-13. A way to estimate the number of theoretical plates with such a peak is to draw a horizontal line across the band at a height equal to one-tenth of the maximum height. The quantities A and

> *Challenge:* If N is constant, show that the width of a chromatographic peak increases with increasing retention time. That is, successive peaks on a chromatogram should be increasingly broad.

B in Figure 22-13 can then be measured. An approximate relation between number of plates and these parameters is[†]

$$N = \frac{41.7(t_r/w_{0.1})^2}{(A/B + 1.25)} \tag{22-38}$$

where $w_{0.1}$ is the width at one-tenth-height $(= A + B)$. All quantities must be measured in the same units, such as minutes or centimeters of chart paper.

Rate Theory of Chromatography

A more realistic description of the movement of solute through a chromatography column takes into account the finite rate at which solute can equilibrate between the mobile and stationary phases. That is, equilibration is not infinitely fast (as plate theory assumes), and the resulting band shape depends on the rate of elution. The band shape is also affected by diffusion of solute along the length of the column and by the availability of different paths for different solute molecules to follow as they travel between particles of the stationary phase.

> The rate, *v*, can be a linear flow rate or a volume flow rate, because each is proportional to the other.

All of the effects just mentioned depend on the rate, *v*, at which the mobile phase passes through the column. Detailed consideration of the various mechanisms by which the solute band is broadened leads to the **van Deemter equation** for plate height:

$$\text{H.E.T.P.} = A + \frac{B}{v} + Cv \tag{22-39}$$

where *A*, *B*, and *C* are constants characteristic of a given column and solvent system. Equation 22-39 says that there will be an optimal velocity for the operation of any column, at which the plate height (H.E.T.P.) reaches its minimum value (Figure 22-14).

> A broad band equilibrates with a longer section of column than does a narrow band. The "section of column" required for equilibration is the theoretical plate.

Let's examine each term of the van Deemter equation to understand the velocity-dependence of H.E.T.P. First, it should be realized that any mechanism that causes a band of solute to broaden will increase the height of a theoretical plate. Each term of Equation 22-39 therefore represents a zone-broadening mechanism.[‡]

> Longitudinal diffusion means diffusion parallel to the direction of flow.

Consider **longitudinal diffusion** of solute as it travels through the column (Figure 22-15 top). *Since the concentration of solute is lower at the edges of the solute zone than at the center, solute is always diffusing toward the edges of the zone.* The more time the solute spends on the column, the greater will be its diffusive spreading. This gives the B/v term of the van Deemter equation. The greater the mobile phase velocity, the less time is spent on the column and the less diffusion occurs.

The term Cv arises from the finite time required for solute to equilibrate between the mobile and stationary phases. Consider an instant at which some solute is in each phase (Figure 22-15 center). If the mobile phase is moving

---

[†] J. P. Foley and J. G. Dorsey, *Anal. Chem.*, **55**, 730 (1983); B. A. Bidlingmeyer and F. V. Warren, Jr., *Anal. Chem.*, **56**, 1583A (1984).

[‡] A thorough discussion of the van Deemter equation is presented by S. J. Hawkes [*J. Chem. Ed.*, **60**, 393 (1983)].

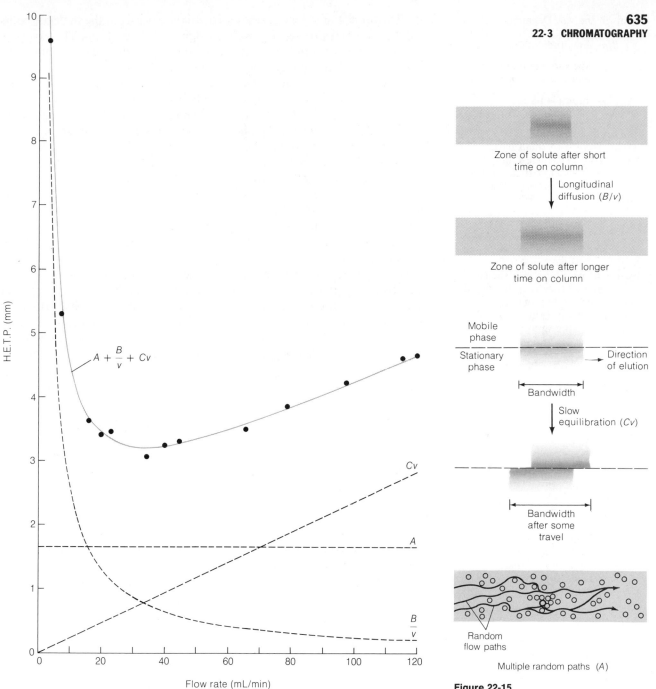

Figure 22-14
Application of van Deemter equation to gas chromatography. Experimental points are data from H. W. Moody, *J. Chem. Ed.*, **59**, 291 (1982). Van Deemter parameters are $A = 1.65$ mm, $B = 25.8$ mm · mL/min, and $C = 0.023\ 6$ mm · min/mL.

Figure 22-15
Mechanisms of band-broadening in chromatography.

rapidly and if solute cannot "escape" from the stationary phase rapidly, then the zone of solute in the mobile phase moves ahead of the zone in the stationary phase. The effect is to broaden the total solute band. The greater the velocity, v, the worse this broadening becomes. The term Cv expresses this effect.

Terms in the van Deemter equation:

B/v—longitudinal diffusion.

Cv—speed of equilibration, also called *mass transfer* term.

A—multiple paths.

Equilibration with small particles is faster, so small particles allow higher flow rates. Smaller particles also give more theoretical plates per length of column.

Because gases diffuse into the stationary phase faster than liquids, the optimum flow rate in gas chromatography is greater than that for liquid chromatography.

The term A in the van Deemter equation arises from the multiple paths of different lengths traveled by solute (Figure 22-15 bottom). The existence of multiple paths of random length tends to broaden the zone of solute. This effect is approximately independent of flow velocity.

The optimum condition occurs when H.E.T.P. is a minimum. At velocities below the optimal velocity, longitudinal diffusion causes the zone to broaden and increases the plate height. At velocities above the optimum, slow equilibration of solute between the phases causes the solute to spread out.

For a given type of stationary phase, *the smaller the particle size the more efficient is the column.* This is because solutes must travel shorter distances to equilibrate with smaller particles of stationary phase. Particle size in high performance chromatography columns is approximately 5 μm. *The smaller the particle diameter, the higher is the optimal flow rate.* Columns packed with tiny particles provide a great resistance to solvent flow. Consequently, high pressures are needed to operate the most efficient liquid chromatography columns.

Because the diffusion coefficients of gases are so much greater than those of liquids, the optimum flow rate in gas chromatography is higher than that in liquid chromatography. Whereas it is possible to run a gas chromatograph below its optimum flow rate, it is rare for the flow rate in liquid chromatography to be below the optimal value. In most cases, lowering the liquid flow rate raises the efficiency (decreasing H.E.T.P.) of a liquid chromatography column.

Resolution

The **resolution** of two peaks from each other is defined as

$$\text{Resolution} = \frac{\Delta t_r}{w_{\text{av}}} = \frac{\Delta V_r}{w_{\text{av}}} \tag{22-40}$$

where Δt_r or ΔV_r is the separation between peaks (in units of time or volume) and w_{av} is the average width of the two peaks in corresponding units. (Peak width, measured at the base, is defined in Figure 22-12.) Figure 22-16 shows how well separated two peaks are at various values of resolution. When resolution = 0.5, the overlap of the two peaks is 16%. When resolution = 1.0, the overlap is 2.3%, and the overlap is reduced to 0.1% at a resolution of 1.5.

It can be shown[†] that the resolution is given by

$$\text{Resolution} = \frac{\sqrt{N}}{4}\left(\frac{\alpha - 1}{\alpha}\right)\left(\frac{k_2}{1 + k_{\text{av}}}\right) \tag{22-41}$$

where N is the number of theoretical plates in the column, α is the relative retention of the two peaks (Equation 22-26), k_2 is the capacity factor for the second (more retained) component (Equation 22-27), and k_{av} is the average capacity factor for both components. If the number of theoretical plates for the two peaks is not the same, replace N by $\sqrt{N_1 N_2}$ and replace k_2 by k_{av}. N_1 and N_2 are the number of theoretical plates for each component.

[†] J. P. Foley, *Anal. Chem.* (1990), in press.

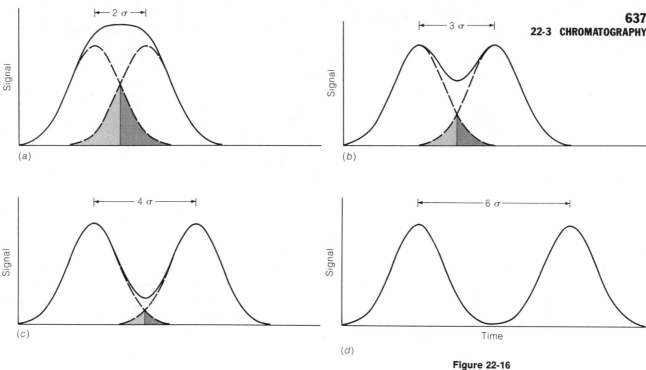

Figure 22-16
Resolution of two Gaussian peaks of equal area and amplitude. Dashed lines show individual peaks, and solid lines are the sum of two peaks. Overlapping area is shaded. (a) Resolution = 0.50. (b) Resolution = 0.75. (c) Resolution = 1.00. (d) Resolution = 1.50.

The most important feature of Equation 22-41 is that resolution is proportional to $\sqrt{N}$. Since the number of theoretical plates of a column is proportional to the column length, *doubling the column length increases the resolution by* $\sqrt{2}$. This is analogous to the behavior of countercurrent distribution and the reasons are the same. Separation of components increases in proportion to N, but band spreading is proportional to $\sqrt{N}$. The net result is that resolution is proportional to $\sqrt{N}$ (or equivalently, $\sqrt{\text{column length}}$). Equation 22-41 also tells us that resolution increases as the relative retention (α) increases, and as the capacity factor k_2 increases. Increasing the capacity factor is equivalent to increasing the fraction of time spent by the solute in the stationary phase. There is a practical limit to increasing the capacity factor, because retention times become too long and peaks become too broad.

Resolution $\propto \sqrt{N} \propto \sqrt{L}$

Open Tubular Columns

The column in Figure 22-9 is called a **packed column** because it is filled with the chromatography medium (the stationary phase) on which separation of solutes occurs. For gas chromatography, an alternative is to coat the inside wall of a narrow capillary column with the stationary phase (Figure 23-2). Such **open tubular columns** offer much higher resolution than packed columns, shorter analysis times, and increased sensitivity to small quantities of analyte. Open tubular columns have small sample capacity, and are not useful for preparative separations in which the object is to purify and isolate significant quantities of a compound.

The advantages of the open tubular design arise from the open pathway in the middle of the column. In a packed column the stationary phase resists

Compared to packed columns, open tubular columns can provide

1. higher resolution
2. shorter analysis time
3. increased sensitivity (discussed in Section 23-1)
4. lower sample capacity

Table 22-3
Comparison of packed and wall-coated open tubular column performance[†]

Property	Packed	Open Tubular
Column length, L	2.4 m	100 m
Linear gas velocity	8 cm/s	16 cm/s
H.E.T.P. for methyl oleate	0.73 mm	0.34 mm
Capacity factor, k, for methyl oleate	58.6	2.7
Theoretical plates, N	3 290	294 000
Resolution of methyl stearate and methyl oleate	1.5	10.6
Retention time of methyl oleate	29.8 min	38.5 min

SOURCE: L. S. Ettre, *Introduction to Open Tubular Columns* (Norwalk, Connecticut: Perkin-Elmer Corp., 1979), p. 26.
[†] Methyl stearate ($CH_3(CH_2)_{16}CO_2CH_3$) and methyl oleate (*cis*-$CH_3(CH_2)_7CH{=}CH$-$(CH_2)_7CO_2CH_3$) were separated on columns with DEGS (Table 23-1) stationary phase at 180°C.

For a given pressure, flow rate is proportional to the cross-sectional area of the column and inversely proportional to column length:

$$\text{Flow} \propto \frac{\text{area}}{\text{length}}$$

Open tubular design permits

1. increased linear flow rate and/or a longer column
2. decreased H.E.T.P., which means higher resolution

flow of the mobile phase, so the linear flow rate cannot be very fast. For the same length of column and applied pressure, the linear flow rate in an open tubular column is much higher than that of a packed column. Therefore the open tubular column can be made very much longer (typically 100 times longer) than the packed column, to give a similar pressure drop and flow rate. If all else is the same, this provides about 100 times more theoretical plates, yielding greatly increased resolution.

In addition, the height equivalent to a theoretical plate (H.E.T.P.) is usually smaller for an open tubular column, giving increased resolution per unit length. The reason for this can be seen in Figure 22-15 and the van Deemter equation (Equation 22-39). H.E.T.P. depends on three factors. The first term (A) arises from the many different paths that are followed by the mobile phase as it moves through a packed column. Since some paths are longer and some are shorter, the width of the solute band broadens as it travels through the column, decreasing the resolution of closely spaced solutes. *In an open tubular column there is no multiple path broadening,* because all solute follows essentially the same path. Now consider the van Deemter curve for a typical packed column in Figure 22-14. At the most efficient flow rate (minimum H.E.T.P.) near 30 mL/min, the A term accounts for half of the plate height. If the A term were deleted, the plate height would be reduced by a factor of 2 and the number of plates on the column would be doubled.

To obtain high performance from an open tubular column, the radius of the column must be small and the stationary phase must be as thin as possible. The small dimensions ensure rapid exchange of solute between the mobile and stationary phases, which is the basis for separating solutes from each other.

Table 22-3 compares actual operating characteristics for a packed column and an open tubular column with the same stationary phase used to separate a pair of solutes by gas chromatography. For similar analysis times (about half an hour), the open tubular column gives seven times as great a resolution (10.6 versus 1.5). The open tubular column is 42 times longer and has 89 times as many theoretical plates. Alternatively, speed can be traded for resolution. If the open tubular column is reduced to 5 m in length, the same two solutes are separated with a resolution of 1.5, but the analysis time is reduced from 38.5 min to 0.83 min.

Band spreading before and after chromatography

In discussing the van Deemter equation, we only mentioned ways that the solute bands are broadened in the chromatographic process. However, the solute cannot be applied to the column in an infinitesimally thin zone, so the band has a finite width even before it begins spreading on the column. After elution, further broadening can occur in a poorly designed column outlet or in the detector used to measure the solute emerging from the column.

In well-designed outlets and detectors, laminar flow is achieved. This means that the velocity of the fluid is greatest in the center of the tube and diminishes gradually until, at the walls of the tube, the velocity of fluid is zero. Because fluid in the center of the chamber moves more rapidly than fluid at the edges, the band spreads out. However, the spreading would be much worse in an outlet or detector where each new drop entering the chamber mixed totally with the contents of the chamber. In the large dead space beneath some crude chromatography columns, this does happen.

To minimize band spreading outside the chromatography column, all dead spaces and tubing lengths should be minimized. The sample should be applied in a narrow zone and allowed to enter the column before mixing with the eluent.

Band shapes

A Gaussian band shape results when the partition coefficient, K $(=C_s/C_m)$, is constant, independent of the quantity of solute on the column. In real columns, the ratio C_s/C_m changes somewhat as the total quantity of solute increases, and the resulting band shapes are skewed. A graph of C_s versus C_m (at a given temperature) is called an *isotherm*. Three common isotherms and their resulting band shapes are shown in Figure 22-17. The center isotherm is the ideal one, leading to a symmetric peak.

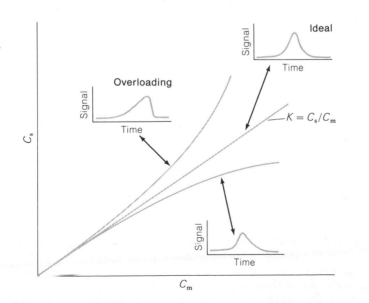

Figure 22-17
Common isotherms and their resulting chromatographic band shapes.

Overloading produces a gradual rise
and an abrupt fall of the chromatographic peak.

A long tail occurs when some sites
retain solute more strongly than
other sites.

The upper isotherm is typical of an *overloaded* column. This is one to which so much solute has been applied that the solute affects the physical properties of the stationary phase. The stationary phase can no longer be considered a dilute solution of solute. An overloaded stationary phase behaves more like a liquid phase made of the solute. Since solute is most soluble in itself ("like dissolves like"), the value C_s/C_m increases with increased solute loading.

The front of an overloaded peak exhibits a normal, gradually increasing concentration. As the concentration increases, the band becomes overloaded and the partition coefficient ($= C_s/C_m$) increases. The solute becomes so soluble in the zone on the stationary phase that there is little solute trailing behind the peak. The band emerges gradually but ends abruptly (Figure 22-17, upper left inset).

The lower isotherm in Figure 22-17 results when small quantities of solute are retained more strongly than large quantities. This is exactly the opposite situation to overloading. It leads to a relatively abrupt rise of concentration at the beginning of the band and a long "tail" of gradually decreasing concentration after the peak.

One cause of tailing is the presence of sites at which the solute is held strongly by the stationary phase. When these sites are saturated with solute, the partition coefficient decreases because no more strong sites are available to hold fresh solute. In gas chromatography, a common solid phase support is made of diatomite, the hydrated silica skeletons of algae. Hydroxyl groups at the surface of the particles can form strong hydrogen bonds with polar solutes, leading to serious tailing. One strategy to reduce tailing is to cover these sites with a polar compound, such as acetic acid, which remains bound indefinitely. Another technique, called **silanization,** is to covalently attach trimethylsilyl groups to the silanol hydroxyl groups:

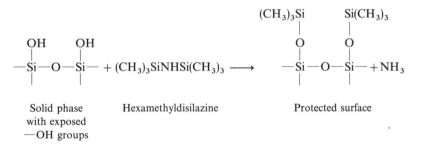

Diatomite so treated produces less tailing than does untreated material. Glass and silica columns used for gas and liquid chromatography can also be silanized to minimize interaction of the solute with active sites on the walls.

Summary

A solute may be extracted from one phase into a second in which it is more soluble. The ratio of solute concentrations in each phase at equilibrium is called the partition coefficient. If more than one form of the solute exists in the solution, we use a distribution coefficient instead of a partition coefficient. In this chapter we derived simple equations relating the fraction of solute extracted to the partition coefficient, volumes of the phases, and pH. A series of small extractions is more effective than a few large extractions. A metal chelator, soluble only in organic solvents, may be used to extract metal ions from aqueous solution, with selectivity achieved by controlling pH. Hydrophobic counterions, ion-pair formation, or crown ethers may also be used to extract ions from water into a hydrophobic phase.

Countercurrent distribution is a serial extraction procedure in which solutes with differing partition coefficients may be separated. The distribution of solutes may be described by a binomial distribution in which the band position is proportional to the number of steps and the bandwidth is proportional to the square root of the number of steps.

In adsorption and partition chromatography, an essentially continuous equilibration of solute between the mobile and stationary phases occurs. In ion-exchange chromatography, the solute is attracted to the stationary phase by coulombic forces. In molecular exclusion chromatography, the fraction of stationary-phase volume available to solute decreases as the size of the solute molecules increases. Affinity chromatography relies on specific, noncovalent interactions between the stationary phase and one solute in a complex mixture.

We can speak of a theoretical plate as the length of column needed for one imaginary equilibration step to occur. The number of theoretical plates in a column pro-

ducing Gaussian peaks is defined as $16t_r^2/w^2$, where t_r is the retention time of a peak and w is its width at the base. The van Deemter equation states that the height equivalent to a theoretical plate is given by $A + B/v + Cv$, where v is the solvent velocity and A, B, and C are constants. The first term represents irregular flow paths, the second longitudinal diffusion, and the third the finite rate of equilibration. An optimum flow rate is one that minimizes the plate height. The optimum flow rate is faster for gas chromatography than for liquid chromatography. Both the number of plates and the optimal flow rate increase as the stationary-phase particle size decreases. Open tubular columns provide higher resolution and shorter analysis times than packed columns. The resolution of a column is proportional to the square root of its length. Bands spread not only on the column, but during injection and detection. Overloading and tailing can be corrected by using smaller samples and by masking strong adsorption sites on the stationary phase.

Terms to Understand

adjusted retention time
adsorption chromatography
affinity chromatography
capacity factor
chromatogram
countercurrent distribution
distribution coefficient
eluate
eluent
elution
extraction
gel filtration chromatography
gel permeation chromatography
H.E.T.P.
ion-exchange chromatography
ionophore
ion pair
linear flow rate

longitudinal diffusion
miscible
mobile phase
molecular exclusion chromatography
open tubular column
packed column
partition chromatography
partition coefficient
phase transfer catalysis
relative retention
resolution
retention time
retention volume
silanization
stationary phase
theoretical plate
van Deemter equation
volume flow rate

Exercises

22-A. Consider a chromatography experiment in which two components with capacity factors $k_1 = 4.00$ and $k_2 = 5.00$ are injected into a column with $N = 1.00 \times 10^3$ theoretical plates. The retention time for the less-retained component is $t_{r1} = 10.0$ min.

(a) Calculate t_m, t_{r2}, and the width at half-height ($w_{1/2}$) for each peak. Also calculate the width, w, at the base of each peak.

(b) Using graph paper, sketch the chromatogram analogous to Figure 22-11, supposing that the two peaks have the same amplitude (height). Draw the half-widths accurately.

(c) Calculate the resolution of the two peaks and compare this value to those drawn in Figure 22-16.

22-B. A solute with a partition coefficient of 4.0 is extracted from 10 mL of phase 1 into phase 2.

(a) What volume of phase 2 is needed to extract 99% of the solute in one extraction?

(b) What will be the total volume of solvent 2 needed to remove 99% of the solute in three equal extractions instead?

22-C. Consider the 100-step countercurrent distribution of solutes A and B, with distribution coefficients $D_A = 1$ and $D_B = 0.6$. Suppose that the volumes of mobile and stationary phases are equal.

(a) Calculate r_{max} and σ for each band.

(b) If the fractions 28–43 and 44–60 are each pooled, find the percent recovery and purity of each solute.

22-D. (a) Find the capacity factors for octane and nonane in Figure 22-11.

(b) Find the ratio

$$\frac{\text{time octane spends in stationary phase}}{\text{total time octane spends on column}}.$$

(c) Find the relative retention for octane and nonane.

(d) If the volume of the stationary phase equals half the volume of the mobile phase, find the partition coefficient for octane.

22-E. A gas chromatogram of a mixture of toluene and ethyl acetate is shown below.

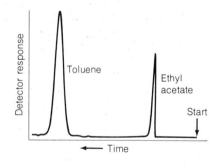

(a) Use the width of each peak (measured at the base) to calculate the number of theoretical plates in the column. Estimate all lengths to the nearest 0.1 mm.

(b) Using the width of the toluene peak at its base, calculate the width expected at half-height. Compare the measured and calculated values. When the thickness of the pen trace is significant compared to the length being measured, it is important to take the pen width into account. It is best to measure from the edge of one trace to the corresponding edge of the other trace, as shown above right.

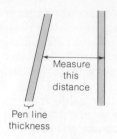

22-F. The three chromatograms below were obtained with 2.5, 1.0, and 0.4 μL of ethyl acetate injected on the same column under the same conditions. Explain why the asymmetry of the peak decreases as the sample size decreases.

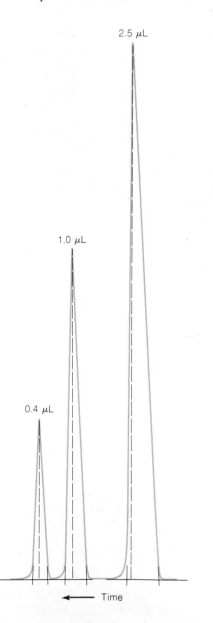

Problems

$K = \dfrac{Phase\ 2}{Phase\ 1} = 4.0$

A22-1. Solute S has a partition coefficient of 4.0 between water (phase 1) and chloroform (phase 2) in Equation 22-1. Consider the equilibration of S in a mixture of water and chloroform.
  (a) Calculate the concentration of S in chloroform if the concentration in water is 0.020 M.
  (b) If the volume of water is 80.0 mL and the volume of chloroform is 10.0 mL, find the quotient (mol S in chloroform)/(mol S in water).

A22-2. The solute in the previous problem is initially dissolved in 80.0 mL of water. It is then extracted six times with 10.0 mL portions of chloroform. Use Equation 22-6 to find the fraction of solute remaining in the aqueous phase.

A22-3. The weak base B $(K_b = 1.0 \times 10^{-5})$ equilibrates between water (phase 1) and benzene (phase 2).
  (a) Define the distribution coefficient, D, for this system.
  (b) Explain the difference between D and K, the partition coefficient.
  (c) Calculate D at pH 8.00 if $K = 50.0$
  (d) Will D be greater or less at pH 10 than at pH 8? Explain why.

A22-4. Why is the extraction of a metal ion into an organic solvent with 8-hydroxyquinoline more complete at higher pH?

A22-5. In a countercurrent extraction, a solute is distributed 37% in the upper phase and 63% in the lower phase in each extraction.
  (a) In which tube will the peak concentration appear after 150 extraction steps?
  (b) How many tubes wide is the standard deviation for this solute?

A22-6. Match the terms in the first list with the characteristics in the second list.

  1. adsorption chromatography C
  2. partition chromatography D
  3. ion-exchange chromatography A
  4. molecular exclusion chromatography E
  5. affinity chromatography B

  A. Ions in mobile phase are attracted to counterions covalently attached to stationary phase.
  B. Solute in mobile phase is attracted to specific groups covalently attached to stationary phase.
  C. Solute equilibrates between mobile phase and surface of stationary phase.

  D. Solute equilibrates between mobile phase and film of liquid attached to stationary phase.
  E. Different-sized solutes penetrate voids in stationary phase to different extents. Largest solutes are eluted first.

A22-7. Solvent passes through a column in 3.0 min but solute requires 9.0 min.
  (a) Calculate the capacity factor, k.
  (b) What fraction of time is the solute in the mobile phase as it passes through the column?
  (c) The volume of stationary phase is one-tenth of the volume of mobile phase in the column $(V_s = 0.10\ V_m)$. Find the partition coefficient, K, for this system.

A22-8. Solvent occupies 15% of the volume of a chromatography column whose inner diameter is 3.0 mm. If the volume flow rate is 0.2 mL/min, find the linear flow rate.

A22-9. Consider a chromatography column in which $V_s = V_m/5$. Find the capacity factor if $K = 3$ and if $K = 30$.

A22-10. Which column is more efficient: (a) H.E.T.P. = 0.1 mm or (b) H.E.T.P. = 1 mm

A22-11. A chromatogram with ideal Gaussian bands has $t_r = 9.0$ min and $w_{1/2} = 2.0$ min.
  (a) How many theoretical plates are present?
  (b) The column is 10 cm long. Find H.E.T.P.

A22-12. The asymmetric chromatogram in Figure 22-13 has a retention time equal to 15 cm of chart paper, and the values of A and B are 3.0 and 1.0 cm, respectively. Find the number of theoretical plates.

A22-13. Explain why one term of the van Deemter equation depends on v and why one term depends on $1/v$.

A22-14. Why is the optimal flow rate greater for a chromatographic column if the stationary-phase particle size is smaller?

A22-15. What is the optimal flow rate in Figure 22-14 for best separation of solutes?

A22-16. Two chromatographic peaks with widths of 6 min are eluted at 24 and 29 min. Which diagram in Figure 22-16 will most closely resemble the chromatogram?

A22-17. Explain how silanization reduces tailing of chromatographic peaks.

22-18. Consider the extraction of M^{n+} from aqueous solution into organic solution by reaction with protonated ligand, HL:

$$M^{n+}(aq) + nHL(org) = ML_n(org) + nH^+(aq)$$

$$K_{extraction} = \frac{[ML_n]_{org}[H^+]_{aq}^n}{[M^{n+}]_{aq}[HL]_{org}^n}$$

Rewrite Equation 22-19 in terms of $K_{extraction}$ and express $K_{extraction}$ in terms of the constants in Equation 22-19. Give a physical reason why each constant increases or decreases $K_{extraction}$.

22-19. Give a physical interpretation of Equations 22-9 and 22-10 in terms of the fractional composition equations for a monoprotic acid in Section 12-2.

22-20. Describe how nonlinear partition isotherms lead to non-Gaussian band shapes. Draw the band shape produced by an overloaded column and a column with tailing.

22-21. Butanoic acid has a partition coefficient of 3 (favoring benzene) when distributed between water and benzene. Find the formal concentration of butanoic acid in each phase when 100 mL of 0.10 M aqueous butanoic acid is extracted with 25 mL of benzene (a) at pH 4.00 and (b) at pH 10.00.

22-22. For a given value of $[HL]_{org}$ in Equation 22-19, over what pH range (how many pH units) will D change from 0.01 to 100 if $n = 2$?

22-23. For the extraction of Cu(II) by dithizone in CCl$_4$, $K_L = 1.1 \times 10^4$, $K_M = 7 \times 10^4$, $K_a = 3 \times 10^{-5}$, $\beta = 5 \times 10^{22}$, and $n = 2$.
 (a) Calculate the distribution coefficient for extraction of 0.1 μM Cu(II) into CCl$_4$ by 0.1 mM dithizone at pH 1.0 and pH 4.0.
 (b) If 100 mL of 0.1 μM aqueous Cu(II) is extracted once with 10 mL of 0.1 mM dithizone at pH 1.0, what fraction of Cu(II) remains in the aqueous phase?

22-24. Calculate the fraction of each solute in Table 22-1 in each phase after seven steps. Make a graph similar to Figure 22-7 showing the relative proportions of A and B in each tube after seven steps.

22-25. Consider a countercurrent distribution in which substances A and B have the following distribution coefficients:

$$\frac{[A]_{upper}}{[A]_{lower}} = 3 \qquad \frac{[B]_{upper}}{[B]_{lower}} = \frac{1}{3}$$

 (a) Suppose that 1 mmol of A and 1 mmol of B are to be separated from each other. Calculate the number of millimoles of each substance in each tube after six steps of countercurrent distribution.
 (b) If tubes 5 and 6 are combined, what would be the percent recovery and purity of solute A?

22-26. Suppose that a solute with $D = 1.5$ is subjected to 100 steps of countercurrent distribution, using 4 mL of solvent (2 mL in each phase) in each tube.
 (a) Calculate the bandwidth (expressed as σ for the band) in milliliters and in tubes.
 (b) Find the bandwidth in milliliters and in tubes if the same countercurrent distribution is carried out with 200 tubes, each containing 1 mL of each phase.
 (c) What can you conclude about the utility of using more and smaller steps to effect a separation?

22-27. Derive a formula giving the number of countercurrent distribution steps (n) needed such that the separation of peaks A and B is $2\sigma_A + 2\sigma_B$. Call the fractional distribution of each solute p_A, q_A, p_B, and q_B. Use the Gaussian distribution as a good approximation for the shape of each band.
 (a) Calculate the value of n if $p_A = 0.40$ and $p_B = 0.30$.
 (b) Find n if $p_A = 0.36$ and $p_B = 0.34$.

22-28. If the volumes of mobile and stationary phases are not equal in countercurrent distribution, show that

$$p = \frac{DV_m}{DV_m + V_s} \qquad q = \frac{V_s}{DV_m + V_s}$$

where D is the distribution coefficient for solute between the two phases: $D = C_m/C_s$. (C is the concentration of solute in each phase.)

22-29. A band having a width of 4.0 mL and a retention volume of 49 mL was eluted from a chromatography column. What width is expected for a band with a retention volume of 127 mL? Assume that the only band spreading occurs on the column itself.

22-30. Two compounds with partition coefficients of 0.15 and 0.18 are to be separated on a column with $V_m/V_s = 3.0$. Calculate the number of theoretical plates needed to produce a resolution of 1.5.

22-31. The retention volume of a solute is 76.2 mL for a column with $V_m = 16.6$ mL and $V_s = 12.7$ mL. Calculate the partition coefficient and capacity factor for this solute.

22-32. Write the relation between α, k_1, and k_2 for two chromatographic peaks. Calculate the number of theoretical plates needed to achieve a resolution of 1.0 if
 (a) $\alpha = 1.05$ and $k_2 = 5.00$
 (b) $\alpha = 1.10$ and $k_2 = 5.00$
 (c) $\alpha = 1.05$ and $k_2 = 10.00$
 (d) How can you increase N, α, and k_2 in a chromatography experiment? In this problem, which has a larger effect on resolution: α or k_2?

23 Chromatographic Methods

Separation of compounds for the purpose of purification, identification, and quantification is a challenging aspect of analytical chemistry. In this chapter we discuss the major types of chromatography, which are among our most powerful methods of separation.

23-1 GAS CHROMATOGRAPHY

In **gas chromatography,** a gaseous solute (or vapor from a volatile liquid) is carried by a gaseous mobile phase. The stationary phase is usually a non-volatile liquid coated on the inside of the column or on a fine solid support, as at the upper right in Figure 22-10. This most common form of gas chromatography is called *gas–liquid partition chromatography.* Occasionally, solid particles on which solute can be adsorbed serve as the stationary phase, as at the upper left in Figure 22-10. In this case, the technique is called *gas–solid adsorption chromatography.*[†]

A schematic diagram of a gas chromatograph is shown in Figure 23-1. A volatile liquid sample is injected through a rubber **septum** (a thin disk) into a heated glass injector port, which vaporizes the sample. Gaseous samples can be injected by gas-tight syringe or through a gas-sampling valve. The sample is swept through the column by an inert carrier gas (usually He, N_2, or H_2), which serves as the mobile phase. After passing through the column containing the stationary phase, the separated solutes flow through a detector, whose response is displayed on a recorder or computer. The column temperature need not be above the boiling point of all solutes. It must only be hot enough for each solute to have sufficient vapor pressure to be eluted in a reasonable time. The detector is maintained at a higher temperature than the column, so that all solutes are gaseous in this chamber.

Gas chromatography:

mobile phase—gas

stationary phase—usually a non-volatile liquid on a solid support, but sometimes a solid

solute—gas or volatile liquid

The choice of carrier gas depends on the detector and desired separation efficiency and speed.

[†] Classroom demonstrations of both types of chromatography have been described by A. Wollrab [*J. Chem. Ed.,* **59,** 1042 (1982)] and C. E. Bricker, M. A. Taylor, and K. E. Kolb [*J. Chem. Ed.,* **58,** 41 (1981)].

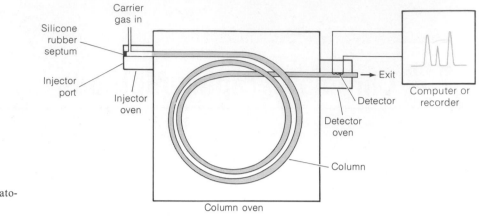

Figure 23-1

Schematic diagram of a gas chromatograph.

Solutes emerging from the gas chromatograph can be collected for identification or can be sent directly into an infrared or mass spectrometer for immediate analysis of each chromatographic peak. To collect microliter quantities of solute, a U-shaped glass tube is inserted into the exit port of the chromatograph. The bottom of the U is cooled with dry ice or liquid nitrogen to condense solute.

The Column

Packed columns made of stainless steel or glass are typically 3–6 mm in diameter and 1–5 m in length. The column is filled with a fine solid support coated with a nonvolatile liquid stationary phase, or the solid may be the stationary phase. For preparative purposes, the thickness of the stationary phase is increased so that more sample can be accommodated, and fatter columns are preferred.

Open tubular columns (Figure 23-2) are much narrower and usually much longer than packed columns. As described in Section 22-3, the open tubular design offers higher resolution, shorter analysis time, and greater sensitivity than packed columns, but can handle much less sample. Resolution is increased because the column is longer and the number of theoretical plates per unit length is increased. Analysis time is decreased because the column contains much less stationary phase than is found in a packed column, so solutes are less retained. The smaller volume of stationary phase means that much less sample can be applied to an open tubular column than to a packed column. Sensitivity is a measure of the detector signal-to-noise ratio for a given concentration of analyte in the injected solution. The sensitivity obtained with an open tubular column is greater than that from a packed column because (1) gas that is optimum for detector performance is added to the sample stream as it enters the detector; (2) contaminants from the septum are reduced by continuous purge of the inlet system; (3) slow bleed of the stationary phase from the column into the detector is reduced because less stationary phase is present; and (4) fluctuations of carrier gas flow that lead to fluctuations of detector response are damped by the long capillary column.

Figure 23-3 compares the separation of a perfume oil on a packed column to the same separation on an open tubular column. Note how much sharper the peaks from the open tubular column are and how many more components are resolved.

Compared to packed columns, open tubular columns offer

1. higher resolution
2. shorter analysis time
3. greater sensitivity
4. less sample capacity

Resolution and analysis time can be traded off against each other, with decreased analysis time at the expense of decreased resolution.

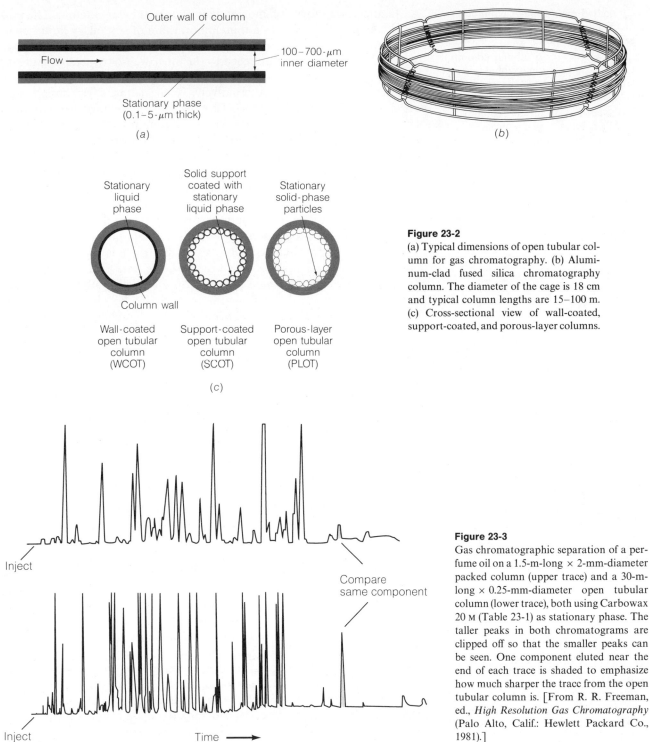

Figure 23-2
(a) Typical dimensions of open tubular column for gas chromatography. (b) Aluminum-clad fused silica chromatography column. The diameter of the cage is 18 cm and typical column lengths are 15–100 m. (c) Cross-sectional view of wall-coated, support-coated, and porous-layer columns.

Figure 23-3
Gas chromatographic separation of a perfume oil on a 1.5-m-long × 2-mm-diameter packed column (upper trace) and a 30-m-long × 0.25-mm-diameter open tubular column (lower trace), both using Carbowax 20 M (Table 23-1) as stationary phase. The taller peaks in both chromatograms are clipped off so that the smaller peaks can be seen. One component eluted near the end of each trace is shaded to emphasize how much sharper the trace from the open tubular column is. [From R. R. Freeman, ed., *High Resolution Gas Chromatography* (Palo Alto, Calif.: Hewlett Packard Co., 1981).]

Figure 23-2c shows common designs of open tubular columns. The **wall-coated** configuration features a thin, uniform film of stationary liquid phase on the inner wall of the column. In the **support-coated** design, a solid support such as silica or alumina is attached to the inner wall of the column, greatly increasing the available surface area. The solid support is coated with a film of stationary liquid phase. Alternatively, in the **porous-layer** design, solid particles attached to the column wall are the active stationary phase. Because of their increased surface area, support-coated columns have a greater volume of stationary phase than wall-coated columns. This greater volume allows larger samples to be analyzed with support-coated columns, which are therefore more suitable for trace analysis, in which the desired analyte is present in very low concentration in the sample. The increased volume of stationary phase gives a support-coated column a greater capacity factor (Equation 22-30), which allows better resolution of poorly retained solutes at the beginning of the chromatogram. Many liquid stationary phases appear to stick more tightly to support-coated columns, so that there is less bleed into the detector and a lower noise level. In general, the performance of support-coated columns is intermediate between those of wall-coated columns and packed columns.

Open tubular columns are generally made of fused silica, whose composition is SiO_2. This material contains <1 ppm of metallic impurities that could lead to active sites producing *tailing* of solutes (Figure 22-17) and decreased resolution. By comparison, glass contains orders-of-magnitude more active sites, leading to poorer resolution. Compared to glass, fused silica also has an inherently low concentration of surface silanol (Si—O—H) groups that bind to solutes and produce tailing. Fused silica is mechanically stronger than glass and the outside of the fused silica column is coated with polyimide (a plastic capable of withstanding 350°C) or aluminum for additional mechanical support and to prevent weakening induced by atmospheric water vapor. As a fused silica column ages and stationary phase is baked off, surface silanol groups are exposed and tailing increases. Exposure of the column to oxygen at high temperatures also leads to degradation and tailing.

Solid Support

Packed columns contain a nonvolatile liquid coated on fine particles of an "inert" solid support. The support should consist of strong, uniformly shaped, small particles with a large surface area. Most supports are made of diatomaceous earth, the silica skeletons of algae. Ideally, the support should not interact with solutes, but no support is totally inert. *Silanization* with hexamethyldisilazane (page 640) or dichlorodimethylsilane ($(CH_3)_2SiCl_2$) is a widely used means of reducing the interaction of silica particles with polar solutes. For tenaciously binding solutes, Teflon can be a useful support.

Uniform particle size decreases the multiple path (A) term in the van Deemter equation (22-39), thereby reducing the plate height and increasing resolution. Small particle size decreases the time required for solute equilibration, thereby improving column efficiency. However, the smaller the particle size, the more pressure is required to force the mobile phase through the column. Chromatographic support particle size is expressed as a **mesh size,** which refers to the size of screens through which the particles are passed or retained (Table 24-2). A 100/200 mesh particle will pass through a 100 mesh screen, but not through a 200 mesh screen. The mesh gives the number of openings per linear inch of screen.

A bewildering assortment of liquid phases is available for gas–liquid partition chromatography. A few are listed in Table 23-1. The choice of liquid phase for a given problem is based on the rule "like dissolves like." The classes of solutes most strongly retained by each liquid phase are given in Table 23-2. In general, a variety of liquid phases might be suitable for a given separation problem.[†]

Tables 23-1 and 23-2 tell us, for example, that the silicone gum rubber SE-30 would be a useful liquid phase for the separation of a mixture of olefins. In separating compounds of different classes, but with similar boiling points, the tables let us predict the order of elution. For example, suppose we wish to separate 1-propanol (b.p. 97°C) from 2-chloropentane (b.p. 97°C). Propanol is a polar compound in class III, while 2-chloropentane is a relatively nonpolar compound in class I. If the polar stationary phase Zonyl E-7 were used, propanol would have a longer retention time than 2-chloropentane. If the nonpolar liquid phase SE-30 were used, propanol would be eluted before 2-chloropentane.

The amount of liquid phase used is expressed as a weight percent of the solid support. Loadings in the range 2–10% are most common. Beyond 30%, column efficiency decreases as pools of liquid phase are formed. Below 1%, the surface of the support may not be entirely covered. In general, a large percent of liquid phase leads to greater capacity for solute, longer retention times, and more separation between peaks. A lower percentage gives more rapid analysis and smaller values of H.E.T.P. High loadings are used for preparative work.

Open tubular columns use liquid phases similar to those in Table 23-1, with modifications that minimize liquid-phase bleeding from the column. There may be covalent chemical attachment of the liquid phase to the solid phase, in which case the liquid is said to be *bonded*. Covalent crosslinking of long liquid-phase molecules to each other can be promoted by free radical reactions. Bonding and crosslinking both reduce the tendency of stationary phase to be lost from the column at elevated temperature.

Several solid stationary phases are used in both packed and open tubular columns. Alumina (Al_2O_3) is a common material that can separate hydrocarbons in gas–solid partition chromatography. **Molecular sieves** (Figure 23-4) are a noteworthy solid phase widely used in chromatography. These are inorganic materials or organic polymers having large cavities into which small molecules can enter and where they are partially retained. Small molecules such as H_2, O_2, N_2, CO_2, and CH_4 can be separated from each other. Gases can be dried by passage through traps containing molecular sieves because water is strongly retained by sieves. Inorganic sieves can be regenerated (freed of water) by heating to 300°C in vacuum or under flowing N_2.

Temperature Programming

Increasing the temperature of a column decreases retention time and resolution. When separating a mixture of compounds with a wide range of boiling points or polarities, it is very useful to be able to change the column temperature *during* the separation. An example is shown in Figure 23-5. At a

Each stationary phase retains solutes in its own class best.

Raising percentage of stationary phase leads to

1. greater capacity for solute
2. longer retention time
3. increased H.E.T.P.

Box 23-1 gives an example of an optically active bonded phase used to separate optical isomers.

Raising column temperature

1. decreases retention time
2. decreases resolution
3. sharpens peaks

[†] A more quantitative basis for selecting a stationary phase is the McReynolds number. See W.O. McReynolds, *J. Chromatogr. Sci.,* **8**, 685 (1970).

Table 23-1
Some common liquid phases used in gas chromatography

Liquid phase	Structure	Class of solutes retained most strongly	Maximum temperature (°C)
Squalane	CH_3 groups on $HC—(CH_2)_3—CH—(CH_2)_3—CH—(CH_2)_4—CH—(CH_2)_3—CH—(CH_2)_3—CH$	I	100
SE-30	$[\,Si(CH_3)_2—O—Si(CH_3)_2—O\,]_r$ (Polymers with the structure $—SiR_2—O—SiR_2—O—$ are called silicones.)	I	350
Apiezon	(Mixed hydrocarbons)	I	275–300 (depending on type of Apiezon)
Dibutyl tetrachlorophthalate	$CO_2(CH_2)_3CH_3$ groups, tetrachloro benzene ring	II	150
Dinonyl phthalate	$CO_2(CH_2)_8CH_3$ groups on benzene ring	II	175
QF-1	$(CH_3)_3Si—O—Si(CH_2CH_2CF_3)(CH_3)—[\,O—Si(CH_3)_2\,]—O—Si(CH_3)_3$	II	250
OV-17	Methyl phenyl silicone	II	300

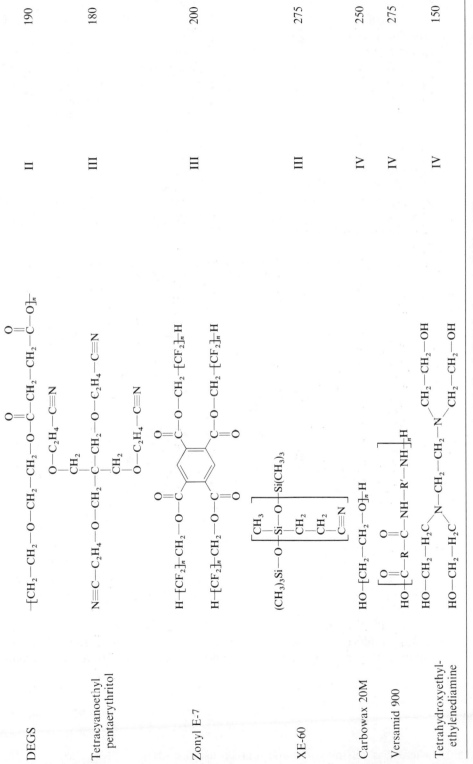

Name	Structure class	Max. temp.
DEGS	II	190
Tetracyanoethyl pentaerythritol	III	180
Zonyl E-7	III	200
XE-60	III	275
Carbowax 20M	IV	250
Versamid 900	IV	275
Tetrahydroxyethyl-ethylenediamine	IV	150

SOURCE: Adapted from H. M. McNair and E. J. Bonelli, *Basic Gas Chromatography* (Palo Alto, Calif.: Varian Instrument Division, 1968).

Table 23-2
Solute classes for Table 23-1

I: Low polarity	II: Intermediate polarity
Saturated hydrocarbons	Ethers
Olefinic hydrocarbons	Ketones
Aromatic hydrocarbons	Aldehydes
Halocarbons	Esters
Mercaptans	Tertiary amines
Sulfides	Nitro compounds (without α-H atoms)
CS_2	Nitriles (without α-H atoms)

III: Polar	IV: Very polar
Alcohols	Polyhydroxyalcohols
Carboxylic acids	Amino alcohols
Phenols	Hydroxy acids
Primary and secondary amines	Polyprotic acids
Oximes	Polyphenols
Nitro compounds (with α-H atoms)	
Nitriles (with α-H atoms)	

SOURCE: Adapted from H. M. McNair and E. J. Bonelli, *Basic Gas Chromatography* (Palo Alto, Calif. Varian Instrument Division, 1968).

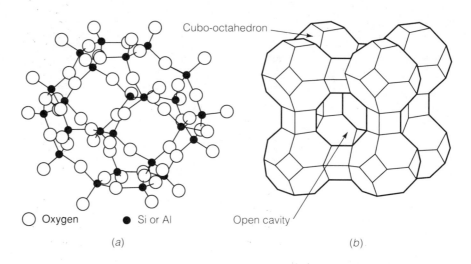

Figure 23-4
Structure of molecular sieves. (a) Aluminosilicate framework of one cubo-octahedron of a mineral class called *zeolites*. (b) Interconnection of eight cubo-octahedra to produce a cavity into which small molecules can enter. The synthetic zeolite $Na_{12}(Al_{12}Si_{12}O_{48}) \cdot 27H_2O$ has this structure.

○ Oxygen   ● Si or Al   Open cavity

Cubo-octahedron

(a)   (b)

constant temperature of 150°C, the more volatile compounds emerge very close together and the less volatile compounds may not even be eluted from the column. If the temperature is increased from 50° to 250°C at a rate of 8°/min, all of the compounds are eluted and the separation between peaks is fairly uniform. For rapid heating, it is desirable to use a small-diameter column. The equation describing the dependence of retention time on temperature is

If the retention time of a compound at two temperatures is known, Equation 23-1 lets us predict the retention time at a third temperature.

$$\log t_r' = \frac{a}{T} + b \tag{23-1}$$

where T is kelvins and a and b are constants.

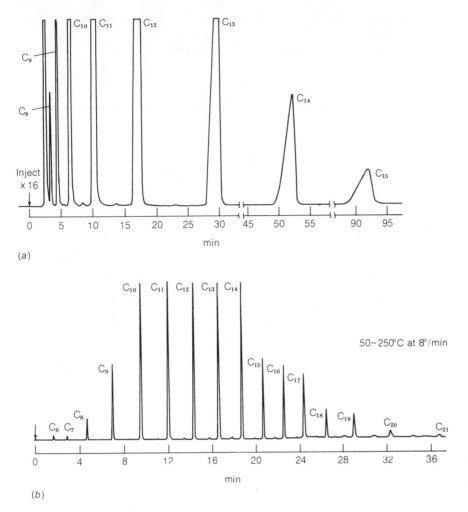

(a)

(b)

50–250°C at 8°/min

Figure 23-5
Comparison of (a) isothermal and (b) programmed temperature chromatography. Each sample contains linear alkanes run on a 1.6-mm × 6-m column containing 3% Apiezon L (liquid phase) on 100/120 mesh VarAport 30 solid support with He flow rate of 10 mL/min. [H. M. McNair and E. J. Bonelli, *Basic Gas Chromatography* (Palo Alto, Calif.: Varian Instrument Division, 1968).]

Carrier Gas

The choice of the mobile phase in gas chromatography affects column and detector performance. Figure 23-6 shows representative van Deemter curves for H_2, He, and N_2, the three most common carrier gases. All three give similar optimal plate height (0.3 mm), but at significantly different flow rates. N_2 requires relatively slow flow for optimum performance, while He and H_2 can be used at higher flow. Furthermore, the performance of H_2 is not degraded very much at flow rates faster than optimum. This means that H_2 can provide resolution similar to the other gases, but at much faster flow.

The reason why H_2 and He give good resolution at high flow rate is that solute gases diffuse more rapidly through H_2 and He than through N_2. Figure 22-14 shows that decreased resolution (increased H.E.T.P) occurs at high flow rates because of the speed of equilibration term, Cv, in the van Deemter equation (22-39). The more rapidly a solute equilibrates between the mobile and stationary phases, the smaller the Cv term. Rapid diffusion of solutes in H_2 and He promotes rapid equilibration between the mobile and stationary phases, decreasing the plate height. If the carrier is run slowly, the same resolution can be obtained with N_2.

H_2 and He give optimal resolution at higher flow rates than N_2.

Box 23-1 SEPARATION OF OPTICAL ISOMERS BY GAS CHROMATOGRAPHY

Gas or liquid chromatography with an optically active stationary phase provides a powerful means of separating mirror image isomers (*enantiomers*). In the chromatogram below, the D and L enantiomers of amino acids are separated from each other in an open tubular column to which is covalently attached an optically active phase containing L-valine:

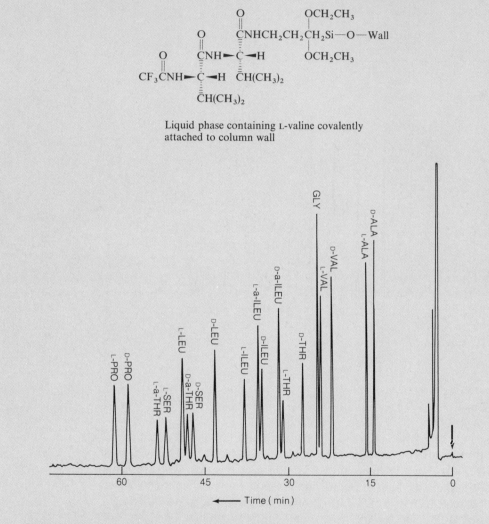

Liquid phase containing L-valine covalently
attached to column wall

Separation of amino acid enantiomers at 110°C on a 0.30-mm (inner diameter) × 39-m column with $N = 100\,000$ plates. [W. A. Koenig and G. J. Nicholson, *Anal. Chem.*, **47**, 951 (1975).]

Amino acids are not volatile enough to be chromatographed directly. Instead, the more volatile *N*-trifluoroacetyl isopropyl esters are prepared. The use of volatile derivatives is commonplace in gas chromatography.

Amino acid   Volatile derivative

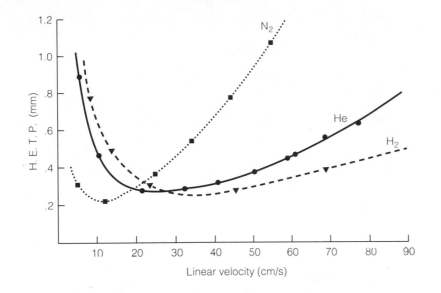

Figure 23-6
Van Deemter curves for gas chromatography of n-$C_{17}H_{36}$ at 175°C using N_2, He, or H_2 in a 0.25-mm-diameter × 25-m-long tubular wall-coated column with OV-101 stationary phase. [From R. R. Freeman, ed., *High Resolution Gas Chromatography* (Palo Alto, Calif.: Hewlett Packard Co., 1981).]

For column efficiency and speed of analysis, H_2 is a good choice of carrier gas. In programmed temperature runs, the viscosity (resistance to flow) of the gas increases as the temperature is raised. If the column is run at a constant pressure, the flow rate decreases as the temperature increases. Therefore, another advantage of H_2 is that resolution does not change very much as the flow decreases, provided that the entire run is conducted in the flat lower region of the curve in Figure 23-6. Since H_2 forms explosive mixtures with air, it must be vented safely. He is frequently used instead of H_2 because of this safety hazard, but He is not a renewable resource.

Different detectors either require, or operate better with, different gases (as will be discussed later in this section under "Detectors"). In capillary column chromatography, the flow through the column may be much slower than that needed for the best detector performance. Therefore the optimum gas for separation can be used in the column, and the best gas for detector operation can be added to the flow between the column and the detector.

Gas added to the stream after the column is called *makeup gas.*

Impurities in the carrier gas slowly degrade the column by chemical reaction with the stationary phase, increasing the number of polar (strongly adsorbing) sites and increasing bleeding of the stationary phase during chromatography. Thus high-quality gases should be used, and even these are often passed through purifiers to further reduce unwanted components. Molecular sieves remove water and small hydrocarbons, while oxygen scrubbers of various types remove O_2, which reacts with the stationary phase at high temperatures.

Sample Injection

Introducing a liquid sample into a packed column requires a sample port, usually containing a glass tube inside a hot metal block. The sample is injected by syringe through a rubber septum and is vaporized in the glass tube. Carrier gas sweeps the vaporized sample through the chromatography column. Analytical columns typically require 0.1–10 μL of liquid sample, whereas preparative columns can handle 20–1 000 μL. Gases can be introduced by gas-tight syringe or by gas-sampling valves. Analytical work usually requires 0.5–10 mL of gas, while preparative columns can handle up to 1 L of gas.

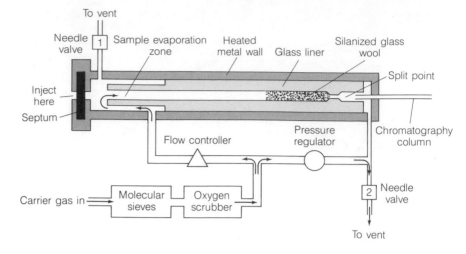

Figure 23-7
Port for split or splitless injection into an open tubular column.

More sophisticated injection ports are required for open tubular columns, which cannot handle large samples. The injector in Figure 23-7 contains a glass liner and glass wool, both of which slowly become contaminated with nonvolatile and decomposed samples. These parts must be replaced periodically.

Split injection is a means of reducing sample volume for open tubular columns. Only 0.1–10% of the 0.1–2 μL of injected sample reaches the column. The remainder is discarded. In Figure 23-7 the sample is injected through the septum into the sample evaporization zone. The needle valve 1 is closed. Carrier gas from the flow controller sweeps the sample through silanized glass wool, where complete vaporization and good mixing occur. At the split point some of the sample enters the chromatography column and the remainder goes through needle valve 2 to a waste vent. The fraction of sample discarded is controlled by the pressure regulator leading to needle valve 2. Fractionation of a sample can occur during injection if some components are not completely vaporized, leading to errors in quantitative analysis. If the injector temperature is too high, decomposition can occur, with loss of some components and creation of new species not present in the original sample.

Split injection is not useful for trace analysis, since most of the sample is discarded. For good quantitative analysis and for trace analysis, **splitless injection** is appropriate. In this case, a dilute solution of sample in a lower boiling solvent is injected into the port in Figure 23-7, with needle valves 1 and 2 closed. The initial column temperature is 20–25° lower than the boiling point of the solvent, which therefore condenses at the beginning of the column. As solutes catch up with the condensed plug of solvent, they are trapped in a small band at the beginning of the column. This *solvent trapping* leads to sharp chromatographic peaks. Some of the solvent (and sample) remains as a vapor near the septum, which would lead to tailing. Therefore after 20–60 s needle valve 1 is opened and vapors near the septum are swept from the port. In splitless injection, most of the sample (perhaps 80%) is applied to the column, and little fractionation of solutes occurs during injection. An alternative means of condensing solutes in a narrow band at the beginning of the column is called *cold trapping*. The initial column temperature is 150° lower than the boiling points of the solutes of interest. Solvent and low-boiling components are eluted rapidly, but high-boiling solutes remain in a

Injection into open tubular columns:

Split—routine means of introducing small sample volume into open tubular column

Splitless—best for trace levels of high-boiling solutes in low-boiling solvents. Better than split injection for quantitative analysis.

On-column—best for thermally unstable solutes and high-boiling solvents narrow band. Upon warming the column, chromatography of the high-boiling solutes occurs.

On-column injection is used for samples that might decompose upon heating above their boiling point during injection. The sample solution is inserted directly into the beginning of the chromatography column, without going through a hot injector. The initial column temperature is near the boiling point of the low-boiling solvent, which therefore condenses and traps solutes in a narrow zone. Chromatography occurs as the column temperature is raised. Samples are subjected to the lowest possible temperatures in this procedure, and little loss of any solute occurs.

Detectors

Thermal Conductivity Detector

Thermal conductivity measures the ability of a substance to transport heat from a hot region to a cold region.[†] The greater the thermal conductivity, the faster heat is carried. Shown in Figure 23-8, the **thermal conductivity detector** usually consists of a hot tungsten–rhenium filament over which is directed the gas emerging from the chromatography column. The electric resistance of the filament increases as the temperature of the filament increases. As long as the carrier gas is flowing at a constant rate over the filament, the resistance is constant and the output signal is constant. When a solute emerges from the column, the thermal conductivity of the gas stream decreases, so the rate at which the filament is cooled by the gas stream decreases proportionally. The filament becomes hotter, its resistance increases, and a change in the output signal is observed.

Since the detector responds to *changes* in thermal conductivity of the gas stream, it is desirable that the conductivities of solute and carrier gas be as different as possible. Table 23-3 shows that H_2 and He have the highest thermal conductivity, which generally decreases as molecular weight increases.

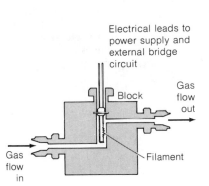

Electrical leads to power supply and external bridge circuit

Block

Gas flow out

Gas flow in

Filament

Figure 23-8
Schematic diagram of a thermal conductivity detector. [Courtesy Varian Associates, Palo Alto, Calif.]

[†] The energy per unit area per unit time flowing from a hot region to a cold region is given by: energy flow $(J/m^2 \cdot s) = -\kappa \, (dT/dx)$, where κ is the thermal conductivity [units $= J/(K \cdot m \cdot s)$] and dT/dx is the temperature gradient (K/m).

Table 23-3
Thermal conductivity at 273 K and 1 atm

Gas	Thermal conductivity $J/(K \cdot m \cdot s)$	Molecular weight
H_2	0.170	2
He	0.141	4
NH_3	0.021 5	17
N_2	0.024 3	28
C_2H_4	0.017 0	28
O_2	0.024 6	32
Ar	0.016 2	40
CO_2	0.014 4	44
C_3H_8	0.015 1	44
Cl_2	0.007 6	71

Thermal conductivity detector:

1. Responds linearly over four orders of magnitude of solute concentration.
2. H_2 and He give best sensitivity.
3. To increase sensitivity:
   (a) Increase filament current (but do not exceed maximum).
   (b) Decrease flow rate in detector.
   (c) Lower detector block temperature (but solutes must remain gaseous).

These two gases are therefore the carriers of choice when a thermal conductivity detector is employed.

It is common practice to split the carrier gas into two streams, sending part of it through the column and part over a reference filament. The resistance of the working filament is measured with respect to that of the reference filament. The sensitivity increases roughly with the square of the filament current. However, the maximum current recommended by the manufacturer should not be exceeded, to avoid burning out the filament. The filament should never be left on without carrier gas flow since it might overheat and be ruined.

The sensitivity of a thermal conductivity detector (but *not* the flame ionization detector described next) is inversely proportional to flow rate. This means that the detector is more sensitive at a lower flow rate and that its sensitivity will change whenever flow rate changes. The sensitivity also increases with greater temperature differences between the filament and the surrounding block in Figure 23-8. The detector block should therefore be maintained at the lowest temperature that will allow all solutes to remain gaseous. Detection limits and linear response ranges for several detectors are given in Table 23-4.

Flame ionization detector

A schematic diagram of the **flame ionization detector** is shown in Figure 23-9. The column eluate is mixed with H_2 and air and burned in a flame inside the detector. Carbon atoms of organic compounds (with the notable exceptions of carbonyl and carboxyl carbons) can produce CH radicals, which go on to produce CHO^+ ions in the hydrogen–oxygen flame.

$$CH + O \rightarrow CHO^+ + e^-$$

Only about one in 10^5 carbon atoms produces an ion. However, this production is steady and is strictly proportional to the number of susceptible carbon atoms entering the flame. The flame ionization detector is insensitive to inorganic compounds such as O_2, CO_2, H_2O, and NH_3.

The CHO^+ produced in the flame carries current to the cathodic collector above the flame. The current flowing between the anode at the base of the

Table 23-4

Detection limits and linear ranges of gas chromatography detectors

Detector	Approximate detection limit	Linear range
Thermal conductivity	400 pg/mL (propane)	$>10^5$
Flame ionization	2 pg/s	$>10^7$
Electron capture	As low as 5 fg/s	10^4
Flame photometric	<1 pg/s (phosphorus)	$>10^4$
	<10 pg/s (sulfur)	$>10^3$
Alkali flame	100 fg/s	10^5
Fourier transform infrared	200 pg to 40 ng	10^4
Mass spectrometric	25 fg to 100 pg	10^5

SOURCE: D. G. Westmorland and G. R. Rhodes, *Pure. Appl. Chem.*, **61**, 1147 (1989).

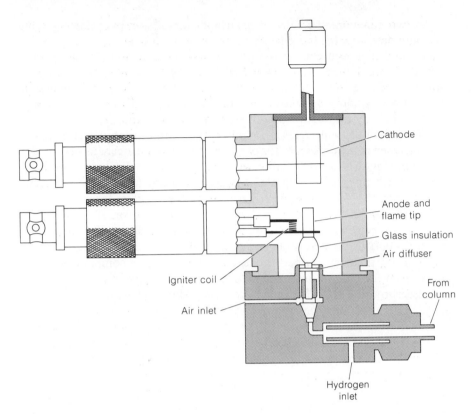

Cathode

Anode and
flame tip

Glass insulation

Air diffuser

From
column

Igniter coil

Air inlet

Hydrogen
inlet

Figure 23-9
Schematic diagram of a flame ionization
detector. [Courtesy Varian Associates,
Palo Alto, Calif.]

flame and the cathodic collector is measured and translated into a signal on a recorder. In the absence of organic solutes, the current is almost zero, and the detector response to organic compounds is directly proportional to solute mass over seven orders of magnitude. This sensitivity is about two orders of magnitude greater than that of the thermal conductivity detector (Table 23-4). The sensitivity of a flame ionization detector is about twice as great when N_2 is the carrier gas as when He is the carrier. With an open tubular column, excellent results are obtained with H_2 or He as carrier and N_2 added as a makeup gas to increase the flow rate as the eluate enters the detector.

Other Detectors

The **electron capture detector** measures a loss of signal when analyte is eluted from the chromatography column. The carrier gas can be either very dry N_2 or 5% methane in Ar. Alternatively, these can be added as makeup gases just before the detector, if H_2 or He is the carrier gas. Nitrogen gas entering the detector is ionized by high-energy electrons ("beta rays") emitted from a foil containing radioactive ^{63}Ni or ^3H. Electrons thus formed are attracted to an anode, producing a small, steady current. When analyte molecules with a high electron affinity enter the detector, they capture some of the electrons and reduce the current to the anode. This detector is particularly sensitive to halogen-containing molecules, conjugated carbonyls, nitriles, nitro compounds, and organometallic compounds. The electron capture detector is insensitive to hydrocarbons, alcohols, and ketones. Carrier and makeup gases must be extremely dry, because water decreases the sensitivity. The ^{63}Ni source can be operated at higher temperature than the ^3H source, and is therefore less prone to buildup of contamination on the surface of the foil.

Flame ionization detector:

1. Use of N_2 is best for detector sensitivity, but H_2 and He can be used.
2. Signal is proportional to number of susceptible carbon atoms.
3. Is more than 10^2 times as sensitive as thermal conductivity.
4. Responds linearly over seven orders of magnitude of solute concentration.
5. Has a very stable background (baseline).

A **flame photometric detector** is an optical emission photometer useful for compounds containing phosphorus or sulfur, such as those in pesticide and air pollution samples. Eluate passes through a hydrogen–oxygen flame, as in the flame ionization detector. Sulfur and phosphorus species are promoted to excited states that emit characteristic light. Phosphorus emission at 536 nm or sulfur emission at 394 nm can be isolated by a narrow-band interference filter and detected with a photomultiplier tube.

Electron capture detector: good for molecules with halogens, conjugated $C=O$, $-C\equiv N$, $-NO_2$

Flame photometric detector: responds to phosphorus and sulfur

Alkali flame detector: responds to phosphorus and nitrogen

The **alkali flame detector** is a modified flame ionization detector that is selectively sensitive to phosphorus or phosphorus and nitrogen. It is especially important for drug analysis. Ions produced by these elements when they contact a Rb_2SO_4-containing glass bead at the burner tip create a current that is measured. N_2, He, or H_2 can be used as carrier gases for phosphorus-containing samples, but N_2 cannot be used for nitrogen-containing samples.

Mass spectrometers and *Fourier transform infrared spectrophotometers* (Section 20-4) are widely used in conjunction with gas chromatography. Modern instruments provide the entire spectrum of each component as it is eluted from the column, and can perform qualitative analysis by comparing the result to a library of spectra. A commercially available *atomic emission detector* directs the chromatography eluate through a helium plasma in a microwave cavity. Every element of the periodic table produces characteristic atomic emission that can be detected by a photodiode array–polychromator (Section 20-2). You can set the detector to observe any element you choose in each analyte as it emerges from the column.

Analytical Methods

Qualitative Analysis

Numerous spectrometric methods of identifying chromatographic peaks are available. As we just mentioned, each peak can be directed into a mass spectrometer or Fourier transform infrared spectrophotometer to record a complete spectrum as the substance is eluted from the column. Alternatively, eluate can be condensed and collected and subjected to any desired analysis, such as nuclear magnetic resonance. For this purpose, a nondestructive detection method, such as thermal conductivity, must be used. Flame detectors burn the sample.

Co-chromatography with authentic sample provides nearly conclusive identification if carried out on two or more different types of columns.

The simplest method of identification of a chromatographic peak is comparison of its retention time with that of an authentic sample of the suspected compound. The most reliable way to do this is by **co-chromatography,** in which authentic sample is added to the unknown. If the added compound is identical to one component of the unknown, the area of that one peak will increase with respect to the others in the chromatogram. This identification is only tentative when it is carried out with a single type of column, but it is nearly conclusive when carried out with two or more different kinds of columns. It is unlikely that two compounds with the same retention time on one stationary phase will have the same retention times on two different stationary phases.

The Kovats index relates the retention time of a solute to the retention times of linear alkanes.

The *retention index* is a measure of relative retention time. One common definition is the **Kovats index,** which is a logarithmic scale on which the adjusted retention time of a peak is compared with those of linear alkanes.

An example was shown in Figure 22-11. The unknown had a retention time between those of octane and nonane. The Kovats index, I, for the unknown is calculated from the formula

$$I = 100 \left[n + (N - n) \frac{\log t'_r(\text{unknown}) - \log t'_r(n)}{\log t'_r(N) - \log t'_r(n)} \right] \quad (23\text{-}2)$$

where n is the number of carbon atoms in the *smaller* alkane

N is the number of carbon atoms in the *larger* alkane

$t'_r(n)$ is the adjusted retention time of the *smaller* alkane

$t'_r(N)$ is the adjusted retention time of the *larger* alkane

This definition makes the Kovats index for a linear alkane equal to 100 times the number of carbon atoms. For octane $I = 800$, and for nonane $I = 900$.

EXAMPLE: Kovats Index

Suppose that the unadjusted retention times in Figure 22-11 are $t_r(\text{air}) = 0.5$ min, $t_r(\text{octane}) = 14.3$ min, $t_r(\text{unknown}) = 15.7$ min, and $t_r(\text{nonane}) = 18.5$ min. The Kovats index for the unknown is

$$I = 100 \left[8 + (9 - 8) \frac{\log 15.2 - \log 13.8}{\log 18.0 - \log 13.8} \right] = 836$$

The retention index of an unknown measured on several different columns is useful for identifying the unknown by comparison with tabulated retention indexes. For a homologous series of compounds (those with similar structures, but differing by the number of CH_2 groups in a chain, such as $CH_3CO(CH_2)_nCO_2CH_2CH_3$) $\log t'_r$ is usually a linear function of the number of carbon atoms.

Quantitative Analysis

When using chromatography for quantitative analysis, the area of each peak is related to the quantity of that component in the mixture. We normally choose conditions under which the response is *linear*, which means that *the area of a peak is proportional to the amount of that component.* Figure 23-10 illustrates linear response to I^- and nonlinear response to Br^- by an electron capture detector. In the discussion that follows, it is assumed that response is linear, which greatly simplifies the analysis.

Common ways to measure chromatographic peak area, in order of preference, are the following:

1. Modern chromatographs usually include a computer that automatically calculates peak area. If the baseline slopes, it is important that the slope be taken into account prior to calculating peak area.[†]

2. Peak areas can be measured by tracing the edges with a mechanical device called a *planimeter*.

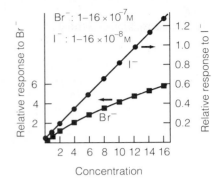

Figure 23-10
Calibration curve for electron capture detector response to iodide and bromide in natural waters. Iodate that may be present is first reduced to I^- with thiosulfate. Then Br^- and I^- are oxidized to Br_2 and I_2 with chromic acid in the presence of acetone, which reacts to give $BrCH_2COCH_3$ and ICH_2COCH_3. The products are extracted into benzene and analyzed by gas chromatography, using a constant concentration of *p*-dichlorobenzene as *internal standard*. The graph compares detector response for each compound to that of the internal standard. [From L. Maros, M. Káldy, and S. Igaz, *Anal. Chem.,* **61**, 733 (1989).]

A plot of $\log t'_r$ versus (number of carbon atoms) is a fairly straight line for a homologous series of compounds.

Linear detector response means that peak area is proportional to analyte concentration. For very sharp peaks, peak height is often substituted for peak area.

[†] Do not place blind faith in an integrator. For a discussion of integration errors, see A. N. Papas and M.F. Delaney, *Anal. Chem.,* **59**, 55A (1987).

3. For a Gaussian peak, the product of peak height times width at half-height equals 84% of the total area.

4. A triangle can be drawn with two sides tangent to the inflection points on each side of the peak. When the third side is drawn along the baseline, the area of the triangle ($=\frac{1}{2}$(base × height)) is 96% of the area of a Gaussian peak.

If detectors responded equally to all solutes, the relative peak areas would be equal to the relative concentrations of each. However, detectors do not respond equally to different compounds. For example, the signal from a thermal conductivity detector depends on the thermal conductivity of the analyte. The smaller the thermal conductivity, the greater the response for a given quantity of compound. An empirical **response factor** for each substance must therefore be measured for quantitative analysis. The data in Figure 23-10 were obtained by measuring the peak area of each analyte with respect to the peak area of the **internal standard,** *p*-dichlorobenzene. That is, a constant concentration of *p*-dichlorobenzene was added to each sample and the area of each analyte peak was compared to that of the standard. The response factor, *F*, is defined by the relation

$$F = \frac{C_{sol}}{C_{st}} \frac{A_{st}}{A_{sol}}$$

Definition of response factor, F

$$\frac{C_{st}}{C_{sol}} \alpha \frac{A_{st}}{A_{sol}} \qquad \frac{\text{Concentration of solute}}{\text{Concentration of standard}} = F\left(\frac{\text{area of solute}}{\text{area of standard}}\right) \qquad (23\text{-}3)$$

$$F = \frac{C_{st}}{C_{sol}} \frac{A_{sol}}{A_{st}}$$

EXAMPLE: Measuring a Response Factor

A mixture containing 8.0×10^{-8} M I^- and 6.4×10^{-8} M *p*-dichlorobenzene gave the relative detector response (I^- peak area)/(*p*-dichlorobenzene peak area) = 0.71. Calculate the response factor for I^-.

Plugging into Equation 23-3 we get

$$\frac{8.0 \times 10^{-8} \text{ M}}{6.4 \times 10^{-8} \text{ M}} = F\left(\frac{0.71}{1}\right) \quad \Rightarrow \quad F = 1.76$$

That is, the detector responds 1.76 times more to *p*-dichlorobenzene than to I^- when both are present at equal concentration.

EXAMPLE: Using an Internal Standard and a Response Factor

Now suppose that a new mixture is made that contains 9.3×10^{-7} M *p*-dichlorobenzene and an unknown concentration of I^-. When chromatography is performed, the quotient of areas is found to be

$$\frac{\text{Area of solute}}{\text{Area of standard}} = 1.21$$

Find the concentration of I^- in the sample.

Once again we use Equation 23-3, with the response factor measured in the previous example.

$$\frac{[I^-]}{9.3 \times 10^{-7} \text{ M}} = 1.76\left(\frac{1.21}{1}\right) \quad \Rightarrow \quad [I^-] = 19.8 \times 10^{-7} \text{ M}$$

Before this result can be relied upon, it is necessary to construct a calibration curve using similar conditions to demonstrate that the response is linear up to this I^- concentration.

Modern chromatography evolved from the type of experiment depicted in Figure 22-9, in which analyte and eluent are applied to the top of an open, gravity-fed column containing stationary phase. This technique is widely used for preparative chemistry and biochemistry, in which relatively large quantities (milligrams to grams) of material are to be isolated. For analytical purposes, closed high-pressure columns capable of exquisite separations of tiny quantities (picograms to micrograms) of analyte are most common. This **high-performance liquid chromatography (HPLC)** will be discussed in the next section. Various types of chromatography (such as ion exchange and molecular exclusion, which are discussed later in this chapter) can be practiced in both gravity-fed and pumped columns for preparative or analytical work. HPLC can also be employed for preparative separations, but open columns are still more commonly used. Liquid chromatography is vital in chemistry, because most compounds are not sufficiently volatile or do not have sufficient thermal stability for gas chromatography.

Columns

Broadening of the analyte zone during chromatography is a natural consequence of diffusion and the finite rate of equilibration of analyte between the mobile and stationary phases. Broadening that occurs because of imperfect application of analyte to the column and mixing of zones during flow to the detector causes unnecessary loss of resolution. The column in Figure 23-11 is designed to minimize such broadening. The stationary phase is supported by a porous nylon net beneath which is a very small dead space leading to the exit tube. An adjustable **flow adaptor** is pressed tightly against the top of the solid phase so that no volume is available for sample and solvent to mix above the column. The inlet and outlet tubing have a 1-mm diameter in order to minimize mixing of liquids within the tubing. Each of these measures decreases band spreading and increases resolution.

Stationary Phase

The most common stationary phase is *silica gel* ($SiO_2 \cdot xH_2O$, also called silicic acid), obtained by acid precipitation of silicate solutions. The pore size, surface area, and surface pH depend on the conditions of preparation. The active adsorption sites are surface Si—O—H (silanol) groups, which can adsorb water from the air and are thereby slowly deactivated. The gel can be re-activated by heating to 200°C to drive off the water. Higher temperatures cause an irreversible dehydration with loss of surface area. Silica is weakly acidic and interacts most strongly with basic solutes.[†] The surface of the silica may be chemically modified to alter its interaction with solutes.

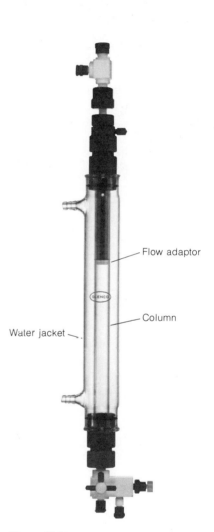

Flow adaptor

Column

Water jacket

Figure 23-11
Glass chromatography column with a flow adaptor at the top. The column is enclosed by a jacket through which water from a constant-temperature bath may be circulated to keep delicate biological samples refrigerated. [Courtesy Glenco Scientific Co., Houston, Tex.]

[†] An order to avoid contamination from previous use, researchers rarely reuse silica gel. However, regeneration and reuse is economical for student work. To a boiling mixture of 500 g of used silica gel in 500 mL of water is added dropwise 50 mL of 5% (wt/wt) aqueous sodium hypochlorite (commercial bleach solution). The pH should be maintained below 7 with HCl during the addition. After 1 h of further boiling, cool, decant, and wash the gel with three 500-mL volumes of water, three 500-mL volumes of 1 M HCl, and sufficient water to bring the pH to 5.5–6. Air dry and then activate by heating to 100–200°C for 1 h. [Procedure from A. E. Guarconi and V. F. Ferriera, *J. Chem. Ed.*, **65**, 891 (1988).]

Alumina grades:

I—anhydrous
II—3% H_2O
III—6% H_2O
IV—10% H_2O
V—15% H_2O

Ion exchange is discussed in Section 23-4.

Alumina ($Al_2O_3 \cdot xH_2O$), the other most common adsorbent, displays increased activity after heating to 1 100°C. Overnight heating to 400°C in air produces highly adsorbent *activity grade I* alumina. Equilibration with various quantities of water (for 1 day with occasional shaking) reduces the activity. Alumina is typically sold in acidic (pH 4), neutral (pH 7), or basic (pH 10) forms. Neutral alumina is used for most nonaqueous separations. It has basic sites that *strongly* adsorb acidic solutes. It also has Lewis acid (electron acceptor, Al^{3+}?) sites that adsorb unsaturated compounds quite well. The relative proportion of acidic sites increases as the activation temperature is increased. Activated alumina may react with esters, anhydrides, aldehydes, and ketones, and may catalyze elimination or isomerization reactions. Basic alumina can also be used for hydrocarbon separations, and it has cation-exchange properties in aqueous solution. Acidic alumina is an anion exchanger and may be used to separate organic anions in aqueous solution.

Cellulose is used as an adsorbent for compounds that are too polar to be eluted from silica gel or alumina. Other adsorbents include activated charcoal, Florisil (a coprecipitate of silica and magnesium oxide), and magnesia ($MgO \cdot xH_2O$). The latter is particularly useful for olefins and aromatic compounds.

Solvents

In adsorption chromatography, the solvent competes with the solute for active adsorption sites on the stationary phase. The relative abilities of different solvents to elute a given solute from the column are nearly independent of the nature of the solute. That is, the elution can be described more as a displacement of solute from the adsorbent, rather than partitioning of the solute between two phases.

An **eluotropic series** (Table 23-5) lists solvents by their relative abilities to displace solutes from a given adsorbent. The **eluent strength** ($\varepsilon°$) is a measure of the solvent adsorption energy, with the value for pentane defined as zero. The eluent strengths in Table 23-5 apply to alumina, but a similar relative order is observed for silica gel. In general, the greater the eluent strength, the more rapidly will solutes be eluted from the column.

In practice, a *gradient* of eluent strength is used for many separations. First, the less highly retained solutes are eluted with a solvent of low eluent strength. Then a second solvent is mixed with the first, either in discrete steps or continuously, increasing eluent strength. By this means, more strongly adsorbed solutes can be eluted from the column. Eluent strength is not a linear function of the relative proportions of each solvent. A small amount of polar solvent will markedly increase the eluent strength of a nonpolar solvent.

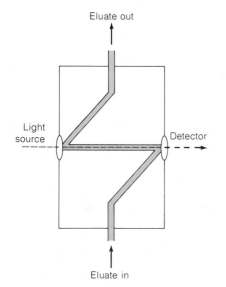

Eluate out

Light source

Detector

Eluate in

Figure 23-12
Light path in a micro flow cell of a spectrophotometric detector. Cells are available having a 0.5-cm path length and containing only 10 μL of liquid.

Detection

The most common means of detecting solutes emerging from a chromatography column is by ultraviolet absorption of the eluate as it passes through a flow cell (Figure 23-12). Other detectors are described in the discussion of high-performance liquid chromatography.

Table 23-5
Eluotropic series

Solvent	ε° (for alumina)	Solvent	ε° (for alumina)
Fluoroalkanes	−0.25	Dichloromethane	0.42
n-Pentane	0.00	Tetrahydrofuran	0.45
i-Octane	0.01	1,2-Dichloroethane	0.49
n-Heptane	0.01	2-Butanone	0.51
n-Decane	0.04	Acetone	0.56
Cyclohexane	0.04	Dioxane	0.56
Cyclopentane	0.05	Ethyl acetate	0.58
Carbon disulfide	0.15	Methyl acetate	0.60
Carbon tetrachloride	0.18	1-Pentanol	0.61
1-Chloropentane	0.26	Dimethyl sulfoxide	0.62
i-Propyl ether	0.28	Aniline	0.62
i-Propyl chloride	0.29	Nitromethane	0.64
Toluene	0.29	Acetonitrile	0.65
1-Chloropropane	0.30	Pyridine	0.71
Chlorobenzene	0.30	2-Propanol	0.82
Benzene	0.32	Ethanol	0.88
Bromoethane	0.37	Methanol	0.95
Diethyl ether	0.38	1,2-Ethanediol	1.11
Chloroform	0.40	Acetic acid	Large

Source: S. G. Perry, R. Amos, and P. I. Brewer, *Practical Liquid Chromatography* (New York; Plenum Press, 1972); and L. R. Snyder, *Principles of Adsorption Chromatography* (New York: Marcel Dekker, 1968).

Choosing Conditions

Two procedures are commonly used to select conditions for a chromatographic separation. One uses **thin-layer chromatography** (TLC) with different solvents and stationary phases to find the best conditions. In TLC, a tiny spot of sample is applied near the bottom of a sheet of glass or plastic coated with a thin layer of stationary phase. When the plate is placed in a shallow pool of solvent in a closed chamber, solvent migrates up into the stationary phase by capillary action, performing a chromatographic separation of the sample along the way (Color Plate 17). When solvent nears the top of the plate, the plate is removed from the solvent and dried. Spots can be made visible by placing the plate in a warm chamber containing a crystal of I_2. The I_2 vapor is reversibly adsorbed on most substances and creates a dark spot wherever a compound is located.

Alternatively, many TLC plates contain a fluorescent material whose emission is *quenched* (reduced) by most solutes. After the solvent evaporates, the plate is viewed under an ultraviolet lamp in the dark. Solute spots appear dark, while the rest of the plate is bright. With TLC, you can experiment rapidly with different solvents and stationary phases to find a suitable combination for chromatography.[†]

Two ways to select chromatographic conditions:

1. thin-layer chromatography
2. column chromatography in Pasteur pipet

[†] Although TLC is usually conducted in the low-resolution manner described above, high-resolution separations can be performed with proper equipment and forced flow of solvent. For a review, see C. F. Poole and S. K. Poole, *Anal. Chem.*, **61**, 1257A (1989).

Alternatively, a disposable Pasteur pipet with a plug of glass wool makes a fine chromatography column with a total volume of about 1 mL. It is possible to run several such columns in a short time to select conditions for a larger-scale separation.

23-3 HIGH-PERFORMANCE LIQUID CHROMATOGRAPHY

A representative HPLC system is shown in Figure 23-13. The column is the short length of steel tubing at the center (typically 10–30 cm in length with an inner diameter of 2–5 mm). Other essential components that will be described in this section include a solvent delivery system, an injection valve for sample, a detector, and a recorder or computer to display results.

The resolution of liquid chromatography benefits from reduced particle size of the stationary phase. If solute can diffuse rapidly between the mobile and stationary phases, the height equivalent to a theoretical plate is decreased, and the efficiency of a given length of column is increased. In capillary gas chromatography, rapid equilibration of solute between the two phases is achieved by decreasing the thickness of the liquid stationary phase coated on the column wall or solid support, and by decreasing the diameter of the column so that molecules can diffuse quickly from the center of the column to the stationary phase at the outer edge. In liquid chromatography, a corresponding increase in the rate of equilibration of solute between phases is promoted by reducing the dimensions of the stationary-phase particles, which reduces the distance through which solute can diffuse in both phases. Also, the migration paths between particles are more uniform with smaller particles, which decreases the A term of the van Deemter equation (22-39).[†]

[†] Reducing the HPLC column diameter to capillary diameters (as small as 20 μm) also increases the efficiency, but such columns are not in common use. See R. T. Kennedy and J. W. Jorgenson, *Anal. Chem.*, **61**, 1128 (1989).

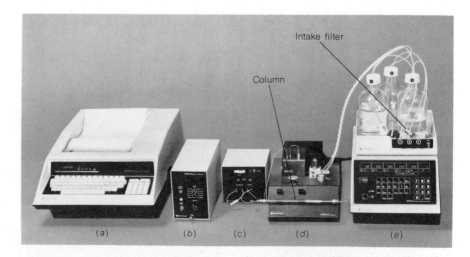

Figure 23-13
Equipment for high-performance liquid chromatography (HPLC). (a) Computing integrator and plotter. (b) Detector control unit. (c) UV detector. (d) Organizer module, including pump, column, and injector. (e) Programmable three-reservoir delivery system. [Courtesy Spectra Physics, Santa Clara, Calif.]

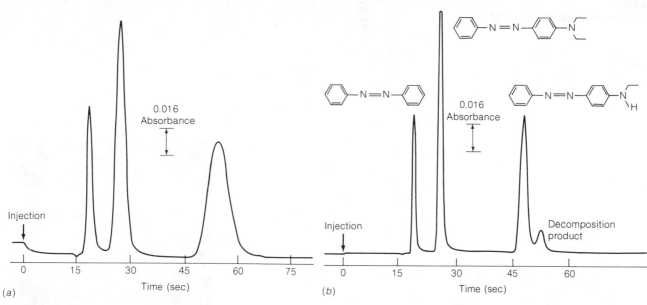

(a)

Injection

0.016
Absorbance

Time (sec)

0   15   30   45   60   75

(b)

Injection

0.016
Absorbance

Decomposition
product

Time (sec)

0   15   30   45   60

Figure 23-14
Chromatograms of the same sample using (a) 10-μm and (b) 5-μm particle diameter silica. [From R. E. Majors, *J. Chromatogr. Sci.*, **11**, 88 (1973).]

Figure 23-14 illustrates the increased resolution afforded by decreasing the stationary-phase particle size from 10 μm to 5 μm. Notice how much sharper the peaks become and how a decomposition product is resolved from the slow-moving component when particle size is reduced. Figure 23-15 shows that smaller particle size decreases plate height, even at higher flow rates.

The penalty for using very fine particles is resistance to solvent flow. It is therefore necessary to use high pressure to force liquid through the column. Pressures of ~7–40 MPa (70–400 atm) are routinely required to attain flow rates of ~0.5–5 mL/min. The column and its associated hardware must be constructed of strong material—usually stainless steel.

Decreased particle size increases chromatographic resolution, but requires high pressure to obtain reasonable flow rate.

Stationary Phase

All types of chromatography can be carried out in the high-performance mode, so stationary phases for every common separation mechanism are available. These includes liquid–solid adsorption, liquid–liquid partition, ion-exchange, molecular-exclusion, and affinity chromatography.

A common support is **microporous particles** of silica with diameters of 5–10 μm (Figure 23-16a). Compared to the irregularly shaped particles, the more expensive spherical variety gives somewhat better separations (10–20% smaller plate height), produces more symmetric peaks, allows about 10% less volume of the mobile phase on the column (called the *void volume*), packs in a more stable manner, and requires 10–30% less pressure for a given flow rate. Compared to both types of microporous particles, spherical **pellicular**

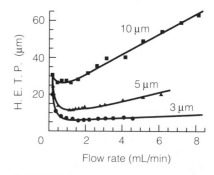

Figure 23-15
Height equivalent to a theoretical plate as a function of flow rate for stationary-phase particle sizes of 10, 5, and 3 μm. [Courtesy Perkin-Elmer Corporation.]

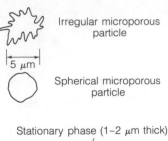

Stationary phase (1–2 μm thick)

Solid glass
core
10–40 μm

Pellicular particle

(a)

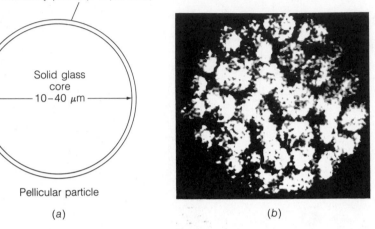

Figure 23-16
(a) Common types of support particles for HPLC. (b) Magnetic resonance imaging of benzene inside a porous 500-μm-diameter polystyrene bead. [From E. Bayer, W. Müller, M. Ilg, and K. Albert, *Angew. Chem. Int. Ed. Engl.*, **28,** 1029 (1989).]

(b)

particles have excellent packing characteristics and lower resistance to solvent flow, but give reduced sample capacity and separation efficiency. Microporous particles are totally permeable to solvent, whereas only the thin coat of a pellicular particle is permeable. Common microporous supports have a surface area of 500 m^2 per gram of silica.

Less rigid supports than silica (such as polystyrene; see Figure 23-25) are also used in HPLC, because of their favorable pore sizes for molecular exclusion separations. Figure 23-16b shows the penetration of a single polystyrene bead by the solvent benzene. White areas in the photograph are rich in benzene. As with other porous particles, most of the inside of the bead is accessible to solvent.

Adsorption chromatography is carried out directly on the surface of the silica particles. More commonly, however, liquid–liquid partition chromatography is conducted with a stationary phase covalently attached to silanol groups on the silica surface. This **bonded phase** is prepared by reactions such as

Residual, unreacted silanol groups on the silica surface are capped with trimethylsilyl groups by reaction with ClSi(CH$_3$)$_3$. This reduces polar adsorption sites that promote peak tailing.

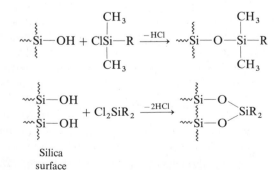

Silica
surface

Common polar phases		Common nonpolar phases	
R = $(CH_2)_3NH_2$	amino	R = $(CH_2)_{17}CH_3$	octadecyl
R = $(CH_2)_3C \equiv N$	cyano	R = $(CH_2)_7CH_3$	octyl
R = $(CH_2)_3OCH(OH)CH_2OH$	diol	R = $(CH_2)_3CH_3$	butyl
		R = CH_2CH_3	ethyl
		R = $(CH_2)_3C_6H_{11}$	hexyl
		R = $(CH_2)_3C_6H_5$	phenyl

The siloxane (Si—O—Si) bond by which the stationary phase is attached to the support is stable over a moderate pH range (typically pH 2–8). Coverage is approximately 4 μmol of R groups per square meter of support surface area, with very little bleeding of the stationary phase from the column during chromatography. Excellent separations are afforded by a bonded stationary phase on 3- to 5-μm-diameter microporous particles. It is common to obtain 50 000–100 000 theoretical plates per meter of column length.

When a polar phase is attached to the support and a less polar solvent is used as the mobile phase, we say that **normal-phase chromatography** is conducted. The eluent strength of the solvent is increased by addition of a more polar solvent. The more important and more common scheme utilizing a nonpolar bonded phase and a polar solvent is called **reverse-phase chromatography.** In this case, the eluent strength is increased by addition of less polar solvent. Reverse-phase chromatography provides excellent separations and eliminates the peak tailing associated with adsorption of polar compounds by polar packings. It is also less sensitive to polar impurities (such as water) in the eluent.

Box 23-2 describes a stationary phase that resembles a cell membrane.

Selecting the Mode of Separation

With so many different kinds of chromatographic stationary phases available, how do you decide which one to use? There are usually several different suitable ways to separate the components of a given mixture. If a reasonable procedure is already described in the literature, most workers will use the known method, rather than develop a new one. If a method has not been described previously, or if the required column is not available, then Figure 23-17 gives a decision tree for choosing a starting point.

If the molecular weight of the analyte is below 2 000, we use the upper part of Figure 23-17; if the molecular weight is greater than 2 000, we use the lower part. In either part, the first question is whether the solutes dissolve in water or in organic solvents. Suppose we have a mixture of small molecules (molecular weight < 2 000) soluble in dichloromethane. Consulting the list of solvents in Table 23-5, we see that nonpolar and weakly polar solvents are in the left half of the table, while moderately polar to polar solvents are in the right half. The eluent strength of dichloromethane (0.42) is closer to that of $CHCl_3$ (0.40) than it is to those of alcohols, acetonitrile, or ethyl acetate (> 0.58). Therefore Figure 23-17 suggests that we try adsorption chromatography, for which microporous silica is by far the most common HPLC stationary phase.

If the solutes are soluble only in hydrocarbons, the decision tree suggests that we try bonded reverse-phase chromatography. Now our choices are many, with available bonded phases containing octadecyl (n-$C_{18}H_{37}$), octyl,

There are no hard and fast rules in Figure 23-17. Methods in either part of the diagram may work perfectly well for molecules whose size belongs to the other part.

Box 23-2 A STATIONARY PHASE THAT MIMICS A CELL MEMBRANE

Animal cells are enclosed by a flexible membrane that limits contact between inside and outside, and in which are embedded numerous proteins, such as transport and receptor molecules. A typical membrane component is a *phospholipid*, which has a highly polar headgroup and a long hydrocarbon tail.

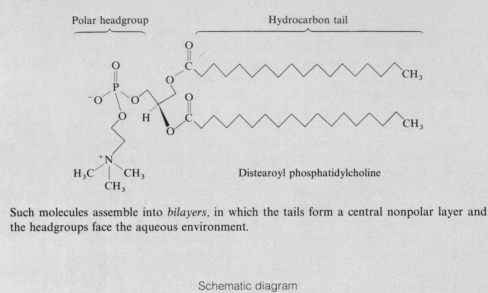

Distearoyl phosphatidylcholine

Such molecules assemble into *bilayers*, in which the tails form a central nonpolar layer and the headgroups face the aqueous environment.

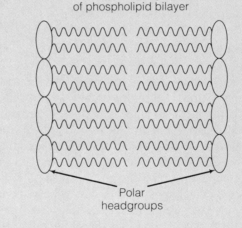

Schematic diagram
of phospholipid bilayer

Polar
headgroups

butyl, ethyl, methyl, phenyl, and cyano groups. The most common reverse-phase supports contain octadecyl groups, and this support may be the only one available in your lab. If several potentially suitable supports are available, you may need to experiment with more than one until you achieve an acceptable separation.

If the molecular weight of the solutes is $> 2\,000$ and they are soluble in

Membrane proteins consist of nonpolar regions that lodge in the central hydrocarbon portion of the membrane and polar regions projecting out of the membrane. Biochemists use *detergents* to extract and dissolve membrane proteins, because detergents have water-soluble ionic headgroups attached to long hydrocarbon tails.

The diagram below shows the chemical structure of a bonded chromatographic stationary phase whose structure resembles half of a lipid bilayer.[†] The membrane protein, cytochrome P-450, can be purified from a crude extract by a single pass through a column containing this stationary phase. After proteins soluble in water and dilute detergent solution are eluted, a more concentrated detergent can be used to elute cytochrome P-450 from the column.

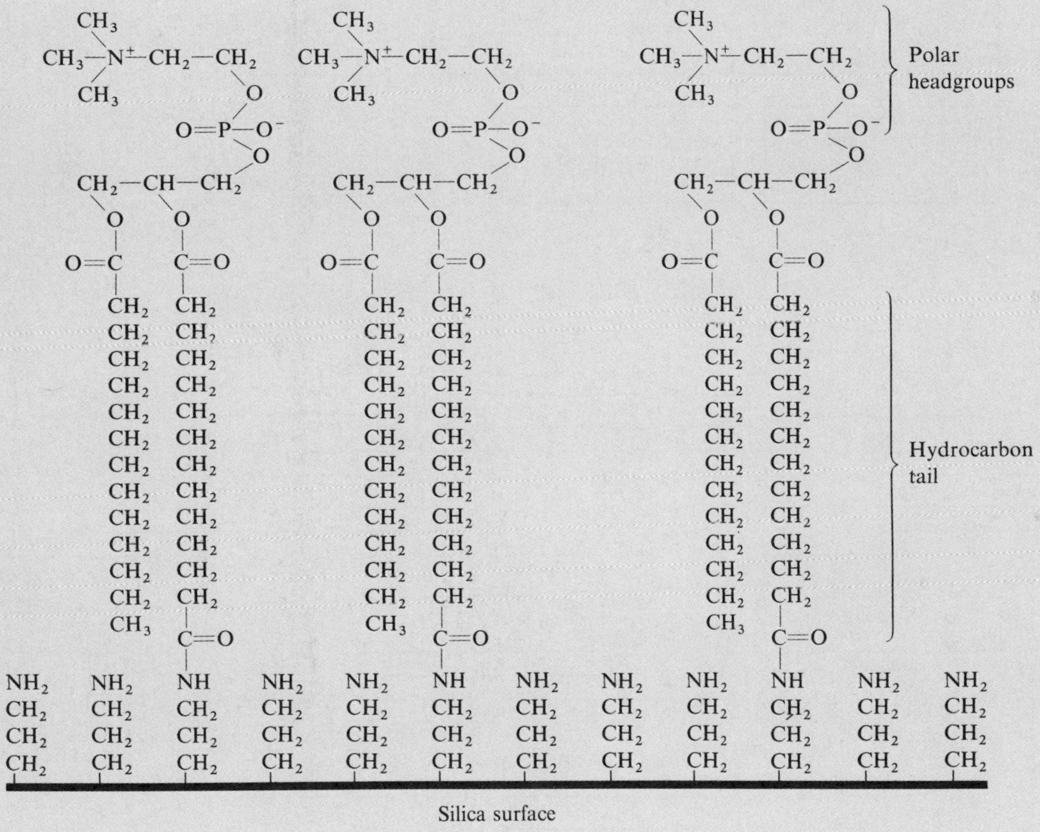

Silica surface

[†] *Chem. Eng. News*, 12 December 1988, p. 23.

organic solvents and their molecular diameter is > 30 nm, Figure 23-17 tells us to try molecular-exclusion chromatography. Stationary phases for this type of separation are described in more detail later. If the molecular weight of solutes is $> 2\,000$, and they are soluble in water, but not ionic, and have diameters < 30 nm, the decision tree says to use bonded reversephase chromatography (such as C_{18}-silica), or *hydrophobic interaction chromatography*.

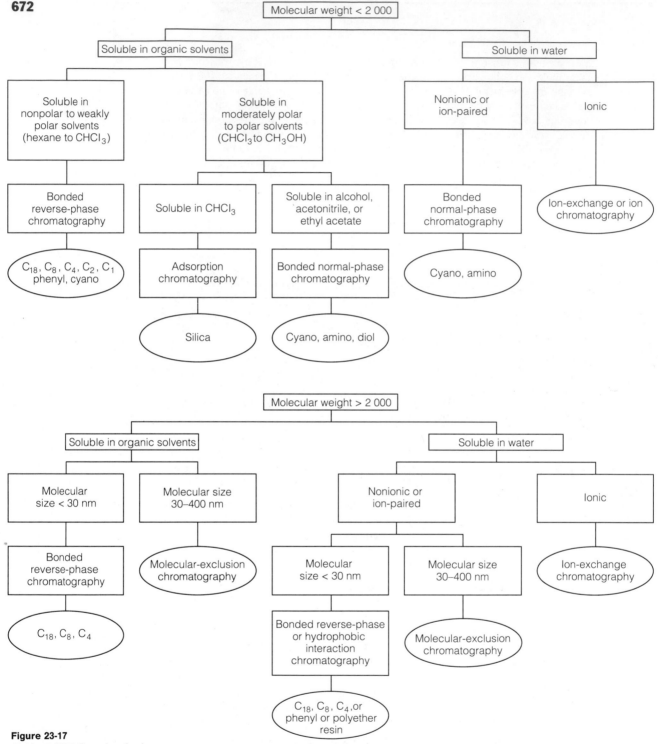

Figure 23-17
Guide to HPLC mode selection.

Glass is an example of a hydrophilic substance that is not soluble in water, but whose surface attracts (is *wetted* by) water. Teflon is an insoluble material whose hydrophobic surface is not wetted by water.

Hydrophobic interaction chromatography is based on the interaction of a hydrophobic stationary phase with a hydrophobic solute. (**Hydrophobic** substances are insoluble in water or do not attract water to their surface. **Hydrophilic** substances are soluble in water or attract water to their surface.) Two commercially available hydrophobic stationary phases are shown in

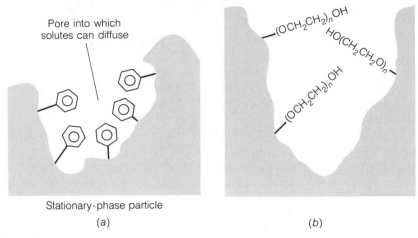

Pore into which
solutes can diffuse

Stationary-phase particle

(a)   (b)

Figure 23-18
Schematic representation of stationary phases for hydrophobic interaction chromatography.

Figure 23-18. The bulk particle is made of styrene–divinylbenzene polymer (Figure 23-25) with pores approximately 100 nm in diameter, through which most molecules can diffuse. The surface of the polymer is coated with many phenyl groups or with poly(ethylene glycol), both of which can interact with hydrophobic solutes and provide a means of chromatographic separation.

Solvents

To avoid degrading expensive HPLC columns and to minimize background signal in detectors, the solvents used for HPLC should be very pure. The intake tubing in the solvent reservoir in Figure 23-13 is protected by a fine filter that rejects dust particles larger than 2–5 μm. Sample and solvent may be passed through a short, expendable **guard column** containing the same stationary phase (often on a pellicular support) as the analytical column, placed between the injection port and the analytical column. Impurities in the solvent and sample that are irreversibly adsorbed by the stationary phase remain in the guard column, which is periodically discarded. Since some impurities absorb ultraviolet light, pure solvents are required to reduce the background absorbance and increase the sensitivity of the detector to analytes. Gas bubbles formed by pressure changes or the mixing of certain solvents interfere with proper functioning of the column and the detector. Therefore solvents are routinely degassed by the chromatographer or the solvent delivery system by evacuation, boiling, or purging with helium, which is very insoluble.

Elution with a single solvent is said to be **isocratic.** If one solvent does not discriminate adequately between several components of a mixture, or if the solvent does not provide sufficiently rapid elution of all components, more than one solvent may be used. Solvent B may simply be substituted for solvent A after an appropriate time. More commonly, a continuous change of solvent composition, called **gradient elution,** is carried out. Figure 23-19 shows an example of reverse-phase chromatography in which a water–acetonitrile gradient was used to elute a series of insecticides from a C_{18}-silica. Increasing the acetonitrile concentration causes solutes that are initially strongly retained by the column to become more soluble and to be eluted.

Elution can be

isocratic—one solvent
gradient—continuous change of solvent composition to increase eluent strength
stepwise—discontinuous change to increase eluent strength

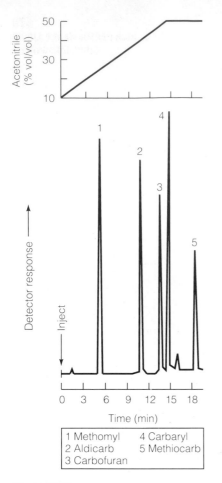

Figure 23-19
Reverse-phase gradient separation of carbamate insecticides using a flow rate of 1.6 mL/min in a 4.6-mm-diameter × 15-cm-long column containing octadecyl groups on spherical 5-μm-diameter microporous particles. Solvent composition was changed linearly from 10% (vol/vol) CH_3CN in water to 50% CH_3CN in water during the first 15 min. [Courtesy Anspec Company.]

Pumps

The quality of a pump is measured by how steady and reproducible a flow it can produce. For some detectors, fluctuations in flow rate give fluctuating response, which is noise that degrades weak signals. Several different pumps are used in HPLC, but we will describe only a piston type designed to produce a programmable, constant flow rate up to 10 mL/min at pressures up to 40 MPa (400 atm).

The pump in Figure 23-20 has two sapphire pistons to give particularly smooth flow. Solvent from the reservoir at the left passes through an electronic inlet valve synchronized with the large piston and designed to minimize the formation of solvent vapor bubbles during the intake stroke. The spring-loaded outlet valve maintains a constant outlet pressure, and the damper is designed to further reduce pressure surges. The midsection of the damper can "breathe" against a constant outside pressure. Surges of pressure from the first piston are decreased by damper volume expansion. Pressure pulses are typically <1% of the operating pressure. As the large piston draws in liquid, the small piston propels liquid into the chromatograph. During the return stroke of the small piston, the large piston delivers solvent into the expanding chamber of the small piston. Part of the solvent fills the chamber, while the remainder flows into the chromatograph. Delivery rate is controlled by decreasing the stroke volumes, not the frequency. Tiny bubbles that may form during operation are passed onto the column. The geometry precludes a buildup of any gas pockets in the pumping mechanism.

Gradients made from up to four solvents are constructed by proportioning the liquids through a four-way valve at low pressure, and then pumping the mixture at high pressure into the column. The gradient-making process is electronically controlled and programmable in 0.1-volume-percent increments from the four channels.

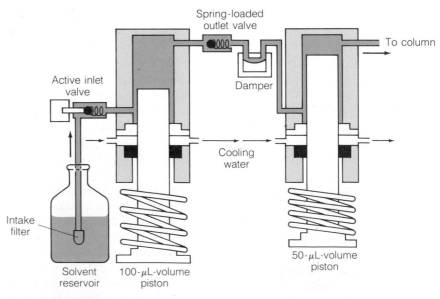

Figure 23-20
High-pressure piston pump for HPLC. [Courtesy Hewlett-Packard Company.]

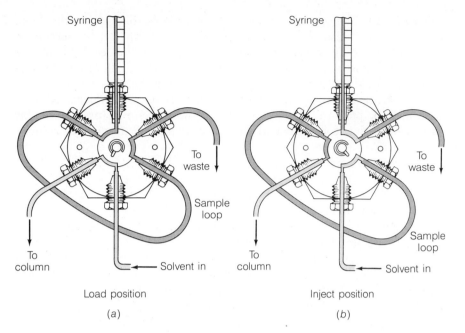

Load position
(a)

Inject position
(b)

Figure 23-21
Injection valve for HPLC. Replaceable sample loop comes in various fixed-volume sizes.

Injection Valve

Although several types of sample injection ports are available, the valve in Figure 23-21 is one of the most popular. The interchangeable sample loop is a steel tube with a narrow bore that holds a fixed volume between 2 and 1 000 μL. In the load position a syringe is used to wash and load the loop with fresh sample at atmospheric pressure. High-pressure flow from the pump to the column passes through the segment of valve at the lower left. When the valve is rotated 60° counterclockwise, the contents of the sample loop are injected into the column at high pressure. The injector, with a handle for rotation, is located on module (d) in Figure 23-13.

Samples should be passed through a 0.5- to 2-μm filter prior to injection, to avoid contaminating the column with particles, plugging the tubing, and damaging the pump.

Column Oven

The chromatography column in Figure 23-13 is kept at ambient temperature. In some systems, the column is enclosed in an oven whose temperature can be regulated between 30° and 100°C. Increased column temperature lowers the viscosity of the solvent, requiring less pressure to produce a given flow rate, or allowing faster flow at a given pressure. Increased temperature also decreases the retention volume of solutes and changes column selectivity for different solutes. The efficiency of the column is increased (smaller plate height) at elevated temperature because the rate of diffusion of solutes in both phases increases, leading to increased equilibration rate between phases.

Linear range: analyte concentration range over which detector response is proportional to concentration

Dynamic range: range over which detector responds in any manner (not necessarily linearly) to changes in analyte concentration

Detection limit: concentration of analyte that gives a signal-to-noise ratio of 2 (Figure 21-19)

Detectors

An ideal detector is sensitive to low concentrations of every analyte, provides linear response (signal proportional to analyte concentration) over a wide range of analyte concentrations, and contributes zero broadening to the eluted peaks. It should also be insensitive to changes in temperature and solvent composition. To avoid peak broadening, the detector flow cell should have a volume that is small (20% or less) compared to the width of the chromatographic peak. Cell volumes of 1–20 μL are common. Gas bubbles in the detector are a serious source of noise, so solvents are usually degassed prior to use. Back pressure may be applied to the detector to prevent bubble formation arising from decreased pressure after passage through the column.

Ultraviolet detector

The ultraviolet detector is the most common HPLC detector, because many solutes absorb ultraviolet light and the detector is fairly sensitive. The flow cell in Figure 23-12 is common, with a typical volume of 10 μL for a pathlength of 1 cm. The simplest systems employ the intense 254-nm emission of a mercury lamp and single-wavelength detection. More versatile instruments use a deuterium lamp and monochromator for variable-wavelength measurements. This allows response to compounds that do not absorb strongly at 254 nm, but absorb in some other region. High-quality detectors provide full-scale absorbance ranges from 0.005 to 3 absorbance units. This means that on the most sensitive scale, an absorbance of 0.005 would give a 100% signal. Noise is around 1% of full scale in the most sensitive range. The linear range extends over five orders of magnitude of solute concentration (which is another way of saying that Beer's law is obeyed over this range). Ultraviolet detectors are good for gradient elution with nonabsorbing solvents. The system in Figure 23-22 uses a photodiode array to record the entire spectrum of each solute as it passes through the detector. It can even search for the absorbance maximum of each peak as it is eluted.

Figure 23-22

Photodiode array ultraviolet detector for HPLC. (a) Dual-beam optical system uses grating polychromator, one diode array for the sample spectrum, and another diode array for the reference spectrum. (b) Reverse-phase chromatography (using C_{18}-silica) of sample containing 0.2 ng of anthracene, with diode array detection at 250 nm. Full-scale absorbance is 0.001. (c) Spectrum of anthracene recorded as it emerged from the column. [Courtesy Perkin-Elmer Corp.] Photodiode arrays are described in Section 20-2.

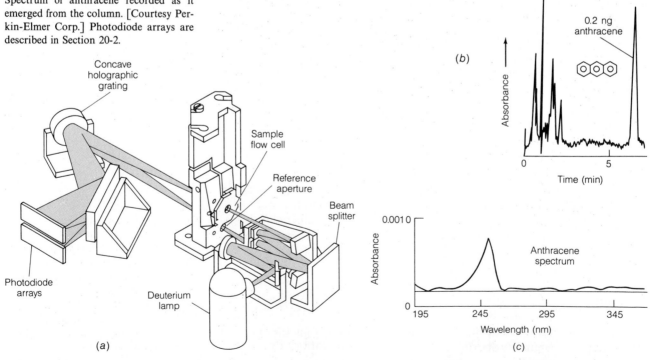

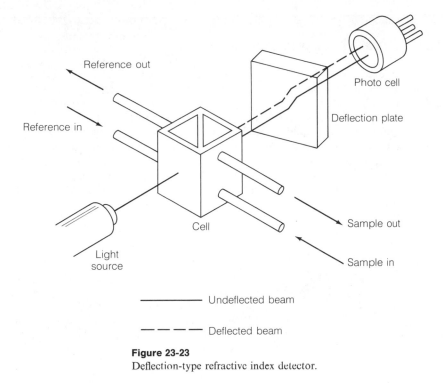

Reference out

Photo cell

Reference in

Deflection plate

Cell

Sample out

Light
source

Sample in

————————— Undeflected beam

— — — — — Deflected beam

Figure 23-23
Deflection-type refractive index detector.

Refractive index detector

The refractive index detector is very nearly a universal detector (responding to almost any solute), but its sensitivity is about 1 000 times poorer than that of the ultraviolet detector, and it is not useful for gradient elution. The principle of operation of a deflection-type refractive index detector is illustrated in Figure 23-23. The cell has two triangular 5- to 10-μL-volume compartments through which are passed a pure solvent reference or column eluate. Light from a tungsten lamp is collimated (made parallel) and filtered to remove infrared radiation, which would heat the sample. The incident light passes through the cell and is directed to the center of the photocell detector by the deflection plate with pure solvent in both compartments. When solute with a different refractive index from that of the solvent is present, the beam is deflected from the center of the photocell and the detector output signal changes.

The relation between angle of refraction and refractive index is described in Section 20-1.

There are more elaborate designs, but all are based on the difference in refractive index of pure solvent and solvent plus solute. In gradient elution it is impossible to match exactly the sample and the reference, so refractive index detectors are useless for this application. These detectors are sensitive to changes in pressure and temperature. Therefore flow rate should be steady (since changes in flow rate represent changes in pressure) and the temperature must be constant within $\sim 0.01°$C. Because of their low sensitivity, they are not useful for trace analysis. They also have a small linear range, spanning only a factor of 500 in solute concentration. The primary appeal of the refractive index detector is its nearly universal response to all solutes, such as carbohydrates and aliphatic polymers that have little ultraviolet absorption.

Box 23-3 SUPERCRITICAL FLUID CHROMATOGRAPHY

This technique fills a gap between gas and liquid chromatography, because the solvent is a fluid whose properties are between those of gas and liquid. Consider the phase diagram of carbon dioxide below. At a temperature of $-78°C$, solid CO_2 (dry ice) is in equilibrium with gaseous CO_2 at 1 atm. We say that the solid *sublimes,* without turning to liquid. Above the triple point temperature of $-56.6°C$, a liquid phase does exist. For each temperature above $-56.6°C$, there is a pressure at which liquid and vapor coexist as separate phases. For example, at $0°C$, liquid is in equilibrium with gas at 34.4 atm. Moving up the liquid–gas dividing line, two phases always exist until the *critical point* is reached at $31.3°C$ and 72.9 atm. *Above this temperature, only one phase exists, no matter what the pressure.* We call this phase a **supercritical fluid.** Its density and viscosity are between those of the gas and liquid, as is its ability to act as a solvent. The table lists critical constants for several compounds.

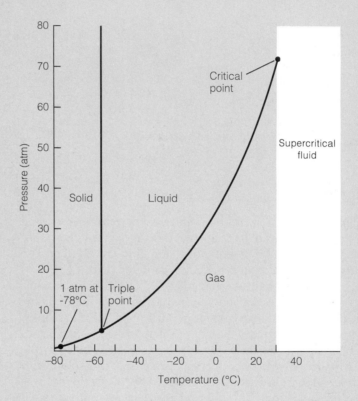

Critical constants

Compound	Critical temperature (°C)	Critical pressure (atm)	Critical density (g/mL)
Carbon dioxide	31.3	72.9	0.448
Ammonia	132.3	111.3	0.24
Water	374.4	226.8	0.344
Methanol	240.5	78.9	0.272
Diethyl ether	193.6	36.3	0.267

Chromatography with a supercritical fluid as the solvent provides increased speed and resolution, compared to liquid chromatography, because of the increased diffusion coefficients of solutes in supercritical fluids, compared to diffusion in liquids. (However, speed and resolution are slower than those of gas chromatography.) Unlike gases, supercritical fluids can dissolve nonvolatile solutes. When the pressure on the supercritical solution is released, the solvent rapidly turns to gas, leaving the solute in the gas phase for easy detection. To date, carbon dioxide has been the supercritical fluid of choice for chromatography because it is compatible with the versatile flame ionization detector of gas chromatography, it has a low critical temperature, and it is nontoxic. Unfortunately, it is not a particularly good solvent for highly polar or high molecular weight solutes.

Equipment for supercritical fluid chromatography is similar to that for HPLC, but columns are more similar to those of gas chromatography. Open tubular columns with a 50-μm diameter and 10 to 20-m lengths are typical. A bonded stationary phase is required, so that it does not dissolve in the solvent. Most detectors compatible with HPLC or gas chromatography are useful, but flame ionization and ultraviolet absorption are most common. Eluent strength is increased in HPLC by gradient elution, and in gas chromatography by raising the temperature. In supercritical fluid chromatography, eluent strength is increased by making the solvent *denser* (by increasing the pressure). The chromatogram below illustrates density gradient elution.

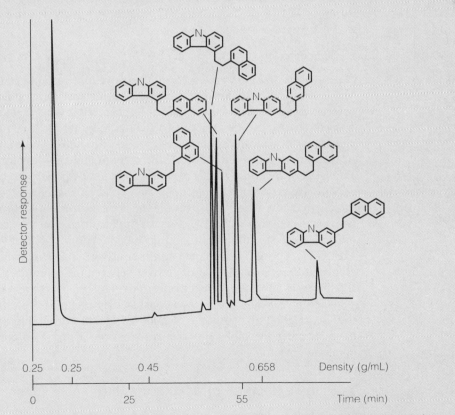

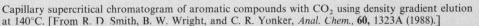

Capillary supercritical chromatogram of aromatic compounds with CO_2 using density gradient elution at 140°C. [From R. D. Smith, B. W. Wright, and C. R. Yonker, *Anal. Chem.,* **60,** 1323A (1988).]

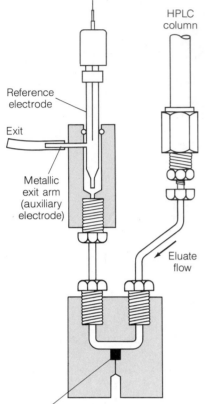

Figure 23-24

Cross-sectional view of an electrochemical detector. A thin layer of eluate passes over the top surface of the working electrode. The auxiliary electrode is the hollow metal sidearm at the exit port on the left. [Courtesy Bioanalytical Systems.]

Anion exchangers contain bound *positive* groups.

Cation exchangers contain bound *negative* groups.

Electrochemical detector

The electrochemical detector is rather selective, because only certain analytes are readily oxidized or reduced. For example, phenols, aromatic amines, peroxides, and mercaptans can be detected by oxidation, and ketones, aldehydes, conjugated nitriles, aromatic halogen compounds, and aromatic nitro compounds can be detected by reduction. Figure 23-24 is a schematic diagram of a three-electrode system in which eluate passes through a narrow channel over the surface of a glassy carbon electrode, which is commonly used for oxidizable solutes. The potential of the electrode is maintained at a selected value with respect to the silver–silver chloride reference electrode, and current is measured between the glassy carbon working electrode and the stainless steel auxiliary electrode. For reducible solutes, a drop of mercury suspended in the eluate stream is a good working electrode. Other common working electrodes include platinum, gold, and carbon paste.

Electrochemical detectors are simple and extremely sensitive for those analytes that can be detected. It is easy to measure nanoampere currents, corresponding to picogram-per-milliliter detection limits. The current is proportional to solute concentration over six orders of magnitude. Aqueous or other polar solvents containing dissolved electrolytes are required, and they must be rigorously free of oxygen. Metal ions that might be picked up from tubing and connections may be rendered electrochemically inactive by addition of EDTA to the solvent. The detector is very sensitive to flow rate and temperature changes.

Other detectors

Fluorescence detectors are especially sensitive, but respond only to the very limited range of analytes that fluoresce. To increase the applicability of fluorescence and electrochemical detectors, chemical groups that are fluorescent or electroactive can be attached to the desired analytes. Such **derivatization** can be performed on the mixture prior to chromatography, or by addition of reagents to the eluate between the column and the detector (called *post-column derivatization*). Other detectors for ionic species are based on *ion-selective electrodes* and *conductivity measurement. Inductively coupled plasma atomic emission, mass spectrometry,* and *Fourier transform infrared spectroscopy* have a variety of HPLC detector applications. The very sensitive *flame ionization detector* of gas chromatography (Section 23-1) can be adapted to HPLC by depositing the eluate continuously on a moving wire that carries eluate into the flame. As the wire moves between the column and the detector the solvent evaporates, allowing only nonvolatile solutes to reach the detector. *Low-angle laser light scattering* is particularly useful for detecting high molecular weight polymers in molecular-exclusion chromatography, and for characterizing the molecular size distribution. Table 23-6 compares some common detectors.

23-4 ION-EXCHANGE CHROMATOGRAPHY

Ion-exchange chromatography is based on the equilibration of solute ions between the solvent and charged sites fixed on the stationary phase (Figure 22-10 and Color Plate 18). In **anion exchangers,** positively charged groups are covalently bound to the packing. Solute anions are attracted to these sites. **Cation exchangers** contain covalently bound, negatively charged sites that bind solute cations.

Table 23-6
Comparison of commercial HPLC detectors

Detector	Approximate limit of detection[†] (ng)	Useful with gradient?
Ultraviolet	0.1–1	yes
Refractive index	100–1 000	no
Electrochemical	0.01–1	no
Fluorescence	0.001–0.01	yes
Conductivity	0.5–1	no
Mass spectrometry	0.1–1	yes
Fourier transform infrared	1 000	yes

[†] Detection limits from E. W. Yeung and R. E. Synovec, *Anal. Chem.*, **58**, 1237A (1986).

Ion Exchangers

Polystyrene and polyacrylic acid resins

Resins are small, amorphous (noncrystalline) particles of organic material. Polystyrene resins for ion exchange are made by the copolymerization of styrene and divinylbenzene (Figure 23-25). The divinylbenzene content is

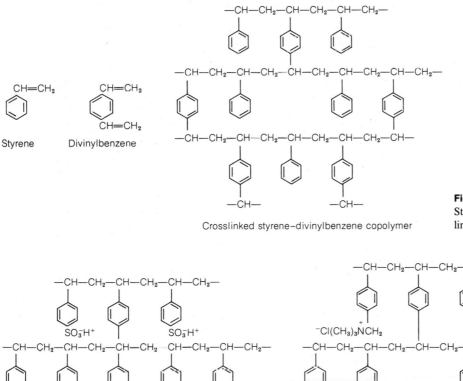

Styrene   Divinylbenzene

Crosslinked styrene–divinylbenzene copolymer

Figure 23-25
Structures of styrene–divinylbenzene cross-linked ion-exchange resins.

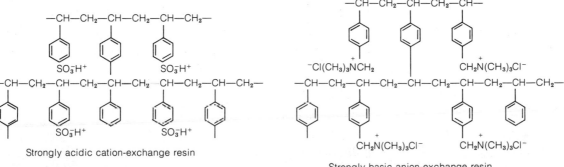

Strongly acidic cation-exchange resin

Strongly basic anion-exchange resin

Table 23-7
Ion-exchange resins

Resin type	Chemical constitution	Usual form as purchased	Common trade names Rohm & Haas	Common trade names Dow Chemical	Selectivity	Thermal stability
Strongly acidic cation exchanger	Sulfonic acid groups attached to styrene and divinyl-benzene copolymer	$\phi-SO_3^-H^+$	Amberlite IR-120	Dowex 50W	$Ag^+ > Rb^+ > Cs^+ > K^+ >$ $NH_4^+ > Na^+ > H^+ >$ Li^+ $Zn^{2+} > Cu^{2+} > Ni^{2+} >$ Co^{2+}	Good up 150°C
Weakly acidic cation exchanger	Carboxylic acid groups attached to acrylic and divinylbenzene copolymer	$R-COO^-Na^+$	Amberlite IRC-50	—	$H^+ \gg Ag^+ > K^+ >$ $Na^+ > Li^+$ $H^+ \gg Fe^{2+} > Ba^{2+}$ $Sr^{2+} > Ca^{2+} > Mg^{2+}$	Good up to 100°C
Strongly basic anion exchanger	Quaternary ammonium groups attached to styrene and divinylbenzene copolymer	$[\phi-CH_2N(CH_3)_3]^+Cl^-$	Amberlite IRA-400	Dowex 1	$I^- >$ phenolate$^- >$ $HSO_4^- > ClO_3^- >$ $NO_3^- > Br^- > CN^- >$ $HSO_3^- > NO_2^- > Cl^- >$ $HCO_3^- > IO_3^- >$ $HCOO^- >$ Acetate$^- >$ $OH^- > F^-$	OH^- form fair up to 50°C Cl^- and other forms good up to 150°C
Weakly basic anion exchanger	Polyalkylamine groups attached to styrene and divinylbenzene copolymer	$[\phi-NH(R)_2]^+Cl^-$	Amberlite IR-45	Dowex 3	$\phi SO_3H > Citric > CrO_3 >$ $H_2SO_4 >$ tartaric $>$ oxalic $> H_3PO_4 >$ $H_3AsO_4 > HNO_3 >$ $HI > HBr > HCl >$ $HF > HCO_2H >$ $CH_3CO_2H > H_2CO_3$	Extensive information not available; tentatively limited to 65°C

SOURCE: Adapted from J. X. Khym, *Analytical Ion-Exchange Procedures in Chemistry and Biology* (Englewood Cliffs, N.J.: Prentice-Hall, 1974).

varied from 1% to 16% to increase the extent of **crosslinking** of the insoluble hydrocarbon polymer. The benzene rings can be modified to produce a cation-exchange resin, containing sulfonate groups ($-SO_3^-$), or an anion-exchange resin, containing ammonium groups ($-NR_3^+$). If methacrylic acid is used in place of styrene, a polymer with carboxyl groups results.

Ion exchangers are commonly classified as being strongly or weakly acidic or basic, as indicated in Table 23-7. The sulfonate groups ($-SO_3^-$) of strongly acidic resins remain ionized even in strongly acidic solutions. The carboxyl groups ($-CO_2^-$) of the weakly acidic resins are protonated at around pH 4 and lose their cation-exchange capacity. "Strongly basic" quaternary ammonium groups ($-CH_2NR_3^+$) remain cationic at all values of pH and function as anion exchangers. The "weakly basic" tertiary ammonium ($-CH_2NHR_2^+$) anion exchangers are deprotonated in moderately basic solution and lose their ability to bind anions.

The extent of crosslinking is indicated by the notation "-XN" after the name of the resin. For example, Dowex 1-X4 contains 4% divinylbenzene, and Bio-Rad AG 50W-X12 contains 12% divinylbenzene. The resin becomes more rigid and less porous as the extent of crosslinking increases. Lightly crosslinked resins permit rapid equilibration of solute between the inside and outside of the particle. On the other hand, resins with little crosslinking swell in water. This decreases both the density of ion-exchange sites and the selectivity of the resin for different ions. More heavily crosslinked resins exhibit less swelling and higher exchange capacity and selectivity, but longer equilibration times.

The pore size of polystyrene and polyacrylic acid resins effectively limits their use to molecules weighing less than 500. Larger molecules cannot penetrate these resins. Also, the charge density of these exchangers is so great that large molecules with many charged groups can be irreversibly bound to the resin.

Cellulose, dextran, and related ion exchangers

These derivatized polysaccharides possess much larger pore sizes and lower density of charged groups. They are well suited to ion exchange of macromolecules, such as proteins. Cellulose and dextran are polymers of the sugar glucose. Dextran, crosslinked by glycerin, is sold under the name Sephadex. (Figure 23-26). Other macroporous ion exchangers are based on the polysaccharide agarose and on polyacrylamide.

The common charged functional groups that are synthetically bound to occasional hydroxyl groups of the polysaccharides are listed in Table 23-8. DEAE-Sephadex, for example, refers to an anion-exchange Sephadex containing diethylaminoethyl groups.

HPLC ion-exchange columns

The ion-exchange groups in Table 23-8 can be attached to silica particles, polystyrene resins, and hydroxylated polyether resins for use in HPLC. Silica is generally only useful in the pH range 2 8, but the synthetic polymers can be used over the pH range 2–12. HPLC ion-exchange stationary phases are further classified by their pore size, which determines the maximum size of solute that can enter the porous particle.

$$H_2C=C\begin{array}{c} CO_2H \\ \\ CH_3 \end{array}$$

Methacrylic acid

Strongly acidic cation exchangers: RSO_3^-

Weakly acidic cation exchangers: RCO_2^-

"Strongly basic" anion exchangers: $RNR_3'^+$

Weakly basic anion exchangers: $RNR_2'H^+$

Because they are much softer than polystyrene *resins*, dextran and its relatives are called **gels.**

See Figure 23-32 for the structure of polyacrylamide.

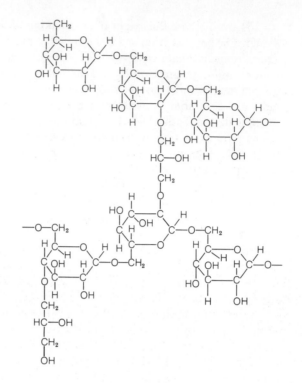

Figure 23-26
Structure of Sephadex, a crosslinked dextran sold by Pharmacia Fine Chemicals, Piscataway, New Jersey.

Table 23-8
Some active groups of ion-exchange gels

Type	Abbreviation	Name	Structure
Cation exchangers			
Strong acid	SP	Sulfopropyl	$-OCH_2CH_2CH_2SO_3H$
	SE	Sulfoethyl	$-OCH_2CH_2SO_3H$
Intermediate acid	P	Phosphate	$-OPO_3H_2$
Weak acid	CM	Carboxymethyl	$-OCH_2CO_2H$
Anion exchangers			
Strong base	TEAE	Triethylaminoethyl	$-OCH_2CH_2\overset{+}{N}(CH_2CH_3)_3$
	QAE	Diethyl(2-hydroxypropyl) quaternary amino	$-OCH_2CH_2\overset{+}{N}(CH_2CH_3)_2$ $CH_2CHOHCH_3$
Intermediate base	DEAE	Diethylaminoethyl	$-OCH_2CH_2N(CH_2CH_3)_2$
	ECTEOLA	Triethanolamine coupled to cellulose through glyceryl chains	
	BD	Benzoylated DEAE groups	
Weak base	PAB	*p*-Aminobenzyl	$-O-CH_2-$⬡$-NH_2$

Table 23-9
Relative selectivity coefficients of ion-exchange resins

	Sulfonic acid cation-exchange resin			Quaternary ammonium anion-exchange resin	
	% divinylbenzene				Relative
Cation	4	8	10	Anion	selectivity
Li^+	1.00	1.00	1.00	F^-	0.09
H^+	1.30	1.26	1.45	OH^-	0.09
Na^+	1.49	1.88	2.23	Cl^-	1.0
NH_4^+	1.75	2.22	3.07	Br^-	2.8
K^+	2.09	2.63	4.15	NO_3^-	3.8
Rb^+	2.22	2.89	4.19	I^-	8.7
Cs^+	2.37	2.91	4.15	ClO_4^-	10.0
Ag^+	4.00	7.36	19.4		
Tl^+	5.20	9.66	22.2		

SOURCE: *Amberlite Ion Exchange Resins—Laboratory Guide* (Rohm & Haas Co., 1979).

Inorganic ion exchangers

The resins and gels mentioned above are unstable at high temperature or in the presence of intense radiation. A number of inorganic substances are stable under these conditions and exhibit useful ion-exchange properties. Hydrous oxides of Zr, Th, Ti, Sn, and W exhibit both anion- and cation-exchange properties. These same metals form insoluble phosphate, molybdate, tungstate, and arsenate salts with cation-exchange properties.

Ion-Exchange Equilibria

Selectivity

Consider the competition of Na^+ and Li^+ for sites on the cation-exchange resin, R^-:

$$R^-Na^+ + Li^+ \rightleftharpoons R^-Li^+ + Na^+ \qquad K = \frac{[R^-Li^+][Na^+]}{[R^-Na^+][Li^+]} \quad (23\text{-}4)$$

The equilibrium constant for Reaction 23-4 is called the **selectivity coefficient,** because it describes the relative selectivity of the resin for Li^+ and Na^+.

The relative selectivities of some strongly acidic and basic polystyrene resins are shown in Table 23-9. Note that the relative selectivities for certain ions increase with the extent of crosslinking. This is because the pore size of the resin shrinks as crosslinking increases. Ions such as Li^+, with a large hydrated radius (Table 6-1), do not have as much access to the resin as smaller ions, such as Cs^+.

In general, ion exchangers favor the binding of ions of higher charge, decreased hydrated radius, and increased **polarizability.** A fairly general order

Polarizability refers to the ability of an ion's electron cloud to be deformed by nearby charges. Deformation of the electron cloud induces a dipole in the ion. The attraction between the induced dipole and the nearby charge increases the binding of the ion to the resin.

of selectivity for cations is the following:

$$Pu^{4+} \gg La^{3+} > Ce^{3+} > Pr^{3+} > Eu^{3+} > Y^{3+} > Sc^{3+} > Al^{3+} \gg$$

$$Ba^{2+} > Pb^{2+} > Sr^{2+} > Ca^{2+} > Ni^{2+} > Cd^{2+} > Cu^{2+} >$$

$$Co^{2+} > Zn^{2+} > Mg^{2+} > UO_2^{2+} \gg Tl^+ > Ag^+ > Rb^+ > K^+ >$$

$$NH_4^+ > Na^+ > H^+ > Li^+$$

This is simply an application of Le Châtelier's principle.

Reaction 23-4 can be driven in either direction, even though Na^+ is bound more tightly than Li^+. Washing a column containing Na^+ with a substantial excess of a Li^+ salt will displace Na^+ and replace it with Li^+. Washing a column in the Li^+ form with Na^+ will convert it to the Na^+ form.

Ion exchangers loaded with one kind of ion will generally bind small amounts of a different ion nearly quantitatively. Thus an Na^+-loaded resin will bind small amounts of Li^+ nearly quantitatively, even though the selectivity is greater for Na^+. The same column will tightly bind large quantities of, for example, Ni^{2+} or Fe^{3+}, because the resin has greater selectivity for these ions than for Na^+. Even though Fe^{3+} is bound more tightly than H^+, Fe^{3+} can be quantitatively removed from the resin by washing with a large excess of acid.

Donnan equilibrium

A phase containing bound charges tends to exclude electrolyte.

When an ion exchanger is placed in an electrolyte solution, *the concentration of electrolyte at equilibrium is higher outside the resin than inside it.* The equilibrium between charged species in solution and charged species inside the resin is called the **Donnan equilibrium.**

Consider a quaternary ammonium anion-exchange resin (R^+) in its Cl^- form immersed in a solution of KCl. Let the concentration of an ion inside the membrane be $[X]_i$ and the concentration outside the membrane be $[X]_o$. It can be shown from thermodynamics that the ion product inside the resin is approximately equal to the product outside the resin. In this example, we can write

$$[K^+]_i[Cl^-]_i = [K^+]_o[Cl^-]_o \tag{23-5}$$

From considerations of charge balance, we know that

We are ignoring H^+ and OH^-, which are assumed to be negligible.

$$[K^+]_o = [Cl^-]_o \tag{23-6}$$

Inside the resin, there are three charged species, and the charge balance is

$$[R^+]_i + [K^+]_i = [Cl^-]_i \tag{23-7}$$

where $[R^+]$ is the concentration of quaternary ammonium ions attached to the resin. Substituting Equations 23-6 and 23-7 into Equation 23-5 gives

$$[K^+]_i([K^+]_i + [R^+]_i) = [K^+]_o^2 \tag{23-8}$$

Equation 23-8 says that $[K^+]_o$ must be greater than $[K^+]_i$.

EXAMPLE: Exclusion of Cations by an Anion Exchanger

Suppose that the concentration of cationic sites in the resin is 6.0 M. When the Cl^- form of this resin is immersed in 0.050 M KCl, what will be the ratio $[K^+]_o/[K^+]_i$?

Let us assume that $[K^+]_o$ remains 0.050 M. Putting values into Equation 23-8 gives

$$[K^+]_i([K^+]_i + 6.0) = (0.050)^2$$

from which we find $[K^+]_i = 0.000\ 42$ M. The ratio $[K^+]_o/[K^+]_i$ is 120. The concentration of electrolyte inside the resin is less than 1% of that outside the resin.

Note that the Donnan theory predicts that ions with the *same* charge as the resin are excluded. (The quaternary ammonium resin excludes K^+.) The counterion, Cl^- in the above example, is *not excluded* from the resin. There is no electrostatic barrier to penetration of any anion into the resin. Anion exchange takes place freely in the quaternary ammonium resin even though cations are repelled from the resin.

> The high concentration of positive charges within the resin repels cations from the resin.

The Donnan equilibrium is the basis of **ion-exclusion chromatography.** Since dilute electrolytes are effectively excluded from the resin, they will pass through a column when the volume of mobile phase (V_m) has been eluted. Nonelectrolytes, such as sugar, freely penetrate the resin. They will not be eluted until a volume $V_m + V_s$ (where V_s is the volume of liquid inside the resin) has passed. Thus, if a solution of NaCl and sugar is applied to an ion-exchange column, the NaCl will emerge from the column *before* the sugar will.

> The volume V_m is available to an electrolyte. The volume $V_m + V_s$ is available to a nonelectrolyte.

Conducting Ion-Exchange Chromatography

Choosing an ion exchanger

In general, ion-exchange *resins* are used for applications involving small molecules (M.W. $\lesssim 500$), which can penetrate the small pores of the resin. Ion-exchange *gels* are used for large molecules (such as proteins and nucleic acids), which could not penetrate the pores of resins. The large molecules often have such great charge that if they could penetrate the resin they might be held too tightly to be eluted. The density of ion-exchange sites in gels is much lower than it is inside resins. For separations involving harsh chemical conditions (high temperature, high radiation levels, strongly basic solution, powerful oxidizing agents), the resins and gels are unsuitable. *Inorganic exchangers* should be tried in such cases.

> Three classes of ion exchangers:
> 1. resins
> 2. gels
> 3. inorganic exchangers

For resins, a mesh size of 100/200 is suitable for most work. Higher mesh numbers (smaller particle size) lead to finer separations, but slower column operation. Very coarse particles are useful for gross separations or batch processes where speed or rapid settling of the resin from a suspension is desired. The selectivity of a resin increases with the degree of crosslinking, but the speed of equilibration decreases. For gels, the size of a macromolecule dictates the minimum pore size that can be used.

> The selectivity of a resin increases with crosslinking.

The choice between strong and weak ion exchangers depends on the operating pH and relative selectivity needed for a separation. A weak-acid (RCO_2^-) cation exchanger becomes protonated below pH ≈ 4 and loses its ion-exchange capacity. Clearly, it would not be useful for a column eluted with a medium containing 1 M HCl. Similarly, a weak-base anion exchanger (R_3NH^+) loses its charge in strongly basic solution. The strongly acidic (RSO_3^-) or strongly basic (R_4N^+) ion exchangers are useful over a greater range of pH. The order of ion selectivites of different ion exchangers is not the same. Manufacturers' catalogs usually contain some selectivity information for each resin.

Gradients

Elution of a column with a **gradient** of ionic strength or pH is extremely valuable in ion-exchange chromatography. The mechanics of creating a gradient in classical chromatography are described in Section 23-8. Consider a column to which the anions A^- and B^- are bound. Suppose the binding of A^- is stronger than that of B^-. One good way to separate A^- from B^- is to elute the column with solution containing the anion C^- (which is less tightly bound than A^- or B^-). At low concentrations of C^-, neither A^- nor B^- is displaced. As the concentration of C^- is increased, B^- will eventually be displaced and move down the column. At a still higher concentration of C^-, the anion A^- will also be eluted. A nonlinear gradient may be used to increase the separation of some components (with a shallow gradient) or to decrease the separation of well-resolved components (with a steep gradient).

An ionic strength gradient is analogous to a solvent or temperature gradient.

Figure 23-27 shows a chromatogram in which discontinuous HCl concentrations were used for a cation-exchange separation. Low concentrations of H^+ displaced Li^+ and Na^+ from the resin, but high concentrations were required for Ca^{2+}. The position of Fe^{3+} is anomalous because at sufficiently high Cl^- concentration Fe(III) is converted to an anionic species that is not retained by a cation exchanger:

$$Fe^{3+} + 4Cl^- \rightleftharpoons FeCl_4^-$$

Column dimensions

A column length-to-diameter ratio of 10 or 20 is adequate for most purposes. Difficult separations might require a greater length. In scaling up a procedure

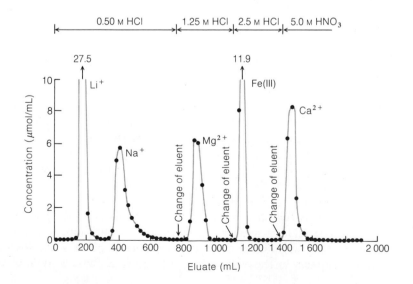

Figure 23-27
Elution of a mixture containing 0.5 mmol Na^+, Mg^{2+}, Ca^{2+}, and Fe^{3+} plus 1 mmol Li^+ from a 2.1 × 16 cm column of Bio-Rad AG MP-50 resin (200/400 mesh) at a rate of 120 mL/h. [From F. W. E. Strelow, *Anal. Chem.*, **56**, 1053 (1984).]

for preparative purposes, the *cross-sectional area* of the column should be increased in proportion to sample size. The column length need not be increased. For preparative separations, the sample may occupy 10–20% of the exchanger volume when the sample is first applied to the column.

Applications

In addition to its considerable value as a chromatographic separation technique, ion exchange has many other applications.

Deionization

Ionic impurities in water can be removed by passing the liquid through an anion-exchange resin in its OH^- form and a cation-exchange resin in its H^+ form. Suppose, for example, that $Cu(NO_3)_2$ is present in the solution. The cation-exchange resin binds Cu^{2+} and replaces it with $2H^+$. The anion-exchange resin binds NO_3^- and replaces it with OH^-. The eluate will be pure water:

Water softeners use ion exchange to remove Ca^{2+} and Mg^{2+} from "hard" water.

$$\left.\begin{array}{l} Cu^{2+} \xrightarrow{\ H^+ \text{ ion exchange}\ } 2H^+ \\[2mm] 2NO_3^- \xrightarrow{\ OH^- \text{ ion exchange}\ } 2OH^- \end{array}\right\} \longrightarrow \text{ pure } H_2O$$

Water so treated is called **deionized water.**

Interconversion of salts

One salt can be converted to another by appropriate ion exchange. For example, tetrapropylammonium hydroxide solution can be prepared if a tetrapropylammonium salt of some other anion is available:

$$(CH_3CH_2CH_2)_4N^+I^- \xrightarrow[OH^-\text{ form}]{\text{anion exchanger}} (CH_3CH_2CH_2)_4N^+OH^-$$

Tetrapropylammonium   Tetrapropylammonium
iodide   hydroxide

A neutral organic acid can be prepared from its sodium salt as follows:

$$N(CH_2CO_2^-Na^+)_3 \xrightarrow[H^+\text{ form}]{\text{cation exchanger}} N(CH_2CO_2H)_3$$

Trisodium   Nitrilotriacetic
nitrilotriacetate   acid

If, instead, the trisodium salt had been titrated with HCl, the solution would contain nitrilotriacetic acid plus NaCl.

Concentration of trace species

It is sometimes necessary to concentrate trace components of a solution in order to obtain enough for analysis. This process is called **preconcentration.** For example, a large volume of fresh lake water could be passed through a cation-exchange resin in the H^+ form to concentrate metal ions from the water onto the resin. Chelex 100, a styrene–divinylbenzene resin containing iminodiacetic acid groups, is noteworthy for its ability to bind transition metal ions.

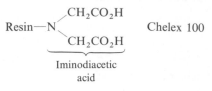

The metals can be eluted in a small volume by 2 M HNO_3, which protonates the iminodiacetate groups.

23-5 ION CHROMATOGRAPHY

Ion chromatography is a high-performance version of ion-exchange chromatography that has become the method of choice for routine anion analysis and has many applications in cation analysis as well. The *suppressed-ion* technique requires two consecutive operations, while the *single-column* version is conducted with standard HPLC equipment. In either technique, ions are usually detected by measuring the conductivity of the eluate.

Suppressed-Ion Anion and Cation Chromatography

Figure 23-28a illustrates the basic idea of **suppressed-ion anion chromatography,** in which a mixture of anions is separated by ion exchange and detected by their electrical conductivity. For the sake of illustration, a sample containing KNO_3 and $CaSO_4$ is injected into the *separator column*—an anion-exchange column in the carbonate form—and eluted with Na_2CO_3. The NO_3^- and SO_4^{2-} equilibrate with the resin and are slowly displaced by the CO_3^{2-} eluent. The K^+ and Ca^{2+} cations are not retained and simply wash through. After a period of time, $NaNO_3$ and Na_2SO_4 are eluted from the separator column, as shown in the upper graph of Figure 23-28a. These species cannot be easily detected because the solvent contains a high concentration of Na_2CO_3, whose high conductivity obscures that of the analyte species.

> The separator column separates the analytes and the suppressor replaces the ionic eluent with a nonionic species.

To remedy this problem, the solution is next passed through a *membrane ion suppressor*, in which all cations are replaced by H^+. The suppressor is a narrow-diameter cylinder or thin, flat channel enclosed by a thin, semipermeable cation-exchange membrane made of sulfonated polyethylene. The sulfonate groups ($-SO_3^-$) in the membrane repel anions, but allow cations to pass freely. The outside of the membrane is bathed in H_2SO_4 solution. When $NaNO_3$ and Na_2CO_3 from the separator column passes through the suppressor, Na^+ is replaced by H^+, making a solution of HNO_3 and H_2CO_3 ($\rightleftharpoons CO_2 + H_2O$). Na_2SO_4 formed in the outside bath is washed away. The process is analogous to kidney dialysis, in which large protein molecules are retained within semipermeable hollow fibers, while small waste products diffuse into the surrounding medium (Demonstration 8-1).

> The Donnan equilibrium tells us that ions with the same charge as those covalently attached to the membrane are excluded from the membrane.

In the absence of analyte, only H_2CO_3, which has very low conductivity, emerges from the suppressor. When analyte is present, HNO_3 or H_2SO_4 with

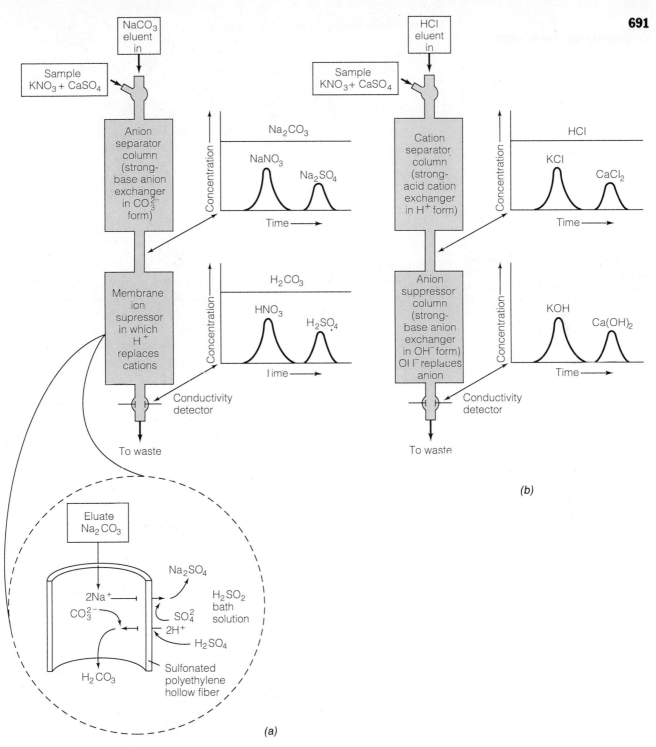

Figure 23-28
Schematic illustrations of (a) suppressed-ion anion chromatography and (b) suppressed-ion cation chromatography. [Adapted from H. Small, *Anal. Chem.,* **55,** 235A (1983).]

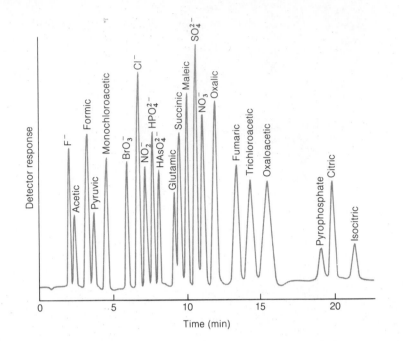

Figure 23-29
Chromatogram illustrating resolving power of anion chromatography. The HPLC apparatus used in this experiment is capable of drawing solvents from four reservoirs, giving gradients of solvent composition, ionic strength, and pH. Sodium carbonate background is suppressed by the hollow-fiber ion-exchange membrane. [Courtesy Dionex Corporation.]

Benzene-1,4-diammonium cation is a stronger eluent that can be used instead of H^+ for suppressed-ion cation chromatography. After passing through the suppressor column, a neutral product is formed:

$$H_3\overset{+}{N}—\!\!\langle\!\langle\!\bigcirc\!\rangle\!\rangle\!\!—\overset{+}{N}H_3 \xrightarrow{OH^-}$$

Benzene-1,4-diammonium ion

$$H_2N—\!\!\langle\!\langle\!\bigcirc\!\rangle\!\rangle\!\!—NH_2$$

high conductivity is produced and detected. Figure 23-29 shows the remarkable resolving power of anion chromatography.

Suppressed-ion cation chromatography is conducted in a similar manner, but the membrane suppressor is replaced by an anion-exchange suppressor column loaded with OH^-. Figure 23-28b illustrates the separation of KNO_3 and $CaSO_4$. With HCl as eluent, KCl and $CaCl_2$ emerge from the cation-exchange separator column, while KOH and $Ca(OH)_2$ emerge from the suppressor column. The HCl eluate is converted to H_2O in the suppressor.

Single-Column Anion and Cation Chromatography

The key feature of suppressed-ion chromatography is removal of the unwanted electrolyte prior to conductivity measurement. Alternatively, if the ion-exchange capacity of the separator column is sufficiently low, and if very dilute eluent is used, the suppressor can be eliminated. For **single-column anion chromatography,** a resin with an exchange capacity near 5 μequiv/g is used, with 10^{-4} M sodium or potassium salts of benzoic, *p*-hydroxybenzoic, or phthalic acid as eluent. These eluents are sufficiently dilute that the constant background conductivity is fairly low, and analyte anions can be detected by a small *change* in conductivity as they emerge from the column. By judicious choice of pH, an average eluent charge between 0 and -2 can be obtained, which allows control of the chromatographic eluent strength. Even the dilute carboxylic acids (which are slightly ionized) are suitable eluents for some separations.

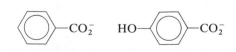

Benzoate   *p*-Hydroxybenzoate

Phthalate

Single-column cation chromatography is conducted in an analogous manner, using dilute HNO_3 as eluent for monovalent ions, and ethylenediammonium salts ($^+H_2NCH_2CH_2NH_2^+$) for divalent ions. No modification of standard HPLC equipment is needed for single-column ion chromatography; the only requirements are a suitable ion-exchange column and detector.

Detectors for Ion Chromatography

The conductivity detector is universal for ions, but somewhat limited in sensitivity. In suppressed-ion chromatography, the eluent conductivity is reduced to near zero by the suppression step, so it is easy to see analyte electrolytes as they emerge. Eluent concentration gradients can be used with suppression. In single-column anion chromatography the analyte anion conductivity is higher than that of the eluent, and an increase in conductivity is observed when analyte emerges from the column. Detection limits are in the low-ppm range. Use of carboxylic acids instead of carboxylate salts as eluent lowers the detection limit by a factor of 10. In single-column cation chromatography the analyte conductivity is usually less than that of the eluent cation. Therefore a *decrease* in conductivity is observed when the analyte emerges.

Even though many common anions have little ultraviolet absorption, use of benzoate or phthalate eluents permits sensitive *indirect spectrophotometric detection.* An example is shown in Figure 23-30. The eluate has a strong, constant ultraviolet absorption. In each emerging peak, nonabsorbing analyte anion replaces an equivalent amount of the absorbing eluent anion. The absorbance therefore *decreases* when the analyte appears. Detection limits are reduced below the ppm range by this method. For cation chromatography, $CuSO_4$ is a suitable ultraviolet-absorbing eluent.

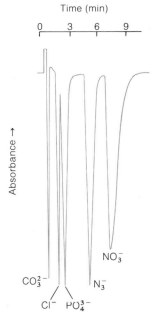

Figure 23-30

Indirect spectrophotometric detection of transparent ions. Column was eluted with 1 mM sodium phthalate plus 1 mM borate buffer, pH 10. [Reproduced from H. Small, *Anal. Chem.,* **54**, 462 (1982).]

Ion-Pair Chromatography

The technique of **ion-pair chromatography** (also called *ion-interaction chromatography*) uses a reverse-phase HPLC column instead of an ion-exchange column. Consider the separation of a mixture of sodium salts of different anions. If the mobile phase contains a hydrophobic cation such as $Fe(phenanthroline)_3^{2+}$, the *ion pairs* [$Fe(phenanthroline)_3^{2+}$][anion] can bind to the stationary phase by interaction of the hydrophobic stationary phase with the hydrophobic cation. The ion pair does not exist in the aqueous solution, but pairing occurs in the nonpolar stationary phase. Since it binds one member of the ion pair, the reverse-phase column effectively behaves as an ion-exchange column, able to carry out efficient separation of ions that can be detected by their conductivity. Alternatively, since $Fe(phenanthroline)_3^{2+}$ has a strong visible absorption, it is useful for indirect spectrophotometric detection of anions. Each anion eluted from the column is accompanied by an equivalent quantity of the colored cation.

Mobile-phase additives for ion-pair chromatography:

For separating anions:

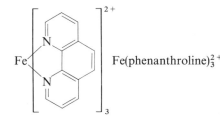

$Fe(phenanthroline)_3^{2+}$

tetraalkylammonium (R_4N^+) with long-chain R group

23-6 MOLECULAR-EXCLUSION CHROMATOGRAPHY

Molecular-exclusion chromatography is also commonly called **gel-filtration** or *gel-permeation chromatography.* This type of chromatography, in which

For separating cations:

alkylsulfonate (RSO_3^-) with long-chain R group

molecules are separated according to their size, is widely used in biochemistry for the resolution of large molecules, such as proteins and carbohydrates. It finds use in polymer chemistry in the separation and characterization of synthetic polymers.

Principles

A schematic representation of the principle of molecular-exclusion chromatography is given in Figure 22-10. The stationary phase contains small pores that can be penetrated by small molecules but not by large molecules. The volume available to small molecules is thus greater than that available to large molecules. Large molecules will therefore be eluted from the column before small molecules (Figure 23-31). Another approach is to consider that large molecules spend all of their time in the mobile phase, whereas small molecules spend only a fraction of their time in the mobile phase. Small molecules are therefore transported more slowly.

> Large molecules pass through the column faster than small molecules.

In molecular-exclusion chromatography, the volume of mobile phase (V_m) is usually called the **void volume**, V_0. In Equation 22-34 we derived the relationship

$$V_r = V_m + KV_s \qquad (22\text{-}34)$$

where V_r is the retention volume, V_s is the volume of stationary phase, and K is the partition coefficient ($=$concentration of solute in stationary phase/concentration in mobile phase). Equation 22-34 can be rearranged to the form

> The void volume is the same as the volume of mobile phase: $V_0 = V_m$.

$$K = \frac{V_r - V_m}{V_s} = \frac{V_r - V_0}{V_s} \qquad (23\text{-}9)$$

The volume of solvent inside the gel particles is V_s. If the gel matrix occupied no volume, then V_s would be $V_t - V_0$, where V_t is the total volume of the

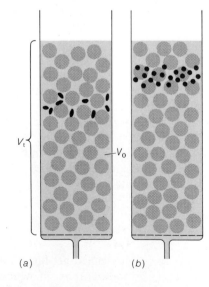

V_t
V_0

(a) (b)

Figure 23-31
(a) Large molecules cannot penetrate the pores of the stationary phase. They are eluted by a volume of solvent equal to the volume of mobile phase. (b) Small molecules that can be found inside or outside the gel require a larger volume for elution. V_t is the total column volume occupied by gel plus solvent. V_0 ($= V_m$) is the volume of the mobile phase. $V_t - V_0$ is the volume occupied by the gel plus its internal liquid phase.

column (Figure 24-19). The gel matrix does occupy some space, so $V_t - V_0$ is greater than V_s. However, V_s is proportional to $V_t - V_0$, since the liquid inside the particles occupies a constant fraction of the particles' volume.

The quantity K_{av} (read "K average") is defined as

$$K_{av} = \frac{V_r - V_0}{V_t - V_0} \tag{23-10}$$

K_{av} is proportional to the partition coefficient, K.

For a large molecule that does not penetrate the gel, $V_r = V_0$, and $K_{av} = 0$. For a small molecule that freely penetrates the gel, $V_r \approx V_t$, and $K_{av} \approx 1$. Molecules of intermediate size can penetrate some gel pores, but not others. For these molecules, K_{av} will be between 0 and 1.

The value of V_0 is measured by passing a large, inert molecule through the column. Its elution volume is defined as V_0. Blue Dextran 2 000, a blue dye of molecular weight 2×10^6, is most commonly used for this purpose. The total volume, V_t, is calculated from the column dimensions ($V_t = \pi r^2 \times$ length).

Ideally, gel penetration is the only mechanism by which molecules are retained in this type of chromatography. In fact, there is always some adsorption. Aromatic molecules, in particular, are known to be adsorbed by some Sephadex ion exchangers and can exhibit $K_{av} > 1$.

Types of Gels

The most widely used gels for molecular exclusion are of the Sephadex variety, whose structure was given in Figure 23-26. The Bio-Gel P gels are made of polyacrylamide crosslinked by N,N'-methylenebisacrylamide (Figure 23-32). The pore size of either gel is controlled by the extent of crosslinking in the manufacturing process. The approximate molecular weight fractionation range of each gel is listed in Table 23-10. The fractionation range applies to "globular" molecules, which are roughly spherical. An elongated molecule, such as a polysaccharide, cannot penetrate the gel as well as a globular molecule of the same weight. The fractionation range for elongated molecules is therefore at a lower molecular weight than for globular molecules. Each gel is available in several particle sizes. The finer the particle size, the greater the resolution and the slower the flow rate of the column.

Fractionation range is more properly defined in terms of molecular size than molecular weight.

Stationary Phases for HPLC

A common stationary phase designated TSK SW, based on microporous silica with a controlled pore size, provides 10 000–16 000 theoretical plates per meter for molecular-exclusion chromatography (Table 23-11). The silica is coated with a proprietary hydrophilic phase that minimizes adsorption of solutes. Another HPLC medium, called TSK PW, is a hydroxylated polyether with a well-defined pore size. It can be used over the pH range 2–12, whereas silica phases generally cannot be used above pH 8. Proprietary hydrophobic polymers of unspecified composition are used for separations of organic polymers in nonaqueous solutions.

Table 23-10

Properties of some gel filtration media

Name	Fractionation range (M.W.) for globular proteins
Sephadex G-10	−700
Sephadex G-15	−1 500
Sephadex G-25	1 000–5 000
Sephadex G-50	1 500–30 000
Sephadex G-75	3 000–80 000
Sephadex G-100	4 000–150 000
Sephadex G-150	5 000–300 000
Sephadex G-200	5 000–600 000
Sephacryl S-200	5 000–250 000
Sephacryl S-300	10 000–1 500 000
Sepharose 2B	70 000–40 000 000
Sepharose 4B	60 000–20 000 000
Sepharose 6B	10 000–4 000 000
Bio-Gel P-2	100–1 800
Bio-Gel P-4	800–4 000
Bio-Gel P-6	1 000–6 000
Bio-Gel P-10	1 500–20 000
Bio-Gel P-30	2 500–40 000
Bio-Gel P-60	3 000–60 000
Bio-Gel P-100	5 000–100 000
Bio-Gel P-150	15 000–150 000
Bio-Gel P-200	30 000–200 000
Bio-Gel P-300	60 000–400 000
Bio-Gel A-0.5 m	<10 000–500 000
Bio-Gel A-1.5 m	<10 000–1 500 000
Bio-Gel A-5 m	10 000–5 000 000
Bio-Gel A-15 m	40 000–15 000 000
Bio-Gel A-50 m	100 000–50 000 000
Bio-Gel A-150 m	1 000 000–150 000 000

Note: Sephadex and Sephacryl are manufactured by Pharmacia Fine Chemical Co., Piscataway, N.J. Bio-Gel is sold by Bio-Rad Laboratories, Richmond, Calif.

Source: The information in this table was taken from the manufacturers' bulletins, which provide a great deal of useful information about these products.

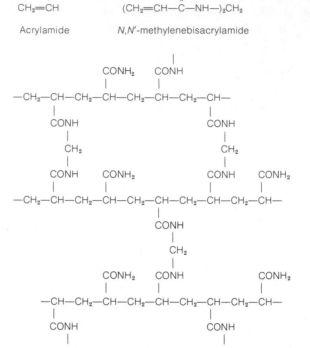

Figure 23-32
Structure of polyacrylamide.

Table 23-11

TSK SW silica for HPLC molecular-exclusion chromatography

Designation	Pore size (nm)	Molecular weight range for globular proteins
G2000SW	13	500–60 000
G3000SW	24	1 000–300 000
G4000SW	45	5 000–1 000 000
G5000SW	100	>1 500 000

Source: Perkin-Elmer liquid chromatography catalog.

Applications

Chromatography

Gel filtration is used mainly to separate mixtures of molecules of different molecular weight. Figure 23-33 shows a protein separation by HPLC using spectrophotometric detection.

For each stationary phase, there is a range over which there is a logarithmic relation between molecular weight and elution volume, as illustrated in Figure 23-34. By comparing K_{av} for an unknown with a series of standards, it is possible to estimate the molecular weight of the unknown. Caution is called for however, because molecules with the same molecular weight but different shapes exhibit different elution characteristics. It is also important to use a high enough ionic strength (>0.05 M) to eliminate extraneous electrostatic adsorption of solute by occasional charged sites on the gel.

Desalting

Salts of low molecular weight (or any small molecule) can be removed from solutions of large molecules by passage through a gel filtration column. This technique, called **desalting,** is useful for changing the buffer composition of a macromolecule solution or for separating small reactants and products from large molecules after a chemical reaction.

23-7 AFFINITY CHROMATOGRAPHY

Affinity chromatography is probably the most rational and powerful development in the field of separations. The principle is illustrated in Figure 22-10. A molecule that has a specific interaction with just one solute of a complex mixture is covalently attached to the stationary phase in a column. When the sample is passed through the column, only one solute binds to the stationary phase. After everything else has washed through, the one adhering solute is eluted by changing conditions to weaken the binding between the solute and the stationary phase.

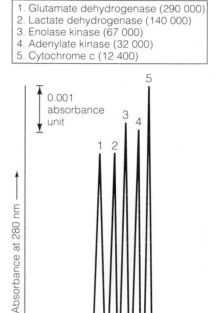

1. Glutamate dehydrogenase (290 000)
2. Lactate dehydrogenase (140 000)
3. Enolase kinase (67 000)
4. Adenylate kinase (32 000)
5. Cytochrome c (12 400)

Figure 23-33
Separation of proteins by molecular-exclusion chromatography with TSK 3000SW column. [Courtesy Varian Associates.]

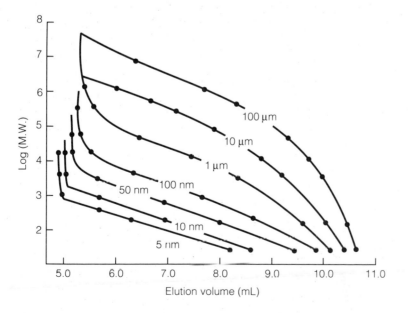

Figure 23-34
Molecular weight calibration graph for polystyrene on Beckman μSpherogel™ molecular-exclusion column (7.7-mm diameter × 300-mm length). Different gels have pore sizes ranging from 5 nm to 100 μm. [Courtesy Anspec Company.]

This is the only form of chromatography in which the chemist *designs* the stationary phase to interact with a specific solute. The technique is especially applicable to biochemistry, where advantage can be taken of such well-known specific interactions as those involving enzymes and substrates or coenzymes, antibodies and antigens, and receptors and hormones. Affinity chromatography has even been used to separate one type of cell from others.

The literature on affinity chromatography in biochemistry is growing explosively.[†] In this section, we mention one application of this technique to the separation of small molecules rather than macromolecules. Molecules with coplanar *cis*–diol groups ($-\overset{\underset{\displaystyle |}{OH}}{C}-\overset{\underset{\displaystyle |}{OH}}{C}-$) can be separated from a complex mixture by passage through a column to which is attached phenylboronic acid. Affi-Gel 601 is a commercially available[‡] form of Bio-Gel P-6 incorporating phenylboronic acid:

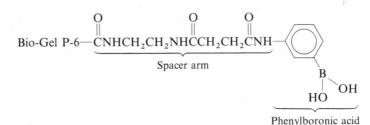

Phenylboronic acid

The boronic acid forms a covalent linkage with coplanar *cis*–diols. By this means, nucleotides (which have a *cis*–diol group) can be separated from deoxynucleotides or cyclic nucleotides, which do not have *cis*–diols. The nucleotide can be displaced from the affinity column by eluting the column with a citrate-containing buffer.

The spacer arm in the structure drawn above is a means of extending the boronic acid away from the bulk of the gel particle. Spacer arms are widely used for affinity chromatography of macromolecules that are sterically hindered from approaching the bulk gel particle closely enough to bind (Figure 23-35).

[†] Readable reviews describing the scope, applications, and chemistry of biochemical affinity chromatography are R. R. Walters, *Anal. Chem.,* **57,** 1099A (1985) and I. Parikh and P. Cuatrecasas, *Chem. Eng. News,* Aug. 26, 1985, pp 17–32. Practical, up-to-date information on affinity chromatography can be found in the following manufacturers' bulletins: *Affinity Chromatography: Principles and Methods* (Piscataway, N.J.: Pharmacia Fine Chemical Co.); and *Materials, Equipment and Systems for Chromatography, Electrophoresis, Immunochemistry and HPLC* (Richmond, Calif.: Bio-Rad Laboratories).

[‡] Bio-Rad Laboratories, Richmond, California.

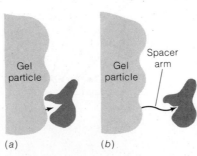

(a)　　　(b)

Figure 23-35
The use of a spacer arm allows solute–stationary phase interactions that might otherwise be sterically forbidden.

There is a certain degree of art to pouring uniform columns, applying samples evenly, and obtaining symmetric elution bands. In this section, we mention some of the practical aspects of liquid column chromatography.

Preparing the Stationary Phase

Before pouring a column, the stationary phase must be equilibrated with the solvent. For alumina and silica gel, this means simply making a slurry of the solid in the appropriate solvent. For ion-exchange and molecular-exclusion gels, this means following the manufacturer's instructions concerning time and temperature of equilibration, as well as what solution might be needed to fully swell and hydrate the gel. When mixing a dry gel with solvent, the gel should be slowly sprinkled on top of the liquid and allowed to settle with gentle stirring, using a glass stirring rod. *Never* use magnetic stirring, which breaks and destroys gel particles.

After the gel has been equilibrated with solvent, it should be suspended in ~2–5 volumes of solvent and allowed to stand until 90–95% has settled. The smaller particles (called **fines**) that are still suspended should be removed by decanting or suction. If left in the gel, the fines significantly reduce the column flow rate. The procedure to remove fines should be repeated several times. Many gels are sold in a pre-equilibrated form from which the fines have already been removed.

Pouring the Column

After it has been equilibrated with solvent, the stationary phase should be suspended in enough solvent so that the solid will occupy from one-half to two-thirds of the total volume when it settles. The chromatography column should contain a few centimeters of liquid before pouring the gel.

The slurry of suspended phase is then poured gently down the wall of the column, preferably directed by a glass rod (Figure 23-36). Foaming and violent convection currents are undesirable. If most of the column is to be filled with stationary phase, additional volume must be available to hold the slurry (only about half of which is stationary phase). For this purpose, a wide-stem funnel held to the top of the column by a rubber stopper is adequate. Reservoirs that screw into the top of some columns are commercially available. It is *not* desirable to pour part of a column, allow it to settle, and then pour the rest of the column. This creates discontinuities in the bed.

When the entire slurry has been added to the column, it should be allowed to settle gently, like a fine snow. After a few centimeters have settled, begin a slow flow of solvent to pack the remainder. The hydrostatic pressure limit set by the manufacturer should not be exceeded during packing or any other phase of chromatography. (Hydrostatic pressure measurement is discussed below.)

Liquid should never be allowed to drain below the top of the gel in the column. Otherwise, air spaces might be introduced that could lead to irregular flow patterns. Solvent should be directed gently down the wall of the column or on top of a few centimeters of liquid above the packing. In *no* case should the solvent be allowed to dig a channel into the gel.

A slurry is a suspension of a solid in a liquid.

Add gel *to* solvent. *Never* use magnetic stirring.

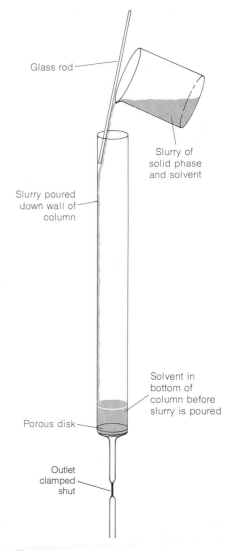

Glass rod

Slurry of solid phase and solvent

Slurry poured down wall of column

Solvent in bottom of column before slurry is poured

Porous disk

Outlet clamped shut

Figure 23-36
A slurry of the stationary phase in starting solvent should be poured gently down the wall of the column.

Applying the Sample

To apply sample, the solvent is drained to the top of the gel, and flow is halted. Then sample is applied gently down the wall of the column by pipet. The sample should be applied evenly around the whole column. Flow is resumed, the sample is drained into the gel, and flow is halted again. Then *small* quantities of solvent are added down the walls by pipet to finish washing the sample into the gel. Finally, a layer of solvent is built up by pipet and maintained during column operation.

Running the Column

Flow is usually maintained by siphoning solvent from a reservoir into the column. The proper flow rate depends on the column diameter and the degree of separation between solutes. Maximum resolution usually demands a very slow flow rate. The manufacturer's bulletin for the particular stationary phase should be consulted to select a flow rate.

Pressure is measured from the inlet level to the outlet level.

The flow rate is governed by the hydrostatic pressure, measured as the distance between the level of liquid in the reservoir and the level of the column outlet (Figure 23-37a). If a **Mariotte flask** is used as a reservoir, the pressure is measured between the bottom of the air inlet and the level of the column outlet (Figure 23-37b). In an ordinary reservoir, the hydrostatic pressure pushing liquid into the outlet tube is equal to the height of liquid above the opening of the tube within the reservoir. This pressure decreases as liquid is drained from the reservoir. In a Mariotte flask, the pressure at the bottom of the air inlet tube must be equal to atmospheric pressure. This pressure remains constant unless the liquid level falls below the bottom of either tube. A Mariotte flask thus maintains a constant pressure for the duration of the chromatography.

A Mariotte flask maintains a constant pressure and therefore a constant flow rate.

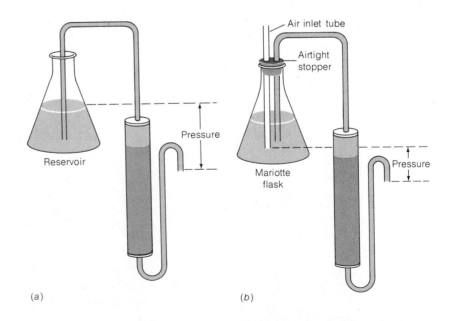

Figure 23-37
Measurement of the hydrostatic pressure with (a) an ordinary solvent reservoir and (b) a Mariotte reservoir.

(a)   (b)

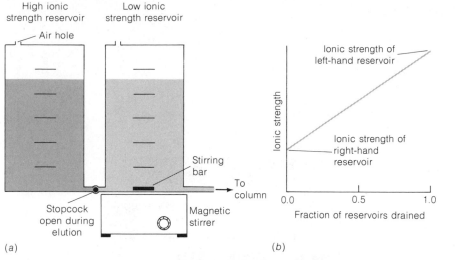

Figure 23-38
A device for gradient elution.

Tubing that comes in contact with any organic solvent is a potential source of contamination because the plasticizers used in tubing manufacture can be leached out by some solvents. For work with nonaqueous solutions, common tubing such as Tygon should be avoided, and fluoroplastic or glass tubing should be used instead.

Gradients

In ion-exchange chromatography, a gradient of eluent ionic strength may be needed to displace the strongly retained solutes from the stationary phase. A device for making a linearly increasing gradient is shown in Figure 23-38. A solution of low ionic strength is placed in the right compartment, and an equal volume at high ionic strength is placed in the left compartment. The first drop of liquid leaving the right side has low ionic strength. As liquid from the left side flows into the right side to keep the levels equal, the ionic strength in the right compartment increases linearly. The last drop to be drained from this system has the ionic strength of the left-hand reservoir.

Accessories

Figure 23-39 shows a rather complete setup for liquid chromatography. Eluent is fed to the column by a peristaltic pump. The flow rate is set by the pump speed, so regulation of hydrostatic pressure is not necessary. The eluate passes through a spectrophotometric flow cell on its way to a fraction collector, which automatically changes test tubes when a preset volume has been collected in each one. The absorbance of the eluate is displayed on a recorder, which also marks each time the fraction collector changes tubes.

All chromatographic devices require a great deal of loving attention, but the automatic features shown in Figure 23-39 allow you to go home and sleep fitfully as the column runs through the night.

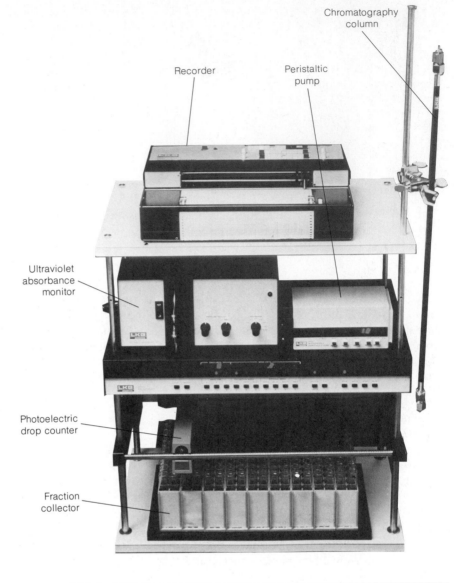

Figure 23-39
Automated apparatus for liquid chromatography. [Courtesy LKB–Produkter AB, Pleasant Hill, Calif.]

Summary

In gas chromatography, a volatile liquid or gaseous solute is carried by a gaseous mobile phase over a stationary liquid coated on a solid support, or over an adsorptive solid surface. Short packed columns provide high capacity for solute, but poor resolution. Long, narrow open tubular columns have low capacity, but give excellent separations. Fused silica open tubular columns may be wall-coated, support-coated, or porous layer. Each liquid stationary phase most strongly retains solutes in its own polarity class. Common solid stationary phases include alumina and molecular sieves. Temperature programming is used to reduce elution times of highly retained components. Without compromising separation efficiency, the linear flow rate may be increased when H_2 or He, instead of N_2, is used as carrier gas. Split injection is routine, whereas splitless injection is better for quantitative analysis. On-column injection is best for thermally unstable solutes. Thermal conductivity detection has universal response, but flame ionization is more sensitive. Electron capture, flame photometry, alkali flame detectors, mass spectrometry, and Fourier transform infrared spectroscopy are also used for detection. A compound can be identified by its retention time on different columns and quantified by the area of its elution peak. Response factors relate the area to the concentration of different compounds.

Classical liquid chromatography utilizes large, packed columns with gravity-fed (or very low-pressure pumped) mobile phase. Common stationary phases for adsorption chromatography are silica gel and alumina. The eluent strength describes the ability of a given solvent to elute solutes from the column. Thin-layer chromatography and small-scale columns are used to select conditions for preparative separations by large columns.

In high-performance liquid chromatography (HPLC), solvent is pumped at high pressure through a small column containing 3- to 10-μm diameter particles of stationary phase. The smaller the particle size, the more efficient the column, but the greater the resistance to flow. Spherical and irregular microporous particles with an adsorptive surface or a bonded liquid phase are most common. In normal-phase chromatography, the stationary phase is polar and a less polar solvent is used. The more common reverse-phase chromatography employs a nonpolar stationary phase and polar solvent. The choice of separation procedure is based on the size, polarity, and ionic nature of the solute. Isocratic or gradient elution may be used, with a piston pump providing relatively pulse-free flow. Injection valves allow rapid, precise sample introduction, preferably through a guard column. Ultraviolet detection is most common; refractive index is more universal but less sensitive. Electrochemical and fluorescense detectors are extremely sensitive, but selective. In supercritical fluid chromatography, nonvolatile solutes are separated by a process whose efficiency, speed, and detectors more closely resemble those of gas chromatography than of liquid chromatography

Ion-exchange resins and gels contain covalently bound charged groups that attract solute counterions (and that exclude ions having the same charge as the resin). Resins of the polystyrene type are useful for separation of small ions. Greater crosslinking of the resin increases the capacity, selectivity, and time needed for equilibration. Ion-exchange gels based on cellulose and dextran have large pore sizes and low charge densities, suitable for the separation of macromolecules. Certain inorganic solids have ion-exchange properties and are useful at extremes of temperature or radiation. HPLC ion exchangers are both silica- and polymer-based. All ion exchangers operate by the principle of mass action, with a gradient of increasing ionic strength most commonly used to effect a separation.

In suppressed-ion chromatography, a separator column separates ions of interest, and a suppressor (membrane or column) converts eluent to a nonionic form so that analytes can be detected by their conductivity. Alternatively, single-column ion chromatography uses an ion-exchange column and low-concentration eluent. If the eluent absorbs light, indirect spectrophotometric detection is convenient and sensitive. Ion-pair chromatography utilizes a hydrophobic ionic additive in the eluent to make a reverse-phase column behave as an ion-exchange column.

Molecular-exclusion chromatography is based on the relative inability of large molecules to enter the pores in the stationary phase. Small molecules can enter these spaces and therefore exhibit longer elution times than large molecules. Exclusion chromatography is used for separations based on size and for approximate molecular weight determinations of macromolecules. In affinity chromatography, the chemist designs a stationary phase to interact with one particular solute in a complex mixture. After all other components have been eluted, the desired species is liberated by a change in conditions.

Terms to Understand

alkali flame detector
anion exchanger
bonded-phase particles
cation exchanger
co-chromatography
crosslinking
deionized water
derivatization
desalting
Donnan equilibrium
electrochemical detector
electron capture detector
eluent strength
eluotropic series
fines
flame ionization detector
flame photometric detector
flow adaptor
gas chromatography

gel
gel filtration
gradient elution
guard column
high-performance liquid chromatography
hydrophilic
hydrophobic
internal standard
ion chromatography
ion-exclusion chromatography
ion-pair chromatography
isocratic elution
Kovats index
Mariotte flask
mesh size
microporous particles
molecular sieve
normal-phase chromatography
on-column injection

open tubular column
packed column
pellicular particles
polarizability
porous-layer column
preconcentration
refractive index detector
resin
response factor
retention index
reverse-phase chromatography
selectivity coefficient
septum

single-column ion chromatography
split injection
splitless injection
supercritical fluid
supercritical fluid chromatography
support-coated column
suppressed-ion chromatography
thermal conductivity detector
thin-layer chromatography
ultraviolet detector
void volume
wall-coated column

Exercises

23-A. When 1.06 mmol of 1-pentanol and 1.53 mmol of 1-hexanol were dissolved together and separated by gas chromatography, they gave relative peak areas of 922 and 1 570 units, respectively.
  (a) Calculate the response factor for hexanol relative to pentanol, the internal standard.
  (b) When 0.57 mmol of pentanol was added to an unknown containing hexanol, the relative chromatographic peak areas were 843:816 (pentanol:hexanol). How much hexanol did the unknown contain?

23-B. A compound was known to be a member of the family $(CH_3)_2CH(CH_2)_nCH_2OSi(CH_3)_3$.
  (a) From the gas chromatographic retention times below, estimate the value of n in the chemical formula.

  $n = 7$:   4.0 min   air: 1.1 min

  $n = 8$:   6.5 min   unknown: 42.5 min

  $n = 14$:   86.9 min

  (b) Calculate the capacity factor for the unknown.

23-C. Vanadyl sulfate ($VOSO_4$), as supplied commercially, is contaminated with H_2SO_4 and H_2O. A solution was prepared by dissolving 0.244 7 g of impure $VOSO_4$ in 50.0 mL of water. Spectrophotometric analysis indicated that the concentration of the blue VO^{2+} ion was 0.024 3 M. A 5.00 mL sample was passed through a cation-exchange column loaded with H^+ to bind VO^{2+} and release H^+, which required 13.03 mL of 0.022 74 M NaOH for titration. Find the weight percent of each component ($VOSO_4$, H_2SO_4, and H_2O) in the vanadyl sulfate.

23-D. Blue Dextran 2 000 was eluted in a volume of 36.4 mL from a 2.0 × 40-cm (diameter × length) column of Sephadex G-50.
  (a) At what retention volume would hemoglobin (M.W. 64 000) be expected?
  (b) At what volume is $^{22}NaCl$ expected?
  (c) What would be the retention volume of a molecule with $K_{av} = 0.65$?

23-E. Make a graph showing the qualitative shape of the ionic-strength gradient that would be produced by each device below.

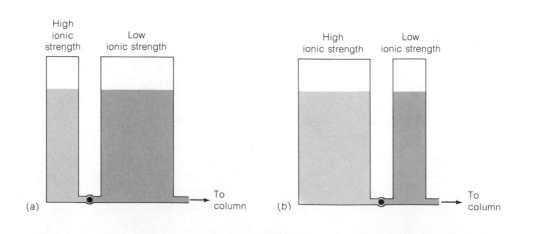

(a)   (b)

Problems

A23-1. Why does the eluent strength increase as solvent becomes less polar in reverse-phase chromatography, whereas the eluent strength increases as solvent becomes more polar in normal-phase chromatography?

A23-2. (a) What is the advantage of temperature programming in gas chromatography?
(b) Why is gradient elution useful in liquid chromatography?
(c) What kind of gradient is used in ion-exchange chromatography?
(d) What kind of gradient is used in supercritical fluid chromatography?

A23-3. (a) What are the relative advantages and disadvantages of packed and open tubular columns in gas chromatography?
(b) Explain the difference between wall-coated, support-coated, and porous-layer open tubular columns.
(c) What is the advantage of a bonded stationary phase in gas chromatography?

A23-4. (a) Why do open tubular columns provide greater resolution than packed columns in gas chromatography?
(b) Why does the efficiency (decreased H.E.T.P.) of liquid chromatography increase as the stationary-phase particle size is reduced?
(c) Why do H_2 and He allow more rapid linear flow rates in gas chromatography than does N_2, without loss of column efficiency (Figure 23-6)?
(d) Why is the curve for 3-μm particles so much flatter than the other curves in Figure 23-15?

A23-5. (a) When would you use split, splitless, or on-column injection in gas chromatography?
(b) Explain how solvent trapping and cold trapping work in splitless injection.

A23-6. To which kinds of analytes do the following gas chromatography detectors respond?
(a) thermal conductivity
(b) flame ionization
(c) electron capture
(d) flame photometric
(e) alkali flame

A23-7. What does the designation 200/400 mesh mean on a bottle of chromatography stationary phase? What is the size range of such particles? (See Table 24-2.) Which particles are smaller, 100/200 or 200/400 mesh?

A23-8. Name two small-scale methods that can be used to select a suitable solvent and stationary phase for large-scale chromatographic separations.

A23-9. Why are the relative eluent strengths of solvents in adsorption chromatography fairly independent of solute?

A23-10. Why is high pressure needed in HPLC? What is the difference between microporous and pellicular particles for chromatography? What is a bonded phase in liquid chromatography?

A23-11. Using Figure 23-17, suggest which type of liquid chromatography you could use to separate compounds in each of the following categories:
(a) low molecular weight ($<2\,000$), soluble in octane
(b) low molecular weight ($<2\,000$), soluble in dioxane
(c) low molecular weight ($<2\,000$), ionic
(d) high molecular weight ($>2\,000$), soluble in water, nonionic, size 50 nm
(e) high molecular weight ($>2\,000$), soluble in water, ionic
(f) high molecular weight ($>2\,000$), soluble in tetrahydrofuran, size 50 nm

A23-12. State the purpose of the separator column and suppressor in suppressed-ion chromatography. For cation chromatography, why does the suppressor contain anion-exchange resin?

A23-13. State the effects of increasing crosslinking on an ion-exchange column.

A23-14. What is deionized water? What kind of impurities are not removed by deionization?

A23-15. (a) How can molecular-exclusion chromatography be used to measure the molecular weight of a protein?
(b) Which pore size in Figure 23-34 is most suitable for chromatography of molecules with molecular weight near 100 000?

A23-16. What is the purpose of a Mariotte flask?

A23-17. This problem reviews some concepts from Chapter 22. An unretained solute passes through a chromatography column in 3.7 min and the analyte requires 8.4 min.
(a) Find the adjusted retention time and capacity factor for the analyte.
(b) The volume of the mobile phase is 1.4 times the volume of the stationary phase. Find the partition coefficient for the analyte.

A23-18. The unadjusted retention times in Figure 22-11 are 1.0 min for air, 12.0 min for octane, 13.0 min for the unknown, and 15.0 min for nonane. Find the Kovats index for the unknown.

A23-19. A compound is eluted from a gas chromatography column at an adjusted retention time $t'_r = 15.0$ min when the column temperature is

373 K. At 363 K, $t'_r = 20.0$ min. Find the parameters a and b in Equation 23-1. Predict t'_r at 353 K.

A23-20. Suppose that a solution containing 6.3×10^{-8} M I^- and 2.0×10^{-7} M p-dichlorobenzene gave peak areas of 395 and 787, respectively, in the experiment in Figure 23-10.

(a) Find the response factor for I^- with respect to the internal standard, p-dichlorobenzene.

(b) A 3.00-mL solution of unknown containing I^- was treated with 0.100 mL of 1.6×10^{-5} M p-dichlorobenzene and the mixture was diluted to 10.00 mL. Gas chromatography gave peak areas of 633 and 520 for I^- and p-dichlorobenzene, respectively. Find the concentration of I^- in the 3.00 mL of original unknown.

A23-21. A gel-filtration column has a radius (r) of 0.80 cm and a length (l) of 20.0 cm.

(a) Calculate the volume (V_t) of the column, which is equal to $\pi r^2 l$.

(b) The void volume (V_0) was found to be 18.1 mL and a solute was eluted at 27.4 mL. Find K_{av} for the solute.

23-22. The exchange capacity of an ion-exchange resin can be defined as the number of moles of charged sites per gram of dry resin. Describe how you would measure the exchange capacity of an anion-exchange resin using standard NaOH, standard HCl, or any other reagent you wish.

23-23. Consider a protein with a net negative charge tightly adsorbed on an anion-exchange gel at pH 8.

(a) How will a gradient of eluent pH (from pH 8 to some lower pH) be useful for eluting the protein? Assume that the ionic strength of the eluent is kept constant.

(b) How would a gradient of ionic strength (at constant pH) be useful for eluting the protein?

23-24. How should Figure 23-38 be modified to produce a linear gradient of decreasing ionic strength?

23-25. Propose a scheme for separating trimethylamine, dimethylamine, methylamine, and ammonia from each other by ion-exchange chromatography.

23-26. An unknown compound was co-chromatographed with heptane and decane. The adjusted retention times were: heptane (12.6 min), decane (22.9 min), unknown (20.0 min). Realizing that the scale of the Kovats index is logarithmic and that the indices for heptane and decane are 700 and 1 000, respectively, find the Kovats index for the unknown.

23-27. The compounds ferritin (M.W. 450 000), transferrin (M.W. 80 000), and ferric citrate were separated by molecular-exclusion chromatography on Bio-Gel P-300. The column had a length of 37 cm and a 1.5-cm diameter. Eluate fractions of 0.65 mL were collected. The maximum of each peak came at the following fractions: ferritin, 22; transferrin, 32; and ferric citrate, 84. (That is, the ferritin peak came at an elution volume of $22 \times 0.65 = 14.3$ mL.) Assuming that ferritin is eluted at the void volume, find K_{av} for transferrin and for ferric citrate.

23-28. (a) The void volume in Figure 23-34 is the volume at which the curves rise vertically at the left. What is the smallest molecular weight of molecules excluded from the 10-nm pore size column?

(b) What is the molecular weight of molecules eluted at 6.5 mL from the 10-nm column?

23-29. Suppose that an ion-exchange resin ($R^- Na^+$) is immersed in a solution of NaCl. Let the concentration of R^- in the resin be 3.0 M.

(a) What will be the ratio $[Cl^-]_o/[Cl^-]_i$ if $[Cl^-]_o$ is 0.10 M?

(b) What will be the ratio $[Cl^-]_o/[Cl^-]_i$ if $[Cl^-]_o$ is 1.0 M?

(c) Will the fraction of electrolyte inside the resin increase or decrease as the outside concentration of electrolyte increases?

23-30. Compounds with $-\overset{\underset{|}{OH}}{C}-\overset{\underset{|}{OH}}{C}-$ or $-\overset{\underset{|}{OH}}{C}-\overset{\underset{|}{NH_2}}{C}-$ linkages can be analyzed by cleavage with periodate. For example, one mole of 1,2-ethanediol consumes one mole of iodate:

$$\begin{matrix} CH_2OH \\ | \\ CH_2OH \end{matrix} \quad + \quad IO_4^- \quad \longrightarrow$$

1,2-Ethanediol   Periodate
M.W. 62.068

$$2CH_2{=}O \; + H_2O + \; IO_3^-$$

Formaldehye   Iodate

To analyze 1,2-ethanediol, oxidation with excess IO_4^- is followed by passage of the whole reaction solution through an anion-exchange resin that binds both IO_4^- and IO_3^-.[†] The IO_3^- is then selectively and quantitatively removed from the resin by elution with NH_4Cl. The absorbance of the eluate is measured at 232 nm ($\varepsilon = 900$ M^{-1} cm^{-1}) to find the quantity of IO_3^- produced by the reaction. In one experiment, 0.213 9 g of aqueous

[†] J. X. Khym and W. E. Cohn, *J. Am. Chem. Soc.*, **82**, 6380 (1960).

1,2-ethanediol was dissolved in 10.00 mL. Then 1.000 mL of the solution was treated with 3 mL of 0.15 M KIO_4 and subjected to ion-exchange separation of IO_3^- from unreacted IO_4^-. The eluate (diluted to 250.0 mL) gave $A_{232} = 0.521$ in a 1.000-cm cell, and a blank gave $A_{232} = 0.049$. Find the weight percent of 1,2-ethanediol in the original sample.

23-31. The substances below were chromatographed on a gel-filtration column. Estimate the molecular weight of the unknown.

Compound	V_r(mL)	Molecular weight
Blue Dextran 2 000	17.7	2×10^6
Aldolase	35.6	158 000
Catalase	32.3	210 000
Ferritin	28.6	440 000
Thyroglobulin	25.1	669 000
Unknown	30.3	?

23-32. (a) Nonpolar aromatic compounds were separated by HPLC using a bonded phase containing octadecyl groups $[—(CH_2)_{17}CH_3]$ covalently attached to silica particles. The eluent was 65% (vol/vol) methanol in water. How would the retention times be affected if 90% methanol were used instead?
(b) Octanoic acid and 1-aminooctane were passed through the same column as in part a using an eluent of 20% (vol/vol) methanol in water adjusted to pH 3.0 with HCl. State which compound is expected to be eluted first and why.

$$CH_3CH_2CH_2CH_2CH_2CH_2CH_2CO_2H$$
Octanoic acid

$$CH_3CH_2CH_2CH_2CH_2CH_2CH_2CH_2NH_2$$
1-Aminooctane

23-33. Suppose the HPLC column produces ideal Gaussian profiles for which

$$\text{Area of peak} = (1.19)hw_{1/2}$$

where h is peak height and $w_{1/2}$ is the width at half the maximum peak height. The detector measures absorbance at 254 nm. A sample containing equal moles of compounds A and B was injected into the column. Suppose that compound A ($\varepsilon_{254} = 2.26 \times 10^4$ M^{-1} cm^{-1}) has $h = 128$ mm and $w_{1/2} = 10.1$ mm. Compound B ($\varepsilon_{254} = 1.68 \times 10^4$ M^{-1} cm^{-1}) has $w_{1/2} = 7.6$ mm. What is the height of peak B in mm?

23-34. A known mixture of compounds C and D gave the following HPLC results:

Compound	Concentration (mg/mL) in mixture	Peak area (cm^2)
C	1.03	10.86
D	1.16	4.37

A solution was prepared by mixing 12.49 mg of D plus 10.00 mL of unknown containing just C, and diluting to 25.00 mL. Peak areas of 5.97 and 6.38 cm^2 were observed for C and D, respectively. Find the concentration of C (mg/mL) in the unknown.

23-35. The graph below shows the dependence of H.E.T.P. on flow rate for the separation of three compounds by liquid chromatography.
(a) Notice that H.E.T.P. is different for each compound. This is a typical observation for different compounds on the same column. Explain why this should be so.
(b) The Blue Dextran data are flat, but for tryptophan and acetone H.E.T.P. rises with increasing flow rate. Rationalize these behaviors in terms of the van Deemter equation, 22-39.

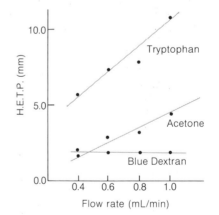

Graph of H.E.T.P. versus flow rate for separation of Blue Dextran (a high molecular weight polymer), acetone, and tryptophan on a 0.4 × 50-cm column of Sephadex G-25. [From J. M. Sosa, *Anal Chem.*, **52**, 910 (1980).]

23-36. *Theoretical performance in gas chromatography.* As the inside radius of a gas chromatography column is decreased, the maximum possible column efficiency increases (i.e., H.E.T.P. decreases) and the sample capacity decreases. For a very thin stationary phase (that equilibrates rapidly with analyte), the minimum theoretical plate height is given by

$$\frac{\text{H.E.T.P.}_{\cdot\text{min}}}{r} = \sqrt{\frac{1 + 6k + 11k^2}{3(1 + k)^2}}$$

where r is the inside radius of the column and k is the capacity ratio.

(a) Find the limit of the square root term as $k \to 0$ (unretained solute) and as $k \to \infty$ (infinitely retained solute).

(b) If the column radius is 0.10 mm, find H.E.T.P.$_{\text{min}}$ for the two cases in part a.

(c) What is the maximum number of theoretical plates in a 50-m-long column with a 0.10-mm radius if $k = 5$?

(d) The relation between the capacity factor, k, and partition coefficient, K (Equation 22-30) can also be written $k = 2dK/r$, where d is the thickness of the stationary phase in a wall-coated column and r is the inside diameter of the column. Find k if $K = 1\,000$, $d = 0.2$ μm, and $r = 0.10$ mm.

23-37. *Separation number.* In chromatography this is defined as

$$\text{Separation number} = TZ = \frac{t_{r2} - t_{r1}}{w_2 + w_1} - 1$$

where t_{ri} is the retention time of peak i and w_i is the width of peak i at half-height. (The symbol TZ comes from the German word *trennzahl*, meaning "separation number.") If all peak widths are equal, TZ tells us how many peaks can be placed between peaks 1 and 2, with a resolution (Figure 22-16) of 1.17 between all peaks. In Figure 22-11, suppose that t_r(octane) = 14.3 min, t_r(nonane) = 18.5 min, and $w = 0.5$ min for both peaks. Calculate the separation number. Does it look as though this many peaks can be placed between octane and nonane in Figure 22-11?

23-38. *Ion-exclusion chromatography.* In this technique, ionic compounds are separated from nonelectrolytes when passed through an ion-exchange column. The nonelectrolytes can penetrate the stationary phase, while the ionic compounds are repelled by the fixed charges. Since electrolytes have access to less of the column volume, they are eluted before the nonelectrolytes. The chromatogram at upper right shows the separation of trichloroacetic acid (TCA, $pK_a = 0.66$), dichloroacetic acid (DCA, 1.30), and monochloroacetic acid (MCA, 2.86) by passage through a cation-exchange resin eluted with 0.01 M HCl. Explain why the three acids are separated and the order of elution.

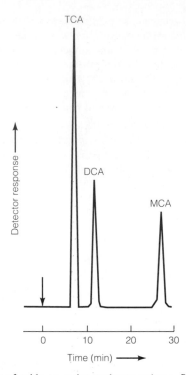

Separation of acids on cation-exchange column. [From V. T. Turkelson and M. Richards, *Anal. Chem.*, **50**, 1420 (1978).]

23-39. The compound norepinephrine (NE) in human urine is conveniently assayed by ion-pair chromatography using an octadecylsilane stationary phase and sodium octyl sulfate as the mobile-phase additive. Electrochemical detection (oxidation at 0.65 V versus Ag|AgCl) is used, with 2,3-dihydroxybenzylamine (DHBA) as internal standard,

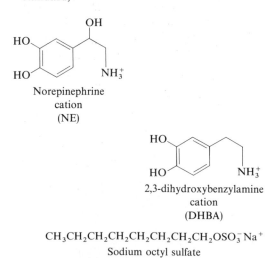

Norepinephrine
cation
(NE)

2,3-dihydroxybenzylamine
cation
(DHBA)

$$CH_3CH_2CH_2CH_2CH_2CH_2CH_2CH_2OSO_3^- Na^+$$
Sodium octyl sulfate

(a) Explain the physical mechanism by which an ion-pair separation works.

(b) A urine sample containing an unknown amount of NE and a fixed, added concentration of DHBA gave a detector peak height ratio NE/DHBA = 0.298. Then small standard additions of NE were made, with the following results:

Added concentration of NE	Peak height ratio NE/DHBA
12 ng/mL	0.414
24 ng/mL	0.554
36 ng/mL	0.664
48 ng/mL	0.792

Using the graphical treatment of standard addition in Figure 21-22, find the original concentration of NE in the specimen.

23-40. *Eluent strength and selectivity in reverse-phase HPLC.* Eluent strength (also called *solvent strength*) is a measure of the ability of a solvent to elute solutes from a chromatography column. The greater the eluent strength, the more easily solutes are eluted and the lower the capacity factors are. *Selectivity* is the ability of a solvent to distinguish between different solutes and separate them. Different mixtures of solvents A or B with water have the same approximate eluent strength when eluent strength = $\phi_A \cdot \varepsilon_A = \phi_B \cdot \varepsilon_B$, where ϕ_A and ϕ_B are the volume fractions of A or B in the mixture, and ε comes from the table below.

Eluent strength factors for reverse-phase chromatography

Solvent	ε	Solvent	ε
methanol	3.0	ethanol	3.6
acetonitrile	3.1	2-propanol	4.2
acetone	3.4	tetrahydrofuran	4.4
dioxane	3.5		

From L. R. Snyder and J. J. Kirkland, *Introduction to Modern Liquid Chromatography,* 2nd ed. (New York: Wiley, 1979), p. 264.

Thus, a 40% (vol/vol) mixture of 2-propanol in water has the same solvent strength as a 50% mixture of acetone in water because $\phi_A \cdot \varepsilon_A = (0.40)(4.2) = \phi_B \cdot \varepsilon_B = (0.50)(3.4)$.

(a) Which mixture has a greater eluent strength for reverse-phase chromatography, 30% tetrahydrofuran in water or 40% methanol in water?

(b) Which solvent is expected to produce lower capacity factors for most solutes in reverse-phase chromatography, 30% tetrahydrofuran in water or 40% methanol in water?

(c) For similar solutes, it is desirable to keep the eluent strength approximately constant during chromatography, so that one component does not require much longer than another to be eluted. However, it is necessary to have some difference in retention, or else the peaks are not separated. If two solutes have a reasonable capacity factor (say, between 1 and 5), but are not resolved from each other, you can try a different solvent mixture that ought to give similar capacity factors, but might yield better resolution. For example, two components not separated by acetonitrile in water might be separated by tetrahydrofuran in water. What volume percent of tetrahydrofuran in water has the same eluent strength as 20% acetonitrile in water?

23-41. An extremely powerful bonded stationary phase for the separation of optical isomers by HPLC has the structure

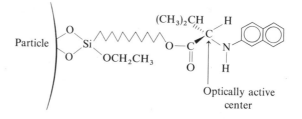

Optically active center

To resolve the optical isomers of amines, alcohols, or thiols the compounds are first attached to a nitroaromatic group that increases their interaction with the bonded phase and makes them observable with a spectrophotometric detector.

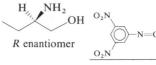

R enantiomer

S enantiomer

Mixture to be resolved (RNH_2)

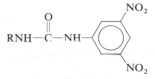

Aromatic derivative (mixture of two enantiomers)

When the mixture is eluted with 20% (vol/vol) 2-propanol in hexane, the *R* enantiomer is eluted before the *S* enantiomer, with the following chromatographic parameters:[†]

$$\text{Resolution} = \frac{\Delta t_r}{\overline{w}} = 7.7 \qquad \text{(Equation 22-40)}$$

[†] W. H. Pirkle and T. C. Pochapsky, *J. Am. Chem. Soc.*, **108**, 352 (1986).

$$\frac{\text{Capacity factor}}{\text{for } R \text{ isomer}} = 1.35 \qquad \text{(Equation 22-27)}$$

$$\text{Relative retention} = 4.53 \qquad \text{(Equation 22-26)}$$

where $\overline{w}$ is the average width of the two peaks at their base. A schematic chromatogram is shown below. Find t_1, t_2, and $\overline{w}$, with units of minutes.

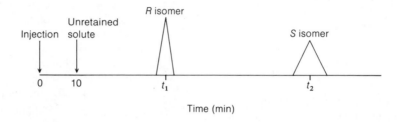

24 Sample Preparation

Real problems in chemical analysis usually begin with a sample that is not suitable for a laboratory experiment. The sample may be a mineral, a chunk of plastic, a 2 000-year-old urn, a lake full of water, or a trainload of coal. There are five steps in most analyses:

1. Obtain a representative specimen of the material you want to analyze.
2. Transform the specimen into a form suitable for analysis, which usually means dissolving the substance.
3. Remove or mask species that will interfere with the chemical analysis.
4. Carry out the analysis.
5. Interpret your results.

Most of this book has dealt with step 4, with some mention of steps 3 and 5. In this chapter we discuss steps 1 and 2, as well as standards for chemical analysis.

24-1 SELECTING THE SAMPLE

In a student laboratory exercise, care is usually taken to ensure that samples are **homogeneous**—that they have the same composition throughout. Variation in analytical results should reflect only the ability of the person doing the analysis and the inherent variability of the analytical method. Samples in the real world are usually **heterogeneous**—their composition varies from one part to another. Examples are a pile of ore, a field of watermelons, or

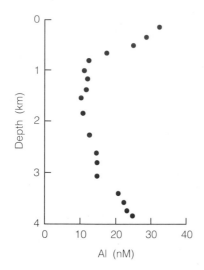

Figure 24-1
Aluminum concentration in seawater as a function of depth in the Atlantic Ocean at 36° 18′ north latitude and 72° 21′ west longitude. [From C. I. Measures and J. M. Edmond, *Anal. Chem.,* **61,** 544 (1989).]

lot
(might be composed of units)
↓
bulk sample
↓
laboratory sample
↓
test portions

sediment under a lake. Figure 24-1 shows the variation in aluminum concentration of ocean water as a function of depth. If you took water from just one depth, your result would not be representative of the entire body of water. It is easy to believe that upper and lower layers of deep oceans do not mix rapidly, but even relatively shallow lakes do not mix well. The topmost layer is probably in equilibrium with the atmosphere and the bottom layer is in equilibrium with the sediment. Temperature and density gradients of the water prevent rapid mixing. When you analyze a heterogeneous material, the result you observe will depend on how you choose samples of the bulk material and how you treat the sample after collection.

Sample composition may change with time after collection because of internal chemical changes, reaction with the air, or interaction of the sample with its container. Glass containers are notorious for ion exchange reactions at the glass surface that can alter the composition of trace ionic species in a solution. For this reason, plastic (especially Teflon) collection bottles are frequently employed. Even plastic containers can be a source of contamination, especially if they are not washed before use. Table 24-1 shows that the manganese content of blood serum samples increased by a factor of 7 when stored in unwashed polyethylene containers prior to analysis. Steel needles are a common source of metal contamination in biochemical analysis. An oft-repeated adage relearned by one analytical chemist after another states that "unless the complete history of any sample is known with certainty, the analyst is well advised not to spend his time in analyzing it."[†] Your laboratory notebook ought to include information on how a sample was collected and stored, as well as on how it was analyzed.

Several quantities are defined in the sampling operation. A **lot** is the total material from which samples are taken. It might be a truckload of reagents, a lake full of water, or a horse spleen. Frequently, lots are composed of *sample units,* such as individual boxes in the truck. A **bulk sample** (also called the *gross sample*) is then taken from the lot for analysis or archiving. The bulk sample is usually chosen to be representative of the lot, and the choice of bulk sample is critical to producing a valid analysis. From the bulk sample, a smaller **laboratory sample** is formed that must have the exact same composition as the bulk sample. For example, this might be done by grinding an entire solid bulk sample to a fine powder, mixing thoroughly, and keeping one bottle of powder for testing. **Test portions** (*aliquots*) of the laboratory sample are then used for individual analyses.

A careful plan is necessary for taking portions of a lot to make a bulk sample. A **random sample** is collected by dividing the lot into many real or imaginary segments. Samples are then taken from some of the segments at random, preferably with the aid of a table of random numbers. For highly **segregated materials** (in which different regions clearly have different compositions), a **composite sample** is constructed to be as representative as possible of the entire lot. For example, if a solid appears to have three zones whose relative volumes are 1:6:3, then a composite sample would be made up of samples of each zone in the volume ratio 1:6:3. Sampling within each zone might be done at random. The composite sample is then homogenized (perhaps by grinding) and a laboratory sample is taken. The hope is that the composition of the composite is representative of the composition of the entire lot.

[†] R. E. Thiers, *Methods of Biochemical Analysis* (D. Glick, ed.), Vol. 5 (New York: Interscience, 1957), p. 274.

For random errors, the overall standard deviation, s_o, is related to the standard deviation of the analytical procedure, s_a, and the standard deviation of the sampling operation, s_s:

$$s_o^2 = s_a^2 + s_s^2 \qquad (24\text{-}1)$$

$$\frac{\text{Total}}{\text{variance}} = \frac{\text{analytical}}{\text{variance}} + \frac{\text{sampling}}{\text{variance}}$$

Since the square of the standard deviation is the **variance**, we see that the variances of sampling and analysis are additive. One consequence of Equation 24-1 is that if either s_a or s_s is sufficiently smaller than the other, there is little point trying to reduce the smaller one. For example, if s_s is 10% and s_a is 5%, the overall standard deviation is 11% ($\sqrt{0.10^2 + 0.05^2} = 0.11$). Using a more expensive and time-consuming analytical procedure that reduces the analytical standard deviation to 1% only improves the overall standard deviation from 11% to 10% ($\sqrt{0.10^2 + 0.01^2} = 0.10$).

Origin of Sampling Variance

Consider a random mixture of two kinds of solid particles. If we select part of the mixture for analysis, the theory of probability allows us to state the likelihood that our sample has the same composition as the bulk sample. It may surprise you to learn how large a sample is required for accurate sampling.

Suppose that the mixture contains n_A particles of type A and n_B particles of type B. The probabilities of drawing A or B from the mixture are

$$p = \text{probability of drawing A} = \frac{n_A}{n_A + n_B} \qquad (24\text{-}2)$$

$$q = \text{probability of drawing B} = \frac{n_B}{n_A + n_B} = 1 - p \qquad (24\text{-}3)$$

If n particles are drawn at random, the expected number of particles of type A are np and the standard deviation of many drawings is known from the binomial distribution to be

$$\sigma_n = \text{standard deviation} = \sqrt{npq} \qquad (24\text{-}4)$$

For example, suppose the mixture contains 1% KCl particles and 99% KNO_3 particles. If 10^4 particles are taken, what is the expected number of KCl particles, and what will be the standard deviation in that number if the experiment is repeated many times? The expected number is just

$$\text{Expected number of KCl particles} = np = (10^4)(0.01) = 100 \qquad (24\text{-}5)$$

and the standard deviation will be

$$\text{Standard deviation} = \sqrt{npq} = \sqrt{(10^4)(0.01)(0.99)} = 9.9 \qquad (24\text{-}6)$$

Table 24-1

Manganese concentration of serum stored in washed[†] and unwashed polyethylene containers

Container	Mn (ng/mL)
Unwashed	0.85
Unwashed	0.55
Unwashed	0.20
Unwashed	0.67
Average	0.57 ± 0.27
Washed	0.096
Washed	0.018
Washed	0.12
Washed	0.10
Average	0.084 ± 0.045

SOURCE: J. Versieck, *Trends in Anal. Chem.*, **2**, 110 (1983).

[†] Washed containers were rinsed with water distilled twice from quartz vessels, which introduce much less contamination into water than does glass.

[†] B. Kratochvil and J. K. Taylor, *Anal. Chem.*, **53**, 924A (1981); H. A. Laitinen and W. E. Harris, *Chemical Analysis*, 2nd ed. (New York: McGraw-Hill, 1975), Chap. 27.

Table 24-2
Standard test sieves

Sieve number	Screen opening (mm)
5	4.00
6	3.35
7	2.80
8	2.36
10	2.00
12	1.70
14	1.40
16	1.18
18	1.00
20	0.850
25	0.710
30	0.600
35	0.500
40	0.425
45	0.355
50	0.300
60	0.250
70	0.212
80	0.180
100	0.150
120	0.125
140	0.106
170	0.090
200	0.075
230	0.063
270	0.053
325	0.045
400	0.038

There is no advantage to reducing the analytical uncertainty if the sampling uncertainty is high, and vice versa.

This is both the standard deviation of the number of KCl particles and the standard deviation of the number of KNO_3 particles. The standard deviation is 9.9% of the expected number of KCl particles, but only 0.1% of the expected number of KNO_3 particles ($nq = 9\,900$). If you want to know how much nitrate is in the mixture, this sample is probably sufficient for most purposes. For chloride, 9.9% uncertainty may not be acceptable.

How much sample corresponds to 10^4 particles? Suppose that the particles are 1-mm-diameter spheres. The volume of a 1-mm-diameter sphere is $\frac{4}{3}\pi r^3 = 0.524\ \mu L$. The density of KCl is 1.984 g/mL and that of KNO_3 is 2.109 g/mL, so the average density of the mixture is $(0.01)(1.984) + (0.99)(2.109) = 2.108$ g/mL. The mass of mixture containing 10^4 particles is $(10^4)(0.525 \times 10^{-3}\ mL)(2.108\ g/mL) = 11.0$ g. *If you take 11.0-g test portions from a larger laboratory sample, the expected sampling standard deviation for chloride is 9.9%!* The sampling standard deviation for nitrate will only be 0.1%.

You might make a mixture of approximately 1-mm-diameter particles by grinding larger particles and passing them through a 16 mesh sieve, whose screen openings are 1.18 mm (Table 24-2). The solid that passes through the screen is then passed through a 20 mesh sieve, whose openings are 0.85 mm, and material that does not pass is retained. This gives particles whose diameters are in the range 0.85–1.18 mm. We refer to the size range as *16/20 mesh*.

Suppose that much finer particles of 80/120 mesh size (average diameter = 150 μm, average volume = 1.77 nL) were used instead. Now the mass that contains 10^4 particles is reduced from 11.0 g to 0.037 2 g. We could analyze a larger sample and reduce the sampling uncertainty for chloride.

EXAMPLE: Reducing Sampling Uncertainty with a Larger Test Portion

How many grams of 80/120 mesh sample are required to reduce the chloride sampling uncertainty to 1%?

We are looking for a standard deviation of 1% of the number of KCl particles:

$$\sigma_n = \sqrt{npq} = (0.01)np$$

Using $p = 0.01$ and $q = 0.99$, we square both sides and solve for n to find $n = 9.9 \times 10^5$ particles. With a particle diameter of 1.77 nL and an average density of 2.108 g/mL, the mass of sample required for 1% chloride sampling uncertainty is

$$\text{Mass} = (9.9 \times 10^5)(1.77 \times 10^{-6}\ mL)(2.108\ g/mL) = 3.69\ g$$

Even with an average particle diameter of 150 μm, we must analyze a full 3.69-g portion to reduce the sampling uncertainty to 1%. There is no point using an analytical method with a precision of 0.1%, because the overall uncertainty will still be 1% based on sampling uncertainty.

You should now see that sampling uncertainty arises from the random nature of drawing particles from a mixture. If the mixture is a liquid and the particles are molecules, there are around 10^{22} particles/mL. It will not require much volume of homogeneous liquid solution to reduce the sampling error to a negligible value. On the other hand, solids must be ground to very fine dimensions and large quantities used to ensure a small sampling variance.

Compromises may be required because many samples cannot be ground indefinitely without changing their composition. Some solids that adsorb negligible water when their surface area is small adsorb significant quantities of water when their surface area is large. The act of grinding usually contaminates the solid being ground with material from the grinding apparatus.

Sampling Materials Whose Composition Varies Randomly

Now we consider heterogeneous materials in which the variation from place to place is random. The mixture of solid KCl and KNO_3 powders is an example of such a material. Two important questions are "How much sample is required to reduce the sampling variance to a desired level?" and "How many samples must be analyzed to produce a desired level of confidence in the answer?"

How much should be analyzed?

Figure 24-2 shows experimental results for the sampling of the radioisotope ^{24}Na in human liver. The tissue has been "homogenized" in a blender, but is not truly homogeneous because it is a suspension of small particles in water. The average number of radioactive counts per second per gram of sample is about 237. When the mass of sample for each analysis was about 0.09 g, the standard deviation (shown by the error bar at the left in the diagram) was ± 31 counts per second per gram of homogenate, which is $\pm 13.1\%$ of the mean value (237). When the sample size was increased to about 1.3 g, the standard deviation was decreased to ± 13 counts per second per gram, or $\pm 5.5\%$ of the mean. For a sample size near 5.8 g, the standard deviation was reduced to ± 5.7 counts per second per gram, or $\pm 2.4\%$ of the mean.

Equation 24-4 told us that when samples are drawn from a mixture of two kinds of particles (such as liver tissue particles and droplets of water), the sampling standard deviation will be $\sigma_n = \sqrt{npq}$, where p and q are the fraction of each kind of particle present. The relative standard deviation is $\sigma_n/n = \sqrt{npq}/n = \sqrt{pq/n}$. The relative variance, $(\sigma_n/n)^2$, is therefore

$$\text{Relative variance} = R^2 = \left(\frac{\sigma_n}{n}\right)^2 = \frac{pq}{n} \quad \Rightarrow \quad nR^2 = pq \qquad (24\text{-}7)$$

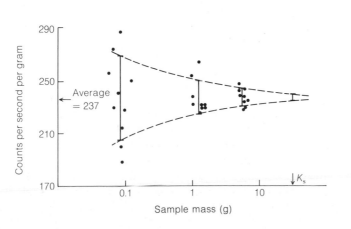

Figure 24-2
Sampling diagram of experimental results for ^{24}Na in liver homogenate. Dots are experimental points, and error bars extend ± 1 standard deviation about the mean. [B. Kratochvil and J. K. Taylor, *Anal. Chem.*, **53**, 925A (1981); National Bureau of Standards Internal Report 80-2164, 1980, p. 66.]

Table 24-3
Calculation of sampling constant for Figure 24-2

Sample mass, m (g)	Relative standard deviation (%)	mR^2 (g)
0.09	13.1	15.2
1.3	5.5	39.1
5.8	2.4	33.5

Noting that the number of particles drawn, n, is proportional to the mass of sample drawn, m, we can rewrite Equation 24-7 in the form

$$mR^2 = K_s \qquad (24\text{-}8)$$

in which m is the mass of sample, R is the relative standard deviation, and K_s is called the *sampling constant*. K_s can also be interpreted as the mass of sample required to reduce the relative sampling standard deviation to 1%.

Let's see if Equation 24-8 describes the data in Figure 24-2. Table 24-3 shows that the product mR^2 is approximately constant for sufficiently large masses of sample, but the agreement is poor for the smallest samples (0.09 g). Attributing the poor agreement at low mass to random sampling variations, we can assign the approximate value 36 g to K_s in Equation 24-8. This is the average obtained from the 1.3- and 5.8-g sample sizes in Table 24-3.

EXAMPLE: Mass of Sample Required to Produce a Given Sampling Variance

With the sampling constant $K_s \approx 36$ g, we can ask the question "What mass of sample in Figure 24-2 will give a sampling standard deviation of $\pm 7\%$?"

The answer is straightforward:

$$m = \frac{36}{7^2} = 0.73 \text{ g}$$

A 0.7-g sample should give approximately a 7% sampling standard deviation. This is strictly a sampling standard deviation. If the analytical procedure also has a significant uncertainty, the net uncertainty will depend on both factors, as described by Equation 24-1.

How many samples should be analyzed?

In the previous section, we saw that a single 0.7-g sample is expected to give a sampling standard deviation of $\pm 7\%$. Now we ask the question "How many 0.7-g samples must be analyzed to give 95% confidence that the mean is known to within $\pm 5\%$?" We are still referring only to sampling uncertainty (and assuming that analytical uncertainty is much smaller than sampling uncertainty). A rearrangement of the Student's t equation 4-6 allows us to answer this question:

$$n = \frac{t^2 s_s^2}{e^2} \qquad (24\text{-}9)$$

The sampling contribution to the overall uncertainty can be reduced by analyzing more samples.

in which n is the number of samples needed, s_s is the standard deviation of the sampling operation, and e is the sought-for uncertainty in the mean value. The quantities s_s and e must both be expressed as absolute uncertainties or both as relative uncertainties. The number t is the value of Student's t for the 95% confidence level and $n - 1$ degrees of freedom. Since n is not yet known, the value of t for $n = \infty$ can be used to estimate n. After a value of n is calculated. The process is repeated a few times until a constant value of n is found.

EXAMPLE: Sampling a Random Bulk Material

Now let's find out how many 0.7-g samples must be analyzed to give 95% confidence that the mean is known to within $\pm 5\%$. A 0.7-g sample gives $s_s = 7\%$, and we are seeking $e = 5\%$. We will express both uncertainties in relative form. Taking $t = 1.960$ (from Table 4-2 for 95% confidence and ∞ degrees of freedom) as a starting value, we find

$$n \approx \frac{(1.960)^2(0.07)^2}{(0.05)^2} = 7.5 \approx 8$$

For $n = 8$ there are 7 degrees of freedom, so a second trial value of Student's t (from Table 4-2) is 2.365. A second cycle of calculation gives

$$n \approx \frac{(2.365)^2(0.07)^2}{(0.05)^2} = 11.0 \approx 11$$

For $n = 11$ there are 10 degrees of freedom and $t = 2.228$, which gives

$$n \approx \frac{(2.228)^2(0.07)^2}{(0.05)^2} = 9.7 \approx 10$$

For $n = 10$ there are 9 degrees of freedom and $t = 2.262$, which gives

$$n \approx \frac{(2.262)^2(0.07)^2}{(0.05)^2} = 10.0 \approx 10$$

The calculations reach a constant value near $n \approx 10$, so it will require about 10 samples of 0.7-g size to measure the mean value to within 5% with 95% confidence.

For the preceding calculations we needed some prior knowledge of the standard deviation. Some preliminary study of the sample must be made before the remainder of the analysis can be planned. If there are many similar samples to be analyzed, a thorough analysis of one sample might allow you to plan less thorough—but adequate—analyses of the remainder.

24-3 PREPARING SAMPLES FOR ANALYSIS[†]

Once a bulk sample is selected, a laboratory sample must be taken and individual test portions analyzed. If the bulk sample is a liquid, a fine powder, or a suspension, it may be homogeneous enough to provide a laboratory sample. A coarse solid must be ground so that the laboratory sample has the same composition as the bulk sample (within an allowed sampling uncertainty). Solids can be ground in a **mortar and pestle** like those shown in Figure 24-3. The steel mortar (also called a percussion mortar or "diamond" mortar) is a hardened steel tool into which the steel sleeve and steel pestle fit snugly. Hard, brittle materials such as ores and minerals can be crushed by striking the pestle lightly with a hammer. The agate mortar (or similar ones made of porcelain, mullite, or alumina) is designed for grinding small

[†] D. C. Bogen, *Treatise on Analytical Chemistry,* 2nd ed. (P. J. Elving, E. Grushka, and I. M. Kolthoff, eds.), Part I, Vol. 5 (New York: Wiley, 1982), Chap. 1; E. C. Dunlop and C. R. Ginnard, *Ibid.,* Chap. 2.

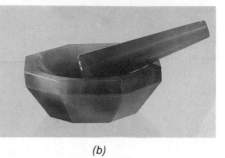

(a)   *(b)*

Figure 24-3
(a) Steel and (b) agate mortars and pestles. The mortar is the base and the pestle is the grinding tool. [Courtesy Thomas Scientific, Swedesboro, NJ.]

Hygroscopic samples pick up water readily from the air. In extreme cases such powders can only be handled in a glove box with a very dry atmosphere.

particles into a fine powder. Less expensive mortars tend to be more porous and more easily scratched, which leads to contamination of the sample with mortar material or portions of previously ground samples. If the ground material comes out easily, a ceramic mortar can be cleaned by wiping with a wet tissue and washing with distilled water. More difficult samples may be cleaned by grinding with 4 M HCl in the mortar or by grinding an abrasive cleaner (such as Ajax®), followed by washing with HCl and water. A **ball mill** is another grinding device in which sample mixed with balls made of a hard ceramic are rotated in a ceramic drum to crush the sample to a fine powder.

Solids are typically dried at 110°C at atmospheric pressure to remove adsorbed water prior to analysis. In some cases water may be retained under these conditions, and in other cases removal of water produces significant changes in the properties of the sample. Heating may cause volatile components of the sample to evaporate or can cause decomposition. Some samples are simply stored in an environment that brings them to a constant, reproducible moisture level.

Once a laboratory sample is obtained, it is dissolved for analysis of desired constituents. It is important to dissolve the entire sample, or else we cannot be sure that some of the intended analyte did not dissolve. If the sample does not dissolve under mild conditions, *acid digestion* or *fusion* may be used. Organic material may be destroyed by *combustion* or *wet ashing* to place inorganic elements in suitable form for analysis.

Dissolving Inorganic Materials with Acids

Table 24-4 lists common acids used for dissolving inorganic materials. The nonoxidizing acids HCl, HBr, HF, H_3PO_4, dilute H_2SO_4, and dilute $HClO_4$ dissolve metals by the redox reaction

$$M + nH^+ \rightarrow M^{n+} + \frac{n}{2}H_2 \qquad (24\text{-}10)$$

Metals with negative reduction potentials should dissolve, although some, such as Al, form a protective oxide coat that inhibits dissolution. Many inorganic compounds also dissolve in hot, concentrated acids. Volatile species formed by protonation of such anions as carbonate ($\rightarrow H_2CO_3 \rightarrow CO_2$), sulfide ($H_2S$), phosphide ($PH_3$), fluoride ($HF$), and borate ($H_3BO_3$) will be lost from hot acids in open vessels. Some metal halides such as $SnCl_4$ and $HgCl_2$ and some molecular oxides such as OsO_4 and RuO_4 are volatile and can also be lost. Hydrofluoric acid is especially useful for dissolving silicates at temperatures near 110°C. Glass or platinum vessels may be used for all but HF, which can be used in Teflon, silver, or platinum vessels. For most purposes the highest quality acids must be used in order to minimize contamination of the sample by the concentrated reagent.

Table 24-4
Acids for sample dissolution

Acid	Typical composition	Notes
HCl	37%, density 1.19 g/mL	Nonoxidizing acid useful for many metals, oxides, sulfides, carbonates, and phosphates. Constant boiling composition at 109°C is 20% HCl (see Problem 11-46).
HBr	48–65%	Similar to HCl in solvent properties. Constant boiling composition at 124°C is 48% HBr.
H_2SO_4	95–98%, density 1.84 g/mL	Good solvent at its boiling point of 338°C. Attacks metals and dehydrates and oxidizes organic compounds.
H_3PO_4	85%, density 1.70 g/mL	Hot acid dissolves refractory oxides insoluble in other acids. Becomes anhydrous above 150°C. Dehydrates to pyrophosphoric acid (HPO_3—O—PO_3H) above 200°C and dehydrates further to metaphosphoric acid ($[HPO_3]_n$) above 300°C.
HF	50%, density 1.16 g/mL	Used primarily to dissolve silicates, making SiF_4. This product and excess HF are removed by adding H_2SO_4 or $HClO_4$ and heating. Constant boiling composition at 112°C is 38% HF. Used in Teflon, silver, or platinum containers. Extremely harmful upon contact or inhalation.
$HClO_4$	60–72%, density 1.54–1.67	Cold and dilute acid are not oxidizing, but hot, concentrated acid is an extremely powerful, explosive oxidant, especially useful for organic matter that has already been partially oxidized by hot HNO_3. Constant boiling composition at 203°C is 72%. **Before using $HClO_4$, evaporate the sample to near dryness several times with hot HNO_3 to destroy as much organic material as possible.**

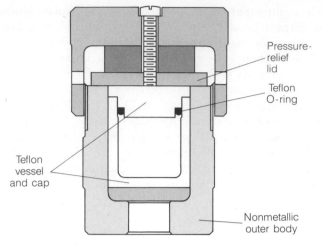

Figure 24-4
Microwave digestion bomb lined with Teflon. The outer container retains strength to 150°C, but rarely reaches 50°C. Deformation of the cap releases pressure exceeding 80 atm from inside the vessel, thereby preventing explosions. [Courtesy Parr Instrument Co., Moline, IL.]

HF is extremely harmful to touch or breathe. If contacted, flood the affected area with water, coat the skin with calcium gluconate (or another calcium salt), and seek medical help.

Substances that do not dissolve in nonoxidizing acids may dissolve in the oxidizing acids HNO_3, hot concentrated H_2SO_4, or hot concentrated $HClO_4$. Nitric acid attacks most metals, but not Au and Pt, which dissolve in the 3:1 (vol/vol) mixture of $HCl:HNO_3$ called **aqua regia**. Strong oxidants such as Cl_2 or $HClO_4$ in HCl dissolve such difficult materials as Ir at elevated temperature. A mixture of HNO_3 and HF attacks the refractory carbides, nitrides, and borides of Ti, Zr, Ta, and W. Hot concentrated $HClO_4$ is a **dangerous,** powerful oxidant whose oxidizing power is increased by adding concentrated H_2SO_4 and catalysts such as V_2O_5 or CrO_3. (Its use is described later in the subsection "Decomposition of Organic Substances.")

A convenient digestion procedure applicable to inorganic and organic materials utilizes a Teflon-lined digestion **bomb** (a sealed vessel) heated in a microwave oven. The apparatus in Figure 24-4 has a volume of 23 mL and can be used to digest up to 1 g of inorganic material (or 0.1 g of organic material, which releases a great deal of $CO_2(g)$) in up to 15 mL of concentrated acid. A household microwave oven will bring the contents of the bomb to 200°C in a minute. A critical safety feature is that the lid assembly deforms and releases gas from the vessel if the internal pressure exceeds 8 MPa (80 atm). The bomb for microwave heating cannot be made of metal, which absorbs microwaves. Teflon-lined metal bombs heated in a furnace are also employed. An advantage of any bomb is that it is cooled prior to opening, thus preventing loss of many volatile products.

Dissolving Inorganic Materials by Fusion

Samples that will not dissolve in acids can usually be dissolved by a hot, molten inorganic **flux** (Table 24-5). The finely powdered unknown is mixed with 5–20 times its mass of solid flux, and **fusion** (melting) is carried out in a platinum crucible at 300–1 200°C in a furnace or over a burner. After a period of minutes to hours, the entire sample dissolves and the melt is allowed to cool slowly. Before the melt solidifies, the container may be swirled to distribute the solidifying mass as a thin coat on the walls. The solid mass is then dissolved in an appropriate solvent, usually dilute aqueous acid. A major disadvantage of a flux is that impurities can be introduced by the large mass of solid reagent. If part of the unknown can be dissolved with acid prior to fusion, it should be dissolved. Then the insoluble component is dissolved with flux and the two portions are combined for analysis.

Fusion is a last resort, because the flux can introduce impurities.

One of the most common fluxes is Na_2CO_3, whose reaction with silicates might be represented by these equations:

Fusion:

$$MSiO_3 + Na_2CO_3 \rightarrow Na_2SiO_3 + MCO_3 \text{ (or } MO + CO_2)$$

$$(24\text{-}11a)$$

Table 24-5
Fluxes for sample dissolution

Flux	Crucible	Uses
Na_2CO_3	Pt	For dissolving silicates (clays, rocks, minerals, glasses), refractory oxides, insoluble phosphates, and sulfates.
$Li_2B_4O_7$ or $LiBO_2$ or $Na_2B_4O_7$	Pt, graphite Au–Pt alloy, Au–Rh–Pt alloy	Individual or mixed borates are used to dissolve aluminosilicates, carbonates, and samples with high concentrations of basic oxides. $B_4O_7^{2-}$ is called tetraborate and BO_2^- is borate.
NaOH or KOH	Au, Ag	Dissolves silicates and SiC. Frothing occurs when H_2O is eliminated from flux, so it is best to prefuse flux and then add sample. Analytical capabilities are limited by impurities in NaOH and KOH.
Na_2O_2	Zr, Ni	Strong base and powerful oxidant good for silicates not dissolved by Na_2CO_3. Useful for iron and chromium alloys. Since it slowly attacks crucibles, a good procedure is to coat the inside of a Ni crucible with molten Na_2CO_3, cool, and add Na_2O_2. The peroxide melts at lower temperature than the carbonate, which shields crucible from the melt.
$K_2S_2O_7$	Porcelain, SiO_2, Au, Pt	Potassium pyrosulfate ($K_2S_2O_7$) is prepared by heating $KHSO_4$ until all water is lost and foaming ceases. Alternatively, potassium persulfate ($K_2S_2O_8$) decomposes to $K_2S_2O_7$ upon heating. Good for refractory oxides, not silicates.
B_2O_3	Pt	Useful for oxides and silicates. Principal advantage is that flux can be completely removed as volatile methyl borate ($[CH_3O]_3B$) by several treatments with HCl in methanol.
$Li_2B_4O_7 + Li_2SO_4$ (2:1 wt/wt)	Pt	Example of a powerful mixture for dissolving refractory silicates and oxides in 10–20 min at 1 000°C. One gram of flux dissolves 0.1 g of sample. The solidified melt dissolves readily in 20 mL of hot 1.2 M HCl.

Dissolution:

$$Na_2SiO_3 + MCO_3 + 4H^+ \rightarrow H_2SiO_3 + 2Na^+ + M^{2+} + H_2CO_3$$

$$(24\text{-}11b)$$

Fluxes are usually classified as acidic, basic, or *amphoteric* (having both acidic and basic properties). Basic fluxes in Table 24-5 are Na_2CO_3, NaOH, KOH, Na_2O_2, and $LiBO_2$. Acidic fluxes are $K_2S_2O_7$, B_2O_3, $Li_2B_4O_7$, and $Na_2B_4O_7$. Basic fluxes are best used to dissolve acidic oxides of Si and P. Acidic fluxes are most suitable for basic oxides of the alkali metals, alkaline earths, lanthanides, and Al. KHF_2 is a useful flux for lanthanide oxides. Sulfides and some oxides, some iron and platinum alloys, and some silicates require an oxidizing flux for dissolution. For this purpose, pure Na_2O_2 may be suitable, or such oxidants as KNO_3, $KClO_3$, or Na_2O_2 may be added to Na_2CO_3. Boric oxide is a special flux that can be converted to volatile form after fusion and removed completely. The solidified flux is treated with 100 mL of methanol saturated with HCl gas and heated gently. The B_2O_3 is converted to $B(OCH_3)_3$, which evaporates. The procedure is repeated several times, if necessary, to remove all of the boron.

Decomposition of Organic Substances

The analysis of elements in organic materials usually begins with chemical destruction of the organic compounds. Procedures are broadly classified as **dry ashing** if they do not involve liquids for the destruction of the organic component, or **wet ashing** if liquid is used. Occasionally, fusion with Na_2O_2 (called Parr oxidation) or alkali metals may be carried out in a sealed bomb. Figure 8-1 illustrates **combustion analysis** (a type of dry ashing), in which organic material is burned in a stream of O_2, usually with the help of catalysts to complete the transformation of carbon to CO_2 and hydrogen to H_2O. Each of these products is trapped separately and the quantity measured. Variations of combustion analysis can also be used to measure nitrogen (Section 8-1), halogens (Problem 9-22), and sulfur in organic compounds. In the latter case, the product stream containing SO_2 and SO_3 is passed through aqueous H_2O_2 to convert sulfur to H_2SO_4 that is titrated with standard base.

The most convenient *wet ashing* procedures involve microwave digestion with acid in a Teflon bomb (Figure 24-4). For example, 0.25 g of animal tissue can be digested for metals analysis by placing it in a 60-mL Teflon vessel containing 1.5 mL of high-purity 70% HNO_3 plus 1.5 mL of high-purity 96% H_2SO_4 and heating in a 700-W kitchen microwave oven for one minute.[†] An extremely important wet ashing process is the *Kjeldahl* digestion with H_2SO_4 used for nitrogen analysis (Experiment 25-6). Digestion with fuming HNO_3 (which contains excess dissolved NO_2) in a sealed heavy-walled glass tube at 200–300°C is called the *Carius* method.

Wet ashing with refluxing HNO_3–$HClO_4$ (Figure 24-5) is a widely applicable, but hazardous, procedure.[‡] **Perchloric acid has caused numerous explosions.** Use a good blast shield between yourself and your experiment,

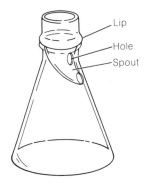

Figure 24-5
Reflux cap for wet ashing. The lip holds the cap at the top of the Erlenmeyer flask in which wet ashing is performed. The hole allows vapor to escape, and the spout is curved to contact the side of the flask. [D. D. Siemer and H. G. Brinkley, *Anal. Chem.*, **53**, 750 (1981).]

Lip
Hole
Spout

$HClO_4$ with organic material is an extreme **explosion hazard.** Always oxidize first with HNO_3. Always use a blast shield for $HClO_4$.

[†] P. Aysola, P. Anderson, and C. H. Langford, *Anal. Chem.*, **59**, 1582 (1987).
[‡] A. A. Schilt, *Perchloric Acid and Perchlorates* (Columbus, OH: G. F. Smith Chemical Co., 1979).

which should only be carried out in a metal-lined fume hood designed for $HClO_4$. The sample is first slowly heated to boiling with HNO_3 in the *absence* of $HClO_4$. The HNO_3 solution should be evaporated to near dryness to complete oxidation of easily oxidized material that might explode if $HClO_4$ were present. Then add fresh HNO_3 and repeat the evaporation several times. After cooling to room temperature, $HClO_4$ is added and heating is begun again. If possible, HNO_3 should be present during the $HClO_4$ treatment. A large excess of HNO_3 should be present when oxidizing inorganic materials. A reviewer of this book once wrote "I have seen someone substitute perchloric acid for sulfuric acid in a Jones reductor experiment with spectacular results— no explosion but the tube melted!" Bottles of $HClO_4$ should not be stored on wooden shelves, because acid spilled on wood can form explosive cellulose perchlorate esters. Perchloric acid also should not be stored near organic reagents or reducing agents.

A very mild wet ashing procedure involving no concentrated reagents was used to oxidize organic material in urine to release traces of mercury for analysis.[†] A 50-mL sample containing <50 ng of Hg was adjusted to pH 3–4 with 0.5 M H_2SO_4. Then 50 μL of saturated aqueous ferrous ammonium sulfate, $Fe(NH_4)_2(SO_4)_2$, was added, followed by 100 μL of 30% H_2O_2. The combination of Fe^{2+} and H_2O_2, called *Fenton's reagent*, is an excellent source of the powerful oxidant hydroxyl radical ($OH\cdot$).

$$Fe^{2+} + H_2O_2 \rightarrow Fe^{3+} + OH^- + OH\cdot \qquad (24\text{-}12)$$

$$OH\cdot + H^+ + e^- \rightarrow H_2O \qquad E° = 2.38 \text{ V} \qquad (24\text{-}13)$$

The $OH\cdot$ radical generated by Fenton's reagent oxidizes organic matter.

After 30 minutes at 50°C the organic matter had been destroyed and the solution contained Hg^{2+}.

24-4 PRECONCENTRATION, CLEANUP, AND DERIVATIZATION

Trace analysis may involve methods that are sensitive to very low concentrations of the desired analyte, or may require **preconcentration** of the analyte to a higher level. A common means of preconcentrating metal ions in natural waters utilizes cation-exchange resin. In one example,[‡] a 500-mL volume of seawater adjusted to pH 6.5 with ammonium acetate and ammonia was passed through 2 g of Chelex-100 in the Mg^{2+} form. Washing with 2 M HNO_3 eluted heavy metals in a total volume of 10 mL, giving a concentration increase of $500/10 = 50$. Metals in the HNO_3 solution were then analyzed by graphite furnace atomic absorption spectroscopy with a typical detection limit for Pb being 15 pg/mL. The detection limit for Pb in the seawater is therefore 50 times lower, or 0.3 pg/mL. Figure 24-6 shows the effect of pH on the recovery of various metals from the seawater. At low pH, H^+ competes with the metal ions for ion-exchange sites and prevents complete recovery.

Ion-exchange resins can capture basic or acidic gases released from dissolved samples. Carbonate from $(ZrO)_2CO_3(OH)_2 \cdot xH_2O$ used in nuclear fuel reprocessing can be measured by placing a known amount of powdered solid in the test tube in Figure 24-7 and treating with 3 M HNO_3. Liberated

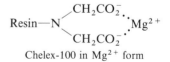

Chelex-100 in Mg^{2+} form

[†] L. Ping and P. K. Dasgupta, *Anal. Chem.,* **61,** 1230 (1989).
[‡] S.-C. Pai, P.-Y. Whung, and R. L. Lai, *Anal. Chim. Acta,* **211,** 257, 271 (1988).

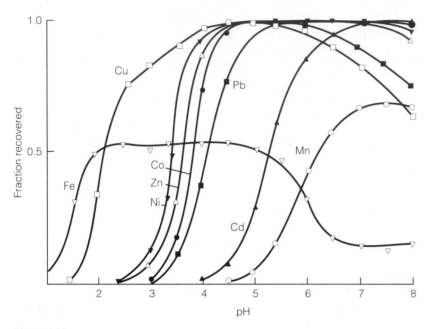
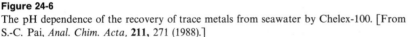

Figure 24-6
The pH dependence of the recovery of trace metals from seawater by Chelex-100. [From S.-C. Pai, *Anal. Chim. Acta,* **211,** 271 (1988).]

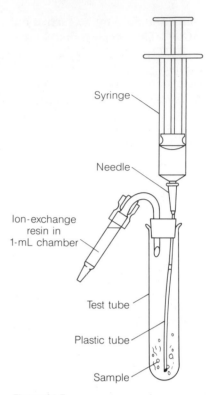

Figure 24-7
Apparatus for trapping basic or acidic gases by ion exchange. [From D. D. Siemer, *Anal. Chem.,* **59,** 2439 (1987).]

CO_2 is captured quantitatively by moist anion-exchange resin in the sidearm, in which the following reactions occur:

$$CO_2 + H_2O \rightarrow H_2CO_3 \qquad (24\text{-}14)$$

$$2\,Resin^+OH^- + H_2CO_3 \rightarrow (resin^+)_2CO_3^{2-} + H_2O \qquad (24\text{-}15)$$

Carbon dioxide removal from the test tube is completed by purging with nitrogen. The carbonate trapped in the sidearm is eluted with 1 M $NaNO_3$ and measured by titration with acid. Table 24-6 gives other applications of this technique.

> The substance used to coprecipitate traces of analyte is called a *gathering agent.*

Coprecipitation with a large excess of highly adsorbent, high-surface-area solid provides significant preconcentration. For example, many metal ions in seawater coprecipitate with $Ga(OH)_3$. To accomplish this, 200 μL of HCl solution containing 50 μg of Ga^{3+} is added to 10.00 mL of the water sample. When the pH is brought to 9.1 with a predetermined volume of NaOH solution, a jellylike precipitate forms, on which is adsorbed a large fraction of the cations in the solution (Table 24-7). After centrifugation to pack the precipitate, the water is removed and the gel washed with water. Then the gel is dissolved in 50 μL of 1 M HNO_3 and aspirated into an inductively coupled plasma for atomic emission analysis. The preconcentration factor is 10 mL/50 μL = 200 and the detection limits for the elements in Table 24-7 in the original seawater range from 15 pg/mL for Zn to 3 ng/mL for Mo.

Table 24-6
Use of ion-exchange resin for trapping gases

Gas	Species trapped	Eluent	Analytical method
CO_2	CO_3^{2-}	1 M $NaNO_3$	Titrate with acid
H_2S	S^{2-}	0.5 M $Na_2CO_3 + H_2O_2$	S^{2-} is oxidizied to SO_4^{2-} by H_2O_2. The sulfate is measured by ion chromatography.
SO_2	SO_3^{2-}	0.5 M $Na_2CO_3 + H_2O_2$	SO_3^{2-} is oxidized to SO_4^{2-} by H_2O_2. The sulfate is measured by ion chromatography.
HCN	CN^-	1 M Na_2SO_4	Titration of CN^- with hypobromite: $$CN^- + OBr^- \rightarrow CNO^- + Br^-$$
NH_3	NH_4^+	1 M $NaNO_3$	Colorimetric assay with Nessler's reagent: $$2K_2HgI_4 + 2NH_3 \rightarrow NH_2Hg_2I_3 + 4KI + NH_4I$$ absorbs strongly at 400–425 nm

SOURCE: From D. D. Siemer, *Anal. Chem.*, **59**, 2439 (1987).

Table 24-7
Recovery of elements by coprecipitation with $Ga(OH)_3$

Element	Recovery (%)	Element	Recovery (%)	Element	Recovery (%)
Al	101.1 ± 2.1	Ti	98.8 ± 0.6	V	33.5 ± 4.2
Cr	100.7 ± 7.9	Mn	78.3 ± 8.9	Fe	95.9 ± 6.3
Co	92.4 ± 4.6	Ni	93.5 ± 3.9	Cu	94.0 ± 4.7
Zn	98.3 ± 5.6	Y	91.0 ± 5.4	Mo	1.3 ± 0.4
Cd	42.4 ± 7.2	Pb	98.8 ± 11.2		

SOURCE: T. Akagi and H. Haraguchi, *Anal. Chem.*, **62**, 81 (1990). Data is average plus standard deviation for five replicate analyses of artificial samples.

Traces of organic analyte are conveniently preconcentrated from a large volume of water by passage through a short, disposable reverse-phase chromatography column containing octyl ($-(CH_2)_7CH_3$) or octadecyl ($-(CH_2)_{17}CH_3$) groups covalently bound to 40-μm-diameter silica particles. Nonpolar molecules retained by the column are then eluted in a small volume of nonpolar solvent, such as hexane or dichloromethane. Other media used to adsorb various analytes include polymers, silica, alumina, and activated charcoal. Volatile analytes may be preconcentrated by passage through a cold trap and nonvolatile analytes may be preconcentrated by evaporation of solvent.

Similar approaches are useful for **sample cleanup,** in which undesirable components of the unknown are removed prior to analysis. If the impurities are more polar than the analyte, the unknown is dissolved in a polar solvent (such as aqueous methanol) and passed through a short, disposable reverse-phase column. The polar impurities pass through the column, while the analyte and nonpolar impurities are retained. Elution with a nonpolar solvent gives a solution of analyte free of polar impurities. If the undesired impurities are less polar than the analyte, sample dissolved in nonpolar solvent is passed through a plain silica column. Impurities pass through while analyte is retained. Finally, the analyte is eluted with a more polar solvent.

The preliminary *cleanup* removes a large fraction of material that might interfere in the analysis.

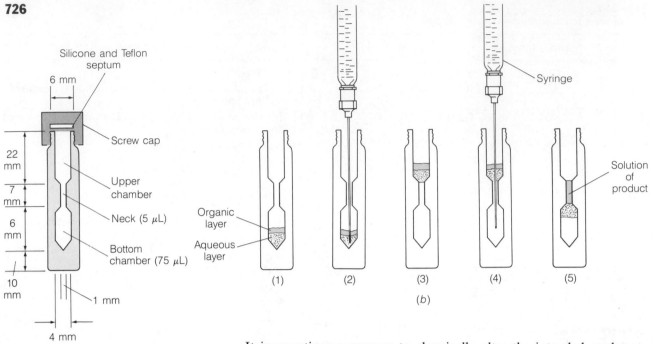

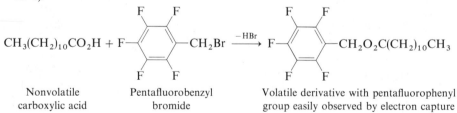

Silicone and Teflon septum

6 mm

22 mm

7 mm

6 mm

10 mm

1 mm

4 mm

Screw cap

Upper chamber

Neck (5 μL)

Bottom chamber (75 μL)

(a)

Organic layer

Aqueous layer

(1) (2) (3) (4) (5)

Syringe

Solution of product

(b)

Figure 24-8

(a) Dimensions of a Keele microreactor. (b) Procedure for derivatization and extraction of product. [From A. B. Attygalle and E. D. Morgan, *Anal. Chem.,* **58,** 3054 (1986).]

It is sometimes necessary to chemically alter the intended analyte to make it more suitable for analysis. An example of such **derivatization** is the attachment of a fluorescent group to a nonfluorescent molecule so that it can be easily detected. In biochemistry, an enzyme that is easily detected by the reaction it catalyzes may be covalently bound to another large molecule that is otherwise difficult to observe. The large molecule is then located wherever the enzyme is seen.

Nanogram quantities of nonvolatile carboxylic acids may be converted to more volatile pentafluorobenzyl esters that are suitable for gas chromatography and readily detected with an electron capture detector (Section 23-1).

$$CH_3(CH_2)_{10}CO_2H + F{-}\underset{F\quad F}{\overset{F\quad F}{\bigcirc}}{-}CH_2Br \xrightarrow{-HBr} F{-}\underset{F\quad F}{\overset{F\quad F}{\bigcirc}}{-}CH_2O_2C(CH_2)_{10}CH_3$$

Nonvolatile carboxylic acid

Pentafluorobenzyl bromide

Volatile derivative with pentafluorophenyl group easily observed by electron capture

A *Keele microreactor* is useful for this micro-scale derivatization. Into the bottom of the reactor in Figure 24-8a is placed 10 μL of 0.1 M aqueous tetrabutylammonium hydroxide, 10 μL of 0.1 M NaOH, 100 ng of carboxylic acid in 10 μL of hexane, and 2 μL of pentafluorobenzyl bromide. After the mixture has been blended by being taken up in a syringe several times, the vial is sealed with its Teflon-lined screw cap and shaken for 20 minutes. Figure 24-8b shows how the sample is manipulated to extract the product for chromatography. First, 20 μL of water and 10 μL of hexane are added to give the two phases in diagram 1. The mixture is then withdrawn by syringe (2) and placed above the thin neck of the vial (3). Withdrawal of air from the lower chamber with the syringe (4) causes the hexane layer containing organic product to move to the thin neck (5), from which it can be removed by syringe.

Analytical methods are validated by demonstrating that accurate results are obtained with known samples. The more complex the unknown, the harder it is to prepare suitable knowns in which the **matrix** (the bulk of the sample) and potential interfering species are similar to those of the unknown. The method of *standard addition*, in which standards are added to the unknown, helps assure that matrix effects and interference are under control. *Standard Reference Materials* from the National Institute of Standards and Technology (Box 9-1) are an extremely important means by which procedures can be tested on samples with known composition. The validity of analytical methods can also be checked by showing that different methods produce similar results with the same unknown.

Validation of analytical method:

1. Analyze standard reference materials.
2. Verify results by different methods.
3. Use standard additions.

Table 24-8 recommends primary standards for many elements. As an *elemental assay standard*, the material must contain a known amount of the desired element. As a *matrix matching standard*, the most important (and demanding) characteristic is that the material must contain extremely low concentrations of undesired impurities. For example, if you want to prepare 10 ppm Fe in 10% aqueous NaCl, the NaCl cannot contain significant Fe impurity, or the Fe impurity would have a higher concentration than the added Fe. Manufacturers frequently indicate elemental purity by some number of 9s. This deceptive nomenclature is based on the measurement of certain impurities. For example, 99.999% (five 9s) pure Al is certified to contain $\leq 0.001\%$ *metallic* impurities, based on the analysis of other metals present. However, C, H, N, and O are not measured. The Al might contain 0.1% Al_2O_3 and still be "five 9s pure." For the most accurate work, the dissolved gas content in solid elements may also be a source of error.

Elemental standard: provides known amount of analyte
Matrix matching standard: must be free of desired analyte

Carbonates, oxides, and other compounds may not possess the exact stoichiometry on the label. For example, TbO_2 will have a high Tb content if some Tb_4O_7 is present. Ignition in an O_2 atmosphere may be helpful, but the final stoichiometry is never guaranteed. Carbonates may contain traces of bicarbonate, oxide, and hydroxide. Firing in a CO_2 atmosphere may improve the stoichiometry. Sulfates may contain some HSO_4^-.

Most metals dissolve in 6 M HCl or HNO_3, or a mixture of the two, possibly with heating. Frothing accompanies dissolution of metals or carbonates in acid, so vessels should be loosely covered by a watchglass or Teflon lid to prevent loss of material. Concentrated (16 M) HNO_3 may *passivate* some metals, forming an insoluble oxide coat that prevents dissolution. If you have a choice between a bulk element and a powder, the bulk form is preferred because it has a smaller surface area on which oxides may form and impurities may be adsorbed. After cutting a pure metal to be used as a standard, it should be etched ("pickled") in a dilute solution of the acid in which it will be dissolved to remove surface oxides and contamination from the cutter. The material should be washed well with water after etching and dried in a vacuum desiccator.

Dilute solutions are best prepared in Teflon or plastic vessels, because glass is a notorious ion exchanger that can replace analyte species. Since volumetric dilutions are rarely more accurate than 0.1%, gravimetric dilutions are required for greater accuracy. Evaporation of standard solutions is an important source of error that is avoided if the mass of the reagent bottle is recorded after each use. If the mass changes between uses, the contents are evaporating.

Use buoyancy corrections (Equation 2-2) for accurate weighing.

Table 24-8
Calibration standards

Element	Source[†]	Purity	Comments[‡]
Li	SRM 924 (Li_2CO_3)	$100.05 \pm 0.02\%$	E; dry at 200°C for 4 h
	Li_2CO_3	five–six 9s	M; purity calculated from impurities. Stoichiometry unknown.
Na	SRM 919 or 2201 (NaCl)	99.9%	E; dry for 24 h over $Mg(ClO_4)_2$.
	Na_2CO_3	three 9s	M; purity based on metallic impurities.
K	SRM 918 (KCl)	99.9%	E; dry for 24 h over $Mg(ClO_4)_2$.
	SRM 999 (KCl)	$52.435 \pm 0.004\%$ K	E; ignite at 500°C for 4 h.
	K_2CO_3	five–six 9s	M; purity based on metallic impurities.
Rb	SRM 984 (RbCl)	$99.90 \pm 0.02\%$	E; hygroscopic. Dry for 24 h over $Mg(ClO_4)_2$.
	Rb_2CO_3		M
Cs	Cs_2CO_3		M
Be	metal	three 9s	E, M; purity based on metallic impurities.
Mg	SRM 929	$100.1 \pm 0.4\%$	E; magnesium gluconate clinical standard.
		$5.403 \pm 0.022\%$ Mg	Dry for 24 h over $Mg(ClO_4)_2$.
	metal	five 9s	E; purity based on metallic impurities.
Ca	SRM 915 ($CaCO_3$)	three 9s	E; use without drying.
	$CaCO_3$	five 9s	E, M; dry at 200°C for 4 h in CO_2. User must determine stoichiometry.
Sr	SRM 987 ($SrCO_3$)	99.8%	E; ignite to establish stoichiometry. Dry at 110° for 1 h.
	$SrCO_3$	five 9s	M; up to 1% off stoichiometry. Ignite to establish stoichiometry. Dry at 200° for 4 h.
Ba	$BaCO_3$	four–five 9s	M; dry at 200°C for 4 h.
B	SRM 951 (H_3BO_3)	100.00 ± 0.01	E; expose to room humidity ($\sim 35\%$) for 30 min before use.
Al	metal	five 9s	E, M; SRM 1257 Al metal available.
Ga	metal	five 9s	E, M; SRM 994 Ga metal available.
In	metal	five 9s	E, M
Tl	metal	five 9s	E, M; SRM 997 Tl metal available.
C			No recommendation
Si	metal	six 9s	E, M; SRM 990 SiO_2 available.
Ge	metal	five 9s	E, M
Sn	metal	six 9s	E, M; SRM 741 Sn metal available.
Pb	metal	five 9s	E, M; several SRM's available.
N	NH_4Cl	six 9s	E; can be prepared from $HCl + NH_3$.
	N_2	>three 9s	E
	HNO_3	six 9s	M; contaminated with NO_x. Purity based on impurities.
P	SRM 194 ($NH_4H_2PO_4$)	three 9s	E
	P_2O_5	five 9s	E, M; difficult to keep dry.
	H_3PO_4	four 9s	E; must titrate 2 hydrogens to be certain of stoichiometry.
As	metal	five 9s	E, M
	SRM 83d (As_2O_3)	$99.9926 \pm 0.0030\%$	Redox standard. As assay not assured.
Sb	metal	four 9s	E, M
Bi	metal	five 9s	E, M
O	H_2O	eight 9s	E, M; contains dissolved gases.
	O_2	>four 9s	E
S	element	six 9s	E, M; difficult to dry. Other sources are H_2SO_4, Na_2SO_4, and K_2SO_4. Stoichiometry must be proven (e.g., no SO_3^{2-} present).

(continued)

Table 24-8 (continued)

729

Element	Source[†]	Purity	Comments[‡]
Se	metal	five 9s	E, M; SRM 726 Se metal available.
Te	metal	five 9s	E, M
F	NaF	four 9s	E, M; no good directions for drying.
Cl	NaCl	four 9s	E, M; dry for 24 h over $Mg(ClO_4)_2$. Several SRM's (NaCl and KCl) available.
Br	KBr	four 9s	E, M; need to dry and demonstrate stoichiometry.
	Br_2	four 9s	E
I	sublimed I_2	six 9s	E
	KI	three 9s	E, M
	KIO_3	three 9s	Stoichiometry not assured.

Transition metals: Use pure metals (usually ≥ four 9s) for elemental and matrix standards. Assays are based on impurities and do not include dissolved gases.

Lanthanides: Use pure metals (usually ≥ four 9s) for elemental standards and oxides as matrix standards. Oxides may be difficult to dry and stoichiometry is not certain.

SOURCE: J. R. Moody, R. R. Greenberg, K. W. Pratt, and T. C. Rains, *Anal. Chem.*, **60**, 1203A (1988).
[†] SRM is the National Institute of Standards and Technology designation for a standard reference material.
[‡] E means elemental assay standard; M means matrix matching standard.

The path to accurate analysis is fraught with obstacles that must be appreciated by those who use analytical results and that challenge the skill and imagination of the hard-core analyst.

Summary

Real samples are heterogeneous. Therefore care is required in selecting a bulk sample from the lot being analyzed. A random sample is appropriate for a random material, whereas a composite sample is appropriate for a segregated material. The bulk sample may be finely ground with a mortar and pestle or ball mill to make a laboratory sample with the same composition. Test portions are then taken for analysis. The variance of an analysis is the sum of the sampling variance and the analytical variance. Sampling variance can be understood in terms of the statistics of selecting particles from a heterogeneous mixture. If the probabilities of selecting two kinds of particles from a two-particle mixture are p and q, the standard deviation in selecting n particles is $\sqrt{npq}$. You should be able to use this relationship to estimate how large a sample is required to reduce the sampling variance to a desired level. Student's t can be used to estimate how repetition of the analysis reduces the uncertainty of the result.

Relatively insoluble inorganic materials may be dissolved in strong acids with heating. Glass vessels are often useful, but Teflon, platinum, or silver are required for HF, which dissolves silicates. If a nonoxidizing acid is insufficient, aqua regia or other oxidizing acids may do the job. A Teflon-lined bomb heated in a microwave oven is a particularly convenient means of dissolving difficult samples. If acid digestion fails, fusion in a molten salt will usually work. Fluxes are classified as acidic, basic, and oxidizing. Organic materials are decomposed by wet ashing with hot concentrated acids or dry ashing with heat. Combustion analysis is commonly used for analysis of C, H, N, S, and halogens.

A dilute analyte may be preconcentrated by ion exchange, chromatographic adsorption, or coprecipitation. Sample cleanup to remove potential interference may be accomplished by chromatography. Derivatization is used to transform the analyte into a more easily detected or separated form. Analytical methods may be validated with standard reference materials, by using different methods to produce the same results, and by standard addition. Elemental standards provide a known amount of analyte, while matrix matching standards must be completely free of the intended analyte. Reagent purity is affected by such factors as oxide coatings and nonstoichiometry. Care and thought are required in every phase of chemical analysis.

Terms to Understand

aqua regia	coprecipitation	homogeneous	random sample
ball mill	derivatization	laboratory sample	sample cleanup
bomb	dry ashing	lot	segregated material
bulk sample	flux	matrix	test portion
combustion analysis	fusion	mortar and pestle	variance
composite sample	heterogeneous	preconcentration	wet ashing

Exercises

24-A. A box contains 120 000 red marbles and 880 000 yellow marbles.

(a) If you draw a random sample of 1 000 marbles from the box, what is the expected number of red and yellow marbles in your sample?

(b) Now put those marbles back in the box and repeat the experiment again. What will be the absolute and relative standard deviations for the numbers in part a after many drawings of 1 000 marbles?

(c) What will be the absolute and relative standard deviations after many drawings of 4 000 marbles?

(d) Fill in the blanks: If you quadruple the size of the sample, you decrease the sampling standard deviation by a factor of _____. If you increase the sample size by a factor of n, you decrease the sampling standard deviation by a factor of _____.

(e) What sample size is required to reduce the sampling standard deviation of red marbles to $\pm 2\%$?

24-B. (a) What mass of sample in Figure 24-2 is expected to give a sampling standard deviation of $\pm 10\%$?

(b) Using the sample size of part a, decide how many samples should be taken to assure 95% confidence that the mean is known to within ± 20 counts per second per gram.

24-C. Because they are used in gasoline to boost the octane rating, alkyl lead compounds find their way into the environment as toxic pollutants. One way to measure alkyl lead compounds in natural waters is by anodic stripping voltammetry, in which the lead is first reduced to the element at a mercury electrode and dissolves in the mercury. Reoxidation occurs when the electrode potential is made sufficiently positive, with current proportional to the concentration of dissolved Pb. An example of this sequence is

Reduction at -1.2 V:

$$(CH_3CH_2)_3PbCl \rightarrow Pb(in\ Hg)$$

Oxidation at -0.5 V:

$$Pb(in\ Hg) \rightarrow Pb^{2+}(aq) \qquad \text{(see Figure 18-18)}$$

Inorganic (ionic) forms of lead are present in much higher concentrations than alkyl lead in natural samples, so inorganic lead must be selectively removed before analysis of alkyl lead. One way to do this is by *coprecipitation* with $BaSO_4$, as shown below.

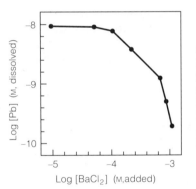

Coprecipitation of Pb^{2+} with $BaSO_4$ from distilled water at pH 2. [Data from N. Mikac and M. Branica, *Anal. Chim. Acta,* **212,** 349 (1988).]

In this experiment, $BaCl_2$ added to Pb^{2+} in distilled water at the concentration on the abscissa was precipitated with a 20% excess of Na_2SO_4 and the residual concentration of lead was measured by stripping voltammetry. Describe control experiments that should be performed to show that coprecipitation may be used to remove a large amount of inorganic lead from a small quantity of $(CH_3CH_2)_3PbCl$ in seawater, without removing the alkyl lead compound.

24-D. A soil sample contains some acid-soluble inorganic matter, some organic material, and some minerals that do not dissolve in any combination of hot acids that you try. Suggest a procedure for dissolving the entire sample.

24-E. What is meant when we say that a standard may not have the expected stoichiometry? Give an example.

Problems

A24-1. (a) What is the difference between homogeneous and heterogeneous samples? Give an example of each.

(b) What is the difference between random heterogeneous material and a segregated heterogeneous material?

A24-2. Explain what is meant by the statement "Unless the complete history of any sample is known with certainty, the analyst is well advised not to spend his time in analyzing it."

A24-3. (a) In the analysis of a barrel of powder, the standard deviation of the sampling operation is $\pm 4\%$ and the standard deviation of the analytical procedure is $\pm 3\%$. What is the overall standard deviation?

(b) To what value must the sampling standard deviation be reduced so that the overall standard deviation is $\pm 4\%$?

A24-4. Describe how you would prepare a composite sample of a segregated material.

A24-5. What mass of sample in Figure 24-2 is expected to give a sampling standard deviation of $\pm 6\%$?

A24-6. Explain how to prepare a powder with an average particle diameter near 100 μm by using the sieves in Table 24-2. How would such a particle mesh size be designated?

A24-7. Based on standard reduction potentials, specify which of the following metals you would expect to dissolve in HCl by the reaction $M + nH^+ \rightarrow M^{n+} + (n/2)H_2$: Zn, Fe, Co, Al, Hg, Cu, Pt, Au. (When the potential predicts that the element will not dissolve, it probably will not. If it is expected to dissolve, it may dissolve if some other process does not interfere. Predictions based on standard reduction potentials at 25°C are only tentative, because the potentials and activities in hot, concentrated solutions vary widely from those in the table of standard potentials.)

A24-8. The following wet ashing procedure was used to measure arsenic in organic soil samples by atomic absorption spectroscopy:[†] A 0.1- to 0.5-g sample was heated in a 150-mL Teflon bomb (Figure 24-4) in a microwave oven for 2.5 min with 3.5 mL of 70% HNO_3. After cooling, a mixture containing 3.5 mL of 70% HNO_3, 1.5 mL of 70% $HClO_4$, and 1.0 mL of H_2SO_4 was added and the sample was reheated for three 2.5-min intervals with 2-min unheated periods in between. The final solution was diluted with

[†] J. Huang, D. Goltz, and F. Smith, *Talanta,* **35,** 907 (1988).

0.2 M HCl for analysis. Explain why $HClO_4$ was not introduced until the second heating.

A24-9. Barbiturate drugs such as barbital can be extracted from urine for analysis by passage of the whole urine sample through a short, reverse-phase octadecyl chromatography column. The barbital is then eluted with 1:1 acetone: chloroform. Explain how this procedure works.

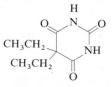

Barbital

A24-10. Explain the difference between an elemental standard and a matrix matching standard.

A24-11. Why is a metal "pickled" before being used as a standard?

A24-12. Referring to Table 24-6, explain how an anion-exchange resin may be used for absorption and analysis of SO_2 released by combustion.

24-13. In the two paragraphs following Equation 24-6, we have an example of a mixture of 1-mm-diameter particles of KCl and KNO_3 in a number ratio 1:99. A sample containing 10^4 particles weighs 11.0 g. What is the expected number and relative standard deviation of KCl particles in a sample weighing 11.0×10^2 g?

24-14. When you flip a coin, the probability of its landing on each side is $p = q = \frac{1}{2}$ in Equations 24-2 and 24-3. If you flip it n times, the expected number of heads equals the expected number of tails $= np = nq = \frac{1}{2} n$. The expected standard deviation for n flips is $\sigma_n = \sqrt{npq}$. From Table 4-1, we expect that 68.3% of the results will lie within $\pm 1\sigma_n$ and 95.5% of the results will lie within $\pm 2\sigma_n$.

(a) Find the expected standard deviation for the number of heads in 1 000 coin flips.

(b) By interpolation in Table 4-1, find the value of z that includes 90% of the area of the Gaussian curve. We expect that 90% of the results will lie within this number of standard deviations from the mean.

(c) If you repeat the 1 000 coin flips many times, what is the expected range for the number of heads that includes 90% of the results? (For example, your answer might be "The range 490 to 510 will be observed 90% of the time.")

24-15. In analyzing a lot with random sample variation, there is a sampling standard deviation of $\pm 5\%$.

Assuming negligible error in the analytical procedure, how many samples must be analyzed to give 95% confidence that the error in the mean is within $\pm 4\%$ of the true value? Answer the same question for a confidence level of 90%.

24-16. In an experiment analogous to that in Figure 24-2, the sampling constant is found to be $K_s = 20$ g.

(a) What mass of sample is required for a $\pm 2\%$ sampling standard deviation?

(b) How many samples of the size in part a are required to produce 90% confidence that the mean is known to within 1.5%?

24-17. Consider a random mixture containing 4.00 g of Na_2CO_3 (F.W. 105.99, density 2.532 g/mL) and 96.00 g of K_2CO_3 (F.W. 138.21, density 2.428 g/mL) with a uniform spherical particle radius of 0.075 μm.

(a) Calculate the mass of a single particle of Na_2CO_3 and the number of particles of Na_2CO_3 in the mixture. Do the same for K_2CO_3.

(b) What is the expected number of particles in 1.00 g of the mixture?

(c) Calculate the relative sampling standard deviation in the number of particles of each type in a 0.100-g sample of the mixture and in a 1.00-g sample.

24-18. The figure below shows elemental concentrations in seawater as a function of depth near hydrothermal vents. Analysis was carried out by the $Ga(OH)_3$ coprecipitation method described in Section 24-4.

(a) What is the atomic ratio (Ga added):(Ni in seawater) for the sample with the highest concentration of Ni?

(b) The results given by dashed lines were obtained with seawater samples that were not filtered prior to coprecipitation. The solid-line results refer to filtered samples. The results for Mn and Ni do not vary between the two procedures, but the results for Fe do vary. Explain what this means.

24-19. Barium titanate, a ceramic used in electronics, was analyzed by the following procedure:[†] Into a Pt crucible was placed 1.2 g of Na_2CO_3 and 0.8 g of $Na_2B_4O_7$ plus 0.314 6 g of unknown. After fusion at 1 000°C in a furnace for 30 min, the cooled solid was extracted with 50 mL of 6 M HCl, transferred to a 100-mL volumetric flask, and diluted to the mark. A 25.00-mL aliquot was treated with 5 mL of 15% tartaric acid (which complexes Ti^{4+} and keeps it in aqueous solution) and 25 mL of ammonia buffer, pH 9.5. The solution was treated with organic reagents that complex Ba^{2+} and the Ba complex extracted into CCl_4. After acidification (to release the Ba^{2+} from its organic complex) the Ba^{2+} was back-extracted into 0.1 M HCl. The final aqueous sample was treated with ammonia buffer and methylthymol blue (a metal ion indicator) and titrated with 32.49 mL of 0.011 44 M EDTA. Find the percent of Ba in the ceramic.

[†] F. P. Gorbenko, A. A. Nadezhda, E. I. Dekhovich, and T. A. Luk'yanénko, *J. Anal. Chem. U.S.S.R.*, **38**, 1231 (1984).

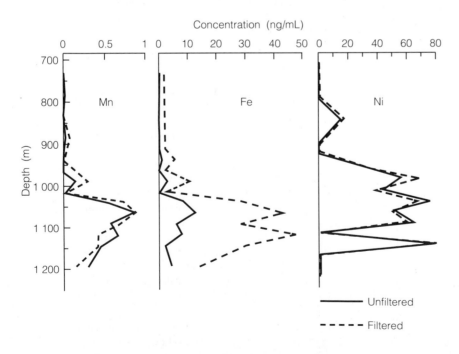

Concentration (ng/mL)

Depth (m)

Depth profile of elements in seawater near hydrothermal vents. [From T. Akagi and H. Haraguchi, *Anal. Chem.*, **62**, 81 (1990).]

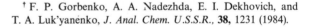

——— Unfiltered

- - - - Filtered

25 Experiments

The experiments described in this chapter are intended to illustrate the major analytical techniques described in the text.[†] These procedures are organized roughly in the same order as the topics in the text.

Although these procedures are safe when carried out with reasonable care, *all chemical experiments are potentially hazardous*. Any solution that fumes (such as concentrated HCl), and all nonaqueous solvents, should be handled in a fume hood. Pipetting should never be done by mouth. Spills on your body should immediately be flooded with water and washed with soap and water; your instructor should be notified for possible further action. Spills on the benchtop should be cleaned immediately. Toxic chemicals should not be flushed down the drain. Your instructor should establish a safe procedure for disposing of each chemical that you use (see Box 2-2).

25-1 CALIBRATION OF VOLUMETRIC GLASSWARE

An important trait of good analysts is the ability to extract the best possible data from their equipment. For this purpose, it is desirable to calibrate your own volumetric glassware (burets, pipets, flasks, etc.) to measure the exact volumes delivered or contained. This experiment also promotes improved technique in handling volumetric glassware. Before beginning this experiment, you should study Section 2-9.

Calibration of 50-mL Buret

In this procedure, we will construct a graph (such as Figure 3-3) needed to convert the measured volume delivered by a buret to the true volume delivered at 20°C.

[†] An excellent source of further experiments involving instrumental methods of analysis is D. T. Sawyer, W. R. Heineman, and J. M. Beebe, *Chemistry Experiments for Instrumental Methods* (New York: Wiley, 1984). For a computerized list of experiments published in the *Journal of Chemical Education*, see *J. Chem. Ed., 64,* 546 (1987).

PROCEDURE

1. Fill the buret with distilled water and force any air bubbles out the tip. See that the buret drains without leaving drops on the walls. If drops are left, clean the buret with soap and water or soak it with peroxydisulfate–sulfuric acid cleaning solution (see footnote on page 21). Adjust the meniscus to be at or slightly below 0.00 mL, and touch the buret tip to a beaker to remove the suspended drop of water. Allow the buret to stand for 5 min while you weigh a 125-mL flask fitted with a rubber stopper. (Do not touch the flask with your hands, to avoid changing its mass with fingerprint residue.) If the level of the liquid in the buret has changed, tighten the stopcock and repeat the procedure.

2. Drain approximately 10 mL of water (at a rate of <20 mL/min) into the weighed flask, and cap it tightly to prevent evaporation. Allow about 30 s for the film of liquid on the walls to descend before you read the buret. Estimate all readings to the nearest 0.01 mL. Weigh the flask again to determine the mass of water delivered.

3. Now drain the buret from 10 to 20 mL, and measure the mass of water delivered. Repeat the procedure for 30, 40, and 50 mL. Then do the entire procedure (10, 20, 30, 40, 50 mL) a second time.

4. Use Table 2-6 to convert the mass of water to the volume delivered. Repeat any set of duplicate buret corrections that do not agree to within 0.04 mL. Prepare a calibration graph such as Figure 3-3, showing the correction factor at each 10 mL interval.

EXAMPLE: Buret Calibration

When draining the buret at 24°C, you observe the following values:

Final reading	10.01	10.08 mL
Initial reading	0.03	0.04
Difference	9.98	10.04 mL
Mass	9.984	10.056 g
Actual volume delivered	10.02	10.09 mL
Correction	+0.04	+0.05 mL
Average correction		+0.045 mL

To calculate the actual volume delivered when 9.984 g of water is delivered at 24°C, look at the column of Table 2-6 headed "Corrected to 20°C." Across from 24°C, you find that 1.000 0 g of water occupies 1.003 8 mL. Therefore, 9.984 g occupies (9.984 g)(1.003 8 mL/g) = 10.02 mL. The average correction for both sets of data is +0.045 mL.

To obtain the correction for a volume greater than 10 mL, add successive masses of water collected in the flask. Suppose that the following masses were measured:

	Volume interval (mL)	Mass delivered (g)
	0.03–10.01	9.984
	10.01–19.90	9.835
	19.90–30.06	10.071
Sum	30.03 mL	29.890 g

The total mass of water delivered corresponds to (29.890 g)(1.003 8 mL/g) = 30.00 mL. Since the indicated volume is 30.03 mL, the buret correction at 30 mL is −0.03 mL.

What does this mean? Suppose that Figure 3-3 applies to your buret. If you begin a titration at 0.04 mL and end at 29.00 mL, you would deliver 28.96 mL if the buret were perfect. In fact, Figure 3-3 tells you that the buret delivers 0.03 mL less than the indicated amount, so only 28.93 mL was actually delivered. To use the calibration curve, either begin all titrations near 0.00 mL or correct both the initial and the final readings. Use the calibration curve whenever you use your buret.

Other Calibrations

Pipets can be calibrated by weighing the water delivered from them. A volumetric flask can be calibrated by weighing it empty and then weighing it filled to the mark with distilled water. Perform each procedure at least twice. Compare your results with the tolerances in Tables 2-2, 2-3, and 2-4.

25-2 GRAVIMETRIC DETERMINATION OF CALCIUM AS CaC₂O₄ · H₂O[†]

Calcium ion can be analyzed by precipitation with oxalate in basic solution to form $CaC_2O_4 \cdot H_2O$ ($K_{sp} = 1.9 \times 10^{-9}$). The precipitate is soluble in acidic solution because the oxalate anion is a weak base (Reactions 8-2 and 8-3). Large, easily filtered, relatively pure cystals of product will be obtained if the precipitation is carried out slowly. This can be done by dissolving Ca^{2+} and $C_2O_4^{2-}$ in acidic solution and gradually raising the pH by thermal decomposition of urea (Reaction 8-4).

REAGENTS

Ammonium oxalate solution: Make one liter of solution containing 40 g of $(NH_4)_2C_2O_4$ plus 25 mL of 12 M HCl. Each student will need 80 mL of this solution.

Unknowns: Prepare one liter of solution containing 15–18 g of $CaCO_3$ plus 38 mL of 12 M HCl. Each student will need 100 ml of this solution. Alternatively, solid unknowns are available from Thorn Smith.[‡]

PROCEDURE

1. Dry three medium-porosity, sintered-glass funnels for 1–2 h at 105°C; cool them in a desiccator for 30 min and weigh them. Repeat the procedure with 30-min heating periods until successive weighings agree to within 0.3 mg. Use a paper towel or tongs, not your fingers, to handle the funnels.

2. Use a few small portions of unknown to rinse a 25-mL transfer pipet, and discard the washings. *Use a rubber bulb, not your mouth, to provide suction.* Transfer exactly 25 mL of unknown to each of three 250–400 mL beakers, and dilute each with ∼75 mL of 0.1 M HCl. Add 5 drops of methyl red indicator solution (Table 11-3) to each beaker. This indicator is red below pH 4.8 and yellow above pH 6.0.

[†] C. H. Hendrickson and P. R. Robinson, *J. Chem. Ed.,* **56,** 341 (1979). For an experiment in thermogravimetric analysis, see J. O. Hill and R. J. Magee, *J. Chem. Ed.,* **65,** 1024 (1988).
[‡] Thorn Smith Inc., 7755 Narrow Gauge Road, Beulah, MI 49617.

3. Add ~25 mL of ammonium oxalate solution to each beaker while stirring with a glass rod. Remove the rod and rinse it into the beaker. Add ~15 g of solid urea to each sample, cover it with a watchglass, and boil gently for ~30 min until the indicator turns yellow.

4. Filter each hot solution through a weighed funnel using suction (Figure 2-15). Add ~3 mL of ice-cold water to the beaker, and use a rubber policeman to help transfer the remaining solid to the funnel. Repeat this procedure with small portions of ice-cold water until all of the precipitate has been transferred. Finally, use two 10-mL portions of ice-cold water to rinse each beaker, and pour the washings over the precipitate.

5. Dry the precipitate, first with aspirator suction for 1 min, then in an oven at 105°C for 1–2 h. Bring each filter to constant weight. The product is somewhat hygroscopic, so only one filter at a time should be removed from the desiccator, and weighings should be done rapidly.

6. Calculate the molarity of Ca^{2+} in the unknown solution or the weight percent of Ca in the unknown solid. Report the standard deviation and relative standard deviation ($s/\overline{x}$ — standard deviation/average).

25-3 GRAVIMETRIC DETERMINATION OF IRON AS Fe_2O_3[†]

A sample containing iron can be analyzed by precipitation of the hydrous oxide from basic solution, followed by ignition to produce Fe_2O_3:

$$Fe^{3+} + (2 + x)H_2O \xrightarrow{\text{base}} FeOOH \cdot xH_2O(s) + 3H^+$$

$$FeOOH \cdot xH_2O \xrightarrow{900°C} Fe_2O_3(s)$$

The hydrous oxide is gelatinous and may occlude impurities. If this is suspected, the initial precipitate is dissolved in acid and reprecipitated. Since the concentration of impurities is lower during the second precipitation, occlusion is diminished. Solid unknowns may be prepared from reagent ferrous ammonium sulfate or purchased from Thorn Smith.

PROCEDURE

1. Bring three porcelain crucibles and caps to constant weight by heating to redness for 15 min over a burner (Figure 2-21). Cool for 30 min in a desiccator and weigh each crucible. Repeat this procedure until successive weighings agree within 0.3 mg. Be sure that all oxidizable substances on the entire surface of each crucible have burned off.

2. Accurately weigh three samples of unknown containing enough Fe to produce ~0.3 g of Fe_2O_3. Dissolve each sample in 10 mL of 3 M HCl (with heating, if necessary). If there are insoluble impurities, filter through qualitative filter paper and wash the filter very well with distilled water. Add 5 mL of 6 M HNO_3 to the filtrate, and boil for a few minutes to ensure that all iron is oxidized to Fe(III).

3. Dilute the sample to 200 mL with distilled water and add 3 M ammonia[‡] with constant stirring until the solution is basic (as determined with litmus paper or

---

[†] Adapted from D. A. Skoog and D. M. West, *Fundamentals of Analytical Chemistry,* 3d ed. (New York: Holt, Rinehart and Winston, 1976).
[‡] Basic reagents should not be stored in glass bottles because they will slowly dissolve the glass. If ammonia from a glass bottle is used, it may contain silica particles and should be freshly filtered.

pH indicator paper). Digest the precipitate by boiling for 5 min and allow the precipitate to settle.

4. Decant the supernatant liquid through coarse, ashless filter paper (Whatman 41 or Schleicher and Schuell Black Ribbon, as in Figures 2-16 and 2-17). Do not pour liquid higher than 1 cm from the top of the funnel. Proceed to step 5 if a reprecipitation is desired. Wash the precipitate repeatedly with hot 1% NH_4NO_3 until little or no Cl^- is detected in the filtered supernate. (Test for Cl^- by acidifying a few milliliters of filtrate with 1 mL of dilute HNO_3 and adding a few drops of 0.1 M $AgNO_3$.) Finally, transfer the solid to the filter with the aid of a rubber policeman and more hot liquid. Proceed to step 6 if a reprecipitation is not used.

5. Wash the gelatinous mass twice with 30 mL of boiling 1% aqueous NH_4NO_3, decanting the supernate through the filter. Then put the filter paper back into the beaker with the precipitate, add 5 mL of 12 M HCl to dissolve the iron, and tear the filter paper into small pieces with a glass rod. Add ammonia with stirring and reprecipitate the iron. Decant through a funnel fitted with a fresh sheet of ashless filter paper. Wash the solid repeatedly with hot 1% NH_4NO_3 until little or no Cl^- is detected in the filtered supernate. Then transfer all the solid to the filter with the aid of a rubber policeman and more hot liquid.

6. Allow the filter to drain overnight—if possible, protected from dust. Carefully lift the paper out of the funnel, fold it as shown in Figure 2-20, and transfer it to a porcelain crucible that has been brought to constant weight.

7. Dry the crucible cautiously with a small flame, as shown in Figure 2-21. The flame should be directed at the top of the container, and the lid should be off. Avoid spattering. After it is dry, *char* the filter paper by increasing the flame temperature. The crucible should have free access to air to avoid reduction of iron by carbon. (The lid should be kept handy to smother the crucible if the paper inflames.) Any carbon left on the crucible or lid should be removed by directing the burner flame at it. Use tongs to manipulate the crucible. Finally, *ignite* the product for 15 min with the full heat of the burner.

8. Cool the crucible briefly in air and then in a desiccator for 30 min. Weigh the crucible and the lid, reignite, and bring to constant weight (within 0.3 mg) with repeated heatings.

9. Calculate the weight percent of iron in each sample, the average, the standard deviation, and the relative standard deviation ($s/\overline{x}$).

25-4 PREPARING STANDARD ACID AND BASE

Section 11-6 provides background information related to the procedures described below. Unknown samples of potassium acid phthalate or sodium carbonate (available from Thorn Smith) may be analyzed by the procedures described in this section.

Standard NaOH

PROCEDURE

1. A 50% (wt/wt) aqueous NaOH solution must be prepared in advance and allowed to settle. Sodium carbonate is insoluble in this solution and precipitates. The solution is stored in a tightly sealed polyethylene bottle and handled gently to avoid stirring the precipitate when supernate is taken. The density is close to 1.50 g of solution per milliliter.

2. Primary standard–grade potassium acid phthalate should be dried for 1 h at 110°C and stored in a desiccator.

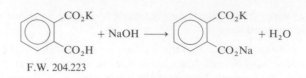

F.W. 204.223

3. Boil 1 L of water for 5 min to expel CO_2. Pour the water into a polyethylene bottle, which should be tightly capped whenever possible. Calculate the volume of aqueous 50% NaOH needed to produce 1 L of ~0.1 M NaOH. Use a graduated cylinder to transfer this much concentrated NaOH to the bottle of water. Mix well and allow the solution to cool to room temperature (preferably overnight).

4. Weigh four samples of solid potassium acid phthalate and dissolve each in ~25 mL of distilled water in a 125-mL flask. Each sample should contain enough solid to react with ~25 mL of 0.1 M NaOH. Add 3 drops of phenolphthalein indicator (Table 11-3) to each, and titrate one of them rapidly to find the approximate endpoint. The buret should have a loosely fitted cap to minimize entry of CO_2.

5. Calculate the volume of NaOH required for each of the other three samples and titrate them carefully. During each titration, you should periodically tilt and rotate the flask to wash all liquid from the walls into the bulk solution. When very near the end, you should deliver less than one drop of titrant at a time. To do this, carefully suspend a fraction of a drop from the buret tip, touch it to the inside wall of the flask, wash it into the bulk solution by careful tilting, and swirl the solution. The end point is the first appearance of faint pink color that persists for 15 s. (The color will slowly fade as CO_2 from the air dissolves in the solution.)

6. Calculate the average molarity, the standard deviation, and the relative standard deviation ($s/\overline{x}$). If you have used some care, the relative standard deviation should be <0.2%.

Standard HCl

PROCEDURE

1. Use the information in the table on the inside cover of this text to calculate the volume of concentrated (~37% wt/wt) HCl that should be added to 1 L of distilled water to produce ~0.1 M HCl, and prepare this solution.

2. Dry primary standard–grade sodium carbonate for 1 h at 110°C and cool it in a desiccator.

3. Weigh four samples containing enough Na_2CO_3 to react with ~25 mL of 0.1 M HCl and place each in a 125-mL flask. As you are ready to titrate each one, dissolve it in ~25 mL of distilled water.

$$2HCl + Na_2CO_3 \rightarrow CO_2 + 2NaCl$$

F.W. 105.989

Add 3 drops of bromocresol green indicator (Table 11-3) to each and titrate one rapidly (to a green color) to find the approximate end point.

4. Carefully titrate each sample until it just turns from blue to green. Then boil the solution to expel CO_2. The solution should return to a blue color.

5. Carefully add HCl from the buret until the solution turns green again. As desired, a blank titration can be performed, by using 3 drops of indicator in 50 mL of 0.05 M NaCl. Subtract the volume of HCl needed for the blank titration from that required to titrate Na_2CO_3.

6. Calculate the mean HCl molarity, standard deviation, and relative standard deviation.

This procedure involves two titrations. First, total alkalinity ($= [HCO_3^-] + 2[CO_3^{2-}]$) is measured by titrating the mixture with standard HCl to a bromocresol green end point:

$$HCO_3^- + H^+ \rightarrow H_2CO_3$$

$$CO_3^{2-} + 2H^+ \rightarrow H_2CO_3$$

A separate aliquot of unknown is treated with excess standard NaOH to convert HCO_3^- to CO_3^{2-}:

$$HCO_3^- + OH^- \rightarrow CO_3^{2-} + H_2O$$

Then all the carbonate is precipitated with $BaCl_2$:

$$Ba^{2+} + CO_3^{2-} \rightarrow BaCO_3$$

The excess NaOH is immediately titrated with standard HCl to determine how much HCO_3^- was present. From the total alkalinity and bicarbonate concentration, we can calculate the original carbonate concentration. Solid unknowns may be prepared from reagent-grade sodium or potassium carbonate and bicarbonate.

PROCEDURE

1. Solid unknown should be stored in a desiccator to keep it dry, but should not be heated. Even mild heating at 50–100°C converts $NaHCO_3$ to Na_2CO_3. Accurately weigh 2.0–2.5 g of unknown into a 250-mL volumetric flask. This is conveniently done by weighing the sample in a capped weighing bottle, delivering some to a funnel in the volumetric flask, and reweighing the bottle. Continue this process until the desired mass of reagent has been transferred to the funnel. Rinse the funnel repeatedly with small portions of freshly boiled and cooled water to dissolve the sample. Remove the funnel, dilute to the mark, and mix well.

2. Pipet a 25.00-mL aliquot of the unknown solution into a 250-mL flask and titrate with standard 0.1 M HCl, using bromocresol green indicator as described in Section 25-4. Repeat this procedure with two more 25.00-mL samples of unknown.

3. Pipet a 25.00-mL aliquot of unknown and 50.00 mL of standard 0.1 M NaOH into a 250-mL flask. Swirl and add 10 mL of 10% (wt/wt) $BaCl_2$, using a graduated cylinder. Swirl again to precipitate $BaCO_3$, add 2 drops of phenolphthalein indicator (Table 11-3), and immediately titrate with standard 0.1 M HCl. Repeat this procedure with two more 25.00-mL samples of unknown.

4. From the results of step 2, calculate the total alkalinity and its standard deviation. From the results of step 3, calculate the bicarbonate concentration and its standard deviation. Using the standard deviations as estimates of uncertainty, calculate the concentration (and uncertainty) of carbonate in the sample. Express the composition of the solid unknown as weight percent ($\pm$uncertainty) for each component. For example, your final result might be written 63.4 ($\pm$0.5) wt% K_2CO_3 and 36.6 ($\pm$0.2) wt% $NaHCO_3$.

[†] For a related experiment, see V. T. Lieu and G. E. Kalbus, "Potentiometric Titration of Acidic and Basic Compounds in Household Cleaners," *J. Chem. Ed.*, **65**, 184 (1988).

25-6 KJELDAHL NITROGEN ANALYSIS

Developed in 1883, the *Kjeldahl procedure* remains one of the most accurate and widely used methods for determining nitrogen. The nitrogen-containing substance is first *digested* in boiling sulfuric acid, which converts the nitrogen to NH_4^+ and oxidizes other elements present:

$$\text{Organic C,H,N} \xrightarrow[\text{H}_2\text{SO}_4]{\text{boiling}} NH_4^+ + CO_2 + H_2O$$

Mercury, copper, and selenium compounds catalyze the digestion process. To speed the rate of reaction, the boiling point of concentrated sulfuric acid (338°C) is raised by adding K_2SO_4. Digestion with H_2SO_4 and H_2O_2 in a microwave bomb (Figure 24-4) is particularly convenient and requires just 15 min. The digestion procedure given below is applicable to amines and amides (proteins). Modifications are necessary for nitro compounds, nitrites, azo compounds, cyanides, and derivatives of hydrazine.[†] (Kjeldahl digestion is also recommended for the destruction of highly toxic organic waste.[‡])

After digestion is complete, the solution containing NH_4^+ is made basic and the NH_3 formed is distilled into a receiver containing a known amount of HCl (see Equations 9-8 to 9-10). The unreacted HCl is titrated with NaOH to determine how much HCl was consumed by NH_3. Since the solution to be titrated contains both HCl and NH_4^+, we must choose an indicator that permits titration of HCl without beginning to titrate NH_4^+. Bromocresol green (transition range 3.8–5.4) fulfils this purpose. (See Problem 11-36.)

Biological samples such as powdered milk, dry cereal, or flour can be analyzed by the procedure described below. Alternatively, unknowns can be prepared from pure acetanilide, N-2-hydroxyethylpiperazine-N'-2-ethanesulfonic acid (HEPES buffer, Table 10-2), tris(hydroxymethyl)aminomethane (TRIS buffer, Table 10-2) or the p-toluenesulfonic acid salts of ammonia, glycine or nicotinic acid.

Digestion

PROCEDURE

1. Dry your unknown at 105°C for 45 min and accurately weigh an amount that will produce 2–3 mmol of NH_3. Place the sample in a *dry* 500-mL Kjeldahl flask (Figure 25-1) so that as little as possible sticks to the walls. Add 10 g of K_2SO_4 and three selenium-coated boiling chips.[§] Pour in 25 mL of 98% (wt/wt) H_2SO_4, washing down any solid from the walls. (*Caution:* Concentrated H_2SO_4 eats people. If you get any on your skin, flood it immediately with water, followed by soap and water.)

Figure 25-1
A Kjeldahl digestion flask. The long neck is designed to prevent loss of sample by spattering.

[†] H. C. Moore, *Ind. Eng. Chem.*, **12**, 669 (1920); *J. Assoc. Offic. Agr. Chemists*, **8**, 411 (1924-1925); I. K. Phelps and H. W. Daudt, *J. Assoc. Offic. Agr. Chemists*, **3**, 306 (1920); **4**, 72 (1921).

[‡] To destroy highly toxic organic compounds, place 10 g of K_2SO_4 and 100 mL of 98% (wt/wt) H_2SO_4 in a 500 mL flask fitted with a reflux condenser. Add up to 5 g of waste and a few glass beads and reflux until the solution is clear. If clarification does not occur in 1 h, cool the flask, slowly add 10 mL of 30% (wt/wt) H_2O_2 with stirring, add a fresh glass bead, and reflux again. Organic nitro, azo, and peroxide compounds should be reduced prior to this destruction procedure. The H_2SO_4 mixture can be reused until it solidifies when cooled.

[§] Hengar selenium-coated granules are available from Scientific Products, 1430 Waukegan Road, McGraw Park, IL 60085. Alternatively, the catalyst may be 0.1 g of Se, 0.2 g of $CuSeO_3$, or a crystal of $CuSO_4$.

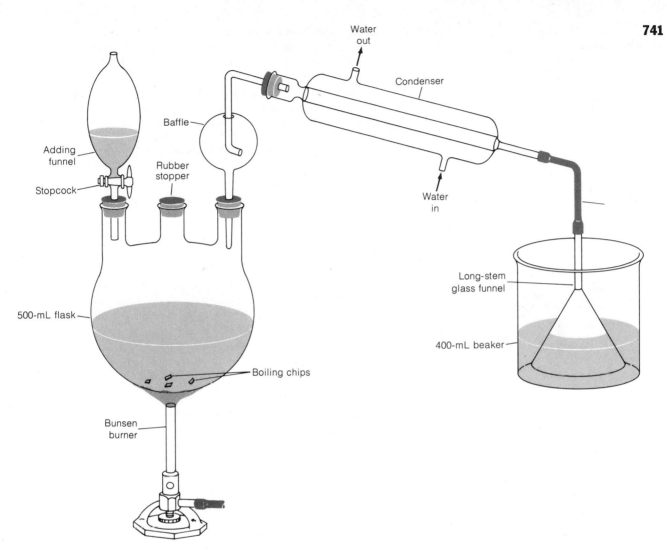

Figure 25-2
Apparatus for the Kjeldahl distillation.

2. In a fume hood, clamp the flask at a 30° angle away from you. Heat gently with a burner until foaming ceases and the solution becomes homogeneous. Gentle boiling should then be continued for an additional 30 min.

3. Cool the flask for 30 min *in the air*, and then in an ice bath for 15 min. Slowly, and with constant stirring, add 50 mL of ice-cold water. Dissolve any solids that have crystallized. Transfer the liquid to a distillation flask (Figure 25-2), wash the Kjeldahl flask five times with 10-mL portions of water, and pour the washings into the distillation flask.

Distillation

PROCEDURE

1. Apparatus such as that in Figure 25-2 should be set up and the connections made airtight. Pipet 50.00 mL of standard 0.1 M HCl into the receiving beaker and clamp the funnel in place below the liquid level.

2. Pour 50 mL of 6 M NaOH into the adding funnel and drip this into the distillation flask over a period of 1 min. Do not let the last 1 mL through the stopcock, so that gas cannot escape from the flask. Close the stopcock and heat the flask gently until two-thirds of the liquid has distilled.

3. Remove the funnel from the beaker *before* removing the burner from the flask (to avoid sucking distillate back into the condenser). Rinse the funnel well with distilled water and catch the rinses in the beaker. Add 6 drops of bromocresol green indicator solution (Table 11-3) to the beaker and carefully titrate to the blue end point with standard 0.1 M NaOH. You are looking for the first appearance of light blue color. (Several practice titrations with HCl and NaOH will familiarize you with the end point.)

4. Calculate the weight percent of nitrogen in the unknown.

25-7   ANALYSIS OF AN ACID–BASE TITRATION CURVE: THE GRAN PLOT[†]

In this experiment, you will titrate a sample of pure potassium acid phthalate (Table 11-4) with standard NaOH. The Gran plot (which you should read about in Section 11-5) will be used to find the equivalence point and K_a. Activity coefficients are used in the calculations of this experiment.

PROCEDURE (AN EASY MATTER)

1. Dry about 1.5 g of potassium acid phthalate at 105° for 1 h and cool it in a desiccator for 20 min. Accurately weigh out ∼1.5 g and dissolve it in water in a 250-mL volumetric flask. Dilute to the mark and mix well.

2. Following the instructions for your particular pH meter, calibrate a meter and glass electrode using buffers with pH values near 7 and 4. Rinse the electrodes well with distilled water and blot them dry with a tissue before immersing in a solution.

3. Pipet 100.0 mL of phthalate solution into a 250-mL beaker containing a magnetic stirring bar. Position the electrode(s) in the liquid so that the stirring bar will not strike the electrode. If a combination electrode is used, the small hole near the bottom on the side must be immersed in the solution. This hole is the reference electrode salt bridge. Allow the electrodes to equilibrate for 1 min (with stirring) and record the pH.

4. Add 1 drop of phenolphthalein indicator (Table 11-3) and titrate the solution with standard ∼0.1 M NaOH. Until you are within 4 mL of the theoretical equivalence point, add base in ∼1.5-mL aliquots, recording the volume and pH 30 s after each addition. Thereafter, use 0.4-mL aliquots until you are within 1 mL of the equivalence point. After that, add base 1 drop at a time until you have passed the pink end point by a few tenths of a milliliter. (Record the volume at which the pink color is observed.) Then add five more 1-mL aliquots.

5. Construct a graph of pH versus V_b (volume of added base). Locate the equivalence volume (V_e) by eye or "feel" as described in Section 11-5. Compare this to the theoretical and phenolphthalein end points.

CALCULATIONS (THE WORK BEGINS!)

1. Construct a Gran plot (a graph of $V_b \cdot 10^{-pH}$ versus V_b) using the data collected between $0.9V_e$ and V_e. Draw a line through the linear portion of this curve and extrapolate it to the abscissa to find V_e. Use this value of V_e in the calculations below. Compare this value to those found with phenolphthalein and estimated from the graph of pH versus V_b.

[†] For a related experiment, see C. A. Castillo S. and A. Jaramillo A., "An Alternative Procedure for Titration Curves of a Mixture of Acids of Different Strengths," *J. Chem. Ed.,* **66,** 341 (1989).

2. Compute the slope of the Gran plot and use Equation 11-51 to find K_a for potassium acid phthalate as follows: The slope of the Gran plot is $-K_a \gamma_{HP^-}/\gamma_{P^{2-}}$. In this equation, P^{2-} is the phthalate anion, and HP^- is monohydrogen phthalate. Because the ionic strength changes slightly as the titration proceeds, so also do the activity coefficients. Calculate the ionic strength at $0.95V_e$ and use this "average" ionic strength to find the activity coefficients.

EXAMPLE: Calculating Activity Coefficients

Find γ_{HP^-} and $\gamma_{P^{2-}}$ at $0.95V_e$ in the titration of 100.0 mL of 0.020 0 M potassium acid phthalate with 0.100 M NaOH.

The equivalence point is 20.0 mL, so $0.95V_e = 19.0$ mL. The concentrations of H^+ and OH^- are negligible compared to those of K^+, Na^+, HP^-, and P^{2-}, whose concentrations are

$$[K^+] = \left(\frac{100}{119}\right)(0.020\ 0) = 0.016\ 8\ \text{M}$$

$$[Na^+] = \left(\frac{19}{119}\right)(0.100) = 0.016\ 0\ \text{M}$$

$$[HP^-] = (0.050)\left(\frac{100}{119}\right)(0.020\ 0) = 0.000\ 84\ \text{M}$$

$$[P^{2-}] = (0.95)\left(\frac{100}{119}\right)(0.020\ 0) = 0.016\ 0\ \text{M}$$

The ionic strength is

$$\mu = \tfrac{1}{2} \sum c_i z_i^2$$
$$= \tfrac{1}{2}[(0.016\ 8) \cdot 1^2 + (0.016\ 0) \cdot 1^2 + (0.000\ 84) \cdot 1^2 + (0.016\ 0) \cdot 2^2]$$
$$= 0.048\ 8\ \text{M}$$

To estimate $\gamma_{P^{2-}}$ and γ_{HP^-} at $\mu = 0.048\ 8$ M, interpolate in Table 6-1. At the bottom of this table, we find that the hydrated radius of P^{2-}—phthalate, $C_6H_4(COO)_2^{2-}$— is 600 pm. The size of HP^- is not listed, but we will suppose that it is also 600 pm. An ion with charge ± 2 and a size of 600 pm has $\gamma = 0.485$ at $\mu = 0.05$ and $\gamma = 0.675$ at $\gamma = 0.01$. Interpolating between these values, we estimate $\gamma_{P^{2-}} = 0.49$ when $\mu = 0.048\ 8$ M. Similarly, we estimate $\gamma_{HP^-} = 0.84$ at this same ionic strength.

---

3. From the measured slope of the Gran plot and the values of γ_{HP^-} and $\gamma_{P^{2-}}$, calculate pK_a. Then choose an experimental point near $\tfrac{1}{3}V_e$ and one near $\tfrac{2}{3}V_e$. Use Equation 10-57 to find pK_a with each point. (You will have to calculate $[P^{2-}]$, $[HP^-]$, $\gamma_{P^{2-}}$, and γ_{HP^-} at each point.) Compare the average value of pK_a from your experiment with pK_2 for phthalic acid listed in Appendix G.

25-8 EDTA TITRATION OF Ca²⁺ AND Mg²⁺ IN NATURAL WATERS

The most common multivalent metal ions in natural waters are Ca^{2+} and Mg^{2+}. In this experiment, we will find the total concentration of metal ions that can react with EDTA, and we will assume that this equals the concentration of Ca^{2+} and Mg^{2+}. In a second experiment, Ca^{2+} is analyzed separately by precipitating $Mg(OH)_2$ with strong base.

REAGENTS

Buffer (pH 10): Add 142 mL of 28% (wt/wt) aqueous NH_3 to 17.5 g of NH_4Cl and dilute to 250 mL with water.

Eriochrome Black T indicator: Dissolve 0.2 g of the solid indicator in 15 mL of triethanolamine plus 5 mL of absolute ethanol.

PROCEDURE

1. Dry $Na_2H_2EDTA \cdot 2H_2O$ (F.W. 372.25) at 80° for 1 h and cool in the desiccator. Accurately weigh out ~0.6 g and dissolve it with heating in 400 mL of water in a 500-mL volumetric flask. Cool to room temperature, dilute to the mark, and mix well.

2. Pipet a sample of unknown into a 250-mL flask. A 1.000-mL sample of seawater or a 50.00-mL sample of tap water is usually reasonable. If you use 1.000 mL of seawater, add 50 mL of distilled water. To each sample, add 3 mL of pH 10 buffer and 6 drops of Eriochrome Black T indicator. Titrate with EDTA from a 50-mL buret and note when the color changes from wine-red to blue. You may need to practice finding the end point several times by adding a little tap water and titrating with more EDTA. Save a solution at the end point to use as a color comparison for other titrations.

3. Repeat the titration with three samples to find an accurate value of the total $Ca^{2+} + Mg^{2+}$ concentration. Perform a blank titration with 50 mL of distilled water and subtract the value of the blank from each result.

4. For the determination of Ca^{2+}, pipet four samples of unknown into clean flasks (adding 50 mL of distilled water if you use 1.000 mL of seawater). Add 30 drops of 50% (wt/wt) NaOH to each solution and swirl for 2 min to precipitate $Mg(OH)_2$ (which may not be visible). Add ~0.1 g of solid hydroxynaphthol blue to each flask. (This indicator is used because it remains blue at higher pH than does Eriochrome Black T.) Titrate one sample rapidly to find the end point; practice finding it several times, if necessary.

5. Titrate the other three samples carefully. After reaching the blue end point, allow each sample to sit for 5 min with occasional swirling so that any $Ca(OH)_2$ precipitate may redissolve. Then titrate back to the blue end point. (Repeat this procedure if the blue color turns to red upon standing.) Perform a blank titration with 50 mL of distilled water.

6. Calculate the total concentration of Ca^{2+} and Mg^{2+}, as well as the individual concentrations of each ion. Calculate the relative standard deviation of replicate titrations.

25-9 SYNTHESIS AND ANALYSIS OF AMMONIUM DECAVANADATE

The balance of species in a solution of vanadium(V) is a delicate function of both pH and concentration.

$$VO_4^{3-} \rightleftharpoons HVO_4^{2-} \rightleftharpoons V_2O_7^{4-} \rightleftharpoons H_2V_2O_7^{2-}$$

In strong
base

$$VO_2^+ \rightleftharpoons H_2V_{10}O_{28}^{4-} \rightleftharpoons \begin{bmatrix} V_4O_{12}^{4-} \\ V_3O_9^{3-} \end{bmatrix}$$

In strong
acid

$$\downarrow \begin{array}{l} NH_4^+ \\ alcohol \end{array}$$

$$(NH_4)_6V_{10}O_{28} \cdot 6H_2O$$
F.W. 1 173.7

The decavanadate ion ($V_{10}O_{28}^{6-}$), which we will isolate in this experiment as the ammonium salt, consists of ten VO_6 octahedra sharing edges with each other (Figure 25-3).

After preparing this salt, we will determine the vanadium content by a redox titration and NH_4^+ by the Kjeldahl method.[†] In the redox titration, V(V) will first be reduced to V(IV) with sulfurous acid and then titrated with standard permanganate (Color Plate 9).

$$V_{10}O_{28}^{6-} + H_2SO_3 \rightarrow VO^{2+} + SO_2$$

$$VO^{2+} + MnO_4^- \rightarrow VO_2^+ + Mn^{2+}$$

Blue   Purple   Yellow   Colorless

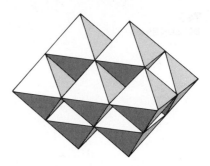

Figure 25-3
Structure of the $V_{10}O_{28}^{6-}$ anion, consisting of VO_6 octahedra sharing edges with each other.

Synthesis

PROCEDURE

1. Heat 3.0 g of ammonium metavanadate (NH_4VO_3) in 100 mL of water with constant stirring (but not boiling) until most or all of the solid has dissolved. Filter the solution and add 4 mL of 50% (vol/vol) aqueous acetic acid with stirring.

2. Add 150 mL of 95% ethanol with stirring and then cool the solution in a refrigerator or ice bath.

3. After maintaining a temperature of 0–10°C for 15 min, filter the orange product with suction and wash with two 15-mL portions of ice-cold 95% ethanol.

4. Dry the product in the air (protected from dust) for two days. Typical yield is 2.0–2.5 g.

Analysis of Vanadium with KMnO₄

Preparation and standardization of KMnO₄[‡]

See Section 16-4.

PROCEDURE

1. Prepare a 0.02 M permanganate solution by dissolving 1.6 g of $KMnO_4$ in 500 mL of distilled water. Boil gently for 1 h, cover, and allow the solution to cool overnight. Filter through a clean, fine sintered glass funnel, discarding the first 20 mL of filtrate. Store the solution in a clean glass amber bottle. Do not let the solution touch the cap.

2. Dry sodium oxalate ($Na_2C_2O_4$) at 105°C for 1 h, cool in a desiccator, and weigh three ~0.25-g samples into 500-mL flasks or 400-mL beakers. To each, add 250 mL of 0.9 M H_2SO_4 that has been recently boiled and cooled to room temperature. Stir with a thermometer to dissolve the sample, and add 90–95% of the theoretical amount of $KMnO_4$ solution needed for the titration. (This can be calculated from the mass of $KMnO_4$ used to prepare the permanganate solution. The chemical reaction is given by Equation 9-1.)

[†] G. G. Long, R. L. Stanfield, and F. C. Hentz, Jr., *J. Chem. Ed.,* **56,** 195 (1979). For related experiments involving synthesis followed by several analyses, see G. H. Searle, G. S. Bull, and D. A. House, "Reinecke's Salt Revisited," *J. Chem. Ed.,* **66,** 605 (1989); and E. P. Dudek, "A Project Lab for an Advanced General Chemistry Course Featuring the Amino Acid Glycine," *J. Chem. Ed.,* **64,** 899 (1987).

[‡] R. M. Fowler and H. A. Bright, *J. Res. National Bureau of Standards,* **15,** 493 (1935).

3. Leave the solution at room temperature until it is colorless. Then heat it to 55–60°C and complete the titration by adding $KMnO_4$ until the first pale pink color persists. Proceed slowly near the end, allowing 30 s for each drop to lose its color. As a blank, titrate 250 mL of 0.9 M H_2SO_4 to the same pale pink color.

Vanadium analysis

PROCEDURE

1. Accurately weigh two 0.3-g samples of ammonium decavanadate into 250-mL flasks and dissolve each in 40 mL of 1.5 M H_2SO_4 (with warming, if necessary).

2. In a fume hood, add 50 mL of water and 1 g of $NaHSO_3$ to each and dissolve with swirling. After 5 min, boil the solution gently for 15 min to remove SO_2.

3. Titrate the warm solution with standard 0.02 M $KMnO_4$ from a 50-mL buret. The end point is taken when the yellow color of VO_2^+ takes on a dark shade (from excess MnO_4^-) that persists for 15 s.

Analysis of Ammonium Ion

Ammonium ion is analyzed by a modification of the Kjeldahl procedure. Transfer 0.6 g of accurately weighed ammonium decavanadate to the distillation flask in Figure 25-2 and add 200 mL of water. Then proceed as described under "Distillation" in Section 25-6.

25-10 IODIMETRIC TITRATION OF VITAMIN C[†]

Ascorbic acid (vitamin C) is a mild reducing agent that reacts rapidly with triiodide (Reaction 16-70). In this experiment, we will generate a known excess of I_3^- by the reaction of iodate with iodide (Reaction 16-65), allow the reaction with ascorbic acid to proceed, and then back-titrate the excess I_3^- with thiosulfate (Reaction 16-66 and Color Plate 10).

Preparation and Standardization of Thiosulfate Solution

PROCEDURE

1. Starch indicator is prepared by making a paste of 5 g of soluble starch and 5 mg of HgI_2 in 50 mL of water. Pour the paste into 500 mL of boiling water and boil until it is clear.

2. Prepare 0.07 M $Na_2S_2O_3$[‡] by dissolving ~8.7 g of $Na_2S_2O_3 \cdot 5H_2O$ in 500 mL of freshly boiled water containing 0.05 g of Na_2CO_3. Store this solution in a tightly capped amber bottle. Prepare ~0.01 M KIO_3 by accurately weighing ~1 g of solid reagent and dissolving it in a 500-mL volumetric flask.

3. Standardize the thiosulfate solution as follows: Pipet 50.00 mL of KIO_3 solution into a flask. Add 2 g of solid KI and 10 mL of 0.5 M H_2SO_4. Immediately titrate

---

[†] D. N. Bailey, *J. Chem. Ed.,* **51,** 488 (1974). For related experiments involving redox titrations, see S. Kaufman and H. DeVoe, "Iron Analysis by Redox Titration, *J. Chem. Ed.,* **65,** 183 (1988); and W. B. Guenther, "Supertitrations: High Precision Methods," *J. Chem. Ed.,* **65,** 1097 (1988). The latter is an extremely precise mass titration that stresses careful laboratory technique.

[‡] An alternative to standardizing $Na_2S_2O_3$ solution is to use anhydrous primary standard $Na_2S_2O_3$, as described on page 412.

with thiosulfate until the solution has lost almost all its color (pale yellow). Then add 2 mL of starch indicator and complete the titration. Repeat the titration with two additional 50.00-mL volumes of KIO_3 solution.

Analysis of Vitamin C

Commercial vitamin C containing 100 mg per tablet may be used. Perform the following analysis three times, and find the mean value (and relative standard deviation) for the number of milligrams of vitamin C per tablet.

PROCEDURE

1. Dissolve two tablets in 60 mL of 0.3 M H_2SO_4, using a glass rod to help break the solid. (Some solid binding material will not dissolve.)
2. Add 2 g of solid KI and 50.00 mL of standard KIO_3. Then titrate with standard thiosulfate as above. Add 2 mL of starch indicator just before the end point.

25-11 PREPARATION AND IODOMETRIC ANALYSIS OF HIGH-TEMPERATURE SUPERCONDUCTOR[†]

In this experiment we will determine the oxygen content of a high-temperature superconductor, yttrium barium copper oxide ($YBa_2Cu_3O_x$). This material is an example of a *nonstoichiometric solid,* in which the value of x is variable, but near 7. Some of the copper in the formula $YBa_2Cu_3O_7$ is in the unusual high-oxidation state, Cu^{3+}. Equations 1–4 of Box 16-1 indicate how Cu^{2+} and Cu^{3+} oxidize iodide to iodine, and therefore how the value of x in the formula $YBa_2Cu_3O_x$ is related to the quantity of iodide oxidized in two separate procedures. The superconductor may be purchased or it can be synthesized if a furnace is available.

Preparation of $YBa_2Cu_3O_x$

PROCEDURE

1. Place in a mortar 0.750 g of Y_2O_3, 2.622 g of $BaCO_3$, and 1.581 g of CuO (atomic ratio Y:Ba:Cu = 1:2:3). Grind the mixture well with a pestle for 20 min and transfer the powder to a porcelain crucible or boat. Heat in the air in a furnace at 920–930°C for 12 h or longer. Turn off the furnace and allow the sample to cool slowly *in the furnace.* This slow cooling step is critical for achieving an oxygen content in the range $x = 6.5$–7 in the formula $YBa_2Cu_3O_x$. The crucible may be removed when the temperature is below 100°C.
2. The black solid mass is gently dislodged from the crucible and ground to a fine powder in a mortar and pestle. It can now be used for this experiment, but better quality material is produced if the powder is heated again to 920–930°C and cooled slowly as in step 1. If the powder from step 1 is green instead of black, raise the temperature of the furnace by 20°C and repeat step 1. The final product must be black, or it is not the correct compound.

[†] D. C. Harris, M. E. Hills, and T. A. Hewston, *J. Chem. Ed.,* **64,** 847 (1987). A more sensitive and elegant procedure for iodometric analysis of superconductors has been described by E. H. Appelman, L. R. Morss, A. M. Kini, U. Geiser, A. Umezawa, G. W. Crabtree, and K. D. Carlson, *Inorg. Chem.,* **26,** 3237 (1987). The procedure of Appelman *et al.* is slightly more difficult to carry out than Experiment 25-11.

Iodometric Analysis

PROCEDURE

1. *Sodium Thiosulfate and Starch Indicator.* Prepare 0.03 M $Na_2S_2O_3$ as described in Experiment 25-10, using 3.7 g of $Na_2S_2O_3 \cdot 5H_2O$ instead of 8.7 g. The starch indicator solution is the same one used in Experiment 25-10.

2. *Standard Cu^{2+}.* Weigh accurately 0.5–0.6 g of reagent Cu wire into a 100-mL volumetric flask. In a fume hood, add 6 mL of distilled water and 3 mL of 70% nitric acid, and boil gently on a hot plate until the solid has dissolved. Add 10 mL of distilled water and boil gently. Add 1.0 g of urea or 0.5 g of sulfamic acid and boil for 1 min to destroy HNO_2 and oxides of nitrogen that would interfere with the iodometric titration. Cool to room temperature and dilute to the mark with 1.0 M HCl.

3. *Standardization of $Na_2S_2O_3$ with Cu^{2+}.* The titration should be carried out as rapidly as possible under a brisk flow of N_2, because I^- is oxidized to I_2 in acid solution by atmospheric oxygen. Use a 180-mL tall-form beaker (or a 150-mL standard beaker) with a loosely fitting two-hole cork at the top. One hole serves as the inert gas inlet and the other is for the buret. Pipet 10.00 mL of standard Cu^{2+} into the beaker and flush with N_2. Remove the cork just long enough to pour in 10 mL of distilled water containing 1.0–1.5 g of KI (freshly dissolved) and begin magnetic stirring. In addition to the dark color of iodine in the solution, suspended solid CuI will be present. Titrate with $Na_2S_2O_3$ solution from a 50-mL buret, adding 2 drops of starch solution just before the last trace of I_2 color disappears. If starch is added too soon, there can be irreversible attachment of I_2 to the starch and the end point is harder to detect. Repeat this standardization two more times and use the average $Na_2S_2O_3$ molarity from the three determinations.

4. *Superconductor Experiment A.* Dissolve an accurately weighed 150- to 200-mg sample of powdered $YBa_2Cu_3O_x$ in 10 mL of 1.0 M $HClO_4$ in a titration beaker in a fume hood.[†] Boil gently for 10 min, so that Reaction 1 in Box 16-1 goes to completion. Cool to room temperature, cap with the two-hole-stopper–buret assembly, and begin N_2 flow. Dissolve 1.0–1.5 g of KI in 10 mL of distilled water and immediately add the solution to the beaker. Titrate rapidly with magnetic stirring as described in step 3. Repeat this procedure twice more.

5. *Superconductor Experiment B.* Place an accurately weighed 150- to 200-mg sample of powdered $YBa_2Cu_3O_x$ in the titration beaker and begin N_2 flow. Dissolve 1.0–1.5 g of KI in 10 mL of 1.0 M $HClO_4$ and immediately add the solution to the titration beaker. Stir magnetically for 1 min, so that the Reactions 3 and 4 of Box 16-1 occur. Add 10 mL of water and rapidly complete the titration. Repeat this procedure twice more.

CALCULATIONS

1. Suppose that mass, m_A, is analyzed in Experiment A and the volume, V_A, of standard thiosulfate is required for titration. Let the corresponding quantities in Experiment B be m_B and V_B. Let the average oxidation state of Cu in the superconductor be $2 + p$. Show that p is given by

$$p = \frac{(V_B/m_B) - (V_A/m_A)}{V_A/m_A} \qquad (25\text{-}1)$$

and x in the formula $YBa_2Cu_3O_x$ is related to p as follows:

[†] Perchloric acid is recommended because it is inert to reaction with superconductor, which might oxidize HCl to Cl_2. We have used HCl instead of $HClO_4$ with no significant interference in the analysis. Solutions of $HClO_4$ should not be boiled to dryness because of their explosion hazard.

$$x = \frac{7}{2} + \frac{3}{2}(2 + p) \qquad (25\text{-}2)$$

For example, if the superconductor contains one Cu^{3+} and two Cu^{2+}, the average oxidation state of copper is $\frac{7}{3}$ and the value of p is $\frac{1}{3}$. Setting $p = \frac{1}{3}$ in Equation 25-2 gives $x = 7$. Equation 25-1 does not depend on the metal stoichiometry being exactly $Y:Ba:Cu = 1:2:3$, but Equation 25-2 does require this exact stoichiometry.

2. Use the average results of Experiments A and B to calculate the values of p and x in Equations 25-1 and 25-2.

3. Suppose that the uncertainty in mass of superconductor analyzed is 1 in the last decimal place. Calculate the standard deviations for steps 3, 4, and 5 of the iodometric analysis. Using these standard deviations as uncertainties in volume, calculate the uncertainties in the values of p and x in Equations 25-1 and 25-2.

25-12 POTENTIOMETRIC HALIDE TITRATION WITH Ag⁺

Mixtures of halides may be titrated with $AgNO_3$ solution as described in Section 9-6. In this experiment, we will use the apparatus in Figure 9-8 to monitor the activity of Ag^+ as the titration proceeds. The theory of the potentiometric measurement is described in Section 15-2.

Each student is given a vial containing 0.22–0.44 g of KCl plus 0.50–1.00 g of KI (both weighed accurately). The object is to determine the quantity of each salt in the mixture. A 0.4 M bisulfate buffer (pH 2) should be available in the lab. This is prepared by titrating 1 M H_2SO_4 with 1 M NaOH to a pH near 2.0.

PROCEDURE

1. Pour your unknown carefully into a 50- or 100-mL beaker. Dissolve the solid in ~20 mL of water and pour it into a 100-mL volumetric flask. Rinse the sample vial and beaker many times with small portions of H_2O and transfer the washings to the flask. Dilute to the mark and mix well.

2. Dry 1.2 g of $AgNO_3$ at 105°C for 1 h and cool in a dessicator for 30 min with minimal exposure to light. Some discoloration is normal (and tolerable in this experiment) but should be minimized. Accurately weigh 1.2 g and dissolve it in a 100-mL volumetric flask.

3. The apparatus in Figure 9-8 should be set up. The silver electrode is simply a 3-cm length of silver wire connected to copper wire. (Fancier electrodes can be prepared by housing the connection in a glass tube sealed with epoxy at the lower end. Only the silver should protrude from the epoxy.) The copper wire is fitted with a jack that goes to the reference socket of a pH meter. The reference electrode for this titration is a glass pH electrode connected to its usual socket on the meter. If a combination pH electrode is employed, the reference jack of the combination electrode is not used. The silver electrode should be taped to the inside of the 100-mL beaker so that the Ag/Cu junction remains dry for the entire titration. The stirring bar should not hit either electrode.

4. Pipet 25.00 mL of unknown into the beaker, add 3 mL of bisulfate buffer, and begin magnetic stirring. Record the initial level of $AgNO_3$ in a 50-mL buret and add ~1 mL of titrant to the beaker. Turn the pH meter to the millivolt scale and record the volume and voltage. It is convenient (but is not essential) to set the initial reading to +800 mV by adjusting the meter.

5. Titrate the solution with ~1-mL aliquots until 50 mL of titrant has been added or until you can see two clear potentiometric end points. You need not allow

more than 15–30 s for each point. Record the volume and voltage at each point. Make a graph of millivolts versus milliliters to find the approximate positions (± 1 mL) of the two end points.

6. Turn the pH meter to standby, remove the beaker, rinse the electrodes well with water, and blot them dry with a tissue.[†] Clean the beaker and set up the titration apparatus again. (The beaker need not be dry.)

7. Now perform an accurate titration, using 1-drop aliquots near the end points (and 1-mL aliquots elsewhere). You need not allow more than 30 s per point for equilibration.

8. Prepare a graph of millivolts versus milliliters and locate the end points as in Figure 9-7. The I^- end point is taken as the intersection of the two dashed lines in the inset of Figure 9-7. The Cl^- end point is the inflection point at the second break. Calculate the mg of KI and mg of KCl in your solid unknown.

25-13 ELECTROGRAVIMETRIC ANALYSIS OF COPPER

Most copper-containing compounds can be electrolyzed in acidic solution, with quantitative deposition of Cu at the cathode. Sections 17-1 to 17-3 discuss the theory of this technique.

Students may analyze a preparation of their own (such as copper acetylsalicylate[‡]) or be given unknowns prepared from $CuSO_4 \cdot 5H_2O$ or metallic Cu. In the latter case, dissolve the metal in 8 M HNO_3, boil to remove HNO_2, neutralize with ammonia, and *barely* acidify the solution with dilute H_2SO_4. Samples must be free of chloride and nitrous acid.[§] Copper oxide unknowns (soluble in acid) are available from Thorn Smith.

Commercial apparatus, such as that in Figure 17-7, is convenient. Alternatively, any 6–12 V direct-current power supply may be set up as in Figure 17-6. A tall-form 150-mL beaker is used as the reaction vessel.

PROCEDURE

1. Handle the Pt gauze cathode with a tissue, touching only the thick stem, not the wire gauze. Immerse the electrode in hot 8 M HNO_3 to remove previous deposits, rinse with water and alcohol, dry at 110°C for 5 min, cool for 5 min, and weigh accurately. If the electrode contains any grease, it can be heated to red heat over a burner after the treatment above.[¶]

2. The sample should contain 0.2–0.3 g of Cu in 100 mL. Add 3 mL of 98% (wt/wt) H_2SO_4 and 2 mL of freshly boiled 8 M HNO_3. Position the cathode so that the top 5 mm are above the liquid level after magnetic stirring is begun. Adjust the current to 2 A, which should require 3–4 V. When the blue color of Cu(II) has disappeared, add some distilled water so that new Pt surface is exposed to the

---

[†] Silver halide adhering to the glass electrode may be removed by soaking in concentrated sodium thiosulfate solution. This thorough cleaning is not necessary between steps 6 and 7 in this experiment. The silver halides in the titration beaker can be saved and converted back to pure $AgNO_3$ using the procedure of E. Thall, *J. Chem. Ed.,* **58,** 561 (1981).

[‡] E. Dudek, *J. Chem. Ed.,* **54,** 329 (1977).

[§] J. F. Owen, C. S. Patterson, and G. S. Rice, *Anal. Chem.,* **55,** 990 (1983), describe simple procedures for the removal of chloride from Cu, Ni, and Co samples prior to electrogravimetric analysis.

[¶] Some metals, such as Zn, Ga, and Bi, form alloys with Pt and should not be deposited directly on the Pt surface. The electrode should be coated first with Cu, dried, and then used. Alternatively, Ag may be used in place of Pt for depositing these metals. Platinum anodes are attacked by Cl_2 formed by electrolysis of Cl^- solutions. To prevent this, 1–3 g of a hydrazinium salt (per 100 mL of solution) may be used as an *anodic depolarizer,* since hydrazine is more readily oxidized than Cl^-: $N_2H_4 \rightarrow N_2 + 4H^+ + 4e^-$.

solution. If no further deposition of Cu occurs on the fresh surface in 15 min at a current of 0.5 A, the electrolysis is complete. If deposition is observed, continue electrolysis and test the reaction for completeness again.

3. *Without* turning off the power, lower the beaker while washing the electrode with a squirt bottle. Then the current can be turned off. (If current is disconnected before removing the cathode from the liquid and rinsing off the acid, some Cu could redissolve.) Wash the cathode gently with water and alcohol, dry at 110°C for 3 min, cool in a desiccator for 5 min, and weigh.

25-14 POLAROGRAPHIC MEASUREMENT OF AN EQUILIBRIUM CONSTANT[†]

In this experiment, we will find the overall formation constant and stoichiometry for the reaction of oxalate with Pb^{2+}:

$$Pb^{2+} + pC_2O_4^{2-} \rightleftharpoons Pb(C_2O_4)_p^{2-2p}$$

$$\beta_p = \frac{[Pb(C_2O_4)_p^{2-2p}]}{[Pb^{2+}][C_2O_4^{2-}]^p}$$

where p is a stoichiometry coefficient. We will do this by measuring the polarographic half-wave potential for solutions containing Pb^{2+} and varying amounts of oxalate. According to Equation 18-21, the change of half-wave potential, $\Delta E_{1/2}$ $[= E_{1/2}(\text{observed}) - E_{1/2}(\text{for } Pb^{2+} \text{ without oxalate})]$ should obey the equation

$$\Delta E_{1/2} = -\frac{RT}{nF} \ln \beta_p - \frac{pRT}{nF} \ln [C_2O_4^{2-}] \qquad (25-3)$$

Here we have converted $0.059\,16 \log x$ in Equation 18-21 to $(RT/F) \ln x$ in Equation 25-3. R is the gas constant, F is the Faraday constant, and T is kelvins. You should measure the lab temperature at the time of the experiment or use a thermostatically controlled cell.

PROCEDURE

1. Pipet 1.00 mL of 0.020 M $Pb(NO_3)_2$ into each of five 50-mL volumetric flasks labeled A–E and add 1 drop of 1% Triton X-100 to each. Then add the following solutions and dilute to the mark with water. The KNO_3 may be delivered carefully with a graduated cylinder. The oxalate should be pipetted.
   A. Add nothing else. Dilute to the mark with 1.20 M KNO_3.

[†] W. C. Hoyle and T. M. Thorpe, *J. Chem. Ed.,* **55,** A229 (1978). For related electrochemistry experiments, see D. Martin and F. Menduciti, "Polarographic Determination of Composition and Thermodynamic Stability Constant of a Complex Metal Ion," *J. Chem. Ed.,* **65,** 916 (1988); J. G. Ibáñez, I. González, and M. A. Cárdenas, "The Effect of Complex Formation upon the Redox Potentials of Metallic Ions," *J. Chem. Ed.,* **65,** 173 (1988); T. J. Farrell, R. J. Laub, and E. P. Wadsworth, Jr., "Anodic Polarography of Cyanide in Foodstuffs," *J. Chem. Ed.,* **64,** 635 (1987); E. Briullas, J. A. Garrido, R. M. Rodríguez, and J. Doménech, "A Cyclic Voltammetry Experiment Using a Mercury Electrode, *J. Chem. Ed.,* **64,** 189 (1987); L. Piszczek, A. Ignatowicz, and K. Kielbasa, "Application of Cyclic Voltammetry for Stoichiometry Determination of Ni(II), Co(II) and Cd(II) Complex Compounds with Polyaminopolycarboxylic Acids," *J. Chem. Ed.,* **65,** 171 (1988); R. S. Pomeroy, M. B. Denton, and N. R. Armstrong, "Voltammetry at the Thin-Film Mercury Electrode," *J. Chem. Ed.,* **66,** 877 (1989); W. H. Chan, M. S. Wong, and C. W. Yip, "Ion-Selective Electrode in Organic Analysis: A Salicylate Electrode," *J. Chem. Ed.,* **63,** 915 (1986).

B. Add 5.00 mL of 1.00 M $K_2C_2O_4$ and 37.5 mL of 1.20 M KNO_3.
C. Add 10.00 mL of 1.00 M $K_2C_2O_4$ and 25.0 mL of 1.20 M KNO_3.
D. Add 15.00 mL of 1.00 M $K_2C_2O_4$ and 12.5 mL of 1.20 M KNO_3.
E. Add 20.00 mL of 1.00 M $K_2C_2O_4$.

2. Transfer each solution to a polarographic cell, deoxygenate for 10 min, and record the polarogram from -0.20 to -0.95 V (versus S.C.E.). Measure the residual current using the same settings and a solution containing just 1.20 M KNO_3 (plus 1 drop of 1% Triton X-100). Record each polarogram on a scale sufficiently expanded to allow accurate measurements. (Use a sweep rate of 0.05 V/min and a chart speed of 2.5 cm/min, with a mercury drop interval of 4 s.)

3. For each polarogram, make a graph of E versus $\log[I/(I_d - I)]$, using 6–8 points for each graph. According to Equation 18-11, $E = E_{1/2}$ when $\log[I/(I_d - I)] = 0$. Use this condition to locate $E_{1/2}$ on each graph. When you measure currents on the polarograms, be sure to subtract the residual current at each potential. Current on the polarographic wave is measured at the top of each oscillation, which corresponds to the maximum size of each mercury drop. Residual current is measured as follows: The electrocapillary maximum is the potential at which the residual current shows no oscillations. If the potential is more *positive* than the electrocapillary maximum, use the top of the oscillation, which corresponds to the maximum mercury drop size. If the potential is more *negative* than the electrocapillary maximum, use the bottom of each oscillation, which also corresponds to the maximum mercury drop size.

4. Make a graph of $\Delta E_{1/2}$ versus $\ln[C_2O_4^{2-}]$. Use Equation 25-3 to find p, the stoichiometry coefficient, from the slope of the graph. Then use the intercept to find the value of β_p. A worthwhile exercise is to use the method of least squares (Section 4-4) to find the standard deviations of the slope and intercept. From the standard deviations, find the uncertainties in p and β_p and express each with the correct number of significant figures.

25-15 COULOMETRIC TITRATION OF CYCLOHEXENE WITH BROMINE[†]

This experiment is described in Section 17-4, and the apparatus is shown in Figure 17-12. A conventional coulometric power supply can be employed, but we use the homemade circuits in Figure 25-4. A stopwatch is manually started as the generator switch is closed. Alternatively, a double-pole, double-throw switch may be used to simultaneously start the generator circuit and an electric clock.

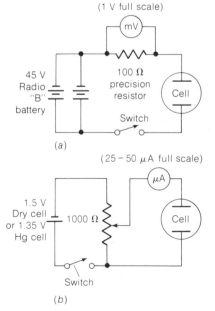

(1 V full scale)

45 V
Radio
"B"
battery

100 Ω
precision
resistor

Cell

Switch

(a)

(25 – 50 μA full scale)

1.5 V
Dry cell
or 1.35 V
Hg cell

1000 Ω

Cell

Switch

(b)

Figure 25-4
Simple circuits for coulometric titrations.
(a) Generator circuit. (b) Detector circuit.

PROCEDURE

1. The electrolyte is a 60:26:14 (vol/vol/vol) mixture of acetic acid, methanol, and water. The solution contains 0.15 M KBr and 0.1 g of mercuric acetate per 100 mL. (The latter catalyzes the reaction between Br_2 and cyclohexene.) The electrodes should be covered with electrolyte. Begin vigorous magnetic stirring (without spattering) and adjust the voltage of the detector circuit to 0.25 V.

2. Generate Br_2 with the generator circuit until the detector current is 20.0 μA. (The generator current is 5–10 mA.)

3. Pipet 2–5 mL of unknown (containing 1–5 mg of cyclohexene in methanol) into

[†] D. H. Evans, *J. Chem. Ed.*, **45,** 88 (1968). A similar experiment in which vitamin C is analyzed has been described by D. G. Marsh, D. L. Jacobs, and H. Veening, *J. Chem. Ed.*, **50,** 626 (1973). Home-built circuits for constant-current or controlled-potential coulometry have been described by E. Grimsrud and J. Amend, *J. Chem. Ed.*, **56,** 131 (1979).

the flask and set the clock or coulometer to zero. The detector current should drop to near zero because the cyclohexene consumes the Br_2.

4. Turn the generator circuit on and simultaneously begin timing. While the reaction is in progress, measure the voltage across the precision resistor (100.0 ± 0.1 Ω) to find the exact current flowing through the cell ($I = E/R$). Continue the electrolysis until the detector current rises to 20.0 µA. Then stop the coulometer and record the time.

5. Repeat the procedure twice more and find the average molarity (and relative standard deviation) of cyclohexene.

6. When you are finished, be sure all switches are off. The generator electrodes should be soaked in 8 M HNO_3 to dissolve Hg that is deposited during the electrolysis.

25-16 SPECTROPHOTOMETRIC DETERMINATION OF IRON IN VITAMIN TABLETS[†]

In this procedure, iron from a vitamin supplement tablet is dissolved in acid, reduced to Fe(II) with hydroquinone, and complexed with o-phenanthroline to form an intensely colored complex (Color Plate 19).

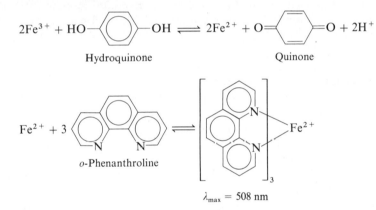

$\lambda_{max} = 508$ nm

REAGENTS

Hydroquinone: Freshly prepared solution containing 10 g/L in water. Store in an amber bottle.

Trisodium citrate: 25 g/L in water.

o-Phenanthroline: Dissolve 2.5 g in 100 mL of ethanol and add 900 mL of water. Store in an amber bottle.

Standard Fe (0.04 mg Fe/mL): Prepare by dissolving 0.281 g of reagent-grade $Fe(NH_4)_2(SO_4)_2 \cdot 6H_2O$ in water in a 1-L volumetric flask containing 1 mL of 98% (wt/wt) H_2SO_4.

[†] R. C. Atkins, *J. Chem. Ed.,* **52,** 550 (1975). For other experiments involving spectrophotometry, see M. A. Grompone, "Determination of Iron in a Bar of Soap," *J. Chem. Ed.,* **64,** 1057 (1987); K. W. Street, "Method Development for Analysis of Aspirin Tablets," *J. Chem. Ed.,* **65,** 915 (1988); T. Matsuo, A. Muromatsu, K. Katayama, and M. Mori, "Construction of a Photoelectric Colorimeter and Application to Students' Experiments," *J. Chem. Ed.,* **66,** 329, 848 (1989); M. A. Williamson, "Determination of Copper by Graphite Furnace Atomic Absorption Spectrophotometry," *J. Chem. Ed.,* **66,** 261 (1989); A. Ríos, M. Dolores, L. de Castro, and M. Valcárcel, "Determination of Reaction Stoichiometries by Flow Injection Analysis," *J. Chem. Ed.,* **63,** 552 (1986); C. L. Stults, A. P. Wade, and S. R. Crouch, "Investigation of Temperature Effects on Dispersion in a Flow Injection Analyzer," *J. Chem. Ed.,* **65,** 645 (1988).

PROCEDURE

1. Place one tablet of the iron-containing vitamin in a 125-mL flask or 100-mL beaker and boil gently (*in a fume hood*) with 25 mL of 6 M HCl for 15 min. Filter the solution directly into a 100-mL volumetric flask. Wash the beaker and filter several times with small portions of water to complete a quantitative transfer. Allow the solution to cool, dilute to the mark, and mix well. Dilute 5.00 mL of this solution to 100.0 mL in a fresh volumetric flask. If the label indicates that the tablet contains <15 mg of Fe, use 10.00 mL instead of 5.00 mL.

2. Pipet 10.00 mL of standard Fe solution into a beaker and measure the pH (with pH paper or a glass electrode). Add sodium citrate solution one drop at a time until a pH of ~3.5 is reached. Count the drops needed. (It will require about 30 drops.)

3. Pipet a fresh 10.00-mL aliquot of Fe standard into a 100-mL volumetric flask and add the same number of drops of citrate solution as required in step 2. Add 2.00 mL of hydroquinone solution and 3.00 mL of *o*-phenanthroline solution, dilute to the mark with water, and mix well.

4. Prepare three more solutions from 5.00, 2.00, and 1.00 mL of Fe standard and prepare a blank containing no Fe. Use sodium citrate solution in proportion to the volume of Fe solution. (If 10 mL of Fe requires 30 drops of citrate solution, 5 mL of Fe requires 15 drops of citrate solution.)

5. Find out how many drops of citrate solution are needed to bring 10.00 mL of the iron supplement tablet solution to pH 3.5. This will require about 3.5 or 7 mL of citrate, depending on whether 5 or 10 mL of unknown was diluted in the second part of step 1.

6. Transfer 10.00 mL of the solution containing the dissolved tablet to a 100-mL volumetric flask. Add the required amount of citrate solution, 2.00 mL of hydroquinone solution, and 3.0 mL of *o*-phenanthroline solution; dilute to the mark and mix well.

7. Allow the solutions to stand for at least 10 min. Then measure the absorbance of each solution at 508 nm. (The color is stable, so all solutions may be prepared and all the absorbances measured at once.) Use distilled water in the reference cuvette and subtract the absorbance of the blank from the absorbance of the Fe standards.

8. Make a graph of absorbance versus micrograms of Fe in the standards. If desired, find the slope and intercept (and standard deviations) by the method of least squares, as described in Section 4-4. Calculate the molarity of Fe (*o*-phenanthroline)$_3^{2+}$ in each solution and find the average molar absorptivity (ε in Beer's law) from the four absorbances. (Remember that all of the iron has been converted to the phenanthroline complex.) If a Spectronic 20 is used for absorbance measurements, assume that the pathlength is 1.1 cm for the sake of this calculation.

9. Using the calibration curve (or its least-squares parameters), find the number of milligrams of Fe in the tablet.

25-17 SPECTROPHOTOMETRIC MEASUREMENT OF AN EQUILIBRIUM CONSTANT[†]

In this experiment, we will use the Scatchard plot described in Section 19-8 to find the equilibrium constant for the formation of a complex between

[†] For related experiments, see J. Lieberman, Jr., and K. J. Yun, "A Semimicro Spectrophotometric Determination of the K_{sp} of Silver Acetate at Various Temperatures," *J. Chem. Ed.,* **65,** 729 (1988); J. J. Cruywagen and J. B. B. Heyns, "Spectrophotometric Determination of the Thermodynamic Parameters for the First Two Protonation Reactions of Molybdate," *J. Chem. Ed.,* **66,** 861 (1989); and H. A. Rowe and M. Brown, "Practical Enzyme Kinetics," *J. Chem.,* **65,** 548 (1988).

iodine and pyridine in cyclohexane:

$$I_2 + N\bigcirc \longrightarrow I_2 \cdot N\bigcirc$$

Both I_2 and $I_2 \cdot$ pyridine absorb visible radiation, but pyridine is colorless. Analysis of the spectral changes associated with variation of pyridine concentration (with a constant total concentration of iodine) will allow us to evaluate K for the reaction. The experiment is best performed with a recording spectrophotometer, but single-wavelength measurements can be used.

PROCEDURE

All operations described below should be carried out *in a fume hood,* including pouring solutions into and out of the spectrophotometer cell. Only a *capped* cuvette containing the solution whose spectrum is to be measured should be taken from the hood. Do not spill solvent on your hands or breathe the vapors. Used solutions should be discarded in a waste container *in the hood,* not down the drain.

1. The following stock solutions should be available in the lab:
    (a) 0.050–0.055 M pyridine in cyclohexane (40 mL for each student, concentration known accurately).
    (b) 0.012 0–0.012 5 M I_2 in cyclohexane (10 mL for each student, concentration known accurately).
2. Pipet the following volumes of stock solutions into six 25-mL volumetric flasks A–F, dilute to the mark with cyclohexane, and mix well.

Flask	Pyridine stock solution (mL)	I_2 stock solution (mL)
A	0	1.00
B	1.00	1.00
C	2.00	1.00
D	4.00	1.00
E	5.00	1.00
F	10.00	1.00

3. Using glass or quartz cells, record a baseline between 350 and 600 nm with solvent in both the sample and the reference cells. Subtract the absorbance of the baseline from all future absorbances. If possible, record all spectra, including the baseline, on one sheet of chart paper. (If a fixed-wavelength instrument is used, first find the positions of the two absorbance maxima in solution E. Then make all measurements at these two wavelengths.)

4. Record the spectrum of each solution A–F or measure the absorbance at each maximum if a fixed-wavelength instrument is used.

DATA ANALYSIS

1. Measure the absorbance at the wavelengths of the two maxima in each spectrum. Be sure to subtract the absorbance of the blank from each.

2. The analysis of this problem follows that of Reaction 19-25, in which P is iodine and X is pyridine. As a first approximation, assume that the concentration of free pyridine equals the total concentration of pyridine in the solution (since $[\text{pyridine}] \gg [I_2]$). Prepare a graph of $\Delta A/[\text{free pyridine}]$ versus ΔA (a Scatchard plot), using the absorbance at the $I_2 \cdot$ pyridine maximum.

3. From the slope of the graph, find the equilibrium constant using Equation 19-35. From the intercept, find $\Delta\varepsilon$ ($= \varepsilon_{PX} - \varepsilon_X$).

4. Now refine the values of K and $\Delta\varepsilon$. Use $\Delta\varepsilon$ to find ε_{PX}. Then use the absorbance at the wavelength of the $I_2 \cdot$ pyridine maximum to find the concentration of bound and free pyridine in each solution. Make a new graph of $\Delta A/[\text{free pyridine}]$ versus ΔA using the new values of [free pyridine]. Find a new value of K and $\Delta\varepsilon$. If you feel it is justified, perform another cycle of refinement.

5. Using the values of free pyridine concentration from your last refinement and the values of absorbance at the I_2 maximum, prepare another Scatchard plot and see if you get the same value of K.

6. Explain why an isosbestic point is observed in this experiment.

25-18 PROPERTIES OF AN ION-EXCHANGE RESIN[†]

In this experiment, we explore the properties of a cation-exchange resin, which is an organic polymer containing many sulfonic acid groups($-SO_3H$). When a cation, such as Cu^{2+}, flows into the resin, the cation is tightly bound by sulfonate groups, and one H^+ is released for each positive charge bound to the resin (Figure 25-5). The bound cation can be displaced from the resin by a large excess of H^+ or by an excess of any other cation for which the resin has some affinity.

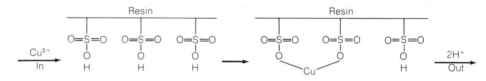

Figure 25-5
Stoichiometry of ion exchange.

First, known quantities of NaCl, $Fe(NO_3)_3$, and NaOH will be passed through the resin in the H^+ form. The H^+ released by each cation will be measured by titration with NaOH.

In the second part of the experiment, we will analyze a sample of impure vanadyl sulfate ($VOSO_4 \cdot 2H_2O$). As supplied commercially, this salt contains $VOSO_4$, H_2SO_4, and H_2O. A solution will be prepared from a known mass of reagent. The VO^{2+} content can be assayed spectrophotometrically, and the total cation (VO^{2+} and H^+) content can be assayed by ion exchange. Together, these measurements enable us to establish the quantities of $VOSO_4$, H_2SO_4, and H_2O in the sample.

REAGENTS

0.3 M NaCl: A bottle containing 5–10 mL per student, with an accurately known concentration.

0.1 M Fe(NO₃)₃ · 6H₂O: A bottle containing 5–10 mL per student, with an accurately known concentration.

[†] Part of this experiment is taken from M. V. Olson and J. M. Crawford, *J. Chem. Ed.*, **52**, 546 (1975).

VOSO₄: The commonly available grade (usually designated "purified") is used for this experiment. Students may make their own solutions and measure the absorbance at 750 nm, or a bottle of stock solution (25 mL per student) can be supplied. The stock should contain 8 g/L (accurately weighed) and state the absorbance.

0.02 M NaOH: Each student should prepare an accurate $\frac{1}{5}$ dilution of standard 0.1 M NaOH.

PROCEDURE

1. Prepare a chromatography column from a 0.7-cm diameter, 15-cm length of glass tubing, fitted at the bottom with a cork having a small hole that serves as the outlet. Place a small ball of glass wool above the cork to retain the resin. Use a small glass rod to plug the outlet and shut off the column. (Alternatively, an inexpensive column such as 0.7×15 cm Econo-Column from Bio-Rad Laboratories[†] works well in this experiment.) Fill the column with water, close it off, and test for leaks. Then drain the water until 2 cm remains and close the column again.

2. Make a slurry of 1.1 g of Bio-Rad Dowex 50W-X2 (100/200 mesh) cation-exchange resin in 5 mL of water and pour it into the column. If the resin cannot be poured all at once, allow some to settle, remove the supernatant liquid with a pipet, and pour in the rest of the resin. If the column is stored between laboratory periods, it should be upright, capped, and contain water above the level of the resin.[‡]

3. The general procedure for analysis of a sample is as follows:
   (a) Generate the H^+-saturated resin by passing ~10 mL of 1 M HCl through the column. Apply the liquid sample to the glass wall so as not to disturb the resin.
   (b) Wash the column with ~15 mL of water. Use the first few milliliters to wash the glass walls and allow the water to soak into the resin before continuing the washing.[§]
   (c) Place a clean 125-mL flask under the outlet and pipet the sample onto the column.
   (d) After the reagent has soaked in, wash it through with 10 mL of H_2O, collecting all eluate.
   (e) Add 3 drops of phenolphthalein indicator (Table 11-3) to the flask and titrate with standard 0.02 M NaOH.

4. Analyze 2.000-mL aliquots of 0.3 M NaCl and 0.1 M $Fe(NO_3)_3$, following the procedures in step 3. Calculate the theoretical volume of NaOH needed for each titration. If you do not come within 2% of this volume, repeat the analysis.

5. Pass 10.0 mL of your 0.02 M NaOH through the column as in step 3, and analyze the eluate. Explain what you observe.

6. Analyze 10.00 mL of $VOSO_4$ solution as described in step 3.

7. Using the molar absorptivity of vanadyl ion ($\varepsilon = 18.0$ $M^{-1} \cdot cm^{-1}$ at 750 nm) and the results of step 6, express the composition of the vanadyl sulfate in the form $(VOSO_4)_{1.00}(H_2SO_4)_x(H_2O)_y$.

[†] Bio-Rad Laboratories, 2200 Wright Avenue, Richmond, CA 94804.

[‡] When the experiment is finished, the resin can be collected, washed with 1 M HCl and water, and reused.

[§] Unlike most other chromatography resins, the one used in this experiment retains water when allowed to run "dry." Ordinarily, you must not let liquid fall below the top of the solid phase in a chromatography column.

25-19 QUANTITATIVE ANALYSIS BY GAS CHROMATOGRAPHY
OR HPLC[†]

We now describe an experiment illustrating the use of internal standards for quantitative analysis. The directions are purposely general, since the specific mixture and analytical conditions depend on the equipment and columns employed. The procedure can be modified for gas or liquid chromatography.

You will receive an unknown solution containing two or three compounds from a specified list of possibilities.[‡] Pure samples of each possible component of the unknown should also be available.

PROCEDURE

1. Identify the components of your mixture. Run a chromatogram of the pure unknown. Then mix pure samples of suspected components with samples of the unknown and run a new chromatogram of each. By observing which peaks grow or whether new peaks appear, you should be able to identify each species in your unknown.

2. Perform a quantitative analysis of one component (designated A) of the unknown, which is well separated from all other components. Find a compound (designated S) that is not a component of the unknown and is separated from all other peaks in the chromatogram. Prepare a mixture of unknown plus S such that the peak areas (or heights, if the peaks are sharp) of A and S are within a factor of two of each other. This is done by trial and error. The mass (or volume) of S and the mass of unknown (not A, but total unknown) in the mixture must be accurately known.

3. Prepare known mixtures of pure A and pure S to establish a calibration curve. If a solvent is used, the mixtures should have the same concentration of S required in step 2, plus variable quantities of unknown. If pure liquids are used without a solvent, the concentrations of both S and A will necessarily vary. Prepare a calibration curve in which you plot the signal ratio (peak area of A/peak area of S) versus the known concentration ratio (concentration of A/concentration of S). The graph should span the range including the signal ratio measured for the unknown mixture plus standard.

4. From your graph, calculate the concentration of species A in the unknown. Calculate the response factor (Equation 23-3) for A relative to S.

[†] For related experiments, see D. E. Goodney, "Analysis of Vitamin C by High-Pressure Liquid Chromatography," *J. Chem. Ed.,* **64,** 187 (1987); R. C. Graham and J. K. Robertson, "Analysis of Trihalomethanes in Soft Drinks," *J. Chem. Ed.,* **65,** 735 (1988); G. W. Rice, "Determination of Impurities in Whiskey Using Internal Standard Techniques," *J. Chem. Ed.,* **64,** 1055 (1987); and S. L. Tackett, "Determination of Methanol in Gasoline by Gas Chromatography," *J. Chem. Ed.,* **64,** 1059 (1987).

[‡] For example, we have done this experiment with Carle student gas chromatographs having columns packed with Carbowax or dinonyl phthalate. We used dichloromethane, chloroform, ethyl acetate, 1-propanol, and toluene as unknowns. We have also done this experiment with a DuPont 841 HPLC instrument containing a 1-m column with an octadecyl-bonded phase eluted with 65% (vol/vol) methanol/water. The unknowns included toluene, biphenyl, naphthalene, 2-naphthol, and 9-fluorenone.

Glossary

Glossary

Absolute Uncertainty   An expression of the margin of uncertainty associated with a measurement. Absolute error could also refer to the difference between a measured value and the "true" value.

Absorbance, A, or **Optical Density, OD**   Defined as $A = \log(P_0/P)$, where P_0 is the radiant power of light striking the sample on one side and P is the radiant power emerging from the other side.

Absorptance, a   Fraction of incident radiant power absorbed by a sample.

Absorption   Occurs when a substance is taken up *inside* another. See also **Adsorption.**

Absorption Coefficient   Light absorbed by a sample is attenuated at the rate $P_2/P_1 = e^{-\alpha b}$, where P_1 is the initial radiant power, P_2 is the power after traversing a pathlength b, and α is called the absorption coefficient.

Absorption Spectrum   A graph of absorbance or transmittance of light versus wavelength, frequency, or wavenumber.

Accuracy   A measure of how close a measured value is to the "true" value.

Acid   A substance that increases the concentration of H^+ when added to water.

Acid–Base Titration   One in which the reaction between analyte and titrant is an acid–base reaction.

Acid Dissociation Constant, K_a   The equilibrium constant for the reaction of an acid, HA, with H_2O:

$$HA + H_2O \rightleftharpoons A^- + H_3O^+ \qquad K_a = \frac{\mathscr{A}_{A^-}\mathscr{A}_{H_3O^+}}{\mathscr{A}_{HA}}$$

Acid Error   Occurs in strongly acidic solutions, where glass electrodes tend to indicate a value of pH that is too high.

Acidic Solution   One in which the activity of H^+ is greater than the activity of OH^-.

Acidity   In natural waters, the quantity of carbonic acid and other dissolved acids that react with strong base when the pH of the sample is raised to 8.3. Expressed as mmol OH^- needed to raise the pH of 1 L to pH 8.3.

Activation Energy   Energy needed for a process to overcome a barrier that otherwise prevents the process from occurring.

Activity, $\mathscr{A}$ The value that replaces concentration in a thermodynamically correct equilibrium expression. The activity of X is given by $\mathscr{A}_X = [X]\gamma_X$, where γ_X is the activity coefficient and $[X]$ is the concentration.

Activity Coefficient, γ   The number by which the concentration must be multiplied to give activity.

Adjusted Retention Time, t_r'   In chromatography, this is given by $t_r' = t_r - t_m$, where t_r is the retention time of a solute and t_m is the time needed for mobile phase to travel the length of the column.

Adsorption   Occurs when a substance becomes attached to the *surface* of another substance. See also **Absorption.**

Adsorption Chromatography   A technique in which the solute equilibrates between the mobile phase and adsorption sites on the stationary phase.

Adsorption Indicator   Used for precipitation titrations, it becomes attached to a precipitate and changes color when the surface charge of the precipitate changes sign at the equivalence point.

Aerosol   A suspension of very fine liquid or solid particles in air or gas. Examples include fog and smoke.

Affinity Chromatography   A technique in which a particular solute is retained by a column by virtue of a specific interaction with a molecule covalently bound to the stationary phase.

Aliquot   Portion.

Alkali Flame Detector   Modified flame ionization detector that responds to N and P, which produce ions when they contact a Rb_2SO_4-containing glass bead in the flame.

Alkalimetric Titration With reference to EDTA titrations, this involves titration of the protons liberated from EDTA upon binding to a metal.

Alkaline Error Occurs when a glass pH electrode is placed in a strongly basic solution containing very little H^+ and a high concentration of Na^+. The electrode begins to respond to Na^+ as if it were H^+, so that the pH reading is lower than the actual pH.

Alkalinity In natural waters, the quantity of base (mainly HCO_3^-, CO_3^{2-}, and OH^-) that reacts with strong acid when the pH of the sample is lowered to 4.5. Expressed as mmol H^+ needed to lower the pH of 1 L to pH 4.5.

Allosteric Interaction An effect at one part of a molecule caused by a chemical reaction or conformational change at another part of the molecule.

Amalgam A solution of anything in mercury.

Amine A compound with the general formula RNH_2, R_2NH, or R_3N, where R is any group of atoms.

Amino Acid One of 20 building blocks of proteins, having the general structure

$$\overset{\displaystyle R}{\underset{\displaystyle}{{}^+H_3NCHCO_2^-}}$$

where R is a different substituent for each acid.

Ammonium Ion The ammonium ion is NH_4^+. An ammonium ion is any ion of the type RNH_3^+, $R_2NH_2^+$, R_3NH^+, or R_4N^+, where R is an organic substituent.

Ampere, A One ampere is the current that will produce a force of exactly 2×10^{-7} N/m when that current flows through two "infinitely" long, parallel conductors of negligible cross section, with a spacing of 1 m, in a vacuum.

Amperometric Titration One in which the end point is determined by monitoring the current passing between two electrodes immersed in the sample solution and maintained at a constant potential difference.

Amperometry The measurement of electric current for analytical purposes.

Amphiprotic Molecule One that can act as both a proton donor and a proton acceptor. The intermediate species of polyprotic acids are amphiprotic.

Analyte The substance being analyzed.

Analytical Concentration See **Formal Concentration.**

Anion Exchanger An ion exchanger with positively charged groups covalently attached to the support. It can reversibly bind anions.

Anode The electrode at which oxidation occurs.

Anodic Depolarizer A molecule that is easily oxidized, thereby preventing the anode potential of an electrochemical cell from becoming too positive.

Anodic Wave In polarography, a flow of current due to oxidation of analyte.

Anolyte The solution present in the anode chamber of an electrochemical cell.

Antibody A protein manufactured by an organism to sequester and mark for destruction foreign molecules.

Antigen A molecule that is foreign to an organism and that causes antibodies to be made.

Antilogarithm The antilogarithm of a is b if $10^a = b$.

Antireflection Coating A coating placed on an optical component to diminish reflection. Ideally, the index of refraction of the coating should be $\sqrt{n_1 n_2}$, where n_1 is the refractive index of the surrounding medium and n_2 is the refractive index of the optical component. The thickness of the coating should be one-fourth of the wavelength of light inside the coating. Antireflection coatings are also made by layers that produce a gradation of refractive index.

Apodization Mathematical treatment of an interferogram prior to computing the spectrum by calculating the Fourier transform. Apodization reduces the wiggly feet that result from the finite extent of the interferogram.

Aprotic Solvent One that cannot donate protons (hydrogen ions) in an acid–base reaction.

Aqua Regia 3:1 (vol/vol) mixture of concentrated (37%) HCl and concentrated (70%) HNO_3.

Aquo Ion The species $M(H_2O)_n^{m+}$, containing just the cation M and its tightly bound water ligands.

Argentometric Titration One using Ag^+ ion.

Arrhenius Equation Empirical relation between the rate constant, k, for a chemical reaction and temperature, T, in kelvins: $k = Be^{-\Delta G^{\ddagger}/RT}$, where R is the gas constant, $\Delta G^{\ddagger}$ has the dimensions of energy and is taken as the free energy of activation for the chemical reaction, and B is a constant called the preexponential factor. This equation is more commonly written in the form $k = Ae^{-E_a/RT}$, where E_a is called the activation energy.

Ashless Filter Paper Specially treated paper that leaves a negligible residue after ignition. It is used for gravimetric analysis.

Asymmetry Potential When the activity of analyte is the same on the inside and outside of an ion-selective electrode, there should be no voltage across the membrane. In fact, the two surfaces are never identical, and some voltage (called the asymmetry potential) is usually observed. The asymmetry potential changes with time and leads to electrode drift.

Atmosphere, atm One atm is defined as a pressure of 101 325 N/m^2. It is also equal to the pressure exerted by a column of Hg 760 mm in height at the earth's surface.

Atomic Absorption Spectroscopy A technique in which the absorption of light by free gaseous atoms in a flame or furnace is used to measure the concentration of atoms.

Atomic Emission Spectroscopy A technique in which the

emission of light by thermally excited atoms in a flame or furnace is used to measure the concentration of atoms.

Atomic Fluorescence Spectroscopy A technique in which atomic electronic transitions are excited by light, and fluorescence is observed at a right angle to the incident beam.

Atomic Weight The number of grams of an element containing Avogadro's number of atoms.

Atomization The process in which a compound is decomposed into its atoms at high temperature.

Autoprotolysis The reaction of a neutral solvent, in which two of the same molecules transfer a proton between each other; e.g., $CH_3OH + CH_3OH \rightleftharpoons CH_3OH_2^+ + CH_3O^-$.

Auxiliary Complexing Agent A species, such as ammonia, that is added to a solution to stabilize another species and keep that other species in solution. It binds loosely enough to be displaced by a titrant.

Auxiliary Electrode The current-carrying partner of the working electrode in an electrolysis.

Azeotrope The distillate produced by two liquids. It is of constant composition, containing both substances.

Back Titration One in which an excess of standard reagent is added to react with analyte. Then the excess reagent is titrated with a second reagent or with a standard solution of analyte.

Ball Mill Ceramic drum in which solid sample is ground to fine powder by tumbling with hard ceramic balls.

Band Gap Energy separating the valence band and conduction band in a semiconductor.

Band Pass Filter A filter that allows a band of wavelengths to pass through, while absorbing or reflecting other wavelengths.

Bandwidth Usually, the range of wavelengths or frequencies of an absorption or emission band at a height equal to half of the peak height. It also refers to the width of radiation emerging from the exit slit of a monochromator.

Base A substance that decreases the concentration of H^+ when added to water.

Base "Dissociation" Constant, K_b The equilibrium constant for the reaction of a base, B, with H_2O:

$$B + H_2O \rightleftharpoons BH^+ + OH^- \qquad K_b = \frac{\mathscr{A}_{BH^+}\mathscr{A}_{OH^-}}{\mathscr{A}_B}$$

Base Hydrolysis Constant Same as base "dissociation" constant, K_b.

Basic Solution One in which the activity of OH^- is greater than the activity of H^+.

Beam Chopper A rotating mirror that directs light alternately through the sample and reference cells of a double-beam spectrophotometer. In atomic absorption, periodic blocking of the beam allows a distinction to be made between light from the source and light from the flame.

Beer's Law Relates the absorbance (A) of a sample to its concentration (c), pathlength (b), and molar absorptivity (ε): $A = \varepsilon bc$.

Biamperometric Titration An amperometric titration conducted with two polarizable electrodes held at a constant potential difference.

Biochemical Oxygen Demand, BOD In a water sample, the quantity of dissolved oxygen consumed by microorganisms during a 5-day incubation in a sealed vessel at 20°C. Oxygen consumption is limited by organic nutrients, so BOD is a measure of pollutant concentration.

Bipotentiometric Titration A potentiometric titration in which a constant current is passed between two polarizable electrodes immersed in the sample solution. An abrupt change in potential characterizes the end point.

Blackbody An ideal surface that absorbs all photons striking it.

Blackbody Radiation Radiation emitted by a blackbody. The energy and spectral distribution of the emission depend only on the temperature of the blackbody.

Blank Titration One in which a solution containing all of the reagents except analyte is titrated. The volume of titrant needed in the blank titration should be subtracted from the volume needed to titrate unknown.

Blocking Occurs when metal ion binds tightly to a metal ion indicator. A blocked indicator is unsuitable for a titration because no color change is observed at the end point.

Bolometer An infrared detector whose electrical resistance changes when it is heated by infrared light.

Boltzmann Distribution The relative population of two states at thermal equilibrium:

$$\frac{N_2}{N_1} = \frac{g_2}{g_1} e^{-(E_2 - E_1)/kT}$$

where N_i is the population of the state, g_i is the degeneracy of the state, E_i is the energy of the state, k is Boltzmann's constant, and T is temperature in kelvins; degeneracy refers to the number of states with the same energy.

Bomb Vessel for conducting high-temperature, high-pressure reactions.

Bonded Phase In HPLC, a stationary liquid phase covalently attached to the solid support.

Boxcar Truncation In Fourier transform spectroscopy, using the entire interferogram up to a predefined limit and then rejecting everything outside that limit.

Brønsted–Lowry Acid A proton (hydrogen ion) donor.

Brønsted–Lowry Base A proton (hydrogen ion) acceptor.

Brester Window A flat optical window tilted at an angle such that light whose electric vector is polarized parallel to the plane of the window is 100% transmitted. Light polarized perpendicular to the window is partially reflected. It is used on the ends of a laser to produce light whose electric field oscillates perpendicular to the long axis of the laser.

Buffer A mixture of an acid and its conjugate base. A buffered solution is one that resists changes in pH when acids or bases are added.

Buffer Capacity or **Buffer Intensity** A measure of the ability of a buffer to resist changes in pH. The larger the buffer capacity, the greater the resistance to pH change. The definition of buffer capacity (β) is $\beta = dC_b/dpH = -dC_a/dpH$, where C_a and C_b are the number of moles of strong acid or base per liter needed to produce a unit change in pH,

Bulk Sample Material taken from lot being analyzed—usually chosen to be representative of the entire lot. Also called *gross sample*.

Buoyancy Occurs when an object is weighed in air and the observed mass is less than the true mass because the object has displaced an equal volume of air from the balance pan.

Buret A calibrated glass tube with a stopcock at the bottom. Used to deliver known volumes of liquid.

Calibration Curve A graph showing the value of some property versus concentration of analyte. When the corresponding property of an unknown is measured, its concentration can be determined from the graph.

Calomel Electrode A common reference electrode based on the half-reaction $Hg_2Cl_2(s) + 2e^- \rightleftharpoons 2Hg(l) + 2Cl^-$.

Candela, cd The basic SI unit of luminous intensity. It is the intensity emitted by $\frac{1}{60}$ cm^2 of Pt at 2 045 K.

Capacity Factor, k In chromatography, the adjusted retention time for a peak divided by the time for the mobile phase to travel through the column. Capacity factor is also equal to the ratio of the time spent by the solute in the stationary phase to the time spent in the mobile phase. Also called *retention factor, capacity ratio, and partition ratio*.

Capacity Ratio See **Capacity factor**.

Capillary Constant The quantity $m^{2/3}t^{1/6}$ characteristic of each dropping Hg electrode. The rate of flow is m (mg/s) and t is the drop interval (s). The capillary constant is proportional to the square root of the Hg height.

Carboxylate Anion The conjugate base (RCO_2^-) of a carboxylic acid.

Carboxylic Acid A molecule with the general structure RCO_2H, where R is any group of atoms.

Carcinogen A cancer-causing agent.

Catalytic Wave One that results when the product of a polarographic reaction is rapidly regenerated by reaction with another species and the polarographic wave height increases.

Cathode The electrode at which reduction occurs.

Cathodic Depolarizer A molecule that is easily reduced, thereby preventing the cathode potential of an electrochemical cell from becoming very low.

Catholyte The solution present in the cathode chamber of an electrochemical cell.

Cation Exchanger An ion exchanger with negatively charged groups covalently attached to the support. It can reversibly bind cations.

Character The part to the left of the decimal point in a logarithm.

Charge Balance A statement that the sum of all positive charge in solution equals the magnitude of the sum of all negative charge in solution.

Charge Coupled Device An extremely sensitive detector in which light creates electrons and holes in a semiconductor material. The electrons are attracted to regions near positive electrodes, where the electrons are "stored" until they are ready to be counted. The number of electrons in each pixel (picture element) is proportional to the number of photons striking the pixel.

Charge Effect With respect to the strength of acids and bases, the repulsion between H^+ and a positive charge in the same molecule. It could also refer to the attraction between H^+ and a negative charge. A positive charge within a molecule increases the acidity, and a negative charge decreases the acidity.

Charring In a gravimetric analysis, the precipitate (and filter paper) are first dried gently. Then the filter paper is *charred* at intermediate temperature to destroy the paper without letting it inflame. Finally, the precipitate is ignited at high temperature to convert it to its analytical form.

Chelate Effect The observation that a single multidentate ligand forms metal complexes that are more stable than those formed by several individual ligands with the same ligand atoms.

Chelator A ligand that binds to a metal through more than one atom.

Chemical Coulometer A device that measures the yield of an electrolysis reaction in order to determine how much electricity has flowed through a circuit.

Chemical Interference In atomic spectroscopy, any chemical reaction that decreases the efficiency of atomization.

Chemical Oxygen Demand In a natural water or industrial effluent sample, the quantity of O_2 equivalent to the quantity of $K_2Cr_2O_7$ consumed by refluxing the sample with a standard dichromate–sulfuric acid solution containing Ag^+ catalyst. Since 1 mol of $K_2Cr_2O_7$ consumes $6e^-$ ($Cr^{6+} \rightarrow Cr^{3+}$), it is equivalent to 1.5 mol of O_2 ($O \rightarrow O^{2-}$).

Chromatogram A graph showing the concentration of solutes emerging from a chromatography column as a function of elution time or volume.

Chromatograph A machine used to perform chromatography.

Chromatography A technique in which molecules in a

mobile phase are separated because of their different affinities for a stationary phase. The greater the affinity for the stationary phase, the longer the molecule is retained.

Chromophore The part of a molecule responsible for absorption of light of a particular frequency.

Chronoamperometry A technique in which the potential of a working electrode in an unstirred solution is varied rapidly while the current between the working and auxilliary electrodes is measured. Suppose that the analyte is reducible and that the potential of the working electrode is made more negative. Initially, no reduction occurs. At a certain potential, the analyte begins to be reduced and the current increases. As the potential becomes more negative, the current increases further until the concentration of analyte at the surface of the electrode is sufficiently depleted. Then the current decreases, even though the potential becomes more negative. The maximum current is proportional to the concentration of analyte in bulk solution.

Chronopotentiometry A technique in which a constant current is forced to flow between two electrodes. The voltage remains fairly steady until the concentration of an electroactive species becomes depleted. Then the voltage changes rapidly as a new redox reaction assumes the burden of current flow. The elapsed time when the voltage suddenly changes is proportional to the concentration of the initial electroactive species in bulk solution.

Clark Electrode One that measures the activity of dissolved oxygen by amperometry.

Coagulation With respect to gravimetric analysis, small crystallites coming together to form larger crystals.

Co-chromatography Simultaneous chromatography of known compounds with an unknown. If a known and an unknown have the same retention time on several columns, they are probably identical.

Coherence The property of laser light that all waves are in phase with each other.

Cold Trapping Splitless gas chromatography injection technique in which solute is condensed far below its boiling point in a narrow band at the start of the column.

Collimated Light Light in which all rays travel in parallel paths.

Colloid A dissolved particle with a diameter in the approximate range 1–100 nm. It is too large to be considered one molecule but too small to simply precipitate.

Combination Electrode Consists of a glass electrode with a concentric reference electrode built on the same body.

Combustion Analysis A technique in which a sample is heated in an atmosphere of O_2 to oxidize it to CO_2 and H_2O, which are collected and weighed or measured by gas chromatography. Modifications permit the simultaneous analysis of N, S, and halogens.

Common Ion Effect Occurs when a salt is dissolved in a solution already containing one of the ions of the salt. The salt is less soluble than it would be in a solution without that ion. An application of Le Châtelier's principle.

Complex Ion Historical name for any ion containing two or more ions or molecules that are each stable by themselves; e.g., $CuCl_2^-$ contains $Cu^+ + 2Cl^-$.

Complexometric Titration One in which the reaction between analyte and titrant involves complex formation.

Composite Sample A representative sample prepared from a heterogeneous material. If the material consists of distinct regions, the composite is made of portions of each region, with relative amounts proportional to the size of each region.

Compound Electrode An ion-selective electrode consisting of a conventional electrode surrounded by a barrier that is selectively permeable to the analyte of interest. Alternatively, the barrier region might convert external analyte into a different species, to which the inner electrode is sensitive.

Concentration An expression of the quantity per unit volume or unit mass of a substance. Common measures of concentration are molarity (mol/L) and molality (mol/kg of solvent).

Concentration Cell A galvanic cell in which both half-reactions are the same, but the concentrations in each half-cell are not identical. The cell reaction increases the concentration of species in one half-cell and decreases the concentration in the other.

Concentration Polarization Occurs when an electrode reaction occurs so rapidly that the concentration of solute near the surface of the electrode is not the same as the concentration in bulk solution.

Conditional Formation Constant See **Effective Formation Constant.**

Conduction Band Energy levels containing conduction electrons in a semiconductor.

Conduction Electron An electron relatively free to move about within a solid and carry electric current. In a semiconductor, the energies of the conduction electrons are above those of the valence electrons that are localized in chemical bonds. The energy separating the valence and conduction bands is called the band gap.

Conductivity, σ Proportionality constant between electric current density, J (A/m^2), and electric field, E (V/m): $J = \sigma E$. Units are $\Omega^{-1} \cdot m^{-1}$. Conductivity is the reciprocal of resistivity.

Confidence Interval The range of values within which there is a specified probability that the true value will occur.

Conjugate Acid–Base Pair An acid and a base that differ only through the gain or loss of a single proton.

Constant-Current Electrolysis Electrolysis in which a constant current flows between the working and auxiliary electrodes. Because an ever-increasing voltage is required to keep the current flowing, this is the least selective form of electrolysis.

Constant Mass In gravimetric analysis, the product is heated and cooled to room temperature in a desiccator until successive weighings are "constant." There is not a standard definition of "constant mass"; but for ordinary work, it is usually taken as about ± 0.3 mg. Constancy is usually limited by the irreproducible regain of moisture picked up by the sample during cooling in the desiccator and during weighing.

Constant-Voltage Electrolysis Electrolysis in which a constant voltage is maintained between the working and auxiliary electrodes. This is less selective than controlled potential electrolysis because the potential of the working electrode becomes steadily more extreme as ohmic potential and overpotential change.

Controlled-Potential Electrolysis A technique for selective reduction (or oxidation), in which the voltage between the working and reference electrodes is held constant.

Convection Process in which solute is carried from one place to another by bulk motion of the solution.

Cooperativity Interaction between two parts of a molecule such that an event at one part affects the behavior of the other. Consider a molecule, M, with two sites to which two identical species, S, can bind to form MS and MS_2. When binding at one site makes binding at the second site more favorable than in the absence of the first binding, there is *positive cooperativity*. When the first binding makes the second binding less favorable, there is *negative cooperativity*. In *noncooperative* binding, neither site affects the other.

Coprecipitation Occurs when a substance whose solubility is not exceeded precipitates along with one whose solubility is exceeded.

Coulomb The amount of charge per second that flows past any point in a circuit when the current is one ampere. There are 96 485.309 coulombs in a mole of electrons.

Coulometric Titration One conducted with a constant current for a measured time.

Coulometry A technique in which the quantity of analyte is determined by measuring the number of coulombs needed for complete electrolysis.

Countercurrent Distribution A technique in which a series of solvent extractions is used to separate solutes from each other.

Counterion Most ionic substances contain two kinds of ions. Often one is of particular interest to us, and the other is said to be the counterion. It is necessary for electrical neutrality, but its nature is unimportant for the purpose at hand.

Critical Temperature Temperature above which a fluid cannot be condensed to two phases (liquid and gas), no matter how great a pressure is applied.

Crosslinking The covalent linkage between different strands of a polymer.

Crystallization Occurs when a substance comes out of solution slowly to form a solid with a regular arrangement of atoms.

Current, *I* Tells how much charge flows through a circuit per unit time.

Current–Overpotential Equation Relation between current and overpotential in a redox process occurring at an electrode.

Cuvet A cell used to hold samples for spectrophotometric measurements.

Cyclic Voltammetry A polarographic technique in which a triangular waveform is applied with a period of a few seconds. Both cathodic and anodic currents are observed for reversible reactions.

Dead Stop End Point The end point of a biamperometric titration.

Debye–Hückel Equation Gives the activity coefficient (γ) as a function of ionic strength (μ). The extended Debye–Hückel equation, applicable to ionic strengths up to about 0.1 M, is $\log \gamma = [-0.51z^2 \sqrt{\mu}] / [1 + (\alpha \sqrt{\mu}/305)]$, where z is the ionic charge and α is the effective hydrated radius in pm.

Decant To pour liquid off a solid or, perhaps, a denser liquid. The denser phase is left behind.

Decomposition Potential In an electrolysis, that voltage at which rapid reaction first begins.

Deionized Water Water that has been passed through a cation exchanger (in the H^+ form) and an anion exchanger (in the OH^- form) to remove ions from the solution.

Deliquescent Substance Like a hygroscopic substance, one that spontaneously picks up water from the air. It can eventually absorb so much water that the substance completely dissolves.

Demasking The removal of a masking agent from the species protected by the masking agent.

Density The mass per unit volume of a substance.

Depolarizer A molecule that is oxidized or reduced at a modest potential. It is added to an electrolytic cell to prevent the cathode or anode potential from becoming too extreme.

Derivatization Chemical alteration to attach a group to a molecule so that it may be detected conveniently. Alternatively, treatment may alter volatility or solubility.

Desalting The removal of salts (or any small molecules) from a solution of macromolecules. Gel filtration is conveniently used for desalting.

Desiccant A drying agent.

Desiccator A sealed chamber in which samples can be dried in the presence of a desiccant and/or vacuum pumping.

Detection Limit That concentration of an element which gives a signal equal to twice the peak-to-peak noise level of the baseline.

Determinant The value of the two-dimensional determinant $\begin{vmatrix} a & b \\ c & d \end{vmatrix}$ is the difference $ad - bc$.

Determinate Error See **Systematic Error.**

Deuterium Arc Lamp Source of broadband ultraviolet light. An electric discharge (a spark) in deuterium gas causes D_2 molecules to dissociate and emit many wavelengths of light.

Dialysis A technique in which solutions are placed on either side of a semipermeable membrane that allows small molecules, but not large molecules, to cross. The small molecules in the two solutions diffuse across and equilibrate with each other. The large molecules are retained on their original side.

Dielectric Constant The electrostatic force, F, between two charged particles is given by $F = kq_1q_2/\varepsilon r^2$, where k is a constant, q_1 and q_2 are the charges, r is the separation between particles, and ε is the dielectric constant of the medium. The higher the dielectric constant, the less force is exerted by one charged particle on another.

Differential Pulse Polarography A technique in which current is measured before and at the end of pulses of potential superimposed on the ordinary wave form. It is more sensitive than ordinary polarography, and the signal closely approximates the derivative of a polarographic wave.

Diffraction Occurs when electromagnetic radiation passes through slits with a spacing comparable to the wavelength. Interference of waves from adjacent slits produces a spectrum of radiation, with each wavelength at a different angle.

Diffuse Reflection Reflection of light in all directions by a rough surface.

Diffusion Random motion of molecules in a liquid or gas (or, very slowly, in a solid).

Diffusion Coefficient, D Defined by Fick's first law of diffusion: $J = -D(dc/dx)$, where J is the rate at which molecules diffuse across a plane of unit area and dc/dx is the concentration gradient in the direction of diffusion.

Diffusion Current In polarography, the current observed when the rate of reaction is limited by the rate of diffusion of analyte to the electrode. See also **Limiting Current.**

Diffusion Layer Region of solution near an electrode surface in which concentrations of ions are not the same as those in bulk solution because of electrostatic attraction or repulsion by the electrode. Concentrations of electroactive species are also different from those in bulk solution because these species are created or destroyed at the electrode.

Digestion The process in which a precipitate is left (usually warm) in the presence of mother liquor to promote particle recrystallization and growth. Purer, more easily filterable crystals result. Also used to describe any chemical treatment in which a substance is decomposed to transform the analyte into a form suitable for analysis.

Dimer A molecule made from two identical units.

Diode A semiconductor device consisting of a p–n junction through which current can pass in only one direction. Current flows when the n-type material is made negative and the p-type material is made positive. A voltage sufficient to overcome the activation energy for carrier movement must be supplied before any current flows. For silicon diodes this is approximately 0.6 V. If a sufficiently large voltage, called the breakdown voltage, is applied in the reverse direction, current will flow in the wrong direction through the diode.

Diprotic Acid One that can donate two protons.

Direct Current Polarography The classical form of polarography, in which a linear voltage ramp is applied to the working electrode.

Direct Titration One in which the analyte is treated with titrant, and the volume required for complete reaction is measured.

Dispersion For a prism, the rate of change of refractive index with wavelength, $dn/d\lambda$. This term also refers to separation of neighboring wavelengths by a monochromator. In this context, it is also called *angular dispersion*.

Displacement Titration An EDTA titration procedure in which analyte is treated with excess $MgEDTA^{2-}$ to displace Mg^{2+}: $M^{n+} + MgEDTA^{2-} \rightleftharpoons MEDTA^{n-4} + Mg^{2+}$. The liberated Mg^{2+} is then titrated with EDTA. This procedure is useful if there is not a suitable indicator for direct titration of M^{n+}.

Disproportionation A reaction in which an element in one oxidation state gives products containing that element in higher and lower oxidation states; e.g., $Cu^+ \rightleftharpoons Cu^{2+} + Cu(s)$.

Distribution Coefficient Describes the distribution of a solute partitioned between two phases. The distribution coefficient is defined as the total concentration of all forms of solute in phase 2 divided by the total concentration in phase 1.

Donnan Equilibrium The phenomenon that ions of the same charge as those fixed on an exchange resin are repelled from the resin. Thus, anions do not readily penetrate a cation-exchange resin, and cations are repelled from an anion-exchange resin.

Dopant When small amounts of substance B are added to substance A, we call B a dopant and say that A is doped with B. Doping is done to alter the properties of A.

Doppler Effect The phenomenon that a molecule moving toward a source of radiation experiences a higher frequency than one moving away from the source.

Double Layer Heterogeneous region consisting of a

charged surface and oppositely charged region of solution adjacent to the surface.

Dropping-Mercury Electrode One that delivers fresh drops of Hg to a polarographic cell.

Dry Ashing Oxidation of organic matter with O_2 at high temperature to leave behind inorganic components for analysis.

Dynamic Range The range of analyte concentration over which a change in concentration gives a change in detector response.

$E°$ The standard reduction potential.

$E°'$ The effective standard reduction potential at pH 7 (or at some other specified conditions).

EDTA (Ethylenediaminetetraacetic Acid) The compound $(HO_2CCH_2)_2NCH_2CH_2N(CH_2CO_2H)_2$—the most widely used reagent for complexometric titrations. It forms 1:1 complexes with virtually all cations with a charge of 2 or more.

Effective Formation Constant or **Conditional Formation Constant** The equilibrium constant for formation of a complex under a particular stated set of conditions, such as pH, ionic strength, and concentration of auxiliary complexing species.

Effervescence Rapid release of gas with bubbling and hissing.

Efflorescence The property in which the outer surface or entire mass of a substance turns to powder from loss of water of crystallization.

Effluent See **Eluate.**

Einstein A mole of photons.

Electric Discharge Emission Spectroscopy A technique in which atomization and excitation are stimulated by an electric arc, a spark, or microwave radiation.

Electric Double Layer The region comprising the charged surface of a particle plus the oppositely charged ionic atmosphere immediately surrounding the particle in solution.

Electric Potential The electric potential (in volts) at a point is the energy (in joules) needed to bring one coulomb of positive charge from infinity to that point. The potential difference between two points is the energy needed to transport one coulomb of positive charge from the negative point to the positive point.

Electroactive Species Any species that can be oxidized or reduced at an electrode.

Electrocapillary Maximum The potential at which the net charge on a mercury drop from a dropping mercury electrode is zero (and the surface tension of the drop is maximal).

Electrochemical Detector Liquid chromatography detector that measures current when an electroactive solute emerges from the column and passes over a working electrode held at a fixed potential with respect to a reference electrode. Also called *amperometric detector.*

Electrochemistry Use of electrical measurements on a chemical system for analytical purposes. Also refers to use of electricity to drive a chemical reaction or use of a chemical reaction to produce electricity.

Electrode A device at which or through which electrons flow into or out of chemical species involved in a redox reaction.

Electrogravimetric Analysis A technique in which the mass of an electrolytic deposit is used to quantify the analyte.

Electrolysis The process in which the passage of electric current causes a chemical reaction to occur.

Electrolyte A substance that produces ions when dissolved.

Electromagnetic Spectrum The spectrum of all electromagnetic radiation (visible light, radio waves, x-rays, etc.).

Electron Capture Detector Gas chromatography detector in which analytes with high electron affinity capture electrons from radioactively ionized N_2, and decrease the ion current.

Electronic Balance A balance that uses an electromagnetic servomotor to balance the load on the pan. The mass of the load is proportional to the current needed to balance it.

Electronic Transition One in which an electron is promoted from one energy level to another.

Eluate or **Effluent** What comes out of a chromatography column.

Eluent The solvent applied to the beginning of a chromatography column.

Eluent Strength A measure of the absorption energy of a solvent on the stationary phase in chromatography. The greater the eluent strength, the more rapidly will the solvent elute solutes from the column. (Also called *solvent strength.*)

Elution The process of passing a liquid or a gas through a chromatography column.

Eluotropic Series Ranks solvents according to their ability to displace solutes from the stationary phase in adsorption chromatography.

Emission Spectrum A graph of luminescence intensity versus luminescence wavelength (or frequency or wavenumber), using a fixed excitation wavelength.

Emissivity A quotient given by the radiant emission from a real object divided by the radiant emission of a blackbody at the same temperature.

Enantiomers Mirror image isomers.

Endergonic Reaction One for which ΔG is positive; it is not spontaneous.

Endothermic Reaction One for which ΔH is positive; heat must be supplied to reactants for them to react.

End Point The point in a titration at which there is a sudden change in a physical property, such as indicator color, pH, conductivity, or absorbance. Used as a measure of the equivalence point.

Enthalpy Change, ΔH The heat absorbed when the reaction occurs at constant pressure.

Enthalpy of Hydration The heat liberated when a gaseous species is transferred to water.

Entropy A measure of the "disorder" of a substance.

Enzyme A protein that catalyzes a chemical reaction.

Equilibrium The state in which the forward and reverse rates of all reactions are equal, so that the concentrations of all species remain constant.

Equilibrium Constant, K For the reaction $a\text{A} + b\text{B} \rightleftharpoons c\text{C} + d\text{D}$, $K = \mathscr{A}_C^c\mathscr{A}_D^d/\mathscr{A}_A^a\mathscr{A}_B^b$, where $\mathscr{A}_i$ is the activity of the ith species.

Equimolar Mixture of Compounds One that contains an equal number of moles of each compound.

Equivalence Point The point in a titration at which the quantity of titrant is exactly sufficient for stoichiometric reaction with the analyte.

Equivalent For a redox reaction, the amount of reagent that can donate or accept one mole of electrons. For an acid–base reaction, the amount of reagent that can donate or accept one mole of protons.

Equivalence Weight The mass of substance containing one equivalent.

Excitation Spectrum A graph of luminescence (measured at a fixed wavelength) versus excitation frequency or wavelength. It closely corresponds to an absorption spectrum because the luminescence is generally proportional to the absorbance.

Excited State Any state of an atom or a molecule having more than the minimum possible energy.

Exergonic Reaction One for which ΔG is negative; it is spontaneous.

Exitance Power per unit area radiating from the surface of an object.

Exothermic Reaction One for which ΔH is negative; heat is liberated when products are formed.

Extended Debye–Hückel Equation See **Debye–Hückel Equation.**

Extensive Property A property of a system or chemical reaction, such as entropy, that depends on the amount of matter in the system; e.g., ΔG, which is twice as large if two moles of product are formed than if one mole is formed. See also **Intensive Property.**

Extinction Coefficient See **Molar Absorptivity.**

Extraction The process in which a solute is allowed to equilibrate between two phases, usually for the purpose of separating solutes from each other.

Fajans Titration A precipitation titration in which the end point is signaled by adsorption of a colored indicator on the precipitate.

Faradaic Current That component of current in an electrochemical cell due to oxidation and reduction reactions.

Faraday Constant $9.648\,530\,9 \times 10^4$ C/mol of charge.

Faraday's Laws These two laws state that the extent of an electrochemical reaction is directly proportional to the quantity of electricity that has passed through the cell. The mass of substance that reacts is proportional to its formula weight and inversely proportional to the number of electrons required in its half-reaction.

Ferroelectric Material A solid with a permanent electric polarization (dipole) in the absence of an external electric field. The polarization results from alignment of molecules within the solid.

Field Effect Transistor A semiconductor device in which the electric field between the gate and base governs the flow of current between the source and drain.

Filtrate The liquid that passes through a filter.

Fines The smallest particles of stationary phase used for chromatography. It is desirable to remove the fines before packing a column because they retard solvent flow.

Fischer Titration See **Karl Fischer Titration.**

Flame Ionization Detector A gas chromatography detector in which solute is burned in a H_2–O_2 flame to produce CHO^+ ions. The current carried through the flame by these ions is proportional to the concentration of susceptible species in the eluate.

Flame Photometer A device that uses flame atomic emission and a filter photometer to quantify Li, Na, K, and Ca in liquid samples. It is widely used in clinical laboratories.

Flame Photometric Detector Gas chromatography detector that measures emission from S and P in H_2–O_2 flame.

Flow Adaptor An adjustable plungerlike device that may be used on either side of a chromatographic bed to support the bed and to minimize the dead space through which liquid can flow outside of the column bed.

Flow Injection Analysis Analytical technique in which sample is injected into a flowing stream. Other reagents can also be injected into the stream, and some type of measurement, such as absorption of light, is made downstream. Because the sample spreads out as it travels, different concentrations of sample are available in different parts of the band when it reaches the detector.

Fluorescence The process in which a molecule emits a photon shortly (10^{-9}–10^{-4} s) after absorbing a photon. It results from a transition between states of the same spin multiplicity.

Flux An agent used as the medium for a fusion.

Formal Concentration or **Analytical Concentration** The molarity of a substance if it did not change its chemical form upon being dissolved. It represents the total number of moles of substance dissolved in a liter of solution, regardless of any reactions that take place when the solute is dissolved.

Formality, F Same as formal concentration.

Formal Potential The potential of a half-reaction (relative to a standard hydrogen electrode) when the formal concentrations of reactants and products are unity. Any other conditions (such as pH, ionic strength, and concentrations of ligands) must also be specified.

Formation Constant or **Stability Constant** The equilibrium constant for the reaction of a metal with its ligands to form a metal–ligand complex.

Formula Weight, F.W. The mass containing one mole of the indicated chemical formula of a substance. For example, the formula weight of $CuSO_4 \cdot 5H_2O$ is the sum of the masses of copper, sulfate, and five water molecules.

Fourier Analysis Process of decomposing a function into an infinite series of sine and cosine terms. Since each term represents a certain frequency or wavelength, Fourier analysis decomposes a function into its component frequencies or wavelengths.

Fourier Series Infinite sum of sine and cosine terms that add to give a particular function in a particular interval.

Fraction of Association, α For the reaction of a base (B) with H_2O, the fraction of base in the form BH^+.

Fraction of Dissociation, α For the dissociation of an acid (HA), the fraction of acid in the form A^-.

Frequency The number of oscillations of a wave per second.

Fugacity The activity of a gas. The activity coefficient for a gas is called the **Fugacity Coefficient.**

Fusion The process in which an otherwise insoluble substance is dissolved in a molten salt such as Na_2CO_3, Na_2O_2, or KOH. Once the substance has dissolved, the melt is cooled, dissolved in aqueous solution, and analyzed.

Galvanic Cell One that produces electricity by means of a spontaneous chemical reaction.

Gas Chromatography A form of chromatography in which the mobile phase is a gas.

Gathering A process in which a trace constituent of a solution is intentionally coprecipitated with a major constituent.

Gaussian Distribution or **Normal Error Curve** This function describes the theoretical bell-shaped distribution of measurements when all error is random. The center of the curve is the mean, and the width is characterized by the standard deviation.

Gel Chromatographic stationary-phase particles, such as Sephadex or polyacrylamide, which are soft and pliable.

Gel Filtration or **Gel Permeation Chromatogrphy** See **Molecular-Exclusion Chromatography.**

Geometric Mean For a series of n measurements with the values x_i, $[\Pi_i(x_i)]^{1/n}$.

Gibbs Free Energy, G The change in Gibbs free energy (ΔG) for any process at constant temperature is related to the change in enthalpy (ΔH) and entropy (ΔS) by the equation $\Delta G = \Delta H - T\Delta S$, where T is temperature in kelvins. A process is spontaneous (thermodynamically favorable) if ΔG is negative.

Glass Electrode One that has a thin glass membrane across which a pH-dependent voltage develops. The voltage (and hence pH) is measured by a pair of reference electrodes on either side of the membrane.

Glassy Carbon Electrode An inert carbon electrode, impermeable to gas, and especially well suited as an anode. The isotropic (same in all directions) structure is thought to consist of tangled ribbons of graphitelike sheets, with some crosslinking.

Globar An infrared radiation source made of a ceramic such as silicon carbide heated by passage of electricity.

Golay Cell Infrared detector that uses expansion of a gas in a blackened chamber to deform a flexible mirror. Deflection of a beam of light by the mirror changes the power impinging on a phototube.

Gooch Crucible A short, cup-shaped container with holes at the bottom, used for filtration and ignition of precipitates. For ignition, the crucible is made of porcelain or platinum and lined with a mat of purified asbestos to retain the precipitate. For precipitates that do not need ignition, the crucible is made of glass and has a porous glass disk instead of holes at the bottom.

Gradient Elution Chromatography in which the composition of the mobile phase is progressively changed to increase the eluent strength of the solvent.

Graduated Cylinder or **Graduate** A tube with volume calibrations along its length.

Gram-Atom The amount of an element containing Avogadro's number of atoms; it is the same as a mole of the element.

Gran Plot A graph such as the plot of $V_b \cdot 10^{-pH}$ versus volume used to find the end point of a titration.

Graphite Furnace A hollow graphite rod that can be heated electrically to about 2 500 K to decompose and atomize a sample for atomic spectroscopy.

Grating Either a reflective or a transmitting surface etched with closely spaced lines; used to disperse light into its component wavelengths.

Gravimetric Analysis Any analytical method that relies on measuring the mass of a substance (such as a precipitate) to complete the analysis.

Gross Sample Same as **Bulk sample.**

Ground State The state of an atom or a molecule with the minimum possible energy.

Guard Column In HPLC, a short column packed with the same material as the main column, placed between the injector and main column. The guard column removes impurities that might irreversibly bind to and degrade the main column. Also called *precolumn*.

Half-Height Half of the maximum amplitude of a signal.

Half-Reaction Any redox reaction can be conceptually broken into two half-reactions, one involving only oxidation and one involving only reduction.

Half-Wave Potential The potential at the midpoint of the rise in the current of a polarographic wave.

Half-Width The width of a signal at its half-height.

Hall Process Electrolytic production of aluminum metal from a molten solution of Al_2O_3 and cryolite (Na_3AlF_6).

Hammett Acidity Function, H_0 Used to measure the acidity of nonaqueous solutions or concentrated aqueous solutions of strong acids.

Hanging-Drop Electrode One with a stationary drop of Hg that is used for stripping analysis.

Hardness Total concentration of alkaline earth ions in natural water. Expressed as mg $CaCO_3$ equivalent to the same number of moles of alkaline earth cations per liter of water.

Henderson–Hasselbalch Equation A logarithmic rearranged form of the acid dissociation equilibrium equation:

$$pH = pK_a + \log \frac{[A^-]}{[HA]}$$

Hertz, Hz The unit of frequency, s^{-1}.

Heterogeneous Not uniform throughout.

HETP (Height Equivalent to a Theoretical Plate) The length of a chromatography column divided by the number of theoretical plates in the column.

Hexadentate Ligand One that binds to a metal atom through six ligand atoms.

Hole Absence of an electron in a semiconductor. When a neighboring electron moves into the hole, a new hole is created where the electron came from. By this means, a hole can move through a solid just as an electron can move through a solid.

Hollow-Cathode Lamp One that emits sharp atomic lines characteristic of the element from which the cathode is made.

Homogeneous Having the same composition everywhere.

Homogeneous Precipitation A technique in which a precipitating agent is generated slowly by a reaction in homogeneous solution, effecting a slow crystallization instead of a rapid precipitation of product.

HPLC (High Performance Liquid Chromatography) A chromatographic technique using very small stationary-phase particles and high pressure to force solvent through the column.

Hydrated Radius The effective size of an ion or a molecule plus its associated water molecules in solution.

Hydrolysis "Reaction with water." The reaction $B + H_2O \rightleftharpoons BH^+ + OH^-$ is often called hydrolysis of a base.

Hydronium Ion, H_3O^+ What we really mean when we write $H^+(aq)$.

Hydrophilic Substance One that is soluble in water or attracts water to its surface.

Hydrophobic Interaction Chromatography Chromatographic separation based on the interaction of a hydrophobic solute with a hydrophobic stationary phase.

Hydrophobic Substance One that is insoluble in water or repels water from its surface.

Hygroscopic Substance One that readily picks up water from the atmosphere.

Ignition The heating to high temperature of some gravimetric precipitates to convert them to a known, constant composition that can be weighed.

Inclusion An impurity that occupies lattice sites in a crystal.

Indeterminate Error See **Random Error.**

Indicator A compound having a physical property (usually color) that changes abruptly near the equivalence point of a chemical reaction.

Indicator Electrode One that develops a potential whose magnitude depends on the activity of one or more species in contact with the electrode.

Indicator Error The difference between the indicator end point of a titration and the true equivalence point.

Indirect Spectrophotometric Detection Ion chromatography detection based on the use of a light-absorbing ionic eluent. A nonabsorbing analyte replaces an equivalent concentration of eluent when it emerges from the column, decreasing the absorption of eluate.

Indirect Titration One that is used when the analyte cannot be directly titrated. For example, analyte A may be precipitated with excess reagent R. The product is filtered, and the excess R washed away. Then AR is dissolved in a new solution, and R can be titrated.

Inductive Effect The attraction of electrons by an electronegative element through the sigma-bonding framework of a molecule.

Inductively Coupled Plasma A high-temperature plasma that derives its energy from an oscillating radio-frequency field. It is used to atomize a sample for atomic emission spectroscopy.

Inflection Point One at which the derivative of the slope is zero: $d^2y/dx^2 = 0$. That is, the slope reaches a maximum or minimum value.

Inner Helmholtz Plane Imaginary plane going through the centers of ions or molecules specifically adsorbed on an electrode.

Inorganic Carbon In a natural water or industrial effluent sample, the quantity of dissolved carbonate and bicarbonate.

Intensity See **Radiant Power.**

Intensive Property A property of a system or chemical reaction that does not depend on the amount of matter in the system; e.g., temperature and electric potential. See also **Extensive Property.**

Intercept For a straight line whose equation is $y = mx + b$, the value of b is the intercept. It is the value of y when $x = 0$.

Interference The effect when the presence of one substance changes the signal in the analysis of another substance.

Interference Filter A filter that transmits a particular band of wavelengths and reflects others. Transmitted light interferes constructively within the filter, whereas light that is reflected interferes destructively.

Interferogram A graph of light intensity versus retardation (or time) for the radiation emerging from an interferometer.

Interferometer A device with a beamsplitter, fixed mirror, and moving mirror that breaks input light into two beams that interfere with each other. The degree of interference depends on the difference in pathlength of the two beams.

Internal Conversion A radiationless isoenergetic electronic transition between states of the same electron-spin multiplicity.

Internal Standard A known quantity of a compound added to a solution containing an unknown quantity of analyte. The concentration of analyte is then measured relative to that of the internal standard.

Interpolation The estimation of the value of a quantity that lies between two known values.

Intersystem Crossing A radiationless isoenergetic electronic transition between states of different electron-spin multiplicity.

Iodimetry The use of triiodide (or iodine) as a titrant.

Iodometry A technique in which an oxidant is treated with I^- to produce I_3^-, which is then titrated (usually with thiosulfate).

Ion Chromatography HPLC ion-exchange separation of ions. See **suppressed-ion chromatography** and **single-column ion chromatography.**

Ion-Exchange Chromatography A technique in which solute ions are retained by oppositely charged sites in the stationary phase.

Ion-Exchange Equilibrium An equilibrium involving replacement of a cation by another cation or replacement of an anion by another anion. Usually the ions in these reactions are bound by electrostatic forces.

Ion-Exchange Membrane Membrane containing covalently bound charged groups. Oppositely charged ions in solution penetrate the membrane freely, but similarly charged ions tend to be excluded from the membrane by the bound charges.

Ion-Exclusion Chromatography A technique in which electrolytes are separated from nonelectrolytes by means of an ion-exchange resin.

Ionic Atmosphere The region of solution around an ion or a charged particle. It contains an excess of oppositely charged ions.

Ionic Radius The effective size of an ion in a crystal.

Ionic Strength, μ Given by $\mu = \frac{1}{2} \sum_i c_i z_i^2$, where c_i is the concentration of the ith ion in solution and z_i is the charge on that ion. The sum extends over all ions in solution, including the ions whose activity coefficients are being calculated.

Ionization Interference In atomic spectroscopy, a lowering of signal intensity due to ionization of analyte atoms.

Ionization Suppressor An element used in atomic spectroscopy to decrease the extent of ionization of the analyte.

Ionophore A molecule with a hydrophobic outside and a polar inside that can engulf an ion and carry the ion through a hydrophobic phase (such as a cell membrane).

Ion Pair A closely associated anion and cation, held together by electrostatic attraction. In solvents less polar than water, ions are usually found as ion pairs.

Ion-Pair Chromatography Separation of ions on reverse-phase HPLC column by adding to the eluent a hydrophobic counterion that pairs with analyte ion and is attracted to stationary phase.

Ion-Selective Electrode One whose potential is selectively dependent on the concentration of one particular ion in solution.

Isocratic Elution Chromatography using a single solvent for the mobile phase.

Isoelectric Focusing A technique in which a sample containing polyprotic molecules is subjected to a strong electric field in a medium with a pH gradient. Each species migrates until it reaches the region of its isoelectric pH. In that region, the molecule has no net charge, ceases to migrate, and remains focused in a narrow band.

Isoelectric pH or **Isoelectric Point** That pH at which the average charge of a polyprotic species is zero.

Isoionic pH or **Isoionic Point** The pH of a pure solution of a neutral, polyprotic molecule. The only ions present are H^+, OH^-, and those derived from the polyprotic species.

Isosbestic Point A wavelength at which the absorbance spectra of two species cross each other. The appearance of isosbestic points in a solution in which a chemical reaction is occurring is evidence that there are only two components present, with a constant total concentration.

Job's Method See **Method of continuous variation.**

Jones Reductor A column packed with zinc amalgam. An oxidized analyte is passed through to reduce the analyte, which is then titrated with an oxidizing agent.

Joule, J The SI unit of energy. One joule is required to heat one milliliter of water by $0.24°C$, to lift a one-kilogram mass 0.98 m at the earth's surface, or to move a charge of one coulomb through a potential difference of one volt.

Junction Potential An electric potential that exists at the junction between two different electrolyte solutions or substances. It arises in solutions from unequal rates of diffusion of different ions.

Karl Fischer Titration A sensitive technique for determining water, based on the reaction of H_2O with pyridine, I_2, SO_2, and methanol.

Kelvin, K The absolute unit of temperature defined such that the temperature of water at its triple point (where water, ice, and water vapor are at equilibrium) is 273.16 K and the absolute zero of temperature is 0 K.

Kieselguhr The German term for diatomaceous earth, which is used as a solid support in gas chromatography.

Kilogram, kg The mass of a particular Pt–Ir cylinder kept at the International Bureau of Weights and Measures, Sèvres, France.

Kinetic current A polarographic wave that is affected by the rate of a chemical reaction involving the analyte and some species in the solution.

Kinetic Polarization Occurs whenever an overpotential is associated with an electrode process.

Kjeldahl Nitrogen Determination Procedure for the analysis of nitrogen in organic compounds. The compound is digested with boiling H_2SO_4 to convert nitrogen to NH_4^+, which is treated with base and distilled as NH_3 into a standard acid solution. The number of moles of acid consumed equals the number of moles of NH_3 liberated from the compound.

Kovats Index In chromatography, a retention index comparing the adjusted retention time of an unknown to those of linear alkanes eluted before and after the unknown.

Laboratory Sample Portion of bulk sample taken to the lab for analysis. Must have the same composition as the bulk sample.

Laser Source of intense, coherent monochromatic radiation. Light is produced by stimulated emission of radiation from a medium in which an excited state has been pumped to a high population. Coherence means that all light exiting the laser has the same phase.

Latimer Diagram One that shows the reduction potentials connecting a series of species containing an element in different oxidation states.

Law of Mass Action States that for the chemical reaction $aA + bB \rightleftharpoons cC + dD$, the condition at equilibrium is $K = \mathscr{A}_C^c \mathscr{A}_D^d / \mathscr{A}_A^a \mathscr{A}_B^b$, where $\mathscr{A}_i$ is the activity of the ith species. The law is usually used in approximate form, in which the activities are replaced by concentrations.

Le Châtelier's Principle States that if a system at equilibrium is disturbed, the direction in which it proceeds back to equilibrium is such that the disturbance is partially offset.

Leveling Effect The strongest acid that can exist in solution is the protonated form of the solvent. Any acid stronger than this will donate its proton to the solvent and be leveled to the acid strength of the protonated solvent. Similarly, the strongest base that can exist in a solvent is the deprotonated form of the solvent.

Lewis Acid One that can form a chemical bond by sharing a pair of electrons donated by another species.

Lewis Base One that can form a chemical bond by sharing a pair of its electrons with another species.

Ligand An atom or a group attached to a central atom in a molecule. The term is often used to mean any group attached to anything else of interest.

Limiting Current In a polarographic experiment, the current that is reached at the plateau of a polarographic wave. See also **Diffusion Current.**

Linear Interpolation A form of interpolation in which it is assumed that the variation in some quantity is linear. For example, to find the value of b when $a = 32.4$ in the table below

a	32	32.4	33
b	12.85	x	17.96

you can set up the proportion

$$\frac{32.4 - 32}{33 - 32} = \frac{x - 12.85}{17.96 - 12.85}$$

which gives $x = 14.89$.

Linear Flow Rate In chromatography, the distance per unit time traveled by the mobile phase.

Linear Range The concentration range over which the change in detector response is proportional to the change in analyte concentration.

Linear Voltage Ramp The linearly increasing potential that is applied to the working electrode in polarography.

Lipid Bilayer Double layer formed by molecules containing hydrophilic headgroup and hydrophobic tail. The tails of the two layers associate with each other, while the headgroups face the aqueous solvent.

Liquid-Based Ion-Selective Electrode One that has a hydrophobic membrane separating an inner reference electrode from the analyte solution. The membrane is saturated with a liquid ion exchanger dissolved in a nonpolar solvent. The ion-exchange equilibrium of analyte between the liquid ion exchanger and the aqueous solution gives rise to the electrode potential.

Liter, L Defined in 1964 as exactly 1 000 cm^3.

Littrow Prism A prism with a reflecting back surface.

Logarithm The logarithm of a is b if $10^b = a$.

Longitudinal Diffusion Diffusion of solute molecules parallel to the direction of travel through a chromatography column.

Lorentzian Refers to an analytical function commonly used to describe the shape of a spectroscopic band: amplitude $= A_{max}\Gamma^2/[\Gamma^2 + (v - v_0)^2]$, where v is the frequency (or wavenumber), v_0 is the frequency (or wavenumber) of the center of the band, Γ is half the width at half-height, and A_{max} is the maximum amplitude.

Lot Entire material that is to be analyzed. Examples are a bottle of reagent, a lake, or a truckload of gravel.

Luminescence Any emission of light by a molecule.

L'vov Platform Platform on which sample is placed in a graphite rod furnace for atomic spectroscopy to prevent sample vaporization before the walls reach constant temperature.

Mantissa The part of a logarithm to the right of the decimal point.

Mariotte Flask A reservoir that maintains a constant hydrostatic pressure for liquid chromatography.

Masking Agent A reagent that selectively reacts with one (or more) component(s) of a solution to prevent the component(s) from interfering in a chemical analysis.

Mass Balance A statement that the sum of the moles of any element in all of its forms in a solution must equal the moles of that element delivered to the solution.

Mass Spectrograph An apparatus in which a sample is bombarded with electrons to produce charged molecular fragments that are then separated according to their mass in a magnetic field.

Mass Titration One in which the mass of titrant (instead of the volume) is measured.

Matrix The medium containing analyte. For many analyses, it is important that standards be prepared in the same matrix as the unknown.

Matrix Modifier Substance added to sample for atomic spectroscopy to retard analyte evaporation until the matrix is fully charred.

Maximum Suppressor A surface-active agent (such as the detergent Triton X-100) used to eliminate current maxima in polarography.

Mean The average of a set of all results.

Mean Activity Coefficient For the salt (cation)$_m$(anion)$_n$, the mean activity coefficient, $\gamma_{\pm}$, is related to the individual ion activity coefficients (γ_+ and γ_-) by the equation $\gamma_{\pm} = (\gamma_+^m \gamma_-^n)^{1/(m+n)}$.

Mechanical Balance A balance with a beam that pivots on a fulcrum and uses standard masses to measure the mass of an unknown.

Median For a set of data, that value above and below which there is an equal number of data.

Mediator With reference to electrolysis, a molecule added to a solution to carry electrons between the electrode and a dissolved species. Used when the target species cannot react directly at the electrode or when the target concentration is so low that other reagents react instead.

Meniscus The curved surface of a liquid.

Mesh Size The number of spacings per linear inch in a standard screen used to sort particles.

Metal Ion Buffer Consists of a metal–ligand complex plus excess free ligand. The two serve to fix the concentration of free metal ion through the reaction $M + nL \rightleftharpoons ML_n$.

Metal Ion Indicator A compound whose color changes when it binds to a metal ion.

Meter, m Defined as the distance light travels in a vacuum during $\frac{1}{299\,792\,458}$ of a second.

Method of Continuous Variation Procedure for finding the stoichiometry of a complex by preparing a series of solutions with different metal-to-ligand ratios. The ratio at which the extreme response (such as spectrophotometric absorbance) occurs corresponds to the stoichiometry of the complex.

Micelle Spherical cluster of molecules with polar or ionic headgroups and nonpolar tails. The tails associate with each other on the inside of the sphere and the headgroups face the aqueous solvent.

Microelectrode A very tiny electrode, with a diameter of perhaps 10 μm. Microelectrodes fit into very small places, such as living cells. Their small current gives rise to little ohmic loss, so they can be used in resistive, nonaqueous media. Small double-layer capacitance allows their voltage to be changed rapidly, permitting short-lived species to be studied.

Microporous Particles A type of stationary phase used in HPLC consisting of porous particles 3–10 μm in diameter with high efficiency and high capacity for solute.

Migration Electrostatically induced motion of ions in a solution under the influence of an electric field.

Miscible Liquids Two liquids that form a single phase when mixed in any ratio.

Mobile Phase In chromatography, the phase that travels through the column.

Mobility The terminal velocity that an ion reaches in a field of 1 V/m. Velocity = mobility × field.

Modulation Amplitude In pulsed polarography, the magnitude of the voltage pulse.

Molality The number of moles of solute per kilogram of solvent.

Molar Absorptivity, ε, or Extinction Coefficient The constant of proportionality in Beer's law: $A = \varepsilon bc$, where A is absorbance, b is pathlength, and c is the molarity of the absorbing species.

Molarity, M The number of moles of solute per liter of solution.

Mole, mol The amount of substance that contains as many molecules as there are atoms in 12 g of ^{12}C. There are approximately $6.022\,136\,7 \times 10^{23}$ molecules per mole.

Molecular Exclusion Chromatography or **Gel Filtration** or **Gel Permeation Chromatography** A technique in which the stationary phase has a porous structure into which small molecules can enter but large molecules cannot. Molecules are separated by size, with larger molecules moving faster than smaller ones.

Molecular Orbital Describes the distribution of an electron within a molecule.

Molecular Sieve A solid particle with pores the size of small molecules. Zeolites (sodium aluminosilicates) are a common type.

Molecular Weight, M.W. The number of grams of a substance that contains Avogadro's number of molecules.

Mole Fraction The number of moles of a substance in a mixture divided by the total number of moles of all components present.

Monochromatic Light Light of a single wavelength (color).

Monochromator A device (usually a prism, grating, or filter) for selecting a single wavelength of light.

Monodentate Ligand One that binds to a metal ion through only one atom.

Mortar and Pestle A mortar is a hard ceramic or steel vessel in which a solid sample is ground with a hard pestle.

Mother Liquor The solution from which a substance has crystallized.

Mull A fine dispersion of a solid in an oil.

Multidentate Ligand One that binds to a metal ion through more than one atom.

Nebulizer In atomic spectroscopy, this device breaks the liquid sample into a mist of fine droplets.

Nephelometry A technique in which the intensity of light scattered by a suspension is measured to determine the concentration of suspended particles.

Nernst Equation Relates the voltage of a cell to the activities of reactants and products:

$$E = E^\circ - \frac{RT}{nF} \ln Q$$

where R is the gas constant, T is temperature in kelvins, F is the Faraday constant, Q is the reaction quotient, and n is the number of electrons transferred in the balanced reaction.

Neutralization The process in which a stoichiometric equivalent of acid (or base) is added to a base (or acid).

Neutron-Activation Analysis A technique in which radiation is observed from a sample bombarded by slow neutrons. The radiation gives both qualitative and quantitative information about the sample composition.

Nonelectrolyte A substance that does not dissociate into ions when dissolved.

Nonpolarizable Electrode One whose potential remains nearly constant, even when current flows; e.g., a saturated calomel electrode.

Normal Error Curve See **Gaussian Curve.**

Normal Hydrogen Electrode (N.H.E.) See **Standard Hydrogen Electrode (S.H.E)**

Normality n times the molarity of a redox reagent, where n is the number of electrons donated or accepted by that species in a particular chemical reaction. For acids and bases, it is also n times the molarity, but n is the number of protons donated or accepted by the species.

Normal-Phase Chromatography A chromatographic separation utilizing a polar stationary phase and a less polar mobile phase.

Nucleation The process whereby molecules in solution come together randomly to form small aggregates.

Occlusion An impurity that becomes trapped (sometimes with solvent) in a pocket within a growing crystal.

Ohm's Law States that the current (I) in a circuit is proportional to voltage (E) and inversely proportional to resistance (R): $I = E/R$.

On-Column Injection Used in gas chromatography to place a thermally unstable sample directly on the column without excessive heating in an injection port. Solute is condensed at the start of the column by low temperature, which is raised to initiate chromatography.

Open Tubular Column In chromatography, a capillary column whose walls are coated with stationary phase.

Optical Density, OD See **Absorbance.**

Optical Fiber Fiber that carries light by total internal reflection because the transparent core has a higher refractive index than the surrounding cladding.

Osmolarity An expression of concentration that gives the total number of particles (ions and molecules) per liter of solution.

Outer Helmholtz Plane Imaginary plane passing through centers of hydrated ions just outside the layer of specifically adsorbed molecules on the surface of an electrode.

Overall Formation Constant, β_n The equilibrium constant for a reaction of the type $M + nX \rightleftharpoons MX_n$.

Overpotential The potential above that expected from the equilibrium potential, concentration polarization, and ohmic potential needed to cause an electrolytic reaction to occur at a given rate. It is zero for a reversible reaction.

Oxidant See **Oxidizing Agent.**

Oxidation A loss of electrons or a raising of the oxidation state.

Oxidation State (Number) A bookkeeping device used to tell how many electrons have been gained or lost by a neutral atom when it forms a compound.

Oxidizing Agent or **Oxidant** A substance that takes electrons in a chemical reaction.

Packed Column A chromatography column filled with stationary phase particles.

Parallax The apparent displacement of an object when the observer changes position. Occurs when the scale of an instrument is viewed from a position that is not perpendicular to the scale, so that the apparent reading is not the true reading.

Particle Growth The process in which molecules become attached to a crystal to form a larger crystal.

Partition Chromatography A technique in which separation is achieved by equilibrium of solute between two phases.

Partition Coefficient The equilibrium constant for the reaction in which a solute is partitioned between two phases: solute (in phase 1) $\rightleftharpoons$ solute (in phase 2).

Partition Ratio See **Capacity factor.**

Pascal, Pa A unit of pressure equal to 1 N/m. There are 101 325 Pa in 1 atm.

Pellicular Particles A type of stationary phase used in liquid chromatography. Contains a thin layer of liquid coated on a spherical bead. It has high efficiency (low HETP), but low capacity.

Peptization Occurs when washing some ionic precipitates with distilled water causes the ions that neutralize the charges of individual particles, and thereby help to hold the particles together, to be washed away. The particles then disintegrate and pass through the filter with the wash liquid.

Permanent Hardness Component of water hardness not due to dissolved alkaline earth bicarbonates. This hardness remains in the water after boiling.

p Function The negative logarithm (base 10) of a quantity: $pX = -\log X$.

pH Defined as $pH = -\log \mathscr{A}_{H^+}$, where $\mathscr{A}_{H^+}$ is the activity of H^+. In most approximate applications, the pH is taken as $-\log[H^+]$.

Phase Correction In Fourier transform spectroscopy, correcting for the slight asymmetry of the interferogram prior to computing the Fourier transform.

Phase-Transfer Catalysis A technique in which a compound such as a crown ether is used to extract a reactant from one phase into another in which a chemical reaction can occur.

pH Meter A very sensitive potentiometer used in conjunction with a glass electrode to measure pH.

Phosphorescence The emission of light during a transition between states of different spin multiplicity (e.g., triplet $\rightarrow$ singlet).

Phospholipid A molecule with a phosphate-containing polar headgroup and long hydrocarbon (lipid) tail.

Photoconductive Detector A detector whose conductivity changes when light is absorbed by the detector material.

Photodiode Array An array of semiconductor diodes used to detect light. The array is normally used to detect light that has been spread into its component wavelengths. One small band of wavelengths falls on each detector.

Photomultiplier Tube One in which the cathode emits electrons when struck by light. The electrons then strike a series of dynodes (plates that are positive with respect to the cathode), and more electrons are released each time a dynode is struck. As a result, more than 10^6 electrons may reach the anode for every photon striking the cathode.

Photon A "particle" of light with energy $h\nu$, where h is Planck's constant and ν is the frequency of the light.

Phototube A vacuum tube with a photoemissive cathode. The electric current flowing between the cathode and the anode is proportional to the intensity of light striking the cathode.

Photovoltaic Detector A photodetector with a junction across which the voltage changes when light is absorbed by the detector material.

pH-Stat A device that maintains a constant pH in a solution by continually injecting (or electrochemically generating) acid or base to counteract pH changes.

Piezoelectric Effect Development of electric charge on the surface of certain crystals when subjected to pressure. Conversely, application of an electric field can cause the crystal to deform.

Pilot Ion In polarography, an internal standard.

Pipet A glass tube calibrated to deliver a fixed or variable volume of liquid.

pK The negative logarithm (base 10) of an equilibrium constant: $pK = -\log K$.

Planck Distribution Equation giving the spectral distribution of blackbody radiation:

$$M_\lambda = \frac{2\pi hc^2}{\lambda^5}\left(\frac{1}{e^{hc/\lambda kT}-1}\right)$$

where h is Planck's constant, c is the speed of light, λ is the wavelength of light, k is Boltzmann's constant, and T is temperature in kelvins. M_λ is the power (watts) per square meter of surface per meter of wavelength radiating from the surface. The integral $\int_{\lambda_1}^{\lambda_2} M_\lambda\, d\lambda$ gives the power emitted per unit area in the wavelength interval from λ_1 to λ_2.

Planck's Constant Fundamental constant of nature equal to the energy of light divided by its frequency: $h = E/\nu = 6.626\,075\,5 \times 10^{-34}$ J·s.

Polarizability The proportionality constant relating the induced dipole to the strength of the electric field. When a molecule is placed in an electric field, a dipole is induced in the molecule by attraction of the electrons toward the positive pole and attraction of the nuclei toward the negative pole.

Polarizable Electrode One whose potential can change readily when a small current flows. Examples are Pt or Ag wires used as indicator electrodes.

Polarogram A graph showing the relation between current and potential during a polarographic experiment.

Polarograph An instrument used to obtain and record a polarogram.

Polarographic Wave The S-shaped increase in current during a redox reaction in polarography.

Polarography A technique in which the current flowing into an electrolysis cell is measured as a function of the applied potential.

Polychromator A device that spreads light into its component wavelengths and directs each small band of wavelengths to a different region.

Polyprotic Acids and Bases Compounds that can donate or accept more than one proton.

Population Inversion A necessary condition for laser operation in which the population of an excited energy level is greater than that of a lower energy level.

Porous-Layer Column Gas chromatography column containing an adsorptive solid phase coated on the inside surface of its wall.

Postprecipitation The adsorption of otherwise soluble impurities on the surface of a precipitate after the precipitation is over.

Potential See **Electric Potential.**

Potentiometer A device that measures electric potential by balancing it with a known potential of the opposite sign. A potentiometer measures the same quantity as that measured by a voltmeter, but is designed to draw much less current from the circuit being measured.

Potentiometry Any analytical method in which electric potential is measured.

Potentiostat An electronic device that maintains a constant voltage between a pair of electrodes.

Power The amount of energy per unit time (J/s) being expended.

ppb (parts per billion) An expression of concentration that refers to nanograms (10^{-9} g) of solute per gram of solution.

ppm (parts per million) An expression of concentration that refers to micrograms (10^{-6} g) of solute per gram of solution.

ppt (parts per thousand) An expression of concentration that refers to milligrams (10^{-3} g) of solute per gram of solution.

Precipitant A substance that precipitates a species from solution.

Precipitation Occurs when a substance leaves solution rapidly (to form either microcrystalline or amorphous solid).

Precipitation Titration One in which the analyte forms a precipitate with the titrant.

Precision A measure of the reproducibility of a measurement.

Precolumn See **Guard column.**

Preconcentration The process of concentrating trace components of a mixture prior to their analysis.

Premix Burner In atomic spectroscopy, one in which the sample is nebulized and simultaneously mixed with fuel and oxidant before being fed into the flame.

Preoxidation In some redox titrations, adjustment of the analyte oxidation state to a higher value so that it can be titrated with a reducing agent.

Prereduction The process of reducing an analyte to a lower oxidation state prior to performing a titration with an oxidizing agent.

Pressure Force per unit area, commonly measured in pascals (N/m) or atmospheres.

Pressure Broadening In spectroscopy, line broadening due to collisions between molecules.

Primary Standard A reagent that is pure enough and stable enough to be used directly after weighing. The entire mass is considered to be pure reagent.

Prism A transparent, triangular solid. Each wavelength of light passing through the prism is bent at a different angle. Therefore, light is dispersed into its component wavelengths by the prism.

Protic Solvent One with an acidic hydrogen atom.

Pyroelectric Effect Variation of electric polarization with temperature of a ferroelectric material.

Q Test Used to decide whether to discard a datum that appears discrepant.

Quaternary Ammonium Ion A cation containing four substituents attached to a nitrogen atom; e.g., $(CH_3CH_2)_4N^+$, the tetraethylammonium ion.

Radiant Power or **Intensity** The energy per unit time per unit area carried by a beam of light.

Random Error or **Indeterminate Error** A type of error, which can be either positive or negative and cannot be eliminated, based on the ultimate limitations on a physical measurement.

Random Sample Bulk sample constructed by taking portions of the entire lot at random.

Range or **Spread** The difference between the highest and the lowest value in a set of data.

Reactant The species that is consumed in a chemical reaction. It appears on the left side of a chemical equation.

Reaction Quotient, Q Has the same form as the equilibrium constant for a reaction. However, the reaction quotient is evaluated for a particular set of existing activities (concentrations), which are generally not the equilibrium values. At equilibrium, $Q = K$.

Reagent Blank A solution prepared from all of the reagents, but no analyte. The blank measures the response of the analytical method to impurities in the reagents or any other effects caused by any component other than the analyte.

Redox Couple A pair of reagents related by electron transfer (e.g., $Fe^{3+}|Fe^{2+}$ or $MnO_4^-|Mn^{2+}$).

Redox Indicator A compound whose different oxidation states have different colors, and which is used to find the end point of a redox titration. The potential of the indicator must be such that its color changes near the equivalence point potential of the titration reaction.

Redox Reaction A chemical reaction involving transfer of electrons from one element to another.

Redox Titration One in which the reaction between analyte and titrant is an oxidation–reduction reaction.

Reduced Plate Height In chromatography, the quotient HETP/d, where the numerator is the height equivalent to a theoretical plate and the denominator is the diameter of stationary-phase particles.

Reducing Agent or **Reductant** A substance that donates electrons in a chemical reaction.

Reduction A gain of electrons or a lowering of the oxidation state.

Reference Electrode One that maintains a constant potential against which the potential of a half-cell may be measured.

Reflectance Fraction of incident radiant power reflected by an object.

Refraction Bending of light when it passes between media with different refractive indexes.

Refractive Index The speed of light in any medium is c/n, where c is the speed of light in vacuum and n is the refractive index of the medium. The refractive index also measures the angle at which a light ray will be bent when it passes from one medium into another. Snell's law states that $n_1 \sin \theta_1 = n_2 \sin \theta_2$, where n_i is the refractive index for each medium and θ_i is the angle of the ray with respect to a normal between the two media.

Refractive Index Detector Liquid chromatography detector that measures the change in the refractive index of a solution as solutes emerge from column.

Relative Supersaturation Defined as $(Q - S)/S$, where S is the concentration of solute in a saturated solution and Q is the concentration in a particular supersaturated solution.

Relative Uncertainty The uncertainty of a quantity divided by the value of the quantity. It is usually expressed as a percent of the measured quantity.

Releasing Agent In atomic spectroscopy, a substance that prevents chemical interference.

Reprecipitation Sometimes a gravimetric precipitate can be freed of impurities only by redissolving it and reprecipitating it. The impurities are present at lower concentration during the second precipitation and are less likely to coprecipitate.

Residual Current The small current that is observed prior to the decomposition potential in an electrolysis.

Resin An ion exchanger, such as polystyrene, which exists as small, hard particles.

Resistance, R A measure of the retarding force opposing the flow of electric current.

Resistivity, ρ A measure of the ability of a material to retard the flow of electric current. $J = E/\rho$, where J is the current density (A/m^2) and E is electric field (V/m). Units of resistivity are $V \cdot m/A = \Omega \cdot m$. The resistance ($\Omega$) of a conductor with a given length and cross-sectional area is given by $R = \rho \cdot \text{length/area}$.

Resolution How close two bands in a spectrum or a chromatogram can be to each other and still be seen as two peaks. In chromatography, it is defined as the difference in retention time between adjacent peaks divided by their width.

Resonance Effect Contribution to a physical property by delocalization of electrons through the pi orbitals of a molecule.

Response Factor An empirically determined factor

measuring the response of a detector to a given compound. It is usually used in gas chromatography to determine the amount of unknown relative to an internal standard.

Retardation The difference in pathlength between light striking the stationary and moving mirrors of an interferometer.

Retention Factor See **Capacity factor.**

Retention Index In gas chromatography, the Kovats retention index is a logarithmic scale that relates the retention time of a compound to those of linear alkanes.

Retention Ratio In chromatography, the time required for solvent to pass through the column divided by the time required for solute to pass through the column.

Retention Time The time, measured from injection, needed for a solute to be eluted from a chromatography column.

Retention Volume The volume of solvent needed to elute a solute from a chromatography column.

Reverse-Phase Chromatography A technique in which the stationary phase is less polar than the mobile phase.

Root-Mean-Square (rms) Noise The quantity

$$\text{rms noise} = \left(\frac{1}{|\lambda_2 - \lambda_1|} \int_{\lambda_1}^{\lambda_2} [N(\lambda) - \overline{N}]^2 \, d\lambda \right)^{1/2}$$

where $N(\lambda)$ is the noise at wavelength λ and $\overline{N}$ is the average noise in the interval λ_1 to λ_2.

Rotating Disk Electrode An motor-driven electrode with a smooth flat face in contact with the solution. Rapid convection created by rotation brings fresh analyte to the surface of the electrode. A Pt electrode is especially suitable for studying anodic processes, in which a mercury electrode would be too easily oxidized.

Rotational Transition Occurs when a molecule changes its rotation energy.

Rubber Policeman A glass rod with a flattened piece of rubber on the tip. The rubber is used to scrape solid particles from glass surfaces in gravimetric analysis.

Salt An ionic solid.

Salt Bridge A conducting ionic medium in contact with two electrolyte solutions. It allows an ionic current to flow without allowing immediate diffusion of one electrolyte solution into the other.

Sample Cleanup Removal of portions of the sample that do not contain analyte and may interfere with analysis.

Sampling Variance The square of the standard deviation associated with the heterogeneity of the sample itself, not the analytical procedure. For inhomogeneous materials, different samples will have different compositions. It is necessary to take larger portions or more portions to reduce the uncertainty of composition due to variation from one region of the sample to another. The total variance of an analysis is the sum of the variances due to sampling and due to the analytical procedure.

Saturated Solution One that contains the maximum amount of a compound that can dissolve at equilibrium.

Scatchard Plot A graph used to find the equilibrium constant for a reaction such as $X + P \rightleftharpoons PX$. It is a graph of $[PX]/[X]$ versus $[PX]$, or any functions proportional to these quantities. The magnitude of the slope of the graph is the equilibrium constant.

S.C.E. (Saturated Calomel Electrode) A calomel electrode saturated with KCl. The electrode half-reaction is $Hg_2Cl_2(s) + 2e^- \rightleftharpoons 2Hg(l) + 2Cl^-$.

Second, s The duration of 9 192 631 770 periods of the radiation corresponding to the transition between two hyperfine levels of the ground state of ^{133}Cs.

Segregated Material A material containing distinct regions of different composition.

Selectivity Coefficient With respect to an ion-selective electrode, a measure of the relative response of the electrode to two different ions. In ion-exchange chromatography, this is the equilibrium constant for displacement of one ion by another from the resin.

Self-Absorption In flame emission atomic spectroscopy, there is a lower concentration of excited-state atoms in the cool, outer part of the flame than in the hot, inner flame. The cool atoms can absorb emission from the hot ones and thereby decrease the observed signal.

Semiconductor A material whose conductivity ($10^{-7} - 10^4 \, \Omega^{-1} \cdot m^{-1}$) is intermediate between that of good conductors ($10^8 \, \Omega^{-1} \cdot m^{-1}$) and that of insulators ($10^{-20}-10^{-12} \, \Omega^{-1} \cdot m^{-1}$).

Sensitivity The response of an instrument or method to a given amount of analyte. In spectrophotometric analysis, sensitivity is the concentration of analyte necessary to produce 99% T (or an absorbance of 0.004 4). For a balance, sensitivity is the deflection of the pointer divided by the difference in mass on the two pans. The greater the sensitivity, the greater the deflection.

Separator Column Ion-exchange column used to separate analyte species in ion chromatography.

Septum A disk, usually made of silicone rubber, covering the injection port of a gas chromatograph. The sample is injected by syringe through the septum.

Signal Averaging Improvement of a signal by averaging successive scans. The signal increase in proportion to the number of scans accumulated. The noise increases in proportion to the square root of the number of scans. Therefore the signal-to-noise ratio improves in proportion to the square root of the number of scans collected.

Significant Figure The number of significant figures in a quantity is the minimum number of figures needed to express the quantity in scientific notation. In experimental data, the first uncertain figure is the last significant figure.

Silanization The treatment of a chromatographic solid support or glass column with silicon compounds that bind to the most reactive Si–OH groups. It reduces irreversible adsorption and tailing of polar solutes.

Silver–Silver Chloride Electrode A common reference electrode containing a silver wire coated with AgCl paste and dipped in a solution saturated with AgCl and (usually) KCl. The half-reaction is $AgCl(s) + e^- \rightleftharpoons Ag(s) + Cl^-$.

Single-Column Ion Chromatography Separation of ions on a low-capacity ion-exchange column using low ionic strength eluent.

Single-Electrode Potential The voltage measured when the electrode of interest is connected to the positive terminal of a potentiometer and a standard hydrogen electrode is connected to the negative terminal.

Singlet State One in which all electron spins are paired.

SI Units The units of an international system of measurement based on the meter, kilogram, second, ampere, kelvin, candela, mole, radian, and steradian.

Slope For a straight line whose equation is $y = mx + b$, the value of m is the slope. It is the ratio $\Delta y / \Delta x$ for any segment of the line.

Slurry A suspension of a solid in a solvent.

Smoothing Use of a mathematical procedure or electrical filtering to improve the quality of a signal.

Snell's Law Relates angle of refraction (θ_2) to angle of incidence (θ_1) for light passing from a medium with refractive index n_1 to a medium of refractive index n_2: $n_1 \sin \theta_1 = n_2 \sin \theta_2$. Angles are measured with respect to the normal to the surface between the two media.

Solid-State Ion-Selective Electrode A type of ion-selective electrode that has a solid membrane made of an inorganic salt crystal. Ion-exchange equilibria between the solution and the surface of the crystal account for the electrode potential.

Solubility Product, K_{sp} The equilibrium constant for the dissolution of a solid salt to give its ions in solution. For the reaction $M_mN_n(s) \rightleftharpoons mM^{n+} + nN^{m-}$, $K_{sp} = \mathscr{A}_{M^{n+}}^m \, \mathscr{A}_{N^{m-}}^n$ where $\mathscr{A}$ is the activity of each species.

Solute A minor component of a solution.

Solvation The interaction of solvent molecules with solute. In general, solvent molecules will orient themselves around a solute to minimize the energy of the solution through dipole and van der Waals forces.

Solvent The major constituent of a solution.

Solvent Extraction A method in which a chemical species is transferred from one liquid phase to another. It is used to separate components of a mixture.

Solvent Trapping Splitless gas chromatography injection technique in which solvent is condensed near its boiling point at the start of the column. Solutes dissolve in a narrow band in the condensed solvent.

Specific Adsorption Process in which molecules are held tightly to a surface by van der Waals forces.

Specific Gravity A dimensionless quantity equal to the mass of a substance divided by the mass of an equal volume of water at 4°C. Since the density of water at

4°C is 1.000 0 g/mL, density and specific gravity are synonymous.

Spectral Interference In atomic spectroscopy, any physical process that affects the light intensity at the analytical wavelength. Created by substances that absorb, scatter, or emit light of the analytical wavelength.

Spectrophotometer A device used to measure absorption of light. It includes a source of light, a wavelength selector (monochromator), and an electrical means of detecting light.

Spectrophotometric Analysis Any method in which light absorption, emission, or scattering is used to measure chemical concentrations.

Spectrophotometric Titrations One in which absorption of light is used to monitor the progress of the chemical reaction.

Spectrophotometry In a broad sense, any method using light to measure chemical concentrations.

Specular Reflection Reflection of light at an angle equal to the angle of incidence.

Split Injection Used in capillary gas chromatography to inject a small fraction of sample onto the column, while the rest of the sample is discarded.

Splitless Injection Used in capillary gas chromatography for trace analysis and quantitative analysis. The entire sample in a low-boiling solvent is directed to the column, where the sample is concentrated by solvent trapping (condensing the solvent below its boiling point) or cold trapping (condensing solutes far below their boiling range). The column is then warmed to initiate separation.

Spontaneous Process One that is energetically favorable. It will eventually occur, but thermodynamics makes no prediction as to how long it will take.

Spread See **Range.**

Square Wave Polarography Form of polarography with a waveform consisting of a square wave superimposed on a staircase wave. The technique is faster and more sensitive than differential pulse polarography.

Stability Constant See **Formation Constant.**

Standard Addition Method A technique in which an analytical signal due to an unknown is first measured. Then a known quantity of analyte is added, and the increase in signal is recorded. Assuming linear response, it is possible to calculate what quantity of analyte must have been present in the unknown.

Standard Curve A graph showing the response of an analytical technique to known quantities of analyte.

Standard Deviation Measures how closely data are clustered about the mean value. For a finite set of data, the standard deviation, s, is computed from the formula

$$s = \sqrt{\frac{\sum(x_i - \overline{x})^2}{n - 1}}$$

where n is the number of results, x_i is an individual result, and $\overline{x}$ is the mean result.

Standard Hydrogen Electrode (S.H.E.) or **Normal Hydrogen Electrode (N.H.E.)** One that contains $H_2(g)$ bubbling over a catalytic Pt surface in contact with aqueous H^+. The activities of H_2 and H^+ are both unity in the hypothetical standard electrode. The cell reaction is $H^+ + e^- \rightleftharpoons \frac{1}{2}H_2(g)$.

Standardization The process whereby the concentration of a reagent is determined by reaction with a known quantity of a second reagent.

Standard Reduction Potential, $E°$ The voltage that would be measured when a hypothetical cell containing the desired half-reaction (with all species present at unit activity) is connected to a standard hydrogen electrode anode.

Standard Reference Materials Certified samples sold by the U.S. National Institute of Standards and Technology containing known concentrations or quantities of particular analytes. Used to standardize testing procedures in different laboratories.

Standard Solution A solution whose composition is known by virtue of the way it was made from a reagent of known purity.

Standard State When writing equilibrium constants, the standard state of a solute is 1 M and the standard state of a gas is 1 atm. Pure solids and liquids are considered to be in their standard states.

Stationary Phase In chromatography, the phase that does not move through the column.

Stepwise Formation Constant, K_n The equilibrium constant for a reaction of the type $ML_{n-1} + L \rightleftharpoons ML_n$.

Stimulated Emission Emission of a photon induced by the passage of another photon of the same wavelength.

Stripping Analysis A very sensitive polarographic technique in which analyte is concentrated from a dilute solution by reduction into a single drop of Hg. It is then analyzed polarographically during an anodic redissolution process.

Strong Acids and Bases Those that are completely dissociated (to H^+ or OH^-) in water.

Strong Electrolyte One that dissociates completely into its ions when dissolved.

Superconductor A material that loses all electric resistance when cooled below a critical temperature.

Supercritical Fluid A fluid above its critical temperature.

Supercritical Fluid Chromatography Chromatography using supercritical fluid as the mobile phase. Capable of highly efficient separations of nonvolatile solutes and able to use detectors suitable for gas or liquid.

Supernatant Liquid Liquid remaining above the solid after a precipitation. Also called *supernate.*

Supersaturated Solution One that contains more dissolved solute than would be present at equilibrium.

Support-Coated Column Open tubular gas chromatographic column in which the stationary phase is coated on solid support particles attached to the inside wall of the column.

Supporting Electrolyte   An unreactive salt added in high concentration to most solutions for voltammetric measurements (such as polarography). The supporting electrolyte carries most of the ion-migration current and therefore decreases the coulombic migration of electroactive species to a negligible level. The electrolyte also decreases the resistance of the solution.

Suppressed-Ion Chromatography   Separation of ions using an ion-exchange column followed by suppressor (membrane or column) to remove ionic eluent.

Suppressor Column   Ion-exchange column used in ion chromatography to transform ionic eluent into a nonionic form.

Surface-Modified Electrode   An electrode whose surface has been changed by a chemical reaction. For example, electroactive materials that react specifically with certain solutes can be attached to the electrode.

Syringe   A device having a calibrated barrel into which liquid is sucked by a plunger. The liquid is expelled through a needle by pushing on the plunger.

Systematic Error or **Determinate Error**   A type of error due to procedural or instrumental factors that cause a measurement to be systematically too large or too small. The error can, in principal, be discovered and corrected.

Systematic Treatment of Equilibrium   A method that uses the charge balance, mass balance(s), and equilibria to completely specify the system's composition.

Tailing   An asymmetric chromatographic elution band in which the later part of the band is drawn out. It often results from adsorption of a solute to a few active adsorption sites on the stationary phase.

Tare   The mass of an empty vessel used to receive a substance to be weighed. Many balances can be tared. That is, with the empty receiver in place, the balance can be set to read zero grams.

Temporary Hardness   Component of water hardness due to dissolved alkaline earth bicarbonates. It is temporary because boiling causes precipitation of the carbonates.

Test Portion   Part of the laboratory sample used for one analysis. Also called *aliquot*.

Theoretical Plate   An imaginary construct in chromatography denoting a segment of a column in which one equilibrium of solute between stationary and mobile phases occurs. The number of theoretical plates on a column with Gaussian bandshapes is defined as $N = t_r^2/\sigma^2$, where t_r is the retention time of a peak and σ is the standard deviation of the band.

Thermal Conductivity   Rate at which a substance transports heat (energy per unit time per unit area) through a temperature gradient (degrees per unit distance). Energy flow $[J/(s \cdot m^2)] = -\kappa\,(dT/dx)$, where κ is the thermal conductivity $[W/(m \cdot K)]$ and dT/dx is the temperature gradient (K/m).

Thermal Conductivity Detector   A device that detects bands eluted from a gas chromatography column by measuring changes in the thermal conductivity of the gas stream.

Thermistor   A device whose electrical resistance changes markedly with changes in temperature.

Thermocouple   An electrical junction across which a temperature-dependent voltage exists. Thermocouples are calibrated for measurement of temperature and usually consist of two dissimilar metals in contact with each other.

Thermogravimetric Analysis   A technique in which the mass of a substance is measured as the substance is heated. Changes in mass reflect decomposition of the substance, often to well-defined products.

Thermometric Titration   One in which the temperature is measured to determine the end point. Most titration reactions are exothermic, so the temperature rises during the reaction and suddenly stops rising when the equivalence point is reached.

Thin-Layer Chromatography   A technique in which the stationary phase is coated on a flat glass or plastic plate. Solute is spotted near the bottom of the plate. The bottom edge of the plate is placed in contact with solvent, which is allowed to creep up the plate by capillary action.

Titer   A measure of concentration, usually defined as how many milligrams of reagent B will react with 1 mL of reagent A. Consider a $AgNO_3$ solution with a titer of 1.28 mg of NaCl per milliliter of $AgNO_3$. The reaction is $Ag^+ + Cl^- \rightarrow AgCl(s)$. Since 1.28 mg of NaCl $= 2.19 \times 10^{-5}$ mol, the concentration of Ag^+ is 2.19×10^{-5} mol/mL $= 0.021\,9$ M. The same solution of $AgNO_3$ has a titer of 0.993 mg of KH_2PO_4, because three moles of Ag^+ react with one mole of PO_4^{3-} (to precipitate Ag_3PO_4), and 0.993 mg of KH_2PO_4 equals $\frac{1}{3}(2.19 \times 10^{-5}$ mol).

Titrant   The substance added to the analyte in a titration.

Titration   A procedure in which one substance (titrant) is carefully added to another (analyte) until complete reaction has occurred. The quantity of titrant required for complete reaction tells how much analyte is present.

Titration Error   Caused by the difference between the observed end point and the true equivalence point of the reaction.

Total Carbon   In a natural water or industrial effluent sample, the quantity of CO_2 produced when the sample is completely oxidized by oxygen at 900°C in the presence of a catalyst.

Total Organic Carbon   In a natural water or industrial effluent sample, the quantity of CO_2 produced when the sample is first acidified and purged to remove carbonate and bicarbonate and then completely oxidized by oxygen at 900°C in the presence of a catalyst.

Total Oxygen Demand   In a natural water or industrial effluent sample, the quantity of O_2 required for complete oxidation of species in the water at 900°C in the presence of a catalyst.

Transmittance, T   Defined as $T = P/P_0$, where P_0 is the radiant power of light striking the sample on one side and P is the radiant power of light emerging from the other side of the sample.

Triangular Apodization   In Fourier transform spectroscopy, multiplying the interferogram by a symmetric function that decreases linearly from 1 at the center to 0 at predefined limits on either side of 0.

Triple Point   The one temperature and pressure at which the solid, liquid, and gaseous forms of a substance are in equilibrium with each other.

Triplet State   An electronic state in which there are two unpaired electrons.

Truncation   The process of cutting off abruptly.

t Test   Used to decide whether the results of two experiments are within experimental uncertainty of each other. The uncertainty must be specified to within a certain probability.

Tungsten Lamp   An ordinary light bulb in which electricity passing through a tungsten filament heats the wire and causes it to emit visible light.

Turbidimetry   A technique in which the decrease in radiant power of light traveling through a turbid solution is measured.

Turbidity   The light-scattering property associated with suspended particles in a liquid. A turbid solution appears cloudy.

Turbidity Coefficient   The transmittance of a turbid solution is given by $P/P_0 = e^{-\tau b}$, where P is the transmitted radiant power, P_0 is the incident radiant power, b is the pathlength, and τ is the turbidity coefficient.

Ultraviolet Detector   Liquid chromatography detector that measures ultraviolet absorbance of solutes emerging from the column.

Valence Band   Energy levels containing valence electrons in a semiconductor. The electrons in these levels are localized in chemical bonds.

van Deemter Equation   Describes the dependence of chromatographic plate height on the velocity of elution: $HETP = A + B/v + Cv$.

Variance   The square of the standard deviation.

Vibrational Transition   Occurs when a molecule changes its vibrational energy.

Void Volume, V_0   The volume of the mobile phase, V_m.

Volatile   Easily vaporized.

Volatilization   The selective removal of a component from a mixture by transforming the component into a volatile (low-boiling) species and removing it by heating, pumping, or bubbling a gas through the mixture.

Volhard Titration   That of Ag^+ with SCN^-, in which the formation of the red complex $Fe(SCN)^{2+}$ marks the end point.

Volt   Unit of electric potential or electric potential difference between two points. If the potential difference between two points is one volt, it requires one joule of energy to move one coulomb of charge between the two points.

Voltammetry   An analytical method in which the relationship between current and voltage is observed during an electrochemical reaction.

Volume Flow Rate   In chromatography, the volume of mobile phase per unit time eluted from the column.

Volume Percent   Defined as (volume of solute/volume of solution) $\times$ 100.

Volumetric Analysis   A technique in which the volume of material needed to react with the analyte is measured.

Volumetric Flask   One having a tall, thin neck with a calibration mark. When the liquid level is at the calibration mark, the flask contains its specified volume of liquid.

von Weimarn Ratio   The quotient $(Q - S)/S$, where Q is the concentration of a solute and S is the concentration at equilibrium. A large value of this ratio means that the solution is highly supersaturated.

Walden Reductor   A column packed with silver and eluted with HCl. An oxidized analyte is reduced upon passage through the column. The reduced product is titrated with an oxidizing agent.

Wall-Coated Column   Hollow chromatographic column in which the stationary phase is coated on the inside surface of its wall.

Watt, W   The SI unit of power, equal to an energy flow of one joule per second. When an electric current of one ampere flows through a potential difference of one volt, the power is one watt.

Wavelength   The distance between consecutive crests of a wave.

Wavenumber, $\tilde{\nu}$ The reciprocal of the wavelength, λ.

Weak Acids and Bases   Those whose dissociation constants are not large.

Weak Electrolyte   One that only partially dissociates into ions when it dissolves.

Weighing Paper   Used as a base on which to place a solid reagent on a balance. The paper has a very smooth surface, from which solids fall easily for transfer to a vessel.

Weight Percent   Defined as (mass of solute/mass of solution) $\times$ 100.

Weight/Volume Percent   Defined as (mass of solute/volume of solution) $\times$ 100.

Weston Cell   An extremely stable voltage source, based on the reaction $Cd(s) + HgSO_4(aq) \rightleftharpoons CdSO_4(aq) + Hg(l)$. It is often used to standardize a potentiometer.

Wet Ashing   The destruction of organic matter in a sample by a liquid reagent (such as boiling aqueous $HClO_4$) prior to analysis of an inorganic component.

White Light   Light of all different wavelengths.

Wien Displacement Law   Approximate formula for the wavelength (λ_{max}) of maximum blackbody emission: $\lambda_{max} \cdot T \approx hc/5k = 2.878 \times 10^{-3}$ m·K, where T is kelvins, h is Planck's constant, c is the speed of light, and k is Boltzmann's constant. Valid for $T > 100$ K.

Working Electrode The one at which the reaction of analytical interest in coulometry or polarography occurs.

Zeeman Background Correction Technique used in atomic spectroscopy in which analyte signals are shifted outside the detector monochromator range by applying a strong magnetic field to the sample. Signal that remains is the background.

Zeeman Effect Shifting of atomic energy levels in a magnetic field.

Zwitterion A molecule with a positive charge localized in one position and a negative charge localized at another position.

Appendixes

A Logarithms and Exponents

The base 10 logarithm of a number, n, is the exponent to which 10 must be raised to produce n:

$$\text{If } n = 10^a, \qquad \log n = a \qquad \text{(A-1)}$$

On a calculator, you can get the log of n by merely pressing the "log" function. If you know $a = \log n$ and you wish to find n, you can calculate the antilog of a or you can raise 10 to the ath power:

$$a = \log n \qquad \text{(A-2)}$$

$$10^a = 10^{\log n} = n \qquad \text{(A-3)}$$

Natural logarithms (ln) work the same way, but are based on the number e ($=2.718\,281\ldots$) instead of the number 10. The ln of a number, n, is the exponent to which e must be raised to produce n:

$$b = \ln n \qquad \text{(A-4)}$$

$$e^b = e^{\ln n} = n \qquad \text{(A-5)}$$

On a calculator, you can find the ln of n with a button marked "ln." To find n when you know $b = \ln n$, use the e^x key.

Logarithms have certain useful properties:

$$\log(a \cdot b) = \log a + \log b \qquad \text{(A-6)}$$

$$\log\left(\frac{a}{b}\right) = \log a - \log b \qquad \text{(A-7)}$$

$$\log(a^b) = b \log a \qquad (A\text{-}8)$$

$$\log 10^a = a \qquad (A\text{-}9)$$

Some useful properties of exponents are

$$a^b \cdot a^c = a^{(b+c)} \qquad (A\text{-}10)$$

$$\frac{a^b}{a^c} = a^{(b-c)} \qquad (A\text{-}11)$$

Problems

Simplify each expression as much as possible. (Answers are given at the end of this appendix.)

(a) $e^{\ln a}$ (b) $10^{\log a}$ (c) $\log 10^a$

(d) $10^{-\log a}$ (e) $e^{-\ln a^3}$ (f) $e^{\ln a^{-3}}$

(g) $\log 10^{1/a^3}$ (h) $\log 10^{-a^2}$ (i) $\log(10^{a^2-b})$

(j) $\log(2a^3 10^{b^2})$ (k) $e^{a+\ln b}$ (l) $10^{(\log 3)-(4\log 2)}$

In working with the Nernst equation or the Henderson–Hasselbalch equation, the case often arises in which an equation of the following form must be solved for the unknown, x:

$$a = b - c \log \frac{d}{gx} \qquad (A\text{-}12)$$

In this equation a, b, c, d, and g are constants. We can isolate x by first bringing the log term to the left and bringing a to the right:

$$c \log \frac{d}{gx} = b - a$$

Dividing by c gives

$$\log \frac{d}{gx} = \frac{b-a}{c}$$

Now raise 10 to the value of each side of the equation:

$$10^{\log(d/gx)} = 10^{(b-a)/c}$$

But $10^{\log(d/gx)}$ is just d/gx, so

$$\frac{d}{gx} = 10^{(b-a)/c}$$

Finally, we rearrange to solve for x:

$$\frac{d}{g10^{(b-a)/c}} = x \tag{A-13}$$

To test yourself on the algebra, solve for the numerical value of x in the equation

$$-0.317 = 0.111 - \frac{0.059\ 16}{2} \log \frac{(x^2)}{0.01}$$

(Answer: $x = 1.72 \times 10^6$.)

Sometimes it is convenient to be able to convert between $\ln x$ and $\log x$. The relation between these is derived by writing

$$x = 10^{\log x}$$

and finding the ln of both sides:

$$\ln x = \ln(10^{\log x}) \tag{A-14}$$

$$= (\log x)(\ln 10) \tag{A-15}$$

Equation A-15 follows from Equation A-14 and the fact that $\ln a^b = b \ln a$.

Answers

(a) a	(b) a	(c) a
(d) $1/a$	(e) $1/a^3$	(f) $1/a^3$
(g) $1/a^3$	(h) $-a^2$	(i) $a^2 - b$
(j) $b^2 + \log(2a^3)$	(k) be^a	(l) $3/16$

B Graphs of Straight Lines

The general form of the equation of a straight line is

$$y = mx + b \qquad \text{(B-1)}$$

where $m = \text{slope} = \dfrac{\Delta y}{\Delta x} = \dfrac{y_2 - y_1}{x_2 - x_1}$

$\quad b = \text{intercept on } y \text{ axis}$

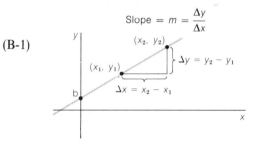

Figure B-1
Parameters of a straight line.

The meanings of slope and intercept are illustrated in Figure B-1.

If you know two points $[(x_1, y_1) \text{ and } (x_2, y_2)]$ that lie on the line, you can generate the equation of the line by noting that the slope is the same for every pair of points on the line. Calling some general point on the line (x, y), we can write

$$\frac{y - y_1}{x - x_1} = \frac{y_2 - y_1}{x_2 - x_1} = m \qquad \text{(B-2)}$$

which can be rearranged to the form

$$y - y_1 = \left(\frac{y_2 - y_1}{x_2 - x_1} \right)(x - x_1)$$

$$y = \underbrace{\left(\frac{y_2 - y_1}{x_2 - x_1} \right)}_{m} x + \underbrace{y_1 - \left(\frac{y_2 - y_1}{x_2 - x_1} \right) x_1}_{b} \qquad \text{(B-3)}$$

When you have a series of experimental points that should lie on a line, the best line is generally obtained by the method of least squares, described

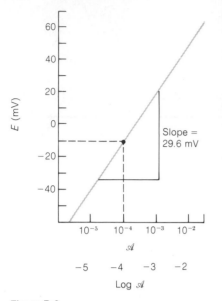

Figure B-2

A linear graph in which one axis is a logarithmic function.

in Chapter 4. This method gives the slope and the intercept directly. If, instead, you wish to draw the "best" line by eye, you can derive the equation of the line by selecting two points *that lie on the line* and applying Equation B-2. In general, your experimental points probably will not lie exactly on the best line. To apply Equation B-2, select two points that do lie on the line in your graph.

Sometimes you are presented with a linear plot in which x and/or y are nonlinear functions. An example is shown in Figure B-2, in which the potential of an electrode is expressed as a function of the activity of analyte. Given that the slope is 29.6 mV and the line passes through the point ($\mathscr{A} = 10^{-4}$, $E = -10.2$), find the equation of the line. To do this, first note that the y axis is linear, but that the x axis is *logarithmic*. That is, the function E versus $\mathscr{A}$ is *not* linear, but E versus $\log \mathscr{A}$ is linear. The form of the straight line should therefore be

$$E = (29.6) \underbrace{\log \mathscr{A}}_{} + b \qquad \text{(B-4)}$$
$$\underset{y}{\uparrow} \qquad \underset{m}{\uparrow} \quad \underset{x}{\uparrow}$$

To find b, we can use the coordinates of the one known point in Equation B-2:

$$\frac{y - y_1}{x - x_1} = \frac{E - E_1}{\log \mathscr{A} - \log \mathscr{A}_1} = \frac{E - (-10.2)}{\log \mathscr{A} - \log(10^{-4})} = m = 29.6$$

or

$$E + 10.2 = 29.6 \log \mathscr{A} + (29.6)(4)$$

$$E \text{ (mV)} = 29.6 \text{ (mV)} \log \mathscr{A} + 108.2 \text{ (mV)} \qquad \text{(B-5)}$$

C A Detailed Look at Propagation of Uncertainty

The rules given for propagation of uncertainty in Equations 3-6 and 3-9 are special cases of a general formula. Suppose you wish to calculate the function, F, of several experimental quantities, x, y, z, ... If the errors in measuring x, y, z, ..., are small, random, and independent of each other, then the uncertainty in F is given approximately by

$$e_F = \sqrt{\left(\frac{\partial F}{\partial x}\right)^2 e_x^2 + \left(\frac{\partial F}{\partial y}\right)^2 e_y^2 + \left(\frac{\partial F}{\partial z}\right)^2 e_z^2 + \cdots} \qquad \text{(C-1)}$$

Here e_F is the uncertainty in F and each e_i is the uncertainty in a measured quantity (x, y, z, etc.). The quantities in parentheses are partial derivatives. A partial derivative is calculated using the same rules for ordinary derivatives, except that all but one variable are treated as constants. For example, if $F = 3xy^2$, $\partial F/\partial x = 3y^2$ and $\partial F/\partial y = (3x)(2y) = 6xy$.

Let's apply Equation C-1 to the function $F = x - y$. For this function,

$$\frac{\partial F}{\partial x} = 1 \qquad \frac{\partial F}{\partial y} = -1$$

$$e_F = \sqrt{(1)^2 e_x^2 + (-1)^2 e_y^2} = \sqrt{e_x^2 + e_y^2}$$

which is just the same as the rule given for addition and subtraction in Chapter 3.

POWERS AND ROOTS

Suppose $F = x^a$, where the exponent a is a constant such as 3 or $-\frac{2}{3}$:

$$\frac{\partial F}{\partial x} = ax^{a-1} = a\frac{x^a}{x} = \left(\frac{a}{x}\right)F$$

$$e_F = \sqrt{\left(\frac{a}{x}\right)^2 F^2 e_x^2} = \left(\frac{a}{x}\right)Fe_x$$

The relative uncertainty will be

$$\frac{e_F}{F} = \frac{a}{x}e_x = a\left(\frac{e_x}{x}\right) \qquad \text{(C-2)}$$

That is, the relative uncertainty in F will be a times the relative uncertainty in x. For example, if $F = \sqrt{x} = x^{1/2}$, a 2% uncertainty in x will yield a $(\frac{1}{2})(2\%) = 1\%$ uncertainty in F. If $F = x^3$, a 2% uncertainty in x leads to a $(3)(2\%) = 6\%$ uncertainty in F.

LOGS AND ANTILOGS

Now consider the function $F = \log x$. This means that

$$x = 10^F$$

But 10 can be rewritten as $10 = e^{\ln 10}$. Putting this into the expression above gives

$$x = 10^F = (e^{\ln 10})^F = e^{F \ln 10}$$

or

$$\ln x = F \ln 10$$

or

$$F = \frac{\ln x}{\ln 10} \approx 0.434\,29 \ln x$$

Recalling that $d(\ln x)/dx = 1/x$ allows us to write

$$\frac{\partial F}{\partial x} = \left(\frac{1}{\ln 10}\right)\frac{\partial \ln x}{\partial x} = \frac{1}{\ln 10}\frac{1}{x}$$

Inserting this derivative into Equation C-1 shows that if $F = \log x$,

$$e_F = \sqrt{\left(\frac{1}{\ln 10}\right)^2\left(\frac{1}{x}\right)^2 e_x^2} = \left(\frac{1}{\ln 10}\right)\frac{e_x}{x} = (0.434\,29)\frac{e_x}{x} \qquad \text{(C-3)}$$

That is, the absolute uncertainty in F is proportional to the relative uncertainty in x.

Finally, consider $F = \text{antilog } x$, which is the same as saying $F = 10^x$. Noting that $d(a^x)/dx = a^x \ln a$, we can write

$$\frac{\partial F}{\partial x} = \frac{\partial(10^x)}{\partial x} = 10^x \ln 10$$

$$e_F = \sqrt{(10^x \ln 10)^2 e_x^2} = (10^x \ln 10)e_x$$

or

$$\frac{e_F}{F} = (\ln 10)e_x = (2.302\ 6)e_x \tag{C-4}$$

Here the relative uncertainty in F is proportional to the absolute uncertainty in x.

Problems

C-1. Verify the following calculations.
   (a) $\sqrt{2.36\ (\pm 0.06)} = 1.53_6 \pm 0.02_0$
   (b) $2.36^{4.39\ (\pm 0.08)} = 43._4 \pm 3._0$
   (c) $2.36\ (\pm 0.06)^{4.39\ (\pm 0.08)} = 43._4 \pm 5._7$
   (d) $\log[3.141\ 5\ (\pm 0.001\ 1)] = 0.497\ 1_4 \pm 0.001_5$
   (e) $\text{antilog}[2.08\ (\pm 0.06)] = 1.2_0\ (\pm 0.1_7) \times 10^2$
   (f) $\ln[3.141\ 5\ (\pm 0.001\ 1)] = 1.144\ 7_0 \pm 0.000\ 3_5$
   (g) $2.08\ (\pm 0.06)^{3.141\ 5\ (\pm 0.001\ 1)} -$
       $10.4\ (\pm 0.1)\log\sqrt{9.28\ (\pm 0.11)} = 4.9_5 \pm 0.9_1$

C-2. Consider the definition of pH to be $\text{pH} = -\log[\text{H}^+]$. If a pH measurement has an uncertainty of ± 0.02, show that the relative uncertainty in $[\text{H}^+]$ is 4.6%.

D Oxidation Numbers and Balancing Redox Equations

The *oxidation number,* or *oxidation state,* is a bookkeeping device used to keep track of the number of electrons formally associated with a particular element. The oxidation number is meant to tell how many electrons have been lost or gained by a neutral atom when it forms a compound. Because oxidation numbers have no real physical meaning, they are somewhat arbitrary, and not all chemists will assign the same oxidation number to a given element in an unusual compound. However, there are some ground rules that provide a useful start.

1. The oxidation number of an element by itself—e.g., $Cu(s)$ or $Cl_2(g)$—is zero.

2. The oxidation number of H is almost always $+1$, except in metal hydrides— e.g., NaH—in which H is -1.

3. The oxidation number of oxygen is almost always -2. The only common exceptions are peroxides, in which two oxygen atoms are connected and each has an oxidation number of -1. Two examples are hydrogen peroxide (H—O—O—H) and its anion (H—O—O$^-$). The oxidation number of oxygen in gaseous O_2 is, of course, zero.

4. The alkali metals (Li, Na, K, Rb, Cs, Fr) almost always have an oxidation number $+1$. The alkaline earth metals (Be, Mg, Ca, Sr, Ba, Ra) are almost always in the $+2$ oxidation state.

5. The halogens (F, Cl, Br, I) are usually in the -1 oxidation state. Exceptions occur when two different halogens are bound to each other or when a halogen is bound to more than one atom. When different halogens are bound to each other, we assign the oxidation number -1 to the more electronegative halogen.

The sum of the oxidation numbers of each atom in a molecule must equal the charge of the molecule. In H_2O, for example, we have

$$
\begin{aligned}
2 \text{ Hydrogen} = 2(+1) &= +2 \\
\text{Oxygen} &= \underline{-2} \\
\text{Net charge} & \quad 0
\end{aligned}
$$

In SO_4^{2-}, sulfur must have an oxidation number of $+6$ in order that the sum of the oxidation numbers be -2:

$$\text{Oxygen} = 4(-2) = -8$$
$$\text{Sulfur} \qquad\quad = \underline{+6}$$
$$\text{Net charge} \qquad\quad -2$$

In benzene (C_6H_6), the oxidation number of carbon must be -1 if hydrogen is assigned the number $+1$. In cyclohexane (C_6H_{12}), the oxidation number of carbon must be -2, for the same reason. Benzene therefore represents a higher oxidation state of carbon than that represented by cyclohexane.

The oxidation number of iodine in ICl_2^- is $+1$. This is unusual since halogens are usually -1. However, since chlorine is more electronegative than iodine, we assign Cl as -1, forcing I to be $+1$.

The oxidation number of As in As_2S_3 is $+3$, and the value for S is -2. This is arbitrary, but reasonable. Since S is more electronegative than As, we make S negative and As positive. Because sulfur is in the same family as oxygen, which is usually -2, we assign S as -2, leaving As as $+3$.

The oxidation number of S in $S_4O_6^{2-}$ (tetrathionate) is $+2.5$. The *fractional oxidation state* comes about since six O atoms contribute -12. Since the charge is -2, the four S atoms must contribute $+10$. The average oxidation number of S must be $+\frac{10}{4} = 2.5$.

The oxidation number of Fe in $K_3Fe(CN)_6$ is $+3$. To make this assignment, we first recognize cyanide (CN^-) as a common ion that carries a charge of -1. Six cyanide ions give -6, and three potassium ions (K^+) give $+3$. Therefore, Fe should have an oxidation number of $+3$ for the whole formula to be neutral. In this approach, it is not necessary to assign individual oxidation numbers to carbon and nitrogen, so long as we recognize that the charge of CN is -1.

Problems

Answers are given at the end of this appendix.

D-1. Write the oxidation state of the boldfaced atom in each of the following species.

(a) **Ag**Br (b) **S**$_2$O$_3^{2-}$ (c) **Se**F$_6$

(d) H**S**$_2$O$_3^-$ (e) H**O**$_2$ (f) **N**O

(g) **Cr**$^{3+}$ (h) **Mn**O$_2$ (i) **Pb**(OH)$_3^-$

(j) **Fe**(OH)$_3$ (k) **Cl**O$^-$ (l) K$_4$**Fe**(CN)$_6$

(m) **Cl**O$_2$ (n) **Cl**O$_2^-$ (o) **Mn**(CN)$_6^{4-}$

(p) **N**$_2$ (q) **N**H$_4^+$ (r) **N**$_2$H$_5^+$

(s) H**As**O$_3^{2-}$ (t) **Co**$_2$(CO)$_8$ (CO group is neutral) (u) (CH$_3$)$_4$**Li**$_4$

(v) **P**$_4$O$_{10}$ (w) **C**$_2$H$_6$O (ethanol, CH$_3$CH$_3$OH) (x) **V**O(SO$_4$)

(y) **Fe**$_3$O$_4$ (z) **C**$_3$H$_3^+$

Structure: H—C $\underset{C}{\overset{C}{\underset{\diagdown}{\overset{\diagup}{\oplus}}}}$ $\begin{array}{c} H \\ | \\ H \end{array}$

D-2. Identify the oxidizing agent and the reducing agent on the left side of each of the following reactions.

(a) $Cr_2O_7^{2-} + 3Sn^{2+} + 14H^+ \rightarrow$
$$2Cr^{3+} + 3Sn^{4+} + 7H_2O$$

(b) $4I^- + O_2 + 4H^+ \rightarrow 2I_2 + 2H_2O$

(c) $\begin{array}{c} O \\ \parallel \\ 5CH_3CH \end{array} + 2MnO_4^- + 6H^+ \rightarrow$

$$\begin{array}{c} O \\ \parallel \\ 5CH_3COH \end{array} + 2Mn^{2+} + 3H_2O$$

(d) $HOCH_2CHOHCH_2OH + 2IO_4^- \rightarrow$
    Glycerol

$$2H_2C{=}O + HCO_2H + 2IO_3^- + H_2O$$
Formaldehyde   Formic
                acid

(e) $C_8H_8 + 2Na \rightarrow C_8H_8^{2-} + 2Na^+$
C_8H_8 is cyclooctatetraene with the structure

(f) $I_2 + OH^- \rightarrow HOI + I^-$
                    Hypoiodite

BALANCING REDOX REACTIONS

To balance a reaction involving oxidation and reduction, we must first identify which element is oxidized and which is reduced. We then break the net reaction into two imaginary *half-reactions,* one of which involves only oxidation and the other only reduction. Although free electrons never appear in a balanced net reaction, they do appear in balanced half-reactions. If we are dealing with aqueous solutions, we proceed to balance each half-reaction, using H_2O and either H^+ or OH^-, as necessary. *A reaction is balanced when the number of atoms of each element is the same on both sides, and the net charge is the same on both sides.*[†]

Acidic Solutions

Here are the steps we will follow:

1. Assign oxidation numbers to the elements that are oxidized or reduced.

2. Break the reaction into two half-reactions, one involving oxidation and the other reduction.

3. For each half-reaction, balance the number of atoms that are oxidized or reduced.

4. Balance the electrons to account for the change in oxidation number by adding electrons to one side of each half-reaction.

5. Balance oxygen atoms by adding H_2O to one side of each half-reaction.

6. Balance the H atoms by adding H^+ to one side of each half-reaction.

7. Multiply each half-reaction by the number of electrons in the other half-reaction so that the number of electrons on each side of the total reaction will cancel. Then add the two half-reactions and simplify to the smallest integral coefficients.

[†] A completely different method for balancing complex redox equations by inspection has been described by D. Kolb, *J. Chem. Ed., 58,* 642 (1981).

EXAMPLE

Balance the following equation using H^+, but not OH^-:

$$\underset{+2}{Fe^{2+}} + \underset{+7}{MnO_4^-} \rightleftharpoons \underset{+3}{Fe^{3+}} + \underset{+2}{Mn^{2+}}$$
$$\text{Permanganate}$$

The steps just listed above apply to this example, as follows:

1. *Assign oxidation numbers.* These are assigned for Fe and Mn in each species in the above reaction.

2. *Break the reaction into two half-reactions.*

Oxidation half-reaction: $$\underset{+2}{Fe^{2+}} \rightleftharpoons \underset{+3}{Fe^{3+}}$$

Reduction half-reaction: $$\underset{+7}{MnO_4^-} \rightleftharpoons \underset{+2}{Mn^{2+}}$$

3. *Balance the atoms that are oxidized or reduced.* Since there is only one Fe or Mn in each species on each side of the equation, the atoms of Fe and Mn are already balanced.

4. *Balance electrons.* Electrons are added to account for the change in each oxidation state.

$$Fe^{2+} \rightleftharpoons Fe^{3+} + e^-$$

$$MnO_4^- + 5e^- \rightleftharpoons Mn^{2+}$$

In the second case, we need $5e^-$ on the left side to take Mn from $+7$ to $+2$.

5. *Balance oxygen atoms.* There are no oxygen atoms in the Fe half-reactions. There are four oxygen atoms on the left side of the Mn reaction, so we add four molecules of H_2O to the right side:

$$MnO_4^- + 5e^- \rightleftharpoons Mn^{2+} + 4H_2O$$

6. *Balance hydrogen atoms.* The Fe equation is already balanced. The Mn equation needs $8H^+$ on the left.

$$MnO_4^- + 5e^- + 8H^+ \rightarrow Mn^{2+} + 4H_2O$$

At this point, each half-reaction must be completely balanced (the same number of atoms and charge on each side), *or you have made a mistake.*

7. *Multiply and add the reactions.* We multiply the Fe equation by 5 and the Mn equation by 1 and add:

$$5Fe^{2+} \rightleftharpoons 5Fe^{3+} + 5e^-$$

$$MnO_4^- + 5e^- + 8H^+ \rightleftharpoons Mn^{2+} + 4H_2O$$

$$\overline{5Fe^{2+} + MnO_4^- + 8H^+ \rightleftharpoons 5Fe^{3+} + Mn^{2+} + 4H_2O}$$

The total charge on each side is $+17$, and we find the same number of each atom on each side. The equation is balanced.

EXAMPLE

Now try the next reaction, which represents the reverse of a *disproportionation*. (In a disproportionation, an element in one oxidation state reacts to give the same element in higher and lower oxidation states.)

$$I_2 \;+\; IO_3^- \;+\; Cl^- \rightleftharpoons ICl_2^-$$
$$\;\;0 \quad\; +5 \qquad -1 \quad\; +1-1$$
$$\text{Iodine}\quad\text{Iodate}$$

1. The oxidation numbers are assigned above. Note that chlorine has an oxidation number of -1 on both sides of the equation. Only iodine is involved in electron transfer.

2. Oxidation half-reaction: $\qquad\qquad I_2 \rightleftharpoons ICl_2^-$
   $$0 \quad\; +1$$

   Reduction half-reaction: $\qquad\qquad IO_3^- \rightleftharpoons ICl_2^-$
   $$+5 \quad\;\; +1$$

3. We need to balance I atoms in the first reaction and add Cl^- to each reaction to balance Cl.

   $$I_2 + 4Cl^- \rightleftharpoons 2ICl_2^-$$

   $$IO_3^- + 2Cl^- \rightleftharpoons ICl_2^-$$

4. Now add electrons to each.

   $$I_2 + 4Cl^- \rightleftharpoons 2ICl_2^- + 2e^-$$

   $$IO_3^- + 2Cl^- + 4e^- \rightleftharpoons ICl_2^-$$

   The first reaction needs $2e^-$ because there are two I atoms, each of which changes from 0 to $+1$.

5. The second reaction needs $3H_2O$ on the right side to balance oxygen atoms.

   $$IO_3^- + 2Cl^- + 4e^- \rightleftharpoons ICl_2^- + 3H_2O$$

6. The first reaction is balanced, but the second needs $6H^+$ on the left.

   $$IO_3^- + 2Cl^- + 4e^- + 6H^+ \rightarrow ICl_2^- + 3H_2O$$

   As a check, the charge on each side of this half-reaction is -1, and all atoms are balanced.

7. Multiply and add.

   $$2(I_2 \;+\; 4Cl^- \qquad\qquad\qquad \rightleftharpoons 2ICl^- + 2\cancel{e}^-)$$
   $$\underline{IO_3^- + 2Cl^- + \; \cancel{4e}^- \; + 6H^+ \rightleftharpoons \; ICl_2^- + 3H_2O}$$
   $$2I_2 \;+\; IO_3^- \;+\; 10Cl^- + 6H^+ \rightleftharpoons 5ICl_2^- + 3H_2O \qquad\qquad \text{(D-1)}$$

   We multiplied the first reaction by 2 so there would be the same number of electrons in each half-reaction. You could have multiplied the first reaction by 4 and the second by 2, but then all coefficients would simply be doubled. We customarily write the smallest coefficients.

Basic Solutions

The method many people prefer for basic solutions is to balance the equation first with H^+. The answer can then be converted to one in which OH^- is used instead. This is done by adding to each side of the equation a number of hydroxide ions equal to the number of H^+ ions appearing in the equation. For example, to balance Equation D-1 with OH^- instead of H^+, proceed as follows:

$$2I_2 + IO_3^- + 10Cl^- + 6H^+ \rightleftharpoons 5ICl_2 + 3H_2O$$

$$+ 6OH^- \qquad\qquad + 6OH^-$$

$$2I_2 + IO_3^- + 10Cl^- + \underbrace{6H^+ + 6OH^-}_{\substack{\cancel{6}H_2O \\ \Downarrow \\ 3H_2O}} \rightleftharpoons 5ICl_2 + \cancel{3H_2O} + 6OH^-$$

Realizing that $6H^+ + 6OH^- = 6H_2O$, and canceling $3H_2O$ on each side, gives the final result:

$$2I_2 + IO_3^- + 10Cl^- + 3H_2O \rightleftharpoons 5ICl_2 + 6OH^-$$

Problems

D-3. Balance the following reactions using H^+, but not OH^-.

(a) $Fe^{3+} + Hg_2^{2+} \rightleftharpoons Fe^{2+} + Hg^{2+}$

(b) $Ag + NO_3^- \rightleftharpoons Ag^+ + NO$

(c) $VO^{2+} + Sn^{2+} \rightleftharpoons V^{3+} + Sn^{4+}$

(d) $SeO_4^{2-} + Hg + Cl^- \rightleftharpoons SeO_3^{2-} + Hg_2Cl_2$

(e) $CuS + NO_3^- \rightleftharpoons Cu^{2+} + SO_4^{2-} + NO$

(f) $S_2O_3^{2-} + I_2 \rightleftharpoons I^- + S_4O_6^{2-}$

(g) $ClO_3^- + As_2S_3 \rightleftharpoons Cl^- + H_2AsO_4^- + SO_4^{2-}$

(h) $Cr_2O_7^{2-} + CH_3\overset{O}{\overset{\|}{C}}H \rightleftharpoons CH_3\overset{O}{\overset{\|}{C}}OH + Cr^{3+}$

(i) $MnO_4^{2-} \rightleftharpoons MnO_2 + MnO_4^-$

(j) $Hg_2SO_4 + Ca^{2+} + S_8 \rightleftharpoons Hg_2^{2+} + CaS_2O_3$

(k) $ClO_3^- \rightleftharpoons Cl_2 + O_2$.

D-4. Balance the following equations using OH^-, but not H^+.

(a) $PbO_2 + Cl^- \rightleftharpoons ClO^- + Pb(OH)_3^-$

(b) $HNO_2 + SbO^+ \rightleftharpoons NO + Sb_2O_5$

(c) $Ag_2S + CN^- + O_2 \rightleftharpoons S + Ag(CN)_2^- + OH^-$

(d) $HO_2^- + Cr(OH)_3^- \rightleftharpoons CrO_4^{2-} + OH^-$

(e) $ClO_2 + OH^- \rightleftharpoons ClO_2^- + ClO_3^-$

(f) $WO_3^- + O_2 \rightleftharpoons HW_6O_{21}^{5-} + OH^-$

(g) $Mn_2O_3 + CN^- \rightleftharpoons Mn(CN)_6^{4-} + (CN)_2$

(h) $Cu^{2+} + H_2 \rightleftharpoons Cu + H_2O$

(i) $BH_4^- + H_2O \rightleftharpoons H_3BO_3 + H_2$

(j) $Mn_2O_3 + Hg + CN^- \rightleftharpoons$
$$Mn(CN)_6^{4-} + Hg(CN)_2$$

(k) $MnO_4^- + H\overset{O}{\overset{\|}{C}}CH_2CH_2OH \rightleftharpoons$
$$CH_2(CO_2^-)_2 + MnO_2$$

(l) $K_3V_5O_{14} + HOCH_2CHOHCH_2OH \rightleftharpoons$
$$VO(OH)_2 + HCO_2^- + K^+$$

Answers

D-1. (a) $+1$ (b) $+2$ (c) $+6$ (d) $+2$
(e) $-\frac{1}{2}$ (f) $+2$ (g) $+3$ (h) $+4$
(i) $+2$ (j) $+3$ (k) $+1$ (l) $+2$
(m) $+4$ (n) $+3$ (o) $+2$ (p) 0
(q) -3 (r) -2 (s) $+3$ (t) 0
(u) -4 (v) $+5$ (w) -2 (x) $+4$
(y) $+8/3$ (z) $-2/3$

D-2.

	Oxidizing agent	Reducing agent
(a)	$Cr_2O_7^{2-}$	Sn^{2+}
(b)	O_2	I^-
(c)	MnO_4^-	CH_3CHO
(d)	IO_4^-	Glycerol
(e)	C_8H_8	Na
(f)	I_2	I_2

Reaction f is called a *disproportionation*, since an element in one oxidation state is transformed into two different oxidation states—one higher and one lower than the original oxidation state.

D-3. (a) $2Fe^{3+} + Hg_2^{2+} \rightleftharpoons 2Fe^{2+} + 2Hg^{2+}$
(b) $3Ag + NO_3^- + 4H^+ \rightleftharpoons 3Ag^+ + NO + 2H_2O$
(c) $4H^+ + 2VO^{2+} + Sn^{2+} \rightleftharpoons$
$2V^{3+} + Sn^{4+} + 2H_2O$
(d) $2Hg + 2Cl^- + SeO_4^{2-} + 2H^+ \rightleftharpoons$
$Hg_2Cl_2 + SeO_3^{2-} + H_2O$
(e) $3CuS + 8NO_3^- + 8H^+ \rightleftharpoons$
$3Cu^{2+} + 3SO_4^{2-} + 8NO + 4H_2O$
(f) $2S_2O_3^{2-} + I_2 \rightleftharpoons S_4O_6^{2-} + 2I^-$
(g) $14ClO_3^- + 3As_2S_3 + 18H_2O \rightleftharpoons$
$14Cl^- + 6H_2AsO_4^- + 9SO_4^{2-} + 24H^+$
(h) $Cr_2O_7^{2-} + 3CH_3CHO + 8H^+ \rightleftharpoons$
$2Cr^{3+} + 3CH_3CO_2H + 4H_2O$
(i) $4H^+ + 3MnO_4^{2-} \rightleftharpoons$
$MnO_2 + 2MnO_4^- + 2H_2O$
(j) $2Hg_2SO_4 + 3Ca^{2+} + \frac{1}{2}S_8 + H_2O \rightleftharpoons$
$2Hg_2^{2+} + 3CaS_2O_3 + 2H^+$
(k) $2H^+ + 2ClO_3^- \rightleftharpoons Cl_2 + \frac{5}{2}O_2 + H_2O$

The balanced half-reaction for As_2S_3 in part g is

$$As_2S_3\ \underset{+3\ -2}{} + 20H_2O \rightleftharpoons$$

$$\underset{+5\qquad +6}{2H_2AsO_4^- + 3SO_4^{2-} + 28e^- + 36H^+}$$

Since As_2S_3 is a single compound, we must consider the $As_2S_3 \rightarrow H_2AsO_4^-$ and $As_2S_3 \rightarrow SO_4^{2-}$ reactions together. The net change in oxidation number for the *two* As atoms is $2(5-3) = +4$. The net change in oxidation number for the *three* S atoms is $3[6-(-2)] = +24$. Therefore, $24 + 4 = 28e^-$ are involved in the half-reaction.

D-4. (a) $H_2O + OH^- + PbO_2 + Cl^- \rightleftharpoons$
$Pb(OH)_3^- + ClO^-$
(b) $4HNO_2 + 2SbO^+ + 2OH^- \rightleftharpoons$
$4NO + Sb_2O_5 + 3H_2O$
(c) $Ag_2S + 4CN^- + \frac{1}{2}O_2 + H_2O \rightleftharpoons$
$S + 2Ag(CN)_2^- + 2OH^-$
(d) $2HO_2^- + Cr(OH)_3^- \rightleftharpoons$
$CrO_4^{2-} + OH^- + 2H_2O$
(e) $2ClO_2 + 2OH^- \rightleftharpoons ClO_2^- + ClO_3^- + H_2O$
(f) $12WO_3^- + 3O_2 + 2H_2O \rightleftharpoons$
$2HW_6O_{21}^{5-} + 2OH^-$
(g) $Mn_2O_3 + 14CN^- + 3H_2O \rightleftharpoons$
$2Mn(CN)_6^{4-} + (CN)_2 + 6OH^-$
(h) $Cu^{2+} + H_2 + 2OH^- \rightleftharpoons Cu + 2H_2O$
(i) $BH_4^- + 4H_2O \rightleftharpoons H_3BO_3 + 4H_2 + OH^-$
(j) $3H_2O + Mn_2O_3 + Hg + 14CN^- \rightleftharpoons$
$2Mn(CN)_6^{4-} + Hg(CN)_2 + 6OH^-$
(k) $2MnO_4^- + \overset{\overset{\textstyle O}{\|}}{H}CCH_2CH_2OH \rightleftharpoons$
$2MnO_2 + 2H_2O + CH(CO_2^-)_2$
(l) $32H_2O + 8K_3V_5O_{14} +$
$5HOCH_2CHOHCH_2OH \rightleftharpoons$
$40VO(OH)_2 + 5HCO_2^- + 9OH^- + 24K^+$

For (k), the organic half-reaction is $8OH^- + C_3H_6O_2 \rightleftharpoons C_3H_2O_4^{2-} + 6e^- + 6H_2O$.

For (l), the two half-reactions are $K_3V_5O_{14} + 9H_2O + 5e^- \rightleftharpoons 5VO(OH)_2 + 8OH^- + 3K^+$ and $C_3H_8O_3 + 11OH^- \rightleftharpoons 3HCO_2^- + 8e^- + 8H_2O$.

E Normality

The *normality*, N, of a redox reagent is n times the molarity, where n is the number of electrons donated or accepted by that species in a chemical reaction.

$$N = n\text{M} \tag{E-1}$$

For example, in the half-reaction

$$MnO_4^- + 8H^+ + 5e^- \rightleftharpoons Mn^{2+} + 4H_2O \tag{E-2}$$

the normality of permanganate ion is five times its molarity, since each MnO_4^- accepts $5e^-$. If the molarity of permanganate is 0.1 M, the normality for the reaction

$$MnO_4^- + 5Fe^{2+} + 8H^+ \rightleftharpoons Mn^{2+} + 5Fe^{3+} + 4H_2O \tag{E-3}$$

is $5 \times 0.1 = 0.5$ N (read "0.5 normal"). In this reaction, each Fe^{2+} ion donates one electron. The normality of ferrous ion *equals* the molarity of ferrous ion, even though it takes five ferrous ions to balance the reaction.

In the half-reaction

$$MnO_4^- + 4H^+ + 3e^- \rightleftharpoons MnO_2 + 2H_2O \tag{E-4}$$

each MnO_4^- ion accepts only *three* electrons. The normality of permanganate for this reaction is equal to three times the molarity of permanganate. A 0.06 N permanganate solution for this reaction contains 0.02 M MnO_4^-.

The normality of a solution is a statement of the moles of "reacting units" per liter. One mole of reacting units is called one *equivalent*. Therefore, the units of normality are equivalents per liter (equiv/L). For redox reagents, *one equivalent is the amount of substance that can donate or accept one electron.*

It is possible to speak of equivalents only with respect to a particular half-reaction. For example, in Reaction E-2 there are five equivalents per mole of MnO_4^-; but in Reaction E-4, there are only three equivalents per mole of MnO_4^-. The mass of substance containing one equivalent is called the *equivalent weight*. The formula weight of $KMnO_4$ is 158.033 9. The equivalent weight of $KMnO_4$ for Reaction E-2 is 158.033 9/5 = 31.606 8 g/equiv. The equivalent weight of $KMnO_4$ for Reaction E-4 is 158.033 9/3 = 52.678 0 g/equiv.

EXAMPLE

Find the normality of a solution containing 6.34 g of ascorbic acid in 250.0 mL if the relevant half-reaction is

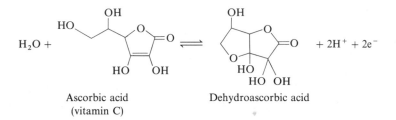

Ascorbic acid
(vitamin C)

Dehydroascorbic acid

The formula weight of ascorbic acid ($C_6H_8O_6$) is 176.126. In 6.34 g, there are (6.34 g)/(176.126 g/mol) = 3.60×10^{-2} mol. Since each mole contains 2 equivalents in this example, 6.34 g = (2 equiv/mol)(3.60×10^{-2} mol) = 7.20×10^{-2} equivalent. The normality is (7.20×10^{-2} equiv)/(0.250 0 L) = 0.288 N.

EXAMPLE

How many grams of potassium oxalate should be dissolved in 500.0 mL to make a 0.100 N solution for titration of MnO_4^-?

$$5H_2C_2O_4 + 2MnO_4^- + 6H^+ \rightleftharpoons 2Mn^{2+} + 10CO_2 + 8H_2O \qquad \text{(E-5)}$$

It is first necessary to write the oxalic acid half-reaction:

$$H_2C_2O_4 \rightleftharpoons 2CO_2 + 2H^+ + 2e^-$$

It is apparent that there are two equivalents per mole of oxalic acid. Hence a 0.100 N solution will be 0.050 0 M:

$$\frac{0.100 \text{ equiv/L}}{2 \text{ equiv/mol}} = 0.050 \ 0 \text{ mol/L} = 0.050 \ 0 \text{ M}$$

Therefore, we must dissolve (0.050 0 mol/L)(0.500 0 L) = 0.025 0 mol in 500.0 mL. Since the formula weight of $K_2C_2O_4$ is 166.216, we should make use of (0.025 0 mol) × (166.216 g/mol) = 4.15 g of potassium oxalate.

The utility of normality in volumetric analysis lies in the equation

$$N_1 V_1 = N_2 V_2 \qquad \text{(E-6)}$$

where N_1 is the normality of reagent 1, V_1 is the volume of reagent 1, N_2 is the normality of reagent 2, and V_2 is the volume of reagent 2. V_1 and V_2 may be expressed in any units, so long as the same units are used for both.

EXAMPLE

A solution containing 25.0 mL of oxalic acid required 13.78 mL of 0.041 62 N $KMnO_4$ for titration, according to Reaction E-5. Find the normality and molarity of the oxalic acid.

Setting up Equation E-6, we write

$$N_1(25.0 \text{ mL}) = (0.041\ 62 \text{ N})(13.78 \text{ mL})$$

$$N_1 = 0.022\ 94 \text{ equiv/L}$$

Since there are two equivalents per mole of oxalic acid in Reaction E-5,

$$M = \frac{N}{n} = \frac{0.022\ 94}{2} = 0.011\ 47 \text{ M}$$

Normality is sometimes used in acid–base or ion-exchange chemistry. With respect to acids and bases, the equivalent weight of a reagent is the amount that can donate or accept one mole of H^+. With respect to ion exchange, the equivalent weight is the mass of reagent containing one mole of charge.

F Solubility Products[†]

Formula	pK_{sp}	K_{sp}	Temperature (°C)	Ionic strength (M)
Azides: L = N_3^-				
CuL	8.31	4.9×10^{-9}	25	0
AgL	8.56	2.8×10^{-9}	25	0
Hg_2L_2	9.15	7.1×10^{-10}	25	0
TlL	3.66	2.2×10^{-4}	25	0
$PdL_2(\alpha)$	8.57	2.7×10^{-9}	25	0
Bromates: L = BrO_3^-				
$BaL \cdot H_2O$	5.11	7.8×10^{-6}	25	0.5
AgL	4.26	5.5×10^{-5}	25	0
TlL	3.78	1.7×10^{-4}	25	0
PbL_2	5.10	7.9×10^{-6}	25	0
Bromides: L = Br^-				
CuL	8.3	5×10^{-9}	25	0
AgL	12.30	5.0×10^{-13}	25	0
Hg_2L_2	22.25	5.6×10^{-23}	25	0
TlL	5.44	3.6×10^{-6}	25	0
HgL_2	18.9	1.3×10^{-19}	25	0.5
PbL_2	5.68	2.1×10^{-6}	25	0
Carbonates: L = CO_3^{2-}				
MgL	7.46	3.5×10^{-8}	25	0
CaL(calcite)	8.35	4.5×10^{-9}	25	0
CaL(aragonite)	8.22	6.0×10^{-9}	25	0
SrL	9.03	9.3×10^{-10}	25	0
BaL	8.30	5.0×10^{-9}	25	0
Y_2L_3	30.6	2.5×10^{-31}	25	0
La_2L_3	33.4	4.0×10^{-34}	25	0
MnL	9.30	5.0×10^{-10}	25	0
FeL	10.68	2.1×10^{-11}	25	0
CoL	9.98	1.0×10^{-10}	25	0
NiL	6.87	1.3×10^{-7}	25	0
CuL	9.63	2.3×10^{-10}	25	0
Ag_2L	11.09	8.1×10^{-12}	25	0
Hg_2L	16.05	8.9×10^{-17}	25	0
ZnL	10.00	1.0×10^{-10}	25	0
CdL	13.74	1.8×10^{-14}	25	0
PbL	13.13	7.4×10^{-14}	25	0

[†] The designations α, β, or γ after some formulas refer to particular crystalline forms (which are customarily identified by Greek letters). Data for salts except oxalates are taken mainly from A. E. Martell and R. M. Smith, *Critical Stability Constants,* Vol. 4 (New York: Plenum Press, 1976). Data for oxalates are from L. G. Sillén and A. E. Martell, *Stability Constants of Metal-Ion Complexes,* Supplement No. 1 (London: The Chemical Society, Special Publication No. 25, 1971). Another source: R. M. H. Verbeeck et al., *Inorg. Chem.,* **23,** 1922 (1984).

Formula	pK_{sp}	K_{sp}	Temperature (°C)	Ionic strength (M)
Chlorides: L = Cl⁻				
CuL	6.73	1.9×10^{-7}	25	0
AgL	9.74	1.8×10^{-10}	25	0
Hg_2L_2	17.91	1.2×10^{-18}	25	0
TlL	3.74	1.8×10^{-4}	25	0
PbL_2	4.78	1.7×10^{-5}	25	0
Chromates: L = CrO_4^{2-}				
BaL	9.67	2.1×10^{-10}	25	0
CuL	5.44	3.6×10^{-6}	25	0
Ag_2L	11.92	1.2×10^{-12}	25	0
Hg_2L	8.70	2.0×10^{-9}	25	0
Tl_2L	12.01	9.8×10^{-13}	25	0
Cobalticyanides: L = $Co(CN)_6^{3-}$				
Ag_3L	25.41	3.9×10^{-26}	25	0
$(Hg_2)_3L_2$	36.72	1.9×10^{-37}	25	0
Cyanides: L = CN⁻				
AgL	15.66	2.2×10^{-16}	25	0
Hg_2L_2	39.3	5×10^{-40}	25	0
ZnL_2	15.5	3×10^{-16}	25	3
Ferrocyanides: L = $Fe(CN)_6^{4-}$				
Ag_4L	44.07	8.5×10^{-45}	25	0
Zn_2L	15.68	2.1×10^{-16}	25	0
Cd_2L	17.38	4.2×10^{-18}	25	0
Pb_2L	18.02	9.5×10^{-19}	25	0
Fluorides: L = F⁻				
LiL	2.77	1.7×10^{-3}	25	0
MgL_2	8.18	6.6×10^{-9}	25	0
CaL_2	10.41	3.9×10^{-11}	25	0
SrL_2	8.54	2.9×10^{-9}	25	0
BaL_2	5.76	1.7×10^{-6}	25	0
ThL_4	28.3	5×10^{-29}	25	3
PbL_2	7.44	3.6×10^{-8}	25	0
Hydroxides: L = OH⁻				
MgL_2	11.15	7.1×10^{-12}	25	0
CaL_2	5.19	6.5×10^{-6}	25	0
$BaL_2 \cdot 8H_2O$	3.6	3×10^{-4}	25	0
YL_3	23.2	6×10^{-24}	25	0
LaL_3	20.7	2×10^{-21}	25	0
CeL_3	21.2	6×10^{-22}	25	0
$UO_2 (\rightleftharpoons U^{4+} + 4OH^-)$	56.2	6×10^{-57}	25	0
$UO_2L_2 (\rightleftharpoons UO_2^{2+} + 2OH^-)$	22.4	4×10^{-23}	25	0
MnL_2	12.8	1.6×10^{-13}	25	0
FeL_2	15.1	7.9×10^{-16}	25	0
CoL_2	14.9	1.3×10^{-15}	25	0
NiL_2	15.2	6×10^{-16}	25	0
CuL_2	19.32	4.8×10^{-20}	25	0
VL_3	34.4	4.0×10^{-35}	25	0
CrL_3	29.8	1.6×10^{-30}	25	0.1
FeL_3	38.8	1.6×10^{-39}	25	0
CoL_3	44.5	3×10^{-45}	19	0
$VOL_2 (\rightleftharpoons VO^{2+} + 2OH^-)$	23.5	3×10^{-24}	25	0
PdL_2	28.5	3×10^{-28}	25	0
ZnL_2(amorphous)	15.52	3.0×10^{-16}	25	0
$CdL_2(\beta)$	14.35	4.5×10^{-15}	25	0
HgO (red) $(\rightleftharpoons Hg^{2+} + 2OH^-)$	25.44	3.6×10^{-26}	25	0

(Continued)

Formula	pK_{sp}	K_{sp}	Temperature (°C)	Ionic strength (M)
$Cu_2O(\rightleftharpoons 2Cu^+ + 2OH^-)$	29.4	4×10^{-30}	25	0
$Ag_2O(\rightleftharpoons 2Ag^+ + 2OH^-)$	15.42	3.8×10^{-16}	25	0
AuL_3	5.5	3×10^{-6}	25	0
$AlL_3(\alpha)$	33.5	3×10^{-34}	25	0
$GaL_3(amorphous)$	37	10^{-37}	25	0
InL_3	36.9	1.3×10^{-37}	25	0
$SnO (\rightleftharpoons Sn^{2+} + 2OH^-)$	26.2	6×10^{-27}	25	0
PbO (yellow) $(\rightleftharpoons Pb^{2+} + 2OH^-)$	15.1	8×10^{-16}	25	0
PbO (red) $(\rightleftharpoons Pb^{2+} + 2OH^-)$	15.3	5×10^{-16}	25	0
Iodates: $L = IO_3^-$				
CaL_2	6.15	7.1×10^{-7}	25	0
SrL_2	6.48	3.3×10^{-7}	25	0
BaL_2	8.81	1.5×10^{-9}	25	0
YL_3	10.15	7.1×10^{-11}	25	0
LaL_3	10.99	1.0×10^{-11}	25	0
CeL_3	10.86	1.4×10^{-11}	25	0
ThL_4	14.62	2.4×10^{-15}	25	0.5
$UO_2L_2(\rightleftharpoons UO_2^{2+} + 2IO_3^-)$	7.01	9.8×10^{-8}	25	0.2
CrL_3	5.3	5×10^{-6}	25	0.5
AgL	7.51	3.1×10^{-8}	25	0
Hg_2L_2	17.89	1.3×10^{-18}	25	0
TlL	5.51	3.1×10^{-6}	25	0
ZnL_2	5.41	3.9×10^{-6}	25	0
CdL_2	7.64	2.3×10^{-8}	25	0
PbL_2	12.61	2.5×10^{-13}	25	0
Iodides: $L = I^-$				
CuL	12.0	1×10^{-12}	25	0
AgL	16.08	8.3×10^{-17}	25	0
$CH_3HgL (\rightleftharpoons CH_3Hg^+ + I^-)$	11.46	3.5×10^{-12}	20	1
$CH_3CH_2HgL (\rightleftharpoons CH_3CH_2Hg^+ + I^-)$	4.11	7.8×10^{-5}	25	1
TlL	7.23	5.9×10^{-8}	25	0
Hg_2L_2	27.95	1.1×10^{-28}	25	0.5
SnL_2	5.08	8.3×10^{-6}	25	4
PbL_2	8.10	7.9×10^{-9}	25	0
Oxalates: $L = C_2O_4^{2-}$				
CaL	7.9	1.3×10^{-8}	20	0.1
SrL	6.4	4×10^{-7}	20	0.1
BaL	6.0	1×10^{-6}	20	0.1
La_2L_3	25.0	1×10^{-25}	20	0.1
ThL_2	21.38	4.2×10^{-22}	25	1
$UO_2L (\rightleftharpoons UO_2^{2+} + C_2O_4^{2-})$	8.66	2.2×10^{-9}	20	0.1
Phosphates: $L = PO_4^{3-}$				
$MgHL \cdot 3H_2O (\rightleftharpoons Mg^{2+} + HL^{2-})$	5.78	1.7×10^{-6}	25	0
$CaHL \cdot 2H_2O (\rightleftharpoons Ca^{2+} + HL^{2-})$	6.58	2.6×10^{-7}	25	0
$SrHL (\rightleftharpoons Sr^{2+} + HL^{2-})$	6.92	1.2×10^{-7}	20	0
$BaHL (\rightleftharpoons Ba^{2+} + HL^{2-})$	7.40	4.0×10^{-8}	20	0
LaL	22.43	3.7×10^{-23}	25	0.5
$Fe_3L_2 \cdot 8H_2O$	36.0	1×10^{-36}	25	0
$FeL \cdot 2H_2O$	26.4	4×10^{-27}	25	0
$(VO)_3L_2 (\rightleftharpoons 3VO^{2+} + 2L^{3-})$	25.1	8×10^{-26}	25	0
Ag_3L	17.55	2.8×10^{-18}	25	0
$Hg_2HL (\rightleftharpoons Hg_2^{2+} + HL^{2-})$	12.40	4.0×10^{-13}	25	0
$Zn_3L_2 \cdot 4H_2O$	35.3	5×10^{-36}	25	0
Pb_3L_2	43.53	3.0×10^{-44}	38	0
GaL	21.0	1×10^{-21}	25	1
InL	21.63	2.3×10^{-22}	25	1

(Continued)

Formula	pK_{sp}	K_{sp}	Temperature (°C)	Ionic strength (M)
Sulfates: $L = SO_4^{2-}$				
CaL	4.62	2.4×10^{-5}	25	0
SrL	6.50	3.2×10^{-7}	25	0
BaL	9.96	1.1×10^{-10}	25	0
RaL	10.37	4.3×10^{-11}	20	0
Ag_2L	4.83	1.5×10^{-5}	25	0
Hg_2L	6.13	7.4×10^{-7}	25	0
PbL	6.20	6.3×10^{-7}	25	0
Sulfides: $L = S^{2-}$				
MnL (pink)	10.5	3×10^{-11}	25	0
MnL (green)	13.5	3×10^{-14}	25	0
FeL	18.1	8×10^{-19}	25	0
CoL(α)	21.3	5×10^{-22}	25	0
CoL(β)	25.6	3×10^{-26}	25	0
NiL(α)	19.4	4×10^{-20}	25	0
NiL(β)	24.9	1.3×10^{-25}	25	0
NiL(γ)	26.6	3×10^{-27}	25	0
CuL	36.1	8×10^{-37}	25	0
Cu_2L	48.5	3×10^{-49}	25	0
Ag_2L	50.1	8×10^{-51}	25	0
Tl_2L	21.2	6×10^{-22}	25	0
ZnL(α)	24.7	2×10^{-25}	25	0
ZnL(β)	22.5	3×10^{-23}	25	0
CdL	27.0	1×10^{-27}	25	0
HgL (black)	52.7	2×10^{-53}	25	0
HgL (red)	53.3	5×10^{-54}	25	0
SnL	25.9	1.3×10^{-26}	25	0
PbL	27.5	3×10^{-28}	25	0
In_2L_3	69.4	4×10^{-70}	25	0
Thiocyanates: $L = SCN^-$				
CuL	13.40	4.0×10^{-14}	25	5
AgL	11.97	1.1×10^{-12}	25	0
Hg_2L_2	19.52	3.0×10^{-20}	25	0
TlL	3.79	1.6×10^{-4}	25	0
HgL_2	19.56	2.8×10^{-20}	25	1

G Acid Dissociation Constants[†]

Name	Structure[†]	pK_a[‡]	K_a
Acetic acid (ethanoic acid)	CH$_3$CO$_2$H	4.757	1.75×10^{-5}
Alanine	$\overset{\overset{\text{NH}_3^+}{\mid}}{\underset{\underset{\text{CO}_2\text{H}}{\mid}}{\text{CHCH}_3}}$	2.348 (CO$_2$H) 9.867 (NH$_3$)	4.49×10^{-3} 1.36×10^{-10}
Aminobenzene (aniline)	C$_6$H$_5$—NH$_3^+$	4.601	2.51×10^{-5}
4-Aminobenzenesulfonic acid (sulfanilic acid)	$^-$O$_3$S—C$_6$H$_4$—NH$_3^+$	3.232	5.86×10^{-4}
2-Aminobenzoic acid (anthranilic acid)	C$_6$H$_4$(NH$_3^+$)(CO$_2$H)	2.08 (CO$_2$H) 4.96 (NH$_3$)	8.3×10^{-3} 1.10×10^{-5}
2-Aminoethanethiol (2-mercaptoethylamine)	HSCH$_2$CH$_2$NH$_3^+$	8.21 (SH) ($\mu = 0.1$) 10.71 (NH$_3$) ($\mu = 0.1$)	6.2×10^{-9} 1.95×10^{-11}
2-Aminophenol (ethanolamine)	HOCH$_2$CH$_2$NH$_3^+$	9.498	3.18×10^{-10}
2-Aminophenol	C$_6$H$_4$(OH)(NH$_3^+$)	4.78 (NH$_3$) (20°) 9.97 (OH) (20°)	1.66×10^{-5} 1.05×10^{-10}
Ammonia	NH$_4^+$	9.244	5.70×10^{-10}
Arginine	$\overset{\overset{\text{NH}_3^+}{\mid}}{\underset{\underset{\text{CO}_2\text{H}}{\mid}}{\text{CHCH}_2\text{CH}_2\text{CH}_2\text{NHC}}}\overset{\text{NH}_2^+}{\underset{\text{NH}_2}{}}$	1.823 (CO$_2$H) 8.991 (NH$_3$) (12.48) (NH$_2$)	1.50×10^{-2} 1.02×10^{-9} 3.3×10^{-13}

[†] Each acid is written in its protonated form. The acidic protons are indicated in bold type.

[‡] pK_a values refer to 25°C and zero ionic strength unless otherwise indicated. Values in parentheses are considered to be less reliable. Data are from A. E. Martell and R. M. Smith, *Critical Stability Constants* (New York: Plenum Press, 1974).

Name	Structure	pK_a	K_a
Arsenic acid (hydrogen arsenate)		2.24 6.96 11.50	5.8×10^{-3} 1.10×10^{-7} 3.2×10^{-12}
Arsenious acid (hydrogen arsenite)	$As(OH)_3$	9.29	5.1×10^{-10}
Asparagine		2.14 (CO_2H) ($\mu = 0.1$) 8.72 (NH_3) ($\mu = 0.1$)	7.2×10^{-3} 1.9×10^{-9}
Aspartic acid		1.990 (α-CO_2H) 3.900 (β-CO_2H) 10.002 (NH_3)	1.02×10^{-2} 1.26×10^{-4} 9.95×10^{-11}
Aziridine (dimethyleneimine)		8.04	9.1×10^{-9}
Benzene-1,2,3-tricarboxylic acid (hemimellitic acid)		2.88 4.75 7.13	1.32×10^{-3} 1.78×10^{-5} 7.4×10^{-8}
Benzoic acid		4.202	6.28×10^{-5}
Benzylamine		9.35	4.5×10^{-10}
2,2'-Bipyridine		4.35	4.5×10^{-5}
Boric acid (hydrogen borate)	$B(OH)_3$	9.236 (12.74) (20°) (13.80) (20°)	5.81×10^{-10} 1.82×10^{-13} 1.58×10^{-14}
Bromoacetic acid	$BrCH_2CO_2H$	2.902	1.25×10^{-3}
Butane-2,3-dione dioxime (dimethylglyoxime)		10.66 12.0	2.2×10^{-11} 1×10^{-12}
Butanoic acid	$CH_3CH_2CH_2CO_2H$	4.981	1.52×10^{-5}
cis-Butenedioc acid (maleic acid)		1.910 6.332	1.23×10^{-2} 4.66×10^{-7}
trans-Butenedioc acid (fumaric acid)		3.053 4.494	8.85×10^{-4} 3.21×10^{-5}
Butylamine	$CH_3CH_2CH_2CH_2NH_3^+$	10.640	2.29×10^{-11}

(Continued)

Name	Structure	pK_a	K_a
Carbonic acid[†] (hydrogen carbonate)	$\underset{\displaystyle \text{HO}-\overset{\displaystyle \text{O}}{\overset{\|}{\text{C}}}-\text{OH}}{}$	6.352 10.329	4.45×10^{-7} 4.69×10^{-11}
Chloroacetic acid	$ClCH_2CO_2H$	2.865	1.36×10^{-3}
3-Chloropropanoic acid	$ClCH_2CH_2CO_2H$	4.11	7.8×10^{-5}
Chlorous acid (hydrogen chlorite)	$HOCl{=}O$	1.95	1.12×10^{-2}
Chromic acid (hydrogen chromate)	$\text{HO}-\overset{\text{O}}{\underset{\text{O}}{\overset{\|}{\underset{\|}{\text{Cr}}}}}-\text{OH}$	-0.2 (20°) 6.51	1.6 3.1×10^{-7}
Citric acid (2-hydroxypropane-1,2,3-tricarboxylic acid)	$\underset{\displaystyle \text{OH}}{\overset{\displaystyle \text{CO}_2\text{H}}{HO_2CCH_2\overset{\|}{\text{C}}CH_2CO_2H}}$	3.128 4.761 6.396	7.44×10^{-4} 1.73×10^{-5} 4.02×10^{-7}
Cyanoacetic acid	$NCCH_2CO_2H$	2.472	3.37×10^{-3}
Cyclohexylamine	$\text{C}_6\text{H}_{11}{-}NH_3^+$	10.64	2.3×10^{-11}
Cysteine	$\underset{\displaystyle \text{CO}_2\text{H}}{\overset{\displaystyle \text{NH}_3^+}{\overset{\|}{\text{CH}}CH_2SH}}$	(1.71) (CO_2H) 8.36 (SH) 10.77 (NH_3)	1.95×10^{-2} 4.4×10^{-9} 1.70×10^{-11}
Dichloroacetic acid	Cl_2CHCO_2H	1.30	5.0×10^{-2}
Diethylamine	$(CH_3CH_2)_2NH_3^+$	10.933	4.7×10^{-10}
1.2-Dihydroxybenzene (catechol)		9.40 12.8	4.0×10^{-10} 1.6×10^{-13}
1,3-Dihydroxybenzene (resorcinol)		9.30 11.06	5.0×10^{-10} 8.7×10^{-12}
D-2,3-Dihydroxybutanedioc acid (D-tartaric acid)	$\underset{\displaystyle \text{OH}}{\overset{\displaystyle \text{OH}}{HO_2CCHCHCO_2H}}$	3.036 4.366	9.20×10^{-4} 4.31×10^{-5}
2,3-Dimercaptopropanol	$\underset{\displaystyle \text{SH}}{HOCH_2CHCH_2SH}$	8.58 ($\mu = 0.1$) 10.68 ($\mu = 0.1$)	2.6×10^{-9} 2.1×10^{-11}
Dimethylamine	$(CH_3)_2NH_2^+$	10.774	1.68×10^{-11}
Ethane-1,2-dithiol	$HSCH_2CH_2SH$	8.85 (30°, $\mu = 0.1$) 10.43 (30°, $\mu = 0.1$)	1.4×10^{-9} 3.7×10^{-11}
Ethylamine	$CH_2CH_3NH_3^+$	10.636	2.31×10^{-11}

[†] The concentration of "carbonic acid" is considered to be the sum $[H_2CO_3] + [CO_2(aq)]$. See Box 5-5.

(Continued)

Name	Structure	pK_a	K_a
Ethylenediamine (1,2-diaminoethane)	$H_3\overset{+}{N}CH_2CH_2\overset{+}{N}H_3$	6.848 9.928	1.42×10^{-7} 1.18×10^{-10}
Ethylenedinitrilotetraacetic acid (EDTA)	$(HO_2CCH_2)_2\overset{+}{N}HCH_2CH_2\overset{+}{N}H(CH_2CO_2H)_2$	0.0 (CO_2H) $(\mu = 1.0)$ 1.5 (CO_2H) $(\mu = 0.1)$ 2.0 (CO_2H) $(\mu = 0.1)$ 2.68 (CO_2H) $(\mu = 0.1)$ 6.11 (NH) $(\mu = 0.1)$ 10.17 (NH) $(\mu = 0.1)$	1.0 0.032 0.010 0.002 1 7.8×10^{-7} 6.8×10^{-11}
Formic acid (methanoic acid)	HCO_2H	3.745	1.80×10^{-4}
Glutamic acid	$\underset{CO_2H}{\overset{\overset{+}{N}H_3}{CHCH_2CH_2CO_2H}}$	2.23 $(\alpha\text{-}CO_2H)$ 4.42 $(\gamma\text{-}CO_2H)$ 9.95 (NH_3)	5.9×10^{-3} 3.8×10^{-5} 1.12×10^{-10}
Glutamine	$\underset{CO_2H}{\overset{\overset{+}{N}H_3}{CHCH_2CH_2\overset{O}{\overset{\|}{C}}NH_2}}$	2.17 (CO_2H) $(\mu = 0.1)$ 9.01 (NH_3) $(\mu = 0.1)$	6.8×10^{-3} 9.8×10^{-10}
Glycine (aminoacetic acid)	$\underset{CO_2H}{\overset{\overset{+}{N}H_3}{CH_2}}$	2.350 (CO_2H) 9.778 (NH_3)	4.47×10^{-3} 1.67×10^{-10}
Guanidine	$H_2N-\overset{\overset{+}{N}H_2}{\overset{\|}{C}}-NH_2$	13.54 $(27°, \mu = 1.0)$	2.9×10^{-14}
1,6-Hexanedioic acid (adipic acid)	$HO_2CCH_2CH_2CH_2CH_2CO_2H$	4.42 5.42	3.8×10^{-5} 3.8×10^{-6}
Hexane-2,4-dione	$CH_3\overset{O}{\overset{\|}{C}}CH_2\overset{O}{\overset{\|}{C}}CH_2CH_3$	9.38	4.2×10^{-10}
Histidine	$\underset{CO_2H}{\overset{\overset{+}{N}H_3}{CHCH_2}}$ imidazole	1.7 (CO_2H) $(\mu = 0.1)$ 6.02 (NH) $(\mu = 0.1)$ 9.08 (NH_3) $(\mu = 0.1)$	2×10^{-2} 9.5×10^{-7} 8.3×10^{-10}
Hydrazoic acid (hydrogen azide)	$HN{=}\overset{+}{N}{=}\overset{-}{N}$	4.65	2.2×10^{-5}
Hydrogen cyanate	$HOC{\equiv}N$	3.48	3.3×10^{-4}
Hydrogen cyanide	$HC{\equiv}N$	9.21	6.2×10^{-10}
Hydrogen fluoride	HF	3.17	6.8×10^{-4}
Hydrogen peroxide	$HOOH$	11.65	2.2×10^{-12}
Hydrogen sulfide	H_2S	7.02 13.9[†]	9.5×10^{-8} 1.3×10^{-14}[†]
Hydrogen thiocyanate	$HSC{\equiv}N$	0.9	0.13
Hydroxyacetic acid (glycolic acid)	$HOCH_2CO_2H$	3.831	1.48×10^{-4}

[†] A better value of pK_2 for H_2S is probably in the range 17–19 [R. J. Myers, *J. Chem. Ed.*, **63**, 687 (1986); S. Licht, F. Forouzan and K. Longo, *Anal. Chem.*, **62**, 1356 (1990)]. The value $pK_2 = 13.9$ was used for problems in this text.

(Continued)

Name	Structure	pK_a	K_a
Hydroxybenzene (phenol)	⟨benzene⟩—OH	9.98	1.05×10^{-10}
2-Hydroxybenzoic acid (salicylic acid)	⟨benzene with CO_2H and OH⟩	2.97 (CO_2H) 13.74 (OH)	1.07×10^{-3} 1.82×10^{-14}
L-Hydroxybutanedioic acid (malic acid)	$HO_2CCH_2\overset{\overset{OH}{\mid}}{C}HCO_2H$	3.459 5.097	3.48×10^{-4} 8.00×10^{-6}
Hydroxylamine	$HO\overset{+}{N}H_3$	5.96	1.10×10^{-6}
8-Hydroxyquinoline (oxine)	⟨quinoline structure with HO and $\overset{+}{N}H$⟩	4.91 (NH) 9.81 (OH)	1.23×10^{-5} 1.55×10^{-10}
Hypobromous acid (hydrogen hypobromite)	HOBr	8.63	2.3×10^{-9}
Hypochlorous acid (hydrogen hypochlorite)	HOCl	7.53	3.0×10^{-8}
Hypoiodous acid (hydrogen hypoiodite)	HOI	10.64	2.3×10^{-11}
Hypophosphorous acid (hydrogen hypophosphite)	$H_2\overset{\overset{O}{\parallel}}{P}OH$	1.23	5.9×10^{-2}
Imidazole (1,3-diazole)	⟨imidazole structure with $\overset{+}{N}H$ and $\overset{\mid}{N}{H}$⟩	6.993	1.02×10^{-7}
Iminodiacetic acid	$H_2\overset{+}{N}(CH_2CO_2H)_2$	1.82 (CO_2H) ($\mu = 0.1$) 2.84 (CO_2H) 9.79 (NH_2)	1.51×10^{-2} 1.45×10^{-3} 1.62×10^{-10}
Iodic acid (hydrogen iodate)	$HO\overset{\overset{O}{\parallel}}{I}{=}O$	0.77	0.17
Iodoacetic acid	ICH_2CO_2H	3.175	6.68×10^{-4}
Isoleucine	$\overset{\overset{NH_3^+}{\mid}}{\underset{\underset{CO_2H}{\mid}}{C}}HCH(CH_3)CH_2CH_3$	2.319 (CO_2H) 9.754 (NH_3)	4.80×10^{-3} 1.76×10^{-10}
Leucine	$\overset{\overset{NH_3^+}{\mid}}{\underset{\underset{CO_2H}{\mid}}{C}}HCH_2CH(CH_3)_2$	2.329 (CO_2) 9.747 (NH_3)	4.69×10^{-3} 1.79×10^{-10}
Lysine	$\overset{\overset{NH_3^+}{\mid}}{\underset{\underset{CO_2H}{\mid}}{C}}HCH_2CH_2CH_2CH_2NH_3^+$	2.04 (CO_2H) ($\mu = 0.1$) 9.08 (α-NH_3) ($\mu = 0.1$) 10.69 (ε-NH_3) ($\mu = 0.1$)	9.1×10^{-3} 8.3×10^{-10} 2.0×10^{-11}

(Continued)

Name	Structure	pK_a	K_a
Malonic acid (propanedioic acid)	$HO_2CCH_2CO_2H$	2.847 5.696	1.42×10^{-3} 2.01×10^{-6}
Mercaptoacetic acid (thioglycolic acid)	$HSCH_2CO_2H$	(3.60) (CO_2H) 10.55 (SH)	2.5×10^{-4} 2.82×10^{-11}
2-Mercaptoethanol	$HSCH_2CH_2OH$	9.72	1.91×10^{-10}
Methionine	$\overset{\overset{+}{N}H_3}{\underset{CO_2H}{CHCH_2CH_2SCH_3}}$	2.20 ($\mu = 0.1$) 9.05 ($\mu = 0.1$)	6.3×10^{-3} 8.9×10^{-10}
2-Methoxyaniline (o-anisidine)		4.527	2.97×10^{-5}
4-Methoxyaniline (p-anisidine)	$CH_3O\!-\!\langle\rangle\!-\!\overset{+}{N}H_3$	5.357	4.40×10^{-6}
Methylamine	$CH_3\overset{+}{N}H_3$	10.64	2.3×10^{-11}
2-Methylaniline (o-toluidine)		4.447	3.57×10^{-5}
4-Methylaniline (p-toluidine)	$CH_3\!-\!\langle\rangle\!-\!\overset{+}{N}H_3$	5.084	8.24×10^{-6}
2-Methylphenol (o-cresol)		10.09	8.1×10^{-11}
4-Methylphenol (p-cresol)	$CH_3\!-\!\langle\rangle\!-\!OH$	10.26	5.5×10^{-11}
Morpholine (perhydro-1,4-oxazine)	$O\langle\rangle\overset{+}{N}H_2$	8.492	3.22×10^{-9}
1-Naphthoic acid		3.70	2.0×10^{-4}
2-Naphthoic acid		4.16	6.9×10^{-5}
1-Naphthol		9.34	4.6×10^{-10}

(Continued)

Name	Structure	pK_a	K_a
2-Naphthol		9.51	3.1×10^{-10}
Nitrilotriacetic acid	$\overset{+}{H}N(CH_2CO_2H)_3$	1.1 (CO_2H) ($20°$, $\mu = 1.0$) 1.650 (CO_2H) ($20°$) 2.940 (CO_2H) ($20°$) 10.334 (NH) ($20°$)	8×10^{-2} 2.24×10^{-2} 1.15×10^{-3} 4.63×10^{-11}
2-Nitrobenzoic acid		2.179	6.62×10^{-3}
3-Nitrobenzoic acid		3.449	3.56×10^{-4}
4-Nitrobenzoic acid	$O_2N{-}\bigcirc{-}CO_2H$	3.442	3.61×10^{-4}
Nitroethane	$CH_3CH_2NO_2$	8.57	2.7×10^{-9}
2-Nitrophenol		7.21	6.2×10^{-8}
3-Nitrophenol		8.39	4.1×10^{-9}
4-Nitrophenol	$O_2N{-}\bigcirc{-}OH$	7.15	7.1×10^{-8}
2,4-Dinitrophenol	$O_2N{-}\bigcirc{-}OH$	4.11	7.8×10^{-5}
N-Nitrosophenylhydroxylamine (cupferron)		4.16 ($\mu = 0.1$)	6.9×10^{-5}
Nitrous acid	$HON{=}O$	3.15	7.1×10^{-4}
Oxalic acid (ethanedioic acid)	HO_2CCO_2H	1.252 4.266	5.60×10^{-2} 5.42×10^{-5}
Oxoacetic acid (glyoxylic acid)	$H\overset{O}{\overset{\|}{C}}CO_2H$	3.46	3.5×10^{-4}
Oxobutanedioic acid (oxaloacetic acid)	$HO_2CCH_2\overset{O}{\overset{\|}{C}}CO_2H$	2.56 4.37	2.8×10^{-3} 4.3×10^{-5}

(Continued)

Name	Structure	pK_a	K_a
2-Oxopentanedioic (α-ketoglutaric acid)	$\overset{O}{\overset{\|}{HO_2CCH_2CH_2CCO_2H}}$	1.85 ($\mu = 0.5$) 4.44 ($\mu = 0.5$)	1.41×10^{-2} 3.6×10^{-2}
2-Oxopropanoic acid (pyruvic acid)	$\overset{O}{\overset{\|}{CH_3CCO_2H}}$	2.55	2.8×10^{-3}
1,5-Pentanedioic acid (glutaric acid)	$HO_2CCH_2CH_2CH_2CO_2H$	4.34 5.43	4.6×10^{-5} 3.7×10^{-6}
Pentanoic acid (valeric acid)	$CH_3CH_2CH_2CH_2CO_2H$	4.843	1.44×10^{-5}
1,10-Phenanthroline		4.86	1.38×10^{-5}
Phenylacetic acid	$-CH_2CO_2H$	4.310	4.90×10^{-5}
Phenylalanine	$\overset{NH_3^+}{\underset{CO_2H}{CHCH_2-}}$	2.20 (CO_2H) 9.31 (NH_3)	6.3×10^{-3} 4.9×10^{-10}
Phosphoric acid[†] (hydrogen phosphate)	$\overset{O}{\overset{\|}{HO-P-OH}}\underset{OH}{}$	2.148 7.199 12.15	7.11×10^{-3} 6.32×10^{-8} 7.1×10^{-13}
Phosphorous acid (hydrogen phosphite)	$\overset{O}{\overset{\|}{HP-OH}}\underset{OH}{}$	1.5 6.79	3×10^{-2} 1.62×10^{-7}
Phthalic acid (benzene-1,2-dicarboxylic acid)	$\overset{CO_2H}{\underset{CO_2H}{}}$	2.950 5.408	1.12×10^{-3} 3.90×10^{-6}
Piperazine (perhydro-1,4-diazine)	$H_2^+N\quad NH_2^+$	5.333 9.731	4.65×10^{-6} 1.86×10^{-10}
Piperidine	NH_2^+	11.123	7.53×10^{-12}
Proline	$\overset{}{\underset{N}{\underset{+H_2}{}}}-CO_2H$	1.952 (CO_2H) 10.640 (NH_2)	1.12×10^{-2} 2.29×10^{-11}
Propanoic acid	$CH_3CH_2CO_2H$	4.874	1.34×10^{-5}
Propenoic acid (acrylic acid)	$H_2C=CHCO_2H$	4.258	5.52×10^{-5}
Propylamine	$CH_3CH_2CH_2NH_3^+$	10.566	2.72×10^{-11}

[†] pK_3 from A. G. Miller and J. W. Macklin, *Anal. Chem.*, **55**, 684 (1983).

(Continued)

Name	Structure	pK_a	K_a
Pyridine (azine)		5.229	5.90×10^{-6}
Pyridine-2-carboxylic acid (picolinic acid)		1.01 5.39	9.8×10^{-2} 4.1×10^{-6}
Pyridine-3-carboxylic acid (nicotinic acid)		2.05 4.81	8.9×10^{-3} 1.55×10^{-5}
Pyridoxal-5-phosphate		1.4 (POH) ($\mu = 0.1$) 3.44 (OH) ($\mu = 0.1$) 6.01 (POH) ($\mu = 0.1$) 8.45 (NH) ($\mu = 0.1$)	0.04 3.6×10^{-4} 9.8×10^{-7} 3.5×10^{-9}
Pyrophosphoric acid (hydrogen diphosphate)	$(HO)_2POP(OH)_2$	0.8 2.2 6.70 9.40	0.16 6×10^{-3} 2.0×10^{-7} 4.0×10^{-10}
Pyrrolidine		11.305	4.95×10^{-12}
Serine		2.187 9.209	6.50×10^{-3} 6.18×10^{-10}
Succinic acid (butanedioic acid)	$HO_2CCH_2CH_2CO_2H$	4.207 5.636	6.21×10^{-5} 2.31×10^{-6}
Sulfuric acid (hydrogen sulfate)		1.99 (pK_2)	1.02×10^{-2}
Sulfurous acid (hydrogen sulfite)	$HOSOH$	1.91 7.18	1.23×10^{-2} 6.6×10^{-8}

(Continued)

Name	Structure	pK_a	K_a
Thiosulfuric acid (hydrogen thiosulfate)	$\begin{array}{c} O \\ \parallel \\ HOSOH \\ \parallel \\ S \end{array}$	0.6 1.6	0.3 3×10^{-2}
Threonine	$\begin{array}{c} NH_3^+ \\ \mid \\ CHCHOHCH_3 \\ \mid \\ CO_2H \end{array}$	2.088 (CO_2H) 9.100 (NH_3)	8.17×10^{-3} 7.94×10^{-10}
Trichloroacetic acid	Cl_3CCO_2H	0.66 ($\mu = 0.1$)	0.22
Triethanolamine	$(HOCH_2CH_2)_3NH^+$	7.762	1.73×10^{-8}
Triethylamine	$(CH_3CH_2)_3NH^+$	10.715	1.93×10^{-11}
1,2,3-Trihydroxybenzene (pyrogallol)	(benzene ring with OH, OH, OH)	8.94 11.08 (14)	1.15×10^{-9} 8.3×10^{-12} 10^{-14}
Trimethylamine	$(CH_3)_3NH^+$	9.800	1.58×10^{-10}
Tris(hydroxymethyl)aminomethane (TRIS or THAM)	$(HOCH_2)_3CNH_3^+$	8.075	8.41×10^{-9}
Tryptophan	$\begin{array}{c} NH_3^+ \\ \mid \\ CHCH_2 \\ \mid \\ CO_2H \end{array}$ (indole ring)	2.35 (CO_2H) ($\mu = 0.1$) 9.33 (NH_3) ($\mu = 0.1$)	4.5×10^{-3} 4.7×10^{-10}
Tyrosine	$\begin{array}{c} NH_3^+ \\ \mid \\ CHCH_2 \\ \mid \\ CO_2H \end{array}$ (benzene ring with OH)	2.17 (CO_2H) ($\mu = 0.1$) 9.19 (NH_3) 10.47 (OH)	6.8×10^{-3} 6.5×10^{-10} 3.4×10^{-11}
Valine	$\begin{array}{c} NH_3^+ \\ \mid \\ CHCH(CH_3)_2 \\ \mid \\ CO_2H \end{array}$	2.286 (CO_2H) 9.718 (NH_3)	5.18×10^{-3} 1.91×10^{-10}
Water[†]	H_2O	13.996	1.01×10^{-14}

[†] The constant given for water is K_w.

H Standard Reduction Potentials[†]

Reaction	$E°$ (volts)	$dE°/dT$ (mV/K)
Aluminum		
$Al^{3+} + 3e^- \rightleftharpoons Al(s)$	-1.677	0.533
$AlCl^{2+} + 3e^- \rightleftharpoons Al(s) + Cl^-$	-1.802	
$AlF_6^{3-} + 3e^- \rightleftharpoons Al(s) + 6F^-$	-2.069	
$Al(OH)_4^- + 3e^- \rightleftharpoons Al(s) + 4OH^-$	-2.328	-1.13
Antimony		
$SbO^+ + 2H^+ + 3e^- \rightleftharpoons Sb(s) + H_2O$	0.208	
$Sb_2O_3(s) + 6H^+ + 6e^- \rightleftharpoons 2Sb(s) + 3H_2O$	0.147	-0.369
$Sb(s) + 3H^+ + 3c^- \rightleftharpoons SbH_3(g)$	-0.510	-0.030
Arsenic		
$H_3AsO_4 + 2H^+ + 2e^- \rightleftharpoons H_3AsO_3 + H_2O$	0.575	-0.257
$H_3AsO_3 + 3H^+ + 3e^- \rightleftharpoons As(s) + 3H_2O$	$0.247\ 5$	-0.505
$As(s) + 3H^+ + 3e^- \rightleftharpoons AsH_3(g)$	-0.238	-0.029
Barium		
$Ba^{2+} + 2e^- + Hg \rightleftharpoons Ba(\text{in Hg})$	-1.717	
$Ba^{2+} + 2e^- \rightleftharpoons Ba(s)$	-2.906	-0.401
Beryllium		
$Be^{2+} + 2e^- \rightleftharpoons Be(s)$	-1.968	0.60
Bismuth		
$Bi^{3+} + 3e^- \rightleftharpoons Bi(s)$	0.308	0.18
$BiCl_4^- + 3e^- \rightleftharpoons Bi(s) + 4Cl^-$	0.16	
$BiOCl(s) + 2H^+ + 3e^- \rightleftharpoons Bi(s) + H_2O + Cl^-$	0.160	
Boron		
$2B(s) + 6H^+ + 6e^- \rightleftharpoons B_2H_6(g)$	-0.150	-0.296
$B_4O_7^{2-} + 14H^+ + 12e^- \rightleftharpoons 4B(s) + 7H_2O$	-0.792	
$B(OH)_3 + 3H^+ + 3e^- \rightleftharpoons B(s) + 3H_2O$	-0.889	-0.492
Bromine		
$BrO_4^- + 2H^+ + 2e^- \rightleftharpoons BrO_3^- + H_2O$	1.745	-0.511
$HOBr + H^+ + e^- \rightleftharpoons \frac{1}{2}Br_2 + H_2O$	1.584	-0.75
$BrO_3^- + 6H^+ + 5e^- \rightleftharpoons \frac{1}{2}Br_2(l) + 3H_2O$	1.513	-0.419
$Br_2(aq) + 2e^- \rightleftharpoons 2Br^-$	1.098	-0.499
$Br_2(l) + 2e^- \rightleftharpoons 2Br^-$	1.078	-0.611
$Br_3^- + 2e^- \rightleftharpoons 3Br^-$	1.062	-0.512
$BrO^- + H_2O + 2e^- \rightleftharpoons Br^- + 2OH^-$	0.766	-0.94
$BrO_3^- + 3H_2O + 6e^- \rightleftharpoons Br^- + 6OH^-$	0.613	-1.287

[†] All species are aqueous unless otherwise indicated. The reference state for amalgams is an infinitely dilute solution of the element in Hg. The temperature coefficient, $dE°/dT$, allows us to calculate the standard potential, $E°(T)$, at temperature T:
$E°(T) = E° + (dE°/dT) \Delta T$, where ΔT is $T - 298.15$ K. Note the units mV/K for $dE°/dT$.

Source: The most authoritative source is S. G. Bratsch, *J. Phys. Chem. Ref. Data*, **18**, 1 (1989). Additional data come from L. G. Sillen and A. E. Martell, *Stability Constants of Metal-Ion Complexes* (London: The Chemical Society, Special Publications No. 17 and 25, 1964 and 1971); G. Milazzo and S. Caroli, *Tables of Standard Electrode Potentials* (New York: Wiley, 1978); T. Mussini, P. Longhi, and S. Rondinini, *Pure Appl. Chem.*, **57**, 169 (1985). Another good source is A. J. Bard, R. Parsons, and J. Jordan, *Standard Potentials in Aqueous Solution* (New York: Marcel Dekker, 1985).

Reaction	$E°$ (volts)	$dE°/dT$ (mV/K)
Cadmium		
$Cd^{2+} + 2e^- + Hg \rightleftharpoons Cd(\text{in } Hg)$	-0.380	
$Cd^{2+} + 2e^- \rightleftharpoons Cd(s)$	-0.402	-0.029
$Cd(C_2O_4)(s) + 2e^- \rightleftharpoons Cd(s) + C_2O_4^{2-}$	-0.522	
$Cd(C_2O_4)_2^{2-} + 2e^- \rightleftharpoons Cd(s) + 2C_2O_4^{2-}$	-0.572	
$Cd(NH_3)_4^{2+} + 2e^- \rightleftharpoons Cd(s) + 4NH_3$	-0.613	
$CdS(s) + 2e^- \rightleftharpoons Cd(s) + S^{2-}$	-1.175	
Calcium		
$Ca(s) + 2H^+ + 2e^- \rightleftharpoons CaH_2(s)$	0.776	
$Ca^{2+} + 2e^- + Hg \rightleftharpoons Ca(\text{in } Hg)$	-2.003	
$Ca^{2+} + 2e^- \rightleftharpoons Ca(s)$	-2.868	-0.186
$Ca(\text{acetate})^+ + 2e^- \rightleftharpoons Ca(s) + \text{acetate}^-$	-2.891	
$CaSO_4(s) + 2e^- \rightleftharpoons Ca(s) + SO_4^{2-}$	-2.936	
$Ca(\text{malonate})(s) + 2e^- \rightleftharpoons Ca(s) + \text{malonate}^{2-}$	-3.608	
Carbon		
$C_2H_2(g) + 2H^+ + 2e^- \rightleftharpoons C_2H_4(g)$	0.731	
$O{=}\langle\bigcirc\rangle{=}O + 2H^+ + 2e^- \rightleftharpoons HO{-}\langle\bigcirc\rangle{-}OH$	0.700	
$CH_3OH + 2H^+ + 2e^- \rightleftharpoons CH_4(g) + H_2O$	0.583	-0.039
$\text{Dehydroascorbic acid} + 2H^+ + 2e^- \rightleftharpoons \text{ascorbic acid}$	0.390	
$(CN)_2 + 2H^+ + 2e^- \rightleftharpoons 2HCN$	0.373	
$H_2CO + 2H^+ + 2e^- \rightleftharpoons CH_3OH$	0.237	-0.51
$C(s) + 4H^+ + 4e^- \rightleftharpoons CH_4(g)$	$0.131\,5$	$-0.209\,2$
$HCO_2H + 2H^+ + 2e^- \rightleftharpoons H_2CO + H_2O$	-0.029	-0.63
$CO_2(g) + 2H^+ + 2e^- \rightleftharpoons CO(g) + H_2O$	$-0.103\,8$	$-0.397\,7$
$CO_2(g) + 2H^+ + 2e^- \rightleftharpoons HCO_2H$	-0.114	-0.94
$2CO_2(g) + 2H^+ + 2e^- \rightleftharpoons H_2C_2O_4$	-0.432	-1.76
Cerium		
$Ce^{4+} + e^- \rightleftharpoons Ce^{3+}$ $ClO_2 + e^- \rightleftharpoons ClO_2^-$	$\left\{ \begin{array}{ll} 1.72 \\ 1.70 & 1\,\text{F}\,HClO_4 \\ 1.44 & 1\,\text{F}\,H_2SO_4 \\ 1.61 & 1\,\text{F}\,HNO_3 \\ 1.28 & 1\,\text{F}\,HCl \end{array} \right.$	1.54
$Ce^{3+} + 3e^- \rightleftharpoons Ce(s)$	-2.336	0.280
Cesium		
$Cs^+ + e^- + Hg \rightleftharpoons Cs(\text{in } Hg)$	-1.950	
$Cs^+ + e^- \rightleftharpoons Cs(s)$	-3.026	-1.172
Chlorine		
$HClO_2 + 2H^+ + 2e^- \rightleftharpoons HOCl + H_2O$	1.674	0.55
$HClO + H^+ + e^- \rightleftharpoons \frac{1}{2}Cl_2(g) + H_2O$	1.630	-0.27
$ClO_3^- + 6H^+ + 5e^- \rightleftharpoons \frac{1}{2}Cl_2(g) + 3H_2O$	1.458	-0.347
$Cl_2(aq) + 2e^- \rightleftharpoons 2Cl^-$	1.396	-0.72
$Cl_2(g) + 2e^- \rightleftharpoons 2Cl^-$	$1.360\,4$	-1.248
$ClO_4^- + 2H^+ + 2e^- \rightleftharpoons ClO_3^- + H_2O$	1.226	-0.416
$ClO_3^- + 3H^+ + 2e^- \rightleftharpoons HClO_2 + H_2O$	1.157	-0.180
$ClO_3^- + 2H^+ + e^- \rightleftharpoons ClO_2 + H_2O$	1.130	0.074
$ClO_2 + e^- \rightleftharpoons ClO_2^-$	1.068	-1.335
Chromium		
$Cr_2O_7^{2-} + 14H^+ + 6e^- \rightleftharpoons 2Cr^{3+} + 7H_2O$	1.36	-1.32
$CrO_4^{2-} + 4H_2O + 3e^- \rightleftharpoons Cr(OH)_3\ (s, \text{hydrated}) + 5OH^-$	-0.12	-1.62
$Cr^{3+} + e^- \rightleftharpoons Cr^{2+}$	-0.42	1.4
$Cr^{3+} + 3e^- \rightleftharpoons Cr(s)$	-0.74	0.44
$Cr^{2+} + 2e^- \rightleftharpoons Cr(s)$	-0.89	-0.04

(Continued)

Reaction	$E°$ (volts)		$dE°/dT$ (mV/K)
Cobalt			
$Co^{3+} + e^- \rightleftharpoons Co^{2+}$	1.92		1.23
	1.817	8F H_2SO_4	
	1.850	4F HNO_3	
$Co(NH_3)_5(H_2O)^{3+} + e^- \rightleftharpoons Co(NH_3)_5(H_2O)^{2+}$	0.37	1F NH_4NO_3	
$Co(NH_3)_6^{3+} + e^- \rightleftharpoons Co(NH_3)_6^{2+}$	0.1		
$CoOH^+ + H^+ + 2e^- \rightleftharpoons Co(s) + H_2O$	0.003		-0.04
$Co^{2+} + 2e^- \rightleftharpoons Co(s)$	-0.282		0.065
$Co(OH)_2(s) + 2e^- \rightleftharpoons Co(s) + 2OH^-$	-0.746		-1.02
Copper			
$Cu^+ + e^- \rightleftharpoons Cu(s)$	0.518		-0.754
$Cu^{2+} + 2e^- \rightleftharpoons Cu(s)$	0.339		0.011
$Cu^{2+} + e^- \rightleftharpoons Cu^+$	0.161		0.776
$CuCl(s) + e^- \rightleftharpoons Cu(s) + Cl^-$	0.137		
$Cu(IO_3)_2(s) + 2e^- \rightleftharpoons Cu(s) + 2IO_3^-$	-0.079		
$Cu(ethylenediamine)_2^+ + e^- \rightleftharpoons Cu(s) + 2\ ethylenediamine$	-0.119		
$CuI(s) + e^- \rightleftharpoons Cu(s) + I^-$	-0.185		
$Cu(EDTA)^{2-} + 2e^- \rightleftharpoons Cu(s) + EDTA^{4-}$	-0.216		
$Cu(OH)_2(s) + 2e^- \rightleftharpoons Cu(s) + 2OH^-$	-0.222		
$Cu(CN)_2^- + e^- \rightleftharpoons Cu(s) + 2CN^-$	-0.429		
$CuCN(s) + e^- \rightleftharpoons Cu(s) + CN^-$	-0.639		
Dysprosium			
$Dy^{3+} + 3e^- \rightleftharpoons Dy(s)$	-2.295		0.373
Erbium			
$Er^{3+} + 3e^- \rightleftharpoons Er(s)$	-2.331		0.388
Europium			
$Eu^{3+} + e^- \rightleftharpoons Eu^{2+}$	-0.35		1.53
$Eu^{3+} + 3e^- \rightleftharpoons Eu(s)$	-1.991		0.338
$Eu^{2+} + 2e^- \rightleftharpoons Eu(s)$	-2.812		-0.26
Fluorine			
$F_2(g) + 2e^- \rightleftharpoons 2F^-$	2.890		-1.870
$F_2O(g) + 2H^+ + 4e^- \rightleftharpoons 2F^- + H_2O$	2.168		-1.208
Gadolinium			
$Gd^{3+} + 3e^- \rightleftharpoons Gd(s)$	-2.279		0.315
Gallium			
$Ga^{3+} + 3e^- \rightleftharpoons Ga(s)$	-0.549		0.61
$GaOOH(s) + 3H_2O + 3e^- \rightleftharpoons Ga(s) + 3OH^-$	-1.320		-1.08
Germanium			
$Ge^{2+} + 2e^- \rightleftharpoons Ge(s)$	0.1		
$H_4GeO_4 + 4H^+ + 4e^- \rightleftharpoons Ge(s) + 4H_2O$	-0.039		-0.429
Gold			
$Au^+ + e^- \rightleftharpoons Au(s)$	1.69		-1.1
$Au^{3+} + 2e^- \rightleftharpoons Au^+$	1.41		
$AuCl_2^- + e^- \rightleftharpoons Au(s) + 2Cl^-$	1.154		
$AuCl_4^- + 2e^- \rightleftharpoons AuCl_2^- + 2Cl^-$	0.926		
Hafnium			
$Hf^{4+} + 4e^- \rightleftharpoons Hf(s)$	-1.55		0.68
$HfO_2(s) + 4H^+ + 4e^- \rightleftharpoons Hf(s) + 2H_2O$	-1.591		-0.355
Holmium			
$Ho^{3+} + 3e^- \rightleftharpoons Ho(s)$	-2.33		0.371
Hydrogen			
$2H^+ + 2e^- \rightleftharpoons H_2(g)$	0.000 0		0
$H_2O + e^- = \frac{1}{2}H_2(g) + OH^-$	$-0.828\ 0$		$-0.836\ 0$

(Continued)

Reaction	$E°$ (volts)		$dE°/dT$ (mV/K)

Indium

$In^{3+} + 3e^- + Hg \rightleftharpoons In(in\ Hg)$	-0.313		
$In^{3+} + 3e^- \rightleftharpoons In(s)$	-0.338		0.42
$In^{3+} + 2e^- \rightleftharpoons In^+$	-0.444		
$In(OH)_3(s) + 3e^- \rightleftharpoons In(s) + 3OH^-$	-0.99		-0.95

Iodine

$IO_4^- + 2H^+ + 2e^- \rightleftharpoons IO_3^- + H_2O$	1.589		-0.85
$H_5IO_6 + 2H^+ + 2e^- \rightleftharpoons HIO_3 + 3H_2O$	1.567		-0.12
$HOI + H^+ + e^- \rightleftharpoons \frac{1}{2}I_2(s) + H_2O$	1.430		-0.339
$ICl_3(s) + 3e^- \rightleftharpoons \frac{1}{2}I_2(s) + 3Cl^-$	1.28		
$ICl(s) + e^- \rightleftharpoons \frac{1}{2}I_2(s) + Cl^-$	1.22		
$IO_3^- + 6H^+ + 5e^- \rightleftharpoons \frac{1}{2}I_2(s) + 3H_2O$	1.210		-0.367
$IO_3^- + 5H^+ + 4e^- \rightleftharpoons HOI + 2H_2O$	1.154		-0.374
$I_2(aq) + 2e^- \rightleftharpoons 2I^-$	0.620		-0.234
$I_2(s) + 2e^- \rightleftharpoons 2I^-$	0.535		-0.125
$I_3^- + 2e^- \rightleftharpoons 3I^-$	0.535		-0.186
$IO_3^- + 3H_2O + 6e^- \rightleftharpoons I^- + 6OH^-$	0.269		-1.163

Iridium

$IrCl_6^{2-} + e^- \rightleftharpoons IrCl_6^{3-}$	1.026	1F HCl	
$IrBr_6^{2-} + e^- \rightleftharpoons IrBr_6^{3-}$	0.947	2F NaBr	
$IrCl_6^{2-} + 4e^- \rightleftharpoons Ir(s) + 6Cl^-$	0.835		
$IrO_2(s) + 4H^+ + 4e^- \rightleftharpoons Ir(s) + 2H_2O$	0.73		-0.36
$IrI_6^{2-} + e^- \rightleftharpoons IrI_6^{3-}$	0.485	1F KI	

Iron

$Fe(phenanthroline)_3^{3+} + e^- \rightleftharpoons Fe(phenanthroline)_3^{2+}$	1.147		
$Fe(bipyridyl)_2^{3+} + e^- \rightleftharpoons Fe(bipyridyl)_3^{2+}$	1.120		
$FeOH^{2+} + H^+ + e^- \rightleftharpoons Fe^{2+} + H_2O$	0.900		0.096
$FeO_4^{2-} + 3H_2O + 3e^- \rightleftharpoons FeOOH(s) + 5OH^-$	0.80		-1.59
$Fe^{3+} + e^- \rightleftharpoons Fe^{2+}$	$\begin{cases} 0.771 \\ 0.732 \\ 0.767 \\ 0.746 \end{cases}$	1F HCl 1F HClO$_4$ 1F HNO$_3$	1.175
$FeOOH(s) + 3H^+ + e^- \rightleftharpoons Fe^{2+} + 2H_2O$	0.74		-1.05
$ferricinium^+ + e^- \rightleftharpoons ferrocene$	0.400		
$Fe(CN)_6^{3-} + e^- \rightleftharpoons Fe(CN)_6^{4-}$	0.356		
$Fe(glutamate)^{3+} + e^- \rightleftharpoons Fe(glutamate)^{2+}$	0.240		
$FeOH^+ + H^+ + 2e^- \rightleftharpoons Fe(s) + H_2O$	-0.16		0.07
$Fe^{2+} + 2e^- \rightleftharpoons Fe(s)$	-0.44		0.07

Lanthanum

$La^{3+} + 3e^- \rightleftharpoons La(s)$	-2.379		0.242
$La(succinate)^+ + 3e^- \rightleftharpoons La(s) + succinate^{2-}$	-2.601		

Lead

$Pb^{4+} + 2e^- \rightleftharpoons Pb^{2+}$	1.69	1F HNO$_3$	
$PbO_2(s) + 4H^+ + SO_4^{2-} + 2e^- \rightleftharpoons PbSO_4(s) + 2H_2O$	1.685		
$PbO_2(s) + 4H^+ + 2e^- \rightleftharpoons Pb^{2+} + 2H_2O$	1.458		-0.253
$3PbO_2(s) + 2H_2O + 4e^- \rightleftharpoons Pb_3O_4(s) + 4OH^-$	0.269		-1.136
$Pb_3O_4(s) + H_2O + 2e^- \rightleftharpoons 3PbO(s,\ red) + 2OH^-$	0.224		-1.211
$Pb_3O_4(s) + H_2O + 2e^- \rightleftharpoons 3PbO(s,\ yellow) + 2OH^-$	0.207		-1.177
$Pb^{2+} + 2e^- \rightleftharpoons Pb(s)$	-0.126		-0.395
$PbF_2(s) + 2e^- \rightleftharpoons Pb(s) + 2F^-$	-0.350		
$PbSO_4(s) + 2e^- \rightleftharpoons Pb(s) + SO_4^{2-}$	-0.355		

Lithium

$Li^+ + e^- + Hg \rightleftharpoons Li(in\ Hg)$	-2.195		
$Li^+ + e^- \rightleftharpoons Li(s)$	-3.040		-0.514

(Continued)

Reaction	$E°$ (volts)	$dE°/dT$ (mV/K)
Lutetium		
$Lu^{3+} + 3e^- \rightleftharpoons Lu(s)$	-2.28	0.412
Magnesium		
$Mg^{2+} + 2e^- + Hg \rightleftharpoons Mg(in\ Hg)$	-1.980	
$Mg(OH)^+ + H^+ + 2e^- \rightleftharpoons Mg(s) + H_2O$	-2.022	0.25
$Mg^{2+} + 2e^- \rightleftharpoons Mg(s)$	-2.360	0.199
$Mg(C_2O_4)(s) + 2e^- \rightleftharpoons Mg(s) + C_2O_4^{2-}$	-2.493	
$Mg(OH)_2(s) + 2e^- \rightleftharpoons Mg(s) + 2OH^-$	-2.690	-0.946
Manganese		
$MnO_4^- + 4H^+ + 3e^- \rightleftharpoons MnO_2(s) + 2H_2O$	1.692	-0.671
$Mn^{3+} + e^- \rightleftharpoons Mn^{2+}$	1.56	1.8
$MnO_4^- + 8H^+ + 5e^- \rightleftharpoons Mn^{2+} + 4H_2O$	1.507	-0.646
$Mn_2O_3(s) + 6H^+ + 2e^- \rightleftharpoons 2Mn^{2+} + 3H_2O$	1.485	-0.926
$MnO_2(s) + 4H^+ + 2e^- \rightleftharpoons Mn^{2+} + 2H_2O$	1.230	-0.609
$Mn(EDTA)^- + e^- \rightleftharpoons Mn(EDTA)^{2-}$	0.825	-1.10
$MnO_4^- + e^- \rightleftharpoons MnO_4^{2-}$	0.56	-2.05
$3Mn_2O_3(s) + H_2O + 2e^- \rightleftharpoons 2Mn_3O_4(s) + 2OH^-$	0.002	-1.256
$Mn_3O_4(s) + 4H_2O + 2e^- \rightleftharpoons 3Mn(OH)_2(s) + 2OH^-$	-0.352	-1.61
$Mn^{2+} + 2e^- \rightleftharpoons Mn(s)$	-1.182	-1.129
$Mn(OH)_2(s) + 2e^- \rightleftharpoons Mn(s) + 2OH^-$	-1.565	-1.10
Mercury		
$2Hg^{2+} + 2e^- \rightleftharpoons Hg_2^{2+}$	0.908	0.095
$Hg^{2+} + 2e^- \rightleftharpoons Hg(l)$	0.852	-0.116
$Hg_2^{2+} + 2e^- \rightleftharpoons 2Hg(l)$	0.796	-0.327
$Hg_2SO_4(s) + 2e^- \rightleftharpoons 2Hg(l) + SO_4^{2-}$	0.614	
$Hg_2Cl_2(s) + 2e^- \rightleftharpoons 2Hg(l) + 2Cl^-$	$\begin{cases} 0.268 \\ 0.241\ \text{(saturated calomel electrode)} \end{cases}$	
$Hg(OH)_3^- + 2e^- \rightleftharpoons Hg(l) + 3OH^-$	0.231	
$Hg(OH)_2 + 2e^- \rightleftharpoons Hg(l) + 2OH^-$	0.206	-1.24
$Hg_2Br_2(s) + 2e^- \rightleftharpoons 2Hg(l) + 2Br^-$	0.140	
$HgO(s, yellow) + H_2O + 2e^- \rightleftharpoons Hg(l) + 2OH^-$	$0.098\ 3$	-1.125
$HgO(s, red) + H_2O + 2e^- \rightleftharpoons Hg(l) + 2OH^-$	$0.097\ 7$	$-1.120\ 6$
Molybdenum		
$MoO_4^{2-} + 2H_2O + 2e^- \rightleftharpoons MoO_2(s) + 4OH^-$	-0.818	-1.69
$MoO_4^{2-} + 4H_2O + 6e^- \rightleftharpoons Mo(s) + 8OH^-$	-0.926	-1.36
$MoO_2(s) + 2H_2O + 4e^- \rightleftharpoons Mo(s) + 4OH^-$	-0.980	-1.196
Neodymium		
$Nd^{3+} + 3e^- \rightleftharpoons Nd(s)$	-2.323	0.282
Neptunium		
$NpO_3^+ + 2H^+ + e^- \rightleftharpoons NpO_2^{2+} + H_2O$	2.04	
$NpO_2^{2+} + 2e^- \rightleftharpoons NpO_2^+$	1.236	0.058
$NpO_2^+ + 4H^+ + e^- \rightleftharpoons Np^{4+} + 2H_2O$	0.567	-3.30
$Np^{4+} + e^- \rightleftharpoons Np^{3+}$	0.157	1.53
$Np^{3+} + 3e^- \rightleftharpoons Np(s)$	-1.768	0.18
Nickel		
$NiOOH(s) + 3H^+ + e^- \rightleftharpoons Ni^{2+} + 2H_2O$	2.05	-1.17
$Ni^{2+} + 2e^- \rightleftharpoons Ni(s)$	-0.236	0.146
$Ni(CN)_4^{2-} + e^- \rightleftharpoons Ni(CN)_3^{2-} + CN^-$	-0.401	
$Ni(OH)_2(s) + 2e^- \rightleftharpoons Ni(s) + 2OH^-$	-0.714	-1.02
Niobium		
$\frac{1}{2}Nb_2O_5(s) + H^+ + e^- \rightleftharpoons NbO_2(s) + \frac{1}{2}H_2O$	-0.248	-0.460
$\frac{1}{2}Nb_2O_5(s) + 5H^+ + 5e^- \rightleftharpoons Nb(s) + \frac{5}{2}H_2O$	-0.601	-0.381
$NbO_2(s) + 2H^+ + 2e^- \rightleftharpoons NbO(s) + H_2O$	-0.646	-0.347
$NbO_2(s) + 4H^+ + 4e^- \rightleftharpoons Nb(s) + 2H_2O$	-0.690	-0.361

(Continued)

Reaction	$E°$ (volts)	$dE°/dT$ (mV/K)
Nitrogen		
$HN_3 + 3H^+ + 2e^- \rightleftharpoons N_2(g) + NH_4^+$	2.079	0.147
$N_2O(g) + 2H^+ + 2e^- \rightleftharpoons N_2(g) + H_2O$	1.769	-0.461
$2NO(g) + 2H^+ + 2e^- \rightleftharpoons N_2O(g) + H_2O$	1.587	-1.359
$NO^+ + e^- \rightleftharpoons NO(g)$	1.46	
$2NH_3OH^+ + H^+ + 2e^- \rightleftharpoons N_2H_5^+ + 2H_2O$	1.40	-0.60
$NH_3OH^+ + 2H^+ + 2e^- \rightleftharpoons NH_4^+ + H_2O$	1.33	-0.44
$N_2H_5^+ + 3H^+ + 2e^- \rightleftharpoons 2NH_4^+$	1.250	-0.28
$HNO_2 + H^+ + e^- \rightleftharpoons NO(g) + H_2O$	0.984	0.649
$NO_3^- + 4H^+ + 3e^- \rightleftharpoons NO(g) + 2H_2O$	0.955	0.028
$NO_3^- + 3H^+ + 2e^- \rightleftharpoons HNO_2 + H_2O$	0.940	-0.282
$NO_3^- + 2H^+ + e^- \rightleftharpoons \frac{1}{2}N_2O_4(g) + H_2O$	0.798	0.107
$N_2(g) + 8H^+ + 6e^- \rightleftharpoons 2NH_4^+$	0.274	-0.616
$N_2(g) + 5H^+ + 4e^- \rightleftharpoons N_2H_5^+$	-0.214	-0.78
$N_2(g) + 2H_2O + 4H^+ + 2e^- \rightleftharpoons 2NH_3OH^+$	-1.83	-0.96
$\frac{3}{2}N_2(g) + H^+ + e^- \rightleftharpoons HN_3$	-3.334	-2.141
Osmium		
$OsO_4(s) + 8H^+ + 8e^- \rightleftharpoons Os(s) + 4H_2O$	0.834	-0.458
$OsCl_6^{2-} + e^- \rightleftharpoons OsCl_6^{3-}$	0.85 1F HCl	
Oxygen		
$OH + H^+ + e^- \rightleftharpoons H_2O$	2.56	-1.0
$O(g) + 2H^+ + 2e^- \rightleftharpoons H_2O$	2.430 1	-1.148 4
$O_3(g) + 2H^+ + 2e^- \rightleftharpoons O_2(g) + H_2O$	2.075	-0.489
$H_2O_2 + 2H^+ + 2e^- \rightleftharpoons 2H_2O$	1.763	-0.698
$HO_2 + H^+ + e^- \rightleftharpoons H_2O_2$	1.44	-0.7
$\frac{1}{2}O_2(g) + 2H^+ + 2e^- \rightleftharpoons H_2O$	1.229 1	-0.845 6
$O_2(g) + 2H^+ + 2e^- \rightleftharpoons H_2O_2$	0.695	-0.993
$O_2(g) + H^+ + e^- \rightleftharpoons HO_2$	-0.05	-1.3
Palladium		
$Pd^{2+} + 2e^- \rightleftharpoons Pd(s)$	0.915	0.12
$PdO(s) + 2H^+ + 2e^- \rightleftharpoons Pd(s) + H_2O$	0.79	-0.33
$PdCl_6^{4-} + 2e^- \rightleftharpoons Pd(s) + 6Cl^-$	0.615	
$PdO_2(s) + H_2O + 2e^- \rightleftharpoons PdO(s) + 2OH^-$	0.64	-1.2
Phosphorus		
$\frac{1}{4}P_4(s, \text{white}) + 3H^+ + 3e^- \rightleftharpoons PH_3(g)$	-0.046	-0.093
$\frac{1}{4}P_4(s, \text{red}) + 3H^+ + 3e^- \rightleftharpoons PH_3(g)$	-0.088	-0.030
$H_3PO_4 + 2H^+ + 2e^- \rightleftharpoons H_3PO_3 + H_2O$	-0.30	-0.36
$H_3PO_4 + 5H^+ + 5e^- \rightleftharpoons \frac{1}{4}P_4(s, \text{white}) + 4H_2O$	-0.402	-0.340
$H_3PO_3 + 2H^+ + 2e^- \rightleftharpoons H_3PO_2 + H_2O$	-0.48	-0.37
$H_3PO_2 + H^+ + e^- \rightleftharpoons \frac{1}{4}P_4(s) + 2H_2O$	-0.51	
Platinum		
$Pt^{2+} + 2e^- \rightleftharpoons Pt(s)$	1.18	-0.05
$PtO_2(s) + 4H^+ + 4e^- \rightleftharpoons Pt(s) + 2H_2O$	0.92	-0.36
$PtCl_4^{2-} + 2e^- \rightleftharpoons Pt(s) + 4Cl^-$	0.755	
$PtCl_6^{2-} + 2e^- \rightleftharpoons PtCl_4^{2-} + 2Cl^-$	0.68	
Plutonium		
$PuO_2^+ + e^- \rightleftharpoons PuO_2(s)$	1.585	0.39
$PuO_2^{2+} + 4H^+ + 2e^- \rightleftharpoons Pu^{4+} + 2H_2O$	1.000	-1.615
$Pu^{4+} + e^- \rightleftharpoons Pu^{3+}$	1.006	1.441
$PuO_2^{2+} + e^- \rightleftharpoons PuO_2^+$	0.966	0.03
$PuO_2(s) + 4H^+ + 4e^- \rightleftharpoons Pu(s) + 2H_2O$	-1.369	-0.38
$Pu^{2+} + 3e^- \rightleftharpoons Pu(s)$	-1.978	0.23
Potassium		
$K^+ + e^- + Hg \rightleftharpoons K(\text{in Hg})$	-1.975	
$K^+ + e^- \rightleftharpoons K(s)$	-2.936	-1.074

(Continued)

Reaction	$E°$ (volts)		$dE°/dT$ (mV/K)

Praseodymium

$Pr^{4+} + e^- \rightleftharpoons Pr^{3+}$	3.2		1.4
$Pr^{3+} + 3e^- \rightleftharpoons Pr(s)$	-2.353		0.291

Promethium

$Pm^3 + 3e^- \rightleftharpoons Pm(s)$	-2.30		0.29

Radium

$Ra^{2+} + 2e^- \rightleftharpoons Ra(s)$	-2.80		-0.44

Rhenium

$ReO_4^- + 2H^+ + e^- \rightleftharpoons ReO_3(s) + H_2O$	0.72		-1.17
$ReO_4^- + 4H^+ + 3e^- \rightleftharpoons ReO_2(s) + 2H_2O$	0.510		-0.70

Rhodium

$Rh^{6+} + 3e^- \rightleftharpoons Rh^{3+}$	1.48	1F $HClO_4$	
$Rh^{4+} + e^- \rightleftharpoons Rh^{3+}$	1.44	3F H_2SO_4	
$RhCl_6^{2-} + e^- \rightleftharpoons RhCl_6^{3-}$	1.2		
$Rh^{3+} + 3e^- \rightleftharpoons Rh(s)$	0.76		0.4
$2Rh^{3+} + e^- \rightleftharpoons Rh_2^{4+}$	0.7		
$RhCl_6^{3-} + 3e^- \rightleftharpoons Rh(s) + 6Cl^-$	0.44		

Rubidium

$Rb^+ + e^- + Hg \rightleftharpoons Rb(in\ Hg)$	-1.970		
$Rb^+ + e^- \rightleftharpoons Rb(s)$	-2.943		-1.140

Ruthenium

$RuO_4^- + 6H^+ + 3e^- \rightleftharpoons Ru(OH)_2^{2+} + 2H_2O$	1.53		
$Ru(dipyridyl)_3^{3+} + e^- \rightleftharpoons Ru(dipyridyl)_3^{2+}$	1.29		
$RuO_4(s) + 8H^+ + 8e^- \rightleftharpoons Ru(s) + 4H_2O$	1.032		-0.467
$Ru^{2+} + 2e^- \rightleftharpoons Ru(s)$	0.8		
$Ru^{3+} + 3e^- \rightleftharpoons Ru(s)$	0.60		
$Ru^{3+} + e^- \rightleftharpoons Ru^{2+}$	0.24		
$Ru(NH_3)_6^{3+} + e^- \rightleftharpoons Ru(NH_3)_6^{2+}$	0.214		

Samarium

$Sm^{3+} + 3e^- \rightleftharpoons Sm(s)$	-2.304		0.279
$Sm^{2+} + 2e^- \rightleftharpoons Sm(s)$	-2.68		-0.28

Scandium

$Sc^{3+} + 3e^- \rightleftharpoons Sc(s)$	-2.09		0.41

Selenium

$SeO_4^{2-} + 4H^+ + 2e^- \rightleftharpoons H_2SeO_3 + H_2O$	1.150		0.483
$H_2SeO_3 + 4H^+ + 4e^- \rightleftharpoons Se(s) + 3H_2O$	0.739		-0.562
$Se(s) + 2H^+ + 2e^- \rightleftharpoons H_2Se(g)$	-0.082		0.238
$Se(s) + 2e^- \rightleftharpoons Se^{2-}$	-0.67		-1.2

Silicon

$Si(s) + 4H^+ + 4e^- \rightleftharpoons SiH_4(g)$	-0.147		-0.196
$SiO_2(s, quartz) + 4H^+ + 4e^- \rightleftharpoons Si(s) + 2H_2O$	-0.990		-0.374
$SiF_6^{2-} + 4e^- \rightleftharpoons Si(s) + 6F^-$	-1.24		

Silver

$Ag^{2+} + e^- \rightleftharpoons Ag^+$	$\begin{cases} 2.000 \\ 1.989 \\ 1.929 \end{cases}$	4F $HClO_4$ 4F HNO_3	0.99
$Ag^{3+} + 2e^- \rightleftharpoons Ag^+$	1.9		
$AgO(s) + H^+ + e^- \rightleftharpoons \frac{1}{2}Ag_2O(s) + \frac{1}{2}H_2O$	1.40		
$Ag^+ + e^- \rightleftharpoons Ag(s)$	0.799 3		-0.989
$Ag_2C_2O_4(s) + 2e^- \rightleftharpoons 2Ag(s) + C_2O_4^{2-}$	0.465		
$AgN_3(s) + e^- \rightleftharpoons Ag(s) + N_3^-$	0.293		

(Continued)

Reaction	$E°$ (volts)		$dE°/dT$ (mV/K)
$AgCl(s) + e^- \rightleftharpoons Ag(s) + Cl^-$	$\begin{cases} 0.222 \\ 0.197 \end{cases}$	saturated KCl	
$AgBr(s) + e^- \rightleftharpoons Ag(s) + Br^-$	0.071		
$Ag(S_2O_3)_2^{3-} + e^- \rightleftharpoons Ag(s) + 2S_2O_3^{2-}$	0.017		
$AgI(s) + e^- \rightleftharpoons Ag(s) + I^-$	-0.152		
$Ag_2S(s) + H^+ + 2e^- \rightleftharpoons 2Ag(s) + SH^-$	-0.272		
Sodium			
$Na^+ + e^- + Hg \rightleftharpoons Na(in\ Hg)$	-1.959		
$Na^+ + \frac{1}{2}H_2(g) + e^- \rightleftharpoons NaH(s)$	-2.367		-1.550
$Na^+ + e^- \rightleftharpoons Na(s)$	$-2.714\ 3$		-0.757
Strontium			
$Sr^{2+} + 2e^- \rightleftharpoons Sr(s)$	-2.889		-0.237
Sulfur			
$S_2O_8^{2-} + 2e^- \rightleftharpoons 2SO_4^{2-}$	2.01		
$S_2O_6^{2-} + 4H^+ + 2e^- \rightleftharpoons 2H_2SO_3$	0.57		
$4SO_2 + 4H^+ + 6e^- \rightleftharpoons S_4O_6^{2-} + 2H_2O$	0.539		-1.11
$SO_2 + 4H^+ + 4e^- \rightleftharpoons S(s)$	0.450		-0.652
$2H_2SO_3 + 2H^+ + 4e^- \rightleftharpoons S_2O_3^{2-} + 3H_2O$	0.40		
$S(s) + 2H^+ + 2e^- \rightleftharpoons H_2S(g)$	0.174		0.224
$S(s) + 2H^+ + 2e^- \rightleftharpoons H_2S(aq)$	0.144		-0.21
$S_4O_6^{2-} + 2H^+ + 2e^- \rightleftharpoons 2HS_2O_3^-$	0.10		-0.23
$5S(s) + 2e^- \rightleftharpoons S_5^{2-}$	-0.340		
$2S(s) + 2e^- \rightleftharpoons S_2^{2-}$	-0.50		-1.16
$2SO_3^{2-} + 3H_2O + 4e^- \rightleftharpoons S_2O_3^{2-} + 6OH^-$	-0.566		-1.06
$SO_3^{2-} + 3H_2O + 4e^- \rightleftharpoons S(s) + 6OH^-$	-0.659		-1.23
$SO_4^{2-} + 4H_2O + 6e^- \rightleftharpoons S(s) + 8OH^-$	-0.751		-1.288
$SO_4^{2-} + H_2O + 2e^- \rightleftharpoons SO_3^{2-} + 2OH^-$	-0.936		-1.41
$2SO_3^{2-} + 2H_2O + 2e^- \rightleftharpoons S_2O_4^{2-} + 4OH^-$	-1.130		-0.85
$2SO_4^{2-} + 2H_2O + 2e^- \rightleftharpoons S_2O_6^{2-} + 4OH^-$	-1.71		-1.00
Tantalum			
$Ta_2O_5(s) + 10H^+ + 10e^- \rightleftharpoons 2Ta(s) + 5H_2O$	-0.752		-0.377
Technetium			
$TcO_4^- + 2H_2O + 3e^- \rightleftharpoons TcO_2(s) + 4OH^-$	-0.366		-1.82
$TcO_4^- + 4H_2O + 7e^- \rightleftharpoons Tc(s) + 8OH^-$	-0.474		-1.46
Tellurium			
$TeO_3^{2-} + 3H_2O + 4e^- \rightleftharpoons Te(s) + 6OH^-$	-0.47		-1.39
$2Te(s) + 2e^- \rightleftharpoons Te_2^{2-}$	-0.84		
$Te(s) + 2e^- \rightleftharpoons Te^{2-}$	-0.90		-1.0
Terbium			
$Tb^{4+} + e^- \rightleftharpoons Tb^{3+}$	3.1		1.5
$Tb^{3+} + 3e^- \rightleftharpoons Tb(s)$	-2.28		0.350
Thallium			
$Tl^{3+} + 2e^- \rightleftharpoons Tl^+$	$\begin{cases} 1.280 \\ 0.77 \\ 1.22 \\ 1.23 \\ 1.26 \end{cases}$	$\begin{matrix} \\ 1F\ HCl \\ 1F\ H_2SO_4 \\ 1F\ HNO_3 \\ 1F\ HClO_4 \end{matrix}$	0.97
$Tl^+ + e^- + Hg \rightleftharpoons Tl(in\ Hg)$	-0.294		
$Tl^+ + e^- \rightleftharpoons Tl(s)$	-0.336		-1.312
$TlCl(s) + e^- \rightleftharpoons Tl(s) + Cl^-$	-0.557		
Thorium			
$Th^{4+} + 4e^- \rightleftharpoons Th(s)$	-1.826		0.557

(Continued)

Reaction	$E°$ (volts)	$dE°/dT$ (mV/K)
Thullium		
$Tm^{3+} + 3e^- \rightleftharpoons Tm(s)$	-2.319	0.394
Tin		
$Sn(OH)_3^+ + 3H^+ + 2e^- \rightleftharpoons Sn^{2+} + 6H_2O$	0.142	
$Sn^{4+} + 2e^- \rightleftharpoons Sn^{2+}$	0.139 1F HCl	
$SnO_2(s) + 4H^+ \rightleftharpoons Sn^{2+} + 2H_2O$	-0.094	-0.31
$Sn^{2+} + 2e^- \rightleftharpoons Sn(s)$	-0.141	-0.32
$SnF_6^{2-} + 4e^- \rightleftharpoons Sn(s) + 6F^-$	-0.25	
$Sn(OH)_6^{2-} + 2e^- \rightleftharpoons Sn(OH)_3^- + 3OH^-$	-0.93	
$Sn(s) + 4H_2O + 4e^- \rightleftharpoons SnH_4(g) + 4OH^-$	-1.316	-1.057
$SnO_2(s) + H_2O + 2e^- \rightleftharpoons SnO(s) + 2OH^-$	-0.961	-1.129
Titanium		
$TiO^{2+} + 2H^+ + e^- \rightleftharpoons Ti^{3+} + H_2O$	0.1	-0.6
$Ti^{3+} + e^- \rightleftharpoons Ti^{2+}$	-0.9	1.5
$TiO_2(s) + 4H^+ + 4e^- \rightleftharpoons Ti(s) + 2H_2O$	-1.076	0.365
$TiF_6^{2-} + 4e^- \rightleftharpoons Ti(s) + 6F^-$	-1.191	
$Ti^{2+} + 2e^- \rightleftharpoons Ti(s)$	-1.60	-0.16
Tungsten		
$W(CN)_8^{3-} + e^- \rightleftharpoons W(CN)_8^{4-}$	0.457	
$W^{6+} + e^- \rightleftharpoons W^{5+}$	0.26 12F HCl	
$WO_3(s) + 6H^+ + 6e^- \rightleftharpoons W(s) + 3H_2O$	-0.091	-0.389
$W^{5+} + e^- \rightleftharpoons W^{4+}$	-0.3 12F HCl	
$WO_2(s) + 2H_2O + 4e^- \rightleftharpoons W(s) + 4OH^-$	-0.982	-1.197
$WO_4^{2-} + 4H_2O + 6e^- \rightleftharpoons W(s) + 8OH^-$	-1.060	-1.36
Uranium		
$UO_2^+ + 4H^+ + e^- \rightleftharpoons U^{4+} + 2H_2O$	0.39	-3.4
$UO_2^{2+} + 4H^+ + 2e^- \rightleftharpoons U^{4+} + 2H_2O$	0.273	-1.582
$UO_2^{2+} + e^- \rightleftharpoons UO_2^+$	0.16	0.2
$U^{4+} + e^- \rightleftharpoons U^{3+}$	-0.577	1.61
$U^{3+} + 3e^- \rightleftharpoons U(s)$	-1.642	0.16
Vanadium		
$VO_2^+ + 2H^+ + e^- \rightleftharpoons VO^{2+} + H_2O$	1.001	-0.901
$VO^{2+} + 2H^+ + e^- \rightleftharpoons V^{3+} + H_2O$	0.337	-1.6
$V^{3+} + e^- \rightleftharpoons V^{2+}$	-0.255	1.5
$V^{2+} + 2e^- \rightleftharpoons V(s)$	-1.125	-0.11
Xenon		
$H_4XeO_6 + 2H^+ + 2e^- \rightleftharpoons XeO_3 + 3H_2O$	2.38	0.0
$XeF_2 + 2H^+ + 2e^- \rightleftharpoons Xe(g) + 2HF$	2.2	
$XeO_3 + 6H^+ + 6e^- \rightleftharpoons Xe(g) + 3H_2O$	2.1	-0.34
Ytterbium		
$Yb^{3+} + 3e^- \rightleftharpoons Yb(s)$	-2.19	0.363
$Yb^{2+} + 2e^- \rightleftharpoons Yb(s)$	-2.76	-0.16
Yttrium		
$Y^{3+} + 3e^- \rightleftharpoons Y(s)$	-2.38	0.034

(Continued)

Reaction	$E°$ (volts)	$dE°/dT$ (mV/K)
Zinc		
$ZnOH^+ + H^+ + 2e^- \rightleftharpoons Zn(s) + 2H_2O$	-0.497	0.03
$Zn^{2+} + 2e^- \rightleftharpoons Zn(s)$	-0.762	0.119
$Zn^{2+} + 2e^- + Hg \rightleftharpoons Zn(\text{in Hg})$	-0.801	
$Zn(NH_3)_4^{2+} + 2e^- \rightleftharpoons Zn(s) + 4NH_3$	-1.04	
$ZnCO_3(s) + 2e^- \rightleftharpoons Zn(s) + CO_3^{2-}$	-1.06	
$Zn(OH)_3^- + 2e^- \rightleftharpoons Zn(s) + 3OH^-$	-1.183	
$Zn(OH)_4^{2-} + 2e^- \rightleftharpoons Zn(s) + 4OH^-$	-1.199	
$Zn(OH)_2(s) + 2e^- \rightleftharpoons Zn(s) + 2OH^-$	-1.249	-0.999
$ZnO(s) + H_2O + 2e^- \rightleftharpoons Zn(s) + 2OH^-$	-1.260	-1.160
$ZnS(s) + 2e^- \rightleftharpoons Zn(s) + S^{2-}$	-1.405	
Zirconium		
$Zr^{4+} + 4e^- \rightleftharpoons Zr(s)$	-1.45	0.67
$ZrO_2(s) + 4H^+ + 4e^- \rightleftharpoons Zr(s) + 2H_2O$	-1.473	-0.344

Stepwise Formation Constants[†]

	$\log K_1$	$\log K_2$	$\log K_3$	$\log K_4$	Temperature (°C)	Ionic strength (M)
Acetate, $CH_3CO_2^-$						
Ag^+	0.73	−0.09			25	0
Ca^{2+}	1.24				25	0
Cd^{2+}	1.93	1.22			25	0
Cu^{2+}	2.23	1.40			25	0
Fe^{2+}	1.82				25	0.5
Fe^{3+}	3.38	3.7	2.6		20	0.1
Mg^{2+}	1.25				25	0
Mn^{2+}	1.40				25	0
Na^+	−0.18				25	0
Ni^{2+}	1.43				25	0
Zn^{2+}	1.28	0.81			20	0.1
Ammonia, NH_3						
Ag^+	3.31	3.92			25	0
Cd^{2+}	2.51	1.96	1.30	0.79	30	0
Co^{2+}	1.99	1.51	0.93	0.64	30	0
	($\log K_5 = 0.06$, $\log K_6 - −0.74$)					
Cu^{2+}	3.99	3.34	2.73	1.97	30	0
Hg^{2+}	8.8	8.7	1.00	0.78	22	2
Ni^{2+}	2.67	2.12	1.61	1.07	30	0
	($\log K_5 = 0.63$, $\log K_6 = −0.09$)					
Zn^{2+}	2.18	2.25	2.31	1.96	30	0
Cyanide, CN^-						
Ag^+	($\log \beta_2 = 20$)		0.95		20	0
Cd^{2+}	5.18	4.42	4.32	3.19	25	?
Cu^+	($\log \beta_2 = 24$)		4.6	1.7	25	0
Ni^{2+}			($\log \beta_4 = 30$)		25	0
Tl^{3+}	13.21	13.29	8.67	7.44	25	4[†]
Zn^{2+}	($\log \beta_2 = 11.07$)		4.98	3.57	25	0
Ethylenediamine (1,2-diaminoethane), $H_2NCH_2CH_2NH_2$						
Ag^+	4.70	3.00	2.0		20	0.1
Cd^{2+}	5.69	4.67	2.44		25	0.5
Cu^{2+}	10.66	9.33			20	0
Hg^{2+}	14.3	9.0	−0.1		25	0.1
Ni^{2+}	7.52	6.32	4.49		20	0
Zn^{2+}	5.77	5.06	3.28		20	0

[†] Stepwise formation constants, K_i, and overall (or cumulative) formation constants, β_i, are defined in Box 5-3. For example, K_2 for the reaction of Ag^+ with acetate refers to the reaction $Ag^+ + 2CH_3CO_2^- \rightleftharpoons Ag(CH_3CO_2)_2^-$. The overall formation constant is related to the stepwise formation constants by $\beta_n = K_1 K_2 \cdots K_n$. Data from L. G. Sillen and A. E. Martell, *Stability Constants of Metal-Ion Complexes* (London: The Chemical Society, Special Publications No. 17 and 25, 1964 and 1971).

[†] J. Blixt, B. Györi, and J. Glaser, *J. Am. Chem. Soc.*, **111**, 7784 (1989).

	$\log K_1$	$\log K_2$	$\log K_3$	$\log K_4$	Temperature (°C)	Ionic strength (M)

Nitrilotriacetate, $N(CH_2CO_2^-)_3$

	$\log K_1$	$\log K_2$	$\log K_3$	$\log K_4$	Temperature (°C)	Ionic strength (M)
Ag^+	5.16				20	0.1
Al^{3+}	9.5				20	0.1
Ba^{2+}	4.83				20	0.1
Ca^{2+}	6.46				20	0.1
Cd^{2+}	10.0	4.6			20	0.1
Co^{2+}	10.0	3.9			20	0.1
Cu^{2+}	11.5	3.3			20	0.1
Fe^{3+}	15.91	8.70			20	0.1
Ga^{3+}	13.6	8.2			20	0.1
In^{3+}	16.9				20	0.1
Mg^{2+}	5.46				20	0.1
Mn^{2+}	7.4				20	0.1
Ni^{2+}	11.54				20	0.1
Pb^{2+}	11.47				20	0.1
Tl^+	4.75				20	0.1
Zn^{2+}	10.44				20	0.1

Oxalate, $^-O_2CCO_2^-$

	$\log K_1$	$\log K_2$	$\log K_3$	$\log K_4$	Temperature (°C)	Ionic strength (M)
Al^{3+}		($\log \beta_3 = 15.60$)			20	0.1
Ba^{2+}	2.31				18	0
Ca^{2+}	1.66	1.03			25	1
Cd^{2+}	3.71				20	0.1
Co^{2+}	4.69	2.46			25	0
Cu^{2+}	6.23	4.04			25	0
Fe^{3+}	7.54	7.05	5.41		?	0.5
Ni^{2+}	5.16	1.3			25	0
Zn^{2+}	4.85	2.7			25	0

1,10-Phenanthroline,

	$\log K_1$	$\log K_2$	$\log K_3$	$\log K_4$	Temperature (°C)	Ionic strength (M)
Ag^+	5.02	7.05			25	0.1
Ca^{2+}	0.7				20	0.1
Cd^{2+}	5.17	4.83	4.26		25	0.1
Co^{2+}	7.02	6.70	6.38		25	0.1
Cu^{2+}	8.82	6.57	5.02		25	0.1
Fe^{2+}	5.86	5.25	10.03		25	0.1
Fe^{3+}		($\log \beta_3 = 14.10$)			25	0.1
Hg^{2+}	($\log \beta_2 = 19.65$)		3.7		20	0.1
Mn^{2+}	4.50	4.15	4.05		25	0.1
Ni^{2+}	8.0	8.0	7.9		25	0.1
Zn^{2+}	6.30	5.65	5.10		25	0.1

Solutions to Exercises

Chapter 1

1-A. (a) $\dfrac{(25.00 \text{ mL})(0.791\ 4 \text{ g/mL})/(32.042 \text{ g/mol})}{0.500\ 0 \text{ L}}$

$= 1.235 \text{ M}$

(b) 500.0 mL of solution weighs (1.454 g/mL) × (500.0 mL) = 727.0 g and contains 25.00 mL (= 19.78 g) of methanol. The mass of chloroform in 500 mL must be 727.0 − 19.78 = 707.2 g. The molality of methanol is

$$\text{Molality} = \frac{\text{mol methanol}}{\text{kg chloroform}}$$

$$= \frac{(19.78 \text{ g})/(32.042 \text{ g/mol})}{0.707\ 2 \text{ kg}}$$

$$= 0.872\ 9 \text{ m}$$

1-B. (a) $\left(\dfrac{48.0 \text{ g HBr}}{100 \text{ g solution}}\right)\left(1.50 \dfrac{\text{g solution}}{\text{mL solution}}\right)$

$= \dfrac{0.720 \text{ g HBr}}{\text{mL solution}}$

$= \dfrac{720 \text{ g HBr}}{\text{L solution}}$

$= 8.90 \text{ F}$

(b) $\dfrac{36.0 \text{ g HBr}}{0.480 \text{ g HBr/g solution}} = 75.0 \text{ g solution}$

(c) 233 mmol $= 0.233 \text{ mol}$

$\dfrac{0.233 \text{ mol}}{8.90 \text{ mol/L}} = 26.2 \text{ mL}$

(d) $V = 250 \text{ mL}\left(\dfrac{0.160 \text{ M}}{8.90 \text{ M}}\right) = 4.49 \text{ mL}$

1-C. $\dfrac{\text{wt Cl}}{\text{wt MgCl}_2} = \dfrac{(2)(35.453)}{24.305 + 2(35.453)} = 0.745$

If $\text{MgCl}_2 = 12.6$ ppt,

$\text{Cl} = (0.745)(12.6) = 9.38 \text{ ppt}$

Chapter 2

2-A. (a) At 15°C, water density = 0.999 102 6 g/mL

$$m = \frac{(5.397\ 4 \text{ g})\left(1 - \dfrac{0.001\ 2 \text{ g/mL}}{8.0 \text{ g/mL}}\right)}{\left(1 - \dfrac{0.001\ 2 \text{ g/mL}}{0.999\ 102\ 6 \text{ g/mL}}\right)} = 5.403\ 1 \text{ g}$$

(b) At 25°C, water density = 0.997 047 9 g/mL and $m = 5.403\ 1$ g.

2-B. Use Equation 2-2 with $m' = 0.296\ 1$ g, $d_a = 0.001\ 2$ g/mL, $d_w = 8.0$ g/mL, and $d = 5.24$ g/mL $\Rightarrow m = 0.296\ 1$ g.

2-C. $\dfrac{c'}{d'} = \dfrac{c}{d}$.

Let the primes stand for 16°C:

$$\Rightarrow \frac{c' \text{ at } 16°\text{C}}{0.998\ 946\ 0 \text{ g/mL}} = \frac{0.051\ 38 \text{ M}}{0.997\ 299\ 5 \text{ g/mL}}$$

$\Rightarrow c'$ at $16° = 0.051\ 46 \text{ M}$

2-D. Column 3 of Table 2-6 tells us that water occupies 1.003 3 mL/g at 22°C. Therefore (15.569 g) × (1.003 3 mL/g) = 15.620 mL.

3-A. (a) 21.0_9 $(\pm 0.1_6)$ or 21.1 (± 0.2); relative uncertainty $= 0.8\%$

(b) 27.4_3 $(\pm 0.8_6)$; relative uncertainty $= 3._2\%$

(c) $(14._9 \pm 1._3) \times 10^4$ or $(15 \pm 1) \times 10^4$; relative uncertainty $= 9\%$

3-B. (a) 2.000 L of 0.169 M NaOH (F.W. = 39.997 1) requires 0.338 mol = 13.52 g NaOH

$$\frac{13.52 \text{ g NaOH}}{0.534 \text{ g NaOH/g solution}} = 25.32 \text{ g solution}$$

$$\frac{25.32 \text{ g solution}}{1.52 \text{ g solution/mL solution}} = 16.6_6 \text{ mL}$$

(b) Molarity =

$$\frac{[16.66\,(\pm 0.10)\text{ mL}]\left[1.52\,(\pm 0.01)\dfrac{\text{g solution}}{\text{mL}}\right]\times\left[0.534\,(\pm 0.004)\dfrac{\text{g NaOH}}{\text{g solution}}\right]}{\left(39.997\,1\,\dfrac{\text{g NaOH}}{\text{mol}}\right)(2.000\text{ L})}$$

Since the relative errors in molecular weight and final volume are negligible (≈ 0), we can write

$$\begin{array}{l}\text{Relative} \\ \text{error in} \\ \text{molarity}\end{array} = \sqrt{\left(\frac{0.10}{16.66}\right)^2 + \left(\frac{0.01}{1.52}\right)^2 + \left(\frac{0.004}{0.534}\right)^2}$$

$$= 1.16\%$$

Molarity $= 0.169\,(\pm 0.002)$

3-C. $0.050\,0\,(\pm 2\%)$ mol =

$$\frac{[4.18\,(\pm x)\text{ mL}]\left[1.18\,(\pm 0.01)\dfrac{\text{g solution}}{\text{mL}}\right]\times\left[0.370\,(\pm 0.005)\dfrac{\text{g HCl}}{\text{g solution}}\right]}{36.461\,\dfrac{\text{g HCl}}{\text{mol}}}$$

Error analysis:

$$(0.02)^2 = \left(\frac{x}{4.18}\right)^2 + \left(\frac{0.01}{1.18}\right)^2 + \left(\frac{0.005}{0.370}\right)^2$$

$$x = 0.05 \text{ mL}$$

Chapter 4

4-A. Mean $= \frac{1}{5}(116.0 + 97.9 + 114.2 + 106.8 + 108.3)$

$$= 108.6_4$$

Standard deviation

$$= \sqrt{\frac{(116.0 - 108.6_4)^2 + \cdots + (108.3 - 108.6_4)^2}{5 - 1}}$$

$$= 7.1_4$$

Median = 108.3 (the middle value)

Geometric mean

$$= [(116.0)(97.9)(114.2)(106.8)(108.3)]^{1/5} = 108.4_5$$

Range $= 116.0 - 97.9 = 18.1$

90% Confidence interval

$$= 108.6_4 \pm \frac{(2.132)(7.1_4)}{\sqrt{5}} = 108.6_4 \pm 6.8_1$$

$$Q = \frac{106.8 - 97.9}{116.0 - 97.9} = 0.49 < [Q(\text{Table 4-4}) = 0.64]$$

Therefore, 97.9 should be retained.

4-B. (a) $z = \dfrac{(45\,800 - 62\,700)}{10\,400} = -1.625.$

In Table 4-1, we see that the area listed for $|z| = 1.6$ is 0.445 2, and the area for $|z| = 1.7$ is 0.455 4. By linear interpolation, the area for 1.625 is

$$0.445\,2 + \left(\frac{1.625 - 1.6}{1.7 - 1.6}\right)(0.4554 - 0.445\,2)$$

$$= 0.447\,8$$

The are *beyond* $|z| = 1.625$ must be $0.500\,0 - 0.447\,8 = 0.052\,2$. The fraction of brakes expected to wear out in less than 45 800 miles is 0.052 2 (or 5.22%).

(b) 60 000 miles lies at $z = -0.259\,6$. 70 000 miles lies at $z = +0.701\,9$. By linear interpolation, the area from $\bar{x}$ to $z = -0.259\,6$ is 0.102 3. The area from $\bar{x}$ to $z = 0.701\,9$ is 0.258 6. Total area $-$ 0.102 3 + 0.258 6 = 0.360 9. Thus approximately 36% of the brakes are expected to wear out between 60 000 and 70 000 miles.

4-C. Confidence 90%: $\mu = \bar{x} \pm \dfrac{ts}{\sqrt{n}}$

$$= 116._4 \pm \frac{(2.132)(3._{58})}{\sqrt{5}}$$

$$= 112._9 \text{ to } 119._8$$

Confidence 99%: $= 116._4 \pm \dfrac{(4.604)(3._{58})}{\sqrt{5}}$

$$= 109._0 \text{ to } 123._8$$

You can be 90% sure, but not 99% sure, that the result is high.

4-D. Using Equations 4-7 and 4-8, we find

$$t = \frac{0.027\,5_6 - 0.026\,9_0}{0.000\,4_5}\sqrt{\frac{5 \cdot 5}{5 + 5}} = 2._{32}$$

Since t(calculated) = 2.32 > t (Table 4-2) = 1.86, the difference is significant.

4-E. (a)

x_i	y_i	x_iy_i	x_i^2	d_i	d_i^2
0.00	.466	0	0	-0.0046	2.12×10^{-5}
9.36	.676	6.327	87.61	$+0.0016$	2.58×10^{-6}
18.72	.883	16.530	350.44	$+0.0048$	2.31×10^{-5}
28.08	1.086	30.495	788.49	$+0.0040$	1.61×10^{-5}
37.44	1.280	47.923	1 401.75	-0.0058	3.34×10^{-5}
Sum: 93.60	4.391	101.275	2 628.29		9.64×10^{-5}

$$D = \begin{vmatrix} \sum(x_i^2) & \sum x_i \\ \sum x_i & n \end{vmatrix}$$

$$= (2\,628.29)(5) - (93.60)(93.60) = 4\,380.5$$

$$m = \begin{vmatrix} \sum x_iy_i & \sum x_i \\ \sum y_i & n \end{vmatrix} \div D$$

$$= \frac{(101.275)(5) - (93.60)(4.391)}{D}$$

$$= 95.377 \div 4\,380.5 = 0.021\,773$$

$$b = \begin{vmatrix} \sum(x_i^2) & \sum x_iy_i \\ \sum x_i & \sum y_i \end{vmatrix} \div D$$

$$= \frac{(2\,628.29)(4.391) - (101.275)(93.60)}{D}$$

$$= 2\,061.48 \div 4\,380.5 = 0.470\,60$$

$$\sigma_y^2 \approx s_y^2 = \frac{\sum(d_i^2)}{n-2} = \frac{9.64 \times 10^{-5}}{3}$$

$$= 3.21 \times 10^{-5}; \; \sigma_y = 0.005\,7$$

$$\sigma_m = \sqrt{\frac{\sigma_y^2 n}{D}} = \sqrt{\frac{(3.21 \times 10^{-5})5}{4\,380.5}} = 0.000\,191$$

$$\sigma_b = \sqrt{\frac{\sigma_y^2 \sum(x_i^2)}{D}} = \sqrt{\frac{(3.21 \times 10^{-5})(2\,628.29)}{4\,380.5}}$$

$$= 0.004\,39$$

Equation of the best line:

$$y = [0.021\,8\,(\pm 0.000\,2)]x + [0.471\,(\pm 0.004)]$$

(c) $x = \dfrac{y(\pm \sigma_y) - b(\pm \sigma_b)}{m(\pm \sigma_m)}$

$$= \frac{0.973\,(\pm 0.005_7) - 0.471\,(\pm 0.004_4)}{0.021\,8\,(\pm 0.000\,1_9)}$$

$$= 23.0 \pm 0.4 \; \mu g$$

Chapter 5

5-A. (a) $\require{cancel}$
$\cancel{Ag^+} + \cancel{Cl^-} \rightleftharpoons AgCl(aq)$
$AgCl(s) \rightleftharpoons \cancel{Ag^+} + \cancel{Cl^-}$
$\overline{AgCl(s) \rightleftharpoons AgCl(aq)}$

$K_1 = 2.0 \times 10^3$
$K_2 = 1.8 \times 10^{-10}$
$\overline{K_3 = K_1K_2 = 3.6 \times 10^{-7}}$

(b) The answer to part a tells us $[AgCl(aq)] = 3.6 \times 10^{-7}$ M.

(c) $\cancel{AgCl_2^-} \rightleftharpoons AgCl(aq) + Cl^-$
$Ag^+ + \cancel{Cl^-} \rightleftharpoons AgCl(s)$
$\cancel{AgCl(aq)} \rightleftharpoons \cancel{Ag^+} + \cancel{Cl^-}$
$\overline{AgCl_2^- \rightleftharpoons AgCl(s) + Cl^-}$

$K_1 = 1/(9.3 \times 10^1)$
$K_2 = 1/(1.8 \times 10^{-10})$
$K_3 = 1/(2.0 \times 10^3)$
$\overline{K_4 = K_1K_2K_3 = 3.0 \times 10^4}$

5-B. (a)

	BrO_3^-	Cr^{3+}	Br^-	$Cr_2O_7^{2-}$	H^+
Initial concentration:	0.010 0	0.010 0	0	0	1.00
Final concentration:	$0.010\,0 - x$	$0.010\,0 - 2x$	x	x	$1.00 + 8x$

$$\frac{(x)(x)(1.00 + 8x)^8}{(0.010\,0 - x)(0.010\,0 - 2x)^2} = 1 \times 10^{11}$$

(b) $[Br^-]$ and $[Cr_2O_7^{2-}]$ will both be 0.005 00 M because Cr^{3+} is the *limiting reagent*. Reaction 5-12 requires two moles of Cr^{3+} per mole of BrO_3^-. The Cr^{3+} will be used up first, making one mole of Br^- and one mole of $Cr_2O_7^{2-}$ per two moles of Cr^{3+} consumed. To solve the equation above, we set $x = 0.005\,00$ M in all terms except $[Cr^{3+}]$. The concentration of $[Cr^{3+}]$ will be a small, unknown quantity.

$$\frac{(0.005\,00)(0.005\,00)[1.00 + 8(0.005\,00)]^8}{(0.010\,0 - 0.005\,00)[Cr^{3+}]^2}$$
$$= 1 \times 10^{11}$$

$$[Cr^{3+}] = 2.6 \times 10^{-7} \; M$$

$$[BrO_3^-] = 0.010\,0 - 0.005\,00 = 0.005\,00 \; M$$

5-C. (a) $[La^{3+}][IO_3^-]^3 = x(3x)^3 = 1.0 \times 10^{-11}$
$$\Rightarrow x = 7.8 \times 10^{-4} \; M$$
$$= 0.13 \; g \; La(IO_3)_3/250 \; mL$$

(b) $[La^{3+}][IO_3^-]^3 = x(3x + 0.050)^3$
$$\approx x(0.050)^3 = 1.0 \times 10^{-11}$$
$$\Rightarrow x = 8 \times 10^{-8}$$
$$= 1.3 \times 10^{-5} \; g$$

5-D. (a) $Ca(IO_3)_2$ (since it has a larger K_{sp})

(b) The two salts do not have the same stoichiometry. Therefore, K_{sp} values cannot be compared. For $TlIO_3$, $x^2 = K_{sp} \Rightarrow x = 1.8 \times 10^{-3}$. For $Sr(IO_3)_2$, $x(2x)^2 = K_{sp} \Rightarrow x = 4.4 \times 10^{-3}$. $Sr(IO_3)_2$ is more soluble.

5-E. $[Fe^{3-}][OH^-]^3 = (10^{-10})[OH^-]^3 = 1.6 \times 10^{-39}$
$$\rightarrow [OH^-] = 2.5 \times 10^{-10}$$

$[Fe^{2+}][OH^-]^2 = (10^{-10})[OH^-]^2 = 7.9 \times 10^{-16}$
$$\Rightarrow [OH^-] = 2.8 \times 10^{-3}$$

5-F. First we need to find which salt will precipitate at the lowest $[C_2O_4^{2-}]$ concentration:

$$[Ca^{2+}][C_2O_4^{2-}] = x^2 = 1.3 \times 10^{-8}$$
$$\Rightarrow [C_2O_4^{2-}] = x$$
$$= 1.1 \times 10^{-4} \text{ M}$$
$$[Ce^{3+}]^2[C_2O_4^{2-}]^3 = (2x)^2(3x)^3 = 3 \times 10^{-29}$$
$$\Rightarrow [C_2O_4^{2-}] = 3x$$
$$= 7.7 \times 10^{-7} \text{ M}$$

$Ce_2(C_2O_4)_3$ is less soluble than CaC_2O_4. The concentration of $C_2O_4^{2-}$ needed to reduce Ce^{3+} to 1% of 0.010 M is

$$[C_2O_4^{2-}] = \left[\frac{K_{sp}}{(0.000\,10)^2}\right]^{1/3} = 1.4 \times 10^{-7}$$

This concentration of $C_2O_4^{2-}$ will not precipitate Ca^{2+} because

$$Q = [Ca^{2+}][C_2O_4^{2-}] = (0.010)(1.4 \times 10^{-7})$$
$$= 1.4 \times 10^{-9} < K_{sp}$$

The separation is feasible.

5-G. Assuming that all of the Ni is in the form $Ni(en)_3^{2+}$, $[Ni(en)_3^{2+}] = 1.00 \times 10^{-5}$ M. This uses up just 3×10^{-5} mol of en, which leaves the en concentration at 0.100 M. Adding the three equiations gives

$$Ni^{2+} + 3en \rightleftharpoons Ni(en)_3^{2+}$$

$$K = K_1 K_2 K_3 = 2.1_4 \times 10^{18}$$

$$[Ni^{2+}] = \frac{[Ni(en)_3^{2+}]}{K[en]^3}$$
$$= \frac{(1.00 \times 10^{-5})}{(2.1_4 \times 10^{18})(0.100)^3} = 4.7 \times 10^{-21} \text{ M}$$

5-H. (a) Neutral—neither Na^+ nor Br^- has any acidic or basic properties.
(b) Basic—$CH_3CO_2^-$ is the conjugate base of acetic acid, and Na^+ is neither acidic nor basic.
(c) Acidic—NH_4^+ is the conjugate acid of NH_3, and Cl^- is neither acidic nor basic.
(d) Basic—PO_4^{3-} is a base, and K^+ is neither acidic nor basic.
(e) Neutral—Neither ion is acidic or basic.
(f) Basic—The quaternary ammonium ion is neither acidic nor basic, and the $C_6H_5CO_2^-$ anion is the conjugate base of benzoic acid.

5-I. $K_{b1} = K_w/K_{a2} = 4.4 \times 10^{-9}$
$K_{b2} = K_w/K_{a1} = 1.6 \times 10^{-10}$

5-J. $K = K_{b2} = K_w/K_{a2} = 1.0 \times 10^{-8}$

5-K. $[H^+][OH^-] = x^2 = K_w \Rightarrow x = \sqrt{K_w} \Rightarrow pH = -\log\sqrt{K_w} = 7.472$ at 0°C, 7.083 at 20°C, and 6.767 at 40°C.

Chapter 6

6-A. (a) $\mu = \frac{1}{2}([K^+] \cdot 1^2 + [Br^-] \cdot (-1)^2) = 0.02$ M
(b) $\mu = \frac{1}{2}([Cs^+] \cdot 1^2 + [CrO_4^{2-}] \cdot (-2)^2)$
$= \frac{1}{2}([0.04] \cdot 1 + [0.02] \cdot 4) = 0.06$ M
(c) $\mu = \frac{1}{2}([Mg^{2+}] \cdot 2^2 + [Cl^-] \cdot (-1)^2$
$+ [Al^{3+}] \cdot 3^2)$
$= \frac{1}{2}([0.02] \cdot 4 +$
$[\quad 0.04 \quad + \quad 0.09 \quad] \cdot 1 + [0.03] \cdot 9)$
$\qquad \uparrow \qquad\qquad \uparrow$
From $MgCl_2$ From $AlCl_3$
$= 0.24$ M

6-B. For 0.005 0 M $(CH_3CH_2CH_2)_4N^+Br^-$ plus 0.005 0 M $(CH_3)_4N^+Cl^-$, $\mu = 0.010$ M. The size of the ion $(CH_3CH_2CH_2)_4N^+$ is 800 pm. [It is listed at the bottom of Table 6-1 as $(C_3H_7)_4N^+$.] At $\mu = 0.01$ M, $\gamma = 0.912$ for an ion of charge ± 1 with $\alpha = 800$ pm. $\mathscr{A} = (0.005\,0)(0.912) = 0.004\,6$.

6-C. (a) $\mu = 0.060$ M from KNO_3 (assuming that AgSCN has negligible solubility)

$$[Ag^+]\gamma_{Ag^+}[SCN^-]\gamma_{SCN^-} = K_{sp}$$
$$[x](0.79)[x](0.80) = 1.1 \times 10^{-12}$$
$$\Rightarrow x = [Ag^+]$$
$$= 1.3 \times 10^{-6} \text{ M}$$

(b) $\mu = 0.060$ M from KSCN

$$[Ag^+]\gamma_{Ag^+}[SCN^-]\gamma_{SCN^-} = K_{sp}$$
$$[x](0.79)\underbrace{[x + 0.060]}_{\approx 0.060}(0.80) = 1.1 \times 10^{-12}$$
$$\Rightarrow x = [Ag^+]$$
$$= 2.9 \times 10^{-11} \text{ M}$$

6-D. Assuming that $Mn(OH)_2$ gives a negligible concentration of ions, $\mu = 0.075$ from $CaCl_2$.

$$[Mn^{2+}]\gamma_{Mn^{2+}}[OH^-]^2\gamma_{OH^-}^2 = K_{sp}$$
$$[x](0.445)[2x]^2(0.785)^2 = 1.6 \times 10^{-13}$$
$$\Rightarrow 2x = [OH^-]$$
$$= 1.1 \times 10^{-4} \text{ M}$$

6-E. For 0.02 M $MgCl_2$ (which dissociates into 0.020 M Mg^{2+} and 0.040 M Cl^-), $\mu = 0.06$ M. At this ionic strength, interpolation in Table 6-1 gives $\gamma_{Mg^{2+}} = 0.506$ and $\gamma_{Cl^-} = 0.795$.

$$\gamma_\pm = (\gamma_{Mg^{2+}}\gamma_{Cl^-}^2)^{1/(1+2)} = 0.68_4$$

6-F. $[H^+]\gamma_{H^+}[OH^-]\gamma_{OH^-} = (x)(0.86)(x)(0.81) = 1.0 \times 10^{-14} \Rightarrow x = [H^+] = 1.2 \times 10^{-7}$ M.
pH $= -\log(1.2 \times 10^{-7})(0.86) = 6.99$.

7-A. $[H^+] + 2[Ca^{2+}] + [CaF^+] = [OH^-] + [F^-]$
Remember that H^+ and OH^- are present in every aqueous solution.

7-B. (a) $[Cl^-] = 2[Ca^{2+}]$
(b) $\underbrace{[Cl^-] + [CaCl^+]}_{\text{Moles of Cl}} = \underbrace{2([Ca^{2+}] + [CaCl^+])}_{\text{Moles of Ca}}$

7-C. (a) $[F^-] + [HF] = 2[Ca^{2+}]$
(b) $\underbrace{[F^-] + [HF] + 2[HF_2^-]}_{\text{Moles of F}} = 2[Ca^{2+}]$
(One mole of HF_2^- contains two moles of fluorine.)

7-D. $2[Ca^{2+}] = 3\{[PO_4^{3-}] + [HPO_4^{2-}]$
$+ [H_2PO_4^-] + [H_3PO_4]\}$

7-E. (a) Charge balance: Invalid because pH is fixed.

Mass balance: $[Ag^+] = [CN^-] + [HCN]$ (1)

Equilibria: $K_b = \dfrac{[HCN][OH^-]}{[CN^-]}$ (2)

$K_{sp} = [Ag^+][CN^-]$ (3)

$K_w = [H^+][OH^-]$ (4)

Since $[H^+] = 10^{-9.00}$ M, $[OH^-] = 10^{-5.00}$ M. Putting this value $[OH^-]$ into Equation 2 gives

$$[HCN] = \frac{K_b}{[OH^-]}[CN^-] = 1.6[CN^-]$$

Substituting into Equation 1 gives

$$[Ag^+] = [CN^-] + 1.6[CN^-] = 2.6[CN^-]$$

Substituting into Equation 3 gives

$$[Ag^+]\left(\frac{[Ag^+]}{2.6}\right) = K_{sp}$$
$$\Rightarrow [Ag^+] = 2.4 \times 10^{-8} \text{ M}$$
$$[CN^-] = [Ag^+]/2.6 = 9.2 \times 10^{-9} \text{ M}$$
$$[HCN] = 1.6[CN^-] = 1.5 \times 10^{-8} \text{ M}$$

(b) With activities:

$$[Ag^+] = [CN^-] + [HCN] \tag{1'}$$

$$K_b = \frac{[HCN]\gamma_{HCN}[OH^-]\gamma_{OH^-}}{[CN^-]\gamma_{CN^-}} \tag{2'}$$

$$K_{sp} = [Ag^+]\gamma_{Ag^+}[CN^-]\gamma_{CN^-} \tag{3'}$$

$$K_w = [H^+]\gamma_{H^+}[OH^-]\gamma_{OH^-} \tag{4'}$$

Since pH = 9.00, $[OH^-]\gamma_{OH^-} = K_w/[H^+]\gamma_{H^+} = 10^{-5.00}$. Putting this value into Equation 2'

$$[HCN] = \frac{K_b\gamma_{CN^-}[CN^-]}{\gamma_{HCN}[OH^-]\gamma_{OH^-}}$$
$$= \frac{(1.6 \times 10^{-5})(0.755)[CN^-]}{1 \cdot 10^{-5.00}}$$
$$= 1.208[CN^-]$$

Here we have assumed $\gamma_{HCN} = 1$. Using the above relation in the mass balance (Equation 1') gives

$$[Ag^+] = 2.208[CN^-]$$

Substituting into Equation 3' gives

$$K_{sp} = [Ag^+](0.75)\left(\frac{Ag^+}{2.208}\right)(0.755)$$
$$\Rightarrow [Ag^+] = 2.9 \times 10^{-8} \text{ M}$$

$$[CN^-] = \frac{[Ag^+]}{2.208} = 1.3 \times 10^{-8} \text{ M}$$

$$[HCN] = 1.208[CN^-] = 1.6 \times 10^{-8} \text{ M}$$

7-F. Charge balance: Invalid because pH is fixed.

Mass balance:

$$[Zn^{2+}] = [C_2O_4^{2-}] + [HC_2O_4^-] \tag{1}$$
$$+ [H_2C_2O_4]$$

Equilibria:

$$K_{sp} = [Zn^{2+}][C_2O_4^{2-}] \tag{2}$$

$$K_{b1} = \frac{[HC_2O_4^-][OH^-]}{[C_2O_4^{2-}]} \tag{3}$$

$$K_{b2} = \frac{[H_2C_2O_4][OH^-]}{[HC_2O_4^-]} \tag{4}$$

$$K_w = [H^+][OH^-] \tag{5}$$

If pH = 3.0, $[OH^-] = 1.0 \times 10^{-11}$. Putting this into Equation 3 gives

$$[HC_2O_4^-] = \frac{K_{b1}}{[OH^-]}[C_2O_4^{2-}] = 18[C_2O_4^{2-}]$$

Using this result in equation 4 gives

$$[H_2C_2O_4] = \frac{K_{b2}}{[OH^-]}[HC_2O_4^-]$$
$$= \frac{K_{b2}}{[OH^-]}\frac{K_{b1}}{[OH^-]}[C_2O_4^{2-}]$$
$$= 0.324[C_2O_4^{2-}]$$

Using these values of $[H_2C_2O_4]$ and $[HC_2O_4^-]$ in Equation 1 gives

$$[Zn^{2+}] = [C_2O_4^{2-}](1 + 18 + 0.324)$$
$$= [C_2O_4^{2-}](19.324)$$

Putting this last equality into Equation 2 produces the result

$$[Zn^{2+}]\left(\frac{[Zn^{2+}]}{19.324}\right) = K_{sp}$$

$$\Rightarrow [Zn^{2+}] = 3.8 \times 10^{-4} \text{ M}$$

If $[Zn^{2+}] = 3.8 \times 10^{-4}$ M, this many moles of ZnC_2O_4 (= 0.058 g) must be dissolved in each liter.

Chapter 8

8-A. One mole of ethoxyl groups produces one mole of AgI. 29.03 mg of AgI = 0.123 65 mmol. The amount of compound analyzed is 25.42 mg/(417 g/mol) = 0.060 96 mmol. There are

$$\frac{0.123\,65 \text{ mmol ethoxyl groups}}{0.060\,96 \text{ mmol compound}}$$

$$= 2.03 \ (=2) \text{ ethoxyl groups/molecule.}$$

8-B. There is one mole of SO_4^{2-} in each mole of each reactant and of the product. Let x = g of K_2SO_4 and y = g of $(NH_4)_2SO_4$.

$$x + y = 0.649 \text{ g} \qquad (1)$$

$$\underbrace{\frac{x}{174.27}}_{\substack{\text{Moles of} \\ K_2SO_4}} + \underbrace{\frac{y}{132.14}}_{\substack{\text{Moles of} \\ (NH_4)_2SO_4}} = \underbrace{\frac{0.977}{233.40}}_{\substack{\text{Moles of} \\ BaSO_4}} \qquad (2)$$

Making the substitution $y = 0.649 - x$ in Equation 2 gives $x = 0.397$ g = 61.1% of the sample.

8-C. Formula and atomic weights: Ba(137.34), Cl(35.453), K(39.102), H_2O(18.015), KCl(74.555), $BaCl_2 \cdot 2H_2O$(244.28). H_2O lost = 1.783 9 − 1.562 3 = 0.221 6 g = 1.230 1 × 10^{-2} mol of H_2O. For every two moles of H_2O lost, one mole of $BaCl_2 \cdot 2H_2O$ must have been present. 1.230 1 × 10^{-2} mol of H_2O implies that 6.150 4 × 10^{-3} mol of $BaCl_2 \cdot 2H_2O$ must have been present. This much $BaCl_2 \cdot 2H_2O$ equals 1.502 4 g. The Ba and Cl contents of the $BaCl_2 \cdot 2H_2O$ are

$$Ba = \left(\frac{137.34}{244.28}\right)(1.502\,4 \text{ g}) = 0.844\,68 \text{ g}$$

$$Cl = \left(\frac{2(35.453)}{244.28}\right)(1.502\,4 \text{ g}) = 0.436\,09 \text{ g}$$

Since the total sample weighs 1.783 9 g and contains 1.502 4 g of $BaCl_2 \cdot 2H_2O$, the sample must contain 1.783 9 − 1.502 4 = 0.281 5 g of KCl, which contains

$$K = \left(\frac{39.102}{74.555}\right)(0.281\,5) = 0.147\,64 \text{ g}$$

$$Cl = \left(\frac{35.453}{74.555}\right)(0.281\,5) = 0.133\,86 \text{ g}$$

Weight percent of each element:

$$Ba = \frac{0.844\,68}{1.783\,9} = 47.35\%$$

$$K = \frac{0.147\,64}{1.783\,9} = 8.276\%$$

$$Cl = \frac{0.436\,09 + 0.133\,86}{1.783\,9} = 31.95\%$$

8-D. Let x = mass of $Al(BF_4)_3$ and y = mass of $Mg(NO_3)_3$. We can say that $x + y = 0.282\,8$ g. We also know that

$$\begin{array}{l}\text{Moles of nitron} \\ \text{tetrafluoroborate}\end{array} = 3(\text{moles of } Al(BF_4)_3) = \frac{3x}{287.39}$$

$$\begin{array}{l}\text{Moles of nitron} \\ \text{nitrate}\end{array} = 2(\text{moles of } Mg(NO_3)_2) = \frac{2y}{148.31}$$

Equating the mass of product to the mass of nitron tetrafluoroborate plus the mass of nitron nitrate, we can write

$$\underbrace{1.322}_{\substack{\text{Mass of} \\ \text{product}}} = \underbrace{\left(\frac{3x}{287.39}\right)(400.18)}_{\substack{\text{Mass of nitron} \\ \text{tetrafluoroborate}}} + \underbrace{\left(\frac{2y}{148.31}\right)(375.39)}_{\substack{\text{Mass of nitron} \\ \text{nitrate}}}$$

Making the substitution $x = 0.282\,8 - y$ in the above equation allows us to find $y = 0.158\,9$ g of $Mg(NO_3)_2 = 1.072$ mmol of Mg = 0.026 05 g of Mg = 9.210% of the original solid sample.

Chapter 9

9-A. (a) Formula weight of ascorbic acid = 176.126

$$0.197\,0 \text{ g of ascorbic acid} = 1.118\,5 \text{ mmol}$$

$$\text{Molarity of } I_3^- = 1.118\,5 \text{ mmol/29.41 mL}$$

$$= 0.038\,03 \text{ M}$$

(b) 31.63 mL of $I_3^- = 1.203$ mmol of $I_3^- = 1.203$ mmol of ascorbic acid = 0.211 9 g

$$= 49.94\% \text{ of the tablet}$$

9-B. 34.02 mL of 0.087 71 M NaOH = 2.983 9 mmol of OH^-. Let x be the mass of malonic acid and y be the mass of anilinium chloride. Then $x + y = 0.237\,6$ g and

$$\begin{array}{l}\text{(Moles of anilinium chloride)} \\ + \ 2(\text{moles of malonic acid})\end{array} = 0.002\,983\,9$$

$$\frac{y}{129.59} + 2\left(\frac{x}{104.06}\right) = 0.002\,983\,9$$

Substituting $y = 0.237\,6 - x$ gives $x = 0.099\,97$ g $= 42.07\%$ malonic acid. Anilinium chloride $= 57.93\%$.

9-C. (a) $\dfrac{0.824 \text{ g acid}}{204.233 \text{ g/mol}} = 4.03$ mmol. This many mmol of NaOH is contained in 0.038 314 kg of NaOH solution

$$\Rightarrow \text{concentration} = \frac{4.03 \times 10^{-3} \text{ mol NaOH}}{0.038\,314 \text{ kg solution}}$$

$$= 0.105_3 \text{ mol/kg solution.}$$

(b) mol NaOH $= (0.057\,911 \text{ kg})(0.105\,3 \text{ mol/kg}) = 6.10$ mmol. Since 2 mol NaOH reacts with 1 mol H_2SO_4,

$$[H_2SO_4] = \frac{3.05 \text{ mmol}}{10.00 \text{ mL}} = 0.305 \text{ M}$$

9-D. The reaction is $SCN^- + Cu^+ \to CuSCN(s)$. The equivalence point occurs when moles of $Cu^+ =$ moles of $SCN^- \Rightarrow V_e = 100.0$ mL. Before the equivalence point, there is excess SCN^- remaining in the solution. We calculate the molarity of SCN^- and then find $[Cu^+]$ from the relation $[Cu^+] = K_{sp}/[SCN^-]$. For example, when 0.10 mL of Cu^+ has been added,

$$[SCN^-] = \left(\frac{100.0 - 0.10}{100.0}\right)(0.080\,0)\left(\frac{50.0}{50.1}\right)$$

$$= 7.98 \times 10^{-2} \text{ M}$$

$$[Cu^+] = 4.8 \times 10^{-15}/7.98 \times 10^{-2}$$

$$= 6.0 \times 10^{-14}$$

$$pCu^+ = 13.22$$

At the equivalence point, $[Cu^+][SCN^-] = x^2 = K_{sp} \Rightarrow x = [Cu^+] = 6.9 \times 10^{-8} \Rightarrow pCu^+ = 7.16$.
    Past the equivalence point, there is excess Cu^+. For example, when $V = 101.0$ mL,

$$[Cu^+] = (0.040\,0)\left(\frac{101.0 - 100.0}{151.0}\right) = 2.6 \times 10^{-4} \text{ M}$$

$$pCu^+ = 3.58$$

mL	pCu	mL	pCu	mL	pCu
0.10	13.22	75.0	12.22	100.0	7.16
10.0	13.10	95.0	11.46	100.1	4.57
25.0	12.92	99.0	10.75	101.0	3.58
50.0	12.62	99.9	9.75	110.0	2.60

9-E. (a) Moles of $I^- = 2$(moles of Hg_2^{2+})

$$(V_e)(0.100 \text{ M}) = 2(40.0 \text{ mL})(0.040\,0 \text{ M})$$

$$\Rightarrow V_e = 32.0 \text{ mL}$$

(b) Virtually all of the Hg_2^{2+} has precipitated, along with 3.20 mmol of I^-. The ions remaining in solution are

$$[NO_3^-] = \frac{3.20 \text{ mmol}}{1.00 \text{ mL}} = 0.032\,0 \text{ M}$$

$$[I^-] = \frac{2.80 \text{ mmol}}{100.0 \text{ mL}} = 0.028\,0 \text{ M}$$

$$[K^+] = \frac{6.00 \text{ mmol}}{100.0 \text{ mL}} = 0.060\,0 \text{ M}$$

$$\mu = \tfrac{1}{2}\sum c_i z_i^2 = 0.060\,0 \text{ M}$$

(c) $\mathscr{A}_{Hg_2^{2+}} = K_{sp}/\mathscr{A}_{I^-}^2$

$$= 4.5 \times 10^{-29}/(0.028\,0)^2(0.795)^2$$

$$= 9.1 \times 10^{-26} \Rightarrow pHg_2^{2+} = 25.04$$

9-F. $V_e = 23.66$ mL for AgBr. At 2.00, 10.00, 22.00, and 23.00 mL, AgBr is partially precipitated and excess Br^- remains.

2.00 mL: $[Ag^+] = \dfrac{K_{sp}(\text{for AgBr})}{[Br^-]}$

$$= \frac{5.0 \times 10^{-13}}{\underbrace{\left(\dfrac{23.66 - 2.00}{23.66}\right)}_{\substack{\text{Fraction} \\ \text{remaining}}} \underbrace{(0.050\,00)}_{\substack{\text{Original} \\ \text{molarity} \\ \text{of } Br^-}} \underbrace{\left(\dfrac{40.00}{42.00}\right)}_{\substack{\text{Dilution} \\ \text{factor}}}}$$

$$= 1.15 \times 10^{-11} \text{ M} \Rightarrow pAg^+ = 10.94$$

By similar reasoning, we find

$$10.00 \text{ mL: } pAg^+ = 10.66$$

$$22.00 \text{ mL: } pAg^+ = 9.66$$

$$23.00 \text{ mL: } pAg^+ = 9.25$$

At 24.00, 30.00 and 40.00 mL, AgCl is precipitating and excess Cl^- remains in solution.

24.00 mL:

$$[Ag^+] = \frac{K_{sp}(\text{for AgCl})}{[Cl^-]}$$

$$= \frac{1.8 \times 10^{-10}}{\left(\dfrac{47.32 - 24.00}{23.66}\right)(0.050\,00)\left(\dfrac{40.00}{64.00}\right)}$$

$$= 5.8 \times 10^{-9} \text{ M} \Rightarrow pAg^+ = 8.23$$

By similar reasoning, we find

$$30.00 \text{ mL: } pAg^+ = 8.07$$

$$40.00 \text{ mL: } pAg^+ = 7.63$$

At the second equivalence point (47.32 mL), $[Ag^+] = [Cl^-]$, and we can write

$$[Ag^+][Cl^-] = x^2 = K_{sp}\text{(for AgCl)}$$
$$\Rightarrow [Ag^+] = 1.34 \times 10^{-5} \text{ M}$$
$$pAg^+ = 4.87$$

At 50.00 mL, there is an excess of $(50.00 - 47.32) = 2.68$ mL of Ag^+.

$$[Ag^+] = \left(\frac{2.68}{90.00}\right)(0.084\ 54 \text{ M}) = 2.5 \times 10^{-3} \text{ M}$$

$$pAg^+ = 2.60$$

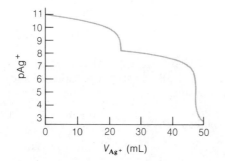

9-G. (a) 12.6 mL of Ag^+ is required to precipitate I^-. $(27.7 - 12.6) = 15.1$ mL is required to precipitate SCN^-.

$$[SCN^-] =$$

$$\frac{\text{moles of } Ag^+ \text{ needed to react with } SCN^-}{\text{original volume of } SCN^-}$$

$$= \frac{[27.7\ (\pm 0.3) - 12.6\ (\pm 0.4)][0.068\ 3\ (\pm 0.000\ 1)]}{50.00\ (\pm 0.05)}$$

$$= \frac{[15.1\ (\pm 0.5)][0.068\ 3\ (\pm 0.000\ 1)]}{50.00\ (\pm 0.05)}$$

$$= \frac{[15.1\ (\pm 3.31\%)][0.068\ 3\ (\pm 0.146\%)]}{50.00\ (\pm 0.100\%)}$$

$$= 0.020\ 6\ (\pm 0.000\ 7) \text{ M}$$

(b) $[SCN]\ (\pm 4.0\%) =$

$$\frac{[27.7\ (\pm 0.3) - 12.6\ (\pm ?)][0.068\ 3\ (\pm 0.000\ 1)]}{50.00\ (\pm 0.05)}$$

Let the error in 15.1 mL be $y\%$:

$$(4.0\%)^2 = (y\%)^2 + (0.146\%)^2 + (0.100\%)^2$$
$$\Rightarrow y = 4.00\% = 0.603 \text{ mL}$$

$$27.7\ (\pm 0.3) - 12.6\ (\pm ?) = 15.1\ (\pm 0.603)$$
$$\Rightarrow 0.3^2 + ?^2 = 0.603^2 \Rightarrow ? = 0.5 \text{ mL}$$

9-H. The transmittance should decrease, and the scattering should increase until the equivalence point is reached. Note that the transmittance is related logarithmically to the absorbance, and hence logarithmically to the particle concentration.

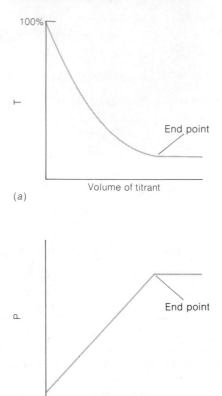

(a)

(b)

Chapter 10

10-A. $pH = -\log \mathscr{A}_{H^+}$. But $\mathscr{A}_{H^+}\mathscr{A}_{OH^-} = K_w \Rightarrow$ $\mathscr{A}_{H^+} = K_w/\mathscr{A}_{OH^-}$. For 1.0×10^{-2} M NaOH, $[OH^-] = 1.0 \times 10^{-2}$ M and $\gamma_{OH^-} = 0.900$ (using Table 6-1, with ionic strength $= 0.010$ M).

$$\mathscr{A}_{H^+} = \frac{K_w}{[OH^-]\gamma_{OH^-}}$$
$$= \frac{1.0 \times 10^{-14}}{(1.0 \times 10^{-2})(0.900)}$$
$$= 1.11 \times 10^{-12}$$
$$\Rightarrow pH = -\log \mathscr{A}_{H^+} = 11.95$$

10-B. (a) Charge balance: $[H^+] = [OH^-] + [Br^-]$

Mass balance: $[Br^-] = 1.0 \times 10^{-8}$ M

Equilibrium: $[H^+][OH^-] = K_w$

Setting $[H^+] = x$ and $[Br^-] = 1.0 \times 10^{-8}$ M, the charge balance tells us that $[OH^-] = x - 1.0 \times 10^{-8}$. Putting this into the K_w equilibrium gives

$(x)(x - 1.0 \times 10^{-8}) = 1.0 \times 10^{-14}$

$$\Rightarrow x = 1.0_5 \times 10^{-7} \text{ M}$$

$$\Rightarrow pH = 6.98$$

(b) Charge balance: $[H^+] = [OH^-] + 2[SO_4^{2-}]$

Mass balance: $[SO_4^{2-}] = 1.0 \times 10^{-8}$ M

Equilibrium: $[H^+][OH^-] = K_w$

As above, writing $[H^+] = x$ and $[SO_4^{2-}] = 1.0 \times 10^{-8}$ M gives $[OH^-] = x - 2.0 \times 10^{-8}$ and $[H^+][OH^-] = (x)[x - (2.0 \times 10^{-8})] = 1.0 \times 10^{-14} \Rightarrow x = 1.10 \times 10^{-7}$ M $\Rightarrow$ pH $= 6.96$.

10-C.

NO$_2$ 2-Nitrophenol F.W. = 139.110

$C_6H_5NO_3$ $K_a = 6.2 \times 10^{-8}$

OH

F_{HA}(formal concentration)

$$= \frac{1.23 \text{ g}/(139.110 \text{ g/mol})}{0.250 \text{ L}}$$

$$= 0.035 \, 4 \text{ M}$$

$$HA \rightleftharpoons H^+ + A^-$$
$$F - x \quad x \quad x$$

$$\frac{x^2}{0.035 \, 4 - x} = 6.2 \times 10^{-8} \Rightarrow$$

$$x = 4.7 \times 10^{-5} \text{ M} \Rightarrow$$

$$pH = -\log x = 4.33$$

10-D.

CH$_3$ CH$_3$

$\rightleftharpoons H^+ +$

OH   O$^-$

$F - x$ x x

But $[H^+] = 10^{-pH} = 8.9 \times 10^{-7}$ M $\Rightarrow [A^-] = 8.9 \times 10^{-7}$ M and $[HA] = 0.010 - [H^+] = 0.010$.

$$K_a = \frac{[H^+][A^-]}{[HA]} = \frac{(8.9 \times 10^{-7})^2}{0.010}$$

$$= 7.9 \times 10^{-11} \Rightarrow pK_a = 10.10$$

10-E. As $[HA] \rightarrow 0$, pH $\rightarrow 7$. If pH = 7,

$$\frac{[H^+][A^-]}{[HA]} = K_a \Rightarrow [A^-] = \frac{K_a}{[H^+]}[HA]$$

$$= \frac{10^{-5.00}}{10^{-7.00}} = 100[HA]$$

$$\alpha = \frac{[A^-]}{[HA] + [A^-]} = \frac{100[HA]}{[HA] + 100[HA]}$$

$$= \frac{100}{101} = 99\%$$

If $pK_a = 9.00$, we find $\alpha = 0.99\%$.

10-F. $CH_3CH_2CH_2CO_2^- + H_2O \rightleftharpoons$
$F - x$

$$CH_3CH_2CH_2CO_2H + OH^-$$
$$x \qquad\qquad x$$

$$K_b = \frac{K_w}{K_a} = 6.58 \times 10^{-10}$$

$$\frac{x^2}{F - x} = K_b \Rightarrow x = 5.7_4 \times 10^{-6} \text{ M}$$

$$pH = -\log\left(\frac{K_w}{x}\right) = 8.76$$

10-G. (a) $CH_3CH_2NH_2 + H_2O \rightleftharpoons$
$\quad F - x$

$$CH_3CH_2NH_3^+ + OH^-$$
$$x \qquad\qquad x$$

Since pH = 11.80, $[OH^-] = K_w/10^{-pH} = 6.3 \times 10^{-3}$ M $= [BH^+]$. $[B] = F - x = 0.094$ M.

$$K_b = \frac{[BH^+][OH^-]}{[B]} = \frac{(6.3 \times 10^{-3})^2}{0.094}$$

$$= 4.2 \times 10^{-4}$$

(b) $CH_3CH_2NH_3^+ \rightleftharpoons CH_3CH_2NH_2 + H^+$
$\quad F - x \qquad\qquad x \qquad\qquad x$

$$K_a = \frac{K_w}{K_b} = 2.4 \times 10^{-11}$$

$$\frac{x^2}{F - x} = K_a \Rightarrow x = 1.5_5 \times 10^{-6} \text{ M}$$

$$\Rightarrow pH = 5.81$$

10-H.

Compound	pK_a (for conjugate acid)
ammonia	9.24 ← Most suitable, since pK_a
aniline	4.60 is closest to pH.
hydrazine	8.48
pyridine	5.23

10-I. pH $= 4.25 + \log 0.75 = 4.13$

10-J. (a) $pH = pK_2$ (for H_2CO_3) $+ \log \dfrac{[CO_3^{2-}]}{[HCO_3^-]}$

$$10.80 = 10.329 + \log \frac{(4.00/138.206)}{(x/84.007)}$$

$$\Rightarrow x = 0.822 \text{ g}$$

(b)

	CO_3^{2-}	$+$ H^+	$\rightarrow$ HCO_3^-
Initial moles:	$0.028\,9_4$	$0.010\,0$	$0.009\,78$
Final moles:	$0.018\,9_4$	—	$0.019\,7_8$

$$pH = 10.329 + \log \frac{0.018\,9_4}{0.019\,7_8} = 10.31$$

(c)

$$CO_3^{2-} + H^+ \rightarrow HCO_3^-$$

Initial moles: $0.028\,9_4$ x —

Final moles: $0.028\,9_4 - x$ — x

$$10.00 = 10.329 + \log\frac{0.028\,9_4 - x}{x}$$

$$\Rightarrow x = 0.019\,7 \text{ mol}$$

$$\Rightarrow \text{volume} = \frac{0.019\,7 \text{ mol}}{0.320 \text{ M}} = 61.6 \text{ mL}$$

10-K. $H_2C_2O_4$ — F.W. = 90.035. 3.38 g = 0.037 5 mol

$$H_2Ox \xrightarrow{OH^-} HOx^- \xrightarrow{OH^-} Ox^{2-}$$

The pH of HOx^- is approximately $\frac{1}{2}(pK_1 + pK_2) = 2.76$. At pH = 2.40, the predominant species will be H_2Ox and HOx^-.

$$H_2Ox + OH^- \rightarrow HOx^- + H_2O$$

Initial
  moles: 0.037 5 x —

Final
  moles: $0.037\,5 - x$ — x

$$pH = pK_1 + \log\frac{[HOx^-]}{[H_2Ox]}$$

$$2.40 = 1.252 + \log\frac{x}{0.037\,5 - x}$$

$$x = 0.035\,0 \text{ mol} \Rightarrow \text{volume} = \frac{0.035\,0 \text{ mol}}{0.800 \text{ M}}$$

$$= 43.8 \text{ mL}$$

(Note that this problem involves a rather low pH, at which Equations 10-64 and 10-65 would be more appropriate to use.)

10-L. (a) $\quad H_2SO_3 \rightleftharpoons HSO_3^- + H^+$

     $0.050 - x$ x x

$$\frac{x^2}{0.050 - x} = K_1 = 1.23 \times 10^{-2}$$

$$\Rightarrow x = 1.94 \times 10^{-2}$$

$$[HSO_3^-] = [H^+] = 1.94 \times 10^{-2} \text{ M}$$

$$\Rightarrow pH = 1.71$$

$$[H_2SO_3] = 0.050 - x = 0.031 \text{ M}$$

$$[SO_3^{2-}] = \frac{K_2[HSO_3^-]}{[H^+]} = K_2$$

$$= 6.6 \times 10^{-8} \text{ M}$$

(b) $\quad [H^+] = \sqrt{\dfrac{K_1K_2(0.050) + K_1K_w}{K_1 + (0.050)}}$

$$= 2.55 \times 10^{-5} \Rightarrow pH = 4.59$$

$$[H_2SO_3] = \frac{[H^+][HSO_3^-]}{K_1}$$

$$= \frac{(2.55 \times 10^{-5})(0.050)}{1.23 \times 10^{-2}}$$

$$= 1.0 \times 10^{-4} \text{ M}$$

$$[SO_3^{2-}] = \frac{K_2[HSO_3^-]}{[H^+]} = 1.3 \times 10^{-4} \text{ M}$$

$$[HSO_3^-] = 0.050 \text{ M}$$

(c) $\quad SO_3^{2-} + H_2O \rightleftharpoons HSO_3^- + OH^-$

     $0.050 - x$ x x

$$\frac{x^2}{0.050 - x} = K_{b1} = \frac{K_w}{K_{a2}} = 1.52 \times 10^{-7}$$

$$[HSO_3^-] = x = 8.7 \times 10^{-5} \text{ M}$$

$$[H^+] = \frac{K_w}{x} = 1.15 \times 10^{-10} \text{ M}$$

$$\Rightarrow pH = 9.94$$

$$[SO_3^{2-}] = 0.050 - x = 0.050 \text{ M}$$

$$[H_2SO_3] = \frac{[H^+][HSO_3^-]}{K_1} = 8.1 \times 10^{-13}$$

10-M. (a) Calling the three forms of glutamine H_2G^+, HG, and G^-, the form shown is HG.

$$[H^+] = \sqrt{\frac{K_1K_2(0.010) + K_1K_w}{K_1 + 0.010}}$$

$$= 1.9_9 \times 10^{-6} \Rightarrow pH = 5.70$$

(b) Calling the four forms of cysteine H_3C^+, H_2C, HC^-, and C^{2-}, the form shown is HC^-.

$$[H^+] = \sqrt{\frac{K_2K_3(0.010) + K_2K_w}{K_2 + 0.010}}$$

$$= 2.8_1 \times 10^{-10} \Rightarrow pH = 9.55$$

(c) Calling the four forms of arginine H_3A^{2+}, H_2A^+, HA, and A^-, the form shown is HA.

$$[H^+] = \sqrt{\frac{K_2K_3(0.010) + K_2K_w}{K_2 + 0.010}}$$

$$= 3.6_8 \times 10^{-11} \Rightarrow pH = 10.43$$

10-N. The reaction of phenylhydrazine with water is

$$B + H_2O \rightleftharpoons BH^+ + OH^- \qquad K_b$$

We know that pH = 8.13, so we can find $[OH^-]$.

$$[OH^-] = \frac{\mathscr{A}_{OH^-}}{\gamma_{OH^-}} = \frac{K_w/10^{-pH}}{\gamma_{OH^-}} = 1.78 \times 10^{-6} \text{ M}$$

(using $\gamma_{OH^-} = 0.76$ for $\mu = 0.10$ M).

$$K_b = \frac{[BH^+]\gamma_{BH^+}[OH^-]\gamma_{OH^-}}{[B]\gamma_B}$$

$$= \frac{(1.78 \times 10^{-6})(0.80)(1.78 \times 10^{-6})(0.76)}{[0.010 - (1.78 \times 10^{-6})](1.00)}$$

$$= 1.93 \times 10^{-10}.$$

$$K_a = \frac{K_w}{K_b} = 5.19 \times 10^{-5} \Rightarrow pK_a = 4.28$$

To find K_b, we made use of the equality $[BH^+] = [OH^-]$.

Chapter 11

11-A. The titration reaction is $H^+ + OH^- \rightarrow H_2O$ and $V_e = 5.00$ mL. Three representative calculations are given:

$$1.00 \text{ mL: } [OH^-] = \left(\frac{4.00}{5.00}\right)(0.010\ 0)\left(\frac{50.00}{51.00}\right)$$

$$= 0.007\ 84 \text{ M}$$

$$pH = -\log\left(\frac{K_w}{[OH^-]}\right) = 11.89$$

5.00 mL: $H_2O \rightleftharpoons H^+ + OH^-$
$\qquad\qquad\qquad x \quad\quad x$

$$x^2 = K_w \Rightarrow x = 1.0 \times 10^{-7} \text{ M}$$

$$pH = -\log x = 7.00$$

$$5.01 \text{ mL: } [H^+] = \left(\frac{0.01}{55.01}\right)(0.100)$$

$$= 1.82 \times 10^{-5} \text{ M} \Rightarrow pH = 4.74$$

V_a (mL)	pH	V_a	pH	V_a	pH
0.00	12.00	4.50	10.96	5.10	3.74
1.00	11.89	4.90	10.26	5.50	3.05
2.00	11.76	4.99	9.26	6.00	2.75
3.00	11.58	5.00	7.00	8.00	2.29
4.00	11.27	5.01	4.74	10.00	2.08

11-B. The titration reaction is $HCO_2H + OH^- \rightarrow HCO_2^- + H_2O$ and $V_e = 50.0$ mL. For formic acid, $K_a = 1.80 \times 10^{-4}$. Four representative calculations are given:

0.0 mL: $HA \rightleftharpoons H^+ + A^-$
$\qquad 0.050\ 0 - x \quad x \quad x$

$$\frac{x^2}{0.050\ 0 - x} = K_a \Rightarrow x = 2.91 \times 10^{-3}$$

$$\Rightarrow pH = 2.54$$

48.0 mL:
$$HA + OH^- \rightarrow A^- + H_2O$$

Initial:	50	48	—	—
Final:	2	—	48	48

$$pH = pK_a + \log\frac{[A^-]}{[HA]} = 3.745 + \log\frac{48.0}{2.0} = 5.13$$

50.0 mL: $A^- + H_2O \overset{K_b}{\rightleftharpoons} HA + OH^-$
$\qquad\quad F - x \qquad\qquad\quad x \quad x$

$$K_b = \frac{K_w}{K_a} \quad \text{and} \quad F = \left(\frac{50}{100}\right)(0.05)$$

$$\frac{x^2}{0.025\ 0 - x} = 5.56 \times 10^{-11} \Rightarrow x = 1.18 \times 10^{-6} \text{ M}$$

$$pH = -\log\left(\frac{K_w}{x}\right) = 8.07$$

$$60.0 \text{ mL: } [OH^-] = \left(\frac{10.0}{110.0}\right)(0.050\ 0)$$

$$= 4.55 \times 10^{-3} \text{ M} \Rightarrow pH = 11.66$$

V_b (mL)	pH	V_b	pH	V_b	pH
0.0	2.54	45.0	4.70	50.5	10.40
10.0	3.14	48.0	5.13	51.0	10.69
20.0	3.57	49.0	5.44	52.0	10.99
25.0	3.74	49.5	5.74	55.0	11.38
30.0	3.92	50.0	8.07	60.0	11.66
40.0	4.35				

11-C. The titration reaction is $B + H^+ \rightarrow BH^+$ and $V_e = 50.00$ mL. Representative calculations:

$V_a = 0.0$ mL: $\quad B + H_2O \rightleftharpoons BH^+ + OH^-$
$\qquad\qquad\qquad 0.100 - x \qquad\quad x \quad\quad x$

$$\frac{x^2}{0.100 - x} = 2.6 \times 10^{-6} \Rightarrow x = 5.09 \times 10^{-4}$$

$$pH = -\log\left(\frac{K_w}{x}\right) = 10.71$$

$V_a = 20.0$ mL:
$$B + H^+ \rightarrow BH^+$$

Initial:	50.00	20.0	—
Final:	30.0	—	20.0

$$pH = pK_a \text{ (for } BH^+) + \log\frac{[B]}{[BH^+]}$$

$$= 8.41 + \log\frac{30.0}{20.0} = 8.59$$

$V_a = V_e = 50.0$ mL: All B has been converted to the conjugate acid, BH^+. The formal concentration of BH^+ is $(\frac{100}{150})(0.100) = 0.066\ 7$ M. The pH is determined by the reaction

$$BH^+ \rightleftharpoons B + H^+$$
$$0.066\ 7 - x \quad x \quad x$$

$$\frac{x^2}{0.066\ 7 - x} = K_a = \frac{K_w}{K_b} \Rightarrow x$$

$$= 1.60 \times 10^{-5} \Rightarrow pH = 4.80$$

$V_a = 51.0$ mL: There is excess H$^+$:

$$[H^+] = \left(\frac{1.0}{151.0}\right)(0.200) = 1.32 \times 10^{-3}$$

$$\Rightarrow pH = 2.88$$

V_a (mL)	pH	V_a	pH	V_a	pH
0.0	10.71	30.0	8.23	50.0	4.80
10.0	9.01	40.0	7.81	50.1	3.88
20.0	8.59	49.0	6.72	51.0	2.88
25.0	8.41	49.9	5.71	60.0	1.90

11-D. The titration reactions are

$$HO_2CCH_2CO_2H + OH^- \rightarrow$$
$$^-O_2CCH_2CO_2H + H_2O$$

$$^-O_2CCH_2CO_2H + OH^- \rightarrow$$
$$^-O_2CCH_2CO_2^- + H_2O$$

and the equivalence points occur at 25.0 and 50.0 mL. We will designate malonic acid as H$_2$M.

0.0 mL: $H_2M \rightleftharpoons H^+ + HM^-$
 $0.0500 - x \quad x \quad x$

$$\frac{x^2}{0.0500 - x} = K_1 \Rightarrow x = 7.75 \times 10^{-3}$$

$$\Rightarrow pH = 2.11$$

8.0 mL:

	H_2M	+ OH^-	$\rightarrow HM^-$	+ H_2O
Initial:	25	8	—	
Final:	20	—	8	

$$pH = pK_1 + \log\frac{[HM^-]}{[H_2M]} = 2.847 + \log\frac{8}{17}$$
$$= 2.52$$

12.5 mL: $V_b = \frac{1}{2}V_e \Rightarrow pH = pK_1 = 2.85$

19.3 mL:

	H_2M	+ OH^-	$\rightarrow HM^-$	+ H_2O
Initial:	25	19.3	—	
Final:	5.7	—	19.3	

$$pH = pK_1 + \log\frac{19.3}{5.7} = 3.38$$

25.0 mL: At the first equivalence point, H$_2$M has been converted to HM$^-$.

$$[H^+] = \sqrt{\frac{K_1K_2F + K_1K_w}{K_1 + F}}$$

where $F = \left(\frac{50}{75}\right)(0.0500)$

$$[H^+] = 5.23 \times 10^{-5} \text{ M} \Rightarrow pH = 4.28$$

37.5 mL: $V_b = \frac{3}{2}V_e \Rightarrow pH = pK_2 = 5.70$

50.0 mL: At the second equivalence point H$_2$M has been converted to M^{2-}

$$M^{2-} + H_2O \rightleftharpoons HM^- + OH^-$$
$$\left(\frac{50}{100}\right)(0.0500) - x \qquad x \qquad x$$

$$\frac{x^2}{0.0250 - x} = K_{b1} = \frac{K_w}{K_{a2}}$$

$$\Rightarrow x = 1.12 \times 10^{-5} \text{ M}$$

$$\Rightarrow pH = -\log\left(\frac{K_w}{x}\right) = 9.05$$

56.3 mL: There is 6.3 mL of excess NaOH.

$$[OH^-] = \left(\frac{6.3}{106.3}\right)(0.100) = 5.93 \times 10^{-3} \text{ M}$$

$$\Rightarrow pH = 11.77$$

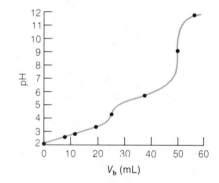

11-E.

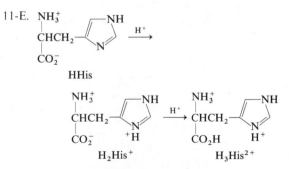

HHis

The equivalence points occur at 25.0 and 50.0 mL.

0 mL: HHis is the second intermediate form derived from the triprotic acid, H$_3$His^{2+}.

$$[H^+] = \sqrt{\frac{K_2K_3(0.0500) + K_2K_w}{K_2 + (0.0500)}}$$
$$= 2.81 \times 10^{-8} \text{ M} \Rightarrow pH = 7.55$$

4.0 mL:

$$HHis + H^+ \rightarrow H_2His^+$$

Initial:	25	4	—
Final:	21	—	4

$$pH = pK_2 + \log \frac{21}{4} = 6.74$$

12.5 mL: $pH = pK_2 = 6.02$

25.0 mL: The histidine has been converted to H_2His^+ at the formal concentration $F = \left(\frac{25}{50}\right) \times (0.050\ 0) = 0.025\ 0$ M.

$$[H^+] = \sqrt{\frac{K_1 K_2 F + K_1 K_w}{K_1 + F}}$$
$$= 1.03 \times 10^{-4} \Rightarrow pH = 3.99$$

26.0 mL:

$$H_2His^+ + H^+ \rightarrow H_3His^{2+}$$

Initial:	25	1	—
Final:	24	—	1

$$pH = pK_1 + \log \frac{24}{1} = 3.08$$

50.0 mL: The histidine has been converted to H_3His at the formal concentration $F = \left(\frac{25}{75}\right)(0.050\ 0) = 0.016\ 7$ M.

$$H_3His^{2+} \rightleftharpoons H_2His^+ + H^+$$
$$0.016\ 7 - x \qquad x \qquad x$$

$$\frac{x^2}{0.016\ 7 - x} = K_1 \rightarrow x = 0.010\ 8\ \text{M}$$

$$\Rightarrow pH = 1.97$$

11-F. Figure 11-2: bromothymol blue: blue → yellow
Figure 11-3: thymol blue: yellow → blue
Figure 11-4: thymolphthalein: colorless → blue

11-G. (a) $A = 2\ 080[HIn] + 14\ 200[In^-]$
(b) $[HIn] = x; [In^-] = 1.84 \times 10^{-4} - x$

$$A = 0.868$$
$$= 2\ 080x + 14\ 200(1.84 \times 10^{-4} - x)$$
$$\Rightarrow x = 1.44 \times 10^{-4}\ \text{M}$$

$$pK_a = pH - \log \frac{[In^-]}{[HIn]}$$
$$= 6.23$$
$$- \log \frac{(1.84 \times 10^{-4}) - (1.44 \times 10^{-4})}{1.44 \times 10^{-4}}$$
$$= 6.79$$

11-H. The titration reaction is $HA + OH^- \rightarrow A^- + H_2O$. It requires one mole of NaOH to react with one mole of HA. Therefore, the formal concentra-

tion of A^- at the equivalent point is

$$\underbrace{\left(\frac{27.63}{127.63}\right)}_{\substack{\text{Dilution factor} \\ \text{for NaOH}}} \times \underbrace{(0.093\ 81)}_{\substack{\text{Initial concentration} \\ \text{of NaOH}}} = 0.020\ 31\ \text{M}$$

Since the pH is 10.99, $[OH^-] = 9.77 \times 10^{-4}$ and we can write

$$A^- + H_2O \rightleftharpoons HA + OH^-$$

$$K_b = \frac{[HA][OH^-]}{[A^-]} = \frac{(9.77 \times 10^{-4})^2}{0.020\ 31 - (9.77 \times 10^{-4})}$$
$$= 4.94 \times 10^{-5}$$

$$K_a = \frac{K_w}{K_b} = 2.03 \times 10^{-10} \Rightarrow pK_a = 9.69$$

For the 19.47 mL point, we have

$$HA + OH^- \rightarrow A^- + H_2O$$

Initial:	27.63	19.47	—
Final:	8.16	—	19.47

$$pH = pK_a + \log \frac{[A^-]}{[HA]} = 9.69 + \log \frac{19.47}{8.16}$$
$$= 10.07$$

11-I. When $V_b = \frac{1}{2}V_e$, $[HA] = [A^-] = 0.033\ 3$ M (using a correction for dilution by NaOH). $[Na^+] = 0.033\ 3$ M, as well. Ionic strength = $0.033\ 3$ M.

$$pK_a = pH - \log \frac{[A^-]\gamma_{A^-}}{[HA]\gamma_{HA}}$$

(from Equation 10-57)

$$= 4.62 - \log \frac{(0.033\ 3)(0.854)}{(0.033\ 3)(1.00)} = 4.69$$

The activity coefficient of A^- was found by interpolation in Table 6-1.

Chapter 12

12-A.

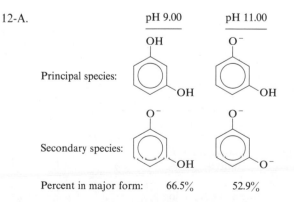

	pH 9.00	pH 11.00
Principal species:	OH	O⁻
Secondary species:		
Percent in major form:	66.5%	52.9%

The percent in the major form was calculated with the formulas for α_0 (Equation 12-19 at pH 9.00) and α_1 (Equation 12-20 at pH 11.00).

12-B.

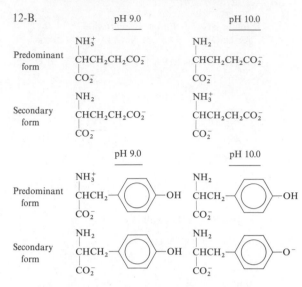

12-C. The isoionic pH is the pH of a solution of pure neutral lysine, which is

$$\underset{\substack{| \\ CO_2^-}}{\overset{\substack{NH_2 \\ |}}{CHCH_2CH_2CH_2CH_2NH_3^+}}$$

$$[H^+] = \sqrt{\frac{K_2K_3F + K_2K_w}{K_2 + F}} \Rightarrow pH = 9.88$$

12-D. We know that the isoelectric point will be near $\frac{1}{2}(pK_2 + pK_3) \approx 9.88$. At this pH, the fraction of lysine in the form H_3L^{2+} is negligible. Therefore, the electroneutrality condition reduces to $[H_2L^+] = [L^-]$, for which the expression, isoelectric pH $= \frac{1}{2}(pK_2 + pK_3) = 9.88$, applies.

12-E.
$$HF + NH_3 \overset{K}{\rightleftharpoons} F^- + NH_4^+$$

$$K = \frac{K_aK_b}{K_w} = \frac{(6.8 \times 10^{-4})(K_w/5.70 \times 10^{-10})}{K_w}$$

$$= 1.2 \times 10^6$$

Since $K \gg 1$, this falls under Case 1 of Section 12-4.

(a)

	HF	$+ NH_3 \rightleftharpoons$	F^-	$+ NH_4^+$
Relative initial amount:	20.0	14.0	—	—
Relative final amount:	6.0	—	14.0	14.0

$$pH = pK_a + \log\frac{[F^-]}{[HF]} = 3.17 + \log\frac{14.0}{6.0} = 3.54$$

(b)

	HF	$+ NH_3 \rightleftharpoons$	F^-	$+ NH_4^+$
Relative initial amount:	14.0	20.0	—	—
Relative final amount:	—	6.0	14.0	14.0

$$pH = pK_{BH^+} + \log\frac{[NH_3]}{[NH_4^+]}$$
$$= 9.244 + \log\frac{6.0}{14.0} = 8.88$$

12-F. (a) $[(CH_3CH_2)_4N^+][HCO_2^-]$ is simply a monoprotic weak base.

$$HCO_2^- + H_2O \rightleftharpoons HCO_2H + OH^-$$
$$\quad 0.100 - x \qquad\qquad x \qquad x$$

$$\frac{x^2}{0.100 - x} = K_b = \frac{K_w}{K_a} \Rightarrow x = 2.36 \times 10^{-6} \text{ M}$$
$$\Rightarrow pH = 8.37$$

(b) $[(CH_3CH_2)_3NH^+][HCO_2^-]$ behaves as the intermediate form of a diprotic acid with $pK_1 = 3.745$ (HCO_2H) and $pK_2 = 10.715$ (($CH_3CH_2)_3NH^+$).

$$[H^+] = \sqrt{\frac{K_1K_2(0.100) + K_1K_w}{K_1 + 0.100}} \Rightarrow pH = 7.23$$

12-G.

$$\underset{\substack{| \\ CO_2^-}}{\overset{\substack{NH_3^+ \\ |}}{CH{-}CH_2}}\!\!-\!\!\bigcirc\!\!-\!OH + NH_3 \overset{K}{\rightleftharpoons}$$

HA $\qquad\qquad$ B

$$\underset{\substack{| \\ CO_2^-}}{\overset{\substack{NH_2 \\ |}}{CH{-}CH_2}}\!\!-\!\!\bigcirc\!\!-\!OH + NH_4^+$$

A$^-$ $\qquad\qquad$ BH$^+$

$$K = \frac{K_aK_b}{K_w} = \frac{K_2 \text{ (for tyrosine)} \cdot K_b \text{(for NH}_3)}{K_w}$$
$$= 1.14$$

Since K is not large, this falls under Case 2 of Section 12-4.

	HA	$+$ B	$\rightleftharpoons A^- +$	BH$^+$
Initial millimoles:	0.250	0.225	—	—
Final millimoles:	0.250$-x$	0.225$-x$	x	x

$$\frac{x^2}{(0.250 - x)(0.225 - x)} = 1.14 \Rightarrow x = 0.122$$

$$pH = pK_a + \log\frac{[A^-]}{[HA]} = 9.19 + \log\frac{0.122}{0.250 - 0.122}$$
$$= 9.17$$

12-H. The answer is K_2HPO_4, since $\frac{1}{2}(pK_2 + pK_3) = 9.77$ for phosphoric acid. The pH values of the other solutions are estimated to be: potassium acid phthalate—$\frac{1}{2}(pK_1 + pK_2) = 4.18$; monosodium malonate—$\frac{1}{2}(pK_1 + pK_2) = 4.27$; glycine—$\frac{1}{2}(pK_1 + pK_2) = 6.06$; ammonium bicarbonate—$\frac{1}{2}[K_1 \text{ (of } H_2CO_3) + K_a \text{ (for } NH_4^+)] = 7.80$.

12-I. The pH of the solution is 7.50, and the total concentration of indicator is 5.00×10^{-5} M. At pH 7.50, there is a negligible amount of H_2In, since $pK_1 = 1.00$. We can write

$$[HIn^-] + [In^{2-}] = 5.0 \times 10^{-5}$$

$$pH = pK_2 + \log \frac{[In^{2-}]}{[HIn^-]}$$

$$7.50 = 7.95 + \log \frac{[In^{2-}]}{5.00 \times 10^{-5} - [In^{2-}]}$$

$$\Rightarrow [In^{2-}] = 1.31 \times 10^{-5} \text{ M}$$

$$[HIn^-] = 3.69 \times 10^{-5} \text{ M}$$

$$A_{435} = \varepsilon_{435}[HIn^-] + \varepsilon_{435}[In^{2-}]$$
$$= (1.80 \times 10^4)(3.69 \times 10^5)$$
$$+ (1.15 \times 10^4)(1.31 \times 10^{-5}) = 0.815$$

Chapter 13

13-A. For every mole of K^+ entering the first reaction, four moles of EDTA are produced in the second reaction.

Moles of EDTA = moles of Zn^{2+} used in titration

$$[K^+] = \frac{(\frac{1}{4})(\text{moles of } Zn^{2+})}{\text{volume of original sample}}$$

$$= \frac{(\frac{1}{4})[28.73 \,(\pm 0.03)][0.043 \,7 \,(\pm 0.000 \,1)]}{250.0 \,(\pm 0.1)}$$

$$= \frac{[\frac{1}{4}(\pm 0\%)][28.73 \,(\pm 0.104\%)][0.043 \,7(\pm 0.229\%)]}{250.0 \,(\pm 0.040 \,0\%)}$$

$$= 1.256 \,(\pm 0.255\%) \times 10^{-3} \text{ M}$$

$$= 1.256 \,(\pm 0.003) \text{ mM}$$

13-B. Total $Fe^{3+} + Cu^{2+}$ in 25.00 mL = (16.06 mL) × (0.050 83 M) = 0.816 3 mmol.

Second titration:

Millimoles EDTA used: $(25.00)(0.050\,83) = 1.270\,8$
Millimoles Pb^{2+} needed: $(19.77)(0.018\,83) = \underline{0.372\,3}$
Millimoles Fe^{3+} present: (difference) $\quad 0.898\,5$

Since 50.00 mL of unknown was used in the second titration, the millimoles of Fe^{3+} in 25.00 mL are 0.449 2. The millimoles of Cu^{2+} in 25.00 mL are $0.816\,3 - 0.449\,2 = 0.367\,1$ mmol/25.00 mL = 0.014 68 M.

13-C. Designating the total concentration of free EDTA as [EDTA], we can write

$$\frac{[GaY^-]}{[Ga^{3+}][EDTA]} = \alpha_{Y^{4-}} K_f = 7.58 \times 10^{11}$$

Some representative calculations are shown below:

$$0.1 \text{ mL: } [EDTA] = \left(\frac{25.0 - 0.1}{25.0}\right)(0.040\,0)\left(\frac{50.0}{50.1}\right)$$
$$= 0.039\,8 \text{ M}$$

$$[GaY^-] = \left(\frac{0.1}{50.1}\right)(0.080\,0)$$
$$= 1.60 \times 10^{-4} \text{ M}$$

$$[Ga^{3+}] = \frac{(1.32 \times 10^{-12})[GaY^-]}{[EDTA]}$$
$$= 5.31 \times 10^{-15}$$
$$\Rightarrow pGa^{3+} = 14.28$$

25.0 mL: Formal concentration of GaY^-

$$= \left(\frac{25.0}{75.0}\right)(0.080\,0) = 0.026\,7M$$

$$Ga^{3+} \ + \ EDTA \rightleftharpoons \ GaY^-$$

Initial concentration:	—	—	0.026 7
Final concentration:	x	x	0.026 7 − x

$$\frac{0.026\,7 - x}{x^2} = 7.58 \times 10^{11}$$

$$\Rightarrow [Ga^{2+}] = 1.88 \times 10^{-7} \text{ M}$$
$$\Rightarrow pGa^{3+} = 6.73$$

$$26.0 \text{ mL: } Ga^{3+} = \left(\frac{1.0}{76.0}\right)(0.080\,0)$$
$$= 1.05 \times 10^{-3} \text{ M}$$
$$\Rightarrow pGa^{3+} = 2.98$$

Summary:

Volume (mL)	pGa^{3+}	Volume	pGa^{3+}
0.1	14.28	24.0	10.50
5.0	12.48	25.0	6.73
10.0	12.06	26.0	2.98
15.0	11.70	30.0	2.30
20.0	11.28		

13-D.
$$HY^{3-} \rightleftharpoons H^+ + Y^{4-} \qquad K_6$$
$$H_2Y^{2-} \rightleftharpoons H^+ + HY^{3-} \qquad K_5$$
$$\overline{H_2Y^{2-} \rightleftharpoons 2H^+ + Y^{4-}} \qquad K = K_5 K_6$$

$$= \frac{[H^+]^2[Y^{4-}]}{[H_2Y^{2-}]}$$

$$[H_2Y^{2-}] = \frac{[H^+]^2[Y^{4-}]}{K_5K_6} = \frac{[H^+]^2\alpha_{Y^{4-}}[EDTA]}{K_5K_6}$$

Using the values $[H^+] = 10^{-4.00}$, $\alpha_{Y^{4-}} = 3.8 \times 10^{-9}$, and $[EDTA] = 1.88 \times 10^{-7}$ gives $[H_2Y^{2-}] = 1.8 \times 10^{-7}$ M.

13-E. (a) One volume of Fe^{3+} will require two volumes of EDTA to reach the equivalence point. The formal concentration of FeY^- at the equivalence point is $(\frac{1}{3})(0.010\ 0) = 0.003\ 33$ M.

$$Fe^{3+} + EDTA \rightleftharpoons FeY^-$$
$$x \qquad x \qquad 0.003\ 33 - x$$

$$\frac{0.003\ 33 - x}{x^2} = \alpha_{Y^{4-}}K_f$$

$$\Rightarrow x = [Fe^{3+}] = 8.8 \times 10^{-8}\ M$$

(b) Since the pH is 2.00, the *ratio* $[H_3Y^-]/[H_2Y^{2-}]$ is constant throughout the *entire* titration.

$$\frac{[H_2Y^{2-}][H^+]}{[H_3Y^-]} = K_4 \Rightarrow \frac{[H_3Y^-]}{[H_2Y^{2-}]} = \frac{[H^+]}{K_4}$$
$$= 4.6$$

13-F. K_f for $CoY^{2-} = 10^{16.31} = 2.0 \times 10^{16}$.

$$\alpha_{Y^{4-}} = 0.054 \text{ at pH } 9.00$$

$$\alpha_{Co^{2+}} = \frac{1}{1 + \beta_1[C_2O_4^{2-}] + \beta_2[C_2O_4^{2-}]^2}$$
$$= 6.8 \times 10^{-6}$$

(using $\beta_1 = K_1 = 10^{4.69}$ and $\beta_2 = K_1K_2 = 10^{4.69}\ 10^{2.46}$)

$$K_f' = \alpha_{Y^{4-}}K_f = 1.1 \times 10^{15}$$

$$K_f'' = \alpha_{Co^{2+}}\alpha_{Y^{4-}}K_f = 7.5 \times 10^9$$

0 mL: $[Co^{2+}] = \alpha_{Co^{2+}}(1.00 \times 10^{-3})$
$$= 6.8 \times 10^{-9}\ M \Rightarrow pCo^{2+} = 8.17$$

1.00 mL: $C_{Co^{2+}} = \left(\frac{1.00}{2.00}\right)(1.00 \times 10^{-3})\left(\frac{20.00}{21.00}\right)$

fraction   initial   dilution
remaining concentration factor

$$= 4.76 \times 10^{-4}\ M$$

$$[Co^{2+}] = \alpha_{Co^{2+}}C_{Co^{2+}}$$
$$= 3.2 \times 10^{-9}\ M \Rightarrow pCo^{2+} = 8.49$$

2.00 mL: This is the equivalence point.

$$C_{Co^{2+}} + EDTA \overset{K_f''}{\rightleftharpoons} CoY^{2-}$$
$$x \qquad x \qquad \frac{20.00}{22.00}(1.00 \times 10^{-3}) - x$$

$$K_f'' = \frac{9.09 \times 10^{-4} - x}{x^2}$$

$$\Rightarrow x = 3.5 \times 10^{-7}\ M = C_{Co^{2+}}$$

$$[Co^{2+}] = \alpha_{Co^{2+}}C_{Co^{2+}}$$
$$= 2.4 \times 10^{-12}\ M \Rightarrow pCo^{2+} = 11.63$$

3.00 mL:

Concentration of excess EDTA $= \frac{1.00}{23.00}(1.00 \times 10^{-2})$
$$= 4.35 \times 10^{-4}\ M$$

Concentration of $CoY^{2-} = \frac{20.00}{23.00}(1.00 \times 10^{-3})$
$$= 8.70 \times 10^{-4}\ M$$

Knowing [EDTA] and $[CoY^{2-}]$, we can use the K_f' equilibrium to find $[Co^{2+}]$:

$$K_f' = \frac{[CoY^{2-}]}{[Co^{2+}][EDTA]} = \frac{[8.70 \times 10^{-4}]}{[Co^{2+}][4.35 \times 10^{-4}]}$$

$$\Rightarrow [Co^{2+}] = 1.8 \times 10^{-15}\ M \Rightarrow pCo^{2+} = 14.74$$

13-G. 25.0 mL of 0.120 M iminodiacetic acid = 3.00 mmol

25.0 mL of 0.050 0 M $Cu^{2+} = 1.25$ mmol

$$Cu^{2+} + \overset{2 \text{ iminodiacetic}}{\underset{\text{acid}}{}} \rightleftharpoons CuX_2^{2-}$$

	Cu^{2+}	2 iminodiacetic acid	CuX_2^{2-}
Initial millimoles:	1.25	3.00	—
Final millimoles:	—	0.50	1.25

$$\frac{[CuX_2^{2-}]}{[Cu^{2+}][X^{2-}]^2} = K_f$$

$$\frac{[1.25/50.0]}{[Cu^{2+}][(0.50/50.0)(4.6 \times 10^{-3})]^2} = 3.5 \times 10^{16}$$

$$\Rightarrow [Cu^{2+}] = 3.4 \times 10^{-10}\ M$$

Chapter 14

14-A. The cell voltage will be 1.35 V because all activities are unity.

$$I = P/E = 0.010\ 0\ W/1.35\ V = 7.41 \times 10^{-3}\ C/s$$
$$= 7.68 \times 10^{-8}\ mol\ e^-/s = 2.42\ mol\ e^-/365\ days$$
$$= 1.21\ mol\ HgO/365\ days = 0.262\ kg\ HgO$$
$$= 0.578\ lb$$

14-B. (a) $I_2(s) + 6H_2O \rightleftharpoons$

$$2IO_3^- + 12H^+ + 10e^- \qquad E° = -1.210\ V$$
$$5Br_2(aq) + 10e^- \rightleftharpoons 10Br^- \qquad E° = 1.098\ V$$
$$\overline{I_2(s) + 5Br_2(aq) + 6H_2O \rightleftharpoons} \qquad E° = -0.112\ V$$
$$2IO_3^- + 10Br^- + 12H^+$$

$$K = 10^{10(-0.112)/0.059\ 16} = 1 \times 10^{-19}$$

(b) $\quad Cr^{2+} + 2e^- \rightleftharpoons Cr(s) \qquad E° = -0.89$ V

$\underline{Fe(s) \rightleftharpoons Fe^{2+} + 2e^- \qquad E° = \quad 0.44}$ V

$Cr^{2+} + Fe(s) \rightleftharpoons \qquad\qquad E° = -0.45$ V
$\qquad Cr(s) + Fe^{2+}$

$$K = 10^{2(-0.45)/0.059\ 16} = 10^{-17}$$

(c) $\quad Mg(s) \rightleftharpoons Mg^{2+} + 2e^- \qquad E° = 2.360$ V

$\underline{Cl_2(g) + 2e^- \rightleftharpoons 2Cl^- \qquad E° = 1.360}$ V

$Mg(s) + Cl_2(g) \rightleftharpoons \qquad\quad E° = 3.720$ V
$\qquad Mg^{2+} + 2Cl^-$

$$K = 10^{2(3.720)/0.059\ 16} = 6 \times 10^{125}$$

(d) $\quad 2[MnO_2(s) + 2H_2O \rightleftharpoons$
$\qquad MnO_4^- + 4H^+ + 3e^-] \quad E° = -1.692$ V
$\underline{3[MnO_2(s) + 4H^+ + 2e^- \rightleftharpoons}$
$\underline{\qquad Mn^{2+} + 2H_2O] \quad E° = \quad 1.230}$ V
$5MnO_2(s) + 4H^+ \rightleftharpoons \qquad E° = -0.462$ V
$\quad 2MnO_4^- + 3Mn^{2+} + 2H_2O$

$$K = 10^{6(-0.462)/0.059\ 16} = 1 \times 10^{-47}$$

An alternative way to answer d is the following:

$5[MnO_2(s) + 4H^+ + 2e^- \rightleftharpoons$
$\qquad Mn^{2+} + 2H_2O \quad E° = \quad 1.230$ V
$\underline{2[Mn^{2+} + 4H_2O \rightleftharpoons}$
$\underline{\qquad MnO_4^- + 8H^+ + 5e^-] \quad E° = -1.507}$ V
$5MnO_2(s) + 4H^+ \rightleftharpoons \qquad E° = -0.277$ V
$\quad 2MnO_4^- + 3Mn^{2+} + 2H_2O$

$$K = 10^{10(-0.277)/0.059} = 2 \times 10^{-47}$$

(e) $\quad Ag^+ + e^- \rightleftharpoons Ag(s) \qquad E° = \quad 0.799$ V
$Ag(s) + 2S_2O_3^{2-} \rightleftharpoons$
$\underline{\quad Ag(S_2O_3)_2^{3-} + e^- \qquad E° = -0.017}$ V
$Ag^+ + 2S_2O_3^{2-} \rightleftharpoons \qquad E° = \quad 0.782$ V
$\quad Ag(S_2O_3)_2^{3-}$

$$K = 10^{0.782/0.059\ 16} = 2 \times 10^{13}$$

(f) $\quad Cu(s) \rightleftharpoons Cu^+ + e^- \qquad\quad E° = -0.518$ V
$\underline{CuI(s) + e^- \rightleftharpoons Cu(s) + I^- \quad E° = -0.185}$ V
$CuI \rightleftharpoons Cu^+ + I^- \qquad\qquad E° = -0.703$ V

$$K = 10^{-0.703/0.059\ 16} = 1 \times 10^{-12}$$

14-C. (a) Anode: $Fe(s) \rightleftharpoons Fe^{2+} + 2e^- \quad E° = 0.44$ V
Cathode: $Br_2(l) + 2e^- \rightleftharpoons 2Br^- \quad E° = 1.078$ V

$$E = 1.51_8 - \frac{0.059\ 16}{2} \log (0.010)(0.050)^2$$

$$= 1.65 \text{ V}$$

(b) Anode: $Cu(s) \rightleftharpoons Cu^{2+} + 2e^- \quad E° = -0.339$ V
Cathode: $Fe^{2+} + 2e^- \rightleftharpoons Fe(s) \quad E° = -0.44$ V

$$E = -0.77_9 - \frac{0.059\ 16}{2} \log \frac{0.020}{0.050}$$

$$= -0.77 \text{ V}$$

(c) Anode: $2Hg(l) + 2Cl^- \rightleftharpoons \qquad E° = -0.268$ V
$\qquad Hg_2Cl_2(s) + 2e^-$
Cathode: $Cl_2(g) + 2e^- \rightleftharpoons 2Cl^- \quad E° = 1.360$ V

$$E = 1.092 - \frac{0.059\ 16}{2} \log \frac{(0.040)^2}{(0.060)^2(0.50)}$$

$$= 1.094 \text{ V}$$

14-D. (a) $E = -0.799 - 0.059\ 16 \log \dfrac{[Ag^+]P_{H_2}^{1/2}}{[H^+]}$

(b) $[Ag^+] = \dfrac{K_{sp}}{[I^-]} = 8.3 \times 10^{-17}/0.10$

$$= 8.3 \times 10^{-16} \text{ M}$$

$$E = -0.799 - 0.059\ 16 \log \frac{(8.3 \times 10^{-16})\sqrt{0.20}}{(0.10)}$$

$$= 0.055 \text{ V}$$

(c) $\qquad E° = -E°$ (for Reaction 1)

$$E = E° - 0.059\ 16 \log \frac{P_{H_2}^{1/2}}{[H^+][I^-]}$$

$0.055 = -E°$ (for Reaction 1) $- 0.059\ 16$

$$\times \log \frac{\sqrt{0.20}}{(0.10)(0.10)}$$

$E°$ (for Reaction 1) $= -0.153$ V

14-E. Left half-cell:
$\qquad Cu(s) \rightleftharpoons Cu^{2+} + 2e^- \qquad E° = -0.339$ V
Right half-cell:
$\qquad 2Ag(CN)_2^- + 2e^- \rightleftharpoons$
$\qquad\qquad 2Ag(s) + 4CN^- \qquad E° = -0.310$ V

$$E = -0.649 - \frac{0.059\ 16}{2} \log \frac{[Cu^{2+}][CN^-]^4}{[Ag(CN)_2^-]^2}$$

At pH 8.21, $\dfrac{[CN^-]}{[HCN]} = \dfrac{K_a}{[H^+]} = 0.10 \Rightarrow$

$$[CN^-] = 0.10\,[HCN]$$

But since $[CN^-] + [HCN] = 0.10$ M, $[CN^-] = 0.009\ 1$ M

$$E = -0.649 - \frac{0.059\ 16}{2} \log \frac{(0.030)(0.009\ 1)^4}{(0.010)^2}$$

$$= -0.481 \text{ V}$$

14-F. (a) $PuO_2^+ + e^- + 4H^+ \rightleftharpoons Pu^{4+} + 2H_2O$

$PuO_2^{2+} \to PuO_2^+ \qquad \Delta G = -1F\,(0.966)$
$PuO_2^+ \to Pu^{4+} \qquad \Delta G = -1F\,E°$
$\underline{Pu^{4+} \to Pu^{3+} \qquad \Delta G = -1F\,(1.006)}$
$PuO_2^{2+} \to Pu^{3+} \qquad \Delta G = -3F\,(1.021)$

$-3F\,(1.021) = -1F\,(0.966) - 1F\,E°$
$\qquad\qquad -1F\,(1.006) \Rightarrow E° = 1.091$ V

(b) $2PuO_2^{2+} + 2e^- \rightleftharpoons 2PuO_2^+$ $\qquad E° = \quad 0.966$ V

$\underline{H_2O \rightleftharpoons \frac{1}{2}O_2(g) + 2H^+ + 2e^- \qquad E° = -1.229 \text{ V}}$

$2PuO_2^{2+} + H_2O \rightleftharpoons$
$\qquad 2PuO_2^+ + \frac{1}{2}O_2(g) + 2H^+ \qquad E° = -0.263$ V

$$E = -0.263 - \frac{0.059\ 16}{2} \log \frac{[\cancel{PuO_2^+}]^2 P_{O_2}^{1/2}[H^+]^2}{[\cancel{PuO_2^{2+}}]^2}$$

$$\text{(since } [PuO_2^+] = [PuO_2^{2+}])$$

$$= -0.263 - \frac{0.059\ 16}{2} \log \sqrt{0.20}\ (0.010)^2$$

$$= -0.134 \text{ V}$$

Water will not be oxidized at pH 2.00. At pH 7.00, $E = +0.16$ V. Water will be oxidized.

14-G. Left half-reaction:

$$2Hg(l) + 2Cl^- \rightleftharpoons Hg_2Cl_2(s) + 2e^- \quad E° = -0.268 \text{ V}$$

Right half-reaction:

$$2H^+ + 2e^- \rightleftharpoons H_2(g) \quad E° = 0 \text{ V}$$

$$E = -0.268 - \frac{0.059\ 16}{2} \log \frac{P_{H_2}}{[H^+]^2[Cl^-]^2}$$

We find $[H^+]$ in the right half-cell by considering the acid–base chemistry of KHP, the intermediate form of a diprotic acid.

$$[H^+] = \sqrt{\frac{K_1 K_2(0.050) + K_1 K_w}{K_1 + 0.050}} = 6.5 \times 10^{-5} \text{ M}$$

$$E = -0.268 - \frac{0.059\ 16}{2} \log \frac{1}{(6.5 \times 10^{-5})^2(0.10)^2}$$

$$= -0.575 \text{ V}$$

14-H. Left half-reaction:

$$Hg(l) \rightleftharpoons Hg^{2+} + 2e^- \qquad E° = -0.852 \text{ V}$$

Right half-reaction:

$$2H^+ + 2e^- \rightleftharpoons H_2(g) \qquad E° = 0 \text{ V}$$

$$E = -0.852 - \frac{0.059\ 16}{2} \log \frac{[Hg^{2+}]P_{H_2}}{[H^+]^2}$$

$$-0.321 = -0.852 - \frac{0.059\ 16}{2} \log \frac{[Hg^{2+}](1)}{(1)^2}$$

$$\Rightarrow [Hg^{2+}] = 1.1 \times 10^{-18} \text{ M}$$

Since $[Hg^{2+}]$ is so small, $[HgI_4^{2-}] = 0.001\ 0$ M. To make this much HgI_4^{2-}, the concentration of I^- must have been reduced from 0.010 M to 0.006 0 M, since one Hg^{2+} ion reacts with four I^- ions.

$$K = \frac{[HgI_4^{2-}]}{[Hg^{2+}][I^-]^4} = \frac{(0.001\ 0)}{(1.1 \times 10^{-18})(0.006\ 0)^4}$$
$$= 7 \times 10^{23}$$

14-I. $CuY^{2-} + 2e^- \rightleftharpoons Cu(s) + Y^{4-} \quad E_1°$

$\underline{Cu(s) \rightleftharpoons Cu^{2+} + 2e^- \qquad E_2° = -0.339 \text{ V}}$

$CuY^{2-} \overset{1/K_f}{\rightleftharpoons} Cu^{2+} + Y^{4-} \qquad E_3°$

$$E_3° = \frac{0.059\ 16}{2} \log \frac{1}{K_f} = -0.556 \text{ V}$$

$$E_1° = E_3° - E_2° = -0.217 \text{ V}$$

14-J. To compare glucose and H_2 at pH = 0, we need to know $E°$ for each. For H_2, $E° = 0$ V. For glucose, we find $E°$ from $E°'$:

$$HA \qquad + 2H^+ + 2e^- \rightleftharpoons \quad G \quad + H_2O$$
Gluconic acid $\qquad\qquad\qquad\qquad$ Glucose

$$E = E° - \frac{0.059\ 16}{2} \log \frac{[G]}{[HA][H^+]^2}$$

But $F_G = [G]$ and $[HA] = \dfrac{[H^+]F_{HA}}{[H^+] + K_a}$. Putting these into the Nernst equation gives

$$E = E° - \frac{0.059\ 16}{2} \log \frac{F_G}{\left(\dfrac{[H^+]F_{HA}}{[H^+] + K_a}\right)[H^+]^2}$$

$$= \underbrace{E° - \frac{0.059\ 16}{2} \log \frac{[H^+] + K_a}{[H^+]^3}}_{\text{This is } E°' = -0.45 \text{ V}}$$

$$- \frac{0.059\ 16}{2} \log \frac{F_G}{F_{HA}}$$

$$-0.45 \text{ V} = E° - \frac{0.059\ 16}{2} \log \frac{10^{-7.00} + 10^{-3.56}}{(10^{-7.00})^3}$$

$$\Rightarrow E° = +0.066 \text{ V for glucose.}$$

Since $E°$ for H_2 is more negative than $E°$ for glucose, H_2 is the stronger reducing agent at pH 0.00.

14-K. (a) Each H^+ must provide $\frac{1}{2}(34.5$ kJ$)$ when it passes from outside to inside.

$$\Delta G = -(\tfrac{1}{2})(34.5 \times 10^3 \text{ J}) = -RT \ln \frac{\mathscr{A}_{\text{high}}}{\mathscr{A}_{\text{low}}}$$

$$\frac{\mathscr{A}_{\text{high}}}{\mathscr{A}_{\text{low}}} = 1.05 \times 10^3$$

$$\Rightarrow \Delta pH = \log (1.05 \times 10^3)$$
$$= 3.02 \text{ pH units}$$

(b) $\Delta G = -nFE$ (where n = charge of $H^+ = 1$)

$$-(\tfrac{1}{2})(34.5 \times 10^3 \text{ J}) = -1FE \Rightarrow E = 0.179 \text{ V}$$

(c) If $\Delta pH = 1.00$, $\mathscr{A}_{\text{high}}/\mathscr{A}_{\text{low}} = 10$.

$$\Delta G(pH) = -RT \ln 10 = -5.7 \times 10^3 \text{ J}$$

$$\Delta G(\text{electric}) = [\tfrac{1}{2}(34.5) - 5.7] \text{ kJ} = 11.5 \text{ kJ}$$

$$E = \frac{\Delta G(\text{electric})}{F} = 0.120 \text{ V}$$

15-A. The reaction at the silver electrode (written as a reduction) is $Ag^+ + e^- \rightleftharpoons Ag(s)$, and the cell voltage is written as

$$E = E_+ - E_-$$

$$= (+0.200) - \left(0.799 - 0.059\,16 \log \frac{1}{[Ag^+]}\right)$$

$$= -0.599 - 0.059\,16 \log[Ag^+]$$

Titration reactions:

$$Br^- + Ag^+ \rightarrow AgBr(s) \qquad (K_{sp} = 5.0 \times 10^{-13})$$

$$Br^- + Tl^+ \rightarrow TlBr(s) \qquad (K_{sp} = 3.6 \times 10^{-6})$$

The two equivalence points are at 25.0 and 50.0 mL. Between 0 and 25 mL, there is unreacted Ag^+ in the solution.

1.0 mL: $[Ag^+] = \left(\frac{24.0}{25.0}\right)(0.050\,0)\left(\frac{100.0}{101.0}\right)$

$$\underset{\text{of } Ag^+}{\underset{\text{Initial concentration}}{}}$$

$$= 0.047\,5 \text{ M} \Rightarrow E = -0.521 \text{ V}$$

15.0 mL: $[Ag^+] = \left(\frac{10.0}{25.0}\right)(0.050\,0)\left(\frac{100.0}{115.0}\right)$

$$= 0.017\,4 \text{ M} \Rightarrow E = -0.495 \text{ V}$$

24.0 mL: $[Ag^+] = \left(\frac{1.0}{25.0}\right)(0.050\,0)\left(\frac{100.0}{124.0}\right)$

$$= 0.001\,61 \text{ M} \Rightarrow E = -0.434 \text{ V}$$

24.9 mL: $[Ag^+] = \left(\frac{0.10}{25.0}\right)(0.050\,0)\left(\frac{100.0}{124.9}\right)$

$$= 1.60 \times 10^{-4} \text{ M}$$

$$\Rightarrow E = -0.374 \text{ V}$$

Between 25 mL and 50 mL, all AgBr has precipitated and TlBr is in the process of precipitating. There is some unreacted Tl^+ left in solution in this region.

25.2 mL:

$$[Tl^+] = \left(\frac{24.8}{25.0}\right)(0.050\,0)\left(\frac{100.0}{125.2}\right)$$

$$= 3.96 \times 10^{-2} \text{ M}$$

$$[Br^-] = K_{sp} \text{ (for TlBr)}/[Tl^+] = 9.0_9 \times 10^{-5} \text{ M}$$

$$[Ag^+] = K_{sp} \text{ (for AgBr)}/[Br^-] = 5.5 \times 10^{-9} \text{ M}$$

$$E = -0.599 - 0.059\,16 \log[Ag^+]$$

$$= -0.110 \text{ V}$$

35.0 mL: $[Tl^+] = \left(\frac{15.0}{25.0}\right)(0.050\,0)\left(\frac{100.0}{135.0}\right)$

$$= 0.022\,2 \text{ M} \Rightarrow$$

$$[Br^-] = 1.62 \times 10^{-4} \text{ M} \Rightarrow$$

$$[Ag^+] = 3.08 \times 10^{-9} \text{ M} \Rightarrow$$

$$E = -0.095 \text{ V}$$

50.0 mL is the second equivalence point, at which $[Tl^+] = [Br^-]$.

50.0 mL:

$$[Tl^+][Br^-] = K_{sp} \text{ (for TlBr)} \Rightarrow [Tl^+] = \sqrt{K_{sp}}$$

$$= 1.90 \times 10^{-3} \text{ M} \Rightarrow$$

$$[Br^-] = 1.90 \times 10^{-3} \text{ M} \Rightarrow$$

$$[Ag^+] = 2.64 \times 10^{-10} \text{ M} \Rightarrow E = -0.032 \text{ V}$$

At 60.0 mL, there is excess Br^- in the solution.

60.0 mL:

$$[Br^-] = \left(\frac{10.0}{160.0}\right)(0.200) = 0.012\,5 \text{ M} \Rightarrow$$

$$[Ag^+] = 4.00 \times 10^{-11} \text{ M} \Rightarrow E = +0.016 \text{ V}$$

15-B. The cell voltage is given by Equation C, in which K_f is the formation constant for $Hg(EDTA)^{2-}$ $(=5.0 \times 10^{21})$. To find the voltage, we must calculate $[HgY^{2-}]$ and $[Y^{4-}]$ at each point. The concentration of HgY^{2-} is 1.0×10^{-4} M when $V = 0$, and is thereafter affected only by dilution because $K_f(HgY^{2-}) \gg K_f(MgY^{2-})$. The concentration of Y^{4-} is found from the Mg–EDTA equilibrium at all but the first point. At $V = 0$ mL, the Hg–EDTA equilibrium determines $[Y^{4-}]$.

0 mL:

$$\frac{[HgY^{2-}]}{[Hg^{2+}][EDTA]} = \alpha_{Y^{4-}} K_f \text{ (for } HgY^{4-})$$

$$\frac{1.0 \times 10^{-4} - x}{(x)(x)} = 1.8 \times 10^{21} \Rightarrow x = [EDTA]$$

$$= 2.36 \times 10^{-13} \text{ M}$$

$$[Y^{4-}] = \alpha_{Y^{4-}}[EDTA]$$

$$= 8.49 \times 10^{-14} \text{ M}$$

Using Equation C, we write

$$E = 0.852 - 0.241$$

$$- \frac{0.059\,16}{2} \log \frac{5.0 \times 10^{21}}{1.0 \times 10^{-4}}$$

$$- \frac{0.059\,16}{2} \log(8.49 \times 10^{-14})$$

$$= 0.237 \text{ V}$$

10.0 mL: Since $V_e = 25.0$ mL, $\frac{10}{25}$ of the Mg^{2+} is in the form MgY^{2-}, and $\frac{15}{25}$ is in the form Mg^{2+}.

$$[Y^{4-}] = \frac{[MgY^{2-}]}{[Mg^{2+}]} \bigg/ K_f \text{ (for } MgY^{2-})$$

$$= \left(\frac{10}{15}\right) \bigg/ 6.2 \times 10^8 = 1.08 \times 10^{-9} \text{ M}$$

$$[HgY^{2-}] = \left(\frac{50.0}{60.0}\right)(1.0 \times 10^{-4}) = 8.33 \times 10^{-5} \text{ M}$$

Dilution
factor

$$E = 0.852 - 0.241$$
$$- \frac{0.059\,16}{2} \log \frac{5.0 \times 10^{21}}{8.33 \times 10^{-5}}$$
$$- \frac{0.059\,16}{2} \log(1.08 \times 10^{-9})$$
$$= 0.114 \text{ V}$$

20.0 mL:

$$[Y^{4-}] = \left(\frac{20}{5}\right) \bigg/ 6.2 \times 10^8 = 6.45 \times 10^{-9} \text{ M}$$

$$[HgY^{2-}] = \left(\frac{50.0}{70.0}\right)(1.0 \times 10^{-4}) = 7.14 \times 10^{-5} \text{ M}$$

$$\Rightarrow E = 0.089 \text{ V}$$

24.9 mL: $[Y^{4-}] = \left(\frac{24.9}{0.1}\right) \bigg/ 6.2 \times 10^8$

$$= 4.02 \times 10^{-7} \text{ M}$$

$$[HgY^{2-}] = \left(\frac{50.0}{74.9}\right)(1.0 \times 10^{-4})$$

$$= 6.68 \times 10^{-5} \text{ M}$$

$$\Rightarrow E = 0.035 \text{ V}$$

25.0 mL: This is the equivalence point, at which $[Mg^{2+}] = [EDTA]$.

$$\frac{[MgY^{2-}]}{[Mg^{2+}][EDTA]} = \alpha_{Y^{4-}} K_f \text{ (for } MgY^{2-})$$

$$\frac{\left(\frac{50.0}{75.0}\right)(0.010\,0) - x}{x^2} = 2.22 \times 10^8$$

$$\Rightarrow x = 5.48 \times 10^{-6} \text{ M}$$

$$[Y^{4-}] = \alpha_{Y^{4-}}(5.48 \times 10^{-6})$$

$$= 1.97 \times 10^{-6} \text{ M}$$

$$[HgY^{2-}] = \left(\frac{50.0}{75.0}\right)(1.0 \times 10^{-4})$$

$$= 6.67 \times 10^{-5} \text{ M}$$

$$\Rightarrow E = 0.014 \text{ V}$$

26.0 mL: Now there is excess EDTA in the solution:

$$[Y^{4-}] = \alpha_{Y^{4-}}[EDTA] = (0.36)\left[\left(\frac{1.0}{76.0}\right)(0.020\,0)\right]$$

$$= 9.47 \times 10^{-5} \text{ M}$$

$$[HgY^{2-}] = \left(\frac{50.0}{76.0}\right)(1.0 \times 10^{-4}) = 6.58 \times 10^{-5} \text{ M}$$

$$\Rightarrow E = -0.036 \text{ V}$$

15-C. At intermediate pH, the voltage will be constant at 100 mV. When $[OH^-] \approx [F^-]/10 = 10^{-6}$ M (pH = 8), the electrode begins to respond to OH^- and the voltage will decrease (i.e., the electrode potential will change in the same direction as if more F^- were being added). Near pH = 3.17 ($= pK_a$ for HF), F^- reacts with H^+ and the concentration of free F^- decreases. At pH = 1.17, $[F^-] \approx 1\%$ of 10^{-5} M $= 10^{-7}$ M, and $E \approx 100 + 2(59) = 218$ mV. A qualitative sketch of this behavior is shown below. The slope at high pH is less than 59 mV/pH unit, because the response of the electrode to OH^- is less than the response to F^-.

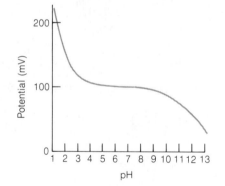

15-D. (a) For 1.00 mM Na^+ at pH 8.00, we can write

$$E = \text{constant} + 0.059\,16 \log([Na^+] + 36[H^+])$$
$$-0.038 = \text{constant} + 0.059\,16$$
$$\times \log[(1.00 \times 10^{-3}) + (36 \times 10^{-8})]$$
$$\Rightarrow \text{constant} = +0.139 \text{ V}$$

For 5.00 mM Na^+ at pH 8.00, we have

$$E = +0.139 + 0.059\,16$$
$$\times \log[(5.00 \times 10^{-3}) + (36 \times 10^{-8})]$$
$$= 0.003 \text{ V}$$

(b) For 1.00 mM Na^+ at pH 3.87, we have

$$E = +0.139 + 0.059\ 16$$
$$\times \log[(1.00 \times 10^{-3}) + (36 \times 10^{-3.87})]$$
$$= 0.007\ V$$

15-E. A graph of $E(mV)$ versus $\log[NH_3(M)]$ gives a straight line whose equation is $E = 563.4 + 59.05 \times \log[NH_3]$. For $E = 339.3$ mV, $[NH_3] = 1.60 \times 10^{-4}$ M. The sample analyzed contains (100 mL) $(1.60 \times 10^{-4}$ M) $= 0.016\ 0$ mmol of nitrogen. But this sample represents just 2.00% (20.0 mL/1.00 L) of the food sample. Therefore, the food contains $0.016/0.020\ 0 = 0.800$ mmol of nitrogen $= 11.2$ mg of N $= 3.59\%$ nitrogen.

15-F. (a) For the original solution, we can write

$$[Ag^+] = \frac{[Ag(CN)_2^-]}{[CN^-]^2(7.1 \times 10^{19})}$$
$$= \frac{(1.00 \times 10^{-5})}{(8.0 \times 10^{-6})^2(7.1 \times 10^{19})}$$
$$= 2.2 \times 10^{-15}\ M$$

$$E = \text{constant} + 0.059\ 16$$
$$\times \log(2.2 \times 10^{-15}) = 206.3\ mV$$
$$\Rightarrow \text{constant} = 1.073\ 4\ V$$

After addition of CN^-, we can say that

$$[Ag^+] = \frac{(1.00 \times 10^{-5})}{(12.0 \times 10^{-6})^2(7.1 \times 10^{19})}$$
$$= 9.8 \times 10^{-16}\ M$$

$$E = 1.073\ 4 + 0.059\ 16 \log(9.8 \times 10^{-16})$$
$$= 185.4\ mV$$

(b) Let there be x mol of CN^- in 50.0 mL of unknown. After the standard addition, the unknown contains

$$x + (1.00 \times 10^{-3}\ L)(2.50 \times 10^{-4}\ M)$$
$$= (x + 2.50 \times 10^{-7})\ \text{mol}\ CN^-$$
$$\Rightarrow [CN^-] = \frac{x + 2.50 \times 10^{-7}\ \text{mol}}{0.051\ 0\ L}$$

$$[Ag^+]_2 = \frac{(1.00 \times 10^{-5})}{\left(\dfrac{x + 2.50 \times 10^{-7}}{0.051\ 0}\right)^2(7.1 \times 10^{19})}$$

Before the addition, we have

$$E = \text{constant} + 0.059\ 16 \log[Ag^+]$$
$$0.134\ 8 = \text{constant} + 0.059\ 16 \log[Ag^+]_1$$

After the addition, we can write

$$0.118\ 6 = \text{constant} + 0.059\ 16 \log[Ag^+]_2$$

Subtracting the second equation from the first gives

$$0.134\ 8 - 0.118\ 6 = 0.059\ 16 \log \frac{[Ag^+]_1}{[Ag^+]_2}$$
$$\Rightarrow \frac{[Ag^+]_1}{[Ag^+]_2} = 1.879$$

But we can also write that

$$1.879 = \frac{[Ag^+]_1}{[Ag^+]_2}$$

$$= \frac{\dfrac{1.00 \times 10^{-5}}{\left(\dfrac{x}{0.050\ 0}\right)^2(7.1 \times 10^{19})}}{\dfrac{(1.00 \times 10^{-5})(50.0/51.0)}{\left(\dfrac{x + 2.50 \times 10^{-7}}{0.051\ 0}\right)^2(7.1 \times 10^{19})}}$$

$$\Rightarrow x = 6.50 \times 10^{-7}\ \text{mol}$$

$$[CN^-] = 6.50 \times 10^{-7}\ \text{mol}/0.050\ 0\ L$$
$$= 1.30 \times 10^{-5}\ M$$

Chapter 16

16-A. Titration reaction: $6Fe^{2+} + Cr_2O_7^{2-} + 14H^+ \rightarrow$

$$6Fe^{3+} + 2Cr^{3+} + 7H_2O \qquad V_e = 10.00\ mL$$

Representative calculations:

0.100 mL:

$$E_+ = 0.771 - 0.059\ 16\ \log\frac{[Fe^{2+}]}{[Fe^{3+}]}$$
$$= 0.771 - 0.059\ 16\ \log\frac{9.90}{0.100}$$
$$= 0.653\ V \Rightarrow E = 0.456\ V$$

10.0 mL:

$$E_+ = +0.771 - 0.059\ 16 \log\frac{[Fe^{2+}]}{[Fe^{3+}]}$$

$$6E_+ = 6(1.36) - (6)\frac{0.059\ 16}{6}\log\frac{[Cr^{3+}]^2}{[Cr_2O_7^{2-}][H^+]^{14}}$$

$$7E_+ = 8.93 - 0.059\ 16 \log\frac{[Fe^{2+}][Cr^{3+}]^2}{[Fe^{3+}][Cr_2O_7^{2-}][H^+]^{14}}$$

At the equivalence point, $[Fe^{3+}] = 3[Cr^{3+}]$ and $[Fe^{2+}] = 6[Cr_2O_7^{2-}]$. Putting in these values gives

$$7E_+ = 8.93 - 0.059\ 16$$
$$\times \log\frac{6[Cr_2O_7^{2-}][Cr^{3+}]^2}{3[Cr^{3+}][Cr_2O_7^{2-}][H^+]^{14}}$$

At the equivalence point, all of the $Cr_2O_7^{2-}$ has been converted to Cr^{3+}. Therefore,

$$[Cr^{3+}] = 2(0.020\ 0)\left(\frac{10.00}{130.00}\right) = 0.003\ 08\ \text{M}$$

Using this value in the log term gives

$$7E_+ = 8.93 - 0.059\ 16 \log \frac{2(0.003\ 08)}{(0.10)^{14}}$$

$$E_+ = 1.18\ \text{V} \Rightarrow E = 0.98\ \text{V}$$

10.10 mL:

$$E_+ = 1.36 - \frac{0.059\ 16}{6} \log \frac{[Cr^{3+}]^2}{[Cr_2O_7^{2-}][H^+]^{14}}$$

$$[Cr^{3+}] = 2(0.020\ 0)\left(\frac{10.00}{130.10}\right) = 0.003\ 07\ \text{M}$$

$$[Cr_2O_7^{2-}] = (0.020\ 0)\left(\frac{0.10}{130.10}\right) = 1.54 \times 10^{-5}\ \text{M}$$

$$E_+ = 1.36 - \frac{0.059\ 16}{6}$$

$$\times \log \frac{(0.003\ 07)^2}{(1.54 \times 10^{-5})(0.10)^{14}}$$

$$= 1.22\ \text{V}$$

The final results are

mL	E(V)	mL	E(V)
0.100	0.456	9.90	0.692
2.00	0.538	10.00	0.98
4.00	0.564	10.10	1.03
6.00	0.584	11.00	1.04
8.00	0.610	12.00	1.04
9.00	0.630		

16-B. Titration reactions:

$$V^{2+} + Ce^{4+} \rightarrow V^{3+} + Ce^{3+}$$

$$V^{3+} + Ce^{4+} + H_2O \rightarrow VO^{2+} + Ce^{3+} + 2H^+$$

$$VO^{2+} + Ce^{4+} + H_2O \rightarrow VO_2^+ + Ce^{3+} + 2H^+$$

5.0 mL:

$$E_+ = -0.255 - 0.059\ 16 \log \frac{[V^{2+}]}{[V^{3+}]}$$

$$\Rightarrow E = -0.496\ \text{V}$$

15.0 mL:

$$E_+ = 0.337 - 0.059\ 16 \log \frac{[V^{3+}]}{[VO^{2+}](1.00)^2}$$

$$\Rightarrow E = 0.096\ \text{V}$$

25.0 mL:

$$E_+ = 1.001 - 0.059\ 16 \log \frac{[VO^{2+}]}{[VO_2^+](1.00)^2}$$

$$\Rightarrow E = 0.760\ \text{V}$$

35.0 mL:

$$E_+ = 1.70 - 0.059\ 16 \log \frac{[Ce^{3+}]}{[Ce^{4+}]}$$

$$= 1.70 - 0.059\ 16 \log \left(\frac{30.0}{5.0}\right)$$

$$\Rightarrow E = 1.41\ \text{V}$$

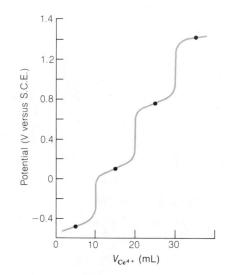

16-C. (a) The second endpoint is halfway between 0.096 and 0.760 = 0.43 V versus S.C.E. (= 0.67 V versus S.H.E.). Diphenylamine would change from colorless to violet at this endpoint.

(b) The third endpoint is approximately midway between 1.41 and 0.760 = 1.08 V versus S.C.E. (= 1.32 V versus S.H.E.). Tris(2,2'-bipyridine)-ruthenium would change from yellow to pale blue at this endpoint.

16-D. (a)

$$\begin{array}{c} \text{HO—CH} \\ | \\ \text{H}_3^+\text{N—CH} \\ | \\ \text{CO}_2^- \end{array} + IO_4^- \rightarrow \begin{array}{c} R \\ | \\ \text{O=CH} + NH_4^+ + IO_3^- \\ + \\ \text{O=CHCO}_2^- \end{array}$$

(For serine, R = H, and for threonine, R = CH_3.) One mole of IO_4^- consumes one mole of serine or one mole of threonine.

(b) $I_3^- + 2S_2O_3^{2-} \rightarrow 3I^- + S_4O_6^{2-}$

823 μL of 0.098 8 M $S_2O_3^{2-}$ = 81.3_1 μmol of $S_2O_3^{2-}$ = 40.6_6 μmol of I_3^-. But one mole of unreacted IO_4^- gives one mole of I_3^- in Reaction 16-72. Therefore, 40.6_6 μmol of IO_4^-

was left from the periodate oxidation of the amino acids. The original amount of IO_4^- was 2.000 mL of 0.048 7 M $IO_4^- = 97.4_0$ μmol. The difference $(97.40 - 40.66 = 56.74)$ is the number of micromoles of serine + threonine in 128.6 mg of protein. But with M.W. $= 58\,600$, 128.6 mg of protein $= 2.195$ μmol. (Serine + threonine)/protein $= 56.74$ μmol/2.195 μmol $= 25.85 \approx 26$ residues/molecule.

(c) 40.66 μmol of I_3^- will react with 40.66 μmol of H_3AsO_3 (Reaction 16-64), which is produced by $\frac{1}{4}(40.66) = 10.16$ μmol of $As_4O_6 = 4.02$ mg.

Chapter 17

17-A. The anode reaction is $Zn(s) \rightarrow Zn^{2+} + 2e^-$

$$5.0 \text{ g Zn} = 7.65 \times 10^{-2} \text{ mol Zn}$$
$$= 1.52 \times 10^{-1} \text{ mol e}^-$$

$(0.152 \text{ mol e}^-)(96\,485 \text{ C/mol}) = 1.48 \times 10^4$ C

The current flowing through the circuit is $I = E/R = 1.02$ V/2.8 $\Omega = 0.364$ A $= 0.364$ C/s.

1.48×10^4 C/(0.364 C/s) $= 4.06 \times 10^4$s $= 11.3$ h

17-B. Anode:

$$H_2O \rightleftharpoons \tfrac{1}{2}O_2(g) + 2H^+ + 2e^- \qquad E° = -1.229 \text{ V}$$

Cathode:

$$\frac{2H^+ + 2e^- \rightleftharpoons H_2(g)}{H_2O \rightleftharpoons \tfrac{1}{2}O_2(g) + H_2(g)} \qquad \frac{E° = \quad 0.00 \text{ V}}{E° = -1.229 \text{ V}}$$

$$E_{eq} = -1.229 - \frac{0.059\,16}{2}\log P_{O_2}^{1/2}P_{H_2}$$

$$= -1.229 \text{ V}$$

$$E_{applied} = E_{eq} - IR - \text{overpotential}$$
$$= -1.229 - (0.100 \text{ A})(2.00 \text{ }\Omega)$$
$$\underbrace{- 0.85 \text{ V}}_{\substack{\text{Anode} \\ \text{overpotential}}} - \underbrace{0.068 \text{ V}}_{\substack{\text{Cathode} \\ \text{overpotential}}} = -2.35 \text{ V}$$

From Table 17-1

For Au electrodes, $E_{applied} = -2.78$ V.

17-C. (a) To electrolyze 0.010 M SbO^+ requires a potential of

E(cathode)

$$= 0.208 - \frac{0.059\,16}{3}\log\frac{1}{[SbO^+][H^+]^2}$$
$$= 0.208 - \frac{0.059\,16}{3}\log\frac{1}{(0.010)(1.0)^2}$$
$$= 0.169 \text{ V}$$

The concentration of $[Cu^{2+}]$ that would be in equilibrium with Cu(s) at this potential is found as follows:

$$Cu^{2+} + 2e^- \rightleftharpoons Cu(s) \qquad E° = 0.339$$

$$E(\text{cathode}) = 0.339 - \frac{0.059\,16}{2}\log\frac{1}{[Cu^{2+}]}$$

$$0.169 = 0.339 - \frac{0.059\,16}{2}\log\frac{1}{[Cu^{2+}]}$$

$$\Rightarrow [Cu^{2+}] = 1.8 \times 10^{-6} \text{ M}$$

Percent of Cu^{2+} not reduced

$$= \frac{1.8 \times 10^{-6}}{0.10} \times 100 = 1.8 \times 10^{-3}\%$$

Percent of Cu^{2+} reduced $= 99.998\%$

(b) In part a, E(cathode versus S.H.E.) $= 0.169$ V.

E(cathode versus Ag|AgCl)

$$= E(\text{versus S.H.E.}) - E(\text{Ag}|\text{AgCl})$$
$$= 0.169 - 0.197 = -0.028 \text{ V}$$

17-D. (a) $Co^{2+} + 2e^- \rightleftharpoons Co(s) \qquad E° = -0.282$ V

E(cathode versus S.H.E.)

$$= -0.282 - \frac{0.059\,16}{2}\log\frac{1}{[Co^{2+}]}$$

Putting in $[Co^{2+}] = 1.0 \times 10^{-6}$ M gives $E = -0.459$ V and

E(cathode versus S.C.E.)

$$= -0.459 - \underbrace{0.241}_{E(\text{S.C.E.})} = -0.700 \text{ V}$$

(b) $Co(C_2O_4)_2^{2-} + 2e^- \rightleftharpoons$

$$Co(s) + 2C_2O_4^{2-} \qquad E° = -0.474 \text{ V}$$

E(cathode versus S.C.E.)

$$= -0.474 - \frac{0.059\,16}{2}$$
$$\times \log\frac{[C_2O_4^{2-}]^2}{[Co(C_2O_4)_2^{2-}]} - 0.241$$

Putting in $[C_2O_4^{2-}] = 0.10$ M and $[Co(C_2O_4)_2^{2-}] = 1.0 \times 10^{-6}$ M gives $E = -0.833$ V.

(c) We can think of the reduction as $Co^{2+} + 2e^- \rightleftharpoons Co(s)$, for which $E° = -0.282$ V. But the concentration of Co^{2+} is that tiny amount in equilibrium with 0.10 M EDTA plus 1.0×10^{-6} M $Co(EDTA)^{2-}$. In Table 13-2, we find that the formation constant for $Co(EDTA)^{2-}$ is $10^{16.31} = 2.0 \times 10^{16}$.

$$K_f = \frac{[Co(EDTA)^{2-}]}{[Co^{2+}][EDTA^{4-}]} = \frac{[Co(EDTA)^{2-}]}{[Co^{2+}]\alpha_{Y4-}F}$$

where F is the formal concentration of EDTA ($=0.10$ M) and $\alpha_{Y4-} = 5.0 \times 10^{-4}$ at pH 7.00 (Table 13-1). Putting in $[Co(EDTA)^{2-}] = 1.0 \times 10^{-6}$ M and solving for $[Co^{2+}]$ gives $[Co^{2+}] = 1.0 \times 10^{-18}$ M.

$$= -0.282 - \frac{0.059\,16}{2}\log\frac{1}{1.0\times10^{-18}}$$

$$- 0.241$$

$$= -1.055 \text{ V}$$

17-E. (a) 75.00 mL of 0.023 80 M KSCN = 1.785 mmol of SCN^-, which gives 1.785 mmol of AgSCN, containing 0.103 7 g of SCN. Final mass = $12.463\,8 + 0.103\,7 = 12.567\,5$ g.

(b) The concentration of Ag^+ in equilibrium with 0.10 M Br^- is $[Ag^+] = K_{sp}/[Br^-] = (5.0 \times 10^{-13})/(0.10) = 5.0 \times 10^{-12}$ M. Writing both cell reactions as reductions

Anode: $Ag^+ + e^- \rightleftharpoons Ag(s)$

$$E° = 0.799 \text{ V}$$

Cathode: $Hg_2Cl_2(s) + 2e^- \rightleftharpoons 2Hg(l) + 2Cl^-$

$$E(S.C.E.) = 0.241 \text{ V}$$

we can say

$$E_{eq} = E(\text{cathode}) - E(\text{anode})$$

$$= 0.241 - \left(0.799 - 0.059\,16 \log\frac{1}{[Ag^+]}\right)$$

Putting in $[Ag^+] = 5.0 \times 10^{-12}$ M gives $E_{eq} = 0.111$ V.

(c) To remove 99.99% of 0.10 M KI will leave $[I^-] = 1.0 \times 10^{-5}$ M. The concentration of Ag^+ in equilibrium with this much I^- is $[Ag^+] = K_{sp}/[I^-] = 8.3 \times 10^{-17}/1.0 \times 10^{-5} = 8.3 \times 10^{-12}$ M. In part b, we found that it takes only 5.0×10^{-12} M Ag^+ to precipitate 0.10 M Br^-. Therefore, the separation is not possible.

17-F. 1.00 ppt corresponds to $30.0/1\,000 = 0.030\,0$ mL of O_2/min = 5.00×10^{-4} mL of O_2/s. The moles of oxygen in this volume are

$$n = \frac{PV}{RT} = \frac{(1.00 \text{ atm})(5.00 \times 10^{-7} \text{ L})}{(0.082\,06 \text{ L atm K}^{-1}\text{ mol}^{-1})(293 \text{ K})}$$

$$= 2.080 \times 10^{-8} \text{ mol}$$

For each mole of O_2, four moles of e^- flow through the circuit, so $e^- = 8.320 \times 10^{-8}$ mol/s = 8.03×10^{-3} C/s = 8.03 mA. An oxygen content of 1.00 ppm would give a current of 8.03 μA instead.

17-G. The Zn^{2+} reacts first with PDTA freed by the reduction of $Hg(PDTA)^{2-}$ in the region BC. Then additional Zn^{2+} goes on to liberate Hg^{2+} from $Hg(PDTA)^{2-}$. This additional Hg^{2+} is reduced in the region DE. The tota Hg^{2+}, equivalent to the added Zn^{2+}, equals one-half the coulombs measured in regions BC and DE (since $2e^-$ reacts with $1\,Hg^{2+}$). Coulombs = $3.89 + 14.47 = 18.36$. Moles of Hg^{2+} reduced = $0.5(18.36\text{C})/(96\,485 \text{ C/mol}) = 9.514 \times 10^{-5}$ mol. $[Zn^{2+}] = 9.514 \times 10^{-5}$ mol/2.00×10^{-3} L = 0.047 57 M.

Chapter 18

18-A. The standard curve is moderately linear with slope = 0.004 19 μA/ppb and intercept of 0.019 8. The concentration of Ni(II) when 54.0 μL of 10.0 ppm solution is added to 5.00 mL is

$$\left(\frac{0.054\,0 \text{ mL}}{5.054\,0 \text{ mL}}\right)(10.0 \text{ ppm}) = 0.107 \text{ ppm}$$

$$= 107 \text{ ppb}$$

The expected current is

$$I = m[Ni(II)] + b$$

$$= (0.004\,19)(107) + 0.019\,8 = 0.468 \text{ } \mu\text{A}$$

A careful examination of the standard curve shows that a better fit might be obtained if the slope and intercept of just the first seven points are calculated. The curve appears to be starting to level off at the higher concentrations in this experiment.

18-B. (a) $[Cd^{2+}]$ in unknown $= \left(\frac{10.00}{50.00}\right)(3.23 \times 10^{-4}$ M$) = 6.46 \times 10^{-5}$ M

$$\frac{(Cd^{2+} \text{ signal}/Pb^{2+} \text{ signal}) \text{ in known}}{(Cd^{2+} \text{ signal}/Pb^{2+} \text{ signal}) \text{ in unknown}}$$

$$= \frac{([Cd^{2+}]/[Pb^{2+}]) \text{ in known}}{([Cd^{2+}]/[Pb^{2+}]) \text{ in unknown}}$$

$$\frac{1.64/1.58}{2.00/3.00} = \frac{3.23/4.18}{6.46 \times 10^{-5}/x}$$

$$\Rightarrow x = 1.30 \times 10^{-4} \text{ M}$$

The concentration of Pb^{2+} in the diluted unknown is 1.30×10^{-4} M. In the undiluted sample, the concentration is $\left(\frac{50.00}{25.00}\right)(1.30 \times 10^{-4}) = 2.60 \times 10^{-4}$ M.

(b) $\dfrac{(1.64 \pm 0.03)/(1.58 \pm 0.03)}{(2.00 \pm 0.03)/(3.00 \pm 0.03)}$

$$= \frac{(3.23 \pm 0.01)/(4.18 \pm 0.01)}{\left[\frac{(10.00 \pm 0.05)}{(50.00 \pm 0.05)}(3.23 \pm 0.01) \times 10^{-4}\right] \Big/ x}$$

$$\Rightarrow x = 1.30_2 (\pm 3._{28}\%) \times 10^{-4} \text{ M}$$

$$[Pb^{2+}]$$

$$= \left(\frac{50.00 \pm 0.05}{25.00 \pm 0.05}\right)(1.30_2 \pm 3._{28}\%) \times 10^{-4} \text{ M}$$

$$= 2.60(\pm 3._{29}\%) \times 10^{-4} \text{ M}$$

$$= 2.60(\pm 0.09) \times 10^{-4} \text{ M}$$

18-C. Sample height (mm) = 26.8 − 2.4 = 24.4
Sample + 1 ppm Cu = 42.2 − 5.6 = 36.6
Sample + 2 ppm Cu = 57.8 − 8.7 = 49.1.

The average response to added Cu is

$$\frac{(36.6 - 24.4) + (49.1 - 24.4)}{3} = 12.3 \frac{\text{mm}}{\text{ppm Cu}}$$

The initial sample must have contained

$$\frac{24.4}{12.03} = 1.98 \text{ ppm Cu}$$

18-D. A graph of $E_{1/2}$ versus $\log[OH^-]$ is shown below. All but the lowest two points appear to lie on a line whose equation is

$$E_{1/2} = -0.080\,6 \log[OH^-] - 0.763$$

According to Equation 18-21, the slope of the graph is $-0.059\,16\ p/n$. Assuming that $n = 2$, we calculate p as follows:

$$p = \frac{(n)(\text{slope})}{-0.059\,16} = 2.72 \approx 3$$

The intercept of Equation 18-21 is given by

$$\text{Intercept} = E_{1/2}(\text{for free } Pb^{2+}) - \frac{0.059\,16}{n} \log \beta_3$$

$$-0.763 = -0.41 - \frac{0.059\,16}{2} \log \beta_3$$

$$\Rightarrow \beta_3 = 9 \times 10^{11}$$

18-E. We see two consecutive reductions. From the value of $E_{pa} - E_{pc}$, we find that one electron is involved in each reduction (using Equation 18-22). A possible sequence of reaction is

$$Co(III)(B_9C_2H_{11})_2^- \rightarrow Co(II)(B_9C_2H_{11})_2^{2-}$$
$$\rightarrow Co(I)(B_9C_2H_{11})_2^{3-}$$

The equality of the anodic and cathodic peak heights suggests that the reactions are reversible. The expected DC (a) and DPP (b) polarograms are sketched below.

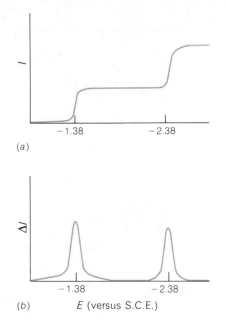

(a)

(b) E (versus S.C.E.)

18-F. In curve a, the current decreases prior to the endpoint because Pb^{2+} is reduced to Pb(Hg) at -0.8 V. Beyond the endpoint, excess $Cr_2O_7^{2-}$ can be reduced and the current increases again. Curve b is level near zero current prior to the equivalence point because Pb^{2+} is not reduced at 0 V (versus S.C.E.). Beyond the endpoint, excess $Cr_2O_7^{2-}$ is reduced, even at 0 V (versus S.C.E.).

18-G. Initially there is no redox couple to carry current, so the potential will be high. As Ce(IV) is added, Fe(II) is converted to Fe(III), the mixture of which can support current flow by the reactions

Anode: $Fe(II) \rightleftharpoons Fe(III) + e^-$

Cathode: $Fe(III) + e^- \rightleftharpoons Fe(II)$

The potential will therefore decrease. At the equivalence point, all of the Fe(II) and all of the Ce(IV) are consumed, so the potential is very high. Beyond the equivalence point, the redox couple $Ce(IV)|Ce(III)$ can support a current and the potential will be low again. The expected curve is shown below.

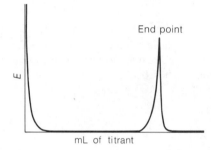

18-H. 34.61 mL of methanol with 4.163 mg of H_2O/mL contains 144.08 mg H_2O = 7.997 8 mmol of H_2O. The titration of "dry" methanol tells us that 25.00 mL of methanol reacts with 3.18 mL of reagent. Therefore, 34.61 mL of methanol will react with (34.61/25.00)(3.18) = 4.40 mL of Karl Fischer reagent. The titer of the reagent is

$$\frac{7.997\ 8 \text{ mmol } H_2O}{(25.00 - 4.40) \text{ mL reagent}}$$

$$= 0.388\ 24\ \frac{\text{mmol } H_2O}{\text{mL reagent}}$$

Reagent needed to react with 1.000 g of salt in 25.00 mL of methanol $= (38.12 - 3.18) = 34.94$ mL. H_2O in 1.000 g of salt $= (0.388\ 24)(34.94) = 13.565$ mmol $= 244.38$ mg of $H_2O = 24.44\%$ (wt/wt) of the crystal.

Chapter 19

19-A. (a) $A = -\log P/P_0 = -\log T = -\log(0.45) = 0.347$

(b) The absorbance will double to 0.694, giving $T = 10^{-A} = 10^{-0.694} = 0.202 \Rightarrow \%T = 20.2\%$.

19-B. (a) 1.00×10^{-2} g of NH_4Cl in 1.00 L $= 1.869 \times 10^{-4}$ M. In the colored solution, the concentration is $(\frac{10}{50})(1.869 \times 10^{-4}$ M$) = 3.739 \times 10^{-5}$ M. $\varepsilon = A/bc = (0.308 - 0.140)/(1.00)(3.739 \times 10^{-5}) = 4.49_3 \times 10^3$ M$^{-1} \cdot$cm^{-1}.

(b) $\dfrac{\text{Absorbance of unknown}}{\text{Absorbance of reference}}$

$$= \frac{0.592 - 0.140}{0.308 - 0.140}$$

$$= \frac{\text{concentration of unknown}}{\text{concentration of reference}}$$

$\Rightarrow$ concentration of NH_3 in unknown

$$= \left(\frac{0.452}{0.168}\right)(1.869 \times 10^{-4})$$

$$= 5.028 \times 10^{-4} \text{ M}.$$

100.00 mL of unknown

$$= 5.028 \times 10^{-5} \text{ mol of N}$$

$$= 7.043 \times 10^{-4} \text{ g of N}$$

$\Rightarrow$ weight % of N

$$= (7.043 \times 10^{-4} \text{ g})/(4.37 \times 10^{-3} \text{ g})$$

$$= 16.1\%.$$

19-C. (a) Milligrams of Cu in flask C $= (1.00)(\frac{10}{250})(\frac{15}{30}) = 0.020\ 0$ mg. This entire quantity is in the isoamyl alcohol (20.00 mL), so the concentration is $(2.00 \times 10^{-5} \text{ g})/(0.020\ 0 \text{ L})(63.546 \text{ g/mol}) = 1.57 \times 10^{-5}$ M.

(b) Observed absorbance

= absorbance due to Cu in rock

+ blank absorbance

$= \varepsilon bc + 0.056$

$= (7.90 + 10^3)(1.00)(1.574 \times 10^{-5}) + 0.056$

$= 0.180$

Note that the observed absorbance is equal to the absorbance from Cu in the rock *plus* the blank absorbance. In the lab we measure the observed absorbance and subtract the blank absorbance from it to find the absorbance due to copper.

(c) $\dfrac{\text{Cu in unknown}}{\text{Cu in known}} = \dfrac{A \text{ of unknown}}{A \text{ of known}}$

$$\frac{x \text{ mg}}{1.00 \text{ mg}} = \frac{0.874 - 0.056}{0.180 - 0.056} \Rightarrow x = 6.60 \text{ mg Cu}$$

19-D. (a) $c = A/\varepsilon b = 0.463/(4\ 170)(1.00) = 1.110 \times 10^{-4}$ M $= 8.99$ g/L $= 8.99$ mg of transferrin/mL. The Fe concentration is 2.220×10^{-4} M $= 0.012\ 4$ g/L $= 12.4$ μg/mL.

(b)

$$A_\lambda = \sum \varepsilon bc$$

$$0.424 = 4\ 170[T] + 2\ 290[D]$$

$$0.401 = 3\ 540[T] + 2\ 730[D]$$

where [T] and [D] are the concentrations of transferrin and desferrioxamine, respectively. Solving for [T] and [D] gives $[T] = 7.30 \times 10^{-5}$ M and $[D] = 5.22 \times 10^{-5}$ M. The fraction of iron in transferrin (which binds two ferrin ions) is $2[T]/(2[T] + [D]) = 73.7\%$.

19-E. The absorbance must be corrected by multiplying each observed absorbance by (total volume/initial volume). For example, at 36.0 μL, A(corrected) = $(0.399)[(2\,025 + 36)/2\,025] = 0.406$. A graph of corrected absorbance versus volume of Pb^{2+} (μL) is similar to Figure 19-17, with the endpoint at 46.7 μL. The moles of Pb^{2+} in this volume are $(46.7 \times 10^{-6}$ L$)(7.515 \times 10^{-4}$ M$) = 3.510 \times 10^{-8}$ mol. The concentration of semi-xylenol orange is $(3.510 \times 10^{-8}$ mol$)/(2.025 \times 10^{-3}$ L$) = 1.73 \times 10^{-5}$ M.

19-F. The appropriate Scatchard plot is a graph of $\Delta A/[X]$ versus ΔA (Equation 19-35).

Experiment	ΔA	$\Delta A/[X]$
1	0.090	20 360
2	0.181	19 890
3	0.271	16 940
4	0.361	14 620
5	0.450	12 610
6	0.539	9 764
7	0.627	7 646
8	0.713	5 021
9	0.793	2 948
10	0.853	1 453
11	0.904	93.6

Points 2–10 lie on a reasonably straight line whose slope is -2.72×10^4 M^{-1}, giving $K = 2.72 \times 10^4$ M^{-1}.

19-G. If self-absorption can be neglected, Equation 19-46 reduces to

$$I = k'P_0(1 - 10^{-\varepsilon_{ex}b_2c}) \qquad (a)$$

At low concentrations, this expression reduces to

$$I = k'P_0(\varepsilon_{ex}b_2c \ln 10) \qquad (b)$$

(using the first term of a power series expansion). As the concentration is increased. Expression b becomes greater than a. When Expression a is 5% below b, we can say

$$k'P_0(1 - 10^{-\varepsilon_{ex}b_2c}) = 0.95k'P_0\varepsilon_{ex}b_2c \ln 10$$

$$1 - 10^{-A} = 0.95A \ln 10$$

By trial and error, this equation can be solved to find that when $A = 0.045$, $1 - 10^{-A} = 0.95A \ln 10$.

(Alternatively, you could make a graph of $1 - 10^{-A}$ versus A and $0.95A \ln 10$ versus A. The solution is the intersection of the two curves.)

Chapter 20

20-A. $T = 0.820 = \dfrac{1 - R}{1 + R}$

$\Rightarrow R = 0.098\,9$ (Equation 20-9)

Now put this value of R into Equation 20-8 and use $n_1 = 1$ for air:

$$0.098\,9 = \left(\frac{1 - n_2}{1 + n_2}\right)^2$$

$$\pm 0.314\,5 = \frac{1 - n_2}{1 + n_2} \Rightarrow n_2 = 1.92 \text{ or } 0.522$$

20-B. (a) $R = \left(\dfrac{1 - 2.17}{1 + 2.17}\right)^2 = 0.136\,2$

$T = \dfrac{(1 - R)^2 e^{-\alpha b}}{1 - R^2 e^{-2\alpha b}}$

$= \dfrac{(1 - 0.136\,2)^2 e^{-(0.47\text{ cm}^{-1})(1.20\text{ cm})}}{1 - (0.136\,2)^2 e^{-2(0.47\text{ cm}^{-1})(1.20\text{ cm})}}$

$= 0.427$

(b) $T = \dfrac{(1 - 0.136\,2)^2 e^{-(0.47\text{ cm}^{-1})(0.120\text{ cm})}}{1 - (0.136\,2)^2 e^{-2(0.47\text{ cm}^{-1})(0.120\text{ cm})}}$

$= 0.717$

(c) $T = \dfrac{(1 - 0.136\,2)^2 e^{-[(0.47 + 0.20)\text{ cm}^{-1}](1.20\text{ cm})}}{1 - (0.136\,2)^2 e^{-2[(0.47 + 0.20)\text{ cm}^{-1}](1.20\text{ cm})}}$

$= 0.335$

20-C. (a) $\dfrac{dM_\lambda}{d\lambda} = 0 = 2\pi hc^2\big[-5\lambda^{-6}e^{-hc/\lambda kT}$

$\qquad + \lambda^{-5}e^{-hc/\lambda kT}\left(\dfrac{hc}{kT}\right)\lambda^{-2}\big]$

$\Rightarrow \lambda_{\max} \cdot T = \dfrac{hc}{5k} = 2.878 \times 10^{-3}$ m$\cdot$K

(b)

$T(K)$	$\lambda_{\max}(\mu m)$
100	28.8
500	5.76
5 000	0.576

20-D. The light ray travels the path $ABCDE$ in the diagram below. At point A the incident angle is 40° and the angle of refraction is θ. From Snell's law, we can say $n_{air} \sin 40° = n_{NaCl} \sin \theta$. Substituting $n_{air} = 1.000\,26$ and $n_{NaCl} = 1.494\,73$ gives $\theta = 25.476\,9°$. In triangle HBC, we know that angle HBC is $90 - \theta$ and angle BHC is 30°. Therefore angle *BCH* is $180 - (90 - \theta) - 30 = 60 + \theta$. Therefore the angle of incidence *BCF* on the mirrored surface is $90 - (60 + \theta) = 30 - \theta$ and the angle of reflection *FCG* is also $30 - \theta$. In triangle *BCF* we know two angles already, so the third angle *BFC* is $150°$. Therefore angle *CFG* is $30°$. In the triangle *CFG* we know two angles already, so angle *FGC* must be $120 + \theta$. Therefore angle *CGJ* is $60 - \theta$, as is angle *GDK*. Finally, we use Snell's law to say $n_{air} \sin \phi = n_{NaCl} \sin(60 - \theta)$. With $\theta = 25.476\,9°$ we find $\phi = 57.876\,2°$.

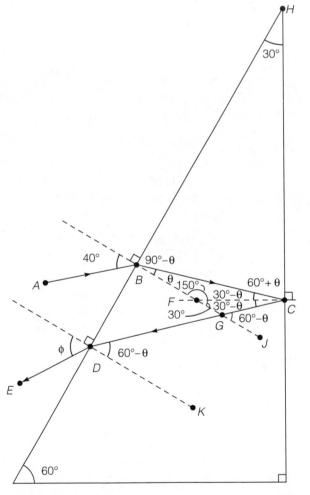

20-E. The ordinate of Figure 20-28 is logarithmic. Replacing the numbers 0.1 and 0.01 by 10^{-1} and 10^{-2}, respectively, we see that for true absorbance $= 1.0$, absorbance error $= 10^{-1.5_3} = 0.03_0$. The relative error in absorbance, and hence concentration, is $0.03_0/1.0 = 3\%$. The measured concentration is 3% low.

20-F. (a) $\Delta\tilde{v} = 1/2\delta = 1/(2 \cdot 1.266\,0 \times 10^{-4}\ \text{cm})$
$$= 3\,949\ \text{cm}^{-1}$$

(b) Each interval is $1.266\,0 \times 10^{-4}$ cm. $409\,6$ intervals $= (409\,6)(1.266\,0 \times 10^{-4}\ \text{cm}) = 0.518\,6$ cm. This is a range of $\pm\Delta$, so $\Delta = 0.259\,3$ cm.

(c) Resolution $\approx 1/\Delta = 1/(0.259\,3\ \text{cm})$
$$= 3.86\ \text{cm}^{-1}$$

(d) Mirror velocity $= 0.693$ cm/s
$$\text{Interval} = \frac{1.266\,0 \times 10^{-4}\ \text{cm}}{0.693\ \text{cm/s}} = 183\ \mu\text{s}$$

(e) $(4\,096\ \text{points})(183\ \mu\text{s/point}) = 0.748$ s

(f) The beamsplitter is germanium on KBr. The KBr absorbs light below 400 cm^{-1}, which the background transform shows clearly.

20-G. The graphs show that signal-to-noise ratio is proportional to $\sqrt{n}$.

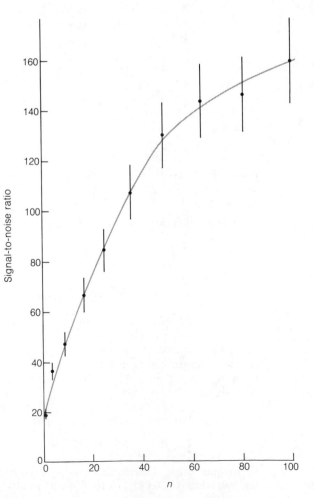

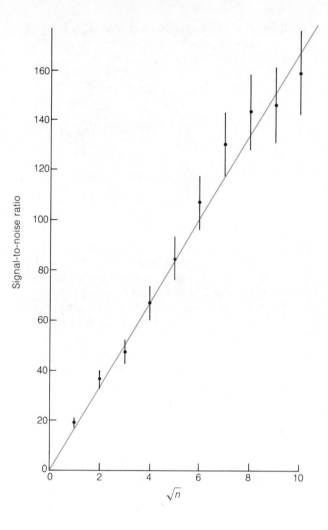

Signal-to-noise ratio vs $\sqrt{n}$

$$\frac{\text{Concentration ratio in unknown}}{\text{Concentration ratio in standard}}$$

$$= \frac{\text{signal ratio in unknown}}{\text{signal ratio in standard}}$$

$$\frac{[\text{Fe}]/2.25}{2.50/2.00} = \frac{0.185/0.128}{1.05}$$

$$[\text{Fe}] = 3.87 \; \mu\text{g/mL}$$

The original concentration of Fe must have been

$$\frac{6.00}{5.00}(3.87) = 4.65 \; \mu\text{g/mL} = 8.33 \times 10^{-5} \; \text{M}.$$

21-C. The ratio of signal to peak-to-peak noise level is measured to be 17 in the figure. The concentration of Fe needed to give a signal-to-noise ratio of 2 is $(\frac{2}{17})(0.048\,5 \; \mu\text{g/mL}) = 0.005\,7 \; \mu\text{g/mL}$ ($=5.7$ ppb).

21-D. In the excitation spectrum, we are looking at emission over a band of wavelengths 1.6 nm wide, while exciting the sample with different narrow bands (0.03 nm) of laser light. The sample can absorb the laser light only when the laser frequency coincides with the atomic frequency. Therefore, emission is observed only when the narrow laser line is in resonance with the atomic levels. In the emission spectrum, the sample is excited by a fixed laser frequency and then emits radiation. The monochromator bandwidth is not narrow enough to discriminate between emission at different wavelengths, so a broad envelope is observed.

Chapter 21

21-A.

Emission intensity	Concentration of added standard (μg/mL)
309	0
452	0.081
600	0.162
765	0.243
906	0.324

A graph of intensity versus concentration of added standard intercepts the x axis at $-0.164 \; \mu$g/mL. Since the sample was diluted by a factor of 10, the original sample concentration is 1.64 μg/mL.

21-B. The concentration of Mn in the unknown mixture is $(13.5)(1.00/6.00) = 2.25 \; \mu$g/mL.

Chapter 22

22-A. (a) $k_1 = \dfrac{t_{r1} - t_m}{t_m}$

$$\Rightarrow t_m = \frac{t_{r1}}{k_1 + 1} = \frac{10.0}{5.00} = 2.00 \; \text{min}$$

$$t_{r2} = t_m(k_2 + 1) = 2.00\,(5.00 + 1) = 12.0 \; \text{min}$$

$$\sigma_1 = \frac{t_{r1}}{\sqrt{N}} = \frac{10.0}{\sqrt{1\,000}} = 0.316 \; \text{min}$$

$$\Rightarrow w_{1/2} \text{ (peak 1)} = 2.35\sigma_1 = 0.74 \; \text{min}$$

$$w_1 = 4\sigma_1 = 1.26 \; \text{min}$$

$$\sigma_2 = \frac{t_{r2}}{\sqrt{N}} = \frac{12.0}{\sqrt{1\,000}} = 0.379 \; \text{min}$$

$$\Rightarrow w_{1/2} \text{ (peak 2)} = 2.35\sigma_2 = 0.89 \; \text{min}$$

$$w_2 = 4\sigma_2 = 1.52 \; \text{min}$$

(b)

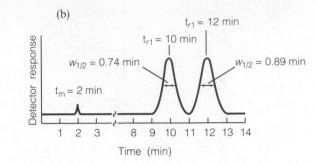

(c) Resolution $= \dfrac{\Delta t_r}{w_{av}} = \dfrac{2}{[(1.26 + 1.52)/2]} = 1.44$

22-B. (a) Fraction remaining $= q = \dfrac{V_1}{V_1 + KV_2}$

$$0.01 = \dfrac{10}{10 + 4.0V_2} \Rightarrow V_2 = 248 \text{ mL}$$

(b) $q^3 = 0.01 = \left(\dfrac{10}{10 + 4.0V_2}\right)^3 \Rightarrow V_2 = 9.1 \text{ mL},$

and total volume $= 27.3$ mL.

22-C. (a) For A, $p = D/(D + 1) = 0.5$. For B, $p = 0.375$. For A, $r_{max} = 100p = 50$, and for B, $r_{max} = 37.5$. $\sigma = \sqrt{npq} = \sqrt{100(0.5)(0.5)} = 5$ for A and $\sqrt{100(0.375)(0.625)} = 4.84$ for B.

(b) The center of A is at tube 50, and $\sigma = 5$ tubes. Tube 28 lies $(50 - 28)/5 = 4.4\sigma$ from the center, and tube 43 lies $(50 - 43)/5 = 1.4\sigma$ from the center. In Table 4-1, we find that the fraction of the area of A to the left of tube 43 is $0.5 - 0.419\,2 = 0.080\,8$. A negligible fraction of A lies to the left of tube 28. Therefore, the fraction of A in tubes 28–43 is 0.080 8. For component B, centered at tube 37.5 and with a standard deviation of 4.84, tube 28 lies $(37.5 - 28)/4.84 = 1.96\sigma$ to the left of center, and tube 43 lies 1.14σ to the right of center. The area between these limits is found to be $0.474\,9 + 0.372\,5 = 0.847\,4$ (by interpolation in Table 4-1). Percent recovery of B $= 84.74\%$, and purity of B $= 0.847\,4/(0.080\,8 + 0.847\,4) = 91.29\%$. Similarly, tubes 44–60 contain 86.22% of A with a purity of $0.862\,2/(0.090\,4 + 0.862\,2) = 90.51\%$.

22-D. (a) Relative distances measured from Figure 22-11:

$$t_m = 10._4$$

$$t'_r = 39._8 \text{ for octane}$$

$$t'_r = 76._0 \text{ for nonane}$$

$k = t'_r/t_m = 3.8_3$ for octane and 7.3_1 for nonane.

(b) Let t_{stat} = time in stationary phase, t_{mob} = time in mobile phase and t be total time on column. We know that $k = t_{stat}/t_{mob}$. But

$$t = t_{stat} + t_{mob} = t_{stat} + \dfrac{t_{stat}}{k}$$

$$= t_{stat}\left(1 + \dfrac{1}{k}\right) = t_{stat}\left(\dfrac{k + 1}{k}\right).$$

Therefore

$$t_{stat}/t = \dfrac{k}{k + 1} = 3.8_3/4.8_3 = 0.79_3.$$

(c) $a = t'_r \text{(nonane)}/t'_r\text{(octane)} = 76._0/39._8 = 1.9_1.$

(d) $K = kV_m/V_s = 3.8_3 (V_m/\tfrac{1}{2}V_m) = 7.6_6.$

22-E. (a) For ethyl acetate, we measure $t_r = 11.3$ and $w = 1.5$ millimeters. Therefore, $N = 16\,t_r^2/w^2 = 910$ plates. For toluene, the figures are $t_r = 36.2$, $w = 4.2$, and $N = 1\,200$ plates.

(b) We expect $w_{1/2} = (2.35/4)w$. The measured value of $w_{1/2}$ is in good agreement with the calculated value.

22-F. The column is overloaded, causing a gradual rise and an abrupt fall of the peak. As the sample size is decreased, the overloading decreases and the peak becomes more symmetric.

Chapter 23

23-A. (a) $\dfrac{[\text{Solute}]}{[\text{Standard}]} = F \dfrac{\text{solute area}}{\text{standard area}}$

$$\dfrac{1.53}{1.06} = F \dfrac{1\,570}{922} \Rightarrow F = 0.848$$

(b) $\dfrac{\text{mmol hexanol}}{\text{mmol pentanol}} = F \dfrac{\text{hexanol area}}{\text{pentanol area}}$

$$\Rightarrow \dfrac{\text{mmol hexanol}}{0.57 \text{ mmol}} = 0.848 \dfrac{816}{843}$$

$$\Rightarrow \text{mmol hexanol} = 0.47$$

23-B. (a) A plot of $\log t'_r$ versus (number of carbon atoms) should be a fairly straight line for a homologous series of compounds.

Peak	t'_r	$\log t'_r$
$n = 7$	2.9	0.46
$n = 8$	5.4	0.73
$n = 14$	85.8	1.93
unknown	41.4	1.62

From a graph of $\log t'_r$ versus n, it appears that $n = 12$ for the unknown.

(b) $k' = t'_r/t_m = 41.4/1.1 = 38$

23-C. 13.03 mL of 0.022 74 M NaOH = 0.296 3 mmol of OH^-, which must equal the total cation charge

$(= 2[VO^{2+}] + 2[H_2SO_4])$ in the 5.00-mL aliquot. 50.0 mL therefore contains 2.963 mmol of cation charge. The VO^{2+} content is (50.0 mL) × (0.024 3 M) = 1.215 mmol = 2.43 mmol of charge. The H_2SO_4 must therefore be $(2.963 - 2.43)/2 = 0.267$ mmol.

1.215 mmol $VOSO_4$ = 0.198 g $VOSO_4$ in 0.244 7 g sample = 80.9%.

0.267 mmol H_2SO_4 = 0.026 2 g H_2SO_4 in 0.244 7 g sample = 10.7%.

H_2O (by difference) = 8.4%.

23-D. (a) Since the fractionation range of Sephadex G-50 is 1 500–30 000, hemoglobin should not be retained and ought to be eluted in a volume of 36.4 mL.

(b) $^{22}NaCl$ ought to require one column volume to pass through. Neglecting the volume occupied by gel, we expect NaCl to require $\pi r^2 \times$ length = $\pi(1.0 \text{ cm})^2(40 \text{ cm})$ = 126 mL of solvent.

(c) $K_{av} = \dfrac{V_r - V_0}{V_t - V_0} \Rightarrow V_r = K_{av}(V_t - V_0) + V_0$

$= 0.65(126 - 36.4) + 36.4 = 95$ mL

23-E. In a, the low ionic strength reservoir drains more rapidly than the high ionic strength reservoir, since the levels must remain equal. In b, the high ionic strength reservoir drains faster than the low ionic strength reservoir.

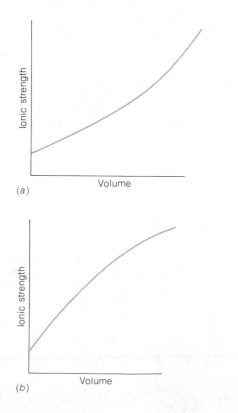

(a)

(b)

24-A. (a) Expected number of red marbles = np_{red} = $(1\,000)(0.12) = 120$. Expected number of yellow = $nq_{yellow} = (1\,000)(0.88) = 880$.

(b) Absolute: $\sigma_{red} = \sigma_{yellow} = \sqrt{npq}$

$= \sqrt{(1\,000)(0.12)(0.88)} = 10.28$

Relative: $\sigma_{red}/n_{red} = 10.28/120 = 8.56\%$.

$\sigma_{yellow}/n_{yellow} = 10.28/880 = 1.17\%$.

(c) For 4 000 marbles, $n_{red} = 480$ and $n_{yellow} = 3\,520$.

$\sigma_{red} = \sigma_{yellow} = \sqrt{npq} = \sqrt{(4\,000)(0.12)(0.88)} = 20.55$. $\sigma_{red}/n_{red} = 4.28\%$. $\sigma_{yellow}/n_{yellow} = 0.58\%$.

(d) 2, $\sqrt{n}$

(e) $\dfrac{\sigma_{red}}{n_{red}} = 0.02 = \dfrac{\sqrt{n(0.12)(0.88)}}{(0.12)n} \Rightarrow$

$n = 1.83 \times 10^4$.

24-B. (a) $mR^2 = K_s \Rightarrow m(10)^2 = 36 \Rightarrow m = 0.36$ g

(b) An uncertainty of ± 20 counts per second per gram is $100 \times 20/237 = 8.4\%$.

$n = \dfrac{t^2 s_s^2}{e^2} \approx \dfrac{(1.96)^2(0.10)^2}{(0.084)^2} = 5.4 \approx 5$

$\Rightarrow t = 2.776$ (4 degrees of freedom)

$n \approx \dfrac{(2.776)^2(0.10)^2}{(0.084)^2} = 10.9 \approx 11 \Rightarrow t = 2.228$

$n \approx \dfrac{(2.228)^2(0.10)^2}{(0.084)^2} = 7.0 \approx 7 \Rightarrow t = 2.447$

$n \approx \dfrac{(2.447)^2(0.10)^2}{(0.084)^2} = 8.5 \approx 8 \Rightarrow t = 2.365$

$n \approx \dfrac{(2.365)^2(0.10)^2}{(0.084)^2} = 7.9 \approx 8$

24-C. 1. The figure shows that coprecipitation occurs in distilled water, so it should first be shown that coprecipitation occurs in seawater, which has a high salt concentration. This can be done by adding standard Pb^{2+} to real seawater and to artificial seawater (NaCl solution) and repeating the experiment in the figure.

2. To demonstrate that coprecipitation of Pb^{2+} does not decrease the concentration of alkyl lead compounds, artificial seawater samples containing known quantities of alkyl lead compounds (but no Pb^{2+}) should be prepared and stripping analysis performed. Then a desired excess of Pb^{2+} would be added to a similar solution containing an alkyl lead compound and coprecipitation with $BaSO_4$ would be carried out. The remaining solution would then be analyzed by stripping voltammetry. If the signal is the same as it was without Pb^{2+} coprecipitation, then it is safe to say that coprecipitation

does not reduce the concentration of alkyl lead in the seawater.

24-D. The acid-soluble inorganic matter and the organic material can probably be dissolved (and oxidized) together by wet ashing with $HNO_3 + H_2SO_4$ in a Teflon-lined bomb in a microwave oven. The insoluble residue should be washed well with water and the washings combined with the acid solution. After the residue has been dried, it can be fused with one of the fluxes in Table 24-5, dissolved in dilute acid, and combined with the previous solution.

24-E. Nonstoichiometry means that the actual composition of a compound is not exactly what the label says. For example, $CaCO_3$ might contain some $Ca(HCO_3)_2$.

Answers to Problems

Chapter 1

4. 1.47×10^3 J/s, 1.47×10^3 W
5. 1.10 M
6. 26.5 g $HClO_4$, 11.1 g H_2O
7. 10^{-3} g/L, 10^3 μg/L, 1 μg/mL, 1 mg/L
8. (a) 1 670 g solution   (b) 1.18×10^3 g $HClO_4$
   (c) 11.7 mol
9. 0.445 F
10. 0.138 M
11. 1.6 kg F^-, 3.54 kg NaF
12. 5.48 g
13. 3.2 L
15. 0.054 8 ppm, 54.8 ppb
16. (a) 55.6 mL   (b) 1.80 g/mL
17. 0.119 osmolar
18. 1.51 m
19. 1.51 g/mL

Chapter 2

2. 9.959 g
11. 0.2%; 0.499 0 M
12. 9.980 mL
13. 49.947 g in vacuum; 49.892 g in air
14. true mass — 50.516 g; mass in air — 50.484 g
15. PbO_2
16. (a) 7.4 torr   (b) 0.001 1 g/mL   (c) 1.001 0 g

Chapter 3

1. (a) 5   (b) 4   (c) 3
2. (a) 1.237   (b) 1.238   (c) 0.135   (d) 2.1   (e) 2.00
3. (a) 0.217   (b) 0.216   (c) 0.217
4. (a) 3.71   (b) 10.7   (c) 4.0×10^1 (d) 2.85×10^{-6}
5. (a) 12.625 1   (b) 6.0×10^{-4} (c) 242
6. (a) 208.232   (b) 560.604
7. 3.124 (± 0.005), 3.124 ($\pm 0.2\%$)
8. (a) c   (b) b   (c) d   (d) a
9. (a) 2.1 (± 0.2 or $\pm 11\%$)   (b) 0.151 (± 0.009 or $\pm 6\%$)
10. (a) 12.3   (b) 75.5   (c) 5.520×10^3 (d) 3.04
    (e) 3.04×10^{-10} (f) 11.9   (g) 4.600
    (h) 4.9×10^{-7}
11. (a) 10.18 (± 0.07 or $\pm 0.7\%$)
    (b) 174 (± 3 or $\pm 2\%$)
    (c) 0.147 (± 0.003 or $\pm 2\%$)
    (d) 7.86 (± 0.01 or $\pm 0.1\%$)
    (e) 2 185.8 (± 0.8 or $\pm 0.04\%$)
12. (b) 0.450 7 (± 0.000 5) M
13. 1.035 7 (± 0.000 2) g
14. 1.054 572 67 (± 0.000 000 64) $\times 10^{-34}$ J·s
15. 0.000 012

Chapter 4

1. (a) 0.682 6   (b) 0.954 6   (c) 0.341 3   (d) 0.191 5
   (c) 0.149 8
2. (a) 1.527 67   (b) 0.001 26   (c) 1.59×10^{-6}
   (d) 1.527 62   (e) 0.003 82

3. (a) 99.7% (b) real differences

4. 90%: $0.14_8 \pm 0.02_8$; 99%: $0.14_8 \pm 0.05_6$

5. yes

6. no

7. no

8. slope $= -1.299(\pm 0.001) \times 10^4$,
   intercept $= 3(\pm 3) \times 10^2$

9. (a) 0.050 4 (b) 0.403 7

10. 104.7

11. (a) no (b) 0.141 1

12. yes

13. no, yes

14. 88

15. $y = [-0.7_5(\pm 0.1_4)]x + [3.9_2(\pm 0.4_9)]$

16. $5.6(\pm 0.8) \times 10^3$

17. (a) $2.0_0 \pm 0.5_0$ (b) 0.3_8 (c) 0.2_6

Chapter 5

2. 1.2×10^{10}

3. 2×10^{-9}

4. 3.0×10^{-6}

5. (a) decrease (b) give off (c) negative

6. 5×10^{-11}

7. (a) right (b) right (c) neither (d) right (e) left

8. unchanged

9. left

10. H_2O

11. (a) $7._1 \times 10^{-5}$ M (b) $1._0 \times 10^{-3}$ mg/100 mL

12. (a) 6.7×10^{-5} M (b) 2.2×10^{-3} mg/100 mL

13. (a) 0.29 g (b) 4.5×10^{-4} g

14. 1.0×10^{-6} M

20. (a) BF_3 (b) AsF_5

21. (a) HSO_3^- (b) HI (c) H_2O

28. a, c

29. 1.6×10^{-5}

31. $K_{a1} = 7.04 \times 10^{-3}$, $K_{a2} = 6.29 \times 10^{-8}$,
    $K_{a3} = 7.1 \times 10^{-13}$

32. 0.096 M

33. 2.7×10^7

35. 1.0×10^{-56}

36. (a) positive (b) endothermic

37. (a) 4.7×10^{-4} atm (b) 153°C

38. (a) 7.82 kJ/mol

39. (a) right (b) $P_{H_2} = 1\,366$ Pa, $P_{Br_2} = 3\,306$ Pa,
    $P_{HBr} = 57.0$ Pa (c) neither (d) formed

40. 0.018

41. AgCl: 1 400 ppb; AgBr: 76 ppb; AgI: 0.98 ppb

42. $I^- < Br^- < Cl^- < CrO_4^{2-}$

43. no, 0.001 4 M

44. (a) 1.3 mg/L (b) 1.3×10^{-13} M (c) 8.4×10^{-4} M

45. $[Zn^{2+}] = 2.9 \times 10^{-3}$ M, $[ZnOH^+] = 2.3 \times 10^{-5}$ M,
    $[Zn(OH)_3^-] = 6.9 \times 10^{-7}$ M, $[Zn(OH)_4^{2-}] = 8.6 \times 10^{-14}$ M

46. 5.2×10^{-7} M

47. (a) 1.5×10^{-4} M (b) 3.3×10^{-3} M (c) 0.059 M

49. (a) 12 mmol (b) solubility increases

50. 15%

52. 0.22 g

53. true

54. (a) endothermic (b) endothermic (c) exothermic

Chapter 6

1. (a) true (b) true (c) true

2. (a) 0.001 2 M (b) 0.08 M (c) 0.30 M

3. (a) 0.660 (b) 0.54 (c) 0.18 (d) 0.83

4. 0.88_7

5. (a) 2.6×10^{-8} M (b) 3.0×10^{-8} M
   (c) 4.2×10^{-8} M

6. $\gamma_{H^+} = 0.86$, pH = 2.07

8. (a) 0.42_2 (b) 0.43_2

9. 0.929

10. 1.5×10^{-6} M

11. 2.5×10^{-3}

12. 11.94

13. increase

14. 0.33

15. 0.80, 0.56, 0.29, 0.19, 0.20, 0.22, 0.26, 0.37

Chapter 7

4. $[H^+] + 2[Ca^{2+}] + [Ca(HCO_3)^+] + [Ca(OH)^+] + [K^+] = [OH^-] + [HCO_3^-] + 2[CO_3^{2-}] + [ClO_4^-]$

5. $[H^+] = [OH^-] + [HSO_4^-] + 2[SO_4^{2-}]$

6. $[H^+] = [OH^-] + [H_2AsO_4^-] + 2[HAsO_4^{2-}] + 3[AsO_4^{3-}]$

7. (a) 0.20 M $= [Mg^{2+}]$ (b) 0.40 M $= [Br^-]$
   (c) 0.20 M $= [Mg^{2+}] + [MgBr^+]$
   (d) 0.40 M $= [Br^-] + [MgBr^+]$

8. (a) $2[Mg^{2+}] + [H^+] = [Br^-] + [OH^-]$
   (b) $2[Mg^{2+}] + [H^+] + [MgBr^+] = [Br^-] + [OH^-]$

9. $[CH_3CO_2^-] + [CH_3CO_2H] = 0.1$ M

10. 7.5×10^{-6}

11. $[Ca^{2+}] = 1.4 \times 10^{-3}$ M, $[F^-] = 1.7 \times 10^{-4}$ M,
    $[HF] = 2.5 \times 10^{-3}$ M

12. $[Y^{2-}] = [X_2Y_2^{2+}] + 2[X_2Y^{4+}]$

13. $[OH^-] = 1.5 \times 10^{-7}$ M

14. 4.0×10^{-5} M

15. 5.8×10^{-4} M

16. (a) 5.0×10^{-9} M (b) 2.1×10^{-6} M
    (c) 1.1×10^{-8} M

17. 4.3×10^{-3} M

18. (a) $12._4$ (b) $[Sr^{2+}] = 1.6 \times 10^{-4}$ M,
    $[F^-] = 4.3 \times 10^{-3}$ M, $[Pb^{2+}] = 2.0 \times 10^{-3}$ M

19. (a) $[NH_4^+] + [H^+] = 2[SO_4^{2-}] + [HSO_4^-] + [OH^-]$
    (b) $[NH_3] = [NH_4^+] = 2\{[SO_4^{2-}] + [HSO_4^-]\}$
    (c) 6.59 M

20. (a) $[FeG^+] + [H^+] = [G^-] + [OH^-]$
 (b) $[FeG_2] + [FeG^+] = 0.050\,0$ M,
  $[FeG^+] + 2[FeG_2] + [G^-] + [HG] = 0.100$ M
 (c) 0.016_4 M

21. (a) $0.10 = [M^{2+}] + \dfrac{2K_1[M^{2+}]^2}{1 + K_1[M^{2+}]} + \dfrac{4K_1K_2[M^{2+}]^3}{(1 + K_1[M^{2+}])^2}$

22. (a) $[Zn^{2+}] = 5.00 \times 10^{-3}$ M
 (b) $[Zn^{2+}] = 6.64 \times 10^{-3}$ M
 (c) $[Zn^{2+}] = 6.85 \times 10^{-3}$ M, 32% ion-paired
 (d) $\gamma_\pm = 0.579$

23. 5.2×10^5 lb

Chapter 8

7. 0.022 86 M
8. 1.94%
9. 11.69 mg CO_2, 2.051 mg H_2O
10. 0.086 57 g
13. 0.508 00 g
14. 0.191 4 g, 0.107 3 g
15. 104.1 ppm
16. 7.22 mL
17. 0.919 g
18. 10.5%
19. 14.5% K_2CO_3, 14.6% NH_4Cl
20. 40.4%
21. 22.65%
22. 13.99%
23. (a) 40.05% (b) 39%
24. (b) 0.204 (± 0.004)
25. (a) 5.5 mg/100 mL (b) 5.834 mg, yes

Chapter 9

4. negative
9. 32.0 mL
10. 43.2 mL $KMnO_4$, 270.0 mL $H_2C_2O_4$
11. 1.72 mg
12. 0.043 88 m
13. 13.08, 8.04, 2.53
14. 12.4%
15. 15.1%
16. 0.149 M
17. (a) 0.020 34 M (b) 0.125 7 g (c) 0.019 83 M
18. 0.092 54 M
19. 56.28%
20. 947 mg
21. 11.04 g, 22.08 g
22. Br = 35.9%, Cl = 30.9%, Br/Cl atomic ratio = 0.514
23. 0.037 3 M
24. 90% T ⇒ $\tau = 0.105$ cm^{-1},
  10% T ⇒ $\tau = 2.30$ cm^{-1},
  $A = 1$ ⇒ $\tau = 2.30$ cm^{-1}
25. (a) 20.17 mL (b) 9.57

26. (a) 19.00, 18.85, 18.65, 17.76, 14.17, 13.81, 7.83, 1.95
 (b) no
27. (a) 25.0 mL (b) 0.090 4 M (c) 24.07
28. upper = 9.6×10^3 M (which means there is no upper limit), lower = 6.2×10^{-4} M
29. 13.61

Chapter 10

1. (a) 3.00 (b) 12.0
3. 4-aminobenzenesulfonic acid
8. 4.37×10^{-4}, 8.93×10^{-13}
9. pH = 3.00, $\alpha = 0.995\%$
10. pH = 11.00, $\alpha = 0.995\%$
11. 4.70
12. (a) pH = 2.51, $[H_2A] = 9.69 \times 10^{-2}$ M,
   $[HA^-] = 3.11 \times 10^{-3}$ M,
   $[A^{2-}] = 1.00 \times 10^{-8}$ M
 (b) 6.00, 1.00×10^{-1} M, 1.00×10^{-3} M, 1.00×10^{-3} M
 (c) 10.50, 1.00×10^{-10} M, 3.16×10^{-4} M, 9.97×10^{-2} M
13. 5.50
14. See Equation 10-57
15. pH = 11.28, $[B] = 0.058$ M, $[BH^+] = 1.9 \times 10^{-3}$ M
16. pH = 5.51, $[B] = 3.1 \times 10^{-6}$ M, $[BH^+] = 0.060$ M
17. pH = 10.95
18. pH = 3.15
19. 0.007 56%, 0.023 9%, 0.568%
20. F = $(0.010\,2)K_a$
21. 0.180, 1.00, 1.80
22. (a) 14 (b) 1.4×10^{-7}
23. (a) pH = 1.95, $[H_2M] = 0.089$ M
 (b) pH = 4.28, $[H_2M] = 3.7 \times 10^{-3}$ M
 (c) pH = 9.35 $[H_2M] = 7.04 \times 10^{-12}$ M
24. $pK_a = 4.19$
25. 3.6×10^{-9}
26. 3.55
27. (b) NaOH
28. (a) 8.19 (b) 8.25
29. 3.27 mL
30. (b) 7.18 (c) 7.00 (d) 6.86 mL
31. (a) 2.56 (b) 2.61 (c) 2.86
32. 11.70, 11.48
33. pH = 11.60. $[B] = 0.296$ M, $[BH^+] = 3.99 \times 10^{-3}$ M, $[BH_2^{2+}] = 2.15 \times 10^{-9}$ M
34. 2.8×10^{-3}, 3.4×10^{-8}
35. (a) 3.03, 9.4% (b) 7.00, 99.9% (c) 8.00, 0.010%
36. 13.7 mL
37. $[A^-] = 0.004\,01$ M
38. pH = 5.70
39. (a) 2.95 g, 92.7 mL HCl
40. 5°C: endothermic; 45°C: exothermic
41. (a) $\Delta H° = -12\,(\pm 1)$ kJ/mol,
   $\Delta S° = -27\,(\pm 4)$ J/(mol·K)
 (b) 5.0 (± 3.6)

42. (a) 1.37   (b) 12.61
43. 0.005 40%
44. (a) 5.88   (b) 5.59
45. pH = 3.69
46. 3.96

Chapter 11

4. H_2SO_4
6. 13.00, 12.95, 12.68, 11.96, 10.96, 7.00, 3.04, 1.75
7. 3.00, 4.05, 5.00, 5.95, 7.00, 8.98, 10.96, 12.25
8. 11.00, 9.95, 9.00, 8.05, 7.00, 5.02, 3.04, 1.75
9. 11.49, 10.95, 10.00, 9.05, 8.00, 6.95, 6.00, 5.05, 3.54, 1.79
10. 2.51, 3.05, 4.00, 4.95, 6.00, 7.05, 8.00, 8.95, 10.46, 12.21
11. 0.079 34 mol/kg
12. yellow, green, blue
13. red, orange, yellow
14. red, orange, yellow, red
17. 2.2×10^9
18. Products are $NaOH + CH_3OH$ and $NaOH + CH_3CH_2OH$.
19. 10.92, 9.57, 9.35, 8.15, 5.53, 2.74
20. $V_e/11$; $10V_e/11$; $V_e = 0$: pH = 2.80;
    $V_e/11$: pH = 3.60; $V_e/2$: pH = 4.60;
    $10V_e/11$: pH = 5.60; V_e: pH = 8.65;
    $1.2V_e$: pH = 12.07
21. 8.74, 5.35, 4.87, 4.40, 3.22, 2.57
22. 8.18
23. 5.09
24. 11.36, 10.21, 9.73, 9.25, 7.53, 5.81, 5.33, 4.85, 3.41, 2.11, 1.85
25. (a) 1.99
26. (b) 7.18
27. (b) 9.81
28. 2.66
29. (a) 9.56   (b) 7.4×10^{-10}
30. 2.97 mL
31. no
33. 2.47
34. violet, blue, yellow
35. $0.99V_e$: pH = 7.22; V_e: pH = 8.96;
    $1.01V_e$: pH = 10.70;
    phenolphthalein (colorless → red)
36. 5.62
37. 2.859%
38. 0.100 0 M
39. 0.30 g
40. 7.1×10^7
41. (a) 9.45   (b) 2.55   (c) 5.15
42. 0.091 78 M
43. 6.28 g
44. 23.40 mL
45. 0.063 56 M

46. (a) 20.254%   (b) 17.985 g
47. (a) acetic acid   (b) pyridine
48. Products are $NaOH + NH_3$ and $LiOH + C_6H_6$.
49. 0.139 M
50. -1.15
51. (a) 3.46, 8.42   (b) 2nd
    (c) thymolphthalein—first trace of blue
52. 20.95 mL
53. 9.72
54. $pK_2 = 9.84$

Chapter 12

1. (a) HA   (b) A^- (c) 1.0, 0.10
2. (a) 4.00   (b) 8.00   (c) H_2A (d) HA^- (e) A^{2-}
3. (a) 9.00   (b) 9.00   (c) BH^+ (d) 1.0×10^3
4. $\alpha_0 = 0.091$, $\alpha_1 = 0.909$, $[A^-]/[HA] = 10$
5. 0.91
7. (a) $10^{4.00}$ (b) 5.00
8. (a) 1.0   (b) 4.70
9. (a) BH^+ (b) 4.50
10. isoelectric pH 5.59, isoionic pH 5.72
12. $\alpha_1 = 0.123, 0.694$
13. $\alpha_1 = 0.110, 0.500, 0.682, 0.500, 2.15 \times 10^{-4}$
14. (b) $8.6 \times 10^{-6}, 0.61, 0.39, 2.7 \times 10^{-6}$
15. 1.07×10^5
16. 9.44
17. 4.16
18. 4.48
19. 4.82
20. 4.44
21. 7.73
22. 3.62
23. 18%
24. (a) 8.34   (b) 7.00
25. 2.21
26. -2
27. 6.95
28. isoionic
29. positive
31. (b) 8.85   (c) chlorophenol red: orange → red
32. 0.37
33. 4.00

Chapter 13

4. (a) 3.4×10^{-10} (b) 0.64
5. HIn^{2-}, wine-red, blue
9. 10.0 mL, 10.0 mL
10. 0.020 0 M
11. (a) 100.0 mL   (b) 0.016 7 M   (c) 0.054
    (d) 5.4×10^{10} (e) 6.8×10^{-7} (f) 1.9×10^{-10} M
12. (a) 4.3×10^3 (b) 0.017

13. (a) 2.93 (b) 6.77 (c) 10.49

14. 0.995 mg

15. 21.45 mL

16. 49.9: 4.87; 50.0: 6.90; 50.1: 8.92

17. 49.9: 7.55; 50.0: 6.21; 50.1: 4.88

18. 2.7×10^{-11} M

19. (a) $\alpha_{ML} = \beta_1[L]/(1 + \beta_1[L] + \beta_2[L]^2)$,
    $\alpha_{ML_2} = \beta_2[L]^2/(1 + \beta_1[L] + \beta_2[L]^2)$
    (b) $\alpha_{ML} = 0.28$, $\alpha_{ML_2} = 0.70$

20. 0 mL − pCu^{2+} = 11.08;
    1.00 mL − pCu^{2+} = 11.09;
    45.00 mL − pCu^{2+} = 12.35;
    50.00 mL − pCu^{2+} = 15.06;
    55.00 mL − pCu^{2+} = 17.73

21. $[Ni^{2+}] = 0.012\,4$ M, $[Zn^{2+}] = 0.007\,18$ M

22. 0.092 6 M

23. 0.024 30 M

24. Mn: 69.64; Mg: 5.150; Zn: 20.89

25. Buffer (a)

26. 0.092 28 M

27. 5.00 g

29. 2 600

Chapter 14

1. (a) $6.241\,506\,3 \times 10^{18}$ (b) 96 485.309

2. (a) $71._5$ A (b) 4.35 A (c) 79 W

3. (a) 1.87×10^{16} e^-/s (b) 9.63×10^{-19} J/e^-
   (c) 5.60×10^{-5} mol (d) 447 V

4. (a) I_2 (b) $S_2O_3^{2-}$ (c) 861 C (d) 14.3 A

5. (a) 1.33 V (b) 1×10^{45}

6. Cl_2

7. Cr^{2+}

8. (a) Fe(s)│FeO(s)│KOH(aq)│Ag_2O(s)│Ag(s);
       Fe(s) + $2OH^-$ ⇌ FeO(s) + H_2O + $2e^-$;
       Ag_2O(s) + H_2O + $2e^-$ ⇌ 2Ag(s) + $2OH^-$
   (b) Pb(s)│$PbSO_4$(s)│K_2SO_4(aq)║H_2SO_4(aq)
       │$PbSO_4$(s)│PbO_2(s)│Pb(s);
       Pb(s) + SO_4^{2-} ⇌ $PbSO_4$(s) + $2e^-$;
       PbO_2(s) + $4H^+$ + $2SO_4^{2-}$ ⇌ $PbSO_4$(s) + $2H_2O$

11. (a) $Fe(CN)_6^{4-}$ ⇌ $Fe(CN)_6^{3-}$ + e^-;
       $Ag(CN)_2^-$ + e^- ⇌ Ag(s) + $2CN^-$
    (b) 2Hg(l) + $2Cl^-$ ⇌ Hg_2Cl_2(s) + $2e^-$;
       Zn^{2+} + $2e^-$ ⇌ Zn(s)

12. (b) −0.359 V

13. (b) 0.430 V

14. $E° = -0.330$ V ⇒ $K_{sp} = 7 \times 10^{-12}$

15. (d) 0.14_3 M

16. (c) 0.317 V

17. 0.054 20, 0.061 54

18. 1.664 V

20. (a) $K = 10^{47}$ (b) $K - 9.2 \times 10^{-7}$
    (c) $K = 1.9 \times 10^{-6}$ (d) $K = 3.2 \times 10^5$

21. (a) Pt(s)│Cr^{2+}, Cr^{3+}║Tl^+│Tl(s)
    (b) 0.08_4 V (d) Pt

22. (b) $K = 2 \times 10^{16}$ (c) -0.02_0 V
    (d) 10 kJ (e) 0.21

23. 0.799 3 V

24. (a) Fe^{3+} (b) Fe^{2+}

25. (a) 0.053 V (b) $2._5 \times 10^{-13}$ (c) $1._6 \times 10^{13}$

26. (b) −2.854 V (c) Br_2 (d) 1.31 kJ
    (e) 2.69×10^{-8} g/s

27. (a) 0.572 V (b) 0.569 V

28. (a) $A = -0.414$ V, $B = 0.059\,16$ V (c) Hg → Pt

29. 9.6×10^{-7}

30. 0.101 V

31. 34 g/L

32. 1.396 V

33. $K = 1.0 \times 10^{-9}$

34. 1.7×10^{-4}

35. −0.447 V

36. 0.117 V

37. 1.341 V

38. 0.984 V

39. $E_2° > E_1°$

40. −0.098 V

41. −0.184 V

42. −0.036 V

43. 7.2×10^{-4}

44. 5.4×10^{13}

45. (a) 1.95×10^{-4} M (b) 2.7×10^{-8}

46. 0.76

47. (a) $[Ox] = 3.82 \times 10^{-5}$ M, $[Red] = 1.88 \times 10^{-5}$ M
    (b) $[S^-] = [Ox]$, $[S] = [Red]$ (c) −0.092 V

48. 6×10^{18}

49. 0.580 V

50. 7.5×10^{-8}

Chapter 15

1. (b) 0.044 V

2. 0.684 V

3. (c) 0.068 V

5. left

8. 10.67

10. (a) −0.407 V (b) $1.5_5 \times 10^{-2}$ M
    (c) $1.5_2 \times 10^{-2}$ M

11. +0.029 6 V

12. (a) K^+ (b) Group I > Group II

13. 3.8×10^{-9} M

14. 0.243 V

15. 0.627

16. H^+: 42.4 s; NO_3^-: 208 s

17. 0.10 pH unit

18. (a) $3._2 \times 10^{13}$ (b) 8% (c) 49.0, 8%

19. (a) 274 mV (b) 285 mV

20. (a) K_I/0.033 3 (b) K_{Cl}/0.020 0 (d) 2.2×10^6

21. $5._2 \times 10^{21}$
22. 1.364 V
23. (b) $1._2 \times 10^{11}$
24. 0.29_6 M
26. (a) 2.4×10^{-3} M (b) 0.951 3
27. (b) $2.43 (\pm 0.09) \times 10^{-3}$ M
28. −0.332 V
29. 0.301 2 V, Zn^{2+}
30. (a) 2.32×10^{-4} M (c) Divide each ΔE by 2.
32. $E = 120.2 + 28.80 \log ([Ca^{2+}] + (6.0 \times 10^{-4})[Mg^{2+}])$
33. 1.4×10^{-4} M
34. (a) 8.9×10^{-8} M (b) 1.90 mmol
35. 1.49×10^{-8} M, 1.25×10^{-13} M
36. (a) 1.13×10^{-4} (b) 4.8×10^{4}
38. (a) 26.9 mV (b) 5.0 mV

Chapter 16

2. (d) 0.490, 0.516, 0.526, 0.536, 0.562, 0.626, 0.99, 1.36, 1.42 V
3. −0.107, −0.102, 0.51, 0.53, 0.84, 0.98 V
4. diphenylamine sulfonic acid: colorless → red-violet; diphenylbenzidine sulfonic acid: colorless → violet; tris(2,2′-bipyridine)iron: red → pale blue; ferroin: red → pale blue
5. (a) 0.029 14 M (b) no
8. products: (a) Mn^{2+} (b) MnO_2 (c) MnO_4^{2-}
10. −0.72, −0.66 −0.06, 0.530 V
11. (c) 0.095 V (d) 0.237 V
12. 0.494, 0.520, 0.530, 0.540, 0.566, 0.586, 0.648, 1.07, 1.15, 1.16, 1.17 V
13. (a) −0.046 V (b) 0.009 V
14. −0.102, 0.53, 0.98, 1.13 V
15. (b) 0.025 1 M UO_2^+, 0.036 5 M Fe^{3+}
    (c) −0.08, 0.526, 1.143, 1.248 V
16. 0.21 V
17. −0.104, 0.413, 0.862 V
18. no
19. no
20. 0.390, 0.450, 0.526, 0.602, 0.662, 0.850, 0.906, 0.962, 1.14, 1.32, 1.38 V
21. (b) $YBa_2Cu_3O_{6.875 \pm 0.038}$
22. −0.207 V
23. (a) 7×10^{2} (b) 1.0 (c) 0.34 g/L
24. 0.011 29 M
25. 41.9%
26. 40.3%
27. 5.730 mg
28. 78.67%
29. 3.826×10^{-3} M
30. (a) $5IO_4^- + C_6H_{12}O_6 \rightarrow 5HCO_2H + H_2CO + 5IO_3^-$
    (b) $C_6H_{13}O_5N + H_2O + 5IO_4^- \rightarrow 5HCO_2H + H_2CO + NH_3 + 5IO_3^-$

(c) $C_3H_6O_3 + IO_4^- \rightarrow H_2CO + HO_2CCH_2OH + IO_3^-$
31. 9.107 mg
33. (a) 1.442 V (b) 0.97 V
34. −0.784 V

Chapter 17

4. constant voltage: V_1; controlled potential: V_2
5. cathodic depolarizer
7. 2.68 h
8. (b) 0.000 26 mL
9. (a) −1.906 V (b) 0.20 V (c) −2.71 V
   (d) −2.82 V
10. 54.77%
11. −0.619 V, negative
12. (a) 5.32×10^{-5} mol (b) 2.66×10^{-5} mol
    (c) 5.32×10^{-3} M
13. (a) anode (b) 52.0 g/mol (c) 0.039 6 M
14. (a) 6.64×10^{3} J (b) 0.012 4 g/h
15. (a) 1.946 mmol (b) 0.041 09 M (c) 4.75 h
16. (a) 0.84 V (b) −1.06 V
17. 0.53 V
18. −2.183 V
19. −1.23 V
20. 26.3% trichloroacetic acid, 49.5% dichloroacetic acid
21. 1.51×10^{2} μg/mL
22. −0.744 V
23. yes
24. (a) current density = 1.00×10^{2} A/m^2, overpotential = 0.85 V
    (b) −0.036 V (c) 1.160 V (d) −2.57 V

Chapter 18

4. 1.03×10^{-9} m^2/s
9. 1.7 mM
10. 1.21 mM
11. 0.81 mM
12. (a) 0.000 35 min^{-1} (b) 3.4 min (c) 0.118%
13. $E_{3/4} - E_{1/4} = 0.056 5/n$ V
14. 0.076 9 ($\pm$0.003 2) mM
15. 2.371 ($\pm$0.024) mM
16. 0.760 ppm
17. similar to Exercise 18-F, lower curve
18. similar to Demonstration 18-1
19. 2.95 mM
20. 0.096 mM
21. −0.01 V
22. peak B: $RNHOH \rightarrow RNO + 2H^+ + 2e^-$; peak C: $RNO + 2H^+ + 2e^- \rightarrow RNHOH$
24. $[Cd^{2+}] = 0.82$ mM, $[Zn^{2+}] = 1.66$ mM
25. (c) $E' = 0.853$ V (versus S.H.E.) (d) $n \approx 2$

Chapter 19

1. double, halve, double
2. (a) 184 kJ/mol   (b) 299 kJ/mol
3. 5.33×10^{14} Hz, 1.78×10^4 cm^{-1}, 213 kJ/mol
4. 3.56×10^4 M$^{-1}\cdot$cm^{-1}
5. violet-blue
7. (a) double   (b) double
8. (a) 7.80×10^{-3} M   (b) 7.80×10^{-4} M
   (c) 1.63×10^3 M$^{-1}\cdot$cm^{-1}
9. (a) 6.97×10^{-5} M   (b) 6.97×10^{-4} M   (c) 1.02 mg
10. $[X] = 8.03 \times 10^{-5}$ M, $[Y] = 2.62 \times 10^{-4}$ M
12. (a) 2.33×10^{-7} mol Fe(III)   (b) 5.83×10^{-5} M
13. (b) $K = 88.2$
14. wavenumber = 1.697 834 5 and
    1.696 114 4 $\times 10^4$ cm^{-1}
15. 0.288
16. (a) 4.97×10^4 M$^{-1}\cdot$cm^{-1} (b) 4.69 μg
17. (a) $2.42_5 \times 10^4$ M$^{-1}\cdot$cm^{-1} (b) 1.26%
18. yes
19. 2.19×10^{-4} M
20. $[\mathrm{MnO_4^-}] = 84.2$ μM; $[\mathrm{Cr_2O_7^{2-}}] = 178$ μM
21. $[A] = 9.11 \times 10^{-3}$ M; $[B] = 4.68 \times 10^{-3}$ M
22. $78\ (\pm 3)$ μL
24. (a) $n = 1$
28. (a) 228 μM   (b) 357 μM
29. (a) $k = 0.485$, $b = 0.084$ (b) 4.6 ng/mL, 83 nM

Chapter 20

3. 0.78
4. 0.970
5. (a) 34°   (b) 0°
6. 77 K − 1.99 W/m^2; 298 K − 447 W/m^2
7. $n_{\text{prism}} > \sqrt{2}$
11. D_2
13. Grating
16. (a) ± 2 cm   (c) 0.5 cm^{-1} (d) 2.5 μm
18. 7
19. (a) 4.2×10^2 cm^{-1} (b) 0.036   (c) 0.86   (d) 0.800
20. 0.051 cm^{-1}
21. 2×10^{-4}
22. (a) (λ, n): (0.2, 1.550 5),   (0.4, 1.470 1),   (0.6, 1.458 0),
    (0.8, 1.453 3), (1, 1.430 4), (2, 1.438 1), (3, 1.419 2),
    (4, 1.389 0), (5, 1.340 4), (6, 1.258 0)   (b) blue
23. (b) 48.59°
24. (a) 61.04°   (b) 51.06°
25. quartz or MgF$_2$
26. (a) $M_\lambda = 8.789 \times 10^9$ W/m^3 at 2.00 μm;
    $M_\lambda = 1.164 \times 10^9$ W/m^3 at 10.00 μm
    (b) 1.8×10^2 W/m^2 (c) 2.3×10^1 W/m^2
    (d) $\dfrac{M_{2.00\ \mu m}}{M_{10.00\ \mu m}} = 7.551$ at 1 000 K
    $\dfrac{M_{2.00\ \mu m}}{M_{10.00\ \mu m}} = 3.163 \times 10^{-22}$ at 100 K

27. (a) 8.0×10^4 W/m^2 (b) 1.2×10^5 W/m^2
28. (a) 2.38×10^3 (b) 143
29. 0.124 2 mm

Chapter 21

8. (b) 204 μg/mL
9. (a) 7.49 μg/mL (b) 25.6 μg/mL
10. 589.3 nm
11. 0.025
12. 17.4 μg/mL
13. Na: 3.3 GHz; Hg: 2.6 GHz
14. (a) 283.0 kJ/mol   (b) 3.67×10^{-6} (c) 8.4%
    (d) 1.03×10^{-2}
15. (a) 1.660   (b) 1.84 μg/mL
16. 0.079 μg/mL

Chapter 22

1. (a) 0.080 M   (b) 0.50
2. 0.088
3. (c) 4.5   (d) greater
5. (a) 55 and 56   (b) 5.9
6. 1-C, 2-D, 3-A, 4-E, 5-B
7. (a) 2.0   (b) 0.33   (c) 20
8. 19 cm/min
9. 0.6, 6
10. 0.1 mm
11. (a) 1.1×10^2 (b) 0.89 mm
12. 138
15. 33 mL/min
16. b
18. $K_{\text{extraction}} = K_M \beta K_a^n / K_L^n$
21. (a) 0.158 M in benzene   (b) 2×10^{-6} M in benzene
22. 2
23. (a) 2.6×10^4 at pH 1 and 2.6×10^{10} at pH 4
    (b) 3.8×10^{-4}
24. fraction of A in each tube: 0.000 012 8, 0.000 358,
    0.004 30, 0.028 7, 0.115, 0.275, 0.367, 0.210;
    fraction of B in each tube: 0.007 81, 0.054 7, 0.164,
    0.273, 0.273, 0.164, 0.054 7, 0.007 81
25. (b) recovery = 53.4%, purity = 99.1%
26. (a) 19.6 mL   (b) 13.9 mL
27. (a) 360   (b) 909 6
29. 10.4 mL
30. 4.0×10^5
31. $K = 4.69$, $k = 3.59$
32. (a) 9.8×10^3 (b) 2.6×10^3 (c) 8.2×10^3

Chapter 23

7. 200/400
17. (a) 4.7 min, 1.3   (b) 1.8
18. 836
19. 27.1 min

20. (a) 0.62_8 (b) 4.3×10^{-7} m
21. (a) $40._2$ mL (b) 0.42
24. Exchange position of buffers.
26. 932
27. transferrin: 0.127; ferric citrate: 0.789
28. (a) 4 000 (b) 1 000
29. (a) 30 (b) 3.3 (c) decrease
30. 38.0%
31. 320 00
32. (a) shorter (b) amine
33. 126 mm
34. 0.418 mg/mL
36. (a) 0.58, 1.9 (b) 0.058 mm, 0.19 mm
    (c) 3.0×10^5 (d) 4
37. 4.2
39. (b) 29 ng/mL
40. (a) 30% tetrahydrofuran (b) 30% tetrahydrofuran
    (c) 14%
41. 23.5, 71.2, 6.2

Chapter 24

3. (a) 5% (b) 2.6%
5. 1.0 g
6. 120/170 mesh
7. Zn, Fe, Co, Al
8. Avoids possible explosion
13. $10^4 \pm 0.99\%$
14. (a) 15.8 (b) 1.647 (c) 474–526
15. 95%: 8; 90%: 6
16. (a) 5.0 g (b) 7
17. (a) Na_2CO_3 mass $= 4.474 \times 10^{-15}$ g;
       number $= 8.941 \times 10^{14}$
       K_2CO_3 mass $= 4.291 \times 10^{-15}$ g;
       number $= 2.237 \times 10^{16}$
    (b) 2.326×10^{14}
    (c) Na_2CO_3: 1.04×10^{-6}; K_2CO_3: 4.14×10^{-8}
18. (a) 53
19. 64.90%

Index